Crustal Evolution and Metallogeny in India

Sanjib Chandra Sarkar
Anupendu Gupta

CAMBRIDGE UNIVERSITY PRESS
Cambridge, New York, Melbourne, Madrid, Cape Town,
Singapore, São Paulo, Delhi, Mexico City

Cambridge University Press
4381/4, Ansari Road, Daryaganj, Delhi 110002, India

Published in the United States of America by Cambridge University Press, New York

www.cambridge.org
Information on this title: www.cambridge.org/9781107007154

First published 2012

Printed in India at Replika Press Pvt. Ltd., Kundli, Haryana

A catalogue record for this publication is available from the British Library

Library of Congress Cataloguing in Publication data
Sarkar, Sanjib Chandra.
Crustal evolution and metallogeny in India / Sanjib Chandra Sarkar, Anupendu Gupta.
p. cm.
Summary: "Documents in detail the nature, origin and evolution of the Indian
mineral deposits in the context of local and regional geology"--Provided by publisher.
Includes bibliographical references and index.
ISBN 978-1-107-00715-4 (hardback)
1. Mineralogy--India. 2. Metallogeny--India. 3. Geology--India. I. Gupta, Anupendu, 1942- II. Title.

QE382.I4S27 2011
549.954--dc22 2010040401

ISBN 978-1-107-00715-4 Hardback

Dedicated
to
the Students of Indian Geology

Contents

List of Figures

Chapter 1 Southern India *1*

1.1 Geology and Crustal Evolution *1*

2.2 Metallogeny *233*

4.2 Metallogeny *423*

Chapter 7 The Himalaya *679*

Chapter 8 Crustal Evolution and Metallogeny in India:A Brief Review in the Context of the World-Scenario *717*

LIST OF TABLES

List of Plates

Chapter 5 North-East India *526*

Chapter 6 Western India *543*

Chapter 7 The Himalaya *679*

Chapter 8 Crustal Evolution and Metallogeny in India: A Brief Review in the Context of the World Scenario *717*

PREFACE

As the title of the book suggests, the two key areas in focus here are crustal evolution and metallogeny. Crustal evolution refers to the changes that the Earth's crust has gone through the geological past. These are due to the effects of the interaction of and changes in the mantle-crust system, the atmosphere, the hydrosphere and the biosphere, or even otherwise. The changes may be chemical or physical, or both. Metallogeny is the genesis of metallic mineral deposits. This term has been used in the book in its conventional sense, although the phosphatic, sulphatic (anhydrite/gypsum, barite), carbonate (magnesite) deposits and even precious stones like diamond are included for the sake of relevance.

Indian geology is now over 150 years old. During the first 100 years or so, the colonial government took interest in exploring mineral resources in India. During the post-independence time, substantial attention was paid to increase the resource (material) base of the country that would ultimately improve the quality of life of the newly liberated people. Interestingly, Pandit Jawaharlal Nehru, the first Prime Minister of India, was a student of Geology at Cambridge. By now a lot of information, but by no means adequate in all respects, on Indian geology has been accumulated. These need to be analysed and synthesised from time to time for whatever they are worth, besides being augmented.

Quite a few texts on modern ore geology are in circulation now. Very few amongst them have any reference to the Indian deposits, or even Indian geology. As a result, an Indian reader finds it difficult to relate to the theories, or the treatment of a particular case in the Indian context. There are many memoirs, records, bulletins and individual scientific papers published by the government agencies and academia. Some of them contain excellent descriptions, but many are not supported by in-depth analysis of the local or regional geology, or regional metallogeny. Moreover, the existing literature is also not always easy to access. Advanced students of Indian geology, Earth science professionals, researchers, teachers, planners and global metallogeneticists who want to be acquainted with the metallogeny in India in the context of crustal evolution therefore, feel disadvantaged, even disappointed. Many of them at different points of time expressed the desire that a volume be brought out where the nature, origin and evolution of the Indian deposits are discussed in the context of local as well as regional geology. Some of the concerned people even personally suggested to the authors to undertake such an endeavour themselves. Looking at the nature and dimensions of the project, we were initially hesitant, but were ultimately obliged to accede. If the book ultimately becomes useful, even partially, we shall feel rewarded.

As will be obvious to a discerning reader, the database for the book is not adequate, or even substantial in every case. However, a synthesis of the available data should not have waited until our database was adequate all over. Collating data and their utilisation is a dynamic process. We from our part have been candid to point out where we thought there existed a cavernous gap. This may help the future researchers in choosing their projects.

We initially were unsure about the best way of presenting the text. Should we start with the oldest rocks and their mineral contents, wherever present, and end with the youngest ones? An alternative was to divide the whole text on a regional basis into several chapters and then synthesise, which we ultimately adopted. Of the eight chapters in total, the first seven chapters deal with the seven regions of the country. These are Southern India (Chapter 1), Central India (Chapter 2), Eastern Ghat Belt (Chapter 3), Eastern India (Chapter 4), North-East India (Chapter 5), Western India (Chapter 6) and the Himalaya (Chapter 7). Each chapter is in turn divided into two sections, the first dealing with the geologic background (crustal evolution) and the second, the metallogeny. One advantage of this system, in our opinion, is that someone, interested in the geology or mineralisation of a particular segment of the country, can easily access the matter in the present arrangement. On the other hand, one who is interested in comprehending the essentials of crustal evolution and its relation to metallogeny in the country without going into every detail, may do so by carefully going through the final chapter, i.e., Chapter 8. This chapter provides essential details regarding crustal evolution and metallogeny of the world. This is followed by a recapitulation of the broad features of the Indian scenario, noting where and how much do they tally. Attempts have been made in this chapter to answer some of the 'whys' and 'hows' of an inquisitive reader.

The authors are grateful to the Department of Science and Technology, Government of India, for sanctioning a two-year book-writing project to Sanjib Chandra Sarkar in August 2002. In hindsight, the initial project itself was ambitious in its scope. It became more so when we modified it on request and suggestions from our friends, acquaintances and former students. Time extended and the next four years or so we had to go on our own, visiting/revisiting mineral deposits and interesting geological sites, conducting/arranging complementary laboratory studies, attending seminars and symposia on relevant topics besides accessing available literature. Interest, confidence and encouragement of the above people, needless to say, sustained us throughout.

Jadavpur University, particularly its Geological Sciences Department, where the first author worked for more than four decades, was kind enough to ungrudgingly extend the possible logistic support. We are obliged to them. The Department has been a traditional centre of economic geology study in the country. We are also thankful to the Geological Survey of India (GSI), Atomic Minerals Directorate of the Department of Atomic Energy, Steel Authority of India Ltd., Hindustan Copper Ltd., Hindustan Zinc Ltd., Tata Steel, National Mineral Development Corporation, Mineral Exploration Corporation, Orissa Mining Co., Indian Aluminium Co. (now Hindustan Aluminium Co.) for their cooperation. We are obliged to the authors and editors of articles, books and journals who have kindly consented to our incorporation of some figures and tables of data from their publications .

Sanjib Chandra Sarkar appreciatingly remembers in this context Late Professor P. Rouchier (BRGM, France) and Late Dr R. W. Boyle (Geol. Surv. Canada), who initially inspired him to take up such a project. The authors are obliged to many of their friends, ex-colleagues and ex-students, including Professors S. Acharya, C. Leelanandam, A. B. Roy, D. M. Banerjee, M. Deb, S. Dasgupta, H. Bhattacharyya and P. Sengupta and Drs M. Ramakrishnan, A. Ghosh, Sumit Ray, V.V. Rao, K.P. Ghosh, R. N. Singh, B. K. Bandyapadhyay for different helps and cooperation.

Sri B. Datta and Sri D. Ghosh are acknowledged for their help with word-processing and Sri S. Roy Chowdhury for drawing/redrawing some of the diagrams.

We are indebted to our families who encouraged us all through, ignoring their inconveniences. Our preoccupation with the job must have affected the performance of the filial duties, not to speak of the management of family matters at a desirable level.

Last, but not the least, we are obliged to Cambridge University Press for publishing the book.

Sanjib Chandra Sarkar
Anupendu Gupta

1

Southern India

1.1 Geology and Crustal Evolution

Introduction

Southern Indian craton today is divided into two regions: the Dharwar Province (DP) and the Southern Granulite Province (SGP) or Southern Granulite Terrain (SGT) (Fig. 1.1.1). The Dharwar Province, ever since the publication of results of excellent geological studies by the geologists of the erstwhile Mysore Geological Survey during the first half of the last century, has remained an area of great interest for Indian earth-scientists. In the last decades, it has drawn the attention of a number of foreign scientists too. The Dharwar Province is characterised by the same trinity that characterises most other Archean terrains in the world (Fig. 1.1.2). It comprises the following:

1. A middle Archean tonalite-trondhjemite-granodiorite (TTG)-dominant basement ('Peninsular Gneiss'),
2. Two generations of Archean green stone belts ('schist belts' in local geological literature since the time of their earliest description), represented by the older Sargur Group and the younger Dharwar Supergroup rocks,
3. Late Archean granitoids with/without a mantle affiliation.

Radhakrishna and Vaidyanathan (1994) suggested that the generalised term 'Peninsular Gneiss' coined by Smeeth about a century ago (Smeeth, 1915) be abandoned and instead, 'older gneiss', be used for those older than 3.0 Ga and 'younger gneissic complex' for those formed at ~ 2.6 Ga. Chadwick et al. (1997) would, however, prefer retention of the term for the gneisses dated > 3.0 Ga. The opinion of Radhakrishna and Vaidyanathan has a valid logic. But the difficulty is that we do not as yet have enough age data, and the field geologists may not be posted with them. Therefore, for a general use, this term may still be retained. The deformational structures in Peninsular Gneisses and the supracrustal enclaves, as also in the adjacent schist belt rocks, may be similar in style and sequence in both Western and Eastern Dharwar. A few examples of such features from the Eastern Dharwar are shown in Fig. 1.1.3 (a) – (d). This prompted some workers to suggest that the Peninsular Gneiss evolved by synkinematic migmatisation of the protoliths (Bhaskar Rao et al., 1991).

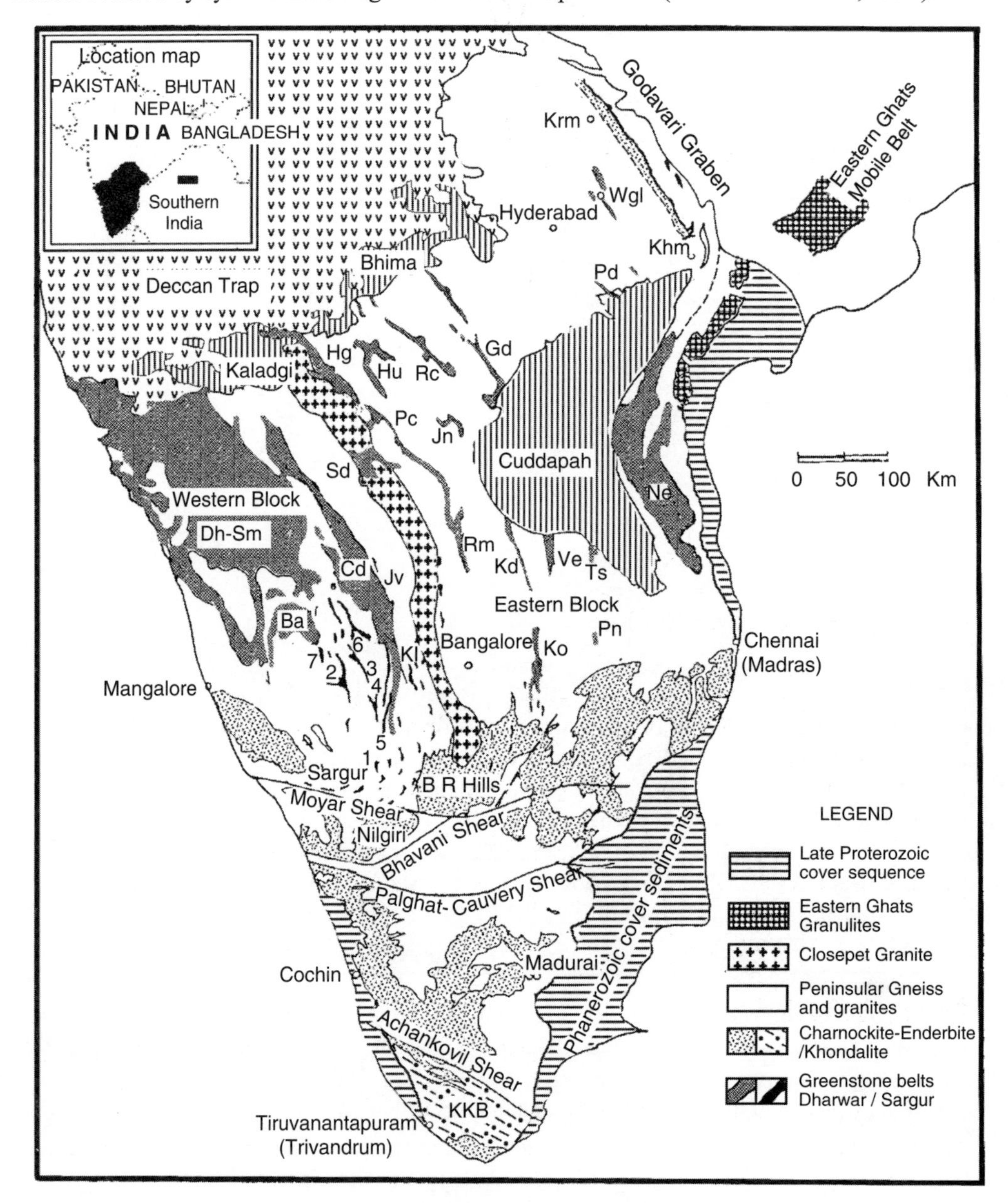

Fig. 1.1.1 Generalised geological map of South India (modified after Ramakrishnan, 1994). Greenstone belts: Western Block (Dharwar and Sargur): Dh-Sm Dharwar-Shimoga; Ba Bababudan; Cd Chitradurga; 1. Sargur; 2. Holenarsipur; 3. Nuggihalli; 4. Nagamangala; 5. K.R.Pet; 6. J.C.Pura; 7. Sigeguda
Eastern Block (Dharwar): Jv. Javanahalli; Kl. Kunigal; Sd. Sandur; Ko. Kolar; Rm. Ramagiri; Pc. Penakacherla; Hu. Hutti-Maski; Hg. Hungund; Kd. Kadiri; Jn. Jonnagiri; Rc. Raichur; Ml. Manglur; Ve. Veligallu; Pn. Penumur; Ts. Tsundapalle; Gd. Gadwal; Pd. Peddavur; Ne. Nellore belt; KKB. Kerala Khondalite belt; Krm. Karimnagar; Wrg. Warangal; Kh. Khammam

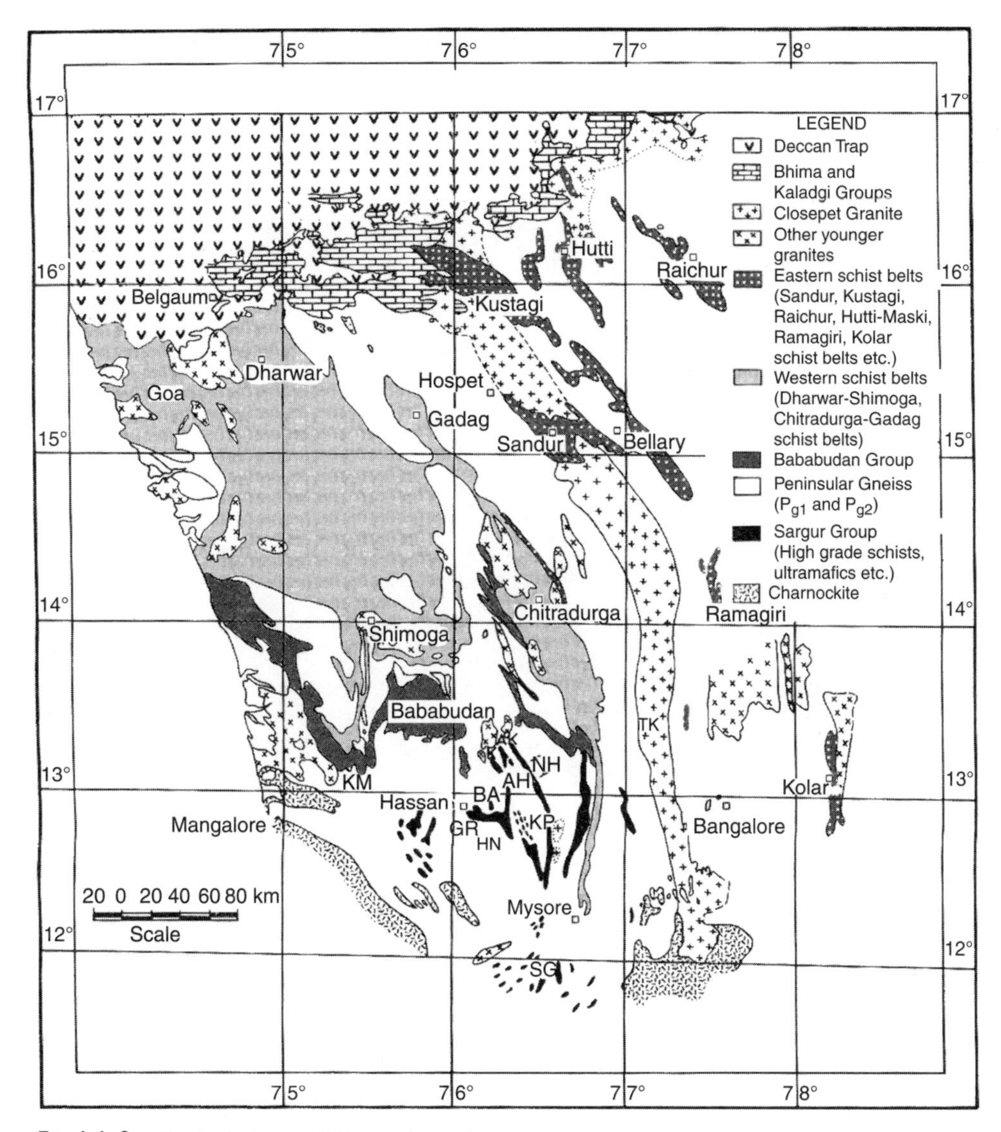

Fig. 1.1.2 Geological map of Karnataka region

Granulite metamorphism is a common feature of many Archean provinces, although all such high-grade metamorphisms need not have been Archean in age. The Dharwar Province is no exception to this generality. The high grade metamorphic rocks of the South Indian Granulite Province skirt the southern boundary of the Dharwar Province. The latter is rimmed by the Eastern Ghat Mobile Belt (EGMB) in the east, and is terminated by the Godavari Rift in the northeast. It has been covered by the Deccan Trap rocks in the north. As we shall see later, the rocks of the Southern Granulite Province differ from those of the Dharwar Province in petrology, structure-tectonics and geochronology.

There are a couple of Proterozoic intracratonic basins within the Dharwar Province. The Dharwar rocks, the schist belts in particular, are rich in some ore deposits. So also are the intracratonic basins, particularly the Cuddapah and the Bhima.

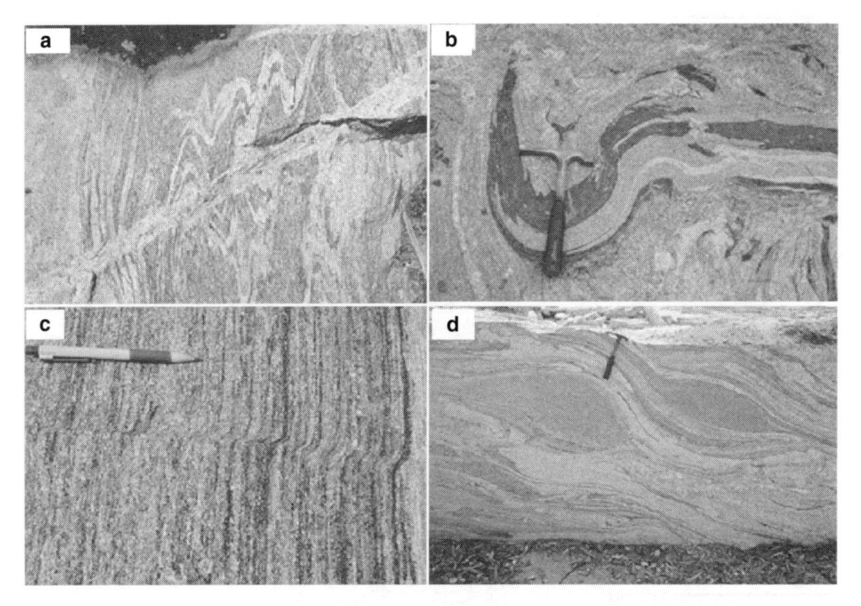

Fig. 1.1.3 (a) Intricate folding in leucosome bands in Peninsular Gneiss, west of Kolar. A later shear zone, filled up by quartzo-feldspathic material, cut across the folded bands. (b) Multiply deformed interbanded gneiss-amphibolite (dark) complex in Peninsular Gneiss, south of Bangalore. A geological hammer added to the natural sickle makes it interesting! (c) Mylonitic (proto-) banding in Peninsular Gneiss, north of Kunigal. (d) Swerving gneissosity around augen like mafic enclaves, NE of Bangalore (Photographs: courtesy V. V. Rao and A. K. Nath, GSI).

Although certain geological characteristics are common through the entire Dharwar Province, differences within it are nonetheless significant. Based on the latter, the Dharwar Province is now divided into two sub-provinces or blocks: the Western Dharwar and the Eastern Dharwar. The divide between them has been suggested to be running along a nearly N-S trending brittle-ductile shear zone that follows the eastern margin of the Gadag–Chitradurga schist belt (Swaminath et al., 1976: Chadwick et al., 2000) (Fig. 1.1.4).

Peninsular Gneiss and Granitoids of Western Dharwar

There have been appreciable studies on the Peninsular Gneiss and the associated granitoids, the salient aspects of which are discussed here.

Gneisses of the Gorur – Hassan – Holenarsipur Area Banded grey gneisses, varying in composition from tonalite to trondhjemite and granodiorite dominate. Patches and bands of metabasic rocks are common. The rocks have been metamorphosed in the amphibolite facies. The rocks may locally be migmatitic with a high K_2O/Na_2O ratio, as in the Hassan district (Naqvi et al., 1983).

The Gorur Gneisses are high-SiO_2 (75–80 wt%), low-Al_2O_3 (commonly 12–13 wt%), and REE-enriched trondhjemite with poorly fractionated REE and negative Eu-anomalies (Bhaskar Rao et al., 1991). Hassan gneiss is chemically somewhat varied.

The Rb-Sr (WR) isochron age of Gorur Gneisses, obtained by Beckinsale et al., (1980), is 3358 ± 66 Ma. Pb-Pb isotope age obtained from feldspar in a gneiss from near Holenarsipur is 3.4 Ga and the content of radiogenic Pb and Sr in the rock led Meen et al. (1992) to suggest that the older crustal protoliths of those gneisses were accreted at or even before 3.8 Ga. Some of the early gneisses at

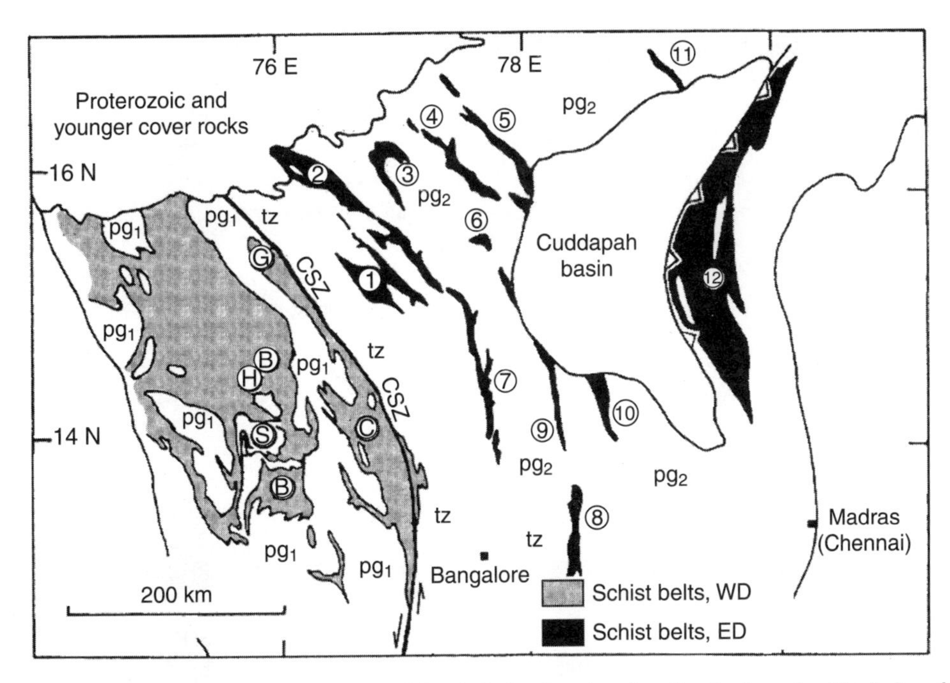

Fig. 1.1.4 Geological map of a part of South India showing the distribution of schist belts of Western Dharwar (WD) and Eastern Dharwar (ED) Blocks (after Chadwick et al., 1996). Legend: pg_1–older Peninsular Gneiss, pg_2 –younger Peninsular Gneiss ('Dharwar batholith'), tz–transition zone between Dharwar and SGP. Supposed interface between WD and ED marked by a brittle-ductile shear zone (CSZ). (B) Bababudan, (C) Chitradurga belt, (G) Gadag belt, (H) Honnali dome, (R) Ranibennur, (S) Shimoga; (1) Sandur, (2) Kustagi, (3) Hutti, (4) Raichur, (5) Gadwal, (6) Jonnagiri, (7) Ramagiri-Penakacherla, (8) Kolar, (9) Kadiri, (10) Veligallu, (11) Peddavur, (12) Nellore

Kabbaldurga also provided a U-Pb concordia upper intercept age of ~3.4 Ga (Buhl, 1987). Hassan gneiss with low Sr_i yielded Rb-Sr (WR) ages also varying from 3.16 Ga to 3.07 Ga (Monrad, 1983). Notable also at the same time is that a U-Pb SHRIMP age of zircon from the gneiss of Kabbaldurga gave an age of 2965 ± 5 Ma (Friend and Nutman, 1992).

Gundlupet Gneisses and Granitoid These rocks occur at the southern end of the high grade Sargur, or older schist belt. The gneisses have two facies: the earlier one is tonalitic and the later one is trondjhemitic. Migmatisation with attendant rise in K_2O/Na_2O ratio is a common feature. REE-distribution shows a highly fractionated (La/Yb = 30–40) pattern. U-Pb (zircon) age obtained by Buhl (1987) is 3.3 Ga. Rb-Sr (WR) age obtained by Janardhan and Vidal (1982), on the other hand, is 2.85 Ga. Granitic plutons intruded the migmatised gneiss at places. It has a high REE content and moderately fractionated REE pattern (Shadaksara Swami et al., 1995).

Honnali Gneiss and Granitoid Structurally they make a composite dome of migmatite and a granitoid rock (Bellur Granite). Both are tonalitic in character. A gneissic sample yielded an age of 3260 Ma (Rb-Sr/WR) (Crawford, 1969).

Anmod Ghat Gneisses The Anmod Ghat Gneisses that form the basement of the Goa Group of the Archean supracrustals are strongly deformed trondhjemitic gneisses with low REE and depleted HREE. The Rb-Sr (WR) isochron age obtained for a sample of these gneisses is 3400 ± 140 Ma (Dhoundial et al., 1987).

Halekote – Kuncha – Tipur and Belur plutons Several trondhjemitic plutons, generally massive with little or no thermotectonic modification, are reported from the region. The Tipur body yielded Rb-Sr (WR) age of 3296 ± 158 Ma, but a composite isochron for the plutons yield Rb-Sr (WR) age of 3095 ± 82 Ma (Rogers and Callahan, 1989).

Chikmagalur Gneisses and 'Granite' The gneisses in the area are tonalite–trondhjemite in composition, whereas the Chikmagalur 'Granite' is in fact a granodiorite. Unlike the gneisses, the granodiorite is massive and shows no deformation structure, except locally (Taylor et al., 1984). The Rb-Sr (WR) analysis of the granodiorite yielded a well-fitted isochron age of 3083 ± 110 Ma with a Sr_i of 0.7013. Pb-isotope analysis of the same samples also yielded a well-fitted $^{207}Pb/^{206}Pb$ isochron age of 3175 ± 45 Ma. Ages obtained for the gneisses are, however, poorly constrained. Single zircon Kober evaporation $^{207}Pb/^{206}Pb$ data obtained for the granodioritic rocks of Chikmagalur, Holenarsipur and Bangalore area are in the range of 3.3–3.2 Ga (Peucat et al., 1993; Jayananda and Peucat, 1996).

Chitradurga Gneisses and Granite The Chitradurga Gneisses are well-exposed, west of the Chitradurga supracrustal belt. They form the basement to the Dharwar Supergroup at the Chitradurga area. In character, they may be banded or homogenous grey gneisses and locally, even migmatites. The gneisses are tonalite to granodiorite and the 'granite' is granodiorite to typical granite, i.e. granite *sensu stricto*. A well-fitted $^{207}Pb/^{206}Pb$ model age ($\mu = 7.62$) obtained for a gneissic rock is 3028 ± 28 Ma. A similar study with the Chitradurga Granite yielded a date of 2605 ± 18 Ma ($\mu = 7.68$) (Taylor et al., 1984).

'Fertile Granite' and Associated Pegmatites of Allapatna – Marlagalla – Chandra Area Pegmatites, rare metal bearing or barren, occur in association with granitoids at places of Holenarsipur, Krishnarajpet and Nagamangala schist belts. These granitoids particularly studied in the Allapatna – Marlagalla – Chandra region, are 2-mica peraluminous, high-K and low-Ca type, and enriched in Rb, Ga, Cs, Nb, Ta, U and Pb. These are some of the chemical characteristics of fertility in the context of production of rare metal ores and hence the attribute 'fertile'.

A composite Rb-Sr isochron age obtained with samples from Allapatna, Pandavpur and Hassanpura is 3098 ± 75 Ma (Krishnamurthy and Sarbajna, 1999).

Kanara Batholith The Kanara batholith is a composite massif, consisting of granodiorite, adamellite and granite, together with the late intrusives of pegmatite and aplite. The obtained Rb-Sr (WR) isochron age of the massif is 2669 ± 60 Ma, with a $Sr_i = 0.7056 \pm 0.0015$ (Balasubrahmanyan, 1978).

Arsikere – Banavara Granite The Arsikere–Banavara 'Granite' is adamellite to granodiorite in composition, meta- to perluminous in nature, with high Zr, Nb, V, Ti, REE and Rb/Sr ratio (Jahangiri et al., 2000). REE distribution pattern is moderately fractionated, with a negative Eu anomaly. The Arsikere Granite analysed to a Rb-Sr (WR) age of 2563 Ma, with $Sr_i = 0.7021$ (Rogers, 1988).

Londa Gneiss In the Londa area, north Kanara district, Karnataka, tonalite-diorite paleosomes occur in granitic neosomes. Migmatitic features are obvious. Rb-Sr (WR) isochron age of the neosome is 2650 ± 130 Ma (Dhoundial et al., 1987).

Melukote – Katteri Gneisses Relics of sedimentary protoliths in the Peninsular Gneiss around Melukote–Katteri area, Karnataka, have been established on the basis of field, petrological and geochemical observations. The Melukote Gneisses are not interlayered with carbonates, quartzites, BIF and para-ultramafic xenoliths that are so common elsewhere, i.e. in the TTG varieties. Instead, these gneisses contain kyanite, staurolite and garnet. HREE pattern and very low negative Eu anomalies distinguish these gneisses from the TTG in the neighbourhood (Naqvi et al., 1983; Udai Raj and Naqvi, 1995).

Summary The gneisses and 'granites' of Western Dharwar may be divided into the following groups based on their characteristics outlined in the previous pages.

1. Banded and foliated grey gneisses generally of TTG composition. They constitute the bulk of the Peninsular Gneiss.
2. Granodiorite–granite plutons, particularly occurring at places like Allapatna–Chandra, Arshikere–Banavara and Chitradurga.
3. Paragneisses, such as of Melukote – Katteri area.

The members of Group–1 vary in age from ~3.4 Ga to 3.0 Ga. They are generally, but not invariably, REE-enriched with varying degree of REE fractionation. Available data indicate negative Eu-anomalies. Depletion of Rb is noted in some old members. Meen et al. (1992) on the basis of radiogenic Pb and Sr isotopic composition of some of the oldest members of this group, suggested that they were derived from other crustal protoliths that possibly accreted ≥ 3.8 Ga ago. The ~ 3.0 Ga event of the magmatic activity, particularly in the central part, locally introduced small trondhjemitic diapirs formed by melting of the older crust. Metabasic enclaves and locally even interbanding of gneisses and metabasic rocks, are not rare. The gneisses with their enclaves have been metamorphosed in the amphibolite facies, and are isostructural with the neighbouring schist belt rocks.

The Group–2 rocks vary in age from ~3.0 Ga (Allapatna–Chandra) to 2.6–2.5 Ga (Chitradurga, Arsikere–Banavara). They are intrusive into the gneisses and schists. Chemically they are characterised by $K_2O/Na_2O \geq 1$, and are enriched in incompatible elements. Arsikere Granite has a highly fractionated REE pattern, with a negative Eu-anomaly.

Group–3 gneisses, reported only from a very few areas, contain relics of metasediments and no mafic–ultramafic enclaves. Enriched HREE patterns and very low negative Eu-anomalies are characteristics of these gneisses. This may be taken to reflect the REE geochemistry of their provenance as REE do not show much fractionation during weathering and sedimentation (McLennan, 1988).

Western Dharwar Schist Belts

The schist belts or the greenstone belts of western Dharwar block collectively occupy a fairly large area. They vary in size, shape, lithologic composition, and in age. The older schist belts constitute the Sargur Group and the younger, the Dharwar Supergroup. All workers do not agree to this straight-jacket classification, as there are some cases where the characteristics are not tell-tale. Even then in the given situation, this division seems reasonably well-founded and useful in describing the geology of the region.

Sargur Group The Sargur Group of rocks are now better exposed in Sargur, Holenarsipur, Nuggihalli and Krishnarajpet areas as well as at other places south of Mysore (Fig. 1.1.2). In the Sargur-Dodhkanya area, pelites are alumina-rich, and produced kyanite, sillimanite, K-feldspar ± corundum (± graphite) on metamorphism. At Gundlupet –Terakanambi, the Al-rich minerals are absent. Quartzite contains zircon and may be fuchsite-bearing. Some, showing $\delta^{18}O \sim +12.5$ per mil, could be of volcanic origin (Naqvi, 1981). The banded iron formation (BIF) consists of alternate layers of quartz and magnetite, commonly with minor orthopyroxene and ortho-amphibole ± garnet. ΣREE and LREE/HREE in the BIF tally with those of the Archean BIFs. Mn-content in the manganiferous horizons ranges 7–19 wt%, and the metal is accommodated mainly in the spessertine garnet. Bedded barite is locally present. The amphibolites associated with the metasediments are oceanic tholeiite. Felsic volcanic rocks are present but not in profusion. The chondrite-normalised REE-distribution in the pelitic sediments of the Sargur Group suggests a silicic-mafic provenance (Janardhan et al., 1990). The grade of metamorphism increases from greenschist facies in the north to granulite facies in the south. Dunite, peridotite, pyroxenite and gabbro–anorthosite bodies are intrusive into the Sargur Group.

Radiometric determinations of age of the Sargur rocks are very few and variable, depending on the stratigraphic positions of the samples chosen, the precision of the methods used and geologic histories of the samples analysed. For example, the detrital zircon, separated from the Sargur Group metapelites from Holenarsipur and Banavara, produced ages in the range of 3.58–3.13 Ga on analysis by U-Pb SHRIMP method (Nutman et al., 1992). Zircon from the Holenarsipur belt, in contrast yielded a U-Pb SHRIMP age of 3.3 Ga (Peucat et al., 1995). The oldest age from the detrital zircon corresponds to the age of the source rock(s). The 3.3 Ga age, obtained from rhyolite should, therefore, be more representative of the Sargur Group itself.

Dharwar Supergroup The Dharwar Supergroup is younger, much more extensive and voluminous and therefore, constitutes much larger 'schist belts' (Figs. 1.1.2 and 1.1.5). The relative age of the Dharwar Supergroup with respect to the Sargur Group has been determined from field observations at a number of places, such as at the south of Bababudan, north of Sigegudda, west of Chitradurga and on both sides of the Kibbanhalli arm of the Chitradurga schist belt (Chadwick et al., 1981; Viswanatha et al., 1982; Ramakrishna and Viswanatha, 1987; Venkata Dasu et al., 1991). The rocks constituting this Supergroup are metapelites, greywackes, metamorphosed basic volcanics, BIF, quartzites, acid volcanic rocks and small volumes of carbonate rocks.

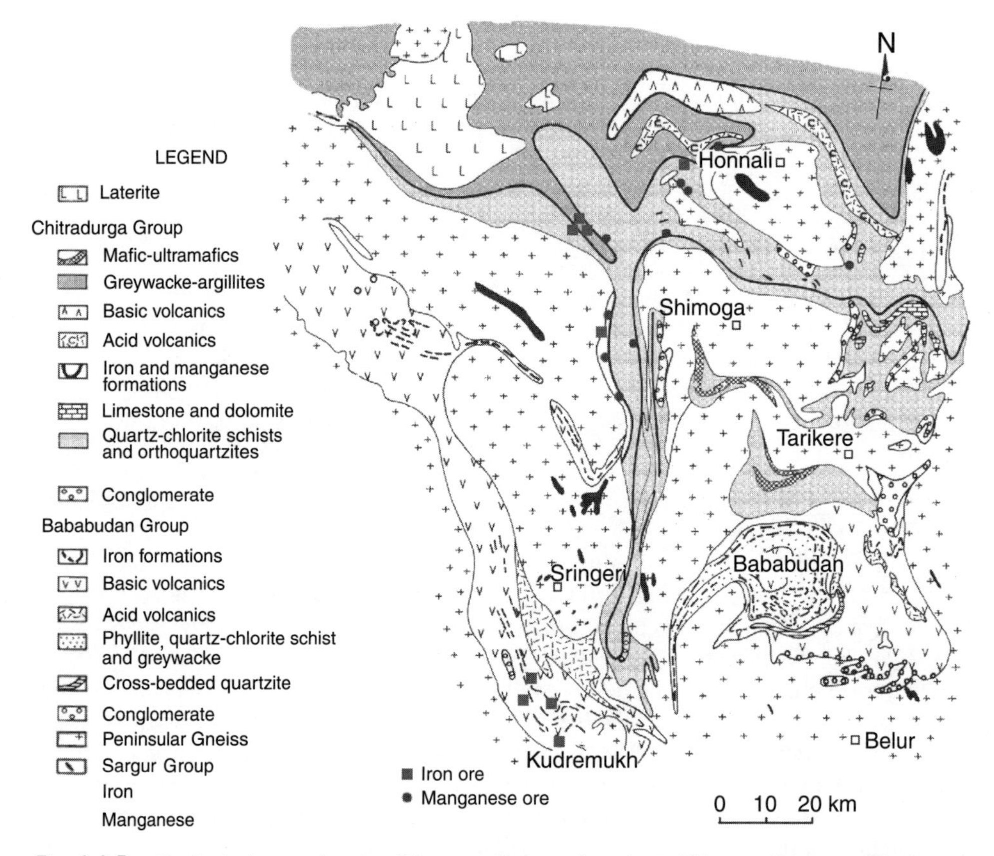

Fig. 1.1.5 Geological map showing Dharwar Group of rocks in Western Dharwar Block and important ore deposits contained by them (Sarkar, 1988)

The Dharwar Supergroup is divided into two Groups: the lower Bababudan Group and the upper Chitradurga Group. Mafic volcanics from Bababudan yielded a Sm-Nd (WR) isochron age of 3.03 ± 0.23 Ga (Drury, 1983) and Rb-Sr (WR) isochron age of 2.58 ± 0.27 Ga (Bhaskar Rao et al., 1992). Trendall et al. (1997a) determined the SHRIMP U-Pb age for zircon from the tuffitic rock, at the base of the Mulaingiri Formation of the Bababudan Group to be ~2.72 Ga. Nutman et al. (1996) obtained a SHRIMP U-Pb date of zircon from acid volcanic rocks from the lower Chitradurga Group, to be ~2.61 Ga. A similar age (2.6 Ga) was obtained by Trendall et al., (1997b) for the same rock using the same analytical method.

Bababudan Group The Bababudan Group was divided into four Formations in the type area (Viswanatha and Ramakrishnan, 1981). These are, from bottom to top, Kalaspura, Allampur, Santaveri and Mulaingiri. Later, Chadwick et al. (1985) subdivided the original Mulaingiri Formation into Mulainigri, Jagar and Mundre Formations as shown in Table 1.1.1 and Fig. 1.1.6. Each of the formations starts with a well-sorted quartzite with preserved shallow water structures. The sequence at the Bababudan belt started with the deposition of a quartz-pebble conglomerate (QPC) that reportedly contains detrital pyrite, uraninite and gold in sub-economic proportions.

Table 1.1.1 *Lithostratigraphy of the Bababudan Group at Bababudan (after Chadwick et al., 1985)*

Formation	Important lithology
Mundre	Kaldurga Conglomerate, quartzite and phyllite
Jagar	Metavolcanic rocks, granite and phyllite
Mulaingiri	BIF, quartzites and phyllites
Santaveri	Metabasalts, ultramafics, quartzites and phyllites
Allampur	Metabasalts, schists and quartzites
Kalaspura	QPC, phyllites, schists and metagabbro
	-----------------*Unconformity*---------------
	Peninsular Gneiss, Sargur Group

Bababudan basalts are generally tholeiitic. Drury (1983a) also reported komatiitic rocks from the Kudremukh area.

Banded Iron Formation (BIF)[1] The iron formation of Bababudan, which is important from both the scientific and economic point of view, requires discussion in some details. It contains the thickest, and the most extensive Banded Iron Formation in the Dharwar shield. The BIF of the Bababudan schist belt overlies the platformal sedimentary facies of conglomerate, quartzite etc. However, it was deposited in a shallow shelf condition, but below the wave base, a condition that ensured its preservation. The BIF in the region is distinguished from that of Sandur, Chitradurga and Shimoga belts by the absence of carbonate, trace fossils like stromatolites, and Fe-Mn formations. The iron formations of Bababudan are divided into two broad types: the cherty banded iron formation and the shaly banded iron formation, the former being dominant. The former on amphibolite facies metamorphism produced quartz, magnetite (surface samples variously oxidised) as the principal phases, together with grunerite, Mg-riebeckite, hornblende, actinolite and stilpnomelane as minor phases. In the shaly variety green to brown biotite, amphiboles, carbonates and talc are the minerals

[1] In this treatment we have considered Banded Iron Formation (BIF) as a distinct litho-unit and hence its discussion has been included in Part 1 of the Chapter. Related ores have been included in the, second part (Metallogeny). Moreover, general aspects of a problem once discussed are not repeated.

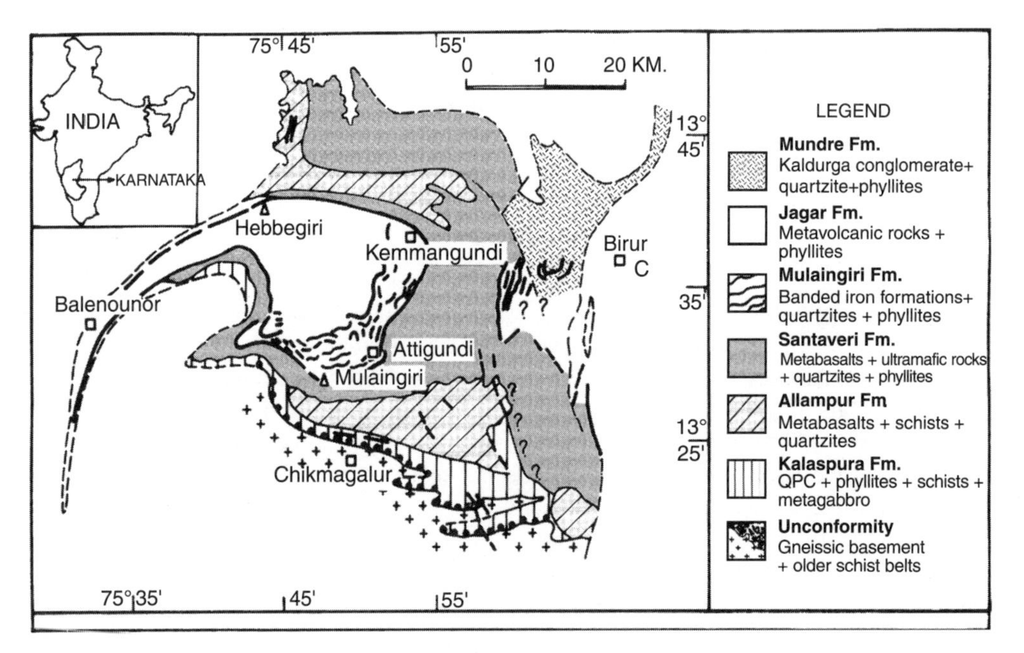

Fig. 1.1.6 Detailed geological map of Bababudan area showing the distribution of its different formations, and their relationship with the gneissic basement and earlier schist belts (modified after Arora et al., 1995)

next in abundance to quartz and magnetite. The once reported mineral 'bababudanite' from these rocks has ultimately been proven to be magnesio-riebeckite (Narayana et al., 1974). Several parallel amositic asbestos seams occur locally, and are even mined. In a crop of 45 samples of Bababudan BIF, SiO_2 ranges 32.39–87.17% (wt), Fe_2O_3 2.45–63.80%, Al_2O_3 0.14–11.24%, MgO 0.91–13.20%, CaO 0.13–22.64%, Na_2O 0.03–1.83%, K_2O 0.01–7.30% (Arora et al., 1995). Bands are generally measurable in terms of centimeters and millimeters.

REE-behaviour of the Bababudan BIF (Fig. 1.1.7) may be discussed in the above context. Arora et al., (1995) divided these BIFs into four groups on the basis of their REE-patterns. 15 samples of low-Mg cherty BIFs are characterised by depleted ΣREE (2.57–19.02 ppm), sloping LREE patterns, positive Eu-anomalies (some weak) and flat HREE. High-Mg cherty BIFs (19 samples) are nearly similar to the above in REE distribution patterns. The shaly BIFs have moderate to enriched ΣREE (9.36–107.41 ppm), sloping REE patterns and positive or negative Eu -anomalies. It may be mentioned here that the shales and the shaly BIFs, now metamorphosed, are interbedded with cherty BIF, and that explains the REE patterns in the shaly BIFs. There appears to have been mixing of Fe-Si-bearing hydrothermal fluids and continent-derived constituents carried by fluvial fluxes. Negative anomaly is a common feature of shales of all geological ages. Recently, Kato et al., (2002) made a study of the content, and pattern of distribution of REE in some Bababudan BIFs. They apparently missed the shaly variety of the BIFs, and reported low ΣREE (5.2–65.3 ppm), a very large negative Ce-anomaly (Ce/Ce* = 0.13–0.83) and a positive Eu-anomaly (Eu/Eu* = 0.96–2.45).

Clastic sediments, volcanic rocks and chemical sediments like the BIFs occur at different levels in the stratigraphic sequence, suggesting fluctuation in geological conditions of the basin. Volcanic activities took place several times; and between the breaks took place deposition of sediments, clastic or chemical (mainly BIF) interbanding of cherty BIF, shales, as reported from the Transvaal

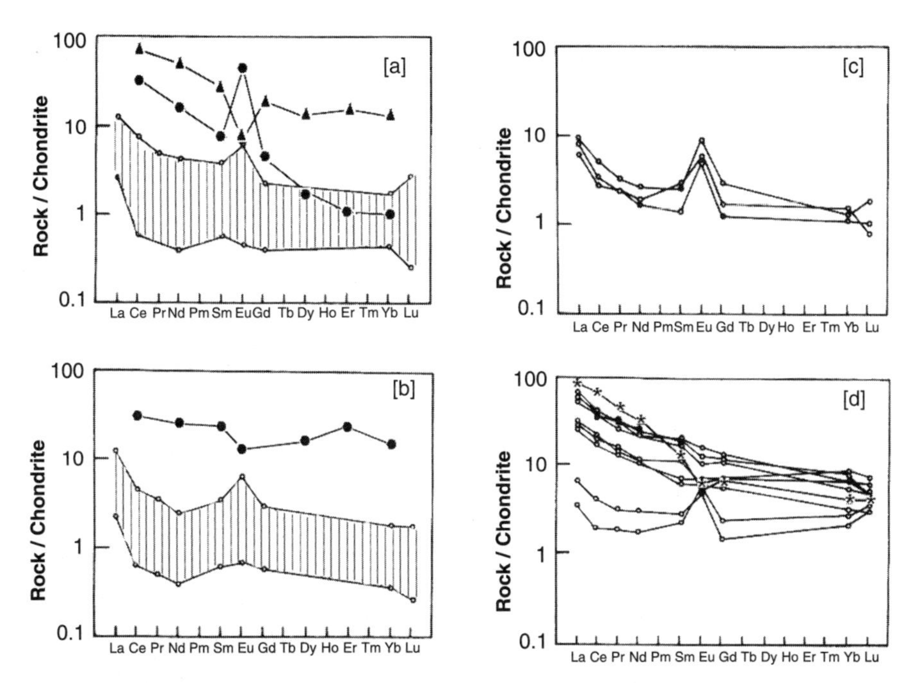

Fig. 1.1.7 (A) Range of REE patterns of low Al_2O_3-MgO cherty BIFs (with +ve Eu anomalies) and its similarity with hydrothermal fluids; closed triangles: average dissolved load in river water, closed circles: mid-ocean ridge hydrothermal fluid. (b) Range of REE pattern of cherty BIFs with low Al_2O_3 but moderately enriched MgO (with +ve Eu anomalies and La-enrichment in REE-depleted samples; non-matching REE pattern (closed circles) of average sea water for comparison. (c) REE pattern of three cherty BIFs of high Na_2O content with strong +ve Eu anomalies. (d) Eight samples of shaly BIFs (open circles) with high Al_2O_3 (~ 10%), MgO, Na_2O and K_2O show absolutely different REE-pattern and abundances; REE pattern of NASC (stars) is superposed for comparison and to illustrate influence of terrigenous components (after Arora et al., 1995).

Supergroup (Klein and Beukes, 1989). Expectedly, during transgression chemical precipitation of rock types like BIF will take place at shallow levels, of course, below the wave base, as both the wave base and the photic zone increased. During the succeeding regression these chemical sediments would be covered by clastic sediments. Such a situation could have prevailed during the evolution of the Bababudan basin. As mentioned above, Kato et al., (2002) reported a large negative Ce-anomaly from the Bababudan BIFs. With the help of this observation, they suggest that an oxidised deep-sea environment prevailed during the precipitation of those BIFs.

Depending on the parameters assumed (utilising the geochemistry of the volcanic rocks), more than one tectonic setting for the deposition of Bababudan Group of rocks may be suggested. This may appear confusing at the first sight. However, a plausible synthesis of the diverse pictures is offered by Arora et al., (1995). They suggest that the Bababudan basin apparently opened as an intracratonic or pericratonic rift, and at this initial stage WPB type volcanic rocks erupted. With time, the basin thus formed widened and deepened, ultimately developing a mid-oceanic ridge. At this stage, MORB type volcanic rocks were extruded, followed by letting out of FeO-SiO_2-bearing hydrothermal flux that ultimately precipitated BIFs. During inversion, the tectonic regime changed, depositing turbidites including polymictic conglomerates at active plate margin, and extruding island-arc type tholeiite, andesite and rhyolite.

Definition and Classification of BIF It may be worthwhile to discuss here the definition, types and classification of the BIFs as also their genetic aspects in general, which would be relevant to such rocks irrespective of their areas of occurrence in the country as well as globally. The BIFs may be defined as thinly laminated or thickly banded rocks made up of layers dominated by iron oxides (hematite or magnetite)/iron sulphides (pyrite-pyrrhotite)/iron carbonates (siderite), alternating with quartz/jasper/chert/shales/wackes. Our present attention would be on the hematitic BIFs (BHQ and BHJ) and magnetitic BIFs (BMQ), which are found closely associated with the iron ores of economic significance. The discussions below are largely restricted to this type of BIF.

Classification of BIFs by Gross (1965) into two broad types, viz, Algoma and the Lake Superior types, was fascinating to many workers on BIF and the related iron ore deposits. The Algoma type BIFs are limited to the Archean and occur in greenstone belts. They have low Fe-content and smaller reserves. The Lake Superior type, on the other hand, would be 'oolitic to granular', extensive over strike-distances of up to several hundred kilometers. Their tectonic setting is generally stable continental platform, the typical stratigraphic sequence of which consists of quartzite, black carbonaceous shales, conglomerate, massive chert and argillite. The bulk of this type of BIF was formed in the Early Proterozoic (2500–1900 Ma). There are some countries including India, however, where such deposits are of Archean age. The Late Proterozoic variety, commonly known as the Rapitan type, is not very common, though locally, as at Urucum, Brazil, this type of BIF gave rise to a super-giant iron ore deposit (Klein and Ladeira, 2004). In reality, however, some BIFs and their related iron ores do not fully conform to this division, and hence differently interpreted. For example, while Gross (1973, 1980) included the Hamersley deposit of Australia under the Superior type, Dimroth (1975) cited the same deposit as an example of the Algoma type. Kimberly (1978) and Barley et al., (1977) pointed out the presence of tangible volumes of basic rocks at Hamersley. The presence of substantial volumes of basic rocks in platforms is not unnatural. Moreover, the so-called Superior type deposits are not necessarily oolitic or granular, as suggested by Gross.

Genesis of BIF Of late, there has been definite progress in our understanding of the genetic aspects of the BIFs through recent researches. However, many basic problems in this context still remain to be satisfactorily solved, such as

1. source of Fe and Si in the BIFs,
2. transport of Fe and Si,
3. repetitive precipitation of Fe-rich and Si-rich layers, and
4. BIF-deposition in finite time slots in the Earth-history.

The suggestion of continental source of iron found favour with many workers for a long time (Cloud, 1973; Drever, 1974; Holland, 1973, 1984). Holland estimated that the annual continental contribution of particulate iron to the oceans is of the order of 10^{15} grams. If $<1\%$ of this volume of iron is mobilised and precipitated as the Fe-mineral in the BIFs, then the problem about the source of iron is solved. But that would require a large production of organic matter globally to mobilise the required amount of iron. This does not appear to be a feasible proposition for that part of the Earth's history (Isley, 1994). Rather on the basis of REE contents of BIFs and relevant research at the laboratory, it is now commonly believed that the bulk of Fe (and also Mn and SiO_2) was obtained through sub-sea floor hydrothermal leaching and possibly also exhalation. The $(Eu/Sm)_{CN} > 1$, similar to what is obtained from the present day volcanic hydrothermal system, is accepted as a corroboration of the above contention (Bau and Moller, 1993; Kato et al., 1998, 2002). Jacobson and Pimental–Klose (1988) reported Fe/Nd ratios of the order 10^5 from the Archean-Paleoproterozoic BIFs, similar to what is obtained from the modern hydrothermal systems. In the early Earth the sea water temperature might have been significantly higher, perhaps as high as 75 °C (Ohmoto and Felder, 1987), i.e., a low temperature hydrothermal system. The above discussion notwithstanding, a possibility remains that there was partial terrestrial contribution to the system through the existent drainage system, without grossly affecting its hydrothermal character (cf. Canfield, 1998).

Iron in the early oceans must have been accumulating as Fe^{2+} in a state of dissolution; and for precipitation into oxidic minerals, it had to be oxidised. But in a generally anoxic Archean–Paleoproterozoic atmosphere, interactions with a similarly anoxic sea could not easily supply the oxygen needed. Preston Cloud (1972, 1973) came forward with a marvellous suggestion to solve this problem. He suggested that photosynthesis by anaerobic bacteria could solve the problem.

$$CO_2 + H_2O\ (+\text{light}) \rightarrow CH_2O + O_2$$

The biota that caused the above reaction did not have O^{2-} mediating enzymes; and so the product oxygen was lethal to these early lives, unless removed immediately from the system. This chemical scavenger was Fe^{2+}, which reacted with this newly produced oxygen and precipitated in the form of ferric oxide or rather oxyhydroxide. The process was prolific during the period 2500–1900 Ma. With time, however, appeared other models in the literature that contested Cloud's.

Widel et al. (1993), on the other hand, suggested a reduction model in which Fe^{2+} oxidised to Fe^{3+}, through reduction of the ambient CO_2 to organic molecules.

$$4Fe^{2+} + CO_2 + 11H_2O \rightarrow 4Fe\ (OH)_3 + CH_2O + 8H^+$$

This model is theoretically sound and could explain the deposition of BIF at least before the generation of oxygen by photosynthesis by bacteria. The scale of operation of the process will, however, remain a question.

Braterman et al. (1983), based on the result of experimental studies suggested that photo-oxidation was possibly a major factor in the precipitation of BIF before the development of the ozone layer (penecontemporaneous with oxyatmoversion). In anoxic water and at pH > 6.5, Fe^{2+} hydrolysed to $Fe(OH)^+$, absorbed ultraviolet rays and some visual radiations, and oxidised to Fe^{3+}. The precipitated phase was FeO(OH). This seems to be another plausible proposition for the development of BIF, when photosynthesis by anaerobic bacteria was either absent or weak (Anbar and Holland, 1992).

Effects of bacterial activity in those remote days have been viewed as important in the genesis of BIFs by some recent workers (Konhauser et al., 2002) also. But their model differs from that of Cloud (op. cit.) in that here the bacteria is given the credit of directly generating ferric iron either by chemolithoautotropic iron oxidation, or by photoferrotrophy. With the help of chemical analysis of the BIF from the Hamersley Group, Australia, they have tried to show that even during the period of maximum iron precipitation, most if not all, of the iron in the BIFs could be precipitated by iron-oxidising bacteria in cell densities much lower than those found in modern Fe-rich aqueous environments. Phosphorous could be an easily available nutrient for these bacteria. Trace metals V, Mn, Co, Zn and Mo, found within some iron-rich bands, could also act as nutrients for the biota.

Oxidation of ocean water, is more likely to be common in a zone just close to the surface. If Fe^{2+} can be transported to this zone from the deeper anoxic part of the ocean, it will have a chance to oxidise to Fe^{3+}. This is now believed to be a plausible mechanism of Fe (III) precipitation when reduced iron-rich deep ocean water is up-welled on to the continental shelf. The initial precipitate, given the chemical parameters of the environment, was possibly $Fe(OH)_3$ (Fig. 1.1.8). It is needless to mention that the carbonate and sulphide facies iron precipitation would be controlled by high f_{CO_2} and high f_{H_2S} or f_{HS^-}, depending on the pH.

Postulations on redox states of Fe in nature in the geologic past and their fixation into large scale sedimentary iron formations do not stop here. Bekker et al., (2010) have recently suggested yet another one. They contend that atmospheric oxidation did not necessarily constrain the entire situation. Rather submarine volcanism was directly responsible for generating extensive ocean and basin-scale anoxea through venting of hydrothermal fluids containing significant volumes of H_2, H_2S, Fe(II) and Mn(II).

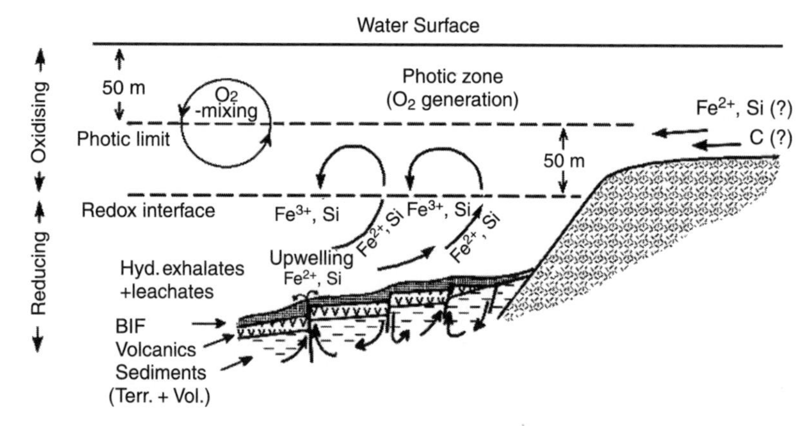

Fig. 1.1.8 A schematic diagram of probable geological situation of BIF precipitation in Indian deposits (from Sarkar, and Gupta, 2005 – adapted and modified from a generalised model of Klein and Beukes, 1989)

Magnetite is present in varying proportions; and as an early phase in oxide layers in the Archean and Paleoproterozoic BIFs throughout the world. It will be facile to interpret this fact, as has been done by some early workers, as indicative of a situation where magnetite was the first phase to precipitate from an aqueous solution. The stability-field of magnetite in the Eh-pH diagram of the Fe-O-C-S system at low T and P (Garrels and Christ, 1965) is so constrained that it will hardly be available in natural sedimentary environments except in a small scale (Fig. 1.1.9). It is probable that the initial precipitate, ferric oxy-hydroxide [$Fe(OH)_3$] is later reduced to magnetite ($FeO \cdot Fe_2O_3$).

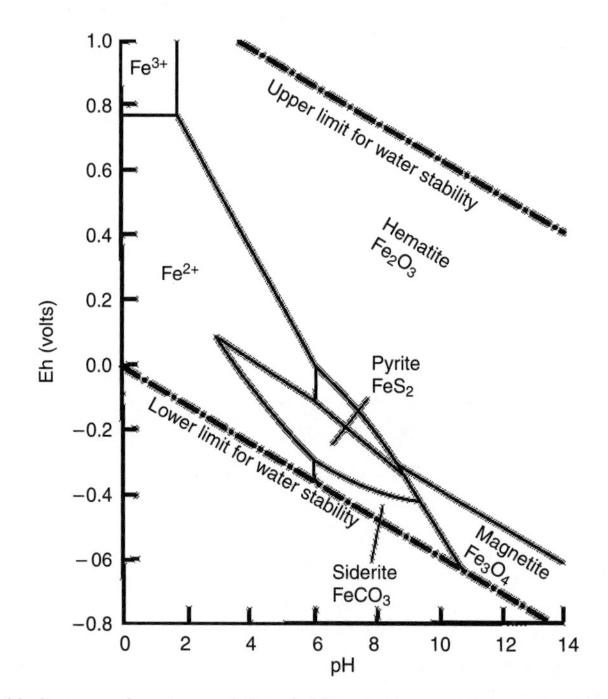

Fig. 1.1.9 Eh-pH diagram showing stability fields of common iron minerals at 25°C and 1b. Molarities of F, S and CO_2 are 10^{-6}, 10^{-6} and 1 respectively (from Sarkar and Gupta, 2005 – adapted from Garrels and Christ, 1965)

The generally preferred and commonly accepted explanation for this conversion is reduction in the sub-oxic environment during diagenesis, commonly in the presence of organic matter. Higher valency Fe, also Mn, if present, would be electron acceptors during the oxidation of organic matter. The varying proportion of martite and blady hematite that we commonly find in the Archean-Proterozoic BIFs are later developments due to atmospheric oxidation or action of oxidising meteoric hydrothermal fluids.

$Fe(OH)_3$	$\rightarrow$	Fe_2O_3	$\rightarrow$	$FeO \cdot Fe_2O_3$	$\rightarrow$	Fe_2O_3
$Fe^{3\pm}$-hydroxide	*Dehydration*	Ht	*Reduction*	Mt	*Oxidation*	Ht

[Ht – hematite, Mt – magnetite]

These processes must have been effective in precipitation of the earliest phase of iron and its subsequent evolution. However, this discussion will remain grossly incomplete without mentioning a recent interesting suggestion of Ohmoto (2003) on the hematite → magnetite conversion. The magnetite ↔ hematite that we briefly discussed above is a metal conservative redox reaction. Ohmoto (op. cit.) suggested that this conversion instead of being a redox reaction could be an acid-base reaction in which Fe^{2+} is added:

$$\underset{\text{Ht}}{Fe_2O_3} + Fe^{2+} + H_2O \rightarrow \underset{\text{Mt}}{FeO.\,Fe_2O_3} + 2H^+$$

According to this model transformation of hematite to magnetite, in many, if not all, BIFs took place during its growth, and not during metamorphism. Superior type BIFs have accumulated in brine pools, where ferric (hydro-) oxide grains nucleated at/near the interface between brine and the overlying O_2-bearing sea water. It settled down on the sea-floor and subsequently converted into magnetite during diagenesis by the above reaction, using Fe^{2+} in the brine pool. This model expects enough Fe^{2+} in the brine pool, as the reduction model outlined above, expects enough reductants to convert all the hematite grains into magnetite. How much of these expectations could be realised in the Archean–Paleoproterozoic time, is a matter of further study.

The development of banding in BIFs is another field of ambiguity in understanding its genesis and evolution. The suggested models in this respect range from diurnal cycles of day (promoting photo-oxidation) and night (nil photo-oxidation), through seasonal repetitions (summer time promoting organic growth, generating oxygen and depositing iron oxide, winter doing just the opposite), to differential flocculation of Fe and Si colloids as a function of changing water-chemistry and may be, some kind of bio-mediation (Robb, 2005).

The question, why the BIFs were destined to some time-slots in the Earth's history has been considered by earth-scientists for too long and yet there is no firm answer. A no-nonsense yet somewhat evasive explanation is that the BIFs were precipitated in profusion when the lithospheric, atmospheric and biospheric situations were favourable. Tectonically, most of the Superior type BIFs were deposited on stable continental shelves at times of marine transgression. From tectonic point of view, these transgressions or sea-level rises would be better explained by continental break up and increase in the volume of mid-ocean ridges (in case of the development of Rapitan type BIFs of Neoproterozoic age, this high sea-stand would be more reasonably attributed to interglacial snow-melting). Prolific growth of micro-biota, a favorable state of ocean-atmosphere evolution, and introduction of large volumes of $Fe^{2\pm}$-and Si-bearing hydrotherms in the Paleoproterozic period must have been conducive to the deposition of BIFs. For the deposition of the Archean BIFs, which are not insignificant in number of occurrence or total volume, models such as those of Widel et al. (1993) and Braterman et al. (op. cit.) assume considerable importance. However, mounting paleontological and isotopic evidence for oxygenic photosynthesis from 3.5 Ga (Schopf and Packer, 1987; Buick, 1992) may make abiotic models redundant, or at least less attractive one day.

Precambrian sea is likely to have been saturated with volcanogenic silica [Si $(OH)_4$] in the absence of silica-mediating organisms, providing a ready reservoir of silica and effecting its precipitation in the BIFs when its solubility limit was exceeded. We revert to the regional geology of South India.

Chitradurga Group The Chitradurga Group of rocks overlies those of the Bababudan Group and the Peninsular Gneiss. It is much more extensive than the Bababudan Group (Fig. 1.1.2), more sediment-rich and also thicker (~8 km thick towards the north). Of course, generally sediments are more prevalent than volcanic rocks in the schist belts of West Dharwar.

The Chitradurga Group at the type area, i.e. at Chitradurga region, is divided into Hiriyur Formation (top), Ingaldhal Formation and Vanivilas Formation (Shesadri et al., 1981). The Vanivilas Formation starts with polymictic conglomerate at Talya and oligomictic conglomerate west of Dodugni, chlorite-biotite schist, quartzite, limestone and dolomite, Mn-formation, banded ferruginous chert, followed again by chloritic schist. The Vanivilas Formation is overlain by the Ingaldhal Formation which is composed of basic, intermediate and acid lavas and pyroclastic rocks, interbanded with layers of ferruginous chert and polymictic conglomerate. A small copper deposit within this formation near Chitradurga (vide Chapter 1.2) was mined for some years in the recent past. The Chitradurga basalts are generally tholeiitic with rare instances of high-Mg varieties (Chaudhuri, 1984). The rocks of the Ingaldahl Formation are overlain by the greywacke-argillites (interspersed with minor volcanic flows and conglomerate) of the Hiriyur Formation. The Chitradurga Greywackes are enriched in Zn, Cr, Ni and suggest a mixed felsic-mafic source. Plotting in the QFL (Quartz-felspar-lithic particle) diagram locates the provenance of these rocks in the dissected arc and/or recycled orogenic history. CIA (Chemical Index of Alteration) values of 58–63 and Al_2O_3/Na_2O ratios of < 6, suggest a poorly weathered nature of the provenance and chemical immaturity of the sediments. The well-defined negative Eu anomaly suggests the same in the provenance in the upper crust (Srinivasan et al., 1989). This supports the contention of Gao and Wedepohl (1995) that a negative Eu-anomaly is a characteristic of the clastic sediments of all geological ages, and not confined only to the post-Archean sediments. Like most Archean greywackes, the Chitradurga Greywackes are also turbidites.

The Chitradurga Group in the Shimoga region is divided into four Formations: Ranibennur (top), Medur, Joldhal and Jhandimatti (Shesadri et al., 1981). Jhandimatti is the most widespread formation in the Shimoga belt. It consists of chlorite schist, quartzite, arkose, greywacke, volcanic rocks and polymictic conglomerate. The gold mineralisations at Honally, Kadur and Lakkavali are located within this formation. The shallow water sediments of Jhandimatti Formation gradually pass on to a predominantly chemogenic sedimentary unit consisting of limestone and dolomite, manganiferous and ferruginous cherts and phyllites and closely interleaved with thin sheets of volcanic rocks and chlorite schists. This is Joldhal Formation, and is important from the economic point of view, as the major manganese mines (and some of limestone) are located within this unit. The Joldhal and Jhandimatti Formations are together equivalent to Vanivilas Formation of Chitradurga belt (Babu et al., 1981). The Medur Formation is a thick pile of volcanic rocks such as quartz porphyry, rhyolite, andesite flows and tuff, pillowed andesite and basalt. Pyritic flows and graphitic cherts are locally present. This formation is correlatable with Ingaldahl Formation of the Chitradurga belt. Ranibennur Formation, correlatable with the Hiriyur Formation of the Chitradurga belt, is essentially a greywacke-argillite-chert-volcanic rock assemblage. The Upper Chitradurga greywackes, essentially turbidites, accumulated in an arc environment and the sediments were probably derived from a dissected arc provenance itself (Srinivasan et al., 1991).

The lithostratigraphy of the Chitradurga Group discussed above, is presented in a synoptic form in Table 1.1.2.

Table 1.1.2 *Lithostratigraphy of the Chitradurga Group*

A. Chitradurga belt (Seshadri et al., 1981)

Formation	Litho-units
Hiriyur	Greywacke – argillite association with mafic to intermediate mafic volcanics, banded ferruginous chert and polymictic conglomerates
	--------------*Disconformity*---------------
Ingaldahl	Mafic (dominant), intermediate and acid lavas and pyroclastic rocks with interbeds of banded pyritiferous chert and argillite
Vanivilas	Chloritic phyllite Banded ferruginous chert Banded manganiferous chert and phyllite Limestone and dolomite Chlorite-biotite ± garnet-bearing phyllite Quartzite and quartz schist Talya Conglomerate
	--------------- *Unconformity* --------------
	Bababudan Group

B. Shimoga belt (Babu et al., 1981)

Ranibennur (~Hiriyur)	Greywacke – argillite schist, volcanic rocks
Medur (~Ingaldahl)	Quartz porphyry, rhyolite-andesitic flows and tuffs, pillowed basalt and andesite
Joldahl	Limestone, dolomite, Mn-, Fe-bearing cherts and phyllites
Jhandimatti	Clastic sediments (dominated by greywacke), volcanic rocks and conglomerate
	---------------*Unconformity*-------------

Structure and Tectonics There is a similarity of the structural patterns in the different schist belts. Three generations of folds are recognisable in most parts of a schist belt and the regional attitudes of the first and second generation of folds are nearly the same in all the schist belts. In fact the complex fold patterns in the schist belts are the results of superposed folding, in many cases with high degree shortening (Mukhopadhyay and Baral, 1985; and Naha et al., 1995). Chadwick and his co-workers also favoured lateral compression as the cause of complex deformation structures in the region. They tried to explain it by subducting an easterly lithospheric plate (their 'Dharwar craton') below a westerly located lithospheric plate, i.e. the West Dharwar block in this discussion (Fig. 1.1.10). The nearly N-S transcurrent shearing (Fig. 1.1.11) could be a result of the oblique convergence (Chadwick et al., 1996, 1997, 2000).

There is another school of thought regarding the development of these structures. Bouhallier et al. (1993) and Chardon et al. (1996) preferred a solid-state diapir model to explain the so-called 'dome and basin' structures of Western Dharwar.

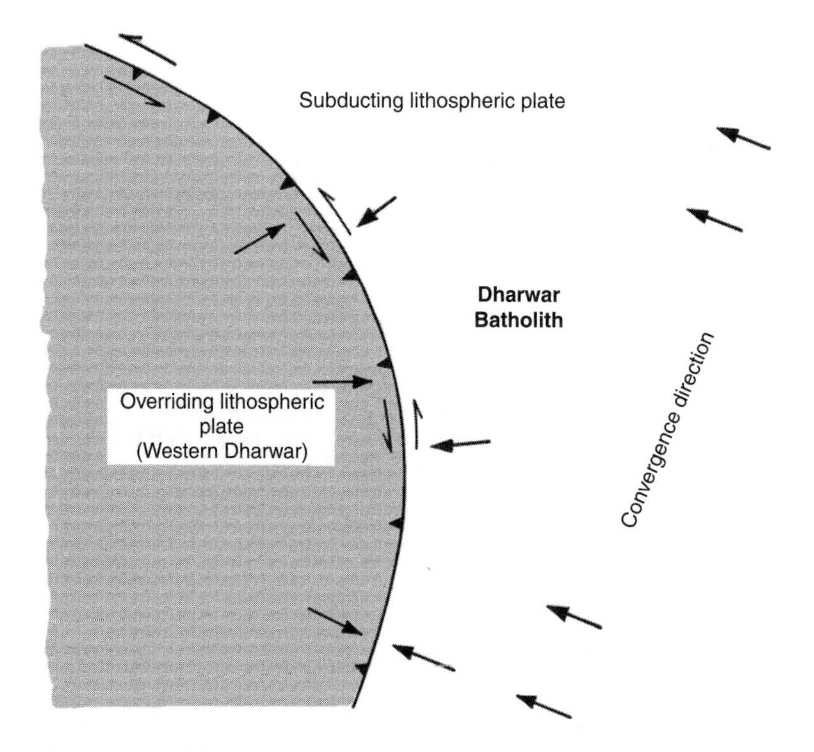

Fig. 1.1.10 The model of Eastern Dharwar Block ('Batholith') subducting below western lithospheric plate, i.e., Western Dharwar Block (Chadwick et al., 2000)

Choukrane et al. (1997), supporting this view concluded that the Dharwar craton (meaning the Western Dharwar block) was dominated by dome and basin tectonics related to the diapiric movement in the middle lower crust. This was followed by E-W shortening, affecting young greenstone belts in particular. This model, drawing from the results of experiments conducted by Ramberg (1973), Mareschal and West (1980), suggests that if a granitic crust is covered with a thick layer of basaltic rocks, and the granite is heated from below through magmatic underplating, mechanical instability will be initiated and granitic semisolid diapirs will be generated/enhanced, sagging the overlaying rocks between the diapirs (Fig. 1.1.12).

Emplacement of Precambrian granite-gneiss massifs in the form of solid/semisolid diapirs has been recommended for such belts elsewhere also (Collins et al., 1998). The model has an advantage, which if found is widely applicable in early Precambrian terrains, the phenomenon will lend support to a concept that the early tectonic developments owe much to mantle plumes and has little to do with the present day plate tectonic model(s) (Fyfe, 1978; Campbell and Griffiths, 1992; Vlaar et al., 1994). But a reasonable solution to the problem possibly lies in the observations in the field. A few diapiric structures here and there notwithstanding, the sagduction model seems generally inappropriate in Western Dharwar as it fails to satisfactory explain the complex fold structures, thrusts, transcurrent shears as well as the sedimentary facies in the schist belts. Moreover, the emplacement of the granitoid bodies in a solid/semisolid state would be accompanied by the development of radial stretching lineations and a 'top up' sense of shear (Collins et al., 1998; Kloppenberg et al., 2001).

Drury et al. (1984) suggested that the early structures in the Dharwar rocks were E-W trending recumbent folds and thrusts, which resulted from a south to north tectonic movement, later modified by N-S trending shear zones. However, later works did not find any structure in the Dharwar schist belts that convincingly supported the model of northward or southward verging, i.e., E-W trending recumbent folds and thrusts.

The suggested tectonic models for the Western Dharwar varied from intracratonic rift or passive margin environment to active continental margin.

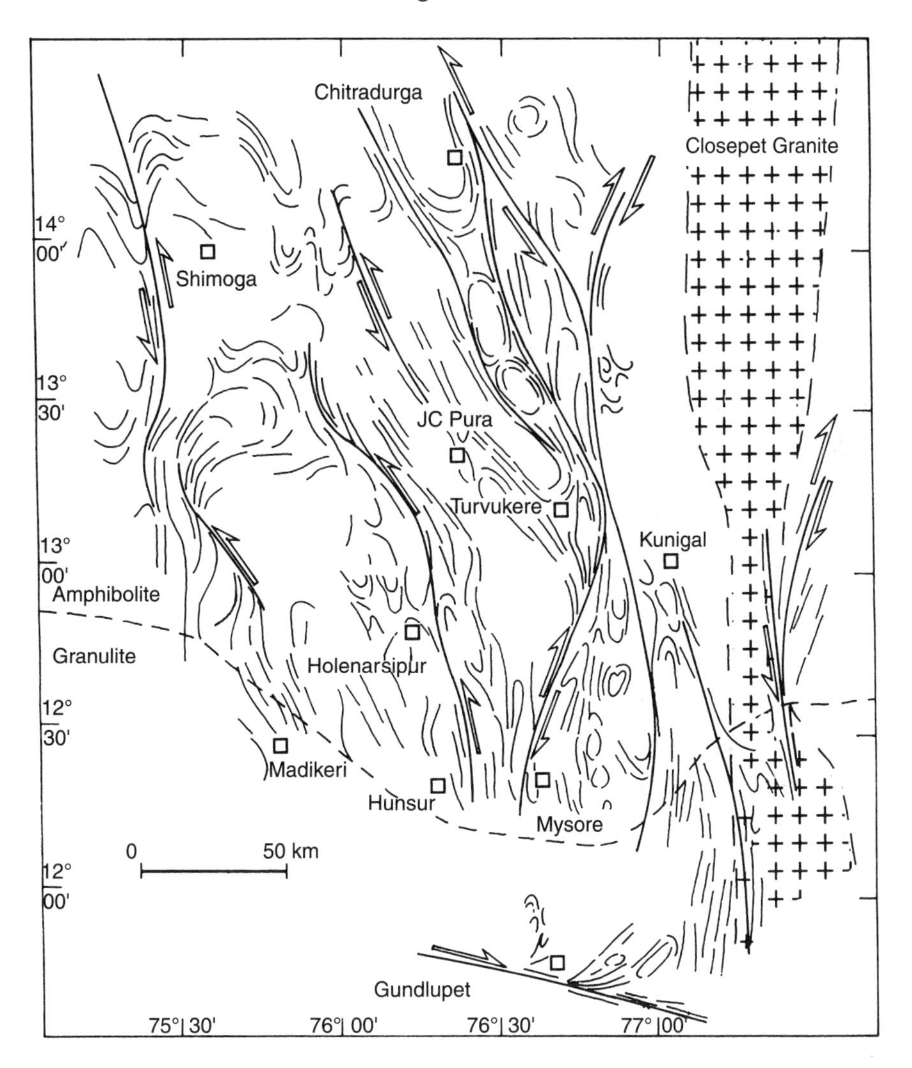

Fig. 1.1.11 Structural map of Western Dharwar with suggested transcurrent shears (Choukrane et al., 1997)

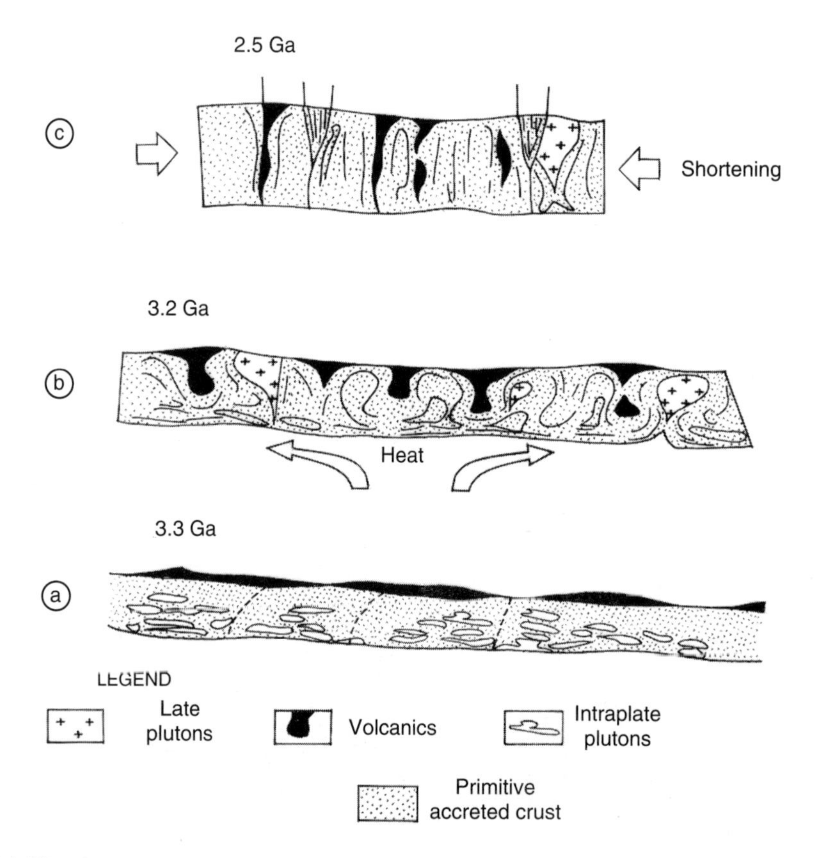

Fig. 1.1.12 Stages of sagduction in Western Dharwar as conceived by Choukrane et al., 1997

Peninsular Gneiss and the Granitoids of Eastern Dharwar

Both the Peninsular Gneiss and granitoids are present in the Eastern Dharwar. But the ratios of granitoids to gneisses are much larger in the Eastern Dharwar than in the Western Dharwar. Again, the ratio of Peninsular Gneiss and granitoids to supracrustals (schist belts or greenstone belts) is also appreciably larger in the Eastern Dharwar. Jayananda et al. (2000) made a detailed study of the region from Closepet 'batholith' to some distance east of Kolar schist belts (KSB). They reported that the outcrops of the basement gneisses decrease from the Closepet batholith (400 km × 50–20 km) to the KSB. Further east, TTG basement gneisses, which occur as enclaves within granitoids, are rarer. The intrusive suite of Closepet batholith outcrops at different structural levels from deep (corresponding to paleopressures of 7–8 kb) to shallow (2–3 kb) crust (Fig. 1.1.13).

The root zone in the south display the strong crust-mantle interaction that resulted in highly heterogeneous, enclave-rich monzonitic to granitic magmas; the transfer zone in the middle with inferred upward movement of these magmas; and a rheological interface at shallow depth in the northern part at which the assent of the magmas seem to have been arrested (Moyen et al., 2003).

The Eastern Dharwar granitoids in general fall in the granodiorite-granite field in an Ab-Or-An diagram (Fig. 1.1.14). The intrusive facies is LREE-enriched. These rocks have been linked to 'sanukitoid' (Jayananda et al., 2000). The granodiorite-granite plutons contain co-magmatic and older mafic enclaves, besides those of TTG, generally migmatised. The three plutonic bodies commonly elongate on plan and have steep sides. They may have penetrative subvertical foliation accordant with the regional fabric (N 10°–15° E).

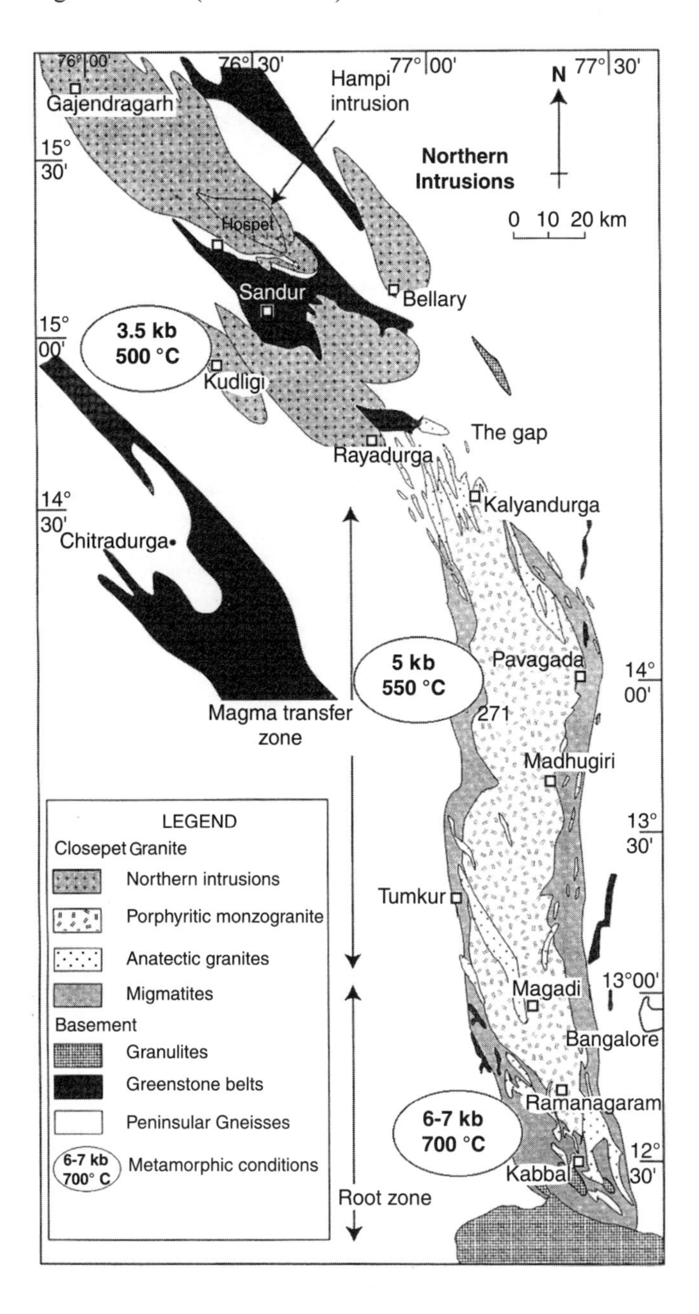

Fig. 1.1.13 Geological map of Closepet 'batholith' (after Moyen et al., 2003)

Near the southern extremity of Closepet batholith, the basement Peninsular Gneiss yielded ages of 3.4–2.96 Ga (U-Pb SHRIMP age in zircon). These basement gneisses underwent migmatisation and locally granulitisation 2.53–2.51 Ga ago. This was penecontemporaneous with the emplacement of the Closepet 'batholith' (Friend and Nutman, 1991, 1992; Mahabaleswar et al., 1995). Zircon from the basement TTG at the western margin of KSB gave a Pb-Pb minimum age of 3.14 Ga. Krogstad et al. (1991) obtained U-Pb zircon age of ~2.63 Ga for Dod Gneiss, ~2.61 Ga for the Dosa Gneiss, ~2.55 Ga for the Patna Granite, and ~2.53 Ga for the Kamba Gneiss. U-Pb SHRIMP zircon age of gneisses close to the eastern boundary of the Sandur schist belt (Belagallu Tanda) is ~2.72 Ga (Nutman et al., 1996).

Subba Rao et al. (1998) outlined the characteristic field-features, petrography, geochemistry and Rb-Sr geochronology of some well-known granitoid bodies of Karnataka and Andhra Pradesh such as Lepakshi, Kadiri, Tanakallu, Hampi, Hyderabad and Perur. In general, they are SiO_2-rich with high (4–5%) Al_2O_3, CaO, MgO, MgO, FeO_{TOTAL} TiO_2 and K_2O-content. The Hyderabad Granite Complex is as large as the Closepet 'batholith' itself. Compositionally these granitoids cluster more in the quartz-monzonite field, spilling over to those of granite and granodiorite and, are characterised by Σ LREE > Σ HREE. These are suggested to have been formed by the anatexis of TTG-type protoliths that had short crustal residence. Their tectonic settings as suggested by plotting in the Rb-(Y+Nb) and Nb-Y fields are not clear (Fig. 1.1.15). Rb-Sr isochron ages for these intrusives are within the range of 2.6–2.5 Ga.

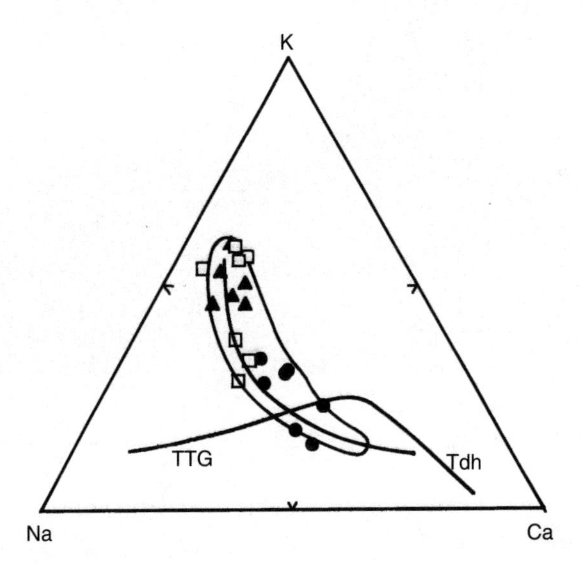

Fig. 1.1.14 Calc-alkaline differentiation trend of the Eastern Dharwar granitoids, in contrast with the TTG trend (after Jayananda et al., 2000)
TTG-Tdh: Tonalite – trodhjemite; filled triangles: Bangalore area; filled circles: Western margin of KSB; open squares: Hoskote – Kolar area

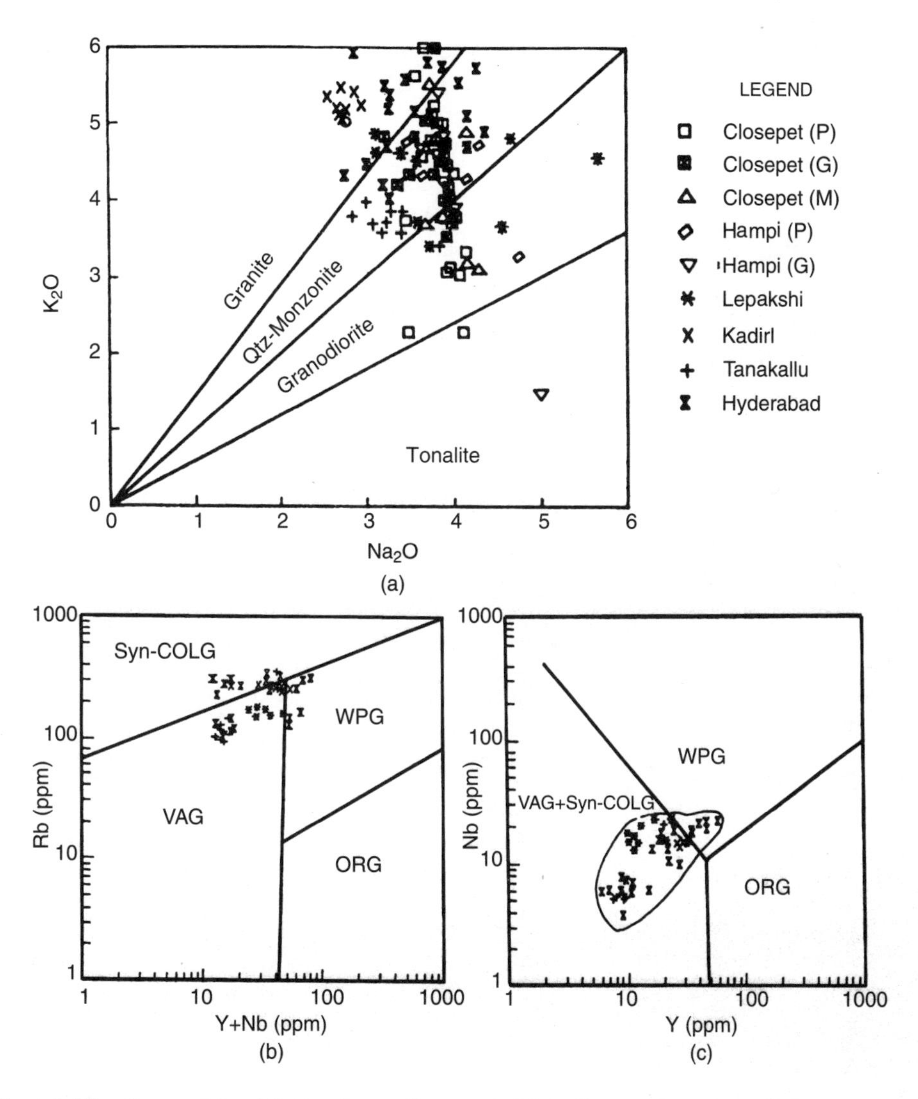

Fig. 1.1.15 Compositional fields of some late granitoids in Eastern Dharwar and their projected tectonic settings (after Subba Rao et al., 1998)

Summing up, the gneissic and granitoid rocks of the Eastern Dharwar block range in age from 3.4 to 2.5 Ga, but cluster around the latter. The rocks vary in composition from tonalite through granodiorite and adamellite to granite.

Eastern Dharwar Schist Belts

The major schist belts in East Dharwar comprise Sandur, Kustagi, Kolar, Ramagiri–Penakacherla, Hutti–Maski, Gadwal, Veligallu, Mangalur, Jonnagiri, Kadiri and Nellore-Khammam. Barring Sandur, the belts are linear with a high length/width ratio. They trend NNW–SSE or N–S (Fig. 1.1.4). Some of them have relatively been better studied than others, apparently due to their proven ore-context. These include Sandur, Kolar, Ramagiri–Penakacherla and Hutti–Maski belts.

Sandur Schist Belt Sandur schist belt occurs near the northern tip of the Closepet Granite body and is apparently intruded by the latter (Fig. 1.1.16).

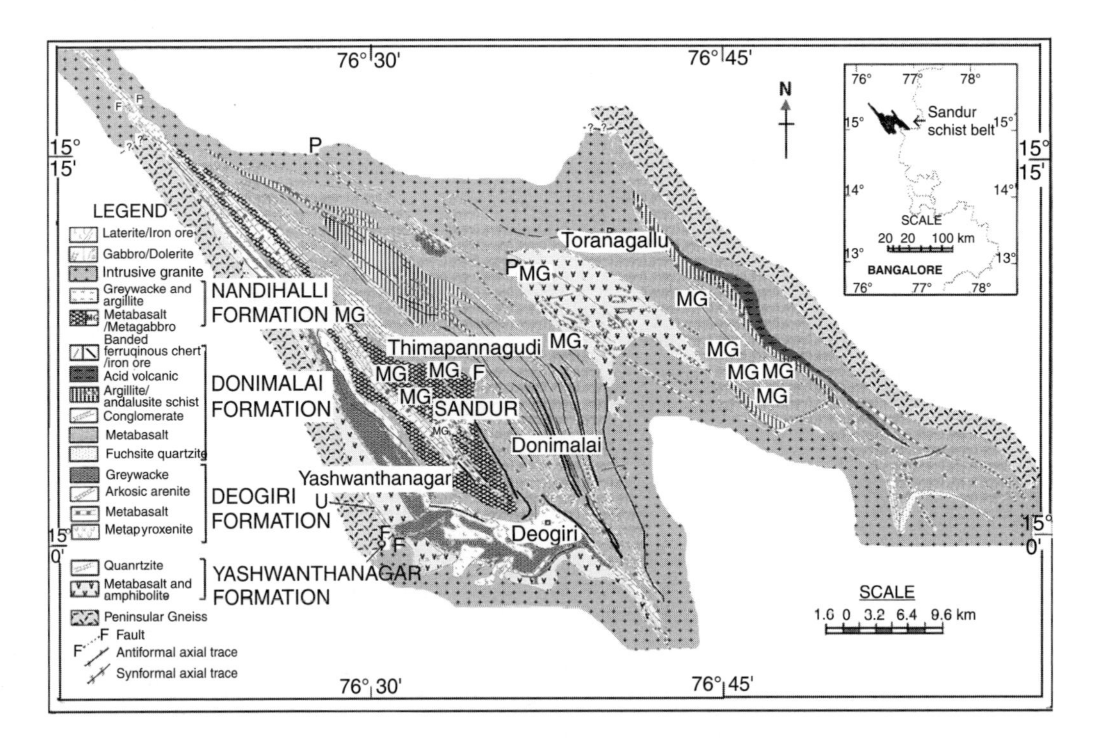

Fig. 1.1.16 Geology of the Sandur schist belt (modified after Roy and Biswas, 1983)

Stratigraphy of the belt has been divided into four major formations by Roy and Biswas (1979, 1983). These are Yeshwantanagar, Deogiri, Donimalai and Nandihalli formations arranged in ascending order. Chadwick et al. (1996), however, divided the Donimalai Formation into three formations, with Ramanmala at the bottom, Donimalai in the middle, and Talur at the top (Table 1.1.3). Again, Mukhopadhyay and Matin (1993) and Manikyamba and Naqvi (1995) are of the view that the rocks of the Deogiri Formation and Donimalai Formation are time-equivalents, i.e. facies variants, rather than one older than the other. This controversy may be resolved in future.

The Yeshwantnagar Formation consists mainly of mafic and ultramafic rocks, interlayered with current-bedded quartzite which is in parts fuchsitic. This formation is followed upward by Deogiri Formation which starts with conglomerate, current bedded quartzite, stromatolitic carbonates interbedded with argillite. Mn-rich layers are repeatedly interlayered with shales. In some layers, Mn-content may be as high as 60% or more. The Donimalai Formation (Roy and Biswas, 1983) consists of volcanic rocks, cherts, carbonates, greywackes, shales, and of course important units of BIF. The top most Nandihalli Formation is made of several rock types of which greywacke (turbidites) and volcanic members deserve special mention. The composition of the volcanic rocks changed with time. In the Yeshwanthnagar Formation, these are mainly ultramafic; in the Deogiri

Table 1.1.3 *Lithostratigraphy of the Sandur schist belt*

Roy and Biswas (1983)	Chadwick et al. (1996)	Lithology
Nandihalli Formation	Vibhutigudda Formation	Conglomerate, greywackes, phyllites-shales, quartzite, acid volcanic, BIFs, metabasalts
Donimalai Formation	Talur Formation	Metabasalts, cherts, carbonates, ultramafics, metagabbro, BIF
	Donimalai Formation	Ferruginous cherts, conglomerates, intermediate-acid volcanic rocks
	Ramanmala Formation	Metabasalts, greywackes, BIF, shales
Deogiri Formation	Deogiri Formation	Argillites, greywackes, Mn-bearing argillites, quartzite, stromatolite-bearing carbonates and Banded Manganese Formation (BMF)
Yeshwantnagar Formation	Yeshwantnagar Formation	Fuchsite quartzite, amphibolite, ultramafic cumulates and ultramafic schists

and Donimalai Formations, these are high-Mg basalt and tholeiite; and in the Nandihalli Formation, these are tholeiite to intermediate and acid volcanic rocks. Hanuma Prasad et al. (1997) took up the mafic-felsic volcanic rocks exposed in the eastern part (Copper Mountain region) for a detailed geochemical study, and a possible tectonic interpretation of the obtained results. They found the basic volcanic rocks divisible into two broad groups, viz, Al-depleted picritic basalts and tholeiitic basalts. Picritic basalt has a HREE-depleted pattern. The tholeiitic basalts have two distinct REE patterns: (i) slightly depleted-flat REE pattern and no HREE fractionation, and (ii) LREE-enriched and HREE-depleted patterns. The felsic volcanics are Na-rhyolites, showing calc-alkaline affinity, and LREE-enriched and HREE-depleted patterns. According to these groups of workers, field geochemical and petrogenetic aspects suggested that the magmatic rocks studied represented an active plate margin environment. Manikyamba et al. (1997) and Krishna Rao and Hanuma Prasad (1995), found no negative Eu-anomaly in the REE-patterns of the clastic sediments studied from this belt, and concluded that the sediments had a mixed mafic source(s) from the intrabasinal arc volcanics.

The BIFs of the Sandur belt show all gradations from chert, through ferruginous chert to cherty BIF and shaly BIF. The BIFs occur above a greywacke bed and a zone of granular iron formation (GIF). They may be interbedded with thinly laminated slate, pillowed lava and pyroclastic rocks. Iron occurs mainly as oxide, but phases like carbonates, sulphides and silicates may also be present. Chert bands from Donimalai Formation contain *Cyanobacteria* fossils (Naqvi et al., 1987; Venkatachala et al., 1990).The gross mineralogic differences in Sandur BIFs are reflected not only in the Fe_2O_3 / SiO_2 ratios, but also in the contents of Al_2O_3 and TiO_2, as well those of the trace elements and REE (Table 1.1.4). REE distribution in mixed oxide-silicate facies BIF of Sandur shows LREE enrichment and a Eu-peak (Fig. 1.1.17). ΣREE increased with the shale-content. Positive Eu-anomaly becomes suppressed under the circumstance. This suggests that the Sandur BIFs had the supply of their constituents from both a hydrothermal (vents at MORB?) and terrestrial sources in the Archean ocean systems. Some workers suggested biogenically mediated precipitation of iron from solution, based on the stromatolytic fossil remains (Manikyamba et al., 1991). But this may not be the unique solution of the problem, as discussed in an earlier section.

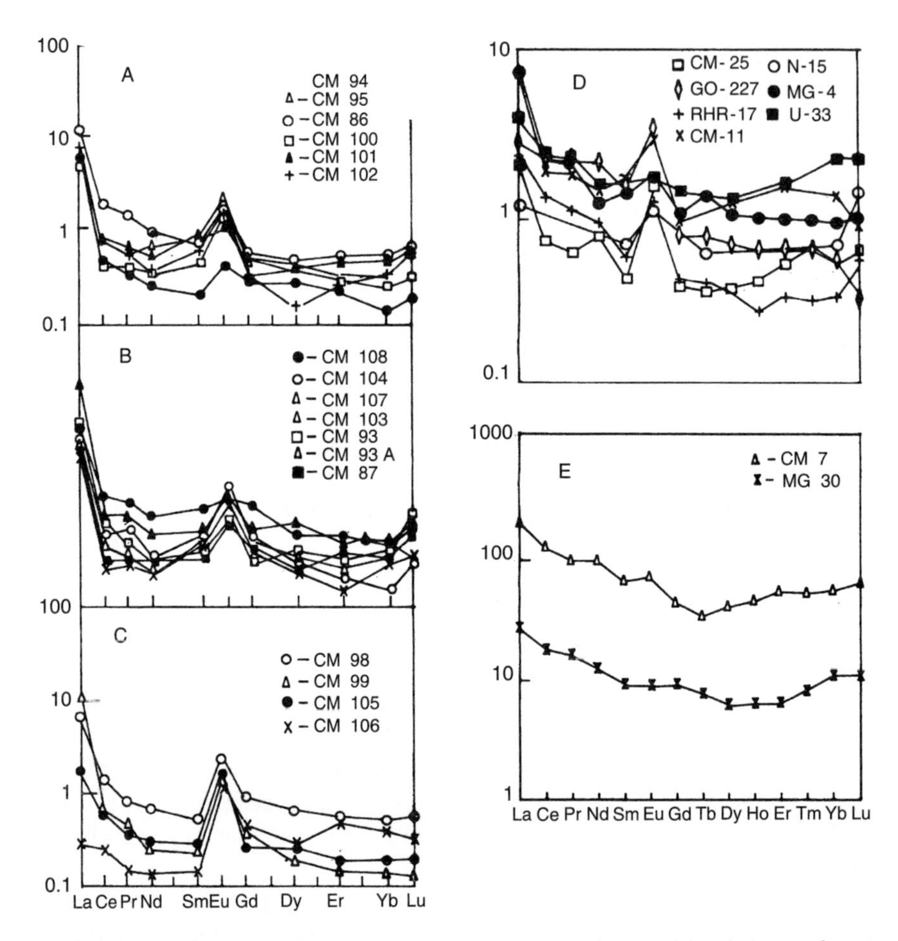

Fig. 1.1.17 REE distribution (A–C) in mixed oxide-silicate facies of banded iron formations (BIF), Sandur. D shows comparison of REE contents in cherty BIFs at Sandur, Kudremukh, Kustagi, Sargur, Bababudan and Kauthimalai. E shows REE pattern in shaly BIFs of Sandur and Bababudan (Manikyamba, 1998)

Table 1.1.4 *Average chemical composition of BIF from Sandur schist belt (Manikyamba et al., 1991)*

Major oxides (wt%), Trace & REE (ppm)	Carbonate	Oxide	Sulphide	Chert	FC	CBIF	SBIF	Shale	Iron ore
	(22)	(27)	(24)	(7)	(8)	(51)	(27)	(9)	(3)
SiO_2	46.19	66.59	38.10	97.71	86.36	51.80	41.84	62.18	8.77
TiO_2	0.01	0.02	0.00	0.01	0.01	0.03	1.17	0.88	0.09
Al_2O_3	0.14	0.33	0.21	0.02	0.05	0.23	17.47	17.63	4.32
Fe_2O_3	27.34	30.17	47.88	2.50	11.01	44.30	29.30	13.93	81.79
CaO	6.74	1.21	0.46	0.14	0.12	0.09	1.11	1.31	1.10
MgO	2.51	0.23	1.79	0.27	0.16	0.50	2.03	1.46	0.25

(Continued)

(Continued)

Major oxides (wt%), Trace & REE (ppm)	Carbonate	Oxide	Sulphide	Chert	FC	CBIF	SBIF	Shale	Iron ore
Na_2O	0.04	0.04	0.05	0.02	0.05	0.19	0.28	0.75	0.15
K_2O	0.13	0.06	0.23	-	0.01	0.07	1.08	2.16	0.01
MnO	0.48	0.31	0.80	0.04	0.10	0.09	0.06	0.08	0.04
P_2O_5	0.08	0.42	0.03	0.11	0.11	0.08	0.14	0.06	0.18
Sc	1.22	1.45	1.29	0.16	0.38	1.30	20.58	23.82	2.64
V	5.58	8.60	9.11	7.53	2.36	1128	210.20	307.79	19.00
Cr	8.92	20.04	10.26	11.40	5.48	55.74	276.01	354 .94	57.67
Co	16.84	56.87	21.68	98.56	73.01	33.72	16.79	24.70	2.06
Ni	8.36	12.08	11.59	7.53	10.73	10.23	82.91	96.02	14.67
Cu	10.10	5.12	32.83	23.75	35.54	12.51	78.13	149.66	14.00
Zn	42.62	33.31	180.34	3.76	3.55	11.21	59.89	98.82	22.82
Rb	1.94	3.60	10.14	0.07	0.18	2.66	40.89	86.52	0.14
Sr	16.16	26.89	14.64	1.61	0.54	3.78	43.09	101.53	27.63
Y	3.35	4.15	3.99	0.46	0.67	1.78	29.21	19.50	2.69
Zr	3.22	4.58	4.08	0.71	0.92	4.01	181.16	258.73	4.52
Nb	0.30	0.57	0.34	0.15	0.22	0.29	62:1	9.73	0.48
Ba	5.12	25.98	19.11	1.08	3.61	14.55	172.20	249.73	6.30
Hf	0.23	0.22	0.20	0.09	0.12	0.18	3.64	3.75	0.18
La	6.79	10.71	9.75	3.57	3.66	5.97	20.29	14.49	37.39
Ce	1.54	2.60	5.77	1.14	0.97	2.10	39.11	38.33	5124
Pr	0.19	0.29	0.58	0.66	0.10	0.28	3.78	2.73	4.39
Nd	1.63	1.47	2.08	5.79	0.24	1.30	20.39	9.35	20.48
Sm	0.29	0.39	0.65	0.49	0.14	0.34	4.13	2.20	2.34
Eu	0.17	0.25	0.46	0.16	0.00	0.18	1.51	0.73	0.68
Gd	0.29	0.47	0.66	0.16	0.08	0.28	3.77	2.22	1.76
Tb	0.06	0.12	0.11	0.03	0.04	0.14	0.64	0.40	0.13
Dy	0.39	0.54	0.74	0.53	0.19	0.37	4.07	2.22	0.77
Ho	0.09	0.12	0.15	0.10	0.04	0.07	0.89	0.47	0.14
Er	0.28	0.35	0.45	0.29	0.15	0.25	2.73	1.49	0.34
Tm	0.04	0.05	0.06	0.03	0.02	0.03	0.38	0.21	0.04
Yb	0.31	0.36	0.41	0.15	0.14	0.19	2.88	1.48	0.21
Lu	0.07	0..09	0.08	0.02	0.02	0.03	0.52	0.24	0.01
Σ REE	11.56	17.19	21.94	13.32	6.11	1.72	96.78	74.82	180.09

(1) Number of samples for each type analysed is given in brackets
(2) Fe_2O_3 is given as Fe_2O_3 (T)
(3) FC: Ferruginous chert, CBIF: Cherty banded iron formation, SBIF: Shaly banded iron formation

Age data on the Sandur rocks are meager. Radhakrishna (1983) correlated the Sandur schist belt rocks with those of the Bababudan Group of rocks in Western Dharwar. But a SHRIMP U-Pb isotopic imprecise concordia intercept ages for zircon grains from the acid volcanic rocks in the Vibhutigudda Formation (Table 1.1.3) were found to be 2658 ± 14 and 2691 ± 18 Ma. These ages of the topmost formation of Sandur belt broadly conform with the topmost formation (Daginkatte Formation) of the Chitradurga Group instead. High strain grey gneisses adjacent to Sandur schist belt and a young granite adjacent to the belt yielded imprecise U-Pb zircon ages of 2719 ± 40 Ma and 2570 ± 62 Ma, respectively (Nutman et al., 1996).

Kustagi Schist Belt It is another NW–SE trending schist belt and it could be as long as 200 km, if it really is a dismembered unit of a large entity, Hungund–Kustagi–Hagari (HKH) schist belt as envisaged by some workers (Bhat et al., 1996). It could even be a part of a larger system that included Ramagiri–Penakacherla belt also.

Lithologic association of the belt consists of metamorphosed basalt, basaltic andesite (pillowed), mafic schists, metamorphosed acid volcanics and subordinate metasediments that comprise phyllite, carbonate and BIFs. Intrusion by younger granites equivalent to Closepet Granite is common (Khan and Naqvi, 1996). The stratigraphic sequence suggested by Bhat et al. (1996) is as follows (Table 1.1.5).

Table 1.1.5 *Lithostratigraphy of the Kustagi schist belt (Bhat et al., 1996)*

Intrusives	Dolerite, gabbro, anorthosite-gabbro intrusions Closepet Granite Anorthosite, anorthositic gabbro, pyroxenite and peridotite
Meta-sediments	BIFs and chert with tuff and phyllite Quartz-chlorite schist, meta-argillite and sub-greywackes
Metamorphosed volcanic and intrusive rocks	Acid volcanic rocks (rhyolite, tuff etc.) Metamorphosed basic volcanic rocks, meta-gabbro and amphibolites
------------------ *Uncomformity* ------------------	
Peninsular Gneiss	Migmatite and banded biotite gneiss with enclaves of amphibolite

There is no direct radiometric dating on the rocks of this belt. Khan and Naqvi suggested an age between 3.0 and 2.6 Ga (Khan and Naqvi, 1996) on indirect observations. It may be pointed out that Bhat et al. (1996) classified the volcanisedimentary rocks of the Sandur belt under Chitradurga Group. We have already seen that the Sandur schist belt and the Chitradurga schist belt rocks are time equivalents. Physical contiguity and geologic similarity suggest that the Sandur and Kustagi belts could initially be parts of the same basin.

Metabasalt is a major member of the rock associations in the Kustagi belt. BIFs are both shaly and cherty. Iron is present mainly in oxidic phases, dominated by hematite. The $Fe_2O_{3\ total}$ / SiO_2 in the BIF vary widely as elsewhere. The REE distribution pattern in chert-rich and hematite-rich bands (Table 1.1.6) show positive Eu-anomalies (Eu / Eu*= 2.27–5.90 and 1.56–2.52) (Fig 1.1.17) and La_N / Yb_N = 0.75 and 0.87, respectively in cherty BIF, suggesting a hydrothermal source of the BIF material.

Table 1.1.6 *Trace and REE composition of Kustagi BIF (Khan and Naqvi, 1996)*

Elements (ppm)	Chert-rich bands (*n* = 10)	Hematite-rich bands (*n* = 10)
Sc	0.37–0.80	1.47–3.11
Cr	0.52–3.89	60.23–147.05
Co	0.51–2.90	0.93–2.22
Ni	1.00–7.04	6.29–62.72
Zn	7.10–12.22	13.60–47.96
Rb	0.24–0.72	0.49–1.28
Sr	2.89–11.67	6.52–22.53
Y	0.77–3.37	2.80–9.05
Zr	1.99–9.67	4.41–16.85
Hf	0.08–0.60	0.11–0.54
Cs	0.05–0.14	0.09–0.24
Pb	5.06–9.22	7.04–19.26
La	0.44–3.18	3.20–9.85
Ce	0.59–4.20	5.75–14.64
Pr	0.11–0.66	0.65–1.52
Sm	0.14–0.90	------------
Eu	0.12–0.45	0.35–0.85
Gd	0.13–0.58	0.73–1.49
Dy	0.17–0.77	0.05–1.55
Er	0.09–0.44	0.38–1.02
Yb	0.08–0.41	0.34–1.06
Lu	0.01–0.06	0.06–0.18
Co/Zn	0.04–0.34	0.03–0.08
Ni/Zn	0.11–0.79	0.24–2.47
La_N/Yb_N	0.40–1.59	0.41–1.81
Nd_N/Yb_N	0.47–1.08	0.41–0.94
Eu/Eu*	2.27–5.90	1.56–2.52

Manglur Schist Belt It is a N-S trending schist belt in Gulbarga district, northern Karnataka with an aerial coverage of about 25 × 5 sq km. The rocks comprising the belt include migmatised mica schists, quartz-chlorite-sericite schists, amphibolite, actinolite-albite-chlorites, and dolerite dykes. The amphibolites vary in petrography from massive through banded to the schistose variety. The quartz-chlorite-sericite schist and mica schists are believed to be metamorphic products of dominantly pyroclastic acid volcanic rocks (Kollapuri, 1990). The belt bears gold mineralisation.

Deodurg Schist Belt It is a discontinuous, long, narrow belt; consisting of pillow-bearing amphibolite, metamorphosed acid volcanic rocks, and quartz-sericite schist. Acid volcanic rocks constitute an important proportion in this rock association. It is possible that the quartz-sericite schists were originally acid volcanics. The rocks have been affected by three phases of deformation.

Highly compressed folds of the first two phases are coaxial. Metamorphism is of amphibolite or lower facies.

Several phases of late granites, such as pink granite, hornblende granite, hornblende biotite granite and grey biotite granite have intruded the supracrustal rocks (Ahmed et al., 1993).

Hutti – Maski Schist Belt Hutti – Maski schist belt is located in the Raichur district of Northern Karnataka. This 100 km long belt is somewhat different in disposition and shape from other schist belts in Eastern Dharwar. The lithologic composition of the supracrustal rocks in the belt is dominated by basic volcanic rocks, now represented by amphibolites and chloritic schists. Meta-sediments like quartzite, phyllite (locally carbonaceous), limestone, BIF are present, particularly in the northern part; but are very much subordinate to metabasic rocks (Fig. 1.1.18). Srikantia (1995) proposed the lithostratigraphy for the Hutti – Maski schist belt (Table 1.1.7).

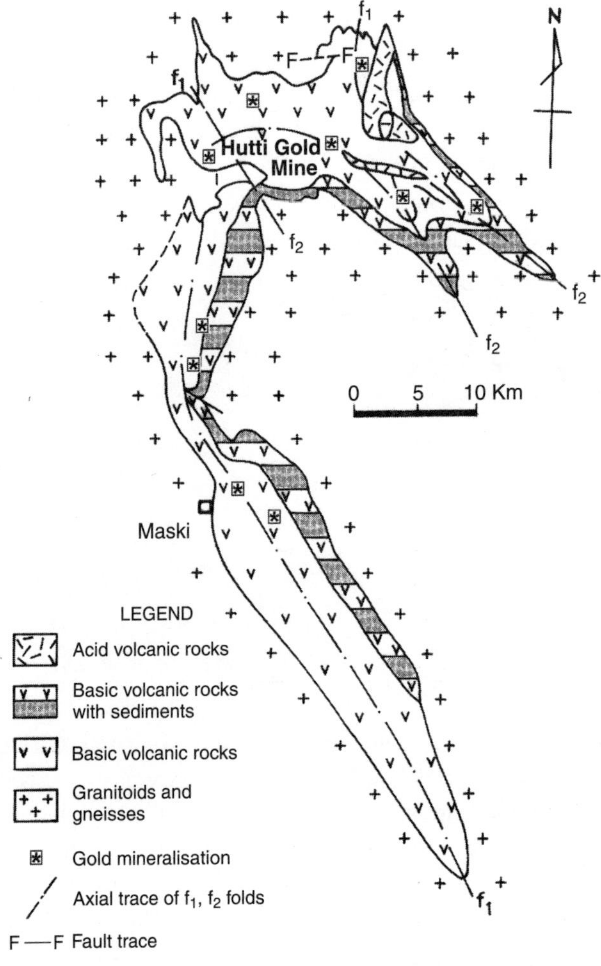

Fig. 1.1.18 Simplified geological map of Hutti–Maski schist belt, also showing locations of gold deposits (after Biswas, 1990a).

Table 1.1.7 *Lithostratigraphy of Hutti–Maski schist belt (Srikantia, 1995)*

Intrusive rocks	Gabbro and dolerite Granitoids
Budini Formation	Massive vesicular and pillowed metabasalt Coarse grained metabasalts Amphibolite, diorite Phyllite (locally carbonaceous), limestone, gritty calcareous quartzite Banded iron formation (BIF) Quartzite, garnetiferous mica schists Meta-acid rocks, Palkamaradi mixite
Bullarpur Formation	Amphibolite Cross-bedded tuffite Diorite porphyry
530 Hill Formation	Pillowed metabasalt
Hussainpur Formation	Banded amphibolite
Peninsular Gneiss	

Metabasalts are generally high Fe-tholeiite (Table 1.1.8), enriched in LIL and show marginal enrichment in LREE. The metamorphic products of pelitic sediments are mica schist, andalusite schist, or garnet-cordierite schist.

Table 1.1.8 *Major element composition of metabasalts of Hutti–Maski schist belt (Srikantia, 1995).*

Major oxides (wt%)	1	2	3	4	5
SiO_2	48.82	52.49	50.53	48.31	50.50
TiO_2	0.97	0.73	0.80	1.18	1.80
Al_2O_3	13.65	14.17	15.75	14.01	12.40
Fe_2O_3+	13.32	-	2.16	12.58	15.55
FeO	-	11.07	10.46	-	-
MgO	6.20	5.95	2.98.	6.51	2.65
CaO	10.77	11.60	12.11	9.48	7.95
Na_2O	1.21	2.32	2.85	1.83	3.38
K_2O	0.31	0.30	0.18	0.21	0.30
P_2O_5	0.09	0.13	0.07	-	-
MnO	0.19	0.14	0.32	0.18	0.10
S	0.09	-	-	0.03	0.10
Cr_2O_3	0.03	-	-	-	-
LOI	2.67	1.50	1.25	6.35	3.10
Total	99.32	100.40	99.46	100.62	97.98

Note: 1. Metabasalt from Hutti (average of 13 samples): Roy, 1991; 2. Metabasalt from Hutti (average of 24 samples): Anantha Iyer and Vasudev, 1979; 3. Metabasalt from Hutti mines: Raju and Sharma, 1991; 4. Metabasalt from Udbal–Maski sector (average of 8 samples): Biswas, 1990a; 5. Metabasalt from Uti (coarse grained – average of 3 samples): Biswas, 1990b.

The country rocks that surround the Hutti–Maski schist belt are a collage of different granitoid intrusives, and patches of gneisses containing inclusions of mafic-ultramafic rocks. Kavital Granite (granodiorite) was followed by Yelagatti Granite as intrusives into the schist belt boundaries with the country rocks. Here again the schist belt rocks underwent three major phases of deformations (D_1, D_2, and D_3) (Roy, 1979). Highly compressed folds of the first two phases are co-axial or nearly so. F_1 and F_2 related to D_1 and D_2 control the map pattern. A pronounced zone of shearing developed during D_2 (Roy, 1979). Kolb et al. (2004a, 2004b) identified five deformational events (D_1–D_5), in which mine-scale structures have been included. General enrichment in LIL elements and slight enrichment in LREE patterns are more suggestive of marginal basin basalts (MBB), rather than mid-oceanic ridge basalt (MORB).

The Kavital Granite has a U-Pb (zircon) age of 2545 ± 7 Ma (Srinivasa Sharma, et al., 2008) and the Yelagatti Granite has a U-Pb (titanite) age of 2532 ± 3 Ma (Anand et al., 2005 in Misra and Pal, 2008).

Ramagiri (–Penakacherla) Schist Belt The Ramagiri – Penakacherla schist belt, with a number of old workings for gold and a recently closed mine at Yeppamana, extends for over 180 km with NNW–SSE trend and <0.5 km to 5 km width from Karnataka border in the south to Anantapur district, Andhra Pradesh in the north.

The Ramagiri belt proper is a N–S trending trident shaped, 2–3 km wide and 60 km long supracrustal belt, surrounded on all sides by granite gneisses and intrusive granites. The belt attains a maximum width of about 3 km near Ramagiri village; and consists of three metavolcanic dominated arms, the eastern, central and western that spread out in the north (Mishra and Rajamani, 1999). These arms are separated from each other and surrounded by granitic rocks, viz, the Chenna Gneiss occurring in the east, Gangam Complex occurring in the west, and the Ramagiri Complex occurring between the central and western limb (Zachariah et al., 1996). We would concentrate on the Ramagiri part of the belt in the following discussion (Fig. 1.1.19).

The eastern block consists of foliated amphibolites and a thin unit of BIF, locally discontinuous. Along the contact zone with Chenna Gneiss there is interleaving of granite, amphibolite and BIF, which is intruded by younger granitic dykes. Deformation in the granite-gneiss is a conspicuous feature rendering it mylonitic-protomylonitic over a wide zone beyond the contact. Compositional bands in the gneisses and the amphibolite-quartzite are co-folded (Zachariah et al., 1996).

The central block contains basic and acid volcanic rocks and turbidites comprising greywacke and argillites and siliceous shales in unequal proportions. The siliceous shales contain organic matter. The major, trace and REE geochemistry of the Ramagiri greywacke suggests its probable association with active island arc volcanism, where rapid erosion had given rise to the metasediments with a very low degree of chemical weathering.

The western block consists of massive and pillowed matabasalts with intercalated schists/chlorite-actinolite schist, BIF, quartzite and discontinuous bodies of serpentinite. The metabasalts are sheared and degraded (into chlorite schist, chlorite-actinolite schists) and differ in REE-content. They bear signature of island arc volcanism (Zachariah et al., 1996). The highest grade of metamorphism of the rocks in the belt is amphibolite facies.

The Chenna Gneiss occurring to the east of the belt is granodioritic with high contents of Sr and moderate contents of REE. The granitoid rocks of the Ramagiri Complex, though varied in composition, are dominated by granite with high Ba-contents. The Gangam Complex has as its members most primitive granitic rocks of quartz-monzodiorite composition with a sanukitoid affinity alike Dod Gneiss occurring on the west side of the Kolar belt (Balakrishnan and Rajamani, 1987: Zachariah et al., 1996). U-Pb zircon age of volcanogenic metagreywacke from the Central block has been determined as 2707 ± 18 Ma. Gangam Complex and Chenna Gneiss yielded U-Pb

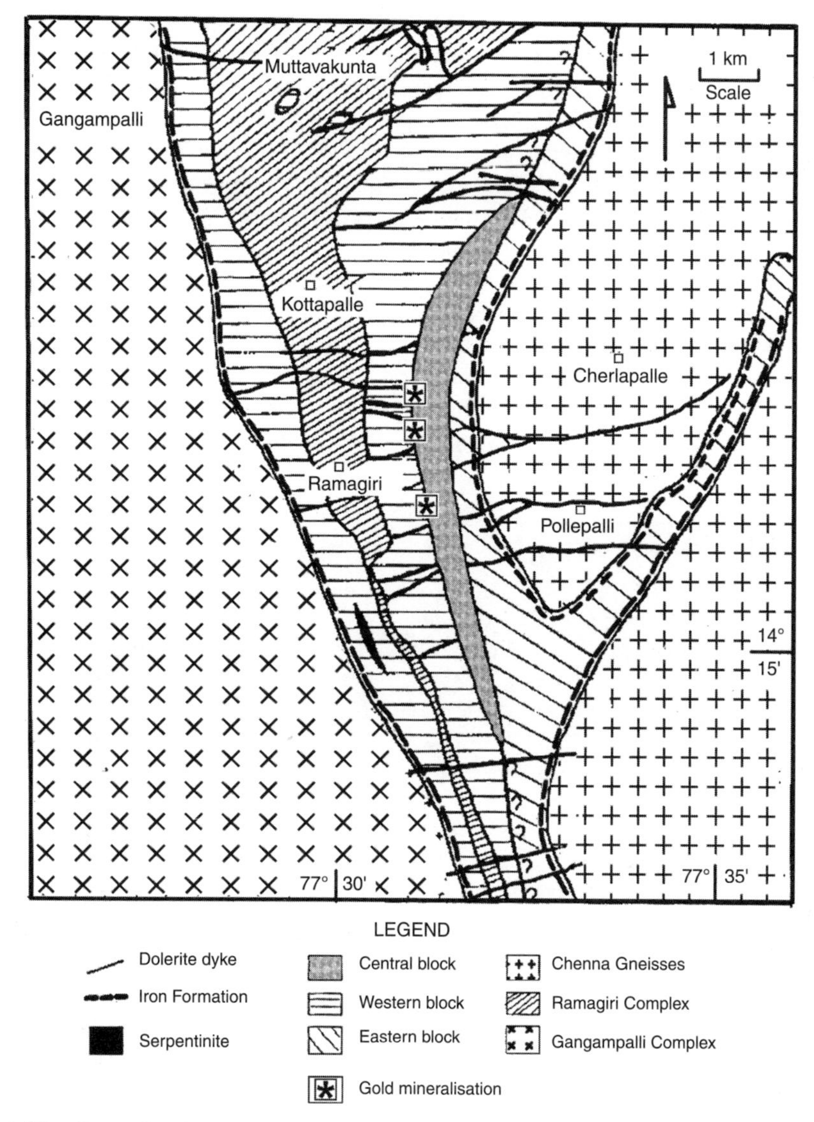

Fig. 1.1.19 Generalised geological map of the Ramagiri area. Principal gold occurrences also indicated. Persistent occurrence of an iron formation towards the base of the greenstone sequence is notable (modified after Zachariah et al., 1996)

sphene ages of 2523 Ma and 2553 Ma, respectively (Krogstad et al., 1992). A swarm of dolerite dykes cut across the belt. One of them was dated to be 2450 Ma in age (Sm-Nd whole rock-mineral isochron age) by Zachariah et al. (1995 a).

Zachariah et al., (1996), based on geochemical data from the basic volcanic rocks and greywackes, postulated an island arc tectonic environment for their emplacement and deposition. The belt overall is suggested to represent a suture zone between two discrete terrains (Balakrishnan et al., 1999).

Kolar Schist Belt The Kolar schist belt is N–S trending, about 80 km in length and 4–8 km in width. The belt is more intensely studied than most others in the region, possibly because of its being the most important gold-mineralised belt in the country, as well as easy accessibility owing to development due to mining, or even otherwise.

The belt is composed of several rock units (Fig. 1.1.20). At the westernmost margin of the belt, close to its contact with the granite-gneiss country rocks, occur banded ferruginous chert/quartzite, interlayered with graphitic and sulphidic schist, schistose metabasalt and other metabasics; apparently originally intrusive. The silicate phases in the BIF, besides the major phases of chert/quartz, magnetite and hematite, are cummingtonite, grunerite and actinolite. Pyrite, pyrrhotite and chalcopyrite are fairly abundant in the sulphidic bands. This BIF (– graphite) sequence was named the Yerrakonda Formation above the lowermost formation in the whole sequence by Viswanatha and Ramakrishnan (1981). This followed in the east, i.e. in the core region of the belt by amphibolitic rocks of complex petrology, followed further by an uncommon type of rock named Champion Gneiss, which again is bordered by amphibolite along the eastern boundary.

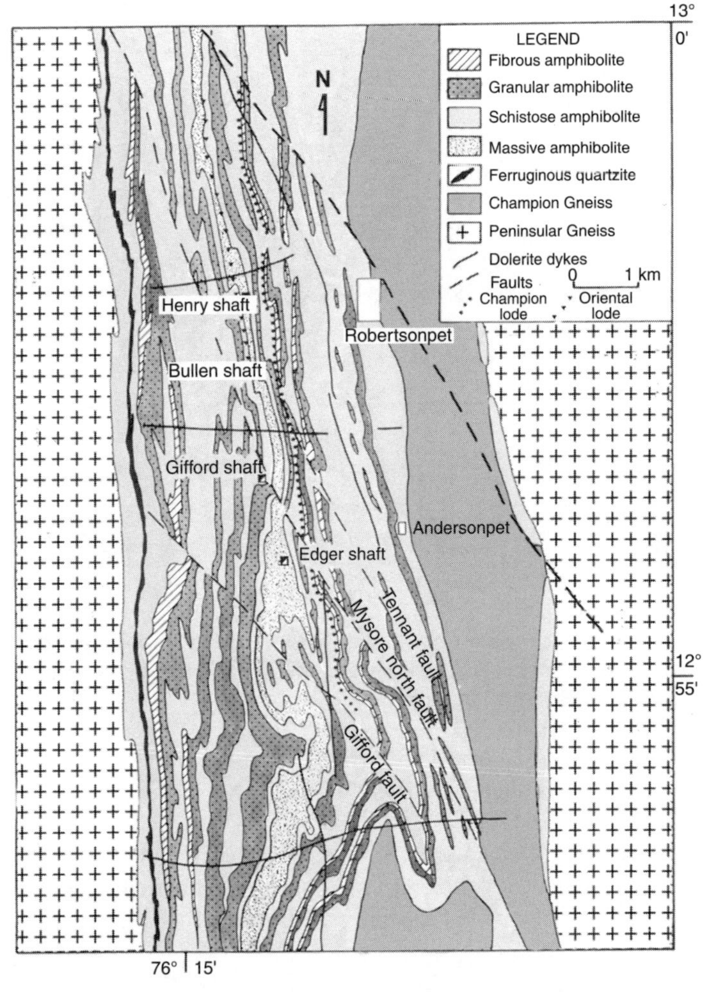

Fig. 1.1.20 Geological map of Kolar schist belt, Karnataka (after Narayanswami et al., 1960)

High-Fe tholeiitic amphibolite is the major member of the central amphibolites of the Kolar belt. Others, sub-ordinate though, are high-Mg basalt to komatiites (MgO 14–21.3 wt%). The latter varieties occur both at the west-central and east-central domains. These high-Mg mafic volcanic rocks show two distribution patterns of REE: (i) Enriched in LREE–MREE, but depleted in Ce vis-à-vis Nd, (ii) depleted in both LREE and HREE .The komatiitic amphibolites from the west central part appear to have been generated from LREE depleted mantle, whereas those from the eastern part formed from LREE enriched mantle. The tholeiitic amphibolites formed from a mantle source with a higher FeO/MgO ratio (Rajamani et al., 1985). Balakrishnan et al. (1990) have shown that the tholeiitic and the high-Mg volcanic rocks have different $^{143}Nd/^{144}Nd$ ratios and Pb-isotope characteristics. These data suggest three or more sources for those rocks in the mantle.

Champion Gneiss is an uncommon type of rock that still eludes a satisfactory model of origin and evolution. It is a fine to medium grained micaceous gneiss with included sheets of granite, pegmatite and aplite. Polymictic conglomerates with pebbles of gneisses and amphibolites are present at places. The main mass appears to be an acid volcanic rock as testified by the bulk composition, and the presence of euhedral grains of feldspar and megacrysts of blue opalescent quartz. Local presence of cordierite–garnet–anthophyllite and sillimanite–andalusite–garnet–muscovite associations suggests interflow of sedimentary material of appropriate chemical composition along with the volcanic material.

Amphibolitic rocks and also a little BIF occur again to the east of Champion Gneiss; and contain pyroxenite, gabbro and BIF, besides the amphibolite. Vishwanatha and Ramakrishnan (1981) called it Kalhalli Amphibolite (vide Table 1.1.9).

Table 1.1.9 *Lithostratigraphy of the Kolar schist belt (Viswanathra and Ramakrishnan, 1981)*

Formations		Litho-associations
Basic intrusives		
Granitic intrusive (tonalite to potassic granite)		
Kolar Group	Goldfield Volcanics	Metabasalt, hypersthene basalt, basaltic andesite, pyroxenite, gabbro, BIF and black shales.
	Champion Gneiss	Wackes, felsic volcanics, BIF and polymictic conglomerates
	Yerrakonda Formation	BIF, graphitic schists, gabbro, pyroxenites
	Kalhalli Amphibolite	Amphibolite with pyroxenite, gabbro and BIF
	Gneisses	

Determination of stratigraphic sequence of a region that was subjected to intense deformation and considerable metamorphism, as has been the fate of many greenstone belts, is fraught with uncertainties in different measures. If in addition to that, the belt is an amalgam of two disparate tectonic blocks, as has been proposed by some workers for the Kolar belt (discussed below), then any proposed stratigraphic sequence, however intelligently constructed, remains a suspect.

This belt is also multiply deformed. The F_1 folds produced by D_1 deformation are isoclinal, recumbent and slightly reclined. F_2, supposedly produced by D_2 deformation, is co-folded with F_1. Numerous brittle-ductile shear zones that characterise the belt and the adjacent country rocks, could be related to D_2. Several cross-faults dissected the belt. The interfaces between the schist belt and the adjacent country rocks are invariably tectonised.

Metabasalts from the western part of the Kolar schist belt have been dated by Balakrishnan et al. (1990) to be 2.7 (±0.15) Ga (Pb-Pb isochron, WR). The same from the eastern part has also been dated 2.7 (±0.04) Ga by Walker et al. (1989). Dod Gneiss, Dosa Gneiss and Patna Granite, all intrusive masses along the western margin of the belt, have been dated respectively to be 2631 (± 6.5), 2610 (±10) and 2551 (±2.5) Ma by Krogstad et al. (1991) using zircon as the probe-material. $^{207}Pb/^{206}Pb$-zircon age of a banded gneiss sample was, however found to be 3140 Ma (minimum). Granitic inclusion in the Champion Gneiss yielded ages of the order of 2532(±3) Ma, 2517(±2) Ma only. A granitic pluton within the Kolar belt was dated 2420(±2) Ma (minimum) (Krogstad et al., 1991).

Krogstad et al. (1989) suggested that the Kolar schist belt is a suture joining the two crustal blocks. At ~2700 Ma, the volcanic rocks of the west and east Kolar developed in two separate segments of the oceanic crust. At that time the middle Archean Peninsular Gneiss existed beyond the western margin but no continent to the east. During the period 2530–2420 Ma the two disparate blocks united, producing compressive and shear structures and also granitic plutons. Later they modified their tectonic model, suggesting that the west side of the Kolar schist belt is an Andean type continental magmatic arc; and the east side is a possible Phanerozoic analog of an evolved island arc, such as Japan (Krogstad et al., 1995).

Table: 1.1.10 *Composition of Kolar Amphibolites (Rajamani et al., 1985)*

Major oxides	Min (wt%)	Max (wt%)
SiO_2	44.74	56.07
TiO_2	0.66	1.50
Al_2O_3	6.34	16.60
FeO_T	9.30	15.88
MnO	0.16	0.34
MgO	5.51	21.34
CaO	6.30	13.96
Na_2O	0.04	3.38
K_2O	0.00	0.73
Trace elements	Min (ppm)	Max (ppm)
V	180	336
Cr	114	3492
Co	27	122
Ni	81	1361
Rb	1	12
Sr	13	315
Ba	10	310
Y	13	28
Zr	31	93
Nb	0.8	14

Gadwal Schist Belt The 85 km long and 1–15 km wide schist belt, shared geographically by the Mehabubnagar and Kurnool districts of Andhra Pradesh, is covered in the north by the Deccan Trap rocks and in the south by the sediments of the Cuddapah basin. It could be continous with the Raichur belt of Karnataka. The basal unit is composed of basalt–andesite, overlain by dacite–rhyolite, volcano-clastic rocks and BIF (Ramam and Murthy, 1997). The granitoid rocks intrusive into the Gadwal shist belt rocks have been grouped by Gopala Reddy and Rao (1992) in to quartz diorite-tonalite, tonalite–granodiorite–adamellite, adamellite–granite. Besides, fine-grained granite and banded gneisses, occurring as small enclaves within the intrusive granite, are tonalite–granodiorite in composition. Three phases of deformation have been noted in the rocks of the belt. The first phase folds F_1 are isoclinals, and the second phase folds F_2 are co-axially folded with F_1. The third phase fold axis makes high angles with those of F_1 and F_2.

Metamorphic grade reached upto amphibolite facies. No radiometric age data of the rocks of the belt are known.

Jonnagiri Schist Belt This rather small (25 km × 0.5–5 km) dolphin-shaped schist belt is located partly in Ananthapur and partly in the Kurnool district of Andhra Pradesh.

The rocks composing the Jonnagiri schist belt are divided into the lower Dona Formation, consisting of metabasalt, meta-felsic volcanics (dacite-rhyolite) and the upper Gavinikonda Formation containing meta-tuff, now represented by quartz-sericite/chlorite schists.

Table 1.1.11 *Lithostratigraphy in the Jonnagiri schist belt (Jairam et al., 2001)*

Formations	Rock types
	Mafic dykes
	Chennampalli Granite (Adamellite – granite)
	Pagadarayi Granodiorite (Granodiorite – tonalite)
Gavanikonda Formation	Metatuff
	Quartz–sericite / chlorite schist
Dona Formation	Rhyolite , dacite
	Metabasalt

Three phases of deformation affected the rocks of the belt and produced the structures including folds, axial cleavages and shear zones. Syn- to late- tectonic intrusive granitoids vary in composition from tonalite through granodiorite and adamellite to granite (Jayram et al., 2001). Post-granitoid gabbro-dolerite intrusives are common.

Veligallu Schist Belt It is another N–S trending schist belt having a length of ~60 km. The northern part of the belt is covered by the Cuddapah Supergroup rocks. On all other sides it is bordered by granites and gneisses. Srinivasan (1990) suggested the following stratigraphic sequence for the Veligallu belt (Table 1.1.12).

Table 1.1.12 *Stratigraphic sequence of rocks in Veligallu schist belt (Srinivasan, 1990)*

Formations	Rock types
	Mafic dykes
	Younger granitoids
Malyankonda	Amphibolite dyke
	BIF
Shivapuram	Rhyolite, rhyodacites
	Quartz-muscovite schist and agglomerate
Tamballapalle	Agglomerate, meta-basalt
	Ultramafic bands and amphibolite

Kadiri Schist Belt This belt is ~80 km long and trend NNW–SSE to N–S in parts of Anantapur and Chittoor districts of Andhra Pradesh. In its northern extremity, it is covered by the Cuddapah Supergroup of rocks.

The stratigraphic sequence consists of metamorphosed basic volcanic rocks in the bottom, overlain by felsic rocks of varying petrography and chemical composition. The meta-basalts vary from tholeiitic to high-Mg (upto 14.6 wt%) varieties in composition. The felsic group contains varieties, ranging from andesite through dacite to rhyolite. Adoption of different chemical parameters plot these rocks in (i) continental type, (ii) island arc type, (iii) AB (LKT type), (iv) active continental margin type fields (Satyanarayana et al., 2000). Three generations of deformational structures, as observed in other schist belts, are also the characteristics of this belt. The interface between the schist belt and the granite–gneiss country rocks is highly tectonised (sheared). Metamorphism did not exceed greenschist facies.

Nellore Schist Belt The Nellore schist belt (NSB) is >200 km in length, and occurs in the Nellore and Prakasham districts of Andhra Pradesh. Conformant to the eastern margin of the Cuddapah basin, the belt is convex towords the west (Fig. 1.1.4) along the western boundary, the NSB has a tectonic (thrust) boundary with the Cuddapah Supergroup rocks. In the north, it is juxtaposed against the Eastern Ghat granulite belt. The boundary is a ductile shear zone, marking out regions of contrasted lithologic composition and metamorphic grade. The southern part of the NSB is covered by Phanerozoic rocks.

The rock types composing the belt comprise both the metasediments including metapelites, quartzites and calc-silicate rocks, and metamorphosed volcanic rocks that vary in composition from basic to intermediate to acidic varieties (Vasudevan and Rao, 1975). The lithostratigraphy of the Nellore schist belt after Srinivasan et al. (1994) is presented in Table 1.1.13.

Table 1.1.13 *Lithostratigraphy of the Nellore schist belt (Srinivasan, et al., 1994)*

Groups & Formations		Rock type
Intrusives		Mafic dykes Granite and pegmatites Gabbro–gabbroic anorthosite
Udaigiri Group		Arenaceous and argillaceous metasediments with interbedded mafic–acid volcanics
Gudur Group	Malakonda Formation	Pelitic metasediments, minor acid volcanics, sideritic limestone and BIF
	Chaganam Formation	Amphibolites, much migmatised
~~*Unconformity*~~~~~~~~~~		
Banded gneissic basement		

The amphibolites of the Chaganam Formation have locally been migmatised to such an extent that relict amphibolite occurs as enclaves within the migmatites. Oxide facies BIF may be extensive over 20 km or more. Pelitic sediments have been metamorphosed to kyanite-staurolite-bearing schists/gedrite-cordierite-staurolite-anthophyllite assemblage. A massif-type anorthosite complex occurs at Inukurti and an ophiolite complex containing sheeted dykes and plagiogranite occurs at Kandra (Leelanandam and Ashwal, 2004). These complexes were emplaced at least before the metamorphism reached its acmate T = 650 °C, P = 7.5 kb (Babu, 1970).

The structural history of the Nellore schist belt is not uniform all over. While the southern part bears evidence of three phases of deformation, the central and northern parts bear evidence of two additional phases of deformation (Srinivasan and Roopkumar, 1995)

The Nellore schist belt is known for containing mica-pegmatites. A barite deposit, extensive for a distance of 2 km and up to 10 m in thickness occurs near Kodandarama, and was being exploited (Vasudevan, 1989).

Peninsular Gneiss in the neighbourhood is generally tonalite to trondhjemite in composition.

Khammam Schist Belt Khammam belt is located north of the Nellore schist belt with EGMB in the east and East Dharwar granite–gneiss in the west. It is considered as the northern extension of the Nellore schist belt. The outermost unit in the west consist of conglomerate, quartzite, carbonates and barites. This is followed in the east by a zone that contains garnet-bearing granodiorite and granodiorite and adamellitic gneiss, having inclusions of granulitic rocks. The rocks of the Khammam schist belt were originally both sediments and volcanic rocks. The metabasalts are hypersthene-normative Fe-tholeiites (Sarvothoman, 1995). A number of anorthositic bodies are reported from this belt, the most important of which is the Chimalpahad anorthosite-gabbro layered complex (~200 sq km) in Khammam district (Ramam and Murthy, 1997), the largest of its kind in Peninsular India. Ultramafic rocks with chromite seams are common members of these complexes. The setting is suggestive of an oceanic crustal environment and the Chimalpahad layered complex may be interpreted as representative of Archean (?) ophiolite assemblage (Leelanandam and Ashwal, 2004).

The rocks have undergone three phases of deformation. The first phase deformation, prograde metamorphism and magmatism are nearly coeval.

We have a few radiometric age data for the rocks of Nellore–Khammam schist belts. The situation is better here than in some other belts of Eastern Dharwar for which, as our foregoing discussion shows, we have as yet little or no data.

K-Ar analyses of whole rock and muscovite yielded ages of 989 ± 23 Ma and 806 Ma, respectively (Ghosh et al., 1994). The Sm-Nd (WR) isochron age of garnet-clinopyroxene bearing amphibolite has been determined to be 824 ± 43 Ma, wheras the Rb-Sr isochron age of the same rock was 481±10 Ma (Okudaira et al., 2001). The first one could represent the peak of metamorphism, while the second one tally with the last Pan-African event. Ramakrishnan (2003) is of the view that the Nellore-Khamam belt developed as a rifted volcanic margin during the post-Cuddapah time. However, the picture is complicated by a U-Th-Pb age of 1630 ±100 Ma obtained from pegmatites intrusive into Nellore schist belt rocks (Ramam and Murthy, 1997). Okudaira et al. (op. cit.) proposed that the Nellore–Khammam schist belt probably represented the foreland of an orogen during Neoproterozoic when the Eastern Ghat terrain accreted to the Dharwar-Bastar cratons. During this orogeny, the late Archean protoliths of Nellore–Khammam belts were thermotectonically reworked and metamorphosed to high-pressure amphibolite facies. Effects of another phase of tectonic movement were superimposed on these rocks during the Early Paleozoic. The above data, however, raise a strong doubt against the rationality of including Nellore–Khammam belts in the group of Dharwar schist belts.

Kimberlites and Lamproites

Kimberlitic and lamproitic rocks have been known to occur at a number of places in Eastern Dharwar, mostly in Andhra Pradesh and a few in the neighbouring districts of Karnataka. Many more occurrences of such rocks have been discovered in this region in the recent past (Fig.1.1.21). These rocks are very uncommon petrochemically, in modes of occurrence, tectonic settings, as well as proneness to weathering and erosion. As a result, in spite of concerted efforts made during the last several decades, certain aspects of these rocks still remain controversial. We outline to-date general knowledge of these rocks, before introducing their Indian occurrences. Needless to mention that these rocks are home to 'primary' diamonds, which will be discussed in a following section.

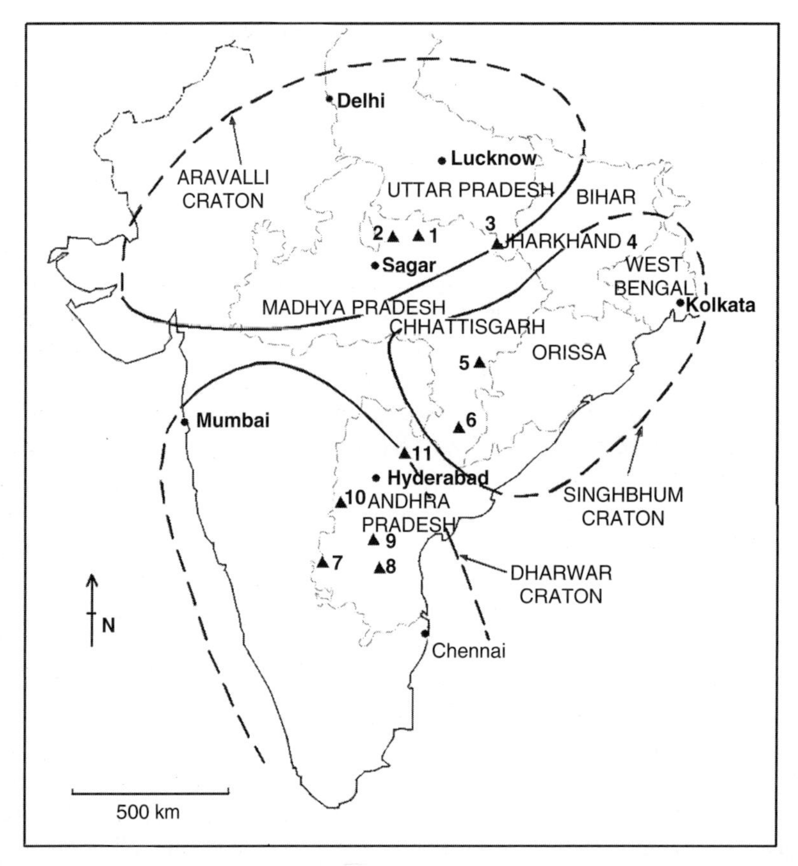

Fig. 1.1.21 Reported locations of kimberlite and lamproite in India (cratons shown after Naqvi, et al., 1974). 1. Majhgaon and Hinota, 2. Angor, 3. Jungel, 4. Gondwana coalfields, 5. Raipur, 6. Tokapal–Bejripadar, 7. Wajrakarur–Lattavaram, 8. Chelima, 9. Zangamrajupalle, 10. Maddur–Narayanpet, 11. Warangal (modified after Scott Smith, 2007)

General Characteristics Kimberlites, first reported from Kimberly, South Africa, are K-enriched ultramafic rocks, commonly containing a sizeable proportion of mantle-derived (not rarely even crust-derived) xenoliths. They commonly occur in the form of narrow dikes, craters and diatremes in Archean cratons and adjacent Proterozoic belts, particularly in the African and Siberian shields, as well as those in Brazil, Canada, Australia and India. In many of these regions, kimberlite emplacement barring a few exceptions has been repeated globally and locally over a period of time that spans from Proterozoic to Cenozoic. Regional and local structures, aided by fluid pressure, determine their localisation. Chemically, the kimberlites differ from the common ultramafic rocks by high contents of volatiles, TiO_2, K_2O, P_2O_5 and other incompatible elements.

Kimberlites are divisible into two textural varieties: macrocrystic and aphanitic. The former is far more common and is dominated by large (> 2mm) anhedral grains of olivine. Subordinate phases include phlogopite, clinopyroxene, garnet, chromite, ilmenite, apatite, monticellite, perovskite, melilite. Deuteric alteration, modifying both the mineral composition and texture is a common feature of this rock. From mineralogical and geochemical points of view, the kimberlites are divided into two types: Group-I, or the archetypal variety, and Group-II (orangeite) (Smith, 1983), the latter being much less common.

In contrast to the common variety of kimberlite, lamproites are characterised by conspicuous difference in petrography and geochemistry, including isotope composition (Mitchell, 2006). However, they may have some petrochemical difference amongst themselves, depending on which they may be named as olivine lamproite, phlogopite lamproite, leucite lamproite etc. Lamproites are generally enriched in LREE and Rb. Although glass and in particular scoriaceous juvenile lapilli have not been found in kimberlites, these are common in lamproites. Complex shapes of olivine macrocrysts and phenocrysts are typical of lamproites. Polysynthetically twinned phlogopites also are more common in lamproite (Scott Smith, 2007). Other minor but typomorphic phases include K-Ti richterite, wadeite and priderite. Lamproites resemble kimberlite-II in some aspects, while their differences are none too negligible (Mitchell, 1995).

The mantle-derived xenoliths in kimberlites fall into two types: ultramafic and eclogitic, the former being dominant or more common. The ultramafic xenoliths or nodules in turn are dominated by garnet lherzolite. Other members are garnet harzburgite, garnet pyroxenite, spinel peridotite and dunite. These come from varying depths as suggested by the presence of spinel vs garnet. Eclogite nodules are composed essentially of omphacite, clinopyroxene and pyrope garnet. Chemically they have the bulk chemistry of basalt (Philpotts, 1994). These are sort of 'black boxes' in guessing, if not establishing, the source of kimberlitic melts.

Archetypal or Group-I kimberlites are geochemically and mineralogically similar in the global scale, irrespective of whether they are emplaced within cratons or their periphery, i.e. accreted mobile belts (Mitchell, 1986, 2006), or the geological time of emplacement, confirming that, they are a distinct magma type formed by a specific reproducible petrogenetic process. Available information from the mantle-derived xenoliths in kimberlites, together with the results of experimental studies and isotopic analyses of kimberlites and the mineral inclusions in the associated diamonds suggested that the archetypal kimberlites were derived from the asthenospheric mantle, which may range up to and even beyond those of the transition zone (400–700 km) (Stachel et. al., 2000). It has also been suggested that the archetypal kimberlites formed by a small degree of partial melting (<1 volume%) of carbonated garnet lherzolite at $P > 5$ G Pa (Dalton and Presnall, 1998; Wyllie and Lee, 1999), or from a larger volume of melts (10 volume%) of metasomatised protoliths (Mitchell, 2004). Another model suggests that the Group-I kimberlite magmatism could be triggered by metasomatism of previous refractory sub-continental lithosphere by fluids yielded by upwelling mantle plume (le Roux et al. 2003). The highly radiogenic nature and the trace element contents of Group-II kimberlites suggest that they are direct melts of the ancient metasomatised sub-continental lithosphere (Smith, 1983). The trace element content, particularly low Nb/La ratio suggests that the metasomatism was probably related to subduction processes. Lamproites and orangeites are thought to have formed by partial melting of metasomatised lithospheric mantle (MLM) of harzburgitic composition, at or above the lithosphere asthenosphere boundary below 150–200 km, in contrast to depleted asthenospheric mantle source in case of kimberlites (Mitchell, 2006). If indeed the kimberlitic (archetypal) and lamproitic/orangeite magmas develop in two different situations, then there is no scope for the use of such expressions as 'transitional' or the like (Mitchell, 2006). Gurney et al. (2005) view the situation in a somewhat different way. According to them both the kimberlites and the lamproites have an asthenospheric source, and they interact with lithosphere enroute to the surface. There are some rocks found around the world which though resemble, do not smugly fit into the definition of kimberlite or lamproite. They are in all possibility products of hybridisation or contamination.

How did the kimberlite/lamproite bodies get themselves emplaced into the crustal domain where we find them today? There is no unique answer to this question, although there is no disagreement over the fact that the explosive breakthrough to the surface takes place when the fluid (volatile) pressure exceeds the confining pressure. According to Clement and Reid (1989) the explosive fluid pressure develops in a closed system as the volatile-charged magma rises upward. Another opinion holds that the diatremes form by multiple downward migrating phreatomagmatic (hydrovolcanic) eruptions resulting from successively deeper melt-groundwater interactions with time (Lorenz, 1985; Lorenz et al., 1999). Mitchell (1986) in a slightly modified version stated that the pipe development may be initiated by the explosive exsolution of juvenile gases, but the diatreme deepens by the downward moving phreatomagmatic explosion. Field and Scott Smith (1999) asserted, citing many examples, that phreatomagmatic activity had no role in the pipe formation. It is rather due to intermittent rise of the gas-charged magma by stoping and wedging until it breaks open to the surface, all through behaving as a closed system. Possibly both the mechanisms operate depending on the 'ground reality'.

In the context of time-space distribution of kimberlitic intrusions, it is pertinent to believe that they should not be unrelated to global tectonics. England and Houseman (1984) believed that the increase in the kimberlite intrusions could be related to retarded plate velocity when uninterrupted mantle convection was the cause of partial melting (and volatile production) in the lithosphere and the subsequent plume activity. Accompanying epeirogenic uplift would create fractures that facilitated a rapid migration of the kimberlite magma upward. In the context of African examples, Robb (2005) thinks that it is a plausible explanation.

Major Occurrences of Kimberlites and Lamproites in Eastern Dharwar (Cuddapah Basin Included) In the recent years, a large number of kimberlite and lamproite bodies have been discovered in South India particularly in Eastern Dharwar. These are generally divided into Wajrakarur kimberlite field (WKF), Narayanpet kimberlite field (NKF), Nallamalai lamproite field (NLF) and Krishna lamproite field (KLF) (cf. Neelakantam, 2000).

Wajrakarur (– Lattavaram) Kimberlite Field (WKF) The Wajrakarur (–Lattavaram) field (WKF) occupies the western part of Anantapur district and the southern parts of Kurnool district of Andhra Pradesh. The field, measuring about 120×60 km, occupies the area skirting the western margin of the Cuddapah basin. The country rocks are represented by Peninsular Gneiss, younger granitoids and patches of Ramagiri-Penakacherla and Jonnagiri greenstone belts. The kimberlites are emplaced into the gneissic basement rocks, as well as the schist belt (greenstone) rocks. On the basis of spatial relationships the WKF kimberlites are generally grouped into three clusters: Wajrakarur–Lattavaram cluster (largest), Chigicherla cluster and the Kalyandurg cluster. Recently discovered Timmasamudram cluster is also referred to this field. Scott Smith (2007) studied some of the kimberlites of WKF. The studied material contained two generations of olivine. Macrocrysts are anhedral grains that are commonly rounded, while the phenocrysts exhibit simple euhedral forms. This is a characteristic of kimberlites. The main groundmass is made up of phlogopite, monticellite, spinel, perovskite, serpentine, carbonate, clinopyroxene, pectolite and probably melilite. In Scott Smith's nomenclature, most of these rocks are mica-bearing montecellite kimberlite.

The various forms of xenoliths are garnet–harzburgite, lherzolite, wehrlite and olivine-clinopyroxene eclogite. Thermobarometric analysis showed the ultrabasic xenoliths to have formed at P = 70–32.5 kb and T = 1490° – 930°C, and the eclogitic xenoliths at P = 53–24kb and T = 1240°–800°C. The PT estimates of these xenoliths suggested that the related lithosphere during the Mid-Proterozoic time was at least 185 km thick and that the peridotitic and eclogitic xenoliths were picked up from the depths of about 100–180 km and 75 – 150 km, respectively (Ganguly and Bhattacharya, 1987).

Narayanpet Kimberlite Field (NKF) NKF is located at about 200 km north of WKF and about 150 km southwest of Hyderabad. It measures 60 km × 40 km, covering the western parts of Mehabubnagar district in Andhra Pradesh and spilling over to the eastern parts of Gulbarga district in Karnataka. The country rocks are the Peninsular Gneiss, granitoid intrusives and greenstone belt (Gadwal) rocks. The kimberlites are emplaced in four clusters: Kotakonda, Maddur, Narayanpet and Bhima. Of these the Maddur and the Narayanpet clusters are larger.

Nallamalai Lamproite Field (NLF) The NLF is located within the intracratonic Cuddapah basin in the Eastern Dharwar, and extends from Tirupati in the south to Jaggayapeta in the north via Chelima in the middle. The area is occupied by the rocks of the Nallamalai Group. The Cuddapah sediments are intruded by lamproite dikes near Chelima, Pachcherla and Zangamrajupalle, the intrusive trends being in the WNW–ESE to NW–SE directions. Diamonds in the gravels of the Sagileru and Kundair rivers, which were extensively worked out in the past, must have come from these lamproite bodies.

Scott Smith (2007) based on limited studies on the Chelima rock reported that it contained abundant phenocrystic phlogopite, a few probable olivine macrocrysts and phenocrysts in a matrix consisting of perovskite, apatite, probable spinel and clinopyroxene. Chalapati Rao (2007) reported the rocks to be SiO_2-deficient. Assuming that the reported quartz and carbonate are secondary, Scott Smith thinks that the rock could be a lamprophyre / minette / lamproite. Chalapati Rao (2007) after a detailed study is, however, convinced that the rocks are lamproites *sensu stricto* and having the distinction of being the oldest amongst the world's known occurrences of the rock. He further concludes, taking into consideration the silica undersaturation, potassic-ultrapotassic nature, enrichment in trace element (including REE), that these dyke rocks were derived essentially from deeper parts of ancient (older than 2 Ga) and anomalously enriched lithospheric mantle in the garnet stability field. The Chelima dykes are unrelated to the sills, dykes and lavas of the Lower Cuddapah, as well as the kimberlite magmatism around. There is no convincing evidence of subduction tectonics and the mantle metasomatism is attributed to volatile and K-rich, extremely low viscosity melts that leak with or without breaks from the asthenosphere, and accumulate in the overlying lithosphere or upper asthenosphere (Chalapati Rao, 2007).

Krishna Lamproite Field (KLF) KLF, located at the northern side of the Krishna river along the eastern margin of the Dharwar craton in Krishna and Nalgonda districts is spread over an area of about 160 sq km, containing more than 25 lamproite bodies. The lamproite bodies occur in clusters in the following localities: Ramannapeta, Vedadri, Pochampalli, Gopinenipalem, Tirumalagiri Anumanchipalli, Polleru, Nallabandagudem and Reddikunta. Recent studies by Paul et al. (2007) shed some light on the potassic ultramafic rocks of the Krishna district. The principal minerals comprising the rocks include olivine, pyroxene, amphibole and phlogopite, the last mineral tangibly varying in composition. There is enrichment in LILE and LREE (La/Yb: 24–104; Gd/Yb: 3.8–8.02). Abundances of Zr (> 500 ppm), Sr (> 1000 ppm) and La (> 200 ppm) support its lamproitic composition, although the Ba content is rather low. In final analysis the authors call it a lamproite, formed by partial melting of a mantle that was metasomatically enriched in Ti and Fe.

Available Rb-Sr isochron ages of kimberlites and lamproites from South India range from ~1400 Ma to ~1000 Ma, amongst which the kimberlites of Wajrakarur and Narayanpet are generally (not strictly) younger (Table 1.1.14; Fig. 1.1.22).

Table 1.1.14 *Rb-Sr isochron ages of kimberlites and lamproites of the Dharwar craton (Neelakantam, 2000)*

Kimberlite / lamproite field	Pipe No.	Age (Ma)
Wajrakarur kimberlite	P-1	1090
	P-2	1092±15
	P-3	966±38
	P-4	1023±40
	P-5	1145±88
	P-7	1091±10
Narayanpet kimberlite	KK-1	1084±13
	NK-3	1099±12
Nallamalai lamproites	Chelima	1354±17
	Zangamrajupalle	1070±22
Krishna lamproites	Ramanapeta	1221±18

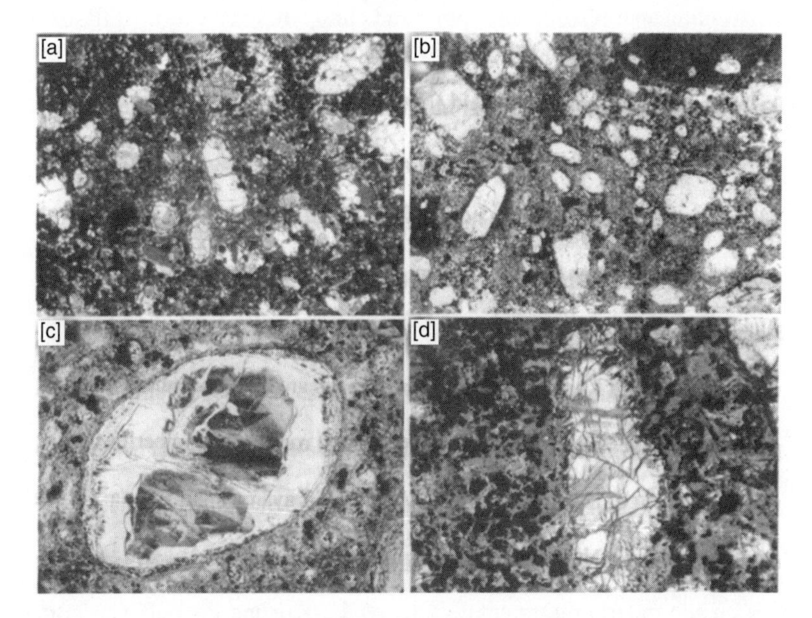

Fig. 1.1.22 Photomicrographs of hypabyssal-facies of (a) kimberlite, Pipe-2, Wajrakarur cluster, (b) carbonate-phlogopite kimberlite, Pipe-7, Wajrakarur cluster, (c) kimberlite, Chigicherla cluster, WKF, (d) pervoskite-phlogopite kimberlite, Pipe-3, Lattavaram cluster (Fareeduddin et al., 2007)

Western and Eastern Dharwar Blocks: Similarity, Contrast and Relationship

The so-called Western Dharwar block is smaller than the Eastern Dharwar block (~45000 sq km vis-à-vis 60,000 sq km). Granite-gneisses mainly belong to the TTG group in West Dharwar, whereas these vary from tonalite, through granodiorite and adamellite to granite (TTG vs calc-alkaline trend) in the Eastern Dharwar. The major events of accretion of granite-gneisses in Western Dharwar correspond to ~ 3.4, ~3., 3.0–2.9 Ga and the last to 2.6 Ga. Barring some banded gneissic rocks near the southern end of the Closepet batholith and west of the Kolar schist belt that have been dated 3.4– 2.96 Ga, most of such rocks in East Dharwar were emplaced in the time slot, 2.6– 2.55 Ga (Fig. 1.1.23).

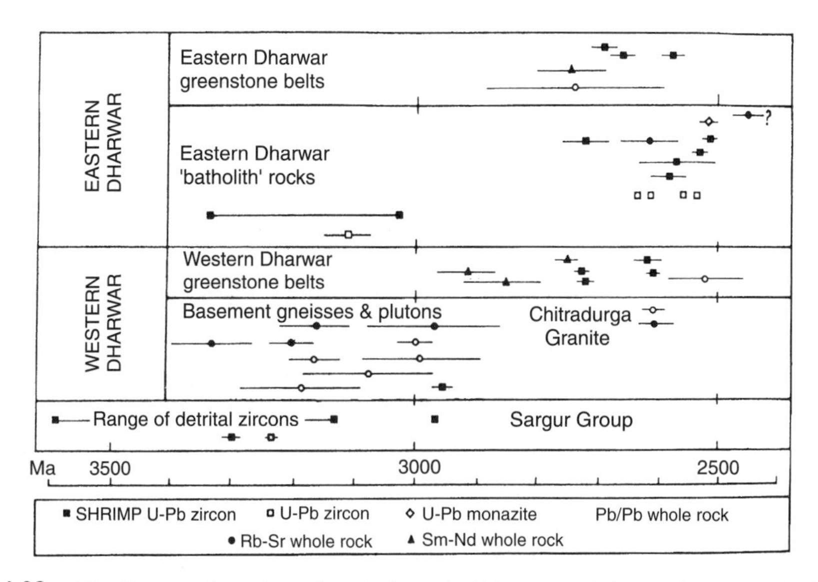

Fig. 1.1.23 Distribution of age data of rocks from the Western and Eastern Dharwar Blocks

In the Western Dharwar, the schist belts are of two ages (~3.3 Ga and 2.7–2.6 Ga). The older type (Sargur) occurs as small belts and even as screens, whereas the younger type (Dharwar) is much more extensive. The volcani-sedimentary package filling the latter may be as thick as 8 km. Sedimentary rocks/ volcanic rock ratios ≥ 1. Sediments may be clastic, such as greywacke, pelites, quartzites and chemical, such as BIF and carbonates. The older schist belts (Sargur type) also contain metapelites, fuchsite quartzite, metamorphosed basic and subordinate acid volcanic rocks and some BIFs. The above age data bring out difference in the temporal relationship between the granite-gneisses and the schist belts in the two blocks.The schist belt rocks in Western Dharwar are generally younger than the surrounding granite-gneisses, whereas reverse is the relationship in Eastern Dharwar, with inconsiderable age-difference, however.

The schist belts in Eastern Dharwar are generally linear. They are of one age (2.75–2.6 Ga). Some of the belts are apparently free of sediments. Some others (Sandur, Kustagi, Ramagiri and of course Nellore–Khammam belts) contain sediments in fair proportions. In the Kolar and Hutti–Maski belts sediments are present but only in insignificant proportions.

Clastic sediments from Western Dharwar show strong negative Eu anomaly. In the Eastern Dharwar, it holds good only for the siliceous shales in the Ramagiri belt. This is better explained by assuming the major source of clastic sediments in the Western Dharwar schist belts being Archean intra-crustal granite-gneisses and that for the Eastern Dharwar belts, the arc/ arc-continent mixed provenances. In Eu/Eu-vs-CIA [Al_2O_3/ (Al_2O_3 + CaO + Na_2O + K_2O) x 100] diagram the greywackes of the two blocks plot in two broadly different fields (Fig. 1.1.24).

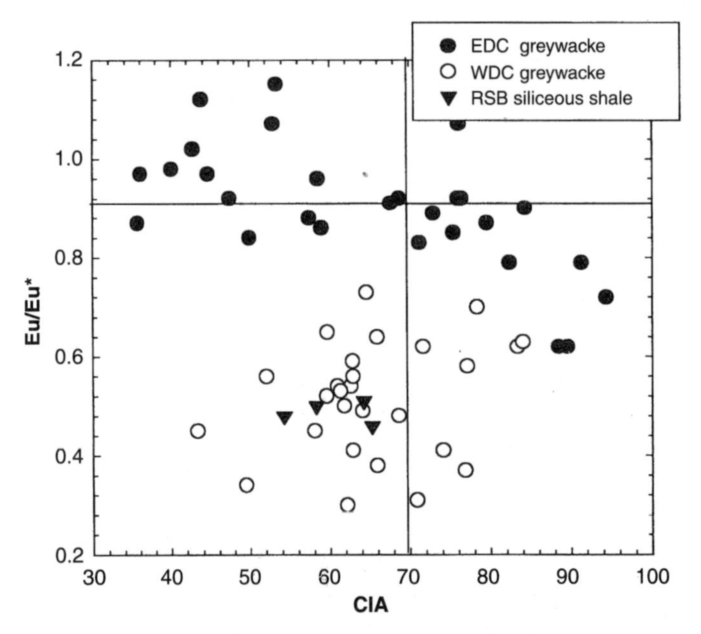

Fig. 1.1.24 Eu/Eu* vs CIA correlations of the greywackes from some schist belts of Western and Eastern Dharwar Blocks (after Mishra and Rajamani, 2003)

Ultramafic and gabbro-anorthosite bodies are present in the schist belts of Western Dharwar. In Eastern Dharwar, they are fewer except in Nellore–Khammam belts where they are more numerous and some are very large. But as has already been alluded to, recent observations raise strong doubt against inclusion of Nellore–Khammam belts in the group of Dharwar schist belts, as we understand them today.

Deformational structures in all the schist belts in both East and West Dharwar blocks have much in common. They are of multiple origin. The first folds are isoclinal and co-folded with the second fold, the third making various angles with the axial trace of the first two. This is, however, a characteristic of most Early Precambrian terrains. Shear zones of various scales are common in the schist belts and adjacent granite-gneisses. The generalised structural pattern of the Dharwar schist belts is apparently the result of E–W compression accompanied by transcurrent shear. Similarity of structures in the schist belts and adjacent gneisses is best explained by mobilisation of the latter during the deformation. Metamorphism varies widely in both the Dharwars. In the Western Dharwar schist belt rocks it ranges from low greenschist to granulite facies and that in the Eastern Dharwar varies from upper greenschist–amphibolite to granulite facies. The tectonic settings for the Eastern Dharwar schist belts have been interpreted to be intra-arc in most cases. In contrast, those for the belts of Western Dharwar are suggested to be intracratonic rifts, or passive-active continental margin settings.

There remains the moot question: How did the two blocks juxtapose? Different models have been proposed to answer this question and these may be divided into (i) collision models, (ii) subduction models, and (iii) plume models. The collision models involve juxtaposition by collision of two terrains with distinct geological histories, producing *inter alia* anatectic granites enriched in compatible elements and alumina. The juvenile inputs characterising the late Archean magmatism, including that in the Eastern Dharwar, are uncommon in collision belts. Moreover, continental collisions lead to great thickness produced by thrusting and is characterised by clockwise isothermal decompressional P-T-t path. This is not the situation in Dharwar (Jayananda et al., 2000). Deep seismic reflection pictures for this region, if available, could be useful in this context. Chadwick et al. (1997, 2000) suggested an interesting model in this context. They suggested the Western Dharwar as the foreland to an accretionary arc in the east represented by a late Archean batholith (their 'Dharwar batholith') with its contained schist belts (Fig. 1.1.10). These schist belts are intra-arc basins. An oblique convergence gave rise to N–S transcurrent shearing. It is indeed an attractive proposition, but as pointed out by Mukhopadhyay and Srinivasan (2003), it does not satisfactorily explain the contemporaneity of the development and evolution of the schist belts in the foreland and the arc. Even if it is true, available radiometric age data suggest that the interface can not be expected to lie along the shear zone along the eastern margin of the Chitradurga belt (Sarkar, 2001). Whatever could be the exact mechanisms, the two blocks were united during 2.6–2.5 Ga.

The simplistic model of the schist belts in Eastern Dharwar being all intra-arc basins caught up in a slightly younger batholith does not satisfactorily explain the findings on the radiogenic isotopes in some of these belts, the Kolar and the Ramagiri belts in particular. According to Balakrishna et al. (1990) the Pb-isotope composition of galena in the Kolar ore veins suggested that the ore fluids might have introduced extraneous Pb from older granitic gneisses to the amphibolitic host rocks, causing $^{207}Pb/^{204}Pb$ ratio in the ores as found today (see 1.2: Gold mineralisation in Kolar belt). Further, Walker et al. (1989) held the view that the Re–Os systematics of many of the rocks of the Kolar belt indicated that osmium was introduced to the area via fluids that carried highly radiogenic osmium ($^{187}Os/^{186}Os = 88 \pm 5$). The Re–Os model age of this sample is around 2.6 Ga. The likely source of this radiogenic osmium was the ancient crust.

Geochemical data (Nd–Sr isotopes and LREE and LIL element contents) suggest highly enriched mantle and ancient TTG crust for the Closepet batholith, enriched mantle and TTG crust for the Bangalore and west of Kolar schist belt areas and a somewhat depleted mantle for the granodiorites to the east of this belt. Jayananda et al. (2000) tried to explain all these observations by a plume model. The centre of this model was an 'enriched hot spot' lying below the Closepet Granite (Fig. 1.1.25).

Such a plume at its hottest zone would produce high temperature magmas and interact with the pre-existing crust causing partial melting. These magmas could cool rather slowly being at the thermal focus, and thus give younger ages to their products. The region away from the thermal focus, in this case the area lying to the west, would cause melting of chondritic or slightly depleted mantle causing production of less enriched magmas. They would cool rather rapidly, and have fewer interactions with the country rocks. This model also has its attraction, but does not explain the origin of the schist belts in Eastern Dharwar. Jayananda (personal communication to SCS, 2000) does it by the following modification of their model. In the modified form, instead of a single model, a two-stage model is proposed. The first stage (2.7–2.6 Ga) was characterised by horizontal movement, i.e. plate tectonics, and the second one (2.55–2.52 Ga) by the impact of a megaplume. Melting of a westward subducting short-lived oceanic crust/mantle during 2.7–2.6 Ga produced the greenstone volcanics and magmatic protoliths of the surrounding granite-gneisses. Closure of the subduction process was followed by sinking of the subducting slab at 660 km seismic discontinuity and its subsequent collapse into lower mantle triggered the generation of the plume. This model also does not satisfactorily explain the origin of the greenstone belts and the presence of 3.4–2.90 Ga old TTG gneisses in Eastern Dharwar.

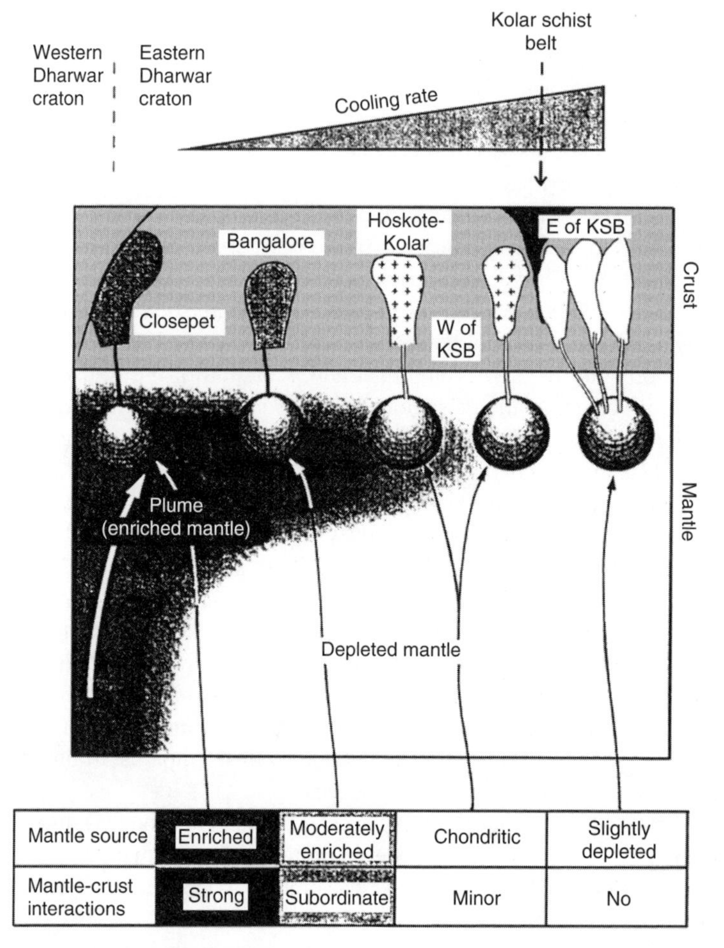

Fig. 1.1.25 'Plume' model of Jayananda et al., 2000 to explain the geology of Eastern Dharwar. The cooling rate increases to the East.

The Western Dharwar block remained quiescent since about 2.5 Ga. The Eastern block on the other hand, was subjected to thermotectonic perturbations in the Proterozoic, producing the intracratonic basins of Cuddapah- Bhima-Kaladgi and possibly also the still less understood Nellore–Khammam schist belts. That the evolution of the Eastern Ghat Mobile Belt (EGMB) must have had its share of mafic alkaline (kimberlite, lamproite and lamprophyre) intrusives within and around the Cuddapah basin during the period 1380–1090 Ma (Chalapathi Rao et al., 1998), also speaks of later thermo-tectonic perturbations, besides other things.

Southern Granulite Province (SGP)

The Peninsular Gneiss and the supracrustal rocks, particularly of the Sargur Group of the Western Dharwar block (WD), and the rocks of the Kolar schist belt (KSB), grade into granulite facies rocks in the south, marking out the uneven *Charnockite line* of Fermor or the *Opx* (orthopyroxene) *isograd* in today's parlance. The region south of this line, sans the area occupied by Phanerozoic cover sediments, is known as the Southern Granulite Terrain (SGT). We would prefer to replace it by 'Southern Granulite Province' (SGP) in consonance with the term Dharwar Province for the region north of it. SGP not only contains granulite facies rocks but also gneisses and supracrustal rocks, all metamorphosed to amphibolite facies in prograde or retrograde regimes of metamorphism. This province has been divided into five blocks, outlined by regional scale shear zones. These blocks are (i) Northern block, (ii) Nilgiri block, (iii) Madras block, (iv) Madurai block, and (v) Trivandrum block/Kerala khondalite belt (Fig. 1.1.26).

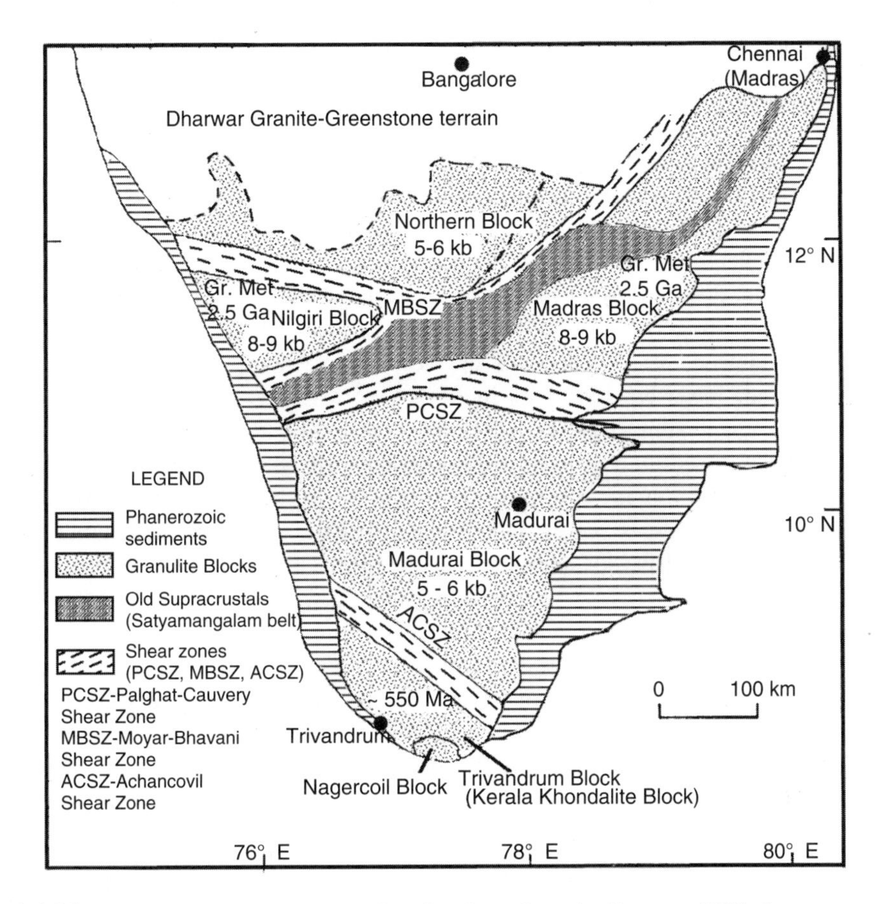

Fig. 1.1.26 Generalised geological map of the Southern Granulite Province (SGP) showing tectonic blocks and intervening regional shear zones

Northern Block The Northern block is situated north of the Moyar shear zone (MSZ). The gradual transition of the amphibolite facies rocks to those of the granulite facies is conspicuous along several sections, such as (i) Kabbaldurga (Satnur) to B.R. Hills (–Sivasamudram), (ii) Kolar–Krishnagiri–Shevroy Hill, (iii) Sakaleshpur (Shiradi) to Coorg. Besides charanockites (Quartz + potash feldspar + plagioclase feldspar + orthopyroxene ± accessores), other granulites and quartzo-feldspathic gneisses are the principal rock types, together with amphibolite, BIF, ultramafic intrusives and granitic rocks (southern part of the Closepet batholith) as subordinate members.

Near the southern tip of the Closepet 'batholith' the maximum P and T of metamorphism are 6–7 kb and 700°C respectively (Moyen et al., 2003) and those in the B.R. Hills in the south these are 7.0–7.5 kb and 730°–800°C respectively (Janardhan et al., 1982; Hansen et al., 1984; Sen and

Bhattacharyya, 1990). The protolith, of amphibolite –granulite facies in the Kabbaldurga–B.R. Hills zone accreted during 3.47–3.0 Ga (Jayananda and Peucat, 1996). The T_{DM} (Depleted Mantle age) analysis of B.R. Hills massif consists of the oldest rocks in the SGP with the protolith ages of charnockite gneises going upto ~3.6Ga and U-Pb zircon ages of ~ 3.4 Ga (Bhaskar et al., 2002). Without denying the possibility that there could be a high grade metamorphic event at 3 Ga (Meen et al., 1992; Jayananda and Peucat, 1996), it may be concluded as per records that a major metamorphic event in the area took place at ~2.5 Ga (Crawford, 1969; Friend and Nutman, 1992; Mahabaleswar et al., 1995). This is true also for the two adjacent blocks, Nilgiri and Madras.

Nilgiri Block The Nlgiri granulite block is an easterly pointing trident shaped massif, extending in strike for about 140 km with maximum width of about 80 km. It is bounded by the eastward converging Moyar and Bhavani shear zones passing along its north and south, respectively (Fig. 1.1.26). Physiographically the terrain forms the highest mountain land of South India, a prominent peak being Dodabetta (2529 m from MSL) near Ootacamond. The Moyar shear zone separates the Nilgiri massif from the B.R. Hills granulites (mainly charnockite) of the Northern block

The Nilgiri highland massif represents the deepest level exposure (c 35 km paleo-depth) of the granulite grade lower crust in South India. This corresponds to a pressure of 9–10 kb and the accompanying temperature of ~700 °C (Raith et al., 1999). The acid granulites of Nilgiris dominantly comprise enderbite with subordinate charno-enderbite and charnockite. The enderbite is both garnetiferous and non-garnetiferous, the former being the dominant variety. There is no significant metasedimentary component in this massif except some sporadic minor bands and lenses of garnet-biotite-feldspar-kyanite-quartz schists/gneisses and banded magnetite quartzite (Srikantappa, 1993).

The garnetiferous enderbitic granulites (quartz + plagioclase + orthopyroxene ± biotite ± garnet) are tonalitic in composition, showing typical calc-alkaline trend. The bulk composition of the rock, its LREE content and pattern of distribution better match a greywacke type protolith (Raith et al., 1999). The U-Pb systematics of composite zircon grains (detrital cores and metamorphic overgrowths), together with the Sm-Nd and Rb-Sr errochron ages and Sm-Nd garnet whole rock data, suggest that the sedimentary precursors and the subsequent high grade tectono-metamorphic processes occurred in the short period between 2600–2500 Ma (Raith et al., op. cit.).

Igneous intrusives compositionally varying from gabbro, anorthositic gabbro to gabbroic anorthosite, occur in the south of Kotagiri. These tabular bodies, a few kilometers along the length and several hundreds of meters in width, are metamorphosed and deformed (Srikantappa, op. cit.).

Madras Block The Madras block, with NE–SW trend is bounded along the northwest by the eastern and northeastern extension and flays of the Moyar–Bhavani shear zone and demarcated in the south from the Madurai block by the Palghat–Cauvery shear zone.

Holland (1893, 1900) was the first to study the granulite rocks of Pallavaram near Madras (now Chennai), later named charnockite after the name of an early (later seventeenth century) English trader of Calcutta (now Kolkata), Job Charnock, whose mausoleum was made of this rock. Since then, of course, quite a few researchers have studied the granulite rocks of this block (Howie, 1955; Weaver, 1980; Bernard–Griffiths et al., 1987; Sen and Bhattacharya, 1990; others). Besides the charnockites other high-grade rocks present are enderbite, leptynite (quartzo-feldspathic granulite), khondalite (quartz + feldspar + sillimanite + garnnet ± graphite bearing granulite) and basic granulites. Granulite metamorphism in the Madras block took place at P = 8–9 kb and T = 750°–900 °C.

Vinogradov et al. (1964) reported a U-Pb zircon age of 2600 Ma for the Madras charnockite. A Rb-Sr (WR) isochron age of 2525± 125 Ma was obtained by Crawford (1969) for the Madras charnockites. A Sm-Nd (WR) isochron age of 2555 ± 140Ma was obtained for the charnockites from south of Madras City. The above data suggest that the magmatic protoliths of Madras granulites accreted during 2.6–2.55 Ga, which was soon involved in granulite metamorphism around 2.5 Ga, followed by a slow cooling upto 2084 Ma (Jayananda and Peucat, 1996).

Madurai Block The Madurai block, outlined by the Palghat–Cauvery shear zone (PCSZ) in the north and Achancovil shear zone (ASZ) in the south (Fig. 1.1.26) is the same as what Ramakrishnan (1994) called 'Pandyan mobile belt'. The granulitic rocks in this block are mainly represented by khondalites, mafic granulites, charnockites and meta-anorthosites. Sapphirine-bearing granulites occur at several places. The original sediments, i.e. the protoliths of khondalites are suggested to have been deposited in shallow marine conditions (Chako et al., 1987). The determined maximum P-T of metamorphism is 5–6 kb and 650°–800 °C. The obtained maximum age of ~2.4 Ga is interpreted to be the age of magmatic crystallisation of the protolith of the metamorphosed rocks analysed (Bartlett et al., 1995). The charnockite and the associated granulites, however, developed at ~550 Ma during the Pan-African event.

There are a number of anorthositic intrusives at Kadavur and Oddanchatram and other places. The Oddanchatram body is the largest amongst them (~1000 sq km) and occurs in the Madurai district in Tamil Nadu. The Kadavur body, which is relatively small, also occurs in Madurai district. They are intrusive into the granulite rocks and were themselves subjected to granulite metamorphism during the late Proterozoic time (Janardhan et al., 1996). Janardhan suggested an extensional thinning of continental lithosphere and invasion of massive anorthosite during Mid- Proterozoic time. In the collisional phase that followed, the crust was thickened and subjected to granulite metamorphism during the Pan-African event.

Trivandrum Block/ Kerala Khondalite and Nagercoil Blocks The Trivandrum block is bordered by the Achancovil shear zone in the north and the Indian Ocean in the south, sans of course the Phanerozoic cover in east. The two units within it are the Kerala Khondalite belt and the Nagercoil block (Fig. 1.1.26). Different rock types present in the former are khondalite, leptynite, incipient-massive charnockite, basic granulite, and calc- rocks, marbles and quartzites as minor members.

The Nagercoil block, located at the southern tip of the Indian Peninsula, is a charnockitic massif. Isotopic studies of carbon in graphite associated with khondalite show that it is of two genetic types: (i) derived from magmatic CO_2, and (ii) derived from organic matter associated with original sediments (Wada and Santosh, 1995). P-T of metamorphism ranged between 5–6 kb and 700–800°C (Harris et al., 1982; Santosh et.al., 1993). Sm-Nd and Rb-Sr ages are of the order of ~500 Ma (Santosh et al., 1992; Chaudhari et al., 1992). Nd model age obtained from Nagercoil is, however, 2100 Ma (Harris et al., 1994; Unnikrishnan–Warrier et al., 1995).

Nature and Geochronology of the Regional Shear Zones and their Interpretations The brittle-ductile shear zones, Moyar, Bhavani, Palghat–Cauvery and Achancovil shear zones are of regional extent. These shear zones, generally occupied by amphibolite facies rocks with relics of granulite rocks here and there, play an important role in understanding the thermo-tectonic evolution of the region. Sm-Nd garnet crystallisation ages and Rb-Sr mica cooling ages from reworked rocks suggest that the Pan-African event caused structurally controlled retrogressive metamorphism, followed by cooling along the Moyar shear zone (MSZ) (garnet 624–591Ma; muscovite 594 Ma, biotite 604–540 Ma). This predates tectonometamorphism in the Bhavani shear zone (garnet 552 Ma, biotite 521–491 Ma) and Palghat shear zone (garnet 521 Ma, biotite 488–485 Ma). A Sm-Nd garnet age of 2355 ± 18 Ma obtained from a granulite remnant in the MSZ recorded an early Paleoproterozoic high-grade metamorphism which may be related to the joining of the Nilgiri block and the Dharwar craton along the Moyar-Bhavani shear zone. A relict charnockite from the southern BSZ yielded a Sm-Nd garnet age of 1705 ± 11 Ma, apparently recording a late Paleoproterozoic metamorphism. Neo-Archean and Paleoproterozoic Nd model ages of 3.0–2.2 Ga suggest that the shear zone rocks are mainly reworked rocks from the adjacent Nilgiri and Madurai crustal domains (Meissner et al., 2002). It may be recalled here that Harris et al. (1994) are of the view that the Nilgiri and Madras blocks are parts of the Dharwar Province. In other words, the Dharwar Province should be extended upto the Palghat–Cauvery lineament. Nd model ages of granites and tonalites that syntectonically intruded the MSZ, provide evidence for crust generation ≤ 1.7 Ga ago. Sm-Nd whole rock data from the unsheared metasediments exposed between the Bhavani and Palghat high strain zones indicate average crustal residence ages of ~ 1.9 Ga, which indicates contribution from an unknown Proterozoic crustal source; and do not confirm their correlation with Sargur or Dharwar supracrustals, as suggested by Nutman et al., (1992).

The magmatic protoliths of the layered anorthosite–gabbro complexes of Sittampundi and Bhavani occurring in the PCSZ are co-eval with granulite metamorphism (Janardhan and Leake, 1975; Ramadurai et al., 1975). The metamorphosed anorthosite of Sittampundi is now composed of calcic plagioclase, amphiboles, garnet, clinozoisite, corundum and anthophyllite. The gabbroic rock in the complex is composed of clinopyroxene, orthopyroxene, garnet, plagioclase and amphibole.

Bartlett et al., (1995), based on P-T and age data came to the conclusion that the Madurai block was not a part of the Archean craton of South India, but was an independent terrain. The addition of Meissner et al. (op. cit.) to this conclusion is that the juxtaposition of the Madurai block and the Dharwar craton occurred less than 1.8 Ga ago, at the latest, during the Pan-African activity in the BSZ and PSZ. After crustal development and metamorphism during 3–2 Ga[2], the block remained practically undeformed until the Pan-African event at ~550 Ma. In this context, it is equally important to mention that the Paleomagnetic data collected by Radhakrishna and Joseph (1996) from the Dharwar Province and the Southern Granulite Province, show that both the provinces have similar pole positions and apparent Polar wandering path since ~2.0 Ga, with the obvious suggestion that the relative position of the Dharwar Province and the Southern Granulite Province remained the same since then. Input from the structural geologists' observations along the PCSZ should also be relevant here. Mukhopadhyay et al. (2001) observed that the granulite facies mineral assemblages and the presence of granoblastic textures in the major part of the area did not suggest that it was a major shear zone, although features of small-scale shears were ubiquitous. The fold pattern is suggestive of shortening across the zone, rather than dextral transcurrent movement. No mylonite belt separates the two blocks. Differences in conclusions based on the differences in approach have, however, to be reconciled.

Causes of Granulite Metamorphism in the Southern Granulite Province It is the general consensus that the P-T conditions of the granulite metamorphism in the region, as inferred from the phase equilibrium data, do not represent a steady-state geotherm, but reflect transient conditions resulting from thermal perturbation. Ganguly et al. (1995) calculated the steady-state geotherm at c 2.5 Ga from the available data and concluded that the granulite facies metamorphism in this region was indeed due to a thermal perturbation which was c 400 °C at 25 km (Fig. 1.1.27).

[2] Chaudhary et al. in a recent publication (JGSI, 77, 227–238, 2011) confirmed crustal accretion and granulite metamorphism in the PCSZ during the Mesoarchean.

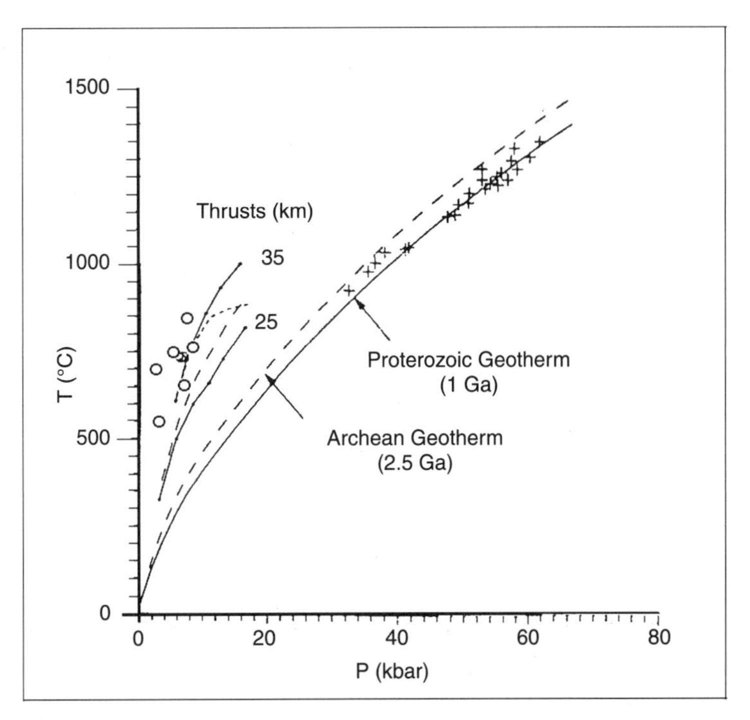

Fig. 1.1.27 Perturbation of the ambient 2.5 Ga geotherm due to suggested overthrusting of blocks, 25–35 km thick, along the horizontal plane (Ganguly et al., 1995)

Suggested Causes of Thermal Perturbation As Ganguly et al. (op. cit.) pointed out, the thermal perturbation causing granulite metamorphism in South India could be due to one or more of the following reasons:

1. A convective heat flux due to CO_2 streaming from a deep source (Harris et al., 1982; Janardhan et al., 1982; Chacko et al., 1987; Ganguly and Singh, 1988; others)
2. Tectonic processes (England and Thompson, 1984)
3. Underplating by basic magmas (Lachenbruch and Sass, 1977; Bohlen and Mezger, 1989)

A number of workers investigated the P-T and fluid conditions related to the development of South Indian granulites, including Janardhan et al., (1982) and Hansen et al. (1984 a, b; 1987) came to the conclusion that the charnockites, rather the granulites, formed under relatively anhydrous conditions ($P_{H_2O} < 0.3P_T$)[3] and that the CO_2 coming from the upper mantle or the lower crust was responsible for the lowering of the P_{H_2O}. Ganguly et al. (op. cit.) calculated the heat flux needed to perturb the steady-state geotherm in South India over a period of geological time and concluded that CO_2 could have played an important role in reducing the P_{H_2O}, needed to produce granulite metamorphism.

[3] However, this may not be invariably true for the entire SGP. Fluid activities during peak metamorphism at Mamandur (Madras block), calculated from mineral-fluid equilibria in granulite facies rocks, show a_{H_2O} varying from 0.16 to 0.69 (Chattopadhyay, 1995). The vein charnockite developed there locally at a retrograde stage could be an effect of canalised CO_2-influx, or due to simple decompression with or without lowering P_{H_2O}. The latter interpretation is preferred in the context of calculated high a_{H_2O} ($\cong 0.7$) during the dehydration reactions leading to the formation of the granulitic rocks and supported by independent evidence for decompression in the area.

Rai et al. (1993), based on the results of teleseismic tomographic studies, concluded that the Southern Granulite Province exhibited crustal thickening coupled with lower velocity in the crust and upper mantle. This contradicts the expected general increase in the average crustal velocity and density beneath a metamorphic terrain, if the latter was underplated by basic magma. Thus the magma underplating is not a satisfactory explanation of thermal perturbation causing granulite metamorphic in South India.

Results of the both seismic tomographic (Rai et al., op. cit.) and gravity studies in South India suggest that crust in the granulite terrain is thicker by 2–4 km over the average thickness of 35–37 km in the adjacent Dharwar Province.

In context of the above, Ganguly et al. (op. cit.) proposed a combination of overthrusting of a 30–35 km thick crustal block and CO_2 fluxing from a deep seated source. This combination would reduce the CO_2 requirement, with respect to situation where CO_2-flux was thought to be the only cause of the necessary thermal perturbation.

Young Granitic Rocks and Migmatitic Gneisses in the Southern Granulite Province In recent years, varieties of granitoid rocks, migmatitic gneisses and pegmatites from several places in Tamil Nadu and Kerala, falling within the Southern Granulite Province have been studied for their general characteristics and emplacement ages. Generally they are high-SiO_2, high-alkalies and low-Ca in composition. They are overall lineament-controlled. Their determined ages are distributed across the Proterozoic–Phanerozoic boundary and correspond with the Pan-African event (Table 1.1.15).

Table 1.1.15 *Geochronology of some late granitic rocks from Southern Granulite Province*

S. No.	Area	Rock/ Mineral studied	Sr_i	Age	Method	Reference
1	Nagamalai, Muttupatti, Mel-Kyuylkudi, Karatupatti, Minakshipurm, Madurai Dist, Tamil Nadu	Granite	0.712	837±34	Rb-Sr	Pandey et al. (1994)
2	An area 12 km west of Madurai, Tamil Nadu	Migmatitic gneiss	0.737	550±15	Rb-Sr	Hansen et al. (1985)
3	a) Kullampatti, Salem Dist, Tamil Nadu b) Suriyamala, Salem Dist, Tamil Nadu	Granite Pegmatite	0.7113 0.7093	534±15 515±9	Rb-Sr Rb-Sr	Pandey et al. (1993)
4	Thaluru, Kerala	Biotite from granite	-	710±20	K-Ar	Nair et al. (1985)
5	Angadimogar, Kerala	Syenite	0.7032	638±24	Rb-Sr	Santosh et al. (1989)
6	Peralimala, Kerala	Alkali granite	0.7031	750±50	Rb-Sr	
7	Ezhimala, Kerala	Granite, granophyre	-	678	Rb-Sr	Nair and Vidyadharan (1982)
8	a) Kalpatta, Kerala b) Kalpatta, Kerala	Biotite from granite Granite	- -	512±20 765	K-Ar U-Pb	Nair et al. (1985)
9	Ambalavayal, Kerala	Alkali granite	0.7171	595±20	Rb-Sr	Santosh et al. (1989)
10	Chengannur, Kerala	Hornblene from granite	-	550	K-Ar	Soman et al. (1983)
11	Munnar, Kerala	Migmatitic gneiss	-	740	U-Pb	Odem, 1982, in Santosh et al. (1989)

Proterozoic–Eocambrian Cover Sequences

Cuddapah Basin The Cuddapah basin, the second largest of the 'Purana basins' (term coined by Holland, 1907) in Peninsular India, hosts perhaps the thickest accumulation of Mesoproterozoic–Neoproterozoic sediments in this part of the world (Kale, 1991). Moreover, no other Purana basin is as rich in mineral deposits as the Cuddapah. Since King's first treatment on the geology of the basin more than a century ago (King, 1872), quite a few interesting studies have been conducted on the geology and mineralisation of this basin, as we will see below. Yet some of these aspects still require further detailed studies.

The crescent-shaped Cuddapah basin, convex towards the west and concave towards the east, presently cover an area of about 4, 40,000 sq km (Fig. 1.1.28).

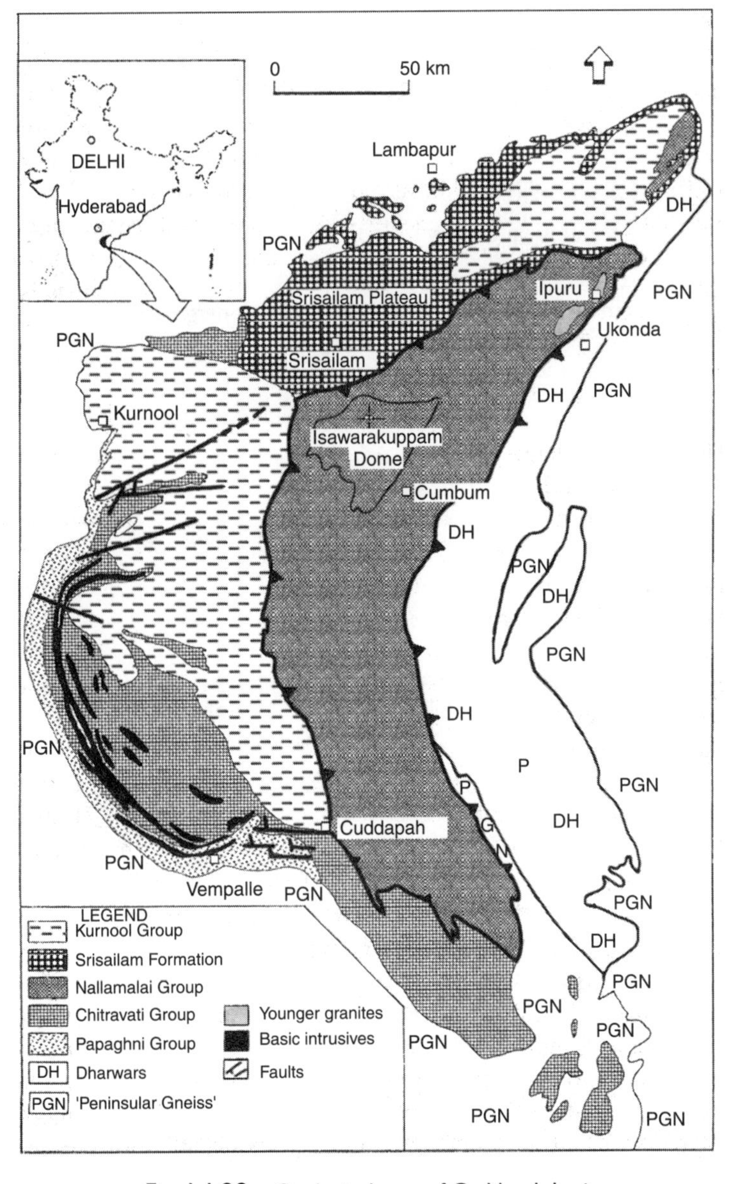

Fig. 1.1.28 Geological map of Cuddapah basin

The volcanic-sedimentary pile which has a cumulative thickness of about 10 km is now assigned the status of a Supergroup, the Cuddapah Supergroup. The Cuddapah Supergroup overlies the granites-gneisses and also schist belts of Eastern Dharwar; and is thrusted over by schists and gneisses in the east, also belonging to the Eastern Dharwar. The Cuddapah basin is generally thought to consist of three sub-basins: Papagni, Nallamalai, and Kurnool; which differed in time of development and details of lithology. Some (Nagaraja et al., 1987) proposed more sub-basins. A west-east geological section of the Cuddapah basin is presented in Fig 1.1.29. The beds in the western part are low-dipping and tilted towards the east through a series of normal faults; and in the east the Cuddapah rocks are intensely deformed, characterised by folding and thrusting with a westerly vergence. The sediments are both clastic and chemical (carbonates). There is intrusion of diatremes of kimberlitic affinity at Chelima that may be related to similar intrusions at Wazrakarur area, west of the Cuddapah basin.

King's stratigraphic classification of the Cuddapah rocks is no more in vogue. Instead several classificatory schemes have been proposed by relatively recent workers (Narayanswami, 1966b; Sen and Narasimha, 1967; Rajurkar and Ramalingaswami, 1975; Geological Survey India, 1981; Meijerink et al., 1984; Nagaraja Rao et al., 1987). Detailed discussions on all these are beyond the scope of the present work. We here choose two: one by the Geological Survey of India (1981) and the other by Meijerink et al. (1984) as they are more widely used (Tables 1.1.16 and 1.1.17). We, however, will follow the Geological Survey of India classification for our discussion further on.

Table 1.1.16 *Stratigraphic classification of the Cuddapah Supergroup (GSI Brochure, 1981)*

Group	Formation	Thickness (m)
Kurnool Group	Nandyal Shale	50–100
	Koilkuntla Limestone	15–50
	Paniam Quartzite	10–35
	Auk Shale	10–35
	Narji Limestone	100–200
	Banganapalle Quartzite and Conglomerate	10–57
	....................*Unconformormity*....................	
Nallamalai Group	Srisailam Quartzite	620+
	(Quartzite, siltstone)	2000+
	Cumbum Formation	1500
	(Shale, phyllite, quartzite, dolomite)	
	Bairenkonda (Nagiri) Quartzite	
	....................*Unconformity*....................	
Chitravati Group	Gandikota Quartzite	1200
	Tadpatri Formation	4600
	(Shales, lava sills)	1–75
	Pulivendla Quartzite	
Papaghni Group	Vempalle Formation	1500
	(Dolomite, shale, flows)	28–250
	Gulcheru Quartzite	
	..................*Unconformity*..................	
	Archean granites, gneisses and schists (Eastern Dharwar)	

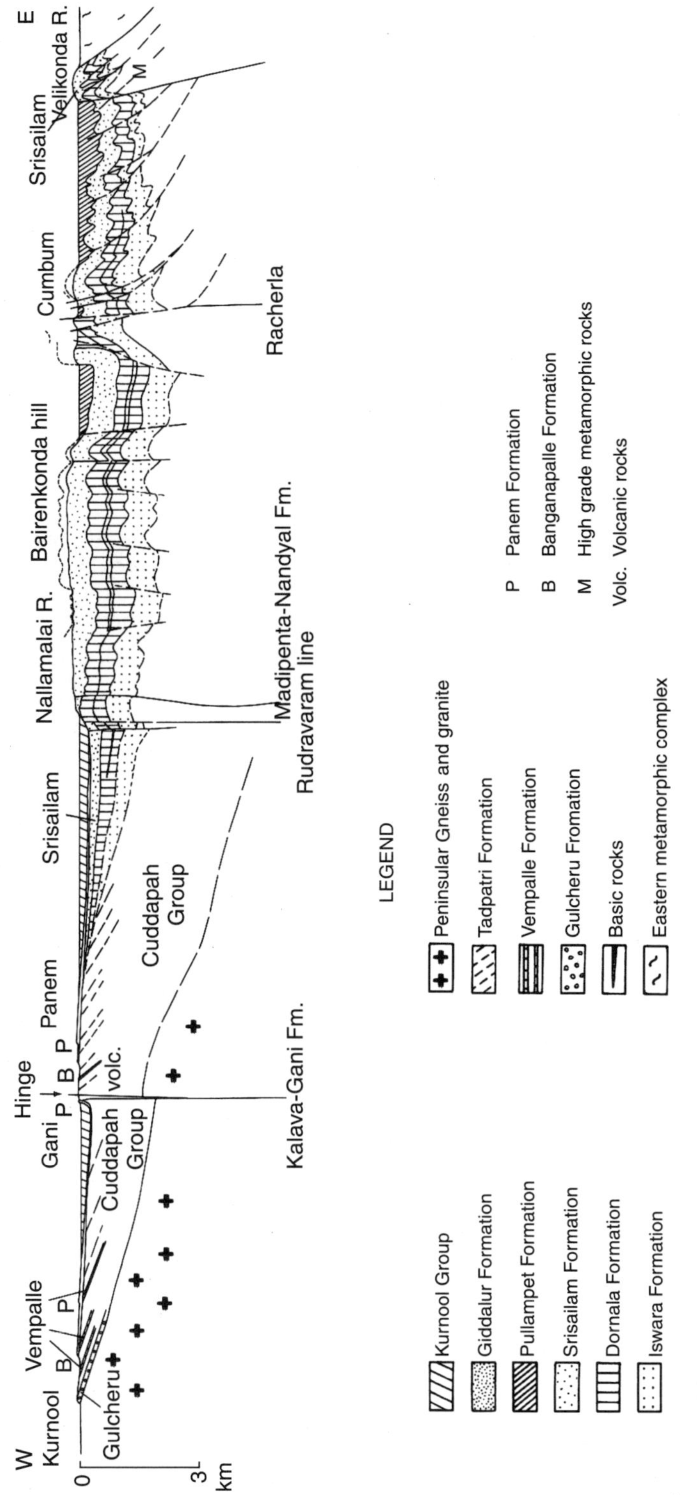

Fig. 1.1.29 A section across the Cuddapah basin showing increasing complexity of structures from west to east (after Meijernik, et al., 1984)

The basic difference between the two classifications presented here are, while the GSI classification divided the whole sequence of the Cuddapah Supergroup into four groups, Meijerink et al., did it into three. The oldest Group in the latter is called Cuddapah. It includes the Papaghni and Chitravati Groups of the former. The latter does not show a thick (1200 m) quartzite unit (Gandikota Quartzite) above the Tadapatri Formation. Another major difference is that the GSI put the Srisailam Formation above Cumbum, i.e. along with the Bairenkonda Subgroup. Within the Kurnool Group also there is conspicuous difference in sub-divisions. Most of the classification schemes proposed by other workers are closer to the GSI scheme than to that of Meijerink, et al.

Table 1.1.17 *Classification of the Cuddapah Supergroup (Meijerink, Rao and Rupke, 1984)*

Group	Subgroup	Formation	Chracteristic lithology
Kurnool Group		Kundair	Shales, limestones
		Panem	Quartzites
		...*Para-conformity*...	Conglomerate, sandstones,
		Banganapalle	shales, Limestones
	....................*Unconformity*....................		
Nallamalai	Cumbum	Giddalur	Quartzite, shales
	Bairenkonda	Pullampet	Shales, phyllites
		Srisailam	Quartzites, sandstones, shales
		Dornala	Quartzites, shales
		Iswara Kuppam	
	..................*Unconformity*....................		
Cuddapah	Mogamu Reru	Tadpatri	Shales, volcanic rocks
	Papagni	Pulivendla	Grits
		...Para-conformity...	Dolomite, shales
		Vempalle	Conglomerate
		Gulcheru	
	....................*Unconformity*....................		
		Archean granites and gneisses	

The Cuddapah basin evolved over an unusually long period of time, e.g., $\geq$ 1 billion years. Crawford and Compston (1973) on the basis of results of Rb-Sr dating of lavas near the base of the sequence concluded that the Cuddapah sedimentation started not later than 1555 Ma, and this could be even earlier (~1700 Ma). The youngest group Kurnool may have started depositing at 850–800 Ma after a major break of sedimentation represented by the large scale igneous activity (acidic, basic and alkaline), and continued till 600–500 Ma (Raha, 1987). Pre-Kurnool dolerites have been dated 980$\pm$110 Ma. Diatremes have been dated at ~1225 Ma and 1140 Ma. Meijerink et al. (op.cit.) suggested the following ages (the basis of which is not elaborated): the lower most group (their Cuddapah Group), 2000–1490/1450 Ma; the middle (Nallamalai Group), 1490/1450–1090/870 Ma; and the Kurnool Group, ~1090/870–500 (?) Ma.

After a critical review of the published geochronological information on the rocks of the Cuddapah basin, Gopalan and Bhaskara Rao (1995) concluded that the basal volcanism in the basin started not later than 1800 Ma and the sedimentation ended before 1600 Ma. The mafic dykes that intruded the basement gneisses, but not the Cuddapah rocks, could even be as old as 2200 Ma.

During 1996, Geological Survey of India launched a project of detailed sedimentological study of the Cuddapah basin along a few transects that started from west to the east, south to north and east to west. Recently Lakshminarayana et al., (2001) reported the salient aspects of the results of this study critically incorporating the results of earlier works. An outline of these findings is given here. Quartz pebble conglomerate and cross-stratified quartzites show easterly paleocurrent in Gulcheru Formation, deposited in an alluvial environment. The overlaying Vempalle Formation comprising laminated shale, stromatolitic dolomite and oolitic chert, was deposited in an intertidal shallow marine environment. The vesicular basalt overlaying the Vempalle Formation and the Pulivendla Conglomerate unconformably overlying the basal unit suggests that the basic volcanism was subaerial. During the deposition of the Pulivendla Conglomerate and Quartzite, the sediment-flow turned northerly. Characteristics of the shale, stromatolitic dolomite, bedded chert and volcaniclastic (?) rocks of the Tadpatri Formation suggest that the basin was relatively deeper during their deposition. This was of course punctuated by volcanism. The shale-siltstone assemblage, succeeded by cross-stratified and ripple-bedded quartzites of the Gandikota Formation indicate shallowing of the basin once again. The paleocurrent was towards NW/N. While earlier works put the Bairenkonda Formation below the Cumbum Formation, Lakshminarayana et al., (2001) proposed a reverse relationship for the two sub-divisions of the Nallamalai Group. According to these investigators, the sediment flow direction during the deposition of the Bairenkonda Quartzite was towards NW/W. Banganapalli Quartzite of Palnad basin (Palnad sediments are equivalent to Kurnool Group) indicate current direction from the NE. Banganapalle Conglomerate contained diamond at places.

The stratigraphic position of Srisailam Quartzite remains a subject of debate. GSI (1981) put it as the topmost unit of the Bairenkonda 'Subgroup' under the Nallamalai Group. Lakshminarayana et al. (op. cit.) make it a facies variant of Bairenkonda/Gandikota Formations. In the recent past, Srisailam Quartzite has assumed a special importance because of its containing uranium mineralisation. In the northern part of the Cuddapah basin, Srisailam Quartzite unconformabily overlies the granite as sub-horizontal sheets. In the eastern part, i.e. towards the inner part of the basin, the unit is folded along with the underlaying slate/phyllite. Near the Nagarjun Sagar dam, a 40–60 m thick pebbly quartzite represents the basal unit of Srisailam Quartzite. The depositional environment has been suggested to be shoreline to tidal flats with fluvial influence (Lakshiminarayana et al., op.cit.)

Several models have been proposed during the last decades to explain the geological evolution of the Cuddapah basin. During the sixties of the last century when the classical geosynclinal model was still in vogue, Narayanaswami (1966b) suggested a progressive evolution of the Cuddapah basin from a platformal stage to geosynclinal stage. Sen and Rao (1967) perceived the development of the basin as a result of the sinking of a basement block along 'Fundamental fractures'. Results of DSS studies led Kaila et al. (1979) to suggest that the Cuddapah basin formed by an echelon type faulting of the Moho-boundary, downthrow being towards the east and the western block remaining rigid (Fig. 1.1.30). The resultant structure was a sort of a half-graben. Bhattacharji and Singh (1984) and Drury (1984) sought the ultimate cause of development of the Cuddapah basin in the thermo-mechanical perturbation of the crust-mantle system at the area and the related phenomena. Bhattacherji and Singh (op.cit.) concluded that mantle perturbation, magmatism, gravitational loading and isostatic subsidence were all interwoven in the process of development and evolution of the Cuddapah basin. Drury concluded that the Archean crust of South India was thermally wrapped into an elliptical dome. Thermal relaxation of this dome, after erosion and crustal thinning produced the initial trough, the deepest part of which lay to the southwest of the present outcrop.

Radhakrishna and Naqvi (1986) are of the view that the Cuddapah basin formation was related to the Eastern Ghat orogeny. We shall get back to this issue in a later section.

Bhima Basin Several Proterozoic–Eocambrian sedimentary basins developed in South India, of which the Bhima basin occurs in the northern Karnataka and in the adjacent parts of Andhra Pradesh. Referring to its location in terms of the Dharwar divisions, the major part of the basin today is located in the Eastern Dharwar.

It is a NE–SW trending basin, about 300 m at its thickest, with an areal spread of about 5000 sq km. It consists of an alternating sequence of clastic and chemical (carbonate) sediments. Carbonates, essentially limestone, dominate over the clastics and the fine clastics (siltstone–shale) over the coarse clastics (sandstone, conglomerate). A lithostratigraphic classification of the sediments in the Bhima basin, i.e. the Bhima Group, has been detailed by Jayaprakash (1999). The latter is presented in Table 1.1.18.

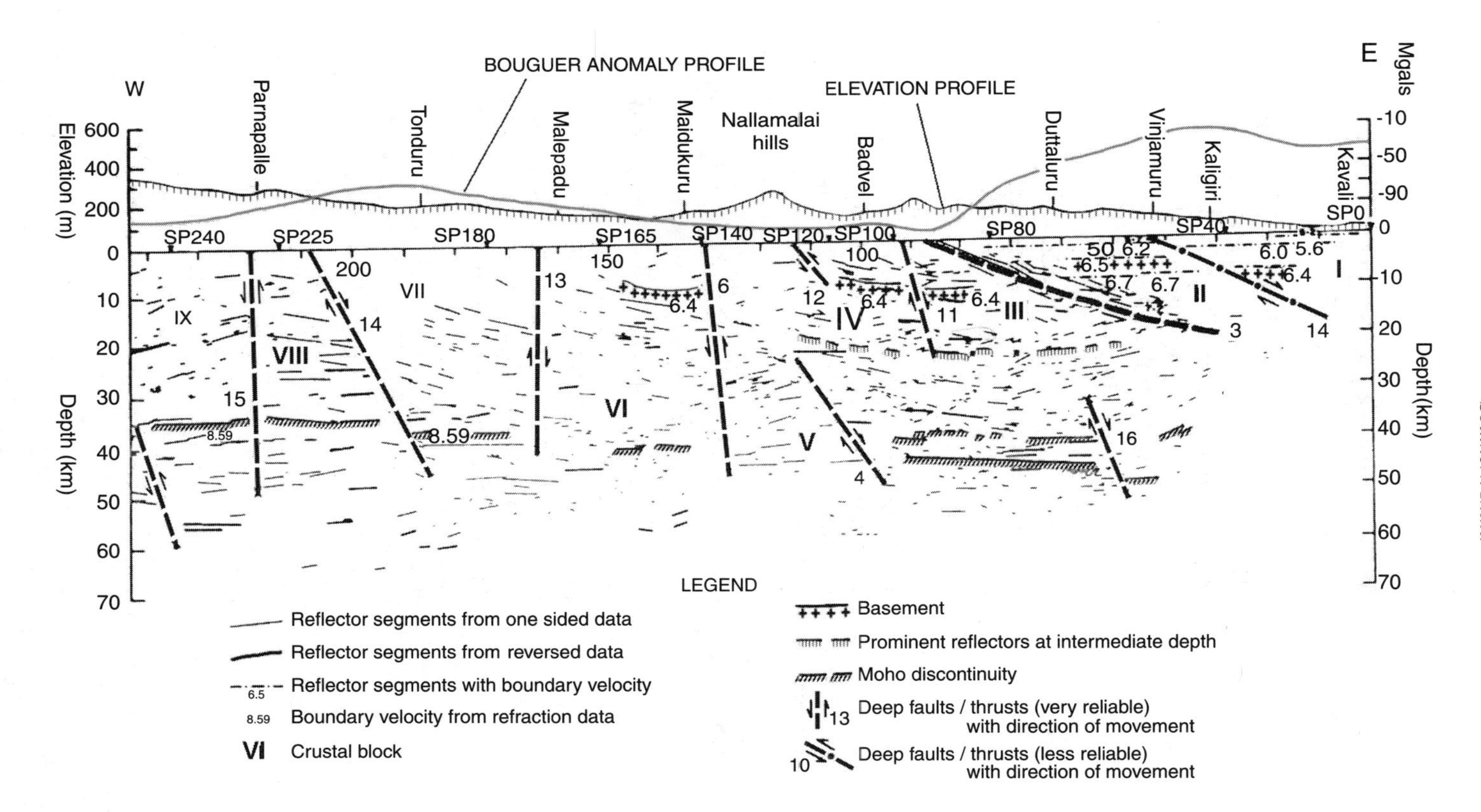

Fig. 1.1.30 DSS profile across the Cuddapah basin (after Kaila et al., 1979)

Table 1.1.18 *Stratigraphic sequence of the Bhima basin (after Jayaprakash, 1999)*

Eon/ Epoch	Group	Formation	Member	Lithology (thickness in m)
Cretaceous to Eocene	Deccan Trap			Basaltic flows with intertrappean sediments
		-------------------- *Unconformity* ————		
Eocambrian	Bhima Group (Total aggregate thickness 297m)	Harwal		Brown, pink to vermilion shale (45)
		Katamade varahalli		Deep grey, occasionally stylolitic, flaggy limestone (40)
		Hulkal		Grey, blackish buff, dull and pale pink shale, occasionally with fine grained, thin, silty beds at the base (30)
		Shahabad	Mulkod Limestone	Deep grey to black flaggy limestone (10)
			Gurdur Limestone	Akin to Wadi limestone, yet slightly inferior in chemical composition (20)
			Sedam Limestone	Variegated medium to thickly bedded, siliceous limestone (60)
			Wadi Limestone	Thickly bedded, stylolitic, relatively superior, cement-grade limestone (15)
			Ravoor Limestone	Flaggy limestone with prominent fissility (Shahabad slabs) (10)
Eocambrian	Bhima Group (Total aggregate thickness 297m)	Rabanpalli	Korla Shale	Fine silty base, grades into green shale, followed by chocolate brown shale with prominent parting (50)
			Kudrapalle Sandstone	Fine silty base, grades into green shale, followed by chocolate brown shale with prominent parting (50)
			Maddebihal Conglomerate	Pebbly conglomerate, locally or at the top, matrix-supported and also granular (2)
		----------Unconformity----------		
Proterozoic to Archean	Basement crystallines			Younger granites, Eastern block greenstone belts, Peninsular Gneiss

Jayaprakash subdivided lowermost Rabanpalli Formation into the Muddebihal Conglomerate at the bottom, followed upward by Kundrapalle Sandstone and Korla Shale, respectively. The conglomerate is matrix supported and contains clasts of quartz, feldspar and some rock fragments. Glauconite is present in all the lithounits of the Formation. Detrital gold in the conglomerate is reported from Balashetehal area (George, 1995). The overlying Shahabad Formation is likewise divided into several Members, which reckoned from below, are: Ravoor Limestone (flaggy), Wadi Limestone (cement grade), Sedam Limestone (siliceous), Gudur Limestone (moderately good grade), and Mulkod Limestone (deep grey and flaggy).

The Hulkal/Halkal Formation is made of grey to pale pink shale. The Katamadevarahalli Formation that overlies the Hulkal/Halkal Formation is composed of stylolitic flaggy limestone. The next upper Formation Harwal is made of brown/red shale. According to Jayaprakash (1999), this is the top-most Formation of the Bhima Group, which is unconformably overlain by the Deccan Traps. However, Janardhana Rao et al., (1975) proposed another formation Halpet above the Harwal Formation and below the Deccan Traps. This Jayaprakash does not recognise.

The sedimentary studies suggest that the sediments were derived from an easterly located provenance and deposited in a gently westward sloping basin. The basin was subjected to vertical tectonics resulting in a number of regional faults. Locally concentrated compression, however, resulted in fold structures at Devan Tegnur and Gogi areas (Jayaprakash, 1999).

Bhima basin has drawn much attention during the recent past because of containing uranium mineralisation particularly in the Ukinal–Gogi area, that will be discussed in a later section.

Kaladgi Basin The Kaladgi basin of Karnataka is one of the Purana basins of Peninsular India, now represented by cycles of orthoquartizites– argillites –carbonates, deposited during the Mesoproterozoic–Neoproterozoic time. Kaladgi Supergroup is sub-divided into two groups, Badani (Neoproterozoic) and Bagalkot (Mesoproterozoic) (Kale and Phansalkar, 1991).

The Bagalkot group in turn is subdivided into Simikeri Subgroup (upper) and and the Lokapur Subgroup (lower).

Relatively undisturbed sequences occur along the basin shoulder and the basin margin, except where affected by marginal faults. However, the Bagalkot Group of rocks underwent intense deformation in the core region of the basin. The younger, Badami Group of rocks, rest unconformably over the Bagalkot Group with a little/no deformation (Viswanathiah, 1977; Jayaprakash et al., 1987).

The only known economic interest in the rocks of the Kaladgi basin is a rather thin (~ 60 cm thick) manganese horizon within the folded sequence of the Simikeri Subgroup (Pillai et al., 1999). Mn content in the manganiferous bed is more than 45%. The Mn-minerals reportedly include tordorokite and lithophorite.

Pranhita-Godavari Basin The Pranhita–Godavari (PG) valley, generally regarded as an intra-cratonic rift basin (Qureshy, et al., 1968; Naqvi, et al., 1974,), marks the dividing line between the Dharwar craton in the south and the Bastar craton in the north. Meso-Neoproterozoic sediments ('Purana') are exposed along two NW–SE trending belts, flanking the margins of the basin, which are traceable for more than 450 km from Wardha in the northwest to Khammam in the southeast. The central part of the valley exposes the Gondwana sediments (220–65 Ma) with significant repository of coal. Two small inliers of Proterozoic rocks (Rajulgutta–Chinnur and Mailaram inliers) occur within the Gondwana rocks. The rift basin extends almost upto the CITZ in the NW and cuts across the EGMB in the SE. It is observed that the Proterozoic sediments of both the flanks are limited to the cratons, while the Gondwana basin-fill sediments extend across the EGMB. The present day Godavari river enters PG rift valley near Ramagundam and flows to the southeast from Rajulgutta along the valley before meeting the Bay of Bengal.

The Proterozoic sediments of the PG valley are flanked on both sides by the cratonic basement complexes of the Peninsular shield, characterised in the immediate vicinity by the Bhopalpatnam granulite belt of Bastar craton along the NE boundary and by the Karimnagar granulite belt (2.6 Ga) of Dharwar craton along the SW margin. Along the SW margin, the Proterozoic rocks unconformably overlie the cratonic basement and are in turn overlain unconformably by the Gondwana sequence. In contrary, these entities are bounded by two sub-parallel faults along the NW boundary (Fig 1.1.31).

After the pioneering works of King (1881), covering large tracts of PG valley, several later workers (Heron, 1949; Rao, 1964; Vinogradov et al., 1964; Basumallick, 1967; Subba Raju et al., 1978; Chaudhury and Howard, 1985; Sreenivasa Rao, 1985, 1987, 2001; Srinivasa Rao et al., 1979; Chaudhuri et al., 1989, 2002; Chaudhuri and Chanda, 1991; Kale, 1991; Saha and Ghosh, 1997;

Chaudhuri and Deb, 2004; Patranabis Deb, 2003; Chaudhuri, 2003) have significantly contributed in building up the data base of the Proterozoics of the PG valley, but most of the studies are sub-basin specific and do not pertain to both the flanks. Some of the sequences exposed in the extreme NW and SE are still undifferentiated as shown in Fig. 1.1.31.

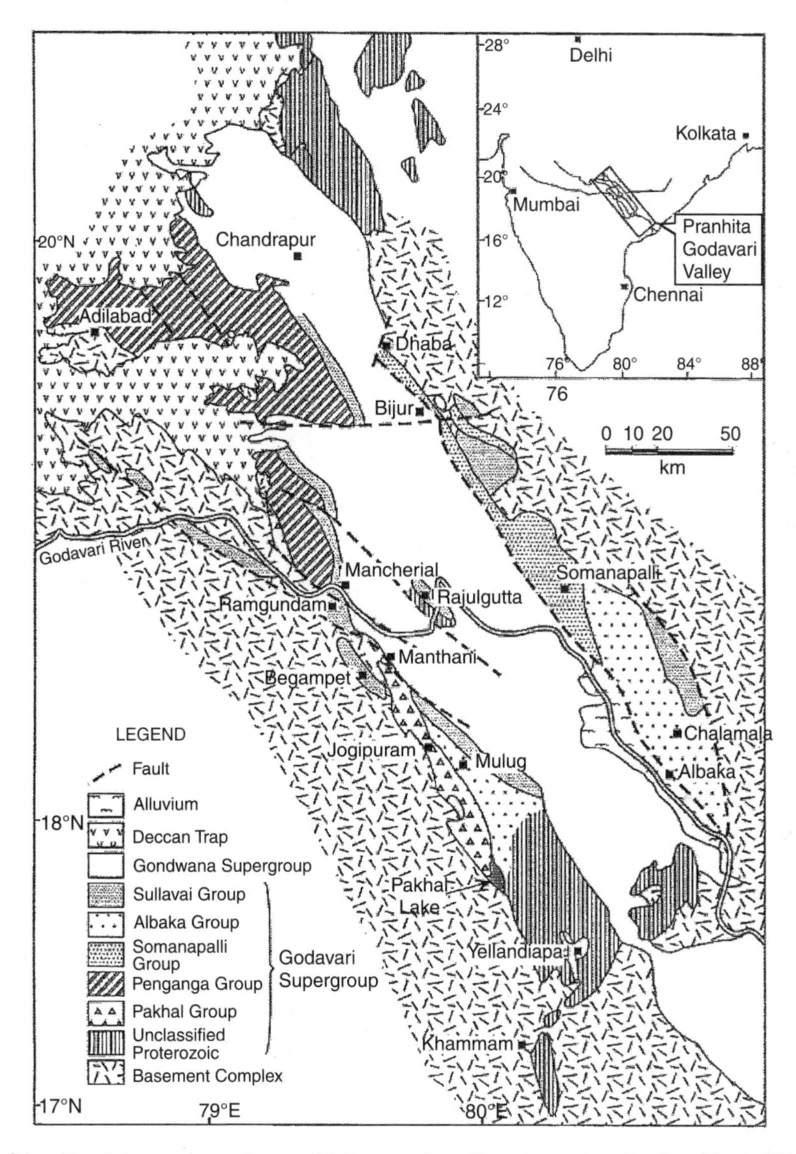

Fig. 1.1.31 Simplified geological map of PG valley (modified after Chaudhuri and Deb, 2004)

Chaudhuri and Chanda (op. cit.) designated the Proterozoic succession of PG valley as a Supergroup, and it was described in detail by Chaudhuri (2003) as several unconformity bounded sequences of 'group' status exhibiting varied depositional environments ranging from unstable rift setting to a stable shelf regime. Signatures of deep marine condition of deposition with attendant volcanicity are also recorded (Patranabis Deb, 2003). The classification and correlation of the sequences exposed in the various segments of the two belts are still fraught with many inconsistencies and doubts. The complexity is 'rooted to a differential response of different parts of the basin to uplift and or subsidence, unequal rates of erosion and deposition, and also the complex history of sea level changes' (Chaudhuri, op. cit.). The fragmentation of the exposures by different generations of faults and the separation of the two belts by the overlying younger Gondwana sediments are considered to have caused further problems in regional correlation.

Before discussing the stratigraphy and related problems of correlation, the various lithological packages occurring in different sectors of PG valley are briefly described below.

Southwestern Outcrop Belt

Pakhal Group The Pakhal sediments are best developed along the southwestern outcrop belt in the southeast of Ramagundam, the type area being the stretch between Pakhal Lake in the SE and Manthani in the NW. The lower part of this package, named as Mallampalli Subgroup, is characterised by basal conglomerate overlain by upward fining sandstone, prominent dolomite and interbeds of thin shales. The upper section, the Mulug Sub-group, has arkose, quartzite and dolomite in the lower part, and a thick shale unit at the top.

Penganga Group The Penganga lithopackage occurring in the northern part of southwestern outcrop belt (Mancherial-Adilabad sector, north of the Godavari river) predominantly comprises argillaceous and carbonate sediments (Satnala Shale and Chanda Limestone) with an arkosic sandstone (Pranhita Sandstone) towards the bottom. The lower contact of the Penganga sequence is faulted against upper Pakhal as well as the basement, whereas the flat lying arenaceous units of Sullavai unconformably overlie it.

Sullavai Group The Sullavai Group, comprising red sandstone-conglomerate and quartzite (Encharani Quartzite, Venkatapur Sandstone and Kapra Sandstone) represents the youngest sediments in the PG valley Proterozoics and overlies with angular unconformity both the Pakhal and Penganga rocks all along this belt as well as overlaps on to the basement at places. Similar successions at comparable stratigraphic level in the northeastern belt, though variously named, are also considered as equivalents of Sullavai.

Northeastern Outcrop Belt

Albaka Group The southern part of the northeastern belt is occupied by the Albaka Group (exposed in the Albaka range and its foothills), which comprises the Somandevara Quartzite, Tippapuram Shale and Albaka (Chalamala) Sandstone in ascending order of superposition. The lower sections have often been correlated with the Pakhals but there is no representation of the highly matured Albaka Sandstone in the southwestern belt.

Somanpalli Group In the northwest of Albaka range, the Somanpalli rocks are traceable up to Dhaba (Fig. 1.1.29), comprising Somnur Sandstone/Shale, Bodlela Vagu Limestone, Kopela Shale and Po Gutta Sandstone in that order.

It may be surmised from the above that the Proterozoic sediments in PG valley are represented by a heterogeneous assemblage of conglomerate, arkose, quartzite, sandstone (glauconitic at places), siltstone, dolomite, limestone, and shale-phyllite. The Pakhal consists of dolomite, while other carbonate-siliciclastic mixed facies (Penganga and Somanpalli) have limestone. The Albaka is made up of quartzite-lithic wacke followed by shale and carbonates. There is no glauconite in Albaka rocks. Sullavai is mainly an arenaceous unit, and represents the youngest in the Proterozoic succession.

Stratigraphy The stratigraphy proposed by the various workers for the southwestern belt and the suggested correlations, as compiled by Chaudhuri (2003), are presented in Table 1.1.19.

Table 1.1.19 *Stratigraphy proposed for the Southwestern Outcrop belt of PG valley by different workers and suggested correlation (after Chaudhuri, 2003)*

King (1881)	Basumallick (1967)	Subba Raju et al.,(1978)	Chaudhuri and Haward (1985)	Sreenivasa Rao (1987)	Chaudhuri (2003)
	Gondwana Supergroup -U-U-U-		Gondwana Supergroup -U-U-U-	Gondwana Supergroup -U-U-U-	Gondwana Supergroup -U-U-U-
Sullavai Series: Kapra Sandstone Venkatpur Sandstone Encherani Quartzite	Sullavai Sandstone	PAKHAL SUPERGROUP — Sullavai Group: Sullavai Sandston and congl.	Sullavai Frm: Venkatapur Sandstone Encharani Quartzite	Sullavai Sandstone	GODAVARI SUPERGROUP — Sullavai Group: Venkatpur Sst. Encharani Fm.
-U-U-U-	-U-U-U-	-U-U-U-	-U-U-U-	-U-U-U-	-U-U-U-
				Penganga Group: Puntur Lst Takhlapalli Arkose	Albaka Group: Tippapuram Shale Somandevara Quartzite
Upper Transition Series: Albaka Sub-division -?-?-?- Pakhal Subdivision	PAKHAL GROUP — Mulug Subgroup: Mulug Shale Mulug Orthqtzite Enchencheruvu Shale -do- Limestone -do- Chert Jakaram Arkose Jakharam Congl.	Mulug Subgroup: Mulug Shale Mulug Orthqtzite Mulug Arkose Jakhram Congl	Mulug Subgroup: Rajaram Limestone Ramgundam Sandstone Damla Gutta Congl.	PAKHAL SUPERGROUP — Mulug Group: Laknavaram Shale Pattipalli Quartzite Enchencheruvu Dolomite Polaram Fm. Jakaram Arkose	Mulug Group: Rajaram Fm. Enchencheruu Chert Ramgundam Sandstone Jakaram Congl.
	-U-U-U-	-U-U-U-	-U-U-U-	-U-U-U-	-U-U-U-
	Mallampalli Subgroup: Pandikunta Shale -do- Lime-stone Mallampalli Congl.	Mallampalli Subgroup: Pandikuta Shale Mallam-palli Dolomite.- -do- Congl.	Mallampalli Subgroup: Pandikunta Limestone Jonalarasi Bodu Fm.	Mallampalli Group: Pandikuna Shale Gunjeda Dolomite Bolapalli Fm.	Mallampall Group: Pandikuta Shale Pandikuta Lst. Bolapalli Fm.
-U-U-U-	-U-U-U-	-U-U-U-	-U-U-U-	-U-U-U-	-U-U-U-
Basement Complex	Basement Complex	Basement Complex	Basement Complex	Basement Complex	Basement Complex

Note: U – Unconformity; Fm – Formation; Lst – Limestone; Sst – Sandstone; Orthqtzite – Orthoquartzite; Congl. – Conglomerate

King (1881) classified the Proterozoic sequence of PG valley into a lower unit named as Upper Transition series (overlying the basement) and an upper unit named as Sullavai Series, with an angular unconformity in between. The Upper Transition series was subdivided into a lower Pakhal Subdivision (sandstone, shale and limestone) and an upper Albaka Subdivision (only siliciclastic rocks). The Sullavai Series comprises red sandstone and conglomerate. In this classical work, the Pakhal subdivision was recorded from both the belts while the Albaka Subdivision was described from only the northeastern belt. A mixed carbonate-siliciclastic sequence identified from the upper parts of the Pahkal in the northern parts of southwestern belt (Ramagundam–Adilabad sector) was named as Penganga Series (Heron, 1949), and was later designated as 'Group' (Sreenivasa Rao, 1985; Chaudhuri et al., 1989). However, its contact relation with Pakhal remained obscure.

Following Basumallick (1967) all the later workers recognised subdivision of Pakhal Group into a lower Mallampalli Subgroup (Group) and an upper Mulug Subgroup (Group), with an intervening unconformity. The Mallampalli Subgroup consists of conglomerate at the base followed upwards by gradually fining sandstone, and topped by limestone and calcareous shale. The Mulug Subgroup also starts with conglomerate, followed upwards by arkose, a supermatured orthoquartzite (Mulug Orthoquartzite) and finally a very thick sequence of shale (Mulug Shale).

Subba Raju et al. (1978) based on the studies in the type area of Mulug–Pakhal Lake (southwestern outcrop belt) and in the Albaka range (northeastern outcrop belt), corroborated the classification of Basumallick (1967), except for designating the subdivisions of Pakhal as Mallampalli and Mulug Groups, and classified the Albaka sequence into lower Tippapuram Shale and upper Chalamala Sandstone with a disconformity in between. The succession of Albaka range was later redefined from bottom to top into Somandevara Quartzite (≡ Mulug Orthoquartzite), Tippapuram Shale(≡ Mulug Shale) and Albaka Sandstone, unconformably overlain by another unit of red arkose and conglomerate (≡ Sullavai Group); and were correlated as shown in parentheses by Srinivasa Rao et al. (1979). Sreenivasa Rao (1987) combined Mallampalli Group, Mulug Group and Albaka Sandstone as Pakhal Supergroup, while Chaudhuri and Chanda (1991) put together the Somandevara Quartzite, Tippapuram Shale and Albaka Sandstone into Albaka Group.

Another assemblage radically different from the Pakhal rocks of type area is the carbonate-siliciclastic succession that occurs around Somanapalli in the northeastern belt. This was named as Somanapalli Group by Saha and Ghosh (1997), and was subdivided into six formations testifying to the initiation of the deposition in a platformal condition followed by deep water accumulation of black shale–chert and volcano-clastic rocks in an unstable basin.

In the north-central parts of the southwestern outcrop belt the Penganga Group unconformably overlies the Pakhal Group or directly on the basement, followed upwards by Sullavai Group after another unconformity. In the southern sector of the same belt, the Pakhal Group is overlain by Sullavai Group, without the development of Penganga sequence. The northern parts of the northeastern outcrop belt is characterised by the thick sequence of Somanpalli mixed carbonate-siliciclastics, while the southern part of the belt comprises the rocks of Albaka Group. Sullavai Group unconformably overlies both Somanpalli and Albaka Groups. The relationship between the Somanpalli, Albaka and Penganga sequences is yet to be unequivocally understood. However, the Albaka sediments having most mature litho-association might represent deposition in the post-rifting stage. Whereas, the sediments of both Penganga and Somanpalli Groups attest to syn-rift platform-slope-deep water deposition in an overall extensional regime. The corollary is that the Albaka Group may be younger than Somanpalli and Penganga Groups, but there are not many clues to resolve the interrelation between the latter two.

Recently widespread occurrences of felsic tuffs have been reported from different stratigraphic levels of the Proterozoics in the PG basin (Patranabis Deb, 2003), which was earlier considered to be largely devoid of any magmatic activity that usually abounds rift valleys. The pyroclastics are noted from both the outcrop belts (in Pakhal, Penganga and Somanpalli Groups) and the Rajulgutta inlier. Thinly bedded welded vitric tuffs, fine ash beds and associated lava occur in association with varied litho-assemblages like limestone, shale, black shale, chert etc.

Paleogeography The longitudinal profile along the PG valley, as reconstructed by Chaudhuri et al. (2003) for the southwestern outcrop belt is presented in Fig. 1.1.32.

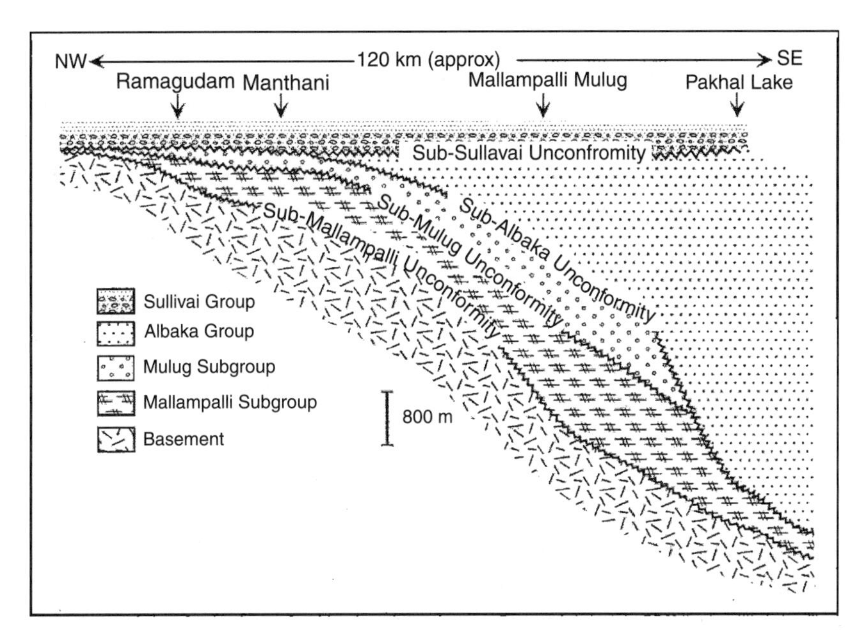

Fig. 1.1.32 Schematic longitudinal profile along SW outcrop belt of PG valley (after Chaudhuri, 2003)

The systematic variation in the thickness of the unconformity bound lithological successions (Mallampalli, Mulug, Albaka and Sullavai Groups) and their interrelationship from the north-west to the south-east along the PG valley indicate episodic uplift and depression of the basin bottom. It also depicts higher rate of subsidence and widening of the valley (more than the present day outcrop width) in the southeast in the pre-Sullavai period, allowing the accumulation of much greater thickness of sediments compared the northeast. It was also inferred by Chaudhuri and Deb (2004) that the PG basin was joined to a major marine basin in the east during Pakhal–Albaka time.

Deformation and Metamorphism The PG valley sequences on both the flanks are affected by a number of steeply dipping to subvertical NW–SE trending faults, which are mostly subparallel to the general strike of the bedding, though transgressive to the lithologic trends at places. These are traceable for hundreds of metres to more than 100 km in some cases, and the displacement along such faults is also of significant order (hundreds of metres to several kms). Besides, there are also some NE–SW trending transverse faults. Uplift of blocks bounded by faults has resulted in the formation of inliers of basement or the Proterozoic rocks within the Gondwana rocks. As mentioned above, along the northern flank of the PG valley the contacts of the Bastar basement complex, the Proterozoics and the Gondwana sequence are marked by regional scale faults running parallel to the length of the basin.

The beds of the southwestern belt, in general, display NW–SE strike and moderate dip towards NE. The rocks of Pakhal and Penganga Groups along this belt show co-folding into a series of NW–SE to NNW–SSE trending anticlines and synclines with northwesterly plunge, whereas the overlying Sullavai beds do not record any folding and are found to transect the folded older sequence along a prominent angular unconformity. The fold deformation in this belt is clearly pre-Sullavai in age.

In the northeastern belt, the Somanpalli rocks are characterised by large scale NW–SE trending folds, imbricate thrusts and shear zones (Saha 1990, 1992; Ghosh and Saha, 2003). Towards the southern sector of this belt (type area of Albaka Group), the Somandevara Quartzite and Tippapuram Shale are folded, while the Albaka Sandstone displays horizontality in the western side but displays tight folding along eastern margin (Sreenivasa Rao, 2001). The sectoral development of tight asymmetric folds and imbricate thrusts along the major faults are regarded as the effects of reactivation of the basin forming faults by strong compression during basin inversion (Patranabis Deb, 2003). Sreenivasa Rao (op.cit.) reports folding in the overlying Usur (≡ Sullavai) sandstone conglomerate sequence, and suggests the fold deformation in this belt to be post-Sullavai (or post-Proterozoic sedimentation).

Data on the metamorphic imprints on the PG valley sequences is scarce. It is reported that the rocks are, in general, very weakly metamorphosed (Patranabis Deb, 2003); and it is almost incipient along the northeastern belt (Sreenivasa Rao, 2001). However, the reported presence of ottrelite (Ramamohan Rao, 1969, 1970; Sreenivasa Rao, 1987), leads us to presume that at least in parts the rocks were subjected to low greenschist facies metamorphism. It is also reported that the grade of metamorphism in the belt increases from NW to SE. Imprints of garnet grade metamorphism are reported from the southeastern extremity close to the Cuddapahs (Patranabis Deb, op.cit.). Near Yellandlapad and in the Mailaram inlier the Pakhal rocks have chloritoid, staurolite, andalusite, garnet and kyanite in pelitic members and talc, tremolite, actinolite and diopside in the dolomitic units (Ramamohan Rao, op.cit.; Sreenivasa Rao, op.cit.). In the vicinity of the EGMB (southeast part of the northern belt) the superposition of NE–SW trending structures over the generally NW–SE trending fabric of the PG belt is prominent at places (Ramamohan Rao, 1969). This deformative episode along with high-grade metamorphism (andalusite–sillimanite grade) is broadly correlatable with the accretion of the EGMB with the Indian cratonic segment (Bastar and Dharwar).

Age Data The radiometric age data are very few from the southwestern belt, and none available from the northeastern belt. K-Ar age of glauconite from Mallampalli sandstone (lower Pakhal) is 1330 ± 53 Ma (Vinogradov et al., 1964). Glauconite from middle part of Pakhal (probably Mulug Group) is reported to have yielded K-Ar ages of 1276 ± 30 Ma, 1188 ± 14 Ma, and 1142 ± 37 Ma and from Sullavai 871 ± 21 Ma (Snelling, 1980 in Mathur, 1982). K-Ar age of glauconite from Sullavai also yielded an age of 871 ± 14 Ma (Chaudhuri and Howard, 1985). Chaudhury et al. (1989) reported Rb-Sr ages of glauconite from sandstone of middle Penganga Group as 775 ± 30 and 790 ± 30 Ma. The available geochronological data show close correspondence with the established stratigraphy except for the ages of Penganga vs Sullavai, which are in contrary to the order of superposition observed in the field.

Chaudhuri and Deb (2004) contemplate that the 1600–1500 Ma may reasonably be presumed as the time of the initiation of Mallampalli sedimentation in view of the available date of 1330 ± 30 Ma for authigenic glauconite from about 200 m above the basement and possible argon loss during deep burial.

Basin Tectonics and Evolution The unconformity and fault bound Proterozoic sequences of the PG basin, the variation in thickness of the basin fill sediments, punctuated by repeated episodes of volcanic effusion amply testify to tectonically controlled uplift and subsidence of the depositional interface, besides the possible sea level changes. The differential response of the different segments of the basin to the tectonic movements is also apparent from the morphology of the depositional interfaces and abrupt sectoral changes in the depositional environment recorded in the sediments. The tectonic behaviour of this linear basin was 'oscillatory', and was accompanied by large number of close spaced basin parallel high angle faults, which are common features of repeated extension, rifting and contraction in an intra-cratonic rift basin. The southern part of the basin appears to have oscillated more vigorously. The most dramatic uplift and erosion of about 4000 m are estimated for the period of sub-Albaka hiatus (Chaudhuri and Deb, op.cit.). Extensive episodic melting at sub-crustal level and resulting emplacement of felsic welded tuffs at various stratigraphic levels further attest to localised stretching and rifting over a protracted period during Meso-Neoproterozoic time (Patranabis Deb, 2003). The limited radiometric age data are indicative of a time span of > 1400 to ~ 700 Ma for the deposition of the Proterozoics of the PG basin.

The above interpretation in regard to the evolution of the PG basin presupposes that the entire width of the basin, developed on the Archean basement, was filled by Proterozoic sediments under riftogenic environment, followed by inversion of the basin under a compressional regime towards the end of Neoproterozoic time. A new rift basin with reduced width formed over the Proterozoic basement at the same locale having identical alignment in which the Gondwana sediments were deposited. The prevailing concept of such two-tier rifting is not acceptable to some workers, who have visualised independent development of two Meso-Neoproterozoic basins at the margins of Dharwar (perhaps continuation of Cuddapah basin) and Bastar cratons, which were tectonically juxtaposed after the deposition of the sediments (Sreenivasa Rao, 2001). According this view, the Godavari rift basin developed much later during Gondwana deposition, which not only dissected the craton, but also the EGMB. The mismatches in the Proterozoic depositional motif and also deformational history between the two flanks of the PG valley and absence of basic volcanics are highlighted in support of this contention. Much younger status (1.6–1.7 Ga) of the Bhopalpatnam granulite (Santosh, et al., 2004) compared to that of Karimnagar granulite (2.6 Ga) may subscribe to the view that the crustal evolutionary trends upto the Mesoproterozoic time was not commonly shared by the Bastar and Dharwar cratons.

1.2 Metallogeny

Introduction

In discussing the metallogeny in South India, we follow the same order as in Section 1.1, i.e. mineralisation in Sargur type schist belts is followed by ore mineralisation in Dharwar type schist belts (and Peninsular Gneiss) and these in turn are followed by ore mineralisation in the Southern Granulite Province and Purana basins, respectively. As will be obvious from the following discussion, the Dharwar schist belts and the Purana basins are better mineralised in South India, and it is a known fact that South India is best endowed with respect to gold in the country. Sandur and Goa–Reddi belts contained rich deposits of BIF and related iron ores. Though low in grade (~ 35% Fe), we have a sizeable reserve of banded quartz–magnetite ores in the Kudremukh area, from which upgradation is neither very difficult nor very expensive. For India, it may be a reserve for the distant future. It has an external market too, but its exploitation is now stopped on grounds of environmental pollution. There is a superlarge deposit of barite in the Cuddapah basin. Other economically/strategically exploitable deposits are of uranium, manganese, chromium, Al-ore (bauxite) and diamond. South India has the distinction of containing the largest reserve of thorium ore (monazite) in beach placers. These and some other less important mineral deposits will be discussed in the following pages.

It may be noted here that in the discussons on the genesis of BIF and the related iron ores, bauxitic ores etc., the status of the subject has been generally reviewed when it came up first, and not repeated in later chapters when similar deposits again come up for discussion.

Mineralisations in Sargur Type Schist Belts

Sargur type schist belts, as we have discussed earlier, occur as small belts and screens. These middle Archean volcanisedimentary sequences do not contain large mineral deposits alike the schist belts comprising the Dharwar Supergroup of rocks. But these are not completely barren of mineral deposits. Mineral deposits reported and even mined from these belts are of chromite, copper, vanadium-bearing titaniferous magnetite, barite and gold.

Chromite Mineralisation Rich but small deposits of chromite ore occur in the Nuggihalli belt, particularly at the Byrapur–Baktharahalli, Tagadur and Jambur. Chromite mineralisation also occurs at Sinduvalli at the Sargur belt. The sheet-like ore body at Bayrapur continues upto a depth of 350 m along the dip (60°–70°). The estimated reserve (including the portion mined) is 3 million tonnes. Host rocks for the mineralisation are dunite, peridotite and pyroxenite, variously altered to serpentinite and talc-chlorite schists. The orebodies are deformed (Fig. 1.2.1), and the ores are metamorphosed.

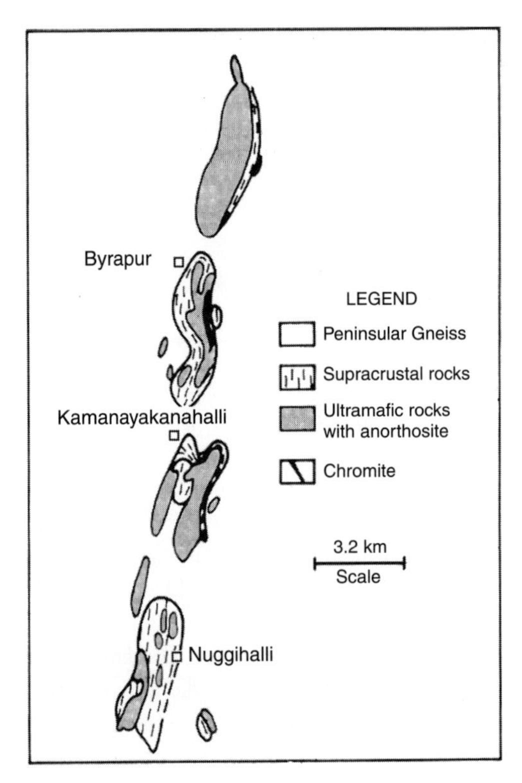

Fig. 1.2.1 A planar view of the Byrapur–Nuggihalli chromite belt

Pitchamuthu (1956) and Varadarajan (1970) regarded the ultramafic bodies (with the chromite ores) as Alpine-type intrusives. Viswanathan (1974), and Naqvi and Hussain (1979), on the other hand, suggested these to be komatiitic. The latter view is not supported, particularly where chromite occurs as rhythmic layers with 'way up' structures and cummulus textures. The chemistry of chromite from these occurrences correlates better with that of stratified chromite deposits, of course with Fe-enrichment in the Byrapur ores (Srikantappa et al., 1980; Nijagunappa and Naganna, 1983). These deposits have many features common with the Archean chromite deposits of Selukwe, Zimbabwe (erstwhile Rhodesia) (Cotterill, 1969).

Copper Mineralisation Copper mineralisation in Sargur-like belts/enclaves within the Peninsular Gneiss occurs at Kalyadi and Aladahalli. Both are small deposits.

The mineralisation at Kalyadi is confined mainly to the quartzite and/or quartz-chlorite schist, but the ore zone is flanked by talc-chlorite-actinolite schists, actinolite-chlorite (-talc) schists, layered amphibolite-quartzite etc. Ores occur as disseminations, veins and locally even laminated (pyrite-chalcopyrite-chert). Veins of quartz-calcite-pyrite-pyrrhotite-chalcopyrite are emplaced within the wall rocks. The ores have been deformed and metamorphosed. However, the temperature determined with the help of sulphur isotope distribution in 'co-existing' sulphide pairs (Menon et al., 1980) appears a little too low.

$\delta^{34}S$ determined on pyrite, pyrrhotite and and chalcopyrite range from +1.2 to +6.7 per mil. A pyrite-bearing magnetite band parallels the ore zone towards the northern end of Block 1. All these are suggestive of a volcanogenic origin of the deposit (Sarkar, 1988). The estimated ore reserve is 1.3 million tonnes (Mt) with an average grade of 0.9% Cu.

Aladahalli copper deposit occurs 15 km to the NNW of Nuggihalli, in another schistose enclave within the gneissic country. The principal rock types comprising this enclave include staurolite and sillimanite bearing schists, quartzites and amphibolites, intruded by lenses of ultramafic dykes, quartz and pegmatite veins. The disseminated ores and streaky veins are dominated by pyrite, chalcopyrite and pyrrhotite. Sphalerite and galena are rare. Magnetite and ilmenite are, however, common. Deformation and metamorphism have obliterated all reliable genetic signatures.

The estimated ore reserve is 3.3 Mt with an average Cu-content of 0.74%.

Vanadium-bearing Titaniferous Magnetite Deposits This type of magnetitic iron ore deposits occur in both the Sargur and Dharwar type schist belts. They are contained within the gabbro-anorthosite members of basic-ultramafic complexes emplaced in quartzite, quartz-carbonate and pelitic rocks. Conformable orebodies are more common, but there are transgressive varieties too. The orebodies are deformed and the ores are metamorphosed. Most of these deposits are reported from Nuggihalli and Shimoga belts. Some of these deposits are fairly large, particularly those at Masanikere and Ubrani (~ 4 Mt each). The ore bands usually contain >50% Fe, >10% TiO_2 and $\leq$ 1% V_2O_5 (Vasudev and Srinivasan, 1979).

Barite Mineralisation Occurrence of barite mineralisation, interstratified with quartzitic rocks in the Ghattihosahalli schist belt, Karnataka, was first reported by Radhakrishna and Sreenivasaiyah (1974). Later, the state Department of Mines and Geology took up systematic exploration in the area for barite. Subsequently, a number of workers including Naqvi (1978), Raase et al., (1983), Hoering (1989), Devaraju et al., (1999), took up studies on this mineralisation, or the rocks that host it.

The barite mineralisation under reference occurs close to the Ghattihosahalli town in the Ghattihosahalli schist belt, considered to belong to the Sargur Group ($\geq$ 3.0 Ga). It is a narrow ~ 25 km long belt, occurring to the immediate west of the main Chitradurga schist belt. The lithologic framework of the belt is constituted of metamorphosed ultramafic-mafic igneous rocks now represented by serpentinite, talc-tremolite schists, amphibolites in the lower part and siliceous meta- sedimentary rocks in the upper. There is even a report of the presence of relict spinifex texture in the serpentinite, suggesting its protolith to be a Mg-rich komatiite. Two metamorphic events are reported. The earlier one that produced an assemblage of staurolite + garnet + kyanite + muscovite in the 'impure quartzite' was overprinted by an assemblage of chloritoid + chlorite + kyanite + muscovite during the deformation and metamorphism of the Dharwar Supergroup rocks (Chadwick et al., 1985).

Barite here occurs as thin (generally < 1m) discontinuous stratiform bands and lenses within quartzite, close to the underlying metabasic and meta-ultramafic rocks (Fig. 1.2.2). The largest body is ~ 120 m long and 1.25 m thick. The quartzitic unit, because of its compositional variations is divided into two types: (i) contains staurolite, kyanite, garnet, chlorite, chloritoid, grunerite, white mica + locally detrital chromite, tourmaline and zircon, besides quartz; (ii) contains Cr- bearing white mica and locally kyanite, besides quartz. Barite is mainly confined to (ii).

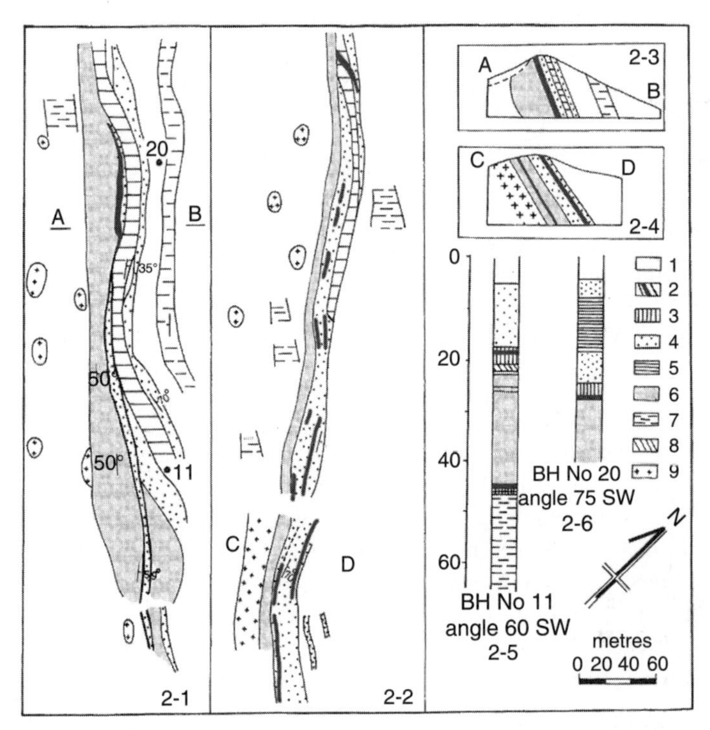

Fig. 1.2.2 Barite mineralisation in the Ghattihosahalli schist belt (2-1 and 2-2 are the geological map of northern and southern blocks; 2-3 and 2-4: Transverse sections across north and south blocks; 2-5 and 2-6: Borehole sections (after Devaraju et al., 1999).
Legend: 1. soil cover, 2. barite seams, 3. chromian muscovite quartzite, 4. grey quartzite, 5. quartz-white mica schist, 6. tremolite-actinolite schist, 7. hornblende schist, 8. mica schist, 9. granite

The barite layers of Ghattihosahalli contain 60–90% (volume) barite and the remainder consists mostly of quartz. Pyrite is a ubiquitous minor phase. Others comprising Ba-Cr-bearing mica, Ba-K-feldspar, tremolite, tourmaline, epidote and Cr-bearing garnet are present in more siliceous and fine grained portions. Fine grained barite shows granoblastic textures, with irregular to polygonal grain boundaries (Devaraju et al., 1999).

Hoering (1989) analysed sulphur isotopes in barite and the coexisting pyrite from Ghattihosahalli, and obtained the following results (Table 1.2.1).

Table 1.2.1 *Isotopic compositions of sulphur in Ghattihosahalli deposit*

Sulphur-isotopes	Barite	Pyrite
$\delta^{33}S$ per mil	+ 1.73 ± 0.17	-3.20 ± 0.16
$\delta^{34}S$ per mil	+ 4.26 ± 0.21	-5.55 ± 0.20
$\delta^{36}S$ per mil	+10.24 ±1.20	-10.25 ±1.50

The value of δ^{34} S obtained from Ghattihosahalli barite agrees with the data obtained from some other ancient barite mineralisations in the world, such as, (i) Waroona Group (3.5 Ga), Pilbora block, Western Australia: δ^{34} S = +3.0 to +4.2 (Lambert et al., 1978); (ii) Fig Tree Group (3.3 Ga), South Africa: δ^{34} S = +2.5 to +4.0 (Perry et al., 1971); (iii) Inegra Series, (3.1 Ga), Aldan Shield, Siberia: δ^{34} S = +3.5 to +8.0 (Vinogradov et al., 1976). The cause of sulphate formation in the (Early-) Middle Archean is not clearly known. It could be volcanogenic, or less possibly oxidation of reduced sulphur by the primitive oxidising atmosphere.

Ideas about the genesis of the Archean bedded barite deposits vary from sedimentary–diagenetic to volcanic-exhalative (Barley, 1992; Buick and Dunlop, 1990). Should we have no objection to imagining that sulphate reducing bacteria of the modern *Desulfovibrio desulfuricans* type were present at that time and place, we may have an explanation for the δ^{34} S in barite and coexisting pyrite. Sulphate reducing bacteria metabolise isotopically light sulphate more easily than the heavier ones to produce sulphides. The residual sulphate will, as a result be enriched in heavier S-isotopes. δ^{34} S of pyrite associated with barite at Ghattihosahalli is unlikely to be volcanogenic. Presence of microbial life is reported from rocks 3.5–3.3 Ga and even older (Schopf and Packer, 1987; Schidlowski, 1998). Because of textural reorganisation of barite and lack of any reported presence of anhydrate in barite layres, it is not possible to suggest if barite formed through the replacement of gypsum. That barium here was introduced to the system as Ba^{2+}, rather than $BaSO_4$, is suggested by the presence of barium in such associated minerals as mica and feldspar.

Gold Mineralisation Gold occurrences have been noted in the following belts of Sargur Group:

1. Nuggihalli (Gollarahalli, Vittalpura, Yelvari and Kempinkote)
2. Nagamangala (Honnebetta)
3. Krishnarajpet (Bellibetta)
4. Sargur (Ambale-Volagere, Porsedyke).

Out of the above occurrences, the most extensive opencast old mine workings are seen at Kempinkote, located at the southern extremity of Nuggihalli belt. Here, the lithoassemblage is made up of mafic-ultramafic flows, with gold mineralisation in the bands of hornblende and biotite bearing quartzo-feldspathic rock of varying width occurring along the interfaces of the flows. The mined out part in the central section of the old working is quite wide (upto 100 m), which presumably constituted a relatively high grade zone. The wall rocks exposed in the surroundings of the old pit analysed 0.1–1.0 g/t Au with occasional 'spot high' values.

In Ambale-Volagere areas, the gold mineralisation is associated with the quartzo-feldspathic and sulphidic quartz-carbonate veins traversing the mafic-ultramafic enclaves occurring in the Peninsular Gneissic Complex. There is a record of 2.9 kg gold production from Volagere in 1906. The onward history is silent, however!

Ore Mineralisation in the Dharwar Schist Belts

The Dharwar schist belts are well-known for their contents of gold. Kolar produced nearly 800 tonnes of the yellow metal, before it closed down a few of years back. There are other deposits of gold too, though most of them are much smaller to sub-economic.

The next important mineral resources of these belts are the iron ores of the Bababudan, Shimoga, Dharwar-Goa, Sandur and Kustagi belts, together containing about 2.6 billion tonnes (Bt) of ores containing $\geq$ 62% Fe and about 5.2 Bt of ores containing 33–47% Fe. There are some mineable manganese deposits also at Chitradurga–Tumkur, Shimoga and Sandur areas.

The Dharwars are not known for containing any important basemetal deposit. The Chitradurga copper deposit with its small reserves (~ 1 Mt at 1.4% Cu) had a mining history until the recent past.

Rare metal deposits occur in pegmatites emplaced without discrimination in the Sargur, Dharwar and the basement Peninsular Gneisses. They are discussed in a separate section.

The Dharwars do not contain any exploitable uranium deposit to-date. However, some lean mineralisations are reported from the basal Dharwar at Kalaspura and Walkunje.

Gold Mineralisation in the Kolar Schist Belt Kolar was a centre of gold mining in the ancient times. In the modern phase, mining resumed during1880–81 in this field under the management of John Taylor and Sons. This company mined the deposits until 1956 when the mines were nationalised. Since then, until the end of the last century when the mines were closed, these were operated by the Bharat Gold Mines Ltd. (BGML), a government undertaking. Kolar gold mines, before closing down a few years back, produced a total of about 800 tonnes of gold from about 48 Mt of ore.

Geology of the Kolar schist belt is discussed in an earlier section. We shall concentrate here on the mineralisation aspects. Based on the geographical positions and characteristics and intensity of mineralisation, the belt is divided into three sectors: North Kolar schist belt, the Central Kolar schist belt ('Kolar Gold Field') and the South Kolar schist belt (Fig. 1.2.3). Concentration of gold mineralisation has, however, taken place in the Central sector or the Kolar Gold Field (KGF). We take it up for discussion first.

Central Kolar Schist Belt (Kolar Gold Field) Gold mineralisation at the Kolar Gold Field (KGF) took place along discrete narrow but laterally continuous zones that correspond with the broadly N–S trending shear zones. Seven such mineralised zones have been identified. However, two of them, i.e. the Champion Reef Lode system along the central zone of the KGF and the Oriental Lode system west of the former, are the most important. The Champion Reef is a free-milling gold quartz lode, whereas the Oriental Lode contains refractory ores of gold quartz–sulphides. The Champion Reef Lode system consists of Champion Reef, Mundy's Lode, Muscom Lode and Pilot Shaft Lode. The Oriental Lode system comprises Oriental Lode, McTaggart West Lode and McTaggart East Lode. The Champion Reef Lode system was responsible for 97% of gold produced from KGF. The Champion Reef Lode system is generally but not invariably hosted by schistose amphibolites, and their contacts are commonly gradational. In contrast, the Oriental Lode system more often has sharp contact with the amphibolite host.

Most of the investigators of gold mineralisation in the belt pointed out that the Champion Reef Lode system was controlled by structures, but they differed with respect to their details. Narayanaswami et al. (1960) concluded that NNW-trending en echelon cross-folds developed on N–S trending shear zones and spatially associated faults controlled the gold mineralisation. Hamilton and Hodgson (1986) on the other hand suggested that there were two major ductile deformations at Kolar due to E–W compression.

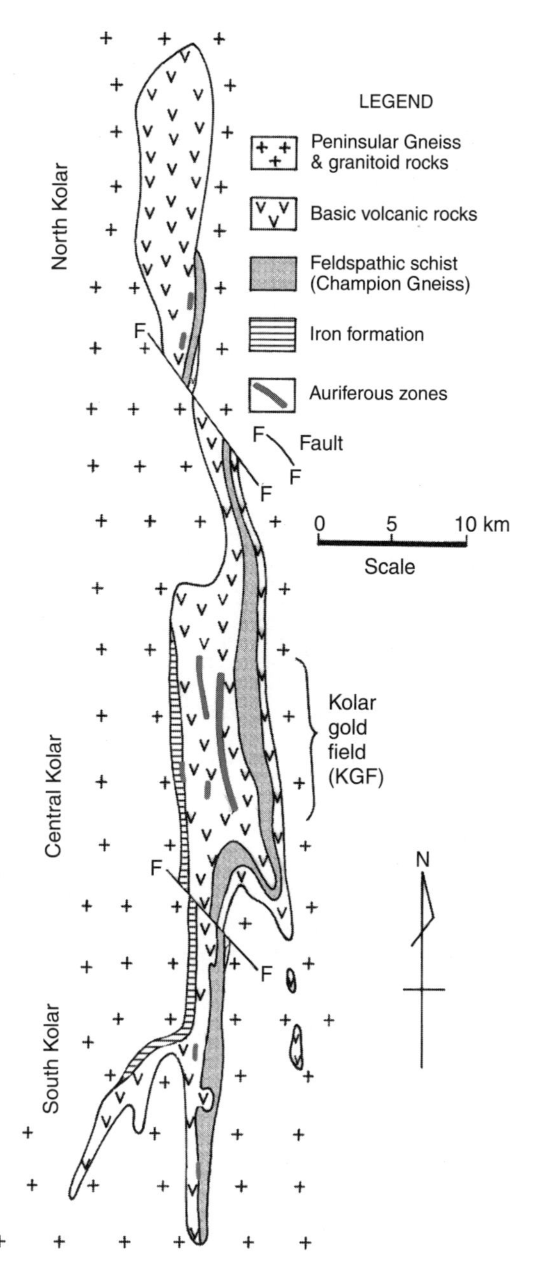

Fig. 1.2.3 A simplified geological map of the Kolar schist belt showing auriferous zones

In each case, the folding culminated in axial cleavage-parallel shearing. The N–S trending steeply west dipping shear planes were of earlier generation. The younger NNW-striking shear planes intersecting the earlier ones, localised the ore mineralisation and produced the NNW-plunging ore shoots (Fig. 1.2.4). Safonov (1988) also agreed that the ore-controlling structures were shear zones, faults and small folds in the shear zones. In some sections of the Champion Reef, several sub-parallel veins are present. These could be members of the Champion Reef system, with some of them being repeated by post-mineralisation faults.

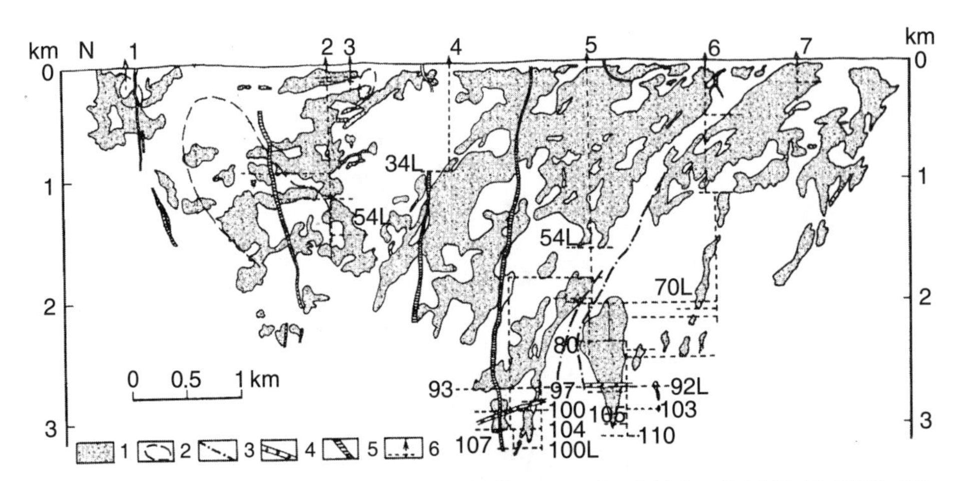

Fig. 1.2.4 Longitudinal vertical section of the Champion Reef, Kolar Gold Field (KGF) (after Genkin, et al., 1988) Legend: 1. Worked out blocks 2. Outline of worked out part of Oriental Lode; 3. Mysore Fault; 4. Pegmatites; 5. Dolerite; 6. Mine shafts. Horizontal dashed lines refer to mine levels.

From petrographic point of view the ores of KGF are divided into two types: (i) Gold-quartz (-calcite) ores, and (ii) Gold-quartz–sulphide ores.

The gold-quartz (- calcite) ores are characteristic of the eastern lode system, comprising (from east to west) Muscoom Reef, Champion Reef and Mundy's Reef. Individual lodes, in an average about 2 m thick except where folded, consisted of a series of sub-parallel veins/stringers of quartz (-calcite). They dip at about 45° due west and become sub-vertical at depth.

Narayanswami et al., (1960) reported wall rock alteration around the quartz lodes in the form of a thin diopside-bearing shell at the immediate vein contact followed outward by the presence of dark hornblende and biotite. Hamilton and Hodgson (1986) reported that the lodes are symmetrically enveloped by a zonal sequence of altered rocks. The outermost alteration zone is characterised by the development of biotite. The next change inward is the development of dark hornblende. This zone is granular in texture, and commonly wears a spotted look due to the presence of plagioclase feldspar and quartz. The zone grades with the increasing development of thin streaks of diopside into a nearly monomineralic selvage of diopside at the quartz vein contact. Lack of good directional fabric in the altered rocks is interpreted as the development of the alteration posterior to the development of the N–S trending shear zones. These authors concluded that the alteration, however, must have been synchronous with the development of the NNW–SSE shears, and both these events preceded the peak metamorphism of the country rocks that corresponded to the middle to upper amphibolite facies.

Four gold-quartz-sulphide lodes were delineated in the western part of the KGF. From west to east, these are West Prospect Lode, Oriental Lode, MacTagaart West Lode and MacTagaart East Lode. Wall rock alteration around the sulphidic lodes is weak, and is represented by a thin (a few inches) zone of biotitisation (Siva Siddaiah and Rajamani., 1989). Lodes consisted of crude layers of cherty quartz, sulphides, mafic silicates and magnetite. However, ore zone alteration in terms of REE redistribution is conspicuous (Fig. 1.2.5).

The gold-quartz (-carbonate) veins consist chiefly of milky white to colourless quartz that may be massive, sheared or laminated. Pyrite, galena, sphalerite, arsenopyrite, pyrrhotite and chalcopyrite occur as fine disseminations or as veinlets in quartz. Gold occurs in free-milling form, usually in the size range of <5 to 50 μm. Scheelite is present with the ores in upper horizons down to 1500 m.

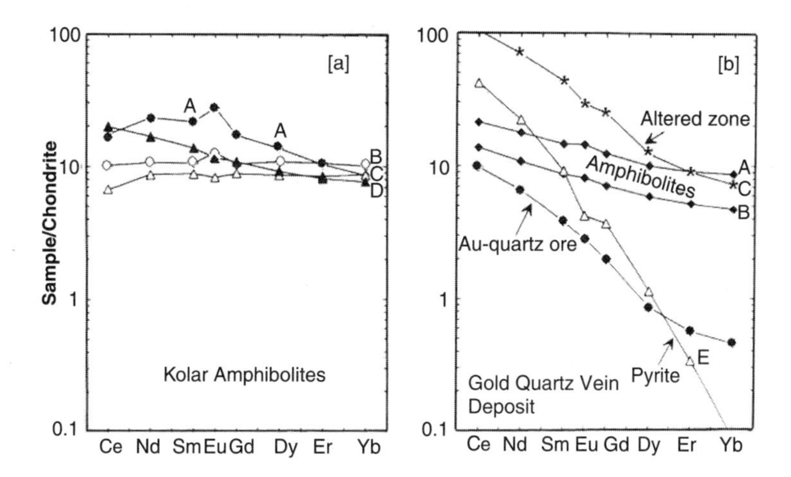

Fig. 1.2.5 REE contents in the Kolar Amphibolites and in the ore zones (after Siva Siddaiah et al., 1994) [a] Kolar Amphibolites: A – komatiitic amphibolite; B – tholeiitic amphibolite (both from the west–central Kolar schist belt); C – tholeiitic amphibolite from the South Kolar schist belt; D – komatiitic amphibolite from the eastern parts of the Central Kolar schist belt. [b] Kolar ore zone rocks: A and B – unaltered amphibolites adjacent to gold-quartz veins; C – altered amphibolite from the immediate contacts of gold-quartz veins; D – gold-quartz ore; E – pyrite from the vein ore.

Gold in the sulphidic lodes is much less in concentration and is associated commonly with sulphides, particularly pyrrhotite and arsenopyrite. Other sulphides present are chalcopyrite, sphalerite and pyrite. Gold in these lodes is present mainly as inclusions and fracture-filling in arsenopyrite. However, a smaller part may occur as free gold in quartz. Siva Siddaiah and Rajamani (1989) are of the view that the sulphide lodes were originally auriferous complex interflow sediments formed from submarine hydrothermal exhalations and clastic sediments. However, interpreting the textures and distribution of gold and sulphides in the lodes, Hamilton and Hodgson (1986) concluded that the gold-bearing sulphides partially replaced a banded and sheared magnetite-bearing ironstone and minor argillaceous and cherty rocks that occur as interflow sediments in volcanic sequences. Besides the ore minerals mentioned above, a host of ore minerals occuring as traces and minor phases are reported from the Kolar ores (Table 1.2.2). Most of them are from the Champion Reef.

The available analytical data show the gold from the Oriental Lode is rich in silver (24.5%–34.83%, $n = 4$). In case of the Champion Reef, the upper levels (comparable with the sample positions in the Oriental lode) are also rich in silver. In the Champion Reef, mercury content increases after some depths (Table 1.2.3). Association of silver with gold is common throughout the world. But locally, as at the Hemlo gold mine, Canada, upto 27% (wt) mercury has been found to have alloyed with gold (Harris, 1989). The system was even richer in mercury, and the excess mercury occurs as cinnabar and aktashite ($Cu_6Hg_3As_4S_{12}$) there. This is an exceptional case altogether. Interestingly the deposit contains large reserves of molybdenite and barite also.

Fluid inclusions in auriferous quartz from the Champion Reef are either aqueous liquid-rich or two-phase CO_2–H_2O type. The fluids evolved from early low salinity (3–5 equivalent wt% NaCl) ones to late moderate salinity (7–12 equivalent wt% NaCl) ones. P-T estimates based on coexisting primary two-phase CO_2–H_2O and aqueous liquid-rich inclusions in vein quartz yeilded P = 1.3 kb and T = 435 °C for the entrapment of immiscible fluids and early sulphide deposition. Bulk of the gold and the associated sulphides and the rare minerals (Table1.2.2), however, was probably deposited at 250°–260 °C (Santosh, 1986).

Ore genesis at the KGF has been studied by several groups of investigators. According to Hamilton and Hodgson (1986) gold mineralisation there took place from the magmatic hydrothermal fluids related to the granitoids in the belt emplaced during 2600–2500 Ma. Siva Siddaiah and Rajamani (1989) have two different models for the origin of the two ore types. According to them, the auriferous quartz-veins formed dominantly if not exclusively, from the magmatic fluids derived from granitic intrusions (2630–2550 Ma). A part of the gold could have been contributed by the footwall komatiitic amphibolite. But the sulphidic ores of the Western lodes are suggested to be complex type of interflow sediments deposited between the volcanic units. These could, however, be enriched in gold by the fluids derived from the granitic intrusives. The model of Safonov et al. (1988) on the genesis of ores in the KGF is also not simple.They envisaged a composite source of the ore fluides.The ore solutions possibly initiated from the upper mantle and then involved the lower crustal basalts and then the metamorphosed tholeiite, quartzite, granites and gneisses of the schist belt that came on the way. They also obtained a mineralisation age (Pb-Pb) of 2600–2500 Ma (Fig. 1.2.6).

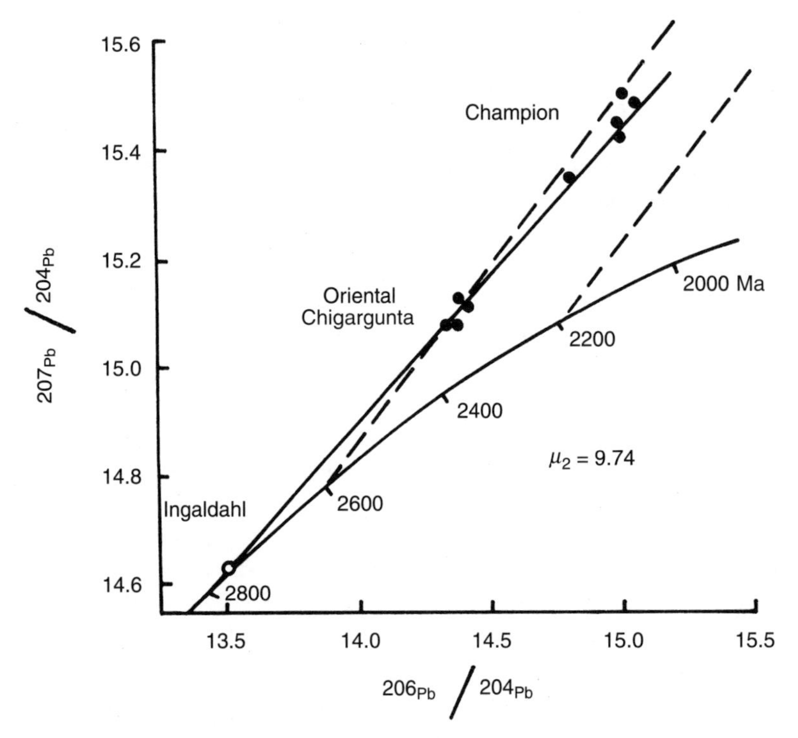

Fig. 1.2.6 Pb-Pb age of gold mineralisation in the Champion and Oriental Lodes and Chigargunta Extension in South Kolar. A single determination of the same from Ingaldhal copper deposit, Chitradurga schist belt (after Genkin et al., 1988).

The model of Chernyshev and Safonov further sharpened, shaping into what is now called the model of 'Orogenic gold' (Groves et al., 1998; Hagemann and Cassidy, 2000; Goldfarb et al., 2001; Robb, 2005). The present authors are of the view that the "Orogenic gold' model explains the origin and evolution of the ores at the KGF reasonably well (Sarkar, 2000).

Old Bisanattam Mines The old Bisanattam Mine is located at about 8 km south of the Kolar mines, in the Chittoor district of Andhra Pradesh. It was mined with large gap of time over a century, finally closing in 2001. The grade was good (5–10 g/t) but proven reserve was rather small.

Gold occurs here in native form in association with quartz. Sulphides, such as pyrite and pyrrhotite are in small concentration. The ore zone has much similarity with the Champion Lode of Kolar.

There were two exploratory mines: Mine I and Mine II, which together produced 45,161 ounces, or 1.4 tonnes of gold before closing down in 2001, along with other mines in Kolar region. At the time of closure the declared reserve was ~ 2 Mt tonnes down to the depth of 300 m with an average grade of 5.8 g/t Au (Seshadri, 2008).

South Kolar Schist Belt South Kolar schist belt contains two deposits: Mallappakonda and Chigargunta (Fig. 1.2.7).

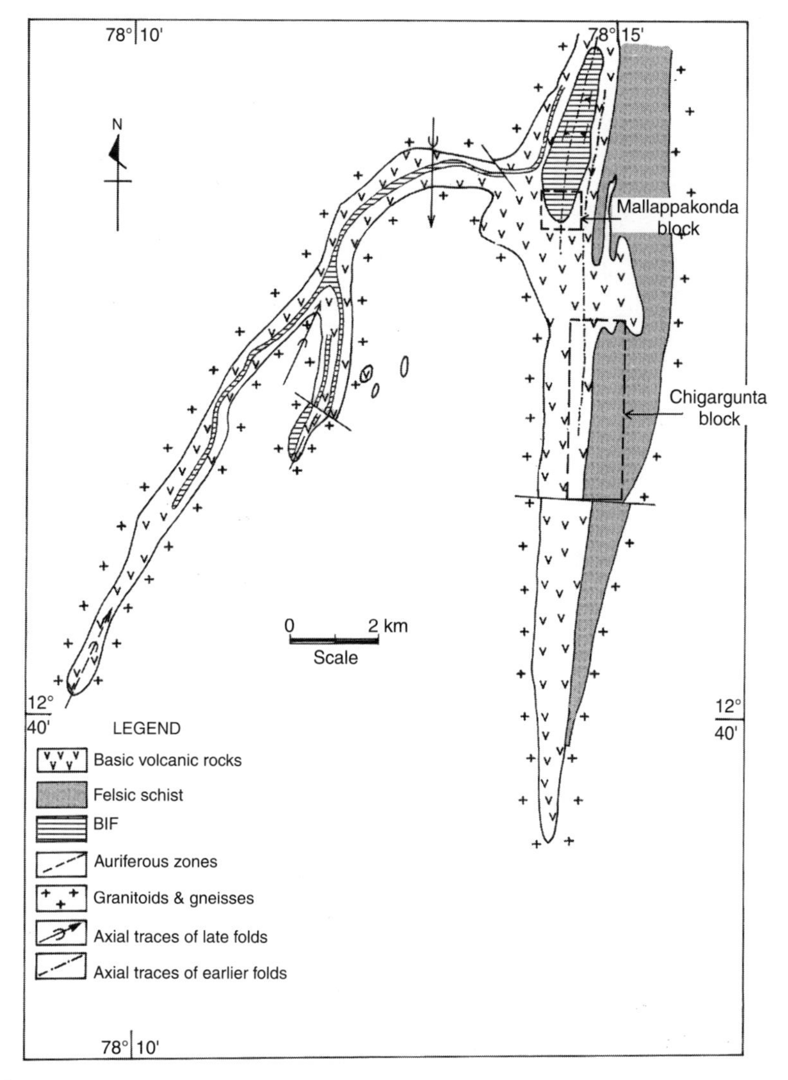

Fig. 1.2.7 Generalised geological map of South Kolar schist belt showing the Mallappakonda and Chigargunta gold prospects

The Mallappakonda deposit consists of chert–sulphide bodies within amphibolite (Fig.1.2.8). The strike length of the ore zone is ~500m, containing N–S trending en echelon lenses. A lode may be upto 40 m in width. The lenses consisted of alternate bands of quartzose chert and sulphidic iron (pyrrhotite and arsenopyrite, with subordinate löllingite). The deposit was mined in the upper 4 levels at ~ 3g/ tonne grade. Gold generally occurs included in arsenopyrite and rarely in pyrite. The ores, as a whole, have a metamorphic fabric and the arsenopyrite yielded an apparently equilibration temperature of 540° ± 10 °C.

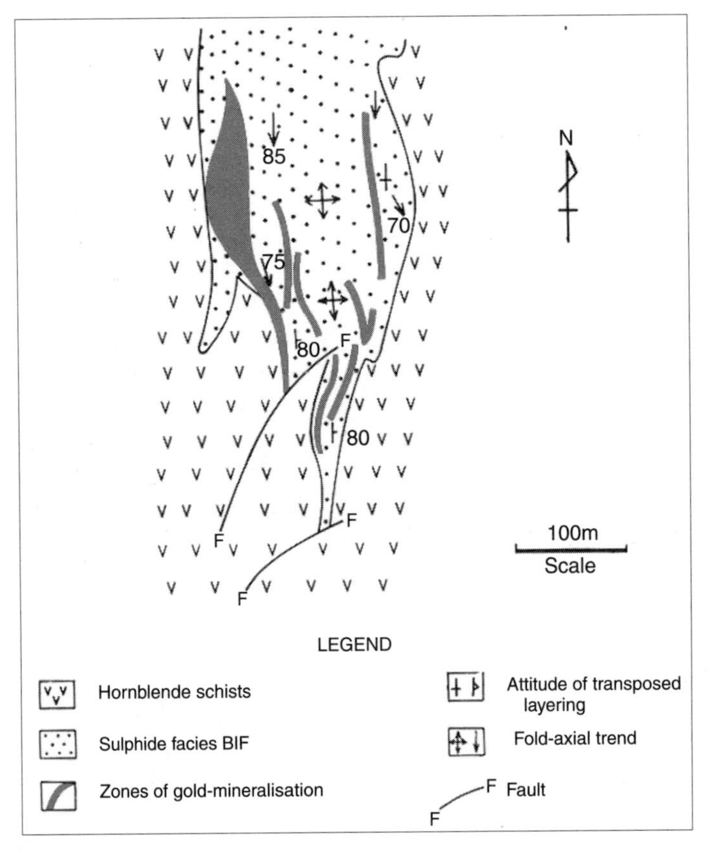

Fig.1.2.8 Map of Mallappakonda deposit (after Mukherjee and Natarajan, 1985)

REE pattern of the Mallappakonda gold sulphide deposits are distinctly different from those of the vein-type gold-quartz deposits in the KGF, and they are similar to those of the associated BIF and to those of many other Archean BIFs in greenstone belts. In some geochemical aspects, particularly on the REE and gold, the deposit is comparable to hydrothermal vent fluids and proximal metalliferous sediments at modern ocean ridges (Fig. 1.2.9) (Siva Siddaiah et al., 1994).

At Chigargunta, south of Mallappakonda, a number of sub-parallel to intersecting zones of gold mineralisation occur over a strike length of about 3 km. On an average, they trend N–S. The mineralised zones are located either in the mafic and felsic schist (Champion Gneiss) or along their interface. They are confined to the zones of ductile/ductile-brittle shearing that intersect the regional schistosity at low angles. These shear zones vary in thickness from a few centimeters to about 10 m (Mukherjee and Natarajan, 1985). South Kolar gold may still be exploited.

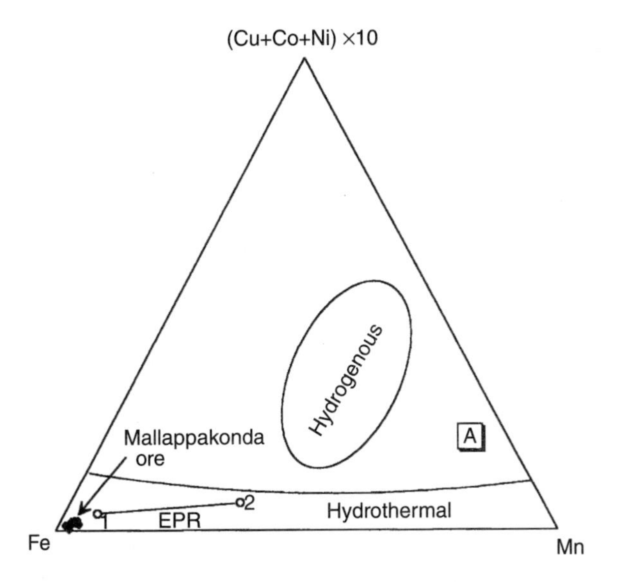

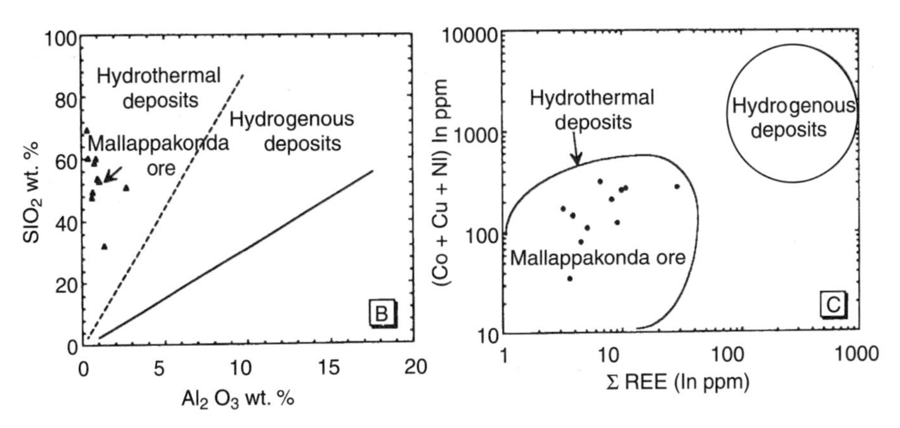

Fig. 1.2.9 Plots of Mallappakonda deposit in (Cu+Co+Ni) x 10 – Fe –Mn; SiO_2 (wt)–Al_2O_3 (wt) and (Co+Cu+Ni) –Σ REE diagrams (Siva Siddaiah et al., 1994) point to the hydrothermal affiliation of the ores

The shear-cum-ore zones in the mafic schist contain calcite, epidote, zoisite and muscovite besides hornblende and plagioclase feldspar. Within this zone there are bands rich in quartz; and shows increased biotitisation, muscovitisation and development of zoisite. The inner alteration zone shows very fine banding and comprises alternate bands of mafic (including diopside) and felsic minerals. The principal sulphide minerals present comprise pyrrhotite, pyrite and sphalerite with minor amounts of galena, chalcopyrite and arsenopyrite. Gold, up to 1 mm in size, occurs both as separate grains and clusters in quartz ± sulphides.

Shear zone-controlled gold mineralisation within the felsic schist is characterised by gold occuring in clusters within vein quartz or migmatitic lenses and show mutual boundary relationship with the sulphides (Mukherjee, 1991, 2002).

Table 1.2.2 *Ore minerals identified in KGF (Genkin et al., 1988)*

Mineral	Formula	Zone
Gold	Au	In all the lodes
Electrum	Au, Ag	Oriental Reef
Scheelite	$CaWO_4$	Champion Reef, Western Reefs
Arsenopyrite	FeAsS	Champion Reef, Western Reefs
Pyrrhotite	$Fe_{1-x}S$	Champion Reef, Western Reefs
Troilite	FeS	Oriental Reef
Pyrite	FeS_2	Champion Reef, Western Reefs
Sphalerite	ZnS	Champion Reef, Western Reefs
Galena	PbS	Champion Reef, Western Reefs
Chalcopyrite	$CuFeS_2$	Champion Reef, Western Reefs
Marcasite	FeS_2	Champion Reef, Western Reefs
Bismuth	Bi	Campion Reef, Oriental Reef
Maldonite	Au_2Bi	Champion Reef
Calaverite	$AuTe_2$	Champion Reef
Hedleyite	Bi_7Te_3	Champion Reef, Oriental Reef
Pilzenite	Bi_4Te_3	Champion Reef
Sumoite	BiTe	Champion Reef
Tellurobismuthite	Bi_2Te_3	Champion Reef
Tetradymite	$Bi_{14}Te_{13}S_8$	Champion Reef
Volynskite	$AgBi_{1.6}Te_2$	Champion Reef
Hessite	Ag_2Te	Champion Reef, Oriental Reef
Altaite	PbTe	Champion Reef
Kolarite*	$PbTeCl_2$	Champion Reef
Radhakrishnaite*	$PbTe_3(Cl,S)_2$	Champion Reef
Kotumite	$PbCl_2$	Champion Reef
Breithauptite	NiSb	Oriental Reef
Ullmannite	NiSbS	Champion Reef, Oriental Reef
Gudmundite	FeSbS	Champion Reef, Oriental Reef
Loellingite	$FeAs_2$	Champion Reef
Hawleyite	CdS	Champion Reef
Tetrahedrite	$Cu_{12}Sb_4S_{13}$	Champion Reef
Pb– sulphoantimonide	$PbSb_8S_{20}$	Champion Reef
Ag-Pb– sulphobismuthite	$AgPbBi_{2.3}S_{3.2}$	Champion Reef
Cubanite	$CuFe_2S_3$	Champion Reef
Pentlandite	$(Fe,Ni)_9S_8$	Champion Reef
Violarite	$FeNi_2S_4$	Champion Reef
Mackinawite	FeS_{1+x}	Champion Reef
Cerussite	$PbCO_3$	Champion Reef

* First report in literature

Table 1.2.3 *Chemical composition (wt%) of mineral gold and electrum from KGF (Genkin et al., 1988)*

Sample No	Depth (m)	Au	Ag	Cu	Hg	Total
			Gold–Champion Reef			
1	150–160	81.09	18.70	Nd	Nd	99.79
2	150–160	77.36	21.68	0.03	0.32	99.39
3	150–160	75.77	22.91	Nd	0.32	99.00
4	150–160	80.87	18.47	0.02	0.38	99.74
5	150–160	81.60	17.98	Nd	0.38	99.96
6	150–160	82.14	15.20	0.12	0.77	98.23
7	~1300	78.19	20.51	0.39	1.35	100.76
8	~2700	94.30	5.40	0.18	Nd	99.88
9	~2700	94.90	4.24	0.06	1.37	100.57
10	~2700	93.52	4.55	0.08	1.55	99.70
11	~2700	95.37	3.70	0.14	1.81	101.02
12	~2700	94.10	3.87	0.16	2.56	100.69
13	~2900	95.69	3.51	0.12	0.68	100.00
14	~2900	94.62	3.66	0.19	0.68	99.15
15	~2900	93.29	5.47	0.11	1.04	99.91
16	~2900	91.12	7.45	0.03	1.05	99.65
17	~2900	92.11	6.43	0.06	1.64	100.24
18	~3000	92.84	6.02	0.35	1.49	100.70
19	~3150	92.27	6.82	0.07	1.56	100.72
20	~3150	90.84	6.97	0.11	1.42	99.35
			Electrum–Oriental lode			
21	~1000	67.30	33.45	Nd	0.95	101.70
22	~1000	65.97	34.83	0.02	0.71	101.53
23	~1000	74.40	24.52	0.05	1.21	100.18
24	~1000	64.60	33.90	Nd	Nd	98.50

Gold Mineralisation in Hutti–Maski Schist Belt Gold mineralisation occurs at a number of places along this 65 km long horse shoe shaped schist belt of north Karnataka. Of these, the mineralisation at Hutti is most intense, which at present sustains the most prominent gold producing mine in the country. Two other satellite mines in this belt are located at Uti and Hira-Buddini. Other prospects of interest in this belt are Udbal, Maski, Buddini, Tuppadhur, Sanbal, Chinchergi, Kadoni, Wandalli and a few others; many of which have old workings, and were explored in the past by GSI and HGML.

Hutti Block The Hutti Gold mine, with annual production of ≥ 3 t of gold is now India's largest operational gold mine (Patil, 2008). Today, the Hutti underground mine is the deepest mine in the country after the closure of Kolar Gold and Mosabani Copper mines. The belt has so far produced ~72 tonnes of gold and the mine is expected to last another 30 years, with an average grade of ~4g/t Au.

There are nine sub-parallel N20°W-trending auriferous quartz reefs in Hutti, which bifurcate and coalesce both along strike and depth. These reefs from west to east are named as Main Reef (western most near granite boundary), Prospect Reef, Oakley's Reef, Middle Reef, Zone-1 Reef, Village Reef, Strike Reef (Hanging wall), Strike Reef (Foot wall) and New East Reef. The host rocks are metabasalts and minor meta-rhyolites, which have undergone peak metamorphism at 3–5 kbar and ~650°C (Mishra and Pal, 2008). Five distinct shear related deformational events (D_1 to D_5), with varied spatial and temporal relation with the mineralisation have been recognised (Kolb et al., 2004a). Disposition of discrete major ore veins at Hutti in plan and section are shown in Fig. 1.2.10 and Fig. 1.2.11. They have nearly sharp boundaries with the host rocks and steeply dip (~70°) towards west. They may be assigned to the 'fault-fill' type veins of Robert and Poulsen (2001). Pre- and/or syn-mineralisation 'dolerite' (dacitic subvolcanic) dykes are exposed in the mine workings.

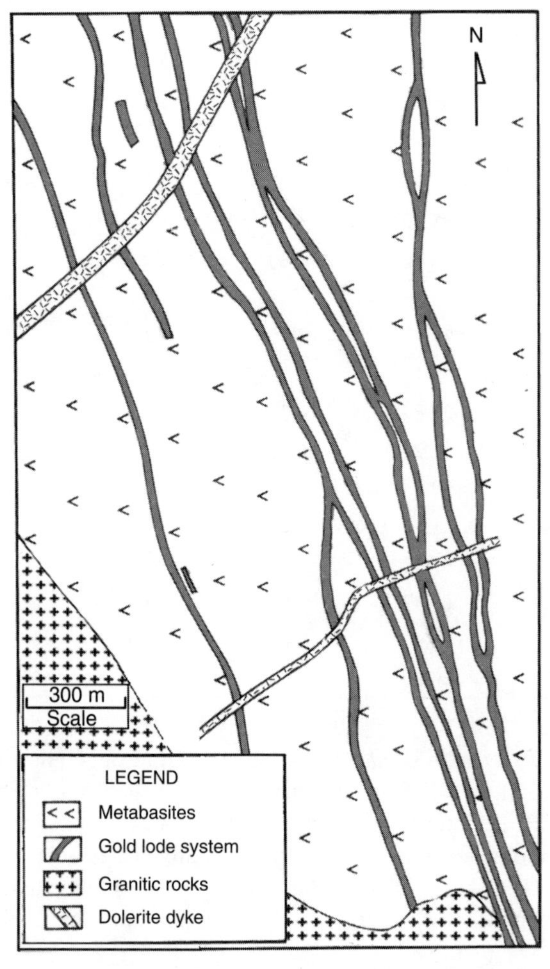

Fig. 1.2.10 Geological map of Hutti mine area, Karnataka, showing the projected disposition of gold lode system and its relationship with country rocks (after Biswas et al., 1985)

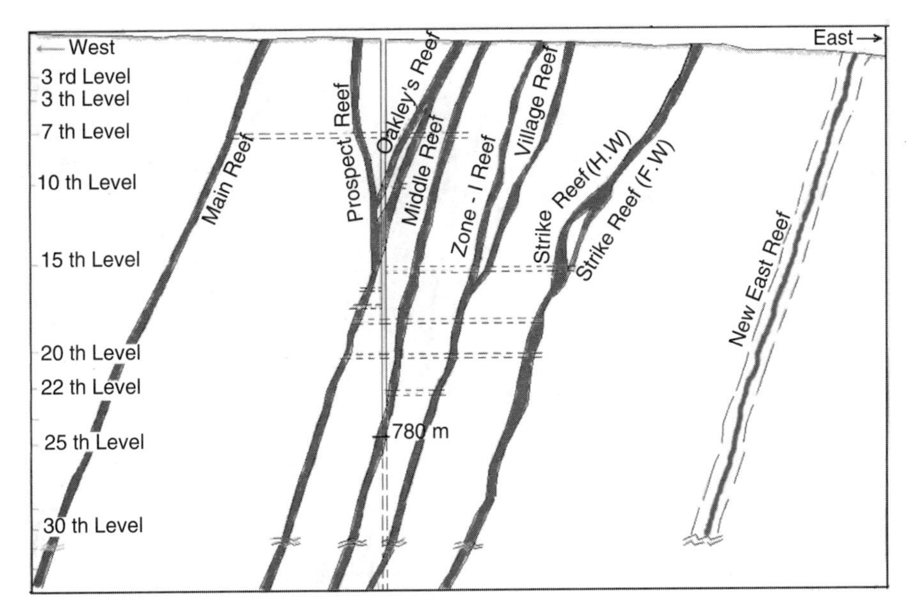

Fig. 1.2.11 Cross-section through the Mallappa shaft, Hutti mines (after Exploration Office, HGML)

There is an increase of SiO_2, Al_2O_3, Na_2O, S and volatiles (H_2O, CO_2) and a general decrease of Fe_2O_3, CaO, TiO_2, and MnO_2 in the ore zone. The minor elements Ba, As, Cu and W show substantial increases in the ore zone, with minor increase of Zn, Ag, Ce and Zr (Biswas, 1990a). Safonov et al. (1988) suggested the presence of the following alteration zones in the wall rocks of the ore bodies, starting from the immediate contacts:

1. Chlorite-biotite zone with calcite and sulphides
2. Chlorite-actinolite zone with rare biotite and sulphides
3. Actinolite-biotite zone
4. Epidote-bearing zone.

Gold at Hutti occurs predominantly with laminated or sheeted quartz veins (D_3) (Fig. 1.2.12, a). The mineralisation is also associated with the proximal biotite-K feldspar (D_2) and sporadic distal chlorite alteration zones. Besides free gold in quartz and proximal wall rock, mineralisation is also in the sulphidic quartz veins in association with arsenopyrite, pyrrhotite, pyrite, chalcopyrite, sphalerite, galena, cobalt-bearing pentlandite, ilmenite and magnetite. Arsenopyrite followed by löllingite and pyrrhotite, are the principal gold-bearing sulphides. In the ore shoots the gold-content varies from 2 to 20 ppm, the average being 4 ppm. Fineness of gold[5] is > 900, the Ag-content not exceeding 8%. Vein quartz is of two generations. The mass of quartz formed early. A later generation of quartz was introduced along with gold and the sulphides, and was emplaced along the fractures in the early quartz. Tectonised vein-quartz shows different degrees of deformation and recrystallisation. The gold-associated, apparently late quartz, contained fluid inclusions that homogenised at 300°C and showed maximum salinity of 13.5% NaCl equivalent. Vein quartz has been found to contain As, Zn and W (up to 10^{-1}%) (Safonov et al., 1988).

Scheelite is a ubiquitous mineral, occurring as fine disseminations (with local concentrations) and veinlets, both in the ore zone and in the wall rocks (Fig. 1.2.12b).

[5] 'Fineness of gold' is defined as the quantity of gold in an alloy, expressed as parts per thousand.

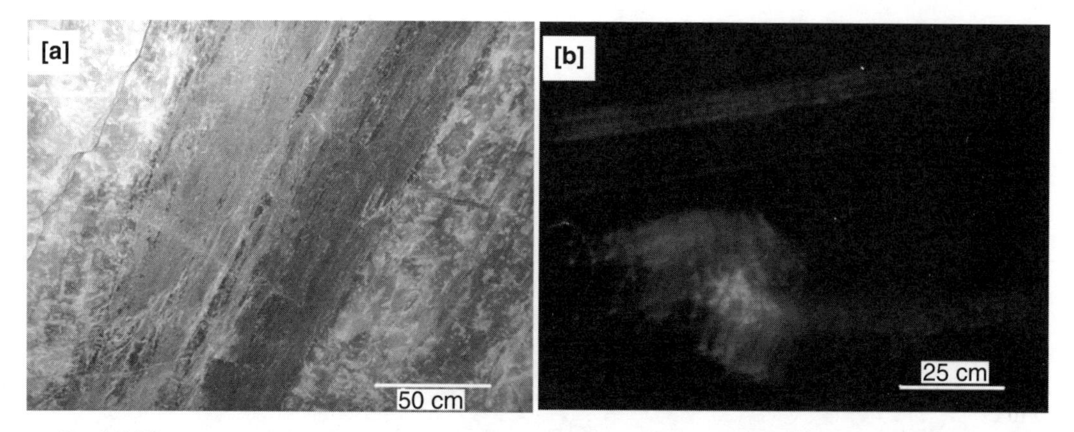

Fig. 1.2.12 Roof-views of Zone-I Reef in the drive at 24th Level (depth–2400 ft), Mallapa Shaft, Hutti mines (a) Laminated (sheeted) auriferous quartz vein with dark biotitic bands; (b) scheelite under UV light, in a biotite-rich band and fold hinge in the wall rock of gold-bearing Reef (photographs by authors)

The gold mineralisations as well as the multiple alteration events in Hutti are post-peak metamorphism. The ore fluid was dominantly metamorphogenic, having low saline aqueous carbonic character, similar to Late Archean orogenic gold deposits world over (Pal and Mishra, 2002; Pandalai et al., 2003; Mishra and Pal, 2008). Mishra (2008) observed that the mineralisation at Hutti took place on the metamorphic retrograde path beginning with initial alteration and sulphidation at upper green schist facies. The ore fluid had a total dissolved sulphur ($m\Sigma_S$) content of ~0.1 m, and decrease in both f_{O_2} and pH caused gold precipitation in the proximal biotite zone from fluid containing Au $(HS)^{2-}$. Fluid inclusions in quartz veins within proximal alteration zone indicate the temperature of precipitation of gold and sufficient CH_4 entrapment at > 400 °C. This stage was followed by phase separation of the initial metamorphogenic fluid during a decompression event, leading to entrapment of gold and the coeval and cogenetic fluids in the laminated quartz veins at 1.0–1.7 kb and 280–320 °C (Pal and Mishra, op. cit.). The gold precipitation in these veins was a result of decrease in $m\Sigma_S$ content of aqueous fluid rather than the wall rock sulphidation and f_{O_2} decrease, suggested for the biotite zone.

Biswas (1990a, b) reports of a syngenetic type lean (up to 400 ppb Au) mineralisation associated with pyritiferous carbonaceous sediments. This is indeed possible, but does not make any contribution to the gold production, or exploitable reserve.

The Hutti–Maski ores share the features of both the Champion and Oriental groups of lodes of Kolar belt in that they are both quartzose and sulphidic. Salinity in the fluid inclusions in quartz is comparable, but the temperatures of homogenisation of the fluid inclusions are lower at Hutti. In the latter, the tellurium- and bismuth-bearing minerals and molybdenite (south Kolar) are conspicuous by their absence. These small differences notwithstanding, it will be reasonable to believe that the Hutti-Maski belt had almost a similar history of metallisation as the Kolar belt.

The felsic volcanic host rocks at Hutti have yielded U-Pb zircon age of 2587 ± 7 Ma, while the age of gold mineralisation determined from hydrothermal monazite is 2547 ± 10 Ma. The syntectonic Kavital granitoid has U-Pb zircon age of 2545 ± 7, which is comparable with the age of gold mineralisation (Sharma et al., 2008).

Uti Block The Uti block is located at 22 km northeast of Hutti. The area is characterised by volcanisedimentary rocks comprising coarse and fine grained amphibolites, metamorphosed felsic volcanics and thin bands of pyritiferous/carbonaceous schists, andalusite-garnet-bearing mica schists and BIF. Dolerite dykes, granite, pegmatite and quartz veins intrude these rocks. The NNE–SSW trending auriferous zone, comprising several sulphide-rich mineralised bands, is mainly confined to the highly deformed, silicified and/or carbonatised amphibolites, though the sheared acid volcanics are also mineralised in some sections. The pinching and swelling of lodes is observed both along the strike and depth. The mineralisation is present in thin stringers and veinlets of quartz associated with pyrite, arsenopyrite and pyrrhotite. Though the mineralised shear zones are discontinuous, they maintain rough parallelism among themselves. The GSI and HGML delineated 17 lodes over a cumulative strike length of ~3.5 km, with strike lengths of 50 to 720 m and widths of 0.3 to 24 m for the individual lodes. 1.027 Mt ore reserve with average grade of + 4 gm/t Au was estimated by GSI for lode 4 down to 75 m depth (Biswas, 1990a,b).This was followed by detailed exploration by the MECL aided by further drilling and exploratory underground mining. An open cast mine was developed in this deposit (Fig. 1.2.13) to work out 300 m of the workable length of lode 4, the ore reserves were recalculated to 8.5 Mt with average width and grade of 12 m and 2.54 g/t Au down to 90 m depth. The mining was commenced in 1996, and is still in progress.

Mishra and Pal (op. cit.) established a clockwise P-T-t path with peak P-T conditions of ~6 kbar and 650–700° C for the metamorphosed host rocks in Uti. They explained the deduced path by invoking a subduction related compressional tectonic setting. Though the mineralisation took place in the metamorphic retrograde path, it differs considerably from that at Hutti, primarily due to lack of any observed link between the initial metamorphogenic fluid and the alteration zone. The virtual absence of carbonic inclusions is another important feature of the auriferous quartz veins of Uti. Mishra et al. (2005) attributed such a fluid chemistry to the proximity of Yelgatti and Kavital granitoids. According to them the formation of the lodes, which are localised along small shear zones, was post metamorphic but possibly synchronous with granite emplacement. In contrast to Hutti, a direct role of metamorphogenic aqueous-carbonic fluid in the formation of the gold-quartz veins is difficult to establish here.

Fig. 1.2.13 Uti open-cast mine showing benches on its western face. Pit length = 360 m, width = 180 m, depth = 70 m; 10 Footwall (west) benches and 11 Hanging wall (east) benches (as in December, 2008) (photograph by the authors)

Hira–Buddini Block The Hira–Buddini prospect, where an underground mine is under development, is located at about 22 km east of Hutti mines. GSI had indicated probable reserve of 0.52 Mt with average grade of 11.9 gm/t Au down to a depth of 110 m. HGML in course of ongoing development of the prospect has indicated a much higher grade at >14 gm/t Au. The strike length of the ore body is 600 m and the width varies from 1–3.5 m. The lodes are characterised by free gold in quartz with nominal sulphides, and are located along the sheared contact of metabasalt and felsic volcanics. The mineralised zone trends at N70°E–S70°W with steep northerly dip of 80°–85° to near vertical. Reversal of dip is also noted at places. The quartz-veined and mineralised shear zone along the contact of the metabasalt and felsic volcanics is called the 'ladder vein', as it is characterised by transverse (sigmoidal) quartz veins and lenses extending from the footwall to the hang wall of the brittle-ductile shear planes. Quartz veins both along the steeply dipping shear planes and shallow dipping sigmoidal extensions are mineralised. Pyrite, chalcopyrite and pyrrhotite are the major sulphides in association with gold-bearing quartz veins. Krienitz, et al. (2008) identified three distinct wall rock alteration zones within and adjacent to Hira-Buddini ore body:

1. Distal chlorite-actinolite zone
2. Proximal actinolite-biotite-calcite-tourmaline zone
3. Inner biotite-K feldspar-quartz zone.

Large density variation in the clusters of closely associated aqueous and carbonic fluid inclusions in the sigmoidal extension veins is suggestive of sudden fluctuation in pressure due to pressure cycling or operation of fault valve. Occurrence of gold within these veins indicates phase separation leading to gold precipitation during sudden pressure drops. The mineralisation at Hira-Buddini took place in shear controlled quartz veins, which developed in the post-peak metamorphic stage and the CO_2-free and high saline nature of the ore fluids point to a granitic source (Mishra and Pal, 2008). Krienitz, et al. (2008) emphasised on the boron isotope dominance in associated tourmaline in concluding on the ore mineralisation in the Hira-Buddini sector. They concluded that the mineralising fluids, which were poor in ^{11}B formed by metamorphic devolatisation and the ^{11}B-rich fluid was probably obtained by degassing of I-type granitic magma(s) that intruded the greenstone belt rocks.

Wandalli Block 25 lodes with individual lengths of 100–600 m and average width of 1.5 m occur within a stretch of ~5 km. Lithological units include coarse grained massive and schistose amphibolite, acid volcanics, carbonaceous phyllite and chert. The lode zones, confined to sheared metavolcanics, contain free gold with minor sulphides. 382 kg gold was produced in the past from this block. The reserve estimate stands at 0.80 Mt with gold values ranging from 2.0–7.5 gm/t (cf. Radhakrishna and Curtis, 1999).

Chinchergi Block This block is situated to the east of Wandalli block. The cumulative strike length and width of the ore zone is 3.45 km and 1 km, respectively. There are more than 16 lodes, individual lode lengths are up to 200 m (width ~ 1 m). The lodes have free gold with pyrite-pyrrhotite dominant sulphides. GSI has estimated about 60,000 tonnes of ore reserve with grade ranging from 4.7–11.5 gm/t Au in this block (GSI, 2000).

Buddini Mines Block This block is situated 25 km south of Hutti gold mine and 12 km NNE of Maski village. The prospective zone in this area is 1.5 to 2 km wide, comprising sheared metabasalts and its variants and thin bands and lenses of metagabbro. The rock units strike in N10°W–S10°E direction with local variation up to N35°W–S35°E and steeply dip at 70°–80° westwards. The rocks are metamorphosed to quartz-albite grade of greenschist facies. There are eight sub-parallel gold-bearing quartz-ankerite veined chlorite schist bands from east to west, of which the thicker and better mineralised bands are named as Mopla Lode, Main Lode and West Lode. Old underground workings are there in all these lodes to limited extents. Out of 1400 m strike length explored by drilling on Mopla and Main Lodes, 200–300 m in each is better mineralised with average 5–6 g/t Au over 1m width. The total reserve estimate for the better mineralised part of Mopla Lode down to 100 m depth is 58,890 tonnes (GSI, 2000).

Tuppadhur Block This prospect is located to the north of Buddini Mines block. Gold mineralisation is associated with quartz veins and quartz-ankerite veins in five parallel chlorite schist bands over a strike length of 1600 m. A probable reserve of 0.089 Mt has been estimated for lode1 (length = 375 m; width = 1.0–1.3 m; depth = 100 m) with average grade of 3.95 g/t Au.

Sanbal Block The Sanbal gold prospect lies to the south east of Buddini block. The area consists of chlorite-actinolite schist and quartz-sericite schist traversed by gabbro dykes. The gold mineralisation is hosted by quartz veins with sporadic sulfides. A total of seven mineralised zones, over cumulative strike length of 935 m have been identified. Very high values of gold (23–305 g/t Au) have been recorded from 0.3–0.75 m wide Lode-1 over a strike legth of 75 m. The HGML has conducted exploratory mining in this block (GSI, 2000).

Gold Mineralisation in the Ramagiri Belt Like most other gold belts of south India, or even the country, Ramagiri belt in Andhra Pradesh has also a history of ancient mining for gold at several places along it. During the first half of the last century (1910–1927 and 1946–1950), some of these sites were taken up for prospecting and even mining, but were soon given up. During the last quarter of the last century mining was resumed by the BGML in the Yeppamana mines, Ramagiri area (Fig. 1.1.19). The mine is now closed again.

Ore mineralisation at the Ramagiri area is confined to grey to greasy quartz veins in schistose metabasites. These rocks, now apparently looking phyllitic, have in reality been transformed from the tholeiitic country rocks and not from andesite as suggested by some earlier workers (Ghosh et al., 1970). Field features and the petrological and petrochemical characters are more suggestive of a tholeiitic protolith (Choudhuri, 1986; Sarkar, 1988). The two principal orebodies at Ramagiri were called Yeppamana and the Main Lode. The ore shoots, better developed in the Yappamana, are folded with the axes plunging 55°–60° towards the north.

Gold occurs within the quartz carbonate veins as discrete grains. Pyrite is the principal sulphide mineral in the ore zone. Other phases are chalcopyrite and arsenopyrite. Comparison of mineralisation with the Kolar belt does not arise. But even compared to the Hutti–Maski belt, mineralisation is poor here.

Gold Mineralisation in the Chitradurga Belt There are following three types of gold occurrences in this belt, controlled by primary and secondary shears as depicted in Fig. 1.2.14. Most of the gold prospects are, however located along the secondary shears, including those in Gadag belt.

1. BIF associated gold prospects (Ajjanahalli, Anesidri, Sasaluhatti, M.N. Halli, Bankrakkanahalli etc.)
2. Auriferous quartz-sulphide lode deposits (G.R. Halli, C.K. Halli, Hosahatti and Paramanahalli)
3. Lode gold deposits associated with intrusive granodiorite-tonalite bodies (Honnamaradi).

Ajjanahalli Deposit Ajjanahalli gold deposit is situated about 80 km to the south of the Chitradurga town. Calculated ore reserve is 1.542 Mt at 2.86 g/t Au (Raghunandan, 1997). Mining started here in March, 1996 by the Hutti Gold Mine Ltd., and was closed down in December, 2000. During this period, the annual gold production was about 200 kg. Mining closed not due to the disappearence of the ore but due to the fact that the ores changed the mineralogy, from oxidised to the sulphidic ores, which necessitated a change of the process of recovery of gold from the ores.

The mineralisation at Ajjanahalli is located about 2.5 km west of the boundary of the Chitradurga schist belt with the Eastern Dharwar granite–granite gneisses (Dharwar batholith of Chadwick et al., 2000). The dominant rock types in the schist belt of this region are quartz–sericite schist with bands of oxide and carbonate facies including BIF. The orebody with economic gold mineralisation at Ajjanahalli, called the Main Reef, occurs in the BIF in the hinge region of an upright antiform (Fig. 1.2.15) (Kolb et al., 2004b).

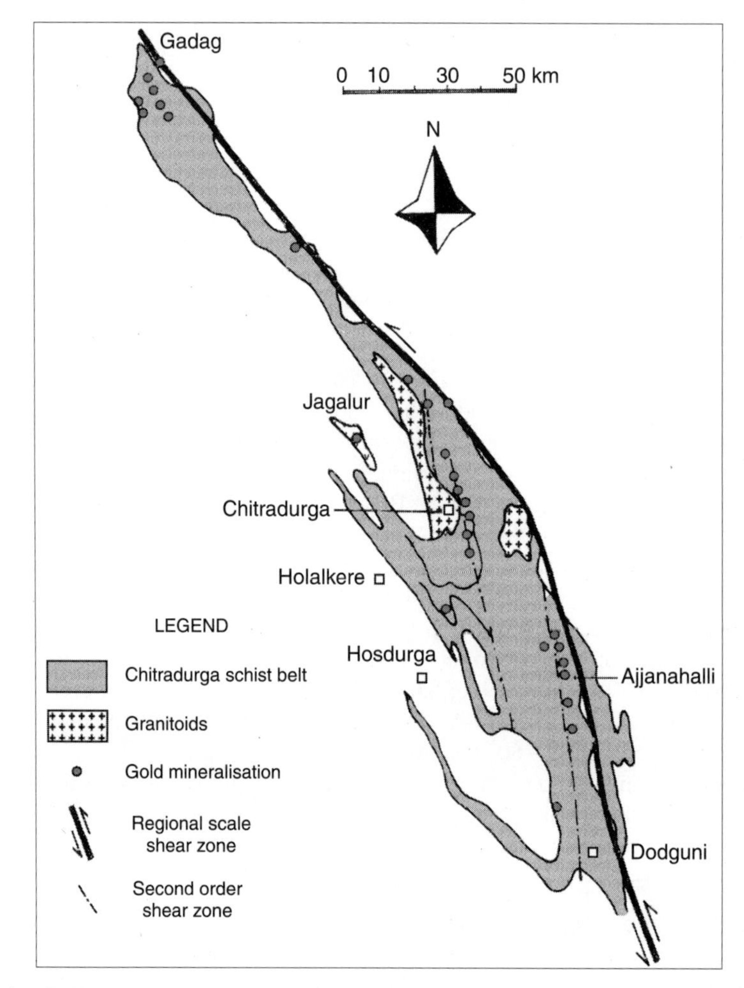

Fig. 1.2.14 Gold occurrences along the first and second order shears in Chitradurga schist belt (modified after Vasudeva in Radhakrishna and Curtis, 1999)

The core of this antiform is occupied by metabasic rock now represented by chlorite–actinolite schist. The mineralised portion of the antiform of the Main Reef is about 100 m wide and consists of six second order antiforms, each 10–15m wide. A barren oxide facies BIF unit, locally referred to as the East Reef, occurs to the east of the Main Reef.

The meta-siliciclastic rocks, phyllite and quartz sericite schists show strong negative Cr-anomaly suggesting a Cr-poor source area. The higher mean Y-content compared to NASC (Kolb et al., 2004) is characteristic of Archean siliciclastic sediments. Further, positive correlation between Co and Ni, V and Sc, U and Th, and Hf and Zr (Kolb et al., 2000b) suggests a source that was made of both mafic and felsic rocks and is characteristic of Archean graywacke and turbidite sequences (McLennan and Taylor, 1991).

The bulk and trace element compositions of the chlorite–actinolite schist confirm it to be a mafic volcanic rock at the initial stage.

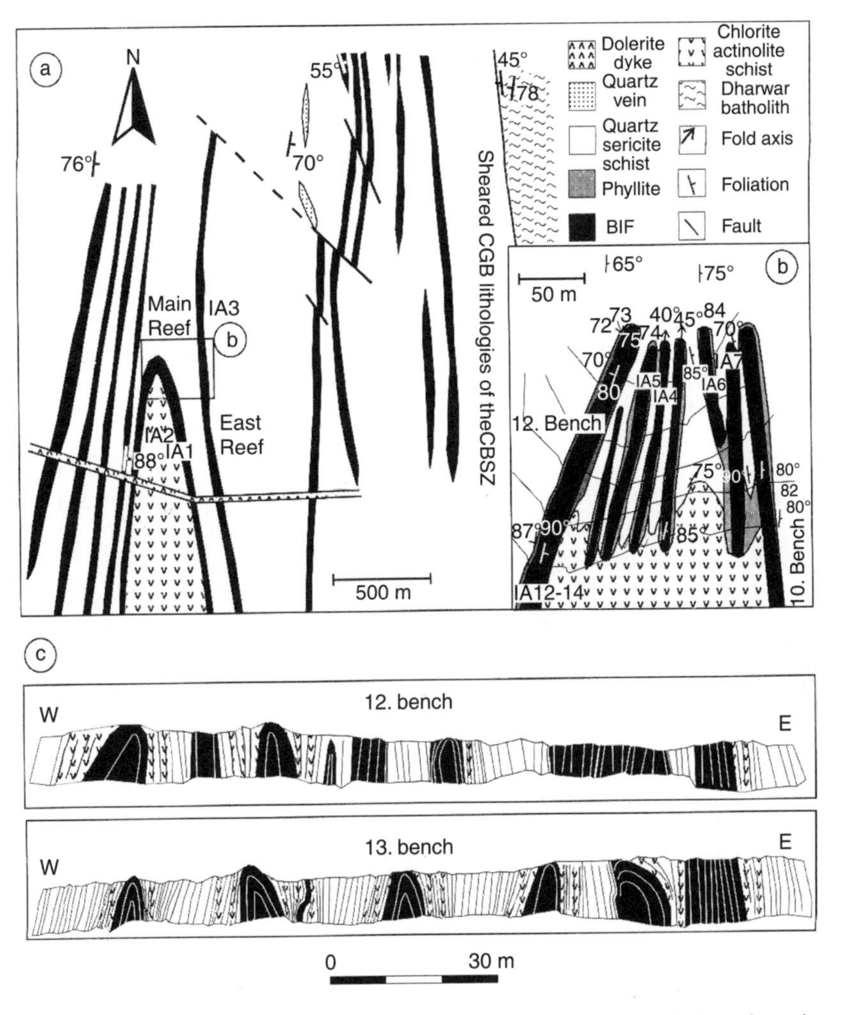

Fig. 1.2.15 (a) Schematic geological map of Ajjanahalli deposit, (b) Detailed geological map of the Main Reef, (c) Bench profiles of Ajjanahalli open pits (12th and 13th Benches) (after Kolb, et al., 2004b)

The BIF is a cherty iron formation containing < 2% Al_2O_3. The Main Reef samples show pronounced positive Eu anomaly (Eu/ Eu* between 1.54 and 3.86). In contrast, the East Reef samples do not show any anomaly (Kolb et al., op.cit.). The situation defies an easy explanation. Gold and sulphide mineralisations in the BIF of Ajjanahalli are correlatable with the increase in the content of Ca, Mg, CO_2, Sr, As, Bi, Cu, Sb, Zn, Pb, Se, Ag and Te content.

Quartz separates from the two mineralised banded iron formation samples yielded $\delta^{18}O$ values of 14.0 and 14.4 per mil. For an estimated temperature of 300–350°C for vein precipitation, the isotopic composition of the water component of the ore fluid in the Au-bearing quartz-ankerite veins is estimated (following Clayton et al., 1972) to have ranged between 6.5 and 8.5 per mil (Kolb et al.,op.cit.). These $\delta^{18}O$ values are within the range characteristic of the orogenic gold deposits involving metamorphic and deep magmatic fluids (Ridley and Diamond, 2000).

There are more than one suggested models for the Ajjanahalli gold mineralisation. A syngenetic model, in which gold was hosted by sulphide facies BIF, has been suggested by early workers (Hussain et al., 1996). Recently, Pal and Misra (2003) suggested an epigenetic model, the essence of which is that the gold-sulphide mineralisation took place from H_2O- CO_2- CH_4- NaCl – bearing fluids. However, their suggestion that the gold mineralisation took place due to phase seperation of the ore fluid during isothermal decompression is considered a matter of conjecture at the moment. Any reasonable genetic model for the Ajjanahalli ores, as suggested by Kolb et al., (2004b), must satisfactorily explain the following observations: (i) Gold mineralsation is hosted by a BIF unit in an antiform, the zone of dialation, (ii) Deformation and mineralisation in the Main Reef corresponding to the peak of metamorphism in the greenschist facis, (iii) Alteration of the BIF, and (iv) Stable isotope data that suggest that the ore-fluids were derived from an external source. These features would rather suggest that mineralisation was epigenetic, formed by syntectonic metamorphic devolatisation that leads to the formation of orogenic gold deposits.

Gold Mineralisation in the Gadag Schist Belt The Gadag schist belt is the northern extremity of the eastern limb of the Chitradurga schist belt in the western Dharwar. The rocks in the Gadag schist belt comprise metavolcanics such as basalt and basaltic andesite and metasediments represented by an argillite-arenite-arkose- greywacke suite, sericite carbonate schist, carbon phyllite, chlorite carbonate schist, BIF and dolomite, belonging to Ingaldahl and Hiriyur Formations of the Chitradurga Group, the upper division of the Dharwar Supergroup. Gabbro and dolerite intrusives are conspicuously present.

The gold mineralisation in the Gadag schist belt is confined to an area of 12 × 10 sq km in the Kappal Gudde block in the northern part of the belt. Gold mining in this region started in the beginning of the last century and continued for about a decade (1905–1911). There have been several subsequent attempts at mining of these deposits, but none of them lasted long. The lodes at the Gadag belt are divided into the following four groups (Bhat and Katti, 1996):

1. Western Group – Lodes of Hosur, Yelishirur, Venkatapur areas located in the acid-intermediate metavolcanic rocks and sericite-carbonate rock
2. Middle (Central) Group – Lodes around Kabulayat Katti, Attikatti, Mysore mine and Sangli mine areas at the contact between metavolcanics and metasediments
3. Eastern Group – Lodes at the south of Beldhadi and Nabhapur at the contact of acid volcanics and meta-sediments
4. Far Eastern Group – Lodes around Sankatodak area in such rocks as talc- chlorite schist $\pm$ BIF $\pm$ sericite schist + sulphides

Of the above, the central lode system, or the Middle Group is best developed with a cumulative strike length of 15 km. The lodes in general, are quartz reefs enclosed on either side by quartz–chlorite/sericite assemblages, permeated by carbonates and sulphides. The gold-bearing quartz-carbonate veins contain arsenopyrite, pyrite, chalcopyrite, sphalerite, pyrrhotite and hematite in order of abundance. Gold generally occurs as fine disseminations.

According to Deb et al. (2008), the commonly noted sericite-carbonate association in the ore zone of Hosur mine may represent metamorphosed marly sediments or pre-/syn-metamorphic alteration zones. They have also documented the microstuctures developed in the Western, Central and Eastern group of lodes in the Gadag schist belt, and the spatial and temporal relation of the ore minerals with the structural elements. A schematic sketch (Fig. 1.2.16a) and photomicrographs (Fig. 1.2.16 b–e) of the microstructures in Hosur ore zone show that arsenopyrite started crystallising in pre- to syn D_2 stage (aligned parallel to folded S_1 and axial planar S_2); and quartz veins containing arsenopyrite and visible gold formed during late to post D_2, as they crosscut the dominant S_2 fabric. It is inferred by the above authors that the gold mineralisation in this lode system took place in the waning phase of deformation and metamorphism.

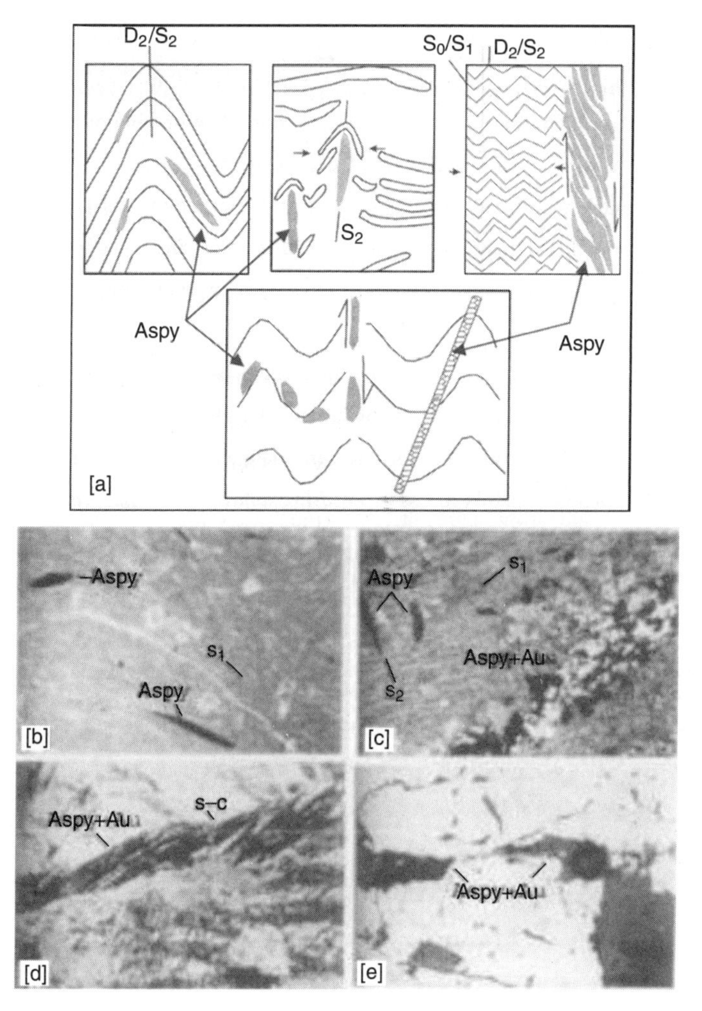

Fig. 1.2.16 (a–e) Small-scale and microstructures in Hosur ore body of Western Lode system, Gadag (a) Schematic diagram showing small-scale structures and their relation with gold-bearing arsenopyrite (Aspy); (b) Elongate Aspy grains parallel to folded (D_2) S_1 foliation; (c) Aspy oriented along S_2 plane; Aspy with visible gold (Au) in a quartz vein crosscutting both S_1 and S_2 surfaces; (d) Aspy + Au displaying S-C like microstructure; (e) Aspy + Au as fracture filling (from Deb et al., 2008).

The Central lode system exposed at Kabuliyatkatti, Attikatti, Mysore mine etc. and Eastern lode system at Beldadih, Nabhapura areas comprise metamorphosed igneous rocks representing andesite, dacite-rhyolite, tonalite-trondhjemite, gabbro and metamorphosed sediments representing greywacke, siltstone, shale, claystone, BIF and carbonate rocks. The gold-bearing quartz veins are dominantly disposed along S_1 planes, which are dissected by strong domainal crenelation cleavages (S_2). Thin quartz veins with significant gold values also occur along the axial planes of tightly

F_2-folded quartz veins (Fig. 1.2.17 a–b). From these observations, it can be inferred that quartz veining took place in this sector throughout the deformation history from early D_1 to syn-D_2 phase, the best gold concentration, however, occuring during syn-D_2 time (Deb et al., 2008).

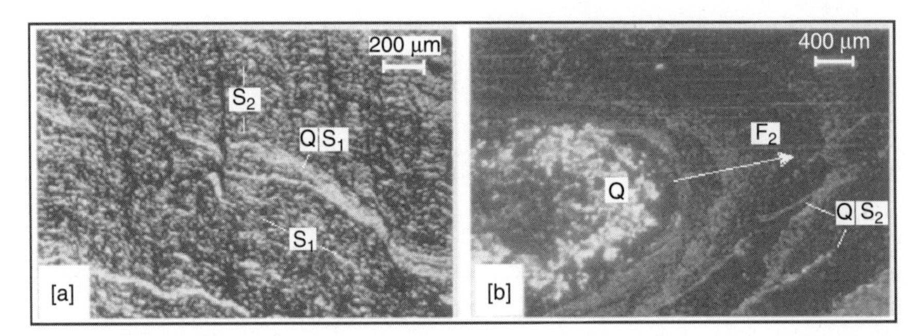

Fig. 1.2.17 (a–b) Microstructures and Au-quartz veins in Kabuliyatkatti ore body. (a) Quartz veins (Q S_1) along S_1 planes, dissected by strong crenulation cleavage (S_2); (b) F_2-folded quartz vein (Q), thin quartz veinlets (Q S_2) with high gold values along S_2 planes (modified after Deb et al., 2008).

The structural relations of the Far Eastern Group of lodes in Sankatodak-Tanda area indicate that the sulfidic gold mineralisation is early to syn-kinematic with respect to $D_2(F_2)$, as documented in Fig. 1.2.18 a–c. The old workings in this area extend parallel to the N–S or NNW–SSE trending axial plane of F_2 folds, which also indicates syn-D_2 mineralisation (Deb et al., op. cit.).

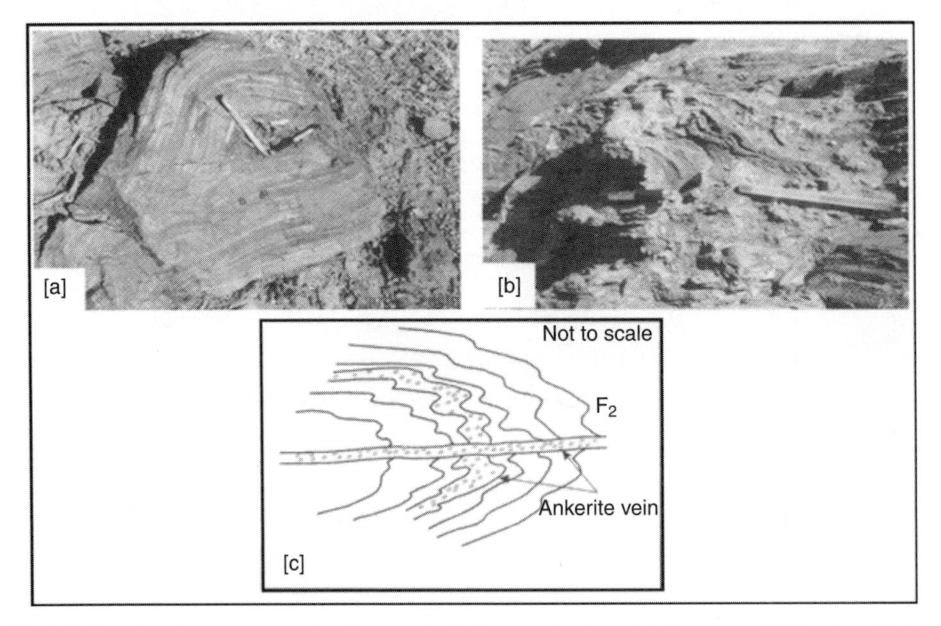

Fig. 1.2.18 (a–c) Structures in Far Eastern Lode System, Sankatodak-Tana area: (a) Reclined F_1 fold overprinted by F_2 fold in BIF; (b–c) ankerite-rich sulphide-gold ore band folded by F_2 and an ore band also occurs along the axial plane (from Deb et al., op. cit.)

Deb et al. (2008) conducted fluid inclusion studies on both mineralised and non-mineralised quartz veins from Western lode system of Hosur and Central lode system of Kabuliyatkatti and Sangli areas. The microthermometric determinations on primary and secondary inclusions in mineralised quartz-carbonate veins of Hosur area revealed a large range of salinity (~6–22 eq wt% NaCl in narrow zone of T_h). This is inferred to be due to the mixing of metamorphogenic low-salinity fluids with magmatogenic high-salinity fluids. The observations on the Central lode system were, however, very different. Here the fluids were entirely metamorphogenic without any mixing, as suggested by narrow salinity spread (0–12 eq wt% NaCl). Though the salinities of the two lode systems are very different, the T_h values are remarkably similar, being 110–330 °C in Western lode and 120–330 °C in the Central lode system, for primary inclusions.

Based on structural and fluid inclusion studies described above, it may be summarised that the orogenic gold mineralisation in Gadag schist belt was induced in contrasting lithologies mainly during D_2 deformation event by metamorphogenic fluids with low salinity (clustering around 4–6 eq wt% NaCl) at temperatures of ~200°C. The western lode at Hosur is an exception in having the mineralisation emplaced mainly during late- to post-D_2 stage, and the effective role of a subjacent magmatic body intruded in the post-tectonic (D_2) stage, in the mineralisation, appears to be a distinct possibility.

U-Pb (EPMA) dating of monazite and xenotime associated with the ore minerals in the Central and Western lode system of Gadag schist belt yielded 2525 ±10 Ma age, which is inferred as the age of gold mineralisation (Sharma et al., 2008).

M/s Ramgad Mines and Minerals Pvt. Ltd. have recently submitted a proposal of open cast mining in the Sangli block (proved reserve: 1.8 Mt with 2.48 g/t Au) at the rate of 1000 tpd for the initial six years and future expansion plan based on the possibility of establishing a total reseve of 27 Mt with + 4 g/t Au in their lease hold (Sawkar and Vasudev, 2009).

Gold Mineralisation in Jonnagiri Schist Belt The Jonnagiri schist belt (JSB), extending for about 25 km from Peravali to Gooty exposes a sigmoid shaped synform in the north trending WNW–ESE for 11 km with ~5 km width. Thereafter, it abruptly changes into a thin linear belt (~0.8 km width) for the rest 14 km with a curvilinear trend from SSE to SSW (Fig. 1.2.19).

The schist belt comprises metabasic and meta-acid volcanics and tuffs with concordant lensoid sills of metagabbro and ultramafites. The belt is surrounded by sheared margins, having tonalite-granodiorite-monzonitic granite (TG) in the west and granite-adamellite (GA) batholithic pluton to the east. The contact between the two accretionery zones is marked by a crustal scale shear zone (Jairam et al., 2001; Rao et al., 2001).

Gold–quartz lodes occur along the shear zones in the southern tip of the sigmoid shaped domain of the JSB, where detailed exploration for assessing economic potential was conducted by GSI and MECL. The assessment results obtained in the blocks explored by the GSI are given below (Rao et al., op.cit.).

Dona Temple Block The mineralised veins are aligned along shear zones cutting across the formational boundaries and dominant schistosity of the metavolcanic sequence and TG suite; strike length of two sub-parallel mineralised zones – 650 m trending WNW–ESE, dip: 60°–80° towards NNE, width: 2.11 m; ore reserve: 0.7 Mt, with average grade of 4.65 g/t Au, down to 150 m vertical depth.

Dona East Block Shear-controlled mineralised veins in altered TG suite. Total strike length of mineralised zone is 650 m, out of which the better mineralised part over 450 m strike length (average width of 19 m) holds an ore reserve of 4.46 Mt with average grade of 2.86 g/t Au, down to 180 m vertical depth. The zone includes a rich ore shoot of 7.5 m width having reserve of 0.073 Mt with average grade of 31.26 g/t Au.

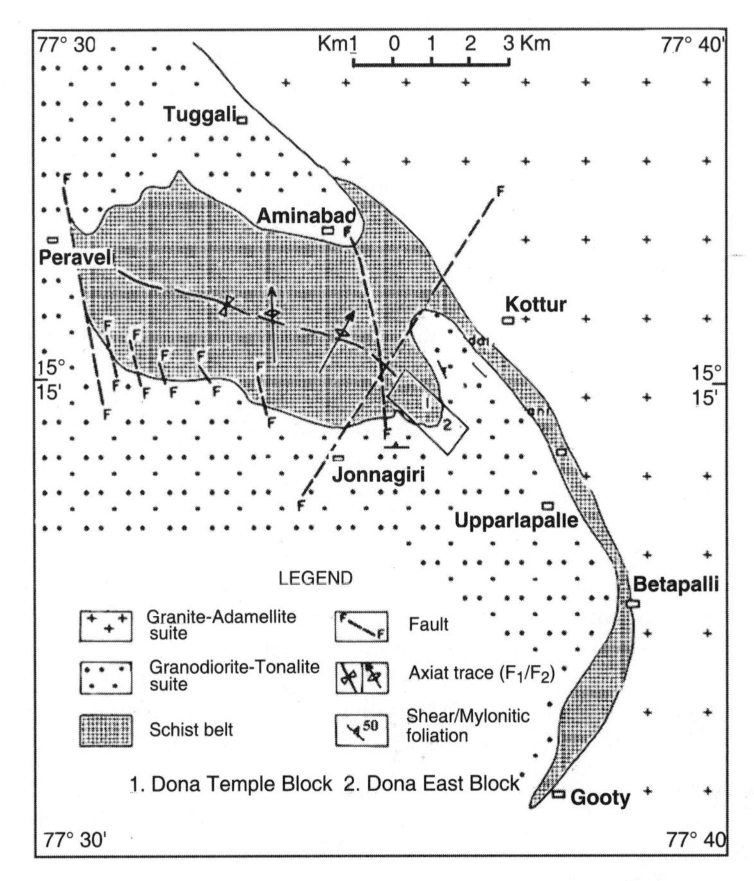

Fig. 1.2.19 Geological map of Jonnagiri schist belt showing locations of gold prospects (composed from various maps of GSI)

Dona South Block Mineralisation in sheared TG, strike length ~ 1.0 km, thickness of ore zone varies from a few meters to 50 m, average grade of 1–3 g/t Au.

Dona North Block A 1 km × 150 m auriferous zone identified in the TG suite is yet to be assessed.

The zones of ductile deformation in both TG suite and metabasalt are marked by mylonite and phyllonite; the auriferous quartz vein/stringer occurring along such zones are remarkably similar, both in having quartz dominant vein systems with total sulphides not exceeding 5% of the bulk, and lack of any positive correlation between the content of gold and sulphide. Gold and bismuth occur in native form; and the sulphides in order of decreasing abundance are pyrite, chalcopyrite, arsenopyrite, pyrrhotite, sphalerite, galena and molybdenite. Minor oxides include ilmenite, magnetite titanite. Sporadically, scheelite is also present. Fluid inclusion studies of gold-quartz veins of Dona Temple and Dona East blocks indicate overlapping transition of low and high salinity inclusions, implying mixing of mineralising fluids, at the temperature range of 200 °C to 600 °C and 0.68 kb pressure. This is indicative of epigenetic mesothermal derivation of the quartz related gold ('orogenic gold'). $\delta\,^{34}S$ values of pyrite from + 1.1 to + 2.33 per mil are attributed to multiple phases of mantle derived and granitic intrusion related mineralising fluids (Rao et al., 2001).

M/s Geomysore Services (India) Pvt. Ltd., in recent years, have conducted detailed exploration and obtained mining lease in the Jonnagiri deposit (Sawkar and Vasudev, 2009).

Gold Mineralisation in Dharwar – Shimoga Belt The Dharwar–Shimoga belt of the Western Dharwar Block, spread over an area of about 25,000 sq km, comprises meta-greywackes and metapelites (sericite chlorite schist) with abundant (1–10 m wide) layers of BIF represented by banded magnetite quartzite/chert and some bands of metabasaltic rocks. The general lithological trend is NNW–SSE, with moderate dips to NE or SW. North of Shimoga, the BIF bands forming hills in an overall flat terrain, show noticeable gold content at several localities, of which the occurrences at Chinmulgund, Karjagi and Kallihalli are more prominent. In the south of Shimoga, auriferous BIF is noted in the Tarikere valley, extending for about 40 km with E–W trend. The BIF hosted gold occurrences of the whole terrain is dotted with ancient workings.

The gold occurrence at Chinmulgund (Karnataka) is observed in the sulphide-rich banded magnetite quartzite (BMQ), interlayerd with greywacke-argillite and tuffaceous rocks. The ore zone extends for about 3.5 km strike length which is marked by several old workings (shafts and inclines). Two parallel lodes (with 1.0 km and 1.2 km strike lengths) were established by extensive trenching and drilling in the prospect area by GSI. The total probable ore reserves calculated (down to 160 m) are 2.93 Mt x 3.39 g/t Au (width up to 9 m) and 4.83 Mt x 2.98 g/t Au (width upto 18 m) under the cut-off grades of 2.0 and 1.0 g/t Au. The possibility of open cast mining and heap leaching of ores were examined by the Mineral Exploration Corpn. Ltd.

A Brief Discussion on Gold Mineralisation in the Dharwar Craton In the foregoing brief treatment of gold mineralisation in the Dharwar craton, an attempt has been made to present the salient features of the deposits and occurrences mined and discovered till date. It is obvious therefore, that most of these are located in metamorphic tectonites, are principally Neoarchean in age and occur in greenstone belts. As of today, it is the best gold-endowed region of the country. Before making some generalisations on this mineralisation in the Dharwar craton, a quick review of the state of the art knowledge on gold mineralisations in metamorphic tectonites and associated granitoid rocks is given.

An Outline of the State of the Art General advancement in Earth Science, as well as fast upward moving unit-price of gold since the seventies of the last century led to both intensive and extensive studies on this metal that has charmed the human society for more than five thousand years for reasons more than one. It is amazing to note that the gold price has increased > 150 times during the last four decades, and yet there is no sign of decrease of interest in this metal. Gold mineralisations in metamorphic tectonites, until relatively recently, were generally referred to as lode gold, mesothermal gold etc. We now know that quite a few deposits belonging to this broad group formed in P-T regimes that do not exactly tally with what was previously understood by the definition of 'mesothermal deposits.' A prevailing inclination these days, therefore, is to call them 'Orogenic gold deposits' (Groves et al., 1998)

The orogenic gold mineralisations are quartz-dominant vein systems in which sulphides (Fe-sulphide dominant) generally constituted 3–5%, carbonates 5–15%. In more common greenschist facies environments, albite, white mica, chlorite, scheelite and tourmaline are present in small but variable proportions. In amphibolite-granulite facies, hornblende, biotite, diopside and even augite may be present. Gold occurs mainly in the veins but may be present in economic concentrations in the wall rocks. Vein systems may continue for kilometers along the strike, 1–2 km along the dip, with a little change in mineralogy and even the gold grade. The ores are interpreted to have formed from low salinity near neutral H_2O-CO_2-CH_4 fluids. The size of a mineral deposit is independent of time and space. However, the Neoarchean has been more prolific in producing such deposits around the world.

The two major controls of gold mineralisation of this type are structures and petrochemistry of the source and host rocks. The first order controlling structures are regional faults/shear zones, several hundred kilometers in cumulative length in case of fragmentation. These disjunctive structures are presumed to be near-vertical major crustal dewatering channels and ultimately flow-paths for the Au-bearing hydrothermal fluids. These major dislocation structures, however, do not localise sites of this type of gold deposits. The preferred sites for the latter are the second and third order structures. The reverse faults are preferred, rather than the normal faults, presumably because of its higher degree of misorientation. Another favoured site is regional fault intersections. Brittle-ductile and ductile shears are common hosts. Lamination is a common feature in quartzose ore veins in metamorphic tectonites (Goldfarb et al., 2005).

From geochemical point of view, the rocks rich in carbon or Fe^{2+}, and occurring along the fluid flow path are conducive to gold deposition through redox reactions. Another possible mechanism of gold precipitation is H_2O-CO_2 phase separation. Experiments showed that depressurisation during say hydrofracturing by an accumulated fluid pressure, may cause gold precipitation (Loucks and Mavrogenes, 1999). Fluid mixing is another possible mechanism.

Emplacement of orebodies vis-a-vis the stage of metamorphism and the source of ore fluids and ore elements, are other relevant issues, all of which as yet are not understood convincingly, at least not at all places. Ore-forming fluids may be introduced for mineralisation at the pre-/syn-peak stages of prograde metamorphism or during retrograde metamorphism. Establishment of gold mineralisation at pre-peak metamorphism is not easy, however. A plausible explanation of common occurrence of gold mineralisation in the retrograde regime is that the deeply sourced ore fluids reach the upper crust when that domain has already entered the retrograde segment of the thermal field. As the ore fluids move up through available channels as an effect of fault-valve activity, thermal equilibrium of these fluids with the ambient rocks may not have been attained. As a result, the marginal parts of the ore veins may show developments of higher temperature minerals. The general consensus with regards to potential fluid sources for such deposits are metamorphic dehydration in a subducted slab, a mantle devolatilisation, magma crystallisation at depths, deep basinal fluids, or even the surface-derived fluids that accessed deep into the crust, i.e. basement (cf. Goldfarb et al., 2005). However, if we accept the non-magmatic genetic model for the supergiant Muruntau deposit of Central Asia, as proposed by Wilde et al. (2001), then a regional deep-level metamorphic (-magmatic) devolatilisation model is confronted by another in which the fluid escaping from the compacting cover sediments can do the job equally well. However, a large resource of reduced sulphur is presumed to be present in the system. The source of gold and the associated metals is neither free from debate. While some authors (Groves et al., 2003; Phillips et al., 1987) are convinced that even giant gold deposits can form from leaching of 'realistic volumes' of any crustal rock type, there are others who would rather go for specific rock types such as iron formations, exhalative sedimentary rocks, komatiites/high-Mg basalts, subducted oceanic crust (including oceanic plateaus), some granitoids, while speculating on the major sources of gold and the associated metals (cf. Keays and Scott, 1976; Bierlein et al. 1998; Bierlein and Pisarevky, 2008). The second view seems more realistic.

In case of typical orogenic gold mineralisation, the metamorphism of the country rocks ± granite emplacement and the mineralisation are not widely separated in time, i.e. they belong to the same geological continuum. But these relationships are not universal. Terrains hosting gold mineralisation in Eastern Asia were metamorphosed to high grades, billons of years in advance of Cretaceous gold mineralisation in them. The situation there is better explained by metamorphic decarbonation and dehydration of the oceanic crust + overlying sediments, subducted below high grade crustal rocks subsequent to at least partial erosion of their Precambrian keel and producing the ore fluids (Goldfarb et al., 2007)

Orogenic gold deposits are less common in higher grade metamorphic (amphibolite-granulite) terrains. There is an apparent contradiction here. In high temperature fluids, the sulphur content could be at least two orders of magnitude higher than that estimated for mesozonal deposits. The gold content likewise should be higher by the same magnitude (Ridley et al., 1996, 2000). If this is correct, then the other factors remaining the same, the higher grade metamorphic terrains should not be as poor in gold mineralisation as they generally appear to be. At least a partial answer to this apparent contradiction lies in the fact that a large scale dehydration is involved during the transition of greenschist to amphibolites facies of metamorphism of rocks. This derived aqueous fluid takes out a substantial proportion of gold and other metals in dissolution and precipitates them in suitable situations.

The occurrence of intrusive granitoid bodies at varying distances from gold deposits is known. Some of them even host gold mineralisations. These granitoid bodies may be pre-, syn-, or post-mineralisation in time of emplacement. In composition they may vary from calcic, through calc-alkaline to Li-, F-, and B-rich S-type granitoids. Many are not agreeable to credit these magmatic bodies with the capacity to carry gold in solution to form mesozonal gold deposits. But there are some convinced supporters of the granitoid model (Sillitoe, 1991; Sillitoe and Thompson, 1998; Thompson et al., 1999; Lang et al., 2000). They believe that even in metamorphic belts, there are some gold deposits in the formation of which granitoid intrusives were involved. The following features are believed to be suggestive of granitic involvement in mesozonal gold deposition:

1. Presence of penecontemporaneous granitoid intrusive(s) of intermediate oxidation state.
2. High salinity cogenetic fluid inclusions in ores and wall rocks.
3. An anomalous metal suite containing all or some of Bi, W, As, Sn, Mo, Te and Sb.
4. Gold ultimately associated with Bi and occurring in low sulphide assemblages.
5. Quartz, alkali feldspars, sericite and carbonates define wall rock alterations. The granitoid body may contribute materially to the genesis of the deposit or may act as the 'heat engine', or both.
6. Presence of gold in skarns and greisens related to the granitoid intrusives.

Some gold deposits contain substantial quantities of radiogenic lead which might have been scavenged by crustal scale hydrothermal systems from the granite-gneiss basement, or young granitoids emplaced in the ore zone.

Dharwar Gold Mineralisation in the Above Context Plots of gold occurrences in a map of India show clustering within the region covered by northern Karnataka, Andhra Pradesh and the adjacent districts of Kerala and Tamil Nadu (Fig. 1.2.20). Some of them belong to 'giant' category (Kolar) and some to 'world class' (Hutti–Maski) (definitions after Laznicka, 1999). They are better explored and exploited and therefore, have better databases. Much of these are lacking in case of other deposits and occurrences, as should be obvious from the foregoing discussions on them.

The Kolar deposit is extraordinary in that the principal ore zone, the Champion Reef, is extensive over a strike length of ~8 km and continues for > 3 km along the dip. The ore shoots pitch at 60°–70° with the north for the greater part of the mineralised zone, sub-parallel with the mineral lineation in the host hornblende schist. Lack of conspicuous tectonic fabric in the immediate wall rocks and the development of neo-amphibole, quartz, feldspar, garnet, diopside, and even augite, at the immediate contacts of ore veins suggest that at least the last phase of mineralisation did not thermally equilibrate with the wall rocks. The ore zone is riven by two sets of shear zones: N–S and NNW–SSE. The control of the first on the ore mineralisation is conspicuous, and that of the second should be controversial for the same reason. A reverse sense in the movement along the N–S shear zone appears to have played a significant role in ore localisation. Summarising, the structure of the Kolar belt is complex and in spite of considerable work having been done on it, some conclusions

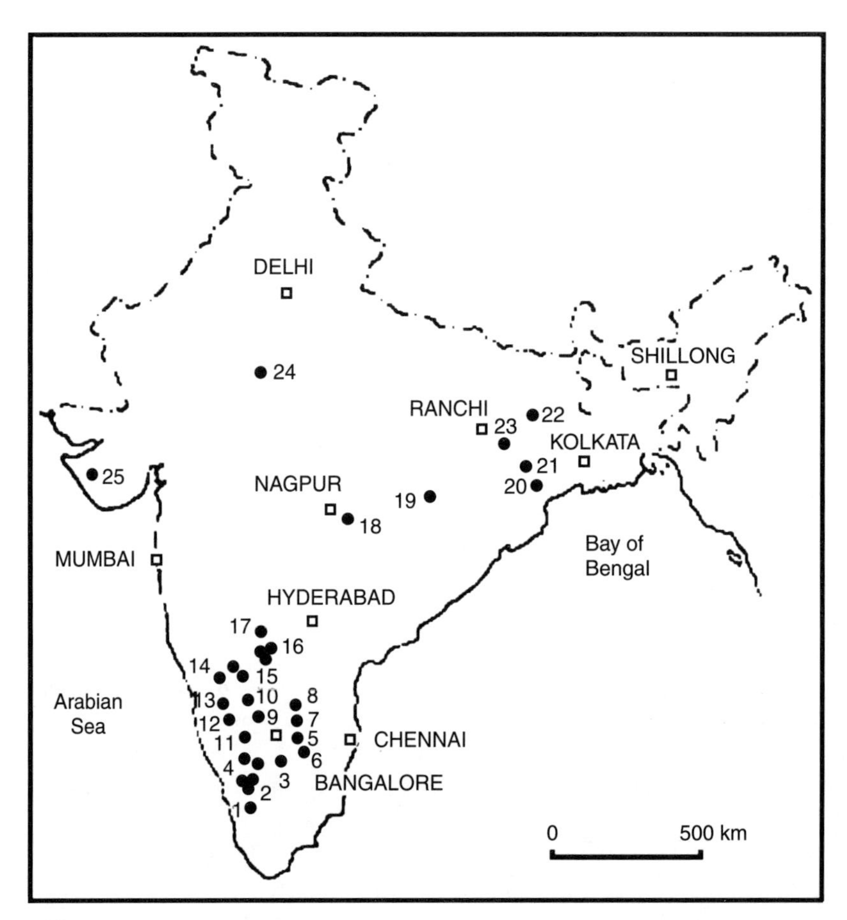

Fig. 1.2.20 Occurrence of primary gold mineralisation in India. Legend: 1. Allapadi, 2. Wynad, 3. Kavudahalli-Porsedyke, 4. Amble–Woolagiri, 5. Kolar, 6. Mallapakonda–Chigargunta, 7. Ramagiri, 8. Penkacharla, 9. Bellara, 10. Ajjanahalli, 11. Kempinkote, 12. Honnuhalli, 13. Kudrekondo–Palavanahalli, 14. Chinmulgund– Haveri, 15. Gadag, 16. Hutti, 17. Manglur, 18. Bhiwapur–Chanderi, 19. Sonakhan, 20. Talkot–Benapal, 21. Kunderkocha, 22. Lawa–Mysera, 23. Parasi, 24. Bhukia, 25. Alech hill.

about it remain controversial as already discussed. Kolar rocks have metamorphosed to amphibolite facies. Peak metamorphism at Hutti took place at 3–5 kb and ~650 °C and that at Uti, ~6 kb and 650°–700 °C (Misra and Pal, 2008). The rock association at the Hutti–Maski belt is almost similar to that of Kolar, except that a rock like Champion Gneiss is absent there. The South Kolar block is geologically very similar to the main Kolar Gold Field (KGF). Only the ore fluid focus there was less sharp.

Regarding ore genesis, more thoughts have been expended on Kolar, followed by Hutti–Maski and Ramagiri–Penakacherla belts. Some recent views on the ore genesis in the KGF are held by Hamilton and Hodgson (1986), Safonov, et al. (1988), Rajamani (1996) and Sarkar, (1988, 2000b, 2010). Hamilton and Hodgson held that the penecontemporaneous granitoid plutons in the belt are the progenitors of the mineralisation in the Kolar belt. Safonov et al. believed that the ore fluids came from a granitic magma chamber, of course 'influenced by mantle processes'. The last two works suggested the mineralisation to be orogenic in nature because of (i) mineralisations being tectonically controlled within the greenstone belts, (ii) low salinity of the ore fluids, and (iii) orientation of the ore shoots accordant with the mineral lineations in the host rocks. On the other hand, presence of radiogenic Pb and also W, As, Bi, Te and also Hg and Mo (the last in south Kolar) in the ores, though in minor/trace quantities, suggest contribution by a contemporaneous granitic pluton(s), or being present as a scavenged product from the granite gneiss basement, or both. However, a more stamping element such as Sn has not been found, Ni being present instead. Gold mineralisation as the petrographic evidence suggests, did not take place by redox reactions involving wall rocks. Instead, it might have taken place by H_2O-CO_2 phase separation caused by fracture related pressure release. An effect, may be partial, of fluid mixing may not be totally discounted either.

Gold in the western lodes, such as the Oriental, is poor in grade, accounting for not more than 3% of the total production. Gold here occurs mainly as inclusions or fracture-filling in arsenopyrite. Two contrasting explanations for the genesis of this kind of ores have been proferred. Siva Siddaiah and Rajamani (1989) held that these sulphidic ores were originally auriferous complex sediments, provided principally by submarine hydrothermal exhalations. Hamilton and Hodgson (1986) explained their genesis by the replacement of a banded and sheared magnetite bearing ironstone and minor argillaceous and cherty rocks, occurring as interflow sediments. However, replacement by both primary (magmatogenous) and secondary (metamorphic) fluids are possible. The Malappakonda deposit is characterised by more sharply banded quartz (chert)-gold bearing sulphides. In physical features and geochemistry (REE-content), they differ from the main KGF ores, although occuring between the KGF and the south Kolar fields. Here also the gold content might have been inherited or introduced before the first phase of metamorphism.

Temporal relationship between magmatism and gold mineralisation at Hutti has been impressively put forward by a recent publication (Sarma et al., 2008). The felsic volcanic rocks yielded a U-zircon age of 2587 ± 7 Ma, about 40 Ma older than the age of gold mineralisation (2547 ± 10 Ma) determined from the hydrothermal monazite. Syntectonic Kavital Granite has a U-Pb zircon age of 2545 ± 7 Ma which overlaps with the age of gold mineralisation. The Patna Granite at Kolar is believed to be contemporaneous with the Kavital Granite. This, as such, makes gold mineralisation at Dharwar 100 my younger to the peak of Late Archean gold mineralisation in most other Archean cratons. This is not unimaginable, as the crustal evolution and metallogeny are not temporally uniform on a global scale.

Eastern Dharwar is much richer in gold than Western Dharwar. A plausible explanation for this is that the gold-rich Kolar and Hutti–Maski schist belts are composed mainly of mafic (-ultramafic) rocks. Strong negative Eu-anomaly in the clastic sediments from the Western Dharwar schist belts suggests that they were derived from intra-crustal granite gneisses and the sediments from the Eastern Dharwar schist belts came from the arc/arc-continent mixed sources. Magmatic arcs are one of the fertile grounds for gold mineralisation in the orogenic gold model of today. But all of them are not equally gold-bearing. Why some of them are very rich in gold and some others contain very little or no gold may possibly be explained by one or more of the following:

1. Size of the hydrothermal cell, larger being the better
2. Composition of the ascending hydrothermal fluids and its reactibility with the wall rocks of the channel
3. Fluid mixing
4. Frequency of the fault-valve operations
5. Aiding and abetting by a contemporaneous granitoid pluton.

Even after accepting the Eastern Dharwar to be an arc accreted to a foreland in the west (Chadwick et al., 1997, 2000), a few details remains to be clarified. One amongst them is where is the probable suture zone? The available information is not clear about it (Sarkar, 2001). Contemporaneity of the development of the schist belts in the 'foreland' and the 'arc' is another point. Does this mean that the accretionary arc and the craton-margin were affected by the same tectonic (tensional) regime via the back arc domain?

The granulite terranes of South India may not be written off as possible host for mineable gold, though they may not hold large accumulations of the metal. Ore-bearing hydrotherms from the upper mantle or lower crust may be locally focused during a later phase of tectonic reactivation.

Copper Mineralisation in Chitradurga Schist Belt Copper mineralisation at Chitradurga, often called Ingaldahl deposit, is located in the 40 km long Ingaldahl– Kunchiganahalu belt, belonging to the Chitradurga Group, the upper division of the Dharwar Supergroup. It is a small deposit, with an ore reserve of 1 million tonne (1.4% Cu). The ores at Ingaldahl–Kunchiganahalu belt occur in two modes: (a) Stratiform, and (b) Vein type.

The stratiform ores occur as a banded sulphide–chert body, 3–20 m thick and extended for about 7 km along the strike. The ores consist principally of pyrite, pyrrhotite and magnetite with chalcopyrite and sphalerite as minor phases. Soft sediment structures and diagenetic features are common. The unit is locally underlain by pyritiferous carbonaceous phyllites (Chaudhuri, 1984).

The vein-type deposits are lensoid bodies, extended for a few hundred metres along the strike; and occur in altered basaltic rocks. The main ore body of Ingaldahl Mine (now closed) is sub-parallel with the dominant secondary structural plane S_2, making a low angle with the formation boundaries (Chaudhuri, 1984; Sarkar, 1988). Mineralogically, chalcopyrite is followed in abundance by pyrite and pyrrhotite. Sphalerite, galena, tennantite, enargite, cobaltite, mackinawite and pentlandite are present as minor to trace minerals. The δ^{34} S from the sulphides cluster around zero. The ores have been metamorphosed in the upper greenschist facies.

Most workers on the deposit are in agreement that the deposit belongs to the 'volcanogenic type' (Naqvi et al., 1977; Chaudhuri, 1984; Anantha Iyer and Vasudev, 1985), although they may differ in various degrees on the probable mechanism of ore deposition–redeposition.

Iron Ores in Dharwar, Karnataka Of the three major iron ore fields in India, one is located in South India, particularly in the Dharwar region, the other two being in Central India (Chhattisgarh) and in Eastern India (Singhbhum – Orissa or Jharkhand – Orissa), respectively. The South Indian deposits under reference are confined to the state of Karnataka.

Major banded iron formations (BIFs) and the related iron ore deposits in Karnataka developed principally in two regions:

1. Bababudan hills, Chikmagalur district
2. Sandur belt comprising Donimalai, Ramandurg and Kumarswami ranges in the Bellary-Hospet region.

In fact the iron-ore field of Goa-Redi, is an extension of the Dharwars through Goa. Besides, there are V-Ti-bearing magnetite ores associated with mafic-ultramafic rocks.

The BIFs of Karnataka are divisible in two major types:

1. Banded magnetite quartzite (BMQ) associated with greenstones
2. Banded hematite quartzite/jasper (BHQ/BHJ).

Banded Magnetite Ores The first type, BMQ itself is considered as an ore although the Fe-content in it, rarely if ever, exceeds 40%. This is because the easy seperability of the iron-oxide mineral by a magnetic separator that gives it an ore-status when no better ores are around. This type of ore occurs in the Chikmagalur district in the Bababudan hills, which forms a magnificent hill range in western Karnataka in the form of a horse-shoe. The highest point, Mulaingiri is 1923 m above MSL. BMQ bands occur all along the 40 km stretch of the horse shoe. The BIF is interbanded with 2.9 Ga old mafic volcanic rocks. Silicate-rich layers with actinolite, cummingtonite, grunerite and garnet are common, suggesting a middle-upper greenschist facies metamorphism. In the proto-ore, iron oxide bands are interlayered with either chert or shale. Some of the silicate-rich bands could be ash beds. These BIFs bear much resemblance to the Algoma type BIF. Estimated reserve of this type of ore in Chikmagalur district is ~ 10 billion tonnes (Bt) (IBM, 2005).

Kudremukh Deposit The Kudremukh deposit occurs in the precipitous hill ranges of the Western Ghats that extend with NNW–SSE trend from near Kudremukh in the south via Agumbe to Kodachadri to further north. The total length and average width of the belt are 140 km and 20 km, respectively (Fig. 1.2.21).

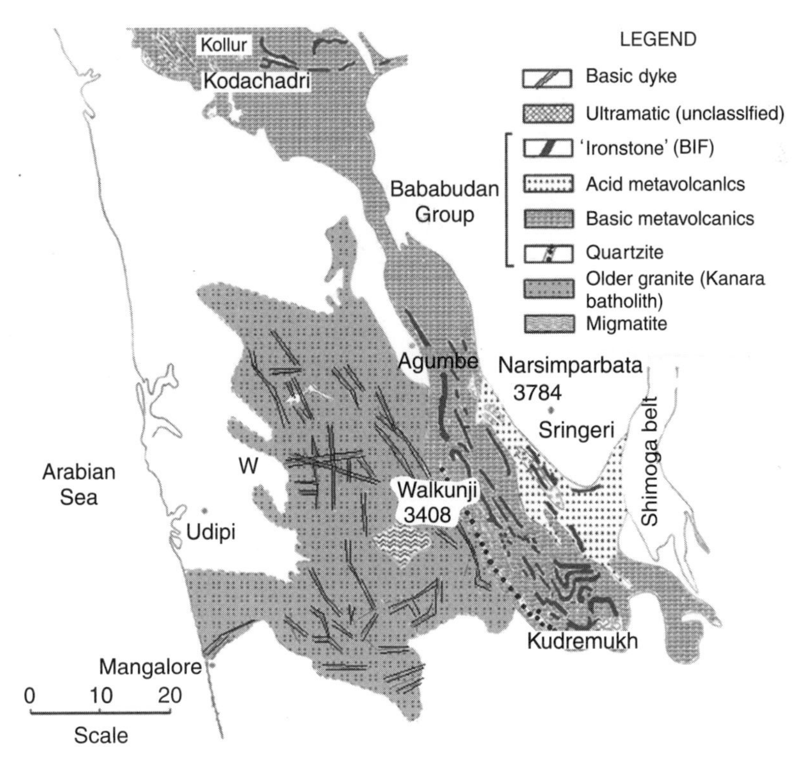

Fig. 1.2.21 Geological map of a part of Western Ghats showing Kudremukh–Agumbe iron ore belt hosted by the Bababudan Group (Modified after Sampat Iyengar, 1910)

About 20 Mt of the ore was annually mined at Kudremukh producing about 6 Mt of concentrate. The mine is located within 100 km of a major sea-port in the west coast. The mining company set up a pelletising plant at Mangalore, capable of producing 3 Mt of Fe-oxide pellets per year. But the whole project is shut down because of a court order involving an environmental issue.

The belt comprises Bababudan Group of rocks which is divisible in the ascending order into Walkunje Formation (conglomerate and arenite), Kudremukh Formation (mafic–ultramafics), Kodachadri Formation (banded magnetite quartzite and associated argillite) and Narasiparvata Formation (acid igneous suite). This sequence overlies the Peninsular Gneissic Complex and granitoids (Kanara Granite), having a series of high-amphibolite grade enclaves of Sargur Group (Balsubrahmanyam, 1978).

Two different types of BIFs are recognised in this belt, viz., (i) the cherty banded iron formation (CBIF), which is exclusively made up of meso- and microbands of quartz and iron-oxide consisting mainly of magnetite, microcrystalline chert/quartz and minor specks of pyrite with or without minor amount of cummingtonite, grunerite and actinolite, and (ii) the shaly banded iron formation (SBIF) comprising amphibole-rich layers in addition to quartz and iron oxide layers. The amphibole-rich layers in the SBIF are made up of actinolite, cummingtonite, grunerite and riebeckite; and are associated with magnetite, quartz, aegirine (minor) and other secondary minerals.

Khan et al. (1992), based on major element, trace element and REE geochemistry of the BIFs suggested a combination of submarine volcanogenic (fumarolic) hydrothermal activity and also the inflow of terrigenous material for their formation. The Cr/Zr, Zr/Cr+Co+Ni, Zr/Y and Nd/Yb ratios reflect that certain fine silicate-rich bands between the iron oxide micro-bands are made up of volcanic ash, whereas others are made up of shales of continental source or a mixture of both. La-enrichment, depletion of total REE, positive Eu anomalies, Nd/Yb, La/Lu and LREE/HREE ratios in Fe-rhythmite indicate that Fe, Si and REEs of the BIF were supplied by hydrothermal and fumarolic activity accompanying submarine volcanism. The enrichment in total REE and obliteration of La signature is probably a consequence of mixing of non-chemogenic components and ambient seawater with the chemical precipitates. Deposition of BIF appears to have taken place below wave base and photic zone of the shallow shelf region. Ocean circulation and upwelling brought FeO and O_2 together resulting in precipitation of iron along with fine-grained argillites (Khan et al., op.cit.).

The geochemistry of the mafic–ultramafic rocks of this belt were studied by Bhaskar Rao and Naqvi (1978) and Drury (1981, 1983). Drury et al. (1983) obtained a Sm-Nd date of 3020 ± 230 Ma (with initial $^{143}Nd/^{144}Nd = 0.50878 \pm 0.00022$) from mafic metavolcanic rocks from near the base of the sequence and also a whole rock Rb-Sr age of 3280 ± 230 Ma (with an initial ratio of 0.7009 ± 12 and MSWD of 1) for the Kanara Granite.

BHQ/ BHJ Related Ores The second type of the BIF and the associated ores are best developed in the Sandur belt, which comprises Donimalai, Ramandurg range and Kumarswami range (Fig. 1.2.22). These were previously bracketed under the Bellary- Hospet deposits.

The *Sandur belt* is a rich repository of high-grade iron ores (hematitic) associated with the BIFs. The major deposits assessed by IBM/GSI and worked under public sector are at Donimalai range (151 Mt ore / + 64% Fe), Ramandurg range (212 Mt ores / + 62% Fe) and Kumarswami range (part) (182 Mt / + 62% Fe) and another 450 Mt were available with the private leaseholds at Ramandurg (45 Mt), NEB range (280 Mt), Kumarswami (90 Mt), Belagal range (12Mt) and other mining blocks (20 Mt).

Besides the above, some other iron ore deposits in this belt which have been exploited/are being exploited, are:

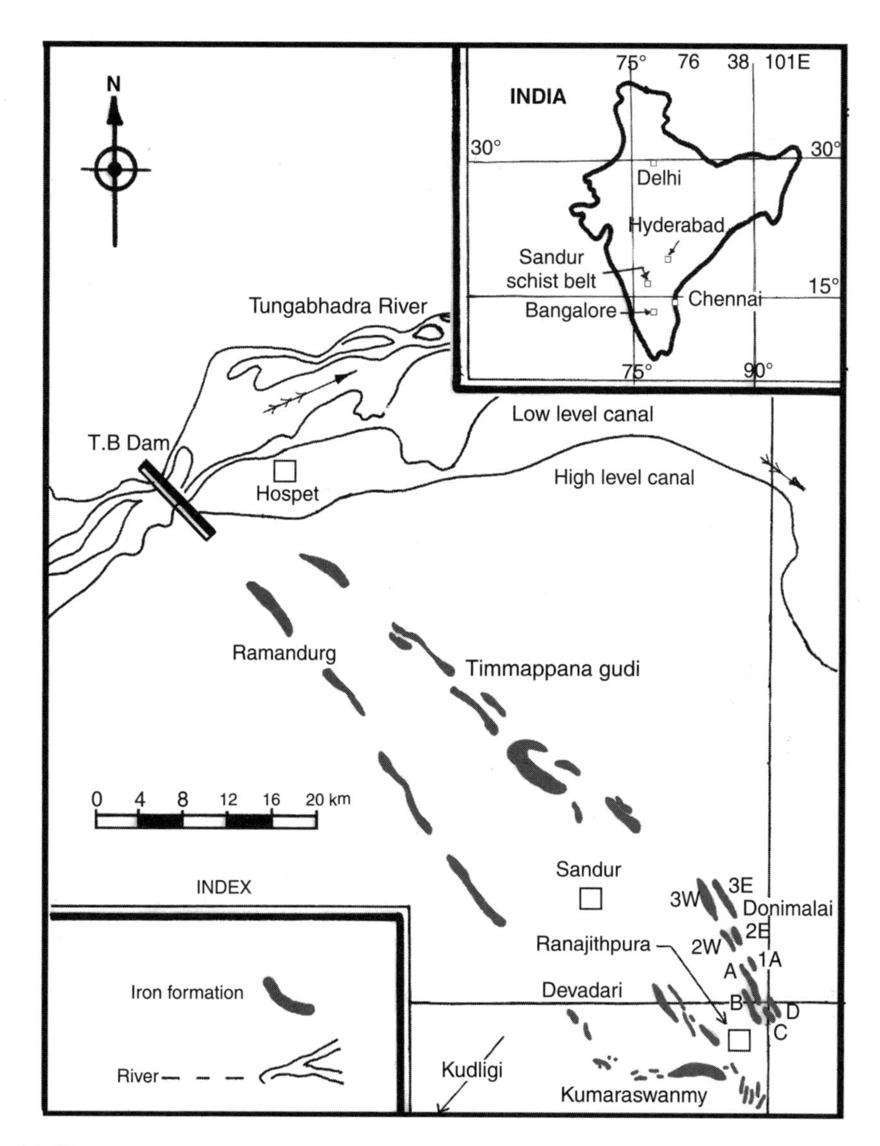

Fig. 1.2.22 Distribution of iron formations and some important iron ore deposits in the Sandur schist belt (after Murthy and Chatterjee, 1995)

1. Copper Mountain Range – Halkundi, Belagal, Vibhutigudde, Hargandona
2. Ettinhatti Range – Ubbalagandi, Rajapuram and Konanharavu
3. Thimmappangudi Range – NEB range, Bharatarayanharavu (Dalmia property), Gogga property, Ingligi, Jambunathalhalli, Sankhapuram
4. Devadari Range

Government reserved an area of about 90 sq. km covering some of the rich deposits of Donimalai, Devadari, Kumarswami and Ramandrug ranges. Within the limits of this reserved area, NMDC has developed a major iron ore mine at Donimalai. Mysore Minerals Ltd, a state owned company, is working in the Taranagar area. Ramandurg and Kumarswami blocks are kept in reserve after detailed exploration.

Ramandurg Deposit It covers an area of about 7 sq km, and forms a part of the southern sector of the Ramandurg Range. Of the 212 Mt ores of + 62% Fe content, 54 Mt are lumpy ores and about 80 Mt are blue dust (65% Fe). High grade lumpy ores have higher demands in the industry.

Kumarswami Deposit The deposit is located in the Kumarswami Range, the SW limb of the Sandur schist belt. It forms a plateau covering an area of 20 × 15 sq km. The estimated reserve of all categories has an average grade of 62.7% Fe, in which lump ores exceed 65% in Fe-content. Blue dust is also present. More than 100 private miners of various denominations are presently engaged in mining of Fe- and Mn- ores (including float ore) outside the leasehold of the government.

Donimalai Deposit Donimalai iron ore deposits are located in SE portion of Sandur belt (Fig. 1.2.22). The BIFs at Donimalai occur in association with shales and tuffs, together constituting the Donimalai Formation. The Donimalai Formation is underlain by the volcanic rocks of the Krishnanagar Formation. The sequence is intersected by dolerite-diabase intrusives. The local stratigraphy is presented the following table (Table 1.2.4).

Table 1.2.4 *Stratigraphy of the Donimalai area (after Murthy and Chatterjee, 1995).*

Formations	Rock types
	Pegmatites
	Dolerite/ diabase
Donimalai Formation	Banded iron formation Shales with tuffs.
Krishnanagar Formation	Metavolcanic rocks (Meta-basalt, meta-andesite, metamorphosed tuffs)

Before entering into a brief discussion on the banded iron formations (BIFs) of this region, it may be mentioned that the definition of BIF given by James (1954) is followed in this work. The definition reads:

> "Banded iron formation is chemical sediment, typically thin bedded or laminated containing 15 percent or more iron of sedimentary origin and commonly but not necessarily containing layers of chert".

The iron minerals present are dominantly oxides (hematite + / magnetite). But they could be one or more of carbonates, sulfides and silicates. Most, if not all, of the large hematitic iron ore deposits around the world are associated with the BIFs. Other names in vogue for the oxidic BIFs are jaspilite, itabirite, taconite, banded hematite quartzite (BHQ), banded hematite jasper (BHJ). The last two names are in vogue in Indian geological literature.

The banded iron formation here is composed of iron oxides and chert/jasper. The chert also contains disseminated iron oxides. The Donimalai BIF banding varies in thickness from less than a millimeter to meter scales, some lensoid ore bodies being 30 m thick. Murthy (1990) has reported some ore bands being as thick as 100 m. The bands measuring less than one millimeter are designated as microband/aftband and thicker bands, usually composite in nature, may be assigned to mesoband, or through calamina to macrobands (meterband).

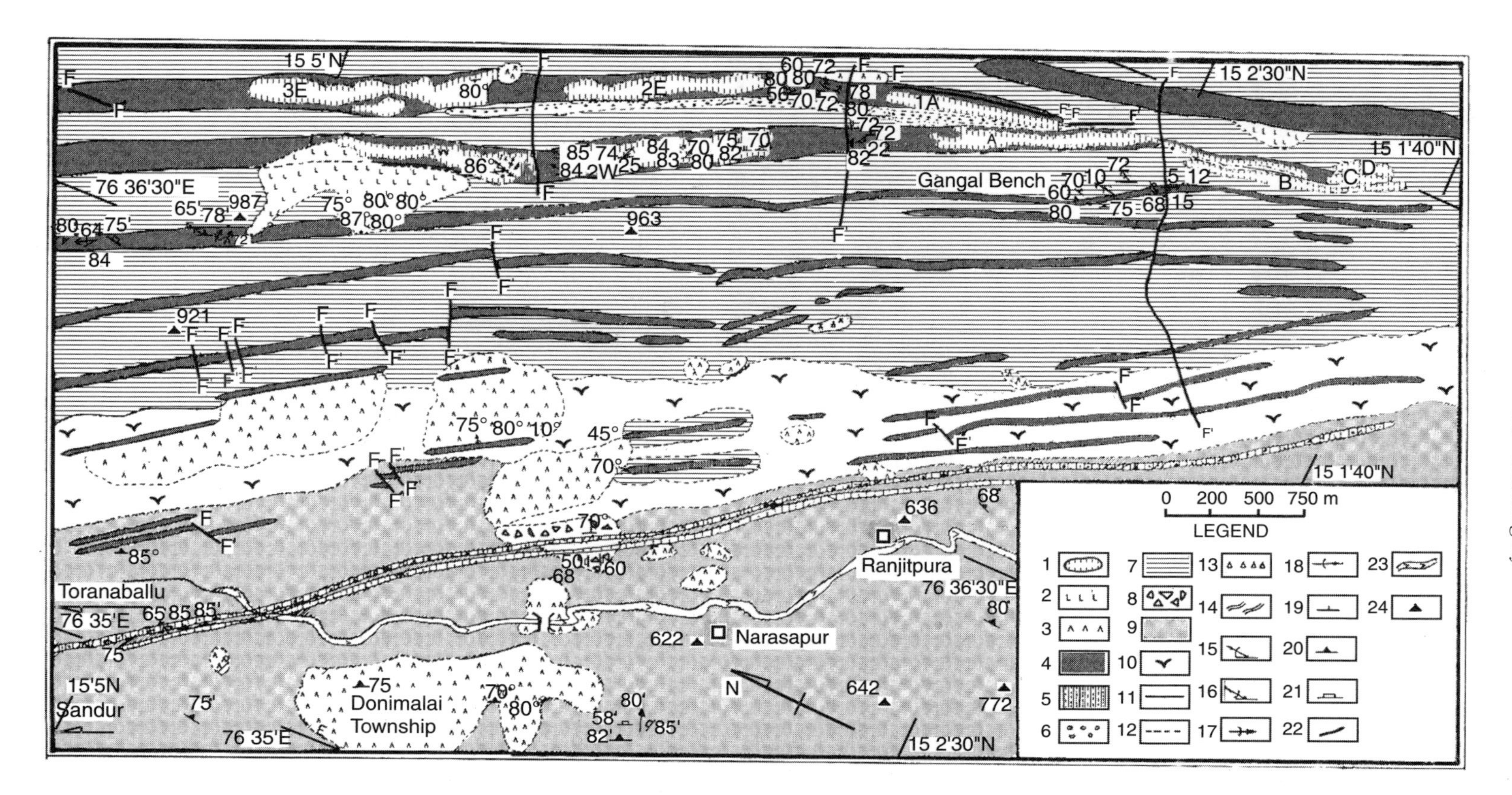

Fig. 1.2.23 Geological map of Donimalai iron ore deposit (modified after Murthy and Chatterjee, 1995). Legend: 1. Soil cover, 2. Laterite, 3. Dolerite, 4. BIF, 5. Banded chert, 6. Tuff, 7. Unclassified shaly formation, 8. Pyroclastic agglomerate, 9. Metavolcanic rocks, 10. Ore deposit boundary, 11. Formation contact, 12. Inferred contact, 13. Breccia, 14. Shear zone, 15. Antiformal early fold, 16. Synformal early fold, 17. Antiformal late fold, 18. Synformal late fold, 19. Strike and dip of lamination, 20. Strike and dip of foliation, 21. Strike and dip of joint, 22. Fault, 23. River, 24. Spot height in metres.

Primary sedimentary structures such as compositional banding, ripple marks, cross-laminations, truncation of the above, synersis cracks, intraformational breccia, slump structures, intraformational faults etc., are reported from the BIF of this region. Murthy et al., (1983) and Murthy and Reddy (1984) respectively reported microbiota and stromatolites from lower sections of Sandur schist belt and suggested a role of micro-organisms in the deposition of the BIFs of Donimalai.

The iron bands in the BIF consist dominantly of hematite (martite), though magnetite is present as relics. Fe-carbonates and sulphides are not reported from this area. Table 1.2.5 shows the average composition of some BIF samples analysed. There is tangible difference in the average analytical results of fresh samples from boreholes and those collected from the surface outcrops. Strangely Fe is partially leached out from the surface samples and SiO_2 added instead. Fresh samples are low in Al_2O_3-content and P-content. The loss of Fe from the surface material is difficult to explain except by dissolution by acidic ground water.

Table 1.2.5 *Average major element composition (wt%) of Donimalai BIFs (Murthy and Chatterjee, 1995)*

Samples	Fe_2O_3	FeO	SiO_2	Al_2O_3	P	LOI
Average of 38 borehole samples of BIF	52.70	0.63	44.23	0.73	0.02	0.57
Average of 8 outcrop samples of BIF	35.44	0.77	60.25	1.57	0.03	1.1

As discussed in an earlier section, the REE distribution in the BIFs of the Sandur belt in general bears the stamp of hydrothermal origin, the Fe and Si being mainly obtained by leaching of volcanic rocks, or by exhalations.

Orebodies and the ores There are nine ore bodies at Donimalai, separated into two blocks, North and South. The ore bodies are intimately associated with the BIF, often with accordant enclaves of the latter within the former. The ore bodies are flanked by shales on either side (Fig. 1.2.24).

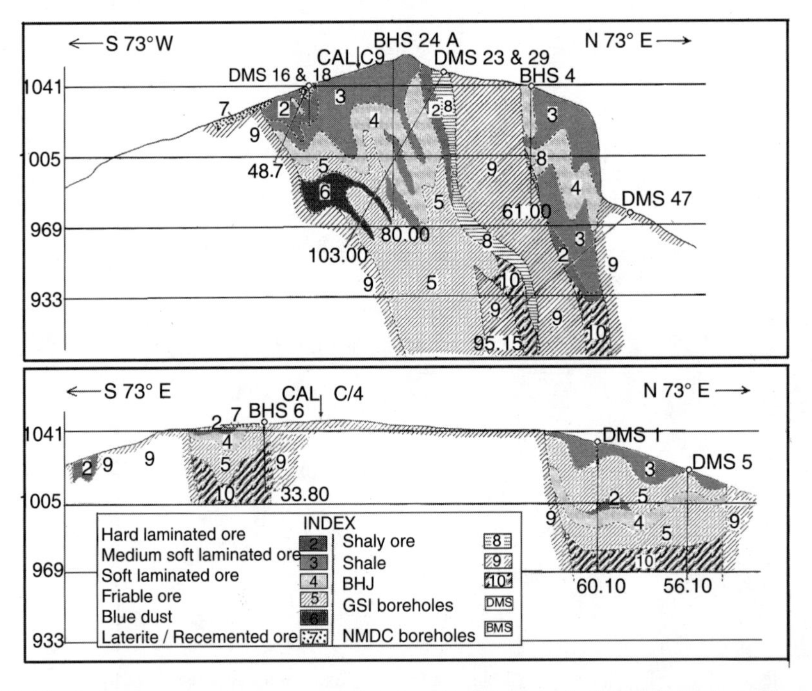

Fig. 1.2.24 Transverse sections across the Donimalai deposit (modified after Murthy and Chatterjee, 1995)

Like most deposits of the type, Donimalai iron ores are divisible into the following types, based on physical properties: very hard ore, hard laminated ore, medium hard laminated ore, laminated ore, soft laminated ore, friable ore and blue dust.

'Very hard ore' is hard, compact and without the obvious lamination and has few or no vugs and voids. It resembles steel grey hematitic ore at Bailadila. The proportion of this type of ore within the whole ore mass is low.

'Hard laminated ore' is hard, but not as hard as the previous type. It is commonly laminated. Voids (few) can be seen along laminations at places. Development of this type of ore decreases with depth (Murthy and Chatterjee, op. cit.). Medium hard laminated ore is not much different from the hard laminated ore, except that the pores occur in greater proportions.

'Soft laminated ore' is same as biscuity ore described from elsewhere. This is well-laminated, but easily breaks into pieces even under light pressure. Friable ore is intermediate between soft laminated one and blue dust.

'Blue dust' or the bluish grey fine grained ($\leq$ 1mm) hematite that occurs in the form of non-cohesive aggregates or pockets.

The above ore types have + 60% Fe in general. There are some other varieties, which are chemically and mineralogically lower in grade but can still be useful, particularly where people do not have higher-grade ores. These are lateritic ore, shaly ore and the recemented ore.

In lateritic ore there is a high proportion of such mineral phases as goethite, lepidocrocite and limonite. Alumina and silica content are also high. The shaly ore occurs in the south block. Iron content in shaly ore varies but may at places exceed + 60% Fe. However, the recovery is poor. Recemented ore is what is called 'Canga' in Brazil and India. It is a product of surficial aggregation of broken ore-pieces, BIF etc. cemented into a hard ore by a mixture of hematite, limonite etc. Canga is rather a rare development in Donimalai deposit.

It is to be noted from Table 1.2.6 that the hard laminated ore is compositionally better in all respects, i.e it has higher Fe and lower contents of SiO_2, Al_2O_3 and P. Other varieties do not vary much in Fe-content but have varying contents of other components such as SiO_2, Al_2O_3 and P. 'Shaly ore' is a misnomer here, because of its very low contents of SiO_2 and Al_2O_3.

Table 1.2.6 *Average chemical composition (wt%) of different ore-types of Donimalai deposit (range in parenthesis) (from Murthy and Chatterjee, 1995)*

Ore type	No of boreholes sampled	No of samples analysed	Fe	SiO_2	Al_2O_3	P
Hard laminated ore	10	16	67.16 (64.00–69.40)	1.21 (0.14–3.30)	1.04 (0.08–2.50)	0.04 (0.01–0.12)
Medium hard laminated ore	16	16	65.23 (60.80–68.60	1.81 (0.12–3.52)	2.72 (0.20–6.44)	0.11 (0.01–0.05)
Soft laminated ore	16	16	65.16 (61.20–68.80)	2.21 (0.24–4.4)	2.30 (0.12–5.50)	0.08 (0.01–0.20)
Friable ore	16	16	65.22 (61.60–69.20)	3.06 (0.34–7.62)	2.10 (0.40–4.20)	0.06 (0.02–0.12)
Shaly ore	6	6	65.76 (63.20–68.54)	1.71 (0.41–3.20)	1.51 (0.26–2.86)	0.07 (0.02–0.20)
Blue dust	13	15	66.20 (62.00–69.60)	2.02 (0.30–7.7)	1.88 (0.08–6.04)	0.05 (0.01–0.11)

There are two distinct views on the origin of the Donimalai ores. Misra (1978) suggested a 'syn-sedimentary origin' of the ores, while Murthy and Chatterjee (1995) uphold an epigenetic origin of the same. The following facts are relevant in this context.

1. The iron ores under discussion are intimately associated with BIF in the field and one passes into the other in all dimensions. Patches and pockets of BIF in the ores constitute a common feature.
2. Tectonic structures in the BIF are present in the iron ores as 'ghosts'.
3. Oriented cavities exist in thinly laminated high-grade ores, for which the permissible explanation should be, unfilled spaces evacuated by the dissolved out silica.
4. Soft ores are much more porous.

We think the view of Murthy and Chatterjee is by and large supportable.

Iron Ores of Goa–Redi Area (Maharashtra) After Goa was integrated with the rest of India in 1961, the Geological Survey of India took up a detailed study of the geology and mineral resources of this former Portugal colony. This study led to the bringing out of a detailed geological map of the region and assessment of the known mineral deposits and discovery of new ore, particularly of bauxite. It was further established that the Precambrian Goan geology was an extension of the geology of the NW Dharwar.

The known mineral deposits in the region are those of iron, manganese, bauxite and clay. In 1970 the iron ore reserve was estimated at 411 million tonnes, comprising 81 million tonnes of lumpy ore with + 60% Fe and the remaining being fines, assaying ~ 62% Fe. About 1.1 million tonnes of black iron ore (containing up to 10% manganese) and also 3.57 million tonnes of manganese ore were assessed.

Till the end of the sixties drilling for exploration of iron ores was limited to the appearance of the underlying phyllite, as that was considered to be the foot wall of the original BIF, and therefore the end of the vertical extension of the ore zone. Drilling was also suspended as soon as the ground water table was met. Subsequently however, deep exploration revealed presence of two or three bands of iron ores separated by beds of manganiferous phyllite. Powdery ore was found below the present water table upto a depth of > 50 m from the MSL. This suggests that another 120–125 million tonnes of low grade (50–55% Fe) lumpy ores may be present in these lower depths.

Government of India has no restriction to the export of iron ores from Goa. The Goan port facility is a boon in this respect. So, bulk of the production of the iron ores from the mines of this region is exported. Now of course, a part of the powdery ores from this region is being used in the newly set up steel plants at Hazira (Gujarat) and Delvi (Maharastra), producing 3 Mt and 2 Mt steel, respectively. The powdery ores from Goa region is made into pellets (± sinter) using natural gas from the Bombay High.

Redi Iron Ore Deposit The deposit is located near Sawantwadi (15° 14′ 30″: 73° 40′) in the Sindhudurg district of Maharashtra, which extends over 3 km in WNW–ESE to E–W direction. Geologically the area represents northward extension of the Dharwars through Goa. The BIF associated iron ore at Redi is massive, compact, which grades into powdery and shaly variety with depth. Total estimated reserve is about 36.66 Mt with Fe content varying from 59.97% to 61.74%. Incidences of gold were reported from the iron ores of this deposit based on chemical analyses. But how the gold occurs here is not known.

Origin of BIF Related Iron Ores and the Dharwar Ores An inescapable conclusion from the above observations is that the BIF related hematitic iron ores formed from the BIFs. Silica leached out with or without the introduction of ferric oxide (oxyhydroxide). Such a conclusion concerned ore geologists since a full hundred years ago (Leith, 1903), if not earlier. But what bothered people specifically, are the answers to the questions:

1. Was the aqueous solution that changed the BIF to iron ores descending or ascending?
2. Was the solution hot (hydrothermal solution), or cold?
3. If ascending, was the water juvenile or modified meteoric?
4. Did the transformation of the BIF take place shortly after the deposition of the BIF (Archean/ Paleoproterozoic, or after a long gap, i.e. Tertiary–Recent?
5. Did the change of BIF to Fe-ore take place by leaching of silica alone, or replacement of silica by Fe-oxide/ oxy-hydroxide, or both in sequence?

There could be other questions too. The debate was initiated by the geologists working on the iron ores of the Mesabi range, located at the northern edge of the Penokean orogen, Minnesota, U.S.A.

Leith and his co-workers for many years asserted that the transformation of BIF to hematitic ores took place due to the action of oxidising surface water percolating downward, i.e. descending water. Oxidation and hydration of primary or diagenetic Fe-oxide to various secondary Fe-oxides resulted in a loss of volume and increasing porosity and permeability of the rock (BIF). This increase in the porosity and permeability in the protolith considerably enhanced the ability of the circulating solutions to further react with the constituent minerals, leaching silica, magnesium, calcium and phosphorus and leading to the final production of the ores (Leith et al., 1935). Another group of workers led by Grunner (1930, 1937) insisted that the geologic features associated with such deposits could be better explained by assuming ascending solutions. This group however, had no convincing proof to justify their conviction. So for several decades thereafter, the descending solution model was supreme, apparently by default. The problem raised its head once again in the seventies and eighties of the last century (Canon, 1976; Morris, 1985). Since then there have been several interesting publications on the problem, which we briefly review here before we conclude on the genesis of the Indian iron ore deposits.

Morris after prolonged study of the iron ores of the Hamersley basin, Australia, presented his view in the form of a model popularly known as the 'Supergene metamorphic model' to explain the origin of the high grade (+60% Fe) iron ores there. In a nutshell, it comprises an early stage of iron addition by processes of deep supergene enrichment in the Proterozoic, caused by deep penetration of oxidising meteoric water. Iron migrated downward. Chert, carbonates and silicates were either replaced by goethite, or leached out of the system. The banded structure of the original BIF was, by and large, still preserved. At the same time magnetite of the original BIF was oxidised to hematite by deep electrochemical processes. Burial and metamorphism of the Proterozoic dehydrated goethite produced micro-platy hematite (Morris, 1980, 1985).

Recently the suggested ore genetic model for the Mesabi Range again came up for a scrutiny. Morey (1999) concluded that the Mesabi ores were not supergene products. They were rather formed in a continental scale ground water flow system associated with the tectonic uplift of the Animikie foreland basin in the Paleoproterozoic. This model is analogous to the model of Garven et al., (1993) in which they suggested a regional flow system resulting from topography- or gravity-driven mechanism associated with the tectonic uplift of the foreland basins.

The model of Powell et al. (1999) with respect to the formation of high grade iron ores of Hamersley is not very different from that of Morey, except that it is a bit more elaborate. They hold that the ore bodies formed during the Opthalmian orogeny (2.2–2.45 Ga). Regional circulation of hydrothermal fluids through the reduced BIFs, causing their iron enrichment, took place during or soon after the Opthalmian orogeny. Oxygen isotopes in quartz veins in the ores and in the ore minerals confirm that hot oxidizing meteoric fluids (200°–400 °C) were involved.

The model of Taylor et al. (2001) for the transformation of BIF to high-grade iron ores of the Hamersley basin (Mt Tom Price deposit in particular) is somewhat more sophisticated. It suggests four stages of transformation. The Stage-1 involves removal of all free silica, not changing the redox state of iron in the minerals and retaining the banding. A hot (150°–250 °C) alkaline, relatively reduced basinal brine is thought to be active at this stage. The Stage-2 involves addition of O_2 by an oxidising low-salinity fluid of elevated temperature, which essentially is deeply circulating water of meteoric origin. Fe^{2+} -bearing minerals oxidise. Stage-3 fully eliminates carbonates by leaching. The fourth or the final stage leaches the ore material of all unwanted constituents and a porous high grade ore is produced. This stage involved oxidising cold fluids of shallow meteoric origin. There is some difference in opinion regarding the particular phase of orogeny involved between the last two models, but that is of little consequence.

The interpretation of the mechanism of formation of the Paleoproterozoic Hotazel iron in Kalahari, South Africa, by Tsikos et al. (2003) offers an interesting thematic variation. They concluded that the genesis of iron ores in this ore field was a two-stage process; the first one represented the initial pre 1.9 Ga meteoric weathering and the second, migration of hydrothermal fluids (meteoric/ diagenetic) along the unconformity surface between the Transvaal Supergroup and the overlying Olifantshak Supergroup, during post-Mapedi times. Oxidation and Fe- enrichment reached down to 60 m below the unconformity.

Rosiere and Rios (2004) studied the origin of the high grade iron ores of the Quadrilatero Ferrifero district, Brazil, particularly applying infra-red microscopy and detailing fluid inclusion characteristics. Their conclusion is that recurrent hydrothermal activities, particularly controlled by the Trans-Amazonian orogeny (3.1–2.0 Ga) and the Brasiliano–Pan African orogeny (0.8–0.6 Ga), was responsible for the development of the giant high grade ore deposit, at structurally preferred sites.

Mc Lellan et al. (2004), also working on the Hamersley iron ores, concluded that isotopic, fluid inclusion, geochemical and structural data suggest that both deep-seated and meteoric fluids were involved in ore genesis and possible upward and downward penetrations of fluids within an extensional environment were important.

A comment of Kazansky (1995) on the ore mineralisation at the Krivoy Rog basin of Ukraine is particularly relevant here. He says some researchers working on the genesis of iron ores there believed that the oxygen rich solutions causing oxidation of the magnetite ores were hypogene in nature. Some even believed in their connection with much deeper sources.

In the context of the above brief discussion and our own observations, we conclude that the high-grade hematitic iron ores of the type under review may have one of the following genetic histories:

1. Wholly supergene origin
2. Wholly hydrothermal or hypogene origin
3. Initial hydrothermal origin overprinted by supergene processes
4. Initial supergene process superceded or overprinted by a hydrothermal event.

Supergene deposits may be of recent or ancient origin. The 'recent' process may, however, be still on. The recently formed deposits can be blanket-like, broadly obeying the topography. Others may be flat bodies that follow bedding or fault zones/shear zones. Ore bodies following faults and fractures of all dimensions are common. Relics of the protolith (commonly BIF) within the ore zone have gradational border with the host. Some supergene deposits may have formed in the distant past (ancient supergene deposits). They would lie below a major erosional unconformity and gradually change to BIF below. The overlying sediments should be red beds that may contain pebbles of iron ore derived from below. Beukes et al. (2002) have cited the Sishen deposit of South Africa as an example of the case in reference. Expectedly, description of such cases will multiply before long.

Iron ore deposits formed by hydrothermal process (i.e. by the action of hot aqueous ascending solutions) formed in the distant past, would be independent of the present day ground water table and would be capped by a BIF cover. However, in limiting conditions, the BIF cap may be removed and the ore body may essentially remain the same that formed by a hypogene or hydrothermal process. Such ores are expected to be free of, or poor in goethite, high lump yielding and moderately porous (up to ~ 30%), expected to have low contents of silica, alumina, alkalies and phosphorus. Hematite crystals should normally be platy and form a dense network. Cited gangue minerals should be suggestive of their not being supergene ores of recent origin. Hydrothermally formed slab-like bodies may be later tectonically deformed, producing folds, faults, cleavages and fractures. Results of fluid inclusion studies in veins that have intersected these ore deposits show that the fluids are generally geothermally heated terrestrial water of meteoric origin. Locally at least the associated shales may bear evidence of having been hydrothermally altered in the process (Haruna et al., 2003). If an igneous body intrusive into such an ore deposit can be successfully dated by a suitable method then that date would fix the minimum age of the deposit. In case of (3) and (4), the criteria generally utilised for establishing overprinting of the products of one geological event by a younger one are followed.

A question is asked by some staunch supporters of the hydrothermal model: If the supergene process is effective in converting BIF to iron ores, then how is it that we do not find iron ore deposits along the entire BIF range of an area? If that is a major weakness of the supergene model, then neither is the other model (hydrothermal) free from the same debilities. Other factors being same, structure induced permeability should be a necessary pre-condition for both the descending and ascending solutions.

Murthy and Chatterjee (1995) hold that the enrichment of BIF to iron ores in the Sandur belt, particularly in the South block of Donimalai deposit, took place by the action of heated water during tectonic movements. Oxidation is deep-going in this part of the belt. In the North block, however, enrichment extends to a shallow depth, and it is of supergene origin. The mechanism of conversion of BIF to iron ore, as suggested by Murthy and Chatterjee for the Southern block, is quite pertinent and it is the forerunner of the models of Morey (op. cit.) and Powell et al. (op. cit.). But such a mechanism, which is orogeny-related fluid activity, should not have normally spared the Northern Block. Cross sections of the South Block (Fig. 1.2.24) do not suggest that the fluids were 'ascending' in the literal sense of the term. The problem needs further study. Some thick ore beds present, could be primary.

Manganese Mineralisation in Dharwar Economic deposits of manganese in South India occur in the Dharwar schist belts, particularly at Chitradurga- Tumkur, Shimoga and Sandur areas. Mn-mineralisation in the Chitradurga–Tumkur and Shimoga areas occur in association with phyllites, carbonates and ferruginous cherts interleaved with thin layers of volcanic rocks. Stratigraphically, the Mn-bearing horizon occurs in the lower part of the Chitradurga Group. In North Kanara and Goa, Mn–mineralisation, however, initially took place at an upper horizon. Iron formations, in all these occurrences, are located at higher stratigraphic positions.

Opinions on the origin of these ores vary from sedimentary–syngenetic (Naganna, 1977) to supergene concentration from Dharwar schists or low grade mineralisation (Fermor, 1909; Roy, 1966). The work of Krishna Rao et al. (1982) on this problem is most exhaustive. They divided these ores into the following types:

1. Reworked metasedimentary ore
2. Sedimentary oolitic ore
3. Cavity–filling replacement type ore
4. Float ore.

The reworked sedimentary ores described by these authors are confined to the manganiferous formations and are found as steeply dipping conformable bodies varying in thickness from 1 m to 8 m. The orebodies, where unaltered, are made up of several thin Mn–rich bands interleaved with phyllitic/quartzitic and ferruginous layers. Extensive reworking (supergene oxidation and enrichment) of manganese has taken place in the manganiferous layers. The associated phyllitic and quartzitic layers are now converted to variegated clay and loosely bound sand. Several mines are/ were working on these deposits.

These reworked metasedimentary ores are biscuity, vuggy, botryoidal, massive and laminated. The ore minerals comprise pyrolusite, psilomelane, cryptomelane, jacobsite, ramsdellite, manganite, hematite, goethite and wad. The ores contain > 50 (Mn + Fe), 4–8% SiO_2, 3–4% Al_2O_3 and about 0.06% P. Mn/Fe ratios vary between 0.75 and 1.20.

'Sedimentary' oolitic/ pisolitic ore occurring as nearly flat, sheet-like or lensoid bodies are overlain by khaki green claystone. They are extensive over a few tens of meters and many are upto 5 m in thickness. The bedded oolitic/ pisolitic ore is hard and massive. The concretions are circular to oval in shape and grew around fragmented ooliths and pisoliths, grains of quartz and patches of clay-like material. Nsutite and cryptomelane are present in ooliths and pisoliths, whereas the cement contains todorokite in addition to cryptomelane. These ores, as a rule, analyse about 45% (Mn + Fe) and 12–17% SiO_2 and the Mn/Fe ratio varies between 1.7 and 2.1. Interstringly, the top 30–50 cm of these ores is free of concretions. It is made up of a massive steel-like Mn and Fe mineral mass with sporadic inclusions of quartz fragments. The present authors are of the view that the features outlined above do not support a 'sedimentary' origin of these ores in the usual sense of the term. They are at best diagenetic, if not produced by guided replacement in the supergene cycle. These should be studied more closely.

The 'cavity- filling replacement' type ore occurs as irregular patches and pockets in lalerite and the weathered phyllite. The ore is made up mainly of psilomelane and wad.

The float ore of the area consists of a heterogenous mixture of boulders, pebbles and fragments of oolitic/ pisolitic ore and reworked- 'metasedimentary' ore.

The lithologic association containing the Mn-horizon in the Sandur belt is characterised by the absence of carbonates. The basalt–andesite volcanics with BIF is overlain by phyllitic rocks containing bedded Mn-ores. The ore horizon is seperated from the volcanic rocks by a pelitic horizon ~150 m thick. The ores are generally low grade (≤ 40% Mn) and consist of pyrolusite, cryptomelane, manganite and lithophorite and a little braunite and manganite (Roy, 1981). Here also an iron formation overlies the Mn horizon. There is no definite indication of the source of manganese of these deposits. However, as in the case of BIF in the Sandur belt, the manganese in this belt could have been derived from the basic volcanic rocks by hydrothermal leaching.

Uranium Mineralisation in the Dharwar Region One of the important fallout of the Second World War was our awareness of the uranium (and by some exptrapolation thorium) atom as the sources of immense energy for war, as well as for peace. Most forward looking nations embarked upon vigorous projects for the exploration of the mineral deposits of these two metals without much delay. Some exploratory gadgets, the Geiger Muller Counter, and later some sophisticated versions of it, came in handy for this purpose. India won freedom in August, 1947, and without much delay started a Department of Atomic Energy (DAE) under the direct supervision of the then Prime Minister. But the real purpose of such a step could be best served if the necessary raw materials could be indigenously found. So the exploratory wing of the Department, the Atomic Minerals

Division (AMD), was soon started. It took up a national programme of search for minerals that were needed for the atomic energy projects. Before long, the effort met its first prize- discovery, the uranium mineralisation along the Singhbhum copper belt in Eastern India.

Many places in South India were searched for uranium mineralisation using the genetic models for uranium deposits in vogue at that time. We knew about the occurrence of monazite deposits in the beach placers of Kerala and Tamil Nadu. But where were the uranium mineralisations? By 1964, the Atomic Minerals Division placed its progress report on the exploration for uranium (-thorium) mineralisations in South India (Bhola, 1965). The report included the following finds:

1. In granites, gneisses and migmatites
   (a) Kashmanclu–Nandigudem–Kasipet (Nalgonda district)
   (b) Places on the Adilabad–Hyderabad road (Karimnagar district)
   (c) Places in Mahboobnagar district, all in Andhra Pradesh
   (d) Parthiapalli (Raichur district) in Karnataka
   (e) Suryamalai Hill and Kullampatti (Salem district) in Tamil Nadu.
2. In granitic pegmatites
   (a) Sankara mine and Parlapally and Darlapally (Nellore district) in Andhra Pradesh,
   (b) Kurumbapalli (Salem district)
   (c) Masinigudi (Nilgiri district) in Tamil Nadu.
3. Associated with basic rocks, Gopalpet (Mahboobnagar district) in Andhra Pradesh.
4. In graphitic schists in the Kolar Gold Field (KFG) in Karnataka.
5. Thorium-bearing conglomerates at
   (a) Gulcheru Conglomerate, Cuddapah basin, Andhra Pradesh
   (b) Kaladgi basal conglomerate, Karnataka.

The most prospective amongst the above occurrences were probed by drilling and necessary petrochemical investigations, but none proved promising.

Quartz-pebble-conglomerate (QPC) type uranium (and associated metals) mineralisation of the Witwatersrand (South Africa) and Blind River (Canada) types was the target of explorers in many countries. Reportedly, samples of Chikmagalur conglomerates were found to contain leachable uranium (Narayan Das et al., 1988). Later Geological Survey of India confirmed the presence of uraninite in the drill core samples obtained during exploration for copper at Kalaspura. Some of these samples analysed up to 0.22% U_3O_8 with negligible thorium. Soon reports on the presence of QPC type uranium mineralisation poured in from a number of places, including Walkunje, occurring at the basal part of the Bababudan Group in Karnataka (Swaminath et al., 1976; Viswanath et al., 1988). But again, none of them was found exploitable on detailed prospectiing.

For a real breakthrough in uranium mineralisation in South India, however, the country had to wait for many years for another chance find, until a few limestone samples collected by the Geological Survey of India for phosphate prospecting in the Vempalle Formation in Cuddapah Supergroup showed tangible uranium content in them (Sundaram et al., 1989). The detailed investigation by the AMD that followed, soon established stratabound uranium mineralisation in the phosphatic dolomites of the Vempalle Formation and the Pulivendla Quartzite that overlies it. This was followed by discovery of uranium mineralisation in the Bhima basin. These are discussed in some details in a later section.

Rare Metal (RM) and Rare Earth Element (REE) Mineralisations The RM and REE mineralisations are generally contained in the pegmatites. The RM–REE bearing pegmatites have been classified into three types: (i) Lithuim-cesium-tantalum (LCT); (ii) Niobium-yttrium-fluorine (NYF); (iii) the 'mixed' type (Cerny, 1991). LCT type is characterised by the abundance of beryl with varying proportions of albite and spodumene and has the geochemical signature of Li, Rb, Cs, Be, Sn, Ga, Ta > Nb (BPF). NYF is REE-important and bears the geochemical signature of Nb > Ta, Tl, Y, Sc, REE, Zr, U, Th and F. The 'mixed' type is a mixture of LCT and NYF.

The ore-bearing pegmatites generally occur in clusters. Such pegmatites are commonly associated with granitoid plutons. These granitic bodies are more commonly late orogenic, although some can be syn-orogenic and post-orogenic and even anorogenic. The LCT granites are silicic, peraluminous to hyperaluminous, S to I types, but poor in Ca, Mg, Fe, Ti. In contrast, the NYF pegmatites are generally somewhat less silicic, largely subaluminous to metaluminous and I to (A + I) types. They are also characterised by LREE>>HREE (Cerny, 1991). Productive pegmatites are dominantly Archean or Proterozoic in age and emplaced in upper greenschist to lower amphibolite facies rocks (Pollard, 1995). Development of rare metal pegmatites is largely a magmatic phenomenon with the ore minerals generally being an integral part of crystallisation (Cerny, 1991; London, 1992).

La Touche is credited with the discovery and reporting of Nb-Ta bearing pegmatites near Maski (12°52′:78°30′) in Karnataka. Investigations by the Mysore Geological Department led to the discovery of a number of mica and beryl-bearing pegmatites in the Holenarsipur schist belt. As a part of comprehensive programme of studies and search for rare metal (RM) and rare earth element (REE) minerals throughout the country, the Atomic Minerals Directorate of the Department of Atomic Energy, Government of India, took up South India also for detailed investigation in the early seventies of the last century. This resulted in the discovery of a number of lithium- cesium-tantalum pegmatites (LCT-type pegmatites of Cerny, 1991) in southern Karnataka, particularly in the older schist belts (Sargur Group) and adjacent gneissic rocks. Important amongst these findings are Arehalli, Doddakadanur and Mundur in the Holenarsipur schist belt, Chinkarli in Krishnarajpet schist belt, Marlagalla and Allapatna in Nagamangala schist belt and Girigoundanadoddi in the Kunigal schist belt. These are all confined to Western Dharwar. Eastern Dharwar is generally barren of this type of mineralisation excepting, of course, such mineralisation in the Nellore schist belt and Kanigiri Granite in Andhra Pradesh.

A number of RM-bearing pegmatites occur within Peninsular Gneisses over an area of ~1000 × 200 sq m near the Arehalli village, Hassan district. The pegmatites are zoned and mineralogically complex. They are Ta-rich pegmatites (40% – 60% Ta_2O_5) and contain niobian tantalite, beryl, spodumene, muscovite and fluor – apatite. In bulk composition these pegmatites are generally peraluminous (A/(C+N+K) = 1.14–1.55) and sodic. Beryl from these pegmatites are alkali beryl, BeO–content being 10.4–10.5%. The Li-content in spodumene vary between 5.2 and 6.1%. Tantalite analyses 15.3%–30.9% Nb_2O_5 and 41.7%–58.1% Ta_2O_5. U_3O_8-content of these pegmatites ranges between 0.03–0.16% (Krishna and Thirupathi, 1999).

A number of RM- and REE-bearing pegmatites are emplaced into the gneissic rocks, close to their contact with the Holenarsipur schist belt. The largest amongst them, a complexly zoned pegmatite at Mandur, Mysore district, has a quartz- microcline core. The intermediate zone consists of perthitic feldspar, quartz, garnet, muscovite, magnetite, monazite, columbite–tantalite, beryl and bismuthite. The outer zone is composed of a quartz feldspar intergrowth. BeO content in beryl is 11.5%.

A number of Nb-Ta bearing pegmatites have intruded into the Krishnarajpet schist belt near Krishnaraj Sagar reservoir and near Chinkurli. Columbite–tantalite (55.9% Nb_2O_5 and 8.6% Ta_2O_5), magnetite and ilmenite are the principal ore minerals.

Complex and zoned pegmatites have been emplaced along the contact between the Peninsular Gneiss and the rocks of the Kunigal schist belt. The pegmatites are peralkaline (A/C_{NK}= 0.43–0.95). In the northern part, the pegmatite is potassic and enriched in tin (Sn = 40–90 ppm). In contrast, the southern part is soda-rich and shows enrichment in Sn, Nb, Ta and Be. Columbite-tantalite composition also is far from simple. Available results of analysis of columbite-tantalite from this deposit give the following composition: Nb_2O_5 58.5–60.9%, Ta_2O_5 5.6–5.7%, TiO_2 0.5–0.6%, SnO_2 0.1–0.14%, WO_4 0.14–0.15% CeO_2 0.01–0.02% and Y_2O_3 upto 0.002% (Krishna and Thirupathi,1999).

The Marlagalla– Allapatna rare metal pegmatite belt is located ~ 5 Km east of Srirangapatna in Mandya district, Karnataka (Fig. 1.2.25). The principal country rocks here are the Sargur supracrustals and Dharwar Supergroup of rocks represented by fuchsite quartzite, kyanite–sillimanite–garnet schist, amphibolite, together with the Peninsular Gneiss and the younger granitoid intrusives. The ore-bearing pegmatites occur in both the schist belt rocks and the Peninsular Gneiss. The pegmatites are generally complex and zoned as follows: (i) an outer zone with blocky feldspar, (ii) an intermediate zone with quartz–feldspar intergrowths, (iii) an inner quartz core. Tantalite, spodumene, beryl, microlite and mica occur along the core and the intermediate zone (Fig. 1.2.26) (Sarbajna and Krishnamurthy, 1996). Marlagalla rare metal pegmatites reportedly contain the largest Li-resource in pegmatite, known in India till date (Banerjee et al., 1994). Beryls from Marlagalla have been assigned to the Na-Li-Cs type by the above authors. Niobian tantalite is rich in Ta_2O_5 (up to 72%).

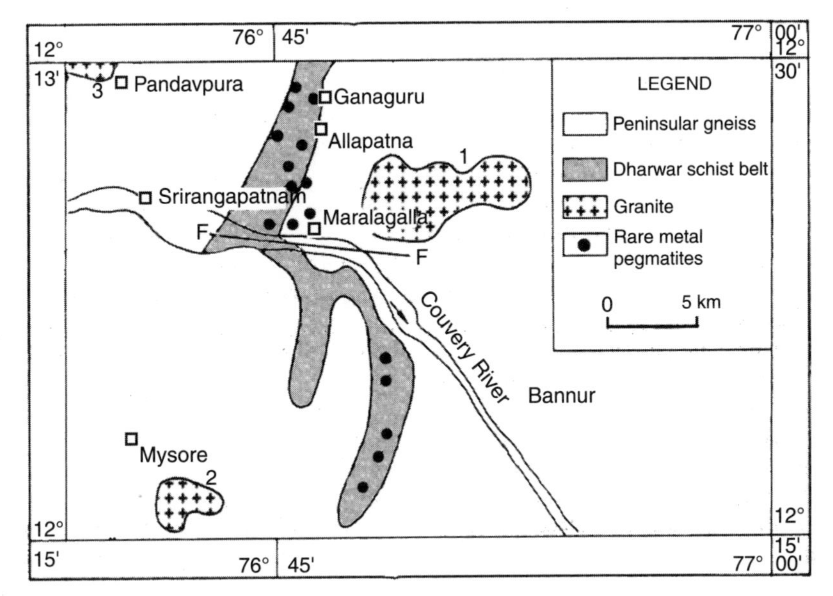

Fig. 1.2.25 Simplified geological map of Allapatna–Marlagalla area. 1. Allapatna Granite, 2. Chamundi Granite, 3. Pandavpura Granite (modified after Sarbajna and Krishnamurthy, 1996)

The Allapatna Granite in the area, on the basis of its peraluminous character, higher Sr_i (0.726 ±0.01) and enrichment in such elements as Li, Be, Rb, Nb, Ta, Ga, Th and U, has been assigned to the S-type. The spatial, temporal and chemical relationship of this granite with the RM-bearing pegmatites of the area, led Sarbajna and Krishnamurthy to assign a parental status to the Allapatna Granite with respect to the pegmatites.

The auriferous pegmatites of Mangalur schist belt are enriched in spodumene (~10% volume) and beryl, suggesting that they belong to albite–spodumene sub-type of pegmatites (Cerny, 1991). These pegmatites are considered a potential source of Li and Be.

The pegmatite deposits discussed above contain rare metal (± rare earth) minerals in economic/sub-economic quantities either within them or in the soil around them (generally both). Mica is hardly of any economic interest for these deposits. But the Nellore schist belt is well known for its mica-bearing pegmatites. Some of them, of course, contain samarskite, fergusonite, sipylite and beryl.

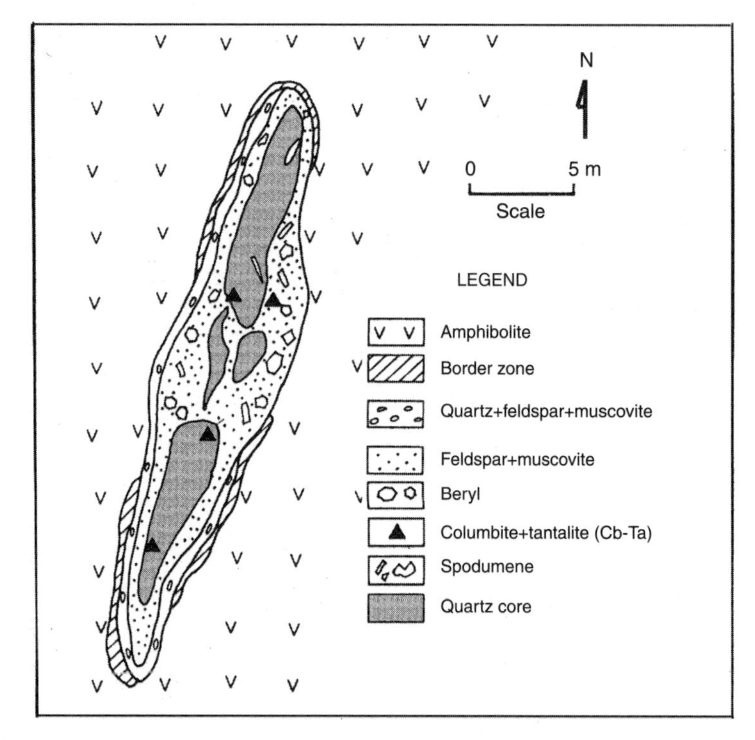

Fig. 1.2.26 Geological sketch map of Marlagalla pegmatite, differentiated into complex zones (after Sarbajna and Krishnamurthy, 1996)

The Kanigiri Granite, occurring in the Prakasam district in Andhra Pradesh is RM-bearing and deserves a brief discussion here. It is not too large a pluton, and is compositionally a biotite-bearing syeno- and monzo-granite with fluorite as an accessory mineral. Other accessory minerals comprise allanite, fergusonite, samarskite, columbite, thorite, monazite, zircon, apatite, rutile, ilmenite, magnetite, hematite and pyrite. Exploration by the Atomic Minerals Directorate indicated: (i) ~ 125 tonnes of $(NbTa)_2O_5$ over an area of 6 × 1.2 sq km, (ii) 300 tonnes of non-magnetic polymineral panned concentrate with an average grade of 3185 g/t.

Ore Mineralisation in Southern Granulitic Province (SGP)

Known ore deposits worth the mention are not many in this region. The most important amongst them are the rare earth (REE) mineral deposits, particularly occurring as beach placers along the sea coast of south Kerala and the adjacent part of Tamil Nadu. Others are of gold, (Wynad–Nilambur, Attapadi), base metals (Mamandur), molybdenite (Harur, Ambalavayal). To these may, of course be added the carbonatites of Samalpatti and Pakkanadu – Mulakkadu, Dharmapuri district, Tamil Nadu, hosting REE mineralisation (Sarkar et al., 1995).

Gold Mineralisation in SGP

Wynad–Nilambur region, Kerala Gold mineralisation in north Kerala, particularly in the Wynad–Nilambur region, has been known to the local people for ages. Even there has been 'modern' mining in the area during the late nineteenth and early twentieth centuries. Gold mineralisation in this region belongs to three types: (i) gold in primary quartz-rich veins, (ii) lateritic gold, and (iii) placer gold.

Nilambur is the eastern extension of Wynad gold field. Relatively recently a primary gold prospect has been located through the exploration by government agencies at Mankada Maruda, NE of Nilambur town (Nair and Suresh Chandran, 1996).

Nilambur–Wynad gold field is in a high grade metamorphic terrain, composed of biotite hornblende gneiss, amphibolite, actinolate schist, charnockite, pyroxene granulite, and quartz–magnetite granulite. Immediate host rocks of the auriferous veins are rich in sericite, chlorite, biotite and carbonates. The mineralised zone is located in the western extension of Bhavani shear zone and the gold-quartz veins are syn- to post- tectonic with respect to shearing. The low- temperature minerals in the mineralised zone are products of retrogression related to shearing, probably augmented by the ore genetic process itself. The density of mixed CO_2-H_2O and H_2O-NaCl inclusions suggest the deposition of gold in the quartz veins at ~2 kb and 300–350°C/400°C (Srikantappa, 2001).

Santosh et al. (1995), based on stable isotope studies, suggested a mantle source of the ore-bearing fluids for the Nilambur–Wynad mineralisation. Binu-Lal et al. (2003) also discussed the mechanism of transport and deposition of gold mineralisation at the Wynad area, based on the results of fluid inclusion studies and gas analysis.

There are a large number of old workings in the laterites near Nilambur town. Unauthorised mining activities on the laterite are still on (Narayanswami and Krishna Kumar, 1996). Some important prospects are Kappil, Anippamadu, Aruvacode, Ponkunnu, Arippukunnu, Chembarassery, Theyyampadikuthu, Maruda, Thannikkadavu and MannucheeNi. A preliminary estimate of gold reserve in the laterite at this place is presented in Table 1.2.7. As commonly happens, many of characteristics of the primary gold are not retained in the lateritic profile.

Table 1.2.7 *Geological reserve and the grade of lateritic gold from Nilambur Valley, Kerala (Narayanaswami and Krishna Kumar, 1996)*

Area	Area of latertic cover (sq km)	Thickness of laterite (m)	Total volume (mill.cub.m)	Grade* (Au g/t)	Quantity (mill. tonnes)
1. Kappil	0.250	20	5.00	2.40	11.50
2. Arippamadu	0.16	15	2.40	2.90	5.54
3. Aruvacode	0.03	10	0.30	0.35	0.69
4. Ponkunnu	0.05	10	0.50	0.10	1.15
5. Arippukunnu	1.00	10	10.00	0.59	23.00
6. Chembarassery	0.0025	12	0.03	0.22	0.07
7. Theyyampadikuthu	0.01	7	0.07	0.05	0.16
8. Maruda	1.00	7	7.00	2.38	16.10
9. Thannikkadavu	0.16	10	1.60	2.00	3.68
10. Mannuchuni	0.16	10	1.60	3.68	1.00
Total					65.57

* Maximum Au-value obtained

In the Nilambur area, hundreds of people are still engaged in recovering placer gold by panning, mostly the top horizons of the gravels of the Chaliyar and Punna Purha rivers. Geological Survey of India in a preliminary study estimated 69,590 ounces of gold in a volume of 8.5 million cubic meters gravel in the Nilambur valley (Sarwarkar, 1969).

Attapadi Valley, Kerala The NE–SW trending high grade supracrustals in Attapadi valley occur between the Nilgiri charnockite hills in the north and Vellingiri granitoid hills in the south, along the western extension of Bhavani shear zone (BSZ). The area mainly comprises amphibolite facies gneisses formed through retrogression of granulite facies rocks, charnockite, migmatitic granite gneiss (PGC) and numerous enclaves of older supracrustals composed of metamorphosed mafic-ultramafic rocks and metasediments (Fig. 1.2.27). Nair (1993) suggested these to be the remnants of an older greenstone belt (possibly equivalent to Sargur supracrustals) within the Archaean granulite terrain. The manifestation of the BSZ in Attapadi valley is represented by a series of NE-SW trending ductile shear zones over a cumulative width of about 8 km.

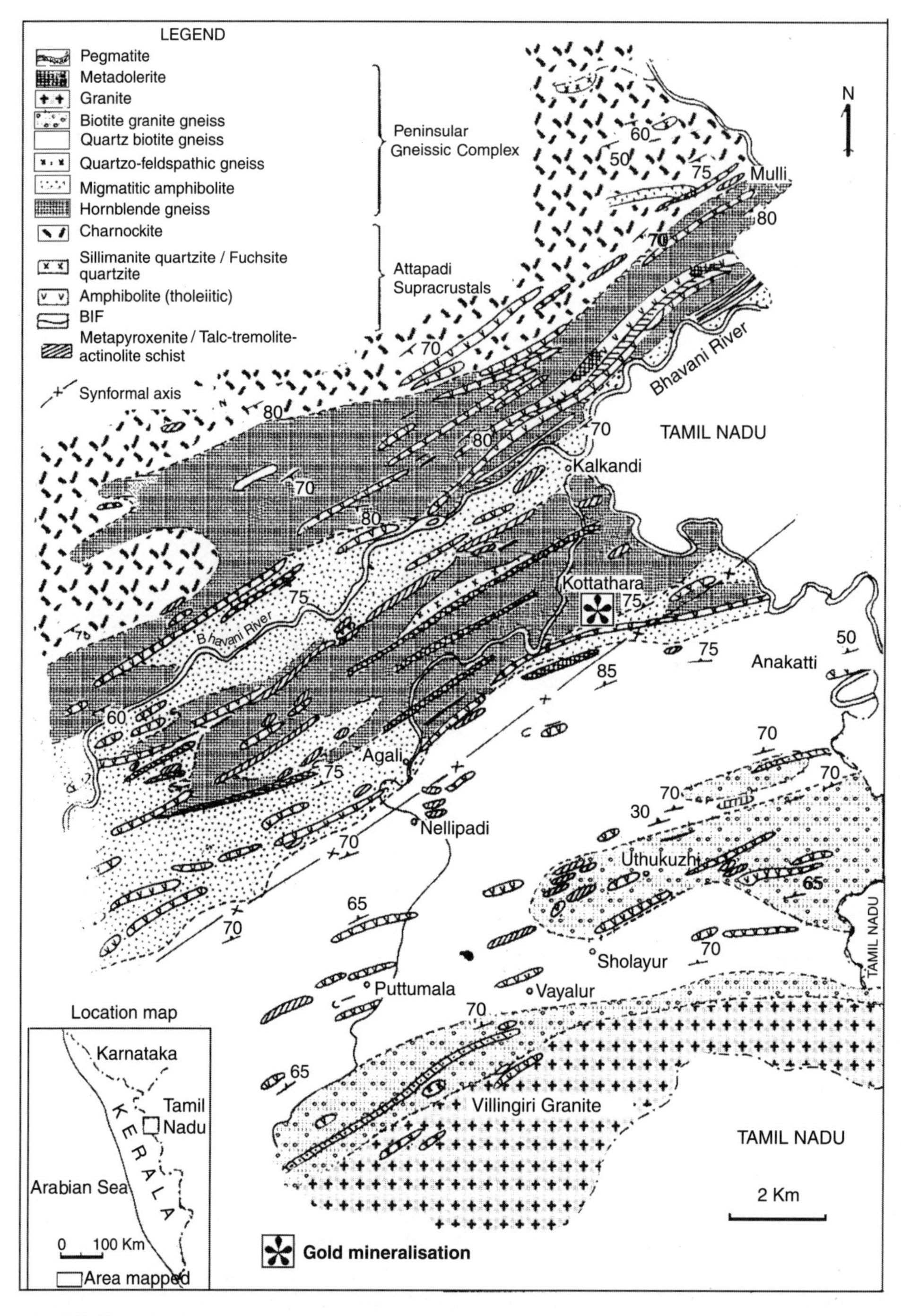

Fig. 1.2.27 Geological map of Attapadi valley showing gold occurrence at Kottathara, Kerala (modified after Nair and Nair, 2001).

Nair and Nair (2001) have worked out a tentative stratigraphic succession of these supracrustals as follows:

Tholeiitic amphibolite
BIF, sillimanite-kyanite quartzite, fuchsite quartzite
Pyroxenite and peridotite (partially or fully altered to talc-tremolite- actinolite schist) and minor anorthosite, pillowed metabasalt and amphibolite, with auriferous veins.

These supracrustals are highly deformed, displaying two phases of near coaxial folding (isoclinal to tight asymmetric F_1 and open asymmetric F_2) with the development of the corresponding axial planar schistosities and their transposition parallel to present NE-SW structural trend. These were followed by regional shearing and wide scale development of mylonitic foliation superposed on the earlier structures. The peak granulite facies metamorphism followed by retrogression and emplacement of auriferous veins (mostly along the sheared contacts of meta-ultramafites and hornblende gneiss) are pre-F_2. The post-F_2 shearing caused remobilisation of gold and associated sulphides (Nair et al., 2005; Nakagawa et al., 2005)

Several auriferous zones have been located in the area, of which the most prominent is the 500 m stretch of discontinuous gold-quartz lode (dipping 65° towards NW) in Kottathara prospect, which is associated with chlorite schist/ biotite quartz schist developed along the contacts of komatiitic pyroxenite/amphibolite and hornblende gneiss. Gold and sulphide mineralisation is hosted by massive (grayish to yellowish white in colour and 20 cm to 2 m in width) quartz veins as well as by the quartz veinlets in the country rock (Naik et al., op. cit.).

The ore minerals in the gold-quartz veins consist of pyrite, chalcopyrite, galena, covellite and native gold-silver alloy in association with pyrite. Arsenopyrite is totally absent; ilmenite, rutile and secondary malachite occur in minor proportion. Quartz is the major gangue, with sparse carbonates. Besides the pyrite associated occurrence, gold also occurs as discrete grain in the quartz gangue. The near surface limonitised portions of the auriferous veins generally have better tenor of gold compared to silver.

Based on the average gold content of 0.20 ppm in the lode-free komatiitic metapyroxenites and close correspondence between Au and MgO values, it is inferred that gold in the quartz veins could have been derived from the proximally occurring Mg-rich rocks or their altered products by scavenging mechanism by hydrothermal solutions (Nair et al., 2005). The timing of quartz veining and primary gold-sulphide mineralisation in the Attapady valley is correlated by these workers with the Paleoproterozoic (2.0–2.5 Ga) tectonic event along the Moyar-Bhavani shear system (locally pre-F_2), and its remobilisation during the Pan-African event (~500 Ma).

Zn-Pb-Cu Mineralisation at Mamandur There is no basemetal deposit in the SGP worth the mention except one at Mamandur in Cuddalore district of Tamil Nadu (Fig. 1.2.28).

Zn-Pb-Cu mineralisation in the area occurs along an elongate hillock, called Shembumalai. In Tamil, it means 'Copper hill', and copper generally produced some spectacularly colourful secondary minerals on being exposed to exogenic chemical processes. It is a small deposit, ~350×17–20 km in aerial extent. It is a steeply dipping sheet-like ore deposit, accordant with the general structural grain in the local geology. The estimated reserve of the ore in the deposit is 700,000 tonnes with an average grade of 0.4% Cu, 3.5% Zn and 2% Pb and 30 ppm Ag and 400 ppm Cd (Terziev, 1971).

The host rocks of mineralisation are pelitic granulites and quartzo-feldspathic gneiss in the footwall side and the same rock or two-pyroxene granulite in the hanging wall side. However, the ore zone is enclosed mostly by the granulitic metapelites and quartzo-feldspathic gneiss.The ore zone is highly sheared, a conclusion that is supported by the conspicuous development of phyllosilicates in the ore zone, C-S fabric in the quartzo-feldspathic gneiss and the fabric of the

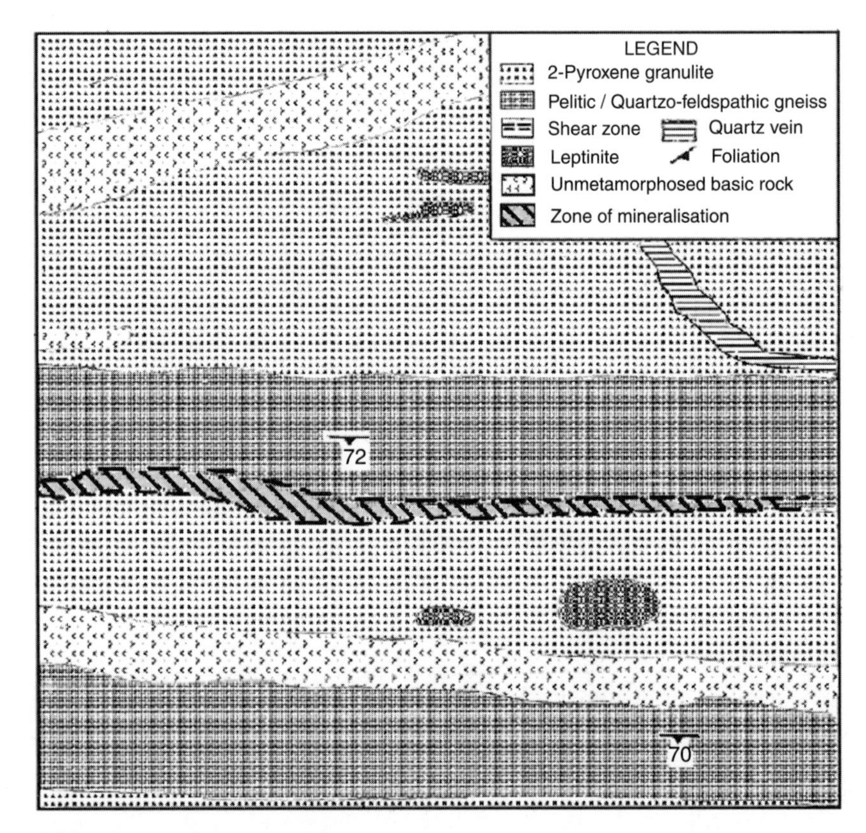

Fig. 1.2.28 Geological map of Mamandur showing zinc-lead-copper mineralisation in granulite rocks, Tamil Nadu (modified after Chattopadhyay, 1995)

ore and the gangue minerals. Within the orebody, the mineralisation is in the form of stringers, streaks and veins. There is a crude metal zoning in the ore body from Cu (bottom) → Pb→ Zn (top) (Sarkar, 1988). The principal sulphide phases in the ores are sphalerite, gelena, pyrite/ pyrrhotite, chalcopyrite. Mackinawite, graphite and arsenopyrite are the minor/trace phases. A small proportion of the ZnS phase is manganoan wurtzite rather than sphalerite. Au was detected in some samples of sphalerite and chalcopyrite (Chattopadhyay, 1995)

A variety of minerals are present as gangue. These comprise quartz, feldspars, biotite, garnet, cordierite, sillimanite, orthopyroxene, clinopyroxene, hornblende, magnetite, Zn-bearing spinel, carbonate and apatite. Strain distribution in the ores is differential, annealing being characteristic of the now low-strain domains.

Deformational features elsewhere suggest that deformation locally outlasted recrystallisation. Presence of roundish sphalerite grains in garnet and cordierite and textural equilibrium of sphalerite and orthopyroxene grains (Fig. 1.2.29a, b) as well as sulphide-silicate equilibria during metamorphism are suggestive of participation of both the ores and the country rocks in the peak of metamorphism. The estimated T_{max} and P_{max} for the Mamandur rocks are 850 °C and 9 kb, respectively. Granulite metamorphism at Mamandur, as already mentioned, was mainly through dehydration melting (Chattopadhyay, 1995).

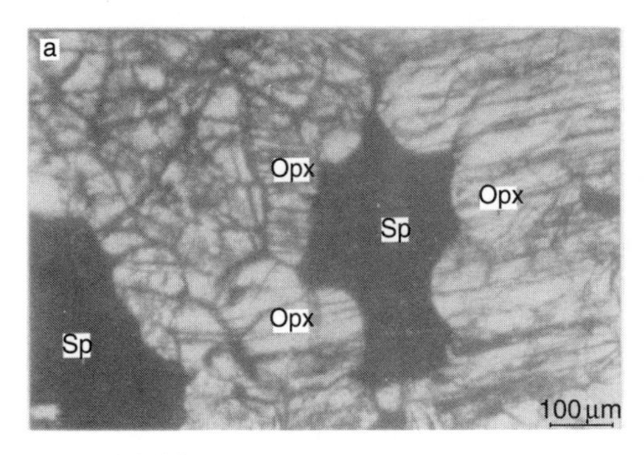

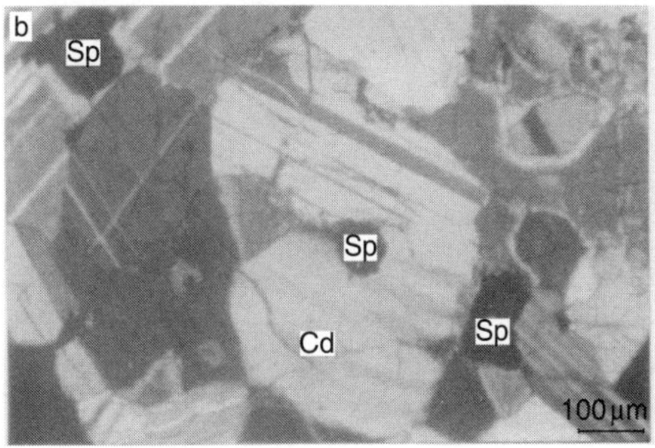

Fig. 1.2.29 Photomicrographs (a) showing equilibrium texture between orthopyroxene (opx) and sphalerite (sp). (b) Spherical inclusions of sphalerite within a cordierite grained in cordierite gneiss. These suggest participation of both silicates and ore minerals in granulite facies metamorphism.

The model Pb-Pb age of Mamandur ores is c 2600 Ma and the lead is less radiogenic than in the Kolar ores but more evolved than the Ingaldahl ores (Chernyshev of IGEM, Moscow –personal communication to SCS, 1989). Chandi (GSI) also obtained a model Pb-age of the Mamandur ores as 2690 Ma, using a solid-source mass spectrometer (personal communication to SCS, 1990).

At this evolved state of the ores and their host rocks, it is difficult to firmly suggest a genetic model for the initial deposition of the ores at Mamandur. But on the basis of paucity of metals (keeping in mind the age of deposition), crude metal zoning, pelitic metasediments as immediate host rocks, it may be permissible to assume that the ores initially formed through sedimentation and diagenesis. Therein may lie the explanation of the deposit being a small one.

Molybdenite Mineralisation in SGP and Transition Zone

Harur–Uttarangi Belt, Dharampuri Dist., Tamil Nadu The most promising molybdenum prospects have been reported from the Harur-Uttarangi belt of Tamil Nadu where an economically viable deposit has been reported for the first time in the country from Velampatti South block.

Between Gudiyattam in the north and Bhavani in the south a litho-tectonic belt is seen bounded by Harur lineament in the east and Dharmapuri lineament in the west. This belt is 40 to 50 m wide and 200 km long. The main lithological unit of the belt is the hornblende gneiss with enclaves of charnockite, pyroxene granulite, etc. A number of alkaline syenite-carbonatite complexes occur within the belt. Within this belt molybdenite occurrences have been identified over 28 km strike length along a prominent shear zone extending from Harur up to Uttarangi. The shear zone is characterised by the development of closely spaced shear foliation with/without prominent vein/reef quartz and is also marked by sericitisation, chloritisation, and phyllonitisation. The mineralisation is confined to the quartz vein emplaced in the shear zone as well as in the sheared country rock with wall rock alterations. Majority of the vein/reef quartz (25 cm–2 m) hosting bulk mineralisation have been emplaced along well defined, low dipping fractures cutting the steeply dipping shear foliation within the shear zone. The mineralised portion in the shear zone have been visualised as low dipping zone, enveloping the vein and reef quartz, and cutting the steeply dipping shear foliation (Nagarajan et al., 2001).

The ore mineragraphic study indicates that the sulphide phase comprises galena, pyrite, chalcopyrite and molybdenite in decreasing order of abundance while rutile represents the oxide phase. The characteristic textures related to deformation, recrystallisation and replacement are commonly observed. Deformation structures are shown by broken pyrite grains in vein quartz as well as gash fillings and mutual boundary textures in chalcopyrite. The recrystallisation structures are indicated by the aggregate habits of the bigger rutile grains and euhedral habit of pyrite. The replacement of chalcopyrite by galena and in turn by molybdenite is indicative of replacement textures. Molybdenite occurs as inclusions in galena and also as independent grains in gangue. Based on these textural characteristics and other mineral habits the following paragenesis has been deduced (Palanisamy et al., 2001):

$$\text{Pyrite}_1 \rightarrow \text{Rutile}_1 \rightarrow \text{Chalcopyrite} \rightarrow \text{Galena} \rightarrow \text{Molybdenite} \rightarrow \text{Rutile}_2 \rightarrow \text{Pyrite}_2$$

In the Velampatti South block, GSI identified reserves of the order of 1.8 million tonnes down to 170 m depth taking 0.10% Mo cut-off grade with 0.13% average grade. Incidence of rhenium has been recorded along with the molybdenum ore. In the Marudipatti Central and North blocks, exploration data indicated 0.484 million tonnes of ore reserves with 0.117% Mo in the Central Block and 0.109 million tonnes of reserves with 0.075% Mo in the North Block. In the Velampatti South block close spaced drilling and exploratory mining has been taken up by MECL.

Mineralisation is hosted by quartz veins emplaced in shear zone and the wall rocks, trending NNE–SSW and dipping 30–40° towards east. The shear zone with intermittent molybdenite mineralisation extends for a strike length of 23 km, of which the southern part is better mineralised. The country rock comprises epidote-hornblende gneiss and quartzo-feldspathic gneiss with enclaves of pyroxene granulite, charnockite, meta-pyroxenite, meta-gabbro.

GSI conducted exploration (1990s) in this belt and established ore reserves in several blocks, of which the Velampatti South block was found to be most prospective, where MECL was engaged by TAMIN for exploratory mining.

1. Marudapatti Central block: 0.074 Mt (0. 14% Mo); strike length = 1.0 km; thickness = 1–5 m; down to 120 m depth
2. Marudapatti North block: 1.17 Mt (0.035% Mo); down to 95 m depth
3. Velampatti South block: 1.61 Mt (0.12% Mo); strike length = 1380 m; thickness = 1–8 m; down to 260 m. depth
4. Velampatti Central block: 2.34 Mt (0.04% Mo); strike length = 1.6 km; thickness = 1–5.6 m; down to 120 m depth.

[Data source: Status of mineral resources in India (Part A) and Profiles of significant prospects (Part B), GSI Professional document, Dec., 2000.]

Ambalavayal, Wayanad District, Northern Kerala An alkali granite pluton of Ambalavayal in the Wayanad district of northern Kerala carries an 800 m wide zone of disseminated molybdenite mineralisation. Molybdenite occurs also as flakes and flaky aggregates in quartz veins. The Ambalavayal Granite is a typical example of within-plate alkaline plutons. The petrochemical characteristics of the granite and the features of ore mineralisation have been synthesised into a tectonic model where the magmatism and the related metallogeny have been linked to deep-seated crustal extension (Santosh, 1988a, b).

The Re-Os dating of two molybdenite samples from the alkali granite and pegmatite of Ambalavayal yielded ages of 567 ± 28 Ma and 566 ± 77 Ma, respectively. These values compare very well with the Rb-Sr (WR) age of 595 ± 20 Ma. It may be considered as a Pan-African metallogenic event, however small.

Rare Metal and Uranium Mineralisation Associated with A-type Granite in Parts of Madurai District, Tamil Nadu Radioactive allanites with uraniferous graphites and also gadolinite-bearing pegmatites were known from the Proterozoic migmatitic terrains of the Madurai district, Tamil Nadu (Perumal, 1974). Further investigation by the Department of Atomic Energy led to more details about these mineralisations.

The country rocks in the area belong to charnockite and khondalite groups in general. But the rocks in the area between Nagamalai and Minakshipuram, comprise a variety consisting of acid and basic granulites, leptynites, granites, granitic gneisses, schists, quartzites, calc- granulites and pegmatites. Graphite is present both in calc-granulites and mica schists (locally exceeding 30% by volume in the latter). The graphitic accumulations are uraniferous.

Granitoid rock occurs as small stock-like bodies (upto 2 sq km). Pegmatites in the area are spatially related to these granites and contain rare earth and rare metal minerals such as gadolinite, fergusonite, and allanite. Chemical analyses, limited though, showed that the granitoids are high silica (>70% SiO_2), metaluminous- peraluminous (A/C +N +K ≈1.0), potassic (K_2O > 5%) and relatively high contents of Rb, Ba, Zr and Ga not only assign them to low-Ca granites but bring them closer to A-type granite. Contents of Rb and Sr also show also similarity with the 'fertile granite' used in the parlance of rare metal metallogeny (Pandey et al., 1994).

Sr_i of the granitoids is rather high. Regression of data on eight samples yields Rb-Sr (WR) isochron age of 837 ± 34 Ma (Pandey et al., op. cit.). Migmatitic gneiss from an area at about 12 km west of Madurai was dated to be 550 ± 15 Ma (Rb-Sr/WR) (Hansen et al., 1985). This phase of granitic activity in the area bears correlation with the Pan-African event.

Carbonatite Type REE–mineralisations Carbonatite bodies are reported from several places in Tamil Nadu and Kerala. Of these, the Samalpatti and Pakkanadu–Mulakkadu carbonatite complexes, located in the Dharmapuri district, Tamil Nadu, in a granulite country, are more important. The carbonatites are sovite, beforsite, Ba-rich benstonite and rauhaugite, associated with dunite, pyroxenite and syenite. The major minerals in these carbonatites are calcite, dolomite/allanite and the minor phases comprise dolomite, siderite, ankerite, ilmenohematite, cerianite, perovskite, pyrochlore, monazite, bastnaesite, barite, sphene, phlogopite and epidote. An ankeritic rauhaugite from Samalpatti analysed as follows (Table 1.2.8).

Table 1.2.8 *Composition of a Samalpatti carbonatite (Subramanium et al., 1978)*

Major element oxides	Wt%	Trace elements	ppm
SiO_2	26.33	Ba	22,000
Al_2O_3	5.18	Sr	1,912
TiO_2	1.13	Zr	243
Fe_2O_3	9.69	La	6,004
FeO	3.00	Ce	7,621
MnO	2.38	Y	200
MgO	7.55	Nb	262
CaO	18.02	U	69
Na_2O	3.60	Th	6,245
K_2O	3.07	(La + Ce) / Y	68.10
P_2O_5	1.20		
CO_2	18.00		
H_2O	1.10		

This carbonatite is much more siliceous and richer in Ba, La, Ce, Th and U, compared to the Ambadongar carbonatites, Rajasthan (Sarkar et al.,1995).

K-Ar dates available from the Pakkanadu –Mulakkadu carbonatites range from 599 ± 60 Ma to 771 ± 2 Ma (Moralev et al., 1975 in Krishnamurthy, 1988; Semenov et al., 1978). These temporally relate them to the Pan- African tectono-magmatic event.

Placer Deposits Containing REE, Th, Ti, Zr, etc. India's REE-resources are mainly exogenic placer type deposits, 70–75% of which occur in the beach sand along the west and east coast of India. The best deposits in the west coast are at the southern tip of Kerala at a place called Chavara and at adjacent Manavalakurichi in Tamil Nadu (Fig 1.2.30). The principal mineral of interest in these deposits is of course monazite (LREE-Th phosphate), but there are also significant concentrations of ilmenite, zircon, sillimanite, kyanite and rutile in the beach placers, which are of economic interest. Of these accompanying minerals, zircon and the Ti-minerals are in great demand. The heavies occur as streaks and inpersistent layers (upto 10 cm thick) in otherwise quartzose sands. Locally, the heavy minerals can be disseminated as well. Old sandstreaks/ dunes near the coast carrying monazite are locally cemented into compact masses by $CaCO_3$ from percolating groundwater. Grain size and shapes of monazite, eroded to different measures, vary.

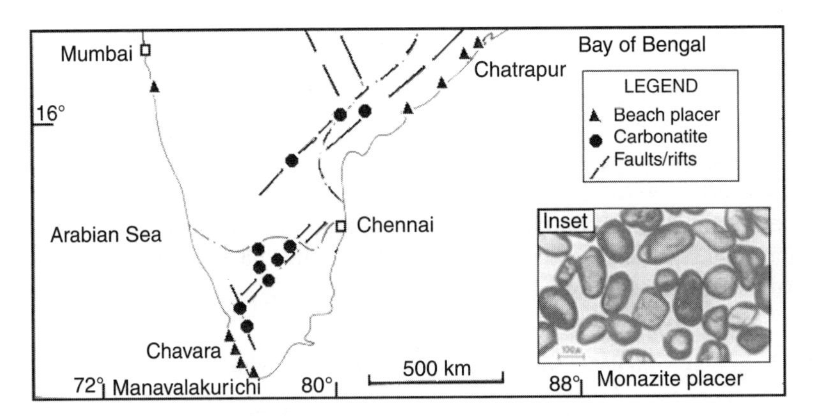

Fig. 1.2.30 Location map of important beach placers containing REE, Th, Ti, Zr etc. in peninsular India. The carbonatite occurrences and faults/rifts are also shown. Inset – Monazite-rich detritals from the placer, Manavalakurichi, Tamil Nadu (modified after Sarkar et al., 1995)

Monazite content of beach sands may be upto 10–11% (by weight). ThO_2 in Indian monazite varies between 1–10.5%. Ce, as usual, is the principal REE. A little uranium (U_3O_8 = 0.25–0.4%) is also present with an average Σ REO of about 60% (wt). The REE composition (wt%) of Indian beach monazite is as follows (Table 1.2.9).

Table 1.2.9 *Average REE composition (wt%) of Indian beach monazite (Sarkar et al., 1995)*

La_2O_3 – 22	Gd_2O_3 – 1.0
CeO_2 – 47	Tb_4O_7 – 0.1
Pr_2O_3 – 6	Dy_2O_3 – 0.2
Nd_2O_3 – 18	Ho_2O_3 – 0.1
Sm_2O_3 – 3	Y_2O_3 – 0.5
Eu_2O_3 – 0.01	

Compared with the average Australian variety, the Indian monazite is a little poor in La and Y, but richer in Ce. The U_3O_8-content is not very much welcome to the users of monazite as a REE-source.

Prospecting for off-shore placers in the Kerala–Tamil Nadu (Chavara–Manavalakurichi sector) did not so far prove to be highly encouraging, particularly with respect to monazite.

Heavy minerals in the beach sands between Mangalore and Cochin could at places be upto 36 wt%. But these are monazite-poor (trace to 0.26%), but can be ilmenite-rich (up to 16.2%) (Rao, 1989).

The estimated reserve of monazite in the country is 8 million tones which is expected to contain $\geq 7 \times 10^5$ tonnes of ThO_2. Bulk of this reserve is located in the Kerala-Tamil Nadu coast.

As is now generally known, the formation of a placer deposit depends on at least the following factors:

1. Presence of a suitable provenance.
2. Resistant physico-chemical properties of the minerals concerned in the context of weathering and erosion process in nature.
3. The duration over which these minerals in their protoliths have been exposed to weathering and erosion.
4. Physico-chemical processes of concentration.

The importance of suitable source rocks can hardly be overemphasised. The best beach placer minerals developed in these sectors particularly where the continental wash pouring out into the sea, was derived from the Precambrian country rocks, comprising granites, granitic pegmatites, granite gneisses, migmatites, and the granulite facies rocks such as charnockites, khondalites and leptinites. They could be recycled Tertiary coastal sediments derived from the same provenance.

The conclusion that monazite, even as accessory mineral is extremely rare in greenschist facies rocks, but is common to abundant in the granulite facies rocks (Overstreet and Olson, 1960) is not generally accepted. Monazite in fact has a large thermal stability field. Hydrothermal monazite, including that formed in low greenschist to low amphibolite facies, are generally poor in ThO_2-content (< 1 wt%), compared to monazite in granitoid and granulite rocks. The latter generally have higher ThO_2-content (upto ~ 10 wt%), provided it was there in the system (Schandl and Gorton, 2004). This upper limit of ThO_2-content is even exceeded in some South Indian monazites occurring in a granulite terrain. Davidson (1956) reported monazite-content of ~ 18% in garnetiferous biotite schist in a migmatite zone at Tadikarakonam, Kerala. Davidson traced this reef (~32 m) for a distance of about 1 mile (1.6 km). Rich zones were discovered during later exploration by the Department of Atomic Energy, Government of India. But nowhere these enrichments have been found good enough for direct mining.

Warm and humid climate with abundant rainfall has been conducive to the decomposition and disintegration of the rocks that originally hosted the minerals that now formed the placer deposits under discussion. Action of waves, tides, current and wind did the necessary winnowing and deposition. Mallik et al. (1987) proposed a three-stage depositional model for the placer sands of Kerala–Tamil Nadu as follows:

1. Initial stage – Supply of heavy minerals through rivers and deposition of sands containing disseminated placers by waves, current etc.
2. Transgressive stage – Erosion and reworking of beach ridges.
3. Regressive stage – Formation of placer deposits on the beach through wave and longshore current activity giving rise to the present configurations.

Iron Formations in SGP Iron formations, composed mainly of banded magnetite quartzite, occur at many places in Southern Granulite Province. They occur as bands <20m thick and upto a few kilometers long in highly metamorphosed supracrustal enclaves in tonalite–trondhjemite gneisses. Some of these deposits, occuring at Kanjamalai, Tirthamalai, Tiruvanamalai and Nainaramalai, have been proved economic to sub-economic in the past (Rao et al., 1977; Rao and Kasipati, 1982). The metal content is generally low (≤40% Fe). The associated supracrustal rocks comprise quartzite–meta-pelites–meta-carbonates with amphibolitic rocks. The grade of metamorphism range between amphibolite and granulite facies (Prasad et al., 1985), corresponding to T = 650°–900°C and P = 7–12 kb. Iron formations within the Sargurs, particularly those occuring close to the granulite province boundary, share many of these characters.

The iron ore formations outlined above have their counterparts in many Precambrian terrains around the world. Prasad et al., (1982) have given a new type name to this kind of iron formations. They have called these 'Tamil Nadu type of iron formation'. This new name does not seem quite warranted as these ores share many features of the common iron formation types, particularly the so called Superior type. Only, these are highly metamorphosed.

Granulite Facies Metamorphism, Ore Deposits and the Southern Granulite Province There is a belief, shared by many, that granulite metamorphic terranes are poor habitats of ore deposits, particularly those of uranium, thorium, gold, basemetals, tin, tungstan etc. Let us examine how far this belief is sustainable. A review of this issue, based on the available information, is not only interesting from a scientific point of view but also for its relevance in exploration of these deposits within such high grade metamorphic terrane, of which we have a fair sharc in India.

Uranium and Thorium in High Grade Metamorphic Rocks It is a general observation that uranium and thorium are relatively insensative to epi- and mesozonal metamorphism, i.e greenschist to amphibolite facis in a closed system. Under these conditions the maximum effect of the regional metamorphism is mobilisation/ remobilisation in millimeter to meter scales. Recrystallisation of pitchblende to uraninite with increased content of Th and REE (depending on availability) is the mineralogical characteristic of this stage. Strata-concordant disseminated type distributions are maintained (Dahlkamp, 1991). On the other hand, the behaviour of these two elements, particularly uranium, is not very clear in granulite facies metamorphism, compared to their contents in low grade metamorphic or igneous rocks (Heirer, 1979; Distal and Capedri, 1978). On the contrary, Adamson (1983 in Dahlkamp, 1993) and Adamson and Parslov (1985) studied the granulite rocks south of the Athabasca basin, Canada, and found no evidence of the depletion of uranium and thorium in these rocks. Pagel and Svab (1985) even found increase in the contents of these elements in the granulite rocks occurring to the west of the area studied by Adamson. This is also supported by the work of Fernandes and Iyer et al., (1984). The general low content of these two elements in granulite rocks may be suggested to be even an inherited feature from the pre-granulite stage. The Indian scenario, as reflected in the published literature, is none the more illuminating. Rao and Narayana (1993) studied the high grade metamorphic rocks of Dharmapuri area, Tamil Nadu and reported that the tonalitic charnockites and mafic charnockites contain less uranium and thorium, compared to the tonalite mafic gneisses of the transition zone. The generally lower Th/U ratios in granulites, compared to the transition zone rocks, as obtained by them, are taken as the evidence of the migration of these elements during granulite metamorphism. They further suggest that uranium was perhaps removed as a CO_2 complex during CO_2-flushing causing granulite metamorphism. However, thorium does not complex with CO_2 as easily as uranium in a given situation and therefore, lowering of the Th/U ratio remains poorly explained. As already discussed, thorium locally accumulated in many granulite terranes of the world, including India. Moreover, not all granulite metamorphism took place due to hot CO_2-flushing from below. Janardhan et al. (1982) studied the composition of vein charnockite of Kabbaldurga and the adjacent tonalite gneiss. They did not find any change in the thorium content (15 ppm), but the uranium content was reduced to 1 ppm in the charnockite from 1.4 ppm in the tonalitic gneiss. Rao et al., (2001) made an extensive study of Southern Granulite Province with respect to the radio elements (U, Th, K), choosing 209 sites in the Northern block (north of Palghat – Cauvery shear zone) and 121 sites in the Southern block. This they did to estimate the crustal contribution to surface heat flow, the crustal heat production being envisaged as a function of the radio elements in the given rocks. Their finding, relevant in this context, is that the sector between Dharmapuri and Namakkal in the Northern block is the lowest heat producing 0.15 μWm^{-3}(n = 38) area in the block. The charnockites and khondalites in the Southern block in contrast, show higher heat production: 0.17–4.0 μWm^{-3} for charnockites (n = 40) and 0.7–15.1 μWm^{-3} for khondalites (n =10). This contrast in the the heat production rates show that the granulite terrains are not necessarily depleted in radio elements. It depends on the details of the metamorphism, location of the material in the granulite column. It might be recalled here that the area north of the Palghat Cauvery shear exposes the deepest part of the granulite crust (corresponding to 9–10 kb). These observations suggest that a granulite terrane need not be axiomatically considered being completely barren of uranium mineralisation. If by chance there was uranium accumulation in an area before it underwent granulite metamorphism, the deposit need not disappear during metamorphism. To do so a very high fluid/rock situation will be needed, which is hardly obtained during granulite facies metamorphism.

Gold Mineralisation in High Grade Rocks Granulite terrains are not known for containing many mineable gold deposits, except possibly a few small ones here and there, the most important amongst which is the Renco deposit in the Limpopo belt, Zimbabwe.

The Renco deposit with a reserve of 26 Mt of ore (Au – 8.7 g/t) occurs in granulites that comprise felsic granulites, enderbite, charnockite (3.56–3.2 Ga). The auriferous reefs range from laminated rocks composed of quartz, sericite, chlorite, magnetite, porphyroblastic feldspar and fine grained sulphides to quartzo-feldspathic pegmatites with Fe-sulphide disseminations. The fine grained gold is accompanied by native bismuth, maldonite, electrum, galena, molybdenite, tellurides and scheelite. The host rocks are dated, but the age of the mineralisation is not established, neither the P-T of the mineralisation. The Chinese picture is clearer in that context. They reported more than 100 deposits/ occurrences of gold in the high grade terrain of North China craton (Gan Shengfei, 1992). Of these several are moderately large. It is believed that the mineralisation took place in the Mesozoic ductile-brittle shear zones at 200°–700° C. A recent paper by Goldfarb et al. (2007) profers a candid explanation of the situation.

Gold mineralisation in the Griffith's Find deposit in the Yilgarn craton, Western Australia, occurs as quartz–pyroxene veins in granulitic rocks (Barnicoat et al., 1991).

There is a common belief that gold moved out during granulite metamorphism of a terrain. No doubt, average Au-content in granulite rocks is lower than few rocks, but it is comparable to that in felsic plutonic and volcanic rocks as well as common sediments. If granulites developed by transformation of such rocks, Au-content will not be high. Sighinolfi and Santosh (1976) obtained an average gold content of 1.51 ppb in the Archean granulites of Bahia, Brazil, with of course a dispersion of <0.4–18 ppb. Low Au-content was correlated with low CaO + MgO and high Au-content with the high CaO + MgO contents of granulites. In the context of the above discussion it may be concluded that gold mineralisation in a granulite terrain may be pre-, syn- or post- with respect to the granulite metamorphism. Because of this high grade metamorphic stamp, a region should not be straight away rejected in any scheme of prospecting for gold.

Mineralisations in and Around the Cuddapah Basin

The Cuddapah basin is a repository of many metallic and non-metallic minerals, of which some like barite at Mangampeta represents one of the largest deposit of its kind in the world, while others like lead- and copper-sulphides (± zinc) occur as small deposits (one of which was worked in the past for a short duration). The uranium prospects, specially some of those located along the western margin of the basin, apparently hold significant economic potential. Fig. 1.2.31 shows the locations of some prominent barite, limestone, asbestos, diamond, basemetal (Pb-Cu-Zn, Zn-Pb-Cu) and uranium mineralisations in and around Cuddapah basin.

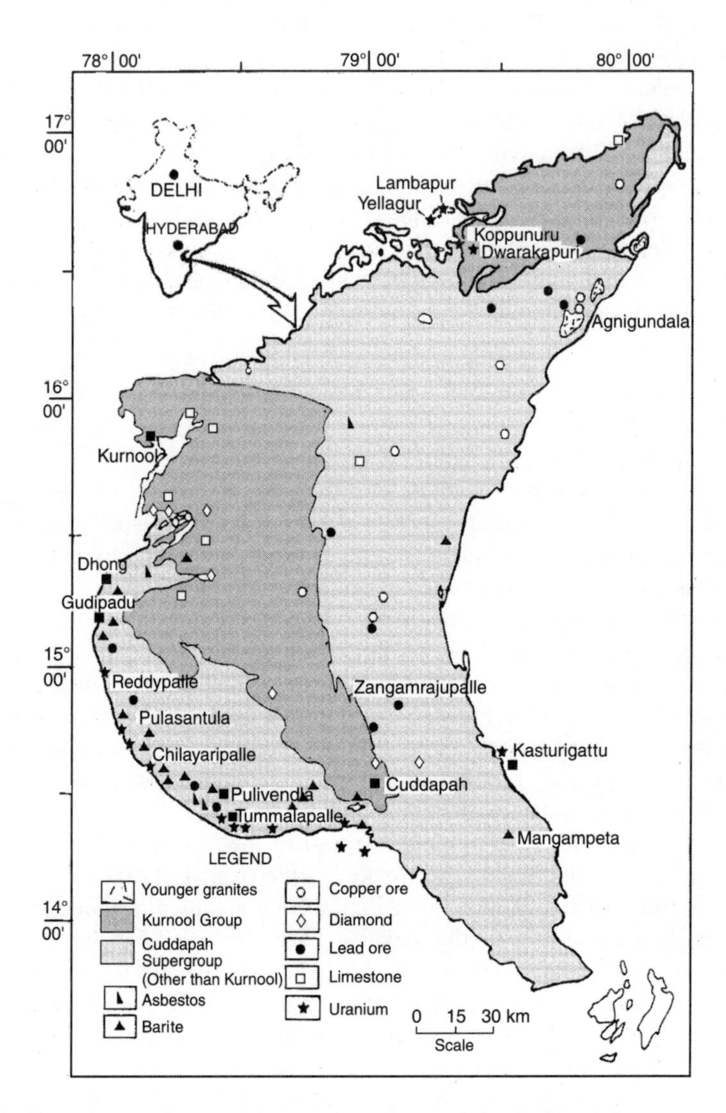

Fig. 1.2.31 Distribution of mineral deposits in and around the Cuddapah basin

Base Metal Mineralisation in Cudappah Basin The basemetal mineralisation in the Cuddapah basin is mainly confined to the rocks of Nallamalai Group in four sectors from north to south as follows (Fig. 1.2.31).

1. Agnigundala (Pb-Cu-Zn)
2. Markapur (Pb,Cu,Zn)
3. Zangamrajupalle–Varikunta– Golapalle (Zn-Pb ± Cu)
4. Rajampet–Mangampeta (very minor)

Apart from the above, Gani–Kalva is another mineralised stretch in Karnool Group of rocks in the west of Nallamalai.

Agnigundala Belt The ore deposits in this area are hosted by the rocks of Cumbum Formation of the Nallamalai Group, comprising brecciated dolomite, dolomitic limestone and calcareous quartzite, interbedded with chloritic and carbonaceous phyllite. The ores are mainly represented by galena, chalcopyrite, bornite, minor sphalerite and subordinate pyrite. Other sulphide phases as accessory and traces are chalcocite, covellite, cobaltite, stibnite, cuprobismuthite etc.

Fine grained carbonaceous dolomite is the dominant host for sulphide mineralisation in this area. The bedding-parallel primary phase of mineralisation is of stratiform 'syngenetic' type, accompanied by a later phase of remobilised ore emplacement along favourable structures by epigenetic processes. However, the remobilisation of ore in general is of very limited extent from the boundaries of stratiform ore bodies.

The reserve and grade of ores in the three main blocks of Agnigundala area, viz, Bandalamottu, Nallakonda and Dhukonda are furnished below (Table 1.2.10).

Table 1.2.10 *Ore reserves and grade in Bandalamottu, Nallakonda and Dhukonda deposits (GSI, 1989)*

Blocks & Grade	Lead		Copper	
	Reserves (million tonnes)	Gross metal content (tonnes)	Reserves (million tonnes)	Gross metal content (tonnes)
Bandalamottu	10.059	6,62,641	0.775	13,300
Grade A: (6.59% Pb, 1.71% Cu)	1.400	43,837	0.261	1,500
Grade B: (3.10% Pb, 0.56% Cu)				
Nallakonda *Grade A:* (1.82% Cu)	....	....	3.144	57,320
Dhukonda Grade: A (8.98% Pb, 1.51% Cu)	0.460	41,300	2.154	32,630

There are several other prospects and occurrences in the Agnigundala area, some of which are located at Karempudi (0.650 mt × 2.34% Pb+Zn), Vummidivaram (0.30 mt × 2.69% Pb), Borrakonda–Lingalakonda, Peddakonda, Ayyanapalem–Koppukonda, etc.

The Bandalamottu deposit (recoverable reserve reestimated as 2.6 Mt × 4.86% Pb) was mined in the past by the Hindustan Zinc Ltd., producing about 200 tonnes of ore per day. The Nallakonda copper deposit was also taken up by the HCL for mine development but was later relinquished before getting into the production stage because of the en echelon pattern and limited strike continuity of the individual orebodies.

Markapur Belt The Markapur belt, occurring in the southwest of Agnigundala, is located in the north eastern parts of Cuddapah basin. The Cumbum Formation in this area dominantly comprises metapelites, with thin intercalations of carbonates and quartzites. The sulphide mineralisation is mainly hosted by the carbonate rock and the mineralised stretch has a cumulative srtike length of about 50 km from Ghantapuram in the south to Gajjelekonda in the north. Structurally, the mineralised belt lies in the eastern limb of a major overturned synform plunging northwards.

There are three mineralised zones, the stratigraphically lowest one (eastern-most) being copper-rich, the middle one is lead-rich, while the upper zone (western-most) is zinc-rich. Apart from the stratigraphic zoning, lateral and depth-wise zoning are also recorded, the latter wherever present shows lead concentration at shallower levels and copper concentration at depth. The dominant sulphide ore minerals are chalcopyrite, galena and sphalerite with ubiquitous association of pyrite. Covellite and bornite are appreciable in copper rich zones.

The mineralisation is essentially 'sedimentary-diagenetic' in origin.

Zangamrajupalli–Varikunta (Z–V) Belt The lenticular dolomite bodies of varied dimension occurring in the rhythmitic shale–dolomite–chert sequence of Cumbum Formation intermittently exposed for 35 km along the central part of Nallamalai basin from Varikunta in the north and Zangamrajupalli in the south are the main host for Zn-Pb-Cu mineralisation in this belt. Mineralisation in the lower stratigrahic levels of the dolomite are dominated by chalcopyrite with minor galena and sphalerite, while the upper cherty dolomite horizon mainly contain sphalerite and galena in varying proportions. Pyrite occurs as fine grains in the mineralised zones. The mineralised horizons are overlain by a carbonaceous shale unit.

The mineralisation is mainly stratiform having sulphidic layers parallel to the bedding lamellae of the host, although there are also disseminations, stringers, fracture fillings and veins of sulphides.

Two small deposits, viz, Golapalli and Zangamrajupalli have been established in this belt with ore reserves of 2.4 Mt (2.73% Pb + 1.97% Zn) and 3.21 Mt (2.55% Zn +1.82% Pb) respectively.

Rajampet–Mangampeta This is the southern most stretch of the Nallamalai belt having continuation of the carbonate–shale sequence of Cumbum (and equivalent Pullampet) Formations, which host basemetal sulphides in the central and northern sectors. However, inspite of the existance of a conducive milieu for mineralisation, the surface shows for sulphides are negligible in the carbonates of this sector. Limited drilling also did not reveal any ore body of significance at depth.

Though poor in sulphides, the unique bedded barite deposit of Mangampeta occures in this sector, which is described separately. The average contents of Zn+Pb+Cu in the barites of Mangampeta do not exceed 0.03%.

Gani–Kalva Copper sulphide dominant mineralisation at Gani and Kalva occurs along a NE–SW trending tectonic zone transecting the Kurnool sub-basin and also transgressing into parts of Nallamalai in the NE and across Papaghni in the SW.

Uranium Mineralisation The Cuddapah basin has in recent times, gained additional importance because of the discovery of uranium mineralisation in and around it. Two principal types of uranium mineralisations have been discovered and established within and around the basin. These are:

1. Stratabound mineralisations in the Cuddapah Supergroup, particularly in the dolomitic rocks of the Vempalle Formation and in the Pulivendla Quartzite that overlies it with a (?) disconformity (Table 1.1.16).
2. Structurally controlled vein-type deposits at the internal or external (present-day) boundary of the basin.

The structurally controlled type (2) is divisible into the following sub-types (Sharma et al., 1996), based mainly on the sites of occurrence and the details of the local geology.

(a) Unconformity-proximal veins in the basement grantoid at Lambapur at the northwestern periphery of the basin.
(b) Veins in the metamorphic tectonites at the eastern margin of the basin, particularly at Kasturigattu, Gudarukuppu or Kullar.
(c) Veins in the basement rocks around Rayachoti at the southern margin of the basin.

Stratabound Mineralisations Stratabound uranium mineralisation occurs in the stromatolite-bearing phosphatic siliceous dolostone (PSD) belonging to the Vempalle Formation, and has been traced discontinuously for a distance of ~140 km from Reddypalle in the northwest to Maddinalugu in the southwest. The important prospects for uranium in this belt are Tummalapalle, Racha Kuntapalle Bakkannagaripalle and Gadankipalle (Fig.1.2.31). The principal characteristics of this type of uranium mineralisation in the Cuddapah belt are furnished in Table 1.2.11.

Table 1.2.11 *Principal characteristics of the stratabound carbonate-hosted uranium mineralisation in the Cuddapah basin*

1	Extension	Extensive discontinuously over a distance of > 130 km
2	Better mineralised zones (deposits)	Tummalapalle, Gadankipalle, Rachakuntapalle, Velamvaripalle and Gudipadu
3	Host rocks	Phosphatic siliceous dolomitic limestone/ dolostone (Vempalle limestone). Basic dykes traversed the carbonate rock.
4	Dimensions of the deposits	1–6.5 km along strike, upto 20m in thickness.
5	Intensity of mineralisation	0.010–0.2% eU_3O_8. Average for the Tummalapalle deposit (best): 0.05% U_3O_8
6	Petrography	Generally alternate bands of light grey medium to fine grained dolomite rich carbonate and dark grey ultrafine grained collophane rich carbonate.
7	Structural features	Generally layered stromatolites; ripple marks and mud cracks present in radioactive outcrops.
8	Chemical composition	Higher SiO_2, Al_2O_3, P_2O_5, MgO and alkalies and lower CaO, compared to common limestones.
9	Uranium minerals	Mainly pitchblende, coffinite and U-Ti (-Si) complex. Th and REE contents low in the U-bearing minerals.
10	Other 'ore' minerals and ore elements	Pyrite, molybdenite, chalcopyrite, bornite, digenite, covellite. Cu (25–670 ppm), Mo (200–260 ppm)
11	Depositional environment	Intertidal and mud flat environment of sedimentation.
12	Wall rock alteration	Not noticed
13	Disequilibrium	18–20% in favour of uranium
14	Probable source of uranium	'Fertile' pre-existing Proterozoic granitoids that surround the basin from the north, west and south.

The uraniferous carbonate rock, well bedded and compact, occurs between the relatively impervious lower middle dolostone and upper shale cherty dolostone (Fig. 1.2.32).

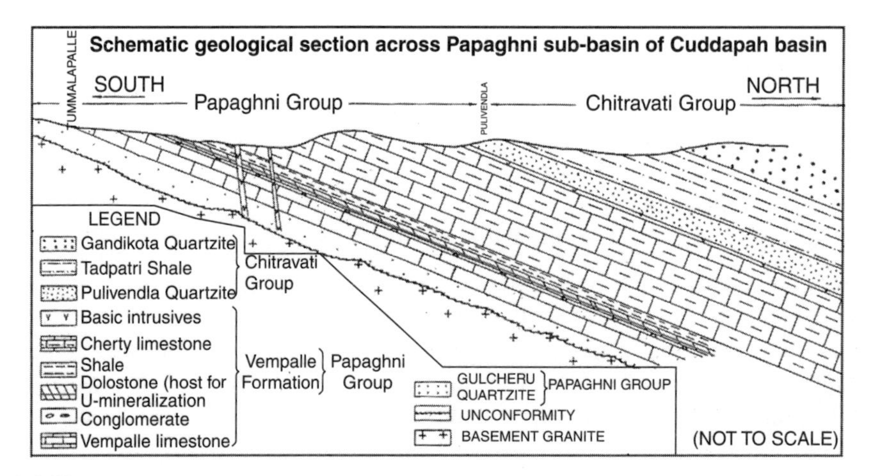

Fig. 1.2.32 Sediment (dolostone)-hosted uranium mineralisation in Papaghni Group, Cuddapah basin (after Dwivedi, 1995)

Primary sedimentary structures like mud-cracks and ripple marks are preserved. This mineralised dolomite unit, upto about 20 m thick, is intercalated with quartz-arenite, shale and mudstone. The uranium mineralisation occurs along the bedding planes, carbonate–phosphate interface, micro-stylolites, grain boundaries of clasts and within pelloids, mainly in the form of ultrafine pitchblende (0.003–0.06 mm) and disseminations in collophane-rich parts. Besides, uranium is present as coffinite, U-Ti (-Si) complex, U-Si and relatively rare uranophane. Associated sulphides as minor phases comprise pyrite, molybdenite, chalcopyrite, bornite, digenite and covellite. Thorium content in all the radioactive mineral phases is very low (0.00–0.24%). So also is the REE- content (Σ REE_2O_3 0.10–1.20%). V_2O_5 analysed to 0.04–0.8%. PbO-content in U-bearing phases vary widely (Roy and Dhanaraju, 1999).

Estimated mineral reserve is ~15,000 tonnes[5] U_3O_8 in the phosphatic dolostone unit. But the carbonate gangue, extremely fine in grain size, and part of the uranium being in the refractory phases pose a serious problem to the desirable recovery of uranium from ores. Preliminary experiments on the metal extraction showed that alkali leaching was preferable to acid leaching and the maximum recoverability of uranium was ~ 70% at 1atmosphere and 700°C (Kaul et al., 1991). Reportedly, experiments on the improvement of the recoverability are still on.

As has been mentioned at an earlier part of this discussion, the basal part of the Gulcheru Formation, represented by impersistent polymictic conglomerate, is known for the paleoplacer type mineralisation of thorium. Recently Umamaheswar et al. (2001) reported discovery of significant concentration of uranium (upto 1.44% U_3O_8) in quartzite, controlled mainly by NE–SW and E–W faults. These faults affected both the Gulcheru Quartzites and the basement granitoids. A part of the uranium occurs as disseminations also.

Uranium is present as pitchblende, uraninite and coffinite. Associated ore minerals comprise hematite (+ limonite), chalcopyrite, arsenopyrite, pyrite, bornite and galena. Contents of trace metals such as Ni (upto 1033 ppm), Mo (upto 643 ppm), Co (upto 77 ppm), Cu (upto 170 ppm), Pb (upto 1515 ppm) and V (upto 107 ppm) are tangible.

Origin of the uranium mineralisation within the rocks of Cuddapah basin is interesting, but not much debated. The fine grain size of the ore minerals, mineral composition of the gangue and low content of Th and REE in the uranium minerals, particularly uraninite and pitchblende and relatively high content of vanadium in the ores in the Vempalle Formation are indicative of low temperature deposition of the ores. These characteristics and the confinement of the mineralisation within a particular stratigraphic unit are suggestive of the mineralisation being sedimentary–diagenetic in

[5] Recent figure is about 3 times larger.

origin. Ore minerals within the Gulcheru Quartzite occur mainly along disjunctive structures with a portion occurring as disseminations within the quartzite. This stratigraphic horizon overlies a 2500–2300 Ma uranium-fertile basement (av. 24 ppm U) and the faults that largely control the mineralisation are also shared by this underlying basement, locally containing similar mineralisation. It, therefore, seems reasonable to assume that the concentration of mineralisation along the disjunctive structures is an effect of episodic reactivation of the basement faults, as suggested by Umamaheswar et al. (2001). The source of uranium in the Vempalle dolostone must have been the same or similar uranium-fertile or 'hot' granitoid rock. Since the mineralisation is post-oxyatmoversion (vide the discussion on the age of the Cuddapah Supergroup in an earlier section), the transfer of uranium (and most other metals) took place in appropriate water-soluble complexes.

Unconformity-proximal Uranium Mineralisation During the early nineties of the last century, uranium mineralisation has been discovered by the Atomic Minerals Division, Department of Atomic Energy at Lambapur–Peddagattu at the north-western fringe of the present day limits of the Cuddapah basin (Fig.1.2.31). The mineralisation occurs along the unconformity between the Srisailam Quartzite belonging to Cuddapah Supergroup and the basement gneisses and younger granites dated 2268 ± 32 Ma to 2482 ± 70 Ma (Pandey et al., 1988). The mineralisation is confined mainly to the basement granite with the overlying Srisailam Quartzite (Sinha et al., 1995). The uranium-bearing ore concentration has taken place in the form of elongate pods or lenticular veins at the intersection of the unconformity and two prominent sets of fractures that trend NNE–SSW and NW–SE within the basement. The depth persistence of the ore is low (≤ 5 m below the surface). Along with uraninite, pitchblende and uranophane, drussy quartz, kasolite [$Pb_2 (UO_2)_2 (SiO_4)_2$ $2H_2O$], galena and a little pyrite and chalcopyrite occur in the ores. This is suggestive of a kind of hydrothermal origin of these veins. The mineralisation has some resemblance to the Proterozoic unconformity-related uranium-bearing mineral deposits at the Athabasca and Thelon basins, Canada. The general consensus about the origin of these deposits is the concentration of the ore elements through pedogenesis in the basement rocks and diagenesis of the overlying sediments and subsequent mobilisation/remobilisation during tectono-thermal evolution of the region (Sarkar, 1995). Sinha et al. (1995) believe that the carbonaceous sediment rich Cumbum Formation, highly tectonised and intruded by a 1575 ± 20 Ma (Rb/Sr) granite should be a more favourable situation for unconformity related uranium mineralisation in the region.

Structurally Controlled Uranium Mineralisation along the Eastern Margin of the Cuddapah Basin Uranium mineralisation at a number of places in the metamorphic tectonites along the eastern outer margin of the Cuddapah basin has been located, the most promising of which are at Kasturigattu, Gudarukoppu and Kullur, all in the Nellore district of Andhra Pradesh.

Mineralisations at all these places occur within feldspathised biotite schists/gneisses. The ore zones are 450–600 m in length and 1.0– 8.0 m in width. U_3O_8 -content in the ores varies between 0.01%–0.53%. Uranium-bearing minerals are uraninite, pitchblende, U-Ti complex and autunite. Associated ore minerals comprise pyrite, chalcopyrite, pyrrhotite, arsenopyrite and apatite (Sharma et al., 1995).

Available information is inadequate to suggest any genetic model for these mineralisations.

Fracture-Controlled Uranium Mineralisation in Southwestern Margin of the Cuddapah Basin Uranium mineralisation in the basement rocks beyond the southwestern margin of the Cuddapah basin and controlled by fractures, have been located around Rayachoti, between Papaghni and Mandavi rivers (Fig. 1.2.31). Mineralisation is associated with those fractures that are restricted to the younger porphyritic pink granites and leucocratic granites that some workers correlated with the Closepet phase of granitic magmatism (Sharma et al., 1995). Of the several fracture- sets, the one oriented ENE–WSW is more frequent and show better grades. A mineralised fracture zone may be upto 16 km in length, 5 m in width and >50m in depth. U_3O_8 ranges from 0.012 – 0.503%. The uraniferous fracture zones show hematitisation, chloritisation, sericitisation and microclinisation, some features of hydrothermal alteration. Some of these mineralised fracture zones continue below the nearby parts of the Cuddapah sedimentary cover. The primary uranium minerals indentified in these ores are ultrafine pitchblende, coffinite and a U-Ti complex.

Occurrence of the mineralisation in fracture zones that are confined to relatively young fertile (av. U ≃ 24 ppm) granites. Hydrothermal type wall rock alteration would suggest that these were deposited from hydrothermal solution(s), related to the host granitoids. We have no radiometric dates of depositon of these ores. But if the field observation that some of these ore-bearing fracture zones continue below the Cuddapah sedimentary cover is not an exception, then it would be a proof of multiphase U-mineralisation in the Cuddapah uranium metallogenic province, as well as a probable contribution of earlier formed fracture controlled deposits to the strata bound deposits within the Cuddapah sequence. The time and mechanism of development of these fracture zones remain a moot question.

Barites in Cuddapah Basin The Cuddapah basin hosts 95% of the country's barite resources, which accounts for about 25% of the world resource of this mineral. Out of the two common types of barite mineralisation, the vein type is spread over most of the formations of Cuddapah sequence while the bedded variety is only restricted to Mangampeta (14°01′: 79°19′) in Rajampeta area. This deposit is the world's largest, having more than 65 Mt of ore reserves. The mineralised horizon occurs within the Pullampet Formation which is principally shaly, with intercalations of dolomite, sandstone and tuff. One important observation is that the black organic-rich pyritic shale is present in association, suggesting anoxic condition during deposition. A stratigraphic section of Mangampeta deposit suggests variation of oxygen levels from oxic to anoxic and again to oxic states (Fig. 1.2.33 a). There are two flat lying lensoid ore bodies separated by about 700 m of barren ground. Out of these, the northern ore body disposed as a doubly plunging synform is 1200 m long, 900 m wide, with an average thickness of 20 m. The other lensoid ore body is much smaller (300 m × 200 m) and of lesser thickness (4-10 m) (Neelkantam and Roy, 1979).

Geochemically, the Mangampeta lapilli barites record higher SiO_2 (~8–11%), R_2O_3 (~3.5%) and Fe_2O_3 (0.6–1.0%) compared to the granular variety, while BaO (59–63%) and SO_3 (31–33%) are higher in the latter. The concentration of Cu, Pb, Zn, Co, Mn and Ni are insignificant, while Ti content (up to 500 ppm) is significant in both the varieties.

The associated tuff at places records more than 1000 ppm Ti and comparably low content of other trace elements like the Ba. Sulphur isotope composition is very significant in the study of barite deposits. Enrichment of heavy isotopes of sulphur in sediment-hosted deposits, particularly in barite rosettes and concretions is attributed to sulphate-reducing bacteria, such as *Desulfovibrio desulfuricans*. These bacteria, thriving in anoxic environments, metabolise isotopically lighter sulphur, leaving the reservoir richer in the heavier sulphur isotope. Enrichment of barite in heavy suphur isotope is therefore interpreted to indicate deposition of barite in a closed basin. Limited data on S-isotopes (n = 7) (Karunakaran, 1976) show that these are much higher than the sulphur isotope content of marine evaporite suggested by Clark et al.(2004) for the Mesoproterozoic (Fig.1.2.33b).

Two principal models are available in the literature to explain the origin of massive (generally bedded) barite deposits. These are:

1. Hydrothermal or exhalative model (Lydon, et al., 1985; Poole, 1988).
2. Biogenic (Jewell and Stallard, 1991; Jewell, 2000).

The hydrothermal or the exhalative model is same as that is utilised to explain many base metal (Zn-Pb deposits), with the difference that the hydrothermal solutions for barite deposition are cooler (< 200°C) and those depositing base metals are hotter (~350°C). The biogenic model is based on coastal upwelling of nutrient-rich cool sea water. Clark et al. (2004) are in support of a biogenic model for the Mangampeta deposit because of the presence of the wide evidence of a reducing environment accompanied with the appropriate organic activities. Neelkantam and Roy, (1979), however, advocated a volcanogenic hydrothermal origin. The actual process could, however, be a combination of both the mechanisms. The origin of the vein type barite occurring at many places of the Cuddapah basin is obviously hydrothermal, the exact origin of the hydrotherms remaining obscure.

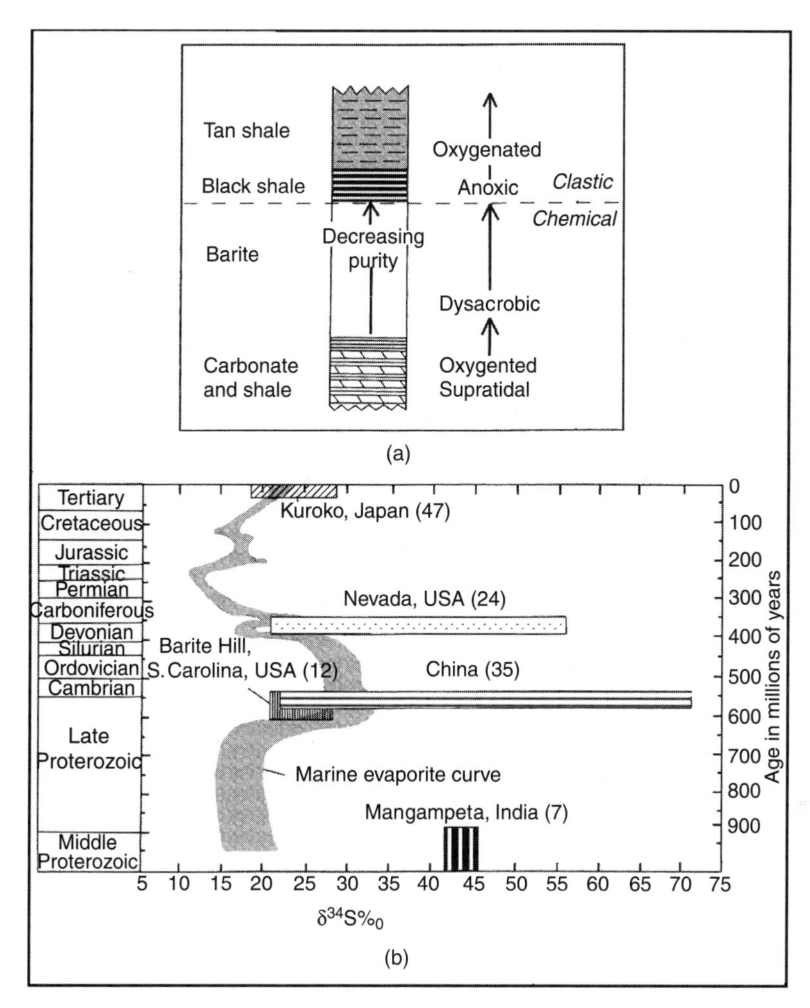

Fig. 1.2.33 Generalised section showing the lithologic association of the Mangampeta barite deposit. (b) $\delta\,^{34}S$ in Mangampeta barite vis-à-vis the same in marine evaporite during the Proterozoic (from Clark et al., 2004).

Mineralisation in the Bhima Basin

Uranium Mineralisation Uranium mineralisation in the Bhima basin, northern Karnataka, is a relatively new find. The discovery is the result of a multi-disciplinary approach in exploration, adopted by the Department of Atomic Energy, Government of India, encouraged apparently by the results of an earlier similar programme on the Cuddapah basin, the salient aspects of which have already been discussed. The guiding principle in this endeavour also must have been – 'the continental Proterozoic basin margins are potential sites of uranium mineralisation(s)' (Fig. 1.2.34).

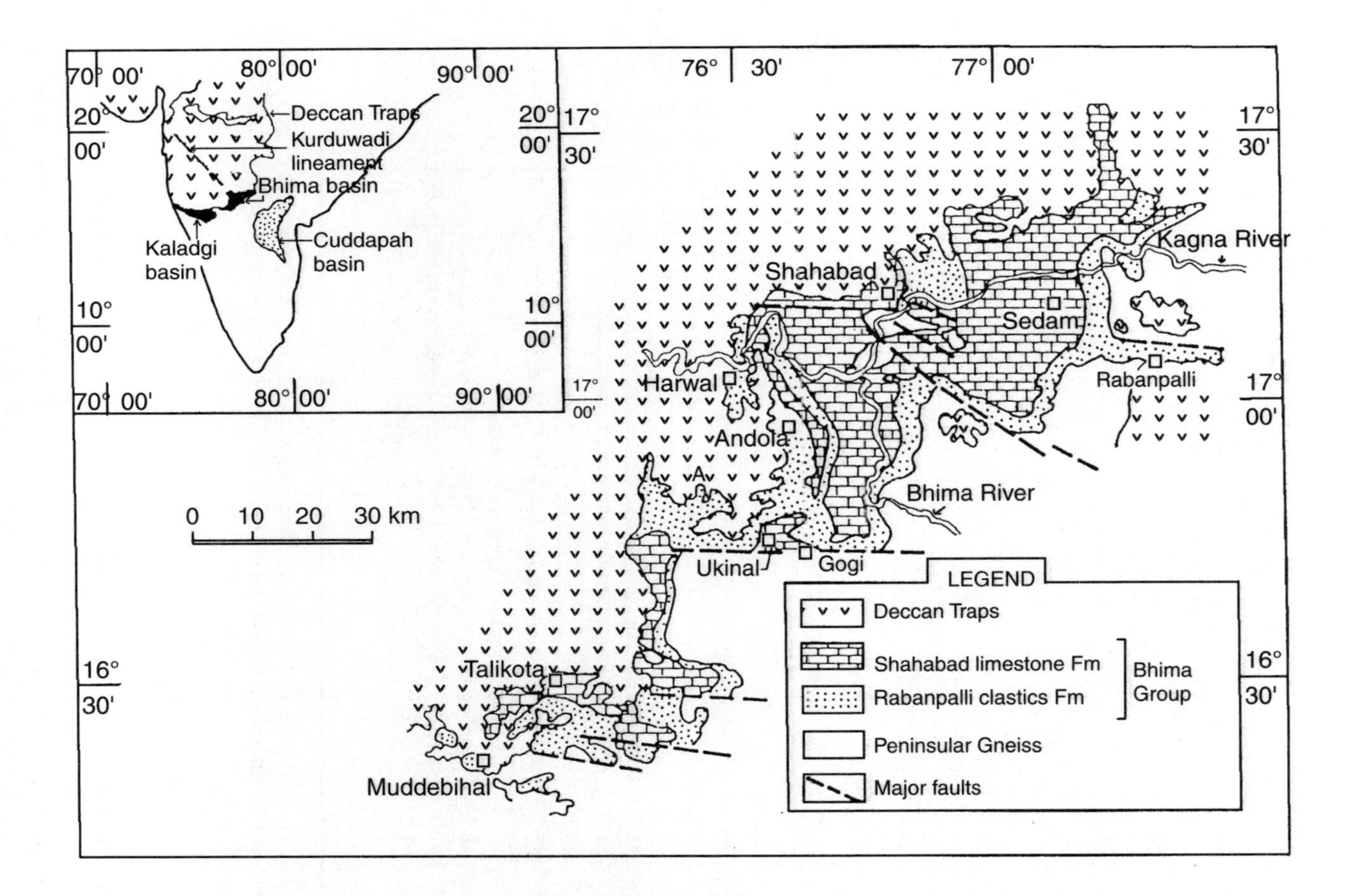

Fig. I.2.34 Geological map of Bhima basin. Ukinal–Gogoi fault zone at the basement–sediment interface is a highly prospective zone of uranium mineralisation (modified after Achar, et al., 1997)

Three distinct types of uranium mineralisation have been identified within the Bhima sequence and the adjacent basement rocks. These are: (i) mineralisation associated with altered phosphatic limestones as at Ukinal, (ii) mineralisation associated with brecciated, non-phosphatic limestones as at Gogi, and (iii) mineralisation within the basement granitoids (Achar et al., 1997).

Ukinal Mineralisation The type (i) mineralisation, best developed near Ukinal, is discontinuously traceable over a distnace of 2 km along cherty limestone–shale boundary. It also occurs at Ramthirth. It is relatively uncommon near the basement granite and along minor faults.

The rock types hosting uranium mineralisation at Ukinal are all phosphatic, some of which are manganiferous and glauconite-bearing. This rock association comprises phosphatic micritic limestone, phosphatic chert, glauconite- bearing calcitic phosphorite and siliceous phosphorite, with P_2O_5 varying in the range of 1.66 –29.5% with an average of 14.8% (Dhanaraju et al., 2002). Uranium in the radioactive carbonate–phosphate rocks of Ukinal occur mostly in collophane. A positive correlation ($r > 0.9$, $n = 140$) exists between U_3O_8 and P_2O_5 in these rocks.

Gogi Mineralisation Uranium mineralisation near Gogi is controlled by a E–W trending Gogi–Kurlagere fault (Fig. 1.2.34) in non-phosphatic brecciated, siliceous limestone as well as in the deformed basement granite, dated 2504 ± 28Ma/ Rb-Sr (Sastry et al., 1999). This fault has been traced over a distance of > 7 km. Exploratory drilling has established a medium grade potentially exploitable uranium deposit at Gogi, mainly in the Shahabad limestone for a vertical depth of ~ 200 m.

The uranium-bearing limestone at Gogi is fine grained, compact, brecciated and contains chert, organic or carbonaceous matter, illitic and smectitic clays, sulphides, limonite, with glauconite and barite present locally. The carbonate is ferroan calcite. Organic material of 'labile' type is present. Available information shows highly varying U-content in the ores (0.013–1.678% U_3O_8). V, Co, Ni, Mo, Cu, Ag and Pb are present as minor/trace metals. Uranium at Gogi ores occurs in coffinite and pitchblende, with a subordinate portion associated with organic matter, clays and anatase. A small part of uranium could exist in labile state also. Both pitchblende and coffinite are intimately associated with organic matter and sulphides and occur as veins, veinlets and fracture fillings (Dhanaraju et al., 2002). A positive correlation of lead with uranium is suggestive of a radiogenic derivation of the lead.

Mineralisation in the Basement Granitoid Rocks The immediate basement is made of granite–granodiorite rocks, characterised by the presence of cataclastic textures and deformational fractures, some of the latter being filled up by calcite and / or fluorite. These granitoids may be called 'fertile' or 'hot' with respect to uranium, since they contain several times more of the element compared to Clarke value. Achar et al. (1997) reported 10–110 ppm U in these rocks. Dhanaraju et al. (2002) reported a range of 14–52 ppm U (average 29 ppm) for 6 of the 11 samples studied and < 5 ppm for the rest.

Uranium mineralisation has been detected in these rocks in the form of lenses that may be upto 150 m in length and 4 m in thickness. The principal source of radioactivity in these lenses is coffinite, followed by pitchblende. A small part may be derived from a U-Ti (- Si) complex and what little uranium is there adsorbed on sericite and chlorite. Also, some sulphide veins containing pyrite, arsenopyrite, chalcopyrite and galena traverse the granitoids.

It is interesting to note that although the U_3O_8 contents in pitchblende from both limestone and granitoid are similar (~ 80%), the PbO in the former is lower than the PbO in the latter. Coffinites from veins in granitoids contain many orders more of all REEs, including Y, than that occurring in coffinites in veins in limestones. This pattern applies to pitchblende also, except for Y (Dhanaraju et al., 2002).

Ore Genesis Uranium mineralisation at Ukinal and Ramthirth appears sedimentary-diagenetic as much of the metal is lodged uniformly in collophane that analysed 0.1–0.2% U_3O_8. Field and textural features suggested this collophane to have been produced by phosphatisation of limestone.

In contrast, uranium mineralisation in the non-phosphatic brecciated siliceous limestone and in the basement granitoids at Gogi, occurs mainly as veins, veinlets and fracture-fillings in tectonised zones. Uranium occurs principally in the form of such discrete phases as coffinite and pitchblende. Pyrite is formed in the ore zone. Thorium is virtually absent. These features, together with the vein-mineralogy as a whole, are suggestive of a low temperature fracture controlled hydrothermal (epithermal) origin of these ores (Achar et al., 1997; Dhanaraju et al., 2002).

The chemical ages of uranium minerals pitchblende, coffinite and U-Ti (-Si) complex have been determined (Dhanaraju et al., 2002) as shown below (Table 1.2.12).

Table 1.2.12 *Chemical ages (Ma) of U-minerals in limestones-hosted and granitoid-hosted veins (Dhanaraju et al., 2002)*

U-Mineral	In limestone		In granitoids	
	Range	Average	Range	Average
Pitchblende	334–862 (*n* = 9)	553	9–798 (*n* = 14)	226
Coffinite	62–1146 (*n* = 16)	330	1–480 (*n* = 20)	126
U-Th (Si) complex			1–12 (*n* = 6)	7

The above data are of little use today, unless these can be satisfactorily explained in terms of the geological process.

The ultimate source of uranium and the associated metals are apparently the 'fertile' granitoid basement. Their accumulation into ore bodies within the basinal sediments must have been through an intermediate stage of transfer to the sediments. From the wet sediments they collected into disjunctive structures in carbonate rocks by hydraulic pumping of the diagenesis related hydrothermal fluids. Mineralisation could have taken place by the reaction of the uranyl complex(es) with reductants, dilution of the ore fluids by mixing with ground water, or any other mechanism. The origin of the uranium-bearing and sulphide-bearing veins in the basement granitoid remains less understood. They as well, are hydrothermal in origin and owe the ore-constituents to the host rocks. But the veins in the carbonate rocks of Gogi do not appear to have formed from the same fluids. A considerable time gap between the emplacement of the basement granitoids and emplacement of veins preclude their formation from hydrothermal solutions produced by an upcoming (depressuring) and cooling granitic magma body. Could this be due to 'reactivation' along pre-existing faults? There also, the exact process remains unclear. Uranium-bearing veins could form from surface water penetrating along faults and forming local hydrothermal systems, with heat provided by the geothermal gradient, or by a magmatic body emplaced during 'reactivation'.

Manganese Mineralisation in the P G Valley

A thin (20–35 cm) laterally extensive high grade manganese mineralisation, interbedded with highly siliceous micritic limestone, occurs in the Chanda Formation of the Penganga Group, exposed mostly in the northwest part of the Godavari valley basin. The grey siliceous limestone immediately below the manganese formation contains framboidal pyrite, now variously altered to hematite. The ore-bearing intervals occur at least in two stratigraphic levels and they have sharp upper and lower intervals (Mukhopadhyay et al., 1997). Bedded chert is associated with the manganese oxides. Small scale mining goes on.

Todorokite and birnessite are the major manganese oxide minerals, with manganite, braunite, bixbyite, psilomelane, pyrolusite and cryptomelane as the subordinate phases. Roy et al. (1990) interpreted the todorokite and birnessite as the primary phases, which subsequently diagenetically converted to other phases. There is little or no metamorphism in the ores and the host sediments. Mukhopadhyay et al. (1997) analysed the facies of the sediments and concluded that the manganese oxide deposition took place at the base of slope of a distally steepened deepwater ramp. The conclusion is based on the close association of these manganese deposits with a variety of mass flow deposits that comprise limestone conglomerates, calc-arenites, plane-bedded micritic limestone and detritus coarser than fine silt. These deposits are components of a major transgressive succession.

This deposit appeared unique, even enigmatic, as all other known deposits in limestone and black shale succession contained Mn-carbonate(s) (Delian et al., 1992; Okita, 1992). But a breakthrough was made by Gutzmer and Beukes (1998, 2000). They discorverd Mn-carbonates, rhodochrosite and kutnohorite below the ore-horizon and decided these to be the protore, which on oxidation later produced the ores. The Mn-carbonates in the form of ooids originated immediately below the sediment-water interface at an early stage of diagenesis. These phases developed by the reduction of Mn^{4+} in the initial oxyhydroxides by organic carbon, accompanied by carbonatisation.

The problem related to the so-called absence of Mn-carbonates in the sequence is solved. There are Mn-carbonates. But there is no unequivocal evidence to prove how did the oregenetic process evolve. Mn oxyhydroxide(s) were the primary Mn-phases which during the process of diagenesis changed to todorokite and birnessite, carbonate development being a product of diagenesis in a somewhat reducing (i.e. disaerobic/suboxic) environment. Alternatively, it was the Mn-carbonates that formed first during early diagenesis, which on supergene oxidation changed to the present state of the ore. Todorokite and birnessite may form as primary phases or secondary phases at low temperature. Solution, therefore, possibly lies in finding out unequivocal replacement features in micro-(microscopic) to macro (field-exposure) scales, observed at numerous places. Gutzmer and Beukes (2000) consider the friable and highly porous nature of the ores to be a convincing feature of supergene oxidation of a Mn-carbonate rock into a Mn-oxide orebody. The logic has its weight, but may not be the sole arbiter of the problem.

None of the workers support a hydrothermal or volcanisedimentary marine origin for this deposit. Gutzmar and Beukes also supported an initial Mn-precipitation on upwelling of Mn-bearing solution from low oxygen zone onto the carbonate platform during a high sea-level stand, or transgression.

Diamonds

India, as it appears from ancient literature, was aware of the diamond as a very hard and useful mineral matter for a few thousand years, but as an outstanding gemstone, for less than a millennium. It became a gemstone when Indians, and not much later, other people learnt to process (to shape and polish) some natural diamonds, bringing out its exquisite intrinsic beauty. It was the attraction of this gem-diamond and some other resources that brought many travellers, traders, and, of course invaders over to this country ever since their stories spread out into the outside world. There were some travellers though, who came purely for intellectual and spiritual pursuits. But that is a different story. However, the scene drastically transformed when large deposits of both primary and secondary diamonds were discovered in other parts of the world such as South Africa, Brazil, Canada and Australia during the nineteenth-twentieth centuries. India's position as a diamond producer became negligible, although it remained in the forefront (>90%) of the diamond-processing industry. Needless to mention, all diamonds do not make the grade of a gemstone. In fact most do not. The non-gem variety is used in the industry because of its high hardness. Diamond being what it is, has been fairly extensively studied on various aspects. It is beyond the scope of the present work to review the results of these studies. However, it will not be out of place to briefly discuss diamond genesis as it is understood today, in the interest of both Earth science and the mining industry.

Origin of Diamond – a Snapshot Diamond is found associated with (i) various igneous rocks (kimberlite, lamproite, lamprophyre, komatiite), (ii) metamorphic rocks (felsic gneisses, calc-silicate rocks, garnet-bearing pyroxenites and peridotite and eclogite), (iii) sediments (alluvia, colluvia, sandstone-conglomerates). Before the discovery of diamond-bearing igneous rock bodies, the sediments supplied all the diamonds. Amongst the igneous rock associated ores, the kimberlitic variety is the most common, followed by the lamproitic. There are now many examples of eclogitic diamonds that have formed in the mantle over a long period of time. Eclogitic diamonds (E-diamonds) have a wide range of ^{13}C (<3.5 to >4 per mil) in contrast to a mantle value of -5 per mil for harzburgitic peridotite diamond (P-diamonds). There is an apparent riddle about the temporal and genetic relationship between diamonds and their host kimberlites and lamproites. Normally economic minerals and their host igneous rocks are nearly cogenetic, or the former develop from the hydrothermal solutions derived from /mainly contributed by the cooling + / depressuring magma. In case of diamonds, economic accumulation takes place in kimberlite and lamproite, particularly when such magmas can access and incorporate diamonds from the lower lithospheric (sub-lithospheric) zone (Fig.1.2.35) enroute to the surface. Information on the timing of diamond generation can be obtained by suitably dating the minerals included within diamonds, coexisting with diamonds or diamond-bearing xenoliths, preferably the first one. A conclusion on the timing of diamond formation states that the diamond generation is episodic and a kimberlite may have multiple paragenetic diamond population that may be of widely varying age and thermal history. For example, in the best studied Kapvaal Province, South Africa, diamonds have been dated from ~3.57 Ga to 1.0 Ga (Gurney et al. 2005). From geothermobarometric studies it is now established that the cratonic areas have cool geothermal gradients that intersect the diamond stability field at about 150 km below the surface in the temperature regime of 950°–1000 °C.

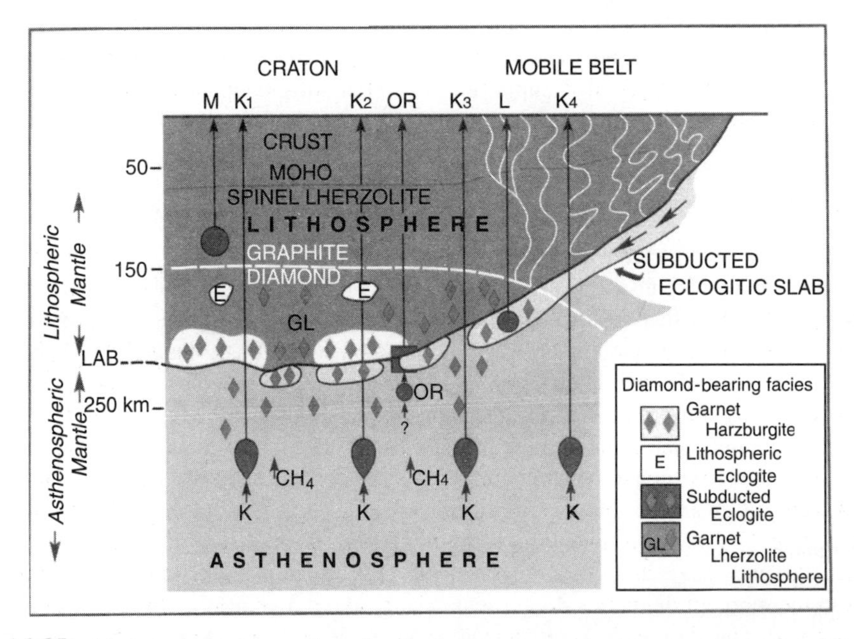

Fig. 1.2.35 Hypothetical cross-section of an Archean craton and adjacent cratonised mobile belt (after Mitchell, 2006)

The origin and distribution of diamond is not a story in itself alone. Its distribution through time, space and composition, particularly mineral and trace element contents, have been of immense help to constrain growth, stabilisation and modification of Archean cratons. Mantle keels and early continental nuclei might have been produced by intense Mid-Archean depletion, accompanied by high degree melting that produced komatiite, followed by metasomatism and early harzburgitic diamond formation even at 3.57 Ga by reaction with subduction related C–O–H–S fluids. Although not necessarily isochronous, characteristic sub-calcic harzburgitic G10 garnets and P-type diamonds have been found, almost globally, in Archean craton roots. Their common presence in economic kimberlites, have made them a valuable tool in diamond exploration. Subducted oceanic lithospheric contributions stabilised the craton and contributed eclogitic diamonds that became an addition to the already existing harzburgite diamond of the Late Archean. Early Proterozoic tectonothermal events modified the composition of the cratonic keel by introducing basaltic material and ultimately adding new generations of lherzolitic and eclogitic diamond to an already existing Archean diamond population. Available Re-Os and Sm-Nd age data for the inclusion in the Udachnaya (Siberian craton) diamonds show that both harzburgitic and lherzolitic diamonds formed at 3.3 Ga and 2.0 Ga events. Note that detrital diamond grains were obtained from the 2.89–2.82 Ga West Rand Group sediments, South Africa. Subduction is emphasized as the most viable mechanism to introduce diamond producing fluids in the Early Precambrian time (Gurney et al., 2005 and the references therein). However, the development of a deep cool refractory cratonic root in the Archean seems to be a necessary prerequisite to the growth of an appreciable inventory of macro-diamonds in that root and in the later underplating events. Haggerty (2000), however, preferred vertical 'foundering' of small thin layers of embryonic komatiitic crust and onward evolution by a plume-like mechanism to a model of subduction, island arc magmatism etc.

Recently diamonds have been recovered from volcaniclastic rocks of komatiitic composition from French Guinea (Kapdevila et al. 1999). Though an unconventional ore type, it may be given the attention it deserves.

Diamond in the South Indian Shield Diamonds in India, South India in particular, may be divided into the following, depending on modes of occurrence:

1. Diamonds in modern sediments
2. Diamonds in ancient sediments
3. Diamonds in kimberlite and lamproite intrusions.

Diamonds in modern sediments in South India refer to those in sands and gravels in river terraces and even beach placers. The two river systems that are particularly well-known in this context are those of Krishna and Pennar. In case of the Krishna river system the most productive area in the past have been the gravels on either side of the river Krishna below the Srisailam gorge and continues up to Vijaywada in Andhra Pradesh. Not many details about the gravels are available although they are known to be large repositories of diamond in the country through the ages, producing some of the historical diamonds of this part of the world, including the Kohinoor, Great Moghul, Pitt etc. These diamonds are believed to have been released on weathering of diamondiferous kimberlites and lamproites in the region. In the recent years, a number of lamproite intrusives have been discovered in the Krishna basin (Reddy et al., 2003). That makes the diamond source rather handy. The gravel beds exposed in the terraces of the Pennar river flowing in the Ananthapur district, Andhra Pradesh, are known for the occurrence of diamonds from very olden days. The more prospective area delineated is about 150 sq km.

Micro-diamond (63–73 microns) was first reported from the beach placers of Krishna-Godavari delta in the East Coast of India (Subrahmanium et al. 2005). Recently both micro- and macro-diamonds have been reported from the Kanyakumari coast at the southern tip of India. The latter diamonds are generally colourless, but may have pink and yellow tints. The larger (macro) ones measured up to 2.05 mm × 1.50 mm. The content is also high, being 0.335 carats from < 10 kg sands (Rau, 2006). The origin and transport of these diamonds is still a matter of speculation.

At Banganapalle, Andhra Pradesh, a sandstone horizon occurs at the base of the Kurnool Group within the Cuddapah Supergroup. It is a 7 m thick unit in which diamonds appear to be confined to lower sections. The primary source of diamond is apparently the diamondiferous plugs in and around the Cuddapah basin.

A report by the Geological Survey of India (Neelkantam, 2000) shows that by the close of the last century, the East Dharwar craton had 54 discovered kimberlite pipes and dykes (this figure must have gone up during the last one decade or so). Among the discovered kimberlite bodies 15 were diamondiferous, 13 non-diamondiferous and the rest were yet to be tested. There are no active mine on any of them till the date. The following table (Table 1.2.13) gives an idea of the diamond recovery from the kimberlite occurrences in the Wajrakarur kimberlite field.

Table 1.2.13 *Pipe-wise record of diamond recovery from Wajrakarur kimberlite field (Neelakantam, 2000)*

Pipe	Area (hectares)	Material processed (tonnes)	Diamond recovery (No/wt in ct)	Diamond incidence (ct /100t)
A. Wajrakarur-Lattavaram cluster				
P/01	19.0	13780	384/80.88	0.56
P/02	2.1	788	Nil	Nil
P/03	0.48	907	12/2.55	0.28
P/04	3.45	2400	30/6.02	0.25
P/05	1.0	482	Nil	Nil
P/06	8.0	8471	1142/119.52	0.67
P/07	1.8	1515	2838/543.26	7.89
Overburden		1221	2838/543.26	44.49
P/08	0.50	1773	40/5.95	0.33
P/09	0.07	229	4/1.17	0.55
P/10	63.00	1434	48/14.71	1.00
P/11	1.1	275	2/2.15	0.78
P/12	0.48	69	Nil	Nil
B. Chigicherla cluster				
CC-1	5.8	436	5/1-31	0.30
CC-2	3.5	302	2/1.05	0.35
CC-4	1.3	88	15/2.25	3.55
CC-5	1.5	175	6/1-38	1.28
C. Kalyandurg cluster				
KL-1	8.7	280	0-08	

Note: From 1053 tonnes of processed material of pipe Nos. MK/2, 3, 4, 7, 8, 9 and KK-1, 2, 3, 4 and NK 3 and 10 no diamonds were recovered. It needs to be remembered that diamond is normally a minute constituent of its host rocks and its distribution is generally non-uniform in any sampling for the diamond-content.

Bauxite Deposits

Bauxite deposits are special types of laterite in which aluminium dominates over iron. Valeton (1981) defined bauxites as economic ores of aluminium, which contain not less than 45% Al_2O_3 and not more than 20% Fe_2O_3 and 3–5% combined SiO_2. As the composition of the ores is never uniform throughout such a deposit, these values may be taken as average for a volume of ore under reference. Bauxite and common laterites are essentially complex mixtures of hydrated alumina and Fe-minerals. The principal Al-mineral in bauxite is gibbsite (Al_2O_3. $3H_2O$), which is easily amenable to leaching by hot NaOH solution (Bayer Process). Boehmite (γ-Al_2O_3.H_2O) comes next in abundance, though gibbsite >> boehmite is the overall picture. Least abundant is diaspore (β-Al_2O_3. H_2O) which, however, may occur in a non-bauxitic environment. Kaolinitic clay may be present in small proportions. This is generally the source of reactive silica, dreaded in industry. Iron occurs as hematite (α-Fe_2O_3), maghemite (γ-Fe_2O_3), ferrihydrite (Fe_2O_3. nH_2O), goethite (α-FeO.OH) and lepidocrocite (γ-FeO·OH), amongst which hematite and goethite are more common. Goethite may be Al-bearing but is unwanted in metallurgy. TiO_2 is generally present upto a few percents. The Ti-mineral phase is commonly referred to as anatase (TiO_2), but for some deposits in electron-probe micro-analyses, Ti has been found to occur in a phase(s) containing Al, Ti and Fe in widely varying proportions, as in the ores from Bagru, Jharkhand (unpublished work, Sarkar and co-workers).

Distribution through Space and Time Five major bauxite provinces are delineated throughout the world. They are:

1. South American Province (Guinea and Central Brazilian shield)
2. Indian Province (East coast; plateau regions of eastern and central India; Western Ghats; numerous other occurrences)
3. West African Province (mainly Guinea, with some deposits in Burkina Faso, Sierra Leone, Ghana, Mali)
4. South-East Asian Province (mainly Vietnam, with other occurrences in Laos and Cambodia)
5. Pre-coastal Province of north and southwest Australia.

Bauxites of Archean age (3100 Ma) are reported from Swaziland, South Africa. The first peak of bauxitisation coincides with the Lower Carboniferous, while a much larger one occured from Late Tertiary to the Present (Bardossy, 1981). Tardy and Roquin (1998) reviewed this aspect of bauxitisation and based on the findings with respect to such deposits in India, Australia and to a lesser extent in Africa, arrived at the conclusion that Eocene was the most favoured period, having a warm and humid climate, conducive to the development of bauxite over a wide zone stretching from the tropics to relatively high latitudes. Cool climate and sea level fall during the middle Oligocene slowed down the rates of weathering. There was a reversal of the climate, as a result of which there was another major phase of bauxitisation during Miocene in the subtropical zones of India, Australia and equatorial Amazonia. In some regions the process is still operative.

Typology of Bauxite Two major types of bauxite are generally distinguished in literature and in the mining industry. These are:

1. Lateritic bauxite
2. Karst bauxite.

Lateritic Bauxite Lateritic bauxite develops in the superficial parts of Al-silicates rich rocks such as granites, granite gneisses, basalts, arkoses, alkaline rocks etc and represents ~ 85% of the World's reserve of ~ 35 Bt (Brimhall and Lewis, 1992; Bardossy, 1981). In most cases it is autochthonous, i.e.

formed virtually in situ by the transformation of the bed rock or protolith to soil. Occasionally, there may be long distance transport of the weathered material, in which case it is called allochthonous.

Karst Bauxite Karst bauxite, as the name suggests, develops in the upper part of carbonate rocks (~ 14%), particularly if the rocks are 'impure', particularly containing clay.

Generalised Weathering Profile Containing Bauxite Deposits In a typical weathering profile containing lateritic bauxite deposit, the protolith at the bottom is overlain by a partially altered (saprolite) zone in which relics of original minerals and rock-textures may be present together with the newly formed minerals, such as kaolinite, halloysite and Fe-hydroxides. Halloysite and goethite are generally more common in the lower part of the saprolite zone, whereas kaolinite and hematite are more common in the upper part. The thickness of the saprolite zone varies from 1–2 m to 100 m or more, depending on several factors such as the age of the profile, characteristics (composition, textures and structures) of the parent rock, annual rainfall and the drainage. The saprolite zone is overlain by the zone of bauxitisation, where kaolin breaks down to gibbsite and silica and the latter moves out of the system, controlled by the pH and circulation of water.

$$\underset{\text{(Kaolinite)}}{Al_2Si_2O_5(OH)_4} + 5H_2O \rightarrow \underset{\text{(Gibbsite)}}{Al_2O_3.3H_2O} + 2H_4SiO_{4\ aq}$$

This stage will, normally, be preceded by an incongruent weathering reaction:

$$\underset{\text{(K-feldspar)}}{2KAlSi_3O_8} + 2H^+_{aq} + 7H_2O \rightarrow \underset{\text{(Kaolinite)}}{Al_2Si_2O_5\,(OH)_4} + 2K^+_{aq} + 4H_4SiO_{4\ aq}$$

Plagioclase feldspar, and even muscovite mica, may be involved in such a reaction, if present in the protolith.

Bauxite is commonly overlain by a duricrust which is an indurated concretionary rock containing more Fe-oxides than alumina and more hematite than gibbsite. Erosion has removed duricrust in some deposits, and in such a case the bauxite zone may be covered by a clay-rich horizon of controversial origin. The deposition of gibbsite and the transfer of SiO_2 to solution is controlled by the pH (4.5–9) of the system. A good drainage is necessary to remove the SiO_2 in solution (H_4SiO_4). A very supportive drainage system is so important that it may help primary silicates, for example feldspar, to skip the intermediate stage of kaolinisation and be directly converted to gibbsite. The gibbsitic bauxite profile without kaolinite is rather rare and may develop in very humid climates with good rain fall and an excellent drainage system, on steep slopes on the top of interfluves. The resultant ore is generally massive (Tardy, 1997).

The possible role of microbiological activity in bauxitisation has apparently not received due attention from investigators.

Further Classifications of Lateritic Bauxites Bardossy and Aleva (1990) divided bauxites into six categories based on the variation in mineral composition and the nature and sequence of the horizons. The first three are lateritic bauxites *sensu stricto*. These are:

1. The zone characterised by the dominance of gibbsite and goethite and a smaller proportion of hematite and kaolinite and low amounts of anatase. Boehmite-content is very low (< 3%).
2. The next variety is still dominated by gibbsite, but may have boehmite content as high as 30%. Iron is present as hematite, rather than goethite.
3. The third type bauxite layers are buried under a kaolinitic cover, the exact mechanism of its formation is still poorly understood.

Tardy (1997) proposed a three fold classification of lateritic bauxite profiles into the following:

1. Orthobauxites
2. Metabauxites
3. Cryptobauxites

'Ortho' means normal or typical. So orthobauxites are typical bauxites. This division corresponds with type-1 of Bardossy and Aleva, briefly discussed above.

The nature of the weathering profile in metabauxites is virtually the same as that of orthobauxites, with the exception that it has higher Al/Fe ratios. Boehmite-content is generally much higher than in the orthobauxites. Metabauxites are much less common than the orthobauxites. Presently it is believed to form by dehydration of bauxitic material in a less humid climate. The contention is confirmed by the co-development of hematite from goethite.

$$\underset{\text{(Gibbsite)}}{Al_2O_3.3H_2O} \rightarrow \underset{\text{(Boehmite)}}{Al_2O_3.H_2O} + 2H_2O$$

$$\underset{\text{(Goethite)}}{2FeOOH} \rightarrow \underset{\text{(Hematite)}}{Fe_2O_3} + H_2O$$

Boehmite, if developed, occurs preferentially at the top of the duricrust, usually in the pisolitic horizon, whereas the goethite dehydration occurs towards the base, forming a massive hematitic Fe-duricrust. This distribution may be due to the reversal of Al and Fe solubilities that occurs in dehydrating environments (Tardy, 1997). This type corresponds to the variety-2 of Bardossy and Aleva.

Cryptobauxites, according to this classification are those bauxites, which are concealed by a thick clay-rich cover. These have been mostly reported from Amazonia.

However, this classification remains silent about the modification of bauxites (orthobauxites) otherwise. Development of kaolinitic veins in bauxites, though usually not prolific, is none the less common. Goethetitisation accompanies the development of kaolin. This observation confirms the fact that the gibbsite-kaolinite reaction is reversible. These developments take place when ground water containing dissolved silica (H_4SiO_4) percolates along fractures in bauxite.

Mc Farlane (1986) proposed another classification, which is not without merit. He divided laterite into Groundwater laterite and Pedogenic laterite. The former is associated with planation surfaces, and other factors being same, is controlled by the groundwater table. The Pedogenic laterite may form on any freely draining low-relief surface at high or low elevation.

Favourable Conditions for Laterite and Bauxite Formation The favourable conditions for laterite and bauxite development are the following.

1. Climate: Tropical monsoon climate with a mean temperature of 23°–26 °C and an annual rainfall of 1200–4000 mm.
2. Drainage: Even in a favourable climatic condition bauxite can form if the drainage through the rock is good. This demands a high permeability of the parent rock. Bauxite formation takes place above the ground water table, or in the zone of its annual fluctuation.
3. Geomorphology: Most laterite–bauxite deposits form as blankets covering the ancient planation surfaces, dissected by valleys. In cases the flat tops may be inherited rather than derived, as exemplified by the Deccan Traps in India. The plateau surface may be gently undulating and the margins sloping at low angles. Bauxite-bearing plateaus are commonly remnants of ancient large scale planation surfaces with an overall dip (surface inclination) 1°–5°. This suggests that, considered on a regional scale (tens or hundreds of kilometers),

the same paleosurface may occur at different altitudes. Of course Freyssinet et al (2005) suggest that in more humid conditions flat lateritic paleosurfaces may be transformed into dome-shaped hills with gentle slopes. This feature is not uncommon in the Indian scenario.

4. Forest cover: A tropical (to Savanna) forest cover generally plays a positive role by decomposing the minerals and mediating the pH of the ground water, increasing the penetration of the rainwater into the rock. Forest cover may help in formation of bauxite on slopes of 20°–30°.
5. Parent rock: Laterites may form from all types of rocks under suitable conditions, but in case of bauxite formation the protolith must have a substantial content of aluminium. Feldspar-bearing igneous, metamorphic and sedimentary (that may also only contain clay), such as granite, basalt, alkaline rocks, granite gneiss, arkose etc are eminently suitable for acting as the parent rocks of bauxites. Schellman (1994) observed that laterite is prevalent over bauxite in acidic rocks in spite of the latter having higher Al/Fe ratios However, the explanations of his observation are not shared by many. On the other hand, it is generally accepted that the physico-chemical conditions being suitable, a rock with a higher Al/Si ratio is a better candidate for being a bauxite protolith. That apparently gives the basic rocks an edge over the so-called acidic rocks. But the latter are also close contenders.
6. Altitude: Laterites and bauxites may form at any altitude between about 2500 m and sea level. Above this altitude the temperature may be cooler, retarding the weathering process. Moderately upper altitudes and well-disected planation surfaces are generally better sites for bauxitisation, because of the better drainage system characterising the latter.
7. Tectonic conditions: Most lateritic bauxitic deposits formed on tectonically quiet continental platforms, whereas karst bauxites develop mainly in orogenic belts. To be precise, most karst bauxites were formed during relatively quiet periods preceding the main phase of orogenesis (Bardossy, 1981).

On the continental platform areas vertical tectonic movements of epeirogenic character expectedly influence bauxite formation by bringing the bauxite area to optimum altitude.

The High and Low Level Bauxites Mention of the altitudinal attribute of the laterite (-bauxite) deposits is common in geological literature, without properly appreciating the geological significance in many cases. However, Bardossy (1983) provided an illuminating discussion on the consequence of topographic variation on bauxitisation. His conclusion that freely draining elevated situations, proximate to, or within the zone of fluctuation of the groundwater table, are favourable for the better development of bauxite, is easy to appreciate. Another observation, particularly based on the results of paleomagnetic investigations of laterite (-bauxite) deposits of peninsular India, says that the high altitude laterites (-bauxites) have undergone a protracted magnetic history during Late Cretaceous and Early Tertiary, while the low altitude laterites appear to be Middle-Late Tertiary in age (Schmidt et al., 1983)

Bauxite Mineralisation in India India is fairly rich in the reserve of Al-ores. The presently estimated reserve is > 3 billion tonnes and this puts India within the first five bauxite-bearing countries of the world. Indian bauxite deposits are confined to Peninsular India, barring some isolated and small deposits in Jammu in the state of Jammu and Kashmir (Fig. 1.2.36). The important bauxite deposits in India occur in the following regions:

1. The Eastern Ghats, extending from Orissa through Andhra Pradesh upto northern Tamil Nadu
2. Plateaus bordering Jharkhand and Madhya Pradesh

3. Maikala Range of Madhya Pradesh, i.e. the Amarkantak deposits
4. The Western Ghats.

(2) and (3) are often clubbed together into the Central Indian bauxite belt.

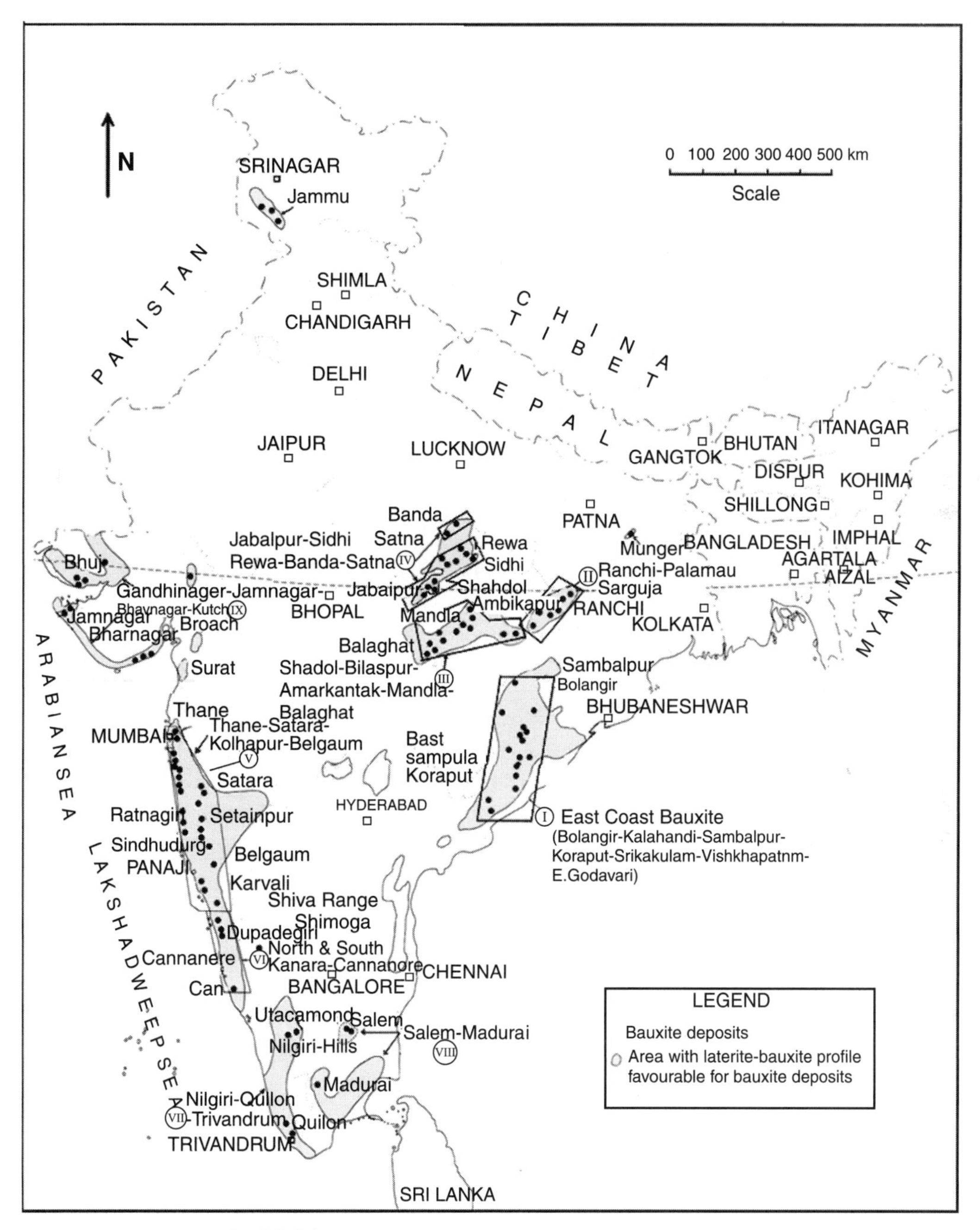

Fig. 1.2.36 Location of bauxite deposits in India (after GSI)

Besides, smaller and isolated deposits occur at a number of places including:

1. Boknur–Navge plateau, Belgaum district and Paduvari plateau, Kanara
2. Shevaroy hills, Kolli and Kodaikanal in Nilgiri-Palni hills in Tamil Nadu
3. Kandangad–Kumbla region, Cannanore district, Kerala; reported also from Trivandrum, Quilon and Allepy districts
4. Vindhyan plateau, particularly one that borders Uttar Pradesh and Madhya Pradesh
5. Riasi tehsil of Jammu
6. Kharagpur hills in Mongher district, Bihar.

We will briefly discuss the bauxite mineralisation in the above regions in the context of local geology, physical environment and to-date knowledge on the subject outlined above. However, these will be taken up in the respective chapters according to the distribution of the deposits.

Bauxite Deposits in Karnataka Bauxite occurs at several places in Karnataka, but the most important of such occurrences are located at Belgaum district, particularly at the Boknur-Navge plateau, and the other is at the Paduvari plateau in South Kanara.

The Boknur-Navge plateau is L-shaped and the laterite-bauxite mineralisation is located at the N–S trending arm (Fig. 1.2.37). The plateau is ~1.5 km long and has a maximum elevation of 1000 m above the Mean Sea Level (MSL). The laterite profile at Boknur–Navge plateau is as follows (Krishna Rao et al., 1989):

1. Bauxite-bearing laterite zone (~10 m thick)
2. Lithomergic clay zone (~2m thick)
3. Altered basalt zone
4. Unaltered basalt.

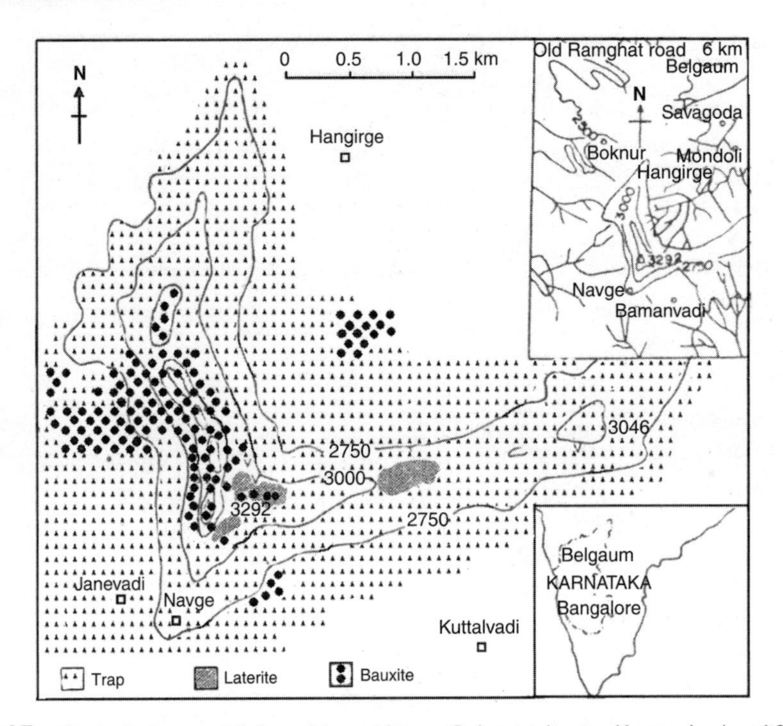

Fig. 1.2.37 Geological map of Boknur-Navge plateau, Belgaum district, Karnataka (modified after Krishna Rao et al., 1989)

Within the bauxite-bearing laterite zone, bauxite and laterite occur as distinct lithounits, sub-horizontal in disposition and lensoid-tabular in shape. In texture bauxite varies from densely pisolitic to almost structureless. Mineralogically the bauxitic ore consisted mostly of gibbsite, with anatase, boehmite, kaolinite and goethite in minor quantities.

Balasubramanium et al. (1987) generalising on Belgaum laterite-bauxite profile, put the height of the bauxite horizon at 900–1100 m (MSL). The bauxitic ores according to them occur as irregular, intermixed pockets within the laterite. The altered zone is 5–15 m thick.

Rao et al. (1989) mention about the later local modification of the bauxite ores by (i) silication and ferrification at the upper horizons, (ii) silication at the intermediate horizon and minor desilication at the lower horizons. These changes are recorded in the modification in texture, mineralogy and chemistry of the ores.

Balasubramanium et al. (op.cit.) reported a generalised chemical variation in the lateritic profiles of Belgaum (Table 1.2.14).

Table 1.2.14 *Chemical variation in lateritic profile, Belgaum (after Balsubramanium et al., 1987).*

Lateritic profile	SiO_2	Al_2O_3	Fe_2O_3	TiO_2	Cu	Pb	Zn
Bauxite	2.68	57.93	5.04	5.81	30	39	13
Lithomerge	42.10	36.22	3.77	5.10	417	70	316
Basalt	49.09	14.01	14.04	2.48	105	8	90

Note: Cu, Pb and Zn are in ppm and the rest are in wt%.

The next important bauxite mineralisation in Karnataka is at the Paduvari plateau in south Kanara. Here the geological background is different from that of the Belgaum region. Here the protolith is granitic gneiss. Expectedly, bauxitic horizon in the lateritic profile is dominated by gibbsite with kaolinite, compared with the goethite hematite in the duricrust and the lithomerge. Gibbsite is believed to have formed by desilication of kaolinite and direct breakdown of feldspar in the protolith. As compared to the basalt-derived bauxite in north Kanara, the Paduvaram deposit is characterised by a relative enrichment in Fe (Khandali and Devaraju, 1987).

Bauxite Deposits in Kerala Bauxite mineralisation in Kerala is no doubt not great. However, the estimated reserve is of the order of 13.20 million tonnes (GSI, 1994). Deposits are reported from Trivandrum, Quilon, Alleppey and Cannanore districts of the state. There are two types of bauxite deposits in Kerala:

1. In-situ bauxite derived from the Precambrian metamorphic and igneous rocks and intimately associated with laterite
2. Detrital and sedimentary bauxites
   Occur as disseminated pebbles, cobbles and boulders in pisolitic laterite in depressions and head water courses of streams.

The first type is common. Rao (1981) suggests lateritisation of bauxite in type (1) through replacement and by fracture filling. There is bauxitic clay also, which may represent partial silication of bauxite, or partial desilication of clay. Kerala bauxites occur below 220 m MSL.

Bauxite Mineralisation in Tamil Nadu The hills capped by laterite (-bauxite) in Tamil Nadu may be referred to theWest coast and East coasts of India. The Nilgiri and Palni hills not far from the West coast have lateritic capping at the altitudes of 2000–2300 m. The other important occurrences at the Shevaroy hills (~1550 m) and the Kollaimalai hills (~1250 m) are commonly thought to be southern extension of the Eastern Ghats. In the Eastern Ghat region the terrain slopes towards the East coast, where lateritic duricrust is known to have developed on Gondwana and Tertiary sediments at altitudes of 9.5–18.5 m above MSL.

Geomorphology of the hill-tops in all the above four hill ranges capped by laterite (-bauxite) are similar. The tops are plateaus with gently rolling topography. These are intersected by U-shaped valleys with meandering streams. The edges of plateaus are steep surfaces that end up at the surrounding plains. Subramanian and Mani (1981) suggest that all the above lateritic hilltops could be part of an extensive peneplane, separated by differential erosion. Altitudinal difference amongst them could be explained even by a small surface gradient, as these sites are located at considerable distances apart.

Lateritic covers have developed on the Mio-Pliocene sandstones along the East Coast at altitudes of 9–18 m above MSL. They occur as discontinuous patches only.

The latertic profile in the high-level occurrences are normal, in which a ferruginous duricrust is followed downward by a bauxite which gives way to kaolinite below and the kaolinite zone is underlain by the altered protoliths and ultimately the protoliths (charnockites, leptinites, granitic gneisses). The major mineral in the high level bauxites is gibbsite, with goethite, hematite, kaolinite and anatase occurring as subordinate/minor phases. The low level laterites are poor in alumina, and iron oxide-hydroxides compared to the high level ones. No gibbsite was detected in these rocks and all of the alumina is distributed in kaolinite and montmorillonite (Subramanian and Mani, 1981). This can be explained by better drainage in elevated situations and a longer period over which the high level plateaus have been subjected to weathering. The high level bauxites under reference have been suggested to be Late Cretaceous–Early Tertiary in age (Valeton, 1972; Subramanian and Mani, 1981; Schmidt et al., 1983). Laterite cappings on Mio–Pliocene sediments along some stretches in the East and West Coasts suggest that their developments are post Mio–Pliocene in time.

The bauxite reserve in Tamil Nadu has been estimated to be ~23 million tonnes (GSI Report, 1994) with Al_2O_3 in the range of 30–60%. Many of these deposits have been under exploitation for quite some time.

2

Central India

2.1 Geology and Crustal Evolution

Introduction

The central segment of the Indian Precambrian Shield comprises two distinct crustal provinces, viz, the Northern (Bundelkhand) Crustal Province and the Southern (Bastar) Crustal Province (Fig. 2.1.1). The Northern Crustal Province again is bifurcated into, the Bundelkhand cratonic area and the ENE–WSW trending wide zone of accretion to its south, known as the Central Indian Tectonic Zone (CITZ).

The Bastar Crustal Province is characterised by an Archean cratonic nucleus represented by widely dispersed older supracrustals (the Sukma Group and its equivalents), regionally deformed and metamorphosed with a TTG crust (> 3.0 Ga) and the younger (Neo–Archean to Meso–Proterozoic) supracrustals occurring as well defined N–S trending volcanisedimentary belts. Younger granitic bodies invaded both the older and younger supracrustals. Till date, the oldest *in situ* rock-age (3.58 Ga) from India has been obtained from the Bastar craton, surprisingly from a typical granite.

A semicircular granite–gneiss massif represents the Archean cratonic nucleus of the Bundelkhand Crustal Province (> 3.0 Ga) with numerous older supracrustal enclaves. The southern and southeastern boundary of Bundelkhand massif is overlain by Paleo– to Meso–Proterozoic Bijawar Group, dominantly composed of metasediments and associated volcanics. There is another metasedimentary package overlying the Bundelkhand massif in its northern part known as the Gwalior Group, which is considered time-equivalent of the Bijawars. Indo–Gangetic alluvial cover marks the northern limit of Bundelkhand massif. The southern, southeastern and western parts of Bundelkhand craton are unconformably overlain by the Vindhyan sediments, causing wide separation at the present exposure level from the Precambrians (BGC–Aravalli–Delhi) of Western India as well as the Mahakoshal belt of the CITZ lying to its south.

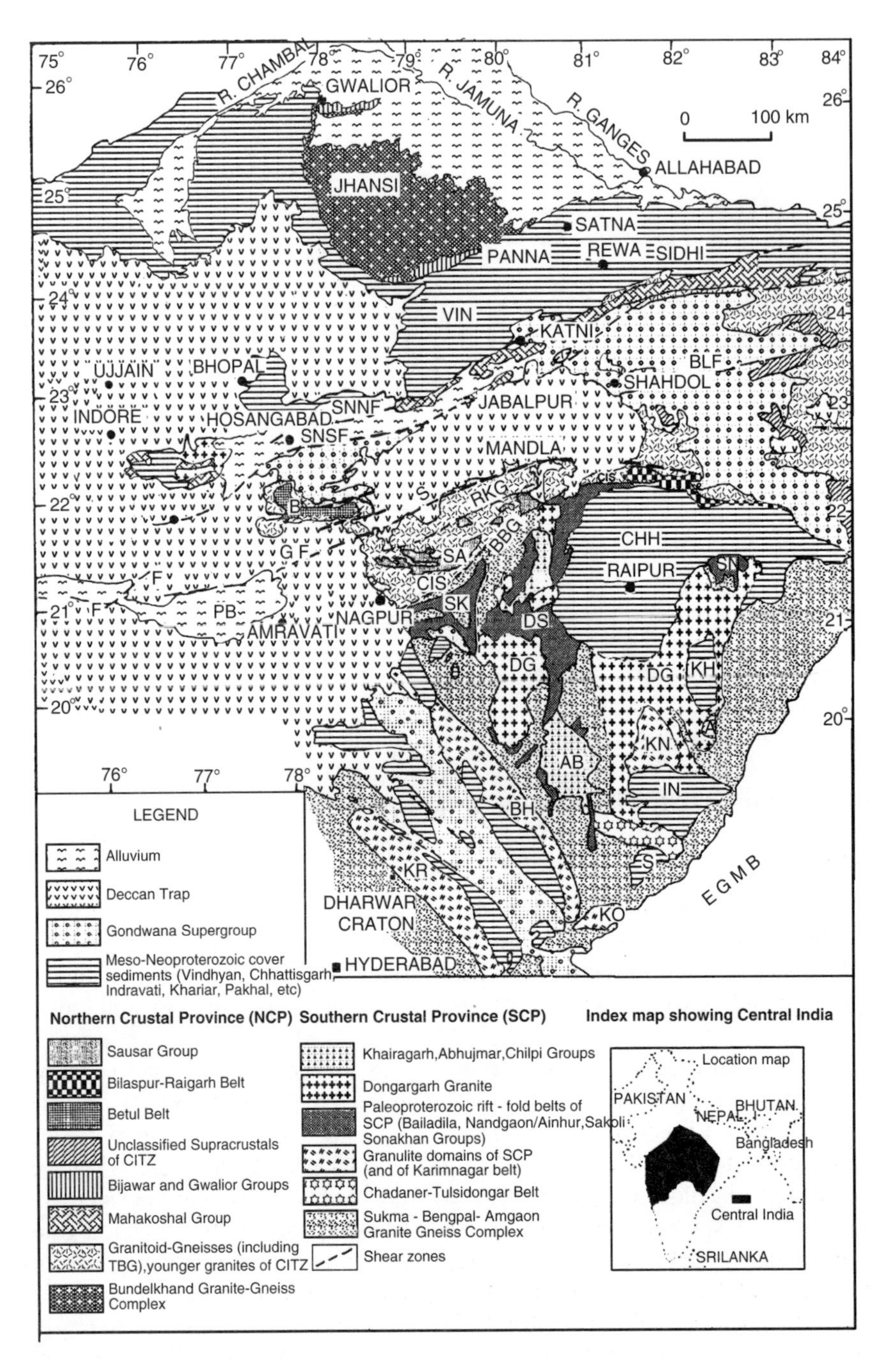

Fig. 2.1.1 Generalised geological map of Central India (compiled from and modified after Roy and Hanuma Prasad, 2001; Ramachandran et al., 2001 and Basu, 2001). Basins: VIN – Vindhyan, CHH – Chhattisgarh, IND – Indravati, K – Khariar, A – Ampani, S – Sabri; Supracrustals: BL – Bailadila, DS – Dongargarh, KH – Khairagarh, AB – Abhujmar, SK – Sakoli, SN – Sonakhan, SA – Sausar, B – Betul; Shear zones: SNNF – Son – Narmada North Fault, SNSF – Son – Narmada South Fault, TS – Tan Shear, GF – Gavilgarh Fault, CIS – Central Indian Shear; Granulite belts: BH – Bhopalpatnam, KN – Kondagaon, K – Konta, KR – Karimnagar, BBG – Balaghat – Bhandara, RKG – Ramakona, EGMB – Eastern Ghat Mobile Belt.

The Central Indian Tectonic Zone (CITZ), referred to as Satpura Province in early literature, is bound by the Son–Narmada North Fault (SNNF) in the north and Central Indian Shear (CIS) in the south (Radhakrishna, 1989; Acharyya and Roy 2000; Roy and Hanuma Prasad, 2003). The CITZ comprises a number of Proterozoic mobile belts (< 2.5 Ga) occuring within largely undifferentiated gneisses, with locally identified TTG members and syn- to post-kinematic K-rich granitic bodies. The Vindhyan and Gondwana sequences and the Deccan Trap rocks, greatly restricting in space the exposure of the Precambrian basement, cover large parts of this region. However, three supracrustal belts of different ages, namely Mahakoshal (2.2–1.8 Ga), Betul (> 1.55–0.85 Ga) and Sausar (1.1–0.95 Ga) are prominent from north to south, each of which is bound by brittle–ductile/ductile shear zones. Besides, there is another belt of supracrustals in the southeastern segment of CITZ situated to the north of Bilaspur. The CITZ is also characterised by three distinct linear tracts exposing granulitic rocks viz, Makrohar Granulites (MG) south of Mahakoshal belt, and Ramakona–Katanai Granulites (RKG) and Balaghat–Bhandara Granulites (BBG) occurring along the northern and southern margins of Sausar supracrustal belt. Besides, granulitic BIF has recently been reported from the northern parts of Betul belt (Roy and Hanuma Prasad, 2003).

As mentioned above, the CITZ is marked by several brittle–ductile to ductile shear zones, the most prominent of which are the Son–Narmada North Fault (SNNF) passing along the northern contact of Mahakoshal belt with the Vindyans, Son–Narmada South Fault (SNNF) marking the southern boundary of Mahakoshal belt, Gavilgarh–Tan Shear (GTS) passing between southern Sausar belt and northern Betul belt, and the Central Indian Shear (CIS) marking the southern limits of the Northern Crustal Province. There are many other subsidiary shear zones close to the contacts of the supracrustal belts, and also within them.

The overall tectonic trend of CITZ is ENE–WSW, which roughly parallels the tectonic trend within the Bundelkhand cratonic massif and the Bijawar supracrustals. This is in sharp contrast with the conspicuous N–S tectonic trend of the Proterozoic supracrustals in the Bastar craton.

A brief description of the various geological components of the crustal provinces enumerated above is presented in the following section, along with brief accounts of tectono–magmatic models proposed for their evolution.

Southern (Bastar) Crustal Province (SCP)

The Bastar Crustal Province is demarcated in the east and northeast by the Eastern Ghat Mobile Belt (EGMB) and Mahanadi Graben, in the southwest by the Godavari Graben and in the north by the Central Indian shear zone (CIS). The possible western extension of the province is obscured by the Deccan Trap cover.

Bose (1898–1899, 1899–1900), Ball (1877), King (1885), Walker (1900), Crookshank (1938) and Ghosh (1941) had pioneering work done on the Bastar craton. Cookshank (1963) divided the Precambrians of southern Bastar into Sukma, Bengpal and Bailadila Iron Ore Series, from the bottom to top. In the northern parts of the province, Sarkar (1957–1958) and Sarkar et al. (1981, 1994) named the oldest gneissic supracrustals as Amgaon Group and the younger N–S trending supracrustals as Dongargarh Supergroup. Following this, all the old gneiss–supracrustals occurring to further northwest and north, surrounding the Sakoli belt, were named as Amgaon Group in the existing literature (Fig. 2.1.1).

Thus, the ancient gneiss–supracrustal complexes in the whole of Bastar cratonic province are named differently as Sukma (or Bengpal) in the southern territory and as Amgaon in the northern parts. These are generally characterised by high-grade (upper amphibolite facies) gneiss–migmatite and supracrustal association, comprising quartzite–BIF–carbonate–pelites and amphibolite–

ultramafites. The supracrustals are interleaved with gneissic rocks containing relics of >3 Ga TTG components and are together intruded and dismembered by younger granites. These are again reworked in the younger granulite and upper amphibolite facies mobile belt. (Fig. 2.1.2 a, b).

Fig. 2.1.2 Bengpal (Sukma) Gneiss exposed in Khardi river section, 3 km south of Bhanupratappur. TTG Complex displays intrusion by younger coarsely crystalline granitoid followed by folding and syn–late tectonic emplacement of amphibolites; (b) TTG with enclaves of metabasics, late aplitic rocks and epidotic veins along fractures and shears, 7 km west of Pithora along NH–6 (photographs by authors).

The Bengpal Series of Crookshank (1963) occurs as a WNW–ESE trending belt from Chandaner (east of Bailadila) in the west to Tulsidongar in the east. But because of an indistinct demarcation between the litho– assemblage of this belt with the wide spread screens of supracustal enclaves within the gneisses, the identification of Bengpal as a separate group has remained controversial. Many later workers opined in favour of a composite Sukma–Bengpal sequence (Ramakrishan, 1973; Chatterjee, 1970; Narayanaswami, 1976). Ramakrishnan (1990), however, reverted to the distinction between Sukma and Bengpal based on exclusive occurrence of diopsidic and calc–silicate rocks in Sukma and basic lava and tuff in Bengpal, with a reported unconformity between the two. Mishra et al. (1988) opted for combining Sukma into one Bengpal Group, but assigned the succession of the Chandaner–Tulsidongar tract a much younger status. Ramachandra et al. (1998) generally endorsed the stratigraphic scheme of Ramakrishnan (1990), and separately delineated the granulite belts of Bhopalpatnam, Kondagaon and Konta, considering the Bengpal sequence of Chandaner–Tulsidongar as representing a mobile belt.

Sukma Gneiss–Supracrustals The southern parts of Bastar Province, in Dantewara, Bastar and Kanker districts of Chhattisgarh, constitute the type area for Sukma gneiss–supracrustals associations, accomodating several occurrences of rock units of mappable scales, besides innumerable small worm-like exposures (Fig. 2.1.3). The gneiss–migmatite complexes have restites of metasedimentary, meta–igneous and TTG components. There is ample evidence to suggest that the hybrid gneissic complex is largely a product of remobilised basement gneiss with varied degrees of assimilation of older supracrustals. However, the restites of orthogneisses of tonalite–trondhjemite composition are recognised at many places, and their involvement in later anatexis, yielding melts of granite–granodiorite composition is suggested (Crookshank, 1963; Ramachandran et al., 2001). The younger granites characterised by coarse to medium grains, lack of banding and weak deformation, are seen to have invaded the basement gneisses at places.

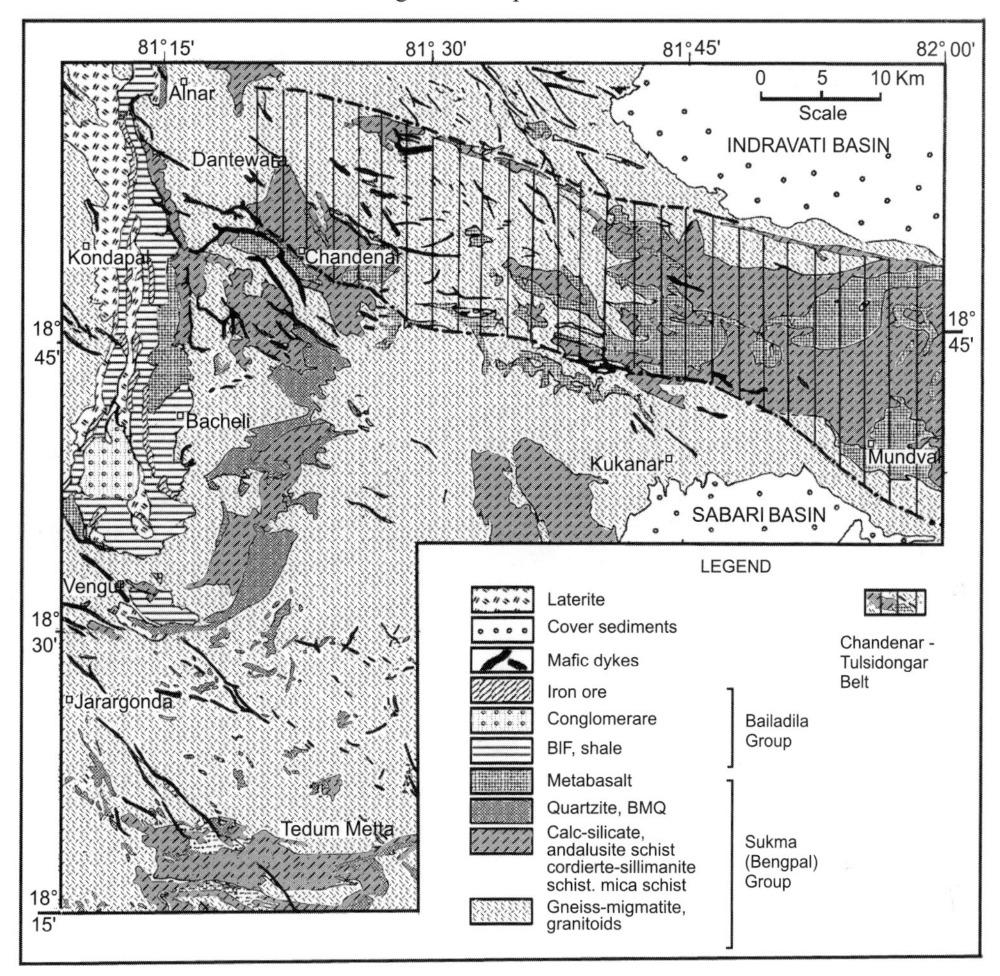

Fig. 2.1.3 Geological map of a part of Bastar region, Chhattisgarh (compiled from and modified after Ramachandran et al., 2001)

The structural trend varies widely in this area, ranging from NW–SE trend (paralleling the trend of Bhopalpatnam granulite belt) in the south western parts, E–W trend in southern parts, NNW–SSE trend in eastern parts and N–S trend (paralleling Bailadila trend) in the north western sector. The WNW–ESE trending sequence of Chandenar–Tulsidongar belt and NE–SW trending Konta granulite belt, also point out discordance in the general trend of Sukma rocks in those sectors.

The supracrustal rocks in the region include quartzite, banded magnetite quartzite (BMQ), calc–silicates (grading into para–amphibolites and calc–gneiss), Al–metapelites (kyanite/sillimanite schists), Al–Mg metapelites (biotite–cordierite gneisses), amphibolite and meta–ultramafics. The quartzite band (interlayered with other metasediments), and BMQ (with locally developed grunerite schist lenses) extend for several kilometers in strike displaying major fold patterns (F_2) in some sectors. The supracrustal restites, in general, indicate a polyphase hydrous upper amphibolite facies metamorphism associated with melt formation (Ramchandran et al., 2001).

The mafic dykes and volcanic rocks are profusely distributed in this terrain showing diverse petrological and geochemical character. These are classsified into two subalkaline and one high-Mg (boninite) swarms, with distinctive trace and REE signatures, indicating three different magma sources for their emplacement at different points of time ranging from Mesoarchaean to Paleoproterozoic (Srivastava and Singh, 2003).

From mineralogical and geochemical considerations, Sarkar et al. (1990) identified six different varieties in the older gneisses of Sukma Group. All those varieties were reported to have relatively high Rb and Sr contents. Though the Rb–Sr and Pb–Pb dates for these gneiss–migmatite rocks showed high scattering of data on isochron plots, at least two Pb–Pb dates with relatively more acceptable accuracy indicated ages of 3018 ± 61 Ma, MSWD 16.8, for Group I (Sukma tonalitic gneiss) and 2573 ± 139 Ma, MSWD 31.03 for Group III (leucogranite plutonic bodies) respectively. Most uniform composition was later reported for the Markampara trondhjemite gneiss from Sukma area (Sarkar et al., 1993), which is characterised by high K/Rb but low Rb/Sr ratios, consistent and highly fractionated REE pattern and moderate negetive Eu anomalies. U–Pb dates of 3509 ± 14 Ma was estimated for the primary crystallisation of trondhjemites and an age of 2480 ± 3 Ma for the late granitic phase surrounding the gneissic enclaves was reported. Rb–Sr whole rock date for the trondhjemite gneiss was determined to be 3610 ± 336 Ma, (Sr_i ~0.7016).

Sarkar et al. (1994) provided two more Rb–Sr age data from the gneisses occuring in the Kakerlanka and Kerulpal, as 2560 + 248 Ma and 2659 + 310 Ma with Sr_i ranging between 0.70899 and 0.70726 respectively. This set of isotopic data is highly scattered, suggestive of largescale reworking of the older gneisses (Sarkar et al., 1994). A summary of geochronological data on Sukma gneissic rocks is presented in Table 2.1.1.

Table 2.1.1 *Geochronological data on Sukma Gneisses and granitic rocks*

S. No.	Area	Rock type	Analytical method	Obtained age (Ma)	References
1	Sukma	Tonalite gneiss	Pb–Pb (WR)	3018± 61	Sarkar et al. (1990)
2	Sukma	Granite	Pb–Pb (WR)	2573± 139	
3	Markampara (near Sukma)	Trondhjemite gneiss	U–Pb (zircon)	3509±14	Sarkar et al. (1993)
			Rb–Sr (WR)	3610± 336	
4	Markampara (near Sukma)	Late granite	U–Pb (zircon)	2480± 3	
5	Kakerlanka	Tonalite gneiss	Rb–Sr (WR)	2560± 248	Sarkar et al. (1994)
		Granite gneiss	Rb–Sr (WR)	2659± 310	

Sarkar et al. (1993) suggested a three-stage evolution with an early period of high U–Pb growth after the emplacement of the 3.5 Ga trondhjemite and U-depletion at 2.5 Ga during the wide spread magmatism and metamorphism. The Sukma Gneisses are considered to have a multistage development and might have been produced by partial melting of amphibolitic to silicic granulitic protoliths with limited plagioclase fractionation.

Longstaffe et al. (2003) reported anomalously low $\delta^{18}O$ values (< + 6 per mil) for all the six groups of granitoids described by Sarkar et al. (1990), inferring crustal recycling and copious low $\delta^{18}O$-fluid activity responsible for the generation of the granite–gneiss complex of Bastar region. Mondal and Hussain (2003), based on detailed geochemical studies of the Bastar granitoids, categorised these as peraluminous, showing a calc–alkaline granodioritic trend, and inferred their generation from sedimentary protoliths by anatexis.

Ghosh (2004) reported U–Pb zircon date of 3561 ± 11 Ga for a tonalite sample that represents basement of Kotri belt near Kapsi in central Bastar. The rock is intruded by biotite gneiss and then by dykes of coarse-grained massive granite. The location is not far from the Khardi river section where the basement gneisses display similar features (vide Fig. 2.1.2). Recently, Rajesh et al. (2009) reported U–Pb zircon (SHRIMP) date of 3582.6 ± 4 Ma (n = 7; MSWD = 0.71) for'true' granite occurring in the immediate east of Dalli–Rajhara BIF (iron ore) belt. Unlike the typical early Archean granitoids, which are commonly deformed into gneisses, this granite is reported to be relatively undeformed. The age and composition of the granite, according to the authors, suggests that the continental crust of the Bastar craton attained sufficient thickness to permit intracrustal melting at 3.6 Ga.

Amgaon Gneiss-Supracrustals Sarkar (1957–1958) included the gneissic supracrustal complex forming the basement of Dongargarh Group in northern parts of Bastar Crustal Province (in Bhandara district) under Amgaon Group. This nomenclature was adopted by later workers for designating the basement complex in the territory lying west and north of the Dongargarh Granite massif, generally limiting uses of the term upto south of the ENE–WSW trending CIS, which demarcates the Bastar Crustal Province from the Satpura Province (CITZ). Nevertheless, there has not been any clear distinction between the Amgaon and the Sukma–Bengpal Group described from southern and eastern Bastar Province, not even along the northeastern flank of Bhopalpatnam belt where the older gneiss–supracrustal complex of Sukma passes uninterruptedly into what is an Amgaon territory.

As per the current use of the term, the Amgaon gneiss–supracrustals occur along more than a 100 km long and 20–30 km wide N–S corridor between the eastern margin of the Sakoli fold belt (SFB) and Dongargarh Granite. Amgaon Gneisses are considered to be the basement of SFB and surround the latter from all sides. Basement inliers/wedges of Amgaon Group are also described from within the Sakoli belt. Due to intense mylonitisation along the eastern and northwestern boundary of SFB, it is difficult to establish basement–cover relation between the Amgaon Gneisses and the Sakoli sequence.

The gneissic component of Amgaon Group ranges in composition from granite to granodiorite with some restites of tonalite–trondhjemite. The gneisses display agmatic and stromatic structures, the latter being more common. Augen structures are well developed in the mylonitised gneiss–migmatite. Granodiorite to tonalilic gneiss occurs as lenticular outcrops, often displaying overprinting of mylonitic fabric on the original gneisic foliation (Roy et al. 1995). Divakara Rao et al. (2000) have described the Amgaon Gneisses under two groups, one dominantly tonalitic and the other adamellitic, most of which exhibit I-type chemical signatures of volcanic arc granite, though a few samples show A-type character.

The supracrustal enclaves in Amgaon Gneisses are of quartzite, BIF, calc–silicate, marble, Al-rich and Al–Mg metapelites, amphibolite and meta–ultramafics. These are strikingly similar to the supracrustals occurring within Sukma gneissic complex. Here also the quartzites, BIF and calc–silicates have large strike extension in several sectors. The presence of kyanite, sillimanite, pyrophyllite, dumortierite characterise the quartzite and metapelites of Amgaon Group. The amphibolites, predominantly composed of hornblende and anthophyllite, vary in strike length from a few metres to several kilometers. Evidence of partial melting is noted at places. Most of the metabasics of Amgaon Group fall within basalt and some within basaltic andesite field and a majority of these are indicative of continental/ocean island arc environment (Sarkar et al. 1994).

The Rb–Sr and K–Ar whole rock and mineral ages for the rocks of Amgaon Group, as reported by Sarkar et al. (1967)and Sarkar et al. (1981), are presented in Table 2.1.2.

Table 2.1.2 *Geochronological data on Amgaon Gneisses (Sarkar et al. 1967, 1981)*

Area	Rock Type	Analytical method	Age (Ma)obtained
Near	Biotite gneiss	Rb–Sr (WR)	2170 ± 60
Amgaon	Amphibolite	K–Ar (WR)	1220, 1150
	Amphibolite	Rb–Sr (biotite)	1640, 1590, 1460
	Amphibolite	K–Ar (hornblende)	1630
		(biotite)	1444, 1490

Bengpal Group Crookshank (1963) proposed and Ramakrishan (1990) and Ramachandran et al. (2001) consolidated the concept of a distinctive belt of low-grade metamorphites (Bengpal Group) comprising a volcanisedimentary assemblage, overlying the Sukma Gneiss-supracrustal rocks of higher metamorphic grade over a stretch of about 100 km and with 10–30 km width from Chandenar in the west to Tulsidongar in the east, with an overall trend of WNW–ESE to NW–SE (Fig. 2.1.3). The lithological sequence of this belt includes metabasalt–gabbro–ultramafic intrusives at the base, followed upwards by conglomerate–sandstone–siltstone–shale and andalusite bearing sericitic quartzite. Ramakrishnan (1990) presented the evidence of distinct unconformable relation between the conglomerate–sandstone–siltstone of Bengpal sequence with the underlying Sukma Gneiss–supracrustals from a place to the south of Bothapara. Ramchandran et al. (2001) considered the Bengpal sequence (Table 2.1.3) to represent a mobile belt, but its contact with the older Sukma basement was described to be mostly tectonic or disconformable.

Table 2.1.3 *Lithostratigraphy in the Chandenar–Tulsidongar mobile belt (Ramchandran et.al., 2001)*

	Mafic dyke swarms
	Granitoids
Bailadila Group	BIF, quartzites, phyllites, conglomerate etc.
	—Tectonic/ unconformable contact—
	Andalusite and rarely fibrolite bearing sericitic quartzite
Bengpal Group	Conglomerate, sandstone, siltstone, shale
	Amygdular metabasalt, gabbro, metaultramafics
	—Tectonic / unconformable /disconformable contact—
	Cordierite–anthophyllite bearing rocks
Sukma Group	Al– and Mg–Al metapelites
	Quartzite, BIF, calc–silicates
	Gneiss–migmatites, amphibolite bands

However, there are contradictory views on the possibility of the supracrustals of south Bastar being a composite Sukama–Bengpal package, as there exists a general structural similarity between the two, despite tectonic, disconformable or unconformable contact relations noted at places. Moreover, the metamorphic grade is also not a distinguishing feature everywhere, and late granites and pegmatites hosting tin mineralisation intrude both the supracrustal sequences.

The amygdaloidal basalt with sheets of gabbro forming the basal formation of Bengpal Group and minor ultramafic intrusives abound to the west of Tulsidongar with a WNW–ESE trend. In the west of Chandenar, the basaltic formation extends along the eastern flank of Bailadila range, paralleling N–S trend of this belt. For this sector, Khan and Bhattacharya (1993) placed these metabasalts (Bhansi Formation) at the base of the Bailadila Group, above the Bengpal Group. The change in the trend of basaltic layer from WNW–ESE in Chandenar–Tulsidongar belt to N–S in the vicinity of the Bailadila belt is explained by the involvement of Bengpal mobile belt in the younger Bailadila orogeny (Ramakrishnan 1990).

Though the precise relation between the various supracrustal components of this terrain is far from uniformly understood, it may be surmised that along Chandenar–Tulsidongar belt, older Sukma gneiss–supracrustals and younger Bengpal metabasalt–supracrustals have been deformed and metamorphosed with the generation of granitic melts at a shallow crustal level resulting in emplacement of younger granites and migmatites (with cassiterite) along this narrow elongated tract. The age data available from the younger granitoids of this region range between 2.1 to 2.5 Ga (Rb–Sr) (Sarkar et al., 1990, 1994).

Granulite Belts There are three granulite belts in the Bastar Province viz, Bhopalpatnam, Kondagaon and Konta belts.

Bhopalpatnam Granulite Belt The NW–SE trending Bhopalpatnam granulite belt extends for about 300 km (with 20–40 km width) along the northeastern edge of the Godavari graben, apparently forming a pair with the Karimnagar granulite belt occurring across the graben along its western shoulder. The belt essentially comprises acid and basic granulites and high-grade metasediments–metaigneous rocks occurring within a gneiss–migmatite complex. Mishra et al. (1988) held that the rocks of Sukma Gneiss–supracrustals are metamorphosed to granulite facies along this belt, accompanied by late-kinematic magmatic inputs. The eastern boundary of the belt is marked by overprinting of granulite facies metamorphism on the original amphibolite facies mineral assemblage with distinct *'opx–in'* isograd in appropriate lithologies. The western margin is characterised by discontinuous mylonitic zones, with reported development of even pseudotachylites at places. Ghosh (1941) held the view that the carbonate rocks (calc–silicates and carbonate facies BIF) in the pre-existing litho-package played a major role in granulite facies metamorphism along the belt. Janardhan et al. (1996) evoked an external source for CO_2 and suggested P–T condition of 6–9 kb at the temperature of 750 °C and an IBC (isobaric cooling) P–T–t path for the granulite facies metamorphism. Due to an overall similarity with the Karimnagar granulite belt, showing overprinting of granulite facies on pre-existing amphibolite grade rocks of Dharwar craton, generation of juvenile magmatic inputs and an IBC evolutionary path (Rajesham et al. 1993), the Bhopalpatnam granulites were considered to be time equivalent with the Karimnagar granulite belt, the upper age limit being marked by the Rb–Sr dates, 2.6–2.7 Ga, for the younger granites in Karimnagar belt. However, the recently available EPMA dates from zircon, monazite and uraninite (Santosh et al., 2004) for the Bhopalpatnam granulites imply a much younger status (1.6–1.7 Ga) for this belt, bringing it closer to EGMB in respect of time equivalence.

Konta Granulite Belt The NW–SE trending Bhopalpatnam granulite belt is truncated and overprinted in the southeast Bastar by the NE–SW trending structures of an apparently younger Konta granulite belt. The trend of Konta belt parallels the EGMB and granulites facies metamorphism is considered coeval with the evolution of EGMB, postdating the development of the Bhopalpatnam and Karimnagar granulite belts. The Konta belt is characterised by the gneiss–supracrustals of Sukma occurring in tectonic contact with orthopyroxene (opx)-bearing BIF, metapelites, two–pyroxene granulites, charnockite, enderbites etc. The granulites are spatially associated with ultramafic and alkaline suite of rocks.

Kondagaon Granulite Belt The Kondagaon granulite belt, extending for 70 km with N–S trend is located to the east of Kondagaon in the east central Bastar Province. Here also the supracrustals of Sukma Group are upgraded to granulite facies rocks and the litho–association is quite similar to that of Bhopalpatnam belt. The western margin of Kondagaon belt shows a diffused, or gradational contact with Sukma Gneiss–supracrustals, while the eastern boundary is characterised by the occurrence of younger granite which separates it from EGMB. Janardhan et al. (1996) have reported migmatisation of calc–silicate rocks in this belt similar to Bhopalpatnam belt. The mineral chemistry of the litho-assemblage indicates a range of 4–6 kb and 700 °C, defining an ITD (isothermal decompression) evolutionary path (Prakash et al.,1996). Rb–Sr dates available for the late granitoid and pegmatite in Kondagaon belt are 2610 ± 143 Ma and 2409 ± 184 Ma (Sarkar et al., 1990), which may be considered as the upper age limit of the granulite metamorphic event.

Bailadila Group The Bailadila Group comprises BIF dominated younger supracrustals developed along an N–S trending chain of belts for about 300 km extending from south of Dantewara to the Durg district, Chhattisgarh. The Bailadila belt (from which the Group name is derived), representing the southern most sequence of this chain, extends for about 60 km from south of Bacheli to the west of Barsur in Dantewara district (Fig. 2.1.4). Chhotadongar, Madenmar and Rowghat belts extending discontinuously for more than 100 km, with their widths ranging from 3 to 8 km, follows this to the north. Further northwards, the belt extends with intermittent gaps for another 120 km (width up to 5 km) through Gatta–Koti, Hahaladi, Sonadehi, Dulki, Mahamaya and Dalli–Rajhara. There are a few subsidiary BIF belts of limited strike extension along both the east and west of the main belt. The chain of BIF belts described above has tectonised contact with the Sukma gneiss– supracrustals, intruded at places by younger granitoids, along its eastern boundary. In the southern segment (most of Bailadila belt), the Sukma Gneiss–supracrustals also occur along the western contact, but further north, Abhujmar cover rocks overlap at many places upto Rowghat belt. Form Hahaladi to Dalli–Rajhara the BIF belt follows the eastern contact of Kotri volcanisedimentary belt, defining the eastern limits of Kotri rift/Kotri lineament.

Bailadila Belt Very large deposits of rich iron ores occur in Bailadila range in association with BIF, shale, quartzite etc. The Bailadila range consists of two near–parallel ridges with a valley in between. The ranges trend N–S and extend for about 50 km in length. The ranges are essentially composed of an imposing sequence of BIF, underlain by ferruginous shale, quartzite and local conglomerates (Fig. 2.1.5). Metabasic rocks are extensively developed along the eastern foothills. A few metadoleritic intrusives cut across the different members of the sequence (Chatterjee, 1970). Along both the ridge axes, a total of 14 clusters of iron ore deposits have been identified (vide Fig. 2.2.10).

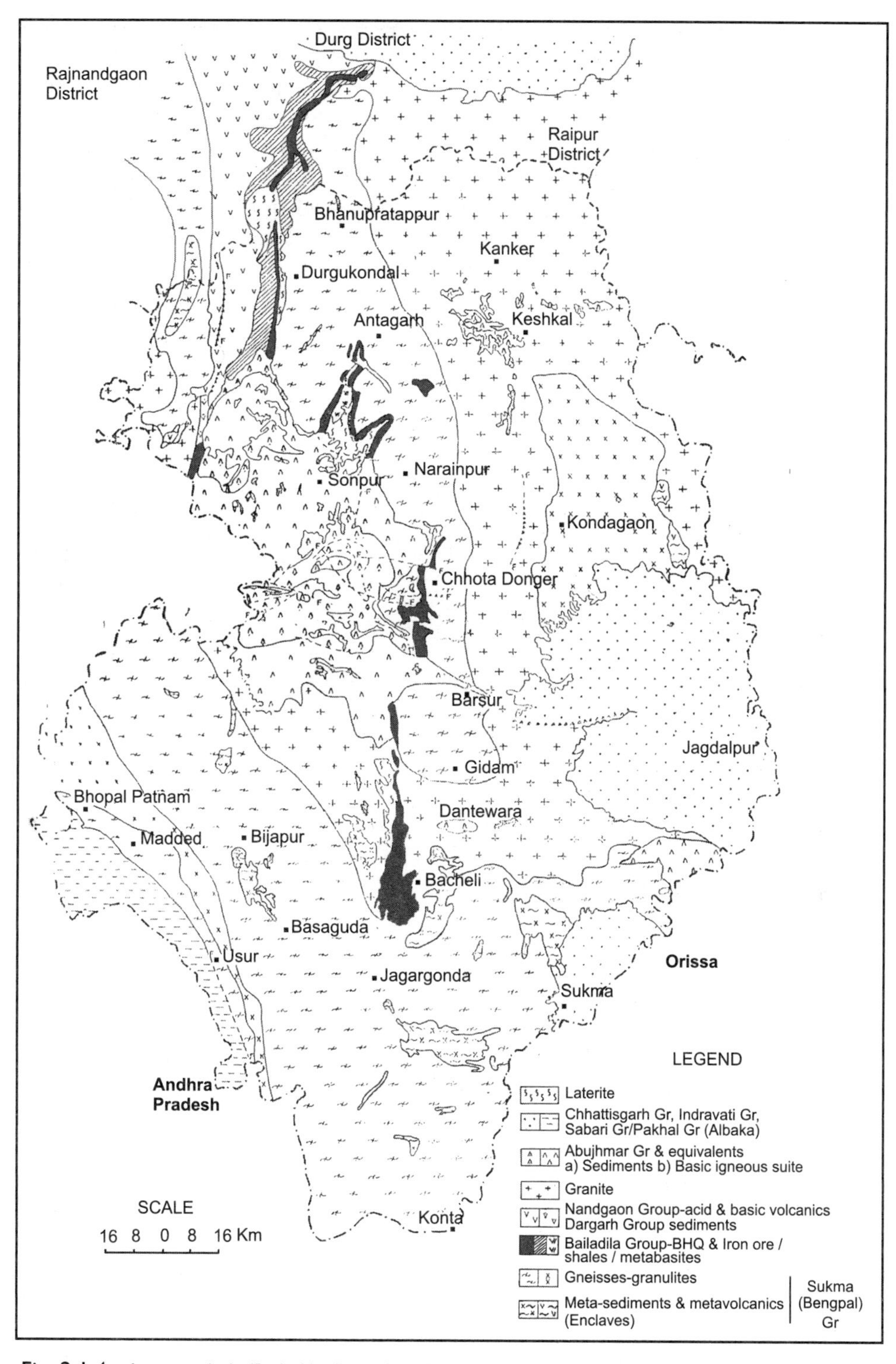

Fig. 2.1.4 Iron ore belt (Bailadila Group) in Chhattisgarh

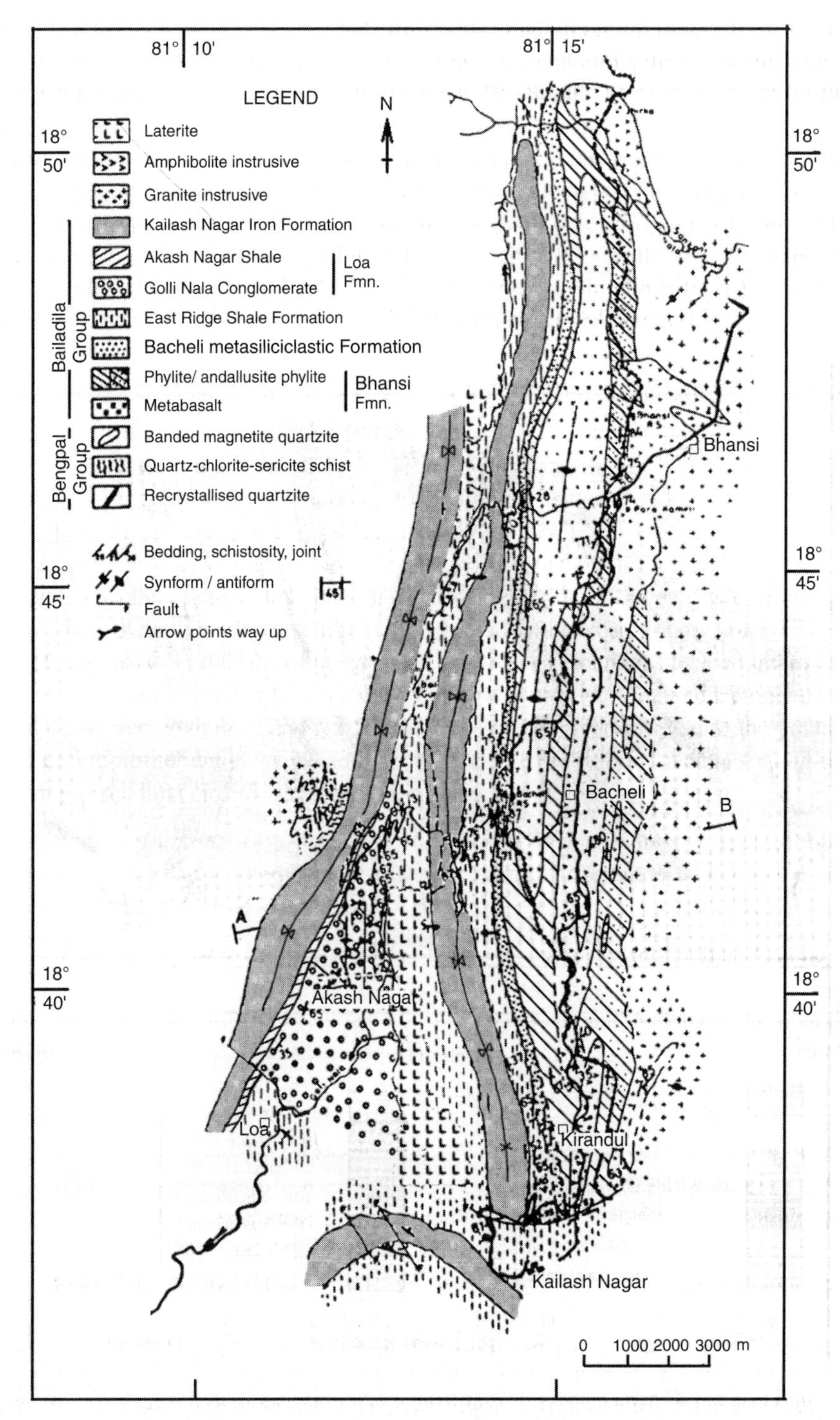

Fig. 2.1.5 Geological map of Bailadila belt (after Khan and Bhattacharya, 1993)

Studies on the geology of the Bailadila belt was first initiated by Bose (1900) in connection with iron ore prospecting. Crookshank (1963) proposed the stratigraphy of Bailadila as presented in Table 2.1.4, and inferred an N–S trending synclinorium with two synclines and one intervening anticline plunging towards north.

Table 2.1.4 *Lithostratigraphy of Bailadila Series (Crookshank, 1963)*

	Cuddapah
	Khondalite series
	......*Position uncertain*......
Bailadila Series	Banded hematite quartzite with associated iron ore
	Grunerite–quartzite, earthy and chloritic hematites, ferruginous phyllite and banded carbonaceous shale
	White quartzite
	.......*Unconformity*.......
	Bengpal Series (Upper Loa Stage, Middle Dantewara Stage and Lower Tulsidongar Stage)
	.......*No clear dividing line*......
	Sukma Series

Later researchers have expressed differing views with respect to both stratigraphy and structure of the Bailadila Group. Both Bandyopadhyay and Hishikar (1977) and Mishra et al. (1988) introduced Loa Stage/Formation (comprising conglomerate, grit, tuffs, carbonaceous shale etc.) in the Bailadila Group, but differed in conjunction to its position below or above the main BIF. Ramakrishnan (1990) kept 'Loa Stage' under the Bengpal Series but considered it to be of only local importance in the Bailadila area (Table 2.1.5).

Table 2.1.5 *Lithostatigraphy of Bailadila Series (Ramakrishnan, 1990)*

Bailadila Series	Banded hematite quartzite with associated iron ore, grunerite–quartzite, chlorite hematite. Ferruginous phyllite, banded carbonaceous shale, white quartzite.
Unconformity	
	Loa Stage
Bengpal Series	Dantewara Stage
	Tulsidongar Stage
No clear dividing line	
Sukma Series	

Khan and Bhattacharyy (1993) proposed another stratigraphic sequence for the area, in which they divided the sequence into three subgroups, the upper two being subdivided into more than one formation (Table 2.1.6). Their upper subgroup has little differences with the upper subgroup of Bandyopadhyay and Hisikar. But the main difference is with respect to the lower part. The Lower 'subgroup' of Khan and Bhattacharyya, which they call Bhansi Formation, contain basic volcanic rocks as an important component. Crookshank placed it at the top of the Bengpal Series (Group), below an unconformity that apparently separates it from the overlying Bailadila Series (Group). Ramakrishnan put it even towards the bottom of the Bengpal Group. In his opinion, the Bengpal Group rich in metabasic volcanic rocks were rotated into concordance with the Bailadila Group during the Bailadila phase of deformation. Field geology has much to contribute towards the resolution of this problem. If the unconformity above the 'Bhansi Formation' and the rocks that overlie it are only local, then its claim to being a part of the Bailadila Group remains a matter of consideration. If on the other hand this conformity is a regional one, then it could as well be a part of the Bengpal Group. A reasonable resolution of this problem has a bearing on the better understanding of the tectonic setting during the deposition of the Bailadila Group and the deposition of BIFs in it.

Table 2.1.6 *Stratigraphic succession of Bailadila Group (Khan and Bhattacharya, 1993)*

Bailadila Group	Upper	Kailasnagar Iron Formation Loa Formation (Akashnagar Shale, Galli nala Conglomerate)
	Middle	East ridge Shale Bacheli metasiliciclastics
	——— local unconformity———	
	Lower	Bhansi Formation (metabasalt, metapelite)
——— Angular unconformity ———		
Bengpal (Sukma) Group		

According to Khan and Bhattacharya the metabasalts of Chandenar–Tulsidongar belt, included under Bengpal Group by Ramakrishnan (1990), are actually the metabasalts of Bhansi Formation belonging to the younger Bailadila Group. From all these controversies, it becomes apparent that the stratigraphy and structure of the rocks in the Bailadila belt are still not understood with the necessary clarity.

However, geological relations in other BIF belts appear to be relatively simple, showing uniformly metasediment-dominant litho-assemblages (BIF, quartzite, and phyllite) with minor metabasalt and meta–aultramafic intrusives. The general strike is N–S and the lithounits have moderate to steep dips towards west to northwest. All the formations are intruded by younger coarse–grained granites, which are correlated with the Dongargarh Granite, suggesting that the upper age limit of Bailadila Group may be placed at 2500–2600 Ma. Overprinting of the Sukma Gneiss–supracrustals by the structures of the Bailadila deformation is also evident all along the belt, the contact between the two often marked by mylonitic zones. Except for locally noted higher metamorphic grades, the rocks of this belt are metamorphosed to greenschist facies, grading upto low amphibolite facies.

The Bhasi metabasalt are massive to layerd. Generally fine–grained, the rock is composed of actinolite – tremolite, chlorite, alkaline plagioclase ± epidote. In the AFM diagram, these plot in the Fe– tholeiite field, although in the Zr/ TiO_2 – Nb/Y diagram some stray into andesitic field. In the discriminant diagrams constructed with trace element and REE data, such as the Th/ Yb – Ta/Yb, Hf/3 –Th–Ta and Zr/Y – Zr, these rocks plot in the arc regions. Read together with the lithologic association of quartz wackes/ lithic wackes–pelite–iron formation, the tectonic setting during the deposition of the Bailadila Group of rocks could be an ensialic rift (Khan and Bhattacharyya, 1994).

The age of the Bailadila Group and hence an estimation of the age of BIF in this sequence is not well constrained. Ramachandran et al. (2001) suggest that the coarse-grained granite intrusive into the Bailadila association can be correlated with Dongargarh Granite, in which case the upper age limit of the Bailadila Group should be limited between 2400–2500 Ma. Such a correlation between the two granitic intrusives without the support of radiometric age data is fraught with a great risk. However, the Rb–Sr age data of 2200–2600 Ma obtained from the basaltic member of the unconformably overlying Abhujmar sequence (Sarkar et al., 1990) may be significant in this context.

Metamorphism in the region is rather low–pressure type and range from sericite schist, through andalusite–biotite schist to sillimanite schist in pelitic metasediments. Generally, the grade of metamorphism increases from west to the east. The entire sequence of Bailadila iron formation together with the underlying sediments shows the lowest grade of metamorphism. Again the grade of metamorphism increases from north to south, as the structurally lower levels are exposed (Chatterjee, 1970).

Chhota Dongar–Mademnar–Rowghat Belts To the north of Bailadila belt, the Chhota Dongar–Mademnar belt extends, with an N–S trend, for about 25 km, with a gap and eastward shift (Fig. 2.1.4). The BIF and associated litho-assemblages of this sector are in direct contact with Sukma gneiss–supracrustals on both east and west. Towards north and south, the basic igneous suite of the Abhujmar Group covers the BIF. The contact of the belt with the older gneisses is tectonic, marked by discontinuous mylonitic zones. This stretch is yet to be explored in detail for its iron ore potential. So far about 48 million tones of ore with + 60% Fe are estimated from parts of Chhota Dongar.

The Rowghat belt, about 28 NNW of Narainpur, is another important stretch of BIF and rich iron ore deposits, which display M-shaped map-scale folds. The litho-association of this belt includes BIF, quartzite, phyllite, chlorite schist and hornblende schist. The BIF is composed of banded hematite quartzite (BHQ), banded hematite jasper (BHJ) and minor banded magnetite quartzite (BMQ). Grunerite–cummingtonite is recorded from some of these bands. The first generation tightly apressed or isoclinal folds on compositional banding (bedding), with axial planes trending NNE–SSW, have resulted in the distinctive map pattern. Cross folds plunging 30°–50° due NW are also prominent at places. The folded sequence of Rowghat BIF belt is in direct contact with Sukma Gneiss– supracrustals all around its outer margin, while the core region is mostly covered by Abhujmar metavolcanics.

Hahaladi–Sonadehi–Dulki–Kanchar–Dalli–Rajhara Belt The belt extends for nearly 120 Km with N–S trend from Hahaladi to Sonadehi–Gurpher, beyond which it has a swerving trend (NE–SW to NS) upto Kanchar–Mahamaya sector, finally assuming nearly E–W trend in Dalli–Rajhara area of Durg district. The belt terminates in the north east of Kokan (east) block, close to the southwestern contact of Chhattisgarh basin (Fig. 2.1.4).

Along this belt the BIF and associated litho-assemblages have a limited width (<1 km to 3 km). The BIF is represented by banded quartz–magnetite–hematite rock in association with BHQ, quartzite, phyllite and shale. Amphibolite and serpentinised pyroxenite are interbanded with BHQ.

This belt is bordered in the west by the younger volcanisedimentary suite of Kotri belt, while the Sukma Gneiss–supracrustals with younger granitoid intrusives mark the eastern contact. Discontinuous stretches of ductile shear zones characterise both the western and eastern contacts of the belt. The younger granitoids (Dongargarh Granite) are seen to intrude the rocks of this belt at many places.

The BIF and BHQ show polyphase deformation at all scales. The F_2 folds are most prominent with N–S axial trend and subvertical axial planes, which are refolded (F_3) around NW–SE axes. The superposed folding has resulted in a series of lenticular BHQ bodies in the ore zone in Dalli–Rajhara sector.

Dongargarh Supergroup (including 'Kotri Supergroup' and Chilpi Group) The NNE–SSW trending volcanisedimentary belt lying to the west of Chhattisgarh basin and east of Sakoli belt and occurring in parts of Durg and Rajnandgaon districts of Chhattisgarh, Balaghat district of Madhya Pradesh and Bhandara district of Maharashtra was included in Dongargarh Supergroup by Sarkar (1957–1958). The southern extension of this belt, continuing along the west of Sukma Gneiss–supracrustals and Bailadila BIF belt in Durg, Rajnandgaon, Kanker and Bastar districts of Chhattisgarh, was named as Kotri Supergroup by Mishra et al., 1988, and Ghosh and Pillai, 1992. Towards the north, the rocks of Dongargarh belt display uninterrupted lithological continuity with those of Chilpi Ghat Series of King (1885), which are also correlated with the Dongargarh Supergroup by some of the later workers (Tripathi et al., 1981; Rao, 1981; Pasayat, 1981; Bhargava and Pal, 2000). Notwithstanding the coinage of varied stratigraphic nomenclature for different parts of the belt, the Dongargarh Supergroup (DSG) is generally understood to include the Paleoproterozoic volcanisedimentary assemblage of continental rift affinity occurring along the N–S trending belt that extends from the north west of Abhujmar basin in the south to the CIS in the north (Fig. 2.1.1).

The DSG essentially comprises an older sequence of acid and basic lava and pyroclasts (Nandgaon Group, Ainhur Group), intruded by Dongargarh Granite (2.2–2.4 Ga) and overlain by the younger volcanisedimentary sequences, named as the Khairagarh, Abhujmar, and Chilpi Groups (Fig. 2.1.6). The stratigraphic successions proposed by the workers for different parts of the belt are presented in Table 2.1.7.

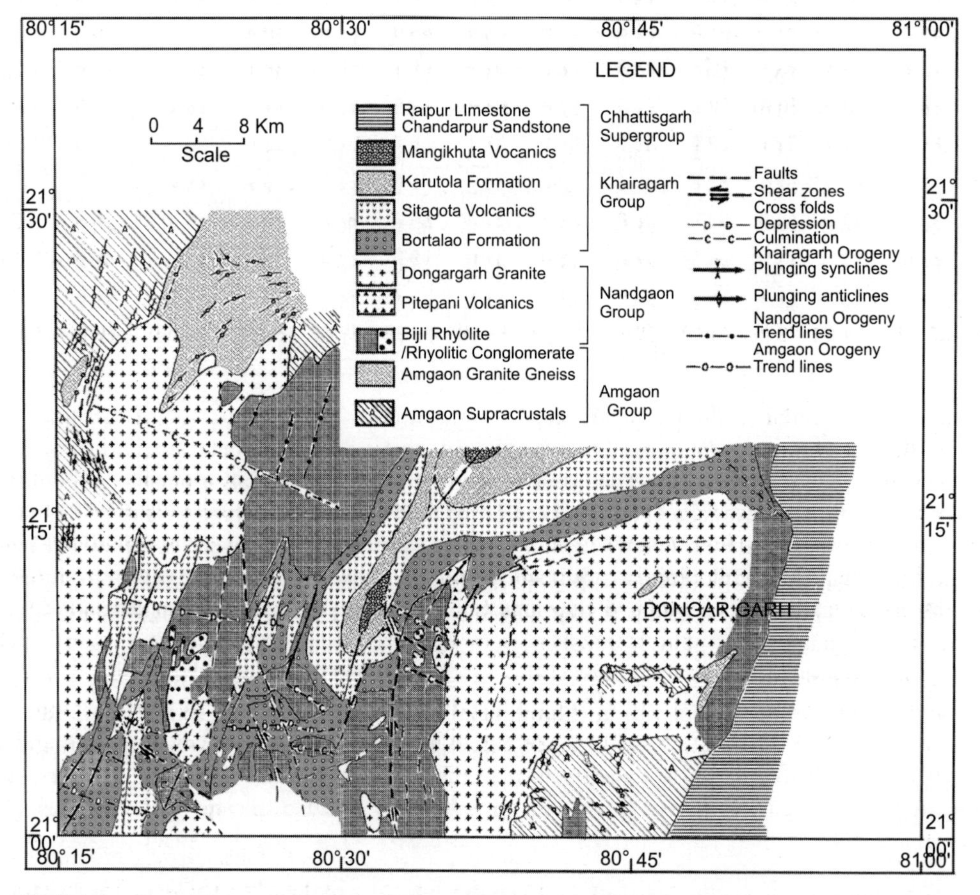

Fig. 2.1.6 Geological map of a part of Dongargarh belt, Central India (modified after Sarkar et al., 1981)

Table 2.1.7 *Stratigraphy of Dongargarh/Kotri Supergroup*

	Sarkar (1957, 1958) Sarkar et al. (1981)	Mishra et al. (1988)	Ghosh and Pillai (1992)
Dongargarh Supergroup	Khairagarh Group (Basal part ~1534 Ma)		
		Abujhmar Group: Gabbro dyke Maspur trap –basalt	
		Abujhmar Group: Gundul Formation conglomerate, sandstone, shale	Kotri Supergroup: Patkasa Formation; Linear gabbro bodies and dykes of various ages
		Abujhmar Group: Intrusives of various ages amphibolite, pegmatite etc.	
	Dongargarh Granite (2270±25Ma)	Sitagaon Granite Suite	Madanbera Granite
	Nandgaon Group	Manghur sediments	
			Ainhur Group: Mendra Formation
	Nandgaon Group: Pitepani volcanics	Ainhur Group: Hammantwahi Basic Suite	
			Ainhur Group: Kurse Kohri Formation
	Nandgaon Group: Bijli Rhyolite (2180 ± 25 Ma)	Ainhur Group: Mahala Rhyolite	Ainhur Group: Pachangi Formation
		Dargarh Group	
	NW–SE trending dyke swarms in the basement		
	Sakoli Group	Bailadila Group	Latemarka Group Hanker Group
	Older Basement	Bengpal Group	

Nandgaon Group (Ainhur Group) The Nandgaon Group comprises two formations. The lower Bijli volcanics are mainly composed of rhyolite associated with sandstone and rhyolite conglomerate that unconformably overlie the older gneisses. This is overlain by the Pitepani volcanics, which include porphyritic and non–porphyritic basalt (with rare pillow structure) of tholeiitic trend. Both komatiitic (Deshpande et al., 1990) and andesitic varieties (Neogi et al. 1996) have been reported from the Pitepani volcanics. The Bijli volcanics are characterised by interbanding of acidic flows, pyroclastics and arenitic sediments.

The equivalent of Nandgaon Group in the southern Kotri belt (Fig. 2.1.7) is named as Ainhur Group which is subdivided into four formations by Ghosh and Pillai (1992) of which the basal Pachangi Formation comprises conglomerate shale, sandstone, rhyolite and rhyolitic pyroclastics. The overlying Kurse Kohri Formation includes the spectacular ignimbrites of Kotri river section. These two formations together correspond to the Bijli volcanic sequence of northern sector, while the overlying Mendra Formation composed of basalt and basaltic tuff, corresponds to the Pitepani volcanics. The youngest Patkasa Formation, comprising conglomerate, siltstone, sandstone etc, occurs only along the eastern margin of the belt as very thin and flat lying beds with pronounced angular unconformity. Though the Patkasa Formation was tentatively included in the Ainhur Group, the undeformed character, prominent unconformity with the underlying volcanic sequence and non-invasion by granitic rocks are suggestive of its equivalence with the Khairagarh (Abhujmar) Group that were deposited after the emplacement of the Dongargarh Granites. Sensarama and Mukhopadhyay (2003), based on field studies of a limited part of the Dongargarh belt and petrochemical considerations, suggested inclusion of the basic volcanics (Sitagota, Karutola and Mangikhuta Formations) of Khairagarh Group into the Nandgaon Group and placement of the Bartalao Formation at their top. Any serious consideration on this proposition will, however, call for much wider ground coverage and attention towards many aspects of stratigraphic correlation established by the earlier workers.

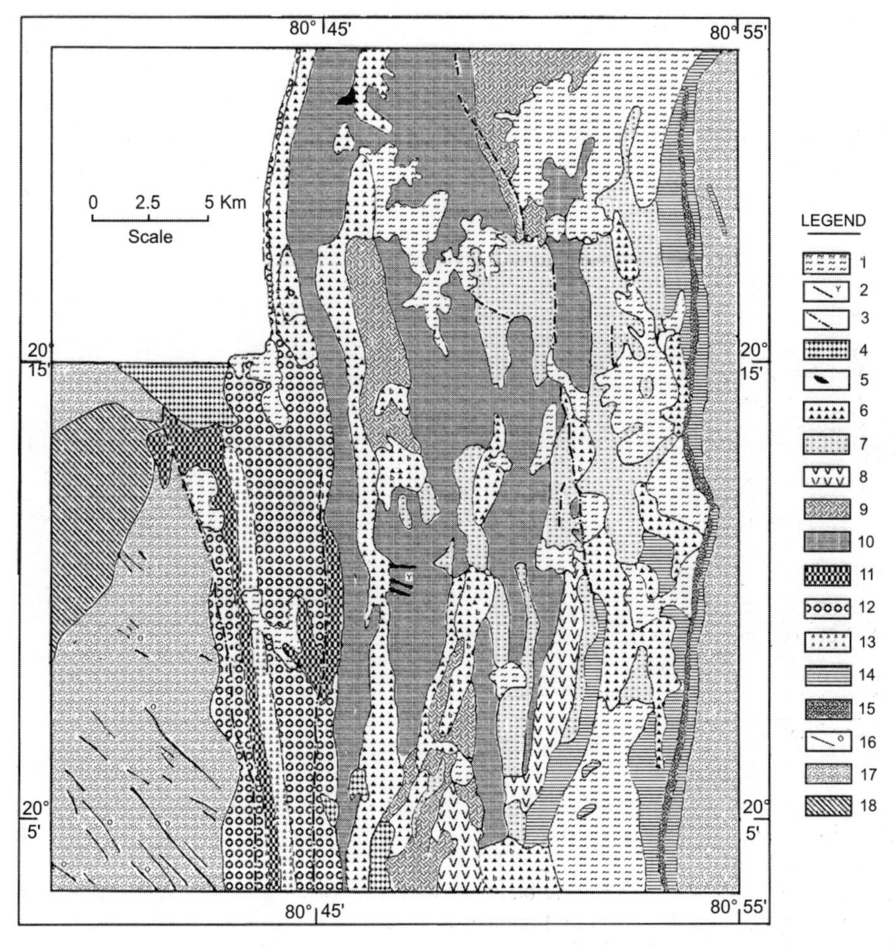

Fig. 2.1.7 Geological map of a part of Kotri belt, Central India (modified after Ramachandran et al., 2001) 1. Laterite, 2. Younger mafic dyke, 3. Silicified shear zone, 4. Granite and granophyre, 5. Ultrmafic intrusives, 6. Basalt and gabbro, 7. Patkasa Sandstone-conglomerate, 8. Basaltic pyroclasts (Mendra Fm.), 9. Ignimbrite and hybrid tuff, 10. Porphyritic rhyolite (Kurse Kohri Fm.), 11. Rhyolite, 12. Volcaniclastic-epiclstic rocks, 13. Agglomerate (Pachangi Fm.), 14. Ferruginous shale, 15. BIF and iron ore (Bailadila Fm.), 16. Older mafic dykes, 17. Basement granite Gneiss, 18. Metasediments (Sukma-Bengpal Gneiss-supracrustals).

Bijli Rhyolite The rhyolites are a volumetrically dominant (> 80%) member of the volcanics of Nandgaon Group. Based on the abundance of the phenocrysts, the Bijli Rhyolites were classified into two (Sarkar et al., 1981) or three (Gangopadhyay and Roy, 1997) types. Chemically, the rhyolites show peraluminous composition with high SiO_2, high K_2O and low MgO, CaO, Fe_2O_3, and A-type granite character (Gangopadhyay and Roy, op. cit.; Divakar Rao et al., 2000). Neogi et al. (1996) reported the Bijli Rhyolites to be dominantly sub-alkaline, characterised by a high but narrow SiO_2 range of 72–75%, low LILE, Ti, Se, V, Ni, and high LREE, Zr, Th, Cr, Rb/Sr (2.48) and K/Rb (306). They display highly fractionated chondrite normalised REE patterns (La_N/Lu_N: 8–14), moderate to strong Eu anomalies and unfrationated HREE (Tb_N/ Lu_N: 0.9–1.5). The major element chemistry of Bijli Rhyolite is presented in the following Table 2.1.8.

Table 2.1.8 *Average major element composition of Bijli Rhyolite, Pachangi Rhyolite and Kurse Kohri Rhyolite/ignimbrite*

Major oxides (wt%)	Bijli Rhyolite				Pachangi Rhyolite	Kurse Kohri Ignimbrite
	I (n = 15)	II (n = 4)	III (n = 5)	IV (n = 25)	V (n = 7)	VI (n = 5)
SiO_2	71.40	71.17	74.35	74.36	72.22	66.92
Al_2O_3	14.28	14.69	12.66	13.17	10.84	12.72
Fe_2O_3	3.49	3.12	–	2.73	3.57	3.01
FeO	–	0.80	2.64*	–	1.30	3.79
CaO	0.98	0.63	0.54	1.05	1.76	3.10
MgO	0.30	1.24	0.23	0.23	0.72	1.45
Na_2O	4.19	2.06	4.23	3.64	3.80	2.92
K_2O	4.06	4.12	4.03	4.49	5.29	3.71
TiO_2	0.23	0.43	0.23	0.27	0.32	2.53
MnO	0.08	Tr	0.04	0.06	0.02	0.04
P_2O_5	0.26	0.73	0.03	0.04	0.44	0.17

Data source: I – Gangopadhyay and Roy, 1997; II – Sarkar et al., 1981; III – Neogi et al., 1996; IV – Divakar Rao et al., 2000; V and VI – Ghosh and Pillai, 2002.

The available Rb–Sr (WR) ages for the rhyolite of Nandgaon Group range from 2466 ± 22 Ma to 2180 ± 25 Ma (Sarkar et al., 1981; Krishnamurthy et al., 1988).

The lower most Pachangi Formation in the Kotri belt occurs along the western boundary as steeply dipping package of alternating epiclastic and pyroclastic sediments with layers of acid volcanics. The acid volcanics are melt phase rhyolite to ignimbrite. The package is dominant in sediments, with 10–25% volcanics. All the variants of acid volcanics in this package fall in the rhyolite–trachite field of the alkali–silica diagram (Ghosh and Pillai, 2002). The overlying Kurse Kohri Formation, exposed to the east of Pachangi Formation, comprises predominantly rhyo–dacite, ignimbrite, and varied types of acid–intermediate tuffs. Subordinate proportion of basic volcanics is also associated, which show mingling and mixing characters with acid volcanics. Major element data of the acid volcanics of Pachangi and Kurse Kohri Formations (Ghosh and Pillai, op. cit.) are presented in Table 2.1.8.

Pitepani Volcanics The Pitepani basic volcanics are represented by basalts and andesite, enriched in Sr, Ba,Zr and Cr, and show progressively more mafic character towards the top. This variation is also reflected in their REE pattern which changes from fractionated with negetive Eu anomaly in the lower levels to flat unfractionated pattern in the upper levels (Neogi et al., 1996).The basic volcanics of Mendra Formation in the Kotri belt, which are stratigraphically equivalent of the Pitepani volcanics, are represented by basaltic flows and volcaniclastics, mostly falling in the 'basaltic field' (Table 2.1.9). Sivkumar et al. (2003), based on large bulk of analytical data described the Pitepani volcanics as basalts and basaltic andesite ranging from sub-alkaline to tholeiitic types. They have also reported ultramafic variants, the composition of which plot in the peridotitic komatiite field but do not show the necessary field characters of komatiites.

Table 2.1.9 *Average chemical composition of Pitepani volcanics (Neogi, 1996)*

	20A	26 BL	60G	26G	60BL	21A	21B	21C	21D	30C	30R	3B
SiO_2	77.02	79.00	71.45	56.71	65.91	46.26	56.59	56.10	56.33	54.87	54.38	51.80
TiO_2	0.18	0.47	0.41	0.60	0.65	1.69	0.85	0.87	0.86	0.55	0.55	0.52
Al_2O_3	11.40	7.30	12.82	8.05	14.28	14.53	13.99	13.90	14.01	12.69	12.69	12.72
FeO^*	1.73	4.26	3.48	9.12	5.40	13.63	7.65	7.87	7,65	8.68	8.70	9.44
MnO	0.02	0.05	0.05	0.14	0.08	0.20	0.12	0.12	0.10	0.17	0.16	0.16
MgO	0.52	2.06	1.06	11.15	1.73	7.10	5.33	5.54	5.07	7.69	7.95	9.51
CaO	0.30	1.53	1.77	6.69	2.66	10.94	6.62	6.12	5.93	10.03	10.06	9.81
Na_2O	0.16	0.00	2.26	0.29	2.40	1.76	2.30	2.56	2.93	2.15	1.95	1.73
K_2O	7.31	3.73	4.85	2.13	4.32	0.65	3.06	3.26	3.09	1.20	1.16	1.41
P_2O_5	0.01	0.12	0.05	0.08	0.08	0.15	0.30	0.30	0.30	0.14	0.15	0.06
H_2O	0.46	0.18	0.25	0.52	0.22	0.26	0.32	0.44	0.64	0.22	0.21	0.23
LOI	0.82	1.50	1.33	3.91	2.07	2.46	2.40	2.65	2.78	1.44	1.47	2.17
Total	99.93	100.20	99.78	99.39	99.80	99.63	99.53	99.73	99.69	99.83	99.43	99.56

Evolutionary Trend in Nandgaon Volcanics Ramachandra (1994) demonstrated magma mingling and mixing characters of the Bijli Rhyolite with Dongargarh granitoids as well as the Pitepani basic volcanics and implied a genetic relation between the three. Sarkar et al. (1994) from major element mixing models opined that both Bijli Rhyolites and Dongargarh Granite were derived in two phases by extensive partial melting of the Amgaon Granite Gneiss basement, while the Pitepani Basalts were produced by partial melting of the ortho–amphibolites of Amgaon Group.

The compositional range of the Dongargarh volcanic suite is shown by the normative An–Ab–Or diagram (Fig. 2.1.8). Based on a more comprehensive geochemical study of the Nandgaon volcanics, Neogi et al. (1996) envisaged that a viable degree of partial melting of metasomatised mantle source led to the formation of initial LREE enriched felsic liquids in low melt fraction, followed by the generation of more and more mafic magma at higher degees of melting. These melts accumulated at the base of lower crust and might have induced some amount of melting of the crustal rocks, partly augmented by upward bending of geotherms and lithospheric thinning in an extensional regime. Though these volcanics show Island arc basalt (IAB) affinity, their genetic connection with a subduction related tectonic setting is ruled out by the authors based on the interlayering of the volcanics with epiclastic sediments (arkosic) and initial predominance of the acid volcanics with evidence of crustal contamination. While proposing a riftogenic origin for these volcanics,

Neogi et al (op cit.) rightly point out the pitfalls in assigning paleotectonic environments based on geochemical 'fingerprints' alone, as these are often complex functions of source characteristics and magmatic processes, not necessarily restricted to a particular tectonic setting. Sivkumar et al.,(2003) also considers the Bijli Rhyolites as tholeiitic rhyolite and dacite having formed in a rift tectonic setting rather than subduction setting for their evolution.

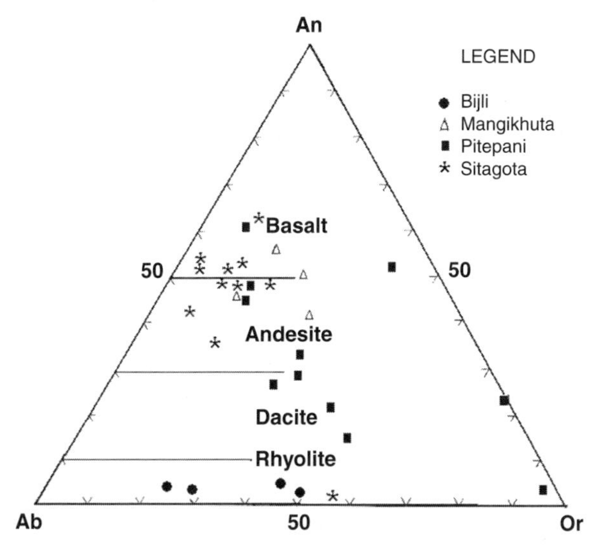

Fig. 2.1.8 Normative An–Ab–Or diagram showing compositional range of the Dongargarh volcanic suite (after Neogi, et al., 1996)

Dongargarh Granite and Equivalent Granitoids The Dongargarh Granite batholith occupies a N–S oriented ovoid area of about 3000 sq km in the southwest and west of Dongargarh (Kotri) Supergroup, comprising three distinct plutons viz, Amgaon Terha pluton in the north, Chichola pluton in the central part and the Dhanora–Manpur pluton (largest) in the south. There are ample evidence of the granite intruding the surrounding older gneiss–supracrustals (Sukma–Bengpal or Amgaon Suite), volcani–sedimentaries of Nandgaon (Ainhur) Group and the BIF bearing Bailadila Group. Enclaves of deformed and metemorphosed members of all the previously mentioned suites of rocks occur within the Dongargarh batholith.

The Dongargarh granitoid plutons display closely comparable mineralogy and texture. Compositionally also there is not much variation in their meta–aluminous character and a distinct tholeiitic lineage, showing post-orogenic rift related A-type and S–I transitional genetic trends (Ramchandra and Roy, 1998; Roy et al., 2000).

The magmatic fabric is generally discernible in the Dongargarh Granite in the form of porphyritic texture, rapakivi texture, occasional magmatic flow orientation of plagioclase phenocryts etc. The presence of enclaves of microgranular gabbro to granodiorite, faint development of north–south trending steeply dipping schistosity/gneissosity are common features. Prominent ductile/ brittle–ductile shear zones, marked by mylonitic stretches of granite, characterise the margins of most of the plutons.

It is important to note that the Dongargarh Granite and its equivalents occur in association with both low–grade metavolcanic assemblages as well as high-grade rocks (upto granulite facies). Sarkar et al. (1994) pointed out the geochemical similarity between spatially associated rhyolites belonging to the Nandgaon Group and the Dongargarh Granite, suggesting shallow level late to post-kinematic emplacement of these granites in the low grade (green schist to lower amphibolite) sequences of Nandgaon Group of rocks and almost syn–emplacement peak of metamorphism. The Malanjkhand Granite in the north, hosting the well-known copper deposit, is near contemporary to the Dongargarh Granite and shares similar shallow level emplacement history (Ramchandra and Roy, 1998). The Madanbera Granite (Ghosh and Pillai, 1992) and Sitagaon Granite (Mishra et al., 1988) exposed in the northern parts of Kotri belt also represent the shallow level intrusive phases of Dongargarh Granite, having comparable relation with the enclosing low grade volcani-sedimentary sequence.

In contrast to the above, the nearly contemporaneous Kanker–Mainpur granitoid batholith, variously termed as Bundeli, Chura, Kanker or Keshkal granitoids, covering an area of about 10,000 sq km in the north eastern parts of Bastar Cratonic Province (Bastar, Kanker and Raipur districts, Chhattisgarh), contains enclaves of high grade metamorphic rocks, including those of granulite facies. These granitoids are clearly intrusive into the rocks of Sukma (Bengpal) Gneiss–supracrustals as well as into the Neo–Archaean/Paleoproterozoic supracrustals of Bailadila and Sonakhan Groups. The batholith is bordered by the Eastern Ghat mobile belt (EGMB) to its east, the Kondgaon ganulites in the southwest, the Sukma (Bengpal) Gneiss–supracrustals along west and the Sonakhan Group rocks in the north. The recently discovered diamondiferous kimberlite bodies of south Raipur occur in the northeastern segment of Kanker–Mainpur batholith. These granitoids are compositionally more varied (granite–granodiorite–monzonite) compared to the type Dongargarh Granite, but both are closely comparable in S–I and A-type characters and REE patterns. Ramchandra and Roy (1998) and Ramchandra et al. (2001), held that the Kanker.Mainpur granitoid represented a deep-seated batholithic massif of the Dongargarh magmatic event, which had some connection with the granulitic crust formation. A possible model of evolution for the Kanker–Mainpur granitoids, as suggested by Ramchandran et al. (2001), contemplates basaltic underplating of the crust causing formation of granulites and associated reworking of sialic crust, contributing infra–crustal addition of granitoids. As for the shallow level intrusive phases of Dongargarh Granite and its equivalents, a direct genetic linkage is suggested between intracratonic rifting, bimodal volcanism and post tectonic granitoid emplacement.

The Dongargarh Granitoids have yielded Rb–Sr (WR) ages of 2270 $\pm$ 90 (Sarkar et al., 1981) and 2466 $\pm$ 22 Ma (Krishnamurthy et al., 1988). The Malanjkhand Granite ranges in Rb–Sr dates from 2100 Ma to 2362 $\pm$ 58 (Ghosh et al., 1986; Panigrahi et al., 1993), but a maximum age of 2.5 Ga was obtained by using U–Pb systematics in zircon (Panigrahi et al., 2002). This is further supported by Re–Os date obtained from molybdenite ore in Malanjkhand Granite (Stein et al., 2004) (vide Section 2.2).

Khairagarh, Abhujmar, Chilpi Groups These Paleo–Mesoproterozoic volcanisedimentary sequences (sediment dominated), post-dating the Dongargarh granitic activity are developed in detached segments along the belt, viz, Abhujmar in southern sector, Khairagarh in the central and Chilpi in the northern parts. The overall history of sedimentation, volcanism and deformation are considered to be near-contemporaneous in these basins despite local variations.

The Khairagarh Group is developed in the central part of Dongargarh belt in two NS to NNE–SSW trending stretches, viz, (i) south of Amgaon pluton (60 × 15 km) and (ii) Khairagarh area north of Dongargarh (50 × 39 km), intervened by Nandgaon Group of rocks. The subdivisions of Khairagarh Group into two formations and the respective litho-assemblages as per Sarkar (1957, 1958, 1981 and 1994) are presented in Table 2.1.10.

Table 2.1.10 *Stratigraphy of the Khairagarh Group (Sarkar 1957–1958, 1981, 1994)*

Chhattishgarh Supergroup	
……………………Unconformity……………………..	
Kotima volcanics	Khairagarh Group
Ghogra Formation	
Mangikhuta volcanics	
Karutola Formation (Sandstone)	
Sitagota volcanics	
Intertrappean shale	
Bartalao Formtion (Sandstone)	
Basal shale	
…………………………Unconformity………………………..	
Dongargarh Granite	
Nandgaon Group	
Bengpal Group/Amgaon Group/Sukma supracrustals	

According to this classification the Bartalao Formation, essentially comprising conglomerate quartz arenite (with arkosic components) siltstone–shale (often ferruginous) is the basal part of the Khairagarh sequence which unconformably overlies the Nandgaon volcanics and Dongargarh Granite. Presence of tuff along with conglomerate and sandstone was recognised in Bartalao Formation by Jain et al., 1995. The Sitagota volcanics overlie the Bartalao Formation and comprises tholeiitic to andesitic basalt with minor tuffs and agglomerates. The basalt at places contains xenoliths of Bartalao sediments. Sarkar et al. (1994) have reported some picritic layers in Sitagota volcanics. The Karutola Formation, comprising ferruginous arenitic units (cross-bedded at places) disconformably overlies the Sitagota volcanic pile. This is again overlain by basaltic flows with minor intratrappeans of Mangikhuta Formation.

The mafic volcanics of the Khairagarh Group are predominantly basaltic, of which the Sitagota low-K tholeiites show an unfractionated primitive character, while the Mangikhuta basalt is slightly fractionated. These volcanics have even closer affinity with volcanic arc basalt than Pitepani volcanics, but overall geological evidences are overwhelmingly in favour of rift related origin (Neogi et al., 1996).

The rocks of Khairagarh Group occurring to the north of Dongargarh show doubly plunging open syncline with axes trending NNE–SSW. Several major faults dissect the Khairagarh sequence, of which the most prominent and extensive are those that are aligned parallel to the eastern margin.

The Abhujmar Group unconformably overlies the Sukma Gneiss–supracrustals, Bailadila and Ainhur (Nandgaon) Group of rocks in the southern extension of Kotri belt, occupying a vast expanse of about 3000 sq km. Mishra et al. (1988) has subdivided the sequence into a lower Gundul Formation comprising sandstone–conglomerate–shale, overlain by Manspur basaltic flows with associated gabbro.

The Chilpi Group (Tripathi et al., 1981), essentially comprising psammopelitic sediments, unconformably overlie the Nandgaon volcanics and Malanjkhand Granite in the northern extension of Dongargarh belt between the northwestern margin of Chhattisgarh basin and Malanjkhand pluton. This sequence is correlated with the Khairgarh Group based on physical continuation of some quartzite units of Chilpi Group with Bartalao Sandstone (Rao, 1981). In Malanjkhand area, Bhargav and Pal (2000) have subdivided the Chilpi Group into a lower phyllite–shale– carbonaceous phyllite unit overlain by sandstone–arkosic grit–conglomerate (with minor phyllite–shale and crystalline limestone) unit.

The K–Ar age data (Sarkar et al., 1964) for the Khairagarh Group ranges from 1771 Ma (silty shale of Sitagota Formation) to 677 Ma (Sitagota Basalt). Apparently, there is some anomaly in the above age data. Sinha et al. (1998a) have provided an Rb–Sr date of 2123 ± 35 Ma for the basaltic rocks of this Group.

Sonakhan Group Another Late Archean to Paleoproterozoic supracrustal belt in the Bastar Cratonic Province is represented by the NW–SE trending Sonakhan greenstones covering an area of about 1200 sq km in the northeastern part. In the south it is surrounded by the older Sukma Gneiss–supracrustals and younger Mainpur–Kanker batholithic suite (locally named as Baya Gneissic Complex), having unconformable or tectonic contact with the latter, whereas in the north, the sequence is unconformably overlain by Chhattisgarh cover sediments (Fig. 2.1.9).

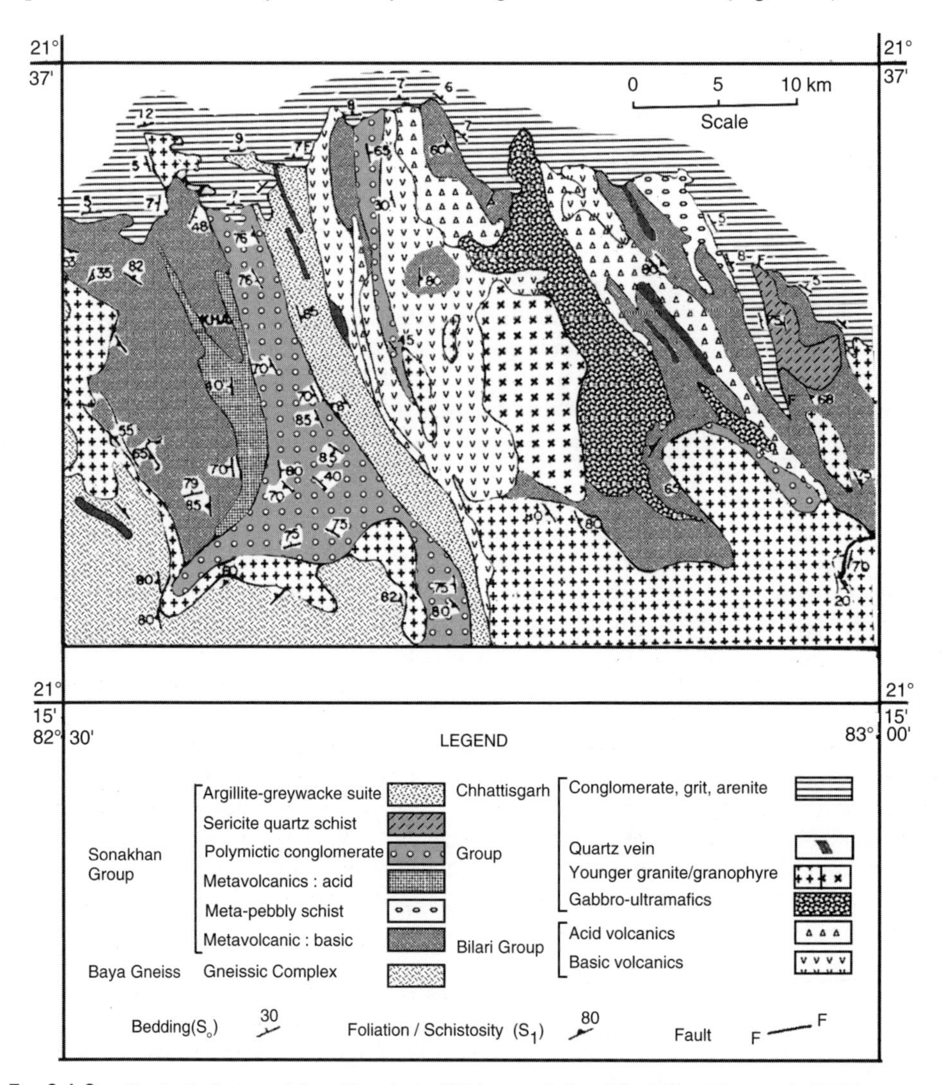

Fig. 2.1.9 Geological map of Sonakhan belt, Chhattisgarh (modified after Das et al., 1990)

Das et al. (1990) identified two groups of rocks in this belt, the lower Sonakhan Group (deformed and metamorphosed volcanisedimentary pile) and the upper Bilari Group (underformed and unmetamorphosed silicic–basic volcanics associated with intrusives).

Table 2.1.11 *Lithostratigraphy of the Sonakhan Group (Das et al., 1990)*

Chhattisgarh Supergroup	Conglomerate, grit, arenite	
....Unconformity..................................		
Basic dykes	Quartz vein, dolerite, gabbro	
Granitoids	Pegmatite, aplite, granophyre, biotite and porphyritic granite	
Gabbro, ultramafic intrusives		
Bilari Group	Lakhdabri acid igneous suite Arangi basic igneous suite	
Disconformity		
Sonakhan Group	Arjuni Formation	Meta–polymict conglomerate, greywacke, argillite, BHJ, arenite with thin bands of basic volcanics
	Baghmara Formation	Metaultramatics, pebbly tremolite/actinolite schist, amphibolite, metabasalt, pillowed metabasalt, pyroclastics, acid volcanics and tuffs BIF, cherts etc..
..................................*Unconformity*................................		
Baya Gneiss Complex	Tonalite, granodiorite gneiss, migmatitic granite, restites of BMQF, metaultramafics, mica schist, fibrolite quartzite etc.	

The Sonakhan Group is further subdivided into the basal Baghmara Formation, comprising pillowed metabasalt (Fig. 2.1.10), acid–basic pyroclasts, ultramafic schists, BIF, chert, etc., and the upper Arjuni Formation, comprising a greywacke–conglomerate–arenite–argillite suite.

Fig. 2.1.10 Pillowed metabasalt in Baghmara Formation, Sonakhan Group, Chhattisgarh (photograph by authors)

These rocks are tightly folded with steep dipping NNW–SSE trending axial planes and superposed cross folds, displaying largescale synformal structures. In contrast to the complex deformation history, the metamorphic grade of these rocks belongs to low green schist facies. The undeformed Bilary Group of rocks has restricted occurrence in the eastern part of the belt, which are profusely penetrated by intrusives of varied composition. Medium to coarse-grained granite with porphyritic and pegmatioid variants intrude the rocks of Sonakhan belt, probably representing a shallow level intrusive phase related to the Mainpur–Kanker granitoid batholith.

The rhyolitic rocks of Baghmara Formation and Bilari Group have yielded Rb–Sr dates of 2390 (± ?) Ma and 2102 ± 74 Ma respectively (Ghosh et al., 1995).

Sakoli Fold Belt The Sakoli fold belt (SFB) is located in the northwestern corner of the Bastar Crustal Province, covering about 3500 sq km. The fold belt is characterised by its subtriangular outcrop pattern (forming a part of the 'Bhandara triangle', vide Pascoe, 1950). Out of the three limbs of the SFB two are marked by prominent ductile shear zones, i.e along the N–S trending eastern limb separating the SFB from the Dongargarh–Kotri belt in the east (with intervening older Amgaon Gneiss–supracrustals) and along ENE–WSW trending north western limb (Central Indian Shear–CIS), demarcating it from the Sausar belt in the north (Roy et al., 1995, 2000; Bandyopadhyay et al., 1995). The southwestern margin of the SFB is partly in tectonic contact with older gneiss–supracrustals and partly covered by the sediments of Gondwana Supergroup. The Amgaon Gneiss–supracrustals also occur as embayments and inliers within SFB and extends in the north of the belt along the CIS (Fig. 2.1.11).

The stratigraphy of the Sakoli and Sausar belts as proposed by pioneer workers (Bhattacharjee, 1928–1932; Chatterjee, 1929 1932; Sarkar, 1957–1958) has undergone major revisions by later workers (Bhoskar, 1983; Roy and Bandyopadhyay, 1989; Roy et al., 1994; Bandyopadhyay et al., 1995). The four-fold classification of the Sakoli Group (Roy et al., 1994 and Bandyopadhyay et al., 1995), presented in Table 2.1.12 below is now widely accepted.

Table 2.1.12 *Lithostratigraphy of Sakoli fold belt (SKFB) (Roy et al., 1994; Bandyopadhyay et al., 1995)*

Gondwana Supergroup		
..................................*Fault contact*..................................		
Quartz veins, reefs and silicified zones. Alkali feldspar granite (Purkabori Granite), Pegmatite, tourmaline granite (Mandhal Granite), gabbro/ dolerite (metamorphosed)		
......................*Igneous intrusions*..........		
Sakoli Group	4.	Pawni Formation: Slate, phyllite, meta–arkose, quartzite, matrix-supported conglomerate
	3.	Bhiwapur Formation: Mainly metapelites (±chloritoid, andalusite, garnet, stautolite) with interbands of metamorphosed acid volcanics/ tuffs of rhyolite/ rhyodacite composition, minor psammites,exhalative sediments (coticules, tourmalinites), banded garnet–amphibole rock (BGA), rare basic volcanics and syngenetic basemetal (Zn, Cu) mineralisation
	2.	Dhabetekri Formation: Mainly metabasalt with subordinate metapelites, chert bands and metaultramafic rocks
	1.	Ghaikuri Formation: Conglomerate, gritty quartzite, meta–arkose, minor pelites (at places carbonaceous) and banded ferruginous quartzite (BIF) and BGA rocks
.........................Tectonised contact....................		
Amgaon Gneissic Complex (AGC) Pre–Sakoli rocks		Gneisses and migmatites, granitoids, amphibolite, chromite bearing metaultramafics and pre-Sakoli supracrustal assemblages of high grade schists including quartzite, kyanite and sillimanite schists, calc–silicate rocks, marble, cordierite–gedrite–anthophyllite schists, garnet–staurolite schists, etc.

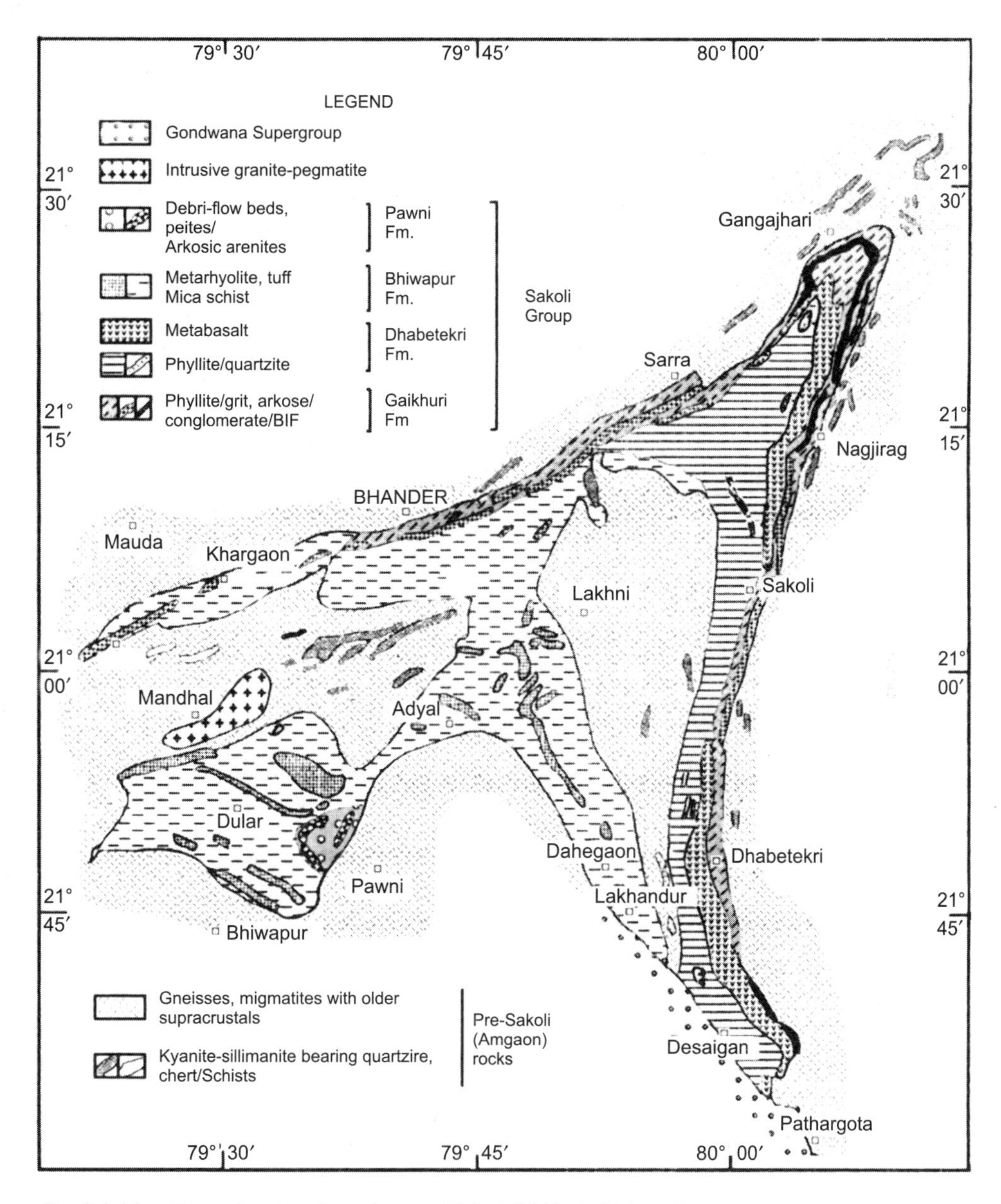

Fig. 2.1.11 Generalised geological map of Sakoli fold belt, Maharashtra

Salient features of the Sakoli fold belt, summarised by Roy et al., 2000, are as follows:

1. The sequence is characterised by a volcanisedimentary assemblage with dominant pelitic sediments and subordinate volcanics. The Amgaon Gneiss–supracrustals are apparently the basement of Sakoli Group, but the basement-cover relationship is obscured in most places due to later tectonic disturbance.
2. The volcanics of Sakoli Group show a bimodal character, represented by an older high-iron tholeiitic unit (Dhabatekri Formation) exposed along the eastern part and younger felsic volcanics and tuffs (Bhiwapur Formation) and late granitic intrusives showing calc–alkaline and tholeiitic trends, occupying the central and western parts.

3. Geochemically, the basic volcanics suggest rift-related emplacement at a craton margin (Roy et al., 1997). A separate magmatic source with contamination by partial melting of sialic crust is suggested for the felsic volcanics (Bandyopadhyay et al., 1992; Roy et al., 1997). An overall ensialic setting, peripheral to a craton margin is envisaged for the SFB.
4. Four episodes of fold deformations are recorded in the rocks of Sakoli Group, each represented by mesoscopic and macroscopic structures. The northeastern apex of Sakoli sub–triangle is interpreted as a largescale reclined to vertical fold closure of first generation (Bandyopadhyay et al., 1995; Roy et al., 1995). Regionally, the Sakoli fold belt represents a doubly plunging synclinorium with NE–SW striking sub-vertical axial surface. The triangular outcrop pattern of SFB is attributed to the combined effect of original basin configuration, superposed folding and the two bounding ductile shear zones along eastern and northwestern margins. Two episodes of low-pressure medium-temperature metamorphism, with the peak outlasting the third deformation, are recorded.
5. Several mineral occurrences are noted in this belt including those of copper, zinc–lead, tungsten, gold and platinoid. Other economic minerals like kyanite, sillimanite, fluorite, barite, corundum and pyrophyllite are reported from the pre-Sakoli basement Amgaon Gneisses. Small pods of chromite occur in ultramafic intrusives in the basement in the southern sector. Bandyopadhyay et al., (1992) recognised SEDEX type stratabound stratiform sub–economic basemetal sulphide and scheelite mineralisations in the Bhiwapur Formation of SFB. Epigenetic gold and platinum occurrences, spatially close to copper sulphide mineralisations, are also noted in the central part of the area.

The age of Sakoli rocks remains largely unresolved. Sarkar (1957–58) considered both Amgaon and Sakoli Groups to be older than 2500 Ma, implying an Archaean age. However, none of the age data for Amgaon Group (oldest Rb–Sr whole rock date of gneissic component is 2170 $\pm$ 60 Ma) or those of Sakoli rocks (in two distinct clusters of Rb–Sr (WR) date, 1290–1340 Ma and 720–923 Ma) made available by Sarkar et al. (1967, 1981) subcribed to this contention. Subsequently, the felsic volcanics of Sakoli Group also yielded Rb–Sr (WR) date of 1295$\pm$ 40 Ma (Bandyopadhyay et al., 1990) and Roy et al. (2000) reported an unpublished Pb–Pb model age of ca 1800 Ma for the Bhiwapur Formation. It was, therefore, presumed that the age range of 1290–1340 Ma represented the main episode of orogenic deformation and metamorphism in the Sakoli belt, while the age range of 720–923 Ma reflected the re–setting of the same during the younger Sausar orogeny. The Rb–Sr (WR) age of 1207 $\pm$ 61 Ma for Purkhbori Granite (Roy et al., 1994), which is intrusive into the SFB, further supported the upper age limit of Sakoli rocks as about 1200 Ma, and the 1800 Ma date of galena vein in acid volcanics could tentatively be considered as close to the emplacement of these volcanics.

Recently furnished new set of geochronological data (Roy, 2002) adds to the enigma of age relationship of the different volcanisedimentary components of SFB vis-a-vis the Amgaon Gneissic Complex, which has been considered as the basement complex based on well-studied geological relations. The metabasalt of Dhabetekri Formation yielded a six-point Sm–Nd isochron age of 2694 $\pm$ 84 Ma (E_{Nd} = 2.80 $\pm$ 0.15 with MSWD = 0.11). The same rock in Rb–Sr systematics yielded a much younger age of 1536$\pm$ 43 Ma (Sr_i = 0.70304 $\pm$ 0.00014, with MSWD = 1.84). On the other hand the acid volcanics of Bhiwapur Formation gave a seven-point whole rock Sm– Nd isochron age of 1769 $\pm$ 110 Ma (E_{Nd} = –5.9 $\pm$ 0.4, MSWD = 0.14). The Rb–Sr isochron age of the same rock is 870 $\pm$ 29 Ma (Sr_i = .0.7838 $\pm$ 0.0035, MSWD = 6.16), which may be taken as the degradational effect of Sausar orogeny. These sets of dates suggest emplacement of Dhabetekri basalt and Bhiwapur acid volcanics at wide apart time slots of about 2.7 Ga and 1.7 Ga. The Rb–Sr date of 1536$\pm$ 43 Ma for Dhabetekri basalt (Roy op. cit.) is interpreted as the effect of Amgaon orogeny, as this date closely compares with Rb–Sr (biotite) age range of 1460–1640 Ma for Amgaon Gneisses (Sarkar, 1994).

As per this interpretation the sedimentation and volcanism (Ghaikuri and Dhabetakni Formations) in Sakoli belt initiated at 2.7 Ga or even earlier (predating Amgaon orogeny) and continued upto 1.7 Ga, or even later (Bhiwapur and Pawni Formations). The eastern part of the belt bears the effects of Amgaon orgeny (1.5 Ga) and the south–central part preserves the effect of Sakoli orogeny (1.3–1.2 Ga), and the whole belt records isotopic resetting during much younger Sausar orgogeny (0.8–0.9 Ga) of the northern crustal province, CITZ. It is extremely difficult to reconcile to such huge span of time (> 1.0 Ga) for sedimentation and volcanism in a single Precambrian supracrustal belt, as the very large time gap between the initiation of basin formation and its inversion. Besides, under the given situation the Amgaon Gneiss–supracrustals also loses its long held status as basement complex with respect to the Sakoli supracrustals.

Northern (Bundelkhand) Crustal Province (NCP)

Bundelkhand Craton The granite and granite–gneiss massif with older supracrustal enclaves constituting the Bundelkhand craton is spread over nearly 26,000 sq km in a semicircular pattern (with southward convexity), in the northern fringe of the Precambrian shield bordering the Indo–Gangetic alluvium (Fig. 2.1.12). The massif is dominantly characterised by multiphase undeformed granitoids emplaced in an intensely deformed basement complex consisting of gneisses, migmatites, metasediments and metavolcanics, resembling an Archaean greenstone–granite complex. The whole sequence is intruded by numerous giant quartz reefs (NE–SW), accompanied by pyrophyllite and mafic, felsic and ultramafic dykes (NW–SE).

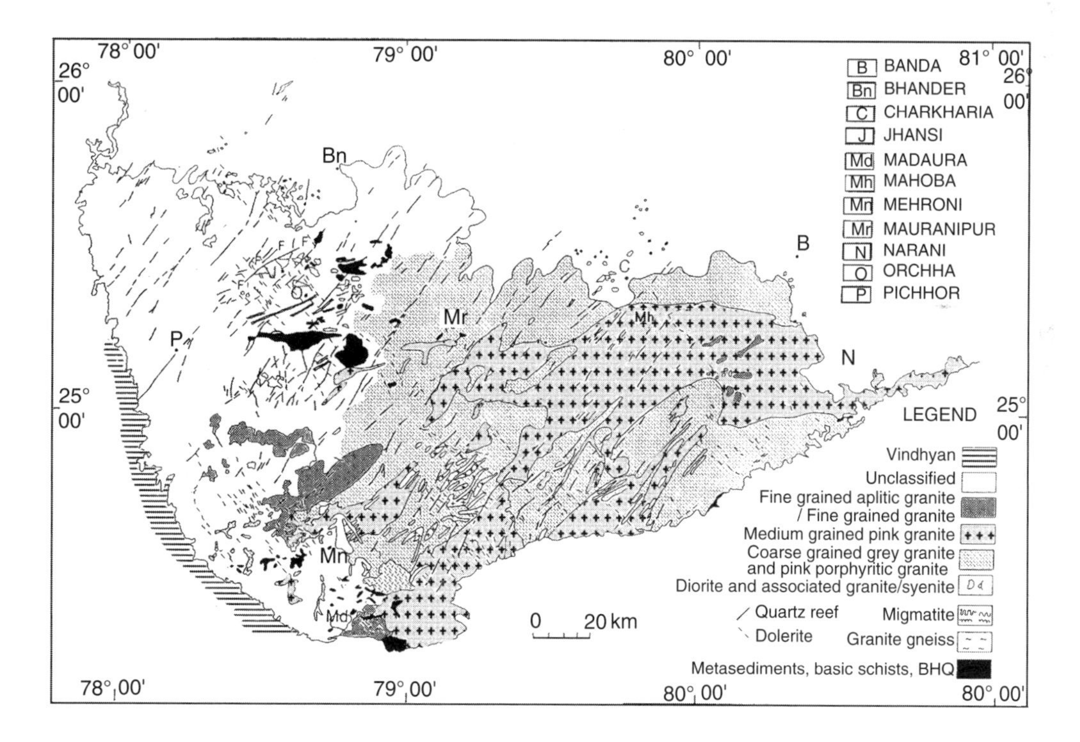

Fig. 2.1.12 Geological map of Bundelkhand Granite Complex (modified after Soni and Jain, 2001)

While working out the stratigraphy of the basement complex, some of the earlier workers (Prakash et al., 1975; Mishra and Sharma, 1975) felt that the older supracrustals represented two distinct lithostratigraphic formations separated by the event of migmatisation. Basu (1986) considered that the migmatisation post-dated both the formations, while Hanuma Prasad et al. (1999) concluded that the older supracrustals represented an earlier lithopackage that was subjected to the migmatitic event, followed by a later phase of felsic volcani-plutonic emplacement, prior to the late intrusion of granitoids.

Thus, the Bundelkhand cratonic massif may be broadly subdivided into three components, viz, (i) highly deformed enclave suite comprising TTG gneisses and old supracrustals (metasediments, meta-igneous rocks), migmatites, felsic volcani-plutonic rocks, (ii) undeformed multi-phase granitoid plutons and associated quartz reefs, (iii) mafic dyke swarm and other late intrusives.

Enclave Suite (Archean Granite–Greenstone Complex) There are three major E–W trending belts exposing the old crustal components in the cratonic massif, Karera–Jhansi and Babina–Kuraicha– Kabrai belts forming the northern tract and Madaura–Rajaula–Girar–Baraitha forming the southern tract. The northern tract has exposures stretching for more than 50 km comprising volcani-sedimentery packages of mafic–ultramafic volcanics, felsic volcanics, pelitic schists, calc-silicates, BIF and fuchsite quartzite. The southern tract is also characterised by a similar volcanisedimentary package with dominance of meta–ultramafics (peridotite and pyroxenite), fuchsite quartzite and basaltic flows and minor pelites.

The mafic volcanics are often interlayered with metasediments including BIF. The major element chemistry of these rocks is suggestive of tholeiitic composition with oceanic affinity (Basu, 1986). The ultramafic intrusives have high Mg-number and some of these show komatiitic affinity (Haldar et al., 1981). The felsic igneous rocks, more prominent in the northern belts, are compositionally akin to rhyolite and rhyodacite and are both fine and coarse grained.

Besides the volcanisedimentary enclaves described above, xenoliths of trondhjemitic gneisses cofolded with the supracrustals are also common within the migmatites and younger granitoids. Sarkar et al. (1996) studied one such occurrence near Baghora (Baghora Trondhjemite Gneiss) which is compositionally low-Al_2O_3 trondhjemite, but with distinctly lower SiO_2 and marginally higher iron and magnesia content. The trace element ratios are indicative of its similarity with average Archean low-Al_2O_3 trondhjemite belonging to the continental trondhjemite field. Further, the suite is moderately LREE enriched, depleted in Yb, and has perceptible negative Eu anomaly similar to the universally noted character of Archean trondjhemite gneisses (Sarkar et al., op. cit.). Another prominent occurrence of trondhjemite, associated with mafic–ultramafic and metasedimentary enclaves within late granitoid, in Loda Pahar hillock (east of Kabraj), has high Al_2O_3, low K_2O/Na_2O and depleted Yb content (Sharma and Rahman, 1995). In the central segment of the northern tract, there are enclaves of meta-aluminous TTG exposures showing intrusive relation with the associated metasediments and mafic-ultramafic components of early supracrustals.

Sarkar et al. (1996) reported Rb–Sr (WR) age of 3503 $\pm$ 99 Ma (Sr_i0.70025; MSWD= 1.65) for the Baghora Trondhjemite, which was interpreted as the emplacement date of the suite from a mantle derived protolith. Pb–Pb dating of single zircon crystals yielded an age of 3270 $\pm$ 3 Ma for the deformed gneissic enclaves within younger hornblende granite pluton near Rampura village. A biotite rich component within the gneissic complex also yielded an age of 3249 $\pm$ 5 Ma (Sharma, 1998). The Loda Pahar Trondhjemite gave a model Sm–Nd date of 3.2 Ga and Rb–Sr age of 2.5 Ga indicating a feeble effect of terminal deformation during late Archean. Mondal et al. (1998) gave age ranges of 3249 $\pm$ 5 to 3270 $\pm$ 3 Ma for Mahoba Gneisses and older supracrustal enclaves, and 3.0 to 3.2 Ga for Lalitpur xenocrystic Gneisses, based on extensive Pb–Pb zircon dating of a large number of samples from different localities.

Undeformed Granitoids Multiple phases of granitoids of varied petrography and composition constitute several plutons, coalesced together to form the composite Bundelkhand batholith. All the phases of granitoids are post-orogenic, as is apparent from their undeformed character. Sarkar et al. (1984) studied the grey granodiorite–granite suite of Jhansi area and the massive/porphyritic pink granites from Babina–Tabehat area. Mondal and Zainuddin (1996) established three compositionally distinct suites of undeformed granitoids with a well-defined sequence of emplacement, as detailed below.

Phase I: Meta-aluminous hornblende granodiorite, best developed in the northwestern parts of the massif; amphibolite enclaves in close proximity.
Age: 2402 ± 70 Ma (Rb–Sr/WR) (Sarkar et al., 1984); 2492 ± 9 Ma (Pb–Pb /zircon) (Mondal et al., 1997–98)

Phase II: Biotite granite, most voluminous with wide variation in texture from fine to porphyritic and colour from grey to pink, is most common.
Age: 2352 ± 30 to 2316 ± 80 Ma (Rb–Sr /WR) (Sarkar et al., 1984, 1990)

Phase III: Leucocratic granitoid, medium to fine grained, white to pinkish in colour and intruding both the older phases; associated with aplitic veins.
Age: 2214 ± 142 to 2271 ± 50 Ma (Rb–Sr/WR) (Sarkar, et al., 1984, 1990)

Rhyolites with large phenocrysts of feldspar and quartz occur in tectonic contact with the granitoids in Jhansi–Lalitpur section. Pb–Pb dating of zircon from the rhyolite yielded an age of 2521 ± 7 Ma, suggesting it to be the extrusive phase of the granitic magmatism.

The entire terrain of Bundelkhand Granite–Gneiss massif is characterised by the presence of spectacular NE–SW trending giant quartz reefs, generally traceable for 30–40 km with an average width of about 50 metres. The largest reef passing through Nivari extends almost uninterrupted for 100 km. Smaller reefs of 10 to 30 m width and 2 to 3 km length are numerous. The field evidences suggest the emplacement of these quartz reefs along brittle–ductile shear zones and associated hydrothermal activity responsible for profuse pyrophyllite–diaspore and sporadic sulphide mineralisation, marking the terminal stage of granite magmatism in the Bundelkhand craton.

Pegmatites are widely distributed as small bodies related to the different phases of granitoid emplacement.

Post-granitoid Intrusives

Mafic Dyke Swarms Mafic dykes (mostly dolerite), showing geochemical affinity toward continental tholeiite, abound in the Bundelkhand Granite–Gneiss massif. The dykes run mostly in NW–SE direction in en–echelon pattern, and show intrusive relation with all the granitoids and the related quartz reefs.

Two age clusters viz, 1825–1787 Ma (mean 1804 Ma) and 1558–1523 Ma (mean 1538 Ma) were established by Sarkar (1997) through K–Ar dating of several mafic dyke bodies. In view of undeformed, unmetamorphosed and unaltered character of the mafic dykes, the above dates are inferred to be their emplacement ages. Sharma (2000) considered the mafic dyke swarm to be related to an extensional event, due to mantle upwells and plume activity, soon after the cratonisation of Bundelkhand massif.

Dykes of other Types Besides the mafic dyke swarm, there are several other varieties of dyke rocks distributed across the terrain. Several varieties of porphyry dykes occur that include diorite porphyry, syenite porphyry and granite porphyry. There are smaller ultramafic intrusives having perdotite–gabbro interbands, confined to the southern margin of the massif. Other types of intrusives are of orbicular granitoid, anorthosite, felsite, keratophyre, lamprophyre, carbonatite and pyroxenite.

Tectono–magmatic Evolution From the available database, it may be surmised that the Bundelkhand craton records polyphase deformation and metamorphism of ancient supracrustal components, attendant with at least three events of syn- to late-kinematic TTG magmatism from middle to late Archean time (3.3 Ga, 3.0 Ga and 2.7 Ga). This was followed by ocean–continent collision related generation of potash–rich magma and emplacement of late to post-kinematic multi-phase granitic plutons in a continental volcanic arc setting (Andean type) during end–Archean to Paleoproterozoic period (2.4 Ga, 2.3 Ga and 2.2 Ga). The subsequent extensional regime resulted in the emplacement of mafic dyke swarm, melt generated by deep seated mantle plumes below a domally uplifted and thickned batholithic complex (Sharma and Rahman, 2000).

Bijawar and Gwalior Groups The Bundelkhand craton is flanked along its southern and northwestern margin by siliciclastic shelf sequence associated with chemogenic sediments and volcanics, and intrusive rocks, which are included in Bijawar and Gwalior Groups. The Bijawar rocks occur along the northern margin of theVindhyan basin in two adjacent ENE–WSW and E–W trending stretches of Bijawar type-area (80 km) and Sonrai area (28 km), besides which there is another occurrence far to the south west in Narmada valley in Barwah–Handia area (Harda inlier) (Fig. 2.1.13). The rocks of Gwalior Group, considered as equivalent of Bijawar Group, occur with E–W trend over about 2000 sq km in parts of Gwalior, Dantia and Bhind districts.

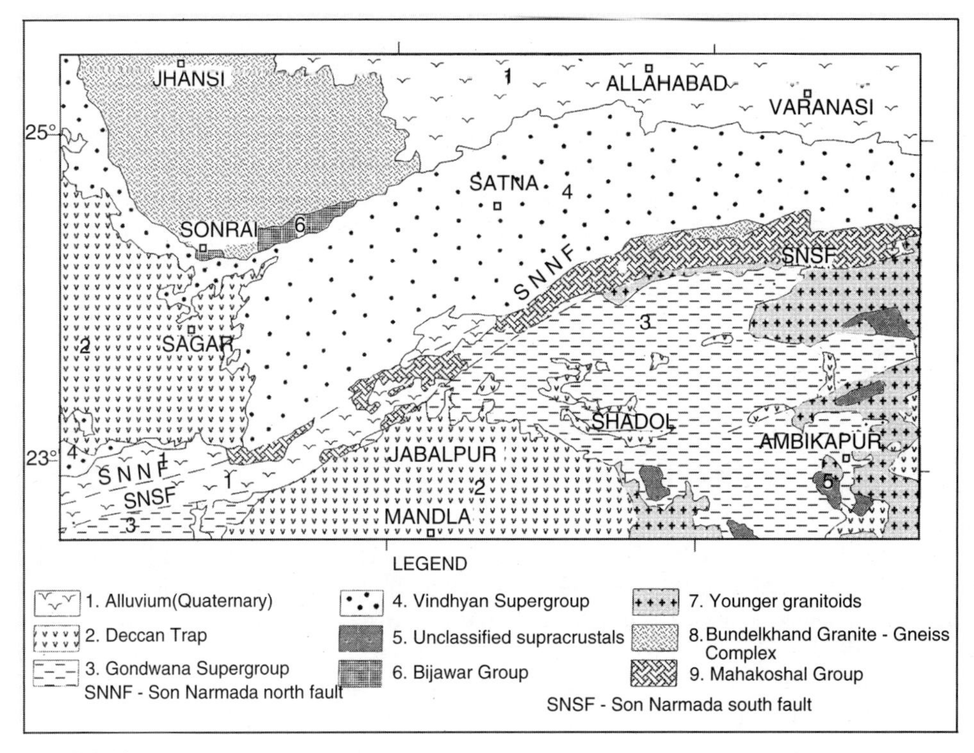

Fig. 2.1.13 Geological setup of Bijawar and Mahakoshal belts in Northern Crustal Province (NCP) (modified after GSI, Spl. Pub. 77, 2004)

The lithostratigraphic classification of Bijawar Group, proposed by Kumar et al., 1990 (Table 2.2.13) gives a detailed account of the various lithounits exposed in the type area.

Table 2.1.13 *Lithostratigraphy of Bijawar Group (Kumar et al., 1990)*

Group	Sub group	Formation	Member	Lithology	Thickness (m)
Semri (Vindhyan Supergroup)					
.... *Erosional and structural unconformity*					
Bijawar Group	Gangau Subgroup	Karri Ferruginous Formation		Basic intrusives and quartz veins. Highly ferruginous sandstone and shales with intraformational conglomerate and breccia	120 m
		Hirapur Phosphorite Formation		Phosphoric shales sandstone, chert, dolomite and breccia	60 m
		.....................*Unconformity*........................			
	Moli Subgroup	Dargawan intrusive Formation		Coarse to medium grained dioritic and gabbroic rocks	100 m
		Bajno dolomite Formation		Stromatolitic crystalline dolomite with chert interbeds	100 m
		Melehra chert breccia Formation	Raidaspura member Sendhpa member	Quarts-arenite, dolomite, conglomerate and basaltic trap	55 m
		Kawar volcanic Formation		Basaltic trap, quartz arenite and basal conglomerate	70 m
.... *Erosional and structural unconformity*					
Bundelkhand Granitoid Complex			Granite, granodiorite and variants with supracrustal enclaves. The sequence intruded by basic and ultrabasic dykes, quartz veins and quartz reefs		

In this sector, volcanic flows of restricted extent interbanded with polymictic conglomerate and arenites unconformably overlie the Bundelkhand basement complex. The succession upwards is dominated by chemogenic sediments like bedded chert, jasperite and dolomite with thin interbands of terrigenous siliciclastics; and some basic intrusive and extrusive units. The upper most part of the sequence is represented by alternations of shale, sandstone and minor dolomite with prominent ferruginous and phosphatic formations. The Vindhyan Supergroup unconformably overlies the whole sequence.

The major and trace element data of the volcanic flows of basal Bijawars indicate compositions ranging from sub–alkaline tholeiitic basalt to basaltic andesite, showing close affinity with continental intraplate volcanics with variable crustal contamination. The REE content suggests a depleted mantle source for the basaltic magma (Rao and Rao 1996; Haldar and Ghosh, 2000). Compositionally, the intrusive bodies of Dargaon Formation are olivine normative sub–alkaline tholeiite, which yielded Rb–Sr whole rock isochron age of 1789 ± 17 Ma (Sr_i – 0.70612 ± 38; MSWD = 0.55). The higher initial Sr ratio (Sr_i) is indicative of significant crustal contamination of mafic parent magma (Sarkar, 1997).

The stratigraphic sequence (Table 2.1.14) of Bijawar Group in the Sonrai sector (Lalitpur District., UP) is not exactly same as that of the type area but here also the basal horizon represents a shallow shelf deposit, passing upwards into shale and carbonate dominated facies. Black shale with syngenetic sulphide and discrete phosphate beds are common. Basic volcanic flows and chloritic shale occur at higher levels, overlain by ferruginous sandstone and shale at the top.

Table 2.1.14 *Stratigraphic succession of Bijawars in Sonrai area (Srivastava, 1989)*

Formation	Member	Lithology
Solda Iron Formation	Solda–Dhorisagar sandstone–chloritic shale	Brown, fine- to medium-grained, ferruginous sandstone, shale and quartzite, with rich pockets and lenses of hematite; dark grey calcareous sandstone with interbedded shale; brown weathering, often hematitic on the surface
		Dark greenish grey, massive to thinly laminated sandy chloritic shale; lenses of sandy calcareous shale are sometimes present (200 m thick)
	Bandai	Grey arenaceous dolomite and calcareous sandstone interbedded with pyritic calcareous shale locally carbonaceous/graphite (40–50 m)
Sonrai Formation	Rohini Carbonate	Kurrat basic flows, grey laminated shaly dolomite interbedded with arenaceous dolomite, often pyritic; discrete phosphate beds, grey sandy dolomite; phosphatic towards base (altered into silicified and phosphatic breccia by supergene process) 280–290 m Off–shore deposit
	Gorakalan Shale	Black carbonaceous shale and calcareous shale with pyrite and graphitic matter (20–30 m). Lagoonal deposit
	Jamuni Member	Grey calcerous shale and brown weathered dolomite grading laterally into phosphatic shale and siltstone, brownish grey thick-bedded dolomite, chert, pebbly sandstone and grit (110–120 m) Shore zone shelf deposit
..............................Bewar Formation/Basement Complex......................		

The stratigraphic classification of the rocks of Gwalior Group (Shrivastava and Shrivastava, 1989) furnished in the Table 2.1.15 below reflects an uneven subdivision of the sequences into lower Par Formation (14–80 m thick) overlain by Morar Subgroup (600 m thick). The Par Formation that unconformably overlies Bundelkhand Granite–Gneiss basement is made up of gritty and flaggy sandstone and quartzite. The overlying Morar Subgroup comprises an interbanded sequence of siliceous and ferruginous shale, chert, limestone, jasperite, with local pockets of clay and hematite. Basic sills (Gwalior Trap) occur at two levels in this sequence, of which the upper one is more prominent, which was dated at 1830 ± 200 Ma (Crawford and Compston, 1979).

The undeformed flat lying Gwalior Group, trending E–W to ENE–WSW, shows steady low dips of 5° towards north.

Table 2.1.15 *Stratigraphic succession of Gwalior Group (Shrivastava and Shrivastava, 1989)*

Age	Group			Lithology
Recent	Alluvium			
Middle Proterozoic	Kaimur Group			Sandstone with breccia and conglomerate at the base
..................................*Unconformity*....................................				
	Gawalior Group	Morar Subgroup	Quartz veins	
			Akbarpur Formation	Green shales with bands of limestone, chert and jasper
			Biraoli Formation	Dioritic sill, ferruginous chert, shale jaspilite, bands of clay and pockets of iron
			Nayagaon Formation	Limestone with bands of green shales with pockets of clays
			Sithanli Formation	Ferruginous shales with bands of chert jasper and limestone at the base
			Sitla Formation	Ferruginous shales with bands and lenses of chert, jasper
Lower Proterozoic			Par Formation	Orthoquartzitic sandstone with alternation of green shales and glauconitic sandstone, flaggy sandstone, gritty sandstone pebbly at base.
..................................Unconformity....................................				
Archean*			Bundelkhand Granite	Pink to greyish white prophyritic granite and granite gneiss with quartz veins and reefs and basic dykes

*Vide the age of Bundelkhand Granite already discussed.

Evolution With regard to the history of sedimetation and volcanism for the Bijawars and equivalents, it is suggested that rifting along the margins of the stabilised Bundelkhand craton, in a subsequent extensional regime, led to the development of a number of small isolated intra-cratonic basins in a graben–horst system, in which there were initial deposition of terrigenous sediments and rift related volcanic products. This was followed by significant chemical precipitation, starting with silicic deposits and ending with the carbonate in broad tidal flats. Stromatolite bioherms and cross–bedded sandstone within carbonate are indicative of local shallowing of the basin and influx of terrigenous material. A period of quiescence is attributed for the deposition of phosphatic sediments and sulphide-bearing black shale facies (Pant and Banerjee, 1990). An overall marginal marine basin environment with deltaic condition and recurrence of reducing environment during regressive phases are contemplated for the sedimentation in Sonrai area (Prakash et al., 1975)

The available age data are suggestive of a close relation between the basement-fracturing rift/graben development, Bijawar sedimentation and mafic igneous activity at about 1.8 Ga (Sharma, 2000). This age corresponds with the older age cluster (1825–1787 Ma) of post-granitoid intrusives within the Bundelkhand craton.

Central Indian Tectonic Zone (CITZ) The Precambrian crust of Central India is characterised by the existence of a very wide (~ 200 km) ENE–WSW trending zone of accretion between the Bundelkhand cratonic nucleus in the north and Bastar craton in the south, named as Central Indian Tectonic Zone (CITZ) (Radhakrishna, 1989; Acharyya and Roy, 2000; Acharyya, 2001; Roy and Hanuma Prasad, 2003). The CITZ, bound by the Son–Narmada North Fault (SNNF) in the north and the Central Indian shear zone (CIS) in the south, predominantly records the history of Proterozoic crustal development (< 2.5– 0.9 Ga) in this region. From north to south, the Mahakoshal, Betul and Sausar represent prominent supracrustal belts of different metamorphic grade within CITZ, which are set in largely undifferentiated gneisses (with TTG components) and syn- to post-kinematic granites. Besides, there is another highly tectonised dismembered supracrustal belt comprising tectonic slivers of metamorphites of different grades and mylonitised granite along southeastern parts of CITZ, bordering the Chhattisgarh basin. The CITZ is further characterised by several brittle–ductile and ductile shear zones, most prominent of which are the northern two sets of Son–Narmada North Fault (SNNF) and Son–Narmada South Fault (SNSF) bounding the Mahakoshal belt, and the southern two sets Gavilgarh–Tan Shear (GTS) and Central Indian Shear (CIS) bounding the Sausar belt. Several other minor shear zones traverse the CITZ, paralleling the general trend of this accretionary zone and often mark the margins of the supracrustal belts. The CITZ also comprises three granulite belts viz, the Makrohar granulite belt (MG) occurring south of the Mahakoshal supracrustals, Ramakona–Katangi granulite belt (RKG) along northern margin of Sausar supracrustals, and Balaghat–Bhandara granulite belt, marking the southern boundary of Sausar belt (Fig. 2.1.14).

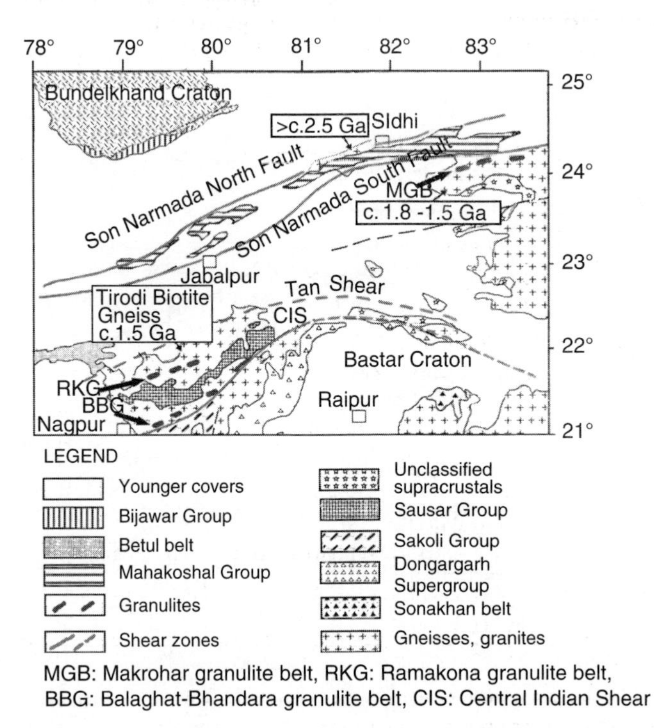

Fig. 2.1.14 Regional tectonic frame of CITZ (after Roy and Hanuma Prasad, 2003)

The CITZ, which forms a part of Satpura Province (King, 1898; Middlemiss, 1915; Pascoe 1928), is largely covered by rocks of the Vindhyan Supergroup, Gondwana Supergroup and the Deccan volcanics. The Meso–Neoproterozoic platformal cover sediments of the Vindhyan Supergroup are restricted to the northern parts of this terrain, while the Paleozoic rift-controlled Gondwana sediments and late Cretaceous Deccan flood basalts form more extensive cover in the central and southern parts. There is ample evidence to suggest that some of the crustal scale shear zones of the CITZ were reactivated in later periods, the latest being the present day prominent seismicity witnessed along the Son–Narmada South Fault.

The various components of the CITZ, i.e the supracrustal belts, granulite belts and shear zones are described in the following pages.

Mahakoshal Supracrustal Belt The ENE–WSW to E–W trending Mahakoshal supracrustal belt extends in strike for about 600 km from south–west of Jabalpur, Madhya Pradesh to Palamau district in Jharkhand, with an average width of 20 km, covering an areal extent of about 9000 sq km. It is a fault-controlled asymmetric rift basin, bound by SNNF and SNSF along the north and south. The Vindhyan Supergroup borders it in the north, except for a limited stretch in Sidhi area, where a linear belt of basement (Archean) gneissic complex intervenes. The southern margin of the belt is marked by the vast expanse of Proterozoic granitic intrusives, or is juxtaposed against the rocks of Gondwana Supergroup, with the prominent SNSF passing in–between (Fig. 2.1.12).

Litho-assemblage of the Mahakoshal belt is represented by quartzite–carbonate pelite–chert–BIF–greywacke, basic metavolcanics– pyroclastics–ultramafic intrusives and several post-tectonic granitic, syenitic and alkaline intrusives. Nair et al., 1995, provided the stratigraphy of the belt (Table 2.1.16).

Table 2.1.16 *Stratigraphic succession of Mahakoshal Group (Nair et al., 1995)*

Vindhyan Supergroup and Jungel Group of sediments	
............Unconformity and faulted contact.........	
Intrusives	Dunite, gabbro, dolerite, quartz porphyry, quartz veins, syenite and associated alkaline dykes, carbonatite, barite veins Lamprophyres, trachyte and associated intrusives. Barambaba granite and equivalents
Parsoi Formation	Tuffaceous and carbonaceous phyllites, feldspathic quartzite and conglomerate, tuffaccous phyllites with metabasalt intercalations
Agori Formation	Banded hematite/magnetite quartzite and jasperiod with associated tuffs and ash beds Impure marble, dolomite and interbedded calc–silicates—chlorite schist with occasional metabasalt lenses, conglomerate
Chitrangi Formation	Basic and ultrabasic plugs and dykes including peridotite and serpentinite, agglomerates, metabasalt and peridotitic pillow lava
———— Faulted contact/local unconformity ————	

As per the scheme of classification presented above, the lower most Chitrangi Formation represents a volcanic assemblege comprising mafic and ultramafic lava with associated dykes and ultrabasic plugs. The middle Agori Formation consists of clastic and non-clastic sediments with minor volcanics. The upper Parsoi Formation comprises thick succession of argillites and turbidites. The end of sedimentation is marked by an array of varied intrusives.

Roy and Devrajan (2000) presented a revised stratigraphic scheme for the Mahakoshal Group, in which the components of Chitrangi and Agori Formations were clubbed together under Sleemanabad Formation, characterised by sediment-dominant lower part and metavolcanics in the upper part. This is followed upwards by phyllite–grewacke–quartz arenite sequence of Parsoi Formation with basal unconformity marked by polymictic conglomerate. This scheme also introduced 'Dudhmaniya Formation' as the youngest horizon, comprising BIF and phyllites, developed mainly in the south eastern parts of the belt (Chitrangi–Gurharpahar–Dudhmaniya area), is characterised by the occurrence of gold bearing quartz–carbonate veins.

Rocks of the Mahakoshal Group are folded into upright to overturned isoclinal folds with ENE–WSW striking and steeply (80°) southward dipping axial planes. The deformation was also accompanied by prominent reverse–slip ductile shearing along SNSF, which led to the emplacement of granitic rocks along this zone (Roy et al., 2002b).

Nair et al. (1995) and Jain et al. (1995) considered the Mahakoshal as a greenstone complex, developed in a typical continental rift setting and the basin representing a spreading centre. Roy and Bandyopadhyay (1990, 1998) with Roy and Devrajan (1999) also subcribed to the concept of continental rift setting for the development of the Mahakosal Group.

According to Roy and Devrajan (2000), a pericratonic basin, developed along the southern margin of Bundelkhand craton, witnessed initiation of Mahakoshal sedimention in a shallow marine environment. This was followed by rifting of the basin aided by thermal doming, resulting in emplacement of tholeiitic lava, pyroclasts and co–magmatic ultramafic intrusives. The rifting aborted and subsequent upliftment of rift shoulders led to the deposition of debri-flow sediments, culminating into the deposition of pelite–BIF during thermal relaxation.

Geochemical Character of Volcanics The geochemical character of mafic volcanic rocks of Mahakoshal belt suggests high degree of melting of shallow mantle source in a rift environment (Kumar, 1993; Nair et al., 1995; Roy and Hanuma Prasad, 2003).

Nair et al. (1995) have shown that the metabasalts and gabbroic intrusives of Mahakoshal belt are MORB type tholeiites, but the plots of ultramafic intrusives fall within komatiitic field (Fig. 2.1.13). The trachytes and syenites on the other hand indicate calc–alkaline affinity. Roy and Bandyopadhyay (1989), based on major and trace element chemistry, also suggested that the metabasalts of Mahakoshal belt represented a Fe-rich, K-poor theoleiitic suite, possibly derived from a considerably evolved magma, emplaced in an intracratonic, mantle activated rift zone (Table 2.1.17).

Table 2.1.17 *Chemical composition of the igneous rocks of the Mahakoshal belt (Nair et al., 1995 and Roy and Bandyopadhyay, 1989)*

	Range of composition (Nair et al., 1995)					Mean (Roy and Bandyopadhyay, 1989)
Major oxides (wt%)	Metabasalt (n = 21)	Ultrabasic intrusives (n = 9)	Trachyte and associated rocks (n = 23)	Syenite and associated alkaline suite (n = 14)	Gabbro (n = 6)	Basalt (n = 27)
SiO_2	41.52–57.68	37.30–49.55	35.73–68.75	55.06–71.07	41.17–52.74	50.68
Al_2O_3	9.35–23.74	2.14–8.19	9.20–19.32	13.21–17.84	9.23–15.37	13.43
FeO	7.56–12.50	4.49–9.25	0.56–10.80	0.90–4.50	6.73–10.98	10.50
Fe_2O_3	1.44–9.37	2.90–8.20	1.40–13.83	0.60–23.00	1.80–8.80	3.29
TiO_2	0.47–2.28	0.33–2.88	0.40–2.08	0.10–1.35	1.05–1.41	1.14
MnO	0.06–0.52	0.12–0.89	0.01–0.25	0.01–3.13	0.11–0.19	0.21
CaO	2.65–20.35	4.60–12.67	0.49–9.87	0.10–5.13	4.90–11.19	0.25
MgO	4.38–9.30	11.13–27.83	0.42–11.21	0.10–5.13	5.64–8.32	6.09
CO_2	0.23–16.0	0.23–8.18	0.57–13.78	0.15–2.15	0.45–13.50	–

(Continued)

(Continued)

	Range of composition (Nair et al., 1995)					Mean (Roy and Bandyopadhyay, 1989)
Na_2O	0.97–4.10	0.02–0.47	0.10–4.40	2.24–6.75	0.87–3.10	2.17
K_2O	0.04–1.50	0.0–13.25	0.25–13.08	0.17–9.41	0.13–2.96	0.25
P_2O_5	0.10–0.48	0.13–2.52	0.01–1.23	0.01–0.89	0.01–0.47	–
$-H_2O$	0.06–0.61	0.04–0.22	0.06–0.41	0.10–0.43	0.02	–
$+H_2O$	1.77–3.82	1.08–4.41	0.31–3.49	0.10–1.00	1.88–2.63	–
CO_2	0.23–16.0	0.23–8.18	0.57–13.78	0.15–2.15	0.45–13.50	–

The Jungel volcanic complex that lies in the eastern part of this belt in Mirzapur District, UP, includes varied magmatic associations, having early tholeiitic phase and a later suite of alkali picritic basalt, alkali olivine basalt and associated tuffs. Besides, dark coloured, zoned lamprophyric dykes are reported from this area, with rims of olivine basalt and an inner core of camptonite (akin to olivine alkali basalt). The pyroclastic rocks are typically undersaturated with rather high alkali. These are further characterised by high CaO, MgO, TiO_2, P_2O_5 and depleted Al_2O_3. Volcanic bombs are relatively more acidic than their host tuff.

Granite Gneiss and Granitic Plutons A narrow sliver of basement gneisses (Bundelkhand Granite–Gneiss) is exposed for a limited strike length along the northern boundary of Mahakoshal supracrustal belt in Sidhi area, while the rest of the northern contact is with the rocks of Vindhyan Supergroup, SNNF marking the contact zone.

Granite gneisses, migmatites and granitic plutons occupy vast areas in the south of Mahakoshal belt, exposed along the southern tectonised contact (SNSF), as well as farther to the southeast, though largely covered by the Gondwana sediments.

The gneissic rocks and the granitic intrusives of this terrain are described under the basket term 'Dudhi Complex' (classified as Dudhi Group in Mirzapur district of UP) and considered as equivalent to Chhotanagpur Gneissic Complex of Jharkhand. The entire complex is characterised by a strongly developed E–W fabric.

The gneissic complex comprises finely foliated leucocratic granodiorite with discontinuous bands of amphibolite and local lenticular patches of massive garnetiferous rock. In the western sector in Sleemanabad area, thin strips of augen gneiss associated with garnetiferous schist occur along the Mahakoshal supracrustals (Devrajan and Hanuma Prasad, 1995–1996). In the Makrohar area, Sidhi district, tonalitic gneiss containing enclaves of garnet–sillimanite–corundum schist, two–pyroxene granulite, marble and calc–silicates have been reported (Pitchai Muthu, 1990). Geochemically the amphibolites in the banded gneissic country are of tholeiitic character, whereas the gneisses can be grouped as calc–alkaline, peraluminous granodiorite–adamellite.

There are several plutons of intrusive granite in the Dudhi Gneissic Complex, which are strongly mylonitised and contain enclaves of Mahakoshal rocks along the sheared contact zone (SNSF). The intrusive phases may be broadly classified as calc–alkaline, per- to metaluminous rocks of granitic to dioritic composition. The granitic intrusives are followed by a series of alkali intrusives (syenite, trachyte, ijolite, minette, tinguite and carbonatite). Sarkar et al. (1998) studied the following four suites of granitic plutons from the eastern segment of the belt: Jhirgadandi pluton, granitoids of Renusagar–Rihand dam area, Muirpur Granite, and Dudhi Granite Gneiss.

Jhirgadandi Pluton This is a 12-km-long lensoid body; trending E–W, with maximum width of 2 km. Medium- to coarse-grained monzo–diorite and monzonitic quartz syenite with phenocrysts of K– feldspar constitute the pluton. It is characterised by enriched LREE, depleted HREE and minor Eu anomaly. The rock yielded six-point Rb–Sr (WR) isochron age of 1813 $\pm$ 65 Ma.

Renusagar–Rihand Dam Area It is a 25-km-long concordant wedge shaped granitoid body. Chemically, the rock is peraluminous, with a calc–alkaline trend. The rock yielded a 11-point Rb–Sr (WR) isochron age of 1731 ± 36 Ma.

Muirpur Granite This granitic body occurs to the south of Rihand, exhibiting leuco- to mesocratic granodiorite–adamellite with strong sub-vertical foliation striking ENE–WSW. It has yielded 5-point Rb–Sr (WR) Rb–Sr isochron age of 1709 ± 102 Ma.

Dudhi Granite Situated to the south of Muirpur, pink granitic gneiss, adamellite to granite in composition, peraluminous. LREE enriched with pronounced negative Eu-anomaly 6-point Rb–Sr(WR) isochron age of 1576 ± 76 Ma.

In addition to the above geochronological data, a few more Rb–Sr whole rock age determinations are available from the granitic and syenite rocks of this area (Tamakhan Granite – 1856 ± 68 Ma by Bandyopadhyay et al., 1990; Sidhi granitoid – 1850 ± 40 Ma by Sarkar et al., 1998; Bari Syenite –1800 Ma by Nair, 1995; Bari–Aniti Syenite –1361 ± 60 Ma by Panday et al., 1998), all of which relate to the post-Mahakoshal thermal reactivation. Though no direct dating is available for the Mahakoshal rocks, the post-tectonic granitic activity of ~1500 to 1850 Ma may be suggestive of Paleoproterozoic age for the former.

Makrohar Granulite Belt In the Rihand area around Makrohar, mappable units of WNW–ESE trending enclaves of amphibolite to granulite facies rocks occur within the sheared granite, close to the southern boundary of the Mahakoshal belt. Pichai Muthu (1990) described tectonic slivers of migmatitic gneiss, sillimanite–corundum schist, mafic granulites and anorthosite from this location. In addition, Roy and Hanuma Prasad (2003) reported rafts of calc–silicate, marble and BIF from this area. Based on the Rb–Sr age range of 1.7–1.5 Ga of the granitic rocks intruding this terrain, the granulite facies metamorphism of the enclave suite may be presumed to be older than 1.70 Ga.

Geophysical Characteristics of Mahakoshal Belt and its Surroundings The Mahakoshal belt is characterised by an ENE–WSW trending gravity high anomaly zone in the Bouger anomaly map of Central India (Venkat Rao et al., 1990; Rama Rao and Srirama, 1995; Das and Mall, 1995). The density model suggests an upwarp of Moho, with emplacement of high-density material into the lower crust along this zone, bounded by two deep-seated faults (Venkat Rao, op. cit.). The DSS studies (Kaila et al., 1988) also revealed fragmented crust along this zone and indicated emplacement of mantle products at a very shallow crustal level (2–3 km depth). The curie–crust is about 3.4 km thick in the Mahakoshal belt, and the Moho depth is interpreted as 39 km. Seismotectonic studies (Acharyya et al., 1998; Devrajan et al., 1997) indicated a steep southerly dipping fault along southern margin of the Mahakoshal belt (SNSF) extending down to a depth of least 33 km, coinciding with the base of the magnetic crust. The SNNF also shows deep crustal penetration, but it is not adequately studied.

Betul Supracrustal Belt The Betul supracrustal belt occurs in the southwestern part of CITZ, situated between the Mahakoshal belt to its northeast and Sausar belt in the southeast. The belt has a strike extension of about 150 km in ENE–WSW direction from Chichola to Chhindwara, its width varying from 20–30 km (Fig. 2.1.15). Volcanisedimentary rocks intruded by mafic–ultramafic and granitic suites constitute this supracrustal assemblage with basement gneisses exposed locally. Rocks of the Betul belt are unconformably overlain by the Gondwana Supergroup and Deccan volcanics in the north and by Deccan volcanics alone in the west, south and east. However, along the Kanhan river section (where the Deccan volcanic cover is eroded), gneissic complex with E–W trending K–granite bodies are exposed between the Betul belt and Sausar supracrustals. There are several ductile shear zones passing through this zone, of which, the Gavilgarh–Tan shear (GTS) is most extensive and prominent. However, the Tapti shear zone (along the Tapti river course, south west of Betul) is closest to the southern margin of the Betul belt but cannot be traced in its eastern extension due to the Deccan trap cover. Brittle-ductile shearing also discontinuously marks the northern contact of the belt with Gondwana rocks. Besides, there are also a few zones of high shear strain within the supracrustal belt, especially along zones of pronounced rheologic contrast.

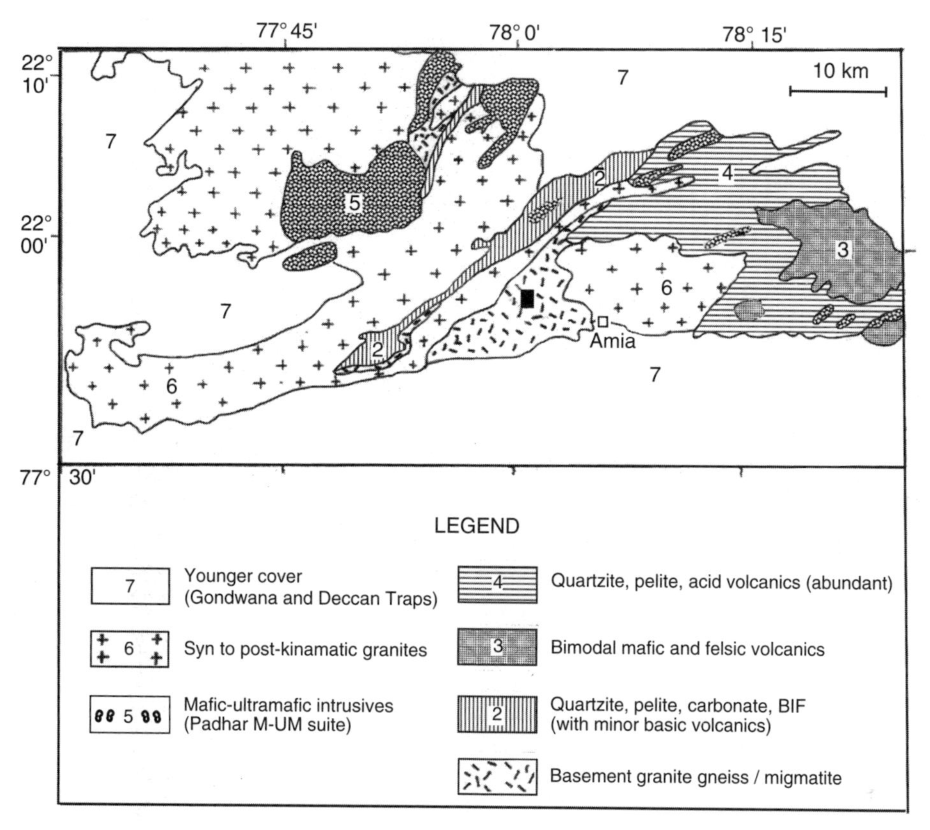

Fig. 2.1.15 Generalised geological map of Betul supracrustal belt, Madhya Pradesh (modified after Roy et al., 2003)

Litho-assemblage of the Betul belt is represented by bimodal acid–basic volcanics (pillowed and non-pillowed metabasalt, rhyolite, felsic pyroclasts), meta–ultramafic volcani-clastics, metasediments (pelite, psammopelite, arenite, BIF, anthophyllite schist, calc silicate, marble), mafic–ultramafic intrusives and syn– to post–tectonic granitic intrusives. The southeastern part of the belt is dominated by a metavolcanic suite interbeded with metasediments, and is characterised by several small Zn–Cu–Pb deposits. About 100 sq km of the northwestern part of the belt exposes a large mafic–ultramafic intrusive complex comprising discrete sheets and plutons of peridotite, pyroxenite, gabbro, norite and diorite. A NE–SW trending sedimentary formation passing along the middle of the belt and separating the volcanic-dominant SE sector from the intrusive-dominant NW sector comprises metapelites (at places carbonaceous, ferruginous), quartzite, and BHQ with minor basic volcanics.

Supracrustals of the Betul area are yet to be properly classified. A stratigraphic succession is presented below, (Table 2.1.18) based on descriptions in available literature and some studies conducted by the authors.

Table 2.1.18 *Generalised stratigraphic succession in Betul belt (modified after Mahakud, 1993)*

Daccan Trap	
Lameta Beds	
Unconformity	
Granitic intrusives Mafic–ultramafic intrusives	
Younger metasediments	Phyllite (carbonaceos, ferruginous), quartzite, quartz–mica schist, BHQ
Volcanisedimentary sequence	Meta–rhyolite and felsic pyroclasts with intercalations of anthophyllite schist, garnetiferous schist, calc–silicates; basic volcanics (pillowed and non-pillowed metabasalt)
Gneissic Complex (Basement)	

The metabasaltic rocks (dominant in eastern sector) are tholeiitic in composition and comparable to low-K tholeiites of near-arc environment (Ramchandra and Pal, 1992; Raut and Mahakud, 2004), while the acid volcanics show calc–alkalline to alkaline character, (Raut and Mahakud, 2004), akin to continental arc setting. Roy and Hanuma Prasad (2003) consider this to be due to non–cogenetic nature of the basic and acid volcanics. Chemical compositions of mafic and felsic volcanics (Mahakud et al., 2001) are presented in Tables 2.1.19 and 2.1.20.

Table 2.1.19 *Chemical analyses of mafic (1–5) and felsic volcanic rocks (6–10), Betul belt (Mahakud et al., 2001). Oxides in wt%, elements in ppm.*

	1	2	3	4	5	6	7	8	9	10
SiO_2	46.79	46.79	46.24	48.3	49.17	73.67	74.65	72.90	70.45	75.40
Al_2O_3	15.00	14.05	13.03	12.47	14.29	11.60	11.93	12.82	13.51	12.50
FeO	5.04	4.32	9.91	9.61	1.38	1.80	2.70	1.89	–	–
Fe_2O_3	5.00	6.00	3.90	5.68	5.60	2.60	1.55	1.93	2.21	2.55
TiO_2	0.66	0.57	0.96	2.19	1.78	0.34	0.24	0.10	0.34	0.17
CaO	16.30	17.3	15.35	8.47	9.37	0.50	0.56	1.76	0.59	1.66
MgO	1.89	8.29	7.32	5.58	8.41	3.27	2.91	0.92	0.85	0.40
K_2O	0.30	0.36	0.70	2.24	1.52	3.38	2.71	4.20	2.11	1.46
Na_2O	1.50	1.5	2.08	1.09	1.82	0.32	2.20	2.85	2.16	4.51
MnO	0.21	0.27	0.27	0.37	0.23	0.04	0.04	0.04	0.12	0.04
P_2O_5	0.22	0.10	0.17	0.53	0.56	0.02	0.02	0.09	0.04	0.022
$-H_2O$	Tr	Tr	0.04	0.47	0.06	0.06	0.07	Tr	0.11	–
$+H_2O$	0.60	0.39	0.36	2.25	1.93	0.91	0.20	0.20	0.72	–
CO_2	Tr	Tr	Tr	Tr	Tr	Tr	–	Tr	–	–
Cr	660	–	500	50	–	110	–	–	–	–
Co	50	–	70	70	–	20	–	–	–	–
Ni	40	–	20	10	–	10	–	–	–	–

The mafic and felsic volcanic rocks are interbanded with lensoid bodies of anthophyllite–cummintonite–talc–tremolite schist, garnetiferous biotite– chlorite–gahnite–staurolite schist and lithic tuffs. These schists are the main hosts of sulphide ores. Chemical composition of the high–Mg schists (Mahakud et al., 2001) is presented in Table 2.1.20.

Table 2.1.20 *Chemical analyses of magnesian schists, Betul belt (Mahakud et al., 2001). Oxides in wt%, elements in ppm.*

	1	2	3	4	5
SiO_2	51.3	55.9	37.18	24.00	17.18
Al_2O_3	2.61	4.00	12.29	4.20	12.29
FeO	4.36	4.68	–	1.40	–
Fe_2O_3	1.29	3.40	0.09	1.9	12.53
TiO_2	0.10	0.11	12.40	0.01	0.09
CaO	13.79	10.25	20.27	28.50	12.40
MgO	24.48	19.37	0.78	18.95	20.27
K_2O	0.13	0.012	Tr	–	0.78
Na_2O	0.09	0.31	0.46	0.09	Tr
MnO	0.81	0.01	0.01	0.55	0.46
P_2O_5	0.04	0.16	–	0.06	0.01
$-H_2O$	–	0.09	–	0.02	0.02
$+H_2O$	0.43	1.39	Tr	1.09	0.88
CO_2	0.10	Tr	50	18.80	Tr
Cr	10	10	50	10	10
Ni	10	10	50	10	10
Co	30	10	–	–	10
Rb)	50	–	–	30	–
Sr	50	–	–	80	–
Zr	50	–		110	–

The extensive plutonic suite of mafic–ultramafic rocks of northwestern part of the belt exhibit cumulus texture and at places coexisting orthopyroxene(opx)–clinopyroxene(cpx)–plagioclase(plag), displaying superposition of granoblastic texture over the pre-existing magmatic texture (Roy and Hanuma Prasad, 2003). Granulite facies BIF enclaves occur within the mafic–ultramafic complex. Roy and Prasad (op. cit.) considered these as evidence of recrystallisation of the suite after emplacement at deeper level of the crust. The mafic–ultramafic intrusives show LREE enrichment and presence of phlogopite, suggesting their generation from an enriched mantle source subsequent to the derivation of low K–tholeiitic melts emplaced in the supracrustal sequence. These rocks are undeformed except along some shear zones, but are intruded by the various phases of granitoids.

The intrusive porphyritic granitoids are emplaced along ductile shear zones, exhibiting steep northerly dipping mylonitic foliation and low to moderately plunging stretching lineation. Intensive shearing and extensive granitic plutonism have dismembered the supracrustals and the basement geneissic complex. Mahakud et al. (2001) have reported two Rb–Sr dates, 1550± 50 and 850 ± 15 Ma for intensely foliated and non-foliated granitoids from Betul belt. The available date for the syn–tectonic intrusive granite may be reasonably taken as the minimum age limit for the Betul supracrustals, making it more or less equivalent to Mahakoshal Group.

Tectonic Setting The Betul belt represents a distinctive supracrustal belt of probable Paleoproterozoic time within the accretionary complex of CITZ. Because of the occurrence of the belt as an inlier of limited extension in the vast Gondwana and Daccan Trap cover, its relationship with the other adjacent supracrustal belts cannot be determined with greater certainity. Available data, however, are suggestive of an arc setting, marginal to the continent, resulting in the emplacement of low–K tholeiitic– calc alkaline felsic volcanics, deposition of continent-derived clastic sediments, followed by emplacement of mafic–ultramafic intrusives at a deeper crustal level from a LREE-enriched magma source. Copious granitic magmatism in the late phase is compatible with the contemplated arc setting (Roy and Hanuma Prasad, 2003). Besides, the metallogenetic environment (Zn–Cu–Pb) also suggests a marginal arc setting.

Sausar Supracrustal Belt The curvilinear Sausar supracrustal belt with southern convexity (trending WNW–ESE to ENE–WSW) occurs along the southern margin of the CITZ for 200 km strike length, bound by CIS along its south and Gavilgarh-Tan shear (GTS) passing along north. The fold belt is constituted of the metasedimentary Sausar supracrustals (SSG) overlying a multicomponent biotite gneiss–migmatite–granulite complex, named as Tirodi Biotite Gneiss (TBG). The granulitic rocks occur as distinct linears along the north and south of SSG, named as Ramakona–Katangi Granulite (RKG) belt and Balaghat–Bhandara Granulite (BBG) belt (Fig. 2.1.16).

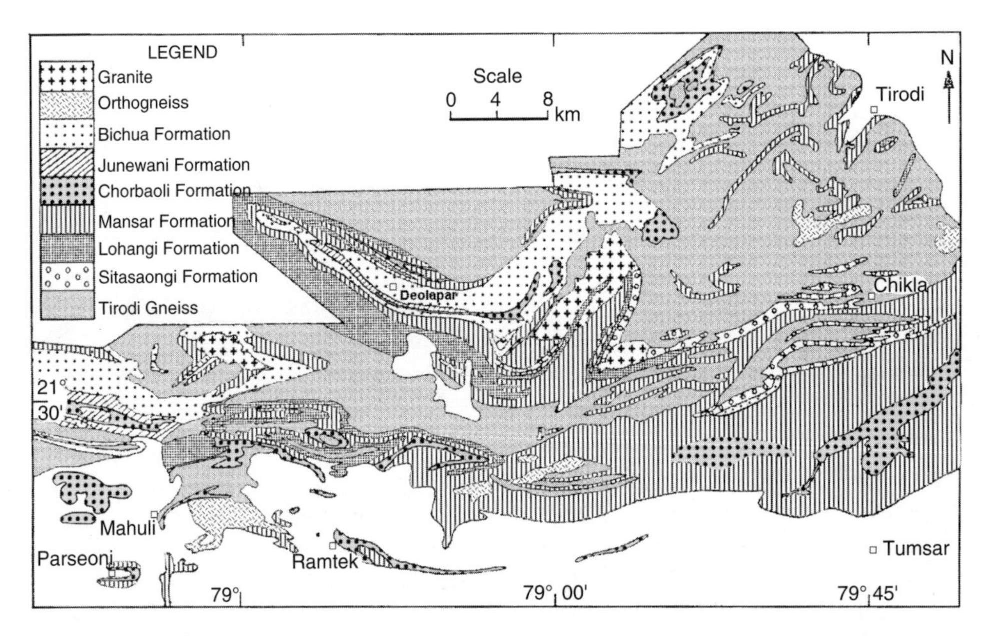

Fig. 2.1.16 Geological map of the central part of Sausar belt, Maharashtra and Madhya Pradesh (modified after Naraynaswami et al., 1963)

Sausar Supracrustal Group (SSG) The litho-assemblage of Sausar supracrustals is characterised by northerly dipping metapelites–quartzite–carbonate and manganese bearing formations and total absence of volcanic rocks. A polymictic conglomerate and grit beds occur along the southern margin of Sausar metasedimentary pile, which have pebbles of gneisses and granite. The Sausar stratigraphy has remained problematic on various counts, viz, a long held debate on its relation with Tirodi Biotite Gneiss (TBG), largescale southerly vergent recumbent folds with thrust sheets causing possible reversal of the sequence, occurrence of dolomitic marble hosted Mn-formations at several stratigraphic levels precluding their utility as definite markers, interleaving of Sausar metasediments with TBG components in macro- to mesoscopic scales, apparent participation of TBG in second folding episode of Sausar supracrustals in contrast to clearly unconformable basement-cover relation between the two along the southern margin, and so on. The stratigraphic successions proposed by different workers are compared in the Table 2.1.21.

Table 2.1.21 *Suggested stratigraphic successions for the Sausar Supracrustals*

	Fermor (in Pascoe, 1928) (Sausar area)	West (1936) (Deolapar area)	Narayanaswami et al., (1963) (mapped area, vide Fig. 2.1.15)	Chattopadhyay et al. (2001) (Deolapar– Ramtek–Chikla area)
Sausar Series / Sausar Group	Ramtek Stage: Sericite quartzite			Intrusives: Massive potassic granite, aplite, pegmatite and quartz vein; foliated granite locally rich in biotite and or fibrolite
	Sapghota Stage: Garnet anthophyllite schist, chlorite schist, magnetite–quartz rock			
	Sitapar Stage: Hornblende schist, garnet amphibolites, pyroxenite	Sitapar Stage: Hornblende schist		
	Bichua Stage: White dolomitic marble, etc.	Bichua Stage: Pure and impure dolomitic marble, diopsidites, etc.	Bichua Formation: Dolomitic marble, calc silicate, etc.	Bichua Formation: Pure and impure dolomitic marble with minor red, yellow and grey chert
		Junewani Stage: Muscovite–biotite schist, autoclastic conglomerate	Junewani Formation: Muscovite–biotite schist, quartz–biotite granulite, etc.	
	Chorbaoli Stage : Feldspathic muscovite quartz schist	Chorbaoli (Ramtek) Stage: Quartzite and quartz muscovite schist	Chorbaoli Formation: Quartzite, quartz schist, quartz muscovite schist	Chorbaoli (Ramtek) Formation: Garnet– staurolite–quartz muscovite schist, micaceous and/or cherty quartzite,locally with garnet and/or magnetite
	Mansar Stage: Gondite type manganese ore bodies	Mansar Stage: Muscovite–biotite–sillimanite schist with Mn-ore	Mansar Formation: Muscovite schists, etc., with Mn ore	Mansar Formation: Biotite–fibrolite quartz muscovite schist, dolomite with Mn ore

(Continued)

(Continued)

Fermor (in Pascoe, 1928) (Sausar area)	West (1936) (Deolapar area)	Narayanaswami et al., (1963) (mapped area, vide Fig. 2.1.15)	Chattopadhyay et al. (2001) (Deolapar– Ramtek–Chikla area)
Lohangi Stage: Calcite marble, calciphyres and manganiferous marbles	Lohangi Stage: Calcitic marble, Mn-ore etc.	Lohangi Formation: Lohangi Substage –Calcitic and dolomitic marble ± Mn– ore	Lohangi Formation: Calc gneiss,calc–silicate rocks and calcic marble,with or without Mn-ore
Utekata Stage: Calc granulite, hornblende–biotite granulite		Utekata Substage– Calc silicate rocks, calc granulite, etc.	
	Kadbikheda Stage: Magnetite biotite rock	Kadbikheda Substage– Quartz –biotite granulite, etc.	
		Sitasaongi Formation: Quartz–muscovite gneiss, conglomerate etc.	Sitasaongi Formation: Gritty quartzite and associated pebbly quartz mica schist
Tirodi Biotite Gneiss		Tirodi Biotite Gneiss: Biotite gneiss and amphibolite etc.	

The stratigraphic scheme of Naraynaswami et al. (1963) is most comprehensive and widely accepted. However, Khan et al. (1999) and Chattopadhyay et al. (2001) have introduced some modifications based on remapping and present day understanding of the poblem. Pal and Bhowmik (1998), Bhowmik et al. (1999), Bhowmik and Pal (2000) and Ramachandra et al. (2001) provided insight into the granulitic rocks of RKG and BBG belts and their tectono–metamorphic evolutionary history.

Narayanaswami et al. (1963) held that the Sausar metasediments were deposited in a shallow marine shelf environment. Based on the polymictic conglomerate (Sitasaongi Formation) along southern margin and lithofacies distribution of the successive younger sequence, Khan et al. (1999) and Chattopadhyay et al. (2001) postulated progressive deepening of the basin to the north and the provenance to the south.

Narayanaswami et al. (1963) have described different rock types of Sausar supracrustals in detail and indicated noticeable change in rock distribution from south to north, across the central E–W trending stretch of TBG. While the psammitic and psammopelitic litho–facies of Sitasaongi, Mansar and Chorbaoli Formations are widespread in the southern part, the calcareous facies of Lohangi and Bichhua Formations are much better developed along northern part of the belt. The stratigraphic classification of Narayanaswami et al. (1963) is largely retained by the later workers (Chattopadhyay et al., Khan et al., op. cit.) except for the abolition of Junewani Formation, which is now considered as tectonic slivers of TBG.

The manganese bearing horizons, including economic deposits, mainly occur within Lohangi and Mansar Formations hosted by marbles/calc.silicates and pelitic schists respectively. The Sausar belt is charecterised by medium to high-grade manganese ore (40–50% Mn, ~0.1% P, ~7% SiO_2), with mostly braunite, pyrolusite, psilomelane, hollandite, jacobsite, hausmanite, spessartine, winchite and hosts of other oxides, hydroxides, silicate and carbonate minerals (for more details vide Section 2.2).

The Sausar supracrustals recorded polyphase deformation, the first phase having been dominated by low angle thrusting leading to the tectonic interleaving of basement TBG and supracrustal rocks (Bhowmik et al., 1999; Roy et al., 2000; Chattopadhyay et al., 2001). The first phase of deformation also resulted in an E–W trending mylonitic foliation and small-scale isoclinal (recumbent to reclined) folds on bedding and mylonitic foliation (Roy and Hanuma Prasad, 2003). The earliest deformation is manifested mainly along the northern contact of Sausar rocks with basement TBG, producing mylonites and layered tectonites with mesoscopic recumbent folds. Development of quartz– fibrolite tabloids and biotite fibrolite schists are very prominent within the TBG in the tectonised zones. This phase of deformation is best recognised in the 'Deolapar nappe area' (West, 1936), the structure corroborated by later workers (Roy et al., op. cit.; Chattopadhyay et al., op. cit.) as product of second deformation with southward tectonic transport. The second phase of deformation folded the earlier structures into upright to steeply inclined non-cylindrical folds. Local warps and cross folds with sinistral asymmetry were produced by the subsequent phases of deformation. The regional schistosity is the product of first deformation, followed by superposition of less pervasive axial planar schistosity related to the subsequent fold deformations. The pattern of deformation in the Sausar supracrustal belt may be comparable with thick skinned fold–thrust model, with foreland in the south and hinterland towards the north (Chattopadhyay, personal comunication, 2001; Roy and Hanuma Prasad, 2003).

The metamorphic grade of Sausar supracrustals varies from low greenschist facies in the south to upper amphibolite facies in the north. In striking contrast to this, the metamorphic grade of TBG rocks is uniformly middle–upper amphibolite facies throughout the belt, irrespective of the metamorphic grade of the overlying Sausar supracrustals. The peak P–T condition of Sausar metamorphism is constrained at 7 kb at 675°C from calc–silicate rocks in Deolapar area (Bhowmik et al., 1999). The Sausar metamorphism did not reach K–feldspar–sillimanite grade and apparently did not cause partial melting.

Two types of granitoids occur within the Sausar belt, a deformed and well-foliated syn-tectonic type and an undeformed post–tectonic variety. The post-tectonic granite has been dated at 960 Ma (Rb–Sr) (quoted in Roy and Hanuma Prasad, 2003), which closely corresponds to the Ar–Ar age of 950 Ma for cryptomelane in the Mn-formation (Lippolt and Hautmann, 1994). The significance of this coincidence is yet to be fully appreciated.

Triodi Biotite Gneiss (TBG) The stratigraphic status of Tirodi Biotite Gneiss has remained a most debated issue. Some of the earlier workers considered TBG as product of intrusion in Sausar sequence (West, 1933), or partial melting of Sausar supracrusts (Phadke, 1990; Sarkar et al., 1986), which were strongly contested by the others (Naraynaswami et al., 1963; Pal and Bhowmik, 1998; Khan et al., 2000), who placed TBG as basement of Sausar Group. Sarkar et al. (op. cit.) interpreted 1525 ± 70 Ma (Rb–Sr/WR) isochron age from Tirodi Biotite Gneiss (TBG) as the main phase of regional amphibolite facies metamorphism, which also involved the Sausar Group metasediments. A younger K–Ar mineral isochron age of 850 Ma from these rocks was inferred to be a later thermal overprint on the Sausar–TBG. These inferences are also contested now based on recent data. Chattopadhyay et al. (2001) considered TBG as a complex medly of biotite gneiss, granite gneiss, migmatitic gneiss and older supracrustal enclaves forming the basement for the Sausar sediments. According to these workers, the basement–cover relation is largely obscured due to Sausar tectonism. Bhawmik et al. (1999, 2000) have established that the Tirodi Biotite Gneiss, predating the deposition of Sausar sediments, were related to a phase of granulite facies metamorphism along RKG belt. It is also noteworthy that a variety of basic intrusives in TBG does not penetrate the Sausar metasediments.

Thus, the monocyclic tectono–metamorphic models for the Sausar Group, TBG and granulites are no longer tenable, as distinct metamorphic and tectonic gaps are now recognised between the Sausar Group at one hand and TBG granulites on the other. Further, a monocyclic evolutionary model is also not applicable to the TBG and granulites, as the occurrence of granulites in amphibolite facies TBG gneisses cannot be explained by such a postulation.

Tectonic Setting Though there is not much difference in opinion about some important geological aspects of the belt, the evolutionary models proposed for the belt by different group of workers vary widely. Yedekar et al. (1990) inferred that the Sausar supracrustals were depostied on a passive margin of Northern Crustal Province (NCP) implying a northern provenance for the sediments. This is contrary to the fact that basal conglomerates of Sausar sediments lie along the south with evidence of northward deepening of the basin, which would suggest a southern provenance for these sediments. Further, the southward vergence of the Sausar supracrustals suggests the Southern Crustal Province (SCP) to be the foreland of the Sausar basin. These evidences, coupled with latest geochronological data, suggest that the Sausar sediments were deposited between 1.5 and 1.0 Ga on the northern fringe of SCP post-dating its collision and suturing with the NCP in pre-Sausar time. Most of the recent workers (Roy and Hanuma Prasad, 2003; Acharyya, 2003; Bhowmik et al., 1999, 2000) share this broad interpretation but differ over details of the mechanism and direction of subduction, location of the suture zone(s), etc. This, however, does not affect the conclusion that the CIS and the BBG belt mark the southern most contact of the accretionary zone (CITZ) between NCP and SCP. Acharyya (2003) is of the view that the Sausar basin was located intermontane over the partly subsided and eroded pre-Sausar collision zone. The reworking and tectonic incorporation of basement gneisses, granulites and pre-Sausar high grade supracrustals into the younger Sausar sediments denote further convergence and renewed spell of fold-thrust teconics along this mobile belt, resulting in the final closure of the collision zone.

Granulite Belts (Flanking the Sausar Supracrustals) There are two granulite belts along north and south of the Sausar supracrustal belt, which depict different P-T-t paths of evolution and different age-ranges for their development.

Ramakona-Katangi Granulite Belt (RKG) The northern granulite belt can be traced as discontinuous exposures along a northerly dipping ductile shear zone for about 240 km from Ramakona in the west to Bichiya in the east through Khawasa and Katangi. The granulite facies rocks, set in felsic migmatitic gneiss are found interleaved with Sausar supracrustals and younger granite. In most of the domains, the rock-association along RKG belt includes felsic migmatitic gneiss, mafic granulites, cordierite gneiss, meta-ultramafites, hornblendite and garnetiferous meta-dolerite. At Ramakona, calcareous rocks (dolomitic marble) represent the Sausar supracrustals. The felsic migmatite gneiss, which is the most dominant constituent of this belt comprises three distinct varieties, viz., 1) biotite ± hornblende bearing tonalite gneiss, 2) granite gneiss migmatite and 3) composite gneiss (Bhowmik and Pal, 2000).

Three stages of metamorphic evolution (PSM_1, PSM_2 and PSM_3), all of them denoting pre-Sausar history, were determined from these granulites, marked by distinctive mineralogical assemblages and P-T stability fields (Bhowmik et al, 1999; Bhowmik and Pal, 2000). The findings were refined and modified based on later researches, depicting four stages of metamorphic evolution (prograde RM_0, peak RM_1 and retrograde RM_2 and RM_3), relating to continent-continent collision tectonism in the late Mesoproterozoic (Grenvillian) time (Table 2.1.22).

Table 2.1.22 *Metamorphic conditions of RKG belt. (Graciously provided by S.K. Bhowmik, 2010, as personal communication)*

Rock name	RM_0 (Prograde)	RM_I (Peak)	RM_2 and RM_3 (Retrograde)
I. Mineral assemblages			
MG	Hbl_I(46-51)+Pl_I(An54-62)+Ilm, Growth zoning in Grt	Grt (8-10, 33-38)+Cpx_I(51-66,2)+Ru±Qtz	Cpx_2(61 -63,2-4)+Pl_2(An_{90}) +Hbl_2(38) (RM_2) Hbl_3(44-46)-HPl_3 (RM_3)
GA	Hbl_I(39-53)+Pl_I(An57) + Ilm_I	Grt(10-12, 22-29)+Qtz+Ttn	Ilm_2+Pl_2(An_{89-94})+Qtz± Hbl_{FOL}/Hbl_2(40-56) (RM_2)
FMG	Bt_I+Pl_I+Qtz	Grt+Antiper+ Per+Qtz	Bt_2+Pl_2 (RM_2)
MPG	St+Qtz, Early Ky, Bt_I(67)+Sil_I+Pl_I (An_{45-54}) +Qtz, Growth zoning in Grt	Grt(42,07)+Kfs+Sil_I	Pl_2(An_{90}), Spl(23-26)+Crd(80-83)+Pl_2(An_{90}) (RM_2) Bt_2(65)+Crd+Qtz, Ath(63)+ Crd(81), Mag+Sil_2, Hög+Mag (RM_3)
II. Mineral reactions			
MG		Hbl_I+Pl_I+Ilm_I=Grt+ Cpx +Ru+Qtz+H2O/ Melt	Grt+Cpx+H_2O=Cpx_{2A}+ Hbl_{2A}+P l_{2A} (RM_2) Grt+Cpx_{2A}+Hbl_{2A} +Ttn=Cpx_{2B}+ Pl_2B+Ilm_{2B}+ H_2O/melt (RM_2) Grt+Qtz=Cpx_{2B}+ Pl_2B (RM_2) Grt+Cpx_{2B}+Qtz+H_2O= Hbl_3+Pl_3 (RM_3)
GA		Hbl_I+Pl_I+Ilm_I= Grt+Ttn+ Qtz+H_2O	Grt+Hbl_{FOL}+Ttn= Ilm_2+Pl_2+ Qtz ±Hbl_2 (RM_2)
FMG		Bt_I+Pl_I+Qtz= Grt+Per+ Melt	Grt+Per+Melt/H_2O=Bt_I+Pl_I+Qt z (RM_2)
MPG	St+Qtz=Grt+Ky+H_2O, Ilm+Ky+Qtz=Grt+Ru, Ilm+Pl_I+Qtz=Grt+Ru, Pl_I=Grt+Ky+Qtz, Ky=Sil_I	Bt+Sil_I+Pl_I+Qtz=Grt+ Kfs+Melt	Grt+Sil_I=Crd+Pl_2+Spl, Grs+ Sil_I+Qtz=An (RM_2) Grt+Kfs+Melt=Crd+Bt_2+Qtz, Splss=Mg-rich Spl+Mag_{ss}, Mag_{ss}+H_2O+O_2=HOg+Mag (RM_3)
III. P-T	**Conditions of metamorphism**		
MG, FMG, MPG		~ 9-10 kbar, ~ 750-850° C	~6-7 kbar, ~750-825°C (RM_2), ~6 kbar, ~650-600°C (RM_3)
GA		~8 kbar, ~675° C	6.4 kbar, ~700°C (RM_2)
IV. Metamorphic P-T Path		CW P-T Path having prograde (RM_0), peak (RM_I), post-peak ITD (RM_2) and post-decompressional IBC (RM_3) segments	
V. Tectonic set-up		Continent-continent collision	
VI. Time		Late Mesoproterozoic (Grenvillian)	

Data Source-: Bhowmik & Pal (2000); Bhowmik & Roy (2003); Bhowmik and Spiering, 2004; Roy et al. (2006); Lippolt & Hautmann (1994).
Mineral abbreviations used in this table are after Kretz (1983) unless otherwise stated; Mineral compositions in Grt and
Pyroxenes: from left to right numbers refer to Prp and Grs mole % in Grt and Mg No. (XMgx 100) and wt. % Al_2O_3 content in pyroxenes respectively; for other ferromagnesian phases, numbers refer to Mg No. values.
Subscripts 1,2 etc. in minerals refer to different generations of its stability; Per/Antiper: Perthite/Antiperthite; MG: Mafic granulite; GA: Garnetiferous amphibolite; FMG: Felsic migmatite gneiss; MPG: Metapelite granulite; RM: Metamorphic stages in the RKG domain; CW: Clockwise; IBC/ITD: Isobaric cooling/Isothermal decompression

Abhijit Roy (pers. comm. to A. Gupta, 2001), has dated metadolerite occurring within mafic granulite and amphibolite occurring within migmatitic gneiss from the RKG belt. The 5-point Sm-Nd (WR) isochron date of 1116 ± 58 Ma for metadolerite is inferred as the pre-Sausar (PSM3) cooling age. The 4-point Rb-Sr (WR) isochron date of 857 ± 47 Ma for metadolerite and 3-point Rb-Sr (WR) isochron date of 929 ± 85 Ma are interpreted as the reflection of Sausar resetting age.

Balaghat- Bhandara Granulite Belt (BBG) The southern BBG belt extends for nearly 190 km along the CIS, and is represented by discontinuously exposed two-pyroxene granulite, charnockite—enderbite, meta-ultramaphites, cordierite granulite, and quartzite-pelite-BIF granulite. Highly deformed and dismembered lenticular bodies of the granulites occur within strongly tectonised and migmatitic Amgaon Gneiss (Ramachandra, 1999), which are bound by two prominent northerly dipping ductile shear zones (Ramchandra and Roy, 2001). The shear sense along the southern shear zone suggests up-dip southerly thrust of BBG rocks over the Amgaon Gneisses, while the Sausar supracrustals appear to be down-thrown, against BBG along northern shear zone (Ramchandra and Roy, 2001).

Two phases of granulite facies mineral assemblages were recorded, the first phase at 7 kbar/900°C and the second phase at 8 kbar/700°C (Ramchandra, 1999; Bhowmik and Pal, 2000), indicating IBC P-T-t path. Textural evidence in charnockite and two-pyroxene granulites were inferred to be suggestive of melt generation at the peak of first phase, which resulted in development of migmatitic gneiss.

Since then, there has been significant flow of new data and varied reinterpretations presented by Bhowmik et al., (2004,2005). Recently, Bhowmik and his co-workers (Bhandari et al., 2010) have reported a pervasive UTH metamorphism in the BBG belt at ~ 8-9 kbar/950-1000°C, with robust monazite dating of the event as 1.6 Ga, followed by fluid induced recrystallisation in isolated segments at ~ 1.47 Ga.

The mineral assemblages and their reactions recorded in the different rock types of BBG belt with respect to the stages of granulite facies metamorphism, as tabulated and communicated by S.K.Bhowmik (2010) are furnished in the Table 2.1.23.

Table 2.1.23 *Metamorphic conditions of the BBG belt (Graciously provided by S.K. Bhowmik, 2010, as personal communication)*

Rock name	RM_0 (Prograde)	BM_1 (Peak)	BM_2 (Retrograde)
I. Mineral assemblage			
FG	Bt+Pl+Qtz	Grt (35,03)+Opx(58-61, 4-6)	
AG	Bt_1Sil_1+Qtz, Bt_1(81)+ Grt_1(35-41, 03)+Sil_1	Spr(73)+Spl_1(25)+Sil_1	Early Grt_2(38,03)+Crn and later $Bt2_+Spl_2$(31-33)+Sil_2
GCG	Bt_1+Sil_1+Qtz	Early Grt(30,03)+Crd (80) and later Hc(15)+ Qtz	Grt_2(18-28,03)+Sil_2, Bt2+Sil_2+ Qtz
BIF		Stability of subcalcic ferroaugite+Pigeonite/ Opx	Pyroxene exsolutions, Coronal/ Lamellar Grt
II. Mineral reactions			
FG		Bt+Pl+Qtz=Grt+Opx +Per +Melt	
AG	Bt_1+Sil_1+Qtz=Grt_1 +Per +Melt	Bt_1+Grt_1+Sil_1±Per =Spr+ Spl_1+Melt	Spr+Spl_1+Sil_1=Grt_2+Crn Grt_2+Crn+Melt=Spl_2+Sil_2+Bt_2± Kfs Grt_2+Per+Melt=Bt_2+Sil
GCG		Bt_1+Sil_1+Qtz=Grt_1+ Crd+Per+MeltGrt$_1$ +Sil_1= Hc+Qtz	Hc+Qtz=Grt_2+Sil_2 Grt+Per+Melt=Bt_2+Sil_2+Qtz
BIF			Ferroaug=Opx_1(L)+Aug(H) Opx_1=Cpx(L)+Opx(H) Aug(H)=Opx2(L)+Cpx(H) Cpx+Pl=Grt(Cor)+Qtz Cpx+CaTs in Cpx+Opx+MgTs in Opx=Grt(Lamellar)
III. P-T Condition of metamorphism		~8-9 kbar, ~950-1000° C	~9 kbar, ~700-750 °C
IV. Metamorphic P-T Path		CCW P-T Path having prograde (BM_0), peak (BM_1), post-peak IBC (BM_2) segments	
V. Time		BM1 - ~1.6 Ga , BM2 – 1.47 Ga	

Data Source - Bhowmik et al. (2005); Bhowmik (2006); Bhandari et al. (2010).
Mineral abbreviations used in this table are after Kretz (1983) unless otherwise stated;
Mineral compositions in Grt and Pyroxenes: from left to right numbers refer to Prp and Grs mole % in Grt and Mg No. (X × Mg × 100) and wt.% Al_2O_3 content in pyroxenes respectively; for other ferromagnesian phases: numbers refer to Mg No. values; subscripts 1,2 etc. in minerals refer to different generations of its stability.
FG: Felsic granulite; AG: Aluminous granulite; GCG: Grt-Crd granulite; BIF: BIF granulite; Per: Perthite; L/H=Lamellae/Host; Cor = Corona; BM: Metamorphic stages in the BBG domain; CCW: Counter clockwise; IBC: Isobaric cooling

Charnockite of BBG belt has yielded Sm-Nd whole rock isochron age of 2672 ±54 Ma (Roy, Abhijit, 2000), which is likely to represent an Archean event of the Southern (Bastar) Crustal Province, rather than the developmental history of CITZ. Both Sm-Nd and Rb-Sr whole rock isochron ages of garnetiferous and non-garnetiferous mafic granulites at around 1400 Ma (Roy, Abhijit, op cit.) may be inferred to represent the pre-Sausar cooling event at mid-crustal level (M_2 BBG), post-dating which, high temperature ductile shearing might have exhumed the granulites to the upper crustal level. The penetrative foliation defined by amphibole along the northern margin of BBG has been interpreted as the imprint of Sausar metamorphic event by some workers (Ramchandra and Roy, 2001; Acharyya, 2003). However, more definitive imprint of this event on the charnockite and mafic granulites of BBG is corroborated by another set of Rb-Sr whole rock isochron dates ranging between 800 ± 16 and 967± 72 Ma (Roy, Abhijit, op cit).

The TBG with granulites exposed in the two ductile shear zones, along north and south flanks of the Sausar belt are considered to be the basement for the overlying Sausar metasediments. The stages of metamorphism in the granulites have been worked out in detail along the mentioned zones but the ambient deformative and metamorphic history of the TBG (with all its components other than granulites occurring along restricted zones) are yet to be adequately discerned. The imprint of amphibolite facies metamorphism along both RKG and BBG are presumed to be the effect of Sausar orogeny, which needs proper evaluation The TBG as a whole must have a pre- Sausar metamorphic history of its own, which might be the cause of amphibolite facies signature on the granulites. It may be pointed out that the metamorphic grade of Sausar supracrustals is of greenschist facies in the southern part of the belt, gradually increasing to upper amphibolite facies in the north, whereas the metamorphic grade of the TBG belongs uniformly to the amphibolite facies.

Tan Shear Zone (TSZ) As described earlier, several regional crustal scale ductile shear zones characterise the Central Indian Tectonic Zone. Most of these shear zones pass along the margins of the Paleo–Mesoproterozoic supracrustal belts, or bind them from both sides. Some of these shear zones have provided avenues for emplacement of granitic sheets/ plutons, which were subsequently subjected to intense ductile deformation. A few bear evidence of reactivation at later periods.

The SNNF and SNSF passing along the north and south of Mahakoshal belt and the CIS demarcating CITZ from SCP have already been described. The ENE–WSW trending TSZ forms another major shear zone, which passes between the Sausar and Betul supracrustal belts. It is described as the Gavilgarh fault, west of Morshi, where it manifests as a major fault in the Deccan Trap country, while named as Tan Shear for more than 900 kms from the Kanhan river section, through Seoni, to south of Ambikapur. The shear zone is recognised by the occurrence of distinctive mylonitised granite wherever basement is exposed along this stretch (Fig 2.1.17).

The Tan shear zone, varying in width from 2–4 km, is characterised by intensely mylonitised quartzo–feldspathic rocks dominated by pink to flesh or brick red coloured potash feldspar, possibly derived from diverse protoliths of granite–gneiss. An ultramylonitic zone passes along the median stretch of Kanhan river valley section for about 15 km. Pseudotachylitic veins are extensively exposed in this part of the zone (Golani et al., 2001). The Kanhan river section exposes spectacular features depicting repeated phases of ductile and brittle deformation along the Tan shear, some of which are presented in the following photographs [Fig. 2.1.18 (a–h)].

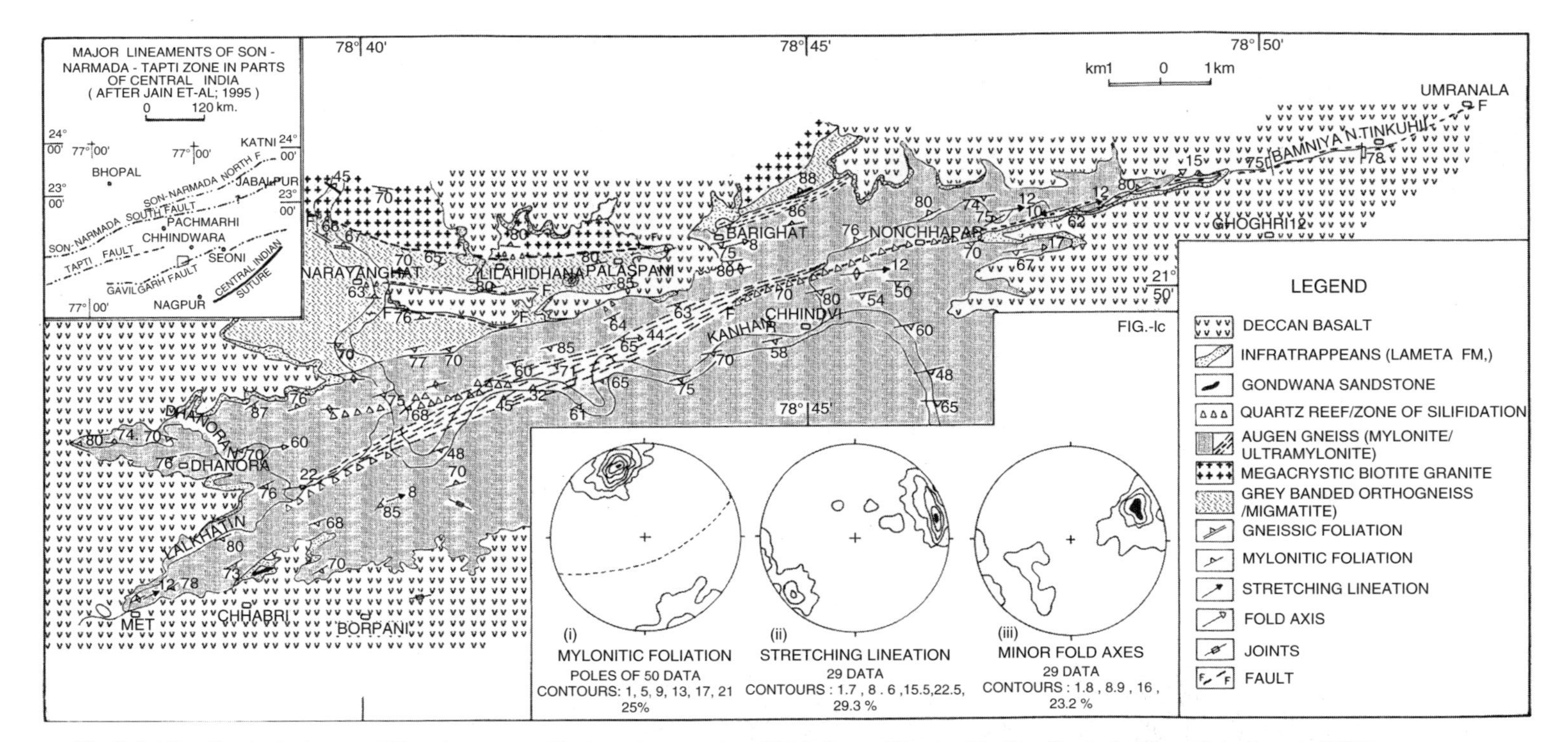

Fig. 2.1.17 Geological map of Tan shear zone, Kanhan river section, Chhindwara District, Madhya Pradesh (after Golani et al., 2001)

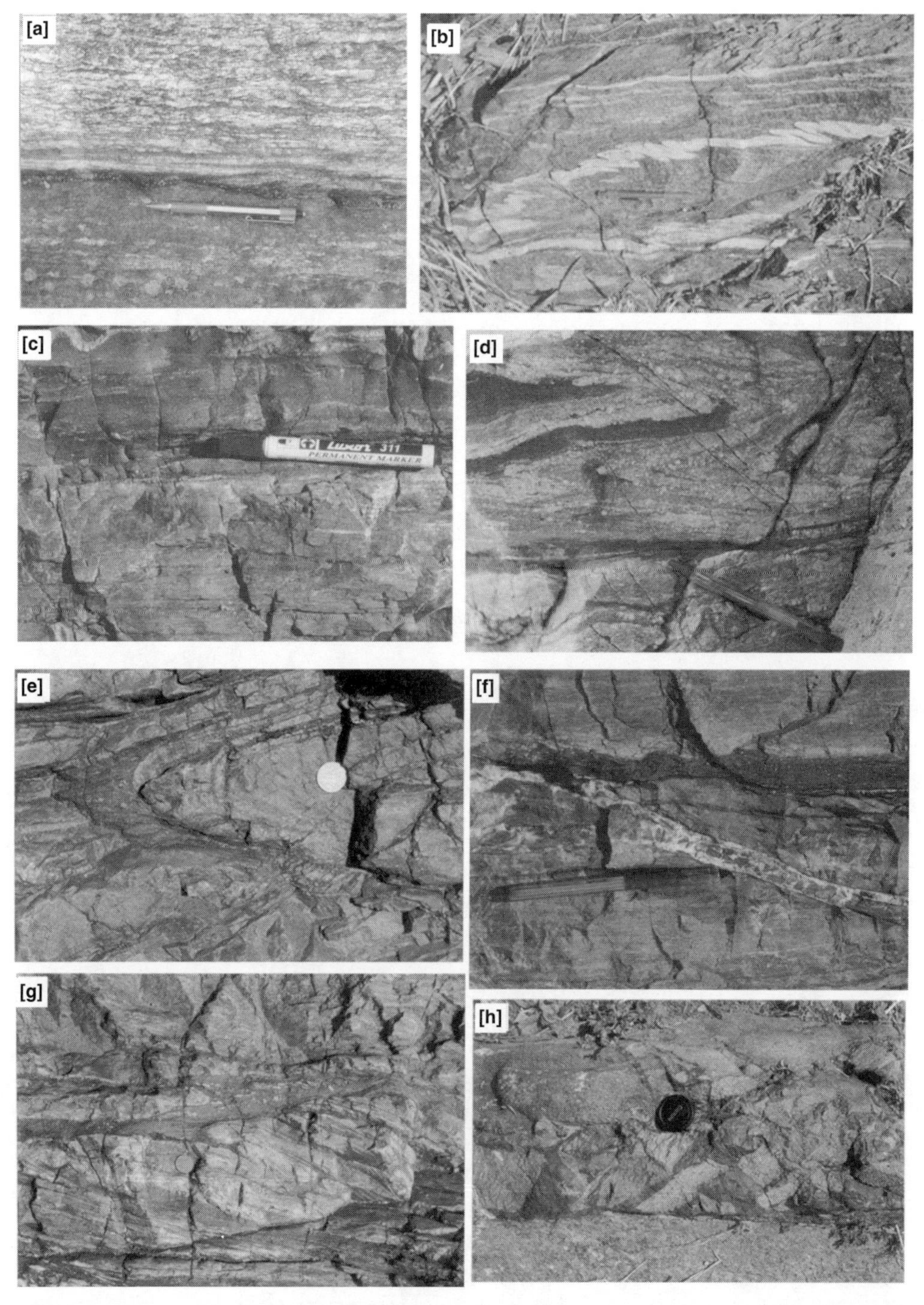

Fig. 2.1.18 (a–h) Field photographs showing (a) strongly mylonitic rock in the upper part and protomylonitic character still retained in the lower part with feldspar porphyroblasts with fish-tails, (b) Axial planes of mesoscopic folds getting nearly aligned parallel to the mylonitic foliation, (c) ultramylonite (brick red) in contact with pseudotachylite (black), (d) folded cataclstite (mottled) and

incipient pseudotachylite bands. (e) folded ultramylonite (brick red) and pseudotachylite (black); note the thickening of pseudotachylite band at the hinge and its flowage along axial planar fracture, (f) ultramylonite (brick red) with a foliation parallel pseudotachylite (black) band and a late discordant quartz vein (white) having included fragments of ultramylonite, (g) the planes depicting a late S–C fabric in mylonitised granite (flesh colour) invaded by pseudotachylite (dark), (h) breccia zone along a fault affecting Deccan Trap at the edge of Tan shear showing fragments of mylonite and basalt (represents last phase of movement along the zone, called Gavilgarh fault, west of Morshi). Location: Tan shear zone, Kanhan river section (Photographs: courtesy B.K.Bandyopadhyay and P.R.Golani, GSI).

The eastern extension of the shear zone, from Seoni to Nainpur (100 km in strike, width 1–3 km) is similarly characterised by distinctive pink coloured granite mylonite. Vast expanse of porphyritic and homophanous granitic rocks are exposed to the south of the shear zone, while except for some granitic inliers the northern terrain is mostly covered by Deccan Trap. Compositionally, the granitoids range from quartz monzodiorite–monzonite–syenogranite–granite, which belong to high–K calc–alkaline series (Roy and Hanuma Prasad, 2001). Some of these granites from Nainpur area yielded Rb–Sr age of 1147 $\pm$ 16 Ma (Pandey et al., 1998). Further east, the Tan shear can be traced along the northern flank of Bilaspur-Raigarh supracrustal belt. Here also the zone is characterised by intensely mylonitised pink to brick red potassic granite. To the west of Kanhan river valley, the Tan shear zone mylonites are exposed as a small inlier within Deccan Trap cover at Salbadi.

The mylonitic fabric in Tan shear zone is aligned E–W to ENE–WSW for most of the length, with steep southerly dip, in contrast to moderate northerly dip of all other major shear zones in the CITZ. The stretching lineations developed on the mylonitic foliations show low plunge ($< 20°$) towards west or east, suggesting the movement vector.

Chattopadhyay et al. (2008) recognised two types of pseudotachylite (Pt–M and Pt–C) in the sheared granites of Tan shear zone. The Pt–M layers, interbanded with mylonite and ultramylonite, display strong internal plastic deformation and buckle folding together with the host rocks, and appear to have formed in the brittle–plastic transitional regime. The Pt–C layers show sharp contacts with the host rock, exhibit abundant coeval cataclasis, preserve no evidence of subsequent plastic deformation, and formed at shallower depths. The two types of pseudotachylites identified by these authors may be seen in the field photographs vide Fig. 2.1.18 (d and e) and Fig. 2.1.18 (g) respectively. Chattopadhyay et al. (op. cit.) further observed that the ductile shear sense in the host mylonites are consistently sinistral, whereas those associated with the deformed pseudotachylite (Pt–M) layers are dextral. It is inferred that the host mylonite/ultramylonite foliation experienced reactivated slip movement in the 'semi-brittle' zone when pseudotachylite was generated and subsequently underwent ductile deformormation. The brittle pseudotachylite (Pt–C) layers were generated later, having been spatially associated with a set of brittle shears with sinistral displacements. These features illustrate a complex history of reactivations along Tan shear zone, leading to the multiple phases of frictional melt generation. These may also facilitate an insight into the seismic behaviour of the upper crust along pre-existing disjunctive structures in the central Indian shield.

Bilaspur–Raigarh Belt (BRB) The WNW–ESE trending Bilaspur–Raigarh belt (BRB) extends for about 100 km along the northern boundary of Chhattisgarh basin, representing a highly tectonised segment of Precambrian supracurstals in the extreme southeastern domain of CITZ (Fig. 2.1.19). The supracrustals are represented by metamorphites of greenschist–amphibolite–granulite facies as tectonic slivers with sheared granites, gneisses, metavolcanics, etc. A number of shear zones with strong mylonitic fabric are developed, bounding the individual mega (tectono–)lithons.

The dismembered supracrustal sequences of the eastern segment of the belt (~ 60 km strike length) are sub-divided into three litho–tectonic assemblages, each separated by prominent shear zones (Bhattacharya and Bhattacharya, 2003).

The Southern supracrustal lithoassemblage (SSL) is bound in the south by the unconformably overlying Chhatishgarh Supergroup, displaying mega-folds in arenitic bands, and in the north by mylonitised Murli Granite Gneiss. The southern part of SSL is represented by metagreywackes and argillites of low-greenschist facies, followed to the north by a prominent horizon of felsic volcanics (Kondra felsic volcanics, KFV) of rhyolite to dacite composition. The felsic volcanic rocks are mostly porphyritic, and display primary flow bands at places. The northern margin of the KFV has developed prominent mylonitic fabric.

The Central supracrustal lithoassemblage (CSL), bound by Murli Granite Gneiss along south and Gondwana sediments along north with a faulted contact, is represented by garnetiferous quartz–biotite (± staurolite) schist, quartz –biotite schist, quartzite and calc–silicate. The metamorphic grade belongs to upper greenschist to lower amphibolite facies.

The Northern supracrustal lithoassemblage (NSL), occurring to the north of Gondwana cover is characterised by a linear zone of granite mylonite and ultramylonites with diagnostic pink to flesh-red potassic feldspar,tectonic slivers of mafic granulte, wide zone of alkali feldspar-bearing megacrystic granite, amphibolite/biotite-bearing granite gneiss and a metasedimentary package of quartz–biotite (± tourmaline) schist–quartzite–calc– silicates.

The tectonic sliver of two–pyroxene granulite (Chhatuabhavna Granulite) is bound on both sides by thin but persistent bands of granite mylonite/ultramylonite and is interleaved with anhydrous garnetiferous granite gneiss. The granulites along with the associated quartzo–feldspathic melt phase display reclined folding. Some relict patches of basement gneissic complex, represented by multiply deformed migmatites, gneisses and folded amphibolite, have been recorded in the northern part of the area near Pipardih and in the southeastern sector around Akhrapol.

The western segment of Bilaspur–Raigarh belt, about a 40 km stretch in Shivtarai–Ratanpur sector), has been described by Sahu et al. (2003), as a highly tectonised belt, similar to the eastern sector. A 4 km wide mylonitised granite (best exposed along Arpa river section between Ratkhandi and Salka) with rafts of quartzite schists–amphibolite, passes through the central part of the area separating two contrastive litho-packages along its south and north. The litho-assemblage occurring to the south of the mylonitised granite essentially comprises low greenschist facies metapelites with minor bands of acid and basic volcanics (Nawagaon Formation) while the northern lithoassemblage (Ratanpur Group) comprises amphibolite facies rocks, quartzite/quartz schist, staurolite–garnet –mica schist/biotite–muscovite–quartz schist, amphibolite, hornblende schist, hornblende gneiss, migmatite, and biotite granite gneiss/pink banded gneiss/fine grained pink granite from south to north.

In order to present a composite picture of the BRB, the authors have compiled three maps of adjacent areas (Bhattacharya and Bhattacharya, 2003 and Sahu et al., 2003) and have attempted a correlation of the lithounits/litho-assemblages, named differently by the two groups of workers (Fig. 2.1.19). It is obvious that the SSL of eastern sector correlates well with Nawagaon Formation of western sector, except for predominance of acid volcanics in the eastern parts. The Murli granite with strong ductile shearing along several discrete zones is correlatable with the mylonitised Salkha Granite. The CSL broadly matches with southern part of Ratanpur Group (Shivpur and Belgaon Formations.) while the NSL is correlated with northern part of Ratanpur Group (Khairjhiti and Sekra Formation).

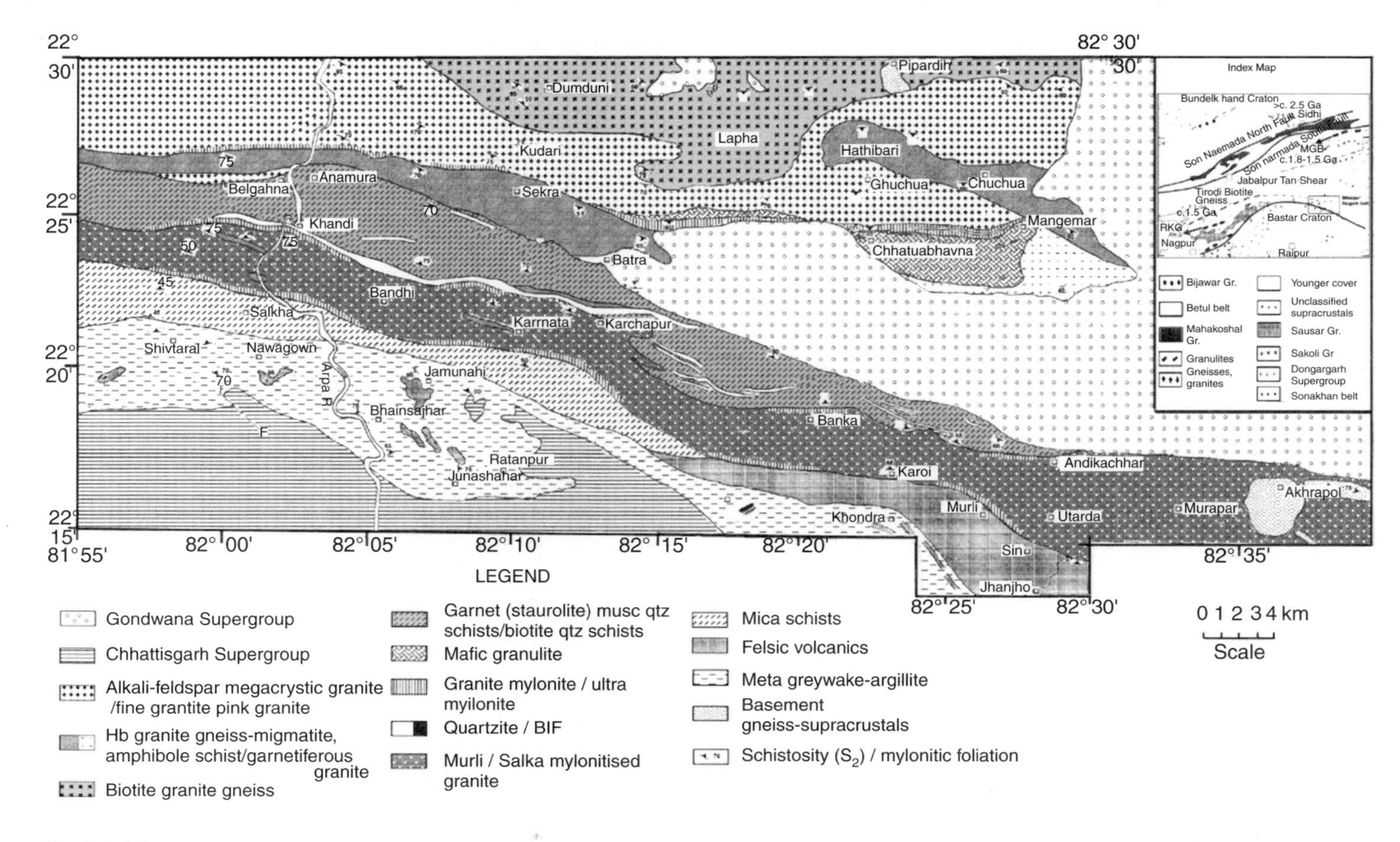

Fig. 2.1.19 Geological map of Bilaspur–Raigarh belt, Chhattisgarh (compiled and modified from Bhattacharya and Bhattacharya, 2003 and Sahu et al., 2003)

The regional schistosity (S_2) and mylonitic foliation, in general, dip to the north in the southern and central part of the belt, while steep southern dip characterises the northern sequence, including the granitic mylonites and tectonic slivers of mafic granulite.

The chemical data and CIPW norms show that the amphibolites and metabasics of BRB are quartz–normative tholeiites while the pyroxenite dykes have calc–alkalic character (Sahu et al., 2003). Granitoids and gneisses (granodiorite to granite in composition) of the BRB plot in continental arc field (Bhattacharya and Bhattacharya, 2003).

Attempts have been made by both the groups of workers to discern the tectonic setting of the Bilaspur–Raigarh belt and to correlate the CIS and Tan shear with the closely spaced ductile shears passing through this belt. Bhattacharya and Bhattacharya (2003) compared it with modern day continental margin arc setting. A compressional tectonic regime is inferred from numerous thrust related ductile shearing and emplacement of continental arc granitoids along the belt. Based on lithological and spatial similarity, the granulitic slivers in BRB are correlated with the RKG belt and the pink feldspar bearing granite mylonite/ ultramylonite passing through its north with the Tan shear zone. Bhattacharya and Bhttacharya (2003) considered the discrete shear zones passing along Murli Granite too insignificant to merit their correlation with CIS, as suggested by some earlier workers (Jain et al., 1995). Sahu et al. (2003), however, prefer to correlate the central mylonitised Salkha Granite with CIS, the granulites with BBG and the amphibolite facies metamorphites of the north (Ratanpur Group) with the Sausar supracrustals, but overlook the need of presenting necessary data in support.

It is evident that the BRB is an integral part of the southern sector of CITZ and is likely to mutually share a crustal development history similar to its western extent. An authentic correlation of the granulitic slivers of BRB with RKG or BBG will depend on properly appreciating the tectono–metamorphic history, aided by a necessary database, which unfortunately is lacking at present. Any such correlation, if successfully made, may not rule out the presence of other pre-Sausar components (TBG) in the belt, even if Sausar supracrustals are not developed. With the available data, however, the eastern extension of Tan shear can reasonably be traced through the north of granulitic slivers in BRB. The southern zone of ductile shear along the margins and within the wide span of Murli/Salkha Granite are certainly not as insignificant as inferred by Bhattacharya and Bhattacharya (op. cit.). From the composite map of the area (Fig. 2.2.19) it is evident that this zone clearly demarcates the northern high grade metamorphites from the southern low grade volcanisedimentary sequence. As such, this zone of ductile shearing, though dissipated compared to the western sector, may be tentatively correlated with the CIS, along the south of which some low grade upper crustal components of SCP might have been involved. A similar situation is established along the BBG belt where lower crustal components of SCP are found involved in a later phase of tectonism along the CIS. It is obvious that neither Tan shear, nor the RKG belt, demarcates the E–W tectonic fabric of NCP from the N–S tectonic fabric of SCP. The demarcating zone between N–S trending Sonakhan belt and E–W trending Bilaspur– Raigarh belt is to be sought below the Chattisgarh cover.

Tectonic Models for Central Indian Precambrian Crust

As expected, varied crustal evolutionary models have been proposed by different groups of researchers for this segment of Precambrian crust in India.

Radhakrishna and Naqvi (1986) and Naqvi and Rogers (1987) proposed a wide tectonised zone transecting Peninsular India, naming it as Middle Proterozoic mobile belt (MPMB). The zone was later termed as Central Indian collisional suture (Yedekar et al., 1990), Satpura mobile belt (Bandyopadhaya et al., 1995) and subsequently as Central Indian Tectonic Zone (CITZ) (Roy et al., 2000; Acharyya and Roy, 2000). The tectonic zone was recognised as a crustal scale feature demarcating the northern Bundelkhand and southern Bastar protocontinents.

Yedekar et al. (1990) proposed an evolutionary model (Fig. 2.1.20) in which the oceanic crust of the Northern Crustal Province was considered to have subducted below the Southern (Bastar) Crustal Province (SCP) at around 2.4 Ga. According to this model, the Central Indian Suture (CIS) passing along the south of Sausar belt marks the suture between the two crustal segments and BBG, represents obducted oceanic crust along the suture. The south directed subduction of the NCP resulted in development of rift basins and volcanic arc related emplacement of Dongargarh volcanics and Granite (and equivalent Malanjkhand Granite) in the SCP. The subduction system culminated into continent–continent collision at ca 1.5 Ga, causing migmatisation of Sausar sediments, deposited on the shelf margin of NCP before the collision. Subsequently, several workers (Jain et al., 1991,1995; Raza et al., 1993; Divakar Rao et al., 1996; Neogi et al., 1996; Yedekar et al., 2000; Reddy, 2001; Mishra, 2000; Acharyya, 2001; Acharyya and Roy, 2000) presented varied tectono–magmatic models. Though these models widely differed on many major issues like the direction of subduction, time of collision, location of the suture zone etc., a basic commonality amongst them being the recognition of an ancient plate tectonic regime involved in the evolution of the Central Indian Shield.

The tectonic model of Yedekar (1990) was subscribed to by many others (Jain et al., op. cit.; Mishra op. cit.; Reddy et al., op. cit.; Divakar Rao, op. cit.) suggesting southward subduction of NCP below SCP, some conclusions being based on seismic reflection and refraction data along the proposed suture zone (CIS). Recently, Yedekar (2003) has reiterated his old model, only modifying the proposed time of subduction to ca 2.7 Ga, based on the recent age data on BBG and accepted the weakness of the model in relating the Dongargarh and Malanjkhand magmatism along N–S trending rift in SCP with the E–W trending subduction zone. However, he neither provided any explanation in this regard, nor on the several geological mismatches pointed out by other groups of workers, and briefly discussed here.

The tectonic model of Yedekar and of others sharing it, was strongly refuted by later workers (Roy et al., 2000; Roy and Hanuma Prasad, 2001; Acharyya et al., 2000), mainly on the basis of new set of petrogenetic data on RKG and BBG made available by Bhowmik et al. (1999), Bhowmik and Pal (2000) and Ramachandra and Roy (2001). The RKG belt, which heretofore escaped the necessary attention, was revealed to represent a collision related tectonic set up, having clockwise ITD P–T–t evolutionary path. On the other hand, the BBG, having a multi-stage metamorphic history with anticlockwise IBC P–T–t path, was found not to qualify for evolution in a collisional set up. Based on these findings and geochemical character of mafic granulite protoliths and overall rock-association, Ramachandra and Roy (2001) suggested an intracontinental tectonic setting for the BBG and negated the possibility of its origin as an obducted oceanic crust along the sutured collision zone. These workers correlated the BBG with other ancient granulite belts in the SCP (implying Archaean origin), which was tectonically cought up in Amgaon Gneisses during Sausar orogeny. This contention was substantiated by a 2672 ± 54 Ma date of charnockite from the BBG belt (Roy, op. cit.). It was further revealed that the Sausar sediments were deposited in a marginal basin of the SCP, progressively younging and deepening towards north with a southern provenance (Chattopadhyay et al., 2001). Recent studies also reveal that the age range of Sausar metamorphism, syntectonic granite emplacement and mineralisation of cryptomelane is 0.8–1.0 Ga (Lippolt and Hautmann, 1994), which postdates the event of granulite facies metamorphism at ca 1.5 Ga.

Roy and Hanuma Prasad (2003) presented an evolutionery model (Fig. 2.1.21), invoking north directed subduction of the oceanic crust of SCP below the NCP along the south of Betul belt at ca 2.2 Ga (or even earlier). This triggered near contemporaneous development of the Mahakoshal belt and Betul belt in back-arc and intra-arc setting respectively. The closure of Mahakoshal basin was marked by extensive calc–alkaline granitic activity at ca 1.8 Ga through reverse slip ductile shear zones in contractional tectonic regime.The magmatism continued in arc-setting upto ca 1.5

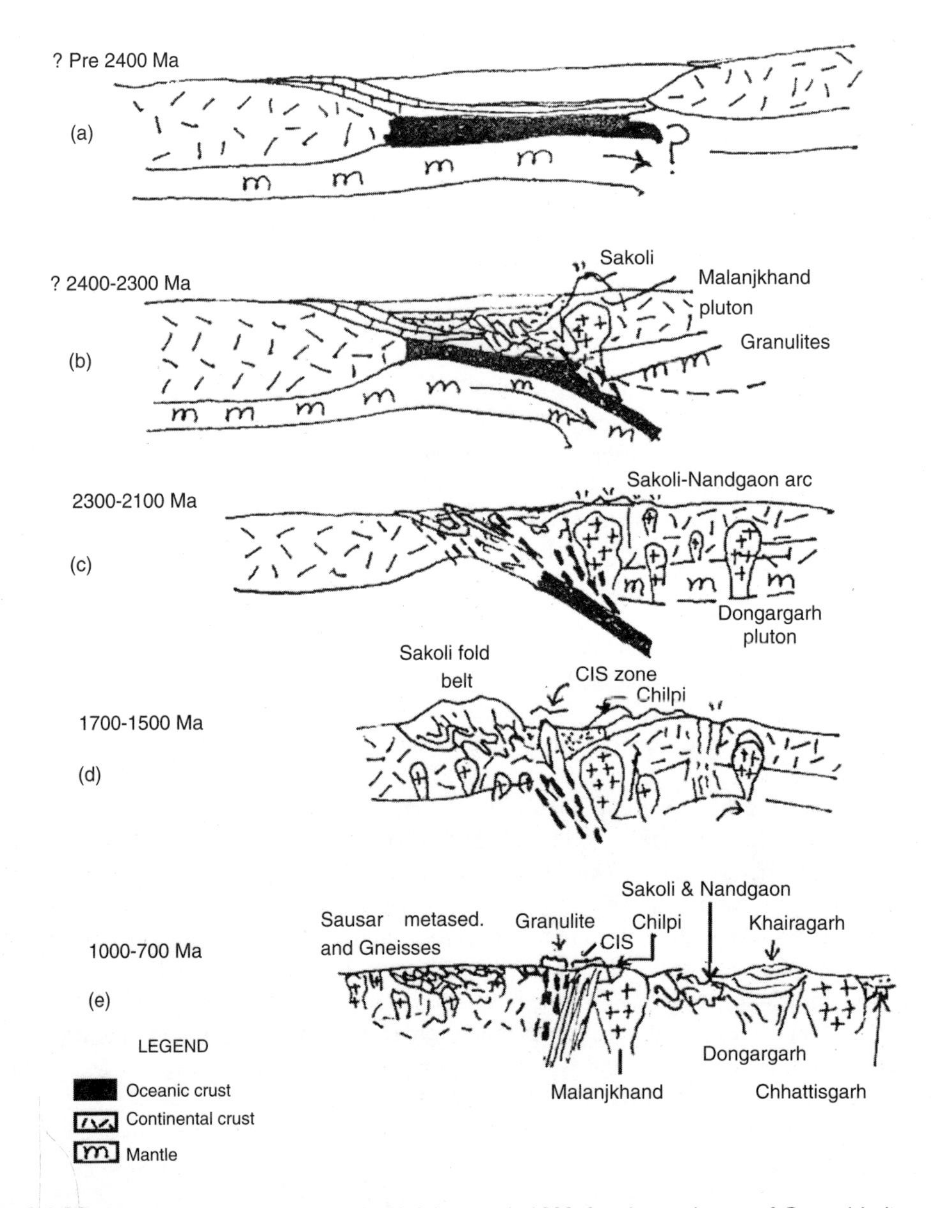

Fig. 2.1.20 Tectonic model proposed by Yedekar et al., 1990, for the evolution of Central Indian Suture (CIS)

Ga, resulting in recycling of the crust with addition of juvenile material, widespread low-pressure medium-temperature metamorphism, and emplacement of tectonic slivers of granulite (Makrohar and Betul Granulites) in the wide expanse of the overriding NCP plate. The Betul basin closed at ca 1.5 Ga, as indicated by the age data of syn-tectonic granitic rocks of this domain, which was also accompanied or slightly preceded by large-scale mantle melting causing copious hydrous ultramafic–mafic magmatic intrusion in the Betul belt. It is further postulated that the active subduction system culminated at ca1.5 Ga with continent–continent collision and suturing along the RKG belt, which reveals a collision related evolutionery signature. The collision also resulted in crustal thickening and attendant migmatisation and lower crustal melting.

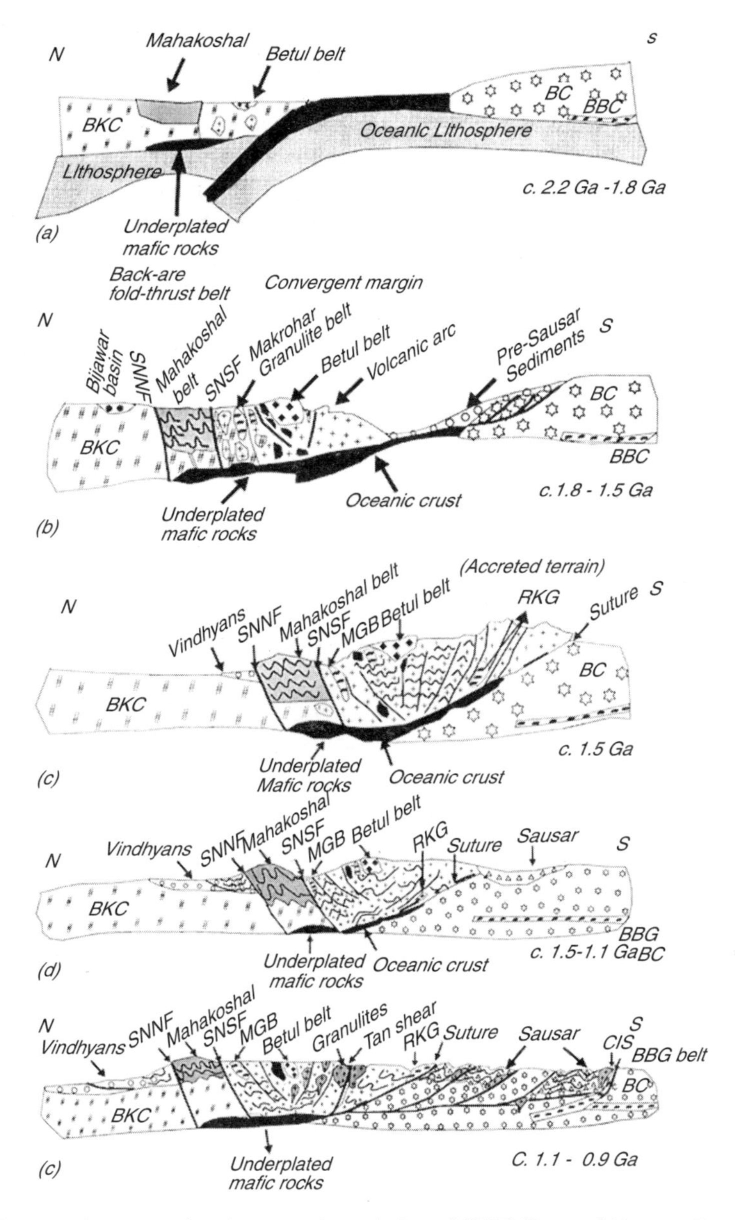

Fig. 2.1.21 Cartoon demonstrating the tectonic evolution of CITZ (Roy and Hanuma Prasad, 2003)

Rapid decompression and exhumation of the RKG granulite to the middle crustal level followed. The sutured northern and southern crust formed the basement for the deposition of the Sausar sediments in a continental shelf setting. The Sausar basin was closed at about 1.1 Ga due to continued south directed thrusting. The emplacement of granitic rocks and probable exhumation of the BBG along CIS also accompanied this event.

In an attempt to conceptualise an uniform evolutionery history for the entire Precambrian crust spanning over the central, eastern and north eastern India, Acharyya (2003), in partial modification of his earlier model (Acharyya, 2001), has suggested dual southward subduction of the NCP below the SCP, one along the south of Mahakoshal belt and the other from the north of Betul belt (Fig. 2.1.22). He further conjectured eastward extention of the SNNF, SNSF and of CIS along the north and south of the Chhotanagpur Granite–Gneiss Complex (CGC), and correlated the CITZ with CGC terrain and further to the east upto the granitic massif of Meghalaya. In context of the central Indian segment, this model also ascertains closure of the Mahakoshal rift as the earliest event triggered by subduction along its south and collision related emplacement of calc–alkaline granitoids (1.8-1.7 Ga) along the SNSF and further south. Simultaneously, or closely following the first event, another southward subduction of NCP is suggested from the north of Betul belt, though not explained explicitly (Acharyya, 2003). This event culminated into another collision of NCP and SCP, after which the Sausar sediments were deposited. According to this model the CITZ witnessed two orogenies, the Satpura orogeny I (1.6-1.5 Ga) during which the pre-Sausar supracrustals including granulites of the RKG belt evolved and Sausar orogeny II (ca 1.0 Ga) which affected the Sausar rocks and left its imprint on the preexisting supracrustals along the southern margin of NCP.The sequence of events relating to the evolution of the BBG belt are rather in conformity with the model of Roy and Hanuma Prasad (2003).

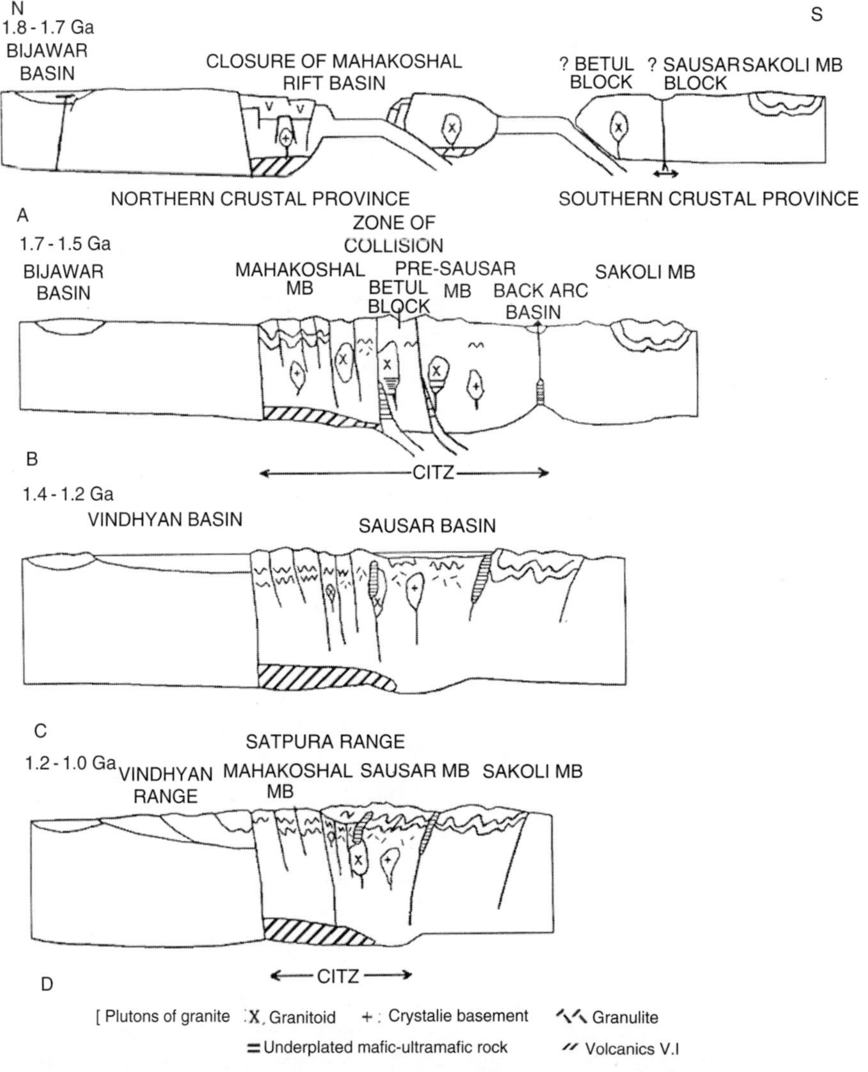

Fig. 2.1.22 Schematic tectonic model of evolution of the CITZ (Acharyya, 2003)

It may be pointed out that none of the models discussed above have attempted any tie-up of the evolutionary history of CITZ with the Archean protocontinental nucleus of Bundelkhand and the Proterozoic Bijawar–Gwalior belts, flanking the craton. The prominent volcanic event at ca 1.8 Ga in both Bijawar–Gwalior belts and within the stabilised Bundelkhand craton seems difficult to isolate from the evolutionary history of CITZ. Another major weakness of the proposed models is the lack of any serious consideration about the Tan shear zone. This is the the most spectacular and extensive zone of ductile deformation and alkali–granite activity (ca 1.1–1.2 Ga) across the the entire width of CITZ, showing low angle transpressive block movement of crustal scale along steep southerly dipping mylonitic surfaces, in contrast to northerly dip and sense of up–dip tectonic transport along all other major shear zones (SNNF, SNSF and CIS).

From the above account, it is evident that the CITZ has a protracted evolutionary history, spanning over the period, > 2.2–0.8 Ga, that culminated in collision and suturing of the NCP and SCP during Mesoproterozoic time. Three prominent tectono–metamorphic events are recorded in this terrain of which the first event (ca 1.8 Ga) is best recorded in the northern parts from south of Mahakoshal belt to Bijawars along cratonic margin and also within the stabilised Bundelkhand craton. The last event (1.1–0.8 Ga) is only recorded in the southern parts of CITZ along the Sausar belt and its vicinity, while the second event (1.6–1.5 Ga) has more or less a pervasive imprint over the entire terrain. As the RKG belt granulites, having collisional evolutionary signature, are dated at ca 1.5 Ga, the second tectono–metamorphic event is considered to be related to the collision of the NCP and SCP. As such, in the context of global orogenic model of Powel et al. (1988) and Gondwanaland reconstructions, the CITZ should have featured as an important tectonic belt in the Rodinian Supercontinent, as the assembly of the NCP and SCP is certainly a pre-Grenvillian event. (vide also Ch. 8).

It may not be out of place to mention some recent contemplation regarding the tectonic setting of the Southern (Bastar) Crustal Province. Steins et al. (2004) pointed out that the Malanjkhand batholith is in tectonic contact with most surrounding units (Dongargarh–Kotri belt, Sakoli, Sausar) and some of those, in turn, are in tectonic contact with each other. The Malanjkhand batholith, having distinct signatures of subduction related arc–volcanism and porphyry type copper mineralisation, cannot be accommodated in any of the adjacent blocks which record continental rift environment in respect of both magmatism and sedimentation. In view of such contrasting character of Malanjkhand block with respect to its surroundings, Steins et al. (op. cit.) suggested that it might be part of a separate microcontinent that grew through subduction-related magmatism in the Late Archean–Early Proterozoic, and was subsequently trapped between the SCP and NCP. The suggestion certainly points towards an often-considered possibility of repeated dispersal and assembly of the ancient cratons. A satisfactory explanation of the apparently anomalous situation will wait till further critical studies.

Proterozoic Cover Sediments in Central India

The Meso–Neoproterozoic cover sediments ('Purana') occur in a number of isolated basins in both Northern and Southern Crustal Provinces of Central India. The Vindhyan basin, skirting the south of Bundelkhand craton, spreads across the northern peninsular India from Rajasthan to Bihar through Madhya Pradesh (MP) and southern Uttar Pradesh (UP). The Vidhyans are essentially constituted of subhorizontal or weakly folded rocks, repeating a sequence of sandstone–shale–limestone, with maximum thickness of ~3 km along southern parts. The Vindhyan rocks unconformably overlie the older Precambrian sequences in most places, though tectonic contacts are also not uncommon. In the southern (Bastar) Province, the cover sediments are developed in several detached basins in Chhattisgarh, which from north to south are Chhattisgarh, Khariar (Pairi), Ampani, Indravati and

Sabari. The constituent sediments of these basins have similar litho–association of siliciclastics, shales, limestone and dolostone, differing, however, in relative proportion and thickness. Some of these sequences are deformed, having tectonic boundary with their basement, while others display prominent angular unconformity.

Vindhyan Basin The Vindhyan basin swerves around the southern crescent shaped margin of Bundelkhand craton (Fig. 2.1.23) .The eastern segment paralleling the ENE–WSW trend of Son–Narmada valley extends from Hoshangabad, MP, in the west to Sasaram, Bihar in the east. The western segment occurring between Bundelkhand region and BGC–Aravalli terrain of Rajashan extends in NE–SW from Chittaurgarh in the south to Dholpur in the north. The curved continuity of the two segments of the basin is only retained along the margin of Bundelkhand craton, while Deccan Trap covers larger parts in its southweast. The northern margin of the basin in the eastern sector is covered by Gangetic alluvium. The Vindhyan rocks are exposed over one hundred thousand sq km, of which about 60% fall in Madhya Pradesh, restricted to the north of Son-Narmada-North Fault (SNNF).

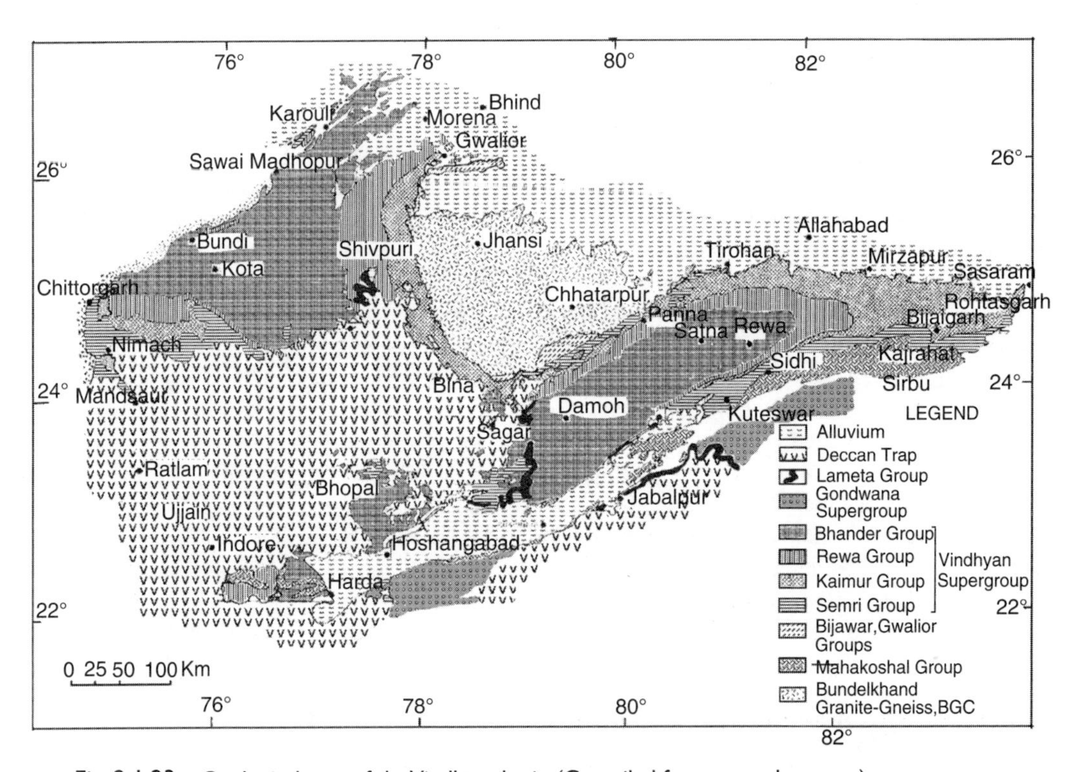

Fig. 2.1.23 Geological map of the Vindhyan basin (Compiled from several sources)

The study of Vindhyans initiated in middle 19th Century, attracting attention of the pioneers like Oldham (1856, 1893), Hacket (1881), Mallet (1869), Vredenberg (1906), and Auden (1933), among many others. Through the subsequent years, scores of workers from GSI and academic institutions have made valuable contributions to various aspects of Vindhyan geology. However, about the stratigraphy, the four-fold classification (Mallet, 1869) of the Vindhyans into Semri, Kaimur, Rewa and Bhander from bottom to top is retained, though the later researchers have suggested many subdivisions and intra-basin variations. Bhattacharya (1996) designated the entire Vindhyan succession as a Supergroup and assigned Group status to the four subdivisions. He also attempted a correlation between the litho-successions developed in different subbasins and proposed a generalised classification applicable to the entire belt following the proposal of Sastry and Moitra (1984).

Semri Group The Semri Group is exposed along the margin of the Vindhyan basin, mainly in Son valley (Madhya Pradesh, Uttar Pradesh and Bihar), along southern margin of Bundelkhand craton (Madhya Pradesh) and Chittorgarh–Mandsaur area (Rajasthan and Madhya Pradesh). The main constituents are alternate beds of shale, sandstone, porcellanite, dolomitic limestone with conglomerate at the base. Many workers consider the porcellanites volcanogenic. Besides, there are occurrences of rhyolitic tuff and volcanic agglomerates in parts of Son valley, indicating synsedimentary volcanism during early Vindhyan history. The limestone beds are very prominent in the eastern part, having some major working deposits.

A generalised succession of the nine Formations of Semri Group and their dominant lithology are presented below (Table 2.1.24).

Table 2.1.24 *Stratigraphic succession of Semri Group (Bhattacharya, 1996)*

Semri Group	Bhagwar Formation	Shale and porcellanite
	Rohtasgarh Formation	Limestone with shale bands
	Rampur Formation	Glauconitic sandstone and shale
	Salkhan/Chorhat Formation	Siliceous and cherty limestone
	Koldaha Formation	Olive green shale
	Deonar Formation	Porcellanite
	Kajrahat Formation	Limestone
	Arangi Formation	Shale
	Deoland Formation	Sandstone with basal conglomerate

Kaimur Group The rocks of Kaimur Group overlie the Semris without any sedimentational break in most places, though some conglomerate–grits are noted in the basal parts of Sasaram Formation in the eastern Son valley (Bihar) and Bundelkhand area (Madhya Pradesh). However, the contact is distinguished in many places by spectacular scarps of Kaimur sandstone. In the margin of Bundelkhand craton, the Kaimurs often transgress over the Semris and come in direct contact with the granitic massif.

The subdivisions of Kaimur Group with respective lithological associations are furnished in the following table (Table 2.1.25).

Table 2.1.25 *Stratigraphic succession of Kaimur Group (Bhattacharya, 1996)*

Kaimur Group	Dhandhraul Formation	Sandstone
	Mangesar Formation	Sandstone with siltstone–shale beds
	Bijaygarh Formation	Shale with glauconitic siltstone
	Ghagar Formation.	Sandstone
	Susnai Formation	Conglomerate–grit–sandstone, breccia
	Sasaram Formation	Sandstone

The eastern Son valley exposes very thick alternating sequence of shale–sandstone mostly belonging to Ghagar and Bijaigarh Formations, attaining a thickness of about 400 m in Rohtasgarh. In Bundekhand region, the Kaimurs are of much lesser thickness. The Kaimur sandstone occurs as blanket like sheets of highly mature quartz arenite over vast areas, showing remarkable lithological homogeneity, suggesting its deposition in high-energy stable marine shelf. The Kaimurs of Rajasthan also suggest a shallow marine shelf condition of deposition, having alternation of mature arenites of different grain size with thin shale partings.

Rewa Group The Rewa Group is exposed in the central Vindhyan basin of Son–Narmada valley as an elliptical girdle with eastern closure around the overlying vast sheet of Bhander Group of rocks. It also occupies substantial area along the west of Bundelkhand craton and near Chittaurgarh. The Rewa Group of rocks is mainly composed of a shale–sandstone sequence, except the dominance of argillaceous units with diamondiferous conglomerate horizons in Satna–Chitrakoot sector.

The Rewa Group is less thick compared to the Semris and Kaimurs, ranging from 100 to 300 m. Among the four Formations of Rewa Group, as detailed in Table 2.1.26, the Jhiri Shales and Govindgarh Sandstone are well-developed in all the sectors, while the lower two Formations of Panna Shale and Asan Sandstone have restricted occurrence.

Table 2.1.26 *Stratigraphic succession of Rewa Group (Bhattacharya, 1996)*

Rewa Group	Govindgarh Formation	Sandstone
	Jhiri Formation*	Shale,basal conglomerate
	Asan Formation*	Sandstone,limestone,chert
	Panna Formation	Shale (calcareous),limestone interbeds

*Note: Rau (2007) named Jhiri Formation as Gahadra Sandstone Formation and Asan Formation as Itawa Sandstone Formation.

Bhander Group The Bhander Group, comprising interbedded sequence of shale–limestone–sandstone, has developed in the central part of the Vindhyan basin in Rewa–Satna–Damoh–Sagar–Bhopal tract of MP and over vast areas of eastern Rajasthan. Stromatolitic limestone at various levels is a characteristic of Bhander sequence. The subdivisions of the Group into five Formations are given in Table 2.1.27.

Table 2.1.27 *Stratigraphic succession of Bhander Group (Bhattacharya, 1996)*

Bhander Group	Balwan Formation	Limestone overlain by shale (only in Rajasthan)
	Shikaonda Formation	Sandstone, limestone and shale
	Sirbu Formation	Shale, siltstone and sandstone
	Bundi Hill Formation	Sandstone
	Lakheri Formation	Limestone with shale partings
	Ganugarh Formation	Shale

Chhattisgarh Basin The saucer shaped intracratonic Chhattisgarh basin in the northern periphery of Southern (Bastar) crustal province covers about 33,000 sq km area, having a E–W stretch of 280 km and maximum N–S extent of 200 km in the western part (Fig. 2.1.24). The basin fills represent multicyclic sedimentation of basal arenaceous and upper shale–carbonate litho-assemblages in a stable shelf condition, attaining a maximum thickness of about 2.5 km. Two main subbasins are recognised in the Chhattisgarh basin, viz, the Hirri or Main subbasin in the west and the Baradwar subbasin in the east, separated by the inferred 'Sonakhan high' trending NNW–SSE. Besides, there is the NE–SW trending Singhora proto-basin, occurring to the south of Baradwar subbasin, which is filled by the oldest sediments (Das et al., 1992).

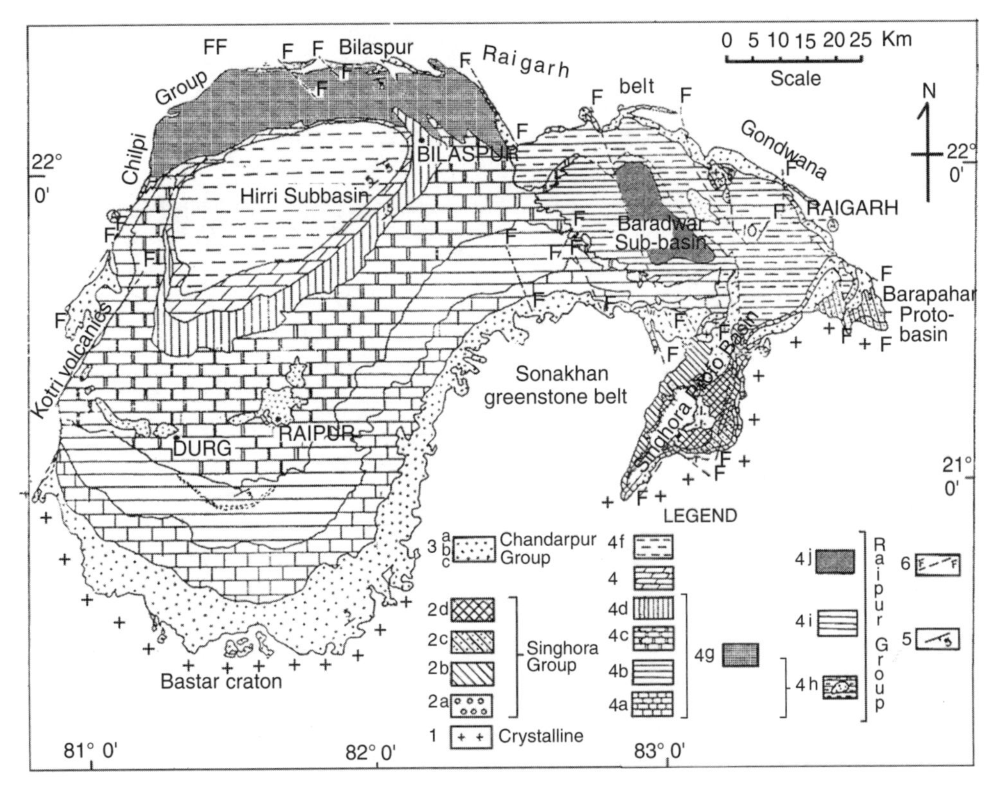

Fig. 2.1.24 Geological map of Chhattisgarh basin (after Das et al., 2001 with some modifications). Legend: 1 – Crystalline Basement; 2 – Singhora Group: 2a – Rehatikhol Formation, 2b – Saraipali Formation, 2c – Bhalukona Formation, 2d – Chhuipalli Formation; 3 – Chandarpur Group: 3a – Lohardih Formation, 3b – Chapadih Formation, 3c – Kansapathar Formation; 4 – Raipur Group: 4a – Chamuria Formation, 4b – Gunderdehi Formation, 4c – Chandi Formation, 4d – Tarenga Formation, 4e – Hirri Formation, 4f – Maniari Formation, 4g – Raigarh Formation, 4h – Pandaria Formation, 4i – Bamandihi Formation, 4j – Saradih Formation; 5 – Bedding; 6 – Fault.

The Chhattisgarh sediments unconformably overlie the Archaean–Proterozoic rocks of the Bastar craton in the south and the supracrustals of Bilaspur–Raigarh belt in the north (Fig. 2.1.25). The rocks are in faulted contact with the Dongargarh Supergroup in the west, while the basin is bound by the Gondwana rocks of Mahanadi graben in the northeast and by the Eastern Ghat Mobile Belt (EGMB) along southeast.

Murti (1987) designated the rocks of Main (Hirri) subbasin and Baradwar subbasin as Chhattisgarh Supergroup and classified them into the lower Chandarpur Group and the upper Raipur Group, which were further subdivided into three and four Formations respectively. Das et al. (1992, 2001) maintaining the overall frame of the aforesaid classification introduced another older sequence as Singhora Group with four formations and described a few facies equivalents developed in different parts of the basin.

The basal arenaceous litho–facies of Chandarpur Group defines the margin of the entire Chhattisgarh basin, except the Singhora proto–basin area in the southeast, and dips towards the centre of the basin. It unconformably overlies the older rocks along the north and south of the basin and the upper formations of Singhora proto-basin. The younger litho–facies of the Chhattisgarh

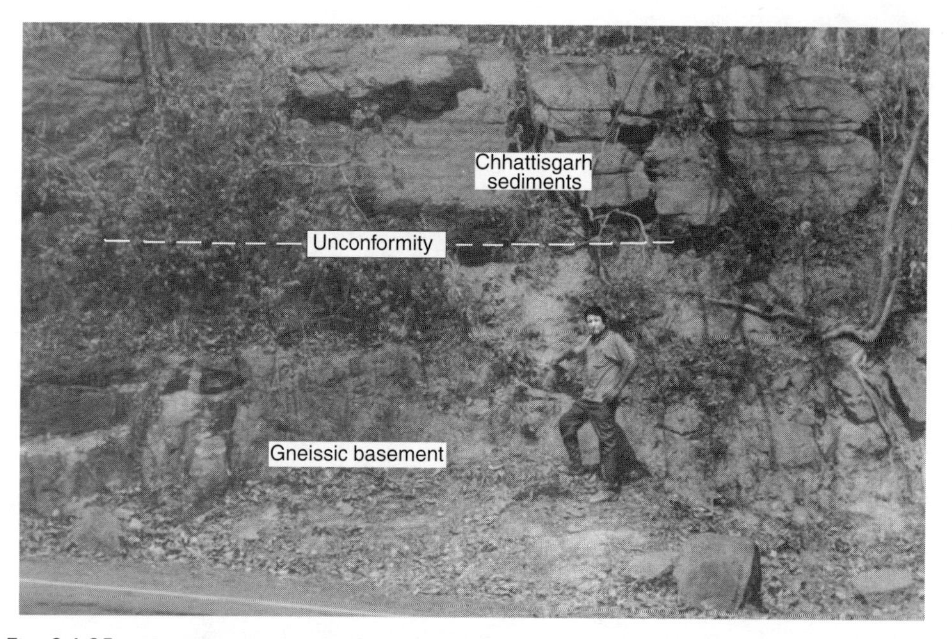

Fig. 2.1.25 Unconformable contact between subhorizontal Chhattisgarh sediments with the underlying basement gneisses (photograph by authors)

basin are developed around the respective depocentres with elliptical outcrop patterns. The central and southern parts of the Hirri subbasin exhibit best development of the various lithounits of the Chhattisgarh Supergroup. The northern part of Hirri subbasin and the northeastern sector of the Baradwar subbasin display wide scale facies variation, prompting Das et al. (op. cit.) to propose different Formation names, indicating their respective equivalence with the Formations in the main areas.

Singhora Group The Singhora Group has restricted occurrence in the southeastern part of the basin and is subdivided into four Formations as depicted in the Table 2.1.28. The basal Rehatikhol Formation, unconformably overlying the crystalline basement, comprises repetitive cycles of conglomerate–arkose, marking the initiation of sedimentation in the Singhora proto–basin as rapid fluvial fan deposits with overbank accumulation of finer sediments (Das et al. 1992; Gupta, 1998). The topmost conglomerate bed of this Formation hosts polymetallic sulphide mineralisation of pyrite–arsenopyrite–galena, associated with uraninite, pitchblende and coffinite (Chakraborti, 1997; Sinha et al. 1998). Saraipali Formation comprising thinly laminated shale–siltstone–chert with bands and pockets of limestone overlies the Rehatikhol arenites and represents shallow shelf sequence under stable platformal condition in a tidal flat environment (Das et al., op. cit.; Gupta, op. cit.). Several volcaniclastic tuffaceous units and thick porcellanite bands occur within this horizon. The overlying Bhalukona Formation is again arenitic (quartz arenite with minor arkose), having mutually exclusive glauconitic and pyritic bands. These sediments represent deposition in coastal to shore line environment in a regressive phase. The top most Chhuipali Fm, characterised by variegated shale–siltstone–chert sequence with stromatolitic limestone/dolomite bands and lenses, occupy the central part of the proto–basin. A marine transgressive phase in a stable offshore environment with frequent storm activity is depicted by the lithoassociation and primary structures preserved in this horizon (Gupta, 1998).

Table 2.1.28 *Stratigraphic succession of Singhora Group (Das et al., 1992)*

Singhora Group	Chhuipali Formation (300 m)	Shale–siltstone with minor chert–limestone (stromatolitic limestone and dolomite at top)
	Bhalukona Formation (20 m)	Quartz arenite–siltstone (minor shale)
	Saraipali Formation (60 m)	Shale–siltstone–limestone (pocellanite, tuff)
	Rehatikhol Formation (20 m)	Feldspathic arenite–arkose conglomerate
	——Unconformity——	
	Crystalline basement (Archaean–Proterozoic)	

By the end of sedimentation in the Singhora proto-basin there was a brief hiatus followed by the development of two large subbasins on either side of the Sonakhan highland, which witnessed subsidence over wide area and inundation by shallow sea.

Chandarpur Group The arenaceous sediments of the Chandarpur Group unconformably overlie the crystalline basement complex along the fringes of major parts of Chhattisgarh basin, except the limits of Singhora proto–basin. The Group is subdivided into three Formations viz, Lohardih, Chaporadih and Kansapathar, as detailed in Table 2.1.29. Along the southern periphery (proximal part of the basin), the Chandarpur Group, represented by all the three Formations is considerably thick (upto 400 m), while it is hardly 20 m thick in the inliers exposed near Durg in the central part and not divisible into Formations. The sequence is relatively thinner in other parts of the basin, not exeeding 200 m. However, not all the formations are developed uniformly showing intrabasinal facies variation.

The Lohardih Formation is constituted of basal conglomerate and arkosic sandstone. The litho-assemblage is interpreted as initial fan deposit related to a braided fluvial system, followed by marine transgression producing wave–tide dominated shallow marine shoal bar, resulting in reworking of the early fluvial sediments (Murti,1987,1996 ; Dutta,1998 ; Das et al., 1992). The overlying Chaporadih Formation, comprising shale–mudstone with fine arenitic bands, is considered to represent maximum extent of transgression depositing sediments in a relatively deeper outer shelf of subtidal to intertidal zone, and also possibly in a lagoonal set up. Signatures of storm generated deposits are also reported from this horizon (Das et al., 1992; Dutta,1998 ; Gupta,1998a). The Kansapathar Formation comprising pinkish glauconitic quartz arenite is inferred to represent a regressional phase in stable shelf condition resulting in the deposition of storm affected tidal sediments and beach deposits (Das et al., op. cit.; Dutta, op. cit.; Murti,1996).

Table 2.1.29 *Stratigraphic succession of Chandarpur Group (Das et al., 1992)*

Chandarpur Group	Kansapathar Formation	White to pinkish glauconitic quartz arenite
	Chaporadih Formation	Green, grey, black shale with fine quartz arenite
	Lohardih Formation	Arkose, gritty wacke and conglomerate

Raipur Group The Raipur Group, subdivided into six Formations and eleven Members (Table 2.1.30), is the thickest and the best developed in the southern and central parts of Hirri subbasin. It is characterised by three cycles of carbonate–shale represented by three pairs of Formations from the bottom to the top, viz, (i) Charmuria Limestone Formation–Gunderdehi Shale Formation, (ii) Chandi Limestone Formation–Tarenga Shale Formation and (iii) Hirri Dolomite Formation–Maniari Shale Formation.

Table 2.1.30 *Stratigraphic succession of Raipur Group (Das et al., 2001)*

Group	Formation	Member	Rock type
Raipur Group	Maniari Formation (70 m)		Purple shale with minor limestone/ dolomite (gypsum)
	Hirri Formation (70 m)		Grey dolomite with argillaceous dolomite
	Tarenga Formation (180 m)	Bilha	Purple dolomitic argillite
		Dagauri	Green clay,chert and shale
		Kusmi	Pink/purple calcareous shale
	Chandi Formation(67 m)	Nipania	Purple limestone/stromatolitic dolomite
		Pendri	Stromatolitic/flaggy limestone with shale–glauconitic arenite bands
	Gunderdehi Formation	Newari	Pink,buff stromatolitic limestone
		Andha	Pink,buff,grey shale with limestone/arenite bands bands
	Charmuria Formation (490 m)	Bagbur	Phosphatic limestone
		Kasdol	Dark grey bedded limestone
		Ranidhar	Cherty limestone/dolomite
		Sirpur	Chert and clay intercalation

There is wide variation in the development of the various lithounits of Raipur Group in different parts of the Chhattisgarh basin, prompting the mappers to coin varied area-specific formation names. Within the Hirri subbasin itself, many of the stratigraphic subdivisions furnished above lose their identity away from the type area in the north, east and west. In the northern parts, the two limestone–argillite cycles from bottom, involving four formations of the type area (i.e. Charmuria–Gunderdehi Formation and Chandi–Tarenga Fms.) coalesce together to form a single limestone –shale pair named as Pandaria Formation. To the east of Hirri subbasin, wide scale facies variation is observed in the Baradwara subbasin where the lithounits of Hirri subbasin cannot be identified. Here the litho–facies are represented by Raigarh shale–limestone, Bamandihi shale–limestone and Saradih limestone–shale in ascending stratigraphic order and each are assigned the status of a 'Formation.'

The tidal to subtidal zone under stable condition caused argillite–carbonate sedimentation in Hirri sub-basin, the cyclicity attributed to repeated fluctuation of sea level due to epeirogenic movements (Murti, 1987, 1996; Das et al., 1992, 2001; Gupta, 1998a, 1998b). The stable isotopic data (δ^{13}C per mil, δ^{18}O per mil) of selected clay and limestone of Raipur Group suggest deposition in marine environment (Murti, 1996). The composition of the rock types and algal content are indicative of both reducing and oxidising state under variable hydrodynamic conditions caused by intermittent regressive and storm agitated phases (Mukherji and Khan, 1996). The phosphate deposit hosted by the chert unit in lower Charmuria Formation suggested a pH value of 8.2 and temperature of the media in the range of 40° to 50° C (Moitra, 1995). Towards the top of the Raipur sequence there is indication of deposition of black shales in tidal flat lagoons and incipient desiccation causing the formation of gypsum pockets.

The sediments of Pandaria Formation in the northern parts of Hirri subbasin show unstable condition depositing argillites with lensoid intercalations of stromatolitic limestone (Das et al., 1992, 2001). The depositional environment of the Baradwar subbasin was also largely dominated by unstable shelf condition leading to the deposition of thick argillaceous sediments of Raigarh Formation (≡ Chamuria–Gunderdehi cycle) and dominant argillites with minor stromatolitic carbonates of Bamandihi Formation (≡ Chandi limestone). The sediments of Saradih Formation (≡ Tarenga Formation) were deposited under restricted circulation with local biogenic activity (Das et al., 1992, 2001).

Igneous Activity The porcellanite and some associated pelites of Saraipali Formation of Singhora Group are reported to be volcaniclastic in origin, displaying texural characters of acidic vitric and crystal tuffs (Chakraborti, 1997). Khan and Mukherjee (1993), based on geochemical attributes, interpreted the lower siliceous clay unit of Chamuria Formation of Raipur Group to be of tuffaceous origin. Various workers (Murti, 1987, 1996; Das et al., 1989, 1992; Rajaiya et al., 1990) also report rocks of tuffaceous lineage from Chandi and Tarenga Formations, but supporting database does not appear convincing in many cases.

Basic dykes and occasional sills are reported from parts of Singhora, Chandarpur and pre-Tarenga Raipur Groups, restricted mostly to the eastern and southeastern parts of the basin. The dykes are mostly doleritic in composition, ranging from a few hundred metres to a few kilometres. One ultramafic body is reported from Rehatikhola Formation (Rajaiya and Ashiya, 1993; Sinha et al., 1998; Tripathi and Murti, 1981).

Structure The E–W trending northern margin of the basin shows tectonic imprints in forms of faults, breccia zones and large scale antiformal and synformal folds, clearly defined by quartzite bands and associated pelites, with E–W axial trend and low easterly plunge. Interestigly, at some places the unconformity surface between the basement (Bilaspur–Raigarh supracrustals) and the cover sequence is also found cofolded. Some broad mesoscopic cross folds with NNW–SSE to NNE–SSW trends are also reported from this sector which are seen to affect both the cover sediments and basement supracrustals (Natarajan et al., 1985; Thorat et al., 1990). These features are indicative of the effect of late phases of tectonism along CITZ.

In eastern parts, the Chhattisgarh rocks display intense folding and faulting especially in the Barapahar area. Near Kharsia, a domal structure is prominent in Chandarpur Sandstone, exposing basement granitoids at the centre (Murti, 1987, 1996). The rocks of Baradwar sub-basin are also deformed into warps with development of long curvilinear faults (NNW–SSE to NNE–SSW), extending from the northern supracrustals of Bilaspur–Raigarh belt to the EGMB in the southeast (Dutta and Dutta, 1990; Das et al., 2001).

Age Based on stromatolitic assemblages two distinct biozones have been established in the Chhattisgarh basin, of which the lower, Biozone-I, is assigned a Mid-Riphean age and the upper Biozone-II is assigned Upper Riphean–Vendian age (Moitra, 1999). The available microbial data suggest early Cambrian as the upper limit of Chhattisgarh sedimentation (Das et al., 2001).

Murti (1996) reported K–Ar date of authigenic glauconite from Chaporadih Formation as 700–750 Ma and based on paleomagnetic studies, estimated the age of Gunderdehi Formation as 1200–1300 Ma.

Recently, Patranabis Deb et al. (2007) reported a stacked set (up to 300m in thickness) of ignimbrites (welded and unwelded tuffs), volcaniclastic sandstone, shale and ash beds near the top of the basin's sedimentary fill in the eastern parts of Chhattisgarh basin (Baradwar subbasin). The mappable tuffaceous horizons (interleaved with shale and sandstone) are located near Sukhda and Sapos villages. Euhedral igneous zircons from three beds of Sukhda tuff unit gave U–Pb SHRIMP ages of 990-1020 Ma, indicating a pre–Neoproterozoic age for the sediments beneath this marker horizon.

Indravati Basin The Indravati basin, occupying about 9000 sq km area in the central part of Southern Crustal Province (SCP) is confined to Bastar district, Chhattisgarh, and Koraput district, Orissa (Fig. 2.1.26). The basin is characterised by late-Proterozoic cover sequence, named as Indravati Group (Sharma, 1975), comprising sandstone–shale–limestone–stromatolitic dolomite, which nonconformably overlies the Archaean basement complex.The eastern margin of the basin is in tectonic contact with the EGMB. The maximum thickness of the sedimentary package is 500 m. It has been subdivided into Tirathgarh Formation, Cherakur Formation, Kanger Formation and Jagdalpur Formation in ascending order (Ramakrishnan, 1987, with minor modifications by Das et al., 2001), as detailed below (Table 2.1.31).

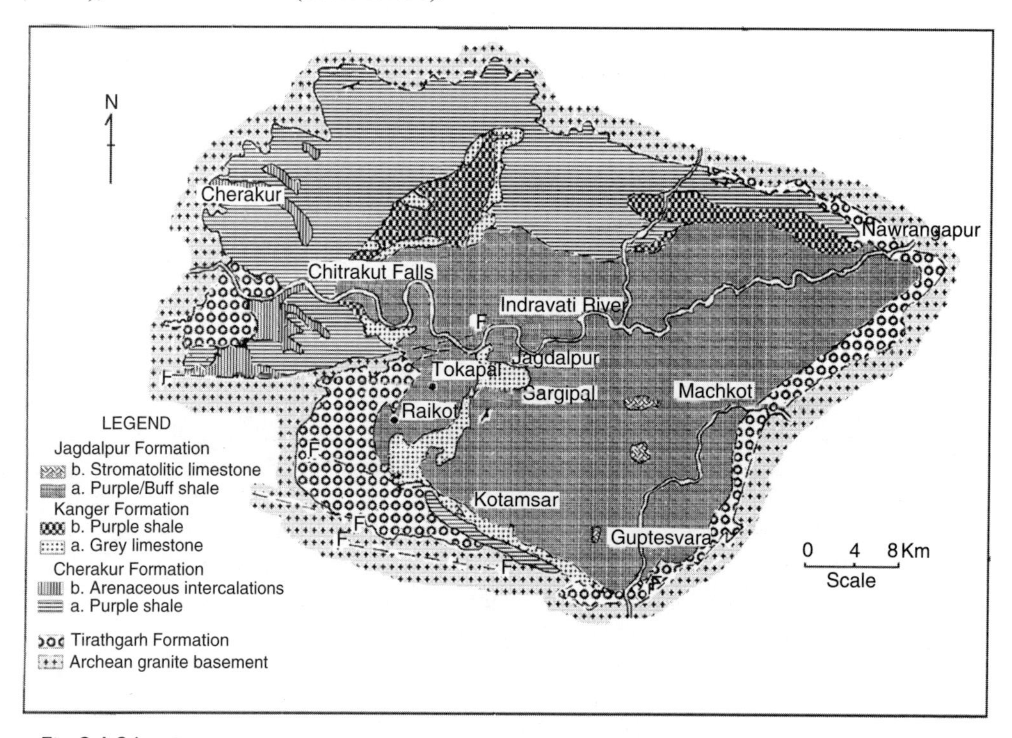

Fig. 2.1.26 Geological map of Indravati Basin (modified after Ramakrishnan, 1987)

Table 2.1.31 *Stratigraphic succession of Indravati Group (Das et al., 2001)*

Intrusive kimberlite		
	Jagdalpur Formation (200 m)	Purple shale, stromatolitic dolomite Purple limestone and shale
Indravati Group	Kanger Limestone Formation (150–200 m)	Grey limestone
	Cherakur Formation (50–60 m)	Purple shale, arkosic sandstone, chert Pebble, conglomerate, grit
	Tirathgarh Formation (50–60 m)	Quartz arenite(Chitrakut Sandstone) Subarkose, conglomerate (Mendri)
	———— Unconformity ————	
Basement Complex		Granite and supracrustals

The roks of Indravati Group lie subhorizontally with very low centrepetal dips. Towards the eastern margin of the basin, the rocks are folded with NE–SW axial trend. The basin is traversed by several E–W to ESE–WNW trending faults, which are more prominent and disposed in en–echelon pattern in the southern part. The major Sirisgura fault has dissected and shifted the western margin of the basin. The other two faults, Tirathgarh and Kanger, have considerably uplifted the quartzite and limestone formations.

Though many earlier workers believed that the Indravati and Chhattisgarh basins were originally connected, the recent workers pointed out that the present basin margins represented their respective shorelines. Ramakrishnan (1987) correlated Chaporadih Formation of Chhattisgarh basin (K–Ar age: 700–750 Ma) with the Chitrakut sandstone.

Igneous Activity Kimberlites and ultramafic intrusives of varied composition occur within the Indravati basin, mainly in the southwestern parts. The multiple kimberlite diatreme at Tokapal (~15 km west of Jagdalpur) is the largest, covering about 2 sq km area, intruding the Tirathgarh Formation and partly penetrating the overlying pelitic formations. Close to the southwestern margin of the Tokapal composite body, another kimberlite pipe of crater facies is exposed at Duganpal, which is surrounded by a ring of pyroclastic rocks containing xenocrysts of olivine, phlogopite and spinel, associated with xenoliths of country rock. At about 4 km northwest of Tokapal, there are two closely spaced kimberlite pipes at Bejripadar, which are compositionally similar to the Tokapal bodies. The kimberlite and associated pyroclastic rocks of Bejripadar show typical clast–matrix texture with pseudomorphs of olivine megacrysts and juvenile lapilli set in fine-grained talc–tremolite–sepentine matrix with locally abundant spinel and sphene. Garnets or diamonds are yet to be located, though the chromites fall in the margin of diamond stability field.

Khariar, Ampani and Sabari Basins There are three more Purana basins along the eastern margin of the SCP bordering EGMB, two of which, Khariar and Ampani(Fig. 2.1.1) are located between Chhattisgarh and Indravati basins while the third, Sabari basin is in the south east of the latter. Of these the Khariar (Pairi) basin is the largest, occupying an irregular oblong area of about 1500 sq km, followed to the immediate south by the Ampani basin of 220 sq km. The Sabri basin with a triangular outcrop pattern covers about 700 sq km area.

Khariar Basin The middle to upper Proterozoic sedimentary package of this basin, exposed on the 150 km long N–S trending tableland rising upto 1000 m height, is named as Pairi Group which overlies the Dongargarh granitoids with a pronounced unconformity.The Group is subdivided into six Formations viz, Devdahara Sandstone (10–80 m), Kulharighat Formation (80–120 m), Neor Formation (40–160 m), Galighat Sandstone (150–300 m), Tarjhar Formation (250 m) and Ling Dongri Sandstone (60–120 m), in ascending stratigraphic order (Fig 2.1.27). The Devdahara Formation is characterised by polymictic conglomerate, sandstone and arkose with intercalations of pyrite bearing black shale. This is conformably overlain by the Kulharighat Formation, which is composed of a thick shaly member with limestone–chert layer at the bottom and stromatolitic dolomite at the top. The Neori, Galighat and Tarjhar Fms. essentially represent alternations of sandstone and shale–siltstone, followed at the top by a thick ferruginous sandstone– orthoquartzite unit of Ling Dongri Formation.

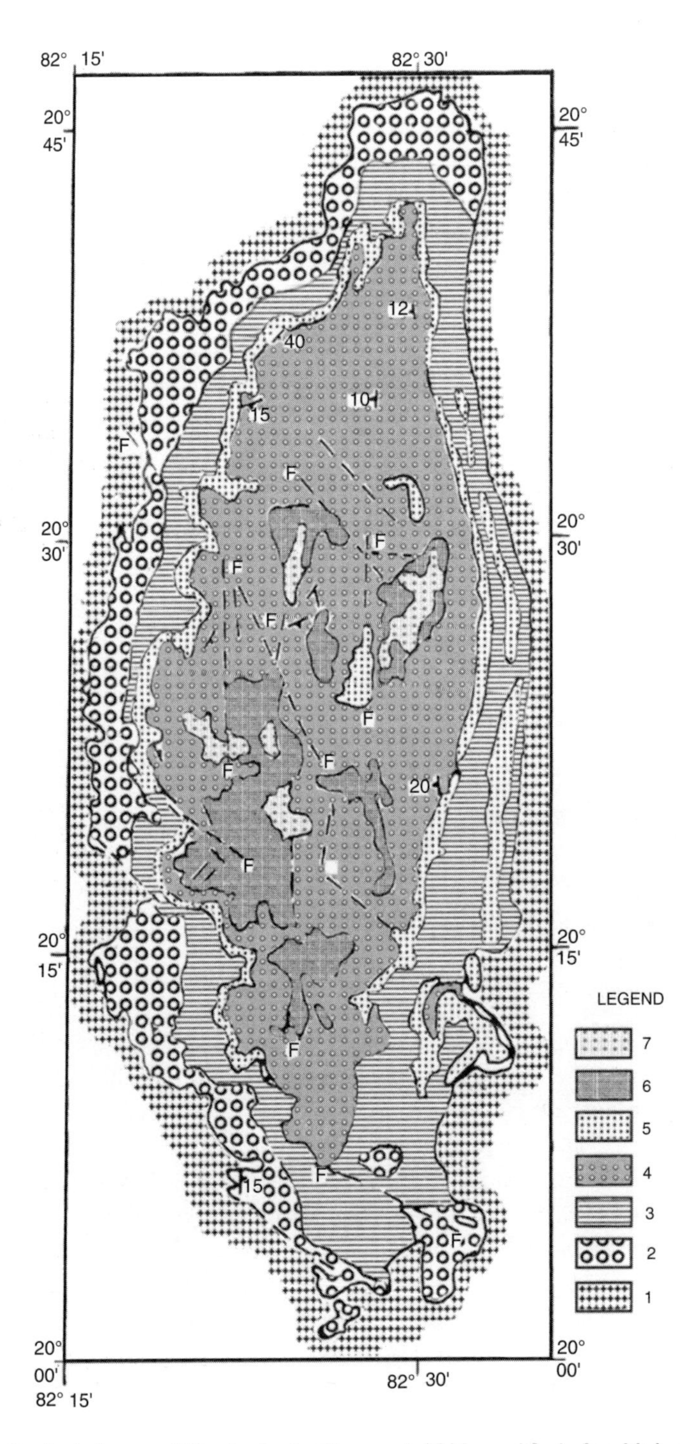

Fig. 2.1.27 Geological map of Khariar basin (Das et al., 2001, modified after Mishra et al., 1989). Legend: 1 –Granitoid, 2 – Devdahara Sandstone, 3 – Kulharighat Formation, 4 – Neor Formation, 5 – Galighat Sandstone, 6 – Tarjhar Formation, 7 – Ling Dongri Sandstone.

In general, the formations of Pairi Group are flat lying but show intense folding and faulting along the eastern part. The folds are open, symmetric to asymmetric with roughly N–S axial trace and 5–25° plunges. The eastern margin of the basin with EGMB basement is highly tectonised along a boundary fault.

Ampani Basin The N–S trending Ampani basin, covering about 220 sq km area is located just to the south of Khariar basin (Fig. 2.1.28). The sedimentary sequence of the basin comprises a 2 m thick and impersistent basal conglomerate–grit on the basement, overlain by a 180 m thick sandstone horizon, followed upwards by 30 m thick silt–shale beds grading into a pyritiferous purple shale with thin calcareous partings. The sandstone and the basement rocks exhibit shearing effects along the eastern margin of the basin, which is marked by a prominent boundary fault. Broad warps and folds are developed with roughly NNE–SSW axial trends, more along the eastern part. The top most shale is often found preseved in the core of synformal warps. The Ampani sediments are considered equivalent of Chadarpur–Tirathgarh sequences of Chhattisgarh and Indravati basins.

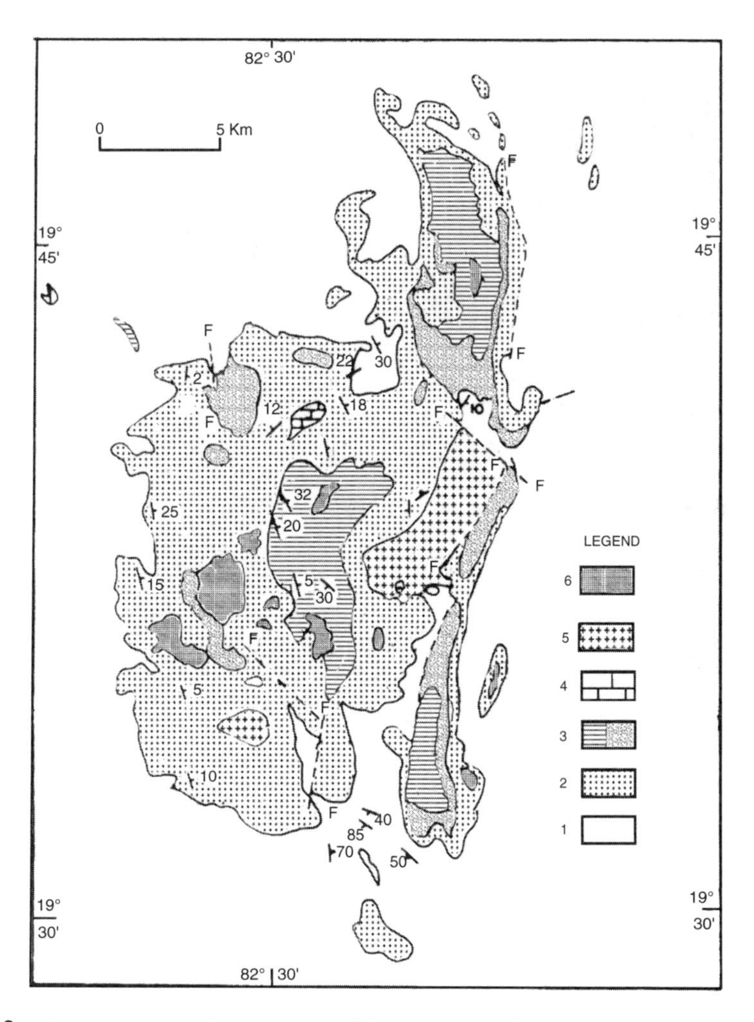

Fig. 2.1.28 Geological map of Ampani basin (after Das et al., 2001).
Legend: 1 – Granite gneiss with older supracrustals, 2 – Conglomerate, 3 – Sandstone, 3 – Purple/ grey shale, 4 – Limestone and calc shale, 5 – Hornblende grandiorite, 6 – Laterite.

Sabari Basin This is the southernmost Purana basin in the Bastar craton, occupying about 700 sq km area in the west bank of Sabari river near Sukma.The sedimentary cover sequence developed in this isolated NE–SW trending triangular basin is quite similar to the stratigraphy of the Indravati basin, though the lithounits are much thinner. The successions of basal conglomerate–quartzite, purple shale, limestone and shale–phyllite of the Sabari basin are correlated with Tirathgarh, Kanger and Jagdalpur Formations of the Indrabati Group. However, no equivalent of Cherakur Formation is found in the Sabri basin.

Deccan Traps and their Relation to Continental Flood Basalts

Deccan Traps About half a million square kilometer of the west–central part of peninsular India is covered by the large Mesozoic continental flood basalts known as Deccan Traps. Blanford (1867) and Crookshank (1935) of the Geological Survey of India conducted the classical studies on these flood basalts. The original spread of this extensive Phanerozoic volcanic suite could be much larger than the present extent, as is obvious from its scattered surface and subsurface occurrences both on land and in the offshore regions around the peninsula (Fig. 2.1.29). Coffin and Eldholm (1993) estimated the original spread of Deccan Traps as 1.5 x 10^6 sq km, that is three time its present extent.

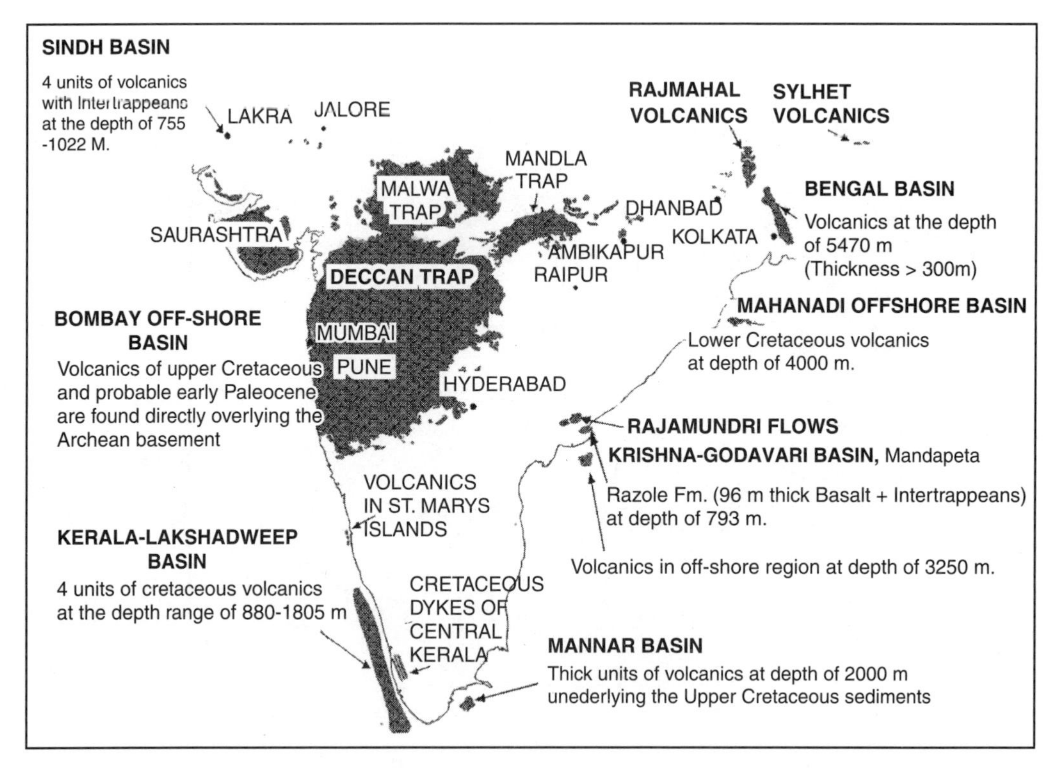

Fig. 2.1.29 Map of peninsular India showing extent of Deccan Traps and correlatable on- and off-shore Mesozoic volcanics (modified after GSI. 2006, Ed. K.S. Misra)

The main mass of the Deccan volcanic succession extends from the northern fringe of Karnataka, through large tracts of western and central Maharashtra to parts of Madhya Pradesh (Malwa and Mandla Traps) in north and northeast and Saurashtra region of Gujarat in the west. The thickness of the volcanic pile is maximum (~3000 m) in the Western Ghats of Maharashtra, while it peters down to a few hundred (up to ~900 m) or even tens of meters in the fringe areas (Beane et al., 1986; Devey and Lightfoot, 1986; Godbole et al., 1996 and Deshmukh et al., 1996). In the main Deccan Province, the older lava flows are in the west and north while the younger flows are emplaced in the south and eastern parts (Mitchel and Cox, 1988; Godbole et al., op. cit.). Apart from these huge volcanic masses, there are time equivalent volcanic piles called the Rajmahal Trap in Jharkhand, the Sylhet Trap in Meghalaya, small outcrops in Kutch, Sirohi and Jhalore in Rajasthan, isolated outcrops between the Mandla and Rajmahal Traps, Rajamundri flows along East Coast, the St. Marys islands along the West Coast and the dykes in central Kerala. Besides, deep drilling for oil exploration has revealed several interbedded volcanic flows within the Cretaceous succession in the Bengal basin, offshore Mahanadi and Krishna–Godavari basins, Mannar basin, Kerala–Lakshadweep basin and on- shore Cambay, Kutch and Sirohi basins.

The main Deccan volcanic province is characterised by linear flat topped hills and ridges with spectacular escarpments interspaced by flat flow-topped valleys, giving rise to a picturesque landscape (Fig. 2.1.30).

Fig. 2.1.30 An escarpment of the Western Ghats showing stacks of compound flows of Deccan Trap, viewed from Mahabaleshwar, Maharashtra

The Deccan flows are largely tholeiitic basalts in composition (Fig. 2.1.31 a–b), except in some restricted areas. Pyroclastic acid volcanics along Panvel flexure near Bombay, rocks of carbonatite affinity in western Narmada valley, alkali basalt flows and plugs with mantle xenoliths in Kutch are some of the prominent variations. Apart from these, the Deccan Traps in Saurashtra display a complete spectrum of differentiated products ranging from oceanite to picrite. Picritic basalt, perhaps generated by the fractionation of tholeiitic basalt, is common along the base of many flows. The picritic basalts in Deccan lava pile are enriched in cumulus olivine and clinopyroxene and do not represent liquid compositions. The parental melts of these picrites are estimated to have contained only ~9–10% MgO (Sheth, 2005b).

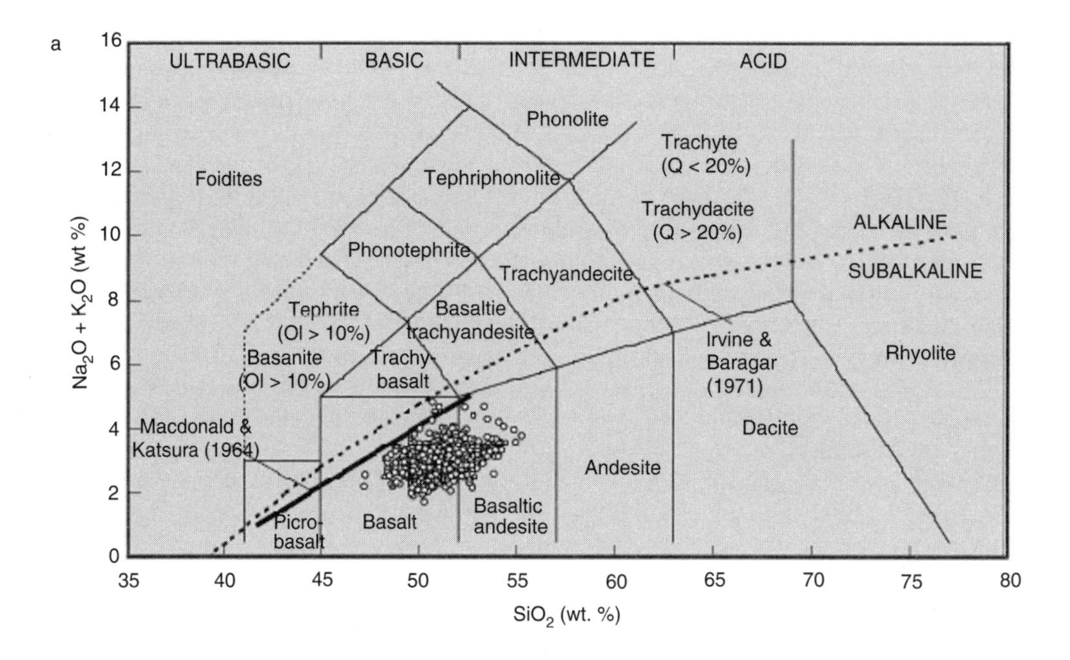

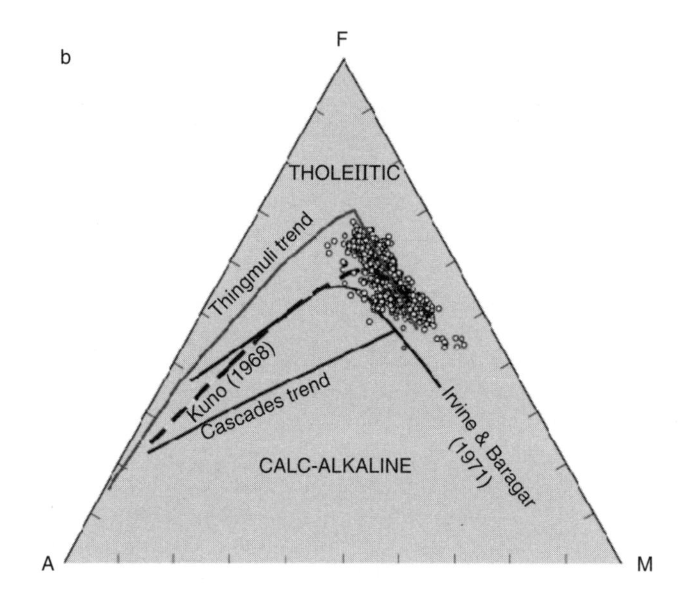

Fig. 2.1.31 (a) A Plot of 624 samples of Deccan Basalts of the Western Ghats (data of Beane et. al., 1986) on the Total Alkali Silica (TAS) diagram. (b) Plot of the same samples in the AFM diagram, showing the Fe-enrichment trend typical of tholeiitic basalts (Sheth, 2005b).

That the Deccan Basalts are overwhelmingly tholeiitic in composition is reflected by the plots in Fig. 2.1.31(a, b). In the TAS diagram, the complete absence of compositions other than basalt and basaltic andesite, and the nearly exclusive subalkalic (tholeiitic) nature is notable. The AFM diagram also shows the Fe-enrichment trend typical of tholeiitic basalts. Interested readers can refer to Sheth (2005b) for an extended petrological discussion.

The Deccan Basalts are mostly of Low-Ti varieties. Sano et al., 2001, based on analysis of 325 samples collected from 27 well-distributed sections across the Deccan Trap Province, subdivided the basalts into two sub-groups. The elemental characteristics of Ba, Sr, TiO_2 and Zr/Nb were used to classify the subgroups; the least-contaminated group has Ba contents < 100 ppm, Sr 190–240 ppm and TiO_2 2.0–4.0 wt%, and the most-contaminated group has TiO_2 contents < 1·5 wt% and Zr/Nb 15. Deccan basalts from south to north along the western Deccan Province, exhibit a rough negative correlation of $_{Nd}$(t) with Mg/Fe and SiO_2 and a positive correlation with Fe, consistent with temperature–controlled assimilation. The basalts with the lowest $_{Nd}$(t) values display distinctive lows for Nb, Ta, P and Ti, and large positive Pb spikes in their primitive–mantle–normalised element patterns, indicative of significant continental lithospheric influence in their petrogenesis (Mahoney et al., 2000).

The Deccan lava pile is composed of both simple (*aa*–) and compound flows (rhythmic alternation of *pahoehoe*–and *aa*–flows). The former type, having limited (20%) aerial extent, erupted at a much faster rate than the slow effusion of the latter, which covers a very large area (80%). The simple flows are thick and massive with prominent columnar joints at places. Ropy structures in pahoehoe flows are characteristic. The megacrystic tholeiitic flows with randomly oriented large plagioclase laths are well-developed in western the Deccan Province. Red and/or green 'bole' beds are often seen along the interface of two flows, which are considered the products of weathering and argillisation along the top of the lower flow (GSI, 1996). Intertrappean beds with terrestrial fossils occur within the lower parts of the volcanic pile. Along the eastern margin, the traps often overlie the basement crystallines with intervening carbonate rich sediments (Lameta) of Upper Cretaceous age. Paleomagnetic studies in some sections have shown N–R–N polarisation events in the course of eruption (GSI, 1996).

The lava piles are often traversed by crosscutting dykes, being most profuse in parts of Narmada–Tapti–Satpura region and in western Maharashtra to the south of Mumbai. The feeder vents and channels of the lava flows are not readily discernible but many workers have considered the dyke swarms along West Coast and Narmada–Tapti to represent in-filled fractures that acted as linear vents, their intersection in northwestern part of Deccan Province being the principal locus of volcanism. However, the horizontal distribution network of the lava flows in form of lava channels, lava tubes etc., have recently been identified over a large area of western Maharashtra, with best documentation from the Nasik–Ahmadnagar–Pune section (Thorat, 1996; Sharma and Vaddadi, 1996; Mishra, 2002). Chatterjee et al. (2001) attempted dynamic and rheologic modeling of the emplacement and migration mechanism of the Deccan flood basalt. The analysis showed that at volumetric eruption rate of 1 km^3/day/km strike length from a 3 m wide conduit could emplace mafic lavas in the order of 10^2 km^3 by volume in matter of weeks over distances as far as 50–100 km.

There are far too many structures and macro- to micro-textures in the lava flows, which are excellently documented by the GSI (cf. Pictorial Atlas of Deccan Flood Basalt, GSI, 1996; Field Guide Book on Deccan Volcanics, *Ed.* K.S.Mishra, GSI, TI, Hyderabad, 2006). Interested readers may also refer to the 'Annals of Deccan Traps Studies and Bibliography…', GSI's Special publication No.38 (1996, 101 p.) for more details.

According to Duncan and Pyle (1988) and Kaneoka et al. (1996), the main bulk of the Deccan Trap volcanics outpoured between 68 and 63 Ma over a short period of time (peak effusion lasting 1 Ma or even less), broadly coinciding with the Cretaceous–Tertiary boundary (KTB – 65.2 ± 0.2 Ma). Based on the subsequently available radiometric and paleomagnetic data from the Deccan Province, Pande (2002) suggested that the volcanism was episodic in nature and probably continued between 69 Ma to 63 Ma, with the most intense pulse of volcanism at 66.9 ± 0.2 Ma. Alexander (1981), however, had suggested longer duration of Deccan and related volcanic activity, starting from the effusion of Rajmahal Traps at the end of lower Cretaceous (116–100 Ma), culminating to the peak activity around 65–60 Ma. He also recorded 62.7 ± 2.05 Ma and 101 ± 3.3 Ma dates for the top and bottom flows (at 488 m depth) from the Dhandhuka borehole drilled in northern Saurashtra. Lisker and Fachmann (2001) have dated the volcanics in the Mahanadi offshore region and the associated onshore mafic dykes near Naraj by step heating analysis and suggested 108 ± 0.8 Ma and 110 ± 3.8 Ma ages respectively. It is observed that the radiometric dates of upper flows recorded from Saurashtra are particularly younger (62–63 Ma) compared to the dates obtained from Panchmarhi (71.8 ± 4 Ma), Mandla Trap (74.9 ± 2.108.5 ± 3 Ma), Dahod (Cambay onshore–72.6 Ma) and St Mary island (85.5 Ma, Pande et al., 2001; 93.1 ± 2.4 Ma, cf Mishra, 2005). The dates of the flows encountered in drill holes in the Kerala–Lakshadweep basin cluster around 100 Ma. The available paleontological data from the sedimentary sequences interlayered with volcanics intersected in the boreholes also suggest a long duration of volcanicity in the Deccan Province and its surroundings (Mishra, 2005).

The origin of Deccan lavas is a matter of debate. One school (Morgan, 1981; Richards et al., 1989; Campbell and Griffiths, 1990) believes that these lavas were erupted when the Indian plate moved over the Réunion hot spot (mantle plume) in the Indian Ocean. However, the long duration of volcanic activity, lack of favourable petrological evidence and the scatter of radiometric dates without testifying a tell-tale northward movement path of the Indian Plate, go against this proposition. According to another school, deep continental fracturing and rifting along the prominent linears in the Deccan Province under an extensional regime led to decompressional melting and outpouring of the flood basalt (Sheth, 1999a-b, 2005a-b; Baksi, 1999; Anderson, 1998; Smith, 1999; Mishra, 2004, 2005). Also a few suggest that the eruption was caused by meteoric impact, relating the main volcanic event of Deccan with the mass extinction of dinosaurs at the K–T boundary. The spectacularly circular Lonar Lake (with wide rim of ejecta) in the Deccan Trap country is considered an 'impact crater' (GSI, 1996).

Continental Flood Basalt Throughout the geological history, both continents and sea floors have experienced periodical eruption of enormous amounts of basalt in different periods, often thousands of cubic kilometers in a matter of a few hundred years, or even days. These may cover areas of as much as 100,000 sq km or more as both flood basalts and intruded sills, the magmas of each major pulse often being remarkably homogenous. At the same time, different units in the same region can differ widely especially in their Ti, P, K, Zr contents, in fact in all residual elements. They are in many respects similar to but are seldom as basic as Ocean Ridge Basalts (ORB). Compositionally, these basalts have 52–56% SiO_2, with high and variable Ti, Fe-contents similar to both the ORB and Ocean Island Basalts (OIB), usually showing marked Fe–Mg fractionation. However, typical continental signatures are also prominent at places, making them indistinguishable from the andesites. In spite of these variations, the bulk of these basalts are usually tholeiitic in composition, with typical cumulates of orthopyroxenite, picrite at places and rarely anorthosite. Late stage alkali basalts, melilites and carbonatites are sometimes associated. Small volumes of residual iron-rich granophyres may also be present.

General Characters Originating in the enriched upper mantle, continental flood basalts have a very different fingerprint than the ORBs. No depleted members are known; the least enriched being close to EMORB composition, but usually with even more elevated Cs, Rb, Ba, U, Th, K but depleted Nb–Ta. Some forms of the most LILE-enriched quartz saturated basalts are known. In fact, they have a typical continental signature, very similar to Andean andesites, though lower in silica and alumina and much more iron-rich in fractionated members.

Because they were extruded through very thick continental plates of lower specific gravity, in most areas covered with up to ~3000 m of detrital sandstone–arkose in the old Gondwana terrains, the basalts are often intruded as sills, many 300–400 m in thickness. The sills in Antarctica are at least 200,000 cu km in volume, though possibly much more is buried under the ice.

Origin Majority of the present researchers tend to assume an origin of flood basalts at great depth near the Earth's metallic core, forming a great mushroom shaped diapir. One might assume that any basaltic rock generated from such depths would be highly alkaline. The salient facts seem to be:

1. Continental flood basalts have a continental or crustal type signature identical to that of the andesite series.
2. All members are highly fluid basalts, not andesites. While some members have high silica (as much as 58%), these are plainly the product of fractionation.
3. Different members may be related by orthopyroxene or OPX + Plagioclase fractionation.
4. In sills especially, Fe and Ti fractionation is well advanced.
5. The overlap of Fe with alumina and the high CaO are quite unlike andesites. High iron is a characteristic of Ferrar and Karoo members especially.
6. All flood basalts are found in a continental environment where the crust has, at sometime in it's history, been subject to subduction and the sub-crustal mantle enrichment.
7. Individual members may vary, but single members of enormous volume (50,000 cu km) are quite homogenous. The post-initial emplacement of huge volumes of orthopyroxenite shows the magma was held for long periods, very slowly cooling in a sub-continental chamber, probably above a spreading axis.

It may be said somewhat tentatively that flood basalts are mantle generated in subcontinental environments where the mantle retains some enrichment from an older subduction phase. Because of the overlying continental roof of low specific gravity, large volumes of magma are generated without penetrating high in the crust. Tensional factures in the crust opening up fissures toward the surface followed by a short compressive phase forcing magma sometimes onto the continental surface would seem to be ideal. An incipient spreading centre developing under continental crust, or a spreading centre being forced under a continent, seem to be a pre-requisite.

Large scale contamination by crustal material is unlikely. The remarkable homogeneity of enormous volumes attests to it. Inspection of contacts, once deep in the crust, shows that granite and sediment are sealed off by a chilled barrier of basalt glass. Xenoliths are not seen to be even partly fused, though often recrystallised.

2.2 Metallogeny

Introduction

Central India is known for containing large deposits of manganese ores, bauxitic ores, BIF and related iron ores. The first two were being exploited from the colonial days, while the iron ores are being mined on commercial scale only after the Independence. Panna diamond is an old name. A large copper (–molybdenum–gold) deposit in calc-alkaline granitoid rocks, uncommon in age (~2.5 Ga) and in some geological characters, occurs at Malanjkhand. Small but economically interesting deposits of tin, rare earth and rare metals occur at several places. Occurrence of gold, uranium, and base metals are reported from several areas. A systematic but concise discussion on them has been lacking for long.

Tin Mineralisation in Bastar

Primary tin mineralisation in the Bastar and Dantewara districts of Chhattisgarh occurs as cassiterite in association with columbite and tantalite, predominantly in the pegmatites traversing the metasedimentary and metabasic assemblages of the Bengpal Group. The other associated minerals are beryl, varlomoffite, sterryte, pandite, kesterite, samarskite and fersmite. Cassiterite also occurs in association with vein quartz. Concentrations of placer tin have been located over wide expanse in the area in the alluvial, colluvial and eluvial material accumulated close to the pegmatite bearing stretches.

The Bastar tin province, which is a part of RM and REE bearing Bastar–Malkangiri pegmatite belt (cf. Ramesh Babu, 1999), extends from near Dantewara-Bacheli, Chhattisgarh, in the WNW to the Koraput district, Orissa, in the ESE through Metapal, Katekalyan, Tongpal and Mundval. The potential blocks with significant tin mineralisation may be grouped into the following five sectors from west to east.

1. Bacheli–Degalras sector
2. Raninala–Bodanar sector
3. Katekalyan sector
4. Tongpal–Mundval sector
5. Mundaguda–Koraput sector

Within these sectors several tin bearing pegmatites have been located over cumulative stretches of 10-15 kms each. Four pegmatite zones with aggregate strike length of 2 km in Bothapara block of Katekalyan area range in grade from 0.06% to 0.14% Sn. GSI estimated primary tin metal reserve of 12,700 tonnes in Bodovada-Katekalyan area. One of the significant secondary concentrations of tin in colluvium is noted over a length of 25 km, width 750 m–1 km, in Tongpal sector. The reserve estimates by GSI of colluvial tin ore is 28.63 Mt, which is equivalent to 5628 tonnes of tin metal.

As per the Mineral Year Book (2003) of Indian Bureau of Mines (IBM) the reserve estimates (as on 1.4.2000) of *in situ* tin ore in the country stands at around 32 Mt (metal 15,989 tonnes), which is entirely contributed by Chhattisgarh (96%) and Orissa (4%). Besides, there are 55.4 Mt of conditional resource in the country (mostly in Tosham, Haryana), of which about 1.45 Mt ore is in the Chhattisgarh-Orissa belt. Tin ore is mined in Govindpal–Tongpal area from the colluvial deposits; and the plants at Raipur and Jagdalpur produce tin metal from this ore.

The granites in this terrain vary from monzo–granite to syeno-granites (Streckeisen, 1976). The granites in the Bastar tin province show higher silica content with increasing tin content upto the critical value of about 50 ppm Sn. With the further concentration of Sn above the critical value there is always a distinct depletion of silica and alkalies in the granite (Babu, 1994). The pegmatites are complex but those related to the two–mica leucogranites, with pronounced greisenisation or wall rock alterations, are better hosts for tin mineralisation. The distribution of cassiterite crystals in the pegmatite is extremely erratic. The major element composition of the pegmatites does not show much variation. However, five types of pegmatite (I–V) have been recognised in the belt based on diagnostic trace element composition of perthite, albite and mica, showing distinct evolutionary trend from type-I to type-V (Ramesh Babu, 1999). The type-I and type-II pegmatites which are depleted in Li, Rb, Cs, are most enriched in Sn and Be (Dwivedi, 1999).

Rare Metals (RM) and Rare Earth Element (REE) Mineralisations

Apart from the most prominent Bastar–Malkangiri Pegmatite Belt (BMPB) in the south Bastar craton, the RM and REE bearing pegmatites are wide spread in Central India, occurring in the Garda –Toya area, west of the Indravati basin, western parts of Chhotanagpur Granite Gneiss Complex in the Sarguja and Jashpur districts of Chhattisgarh, Sausar and Sakoli Groups and the adjoining Precambrian terrain in Nagpur, Bhandara and Chandrapur districts of eastern Maharashtra and Ratnagiri district of western Maharashtra.

The important rare metal (RM) minerals in the pegmatites are beryl, columbite–tantalite, ixiolite, lepidolite and amblygonite and the REE minerals comprise xenotime, monazite and allanite.

As per the classification of Cerny (1991), the RM- and REE- bearing pegmatites of BMPB belong to the petrogenetic family of LCT (lithium, cesium, tantalum), those of Sarguja and Jashpur to NYF (niobium, yttrium, fluorine) and the rest of the occurrences in central India to a mixed family of the aforesaid types (Ramesh Babu, 1999)

RM and REE Pegmatites in BMPB and Garda–Toyar Area As mentioned above, while describing the tin occurrences in Bastar, the BMPB is recognised as a narrow WNW–ESE trending stretch (~ 80 km length × 15–20 km width) of the Archean Bengpal Group of rocks comprising metasediments and metabasics (Fig. 2.2.1).

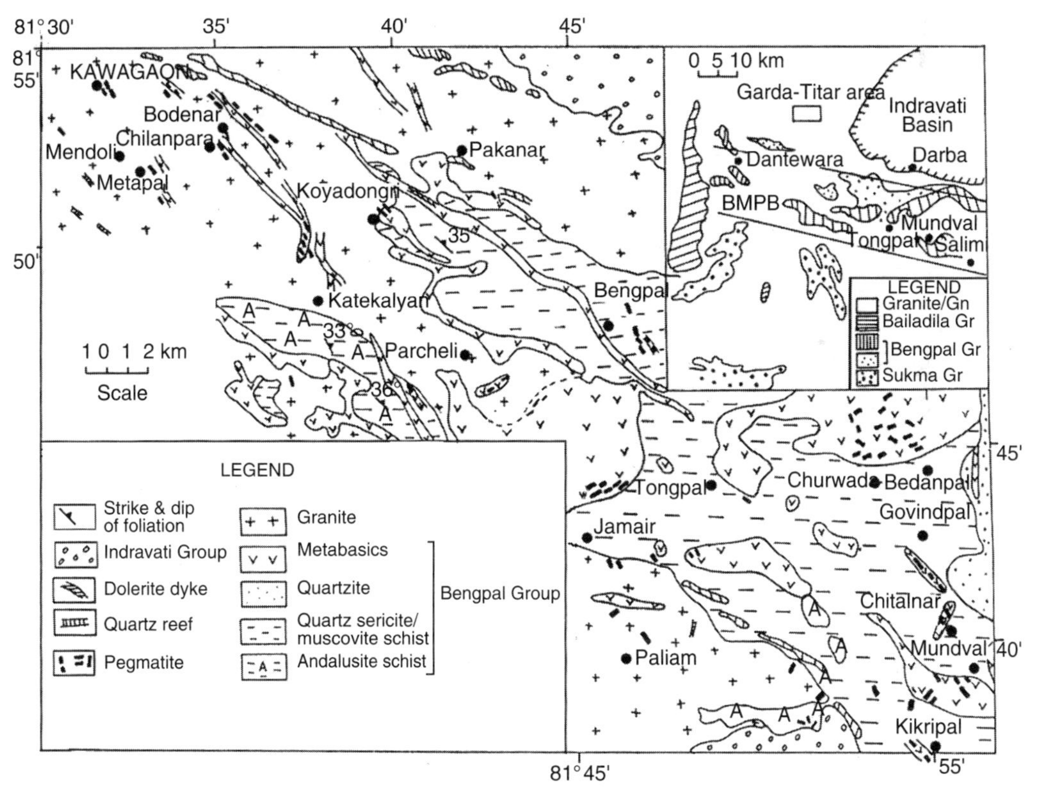

Fig. 2.2.1 Geological map of RM–REE–tin bearing Bastar–Malkangiri pegmatite belt (BMPB) and Garda–Toyar sector, southern Bastar, Chhattisgarh (modified after Ramesh Babu, 1999)

The regional medium temperature–low pressure (Abukuma type) metamorphic event in the belt is dated at 2530 ± 89 Ma (Ramesh Babu op. cit.). Ramesh Babu (1993) classified the pegmatites of BMPB into five types based on their mineralogy, internal zoning and associated ore mineralisation (Table 2.2.1).

Table 2.2.1 *Characteristics of different pegmatite types in BMPB (after Ramesh Babu, 1993)*

Type	Size and shape	Internal zoning	RM, REE mineralogy
I. Simple unzoned biotite–/muscovite–microcline pegmatite	2–20 m long, 0.1–1 m wide; tabular, ribbon-like	Unzoned, intergrowth of quartz + perthite ± biotite ± muscovite with occasional quartz pods	Barren
II. Partly differentiated muscovite–microcline pegmatite	20–50 m long, 2–15 m wide; pod like, crudely lensoid	Partly zoned with quartz pods, blocky microcline perthite and intergrowth zones	Beryl
III. Fully differentiated muscovite–microcline pegmatite	50–300 m long, 10–50 m wide; lensoid	Well-zoned quartz core, blocky perthite intermediate zone, wall zone of finer quartz ± perthite ± muscovite, albitised and greisenised replacement units	Beryl, columbite–tantalite, ixiolite, monazite, microlite, uraninite, ± cassiterite
IV. Replaced lepidolite–albite pegmatite	150–250 m long, 10–20 m wide; lensoid veins	Complex zonation with most of the earlier formed minerals replaced to varying degree	Lepidolite, amblygonite, cassiterite, beryl ± columbite–tantalite
V. Replaced muscovite–albite pegmatite	50–100 m long, 1–2 m wide; tabular veins	Relatively simple quartz + albite + muscovite types with occasional quartz pods; resemble albitised zones in Types III and IV	Cassiterite, beryl

Note: Magnetite, ilmenite, tourmaline, apatite, goethite, zircon, garnet and triplite are the associated minerals in the pegmatites.

Amongst the five paragenetic types of mineralised pegmatites, the fully differentiated, well-zoned, marginal pegmatites, i.e. Type III is the host for significant concentration of Be, Nb, Ta and REE, while the fully replaced Type IV pegmatites carry Li mineralisation (Ramesh Babu op. cit.). The former type is predominant in the northwestern part of the belt while the latter predominates in the southeast.

The geological maps of Bodenar and Metapal Pegmatites, occurring in the BMPB, reflect the details of their morphological and petrological features (Fig. 2.2.2a, b).

There are several plutons of fertile Proterozoic granites in the area, viz, Paliam, Darba–Metapal, etc., (2308 ± 48 Ma), which are geochemically specialised and parental to RM, REE and tin bearing pegmatites (Fig. 2.2.3a, b).Another smaller group of RM and REE mineral-bearing pegmatites occurs in the Garda–Toyar area located at about 20 km north of the BMPB. Here swarms and clusters of columbite–tantalite and beryl-bearing pegmatites are emplaced along the contacts of amphibolite and granite, and along the joints and fractures within these rocks.

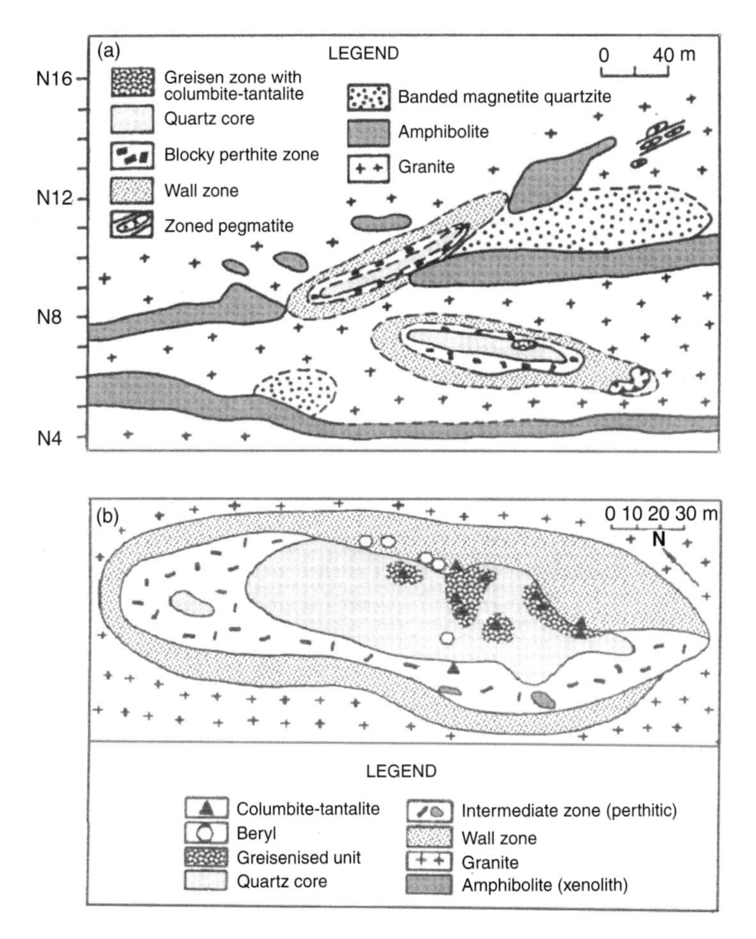

Fig. 2.2.2 Geological maps of (a) Bodenar Pegmatite, (b) Metapal Pegmatite, Bastar–Malkangiri Pegmatite Belt (BMPB) (modified after Ramesh Babu, 1999)

Pegmatites of Bodenar, Challanpara, Metapal, Parcheli and Jamair in BMPB and Adwal in the Garda–Toyar area have in the vicinity, rich columbite–tantalite bearing eluvial, deluvial and colluvial placer deposits, which have been evaluated and substantal reserves have been established.

RM and REE Minerals and their Chemistry The beryl in this belt is alkali-free type, with BeO 12.09–13.40%, total alkali < 0.5%. Columbite–tantalite [(Fe,Mn) $(Nb,Ta)_2O_6$] of BMPB has higher Ta_2O_5 (26.16–50.35%), higher SnO_2 (0.36–6.14%) and lower Nb_2O_5 (12.79–56.26%), while that in Garda–Toyar area has lower Ta_2O_5 (6.65–23.3%), lower SnO_2 (0.07–0.18%) and higher Nb_2O_5 (58.70–70.06%). Ixiolite [(Fe, Mn)(Ta, Nb, Sn)$_2$ O$_6$], the tin-bearing variety of tantalite, occurring in BMPB contains higher Ta_2O_5 (57.10–60.5%) and SnO_2 (8.7–10.2%), but lower Nb_2O_5 (8.5–11.0%) compared to columbite–tantalite of this belt. The other Ta minerals are wodginite (67.8% Ta_2O_5), microlite (89.8% Ta_2O_5) and niobian ixiolite (upto 58.8% Nb_2O_5).

Lepidolite in BMPB analyses 2.89–5.14% Li_2O, with associated Rb (1.98–3.13% Rb_2O) and Cs (0.18–0.58% Cs_2O). Amblygonite has higher Li (8.1–10.1% Li_2O) compared to lepidolite, and 45.3–47.5% P_2O_5.

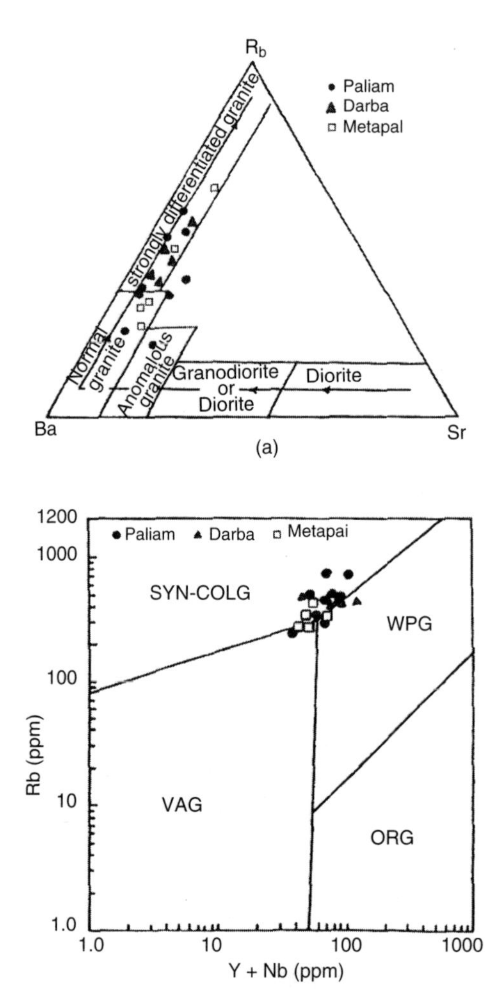

Fig. 2.2.3 (a) Geochemical specialisation of granitoids (parental to RM–REE–tin mineralisation) of Paliam, Darba and Metapal with respect to Rb–Ba–Sr. (b) Discrimination diagram Rb–(Y+Nb) showing the tectonic fields of emplacement of Paliam, Darba and Metapal granitoids (a. modified after Ramesh Babu, 1993; b. after Ramesh Babu, 1999).

The LREE-rich monazite [(LREE, Th) PO_4] is more common in the pegmatites of BMPB and Garda–Toyar area. The HREE–rich mineral xenotime [(Y, HREE) PO_4] occurs in minor proportions in the granites and related placers in BMPB. Some other RM and REE bearing minerals, euxinite and aeschynite are reported from the Metapal pegmatite (Somani et al., 1998).

RM– and REE–pegmatite in Western Chhotanagpur Granite Gneiss Complex The western part of the Chhotanagpur Granite Gneiss Complex in the Sarguja and Jashpur districts of Chhattisgarh is traversed by innumerable pegmatite veins and clusters containing low concentration of columbite–tantalite, beryl, xenotime and monazite and appreciable concentration of riverine placers in the surroundings (Bhatnagar, 1966; Bhola and Bhatnagar, 1969; Rai and Banerjee, 1995; Ramesh Babu et al., 1995).

In the Belangi–Jhapar area of Sarguja, swarms of pegmatite in Precambrian schists and migmatite inliers within Gondwana terrain contain sporadic beryl and columbite–tantalite. Some samples of beryl from zoned pegmatites at Hardibahar analysed 13.0% BeO, and those of tantalite from same locality analysed upto 57.4% Ta_2O_5, 5.86% Nb_2O_5 and U_3O_8 (Ramesh Babu, 1999). The simple unzoned as well as partly zoned pegmatites occurring within the hornblende chlorite schist and granites in parts of Kunkuri area, Jashpur district contain beryl, tourmaline, garnet and magnetite. Some of these pegmatites were mined for beryl.

There are several xenotime and monazite bearing placer deposits in the Mahan and Ib river systems in Sarguja and Jashpur districts (Rai and Banerjee, 1995; Ramesh Babu et al., 1995). The source pegmatites and granites for these placers in the surroundings contain disseminations of both the minerals.

RM– and REE–pegmatites in Sausar and Around Sakoli Belts, Maharashtra Fully differentiated and well–zoned RM–pegmatites related to Pauni Granite are emplaced in Pauni–Salai–Tangla tract, Nagpur district, within the Mansar and Lohangi Formations of Sausar Group and the Tirodi Gneisses. One prominent WNW–ESE trending pegmatite (700 m × 15–20 m) near Salai village, emplaced along the contact of calc–gneiss of Lohangi Formation and mica schist of Mansar Formation, contains amblygonite, lepidolite and fluorite at the core and columbite–tantalite in the outer marginal zone. Eluvial and colluvial placers with appreciable concentration of columbite–tantalite are found around the western parts of Salai Pegmatite. Numerous simple and unzoned type pegmatites in the area are the source for xenotime and monazite bearing riverine placers (with upto 3.59% Y_2O_3 and 13.71% Ce_2O_3) found in Pauni–Tangla area (Ramesh Babu, 1999).

Columbite–tantalite and beryl are reported from lateritic quartzose gravel associated with greisenised pegmatites occurring along the western fringe of Sakoli belt near Shahpur, Bhandara district. Minor columbite–tantalite are also associated with the pegmatites traversing the hornblende gneisses–metabasics–metasediments in Aheri–Allapalli area, Chandrapur district.

RM–pegmatites in Ratnagiri District, Maharashtra Columbite–tantalite [76.74% $(Nb–Ta)_2O_5$], lepidolite (2.8–3.2% Li_2O) and uraninite (65.15% U_3O_8, 10.52% ThO_2) are found in discrete zones in the very large zoned pegmatite body (925 m long and 550 m in maximum width) occurring within the E–W trending Precambrian country rocks near Kadaval, which was worked for mica in the past.

Resource Evaluation and Recovery The placer deposits in the BMPB and other localities hold maximum resources of columbite–tantalite with beryl and REE minerals constituting the byproducts. Sizeable reserves of columbite–tantalite have been established in the pegmatites of Bodenar, Challanpara, Metapal, Parcheli and Jamair in the northwestern parts of BMPB and Adwal in Garda–Toyar area. Salai Pegmatite in Nagpur district also hosts a notable reserve. However, the pegmatites in BMPB are most potential. The Atomic Minerals Directorate has been recovering the RM and REE minerals by setting up 20 TPD pilot scale beneficiation plants at different locations in the BMPB, in both Chhattisgarh and the Jharsuguda district of Orissa. It is feasible to recover lepidolite and amblygonite from several pegmatites in Gobindpal, Chitalnar and Kikripal, including those that were mined in the past, or are presently being worked out for cassiterite.

Iron ores of Dalli–Rajhara–Rowghat–Bailadila Belt

It is a more than 100 km long nearly north-south trending belt containing iron ore deposits of Dalli–Rajhara, Kanchar, Aridongri, Rowghat, Chhotodongar and Bailadila (Fig. 2.1.4). It is interesting to note that the iron ore deposits of the Durg district and the neighbourhood, including those of Dalli–Rajhara were first located by P.N. Bose as early as in 1887 (Bose,1987) and those of Bailadila before 1899 (in Krishnan, 1952). M/s Tata Sons and Co., now Tata Steel Limited, prospected the Dalli Rajhara deposits in the beginning of the last century for setting up an iron and steel plant in that region. But before they began work on that project, the same great Indian geologist of that time, Bose, struck BIF-related hematitic iron ore mineralisation at Gorumahisani ridge in the erstwhile Mayurbhanj State, Eastern India. The Tatas soon shifted the site of the steel plant to a place later named as Jamshedpur or

Tatanagar in Bihar (now in the state of Jharkhand), close to the confluence of Subarnarekha and Kharkai rivers. A number of geologists, including amongst others Crookshank, Ghosh and Heron, studied some of these deposits. Except for Dalli–Rajhara and Bailadila, all other deposits in this belt are yet to be exploited (Table 2.2.2).The Dalli–Rajhara ores are being used in the Bhilai Steel Plant. Bailadila ores until recently were entirely exported. At present, the newly set up steel plant at Vishakhapatnam by Ispat Nigam (PSU) is also drawing its iron-ore feed from the Bailadila mines.

Table 2.2.2 *Some details about the iron ore deposits of the Dalli–Rajhara–Rowghat–Bailadila belt*

Deposit	District	Grade (% Fe)	Reserve (Mt)	Status of mining
Dalli–Rajhara	Durg	66.35	163	Mining on
Kauchar	Durg	NA	82	Unexploited
Aridongri	Kanker	NA	26	Unexploited
Rowghat	Bastar	63.47–68.26	297	Unexploited
Chhotadongar	Bastar	61–66	32	Unexploited
Bailadila	Dantewara	60–69	1135	Mining on
Deposits along the bordes of Durg and Kankar districts	Durg–Kanker	NA	88	Unexploited

It is obvious from the above table that the Bailadila deposits contain the largest reserve of iron ores in this belt, followed by Rowghat and Dalli-Rajhara successively. Rowghat or its neighbourhood is not connected by railways to an existing or forthcoming industrial site/port, a plausible reason for not being taken up for exploitation till date.

Dalli–Rajhara Deposits Dalli–Rajhara iron ore deposits occurring on the Pandarhalli –Rajhara Hills, are located along the eastern part of a 15 km long, nearly ENE–WSW trending iron ore belt, popularly known as the Dalli–Rajhara iron ore belt (subbelt). This belt, rather subbelt, is the northern termination of a longer iron ore belt (mentioned above) that extends northwards from Bailadila (–Kirandul) in the south. The belt is constituted of, besides the BIFs and iron ores, rocks such as phyllite/ mica schists and metabasic rocks (Table 2.2.3). Granitoid rocks occur to the immediate east of the belt.

The BIFs, together with the iron ore deposits, form a distinct ridge/ridges in the backdrop of an otherwise even topography.

Table 2.2.3 *Stratigraphic succession in the Dalli-Rajhara area (after the Mines Office)*

Stratigraphic unit	Rock types
Intrusive and effusive rocks	Dolerite, gabbro, norite, Chikli Granites
Upper ferruginous unit	Upper ferruginous shales with clay, and grit with lean ore pockets at places
Middle ferruginous unit	Banded iron formation, Lower ferruginous shale/schist, Banded ferruginous quartzite (BHQ)
Lower ferruginous unit	Talc, shale/schists, banded quartzites phyllites, shales/ schists
Basement not known	

The iron ore deposits of the Dalli-Rajhara belt are divisible into two sectors:

1. The Rajhara sector
2. The Dalli Sector.

In the Rajhara sector, Rajhara Main block is the principal deposit, flanked by the Konkan East block and the Konkan West block on its two sides. The Dalli sector begins with Jharandalli in the east, followed westward by the Dalli Mechanised Mine (Kondekasa) and the Dalli Manual Mine (Mayurpani) respectively. The two groups of deposits (Dalli and Rajhara) are separated by a fault with limited displacement (Fig. 2.2.4). All these deposits are presently being mined. There is another operative iron ore mine at about 15 km down south along the belt, called the Mahamaya deposit. Mining first started in the Rajhara Main block in 1955. Working on others followed at different points in time.

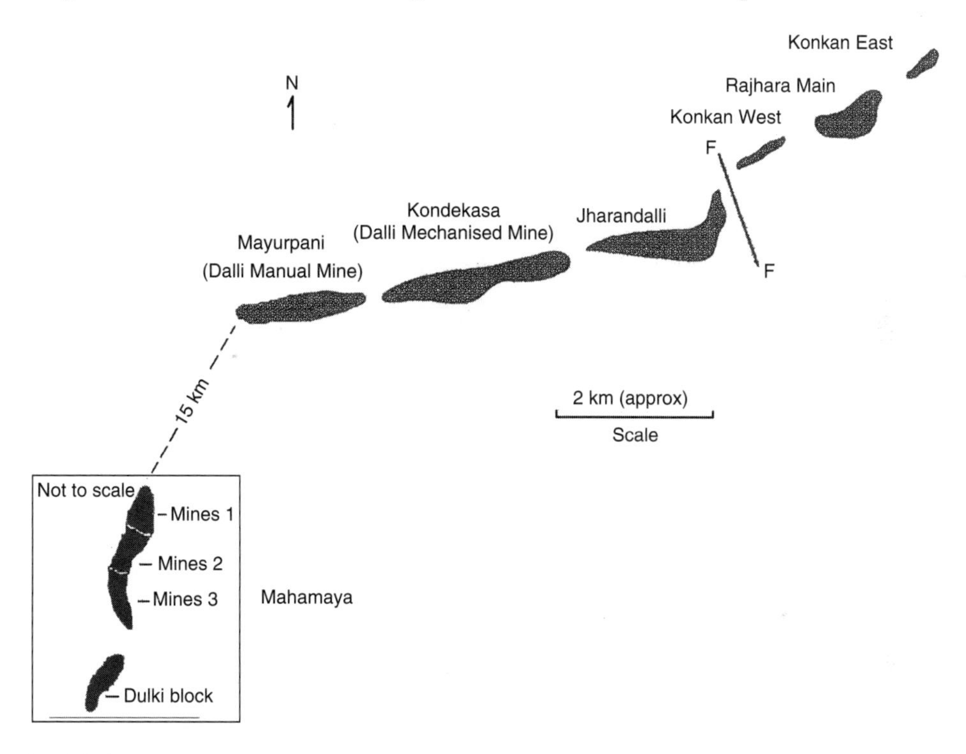

Fig. 2.2.4 Distribution of iron ore deposits in Dalli–Rajhara and Mahamaya sectors

The Rajhara Main block is ~1 km long at 443 m RL and is more than 200 m at its widest and continues downward to a depth of about 150 m from the top (March, 2004). Near the bottom, it shows bifurcation at 333 m RL and finally pinches out at about 250 m RL. The hanging wall of the deposit is constituted of shale, phyllites, basic volcanic rocks; and the footwall by banded hematite quartzite. The variation in the disposition, shape and thickness of the ore body in the Rajhara Main block, as depicted in the cross sections, is interesting (Fig. 2.2.5).

The Rajhara Main block contains mostly massive ore in the eastern and central parts and at depths in the western part. The upper sections of the western part of the Main block and Konkan West block mainly contain laminated ore. However, the Konkan East, the eastern, and the western parts of the Rajhara Main block were capped by laterites, now removed. The massive ores of Rajhara contain + 67% Fe. The Al_2O_3 content is generally low (Table 2.2.4). The Konkan East block is largely mined out, with the remaining balance reserve of 0.28 Mt (with 60% Fe, 5.5% SiO_2, and 4.50% Al_2O_3) as on April 2003, out of original reserve of 27 Mt.

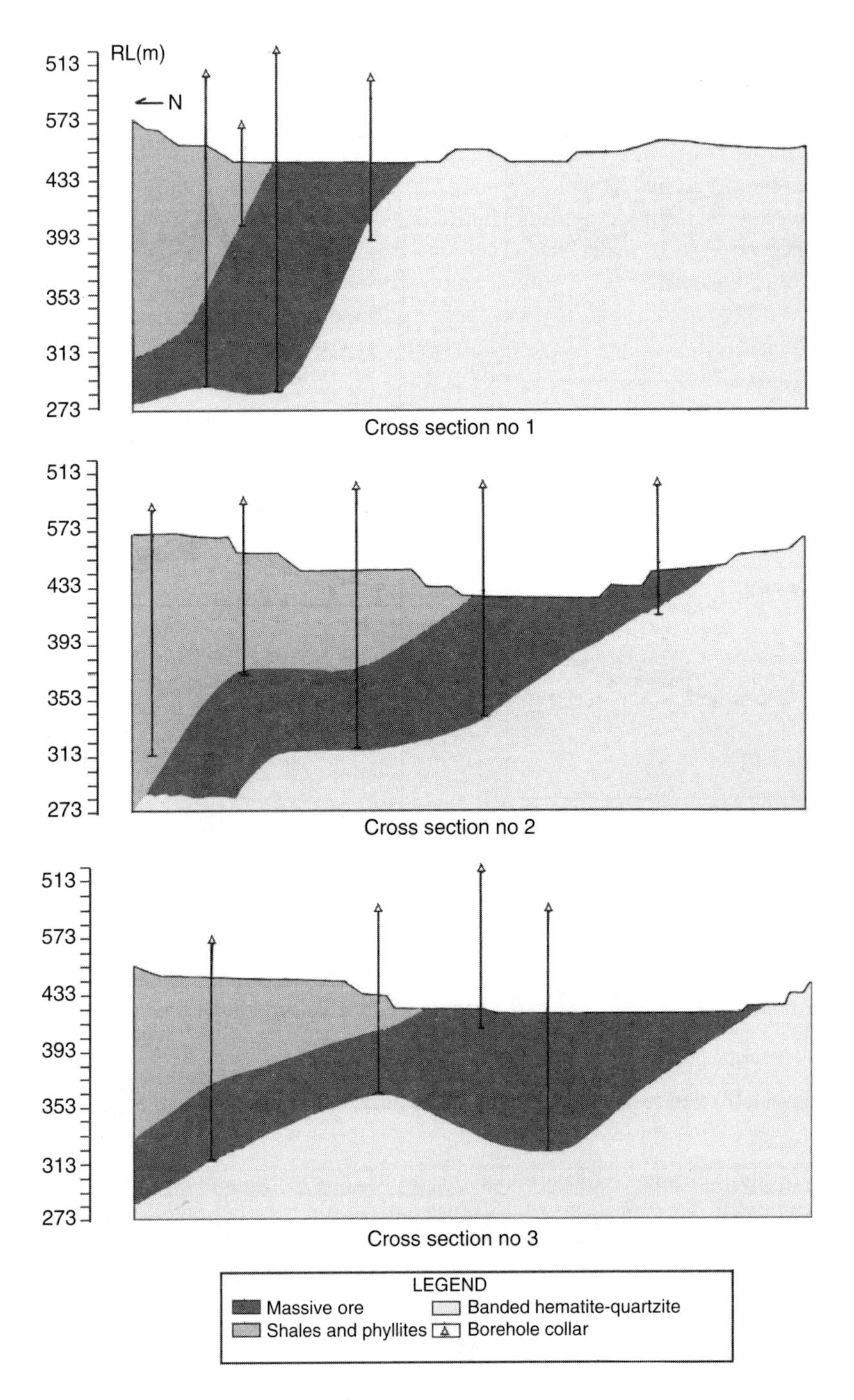

Fig. 2.2.5 Cross-sections showing the variation in disposition, shape and thickness of the iron ore body in Rajhara Main block

Table 2.2.4 *Geological reserves and composition of iron ores in the Rajhara Main Block (Reserves in kilotonnes, Kt) (Data source: Mines office, estimates as on 01.04.2003)*

S No.	Slice mRL	Bench	Ore reserve	Fe(%)	SiO_2 (%)	Al_2O_3(%)	P(%)
1	513–523	513 mRL	26	60.88	6.60	2.38	0.130
2	503–513	503 mRL	23	55.80	13.76	3.27	0.040
3	493–503	493 mRL	12	62.40	6.60	1.56	0.120
4	483–493	483 mRL	5	63.60	4.00	3.40	0.080
5	473–483	473 mRL	18	63.41	3.94	1.84	0.080
6	463–473	463 mRL	26	63.32	6.45	2.69	0.010
7	453–463	453 mRL	145	65.00	3.20	2.80	0.080
8	443–453	443 mRL	482	65.50	2.80	2.60	0.090
9	433–443	433 mRL	1034	66.00	2.40	2.50	0.060
10	423–433	423 mRL	2647	67.10	1.00	1.20	0.050
11	413–423	413 mRL	3712	67.00	1.10	1.30	0.050
12	403–413	403 mRL	3537	68.02	0.73	0.68	0.030
13	393–403	393 mRL	3380	67.68	1.37	1.21	0.040
14	383–393	383 mRL	3152	67.83	1.94	0.68	0.050
15	373–383	373 mRL	2952	67.68	1.38	1.20	0.040
16	363–373	363 mRL	2709	67.60	1.38	1.24	0.040
17	353–363	353 mRL	2407	67.59	2.40	0.65	0.040
18	343–353	343 mRL	2163	68.12	2.10	0.33	0.050
19	333–343	333 mRL	1738	68.05	1.52	0.59	0.050
20	323–333	323 mRL	1499	68.07	1.50	0.57	0.050
21	313–323	313 mRL	1452	67.61	1.74	0.81	0.050
22	303–313	303 mRL	1300	67.68	1.57	0.53	0.040
23	293–303	293 mRL	1126	64.42	3.64	0.56	0.040
24	283–293	283 mRL	956	64.43	3.61	0.55	0.040
25	273–283	273 mRL	788	66.83	2.64	0.56	0.040
26	263–273	263 mRL	348	68.80	0.58	0.56	0.030
27	253–263	253 mRL	178	68.80	0.54	0.49	0.030

Total reserve: 37817 Kt or 38 Mt (approx.)
Average grade: Fe – 67.38%, SiO_2 – 1.66%, Al_2O_3 – 0.95%, P – 0.045%

The geology of the Dalli sector is not much different from that of the Rajhara sector. Two barren zones separate the iron ore deposits in the Dalli hills into three blocks: Jharandalli (east), Kondekasa (middle) and Mayurpani (west). The barren zone between Jharandalli and Kondekasa is constituted of BIF and ferruginous shales, with a cover of laterite. The geological succession and the lithological association in both Jharandalli and Kondekasa are the same. On plan, the ore body in the Jharandalli block is sickle-shaped and occurs mainly over the BIF(BHQ) and locally over ferruginous shale (Fig. 2.2.6 a,b). The mean thickness of the ore body is 31m and its axial length is 1.76 km. The ore reserve of the Jharandalli block, as estimated in April 2003, was 17.62 Mt with + 63% Fe, 4.52% SiO_2 and 2.3% Al_2O_3.

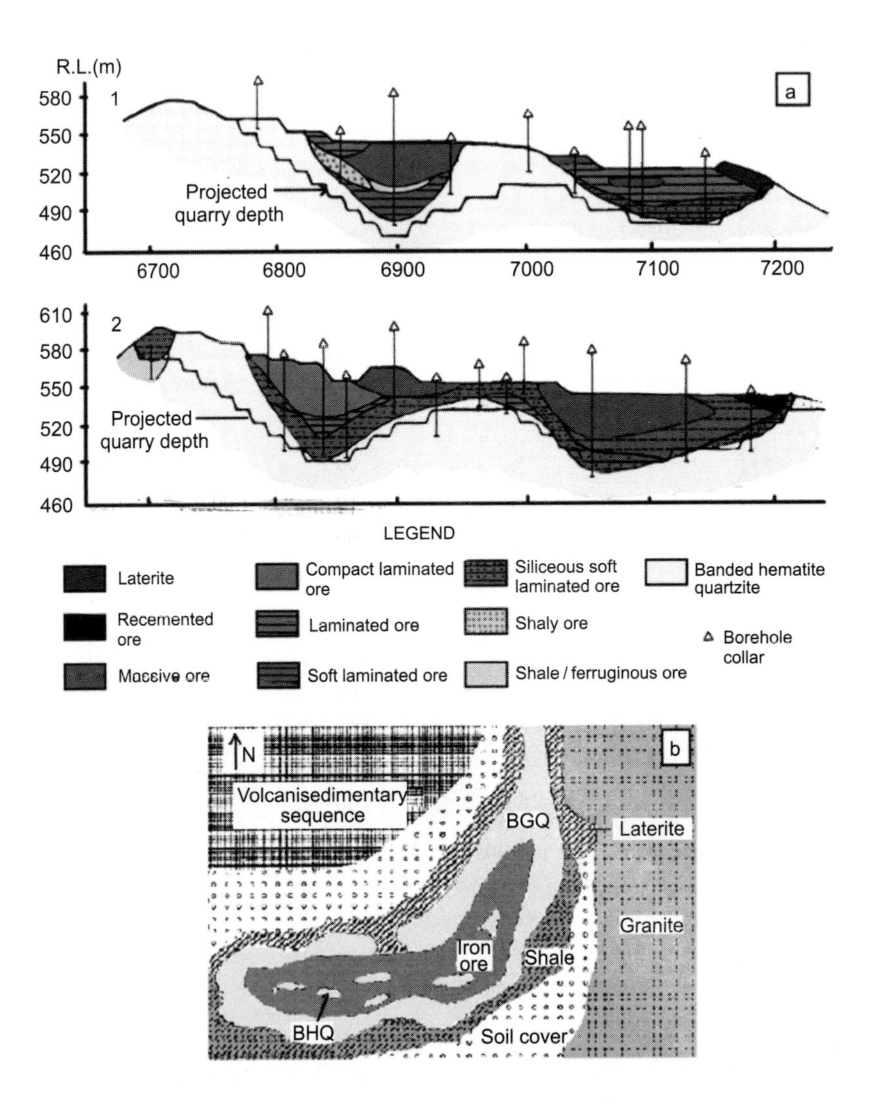

Fig. 2.2.6 (a) Cross-sections showing the disposition of iron ores with respect to BHQ (BIF) and shale in Jharandalli block, Dalli deposit (after Mines office); (b) Sketch map of Jharandalli block at 423 m RL bench level (March, 2004) (Authors).

The Kondekasa block (Dalli Mechanised Mine) has a spindle shaped ore body, with an axial length of 2.43 km. The deposit is folded at the centre, attaining a maximum width of 500 m. The mean thickness of the ore body is 25 m. As in April 2003, the ore reserves of the block stood at 51 Mt with 64.03% Fe, 4.02% SiO_2 and 2.18% Al_2O_3.

The Mayurpani block (Dalli Manual Mine) is the western most block of Dalli sector. The orebody extends for 1.53 km in strike and has a proven depth extension for 150 m (640 to 490 m RL).

The physical properties of the orebodies in the Dalli sector (Jharandalli, Kondekasa and Mayurpani) vary within a short distance from massive to different types of laminated ores (compact laminated, laminated, to soft laminated). However the laminated ores dominate. Table 2.2.5 shows compositional variation of the iron ores in the Dalli Mechanised (Kondekasa) mine. It is interesting to note that the Al_2O_3 and SiO_2 content increase markedly from massive through laminated to the

Table: 2.2.5 *Reserve and composition of ore types in Dalli Mechanised Mine (Data source: Mines office)*

Levels	MO				C.LAM				LAM				S.LAM				SHO			
	QTY(T)	Fe	SiO_2	Al_2O_3	QTY(T)	Fe	SiO_2	Al_2O_3	QTY (T)	Fe	SiO_2	Al_2O_3	QTY(T)	Fe	SiO_2	Al_2O_3	QTY(T)	Fe	SiO_2	Al_2O_3
1	2	3	4	5	6	7	8	9	10	11	12	13	14	15	16	17	18	19	20	21
620–610 m									3717	61.05	4.40	3.92	17200	60.37	5.20	4.24	2763	57.22	7.50	7.48
610–600									6275	61.58	4.20	3.08	7533	59.67	5.73	5.28	47718	57.22	7.25	7.42
600–590					31005	63.80	3.26	3.01	165902	61.59	4.67	4.28	111935	62.03	4.95	4.36	107858	57.22	7.727.91	8.14
590–580	1201	65.40	2.78	1.60	123893	64.70	2.85	3.41	762898	61.11	5.47	4.05	164965	61.37	5.58	3.49	210917	57.22	7.24	8.03
580–570	149063	66.50	2.43	1.75	139120	64.53	3.65	3.12	1283699	62.15	5.91	3.69	275876	61.02	6.52	4.97	154110	57.88	8.45	8.32
570–560	195656	65.41	2.85	2.02	386208	64.47	3.82	3.20	2147481	63.47	4.90	3.36	403931	62.96	5.98	3.42	348749	56.99	7.56	7.94
560–550	544613	67.73	1.45	1.08	985176	65.51	2.89	1.86	3328933	63.36	4.97	3.21	1002403	64.24	4.55	2.65	308979	57.24	4.59	8.01
550–540	1661236	66.68	1.88	1.30	2320167	66.08	3.20	1.02	3834263	64.80	3.80	1.60	1944636	64.60	4.60	1.80	278827	62.54	6.78	4.96
540–530	1708216	66.48	1.97	1.51	23311880	65.36	3.20	1.20	2902980	64.71	3.85	2.29	1166494	64.05	4.40	2.57	257363	60.56	6.78	5.34
530–520	1486152	66.00	1.98	1.25	1686889	65.27	3.67	1.74	1933023	64.01	4.64	1.88	1243323	63.18	4.80	2.60	22056	60.56	6.78	5.34
520–510	964986	67.00	2.00	1.88	1466779	65.92	3.40	1.34	765172	65.13	3.35	1.62	884835	64.27	3.82	2.40	6162	60.56	6.78	5.34
510–500	188976	67.20	1.88	0.97	1176290	66.08	3.15	1.28	426192	64.72	3.39	2.15	545194	66.25	3.44	2.20	6483	60.56	6.78	5.34
5000–490	176889	67.60	1.80	0.52	559192	67.38	1.41	1.00	166779	68.00	1.21	0.85	221730	62.84	4.12	3.63	18388	60.56	6.78	5.34
490–480	47493	67.60	1.80	0.64	399918	66.83	1.60	1.07	49163	67.66	1.40	0.63	54246	59.69	6.61	5.07				
480–470					169390	66.42	1.89	1.27					46897	59.69	6.61	5.07				
470–460					67261	64.33	3.87	1.43					19861	59.69	6.61	5.07				
	7123479	66.62	1.94	1.40	11843168	65.76	3.13	1.43	17766476	63.94	4.42	2.54	8111061	63.89	4.61	2.60	1740372	58.66	7.14	7.06

MO – Massive ore, C.LAM – Compact laminated, LAM – laminated, S.LAM – Soft laminated, SHO – Shaly ore

Table 2.2.6 *Levelwise, gradewise quantity of ore types, Dalli Mechanised Mine (Data source: Mines Office)*

Levels	RO				SISL				LTRO				AVG			
	QTY (T)	Fe	SiO_2	Al_2O_3	QTY (T)	Fe	SiO_2	Al_2O_3	QTY (T)	Fe	SiO_2	Al_2O_3	QTY (T)	Fe	SiO_2	Al_2O_3
620–610 m													23680	60.11	5.34	4.57
610–600													31526	58.67	6.28	6.04
600–590													416700	60.74	5.43	5.21
590–580													1263873	60.85	5.63	4.58
580–570	58715	58.09	9.98	4.20	3414	60.32	8.74	2.12	36218	58.64	7.12	7.24	2100215	60.97	5.83	3.97
570–560	106453	61.34	5.11	3.89	12656	60.06	9.40	2.68	66759	58.17	7.95	6.89	3667893	61.78	5.21	3.66
560–550					93109	62.58	6.67	2.15	44893	58.36	7.54	7.02	6308106	63.45	4.45	2.92
550–540					656005	62.18	9.98	1.30	101564	58.88	7.69	4.77	10796700	64.50	3.95	1.52
540–530	32606	57.40	9.08	5.32	1045519	62.30	10.00	0.25	6243	58.13	0.00	0.00	9451300	64.66	4.19	1.78
530–520					723208	61.44	9.70	1.17					7094650	64.31	4.40	1.78
520–510					894218	60.44	9.85	1.60					4972151	64.72	4.36	1.73
510–500					491049	62.05	7.66	2.23					2834184	65.27	3.95	1.74
500–490					68902	62.60	5.66	2.70					1210880	66.29	2.26	1.55
490–480					59403	62.93	7.52	1.69					610223	65.94	2.62	1.41
480–470													216287	6496	2.91	2.09
470–460													87122	6327	4.49	2.26

RO – Recemented ore (Canga), SISL – Siliceous soft laminated ore (contact zone with BHQ), LTRO – Lateritic ore. Grades-wt% Fe.

shaly ores. At the same time, no pattern in depthwise variation is indicated. Al_2O_3 and SiO_2, as expected, are high in recemented and lateritic ores. However, surprisingly, the SiO_2-contents are abnormally high in the siliceous soft laminated ores. This variety occurs close to the BIF. This could be due to inadequate removal of SiO_2 during the alteration of BIF into Fe-ores. As in the Rajhara mines, there are pockets and patches in the ores where $SiO_2 \pm Al_2O_3$ contents are relatively high. Folds on all scales, defined by the interlayered ore–BIF(BHQ)–shale sequence, are prominant in all the blocks of Dalli–Rajhara belt (Fig.2.2.7).

Fig. 2.2.7 (a) Folded massive iron ore in Rajhara Main block, Bench – 423 m RL; (b) Iron ore in the core of a chevron fold defined by BIF in Kandekasa (Dalli) block, Bench – 550 m RL; (c) Hard ore intimately admixed with soft ore as seen in Rajhara Mine face at 423 m RL (photographs by authors); (d) Sketch of bench faces at Rajhara and Mayurpani Mines.

Mahamaya Deposit An isolated deposit with three working mines M–1, M–2 and M–3, occurring at about 15 km south of the Dall–Rajhara belt proper, is known as Mahamaya deposit (Fig. 2.2.8).

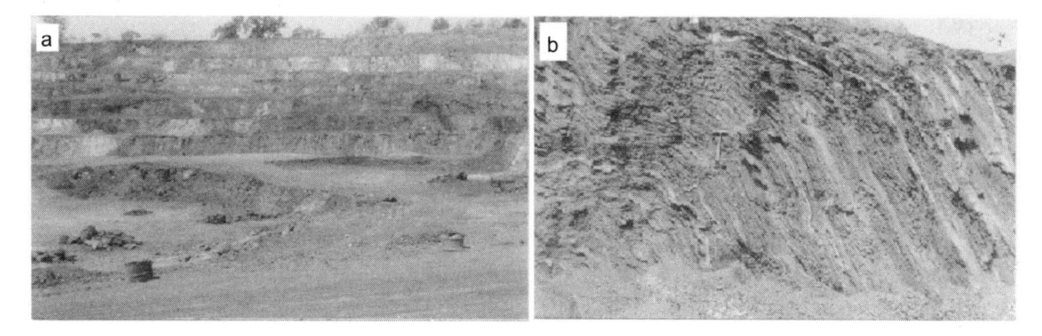

Fig. 2.2.8 (a) Massive ore below laterite capping and flat peneplained surface, Mahamaya deposit (Mine–3); (b) Warping in ore–shale interlaminates in Mahamaya deposit (Mine–3) (photographs by authors).

Further south, close to Mahamaya, there is another block named Dulki, which is also under development. About 20% of ores in the Mahamaya deposit belong to compact (hard) laminated category (Fig.2.2.8a) while the rest are soft laminated ore or blue dust. The orebodies occur variably in the immediate proximity of both BIF and shale. The ore–shale interlaminates display folds and warpings (Fig. 2.2.8b). The ore reserves, as in April 2003, stood at 6.5 Mt with 62% Fe, 5% SiO_2 and 2.90% Al_2O_3. The strike of the ore body varies from N–S to NW–SE, and the dip from subvertical to 60–70 towards east.

Rowghat and Some Smaller Deposits If the Dalli–Rajhara and Bailadila (particularly Kirandul) are considered as the two ends of the Dalli–Rajhara–Bailadila iron ore belt, then Rowghat is located almost at the middle.

The mineralisation at Rowghat is broadly divided into 6 deposits. The outcrops make an M-shaped pattern with an antiform at the middle and two synforms on both the sides. The deposits have been named A to F, the last one being the richest and the largest (Fig. 2.2.9). It is further, sub-divided into several blocks. The ores in the deposit F are massive, massive laminated and lateritic in types. The ore is generally of high-grade, the Fe-content varying from 63–68% and ore reserve estimated at about 521.45 million tonnes.

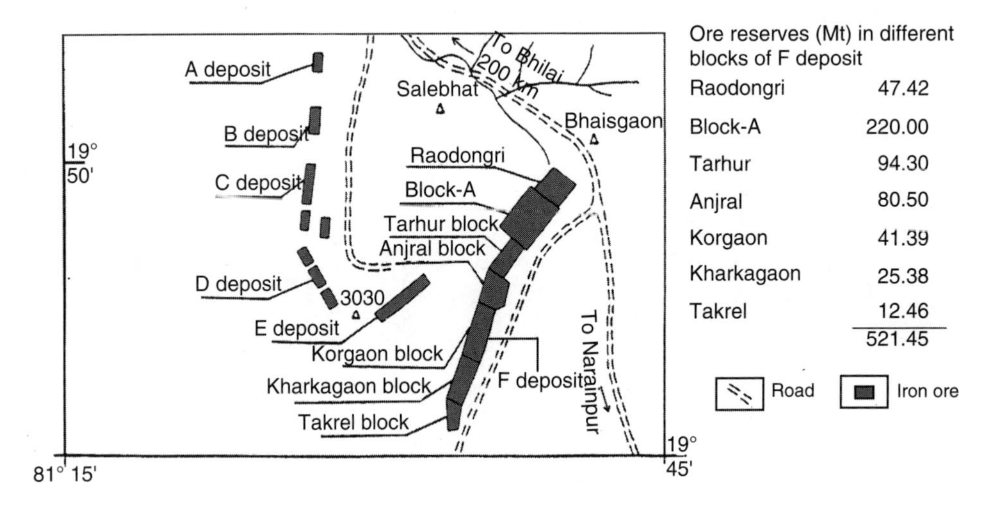

Fig. 2.2.9 A sketch map of the Rowghat iron ore 'belt' showing the disposition of different deposits and blocks and reserve of F-deposit

The Kanchar deposit is located 12 km SW of Rajhara hills. The BIF has been traced over a strike length of 3.2 km. It is folded into a series of antiforms and synforms. Several ore bodies consisting of massive and laminated hematitic ores occur at the upper part of the ridge. The estimated reserve is 82 Mt. However, no information on the ore grade is available.

In the Chhotadongar area, small occurrences of hematitic ores are located on the top of the Bargaon Dongri and Dhanja Dongri hills. The ore is mainly massive and laminated type, with an average ore-grade of 66.35% Fe.

Iron ores occur in the Ari Dongri area in the form of narrow bands or lenses, spread over the ridge-length of about 5 km. The ores in the eastern half of the ore zone is hard and compact, while in the rest it is soft 'micaceous'. The estimated ore reserve is 26 Mt, ore-grade being not known (Bandyopadhyay et al., 2001).

Continuous mining in the Dalli–Rajhara belt since 1955 has largely depleted the ore reserves there. It has been virtually a captive mine (deposit) for the Bhilai Steel Plant. Therefore, when the Steel Authority of India (SAIL) looks for another large deposit to sustain the Plant, the Rowghat belt will be an eligible candidate, or possibly the best candidate, keeping in view its nearness to the Plant.

Bailadila Group of Deposits The Bailadila group of iron ore deposits occur along two N–S trending sub-parallel ridges that extend for about 30 km from Kirandul in the south to Jhirka in the north where they come closer and then extend further northward.

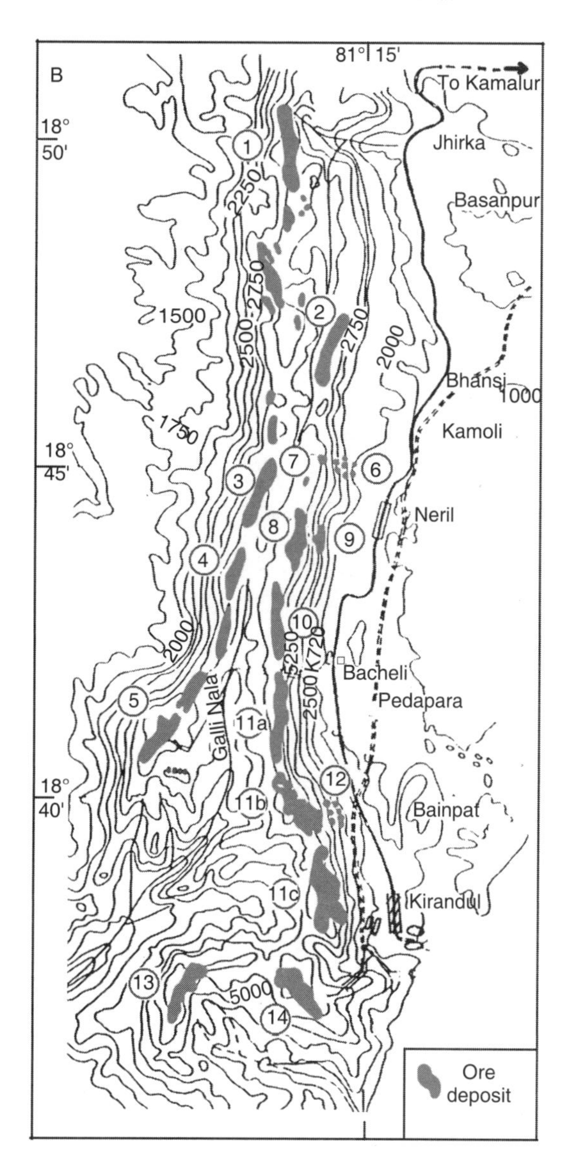

Fig. 2.2.10 Ore deposit distribution map of Bailadila area (compiled from Khan and Bhattacharya, 1993 and Chatterjee, 1993)

The two ridges are in fact synforms and the valley between them is an eroded antiform. BIF (BHQ) commonly occurs along the upper portion of the eastern flank of the eastern ridge and the western flank of the western ridge. Crookshank (1963) identified 14 deposits in both the ridges of this belt and numbered these accordingly, some (for example Deposit No. 11) being further sub-divided (Fig. 2.2.10). Fig.2.1.5 may be referred to for the geological details of Bailadila area.

The scale of mining in some of the deposits in Bailadila are fairly large (Fig.2.2.11a, b).

Fig. 2.2.11 (a) A panoramic view of the Deposit No. 11C, Bailadila; (b) A view of the mine workings at the Deposit No. 5 (photographs by authors).

The balance reserves in these deposits were 1026.33 Mt as on 1st April, 2003 (Table 2.2.7).

Table 2.2.7 *Iron ore reserves (Mt) in the Bailadila group of deposits, Chhattisgarh (Data Source: Mines office)*

S. No.	Deposit No	Estimated reserves	Balance reserves (as on 01.04.2003)	Remarks
1	Bailadila–14	101.50	12.50	Under exploitation since 1968
2	Bailadila–5	207 (mineable)	82.72 (mineable)	Under exploitation since 1968
3	Bailadila–4	107.95	107.95	Estimated by NMDC
4	Bailadila–11 A	29.02	29.02	Estimated by NMDC
5	Bailadila–11B	104.17	104.17	Estimated by NMDC
6	Bailadila–11 C	134.26	40.50	Estimated by NMDC
7	Bailadila–10	219.00	213.69	Estimated by NMDC
8	Bailadila–13	317.00	317.00	Estimated by NMDC
9	Bailadila–1	33.54	33.54	Estimated by GSI
10	Bailadila–2 and 3	74.83	74.83	Estimated by GSI
11	Bailadila–8	10.41	10.41	Estimated by GSI
		1338.68	1026.33	

Note: The deposit nos. 6, 7, 9 and 12, having very small reserves, are not reflected in the above table.

The Fe-rich bands in BIF are composed mainly of hematite and magnetite. Minor phases are minnesotite and grunerite. Magnesio–riebeckite and aegirine are present in some bands. The bands are common in the scale of millimeters to centimeters (meso–bands).

Filamentous and unicellular spheroidal structures regarded as 'possible microfossils' are reported from stromatolytic banded iron formation of the Bailadila Group (Raha et al., 2000). The spheroid bodies have diameters varying between 0.5 and 6.9 μm. The filamentous structures range from < 100 nanometers to 1.2 μm in diameter. Such microfossils are reported from iron formation in Sandur belt, Karnataka and Koira valley (Barbil) ores, Orissa.

The Iron Ores Description of the Bailadila ores given by Chatterjee (1964) still remains the most detailed and reliable. The ore bodies comprising the deposits range in size from thin lenses or beds to large bodies, which extend for about 2.5 km in length and a few hundred meters in thickness. Along the dip, the ore bodies may extend for several hundred meters. Their long dimension on surface is parallel to the strike of compositional lamination in the associated BIFs. The contact of the ore bodies with the unenriched parts of the BIFs is always conformable and usually sharp. Pockets and patches of unenriched or partially enriched BIF are locally present within the ore bodies. The orebodies are not necessarily underlain by BIF. They are at places underlain by shale as well and bear no spatial relation with BIF (Fig. 2.2.12 a–e).

Lamination is a common feature in all types of ores, including the massive variety in places. The laminated ores locally show mesoscopic folds and faults (Fig. 2.2.13 a–b). Joints in the ores, including the massive variety are another common feature. Brecciation is locally visible within the ores, the breccia being cemented by later Fe–oxides /oxyhydroxides.

Based mainly on the physical properties, the Bailadila iron ores are divisible into the following types:

1. Massive ores: A common variety of ores, usually rich in Fe-content and steel grey in colour; hard and tough.
2. Hard laminated ores: (a) Laminated steel – grey variety. Minor-mesoscopic faults and folds are often noticed in them. The lamina thickness varies from a small fraction of millimeter to 5 mm or more. This banding is defined by both the mineralogy of the ores and the volume and disposition of the pore spaces.

   (b) Biscuity ore – easily friable laminated ore, generally brown in colour.

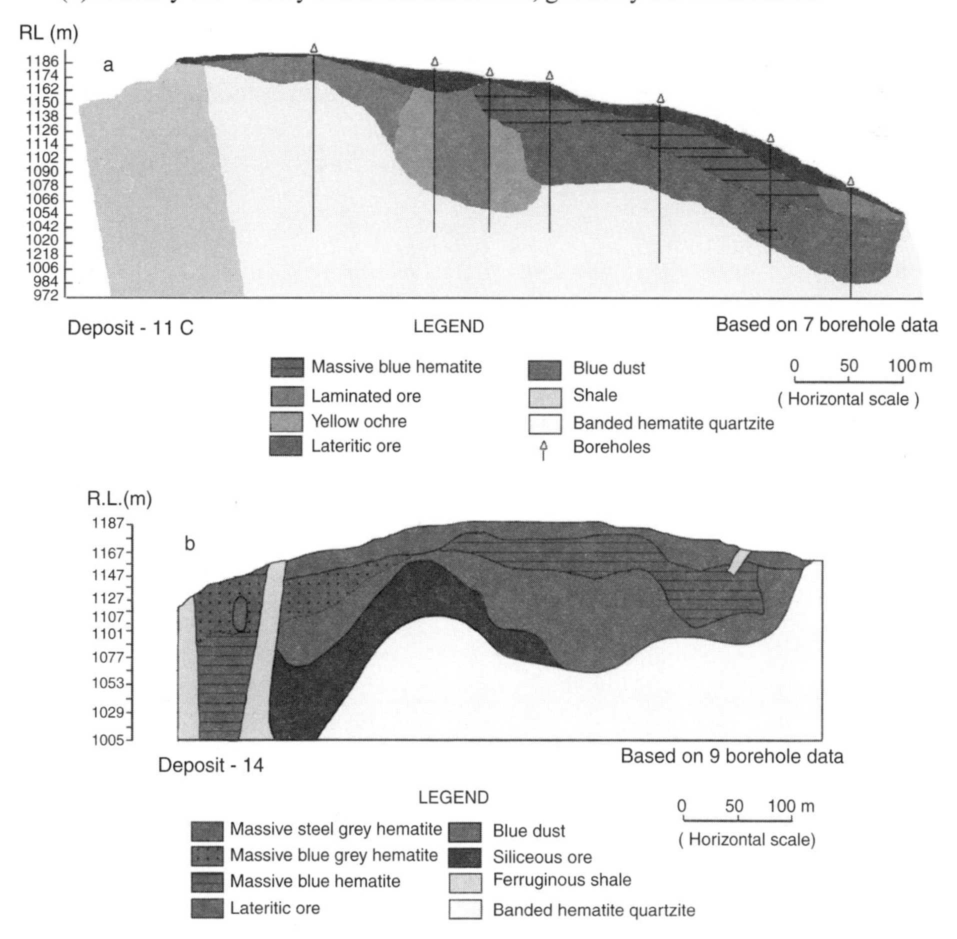

(Continued)

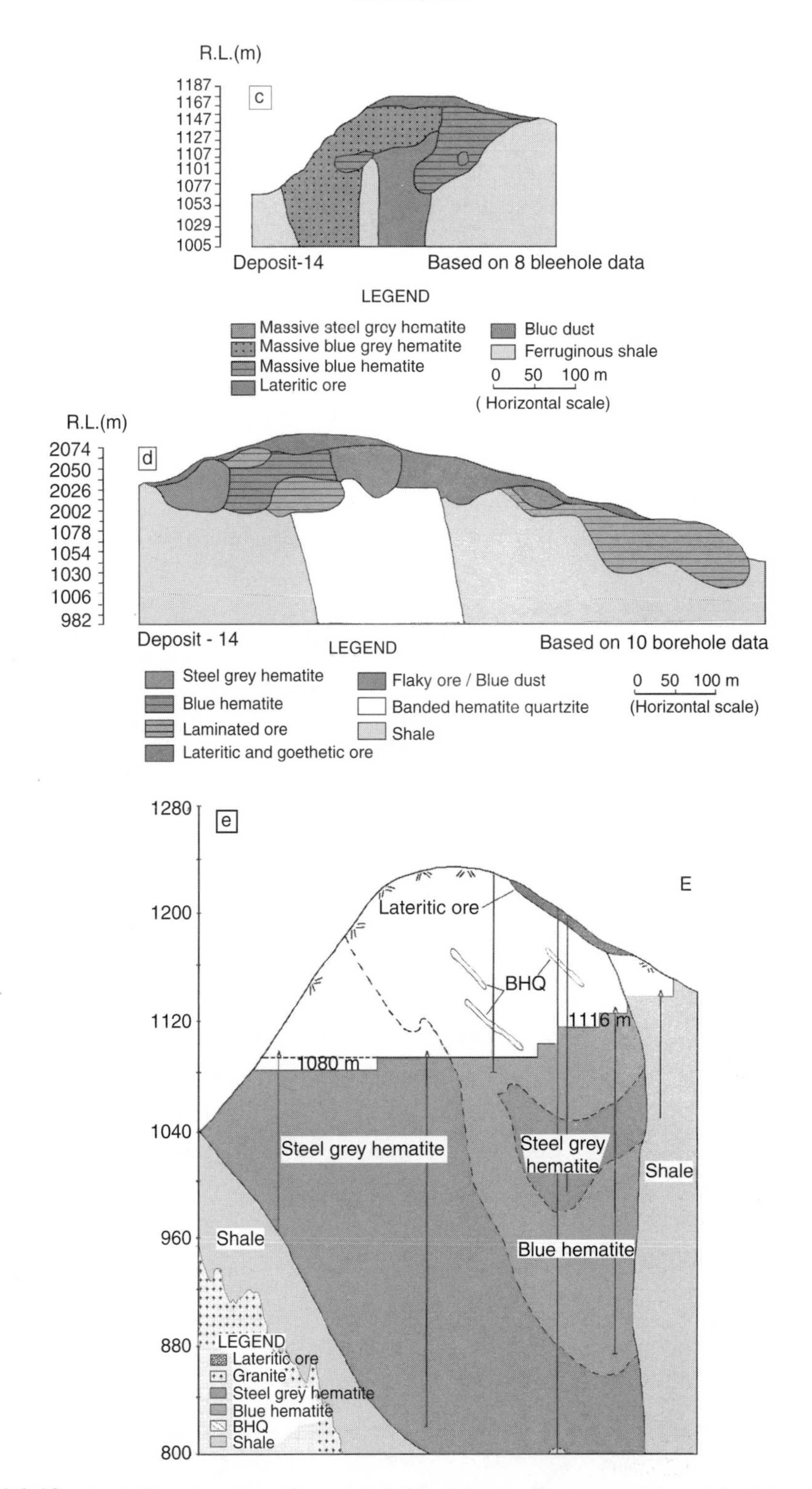

Fig. 2.2.12 (a–e) Cross-sections of some Bailadila deposits showing varied spatial relationship of rich iron ore with BIF (BHQ), shale and lower grade ores (after Mines office)

Fig. 2.2.13 (a) Laminated massive ore, Deposit No. 14, Bailadila (Kirandul); (b) Folding in massive ore, Deposit No. 5, Bailadila (photographs by authors)

3. Soft laminated ore: Bluish grey to dark in appearance and porous. Often contain vestiges of lamination; may grade into blue dust.
4. Blue dust: Fine grained powdery; deep grey to black in colour; association with soft laminated ores is common.
5. Latertic ore: Porous; colour in shades of brown.
6. Canga: Hematite (± BIF) rubbles bonded into a hard conglomeratic iron ore by a cement of hematite + limonite.

The steel grey massive ore (Deposit No. 5) consists mainly of hematite with a little magnetite (hematite:magnetite ≃ 40). Hematite grains are not bladed, neither stubby after magnetite. They are irregular and interlocking with a moderately high porosity (15–20%). Phyllosilicates and quartz represent the scanty gangue minerals.

The high-grade ores of Bailadila are rich in iron and at the same time poor in SiO_2, Al_2O_3 and P (Table 2.2.8). The Al_2O_3/SiO_2 in most cases is ≤1 and $Al_2O_3 + SiO_2 \leq 2$. These properties, together with their hard and tough characters qualify them for use both in the direct reduction route and in the blast furnace. The low H_2O-content (average 1.40%) suggests a low oxy–hydroxide (limonite–goethite) content that is almost invariable.

Locally, quartz veins, cutting across laminations in laminated ores were noticed (Fig. 2.2.14). Quartz grains almost normal to the inner boundary of the vein increase from millimetre to centimetre sizes from the margin to the core. Fluid inclusions (V+L) in quartz from such a vein show maximum salinity of 18.3% NaCl-equivalent and a T_h (P–uncorrected) of upto 400° C (authors' unpublished work).

Table 2.2.8 *Chemical composition of Bailadila iron ore (Samples drawn from rakes during February 2004) (Data source: Mines office)*

S. No.	Date	Rake	Weight of rake sample (tonnes)	Fe%	SiO_2%	Al_2O_3%	P%	Moist	+40	−10
1	01.02.2004	LK–283	0.037	68.20	0.80	0.90	0.023	1.50	0.24	4.92
2	02.02.2004	OK–4	0.035	69.00	0.46	0.20	0.018	1.50	0.32	6.53
3	03.02.2004	OK–14	0.035	68.20	0.40	1.40	0.025	1.60	0.31	7.77
4	03.02.2004	OK–21	0.037	67.70	0.40	2.10	0.035	1.40	0.47	4.24
5	04.02.2004	OK–27	0.035	68.80	0.52	0.60	0.022	1.30	0.33	4.49
6	05.02.2004	OK–34	0.037	68.60	0.68	0.65	0.030	1.50	0.30	7.98
7	05.02.2004	VK–16	0.035	68.40	0.50	0.90	0.031	1.30	0.32	3.07
8	06.02.2004	OK–45	0.035	68.80	0.80	0.20	0.022	1.40	0.41	7.12
9	06.02.2004	OK–52	0.035	68.80	0.68	0.20	0.022	1.30	0.32	4.44
10	07.02.2004	OK–59	0.037	68.70	0.42	0.60	0.024	1.30	0.48	6.53
11	08.02.2004	OK–68	0.037	68.00	0.48	1.40	0.025	1.40	0.24	8.28
12	08.02.2004	VK–27	0.035	69.20	0.18	0.30	0.022	1.30	0.24	7.53
13	09.02.2004	OK–79	0.037	69.10	0.26	0.40	0.022	1.40	0.37	6.88
14	10.02.2004	OK–92	0.035	68.70	0.50	0.60	0.024	1.30	0.41	6.75
15	10.02.2004	VK–31	0.035	68.90	0.38	0.56	0.023	1.40	0.34	5.54
16	11.02.2004	VK–32	0.035	69.30	0.20	0.25	0.016	1.40	0.24	5.06
17	12.02.2004	OK–112	0.037	69.10	0.38	0.20	0.022	1.40	0.65	4.19
18	13.02.2004	OK–124	0.035	69.00	0.46	0.30	0.024	1.30	0.50	4.45
19	14.02.2004	OK–134	0.037	69.20	0.38	0.28	0.016	1.40	0.40	6.53
20	15.02.2004	OK–143	0.035	68.80	0.70	0.40	0.030	1.40	0.76	5.70
21	15.02.2004	OK–144	0.037	68.10	0.80	0.90	0.035	1.30	0.57	4.38
22	16.02.2004	OK–154	0.035	68.50	0.84	0.55	0.022	1.40	0.70	5.67
23	17.02.2004	OK–158	0.037	68.60	0.42	0.60	0.028	1.40	0.56	4.74
24	18.02.2004	VK–51	0.035	68.30	0.72	0.70	0.034	1.30	0.34	4.55
25	18.02.2004	KOT–20	0.035	68.80	0.60	0.50	0.032	1.50	0.32	6.39
26	19.02.2004	OK–178	0.035	68.40	0.70	0.60	0.024	1.40	0.57	3.92
27	19.02.2004	OK–184	0.037	68.10	0.70	1.00	0.031	1.50	0.33	5.36
28	20.02.2004	OK–189	0.035	68.20	0.74	0.90	0.032	1.46	0.32	4.21
29	20.02.2004	OK–197	0.037	68.60	0.90	0.40	0.025	1.50	0.49	6.01
30	23.02.2004	OK–199	0.035	67.60	0.88	1.70	0.029	1.40	0.42	3.64
31	20.02.2004	OK–231	0.035	68.60	0.36	0.70	0.025	1.40	0.25	5.76
32	25.02.2004	OK–243	0.037	69.00	0.42	0.44	0.022	1.30	0.25	5.57
33	26.02.2004	VK–91	0.035	68.70	0.45	0.50	0.024	1.40	0.40	4.19
34	26.02.2004	OK–264	0.035	69.20	0.32	0.30	0.018	1.40	0.17	5.51
35	27.02.2004	OK–273	0.035	68.90	0.60	0.38	0.024	1.40	0.39	3.35
36	28.02.2004	VK–105	0.037	67.80	0.98	0.90	0.030	1.40	0.47	6.95
37	29.02.2004	VK–111	0.035	68.80	0.60	0.50	0.024	1.50	0.24	8.13
38	29.02.2004	OK–297	0.035	69.40	0.26	0.20	0.024	1.40	0.33	5.81
	WTD	AVG	1.36	68.63	0.55	0.64	0.025	1.40	0.39	5.59

Fig. 2.2.14 Quartz vein in laminated iron ore, Deposit No. 11 C, Bailadila. Inset: a sketch of internal texture of quartz in the vein (photograph by authors).

Origin of Iron Ores The origin of hematitic iron ores associated with sedimentary rocks (± basic volcanic rocks) in the sequence is no more axiomatically considered as the products of supergene alteration of BIFs as will be obvious from discussions in an earlier section (Chapter 1.2). More than one mechanism, either singly or in combination may produce iron ores from the protoliths. These are: (i) only a supergene process, (ii) only a hypogene (hydrothermal) process, (iii) hypogene process with supergene overprinting, (iv) supergene alteration followed by hydrothermal overprinting.

Very little is there in print about the origin of the iron ores of Dalli–Rajhara–Rawghat–Bailadila belt. Way back in 1964 Chatterjee outlined the origin of the Bailadila iron ores as follows:

I. Rhythmic chemical precipitation of Fe oxide and silica giving rise to oxide facies of iron formation.
II. Regional metamorphism and diastropism followed in time.
III. Leaching of siliceous materials in BIF 'often accompanied by metamorphic replacement of silica by specularite in structurally, chemically and topographically favourable zones'
IV. Near surface leaching of the members of iron formation giving rise to development of some of the laminated hematitic ores as well as Canga. Near surface leaching also affected the enriched iron ore bodies in altering the physical and mineralogical characters of the ores.

Chatterjee did not spell out whether in his opinion the fluid that leached siliceous material from the BIF was hydrothermal water of formational origin or cool surface water. His reference to near surface leaching in the final stage (IV), suggests that he might have imagined activity of hydrothermal fluids at stage III. If so, this model is not much different from what Beukes et al. (2002) suggested four decades later. According to the latter authors, these are supergene–modified hydrothermal ores, as friable to powdery saprolitic hematite ores are 'invariably associated with bodies of hard hematitic ores'.

The present authors, based on their own observations and existing literature, are of the view that at least the bulk of the ore formation took place penecontemporaneously with late diagenesis–early metamorphism, prior to the peak (buckling) deformation of the iron-bearing sedimentary (–volcanic) package. Wide circulation of the hydrothermal fluids released by the previously mentioned process led to the ore-concentration. The ore materials were not necessarily derived from a BIF body around, as at many places it is absent or only partially present in the immediate vicinity of the ore zones (Fig. 2.2.12 a–e). This contention is supported by (i) folding in the highly enriched ores (Fig. 2.2.13b), (ii) absence of goethite in the rich massive ore, (iii) low Si and Al in the ores, (iv) local presence of hydrothermal quartz veins that showed maximum salinity of 18.3% NaCl-equivalent and T_h (P–uncorrected) of upto 400°C (Fig. 2.2.14) (unpublished work, authors). Effects of supergene overprinting are indeed represented by the development of saprolitic ores, i.e. soft friable ores, blue dust etc.

Future workers should *inter alia* study the trace element, REE and oxygen isotope compositions of the BIFs and associated ores of the region.

Gold in Kotri Belt

The southern stretch of the Dongargarh belt, named as the Kotri belt or Kotri rift,as discussed earlier in this chapter, is a N–S trending nearly 200 km long narrow (~ 40 km width) belt mainly comprising bimodal acid–basic volcanics, bound along east by the Bailadila Group of rocks or the older gneissic complex of Bastar. Incidences of gold have been noted over a wide expanse of this belt, both in the rhyolitic tuffs and ignimbrite of Kotri Supergroup, as well as in the BIFs of the Bailadila Group (Fig. 2.2.15).

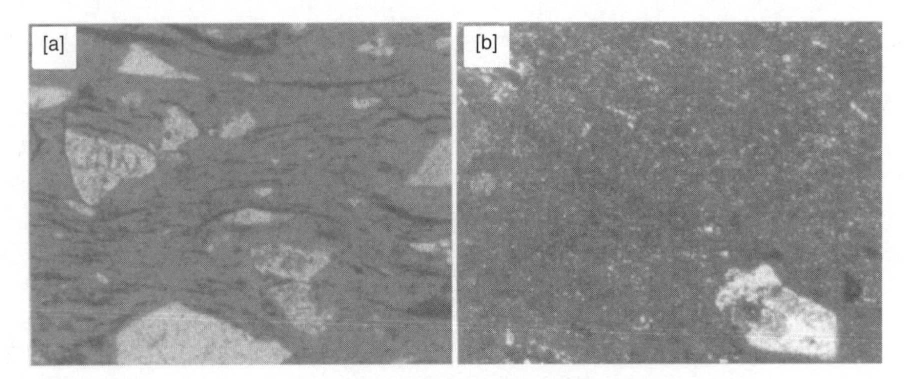

Fig. 2.2.15 Photomicrographs showing (a) auriferous rhyolitic tuff from Gurwandi (x 50), (b) Fine grained crystal tuff (ignimbrite) from Kotri river section (x 100) (unpublished work, A. Gupta).

The prospective auriferous zones identified by GSI in this belt were:

1. Sonadehi prospect (in BIF; strike length: 1200 m; old workings present)
2. Gurwandi prospect (in rhyolitic tuff and flows)
3. Tohe prospect (in ultrabasics; strike length: 500 m)

4. Kumarkatta prospect (in BIF and rhyolitic tuff; strike length: 4 km)
5. Gurpher prospect (in BIF and laterite; strike length: 1.0 km; old pits in laterite)
6. Seetalpur (in rhyolitic tuff and mafic volcanics)

Besides the above, surface sampling, including placers, indicated gold occurrences over a stretch of about 9 km in Belupani area and incidences of Pb, Ag and Au along two shear zones (1200 m and 700 m) in Kattapar are also indicated.

Detailed prospecting by GSI has provided some additional information in respect to the first of the two prospects enlisted above. In Sonadehi, the gold mineralisation is associated with sheared and silicified sulphidic facies of BIF. The 1200 m long and 400 m wide mineralised zone is marked by several old workings (open cast and underground inclines), of which the walls of the two main inclines showed gold values ranging from 0.5 g/t to 11.0 g/t. One borehole intersection confirmed a 4.6 m wide zone with 11.76 g/t gold. The results of further subsurface probe in the prospect suggested steeply pitching shoots with very limited strike extension. At Gurwandi, the mineralisation is mainly within the acid volcanics and tuffs over a wide zone (~30m). The gold tenor is generally low within shallow depths. However, two lodes, 3.0 m × 1.48 g/t Au in the hangwall and 3.0 m x 1.18 g/t Au in the footwall were established at 100 m depth. The footwall lode at the depth of 150 m is 2.0 m wide with an average tenor of 6.0 g/t Au.

Gold in Sonakhan Greenstone Belt

Incidence of gold is wide spread in the Archean Sonakhan greenstone belt, which fringes the southern boundary of the Chhattisgarh basin. Better concentrations of quartz vein hosted gold are reported from Sonakhan and Degon blocks. The auriferous quartz veins are prominent within sulfide–bearing ferruginous chert. Gold with pyrite, galena and often chalcopyrite occur in quartz veins traversing the amphibolites and granite. A mineralised band of black-smoky quartz /recrystallised chert was traced for 630 m strike length with average width of 1.65 m and gold content of 3 g/t Au. The Directorate of Geology and Mining, Govt. of Madhya Pradesh, estimated 2,700 tonnes of gold ore (IBM Mineral Year Book, 1992).

Gold in Raigarh Belt

In the Pandripani–Barjor sector (Lat. 23°30′: Long. 83°30′–83°45′) and surrounding area, auriferous sheared quartz veins are hosted in meta–volcanisedimentary rocks within granite-gneiss country, the latter generally regarded as the western extension of the Chhotanagpur Granite–Gneiss Complex (CGC) of eastern India.

The gold-bearing quartz veins often contain appreciable tourmaline. Besides the quartz veins, gold values were also detected within biotite–chlorite schist in the wall rock and in the recrystallised chert and porphyritic granitoids in contact with the gneisses. Both sulfides (dominated by arsenopyrite) and oxides (magnetite, ilmenite) are associated with the gold–bearing zones. Scheelite and tourmaline are conspicuous in the gangue. The ore bodies are lenticular in shape with limited strike and depth extension.

The GSI explored the Pandripani–Barjor prospect in late eighties and estimated 65,000 tonnes of ore with average grade of 1.4 g/t Au (with a cut–off grade of 0.6 ppm Au). The central Pandripani block hosts 80 kg gold in chert over 125 m strike length and down to 80 m depth with 1.4 g/t Au. The placers in and around the area hosts gold concentration of the order of 0.2–1.0 g/t over extensive areas in gravel bed.

At Bangaon (Lat. 22°32′: Long. 83°54′) crushing and panning of quartz yielded ~1.0 g/t Au. Drilling of a few boreholes in Kansabel–Tapakora area indicated gold values from 0.11–6.84 g/t (IBM Mineral Year Book, 1992).

Copper–Molybdenum (–Gold) Mineralisation at Malanjkhand and Neighbouring Areas

The initial discovery of the Malanjkhand deposit, Balaghat district in the then Central Province was a chance find by a passing civilian officer, sometimes in the nineteenth century. As the exposure (including natural and man made) did not have a telltale story, the initial discovery was not followed up by a detailed investigation until after the middle of the last century, when the Geological Survey of India launched an ambitious programme of mineral exploration in the Central Indian craton. One important outcome of these exploration activities was the proving of Malanjkhand by 1966 as a granite–hosted Cu–Mo (–Au) deposit of great economic value and scientific interest. The deposit has, during the period 1984–1985 to 2001–2002, produced 3, 44, 46, 606 tonnes of ores, varying in average Cu-content from 1.01% to 1.54%, Mo-content of 68–2200 ppm and Au-content of 0.11 ppm to 0.14 ppm (Table 2.2.9). The total ore reserve in the deposit, including mined out tonnage, was 135.5 million tonnes (Mt) with an average grade of 1.32% Cu (at a cut–off grade of 0.45% Cu), down to the depth of –300m RL, as estimated in 2003. A liberal estimate ('geological reserve') at the same time is 248 Mt with an average grade of 1.3% Cu at a cut–off grade of 0.45% Cu. The proved reserve of the metal (1.78 million tonnes) does not qualify it in the category of 'giant' copper deposit (> 2 Mt Cu) of Singer (1995), but brings it very close to it. It falls under the 'Very large' category (1.0–3.162 Mt) of Clark (1993). However, it could be a 'giant' deposit if the geological estimate of 470 Mt with an average of 0.9% Cu at a 0.2% cut off (Bhargava and Pal, 2000) comes true. With respect to grade, it falls under 'high grade' (>0.75% Cu) category of Singer (1995). A number of publications now exist on different aspects of this deposit, workers, as is common, agreeing on some points and differing on others. The list of publications includes those of Sharma and Kumar (1969), Seetharam (1976), Petruk and Sikka (1987), Naik (1989), Sikka (1989) Sarkar et al. (1989, 1996), Rai and Venkatesh (1990), Ramanathan et al. (1990), Panigrahi and Mookherjee (1997), Bhargava et al. (1999), Bhargava and Pal (2000), Vishwakarma (2001), Sikka and Nehru (2002), Panigrahi et al. (2004), Stein et al. (2004), Stein et al. (2006).

'Proved reserve' of ores in Malanjkhand, as of 2003, is 135.5 million tonnes with an average grade of 1.32% Cu at a cut-off of 0.45% Cu. In this estimation, data collected from a depth upto –300m RL have been considered. In a liberal estimation (geological reserve) at the same time, it is 248 Mt, with an averge grade of 1.3% Cu, at a cut-off of 0.45% Cu. Here the estimation is limited to O-meter RL. Malanjkhand produced 2, 233, 523t (1% Cu)ore during 2010–11.

Regional Geology Geographically, the area is located in the Balaghat district of Madhya Pradesh, the nearest railhead (90 km) being Gondia on the Kolkata–Nagpur–Mumbai railway. Geologically, it is located in the Southern (Bastar) Crustal Province of Central India, to the immediate south of the Central Indian Shear Zone (CIS) (Fig. 2.2.16).

The Malanjkhand ore zone occurs within a granitoid pluton of batholithic dimension (~1500 sq km) which is composite in nature, being composed of multiply intruded granitoids (including micro-granitoids) and mafic intrusions. The Malanjkhand granitoid rocks (henceforth be called Malanjkhand Granite), reportedly intruded the Nandgaon Group, composed of volcanic rocks (basalt, rhyolite and andesite) and coarse clastic sediments. The Nandgaon Group overlies the gneisses, schists and amphibolites of the Amgaon Group. The Chilpighat Group or the Chilpi Group overlies the Malanjkhand Granite with an unconformity. Some pebbles in the basal Chilpi at places are composed of granitoid rocks and correspond to the Malanjkhand Granite. The Chilpi Group is sub-divided into two: the Lower Chilpi and the Upper Chilpi. The Lower Chilpi is principally phyllitic, while the Upper Chilpi is composed of fine to coarse clastic sediments and carbonate rocks. The youngest formation is represented by the Deccan Traps (Table 2.2.10).

Table 2.2.9 *Ore production and important metal-contents in ores, ore-concentrates and tailings from Malanjkhand Mine (1984–2002) [Courtesy: Hindusthan Copper Limited (HCL)]*

YEAR	Ores mined and their average metal-content						Concentrates						Tailings				
	Quantity (t)	Cu(%)	Au	Ag	Mo	Co	Quantity (t)	Cu(%)	Au	Ag	Mo	Co	Quantity (t)	Au	Ag	Mo	
1984–85	1306321	1.54	0.13	3.34	68	–	57147	31.3	2.16	55.03	–	–	1249174	0.04	0.9	17	
1984–85	1306321	1.5	0.1	3.34	68	–	57147	31.3	2.16	55.03	–	–	1249174	0.04	0.9	17	–
1985–86	1694196	1.3	0.1	3.3	98	–	72910	29.72	1.95	56.7	1810	–	1621286	0.03	0.9	21	–
1986–87	1537879	1.5	0.1	4.67	157	–	63139	29.1	2.22	65.15	2485	–	1474740	0.04	2.15	51	–
1987–88	1917479	1.3	0.1	4.02	136	–	74727	29.78	2.14	60	2485	–	1842752	0.04	1.6	39	–
1988–89	1851102	1.3	0.1	3.92	120	–	74367	28.93	2.14	60	2403	–	1776735	0.03	1.34	26	–
1989–90	1956000	1.4	0.1	3.97	122	–	85230	27.81	2.03	60	2290	–	1870770	0.03	1.3	27	–
1990–91	1975475	1.4	0.1	3.87	83	–	83299	27.98	2.16	59	1734	–	1892176	0.03	1.37	22	–
1991–92	2024801	1.4	0.1	4.4	101	–	87950	27.3	2.25	63	2135	–	1936851	0.04	1.5	21	–
1992–93	2070930	1.3	0.1	4.25	138	–	85980	27	2.32	61.8	2680	–	1984950	0.04	1.53	26	–
1993–94	1975488	1.3	0.1	4.2	136	–	84699	27.17	2.3	63	2633	–	1890789	0.03	1.3	33	–
1994–95	1682000	1.3	0.1	4.18	143	31	75381	25.54	2.19	60.25	2563	245	1606619	0.03	1.35	39	18
1995–96	2066570	1.2	0.1	3.85	166	27	92259	23.74	2.19	61	2713	218	1974311	0.03	1.3	45	18
1996–97	1761157	1.1	0.1	3.85	131	29	71847	24.57	2.25	60.25	2563	233	1689305	0.02	1.12	47	17
1997–98	2045730	1.1	0.1	3.9	128	32	78703	26.14	2.38	61.55	2149	220	1967027	0.02	1.25	41	25
1998–99	2236076	1	0.1	3.67	114	30	75930	27.14	2.41	63.13	1744	176	2160146	0.02	1.22	45	25
1999–2000	1900223	1.2	0.1	3.97	220	32	74172	27.52	2.4	62	4351	151	1826051	0.03	1.25	71	27
2000–01	2191985	1.1	0.1	3.93	164	41	84483	26.5	2.49	62.55	3015	182	2107502	0.03	1.28	60	34
2001–02	2253194	1.2	0.1	3.95	162	43	89449	26.93	2.28	59.58	2990	225	2163745	0.03	1.14	52	36

Note: Contents of Au, Ag, Mo and Co in ppm.

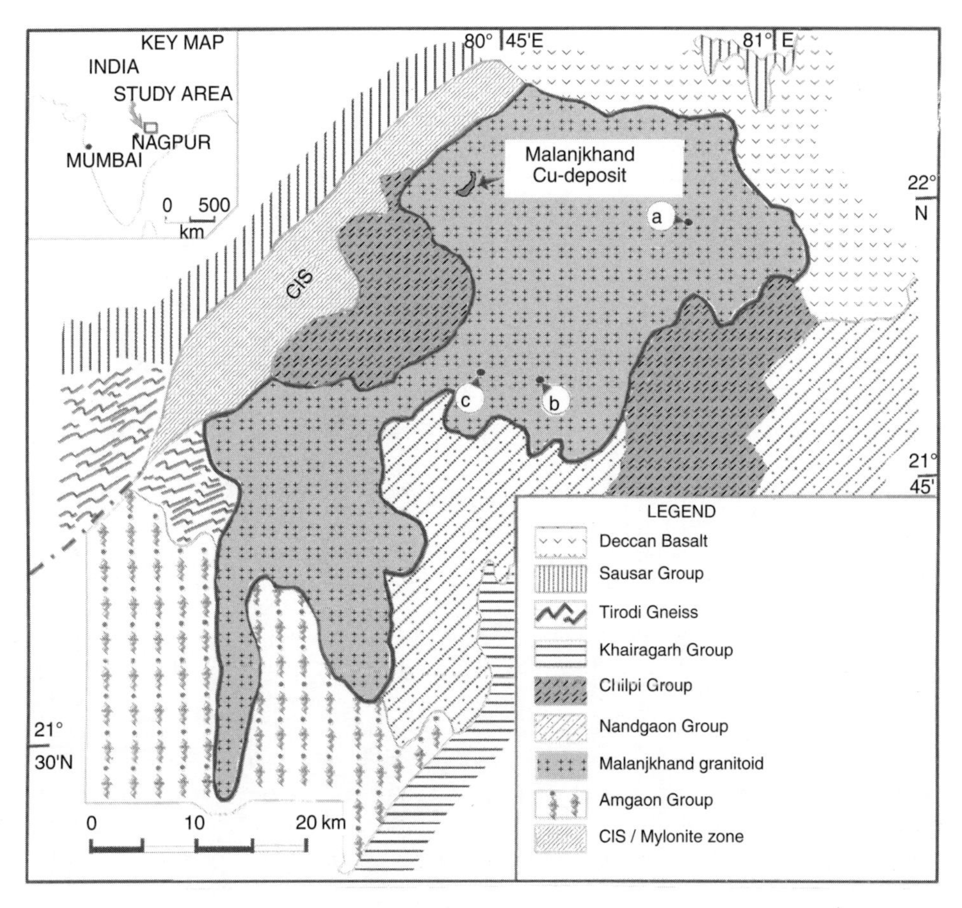

Fig. 2.2.16 Regional geological map of Malanjkhand and surrounding area (modified after Pujari and Shrivastava, 2003).CIS – Central Indian shear zone ; Solid circles – reported evidence of Cu-mineralisation at (a) Patharatola, (b) Dhorli, (c) Pipardhar.

Table 2.2.10 *Stratigraphic succession in the Malanjkhand area, Madhya Pradesh*

	Deccan Traps	
	...............Unconformity...............	
Chilpi Group	Upper Chilpi Formation	Phyllite, grit, limestone arkosic grit, conglomerate
	Lower Chilpi Formation	Phyllite, shale, conglomerate
	...............Unconformity.................	
	Basic dykes Malanjkhand Granite (batholith) (Oldest member ~2490 Ma)	
Nandgaon Group		Basalt, andesite, rhyolite quartzite and conglomerate
Amgaon Group		Gneisses, schists and amphibolites

Ghosh et al. (1986) dated the Malanjkhand granitoid rocks by the Rb–Sr (WR) method. The oldest age obtained by them is 2362 ± 60 Ma. Panigrahi et al. (1993) obtained the following ages: grey granitoid, 2456 ± 38 Ma; pink granitoid, 2243 ± 217 Ma, using the same systematics (Sr_i

= 0.702 ± 0.001). These data are similar to the Rb–Sr ages obtained on the Dongargarh Granite (Sarkar et al., 1981; Krishnamurthy et al., 1990). This led some people to correlate the Malanjkhand Granite and the Dongargarh Granite. However, Panigrahi et al., themselves in a recent publication (Panigrahi et al., 2004), have modified their earlier data. This time they analysed zircon from both the 'grey' granitoids and 'pink' granitoids and obtained ^{207}Pb/ ^{206}Pb ages of 2476 ± 7 and 2477 ± 8 Ma respectively from a pooled set of samples. Some values are of course higher, i.e. 2,490 Ma and some lower, i.e. ~2,450 Ma. The oldest zircons have the highest Th/U ratios suggestive of magmatic values. The youngest zircons in contrast, have notably lower Th/U ratios indicative of trace elements (Y, HREE, Th and P) loss during modification at a later stage. Stein et al. (2006) ascribe metamorphism as the causal factor for this change. However, as we shall see in a later section, barring dynamic recrystallisation in the shear zone(s) within the Malanjkhand Granite, there is virtually no evidence of tangible metamorphism of the latter.

The authors share the opinion of Stein et al. (2006) that the bulk of Malanjkhand Granite crystallised at ~ 2,490 Ma. In the context of field relations outlined in an earlier section and these age data of the Malanjkhand Granite, the Amgaon Group of rocks should be Archean in age. The possibility of the Nandgaon Group rocks being Late Archean in age is also there. This view supports an earlier suggestion in this respect by Sarkar et al. (1990). It is also possible that the Nandgaon Group represents the volcanic phase of a major magmatic event, the plutonic phase of which is represented by the Malanjkhand Granite. Mafic intrusives (gabbro–diabase) are moderately common in the Malanjkhand area. Diabase dykes are poorly exposed on the surface. It is likely that they are of more than one generation. Apparently, the last one has penetrated all the petrographic variants of Malanjkhand Granite (G_2, G_2', and G_4), as well as the ores. Petrographically they are only weakly transformed, proven by the fact that pyroxene is locally replaced by hornblende/actinolite and the plagioclase underwent limited saussuritisation. Hornblende grains have locally been chloritised along the margins, due to increased fluid activity in those domains. Some of these intrusives have been found to contain such sulphides as chalcopyrite and pyrite as fine disseminations. The mafic dykes occur as a broadly N–S trending swarm along extensional disjunctions, presumably due to N–S compression. There are a few intrusive bodies of coarse–grained gabbros, where also pyroxene has been largely replaced by hornblende.

The Chilpi Group of rocks does not show metamorphism higher than the low greenschist facies. Shear zones within the Malanjkhand Granite do not affect these cover sediments. Diastropic structures have poorly developed in the Chilpi Group rocks, the best being large fold warps (Bhargava and Pal, 2000).

The broad stratigraphic sequence in the Malanjkhand area is presented in Table 2.2.10.

Mineralisation at Malanjkhand

Host Rocks(s) The Malanjkhand Granite is the host for mineralisation. Sarkar et al. (1996) divided the Malanjkhand Granite into four petrographic varieties and referred to them as G_1, G_2, G_3 and G_4 (Fig. 2.2.17).

G_1 is moderately coarse-grained porphyritic granite, containing alkali feldspar megacrysts (oikocrysts) upto 2 cm in length in a quartzo–feldspathic matrix, with biotite, epidote–zoisite, sericite, titanite and carbonate as minor phases. Anhedral grain shapes and clustering together suggest that these minor phases could be secondary. Relict plagioclase, epidote quartz and carbonate in alkali feldspar megacrysts may indicate that the latter also are secondary, formed at a sub-solidus stage. Locally, a faint grain orientation is noticed. The distribution of G_1 is limited to the northern part of the Malanjkhand area.

G_2 is the major member of the Malanjkhand Granite in the area. It is an overall coarse-grained rock, varying in composition from tonalite through granodiorite and trondhjemite to granite. The original feldsper in tonalite–granodiorite is sodic andesine (An = 35–36). Magnesio–hornblende [Si = 6.39–6.72 and (Ca+Na+K) =2.19–2.40], and lesser amount of biotite, epidote and titanite are

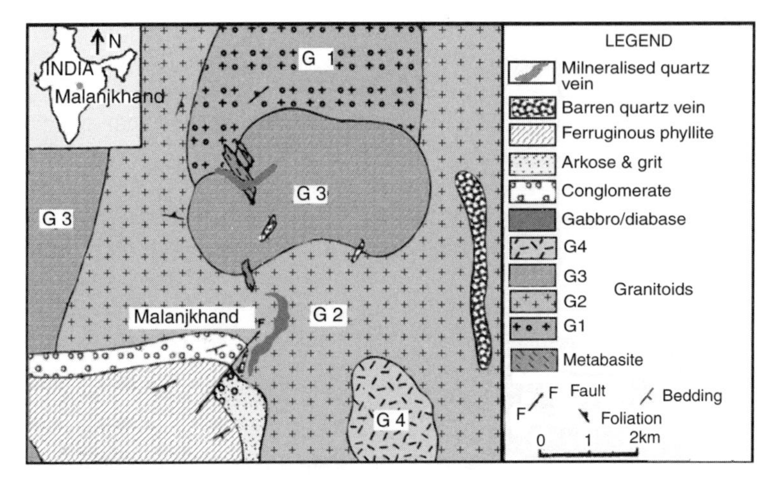

Fig. 2.2.17 Geological map of Malanjkhand region showing the distribution of different petrographic varieties of granite (G_1, G_2, G_3 and G_4) (after Sarkar et al., 1996)

important accessories. The feldspar grains vary in size from 0.25 mm to 6.5 mm, the coarser grains being dominant. The smaller grains of feldspar and quartz occupy the interstices of the feldspar phenocrysts. Hornblende is variously replaced by biotite, epidote and titanite. In and around the ore zone, the G_2 described above is modified to a reddish granitoid rock, G_2'. This colour is partly due to the development of K–feldspar and partly due to being dusted with haematite. Some have mistakenly called it a separate intrusive phase of potassic granite. However, the gradual change over of G_2 to G_2' is noticed in the field. This is confirmed under the microscope by the replacement of andesinic feldspar, first by albite and then by K–feldspar accompanied by dusting of the mineral by hematite (Sarkar et al., 1996) (Fig. 2.2.18). General increase of Fe_2O_3 in the ore zone rocks is also reflected in chemical analyses (Fig. 2.2.19). Panigrahi et al. (1993) and Panigrahi and Mookherjee (1997) referred to the G_2 as the 'grey granite' and G_2' as 'pink granite' and gave them two different ages. However, in their latest age determination of the Malanjkhand rocks (Panigrahi et al., 2004, Table 1) they found no tangible difference of age between these two varieties. This supports the contention of Sarkar et al. (1996) about the relationship of these two varieties of rocks.

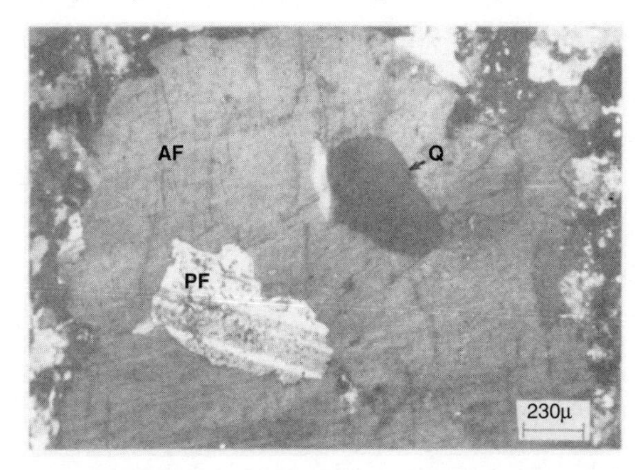

Fig. 2.2.18 Photomicrograph showing relict plagioclase feldspar (PF) in late (secondary) alkali feldspar (AF) in partially altered G_2 granite at the periphery of the mineralised zone. Q – Quartz. Nicols crossed (Sarkar et al., 1996)

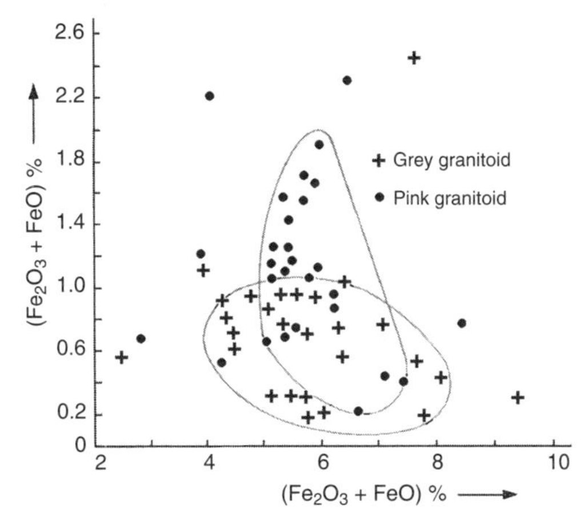

Fig. 2.2.19 Fe_2O_3/FeO variation diagram showing general increase of Fe_2O_3 in the ore bearing grey and pink granitoids (Bhargava and Pal, 2000)

G_3 is similar to G_2, except that it has little or no hornblende. Accessory biotite, epidote and titanite control the colour index and the rock resembles biotite granite or trondhjemite in physical appearance.

G_4 is an aplite that intrudes both G_2 (including G_2' in the mine area) and G_3. Petrographically it is characterised by the presence of the phenocrysts of alkali feldspar (upto 2.5 mm) in a fine-grained (< 0.5 mm) quartzo–feldspathic matrix. Compositionally, it varies from granodiorite to granite.

Magnetite is a common accessory in all the granitoid rocks described above. Accessory pyrrhotite, chalcopyrite and even molybdenite grains reported from a number of samples, particularly belonging to G_2 (Sarkar et al., 1996) are of special interest in the consideration of ore genesis, as we shall see later. It bears mentioning here that chalcopyrite, pyrite and molybdenite occur as accessories also in the micro–granite (G_4) (Bhargava and Pal, 2000). The magnetic susceptibility, measured for these rocks ranges between $2.13x10^{-4}$ to 4.97×10^{-4} emu/g (n = 11), and apparently owes mainly to magnetite (magnetite >> monoclinic pyrrhotite) (Sarkar et al., 1996). Other accessory mineral phases include apatite and more rarely zircon, ilmenite and allanite.

Textually, G_1, G_2 and G_3 are generally granitoid and do not show granoblastic, gneissose or schistose fabrics, except in places of high strain (Fig. 2.2.20 a, b)

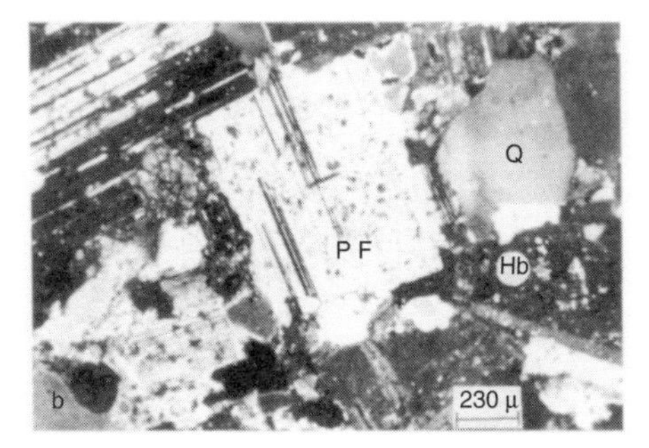

Fig. 2.2.20 (a) Photograph of hand specimen of G_2 granite, (b) Photomicrograph of G_2 granite, showing undeformed texture of the granite hosting the mineralisation. Q – quartz, PF – Plagioclase feldspar, Hb – hornblende (Sarkar et al., 1996)

Table 2.2.11a *Major element composition (wt.%) of granitoid rocks from Malanjkhand (Sarkar et al., 1996)*

	G_1			G_2						G_2'					G_3				G_4				
	(1)	(2)	(3)	(4)	(5)	(6)	(7)	(8)	(9)	(10)	(11)	(12)	(13)	(14)	(15)	(16)	(17)	(18)	(19)	(20)	(21)	(22)	(23)
SiO_2	68.38	73.46	71.00	63.02	67.82	63.04	65.16	70.82	65.14	65.22	67.76	66.68	60.82	72.50	63.58	67.18	72.77	72.36	74.10	70.94	71.30	70.68	75.14
Al_2O_3	13.89	10.92	12.98	14.85	13.34	15.20	15.54	13.73	14.80	14.71	14.00	13.22	16.45	9.20	14.38	14.29	10.49	13.26	11.50	13.85	15.86	13.19	13.40
TiO_2	0.41	0.00	0.00	0.60	0.49	0.66	0.76	0.00	0.68	0.56	0.50	0.37	0.00	0.48	0.70	0.80	0.76	0.49	0.00	0.00	0.18	0.49	0.00
Fe_2O_3	2.40	5.38	4.95	2.77	2.57	2.40	5.36	4.39	3.20	1.50	2.41	2.09	3.14	2.28	3.15	6.67	6.62	3.23	1.38	1.85	0.18	4.85	0.10
FeO	2.16	0.29	0.32	3.25	2.30	3.02	1.35	0.85	2.59	2.88	2.64	2.30	0.85	3.02	2.92	1.29	1.15	0.92	1.29	1.15	0.72	0.68	0.36
MgO	1.08	0.13	0.36	2.24	1.48	2.02	1.66	0.83	1.88	1.74	1.92	1.37	0.68	1.75	2.68	0.11	0.14	0.55	0.31	0.43	0.58	0.15	0.36
MnO	0.04	0.02	0.08	0.04	0.05	0.04	0.07	0.12	0.04	0.05	0.04	0.05	0.03	0.04	0.04	0.05	0.03	0.03	0.03	0.03	0.05	0.02	0.06
CaO	3.02	0.44	0.56	4.87	2.80	4.58	2.20	1.40	4.87	3.06	2.21	0.95	3.24	1.12	3.58	1.05	0.56	0.56	1.06	1.83	2.07	0.77	0.84
Na_2O	4.19	4.22	4.42	4.12	4.35	4.25	4.38	4.35	4.44	6.08	2.85	2.11	1.53	0.92	4.58	4.20	4.80	4.80	2.35	3.64	4.93	6.18	3.62
K_2O	2.93	4.65	4.65	1.93	3.65	3.03	2.89	2.56	1.53	2.79	3.73	9.07	12.24	7.18	2.74	4.05	2.24	2.24	7.50	5.32	2.82	2.84	5.23
P_2O_5	0.18	0.01	0.01	0.31	0.25	0.21	0.02	0.01	0.03	0.25	0.00	0.17	0.61	0.24	0.18	0.02	0.01	0.01	0.16	0.22	0.03	0.01	0.01
H_2O_T	0.92	0.31	0.31	1.80	1.60	1.10	0.26	0.64	1.36	1.08	1.48	1.20	0.50	0.88	1.36	0.26	0.24	0.24	0.36	0.86	0.82	0.23	0.84
Total	99.59	99.83	99.83	99.80	100.70	99.54	99.64	99.70	100.56	99.92	99.54	99.58	100.09	99.60	99.89	99.96	99.81	99.81	100.03	100.12	99.54	99.82	99.95

Analysis by wet chemistry

The granitoid rocks of Malanjkhand (Table 2.2.11a) plot within the fields of tonalite, granodiorite, quartz–monzonite, trondhjemite and granite in a normative albite–orthoclase–anorthite diagram (Fig. 2.2.21) after O' Connor (1965), modified by Barker (1979).

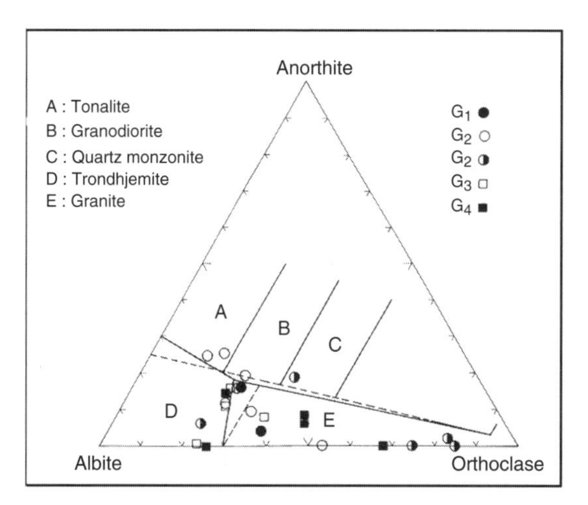

Fig. 2.2.21 Plots of Malanjkhand granitoids in Ab–An–Or triangular diagram (Sarkar et al., 1996)

The so-called granites are in fact, hydrothermally modified versions of the original calc–alkaline granitoids. The Aluminium Saturation Indices (ASI) [mol% $Al_2O_3/(CaO + K_2O + Na_2O)$] are ≤ 1.1 and hence indicative of a meta-aluminous trend. Normative corundum was not expected and it is absent; but diopside is present as a normative phase. Judged in the context of chemical discriminants (ASI index, Na_2O/K_2O ratios) and the petrologic discriminants (presence of magnetite and the absence of cordierite), the Malanjkhand Granite may be assigned to the I-type granites of White and Chappel (1977) and White et al. (1986). Moreover, the high magnetic susceptibility ($> 2 \times 10^{-4}$ emu/g), predominance of magnetite over ilmenite, high bulk Fe_2O_3/FeO ratios (> 1) and titanite as the principal Ti-mineral, place these rocks in the 'Magnetite Series' of Ishihara (1981).

Trace element and REE contents of the Malanjkhand Granitoids are presented in Table 2.2.11b. The chondrite–normalised plots indicate a clear LREE enrichment, an inconspicuous Eu-anomaly and a concavity of the HREE field for these rocks (Fig. 2.2.22).

Table 2.2.11b *Rare earth element (REE) and other trace element contents (ppm) in Malanjkhand rocks (Sarkar et al., 1996)*

		La	Ce	Sm	Eu	Tb	Yb	Lu	Th	Sc	Cu[a]	Co	Cr	W[b]	Hf	Nb[b]	Ta	Ba[b]	Rb[b]	Sr[b]	Rb/Sr
GI	1	46	55	5.8	1.1	0.76	1.7	0.49	9.3	4.5	6	6.3	34	20	4.9	40	1.9	500	160	260	0.62
	2	48	70	5.8	0.97	0.66	2.2	0.34	30.0	5.8	5	5.7	29	30	3.6	30	1.4	470	170	250	0.68
	3	37	61	5.0	0.68	0.32	2.2	0.32	19.0	4.3	3	5.1	28	40	3.4	50	2.7	460	190	220	0.86
GI	4	39	66	4.6	1.10	0.65	1.1	0.21	12.0	6.8	12	9.3	26	20	3.5	90	1.3	570	150	110	1.36
	5	60	86	4.1	1.10	0.81	1.1	0.21	15.0	4.4	16	8.0	23	10	2.3	30	1.0	1200	110	320	0.34
	6	98	150	7.4	1.6	0.41	2.3	0.34	19.0	8.7	12	14.0	36	10	4.0	10	1.2	490	90	680	0.13
GI	7	36	54	3.9	0.84	0.28	1.2	0.23	17.0	3.6	2240	9.6	15	10	3.2	30	1.2	670	120	410	0.29

(Continued)

(Continued)

		La	Ce	Sm	Eu	Tb	Yb	Lu	Th	Sc	Cu[a]	Co	Cr	W[b]	Hf	Nb[b]	Ta	Ba[b]	Rb[b]	Sr[b]	Rb/Sr
	8	14	23	1.5	0.47	0.38	0.74	0.25	7.5	1.3	13	1.5	8.8	80	2.4	50	1.0	540	170	190	0.89
	9	37	62	4.6	1.1	0.85	1.4	0.37	15.0	6.1	30	12.0	24	10	3.7	50	1.1	570	160	150	1.07
	10	43	68	3.6	1.1	0.60	1.6	0.24	7.6	5.4	13	10.0	32	10	3.6	10	0.54	620	130	550	0.24
GI	11	45	74	4.0	1.0	0.72	1.5	0.21	12.0	5.2	51	8.5	38	10	4.5	40	1.0	820	150	250	0.60
	12	54	78	6.0	1.1	1.50	0.74	0.39	25.0	8.3	14	7.0	48	20	4.9	40	1.2	570	190	220	0.86
	13	20	34	3.7	1.1	0.29	1.9	0.17	1.6	12.0	40	16.0	26	10	3.2	40	0.58	460	70	340	0.21
GI	14	30	38	3.1	0.56	0.87	2.7	0.14	15.0	2.7	8	3.0	12	60	3.4	50	1.2	360	160	240	0.67

Note: G^1 – granite; G_2 – tonalite–granite; G_2 – hydrothermally modified G_2; G_3 – trondhjemite–granite; G_4 – trondhjemite–granite.
a: Analysis by AAS; b: Analysis by XRFS; Rest by NAA.

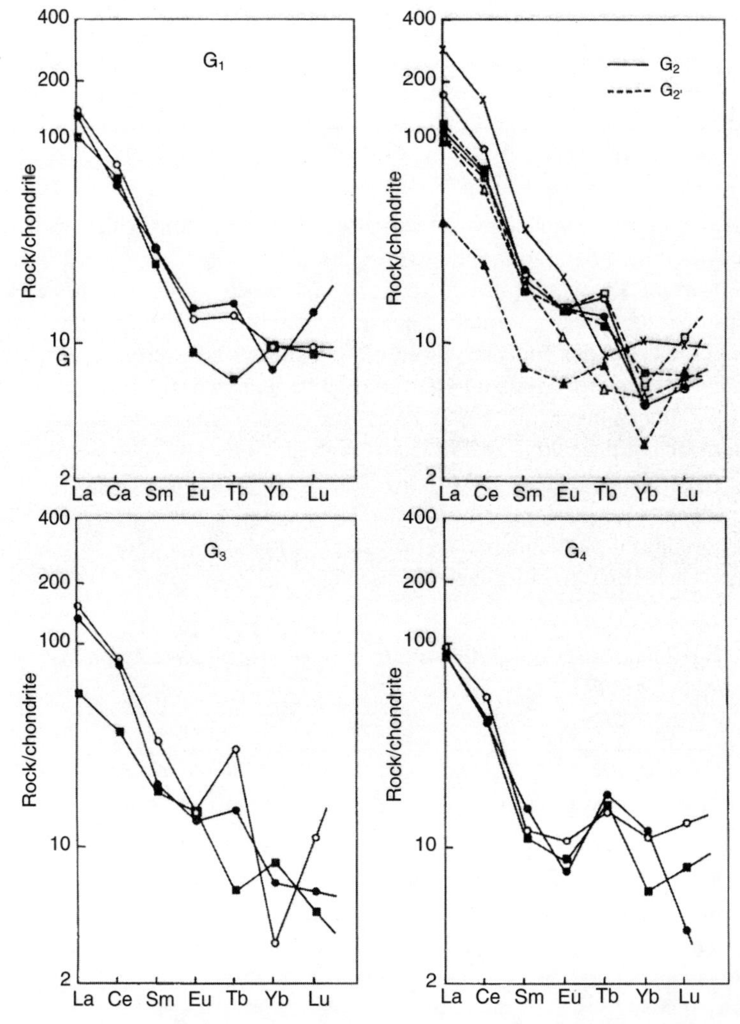

Fig. 2.2.22 Chondrite normalised values of REE in granitoids show LREE-enrichment, inconspicuous Eu-anomaly and concavity of HREE field. Sub-sets refer to G_1, G_2–G'_2, G_3 and G_4 (Sarkar et al., 1996)

The absence of a Eu-anomaly may imply that the system was oxidising, and that the melt was in equilibrium with subequal proportions of a phase with positive Eu-anomaly (plagioclase) and another with a negative Eu-anomaly (amphibole/pyroxene). This REE-distribution pattern is generally characteristic of the calc–alkaline magmas (cf. Cullers and Graf, 1984; Anthony and Titley, 1988). The Rb/Sr ratios exhibit some consistencies in the analyses of G_1 and G_4 rocks, but are highly variable in G_2, G_2' and G_3. Unpublished data of the authors on the geochemistry of the dacite–andesite volcanic rocks of the Malanjkhand area show their close resemblance to the plutonic–hypabyssal rocks of the area and suggest a genetic link.

Fluorine-content of the Malanjkhand granitoid rocks is low, (Table 2.2.12) closely resembling more the high–Ca granites of Turekian and Wedepohl (1961), or the fluorine–poor (mean F = 410 ppm) granitoids of the northwestern Great Basin, that are thought to have intruded a variety of allochthonous oceanic or island arc terrains (Theodore and Menzie, 1984; Christiansen and Lee, 1986). In the Ta–Yb and Rb–(Ta +Yb) discriminant diagrams of Pearce et al. (1984), the Malanjkhand granitoids plot in the 'Volcanic Arc Granite (VAG)' field (Fig. 2.2.23 a, b). Petrography and the bulk chemistry of the rocks already outlined are consistent with an arc model (Sarkar et al., 1996).

Table 2.2.12 *Fluorine-content of Malanjkhand granitoid rocks (Sarkar et al., 1996)*

Type of the granitoid	n	Range (ppm)	Average (ppm)
G_1	4	280–420	350
G_2	8	340–540	460
G_2'	5	100–520	300
G_3	7	260–640	460
G_4	2	320–340	330

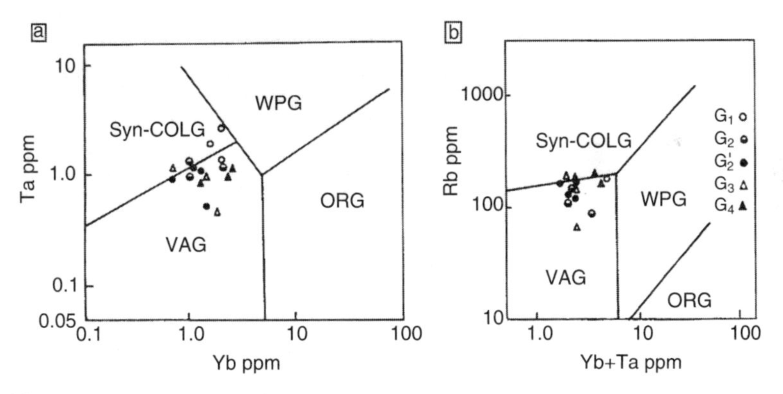

Fig. 2.2.23 (a) Ta–Yb and (b) Rb–(Ta + Yb) discriminant diagrams after Pearce et al. (1984), suggesting possible tectonic setting of Malanjkhand granitoids. VAG – Volcanic Arc Granite, WPG – Within Plate Granite, ORG – Ocean Ridge Granite, Syn-COLG – Syn-collision Granite (Sarkar et al., 1996).

The relatively low $^{87}Sr/^{86}Sr$ initial ratios point to an important contribution by mantle material. It is possible that the calc–alkaline felsic magmas have been derived by partial melting of amphibolitic rocks in the lower crust (the suggestion is supported by the concavity of the REE distribution pattern). Equally plausible is the subduction of submarine ridges, oceanic plateaus and seamount chains, which are thought to cause perturbation to the steady state condition of subduction (cf. Cooke et al., 2005).

In regard to the relative age of the granitoid members of the Malanjkhand Granite, we do not as yet have enough geochronological data to temporally discriminate them and as such, we have to fall back mainly upon the field observations. Patches of G_2 are noticed with G_3, the boundaries being gradational. G_2' ('Pink granite') occurs mainly (though not exclusively) in and around the ore zone and grades into the typical G_2. Worth repeating here is that Panigrahi et al. (2004) did not find any age difference between G_2 (their 'grey granite') and G_2' (their 'pink granite'). G_4 penetrates G_2, G_2' and G_3 (Sarkar et al., 1996).

Orebody: Its External and Internal Features and Disposition In the Malanjkhand deposit, the ore bodies are generally defined by assay contacts as observed in both plan and section (Fig. 2.2.24 and Fig. 2.2.25). However, along the interface with the later intrusives the orebodies display sharp natural contacts.

On plan the ore zone is elongate (> 2 km) with irregular boundaries. In the south, it trends nearly NE, and is more than 200 m thick and has not only been penetrated by the late intrusives (aplite, diabase) but has also been dismembered by them. The ore zone dips 65°–75°easterly. The curvilinear bench developments in the Malanjkhand mine and a bench face showing fractured granite and subhorizontal cover by the rocks of Chilpi Group in the footwall of the ore zone are presented in Fig 2.2.26 a, b.

Most of the ore zone for about 2/3 of its horizontal extent is occupied by sheared vein quartz, with thin slivers of mineralised granite. In the northern 1/3 of the ore zone, however, mineralisations within the granitoid rocks become more important. The mineralised granitic rock is reddish, where the feldspar is alkaline, commonly with hematite–dusts. Hornblende is usually transformed into biotite and chlorite. In this zone, mineralisation occurs along zones of sheeting without or with silicification (Fig. 2.2.26 c, d). Fine to coarse disseminations without any obvious planer controls are prominent in either side of the silicified zone in the northern extremity of the deposit. Locally these ores are coarse-grained approaching pegmatitic character (Fig. 2.2.27).

Small scale features in the mine indicate multiple introduction of vein quartz into the reef, some of which are ore-free and may have slightly modified the host rock at contact. It may be noted that on larger scale too the silicified zone is not entirely ore-bearing (Fig. 2.2.28). Its possible significance will be discussed in a later section.

Wall Rock Alteration Wall rock alterations at Malanjkhand are represented by (i) silicification, (ii) replacement of andesine by alkali feldspar, (iii) replacement of hornblende by biotite, epidote and chlorite, (iv) increase in Mg/Fe ratio in biotite and (v) appearance of hypogence hematite and muscovite as minor phases. These changes may be traced for about 100 m in the hanging wall side and 200m in the footwall side. The dominant feldspar species in this zone is K–feldspar. Sarkar et al. (1996) called these altered rocks G_2'. This is a case of potassic alteration. The higher Mg/(Mg+Fe) ratios (0.70–0.73) in the biotite from this zone, along with the development of hematite suggests a high fo_2 during these alterations, when a part of Fe in biotite was oxidised and removed, resulting in the increase of Mg/(Mg+Fe) ratio. A faint propylitic zone has developed outside of the zone of potassic alteration.

Taylor and Fryer (1982) by way of a general statement concluded that LREE are enriched and some HREE are depleted in potassic alteration zone in hydrothermal ore deposits. Hanson (1980) and Michard (1989) reiterated the same, of course, with the rider that the system should be open. Pirajno (1992) emphasises that such a modification of REE-distribution is a necessary accompaniment of rich hydrothermal mineralisation. These do not well apply to the case of Malanjkhand. Here the distribution of REE in wall rocks (G_2') is not greatly different from that in the unaltered variety of the same rock (Sarkar et al., 1996).

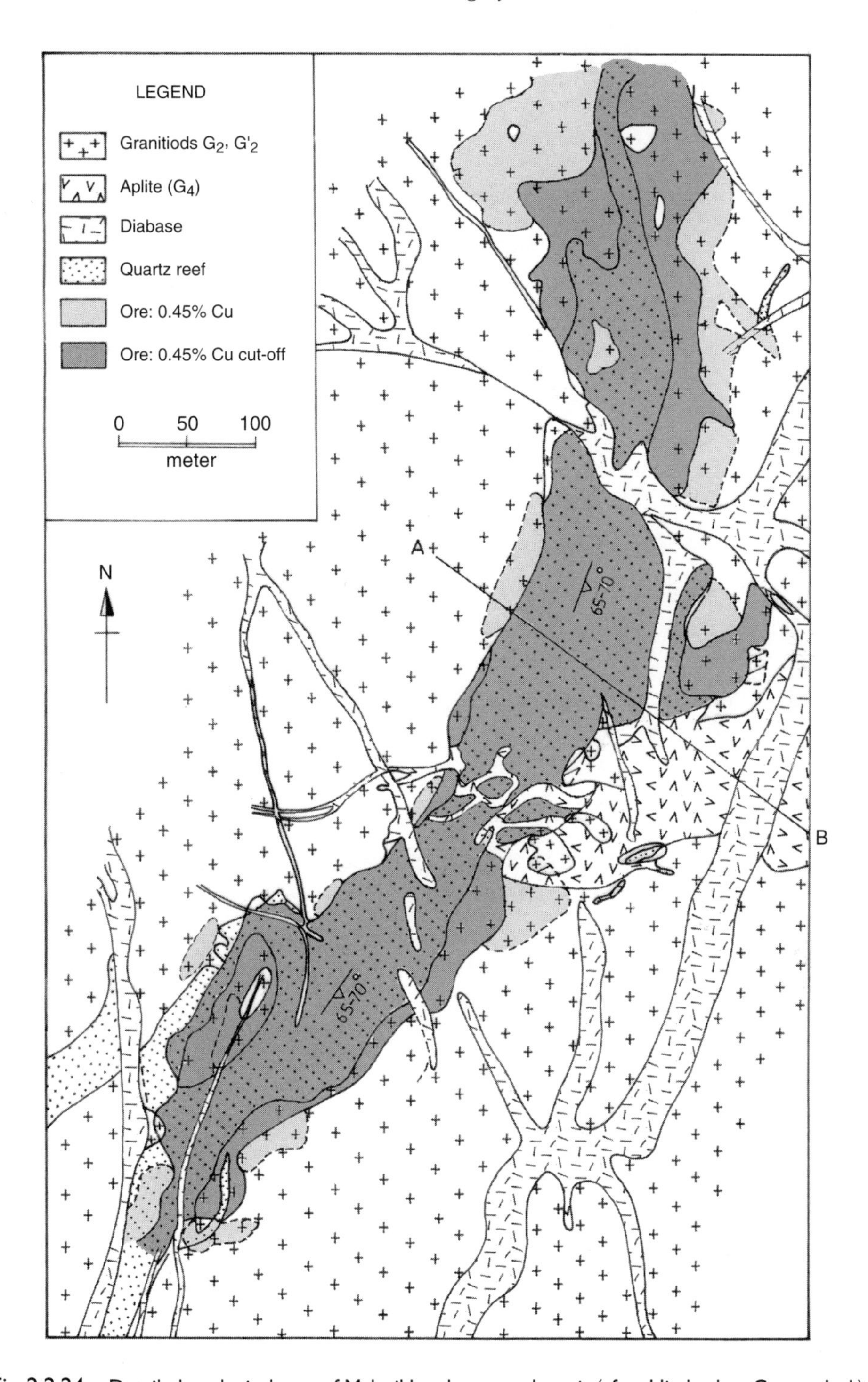

Fig. 2.2.24 Detailed geological map of Malanjkhand copper deposit (after Hindusthan Copper Ltd.)

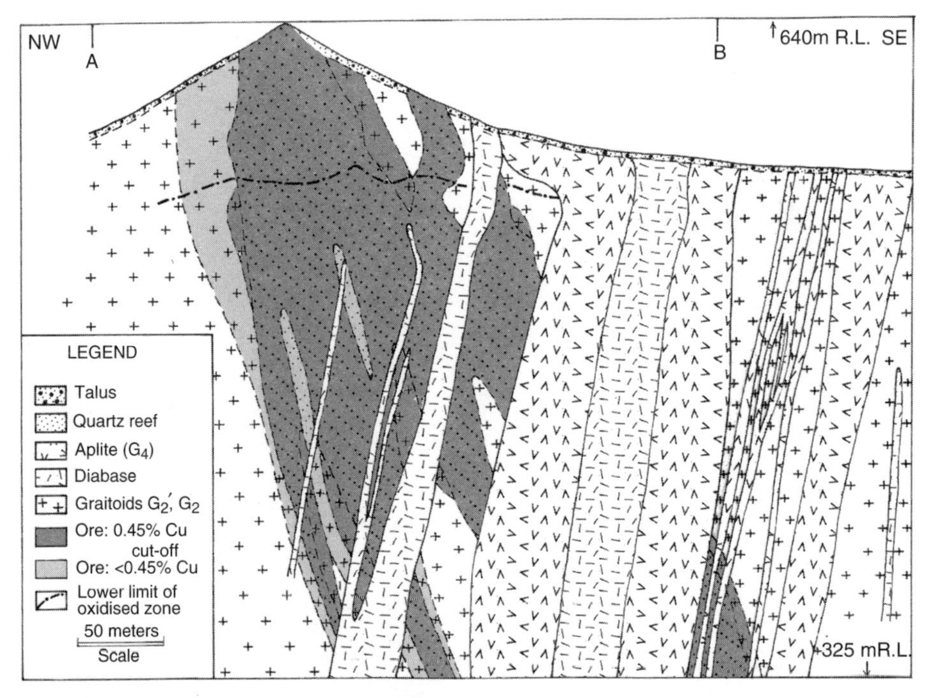

Fig. 2.2.25 Cross-section of ore zone along AB line in Fig. 2.2.24, Malanjkhand copper deposit (after Hindusthan Copper Ltd.)

Fig. 2.2.26 (a) Curvilinear bench developments in Malanjkhand open cast mine; (b) bench face in Malanjkhand mine showing fractured granite in the footwall of ore zone, overlain by subhorizontal formations of Chilpi Group. Field Photographs showing sheeted ore zones; (c) without silicification; (d) with silicification, Malanjkhand copper deposit (photographs by authors).

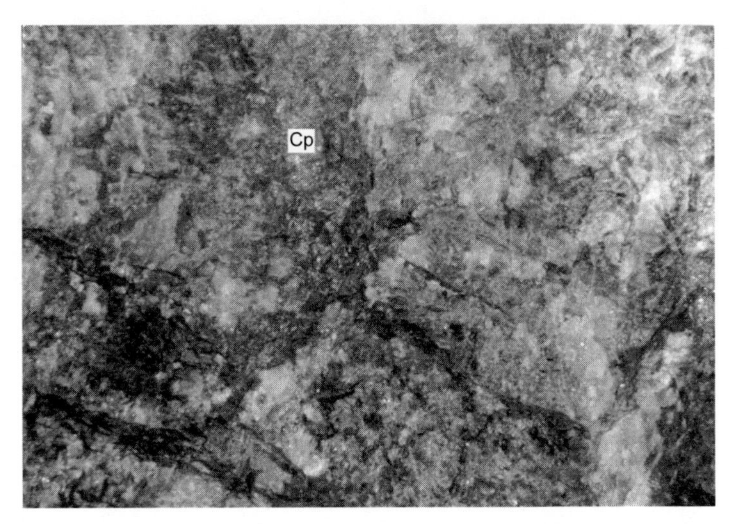

Fig.2.2.27 Field photograph of clots and patches of chalcopyrite-dominant ore in pegmatoid granite in the northern end of Malanjkhand copper deposit (photograph by authors)

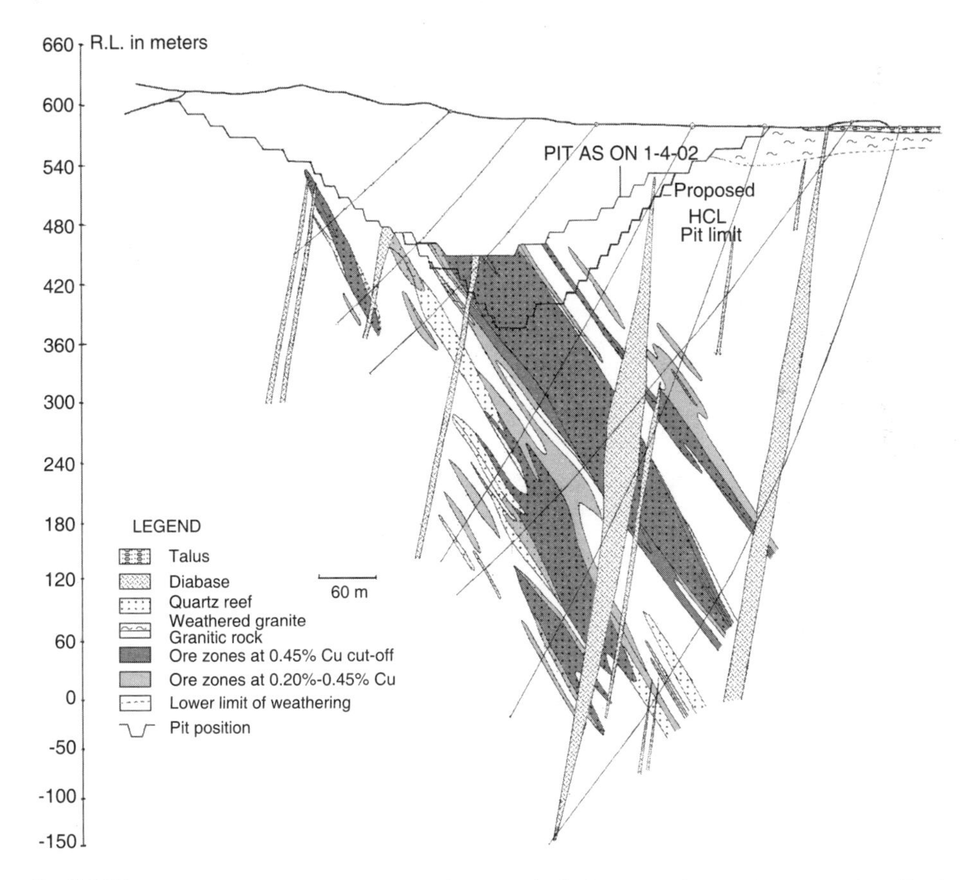

Fig.2.2.28 Cross-section across ore zone showing silicified zones with or without ore mineralisation (after Hindusthan Copper Ltd.)

Mineralogy of the Cu-ores The principal economic metal in the ores is copper, followed by molybdenum and gold. The principal 'ore minerals' in the ore consist of chalcopyrite and pyrite with subordinate magnetite and minor/accessory sphalerite and molybdenite. Trace phases comprise rutile, cassiterite, galena, hessite/stuetzite (Ag_2 Te/Ag_{5-x} Te_2), empressite (Ag Te), clausthalite (Pb Se) and native gold (Sarkar et al., 1996). Quartz is the dominant gangue mineral. Other phases include alkali feldspar. Locally fluorite, gypsum, barite and calcite are present as trace phases.

Quantitative analysis of chalcopyrite composition and X-ray analysis of its structure show that it is compositionally stroichiometric and structurally normal and not otherwise, as suspected by some workers (Panigrahi et al., 1991). Locally chalcopyrite contains Au, Ag, and lesser Sn, Zn, Ni, Co and Se. Gold occurs as minute grains (upto ~3 μm) within chalcopyrite. The range of Au-content in the Malanjkhand area has been furnished at the begining. It may be mentioned that in the concentrates Cu-content broadly ranges between 25% and 30%, where Au-content upgrades to > 2 ppm (Table 2.2.9). The two-layer hexagonal molybdenite – 2H is more common in nature than the three-layer rhombohedral polytype, molybdenite –3R and the Malanjkhand molybdenite is no exception to it. Pyrite grains occur in two size populations in which the coarse grains may be as large as ~1cm, rarely though. They are generally broken with chalcopyrite (± sphalerite) healing the fractures. Cassiterite is present as short prismatic grains in chalcopyrite. Ag-tellurides typically occur together suggesting subsolidus breakdown of a high temperature Ag-telluride.

As much of the ores show deformation and recrystallisation in varying degrees along finite domains of high strain (Figs.2.2.29 a, b), suggestion of a definite sequence of their formation based on mineragraphic criteria will be fraught with much greater uncertainty than what is inherent in the practice. For example, coarse grains of pyrite and magnetite are generally deformed, while the smaller grains are not. Indeed, the coarse grains could be of earlier generation, i.e. pre-deformational. However, at the same time it is also possible that both the coarse and fine grains were pre-deformational, the latter escaped conspicuous deformation due to their small size, i.e. 'Hall Petch relationship' (cf. Müller et al., 1981).

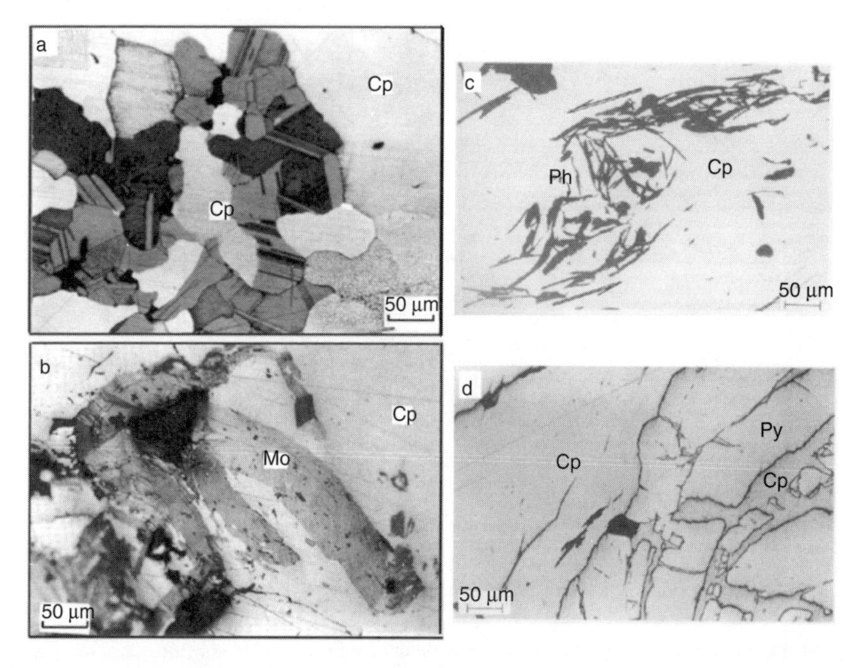

Fig. 2.2.29 (a–d) Textures showing evidences of deformation and recrystallisation in Malanjkhand ore. Cp – chalcopyrite, Mo – molybdenite, Py – pyrite, Ph – phyllosilicates (Sarkar et al., 1996).

The major phases in the secondary ores are Fe–oxides, chalcocite and malachite, followed by brochanthite [$Cu_4SO_4(OH)_6$], bornite, covellite, azurite, anilite ($Cu_{1.75}S$), geerite [$Cu_{1.6}Fe_{0.04}S$], yarrowite ($Cu_{1.13}S$), spionkopite ($Cu_{1.4}S$), native copper, cuprite, delafossite ($CuFeO_{2.3}$), nantokite (CuCl), paraatacamite [$Cu_2(OH)_2Cl$] and claringbullite [$Cu_4Cl(OH)_7$ 0.5 H_2O] (Petruk and Sikka, 1987; Sarkar et al., 1996; D. P. Singh, unpublished work). Electron micro-probe analysis of a chalcocite sample, chosen at random, was found to be composed of 75.4% Cu, 0.51% Fe, 21.6% S with 1000 ppm Zn, 200 ppm Sn, 200 ppm Se, 100 ppm Te, and most interestingly, 1000 ppm Au and 300 ppm Ag (Sarkar et al., 1996).

Sikka and Nehru (2002) reported uraninite and thorite from Malanjkhand, presumably in the ore zone.

Stable Isotope (S, O) Study of the Cu-ores Data on the stable isotope composition of the sulphidic ores are very limited. Sarkar et al. (1996) reported S-isotope composition (δ^{34}S per mil) of a few pyrite and chalcopyrite separates from Malanjkhand ores and these values cluster around zero (Table 2.2.13), similar to those reported in literature from granite associated copper deposits (Ohmoto and Rye, 1979). Oxygen isotope values (δ^{18}O per mil) obtained from the gangue–quartz range from +5.84 to +8.80 (Table 2.2.14).

Table 2.2.13 *δ^{34}S (per mil) of sulphide minerals from the Malanjkhand Mine (Sarkar et al., 1996)*

Sample number	Mineral	n	Average δ^{34}S
1	Pyrite	6	+0.72
	Chalcopyrite	6	+1.74
2	Pyrite	1	–0.38
	Chalcopyrite	4	+1.64
3	Pyrite	2	+0.72
	Chalcopyrite	3	+2.31
4	Pyrite	7	+2.90
	Chalcopyrite	8	+1.92
5	Chalcopyrite	5	+0.05

Note: Samples collected from 556–484 m RL benches. Standard used: Canon Diablo Meteorite.

Table 2.2.14 *δ^{18}O (per mil) of gangue quartz, Malanjkhand deposit (From Sarkar et al., 1996)*

Sample number	n	δ^{18}O
1	10	+ 8.39 to + 8.38
2	8	– 7.54 to + 7.20
3	1	+ 7.21
4	6	+ 7.06 to + 6.99
5	7	+ 5.86 to + 5.81
6	6	+ 8.04 to + 7.75
7	4	+ 8.80 to + 8.73
8	3	+ 8.11 to + 7.31

Note: Standard used – SMOW. Samples collected from 556–484 m RL benches.

Fluid Inclusion Studies Jaireth and Sharma (1986) were the first to make some sustained studies on 339 fluid inclusions (carbonic) from the ore-bearing quartz reef at Malanjkhand. T_h of the fluids ranged between 210°C and 476°C and the salinity, 2.3–18 wt.% NaCl equivalents. They further concluded that 'boiling' was involved and that too recurrently, and this 'boiling' was one of the major causes of ore deposition. If 'boiling' substantially increased the pH of the ore fluid, then ore deposition could indeed be expected. The pressure of entrapment of the fluids from 'boiling' was determined to range between 225 and 445 bars, corresponding to depths of 1–2 km (assuming lithostatic pressure) below the–then surface. Mole% CO_2 ranged between 2 and 5. Naik's (1989) observations are not much too different from those of Jaireth and Sharma.

Ramanathan et al. (1990) studied fluid inclusions in vein quartz 'wherever it shows textual evidence of co–precipitation with chalcopyrite'. According to these authors, these quartz grains were unstrained and undeformed. The fluid inclusions, generally biphase, homogenised at temperatures 310°–360°C (uncorrected for pressure). Salinity varied from 7 to 21 wt.% NaCl equivalent.

Panigrahi and Mookherjee (1997) reported four types of fluid inclusions in quartz in the rocks and ores of Malanjkhand. These are:

1. Type–I: Aqueous biphase (L+V)
   Most common in all samples of ore and the host granitoid. Their homogenisation temperatures (T_h) are as follows: Reef ore – 110° to 362°C; Stringers – 119°to 353°C; Pegmatitic ore – 100° to 258°C; Granitoids – 79° to 222°C. Salinity in most of these inclusions varied from ≤ 3 to ≥ 35 wt.% NaCl equivalent.
2. Type–II: Mixed aqueous–carbonic
   Restricted to stringer and pegmatitic ores. Temperature of total homogenisation $T_{h\ (total)}$ = 240°–329°C
3. Type–III: Pure CO_2
4. Type–IV: L_{aq} + V_{aq} + calcite/gypsum

Estimated pressure ranged from 550–1790 bar during fluid filling in inclusions in the stringer ore. An explanation for such a pressure range would have been illuminative. Exhumation should be the last resort in such a situation for ore deposition that is much too rapid a process. There is no evidence of two time-separated orogenic events either. This pressure range, assuming it is lithostatic, corresponds to a depth of 2–7 km below the then surface. Alternatively, the low pressure corresponds to hydrostatic pressure (developed at a late stage) and the higher one being lithostatic in nature. Change of pressure from lithostatic to hydrostatic over a vertical interval of as small as 100 m is reported from porphyry Cu–Au deposit in the Philippines (Hedenquist et al., 1998). This aspect may be more carefully investigated in future.

Molybdenum Mineralisations Molybdenum as molybdenite (MoS_2) occurs in several modes in the Malanjkhand deposit and its study is as much interesting from scientific point of view, as it is important from its economic worth. It occurs here as:

1. disseminations, and also rarely as pods in the copper ore in the granitic wall rock that underwent potassic alteration,
2. disseminations in the Cu-ore in the quartz reef,
3. discrete veinlets and as 'smears' along fractures in the reef quartz (in cm scales), and
4. discrete veins, meters long and centimetres thick, cutting through the host granite and ore zone (Fig. 2.2.30).

Besides, as mentioned in a previous section, molybdenite occurs as an accessory mineral in granitoid rocks of the Malanjkhand area. Mo–contents in copper ores mined over the years at Malanjkhand have been presented in the Table 2.2.9. It is obvious from this table that the correlation between Cu– and Mo–contents in these ores is poor. This will be further confirmed by the contents of the Table 2.2.15.

Fig. 2.2.30 Discrete Mo-veins cutting across ore zone, Malanjkhand (photograph by authors)

Table 2.2.15 *Mo-contents in ore zones, foot and hanging walls, Malanjkhand deposit (Bhargava et al., 1999)*

Site	Cu (%)	Mo (ppm)
Hanging wall granitoid, close to the ore body	0.09	180
Ore zone at 0.2–0.45% Cu cut–off	0.31	200
Ore zones at 0.45% Cu cut–off	1.47	492

Bhargava et al. (1999) presented a level-wise and block-wise Mo-distribution at the Malanjkhand deposit. Table 2.2.16 shows that there is no finite depth-variation in Mo-content in the deposit, but laterally there is a tangible variation. Limited data from the NNW-block shows that it is the richest zone. Next comes the South-block. Locally the North-block is also rich (> 1000 ppm) in Mo., Estimated Mo-reserve in the Malanjkhand ore zone is presented in Table 2.2.17.

Table 2.2.16 *Vertical and lateral variation of molybdenum (ppm) in the Malanjkhand ore zone (Bhargava et al., 1999)*

Level (m Rl)	South block	Central block	North block	NNW block
Above 460	170	126	462	2194
460–376	387	127		
376–300	251	129	1046	
300–240	204	60	94	2038
240–180	162	68	42	
180–120	405	46	286	NA
120–60	494	80	253	NA
60–0	231	67		
Below 0	504	198	253	NA
Average	307	102	384	2143

Table 2.2.17 *Total estimated molybdenum resource (tonnes) in the Cu-ores of Malanjkhand at 0.4% cut-off (Bhargava et al., 1999)*

Level (mRL)	South block	Central block	North block	NNW block	Total
Upto 376	6449	3374	8667	5031	23542
Between 376 and 0	13438	5917	13104	2280	34734
Below 0	2022	1968	1307	NA	5296
Total molybdenum	21908	11268	23088	7312	63553
Total ore reserve (Mt) (Geological)	71.36	110.48	60.13	3.41	245.38

Re–Os Chronometry Using Malanjkhand Molybdenite Re–Os chronometer in molybdenite is now reasonably well established (Stein et al., 2001). This chronometer provides dependable single mineral ages, given the mineral separates are representative and homogenous. The dating is based on the β-decay of ^{187}Re to ^{187}Os. It may be pointed out that there is insignificant or no measurable initial or common ^{187}Os. The ^{187}Os content in molybdenite is essentially all radiogenic ^{187}Os, whose abundance is a function of Re-concentration in molybdenite and the time for latter's transformation to ^{187}Os.

Malanjkhand molybdenite samples were analysed for Re–Os chronometric studies. Five molybdenite samples were collected by Sarkar and Pal (HCL) from fresh exposures in the mine working, below the zone of supergene alteration. All the samples were collected from the Mo-enriched northern part of the ore zone, from both the reef zone and the granitoid wall rocks that underwent potassic alteration and contain sulphide mineralisation in varying measures. Two of the samples (A–1 and A–2) were from the reef zone and two (A–6 and A–7) from the altered wall rocks. Sample A–8 is from a molybdenite vein without any tangible impact on its wall rocks. The samples were analysed for their ^{187}Re– and ^{187}Os–contents in the AIRIE Program, Colorado State University, Fort Collins, USA. The results of these analyses and the model Re–Os ages for molybdenite are presented in Table 2.2.18. The weighted mean model age for the four samples representing three different mineralisation modes is 2,490 ±2 Ma (without inclusion of the λ ^{187}Re uncertainty). The isochron method is applicable in Re–Os chronometry also. A five-point isochron for the Malanjkhand molybdenite samples provided a model age of 2,489.5 ±1.4 Ma, with a MSWD of 0.5 (Fig. 2.2.31). Restoration of the uncertainty in the ^{187}Re decay constant leads to the conclusion that the Re–Os isochron age of Malanjkhand molybdenite is also 2490 ± 8 Ma. The high Re-content (400–650 ppm) of Malanjkhand molybdenite is indicative of its mantle affiliation.

Table 2.2.18 *Re–Os data for molybdenite from the Malanjkhand deposit (Stein et al., 2004)*

AIRE run	Sample and host	Total Re (ppm)	^{187}Os (ppb)	Age (Ma)
CT – 434 A	A–6 granite	477.1 (4)	12700 (8)	2490 ± 8
CT – 435 A	A–7 granite	458.1 (4)	12195 (8)	2490 ± 8
CT – 456 A	A–2b quartz reef	432.2 (4)	11509 (12)	2491 ± 8
CT – 538 A	A–8 granite	623.3 (6)	16578 (17)	2488 ± 8
Weighted mean age				2490 ± 2
CT – 432 A	A–1 quartz reef	490.0 (3)	12964 (8)	2475 ± 8

(Continued)

(Continued)

AIRE run	Sample and host	Total Re (ppm)	^{187}Os (ppb)	Age (Ma)
CT – 450 A rep	A–1 quartz reef	481.5 (4)	12718 (29)	2471 ± 9
CT – 433 A	A–2 quartz reef	512.5 (3)	13428 (8)	2451 ± 8
CT – 451 A rep	A–2 quartz reef	546.0 (5)	14270 (16)	2446 ± 8

Other Cu–Mo (–Au) Mineralisations (Occurrences) in the Region Malanjkhand and the surrounding areas constitute a sort of a metallogenetic province of Cu–Mo (–Au), the significance of which is yet to be fully understood. Not only are there a couple of nearly similar, though weak, mineralisations elsewhere within the Malanjkhand Granite, but also within the older metamorphics and volcanic rooks (Bhargava and Pal, 2000; Sarkar, unpublished work).

Cu (–Au)–bearing quartz veins occur at Khandapar and Kundikasa within the Amgaon Group and Garhi–Dongri, within Nandgaon andesite and also at Shitalpani/Sitapur at schist–granitoid contacts. Of these, barring Garhi–Dongri, none showed any economic potential at the present level of exploration.

Cu (± Mo ± Au) mineral occurrences within the Malanjkhand Granite, at places other than at Malanjkhand proper, are at Devgaon, Gidori, Pipardhar and Murum. At Devgaon, about 15 km SSE of the Malanjkhand Mine, the sub-economic mineralisation of copper and molybdenum occur in pink and grey granites (granitoids). Here sulphides of Cu, Fe (including Cu–Fe) and Mo occur as disseminations and clots. Stein et al., (2006) conducted Re–Os dating on two fresh molybdenite sample from Devgaon and concluded that the mineral was deposited 2,470–2,465 Ma ago. At Gidori, Pipardhar and Murum Cu–minerals (± Mo ± Au) occur is quartz veins. All these ore occurrences in Malanjkhand Granite are shear zone–controlled. None of these occurrences are thought economically interesting at the present level of exploration. Salient features of these ore occurrences are presented in Table 2.2.19.

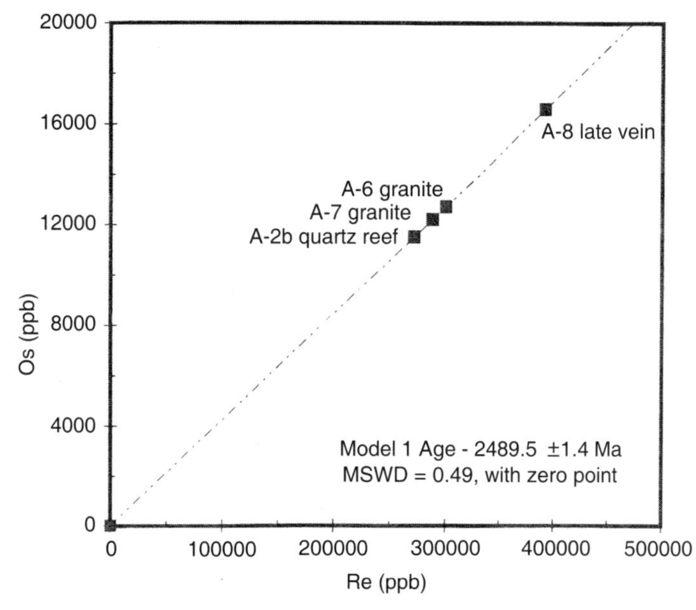

Fig. 2.2.31 Five point Re–Os isochron of molybdenite sample of Malanjkhand (Stein et al., 2004)

Table 2.2.19 *Characteristics of mineralisations in the Malanjkhand Cu–Mo (–Au) metallogenic province (modified after Bhargava and Pal, 2000)*

Location		Characteristic ore minerals	Other observations
A. Mineralisations occurring within the Malanjkhand batholith			
1.Malanjkhand area	In silicified shear zone mainly	Chalcopyrite, pyrite, magnetite, molybdenite. Au, Ag and Co occur in minor/trace quantities.	Superlarge deposit (Geological reserve of 248 Mt at an average grade of 1–3% Cu, with Mo-content varying from 100 to > 2000 ppm and average Au-content of 0.13 ppm). Mineralisation age: 2490 Ma.
2. Devgaon	Disseminated in pink and grey microgranites	Molybdenite, pyrite and chalcopyrite	The host microgranitoid is intrusive into the batholith. Mo-content in the host rock may be as high as 1000 ppm. Soil recorded Cu-content of upto 280 ppm.
3. Gidori	Associated with vein quartz	Chalcopyrite and pyrite. Malachite as a supergene phase	Cu in soil samples > 1000 ppm Au-content in vein quartz = 0.07 ppm.
4. Pipardhar	Associated with vein quartz	Chalcopyrite and pyrite. Malachite and azurite are common as supergene phases	Cu in vein quartz samples > 1000 ppm and Au = 0.37 ppm.
5. Murum	Associated with vein quartz	Chalcopyrite and pyrite	Cu > 200 ppm in soil and Au = 0.23 ppm in schistose rock
B. Mineralisation in the supracrustal rocks in the neighbouring areas			
6. Khandapar	In quartz veins within quartz–sericite schist (Amgaon Group)	Chalcopyrite, pyrite with secondary malachite	
7. Kundikasa	In quartz veins within Amgaon Schists	Chalcopyrite, pyrite with secondary malachite	Weak Cu and Au mineralisation. Cu > 50 ppm in soil and Au = 0.2 ppm in the schistose rock.
8. Garhi–Dongri	In quartz vein in handgaon andesite/ granitoid contact	Chalcopyrite, pyrite with secondary malachite	Cu upto 1000 ppm in soil, and Au upto 3 ppm in quartz veins. Evidence of old mining efforts present.

Origin and Evolution of the Malanjkhand Deposit The Malanjkhand deposit is till date unique in the context of Indian geology. It is for the first time that the Indian geologists got an opportunity of studying a large granite associated Cu deposit. These works varied in details, and in some of the major conclusions. We first discuss the salient aspects of the results of these studies, followed by a logical conclusion in the context of available data. The expressed views on the genesis and evolution of the Malanjkhand deposit may be classified into two broad groups:

1. Pro-Porphyry copper style
2. Anti-Porphyry copper style.

Supporters of the Porphyry copper style include Tripathy (1979, 1983), Ramanathan et al. (1990). Sikka (1989), Sikka and Nehru (2002), Sarkar et al. (1989, 1996), Bhargava and Pal (2000) and Stein et al. (2004). Those who are not supportive of this view include Rai and Venkatesh (1990), Panigrahi and Mookherjee (1997), and Viswakarma (2001). Let us first see what are the main arguments put forward by the Pro-Porphyry style researchers.

Tripathy (op. cit.) favoured a Porphyry style mineralisation for Malanjkhand but did not offer a critical argument. According to Ramanathan et al. (1990), mineralogy, texture and the pattern of distribution of the ores tally with the Porphyry style of Cu-deposits.

Sikka, and his co-workers (1989, 2002) are convinced that the Malanjkhand mineralisation style is typical of a Porphyry Cu-deposit, because (i) it is a low-grade mineralisation, (ii) the deposit is associated with calc–alkaline volcano–plutonic rocks, spatially related to a subduction zone, (iii) weak to well-developed wall rock alterations, (iv) the mineralisation is of stockwork type. According to them, the entire ore-bearing quartz reef is a stockwork. Further, while assuming a subduction in explaining the volcano–plutonic igneous rock association, they involved southward subduction (along Yedekar's 'Central Indian Suture Zone'), but surprisingly in the same paper (Sikka and Nehru, 2002) they conclude that the Malanjkhand mineralisation is in a N–S rift zone and can be associated with the Kotri rift. This is difficult to reconcile. Bhargava and Pal (2000), the mine-geologists, supporting the Porphyry model hold that the stockwork mineralisation is present as an envelope of lower grade mineralisation, surrounding the mineralised quartz reef.

Sarkar et al. (1989, 1996) noted the following features in support of the Porphyry model: (i) Low tenor and high tonnage, (ii) close spatial and temporal relationship with epizonal granitoid intrusions, (iii) calc–alkaline composition of the granitoid host-rock(s), (iv) strong hydrothermal alteration around the orebody. They, however, at the same time stated that the typical stockwork zone is absent, and wondered if it was eroded. Silicification in the ore zone is abnormally high. This last point shall be discussed in a following paragraph in the overall review of the Porphyry model.

Stein et al. (2004) assert that Malanjkhand is a typical Porphyry Cu–Mo–Au deposit, wherein the ore-bearing quartz reef in a 'high–silica stockwork cap' embedded in its primary potassic alteration halo. Its relation with subduction–tectonics is emphasised by high Re-content of molybdenites (400–650 ppm) analysed.

Rai and Venkatesh, on the other hand, believe that the Malanjkhand deposit lacks in some features of Porphyry–Cu deposits and should better be called a vein-type epigenetic deposit, genetically related to the K–rich pink granite. They further hold that Malanjkhand is a peripheral but integral part of the Kotri–Dongargarh rift environment.

Panigrahi and Mookherjee (1997) also concluded that Malanjkhand did not belong to the Porphyry style deposits. Instead, it occurred as a fracture-controlled and quartz–reef controlled mineralisation affiliated to a potassic pink granite that is hosted by a grey granitoid. On finding that the pink and grey granitoids are isochronous, as already suggested by Sarkar et al. (1996) on geologic considerations, they proffer a new suggestion in which 'steam-heated' mineralising fluids from the grey granitoid, triggered by younger aplites and dolerite dikes, were the source of Malanjkhand mineralisation (Panigrahi et al., 2002). They also did not identify a typical stockwork zone at Malanjkhand.

Laznicka (1993) reviewed the geology of the deposit. Based on the literature available until that time, he did not find the 'porphyry copper' model very convincing. A mineralised shear as in the Singhbhum copper belt, appeared to be an alternative genetic explanation for him. The two situations are not alike, however, as will be obvious from a reading of the Cu-mineralisation in the Singhbhum belt (Chapter 4.2).

Viswakarma (2001) is far too radical, if not weakly founded in his model. He believes that the Malanjkhand ores and the host granitoid rock(s) are genetically unrelated, as the Pb in the ores is much more radiogenic than that in the host rock. According to him, the mineralisation is a 'feeder zone filling, developed as a result of hydrothermal exhalation typical of intra–continental rift setting'. The geologic setting of Malanjkhand, as already outlined, has hardly any basis to be called an intra-continental rift. The latter is, in the main, filled by a volcanisedimentary pile. There the intrabasinal fluid, occasionally ore-fluid in composition, is exhaled where the fluid pressure exceeds the hydrostatic pressure. This geological situation is conspicuous at Malanjkhand by its total absence. It bears repetition that the mineralisation there took place in a batholithic massif. Moreover, Pb–Zn, and not Cu dominate a SEDEX deposit. Therefore, Vishwakarma's Model is hardly justifiable on any count. That the ore zone is richer in radiogenic Pb, compared to the granitoid host, may be explained in more than one way. Some investigators have already reported the presence of uraninite in the Malanjkhand ores (Sikka and Nehru, 2002). It could have been a part of the paragenesis with other minerals in the ore, subsequently becoming the ultimate source of the radiogenic Pb. Alternatively, radiogenic Pb could be introduced into the ore zone from an external source during a late recharge, particularly when the endogenic hydrothermal system collapsed, letting in external fluids.

A clear understanding of the genesis and the evolution of the Malanjkhand ore deposit is important from the point of view of both science and industry (exploration). It is the largest Precambrian ore deposit of Cu (–Mo–Au) associated with granitoid rocks, to date. Indeed, the style of mineralisation and the geological environment of the mineralisation are by and large similar to those of Porphyry Cu. But we have to remember at the same time that not all Porphyry copper deposits are exactly the same and no single model of a typical deposit, or of a magma associated with ore, is generally applicable (Hedenquist and Richards, 1998), and there are many 'variations on a theme' (Gustafson and Hunt, 1975; Gustafson et al., 2001). A brief 'state of the art' discussion on Porphyry style Cu-mineralisation follows, in this context.

Porphyry Cu (–Mo–Au) Systems Porphyry Cu (–Mo–Au) systems, are commonly referred to as Porphyry Cu-deposits as Cu is the dominant economic metal in the majority of such deposits. Varieties may be (besides carrying Cu) (i) Mo-rich, Au-poor, (ii) Au-rich, Mo-poor, (iii) carrying Mo and Au, (iv) Au only (rare). Our aim is to characterise the Cu-dominant systems with Mo and Au in varying proportions. It may be mentioned that the Cu-dominant porphyry systems are the source of >50% of the Cu-ores in the world, mined and reserved.

Distribution in Space and Time Porphyry deposits of Cu, Mo and Au occur at many places in the world, but these are numerous, and often very rich around the Pacific rim. Particularly mentionable in this context are the porphyry ore deposits along the southwest border of the USA, western Mexico, the Andean states of Peru, Chile, Argentina and in the Southeast Asian margin, Papua New Guinea–Irian Jaya fold belts, Philippines. Besides, there are fewer but very large deposits in Mongolia (Oyu Tolgoi), Iran (Sar Cheshmeh), Kazakhstan (Aktogay–Aiderly), Uzbekistan (Kal'makyr), China (Yulong, Eastern Tibet), and Pakistan (Reko Diq). In age, they are all Phanerozoic and show a conspicuous polarity towards the Cenozoic.

Known Precambrian deposits are very few and lean, except of course Malanjkhand. Some amongst them are Kopsa, Finland (25 Mt, with 0.18% Cu, 0.57 ppm Au), Ylojarvi, Finland (produced 28,000 t Cu, 427 t W and 50 t Ag), Tallberg, Sweden (45 Mt, with 0.27%Cu) (Sarkar et al., 1996 and references therein). Proterozoic Porphyry Cu-mineralisations are also reported from South Africa (Minnitt, 1986; Viljoen et al., 1986), from Haib river and Lorele areas.

Host Rocks The host rocks for Cu (–Mo–Au) deposits are calc–alkaline to high–K calc–alkaline, biotite– and amphibole–bearing rocks, varying in composition from diorite (SW Pacific rim), through quartz diorite (tonalite), granodiorite, adamellite to varieties of monzonite (upto shoshonite). They occur as medium to small sized intrusive bodies—cupolas, apophyses and dykes thrown out by larger intrusive bodies, generally of batholithic nature. Dacite and andesite (rarely rhyolite and basalt) are present in association, unless removed by erosion. The host rocks are usually porphyritic in texture (quartz and feldspar phenocrysts together constitute 25–50% by volume) and hence the name of the deposit type. In rare cases, as in the Brenda porphyry Cu and Endarko porphyry Mo deposits in British Columbia, Canada, mineralisation occurs in phaneritic granitoids. Both have been interpreted as deep level deposits (Brown, 1969). It is also possible, as suggested by Sillitoe (1972), that these deep level mineralisations were once overlain by typical Porphyry-style mineralisation. These I-type and usually Magnetite Series granitoids are commonly higher in Fe_2O_3/FeO ratios.

Usually host rocks make a complex of multiple intrusions close in space and time. Not all members of the intrusive complex are porphyritic, but some, particularly the early ones are. The magmatism shows an evolutionary trend in terms of increasing SiO_2- and K_2O- contents with time, unless more than one magma batch is involved. Magma mixing, assimilation etc., can make the magmatic trend complex. The REE patterns show strong fractionation with high La/Yb ratios (> 20–25). In case of Andean deposits, this has been taken to suggest that the presence of high-pressure hydrous magma derived from a source with residual amphibole and /or garnet, typical of contractional tectonic regime and continental crust > 45 km thick. Locally Porphyry Cu- mineralisation may spill over to sedimentary country rocks as at San Jorge, Argentina (Williams et al., 1999).

Wall Rock Alteration Wall rock alteration associated with Porphyry ore deposits has been a subject of serious study for many decades now with the results that concurred on some aspects and differed on others. The purpose was as much to find a dependable tool, if not a finger print in exploration, as it was to understand better the genesis of the ores. The model of Lowell and Guilbert (1970) at one point of time found favour with the ore geologists in the industry as well as in the academia. In its simplest form, it suggested a potassic alteration at the core, characterised by the presence of potash feldspar + magnetite + quartz. The 'ore shell' wrapped this. The ore shell is surrounded by a zone of phyllic alteration, characterised by the development of quartz + sericite + pyrite. The next is the zone of intermediate argillic alteration, characterised by the development of clay + chlorite. The outermost, zone is the propylitic zone, having a characteristic mineralogy of epidote + chlorite + albite + calcite.

This scheme of wall rock alterations in Porphyry Cu (–Mo–Au) deposits was ultimately found to be a little too generalised. The phyllic alteration may indeed follow the potassic alteration with the decrease of temperature and a_k/a_{H_2O} ratio. However, in some cases it has developed later or had repetitions, in which case the water involved could be partly/wholly non–magmatic. In cases, the phyllic zone may be absent altogether bringing the potassic alteration zone and the propylitic zone in contact. Argillic alteration by the side of the phyllic zone is also not a regular feature. This scheme does not mention 'advanced argillic alteration' (Meyer and Hemley, 1967), or the 'lithocap' (Sillitoe, 1992) characterised by the development of dickite, kaolinite and /pyrophyllite ± quartz ± alunite ± sericite ± topaz ± zunyite, which is epithermal and occurs at the apical part of the deposit. The ore minerals in this epithermal mineral association comprise pyrite, bornite, and gold. The location of this zone close to the surface makes it more susceptible to weathering, erosion and an ultimate removal.

One thing is common amongst hydrothermal alterations associated with the Porphyry Cu-systems. It is the core-zone of potassic alteration with which the bulk of the mineralisation is commonly associated. The high temperature potassic alteration represented by the development of quartz, K–feldspar, biotite, anhydrite and and magnetite containing veins and veinlets ('A-veins' of Gustafson and Hunt, 1975) may form during the final consolidation of the magma (850°–800° C) (cf. Gustafson and Hunt, 1975; Williams et al.,1999; Tosdal and Richards, 2001; Cannell et al., 2005; Harris et al., 2005).

Base metal sulphide minerals, such as chalcopyrite, bornite, pyrite and molybdenite commonly occur as disseminations in these alteration zones. Of course, the appearance and stable existence of Cu–Fe sulphide phases in these alteration zones would be constrained by their saturation concentrations as well by the stability fields in the Cu–Fe–S system (Vaughan and Craig, 1978). The potassic alteration in the Porphyry Cu–systems takes place over a large temperature range. It may appear at 850°–800° C and may continue upto 400° C or even lower. The high temperature alteration involved a hypersaline liquid (40–60 wt.% NaCl equivalent) commonly co-existing with a low density vapour.

Leaching, transport and deposition of silica are common in any hydrothermal system of ore deposition. In some ore deposits, as in mesothermal gold deposits, quartz is commonly the dominant gangue mineral. In Porphyry Cu–ores, it is present but hardly in such a large proportion. Increasing salinity and increasing pressure push the maximum solubility of silica to higher temperatures (> 400° C). Therefore, the pressure-fall (± fall of temperature) in the hydrothermal system by way of decompression, will be expected to make it saturated with silica. Hydrolysis (H^+ metasomatism) and cation metasomatism of feldspars are commonly held responsible for the release of silica to the hydrothermal systems.

$$3\,KAlSi_3O_8 + 2H^+ \Leftrightarrow KAl_3Si_3O_{10}(OH)_2 + 6SiO_2 + 2K^+$$

(K–feldspar) (Muscovite) (Quartz)

$$2\,NaAlSi_3O_8 + 2H^+ \Leftrightarrow Al_2Si_4O_{10}(OH)_2 + 2SiO_2 + Na^+$$

(Pl–feldspar) (Pyrophyllite) (Quartz)

$$2\,Na\,AlSi_3O_8 + 4\,(Mg, Fe)^{2+} + 2(Fe, Al)^{3+} + 10H_2O \Leftrightarrow$$

(Pl–feldspar)

$$(Mg, Fe)_4\,(Fe,Al)_2\,Si_2\,O_{10}(OH)_8 + 4SiO_2 + 2Na^+ + 12H^+$$

(Chlorite) (Quartz)

The effect of temperature on the solubility of silica is important. Between 600° and 700° C, 0.5–14 g of silica may be deposited per 1000 g of solution, whereas only 0.2–0.3 g of silica can be precipitated from the same weight of solution between 100° C and 200°C (Holland and Malinin, 1979).

Modes of Occurrence Ore bodies may be intrusive (stock/cupola)-centric, or may be tectonically controlled, occurring as linear arrays of veins or fault controlled ore bodies. As is discussed somewhat elaborately in a following section, ores in the Porphyry Cu-systems form from hydrothermal fluids ex-solved from magmas that formed the associated rocks (not necessarily all exposed). The ex-solution is due mainly to depressurisation of the magma as it rises upward due to buoyancy and is emplaced close to the surface. As a granitic magma containing 1–5% water in dissolution ascends, it ex-solves fluids dominated by H_2O. This involves a large positive change in volume, because the volume per unit mass of crystal + vapour > volume of the equal amount of the 'wet' i.e. H_2O-saturated melt. It produces mechanical energy ($P\Delta Vr$, where ΔVr is the change of volume). When the already crystallised rind of the stock can no more accommodate this change of volume by plastic deformation, brecciation and ultimately stock work usually develops at the apical part of the stock (Burnham, 1979). This zone, as a result, is the most favoured volume of the stock for porphyry Cu (–Mo–Au) mineralisation. Strong linear arrays of veins are common in some deposits, which according to the interpretation of Heidrick and Titley (1982), and Tosdal and Richards (2001) formed at depths and therefore, should be more relevant to the deep levels of Porphyry Cu-systems.

Fault control of Porphyry Cu (–Mo–Au) mineralisation is relatively rare for the reasons outlined above. However, a fault or disjunctive structure may develop in an igneous body during a very short-lived episode of high shear strain, even before the complete consolidation of the magma (Fournier, 1999). Older fault structures at the basement may get reactivated and overprint the dislocation on the superjacent granitic body.

The distribution of the Porphyry Cu-systems is very much skewed in favour of the Phanerozoic, or more particularly the late Phanerozoic time.

Mineralogy and Chemistry of the Ores Porphyry Cu-deposits generally contain also other metals in economic to sub-economic proportions. Depending on their content, they may be classified into:

1. Au-rich, Mo-poor
2. Contain both Au and Mo
3. Mo-rich, Au-poor
4. Au-only.

The principal hypogene sulphide minerals in these ores are chalcopyrite, bornite and pyrite. Minor phases include enargite, arsenopyrite, pyrrhotite, tellurides, bismuthinide, selenides, tenantite–tetrahedrite, cassiterite, hypogene chalcocite, djurleite, digenite and gold. In rare cases, as at Butte, Montana, sphalerite and galena may be present in not negligible proportions. Ore petrographic modifications due to post–depositional participation in sulphidation–desulphidation reactions are possibly more common than generally suspected.

Porphyry Cu-ores are rarely metamorphosed in the sense majority of VMS and SEDEX deposits are. Nevertheless, the fault-bound ores subjected to post-mineralisation deformation before complete cooling may undergo dynamic recrystallisation as in the case of Malanjkhand. In some other deposits, for example Chukui Kamata, Chile, recrystallisation of the vein quartz is reported but not in the ores (Ossandon et al., 2001). If vein quartz is recrystallised, the associated Cu-minerals will hardly be spared.

S-isotope composition ($\delta^{34}S$) of sulphide minerals from the porphyry ores do not commonly show a large scatter from the meteoritic pyrrhotite ($\delta^{34}S = 0$ per mil). However, it may deviate significantly from this value depending on the oxidation state (SO_2/H_2S), temperature and fluid/magma ratio (cf. Ohmoto and Rye, 1979).

Tectonic Influence on the Development of Porphyry Cu-systems A close spatial and temporal relationship between the Porphyry Cu (–Mo–Au) deposits and their calc–alkaline/K–calc–alkalic rock–associates, together with their general confinement to arc-domains in tectonic parlance have been noticed and preliminarily explained with the help of the then upcoming plate tectonic model, way back in the early seventies (Sillitoe, 1972; Guild, 1972). However, since then the subject has been studied in much more details.

Many Porphyry–Cu provinces are now known to have been constrained by arc-parallel transcurrent faults, some examples of which are the Andean porphyry belts, North American Cordillera, Southwest Pacific board and Eastern Australia (Richards, 2003 and references therein).

Another observation is that the major Porphyry Cu-deposits of the Andean belts formed late within a given magmatic cycle, the porphyry intrusion typically representing the last intrusive event in the area. In some regions, the localisation of the porphyry systems have been favoured by intersections of transverse lineaments with the arc-parallel transcurrent faults (Lindsay et al., 1995; Richards, 2003; Richards et al., 2001).

What was the role of the evolution, if any, of the subduction zone in the genesis of the arc magmas and the associated porphyry systems? During the time of stable subduction, i.e. the period when the slab subducts without changing the velocity or the angle, slab dehydration and magma generation in the overlying mantle wedge along a narrow strip takes place at a depth of ~100 km. Under the broadly compressional regime, the mafic magma generated accumulates at the base of the crust in the overlying plate (Fig. 2.2.32) and initiates a process envisaged in the MASH (Melting, Assimilation, Storage and Homogenisation) model of Hildreth and Moorbath (1988), along a linear zone beneath the proto-arc.

In course of time, melting, assimilation and homogenisation are advanced, an appreciable volume of magma having lower density and intermediate composition is probably generated, and this magma would tend to move upward due to buoyancy. Advection of the heat into crust, supplied by the upward moving magma, may physically weaken the ambient rocks and may facilitate the development of shear strain, producing new structures and /or reactivating the old ones. Magma ascension may be aided by fault jogs, stepovers and fault intersections, as they will provide with low stress extensional domains. This magma–ascent along zones of weakness explains the confinement of these magmas to narrow arc–parallel tectonic belts. Sillitoe's (1998) brief but succinct comments on the significance of tectonic variations in this context are:

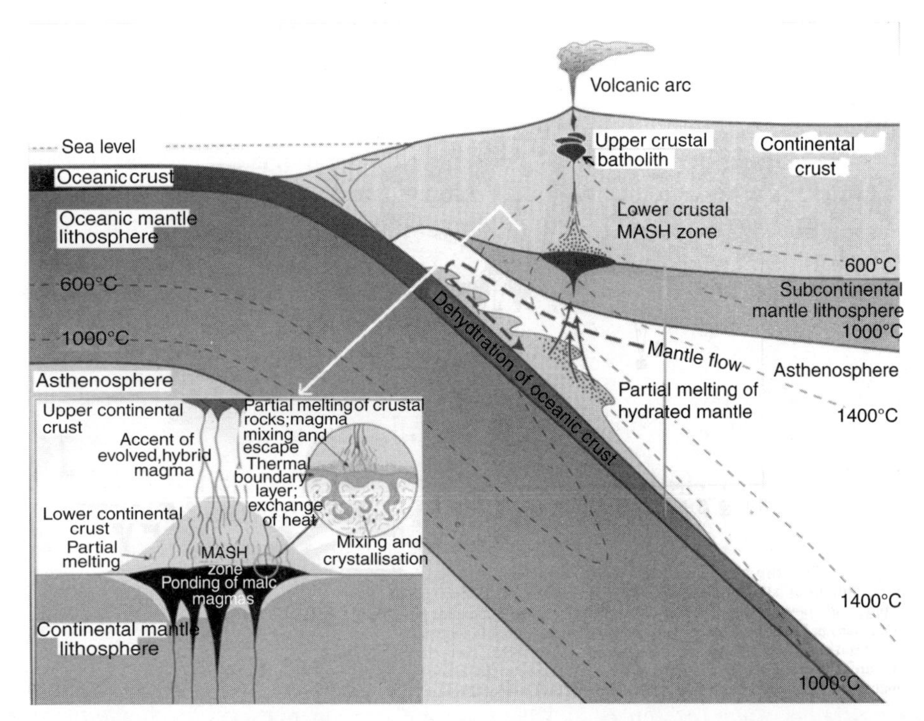

Fig. 2.2.32 A schematic presentation of the MASH model of Hildreth and Moorbath (1988), showing melting, assimilation, storage and homogenisation at lower crust

1. Compression impedes magma ascent through the upper crust, inhibiting volcanism.
2. Resulting shallow magma chambers are larger than those below the extensional arc.
3. Ponding of the magma favours fractionation producing much magmatic–hydrothermal fluids.
4. Compression restricts the number of apophyses on the roof of a large magma chamber. This leads to more efficient hydrothermal fluid focusing.
5. Rapid uplift and erosion favours efficient separation and transport of magmatic hydrothermal fluids due to sudden decompression (Masterman et al., 2005).

Low angle (flat slab) subduction has been made responsible for the development of important supergiant porphyry Cu-deposits (cf. Gutscher et al., 2000). Murphy (2001), studying several (6) supergiant porphyry Cu(–Mo–Au) deposits in SW Arizona and Sonora that were produced during Laramide orogeny, concluded that they were related to flat–slab subduction.

Buoying of a subducting slab by aseismic ridges, seamount chains, and / or oceanic plateaus can provide environments that favour porphyry ore formation. In oceanic island arcs, ridge subduction leads to slab flattening and/or subduction reversal (Cooke et al., 2005; James and Sacks, 1999).

However, Porphyry style Cu-deposit formation has also been discussed without invoking the plate tectonic model. Halter et al. (2002) have suggested that the Porphyry-style Cu-deposits may be related to anatexis of lower crust with the availability of H_2O in the middle crust, ore-bearing fluids may be ex-solved at appropriate depths in the upper crust and give rise to Porphyry-style copper deposits. Such a description and interpretation of a so-called Porphyry-style Cu-deposit are indeed rare in literature.

Origin of the Porphyry Cu (–Mo–Au) Ores The physical, temporal and chemical (compatibility) relationships of these ores with their host rocks suggested to some early workers that these two entities must be genetically closely related in one way or the other. Experimental studies in later times (Candela and Holland, 1986; Cline and Bodnar, 1991) proved that the calc–alkalic magmas had, in general, the potential to form Porphyry Cu-deposits from hydrothermal fluids that would be exsolved when such a magma was depressured, i.e. emplaced close to the surface. Being placed at low load pressures, their saturation vapour (H_2O) pressure would be low, a fact that suggests that their vapour saturation would take place early without allowing crystallisation to advance much. This means that the vapour saturation would be due mainly to ‘first boiling’ (vapour saturation due to decrease of pressure); rather than to ‘second boiling’ (vapour saturation due to crystallisation). This should have a positive effect on the sequestration of Cu into the H_2O-rich exsolved fluid. Cu is a compatible element in the crystallising granitic melt, participating in the rock-forming silicates and/or sulphidic accessory phases (if reduced sulphur is available). So deprived of protracted crystallisation, the magma would release Cu to the aqueous fluid. The vapour phase, highly charged with Cl^-, should scavenge Cu from the silicate melt as a chloride-complex.

Mo, on the other hand is an incompatible element in the given situation. However, because of limited crystallisation, it will have little enrichment in the residual melt. Again, its partition coefficient is small ($D^{Mo}_{fluid/melt} = 2.5$) and it is not known for its affinity for Cl^- that may be an explanation for the relatively lower content of Mo in these ores. Mo (–Cu), or the Climax type deposits are thought to have formed in a somewhat different situation. Here the parental magma was possibly somewhat H_2O-rich and the saturation water content was also higher (Candella and Holland, 1986; Strong, 1988). Results of experimental studies on the solubility and speciation of Mo in water vapour at high T–P shows that Mo may be transported in the vapour–phase as a hydrated complex and then precipitate on reacting with reduced sulphur (H_2S) as follows (Rempel et al., 2006).

$MoO_3.H_2O$ (g) + $2H_2S$ (g) $\Leftrightarrow$ MoS_2(s) + $3H_2O$ (g) + $0.5O_2$ (g)

The enrichment process of porphyry Cu-deposits in Au is less understood. However, in the recent past some research works have pointed out the affinity of Au-rich porphyry systems for high-K calc–alkaline to shosonitic magmas (Sillitoe, 1997; Wilson et al., 2003). These studies may be pursued.

Copper ores in porphyry copper deposits may precipitate at high temperatures along with potassic alteration and 'A' type veins (Gustafson and Hunt, 1975) from highly saline fluids, as in the case of Bingham, Utah and Yerrington, Nevada, both in the USA and the San Jorge deposit, Argentina. A fair portion of these ores, however, formed in medium to low temperatures (400°–200° C, or lower?)

As the solubility of Cu-sulphides in aqueous solution is too low, collection of Cu and reduced sulphur is to be separately done, until they react together to form sulphides. In early high temperature fluids (both vapour and liquid) sulphur occurs mainly as SO_2 and H_2S gases. These gases should partition more strongly into vapour, while Cu as a Cl-complex, should be carried in the brine. Sulphide sulphur for the precipitation of sulphidic ore minerals may be produced by the hydrolysis of sulphur dioxide ($SO_2 \rightarrow H_2S$) (Robb, 2005). A part of H_2S may form by a magnetite-forming reaction like:

$9FeCl_2 + SO_2 + 10H_2O = 3Fe_3O_4 + H_2S + 18\ HCl$

(Ulrich et al., 2001)

When mineralisation is associated with a number of small felsic porphyry stocks and dykes, closely clustered in space and time, the mineralisation is generally most intensely developed with the earliest ore-related intrusive and it becomes weaker with each subsequent porphyry. A plausible explanation of such a situation is the derivation of the magma and volatile fluids for each mineralising porphyry unit from a source magma chamber below, that was being impoverished in volatiles and the contained mineral matter, each time a mass of magma and the associated volatiles are thrown up for a new porphyry system (Proffett, 2003). The situation, however, becomes less simple when one or a set of unmineralised intrusives intervene between sets of mineralised porphyries as at Indio Muerto, El Salvador, Chile and Yerington, Nevada (Gustafson et al., 2001).

In some deposits, such as El Salvador, Chile, and Butte Montana, USA, there is strong ore petrologic evidence to suggest that the sulphidic ores related to an earlier phase of mineralisation have been leached and redeposited at least partially at a upper level, by the fluids given up by an younger intrusive (Gustafson and Hunt, 1975; Gustafson et al., 2001; Brimhall, 1980). The later intrusive could have provided with the reactive hydrothermal fluids, or locally might have played the role of a 'steam engine' only.

Preliminary results on the study of an aspect concerning the ex-solution of fluids from granitoid magmas await confirmation in future. This is about magmatic immiscibility and the formation of a volatile-rich melt as a link between magmas and hydrothermal fluids having potential for ore deposition (cf. Davidson et al., 2005).

The controversy on the relative role of magmatic vs meteoric water in the origin and evolution of Porphyry Cu (–Mo–Au) deposits is not settled yet. However, recent developments in micro-analytical techniques led to determination of significant concentration of metals, including Cu, in vapour-rich fluid inclusions in porphyry ore systems (Hedenquist and Richards, 1998). This suggests that the role of meteoric water, whose presence in varying measures is established in many deposits, has been rather passive, i.e. a fluid-circulator of ore-matter at best, and not its source. Studies on this issue are not yet over.

Distribution of the porphyry Cu-systems is very much skewed in favour of the Phanerozoic, or more particularly, late Phanerozoic time. But why? Is it related to the evolutionary processes of the crust–mantle system, or is it related to something else? A look into the evolutionary trend of the crust–mantle system would show that there could not be a strong geological reason as such to explain this observation. The most plausible reason is the uplift and erosion, which must have destroyed the deposits. The rarely preserved ones are those that have been buried or least disturbed during later orogenic events (Cooke et al., 2005).

Back to the Genesis of Malanjkhand Deposit In the background of the picture of the Porphyry Cu-systems drawn above with a broad brush, one should have least objection in assigning the Malanjkhand deposit to this type, although it may not be a 'typical' one, as claimed by Sikka and Nehru (1998, 2002), or Stein et al. (2004). Its ultimate assignment to porphyry-type may be based on the following: (i) chemical and the mineral composition (Cu, Mo and Au, present mostly as chalcopyrite, molybdenite and gold); (ii) potassic alteration of the immediate host rock (wall rock); (iii) calc–alkaline composition of the host rock, which was originally an oxidized I-type granitic magma; (iv) location in a volcanic arc domain; (v) mantle–affiliation, as indicated by a high (400–650 ppm) Re–content in the Malanjkhand molybdenite (Stein et al., 2004); (vi) a part of the ore is virtually isochronous with the host rock (cf. Stein et al., 2004; Panigrahi et al., 2004). Its difference with many other Porphyry Cu (–Mo–Au) deposits is worth mentioning at the same time.

Malanjkhand deposit differs from most other deposits of porphyry style Cu (–Mo–Au) deposits in not being located in the apical zone of a porphyritic stock/dyke, being located instead in a phaneritic batholithic body. It's being localised in a shear zone-like tectonic structure, is not an exceptional, though a somewhat rare situation. Evidence of very high temperature wall rock alteration, including development of early potassic alteration, i.e. 'A' vein, found in many deposits has not yet been reported from here.

Stockwork is a common feature of pophyry Cu-systems. It develops, as already mentioned, when the solid carapace or rind of a cooling magma body is subjected to increased vapour pressure, in most cases it ultimately yields by fracturing, producing a stockwork mineralisation (Burnham, 1979; Titley, 1995). The principal silicate (gangue) minerals developed in the ores are quartz, K–feldspar and biotite (replaced by chlorite). This is not exactly how Sikka and Nehru (op. cit.) and Stein et al. (2004) perceived of the stockwork they reported from Malanjkhand. They think the fractured and mineralised shear zone rock (the 'quartz reef') is an analogue of stockwork elsewhere. Mineralisation within the host rock and occurring locally as selvage to the main ore zone come closer to the common description of stockwork, but is genetically related to shear mechanics, rather than to fluid dynamics as is generally the case. In authors' opinion, the typical stockwork at Malanjkhand, if present, was eroded away in all possibilities. The top of the orebody has been eroded away before it was overlain by the Chilpi Group sediments (Fig. 2.2.25, b).

Re–Os age–data for molybdenite of Malanjkhand may be divided into three groups (Table 2.2.18):

1. 2490 Ma (from mineralisation in granitic rock and quartz reef)
2. 2475 Ma (from mineralisation in quartz reef)
3. 2451 Ma (from mineralisation in quartz reef).

Age (1) tallies with the Pb–U age of zircon in the host rock (Panigrahi et al., 2004). The younger ages, according to Steins et al. (2006) are due to local remobilisation of earlier 2,490 Ma porphyry-style mineralisation that was subjected to 'intense metamorphic reworking'. However, this explanation is not tenable because of little or no evidence of intense metamorphism (in the classical sense) exists in the country rocks. The major phase (host) of the Malanjkhand Granite is dominantly granitoid in texture (Fig. 2.2.19 a, b), except in areas of local high strain. The altered wall rocks could hardly escape an overprint of this intense metamorphism. For the reworking of Cu (–Fe) sulphide ores the metamorphic fluid had to be a brine and that for molybdenite, water vapour at high T–P. The local geological situation could provide with none. Neither the present authors are very sure about the correct answer. Could it be one (or two) younger granitic intrusive(s) below that did the 'mischief' by throwing up the necessary fluids? The ~2,475 Ma age could be a mixture of 2,490 and 2,450 Ma molybdenite also (Stein et al., 2006). Observations in different scales suggest that the Cu–Mo–Au mineralisation took place in two stages intervened by an event of intense silicification, i.e. reef formation. The initial mineralisation apparently took place along a shear zone in the granitoid rock. This was followed by silicification in the same zone. The silicified zone fractured by repeated movement, acted as loci for the ingress of ore fluids in the second stage. That the ore material and the silica for the reef were not carried in the same solution is suggested by the barren stretches within the reef itself (Fig. 2.2.28).

Uranium Mineralisation in Central India

Uranium mineralisation in the Central Indian craton occurs at a number of places, the most important of which is the Dongargarh uranium province. Other places of occurrence of uranium mineralisation include the Abhujmar basin, the eastern margin of Chhattisgarh basin, and the western parts of Chhotanagpur Granite–Gneiss Complex.

Dongargarh Uranium Province In the Dongargarh region uranium mineralisation is reported from the following three belts (Sen, 1977; Krishnamurthy et al., 1988; Mahadevan, 1992) (Fig. 2.2.33):

1. Bodal–Bhandaritola belt in the east, a little more than 60 km SSW of Rajnandgaon
2. Jangalpur-Baghnadi–Parsori belt in the west, about 50 km west of Dongargarh
3. Dongargaon–Lohra belt.

At Bodal, the uranium mineralisation, apparently hydrothermal, occurs in both metabasalt (now amphibolite) and meta–rhyolite belonging to the Dongargarh Supergroup, close to the eastern margin of Dongargarh Granite. The mineralisation is in the form of veins and veinlets, localised along tectonic structures within the volcanics. The primary U-mineral in the ore is finegrained pitchblende, occurring in association with pyrite, pyrrhotite and minor pentlandite, chalcopyrite and sphalerite together with hematite. The mineralisation is coeval or slightly younger than Dongargarh Granite (Singh et al., 1990). However, the genetic relationship of the uranium mineralisation with the Dongargarh Granite and the felsic volcanic rocks does not seem very straight forward, as neither of them are 'fertile' in the usual sense of the term. Average U-content of Dongargarh Granite (n = 24) is 5.8 ppm and it is same for the felsic volcanic rocks (meta–rhyolite and quartz porphyries) (Mahadevan, 1992). At Bhandaritola, the mineralisation is localised at the interface of meta–rhyolite and metabasalt, in either of the rocks.

Secondary uranium in the form of autunite and uranophane, associated with pyrite and fluorite, is found along shear zones in the rhyolites and felsites at Jangalpur and in the felsic flows and tuffite at Parsori and Baghnadi.

Hansoti and Sinha (1995) reported Iron Oxide–Copper–Gold (IOCG) type metallogeny (with Fe^{+2} +REE ± Cu ± Ni mineralisation) in the Bartalao Sandstones of the Dongargarh–Lohara area, located between the Malanjkhand ore zone within the Malanjkhand Granite Complex and the Chandidongri F+Pb ± Zn ± Cu ± U mineralisation within Dongargarh Granite. The authors are of the view that this occurrence is worth consideration as a first order target area for 'multimetal breccia complex type of deposit'.

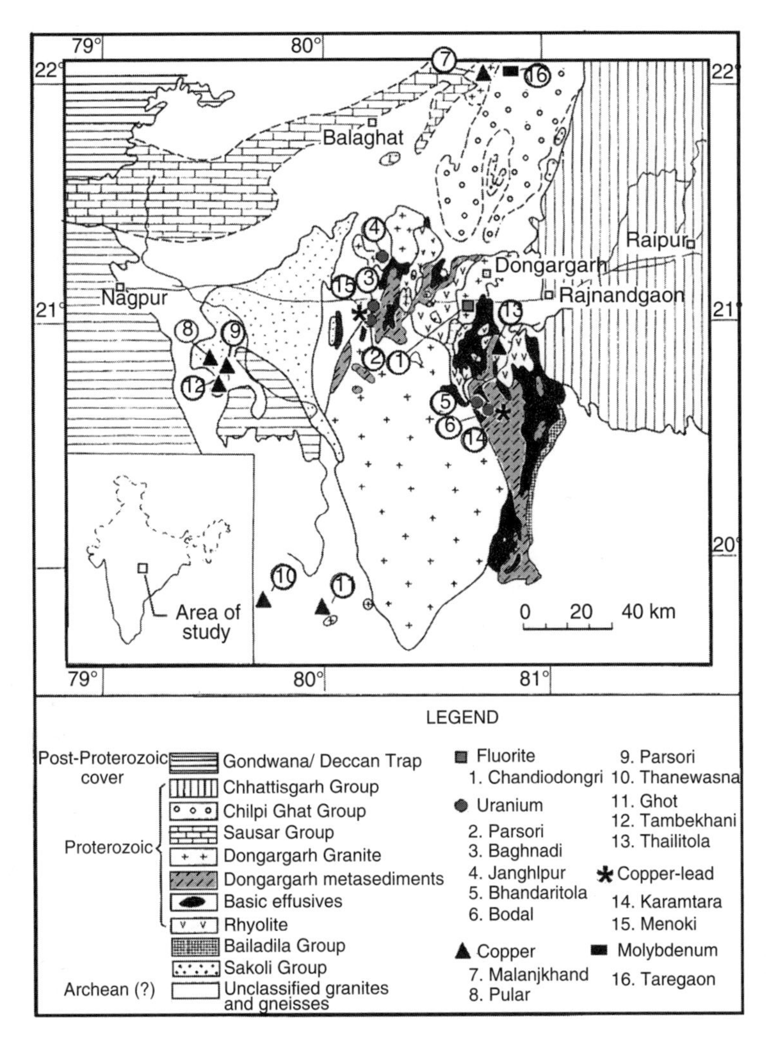

Fig. 2.2.33 Geological map of Dongargarh uranium province showing mineral occurrences (after Mahadevan, 1992)

Uranium Mineralisation in Abhujmar Basin Detailed exploration activity of the Atomic Minerals Directorate (AMD) has led to the identification of two horizons of uranium mineralisation in the Gundal Sandstones in the Abhujmar basin. One of them is located in the upper sandstone/ quartzite below a shale/siltstone cover, particularly at Kerali and Sargipal area. The other is in the pebly quartzite/sandstone, immediately above the basal conglomerate at Bogan and Waler area (Azam Ali et al., 1990). At and about Kerali, about 8 km from basin margin in the southeastern part of the Abhujmar basin, significant uranium anomalies have been located just below a shale/sandstone cover. Grab samples analysed yielded 0.02 to 1.0% U_3O_8 and 63–1827 ppm Cu. Composition and texture of sandstone suggest a mature nature of the sediments. Uranophane and metatorbernite occupy intergranular space and micro-fractures in quartz grains. In the northeastern margin near Bogan, uranium concentration (0.01–0.03% U_3O_8) has been noted in the basal sandstone–conglomerate rocks of suggested fluviatile origin. Older country rocks are apparently the source of uranium in the Abhujmar sediments. However, it is difficult to pinpoint a particular provenance at this stage.

Uranium Mineralisation in and around Chhattisgarh Basin Uranium mineralisation has been located in the southeastern fringe of the Chhattisgarh basin in the following two distinctive geological milieu:

1. The topmost conglomerate-bearing unit of the basal Rehatikhol Formation of Singhora proto-basin hosts polymetallic sulphide mineralisation of pyrite–arsenopyrite–galena, associated with uraninite, pitchblende and coffinite. The mineralisation is closely associated with the cherty/volcani-clastic material near Juba (21°20′15″: 83°15′15″).
2. Uraniferous horizons (upto 1.35% U_3O_8) along NE–SW basement fractures spread over 20 km long and 5 km wide zone (between 21°23′40″and 21°31′45″ : 83°22′35″ and 83°30′35″) close to the southeastern margin of the Chhattisgarh basin, extends from Raigarh district, Chhattisgarh to Bargarh district, Orissa (Fig. 2.2.34).

The second milieu of mineralisation, which has been studied in more detail by Bhairam et al., (1998), reveals some interesting features briefly described here. The host rock of uranium in the linear fracture zone within the basement crystallines is a fine grained brecciated potassic rhyolite (BPR). There is also unbrecciated potassic rhyolite (PR) and quartz veins emplaced along the fracture zone in close proximity of the former, but these are not mineralised.

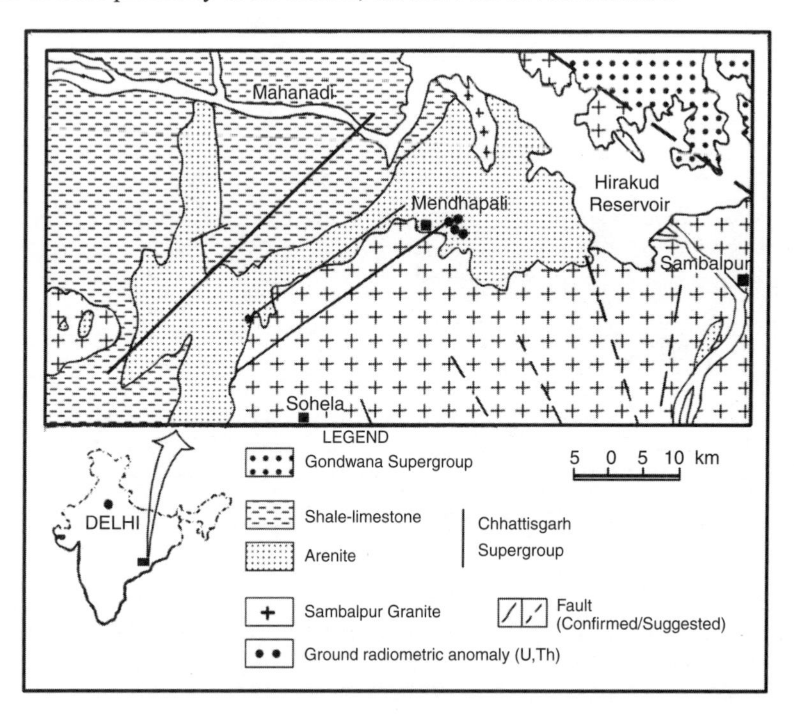

Fig. 2.2.34 Uranium mineralisation at the southeastern margin of Chhattisgarh basin, close to the granitic basement (modified after Bhairam et al., 2000)

The uranium mineralisation in the BPR is in form of 15–400 m long lenses and bands of variable width, displaying a pinch and swell character. The BPR is constituted of angular fragments of potassic rhyolite, granite and glassy material set in microcrystalline quartzose cement. The radioactive minerals are brannerite, torbernite, metatorbernite and secondary uranyl minerals, in association with pyrite, chalcopyrite, anatase, magnetite, fluorapatite, leucoxene, traces of galena, cerrusite, hematite and chlorite. The uraniferous BPR is enriched in K_2O (upto 8.52%), TiO_2 (0.57%), P_2O_5

(up to 1.37%) and depleted in Na_2O (0.5–3.75%) in comparison to the unbrecciated rhyolite. The total iron (~14–17%) and Ba (up to 2298 ppm) in BPR is also much higher than in PR and basement granite. The ore body chemically analysed 0.010–1.35% U_3O_8. Radiometric assay of thorium ranged from < 0.005 to 0.013%.

The unmineralised potassic rhyolite (PR) is geochemically similar to the basement granite. The basement crystallines in this area comprise Sukma Gneiss–supracrustals, invaded by younger Dongargarh Granite (named as Sambalpur Granite in Orissa) dated 2400 Ma (Rb–Sr, WR). The PR comprises fine laths of sanidine, abundant microlites and phenocrysts of embayed quartz with calcite, fluorite, sericite and sphene as accessories. The PR and the host granite are peraluminous adamellites. A few prominent quartz veins are sheared, brecciated and ferruginised giving rise to ferruginous breccia. Secondary uranium minerals occur within the basement granite along the eastern margin of Chhattisgarh basin.

Some workers ascribed the mineralisation to post-Chhattisgarh reactivation of a pre-existing (pre-Singhora) fracture system in the basement, resulting in the formation of BPR and concomitant hydrothermal emplacement of uranium.

Uranium Mineralisation in Western Parts of Chhotanagpur Granite–Gneiss Complex (CGC) The uranium mineralisations in the western parts of the CGC terrain, at the trijunction of the states of Uttar Pradesh (UP), Madhya Pradesh (MP) and Jharkhand, are of interest. The Jajawal occurrences in the Sarguja district of Chhattisgarh and the same in the Rihand valley in the Sonbhadra district, UP, and at the Binda–Nagnaha area of Palamau district, Jharkhand deserve special mention in this context (Fig. 2.2.35). The host rocks trend ENE–WSW with the dominant planar structure dipping northerly. The host rocks are metamorphic crystallines (mica schist, graphitic schists, quartzite and calc–silicate rocks), migmatites and granitic gneisses. They are multiply deformed and affected by several shear zones trending E–W to ENE–WNW. The structures, in general, have a Satpura (≡ CITZ) trend.

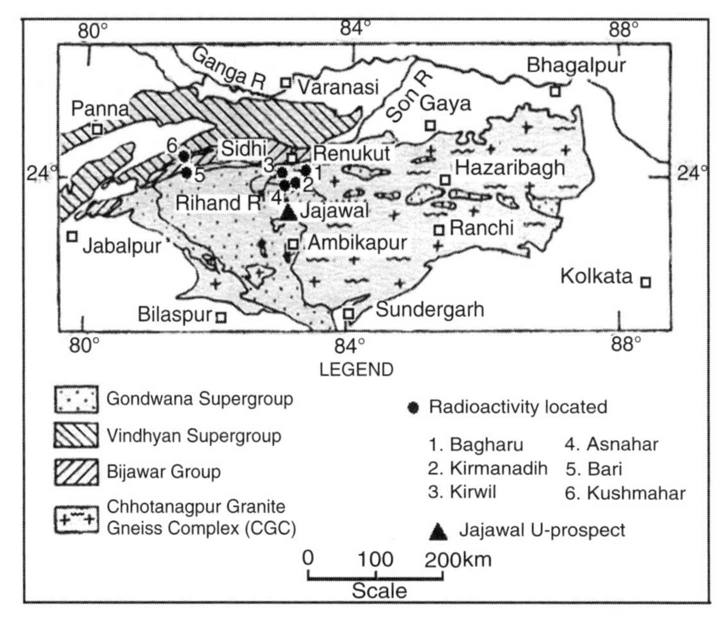

Fig. 2.2.35 Geological map of western CGC in parts of Uttar Pradesh, Madhya Pradesh and Jharkhand showing locations of uranium mineralisation (modified after Bhattacharya et al., 1992)

At Jajawal, the uranium mineralisation occurs as veins of coffinite with minor uraninite along shear zones within granite and associated metamorphic rocks.

At Binda–Nagnaha, the uranium mineralisations, represented by the development of uraninite, brannerite, thorite and allanite, occur in the granitic pegmatites and migmatites (Gupta et al., 1989).

Six samples of granitic rocks from Binda–Nagnaha area and five samples from Jajwal area yielded Rb–Sr (WR) ages of 1242 ± 34 Ma and 1100 ± 20 Ma respectively, both having a $Sr_i = 0.715 \pm 0.002$ (Pande et al., 1986). Mookherjee (1975) obtained a K–Ar age of 880 ± 20 Ma for a biotite from the Dhabi Granite.

In the Rihand valley of Sonbhadra district, U.P, the mineralised zone is fairly extensive (35 x 10 sq km) and is spread from Sidhi in the west to Murratola in the east. Uranium mineralisation in this sector occurs mainly in the migmatites, may be concentrated in the pegmatoid–granitoid leucosomes, and associated biotitic melanosomes. The associated minerals with uraninite in the ore zones are zircon, coffinite, uranophane, rutile, secondary U–minerals: autunite, meta–autunite and the sulphide phases–chalcopyrite, pyrite, molybdenite and rarely pyrrhotite. The highest e U_3O_8 value of 2.5% is reported from Sirsoti. ThO_2 content is usually < 0.1%, except locally higher values (Bhattacharya et al., 1992). Recently, Sengupta et al. (2005) reported an uncommon variety of uraninite containing upto 7.3% CaO from Anjangira area, Sonbhadra district.

The total uranium accumulation in this U-province is large, but it is poorly focused and as a result, is not very attractive for exploitation. It is not clear if all of the uranium and associated 'ore elements' have been contributed by the country rocks, or a part of it came from 'below'. Uranium mineralisation in the western CGC has much resemblance to the same at Rossing, Damara orogenic belt, Namibia.

Polymetallic Mineralisation in Sakoli Fold Belt

The Sakoli fold belt (SFB) is a repository of polymetallic mineralisations comprising stratiform Zn-sulphide, Cu-sulphide ± Au ± PGE and W– mineralisation in the southwestern and south central parts and BIF associated iron ore along the eastern boundary. The older supracrustals in the basement inliers within SFB host local concentrations of sillimanite–kyanite, fluorite, barite, corundum while podiform chromite bodies occur in ultramafic intrusives in the basement at the southern periphery of the belt. Except sillimanite–kyanite and chromite, none of the other prospects has been exploited in the recent past, although exploration data are worth the attention. However, there are quite prominent ancient mining pits in Pular–Parsori and some other locations that were presumably dug for gold. Winning of placer gold still continues in Maru river basin and other streams and rivulets draining the area.

Zinc-sulphide Deposit at Kolari–Bhaonri Stratiform and stratabound zinc-sulphide mineralisation occurs in the metamorphosed and deformed volcanisedimentary sequence of Bhiwapur Formation in the southwestern sector of the Sakoli fold belt. The stratified sphalerite–dominant assemblage is exclusively confined to the bedded layers of spessertine garnet–quartz–biotite–cummingtonite rock, which are interbanded with chert, quartz–chlorite–mica ± andalusite schist, amphibolite (metabasalt), carbonaceous phyllite and tuffs.

In Kolari–Bhaonri area the mineralised horizon extends (strike: N 55° W–S 55°E; dip: 60°–70°due NE) for ~4 km, comprising four parallel ore zones within a length of about 1 km. The other ore minerals associated with predominant sphalerite are minor pyrrhotite, chalcopyrite and sparse galena and pyrite. Magnetite is absent. The ore composition in terms of Zn, Cu and Pb may be expressed as Zn >> Cu > Pb (Zn/Cu ratio upto > 3000 in some cases). Quartz–calcite veinlets containing remobilised sphalerite ± chalcopyrite commonly cut across the stratified sphalerite bands.

The sphalerite–chert rhythmic interbands often display soft sediment structures. The ore bands along with the host rocks are co-folded by the first folds and sphalerite exhibits metamorphic fabric (Bandyopadhaya et al.,1992). Sulphur isotope studies on monomineralic fractions of sphalerite and pyrrhotite separated from the stratified ore have shown average $\delta^{34}S$ values a + 16.99 and + 7.88 per mil respectively (Saha et al., 2001).

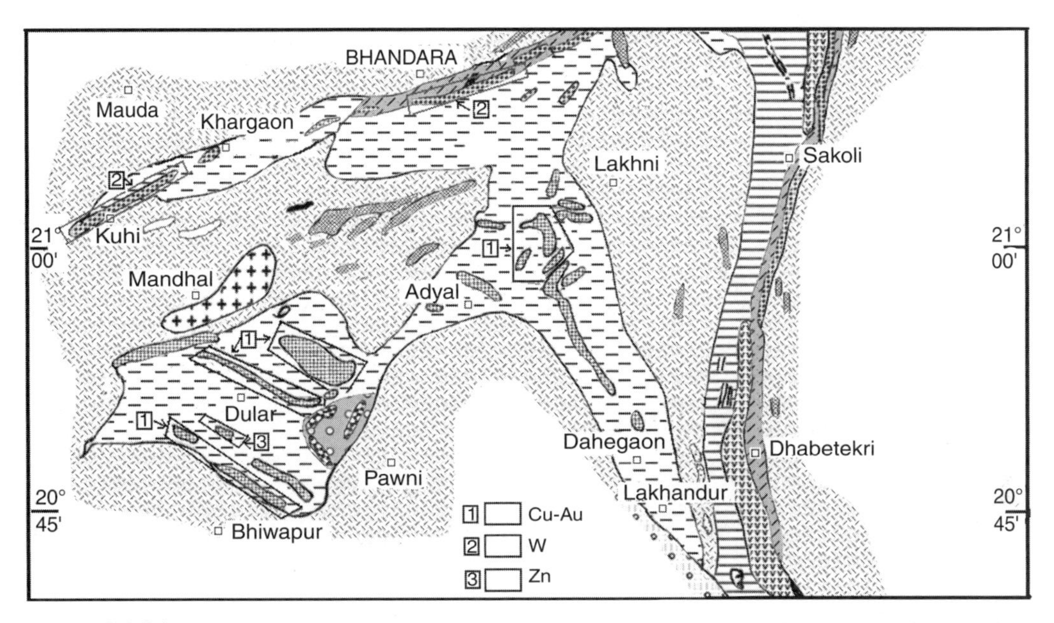

Fig. 2.2.36 The Cu–Au –, W–, and Zn– prospects in Sakoli fold belt (vide Fig. 2.1.11 for legend of lithological units)

The estimated reserve of Kolari–Bhaonri (Zone II) deposit stands at ~4.0 Mt of 5.42% Zn and that of Zone III is 8.3 Mt of 6.79% Zn.

In the Kolari–Bhaonri deposit disseminations and stringers of scheelite in thin quartz veins (1mm to 3 cm) occur parallel to the foliation mainly in the metabasic rocks and Mn garnet-quartz rocks in close proximity of the Zn ore zones. The other ore minerals present in order of abundance in the scheelite bearing veins are arsenopyrite, pyrite and chalcopyrite, all of which occupy the intergranular spaces and fractures in the gangue (Gadadharan and Jog,1990). Saha et al. (2001) reported bedding laminae parallel scheelite mineralisation in the bedded chert. A total reserve of 0.79 Mt of tungsten ore with 0.157% W (equivalent to 2210 tonnes of 65% WO_3 concentrate) over a true width of 1.88 m was estimated over an aggregate strike length of 1380 m down to the depth of 200m in the northwestern part of the area (Gadadharan and Jog op. cit.).

Copper sulphide (± Au, ± PGE) Deposits at Ran Mangli, Pular–Parsori–Thutanbori, Bhimsain Killa and Other Locations Copper sulphide mineralisation, locally associated with gold and/or PGE, are noted at several places in the SFB, which are mostly restricted to the felsic volcanics and to a lesser extent to the adjacent metapelites belonging to the Bhiwapur Formation.

To the north of Bhiwapur, two horizons of felsic volcanics are separated by a ten kilometers wide zone of metapelites and dominantly mafic volcanisedimentary pile. From south to north these are the Ran Mangli-Bhiwapur and the Pular–Parsori–Thutanbori bands, which are considered to be the two limbs of a major west–north–westerly closing synform (Bandyopadhaya et al., op. cit.).

There are two more subsidiary felsic bands at Jogikhera–Pahungaon and Gothangaon. The stratiform Zn-sulphide zone at Kolari –Bhaonri, 3 km away from the southern felsic band of the Ran Mangli–Bhiwapur horizon, occurs at a higher stratigraphic level separated by metapelites and tuffaceous sediments. Similar to the host association of the stratiform Zn-sulphide mineralisation, the Cu-sulphide ore zones and their felsic volcanic hosts are also closely associated with the manganiferous garnet rich rocks (ganetitites and garnet–quartz coticules). Bedded tourmalinites (dravite dominant) are best developed in the metasedimentary partings between the 'stratabound' Cu-sulphide and stratiform Zn-sulphide zones.

Ran Mangli Prospect At Ran Mangli, stratabound copper sulphide mineralisation occurs at the contact of the underlying rhyolite/felsic tuffs and the overlying low-grade metapelites. The mineralisation is in form of impersistent conformable bands, lenses and stringers hosted by chlorite rich top of the felsic volcanics and immediately overlying Mn garnet–quartz rocks. The ore minerals are chalcopyrite with subordinate pyrite and pyrrhotite and minor covellite, sphalerite, cobaltite, bornite, tenorite, cuprite, arsenopyrite and native copper (Bandyopadhaya et al., op. cit.). The main phase of mineralisation is considered as pre-metamorphic by these authors on the basis of deformation and recrystallisation textures displayed by the ore minerals. The effects of later deformation are recorded as crenulation of ore bands along with the enclosing rocks and numerous remobilised sulphide-bearing quartz veins traversing the metapelites. The reserve of Ran Mangli prospect stands at 0.344 Mt of ore with 1.25% Cu.

Pular–Parsori Prospect The Pular–Pasori prospect, marked by old mining pits, is located in the northern felsic horizon, where the copper sulphide dominant mineralisation is hosted by quartz–carbonate veins within sheared rhyolites. Weak and sporadic mineralisation is also noted in the adjacent mica schists. The sulphide assemblage is associated with native gold in the form of minute grains within chalcopyrite and arsenopyrite, elongate to spherical grains within or peripheral to the quartz and other silicate gangue minerals and also in the ramifying carbonate veinlets. The ore assemblage also includes minor quantities of galena, tetradymite, bismuthinite, stannite, cobaltite, wolframite, scheelite, gahnite, cassiterite, franklinite and native platinum (Chattopadhyaya and Saha, 1998; Saha et al., 2001; Mahapatra et al., 2001). The ore reserve of this prospect is estimated at 0.119 Mt with 1.76% Cu. Of late PGE incidences have been noted in the ores samples of Kitari north block, Parsori west and east blocks, Kosari–I block and Marupar blocks of the Pular–Parsori belt.

Thutanbori Prospect The Thutanbori prospect is at the eastern end of the northern felsic horizon, located at about 15 km from Pauni. Here, the copper mineralisation is mainly hosted by chloritic metapelites, often containing magnetite, garnet, andalusite and chloritoid, and also within the subordinate felsic volcanics. Cross cutting quartz–carbonate veins also contain sulphides. The estimated ore reserve of this prospect is 0.845 Mt with 2.24% Cu.

Bhimsain Killa Pahar Prospect In the Bhimsain Killa Pahar prospect, located at about 25 km WSW of Sakoli town and at about 25 km to the northeast of Pular, there is a 900 m long brecciated cherty quartz-vein zone along a brittle shear sub–parallel to the most pervasive foliation (strike NE–SW; dip 60°–70° due SE) within felsic volcanics, which is interbanded with magnetite–andalusite–staurolite bearing mica schist (Mahapatra et al., 2001; Saha et al., 2001a). The individual veins are 5 to 50 m in length and 1 to 3 m in width. The brecciated cherty quartz veins are moderately auriferous, with Au values ranging from 176–1349 ppb (maximum spot value 18 g/t Au) and contain PGE (Pt: 100–590 ppb, Pd: 127–412 ppb, Ir: 3.4–6.2 ppb). In contrast to the wide range of sulphide phases found in the ore assemblages of other felsic horizons of SFB, the associated sulphides in

Bhimsain Killa are very limited and sporadic, dominated by arsenopyrite (often altered to scorodite) with subordinate pyrite, chalcopyrite and chalcocite. SEM–EDX and EPMA studies of the polished sections revealed native gold of upto 10 micron size mainly in quartz gangue and native platinum grains upto 15 microns size dispersed in the gangue and also along the margins or fractures of larger arsenopyrite grains (Mahapatra et al., 2001). Sulphur isotope studies on chalcopyrite and pyrite separates from the ore zone revealed δ^{34} S values in the range of +1.63 and + 2.32 per mil, which is indicative of deep crustal source of sulphur. Good population of both primary and secondary polyphase, biphase and monophase fluid inclusions in the mineralised veins of Bhimsain Killa block II revealed homogenisation temperature (T_h range) of 142.5°–288° C (modal T_h 215°–235° C), moderate to high salinity range of 5.98–36.9 wt.% NaCl equivalent with very low CO_2 (average salinity = 18.68), T_m ice range –78° to –30° C, and mixed salt system Na +K +Ca +Mg with Cl as dominant anion (Saha et al., 2001b).

Ore Genesis Bandyopadhyay et al. (1992) contemplated volcanogenic exhalative origin for both Zn- and Cu- mineralisation along with the associated Au and other elements, in Bhiwapur Formation and categorised it as SEDEX type massive sulphide deposit. Detailed petrology and petrochemistry of the host rocks, sequential evolution of the volcanic products and mode of occurrence and petro-mineralogy of the ore assemblages were taken into account while proposing an intracontinental transtensional tectonic setting for the calc–alkaline felsic volcanism, formation of the protoliths of primary Mn-garnet rich rocks and emplacement of Cu sulphides. The second phase of tholeitic volcanism, after a fairly prolonged interlude of sedimentary accumulations with some exhalative products, was responsible for the generation of the second-generation exhalative Mn-garnet rich rocks and associated stratiform massive Zn-sulphide ores. The sulphide bearing concordant and discordant quartz veins were considered by these workers as the products of remobilisation subsequent to primary ore formation. Most of the other workers (Gadadharan and Jog,1990; Mohan and Bhoskar, 2000; Saha et al., 2001, Mahapatra et al., 2001, Bhoskar et al., 2003) subscribed to a volcanic exhalative origin for the stratiform Zn-sulphide and metabasite-hosted scheelite mineralisation but strongly held that the Cu-sulphide (associated with Au, W, PGE) mineralisation was entirely of hydrothermal origin related to acid volcanism/ acid plutonism, subsequent to the formation of the stratiform Zn-sulphide ores. Disposition of the ore zones parallel to the secondary structural planes, prominent wall rock alteration, the nature of ore mineral assemblage, aided by sulphur isotope and fluid inclusion data on sulphide–gold association in vein quartz are cited in support of this view. On the merit of the data made available by Bandyopadhyay et al. (op. cit.), their views, which proposed the generation of volcanic exhalation related ore fluids during both mafic and felsic volcanism and consequent development of the first generation stratiform and stratabound ores, cannot be outright discarded. However, there is distinct impulse(s) of shear controlled hydrothermal emplacement of Cu–Fe rich sulphides associated with gold, PGE, W in quartz (± carbonate)veins, which may not be merely the products of remobilisation of primary ore during the tectono–metamorphic stage(s), as suggested by Bandyopadhyay et al. (op. cit.). Chattopadhyay and Saha (1998) postulated a possible acid magmatic, mainly plutonic, linkage for the copper–gold and associated mineralisation in at least three stages between 400°C and 271°C (later supported by others based on fluid inclusion studies).The lowest temperature in their opinion could be as low as 175° C, as indicated by the presence of purple fluorite in the mineralised veins.

Tungsten Mineralisation in Kuhi–Khobna–Agargaon Tungsten mineralisation in SFB occurs (i) within tourmalinised and greisenised pelitic schists as concordant bodies, (ii) as concordant and discordant hydrothermal quartz veins in association with basemetal sulphides (± Au ± Ag) hosted by mafic and felsic volcanics and (iii) in skarnoid lenses developed in contact aureoles of carbonate rocks adjacent to intrusive granitic bodies.

Kuhi, Khobna and Agargaon, occurring over a 16 km stretch along the NE–SW trending western limb of the SFB, are the three most prominent tungsten deposits having wolframite and scheelite mineralisations. The associated ore minerals are sphalerite, chalcopyrite and molybdenite. The average chemical composition of the ores is WO_3 – 65.05%, Fe_2O_3 – 18.64%, MnO – 3.73%, SiO_2 – 3.40%, P – 0.04%, MnO/FeO – 0.2%. The total reserve of these deposits is estimated as 11.3 Mt, equivalent to 88,732 tonnes of 65% WO_3. The other occurrences are located at Umrer (only scheelite), Kheripar–Kosamtondi and Pardih–Dahegaon–Pipalgaon areas in different parts of the SFB (Fig. 2.2.36). Fairly prominent scheelite mineralisation also occurs close to the sphalerite rich ore bodies in Kolari–Bhaonri, described above. There are many more incidences of tungsten mineralisation in the area, which are mostly associated with the quartz veins.

At Kuhi–Khobna–Agargaon deposits, the host rocks comprise garnet and andalusite bearing muscovite–chlorite–quartz schist. The wolframite and scheelite mineralisation is mainly confined to tourmaline–quartz– lithium mica greisens, which occur as lenticular bands and lenses, several metres in length and tens of centimeters in width, in the pelitic schists. The greisens show rims of tourmaline around cores of quartz–mica (± topaz, ± fluorite) or only quartz. The tungsten mineralisation is mainly concentrated in the cores and very sparsely in the tourmaline rich outer rims (Mohan and Bhoskar, 1990). The tourmaline in greisens is Fe-rich schorlite and not of Mg-rich variety (dravite) which constitutes the massive tourmalinites of volcanogenic exhalate type described earlier. Minor molybdenite, columbite–tantalite, pyrochlore, struverite, galeno–bismuthite and native bismuth have been reported from the tungsten–bearing greisens and pegmatites in Kuhi–Khobna–Agargaon, Pardih–Dahegaon–Pipalgaon areas and other parts of SFB (Mohan and Bhoskar, 1990; Seetharam, 1990; Saha et al., 2001a).

The workers referred to above are unanimous in relating the wolframite–scheelite mineralisation in greisens with the pneumatolytic–early hydrothermal phase of the post-tectonic granitic activity (Purkabori and Mandhal Granites). The pegmatitic phase of the granite carried tungsten and columbite–tantalite to far away distances (Mohan and Bhoskar, op. cit.). The sulphide associated scheelite mineralisations in quartz veins are considered to be the result of late hydrothermal activity witnessed in different parts of the area. The earliest scheelite mineralisation in the area associated with mafic volcanics, best developed in Kolari–Bhaonri, is assigned an exhalative origin (Gadadharan and Jog, 1990; Bandyopadhyay, et al., 1992). Thus, the multiphased tungsten mineralisation in SFB initiated with the volcanogenic exhalative type found in association with metabasics, followed by acid magmatism related mineralisation in three stages. First was the formation of skarnoid scheelite bodies at 800°–450° C during granite emplacement. The second was the formation of wolframite–scheelite mineralisation related to boron metasomatism and greisen formation at 550°–350°C and the last was the epithermal-mesothermal stage of sulphide–Au–quartz vein associated hydrothermal mineralisation at about 200° C (Saha and Mohan, 2000).

Gold Mineralisation in Mahakoshal Belt

Sporadic gold occurrences are noted at several locations in the volcanisedimentary sequence of Mahakoshal belt in parts of Sonbhadra and Sidhi districts of UP and MP. The mineralisation occurs in different litho-units like phyllites, tuffs, BIF and greywacke belonging to the Agori and Parsoi Formations and is mainly hosted by multiple generations of shear controlled quartz–carbonate veins trending WNW–ESE. At places gold also occurs disseminated in bedding parallel sulphidic arenaceous lamellae alternating with dark grey phyllitic layers. Similar occurrences of gold are also noted further into the west. The gold mineralisation is mostly associated with sulphides (arsenopyrite, pyrite, pyrrhotite, galena and chalcopyrite) and shows positive affinity with some other elements too like Cd, Ag, Ba and Sr. The association of Pt and Pd with gold bearing sulphides have also been recorded at places (Rastogi et al., 2001). Some other subordinate modes of gold mineralisation are noted in the central parts of the belt (north of Jabalpur), where auriferous quartz carbonate veins occupy NNW or NNE trending en echelon fractures in dolomite (Imalia block, Sleemanabad) (Fig. 2.2.37), upper vesicular layer of metabasaltic flows carry arsenopyrite and gold (Shahdar area, Sleemanabad), auriferous quartz veins penetrate the metasediments and post-tectonic granite along the southern flank of the belt, and also the incidences of gold in the polymictic conglomerate of Sleemanabad Formation (Soni and Jha, 2001).

Innumerable old workings (mostly open pits and some underground excavations), huge waste dumps and grinding stones strewn over such locations testify to the ancient mining activity along this belt. A large number of prospects, mostly marked by old workings and/ or positive surface shows, were investigated by surface and sub-surface probe (geophysical survey and drilling) all along Son valley in UP and its western extension in Sidhi district, MP, but none were proved to have much economic significance. In the eastern parts of the Mahakoshal belt, the Gurhar Pahar, Gulaldih and Sona Pahari prospects in the Son valley attracted much attention due to their appreciable strike length (~3 km) and extensive old workings but average tenor of gold in the ore zones (not too wide) did not exceed ~2.0 g/t. Dwivedi et al. (2001) advocated deeper probe in Sona Pahar for locating better gold concentration as the association of As, Pb, Sb and Ag and the presence of quartz–calcite–barite veins indicate that the exploration so far has been limited to shallow depths that represent the supra-ore zone.

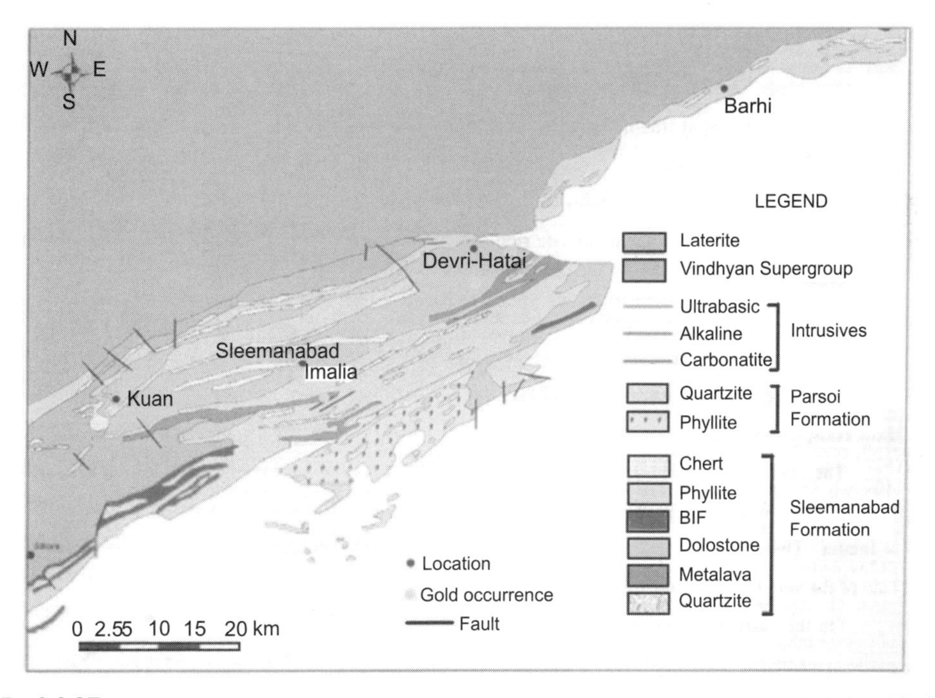

Fig. 2.2.37 Geological map of Imalia gold prospect, district Sleemanabad, Madhya Pradesh (modified after Tripathi, 2008)

Under the microscope gold (native or electrum) is seen to occur as tiny grains (5 μ to 100 μ) in the sulphides (preferably in arsenopyrite, pyrrhotite and pyrite) as well as in the gangue. Other native elements are silver and platinum. Silver commonly occurs as inclusion in galena and sphalerite. Two drill sections in Gurhar Pahar revealed 24 m and 5 m wide Pt +Pd bearing auriferous zones, having average Pt + Pd values of 58 ppb and 109 ppb with associated average gold values of < 2.0 g/t.

Sulphur isotope studies on 15 drill core samples from Gurhar Pahar shows a narrow maximum range of $\delta^{34}S$ values from + 7.81 to + 9.81 per mil in pyrite and +9.16 to +14.45 per mil in arsenopyrite. Rastogi et al. (op. cit.) preferred to relate the mineralisation with a homogenous epigenetic hydrothermal source, rather than considering the possibility of homogenisation of isotopic composition during the low grade metamorphism witnessed here. The fluid inclusion studies conducted on 10 samples, collected from Gurhar Pahar, Gulaldih and Sona Pahari prospects, indicate that the mineralising solution was low to moderately saline (2–14 wt.% NaCl equivalent) and was of moderate density of 0.9 g/cm, and the mineralisation took place at about 3.3 kb pressure and 250–300° C temperature (Prasad et al., 2000; Rastogi et al., op. cit.). Talusain (2001) believes that the mineralisation is closer to the Carlin type. But it appears a little too simplistic a conclusion at the moment.

(Zn–Cu–Pb) Sulphide Mineralisation in Betul Belt

The high magnesian and calc–magnesian schists interleaved with the bimodal acid–basic volcanic suite host the massive stratiform zinc–copper–lead ores occurring over discontinuous stretches in the eastern parts of the Betul belt, between Chhindwara town in the east and Chicholi in the west. The schistose host rocks are represented by anthophyllite–cummingtonite–talc–tremolite schist and garnetiferous biotite–chlorite ± gahnite± staurolite schist. The mineralised schists are closely associated with felsic tuffs and rhyolite. Pillowed and vesicular metabasalts occur as linear bands. The general strike of the belt is ENE–WSW with moderate to steep southerly dip of the formations (Fig. 2.2.38).

A number of small deposits have been located and explored in this belt, viz, Bhawratekra, Kehalpur, Banskhappa–Pipariya, Kherlibazar–Bargaon–Tarora, Ghisi and Mauriya. Some ore bodies, yet to be explored in detail, have also been located in the eastern extremity of the belt at Koparpani and Borkhap. A syn-volcanic hydrothermal alteration (pipe) zone in felsic volcanics with zinc dominated sulphide mineralisation has recently been located in the southeastern parts of the belt near Bhuyari village in Kanhan river section, Chhindwara district (Golani et al., 2006).

The lenticular/tabular ore bodies in the explored blocks are concordant with the regional schistosity, mostly varying in strike length between 250 to 800 m, in thickness from 5–6 m and dip wise extension up to 120 m. The ore reserves and grades estimated for some of these deposits are furnished below (Mahakud et al., 2001).

1. Bhawratekra block–I: 1.58 Mt ore with Zn – 4.04%, Cd – 85 ppm, Ag – 3.5 ppm
2. Bhawratekra block–II: 0.66 Mt ore with Zn – 7.63%
3. Banskhappa–Pipariya block: 3.12 Mt ore with Zn – 2.135%, Cu – 0.55%, Cd – 45 ppm
4. Kherlibazar–Baragaon block: 1.58 Mt ore with Zn – 4.54% (0.66 Mt x 7.63% Zn)

The mineralisation in Bhawratekra-Kehalpur-Banskhappa are stratiform and compact 'massive sulphide' type while at other places it is in the form of schistosity parallel stringers and blebs of ore minerals. The tabular ore bodies range in length from a few metres to tens of metres and in thickness from a few centimetres to tens of metres. Sulphide-bearing quartz veinlets are seen to cut across the massive ore bands at places. The footwall sediments of the stratiform sulphide zones are often marked by silicification. Felsic tuffs of a few centimetres thickness are interbanded with the mineralised schists at all places. Sphalerite is the dominant sulphide associated with chalcopyrite, pyrite, pyrrhotite, and galena, which occur as discrete grains or clusters of grains. The other ore minerals present are ilmenite, hematite, rutile and scheelite. The barium content in the sulphide rich zones varies from 190 ppm to 0.33%. The sulphide phases often display mutual inclusions and deformation twins. The ore minerals also occur as inclusions within staurolite, garnet and gahnite. Zn-rich and Mg-poor gahnite [zincian spinel, $(Zn, Fe, Mg)Al_2O_4$], which is associated with sphalerite-bearing rocks and almost absent in the barren horizons, could be produced by the breakdown of sphalerite during the prograde metamorphic event (Spry, 1987). Ghosh et al. (2006) also believe that gahnite in this metamorphosed massive sulphide (MMS) deposit is the product of desulphidation of sphalerite during the metamorphism. Parveen and Ghosh (2007) reported multiple generations of gahnite from Bhuyari prospect, relating to the pre-metamorphic volcanogenic hydrothermal stage and syn-metamorphic development.

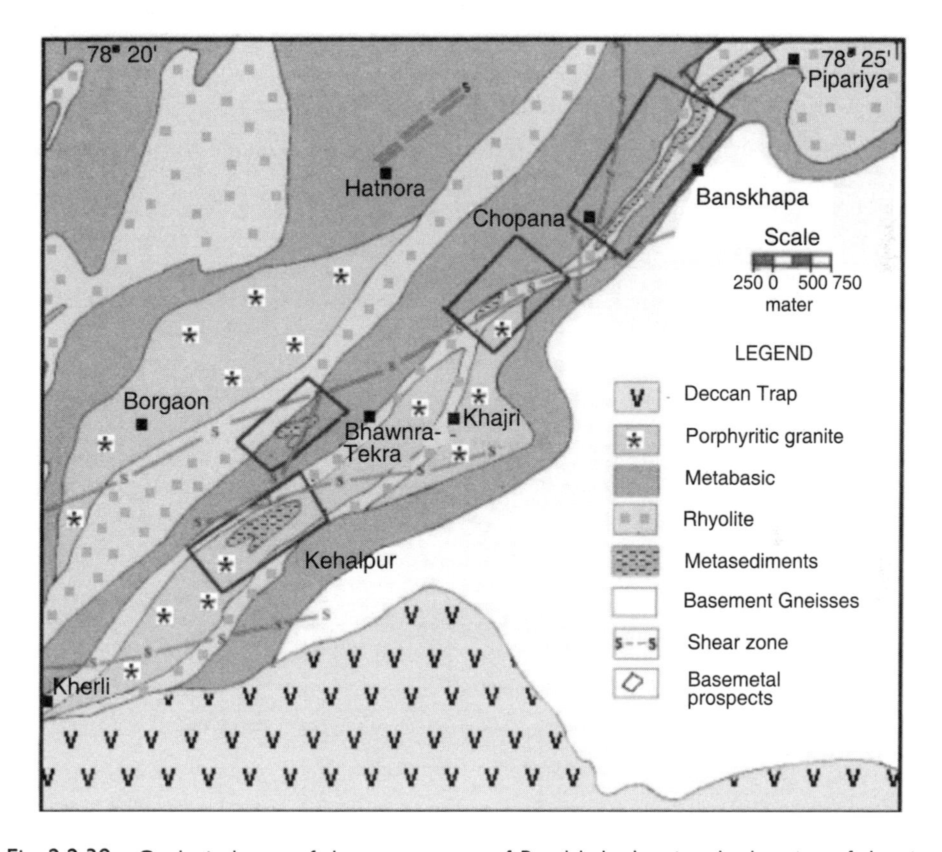

Fig. 2.2.38 Geological map of the eastern part of Betul belt showing the location of the zinc sulphide prospects

Although not much data are available, the overall features of the deposits are suggestive of a submarine volcanic exhalative origin (Mahakud et al., op. cit.). Parveen et al. (2007) have recently classified the prospects of this belt into two groups (Zn–Pb–Cu and Zn–Cu types) of volcanic hosted massive sulphide (VHMS) deposits (Fig. 2.2.39).

Cu
Pb
Zn
Cu type
CR=60
ZR=60
ZR=90
Zn-Pb-Cu type
Zn-Cu type
△ Dehalwara
▲ Mauriya
+ Ghisi
× Bhuyari
○ Bhawnra Tekra
● Bargaon-Tekra
□ Banskhapa-Piparya
■ Koparpani

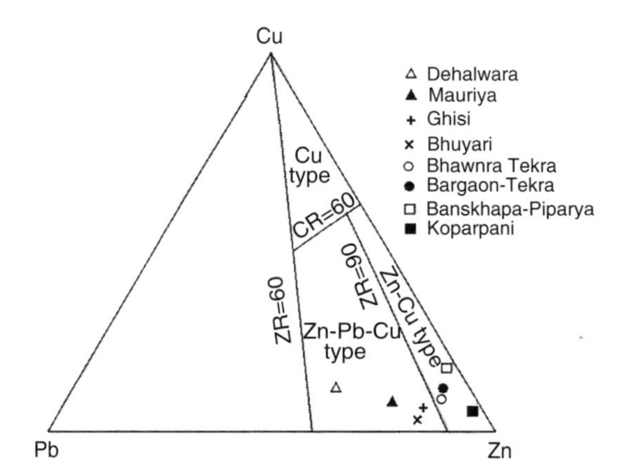

Fig. 2.2.39 Fields with mean grades of eight prospects of VHMS deposits in the Betul belt (after Parveen et al., 2007)

Manganese Mineralisation in the Sausar Group of Rocks

Manganese mineralisation has taken place at different Archean–Proterozoic rocks in India such as in the Dharwar craton (already discussed), Eastern Ghat belt, Sausar belt, Gangpur basin, Mahakoshal belt, Aravalli belt (including Jhabua area) and Pranhita–Godavari Valley. Of these the Sausar Group of rocks constituting the Sausar belt, is the most important. This curvilinear belt is about 200 km long (Fig 2.1.16) and has been the repository of 31.70 Mt of estimated Mn-ore reserves, including 24 Mt of recoverable ore. The inferred reserves of all categories of Mn-ore in the belt total to about 142 Mt (Mulay, 2001).

Within the Sausar Group, Mn-bearing rocks and ores are conformably enclosed in the pelitic Mansar Formation and the calcareous Lohangi Formation in more than one horizon (Fig. 2.2.40). More intense manganese mineralisation took place in the lower two horizons in the Mansar Formation. Pyrite and pyrrhotite bearing graphite schists occur below the manganese ore bodies and near the base in the Mansar Formation. In the Lohangi Formation, the manganese ore deposits present are also conformant with the host carbonate rocks, but the mineralisation as a whole, is weaker than in the Mansar Formation.

Chhind-wara Dst.
Nagpur Dst.
Bhandara Dst.
Balaghat Dst.
Gowari Wadhona Maharku -nd area
Mohagaon area
Junewani area
Chikla-Sita -saongi Dongri Buzurg area
Tirodi Sitapato-re area
Ram -rama -Netra area
Bharweli -Ukwa area
Bichua Fm
June wani Fm
Chorbaoli Fm.
Mansar Fm.
Lohangi Fm.
Sitasaongi Fm.
Tirodi Biotite Gneiss
LEGEND
Bichua & Junewani Fm
Chorbaoli Fm.
Mansar Fm.
Lohangi Fm.
Sitasaongi Fm
Tirodi Biotite Gneiss Fm
Manganese ore

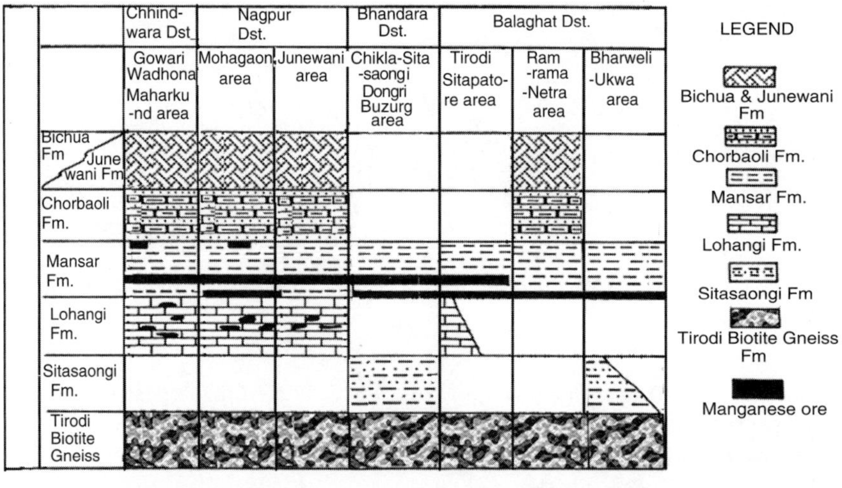

Fig. 2.2.40 Synoptic diagram showing stratigraphic columns of different sectors of the Sausar manganese belt, Maharashtra and Madhya Pradesh (after Dasgupta, 1992a)

Geographically, manganese mineralisation occurs at a number of places, distributed in four districts of Madhya Pradesh and Maharashtra, such as:

1. Chhindwara district, MP (Gowari–Wadhona, Kachi Dhana areas);
2. Balaghat district, MP (Tirodi, Sitapathar–Sukri, Ramrama–Hurki–Netra and Bharweli–Laungur–Ukwa areas);
3. Nagpur district, Maharashtra (Kondegaon, Gumgaon, Ramdongri, Kandri, Mansar areas),
4. Bhandara district, Maharashtra (Chikla, Sitasaongi and Dongri Buzurg areas).

In all these areas manganese deposition took place in one or more horizons (zone I: contact of Mansar with underlying Lohangi /Sitasaongi Formations, zone II: within Mansar Formation and zone III: contact of Mansar and overlying Chorbaoli Formations). In the Chhindwara and Nagpur districts, manganese mineralisation has in addition taken place within the Lohangi Formation. All the Formations of the Sausar Group have not developed at all the places. The manganese bearing Mansar Formation, however, is an exception in being more extensive. The Mn-ores occur as bedding parallel bands of varying thickness and also as schistosity parallel layers, pockets, patches, stringers and lenses interlayered with the Mansar metapelites and psammopelites, gondite bearing quartzite, Sitasaongi feldspathic quartzite and Lohangi carbonates. The Mn-mineralisations in the belt are of primary (sedimentary and diagenetic) bedded type, schistosity controlled metamorphosed type, granite and pegmatite related hydrothermal type and secondary supergene type in oxidation zones. The last type of ore is characterised by the higher oxides. Lippolt and Hautman (1994) reported Ar^{40}/Ar^{39} age of 960 Ma from cryptomelane of Sitapur mine and interpreted it as the time of cooling through cryptomelane closure temperature, following the amphibolite facies metamorphism.

There are more than 200 individual Mn deposits in the Sausar belt, of which about 20 deposits account for nearly 90% of total production. The proved and probable reserve of 17 major deposits is about 20 Mt. The Dongri Buzurg (6.22 Mt) and Chikla (4.71 Mt) are the largest working mines in Maharashtra while Ukhwa in Madhya Pradesh (12.50 Mt) is the single largest Mn deposit in the country, having 5 km strike length, dip wise extension of 600 m and average ore zone thickness of 3 m. All the major deposits in this belt are under the lease hold of Manganese Ore India Ltd. (MOIL). The GSI, besides estimating the ore reserves in all the MOIL leaseholds, also established additional reserves of 4.3 Mt and 4.0 Mt of Mn ores in 9 freehold blocks and 5 private leaseholds (Mulay,2001)

An average Mn-ore of this belt contains 41.50% Mn, 10.26% SiO_2, 4.68% Fe and 0,18% P. Phosphorus content shows a tendency to increase with higher grade of ore. The average composition of high-grade Mn-ore in Balaghat district of MP and Nagpur and Bhandara districts of Maharashtra are tabulated below (Table 2.2.20).

Table 2.2.20 *Average composition of high-grade Mn-ores in Balaghat, Nagpur and Bhandara districts, Madhya Pradesh and Maharashtra (B.K. Bandyopadhyay, personal communication)*

Element/Oxide%	Balaghat	Bhandara	Nagpur
Mn	51.0	50.5	52.5
Fe	7.0	7.5	5.0
SiO_2	6.5	8.0	11.0
P	0.10	0.16	0.19
Manganese peroxide	55.0	28.0	25.0

In rocks characterised by thermo–tectonic modifications, such primary structures as ripple marks, cross laminations and desiccation cracks have locally been noted (Roy, 1981). Penecontemporaneous deformation structures (PCDs), including convolute, slump folds and sedimentary pull-aparts are developed on the primary interbands of chemogenic Mn-oxide ores hosted by the carbonates of Lohangi Formation (Bhowmik et al., 1997; B.K. Bandyopadhyay, personal communication) (Fig. 2.2.41 a,b).

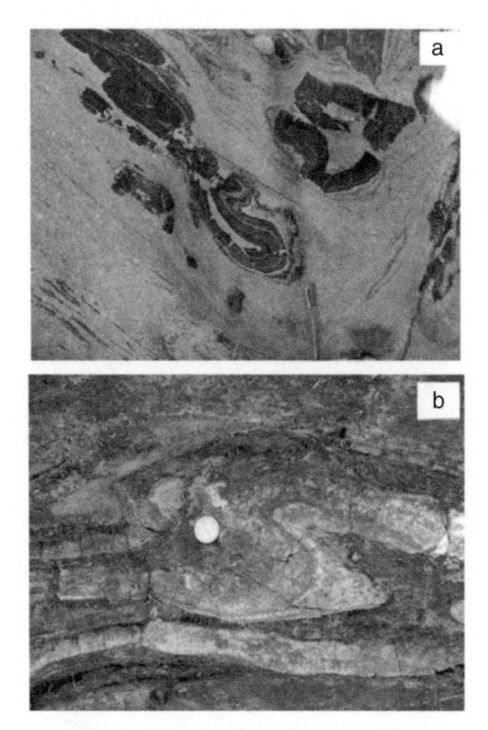

Fig.2.2.41 (a) Penecontemporaneous deformation structures in Mn-ore (dark) bearing carbonate host, Lohangi Formation; (b) Folding in bedded Mn-ore(dark) in Mansar Formation (photographs by B.K. Bandyopadhyay, GSI).

The rocks of Sausar belt hosting the manganese mineralisation are metamorphosed, the grade varying from lower greenschist facies (chlorite–biotite grade) in the south to upper amphibolite facies (sillimanite grade) in the north (Narayanaswami et al., 1963; Sarkar et al., 1977; Roy, 1981). There is also regional lowering of metamorphic grade from west to east along the belt (Mulay, 2001). The ores are deformed and metamorphosed along with the enclosing rocks, except the supergene type.

Five types of manganese rich rocks have been recognised in the Sausar Group, based on their mineralogical associations (Dasgupta et al., 1990). These are:

1. Manganese oxide rocks (Mn-ore bodies)
2. Manganese silicate rocks with minor Mn-oxides
3. Manganese oxide rocks with minor Mn-silicates
4. Manganese silicate–carbonate rocks with minor Mn-oxides
5. Manganese oxide–carbonate rocks

Associations (4) and (5) occur as lenses within manganese oxide ores in random dispositions and as such, do not constitute 'zoned deposits' described by Force and Canon (1988). Interestingly, manganese carbonate-bearing rocks are absent in the Lohangi Formation which is carbonatic. Compositional bandings are commonly present in these rocks. They have been deformed along with the country rocks.

Petrology of the Mn-rich Rocks The banded Mn-oxide ores show alternate bixbyite rich and braunite rich layers separated by silicate/carbonate rich layers. Within the bixbyite-rich bands, the thin laminae comprising coarse aggregates of hematite, often showing deformation twins, represent primary hematite. The braunite rich layers contain the relics of bixbyite. Hausmanite occurs as veins and stringers replacing both bixbyite and braunite. The supergene Mn ore assemblage is marked by the breakdown of the primary braunite and the metamorphic minerals spessartine, rhodonite and other silicates, removal of silica and the development of higher oxides of manganese like pyrolusite, psilomelane, cryptomelane, etc.

Mn/ Fe ratios in the bulk composition of the five ore types described above vary between 5 and 15. Projected in the background of a crustal average of 0.02 and 0.05 in the Mansar metapelites and Lohangi calc–silicate rocks (Dasgupta et al., 1990), it is a case of strong fractionation between Mn and Fe during the deposition of the above-mentioned Mn-rich rocks. The characteristic mineral composition of the five types of Mn- rich rocks mentioned above are given in the following Table 2.2.21.

Table 2.2.21 *Mineralogical compositions of the Mn-bearing rock types in the Sausar Group (Das Gupta et al., 1992)*

Type No.	Type of rock	Mineralogical composition
1	Manganese oxide (ore)	Braunite + bixbyite + hematite + hollandite ± jacobsite + quartz +barite + apatite
2	Manganese silicates with minor oxides	Spessertine + quartz ± braunite + hematite ± pyroxmangite ± jacobsite + Mn-cummingtonite + quartz + barite + apatite
3	Manganese oxide with silicates	Braunite + jacobsite + hematite + quartz ± Mn-cummingtonite ± spessertine + barite + apatite
4	Manganese silicate–carbonate	Mn–Mg clinopyroxene + pyroxmangite + Mn–Mg amphibole + tephroite ± jacobsite ± braunite ± quartz + rhodochrosite + kutnohorite + Mn-calcite +Ca-strontianite + baritocalcite

Type–3 evolved through deoxidation equilibria and the types 4–5 through decarbonation–oxidation equilibria during prograde metamorphism upto the upper amphibolite facies (Bhattacharya et al., 1984; Dasgupta, 1997; Dasgupta et al., 1989; Dasgupta et al., 1992) (Fig. 2.2.42).

The country rocks associated with the types 1–3 above, are commonly referred as 'gondites' a name given by Fermor about a century ago (Fermor, 1909). In Fermor's definition gondites are regionally metamorphosed manganiferous sediments of strictly non-calcareous character, made up essentially of spessertine and quartz with or without other manganese silicates: Later Roy (1966) reported Mn-bearing pyroxene and pyroxenoids (rhodonite, bustamite, blanfordite, Mn-diopside) amphibole (tirodite, winchite), mica (manganophyllite, alurgite and epidote (piedmontite) Because of association of the ores with such rocks, these ores are often referred to as gonditic ores. Metamorphosed manganese ores and gonditic rocks often interband (Fig. 2.2.43).

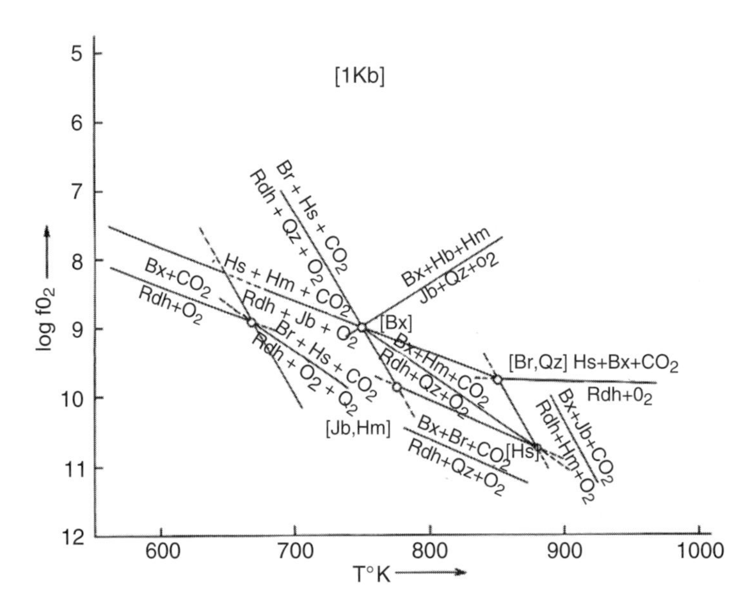

Fig. 2.2.42 Phase relations in Mn–Fe–Si–C–O system at 1 kbar. Qz – quartz, Hs – hausmannite, Br – braunite, Jb – Jacobsite, Bx – bixbyite, Hm – hematite, Rdh-rhodochrosite (Dasgupta, 1997)

Fig. 2.2.43 Bedded manganese oxide ore (Mn) and gondite (Gd), Sausar belt (photograph by B.K. Bandyopadhyay, GSI)

The metamorphosed manganese ore bearing rocks of the Sausar belt are generally granular to schistose in fabric (Fig.2.2.44 a–c). The effects of metamorphic recrystallisation in the manganese ore minerals and their remobilisation are widely recorded in the belt, some of which are illustrated in Fig. 2.2.45 (a–d).

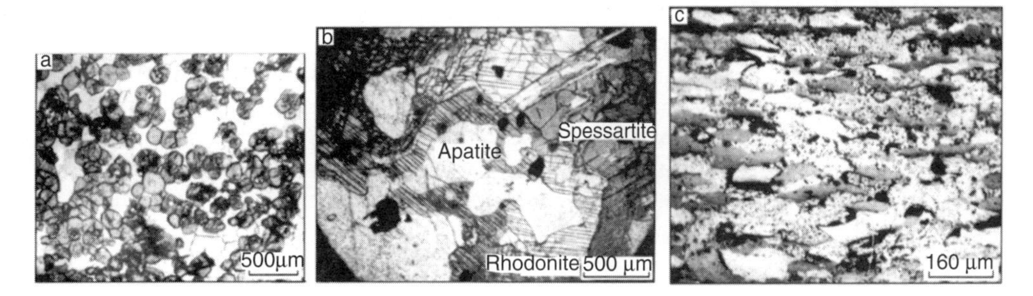

Fig. 2.2.44 Photomicrographs showing (a) granular texture of gondite, (b) Mn-oxide ore with apatite, (c) schistose texture of Mn-oxide ore (Roy, 1966)

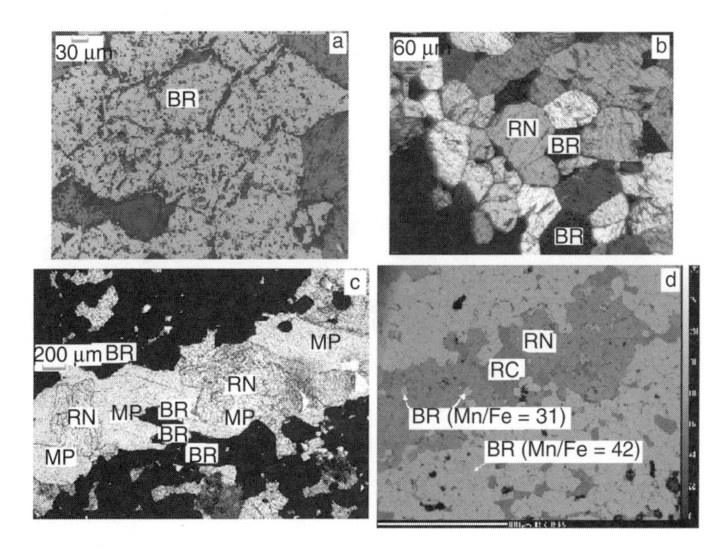

Fig. 2.2.45 (a) Braunite (BR) showing polygonisation and triple–point, Satak Mine, Sausar Belt; (b) Rhodonite (RN) and braunite (BR) showing polygonisation and triple point, Beldongri, Sausar belt; (c) Vein of manganophyllite (MP)–rhodonite(RN)–braunite (Br) in braunite-rich layer, Beldongri, Sausar belt; (d) Variation in Mn/Fe ratio between primary braunite and remobilised braunite in a vein composed of rhodochrosite (RC) and rhodonite (RN) (courtesy: B.K. Bandyopadhyay, GSI)

Fermor (1909) discussed remobilisation of manganese in the weathered crust of the metamorphosed deposits belonging to the Sausar Group. The supergene enrichment is observed down to shallow depths (3–5 m), except locally to deeper extent along shears, faults and joints. Most of it is mined out by now. However, Roy (1981) drew attention to the large manganese deposit at Dongri Buzurg, Maharasthra, composed of low temperature higher oxides of manganese such as, pyrolusite, cryptomelane, manganite and coronadite with relict braunite and jacobsite. The deposit extends E–W for about 1.5 km following the trend of the country rocks. At both the eastern and western ends, the Mn-dioxidic deposit of Dongri Buzurg gradually merges into the usual metamorphosed lower oxide ore bodies (braunite–bixbyite–hollandite–jacobsite–hausmanite) and gondite. Straczek et al. (1956) estimated a depth continuation of the Dongri Buzurg deposit down to 130 m or more from the surface outcrop, which is substantially below the present water table. In contrast, the Chikla and Sitasaongi deposits, which occur only kilometers away, show shallow oxidation only. Straczek et al. (1956)

suggested that the Dungri Buzurg deposit was formed by supergene enrichment of braunite–rich ore and gondite during an ancient weathering cycle. If so, Roy (1981) reasonably wonders why and how the other manganese deposits around Dongri Buzurg could escape this exogenic modification. A hydrothermal alteration process could be a good contender for bringing about the changes that we now see at Dungri Buzurg. One may hold the syntectonic granite–pegmatite intrusion associated with manganese mineralisation along the footwall (Fig. 2.2.46) to be the source of hydrothermal solution. However, this possibility needs to be studied in necessary details in future. Dongri Buzurg is not only a large volume deposit but also a rich deposit containing 50–51% Mn (ROM).

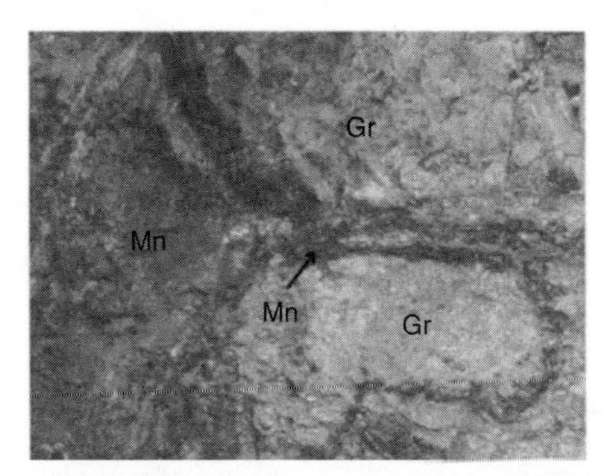

Fig. 2.2.46 Mn-oxide ore (Mn) as veins and patches within pegmatoid granite (Gr), in the footwall of the Main ore zone, Dongri Buzurg mine (photograph by B.K. Bandyopadhyay, GSI)

Deposition of Manganese-rich Sediments Roy (1966, 1981) and Dasgupta et al. (1992) suggested that the metamorphosed manganese oxide ± manganese silicate rocks (Types 1–3 referred to above) are metamorphic products of sedimentary (–diagentic) protoliths that contained Mn-oxide and silica, clay minerals, Fe–oxy–hydroxides in varying proportions. The initial manganese deposition in the carbonate-rich Lohangi Formation was also sedimentary (–diagenetic) in origin. Their depositional models are outlined below.

Manganese Oxide Sediments in the Mansar Formation The type–1 manganese oxide sediments were deposited on oxidised substrate in detritus-starved basin margins. Occasional influx of detritus and its co–precipitation with the manganese from the basin water, produced sediments for the type–2 and type–3 assemblages. Presence of phosphate (apatite) and sulphate (barite) along with the manganese minerals in these assemblages suggest upwelling of reduced manganese from the anoxic low pH deep waters on to the oxidised continental margin, i.e. above the anoxic–oxic boundary and precipitation there (Force and Cannon, 1988). There is no volcanic rock in the sequence. Development of carbonaceous shale (now graphitic schists) with Fe-sulphides (pyrite, pyrrhotite) in dissemination suggests a possible mechanism of partial separation of manganese and iron. Frakes and Bolton (1992) based on the observations on the Eh-control on MnO_2 precipitation, suggest that an Eh value of + 0.60V is critical in determining manganese oxide precipitation in a geological environment. There are two major horizons in the Mansar Formation (Fig. 2.2.40) containing manganese ores. In case of the Kalahari manganese deposit each manganese bed has been correlated with a major transgressive event (Schissel and Aro, 1992). However, later workers (Cornell and Schutte, 1995) have contradicted it. In fact, there is a controversy whether the manganese precipitation will maximise during transgressive or regressive events (Frakes and Bolton, 1984, 1992; Force and Canon, 1988; Schissel and Aro, 1992). Nevertheless, what is important, as

Force and Cannon (op. cit.) emphasised, is that the redox interface intersects the ocean floor on basin margins or shoals, so that sediment does not have to fall through the reduced zone and get redissolved. Transgression or regression whichever will facilitate this situation, will ultimately be the cause of manganese deposition.

Manganese Carbonate Deposits in Mansar Formation Manganese carbonate sediments in the Mansar Formation occur as isolated pockets within the dominant manganese oxide sediments. It is a unique situation. These were not deposited on reduced substrates, nor are they parts of zoned manganese deposits (cf. Force and Cannon, 1988). Dasgupta et al. (1992a) suggested that these manganese carbonate bodies developed by replacement of Mn-oxides, whenever evaporative conditions developed in isolation. Such a conversion of Mn-oxide to Mn-carbonate is reported from number of places. The suggested reaction is:

$$MnO_2 + CaCO_3 + 4H^+ \leftrightarrow MnCO_3 + 2H_2O + Ca^{+2}$$

The reaction is supposed to be very common (Johnson, 1982). The protolith of the type–5 assemblage developed by the above means. The protolith of the type–4 assemblage contained clay minerals and silica in addition to manganese oxide. Otherwise, the diagenetic conversion of $MnO_2 \rightarrow MnCO_3$ was the same.

Manganese Oxide Deposits in the Lohangi Formation Calcitic marbles or calc–silicate rocks constitute the Lohangi Formation dominantly. MnO_2 precipitated on a calcareous substrate when Mn^{+2}-enriched anoxic seawater upwelled over to the anoxic–oxic interface in the shelf zone. The lack of manganese carbonate in these deposits is explained by the ambient high Eh that prevented reduction of Mn-oxides (Dasgupta et al., 1992a).

Phosphorite Deposits in Madhya Pradesh

Jhabua Phosphorite Deposit Phosphorite deposit in Jhabua district, Madhya Pradesh is located in the west of Madrani (23° 00′:74° 27′), at about 22 km from Meghnagar Railway Station. It occurs within theAravalli sequence, may be at a stratigraphically younger horizon than the Jhamarkotra phosphorite deposit in the type Aravallis of southern Rajasthan. Phosphorite in this area is disposed as bands, lenses and irregular bodies within both stromatolitic and non–stromatolitic dolomitic limestone-chert sequence (Munshi et al., 1974; Banerjee and Basu, 1979; Munshi and Khan, 1981).

The phosphorite body is exposed between Amliamal (23°00′:74°25′) in the north and beyond Khatamba (22°58′:74°26′) in the south, over a NNW–SSE trending strike length of 4.5 km with varying dips of 50 to 70 due east and west. The deposit is divided into several blocks, named as Amliamal block, Kelkua block, Khatamba north (strike length –1100 m) and Khatamba south (strike length –1600 m) blocks (Munshi and Khan, op. cit.). Further south of the above area, isolated occurrences of phosphorite are noted upto the south of Rambhapur.

Jhabua phosphorites were probably formed in protected shallow tidal to intertidal waters. While some stromatolitic assemblages have been recognised as ubiquitous, some deposits display specific and restricted algal forms. Geochemical studies and paleogeographic postulations indicate that the water chemistry in various paleodepressions, where phosphorite accumulated, supported a primary algal-induced biochemical origin for the microsphorites in these paleobasins (Banerjee et al., 1980). The phosphorite ore bodies in the Khatamba blocks occur at two levels and are associated with both stromatolitic dolomitic limestone and cherty quartzite. The phosphate in the lower horizon is restricted to the body framework of the stromatolites in dolomitic limestone and as such, the grade is highly variable (5% to +30% P_2O_5), dependant on the concentration of stromatolites. Reserve of 12.5 Mt with average grade of 12% P_2O_5 has been estimated for the dolomite–hosted ore zone, which includes smaller reserves of richer ore with15–20% P_2O_5. The upper phosphatic horizon is cherty and it overlies the dolomitic horizon. The cherty beds are also stromatolitic at places but in this zone, the phosphate is concentrated both in the body framework of stromatolite as well as in the matrix, resulting in richer ore bodies of greater extensions. A reserve of 14 Mt with average 20% P_2O_5 has been estimated for the cherty ore zone. Out of the total reserve of 26.5 Mt, about 4 Mt is with +27% P_2O_5 content (Prasad and Rao, 1999).The deposit is being mined by the MP State Mining Corporation with the present production of 100,000 tonnes of ore /annum, containing + 24% P_2O_5. The beneficiation plant under construction is expected to significantly upgrade the low grade ores and augment the production.

Hirapur and Adjacent Phosphorite Deposits The Hirapur deposit lies at about 2 km southwest of Hirapur village (24°22′:79°12′) in the Sagar district of Madhya Pradesh. The Bijawar Group of rocks comprising ferruginous shale, sandstone, dolomite and phosphatic breccia are exposed in the area. The phosphorite occurs mainly in association with dolomite. The GSI estimated probable reserve of 18.7 Mt with 23% P_2O_5 for the richer part of the deposit and an additional resource of 22 Mt with 7–10% P_2O_5 for the leaner parts. The MP State Mining Corporation is producing about 600,000 tonnes of high-grade ore (+ 30%) per year from this deposit.

Besides Hirapur, there are quite a few other phosphorite deposits with smaller resources located in the surrounding areas of Chhatarpur and Sagar districts. Laminated phosphorite with 20 to 30% P_2O_5 content occurs in Besai (Bessia) (24°22′:79°09′) and Mardeora (24°21′:79°11′).

Sonrai Phosphorite Phosphorite in the Sonrai Formation of Bijawar Group occurs as discrete beds in the siltstone–shale section of Jamuni Member and dolomite sequence of Rohini Member. The phosphate here is an *in situ* chemical precipitate and thus of orthochemical origin with sufficient concentration to form phosphoritic siltstone and phosphoritic dolomite. Phosphate resulting from supergene weathering in the Rohini Member occurs as breccia fillings, forming veins and lenses in silicified breccia, lensoid bodies of precipitated apatite with concentric banding and lenses of aphanitic phosphorite of replacement phase on the hanging wall of silicified breccia and as purely residual mantles of phosphatic sandstone and shale. Near surface, the secondary phosphate bodies have a width of up to 60 m. Fluorapatite is the only phosphate mineral with quartz as the main gangue. Chert, goethite, lepidocrosite and sericite are present in minor amounts.

Shrivastava (1989) was of the opinion that basic volcanism during sedimentation of Sonrai Formation may have contributed to the enrichment of the bottom sea water in phosphorous, and localised upwelling near basement highs brought in phosphorous rich nutrient water to the site of deposition.

Diamonds in Central India

Diamonds in the Panna Belt, Madhya Pradesh The Panna diamond belt occurs along the northern margin of Panna district, Madhya Pradesh (Fig. 2.2.47). It is an 80 km long ENE–WSW trending linear belt, which covers about 4000 sq km and occurs along the northern fringes of the Vindhyan basin. Diamond in this belt occurs in the Majhgaon pipe, located at a distance of 25 km southwest of the Panna town. Discovered in 1827, Majhgaon had to wait one full century for serious investigations (Sinor, 1930). Major exploitation took place during the period of thirties to sixties of the last century. Mining is still going on under the management of the National Mineral Development Corporation (NMDC). The reported grade is 8–15 cts. /100 tonnes. The nearby Hinota pipe, discovered in 1959, is smaller in size and lower in grade. Recently Rio Tinto has estimated a reserve of 40-70 mt (0.3-0.7 ct) from 8 lamproite pipes in the Bunder area, Panna diamond field (SEG News Letter, N. 75, p. 31, 2008). They are interested in its exploitation also.

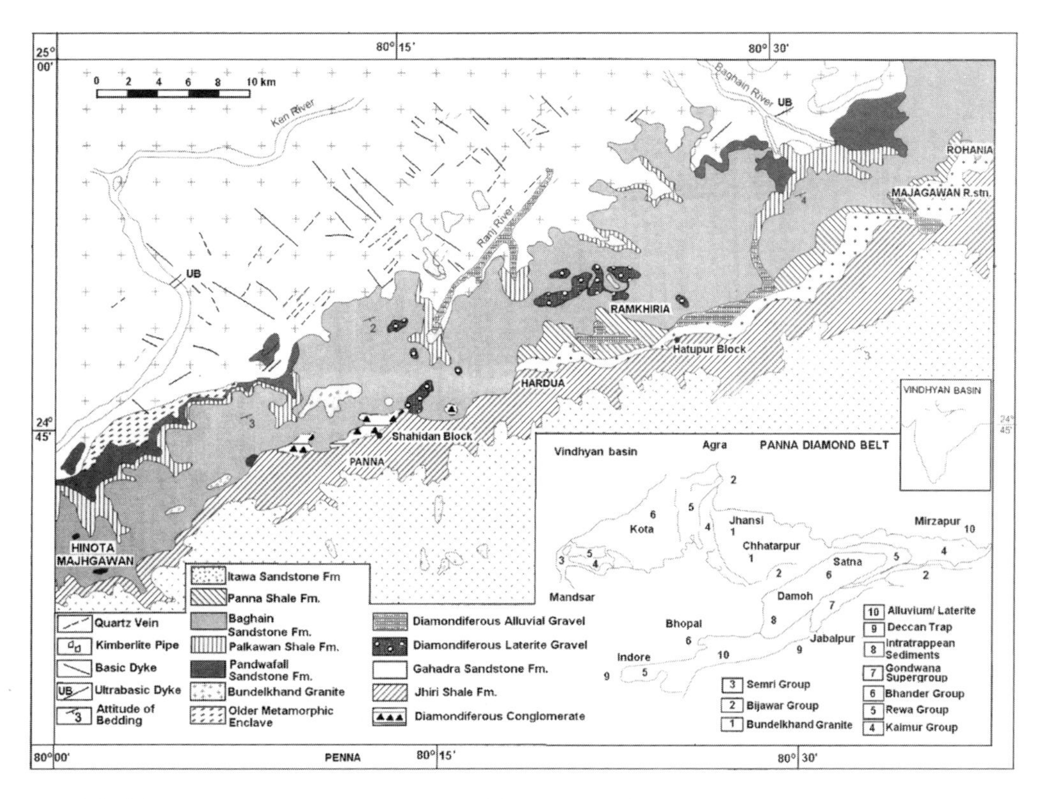

Fig. 2.2.47 Geological map of Panna diamond field, Madhya Pradesh (modified after Rau, 2007)

Majhgaon Pipe Majhgaon and Hinota potassic ultramafic rock bodies intruded the Kaimur Sandstone of the Proterozoic Lower Vindhyan Group that overlies the Archean basement of the central Indian craton. The Majhgaon pipe is near circular on plan and is an inverted cone in shape with the cone walls dipping 70°–75° inward, while Hinota is a shallow crater (Fig. 2.2.48 and Fig. 2.2.49).

The rocks constituting the Majhgaon pipe are generally fragmental (< 25–40 mm) in nature and clast-supported. Fragments are generally of igneous rocks and often the olivine phenocrysts are set up in a fine-grained matrix, forming macrocrystic textures. Occasionally xenoliths are present. Glassy base and plastic deformation (rare) of the fragments suggest that they are juvenile lapilli. Olivine is usually altered to serpentine.

The phenocrystic phlogopite with high TiO_2 and fairly low Al_2O_3 contents present in these rocks are not characteristic of kimberlites. They are rather similar to micas from lamproites (Scott Smith and Skinner, 1984; Mitchell, 1986). They are compositionally different from the groundmass micas in olivine lamproites. Other minerals reported from the Majhgaon rock are melilite, fresh olivine, rare perovskite, rutile, spinel (Mg-chromite) titanomagnetite, barite, pectolite, dolomite, high-Cr and high-Mg ilmenite (Mathur and Singh, 1971; Middlemost and Paul, 1984). Other minerals present are talc, vermiculite, chlorite, smectite and calcite, The Hinota rock is similar to that of Majhgaon. The rocks have been dated at 1100Ma (Rb–Sr).

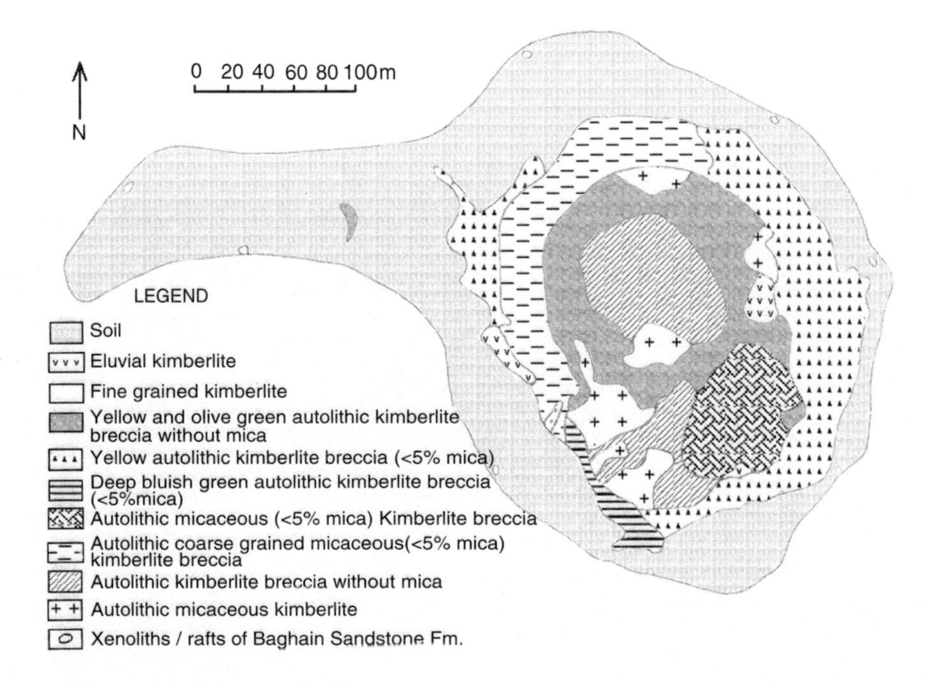

Fig. 2.2.48 Geological map of Majhgaon kimberlite pipe, Madhya Pradesh (modified after Rau, 2007)

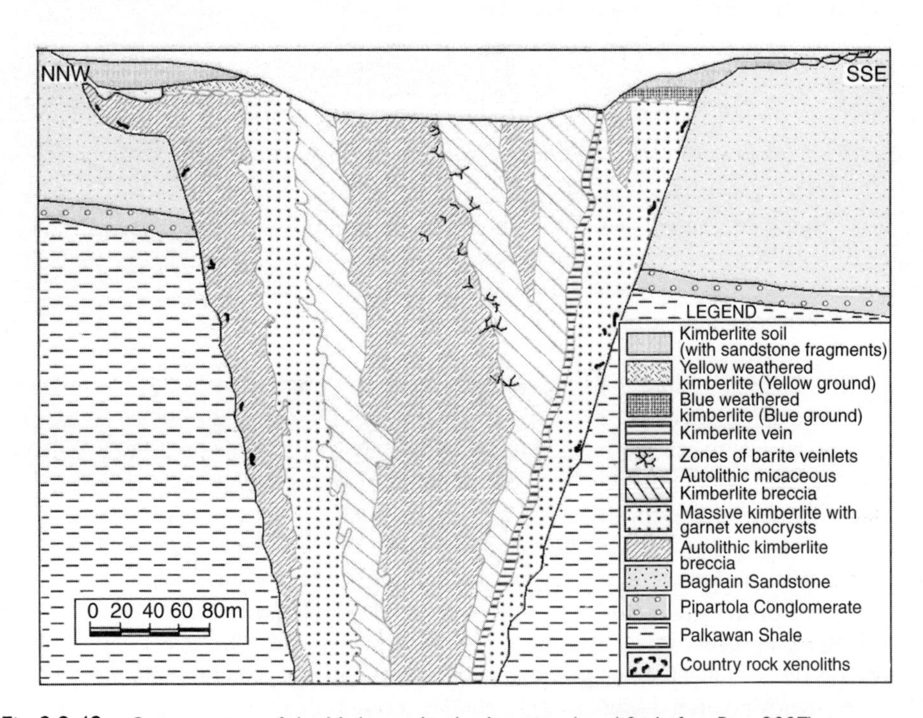

Fig. 2.2.49 Cross-section of the Majhgaon kimberlite pipe (modified after Rau, 2007)

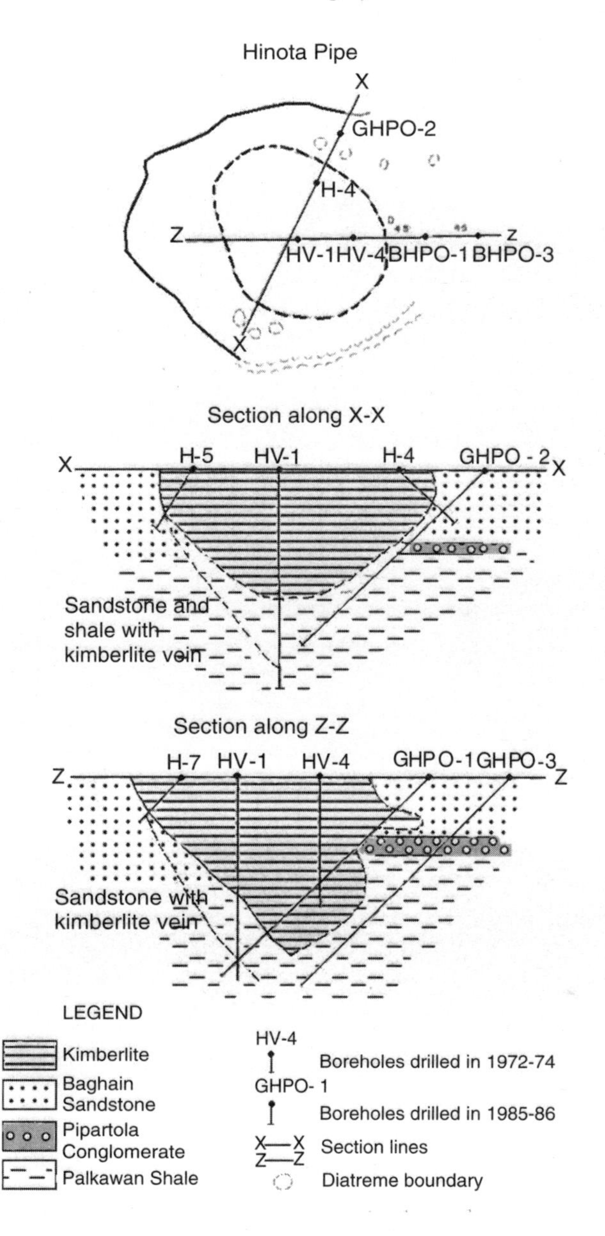

Fig. 2.2.50 Plan and cross-section views of Hinota pipe, Madhya Pradesh (modified after Rau, 2007)

Many workers have called the Majhgaon and Hinota rocks kimberlite. But they differ from the typical kimberlites by the following features: Glass and particularly scoriaceous juvenile lapilli have not been observed in kimberlites, but are typical of lamproites. Generally, in kimberlites the olivine macrocrysts are anhedral and rounded and the phenocrysts have simple euhedral shapes. The complex shapes of the olivine macrocrysts and phenocrysts at Majhgaon and Hinota are typical of lamproites, as also are the polysynthetically twinned mica. The mineral composition and the trace element data also support classification of these rock bodies as lamproites, rather than kimberlites

(Scott Smith, 2007). She also held that none of the igneous intrusive bodies at Angor and Jungel was kimberlite or lamproite and the reported diamond find in these bodies would demand further investigation.

Diamondiferous Vindhyan Sediments Diamonds occur within the sediments of the Vindhyan Supergroup, confined particularly to the Rewa Group. The latter contains diamonds in its upper three Formations (for more discussion on the Vindhyan Supergroup revert to Chapter 2.1).

Rewa Group	Govindgarh/Gahadra Formation	Sandstone with interbeds of diamondiferous conglomerates at the top zone.
	Jhiri Formation	Shale with diamondiferous conglomerate interbeds in its basal part.
	Asan/Itwa Formation	Sandstone with diamondiferous interbeds at its top part
	Panna Formation	Shale with limestone interbeds

The local people have collected diamonds from the above-mentioned geological horizons through ages. However, the difficulties in mining diamonds from these horizons are none too small. The diamondiferous horizons are generally thin and impersistent and covered by overburdens of varying thickness. These were the main sources of diamond before the Majhgaon and Hinota pipes were discovered and accessed.

Alluvial and Colluvial Diamonds There are alluvial and colluvial diamonds in the Panna area. These are mainly (i) alluvial gravel beds along the Baghain and Ranj rivers and their tributaries draining the Baghain plateau, (ii) superficial lateritic gravel spread over the Gahadra and Baghain plateaus. Of special interest are the gravel beds along the Baghain river basin near Ramkheria, where preliminary studies showed a diamond content of about 26 carats/100 tonnes of gravel. National Mineral Development Corporation (NMDC) established a reserve of 1,15,000 carats (ct) of diamond with a tenor of 16ct /100 tonnes. Composition of the gravel beds is suggestive of its derivation from the Vindhyan sediments.

Diamondiferous Kimberlite in the Bastar Craton In the recent decades kimberlites have been discovered at a number of places in the Bastar craton. Some of them have yielded diamonds during preliminary studies; others await detailed investigation. To-date available information on these occurrences is outlined below.

The Mainpur Kimberlite Field (MKF), about 135 km southeast of the Raipur city in the Bastar craton, is relatively better studied amongst the known occurrences. Five diamondiferous kimberlite bodies viz, Payalikhand East and West, Jangra, Bahradih and Kodomali, have so far been reported from the MKF. An $^{40}Ar/^{39}Ar$ (WR) age of 491 ± 11Ma is reported for the diamondiferous Kodomali diatreme (Chalapathi Rao et al., 2007). It may be pointed out that the Siluro–Ordovician age of kimberlites is rare in geological literature, though not absent. If the Pan–African age of the Kodomali Kimberlite corresponds to a collision with a craton then the amalgamation of the Indian shield into a Indo–Antarctic supercontinent took place later than what is commonly assumed (Chalapathi Rao et al., 2007).

The diamondiferous Bahradih kimberlite pipe consists of pelletal textured and olivine macrocrystic diatreme facies. The kimberlite is variously altered. Crustal xenoliths are common. Petrography, bulk chemistry and the trace element composition are typical of a kimberlitic rock (Mainkar and Lehman, 2007)

A kimberlitic rock has recently been reported from the north of Basna town in Mahasamund district, Chhattisgarh. The rocks are altered to various degrees. Emplacement of these rocks appears to be in the form of small plug-like bodies. In one of the three scout boreholes drilled, four pieces of diamond were found from the depth of 15.25–17.05 m (Chellani, 2007).

Kimberlite bodies in Tokapal–Bhejripadar area are located to the west of Jagdalpur. *In situ* U–Pb dating by laser–ablation quadrupole based ICP–MS of an autometasomatic titanite yielded an age of 616 ± 24 Ma. This is the age of the kimberlite volcanism here and the age of chemoclastic sedimentation of the lower Indravati Group (Lehman et al., 2007).

Besides the kimberlites briefly discussed above, diamond is available from the Mahanadi basin and its tributaries such as Ib, Maini and Bhagain. Hirakud, now well known as a large dam-site on the Mahanadi in western Orissa, obtained its name from the presence of mineable diamond in the river-sands of the area in the past. Mahanadi flows into Orissa from the west and flows into the Bay of Bengal near Cuttack.

Bauxite Deposits in Central India

The main bauxite belt of the Central India extends from the Eastern India, totaling nearly 400 km (~ 50 km wide) with roughly ENE–WSW trend from earstwhile Ranchi and Palamau districts in Jharkhand to Balaghat district in Madhya Pradesh, passing through Surguja district of Chhattisgarh and Shadol, Bilaspur and Mandla districts of Madhya Pradesh. The bauxitised mesas along this belt in Chhattisgarh include Jamirapat and Mainpat (Surguja district) and in Madhya Pradesh include Amarkantak (Shadol district), Phutkapahar (Bilaspur district), Kotapahar and Supkhar (Balaghat district), etc. There are two more clusters of bauxite deposits mainly in Madhya Pradesh and partly in Uttar Pradesh, which are located to the north of the previously mentioned belt, viz, Jabalpur–Sidhi and Rewa–Banda–Satna sectors. Before the discovery of the East Coast bauxite deposits in Orissa and Andhra Pradesh, the described region, including those in Jharkhand, was considered as the largest repository of bauxitic ores in the country. Besides the bauxite deposits in the Surguja district, there are also some smaller deposits near Keskal in the northern Bastar region of Chhattisgarh.

Other than Madhya Pradesh and Chhattisgarh, the Central India hosts a number of bauxite deposits along the Western Ghats in Maharashtra, extending from near Thane in the north to Satara and Kolhapur districts in the south and also along the Konkan coast in Raigarh, Ratnagiri and Sindhudurg districts, Maharashtra, extending further south to the coastal areas of Goa, Karnataka and Kerala (vide Fig.1.2.28).

Bauxite Deposits in Madhya Pradesh

Amarkantak Deposits Bauxite deposits occur as lenses and impersistent pockets within the lateritic capping on the dissected plateaus (1050–1100m), particularly in the eastern part of the Maikala Range. The best developed extensive deposits are at the Hazaridadar sections of Amarkantak area, Madhya Pradesh. Fox (1923) studied these deposits and mentioned this as a variety of 'high-level bauxites'. In the recent time Jagannatharao and Krishnamurthy (1981) and Ghosh and McFarlane (1984) have been some of the researchers on these deposits.

The generalised weathering profile in the Amarkantak area are as follows: the basalt (Deccan Trap) at the base gradually gives way upward to altered basalt, underclay, goethitic, ferruginous laterite, aluminous laterite, bauxite, ferruginous laterite and ends up with a top soil (Table 2.2.22). The mineralogical variation in the profile is supplemented by the available data on the chemical variation (Table 2.2.23).

Table 2.2.22 *A generalised lateritic profile of the Amarkantak area, Madhya Pradesh (Jagannatharao and Krishnamurthy, 1981)*

Rock types exposed	Brief description of the weathering zones	Depth (m) from the surface
Top soil	Laleritic soil (red, loamy, ferruginous)	0–1
Ferruginous laterite	Pisolitic and nodular ferruginous laterite	1–3
Bauxite	White or cream coloured hard bauxite	3–5
Aluminous laterite	Buff coloured, non–pisolitic material with brown and white patches	5–6
Ferruginous laterite	Reddish brown to brick red, cavernous ferruginous laterite	6–17
Goethite	Dark brown, compact material	17–18
Underclay	Pink brown, purple, clayey material with gritty bands	18–19
Altered basalt	Dark brownish grey, gritty porous altered basic volcanic rock	19–23

Table 2.2.23 *Chemical analyses (wt%) of different lithounits in the lateritic profile, Amarkantak plateau, Madhya Pradesh (Jagannatharao and Krishnamurthy, 1981)*

Lithounit	Al_2O_3	Fe_2O_3	SiO_2	TiO_2	FeO	ZrO_2	LOI
Pisolitic laterite	51.49	15.53	3.70	8.18	0.07	0.15	19.92
	51.08	14.54	5.20	7.27	0.14	0.14	21.48
	54.49	9.35	3.22	9.05	ND	0.15	21.02
Average	52.35	13.14	4.04	8.17	0.07	0.15	21.47
bauxite	54.14	12.52	0.60	10.91	0.18	0.15	20.61
	52.12	13.64	0.80	9.29	0.07	0.18	23.28
	57.22	6.87	0.84	9.29	0.25	0.16	24.65
	54.69	11.68	1.85	9.06	0.23	0.19	22.08
	63.15	4.65	0.76	9.09	0.07	0.22	21.82
	62.56	4.80	0.43	8.78	0.05	0.18	23.10
	58.68	6.64	2.03	7.29	0.23	0.16	24.19
Average	57.51	8.69	1.04	9.10	0.15	0.18	22.82
aluminous laterite	53.46	11.82	3.68	8.59	0.04	0.12	21.31
	54.10	10.71	1.07	9.60	0.09	0.11	23.73
	43.84	22.02	1.11	9.60	ND	0.14	23.04
Average	50.47	16.80	1.95	9.26	0.04	0.12	22.69
ferruginous laterite	41.22	26.06	1.13	9.09	ND	0.08	21.76
	21.31	42.42	13.03	4.85	0.29	0.08	14.07
Average	32.76	34.24	7.08	6.97	0.15	0.08	17.91

(Continued)

(Continued)

Lithounit	Al_2O_3	Fe_2O_3	SiO_2	TiO_2	FeO	ZrO_2	LOI
Goethetic zone underclay	11.67	61.61	13.60	2.44	–	0.02	9.90
	40.19	24.64	8.54	7.20	0.18	0.12	19.11
	29.05	38.38	11.43	5.25	0.08	0.07	15.36
	24.93	42.11	14.48	4.55	0.16	0.07	12.96
Average	31.39	35.04	11.48	5.67	0.08	0.09	15.81
Altered basalt	20.80	22.60	35.00	3.70	–	–	12.80

Mineralogically, the Amarkantak bauxite consists predominantly of gibbsite with some goethite and hematite. Rutile, magnetite, ilmenite and zircon are trace phases. Kaolinite and cryptocrystalline silica, if present, do not exceed 5% when added together.

Rao and Krishnamurthy (op. cit.) visualised the derivation of bauxitic ores from the Deccan Trap basalts in four stages:

Stage I: breakdown of feldspar and pyroxene and mobilisation of iron
Stage II: desilication of the glass
Stage III: leaching of silica and iron in varying pH
Stage IV: stage of maturity characterised by the development of gibbsite, goethite, hematite and other minor bauxite minerals.

Ghosh and McFarlane (op. cit.) have a reservation about the protolith of bauxite deposits of not only of the Amarkantak plateau but also of the entire Central India. They believe that the bauxite deposits there developed at the cost of hybrid rocks formed by the 'assimilation of mafic magma with acidic derivatives'. Adequate petrochemical evidence is yet to be found out to support this conclusion.

The total reserve of Amarkantak group of deposits in Shahadol district is about 20 million tonnes of bauxite (45–61% Al_2O_3, 1.5–6% SiO_2 and > 3.5% to < 8% TiO_2).

Phutkapahar Deposits Phutkapahar deposits in Bilaspur district, though small in reserve (~3 million tonnes), offer another interesting case for study. They are located in the laterite-capped, plateaus, ~1000 m high (above MSL), starting some distance north of Korba. Mineralogically, the ores consist of both gibbsite and boehmite. Here also the weathering profile, with 20–25 m thick bauxite, is developed on the Deccan Trap according to some workers (GSI Report, 1994). Ghosh and McFarlane (op. cit.) presented a weathering profile in which the Gondwana sediments are shown at the base, rather than the Deccan Trap (Table 2.2.24). This discrepancy should be resolved.

Table 2.2.24 *Phutkapahar weathering profile (after Ghosh and McFarlane, 1984)*

Laterite with alumina-rich patches	0–1.8 m
Massive pisolitic grey/white bauxite	1.8–4.0 m
Highly porous laterite with pisolites	4.0–5.0 m
Differentiated layer of Fe-hydroxides and Al–hydroxides and sandstone/shale	5.0–5.5 m
Coarse sandstone and shale (Gondwana)	5.5–6.4 m

Supkhar and Kotapahar Deposits Bauxite deposits, none very large, are clustered around Supkhar and Kotapahar areas of Balaghat district.

Six separate plateaus around Supkhar have laterite–bauxite capping. The estimated reserve of bauxite ores in this area is 13.5 million tonnes. No more details are available. Kotapahar group has a smaller reserve of 4.2 million tones (40–56% Al_2O_3, 0.85–11.3% SiO_2). In both the areas bauxite occurs as discontinuous lenses and pockets in laterite. Deccan Trap is at the base of the weathering profiles in both the areas (GSI Report, 1994)

Katni Deposit Katni deposits in the Jabalpur district are the most exploited in Madhya Pradesh. Average thickness of the bauxite zone varies from 2 to 12m. Katni bauxite has a good reserve of refractory grade (> 52% < 56% Al_2O_3, 2–11.7% SiO_2, 4.6–7.7% TiO_2) ores. Gibbsite is the principal bauxite mineral.

The Katni bauxite is generally believed to have formed from Vindhyan limestone and shales. Interestingly, in view of the heights of the localities of bauxite capping, Katni bauxite is to be assigned to the 'low level' type.

Semaria Deposits The Semaria area (24°30′–24°58′/81°07′–81°18′) is situated close to Madhya Pradesh Uttar Pradesh border. The lateritic cover is located on the upper Vindhyan sandstone at the Rewa–Satna plateau. Geological Survey of India (1994) estimated a reserve of 10.4 million tonnes of ore (43%–48% Al_2O_3, 3.7%–6% SiO_2).

Rewa–Satna Deposits In Satna district numerous small but high-grade (metallurgical, refractory and chemical grade) bauxite deposits, many of which have been in the state of mining, cap the Vindhyan sandstones. The estimated ore reserve is close to 4 million tonnes.

Bauxite Deposits in Uttar Pradesh A total reserve of 19 million tonnes of bauxitic ores is strewn over several districts (Banda, Lalitpur, Mirzapur and Varanasi) in Uttar Pradesh as small bodies developed on the Vindhyan sandstones. The Al_2O_3 content is moderately high (40%–45%), but the SiO_2 may also be locally high (10%).

Bauxite Deposits in Chhattisgarh

Jamirapat and Mainpat Deposits Bauxite in lateritic cappings occurs in the northern part of the Surguja district, particularly at Jamirapat and Mainpat areas.

At Jamirapat a reserve of 36.6 million tonnes of ore with an average Al_2O_3 content of 52%, has been established by the GSI in a belt that is 40 sq. km in extent. The Department of Geology and Mines, Madhya Pradesh reportedly established a reserve of 33 million tonnes of bauxite ores at Mainpat. In both the areas, the weathering profiles have Deccan Trap at the bottom.

Keskal Deposits Geological Survey of India established occurrences of high-grade bauxite mineralisation over the Bamni Sandstone, Indravati Group, in the Keskal–Amabera and Pirhapal–Tarandul area of Bastar district in the state of Chhattisgarh. A reserve of 9 million tonnes has been estimated.

Bauxite Deposits in Maharashtra Bauxite ore reserves in Maharashtra are of the order of 100 million tonnes (GSI, 1994). Important bauxite deposits in Maharashtra are located at Kolhapur, Kolaba, Ratnagiri and Satara districts, on plateau-tops along the Western Ghats. The bauxites are a part of the lateritic profile developed on the Deccan Trap. Kolhapur and Satara bauxites are so–called 'high level bauxites', having developed at elevations of 1000–1100m above MSL. The laterite–bauxite cappings along the Konkan coastal belt are developed at elevations of 40–350 m above MSL. Bauxite deposits in Kolhapur only are persistent and form blankets. All others are lensoid and pockets within lateritic blankets. The bauxite ores along Konkan coast are generally of high grade (50– >60% Al_2O_3, < 4% SiO_2), though belonging to the so-called 'low–level bauxites'. These bauxites and those at Katni confirm our postulation in earlier relevant discussions that elevation of lateritising surface by itself is not the determinant of the quality of the bauxite. Low-level lateritic profile may develop good quality bauxite if there is adequate rainfall and good drainage.

Dhangarwadi–Penhela deposits A cluster of deposits including those at Dhangarwadi, Girgaon, Rangewadi are located close to Kolhapur–Ratnagiri Highway on a highly dissected 10 x 1.6 sq km plateau. The profile thickness, however, widely varies from about 1m to 14m, with an average of about 3.5m. A major part of the deposit is free of an overburden. The estimated ore reserve is 16 Mt (GSI Report, 1994).

Udgiri Deposit Udgiri deposit, also located within the Kolhapur district, comprises four plateaus. A bauxite zone of 5 m average thickness underlies a lateritic overburden. Estimated ore reserve is 9.5 Mt, with an average composition of 53% Al_2O_3, 1.93% SiO_2, 10.49% Fe_2O_3 and 5.09% TiO_2 (GSI Report, 1994).

Konkan Deposits More than fifty small, pockety but high-grade bauxite deposits with limited reserves, rough estimates totalling to < 20 Mt, have been located along the Konkan coast in Raigarh, Ratnagiri and Sindhudurg districts. None of these have been exploited so far. The largest deposit at Supegaon (18°26′:72°55′; Toposheet 47B/15) has an ore reserve of 3.5 Mt (Bhatia et al., 2001). In most parts of the coastal belt, the laterite–bauxite profiles are developed on the Deccan Trap, except in parts of Sindhudurg district where they form cappings over the northwestern extensions of the Precambrian Dharwar rocks. The bauxite of this belt is mainly composed of gibbsite (modal gibbsite 86.5%), associated with minor amounts of kaolinite, limonite, leucoxene, goethite and rarely quartz and calcite.The coastal bauxites are mainly of metallurgical grade. The composition, as averaged from 50 samples drawn from different parts of the belt, is Al_2O_3 – 51.65%, SiO_2 – 2.42%, Fe_2O_3 – 6.37%, TiO_2 – 2.92% and LOI – 31.42%. The Aronda–Guldev occurrence of bauxite (~1.0 Mt) in Sindhudurg district, overlying the Dharwar rocks, are of better grade: Al_2O_3 – 53.44–62.22%, SiO_2 – 1.33–5.80%, Fe_2O_3 – 2.99–11.98%, TiO_2 – 1.79–3.59%, LOI – 28.44–32.02%. The Geological Survey of India in recent years undertook detailed exploration of some of the high-grade bauxite deposits around Devgarh in Sindhudurg district and Rajpur in Ratnagiri district. Most of the occurrences in this area are at the elevations of 60–100 m from MSL. The potential of exploitability of Konkan bauxite has significantly enhanced due to augmentation of reserves as well as transport facilities offered by the newly built Konkan Railways.

Some recent studies (Joshi et al., 2005a,b) have brought out details on the mineralogical and chemical transformations in the bauxite bearing regoliths at various physiographic levels of western Maharashtra ranging from the hilly tracts to the coastal planes at almost sea level. The representative weathering profiles in the hill section (Girjam –16°45′10″: 73°48′50″), the plateau level (Kumbhawade –16°31′02″:73°27′33″) and coastal level (Vijaydurg –16°33′28″:73°20′07″) show the development of well-differentiated saprolith and pedolith horizons that are unrelated to their respective topograhic position (Fig. 2.2.51).

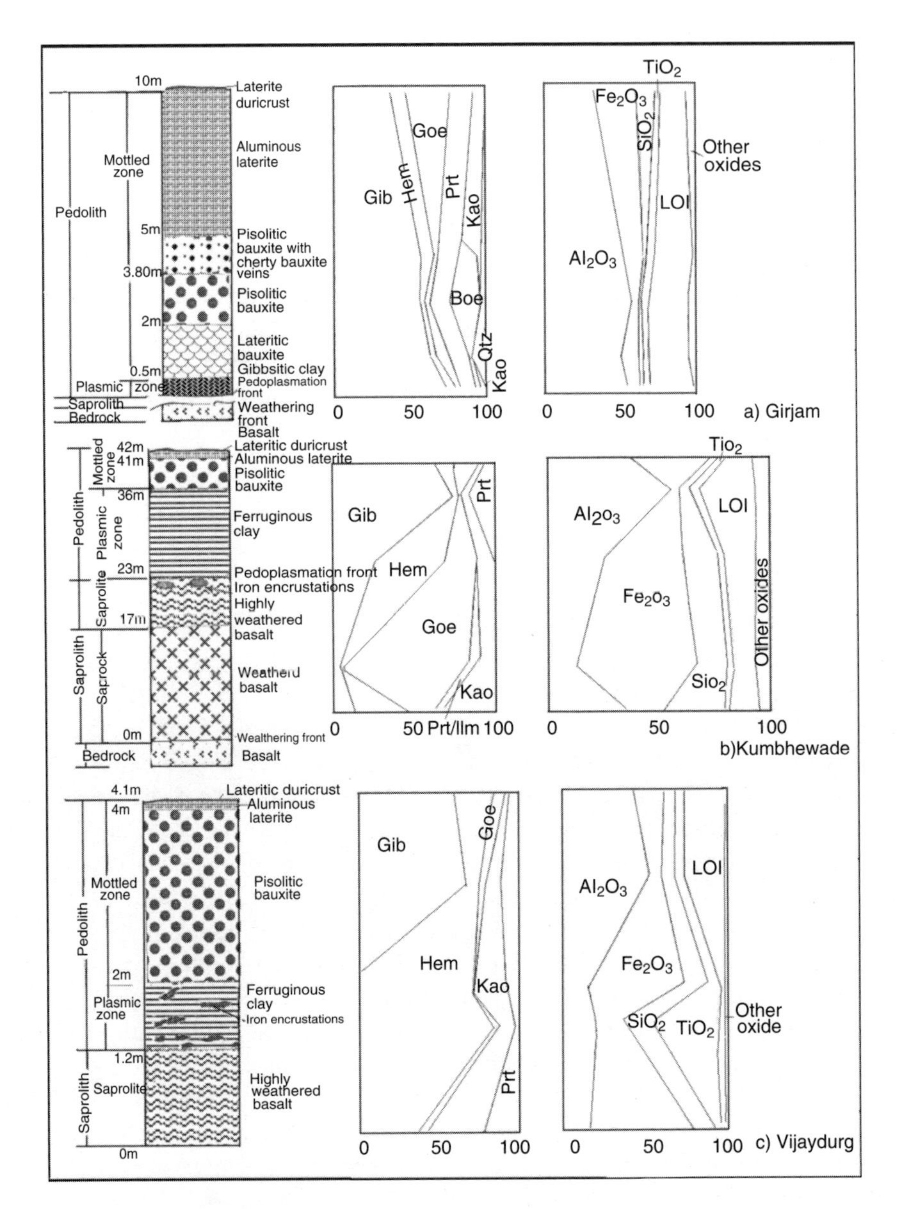

Fig. 2.2.51 Weathering profiles at (a) Girjam, (b) Kumbhawade and (c) Vijaydurg, western Maharashtra. Gib – gibbsite; Boe – boehmite; Hem – hematite; Goe – goethite; Ilm – ilmenite; Kao – kaolinite; Qtz – quartz; Ptr – Pseudorutile (after Joshi et al., 2005a).

The extensive development of an authigenic nodular horizon in the mottled zone of the regolith in this terrain is interesting. The nodules (with > 50% Al_2O_3), studded in a generally indurated ferruginous matrix, are rounded to ovoid in shape and range from pea to boulder size (0.5 mm to 1 m).

3

Eastern Ghat Belt

3.1 Geology and Crustal Evolution

Introduction

The Eastern Ghat belt, geologically commonly referred to as the Eastern Ghat Mobile Belt (EGMB), expose granulite facies rocks over a curvilinear stretch of more than 1000 km length along the eastern coast of India (from Brahmini river basin, Orissa, in the north to Ongole, Andhra Pradesh, in the south). It represents one of the highly deformed and metamorphosed (at Ultra High Temperatures, UHT) Precambrian crustal segments of the Indian Shield. The belt is widest (~ 300 km) in the north and thinner towards south (~50 km), before tapering off beyond Ongole. Two major NW–SE trending rifts (grabens), Mahanadi in the north and Godavari in the south, dissect the belt (Fig. 3.1.1).

The EGMB is bordered by the Bastar and Dharwar cratons along the west and by the East Indian (Singhbhum–North Orissa) craton in the north. The location of the boundary and contact relationship of the granulite facies rocks of the EGMB with the lower grade cratonic components have remained debatable issues. However, most of the workers (Walker, 1902; Fermor, 1936; Pascoe, 1950; Ramakrishnan et al., 1998) identified a crustal scale thrust/shear contact along the western boundary, which roughly coincides with the linear zone of alkaline intrusives (Leelanandam, 1993; Rath et al., 1998) and is marked by a longitudinal 'transition zone' of amphibolite facies rocks (Nanda and Pati, 1989; Ramakrishnan et al., op. cit.). Abrupt variation in Bouger anomalies and deep seismic sounding data across the western boundary are also suggestive of a major thrust contact (Subrahamanyam and Verma, 1986; Kaila and Bhatia, 1981). Chetty and Murthy (1994) identified the crustal scale Kolab-Machkund-Sileru shear zone, based on remote sensing data, and interpreted it as the contact of EGMB with the Bastar craton. Evidence of ductile shearing and thrust sheets over a width of 1 – 2 km in selected sectors of the contact zone are reported from parts of Andhra Pradesh and Orissa (Mukhopadhyay et al., 1995; Biswal and Jena, 1999).

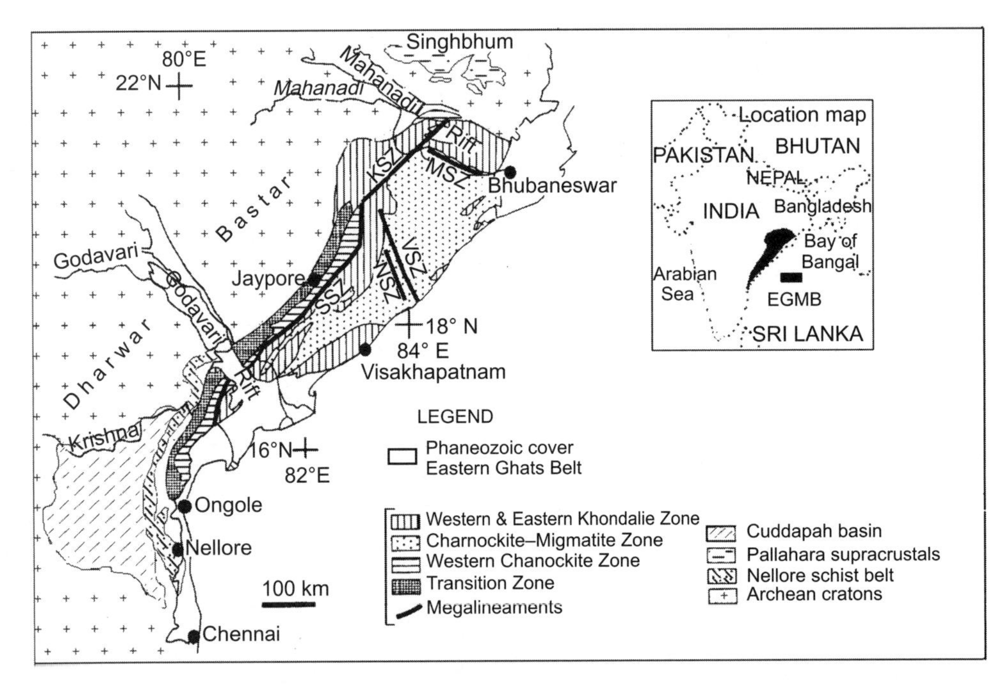

Fig. 3.1.1 Generalised geological map of EGMB and adjacent regions (modified after Ramakrishnan et al., 1998), with prominent megalineaments shown by Chetty (1995, 2001). Megalineaments: KSZ – Koraput-Sonpur shear zone; MSZ – Mahanadi shear zone; NSZ – Nagavalli shear zone; SSZ – Sileru shear zone; VSZ – Vamsadhara shear zone

Gupta et al. (1998) questioned the validity of the 'transition zone' as the physical contact of EGMB with the Bastar craton, and suggested that the contact lay further to the west. The interface of the EGMB with the East Indian Craton in northern Orissa is marked by an ESE–WNW trending megalineament (Gohira–Sukhinda shear/thrust belt), characterised by linear stretches of migmatitic granitoids, several zones of dislocation and ductile shearing (Banerjee, 1990; Bhattacharya et al., 1994; Sarkar and Nanda, 1998). Another redeeming feature is the discontinuous occurrence of a number of late Proterozoic sedimentary basins in the cratonic foreland along the western margin of the EGMB. All of these show variously deformed character along the contact and largely undeformed nature further to the west, i.e. in the cratonic domain.

Lateral Lithologic Zones

The four lateral or longitudinal lithologic zones, described below, are identified across the width of the belt (Nanda and Pati, 1989; Ramakrishnan et al.,1998), which have been depicted in the latest map of the EGMB published by the GSI (1998).

1. Western Charnockite Zone (WCZ)
2. Western Khondalite Zone (WKZ)
3. Central Charnockite Migmatitic Zone (CMZ)
4. Eastern Khondalite Zone (EKZ).

The WCZ (width 20–30 km), which is considered as the basement of EGMB by Ramakrishnan et al.(1998), is dominated by enderbite-charnockite-mafic granulite assemblage with lenses of mafic-ultramafic rocks and minor metasedimentary enclaves. The WCZ is absent in Bolangir–Kalahandi sector of Orissa, where the WKZ is disposed in direct contact with the cratonic foreland.

The WKZ (average width ~ 40 km, at places upto 80 km) mainly comprises khondalite, with intercaleted calc-silicate rocks, intrusive enderbite and charnockites with supracrustal xenoliths and minor marble and quartzite. The zone also exposes minor but significant components of Mg-Al rich cordierite-sapphirine-spinel-orthopyroxene bearing granulites and manganese silicate bearing rocks.There are several occurrences of massif-type anorthosite in this zone, located around Bolangir, Turkel and Jugsaipatna (Dasgupta and Sengupta, 2003).

The CMZ (width 40 – 100 km) is dominantly composed of migmatitic gneisses, anatectic garnetiferous granitoids including leptynite, with enclaves of khondalite, calc-silicates, high Mg-Al granulites, charnockite and mafic granulites. The massif-type anorthosite of Chilka Lake area belongs to this zone.

The eastern most EKZ contains rock assembleges similar to WKZ , the eastern margin of which is covered by the coastal Phanerozoic sediments. Manganiferous horizons form conspicuous markers in parts of this zone. There is no occurrence of anorthosite here.

Though the significance of these subdivisions has been questioned by Dasgupta and Sengupta (2000) because of considerable overlap of the lithological characteristics and lack of isotopic data-support for the delineated zones, the four-fold lateral classification of the EGMB has gained wide acceptence among the later workers as a portrayal of broad lithological framework for the belt (Rickers et al., 2001; Dasgupta and Sengupta, 2003; Dobmeier and Raith, 2003), though not suscribing to all the details.

The lack of unanimity concerns many aspects. The Sileru shear zone, which was earlier considered by Chetty and Murthy (1994) to be the western boundary of EGMB, actually passes between the WCZ and WKZ. The WCZ on the other hand is considered as the basal part of EGMB by Ramakrishnan (1998), and the amphibolite facies 'transition zone' passing along its west as the contact with the craton. As mentioned above, Gupta et al. (1998) do not agree with this view and consider the 'transition zone' as retrograded granulites and prefer to put the EGMB-craton contact further to the west. Dasgupta and Sengupta (op. cit.) further question the validity of restricting the 'migmatites' to the CMZ, as the khondalites of both WKZ and EKZ are also migmatitic at many places. The possible 'terrane' configuration worked out by Rickers et al. (2001) on the basis of isotopic data do not exactly match with the four lateral zones except the WCZ. The crustal architecture proposed by Dobmeier and Raith (2003), however, bears a much better correlation of their 'provinces' with the lateral zones of Ramakrishnan (1998).

Transverse Segments of EGMB

Besides the lateral variations described above, the Mahanadi and Godavari rifts subdivide the belt into three segments across its length, viz, the northern, central and southern. Each of these show perceptible variation in lithotectonic ensemble and tectono-thermal history (Sarkar and Nanda, 1998; Mezger et al., 1996; Dasgupta and Sengupta, 1998). According to Dasgupta and Sengupta (2000), there is evidence of superposed granulite facies metamorphism to the north of Godavari rift, whereas the southern segment distinctly shows only one phase of Ultra High Temperature metamorphism. This view is, however, not shared by many. Mezger and Cosca (1999) suspect that WCZ, both to the north and south of Godavari rift, reflects an older tectono-thermal history. The presence of charnockitic protoliths of Archean age are reported from the north of Godavari (Sarkar and Paul, 1998) and possible existence of similar protoliths are also indicated from the south of Godavari (Sengupta et al., 1999). These observations are supportive of the view of Ramakrishnan et al. (1998) who considered the WCZ as the basement of the EGMB.

In spite of the existing controversies with regard to the metamorphic history of different segments of the belt, there are some distinctive features for each of these. The northern sector to the north of Mahanadi rift is characterised by the dominance of arenaceous facies in the high grade supracrustals, rotation of the NE–SW structural grain of the belt to WNW–ESE, and the presence of late Archean charnockite (Sarkar and Nanda, 1998). In the central sector, lying between Mahanadi and Godavari rifts, khondalites of CKZ are often manganiferous and graphitic. Anorthosite complexes of the belt are exclusively restricted here. Also, the evidences of Mid- to Late-Proterozoic tectono-thermal activity and Pan-African thermal perturbations are most pronounced here (Sarkar, 1981; Mezger et al.,1996; Sarkar and Paul, 1998; Mezger and Cosca, 1999). The southern sector, occurring to the south of Godavari rift, is characterised by abundant occurrence of mafic/ultramafic and alkaline intrusives and Ultra High Temperature metamorphism, presence of banded iron formations in the supracrustals (Sarkar and Nanda, 1998), weak late Proterozoic tectono-thermal impress, and strong effects of the Pan-African event (Sarkar and Nanda, op. cit., Mezger et al., op. cit.).

Geological Framework

Lithologic Assemblage To avoid confusion arising from varied nomenclature used to describe some of the granulite facies rocks of this belt, Dasgupta and Sengupta (2000) suggested the following mineralogical compositions of the major members.

1. Charnockite: Orthopyroxene-quartz-K feldspar-plagioclase, with K feldspar >> plagioclase
2. Enderbite: Orthopyroxene-quartz-plagioclase-K feldspar, with plagioclase >> K feldspar
3. Mafic granulite: Orthopyroxene-clinopyroxene-plagioclase, with or without garnet and quartz
4. Leptynite: Garnet bearing quartzo-feldspathic gneiss, without sillimanite and orthopyroxene
5. Khondalite: Garnet-perthite-quartz-sillimanite-plagioclase gneiss.

Calc-silicate granulites (wollastonite-scapolite-calcite- grandite garnet-clinopyroxene-sphene-plagioclase) and high Mg-Al granulites (sapphirine-spinel-garnet-sillimanite-orthopyroxene-cordierite-corundum/quartz) occur in much smaller proportions as lenses within the khondalites and leptynite or rarely as xenoliths in mafic granulites, but provide most of the information on the P-T trajectory of evolution for the belt (Sengupta et al., 1990, 1999). Quartzites, BIF, and rare marble also occur in the belt in close association with the khondalites.

Other important non-granulitic lithoassemleges occurring in this belt are: Migmatitic granite gneiss and granitoids, mafic-ultramafic rocks, plutonic alkaline complexes, carbonatites, and massif-type anorthosite complexes.

The orthogneisses in the belt are mainly represented by mafic granulites and enderbite, which also provide good constraint on metamorphic P-T regime. A typical charnockite is uncommon, invariably showing intrusive relations with respect to the country rocks. However, two generations of enderbite, the later of pegmatoidal character, have been reported from some areas (Sengupta et al., 1999).

Large bodies of megacrystic granite/granitoid are prominent magmatic rocks in the belt, specially in the central sector (between Mahanadi and Godavari rifts). Massif-type anorthosites are restricted to the northern parts of the belt while the alkaline rocks are prominent along the western margin.

At Kondapalle, located south of the Godavari rift, occurs a discontinuous layered complex, mainly composed of gabbroic and anorthositic rocks (GALS) with subordinate dunite, harzburgite, websterite and pyroxenites, containing sub-economic chromite mineralisation. GALS (Ca 1.7 Ga) preceded the emplacement of ultramafic-chromite bodies. The regional geological setting, rock associations, the geochemistry and mineralogy of the rocks suggest an Andean type continental magmatic arc environment for the emplacement of the complex. The chromites and the ultramafic cumulates of the complex represent a part of the plutonic root of the magmatic arc (Leelanandam and Vijaya Kumar, 2007).

Structural Framework The deformational features bear evidence of polyphase deformation causing folding, faulting and ductile/brittle shearing all along the belt. The overall structural grain has NE–SW trend for about 700 km, swerving to nearly E–W trend to the north of Bolangir, Orissa.

Besides the Godavari and Mahanadi rifts (grabens), there are several lineaments (shear zones) of regional extent which traverse along or across the belt with varied orientation. Of these, the NE–SW trending Kolab-Machkund-Sileru shear zone extends for about 500 km from Koraput, Orissa to Kunavaram, Andhra Pradesh, along the interface of the WCZ and WKZ (Nanda and Pati,1989; Ramakrishnan et al.,1998), but does not demarcate the EGMB from the cratonic terrain as postulated by Chetty and Murthy (1998). Another major lineament, the Vamasdhara–Nagavalli shear zone,extending for 200 km with a NNW–SSE trend, was considered by Chetty and Murthy (op. cit.) to divide the EGMB into two blocks, which has now derived ample support from recent researches (Dobmeier and Raith, 2003; Rickers et al., 2001), though not considered tenable by Sengupta et al. (1999). The tectonic significance of many other major shear zones too, which were not given much credence earlier, are becoming clearer with the generation of more and more isotopic data. However, since quite some time it has been a popular concept to regard the EGMB as a collage of several 'terranes' (Howel, 1989). Varied tectonic models have been proposed for the belt using the structural and some other criteria (Narayanswami, 1966; Murthy et al., 1971; Subrahmanyam and Verma, 1986; Chetty and Murthy, 1998; Bhattacharya, 1997; Ramakrishnan, et al., 1998) but none of these differing views are well-constrained, specially with respect to tectonothermal evolutionary history (Dasgupta and Sengupta, 2000).

Site-specific detailed structural studies in this belt are restricted to small isolated domains, mostly in the north and central segments (Sarkar, 1981; Halden et al., 1982; Park and Dash, 1984; Bhattacharya et al., 1994; Biswal and Jena, 1999; Sanyal and Fukuoka, 1995; Karmakar and Fukuoka, 1998; Bhaumik, 1997; Yoshida et al., 2000). Despite wide difference in interpretation and opinion, these studies reveal that there is a noticeable unifomity in the overall history of progressive fold deformation in the belt. Three phases of major penetrative fold deformations ($D_1 - D_3$) followed by a major phase of shearing characterise most of the studied sectors. The first folds are preserved as tightly appressed isoclinal, often reclined, rootless intrafolial folds on compositional banding (bedding?) in the metapelitic rocks, with the development of strong pervasive NE–SW to N–S trending axial planar schistosity. The second generation folds are near coaxial with the first folds, and are manifested on larger scales as asymmetric to overturned isoclinal folds on the first schistosity. The transposition of the two schistosities is a pervasive feature at many places. The third generation folds have variable geometry and style, with mostly local development of axial planar cleavage or fractures. However, the latter is noted as a pervasive planar element at places with ductile shearing. Dome and basin structures resulting from the interference of the first and third generation folds, are present at many places of the belt.

Metamorphic History and Age Data A spate of researches on thermobarometry over nearly last two decades in several sectors of the belt have largely elucidated the metamorphic history of the belt (Grew and Manton, 1986; Lal et al., 1987; Kamineni and Rao, 1988; Dasgupta, 1993, 1995; Dasgupta and Ehl, 1993; Dasgupta et al., 1991, 1992, 1994; Sengupta et al., 1990, 1991, 1999; Sen et al., 1995; Shaw and Arima, 1996, 1998). In spite of some initial divergence in views, it is generally agreed now that the EGMB records distinct evidence of having undergone Ultra High Temperature (UHT) polymetamorphism in the granulite facies. However, according to Dasgupta et al., 1997 and Sengupta et al., 1999, polymetamorphism is reflected only from the area to the north of Godavari rift, while the southern segment records a single phase of metamorphism. These findings are indicative of both temporal and length-wise spatial variations of the P-T trajectories in the belt, but data density is not sufficient to unequivocally establish the terrain boundaries on the basis of metamorphic criteria, or age and isotopic data available till then.

North of Godavari Rift Based on petrological evidences, Sengupta et al. (1990) first established two discrete events of granulite facies metamorphism, M_1 and M_2, in the northern sector (north of Godavari rift) of EGMB. The peak of M_1 attained ultra-high temperature of ≥ 950° – 1000°C at 9 – 10 kb pressure, with an anti-clockwise prograde P-T path, followed by near isobaric retrograde cooling path (M_{1r}) down to 750° – 800° C. At the peak temperature of M_1, significant partial melting of the pelitic and psammopelitic rocks produced granitic leucosomes, and also resulted in stabilisation of sapphirine + quartz and spinel + quartz assembleges in high Al-Mg granulites and scapolite + wollastonite in calc-granulites (Dasgupta and Sengupta, 2000, and references therein).

The age of M_1 event is largely obscured because of the partial or complete resetting of the isotopic clocks under high temperature metamorphism and intense deformation related to subsequent event(s). The Archean ancestry of the belt is, however, discernible from the Nd model ages (T_{DM}) of 3.9 – 3.2 Ga inferred from orthogneisses (Rickers et al., 2001) and 2.9 – 2.6 Ga deduced from some mafic granulites and charnockites (Paul et al.,1990; Sarkar and Paul, 1998) as well as from more direct determination of 2.6 Ga U/Th-Pb ($^{207}Pb – ^{206}Pb$) age of zircon from khondalite (Sarkar et al., 2000). The next important dated event of the sector is Mesoproterozoic: 1.6 – 1.3 Ga (Sarkar and Paul, 1998), 1.45 Ga (Sm-Nd, WR) as crystallisation age of leptinite and mafic rocks (Shaw et al., 1997), 1.5 Ga zircon date from alkaline rocks along northwest margin (Aftalion et al.,1998). However, these dates are believed to represent a resetting age rather than exactly denoting the M_1 metamorphic event (Dasgupta and Sengupta, 2000). Ricker et al. (2001), however, reported Nd model age of 2.2 – 1.8 Ga from a homogeneous crustal domain, from both orthogneisses and metasediments, to the west of Chilka Lake.

A distinctly later event of granulite facies metamorphism (M_2) recorded in this part of the belt witnessed a lower P-T condition than the M_1, attaining a peak temperature of ~ 850° C at 8 – 8.5 kb pressure, followed by retrograde isothermal decompression (M_{2r}) down to 4.5 – 5 kb. (Sengupta op. cit.; Kamineni and Rao, 1998; Dasgupta et al., 1992, 1995; Mukhopadyay and Bhattacharya, 1997). The M_2 metamorphic event is correlated with the period of structural and metamorphic reworking of the isobarically cooled granulitic crust (post-M_1) and its unroofing and exhumation to the mid-crustal level. The available geochronological data are indicative of an age of 960 – 1000 Ma (Grenvillian event) for the M_2 metamorphism in the EGMB (Grew and Manton, 1986; Mezger et al., 1996; Mezger and Cosca 1999; Ricker et al., 2001; Crowe et al., 2001; Dobmeier and Simmat, 2002).

A late phase of amphibolite facies metamorphism, M_3 (≅ 600°C, 4 – 5 kb), is widely manifested as overprints on the earlier high grade assembleges of M_1 and M_2. This is often associated with hydration,carbonation along ductile shears and emplacement of pegmatites. The M_3 event has been dated at ~ 550 – 500 Ma (Mezger and Cosca, 1999; Dobmeier and Sammit, op.cit.).

South of Godavari Rift In contrast to the polymetamorphic character of the EGMB in the northern sector, the stretch lying to the south of Godavari rift records only one event of Ultra High Temperature metamorphism, corresponding to M_1 described above. Near isobaric heating-cooling P-T trajectories are deduced for the entire sector with peak temperature of ≥ 950°C but peak pressure shows a north-south gradient from ~10 kb in Kondapalli (Sengupta et al., 1999) to 6 kb in Chimakurthy (Dasgupta et al., 1997). The metamorphism at many places in this sector is found to be related to the abundant mafic-ultramafic intrusives.

Ricker et al., (2001) has reported Nd model ages of 2.5 – 2.3 Ga for orthogneisses and 2.8 – 2.6 Ga for metasediments from south of Godavari graben. More direct age data available from this sector are c 1.25 – 1.3 Ga (Rb-Sr/WR) age of alkaline rocks of Elchuru and Uppalapadu (cited in Dasgupta, 1995), c 1.35 – 1.67 Ga (U-Pb/Th-Pb/Pb-Pb) age of allanite and monazite in pegmatites of Kondapalli and Ongole (Mezger and Cosca, 1999). No reliable evidence of later metamorphic events at 1000 Ma and 550 Ma, as noted in the north of Godavari, has so far been obtained from this sector, except for some very local resetting of Ar-Ar radioclock at around 1.10 Ga, reflected by hornblende in basic granulite (Mezger and Cosca, op. cit.).

Crustal Domains/Provinces in EGMB

Recently the terrain concept has been reinvoked in EGMB on a much firmer footing than before. Earlier, the tentative suggestions in this regard were merely based on the shear zones or lithological entities. Based on Nd model ages supported by Rb-Sr and Pb isotope ratios, Rickers et al., (2001) delineated four crustal blocks in EGMB described below (Fig. 3.1.2), which are correlatable in parts with the lithological subdivisions (Ramakrishnan et al.,1998) and demarcated by some of the shear zones (Chetty and Murthy, 1994) identified in the belt.

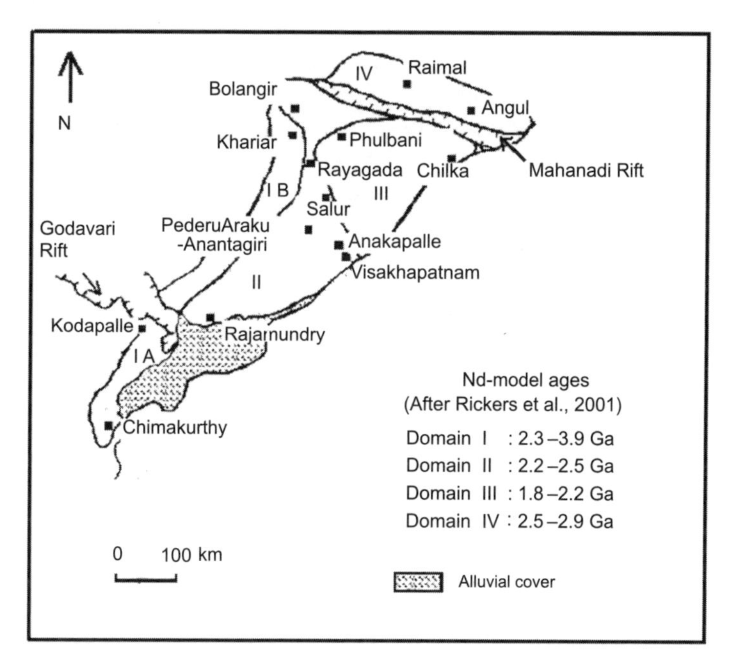

Fig. 3.1.2 Isotopic domains of Rickers et al., 2001 in the EGMB (after Dasgupta and Sengupta, 2003)

1. Domain I: It more or less coincides with WCZ, which is subdivided into the following.
   (IA): It is WCZ in the south of Godavari (Kondapalle–Chimakurthy sector). Nd model age is 2.3 – 2.5 Ga for orthogneisses and 2.6 – 2.8 Ga for metasediments. Primitive Pb-isotopic signature is indicative of reworking of dominantly Archean crust and mixing with minor Proterozoic material. Dasgupta and Sengupta (2003) consider the domain as a distinctly Proterozoic crust formed from juvenile material at 2300 – 1700 Ma, and last metamorphism at 1600 Ma.

(IB): It is WCZ in the north of Godavari (Jaypur–Khariar sector). Nd model age is 3.2–3.9 Ga; Pb isotopes are strongly retarded here (distinct Mid-Archean crustal block).

2. Domain II: It encloses parts of WKZ, CMZ and EKZ in the north of Godavari (Rajamundri-Araku-Vizag-Rayagada-Bolangir area). Nd model age for metasediments is 2.1–2.5 Ga (younger than paragneisses in adjacent WCZ); for orthogneisses, it is 1.8–3.2.(very widespread). Isotopic data are indicative of mixing of Archean and Proterozoic material.

(This is the most heterogeneous but extensively studied crustal block in EGMB. It is sandwiched between two homogeneous blocks of domain IB and III and displays a transitional character. The western and northeastern boundaries are marked by Sileru and Nagavalli-Vamasdhara shear zones, respectively.)

3. Domain III: It is the north eastern part of CMZ and WKZ (Chilka Lake–Phulbani area). It is an almost homogeneous domain with Nd model ages 1.8–2.2 Ga for both orthogneisses and metasediments; Pb isotopes show extremely narrow array.

(Massif-type anorthosite of Chilka Lake area is dated as 790 Ma by zircon U-Pb–`Krause et al., 2001; prominent granulite facies metamorphism at 690–660 Ma – Dobmeier and Raith, 2003).

4. Domain IV: It is the zone lying between East Indian (Singhbhum) craton and Domain III. (Raimal-Angul sector in the north of Mahanadi graben): Nd model age is 2.2–2.8 Ga for metasediments and around 3.2 Ga for orthogneisses.

Based on critical evaluation of the existing geological and isotopic data over large areas of Precambrian peninsular India (including Dharwar-, Bastar-, Singhbhum-cratons, Nellore-Khammam belt and Eastern Ghat belt), Dobmeier and Raith (2003) opined against considering granulite facies metamorphism as the key criterion for defining the extent of EGMB. They recognised four crustal provinces along the eastern coast of India with widely different history of geological evolution, viz, (i) Rengali Province, (ii) Jaypur Province, (iii) Krishna Province, and (iv) Eastern Ghat Province; which are further subdivided into twelve domains characterised by specific features like lithology, structure and metamorphic grade. The Rengali and Jaypur Provinces include domain 1 and 9 respectively, while the Eastern Ghat Province comprises domains 2 to 8 and the Krishna Province domains 10 to 12 (Fig. 3.1.3). Of these, the Eastern Ghat Province, which is broadly correlatable with the domain II and III of Rickers et al. (2001) and includes WKZ, CMZ and EKZ of Ramakrishnan et al. (1998), is considered to have evolved in the Proterozoic times during the assembly of Rodinia. The UHT metamorphism in this Province initiated during Grenvillian orogeny at c. 1.1 Ga and not during the Late Archean as was thought earlier by several workers (Yoshida et al., 1992; Yoshida, 1995; Chetty and Murthy, 1998; Ramakrishnan, et al. 1998; Sarkar and Paul, 1998). The Jaypur Province (≡ domain IB of Rickers et al., 2001/WCZ in the north of Godavari) and part of Rengali Province (≡ part of domain IV of Rickers et al., 2001) are related to pre-Rodinian evolution during Archean time. The Krishna Province comprises the newly defined Ongole domain (≡ southern part of WCZ of Ramakrishnan et al., 1998) and the low- to medium-grade Nellore-Khammam schist belt. The merging of these two apparently unrelated crustal terrains has been justified by these workers on the basis of the congruence of geochronological, magmatic and metamorphic events, pointing towards a common Paleoproterozoic evolution. Dobmeier and Raith (2003) have also substantiated the tectono-metamorphic status of many shear zones described earlier by Chetty, (1995) and Chetty and Murthy (1998), as marking the boundaries of crustal provinces.

Geophysical Attributes of EGMB

The Bouger anomaly map compiled by National Geophysical Research Institute (NGRI) (1975) brings out the EGMB as a prominent curvilinear positive gravity anomaly belt with values ranging between – 40 to 30 mGal, in sharp contrast to the strong negetive values of –10 to –100 mGal in the adjoining Bastar and Dharwar cratons. The conspicuous rise of gravity values to the tune of about 70 mGal over a short distance across the western margin of EGMB has prompted many workers to infer a faulted contact of this belt with the adjacent craton (Subrahmanyan and Verma, 1986; Nayak et al.,1998). Several narrow rectilinear gravity high zones have been delineated within EGMB, most of which trend in NNE–SSW direction, except for the strongest WNW–ESE to WSW–ENE trending Angul gravity high (0–30 mGal) passing through the northern part (Nayak, op. cit.).

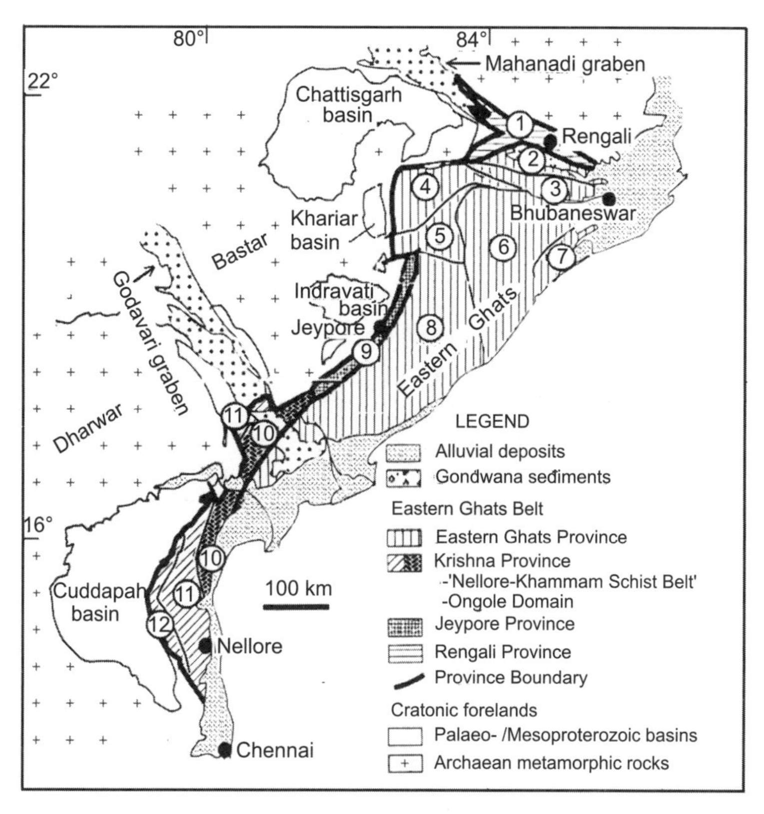

Fig. 3.1.3 Subdivisions of EGMB in four crustal provinces (Rengali, Jaypur, Eastern Ghats and Krishna – as described in the text) based on distinct geological evolution and further divided into twelve domains demarcated by megalineaments and shear zones. The domains indicated by encircled numbers are: 1. Rengali, 2. Angul, 3. Tikarpara, 4. Khariar, 5. Rampur, 6. Phulbani, 7. Chilka Lake, 8. Visakhapatnam, 9. Jaypur, 10. Ongole, 11. Vinjamuru, 12. Udayagiri (slightly modified after Dobmeier and Raith, 2003).

The Deep Seismic Survey (DSS) profiles across the belt have revealed three sub-horizontal discontinuities at the depths of 8 km,12 km and 20 km, followed by the Moho interface at about 34 km depth. The seismic sounding data further suggest ductile nature of the crust below 8 km depth, and the presence of highly disturbed crust in the Ongol sector with the evidence of crust-mantle interaction along a weak zone through a 3.05 g/cm^2 density layer (Kaila and Bhatia., 1981; Nayak et al., op.cit.). The EGMB is also considered to be a moderately active seismic belt, having two prominent axes joining the recorded earthquake epicentres of recent past (magnitude 4–6 on Richter scale), one passing from Kakinada in the south to Bhubaneshwar and beyond in the north, and the other near Ongole. Fault plane solutions of these earthquakes are suggestive of left lateral movement along a NE–SW trending plane parallel to the continental margin for the former and movement along a thrust plane for the latter. All these observations are indicative of a fairly mobile state of the EGMB crust.

Total intensity aeromagnetic and residual gravity data suggest the continuation of both Mahanadi and Godavari grabens into the continental shelf area, thereby providing support to the various models proposed for a possible fit of the eastern margin of Indian Shield with East Antarctica.

Evolutionary History of EGMB

On broad terms, it may be surmised that the EGMB evolved through a protracted period of time from at least Mid-Archean to Late Proterozoic. Besides episodic crustal reactivation leading to polyphase regional metamorphism and deformation, the orogen also witnessed profuse mantle derived magmatism leading to the generation of juvenile crust at different stages of its evolution. This would suggest extremely active mantle-crust interaction, and recurrence of similar magma generating process and tectono-thermal environment along this belt.

We would have expected that the P-T trajectories of UHT metamorphism, worked out in reasonable detail by several workers, would lead to a well-constrained reconstruction of the tectono-thermal history of the EGMB in time and space. But the critical evaluation of the thermo-barometric data in conjuction with petrographic features and available age data presents more problems than solution while attempting to resolve it (Dasgupta and Sengupta, 2000). Irrespective of the age of M_1 event (may be older than 1.6 Ga), extremely high degree of thermal perturbation at crust-mantle boundary at ~11 kb pressure is a prerequisite for this UHT metamorphic event. In view of the stability field of the ubiquitous mineral assemblages, such thermal perturbation cannot be explained either by simple obduction, or 'obduction cum fluid (COH) advection' model of Ganguli et al. (1995). Invoking the classical 'magma-induced thermal pulse' (Wells, 1980; Bohlen, 1987) is also difficult, as the cooling rate here was much slower than the expected fast cooling in case of such short duration thermal pulses. Dasgupta and Sengupta (op. cit.), are tentatively inclined to consider the 'non-extensional lithospheric thinning' of Oxburg (1990) as a possible model, in which, upward migration of ~1400° C isotherm due to some disturbance in the asthenosphere may lead to thinning of the lithosphere without significant stretching. As a consequence, deep crustal rocks are heated to cause UHT metamorphism and simultaneously the partial melting of upper mantle leads to the generation of magmatic products. With the waning of the disturbance, the isotherm slowly subsides back to stable state, allowing the deep crustal granulites and ponded magma to cool down slowly. Under this situation both UHT granulites and the magmatic rocks are considered as the product of the same process, rather than the latter being the cause of the former. However, in the south of Godavari rift, the evidence of mafic magmatic suite- induced regional contact metamorphism is very prominent, which clearly contradicts the contentions of the above model. Moreover, the cooling rate of M_1 is ill-studied all over EGMB and most of the available data of slow cooling of Mezger and Cosca (1999) pertain to M_2 and M_3.

There is considerable gap in knowledge in respect of the timing and interrelationship of the major metamorphic, magmatic and deformational events in the EGMB, thereby greatly restricting our understanding about the tectono-thermal evolution of the belt. Notwithstanding the uncertainties outlined above, Dasgupta and Sengupta (2000) have presented the event stratigraphy for the northern and southern sectors of the belt based on available information, and that is furnished here with minor modification (Table 3.1.1).

Table 3.1.1 *Event stratigraphy of Eastern Ghat Mobile Belt (after Dasgupta and Sengupta, 2000)*

North of Godavari Rift	South of Godavari Rift
Amphibolite facies metamorphism, emplacement of pegmatite at 0.5–0.55 Ga (Pan-African event)	
Massif type anorthosites at 0.8–0.89 Ga	
M_2 granulite facies metamorphism, near isothermal decompression , cooling, emplacement of porphyritic granite/ granitoids at c 1.0 Ga (Grenvillian event)	Local resetting of Ar-Ar radio clock at c 1.0 Ga (geological significance not known)
	Basic dykes Phase II, pegmatite, minor acid intrusives
Emplacement of alkaline and basic magma at 1.5–1.2 Ga	Major phase of crustal reworking at 1.6–1.3 Ga
	Emplacement of alkaline magma at 1.8–1.6 Ga
	Intrusion of enderbitic magma, basic dyke Phase I
Intrusion of basic magma (>>2 Ga) M_1 UTH metamorphism (age unknown)	Intrusion of basic magma,enderbite, and UHT metamorphism

EGMB vis-a-vis East Gondwana Supercontinent

It is widely accepted that the Gondwana supercontinent was divided into two parts, the East and the West Gondwana, of which the East Gondwana, a segment of Rodinia, comprised present day Africa, India, Antarctica, Madagascar and Sri Lanka, and Australia. There are many theories on the possible matching and best fit of these land masses in reconstructing the East Gondwana before its dispersal at about 160 Ma. The basic criteria for the frequently attempted correlation of the eastern boundary of EGMB in peninsular India with the Enderby Land in East Antarctica are the similarity in lithologies and their P-T evolutionery paths (Sengupta et al.,1999), correlation of Mahanadi and Godavari rifts with Lambert and East Enderby rifts and other structural constraints (Hoffman, 1991), and fairly acceptable jigsaw fit of the coast lines between latitude 14°–21° N (Fig. 3.1.4). The recent researches, specially on isotopic signatures and age data of the different crustal provinces of the two supposedly detached counterparts, have cast serious doubts on such a simplified model of correlation.

In order to understand the evolutionary status of EGMB in supercontinental cycles, it would be essential to trace back to the assembly of Rodinia supercontinent (1100–900 Ma), which holds the key to understanding global geodynamic processes from the Mesoproterozoic onwards (Fig. 3.1.5). According to the most accepted model of Rodinia-assembly (Dalziel, 1991; Hoffmann, 1991; Moores,1991), India and Antarctica were sutured along the EGMB. The break-up of Rodinia is considered to have started by late Proterozoic, between 830–750 Ma (Groenewald, 1993; Powell et al., 1994), but some parts remained together untill the Mesozoic fragmentation of Pangea. In all the proposed models, India and East Antarctica remained attached along the EGMB since their suturing during Rodinia assembly, through several subsequent cycles of supercontinent assembly and dispersal including the reassembly of the Gondwana Supercontinent between 500–550 Ma due to Pan-African orogeny (Rogers, 1996). As such, it seems likely that the crustal segments of EGMB and adjacent East Antarctica lived together from Meso-Late Proterozoic–Early Phanerozoic (1100–500 Ma) and were subjected to similar thermo-tectonic events. See also Chapter 8.

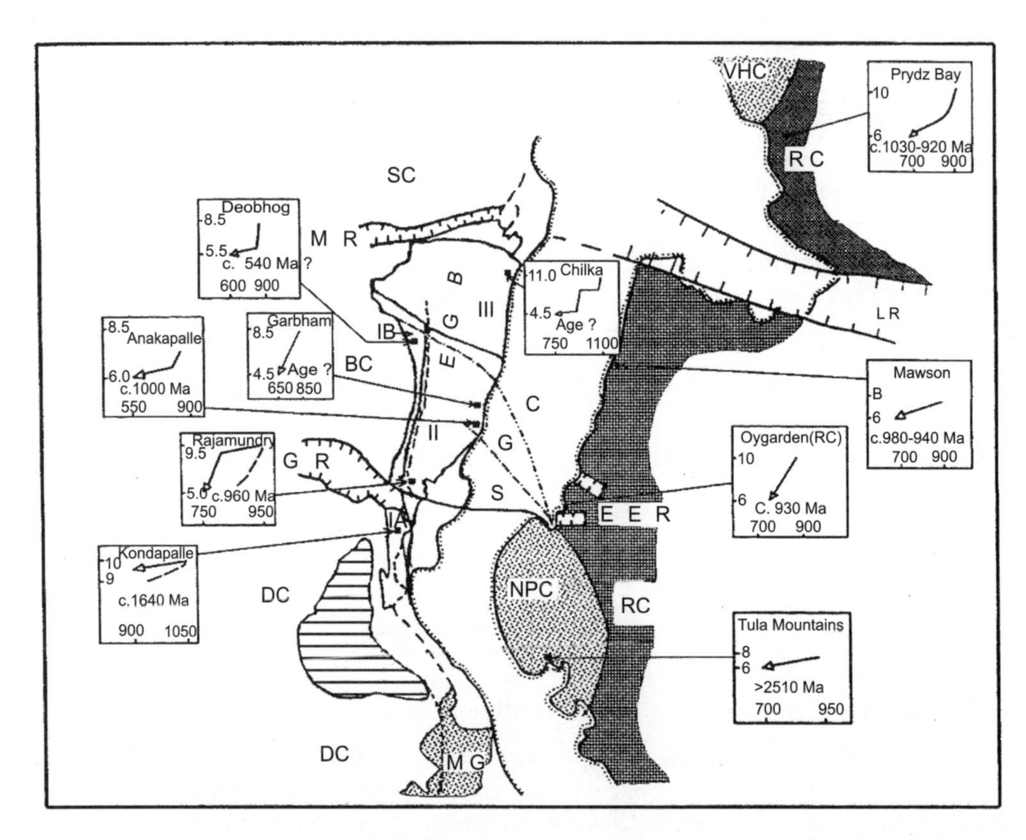

Fig. 3.1.4 Jigsaw fit suggested for the coast lines of the peninsular India and East Antarctica, showing representative P-T paths of metamorphism deduced from the different isotopic domains of Rickers et al. (2001) in Eastern Ghat belt (EGB) and some locations in East Antarctica. S – Grenvillian front in EGB suggested by Dasgupta and Sengupta, SC – Singhbhum craton, MR – Mahanadi rift, BC – Bastar Craton, GR – Godavari rift, DC – Dharwar craton, MG – Madras Granulite; VHC – Vestfold Hills, RG – Rauer Group, LR – Lambert rift, EER – East Enderby rift, NPC – Napier Complex and RC – Rayner Complex (modified after Dasgupta and Sengupta, 2003)

Expectedly, there are many evidences of similar Proterozoic metamorphic imprints having comparable P-T condition(s), reworking of older crust and also juvenile additions of mantle products in isolated sectors of both EGMB and East Antarctica spanning over this time frame. But mere correlation of tectonothermal imprints of mostly post-Rodinia assembly period from isolated sectors did not help much in regional correlation because of a significantly older and diverse crustal evolutionary history recorded in the various domains of both the land masses. The recent attempts made by Rickers et al. (2001) and Dobmeier and Raith (2003) to identify domains/provinces in the EGMB, having characteristic pre-metamorphic crustal evolutionary imprints and dominant Nd model ages (supported by Rb-Sr and Pb-Pb isotope data) have thrown some light on the possible architecture of East Gondwana reassembly. The important inferences arrived at by these workers, supported on many accounts by Dasgupta and Sengupta (2003) on the basis of petrological evidences, are furnished here.

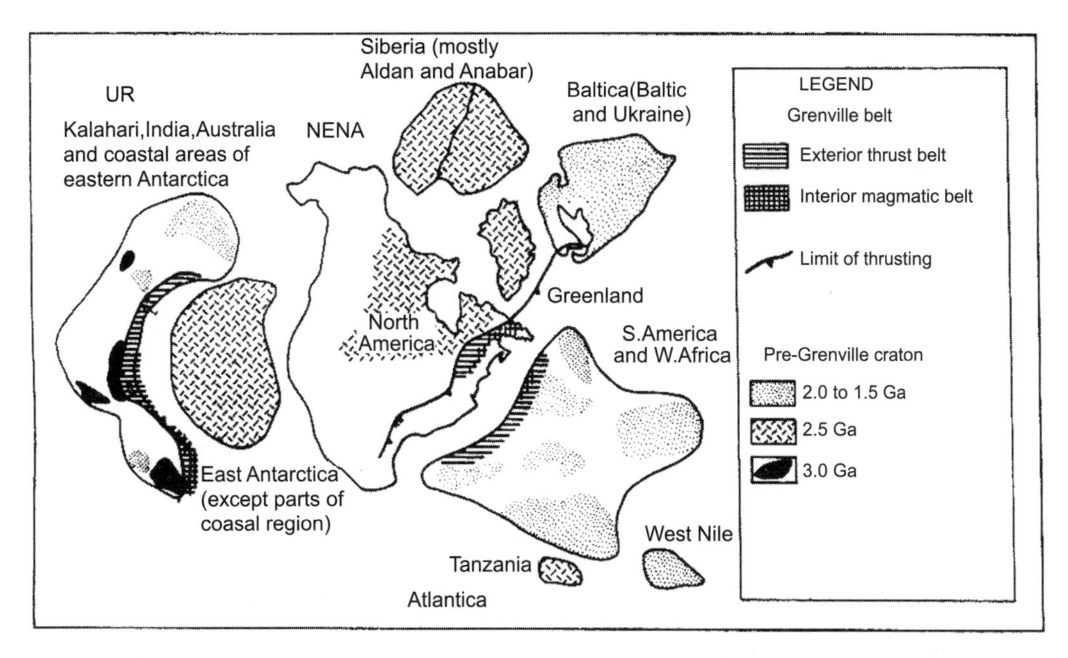

Fig. 3.1.5 Assembly of Rodinia supercontinent as per SW US – East Antarctica (SWEAT) model of Dalziet, 1991; Hoffman, 1991; Moores, 1991 [after schematic diagram of Dasgupta and Sengupta (2003), adopted from Rogers (1996) and Unrug (1996)]

1. There is little correlation in the thermotectonic histories of EGMB and potentially adjacent provinces of East Antarctica before the onset of Grenvillian orogeny (1100–950 Ma).
2. The long held contention of juxtaposing Napier Complex with the WCZ of EGMB, in the south of Godavari for the reconstruction of Gondwana (Hofmann, 1991; Unrug, 1996; Sengupta et al., 1999) is no longer tenable. The Napier Complex (Nd model age > 2.9 Ga) forms a distinct terrain without any counterpart in EGMB. This exotic crustal block could have been welded with the totally independent block of WCZ (Domain 1A of Rickers et al., 2001) at any point of time spanning from assembly of Rodinia to the assembly of Gondwana (Rickers et al., op.cit.).
3. The WCZ in the north of Godavari (Jaypur and part of Rengali province of Dobmeier and Raith, 2003) evolved in Archean time (Kovach et al., 2001) in an active continental margin and was amalgamated with the Bhandara craton previous to or during the Rodinia assembly.
4. The central and eastern domains of EGMB (to the north of Godavari) reveal, a more or less common geological history at ~1000 Ma with the Rayner Complex, Mawson Coast area, Prydz Bay region and Northern Prince Chales Mountains. A general decrease of Nd model age (corroborated by mineral age data of Mezger and Cosca, 1999) from south to north is evidenced both in EGMB and in the mentioned terranes of East Antarctica (Rickers et al., 2001).
5. The model proposed by Boger et al. (2001, in Dasgupta and Sengupta, 2003) suggests continuity of Pan-African orogenesis between Prydz Bay region of Antarctica and Chilka Lake area (Domain III of Rickers et al., op. cit.) of EGMB, which otherwise were quite dissimilar prior to this event. Hence, these two blocks appear to have evolved independently before sharing a common history in the Neoproterozoic (Dasgupta and Sengupta, 2003).

It may be seen that the results of recent researches broadly subscribe to the assembly of EGMB and East Antarctica for reconstruction of the East Gondwana Supercontinent. The thermobarometric information of superposed metamorphic imprints in a composite orogenic belt is often fragmentary, precluding the possibility of discerning the evolutionary history solely on their basis. However, based on new age data and isotope geochemistry, it has been possible now to recognise several crustal domains in both the counterparts, displaying distnctive geological history. Apparent mismatches, quite a few, are due to independent evolution and reworking of a few crustal blocks at different locations and their welding and detatchment at different times over the protracted period of Mesoarchean to Neoproterozoic.

3.2 Metallogeny

Introduction

Eastern Ghats were known for containing occurrences of manganese and graphitic ores at a number of places. Bulk of the graphite is already mined out. But the belt assumed extraordinary importance from the nineteen seventies, on account of the large deposits of bauxitic ores that have been discovered along it. Today it contains about 75% of India's bauxite reserves. Beach sands contain heavy minerals such as ilmenite, rutile, monazite, zircon and sillimanite in the Ganjam coast, particularly close to Chhatrapur where Rare Earth India Ltd., a Government of India undertaking, started a mineral separation plant.

Manganese Mineralisation

Manganese mineralisation occurs discontinuously for more than 200 km in the Eastern Ghat Mobile Belt. In the Orissa sector, it is concentrated over a distance of about 20 km in the Boirani-Kodala zone in Ganjam district, a 30 km zone between Kanaital and Uchhabapalli (Bolangir district), a 15 km long zone near Ambadala (Rayagada district) and another zone of 32 km in the Kutinga-Nishikal area (Rayagada district), close to the Orissa-Andhra Pradesh border. The manganese mineralisation in association with the Eastern Ghat granulites follows on discontinuously into Andhra Pradesh including areas such as Kodur, Garbham and Garividi with other places in Srikakulam district. The rocks associated with the manganese mineralisation are khondalite, quartzite, calc-granulite and 2-pyroxene mafic granulites/gneisses. Granites and granitic rocks occur in profusion (Fig. 3.2.1).

Manganese in the Eastern Ghat Mobile Belt occurs in two mineral associations:

1. Mn-oxide rock, composed of braunite + jaçobsite + hausmanite + quartz. (This is the main source of manganese.)
2. Mn-silicate-carbonate rock, consisting of rhodochrosite + pyroxmangite + rhodonite + bustamite + Mn-clinopyroxene + garnet + olivine.

The manganese of economic interest occurs as:

1. Bedded deposits, co-deformed and co-metamorphosed with the host rocks.
2. Ores concentrated near the surface of the bedded ores, having no relation with deformation or metamorphism.

Ore lenses upto about 150 m in length and 10 m in width are found locally. The pressure and temperature to which the Mn-protoliths together with the associated rocks were subjected to during metamorphism could be as high as 8.5 Kb and 850°C (Dasgupta et al., 1993). The hard lumpy ore, streaky ore and may be with some modifications, the brecciated and friable ores of Bhattacharyya, et al. (2000) had nothing or little to do with the modern exogenic processes. But the cavity filling ores, clayey and wady ores and lateritic ores are obviously the products of modern supergene processes.

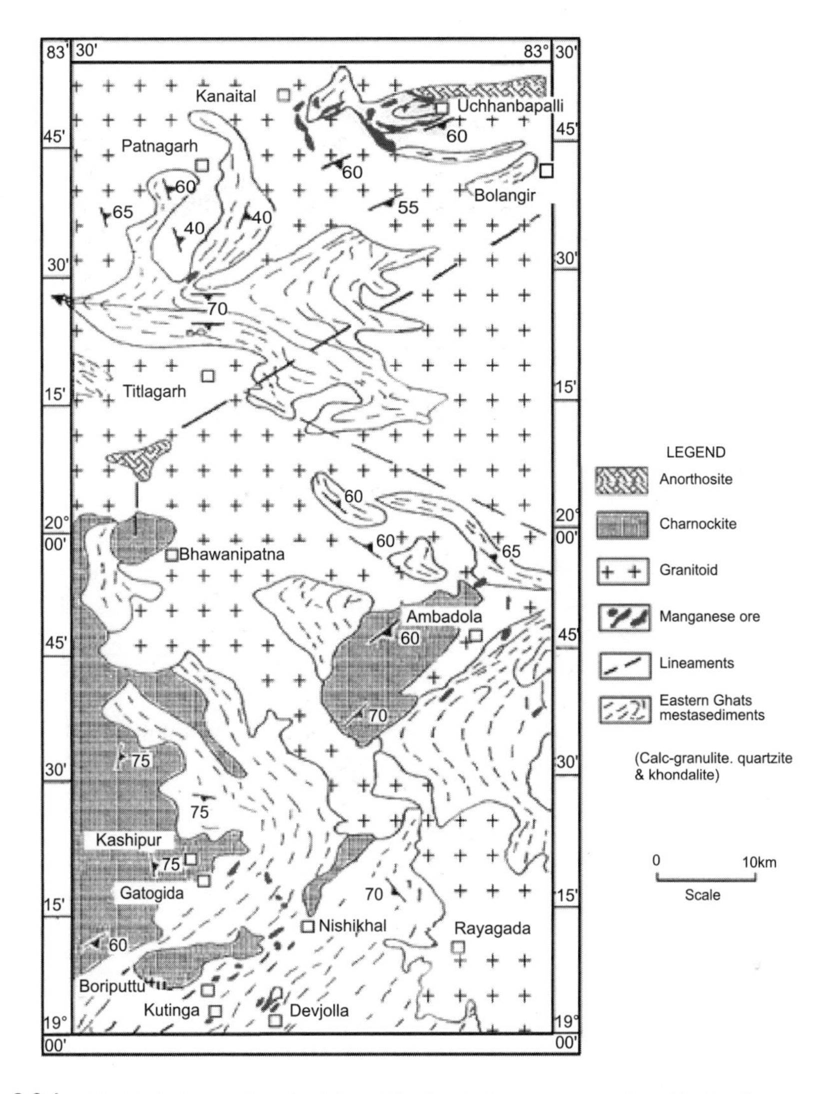

Fig. 3.2.1 Geological map showing khondalite hosted manganese mineralisation in parts of EGMB in Orissa between Kanaital, Bolangir District and Nishikhal–Kutinga, Rayagada District (modified after Bhattacharya et al., 2000)

Primary (metamorphic) Mn (-Fe) oxide minerals present in the ores comprise bixbyite, hollandite, pyrophanite, vredenburgite besides braunite, jacobsite and hausmanite mentioned above. Low temperature, i.e. supergene minerals are cryptomelane, goethite, psilomelane, pyrolusite, lithiophorite, ramsdellite and wad (Sivaprakash, 1980). The silicate-carbonate association in the metamorphosed Mn-bearing silicic-carbonate protolith is represented by rhodochrosite-olivine-garnet-pyroxmangite-rhodonite-bustamite-Mn pyroxene. Olivine is found to contain 65 mol% tephroite, 27 mol% fayalite, and 8 mol% forsterite. Garnet is rich in spessertine (41 – 51 mol%). This might have led Sivaprakash (1980) to assign this rock to the gonditic type, only a little too metamorphosed. Bhattacharyya (1986) reported knebelite [$(Fe, Mn, Mg)_2$ SiO_4] from these rocks. The formation of this mineral is reportedly facilitated by the presence of carbonate in the sediment during metamorphism.

Not enough well-constrained mineralogical data exist about the Mn- bearing rocks metamorphosed to the granulite facies. Dasgupta et al. (1993) found $RSiO_3$ phases (rhodonite-pyroxmangite-bustamite-Mn-clinopyroxene) in textural equilibrium in the Mn-silicate-carbonate rocks of Garbham. They utilised this fact in determining the phase relations in the R (Mn, Mg, Fe, Ca) SiO_3 tetrahedron at the ambient P and T (8.5 Kb and 850°C). Their findings show considerable changes in exchange parameters, Ca-Mn, Mn-Mg and Mn-Fe among manganese pyroxenes and pyroxenoids in granulite facies, compared to the amphibolite facies. Rhodonite accommodates more Ca and clinopyroxene less Mg in the granulite facies. Further, the clinopyroxene field shrinks from Mn apex towards the Fe-apex. In the $MnSiO_3$-$CaSiO_3$-$FeSiO_3$ composition space, rhodonite shows similar Ca-enrichment.

Dash et al. (2005) provide with some data on some Mn-deposits in the Orissa sector of the Eastern Ghat Mobile Belt. These are

1. Kutinga–Nishikal zone: 12.5 Mt/ Av. grade–28.97% Mn, 11.45% Fe and 0.28% P (cut-off grade, 20% Mn).
2. Ambada zone: Mn-content upto 26.26%, analysed from spot samples
3. Kanaital–Uchhabapali zone: 6.5 Mt/ Av. grade–27.38% Mn, 13.74% Fe and 0.31% P
4. Santipur–Sitalpani zone: Reserve data are not available; pit samples from near Sitalpani analysed Mn upto 38.56%, Fe upto 4.33% and P upto 0.238%.

Average grade of Mn-ores from the Srikakulam district in Andhra Pradesh is, Mn = 28–40%, Fe = 5–12% and P upto 0.3% (IBM, 1974).

Fermor (1909) studied the manganese ores from the Kodur area, and concluded that the mineralisation was the supergene product of a rock he named Kodurite. This hypothetical rock was hybrid consisting of garnet (spessertine-andradite), K-feldspar and apatite, the rock having formed by contact metasomatic effect of late granitic plutons on manganiferous country rocks. There is no trace of this rock, however. Moreover, the internal laminations and participation of the Mn-rich rocks in the regional deformation and metamorphism prove that the manganese was originally precipitated as sediment. Only the top parts of the metamorphosed Mn-rocks segregated manganese by supergene enrichment. Of course, the real nature of this manganese mineralisation was brought to light by Mahadevan and Rao (1956), which was supported by Roy (1966). Manganese-silicate-carbonate rocks are reported from several other places in India. As discussed already, carbonate could be a part of the original sediment or diagenetically introduced.

Bauxite Deposits

Lateritisation at the hill-tops in the Eastern Ghats has been reported by a number of geologists, starting from Ball in 1877. But that many of them are repositories of large bauxite deposits came to be known only during the seventies of the last century, thanks to the intensive exploratory activity of the Geological Survey of India, supported by the Government agencies such as the Mineral Exploration Corporation Ltd. (MECL) etc.

The bauxite deposits in the Eastern Ghats, commonly known as East Coast Bauxite are mainly distributed along the coastal regions of Orissa and Andhra Pradesh (the term 'coastal region', however, is not used *sensu stricto*). Orissa has a larger share of these deposits, having a reserve of ~1600 million tonnes, which constitutes ~53% of the country's total reserve. Andhra Pradesh contains 658 million tonnes, or ~22% of the national reserve. In other words, about 75% of India's bauxite deposits are confined to the Eastern Ghats only. The bauxite deposits are scattered over a zone of 400 × 30 sq km (approx.), which is aligned in accordance with the Eastern Ghat orographic trend.

Orissa Sector The major deposits in the Orissa sector of the Eastern Ghats are located at Pottangi, Panchpatmali, Ballada, Maliparbat, Karnapadikonda, Kodingamali and Baphlimali in the Koraput district, and Karlapat, Lanjigarh, Sijimali in the Kalahandi district. The Gandhamardan deposit, in spite of its location in the westerly districts of Bolangir and Sambalpur, is still included under East Coast Bauxites, because of its close similarity with the latter in most respects. It is the second largest bauxite deposit in India with a reserve of 216 million tonnes of ores with + 40% Al_2O_3. The salient features of these deposits are presented in Table 3.2.1.

Table 3.2.1 *Salient features of the major bauxite deposits in the Orissa sector of the Eastern Ghats*

	Deposits	Lat–Long	Elevation (m)	Area (sq.km)	Thickness (m)	Reserve (Mt) and grade
Koraput district	Pottangi	18°34′–18°46′: 82°56–82°58′	1206–1404	4.5	6–32	75 (+40% Al_2O_3)
	Panchpatmali	18°46′–18°55′: 82°57′–83°04′	1180–1355	~13	2.25–31	322 (+40% Al_2O_3)
	Ballada	18°27′–82°00′	1185–1264	0.86	2.1–19.2	12.4 (+40% Al_2O_3; 5% SiO_2)
	Maliparbat	18°39′–18°40′: 82°53′–82°56′	1176–1399	1.1	1–41	9.89 (Met.grade-I)
	Karna-Padikonda	18°46′–18°50′: 83°03′–83°-07′		~2	3.6–19.0	17 (+45%Al_2O_3; 5% SiO_2)
	Kondingamali	19°04′–83°05′	1102–1276	5.3	2.5–32.1	91.4(+40% Al_2O_3; 5% SiO_2)
	Baphlimali	19°18′–19°22′: 82°57′–82°59′	1004–1094	9.6	2.25–25.8	196(+40% Al_2O_3; 5% SiO_2)
Kalahandi district	Karlapat	19°37′–19°41′: 83°08′–83°11′	960–1080	9.73	2–15	51 (Met.grade-I)
	Lanjigarh	19°42′45′–83°22′15′	1016–1306	5.64	2–18	46.4(+40% Al_2O_3; 5% SiO_2)
	Sijimali	19°28′–19°32′: 83°06′–83°10′	991–1233	13	3–9	86 (+40% Al_2O_3)

Preliminary exploration by the State Department of Geology and Mines has revealed the presence of bauxitic deposits that are of economic interest, elsewhere in the districts of Kalahandi, Koraput and Phulbani. They vary from 13 to 30 million tonnes in reserve, and have an average composition of > 40% Al_2O_3, with < 5% SiO_2.

Andhra Pradesh Sector Andhra Pradesh is another state that is rich in bauxitic ores, second only to Orissa. The reserve of bauxitic ore in Andhra Pradesh, occurring along the Eastern Ghats is of the order of 658 million tonnes, making about 22% of the country's total reserve. The major bauxite deposits in the Andhra Pradesh sector of the Eastern Ghats are given as follows:.

1. Gurtedu Group – Katamrajkonda deposit
2. Chintapalli Group – Sapparla deposit, Gudem deposit, Jerala deposit
3. Anantagiri Group – Galikonda deposit, Raktakonda deposit, Katuki deposit, Chittangandi deposit.

For a comparative view of these deposits, their salient features are presented in Table 3.2.2. A comparison with the data in Table 3.2.1 will bring out the similarity and contrast in the characteristics of bauxite deposits in the northern and southern sectors of the Eastern Ghats.

Table 3.2.2 *Salient features of the major bauxite deposits in the Andhra Pradesh sector of the Eastern Ghats*

	Deposit	Lat–Long	Elevation (m)	Area (sq.km)	Thickness (m)	Reserve (Mt) and grade
Gurtedu Group	Katamraj-Konda (2 blocks)	17°50′–17°52′: 81°53′–81°55′	936–1286	1.8	2–39	42.6(46.8–49.9% Al_2O_3; < 3% SiO_2)
Chintapalli Group	Sapparla (12 blocks)	17°53′30″–18°59′30″: 82°05′–82°13′15″	974–1288	15.15	0.75–35.5	184(45–48% Al_2O_3; < 4% SiO_2)
	Gudem (5 blocks)	17°48′31″–17°56′: 82°08′23″–82°13′30″	1020–1232	2.63	3.20–39.5	44.41 (46–49.5% Al_2O_3; < 3% SiO_2)
	Jerala (4 major blocks)	17°57′–18°01′54″: 82°00′–82°17′15″	916–1270	13.5	1.0–31.75	245.74 (45.27–46.84% Al_2O_3; < 3% SiO_2)
Anantgiri Group	Galikonda (3 blocks)	-	1170–1446	0.62	3.5-50.9	15.05 (48.22–49.24% Al_2O_3; 2.78% SiO_2)
	Rakta Konda	-	-	0.42	11.4	5.57 (46.75% Al_2O_3; 2.56% SiO_2)
	Katuki	-	1296	0.14	-	4.43 (49.27–50.58% Al_2O_3; < 3% SiO_2)
	Chittam-gandi	18°22′–82°56′30″	1419	1.52	9.61	25.5(46% Al_2O_3; 2.6% SiO_2)

Gurtedu Group falls under the East Godavari district while the deposits of the Chintapalli and Anantagiri Groups are located within the Visakhapatnam district.

General Characteristics of East Coast Bauxite Deposits in India The isolated hill-tops covered by the lateritic duricrust in the Eastern Ghats are generally elongate and sub-parallel with the Ghat's-trend. Notable, the Eastern Ghats are generally strike-ridges. The monadnocks along the coastal plains are at places capped by laterites. The duricrust or the lateritic cover surprisingly covers a very small proportion (< 1%) of the Eastern Ghat region as a whole.

All the major bauxite deposits along the belt are 'high level' bauxites, being located at elevations greater than 900 m above MSL. Thickness of the mineralised zone widely vary, from < 1.0 m to > 50 m (Tables 3.2.1 and 3.2.2), the two extremes being rare.

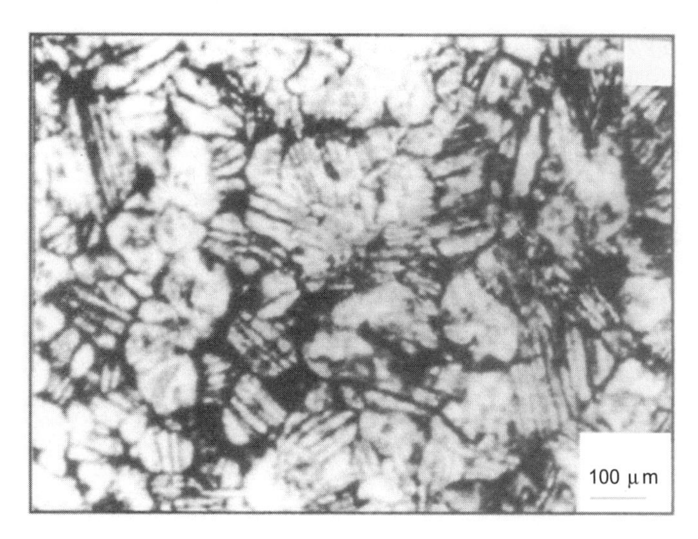

Fig. 3.2.2 Photomicrograph of bauxite showing holocrystalline texture, Panchpatpalli (courtesy: P. Singh)

As already discussed, the major rock types along the Eastern Ghats are khondalites, charnockites, and granite-granitic gneisses with some intrusives of anorthosite, pyroxenite, syenite and carbonatite. Khondalites that occupy the valleys and the coastal plains are generally kaolinised. Bauxitic duricrusts developed on both the khondalites and the charnockites. But the majority of them (> 90%) are localised on the khondalites. The ores are generally spongy and holocrystalline (Fig. 3.2.2), but rarely pisolitic.

One characteristic feature of these ores is the presence of 'pseudofolia' (Sen and Guha, 1987), apparently inherited from the protolith (Fig. 3.2.3). The essential minerals of bauxite (-laterite) are gibbsite, hematite-goethite and kaolinite. TiO_2 is present as anatase. Sillimanite, partially replaced by gibbsite, is not a rare feature under the microscope.

Fig. 3.2.3 Hand specimen showing pseudofolia in bauxite, Panchpatpalli (courtesy: P. Singh)

The weathering susceptibility in the silicate minerals of protoliths is as follows (Sen and Guha, 1987):

Garnet < Na-feldspar < biotite < K-feldspar < muscovite < quartz < sillimanite

Petrographic criteria, supported by theoretical consideration suggest the following genetic relationship of the minerals in the bauxite ores with those in protoliths.

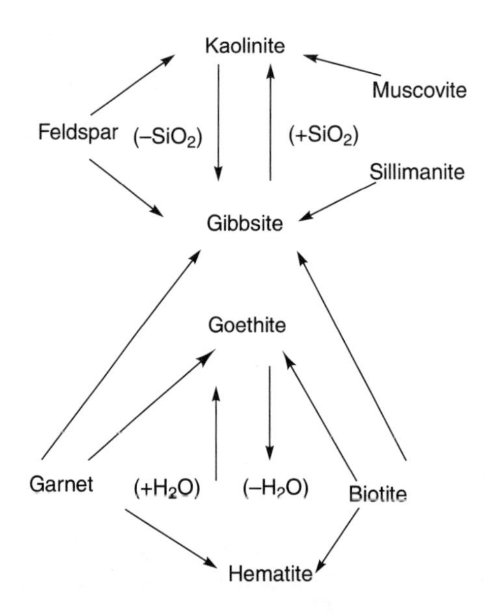

Bauxitic ores in the Andhra Pradesh sector are better in quality compared to those in Orissa, plausibly, due to better drainage. A generalised profile of the weathering crust involving bauxite formation in the East Coast is as shown in Table 3.2.3.

Table 3.2.3 *A generalised weathering profile in the East Coast (GSI, 1994)*

Alteration zones	Thickness (m)
Fe-lateritic duricrust	1–5
Bauxite with lateritic intercalations	2–54
Partially lateritised protolith/lithomerge	3–25
Kaolinitised protoliths	5–8
Unaltered protoliths (khondalite, charnockite)	

Variations within the above generalised picture will be obvious from the weathering profile established by Sen and Guha (1987) across the Panchpatmali bauxite plateau, Koraput (Table 3.2.4). Average chemical composition of khondalitic parent rocks and the weathered rocks derived from them at Panchpatmali and Pottangi bauxite plateaus are presented in Table 3.2.5. Mineralogical variations in the weathering profiles of Pottangi and Galikonda are presented in Fig. 3.2.4.

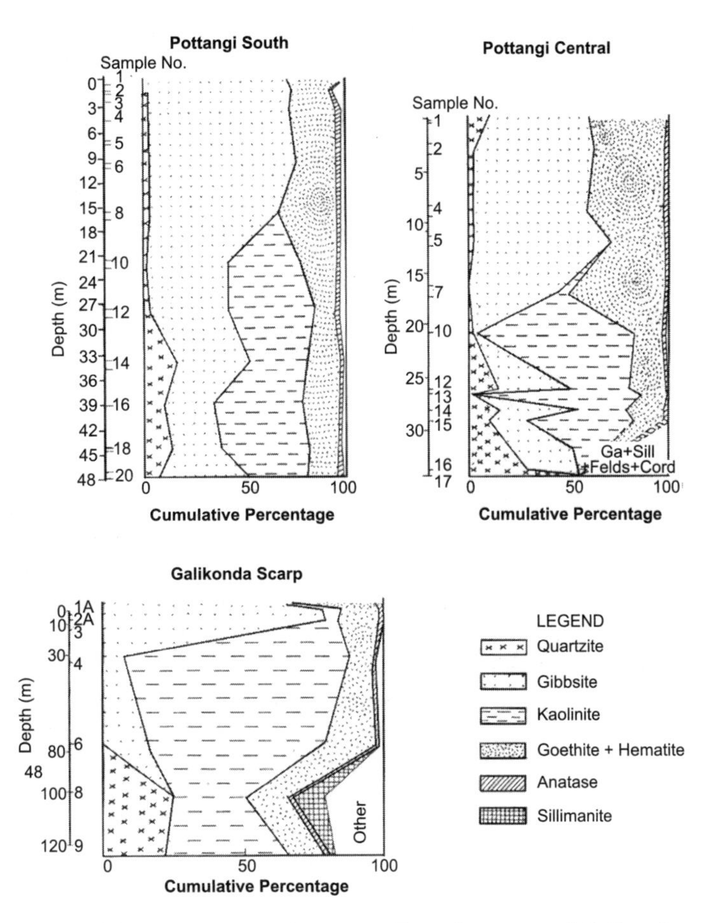

Fig. 3.2.4 Mineralogical variation in weathering profile at Pottangi and Galikonda areas, Andhra Pradesh (after Deb and Joshi, 1984)

Table 3.2.4 *A weathering profile across the Panchpatmali bauxite plateau, Koraput, Eastern Ghats (Sen and Guha, 1987)*

Major Formation	Rock	Depth (m) from surface
Laterite–bauxite zone	Laterite: reddish brown, Fe-rich, pisolitic/nodular, thin crust (duricrust) Lateritic bauxite: low-grade ferruginous, scoriaceous, crystalline, non-pisolitic Bauxite: medium to high grade, spongy, essentially crystalline, non-pisolitic	0.00–0.15 0.15–3.00 3.00–14.00
Saprolite zone	Gibbsitic underclay Highly ferruginous deepbrown clay White/pale to yellow/buff clay	14.00–21.00
	Gibbsitic altered parent rock Zone of neotexture Zone of pseudomorphism	21.00–50.70
Parent rock	Khondalite	Base not exposed

Table 3.2.5 *Average chemical composition of khondalitic parent rocks (protoliths) and the weathered rocks derived from them at the Panchpatmali and Pottangi bauxite plateaus, Koraput (Sen and Guha, 1987)*

Rock type	Number of samples analysed	Analytical results (wt%)									
		SiO_2	Al_2O_3	Fe_2O_3	FeO	TiO_2	CaO	MgO	Na_2O	K_2O	LOI
Laterite-bauxite	65	1.98	48.63	19.60	0.19	1.98	Tr	Tr	0.23	0.05	27.08
Saprolie	26	26.91	34.10	20.40	0.26	1.41	Tr	Tr	0.19	0.02	16.39
Parent rock	4	56.70	28.67	2.22	1.78	0.65	0.48	0.35	2.07	5.68	1.08

Other Minerals

Apatite–Magnetite Zoned apatite–magnetite–vermiculite veins occur in the Kasipatnam area, Visakhapatnam district, Andhra Pradesh. The Kasipatnam hill lies along the foothills of the Eastern Ghats, and is confined within latitudes 18°13′ and 18°13′ and longitudes 83°07′ and 83°11′ in parts of Survey of India toposheet Nos. 65N/3 and 4. The apatite–magnetite–vermiculite veins occupy 'en-echelon' shear fractures.The shear fractures are disposed at angles of 75° to 85° to the foliation of the host rocks (Mahendra, 1975). The veins are characteristically localised in the charnockitic gneisses and spatially associated marginal zones of mafic granulites, calc-granulite, quartzite and leptynite but do not extend into the surrounding khondalitic rocks. The veins that fill the ' en-echelon' tension fractures generally range up to 1 m in width and occasionally reach ~3 m. Individual lenses vary in length from a few metres to about 30 m. Pinching and swelling of veins along both strike and dip are conspicuous. The mineralisation is contemporaneous with or later than the formation of the younger NW–SE fractures. The ore veins are well-zoned and pegmatitic. The various zones recognised are: an outer shell of ferrosalitic clinopyroxene, followed by a shell of vermiculite of variable thickness, a sheath of magnetite, core of apatite and occasionally a central most core of allanite (Rao et al., 1969; Mahendra, 1975). A suggested paragenetic sequence of mineral-formation in the ore veins is as follows: Ferrosalitic clinopyroxene → biotite (hydrated to vermiculite) → zircon → magnetite → apatite → allanite (Mahendra, 1975).

In addition to apatite, magnetite and vermiculite, the veins contain a complex assemblage of accessory minerals, viz, large anhedral crystal of allanite, zircon, scapolite and actinolite. Feldspar occurs in appreciable proportion in some of the veins. Pyrite and chalcopyrite are also noted in some zones occurring as minor fracture-fillings. Pale green to deep red apatite generally occurs in massive, crystalline form. Some major apatite-bearing veins occur at Regulavalasa (1.5–2 m thick) and Sitarampuram (3 m thick) The veins are distinctly radioactive and chemical analyses of apatite from the area suggest these to be fluorapatite with P_2O_5 ranging from 36–42% (Mahendra, 1975). Whitish brown to golden yellow vermiculite is a ubiquitous component in all the veins and occurs as hexagonal crystals forming selvages of variable thickness between the apatite core and the wall rock. At some places the vermiculite books are 30–40 cm thick and in the past were worked as a source of vermiculite with an established export potential (Mahendra, 1975). Magnetite occurs in massive and homogenous form. It is titaniferous with 73% ferric iron and 1% titanium. In some of the veins, pure magnetite occurs close to the surface and gradually gives place to apatite at depth. These veins are believed to have formed in a post-tectonic regime. Zircons from the apatite-magnetite veins have given $^{206}Pb/^{238}U$ age of 502 ± 3 Ma and $^{207}Pb/^{206}Pb$ age of

508 ± 14 Ma suggesting Pan-African thermo-tectonic event (Kovach, et al., 1997). Aggregates of euhedral zircon are confined in the outer margin of the apatite zone suggesting possible crystallisation during the early stages of apatite formation (Rao et al., 1969).

Magnetite dykes also occur in the area. Between 1957 and 1959, about 5000 tonnes of magnetite was mined from the area. Between 1959 and 1970, 34,000 tonnes of apatite, 389 tonnes of vermiculite and 383 tonnes of magnetite were commercially produced. Geologists from GSI estimated 600,000 tonnes of probable (indicated) reserves and 2,000,000 tonnes of possible (inferred) reserves of apatite from the area (Mahendra, 1975).

The apatite–magnetite–vermiculite veins of Kasipatnam area are considered to be products of magmatic injections from a carbonatitic magma by Mahendra (1975). Rao (1969) considered these as metasomatic skarn deposits.

Base Metals The Adash copper prospect (21°23′05″: 84°37′45″) is located in the northern segment of the Eastern Ghats belt at a distance of 120 km from Sambalpur and can be approached from Riamal on the Deogarh–Anugul road (Pattanaik et al., 1998). The copper mineralisation is hosted by silicified mafic granulites, and is controlled by a shear zone along the contact of khondalites and granitic gneisses. The main sulphide minerals are represented by disseminated chalcopyrite and pyrrhotite. GSI drilled a total of 26 boreholes totalling 4811.65 metres to establish copper mineralisation over a strike length of 1600 metres. For operational convenience, the prospect has been divided into three exploration blocks, viz, School, Rampalli and Mundasuni. The copper lodes show pinch and swell structures. The top oxidised zone extends up to 40 m from the surface. Three lodes present, are disposed in an en-echelon fashion. Estimation of reserves and average tenor are based on two cut-off grades, viz, 0.4% Cu and 0.8% Cu. The dimensions and range of Cu% of the three lodes are (after Pattanaik et al., 1998) given below.

(1) Western lode (No.1):	700 × (2 – 21) m	mostly continuous,	0.41 to 2.81% Cu
(2) Central lode (No. 2):	400 × (2 – 11) m	continuous,	0.20 to 2.27% Cu
(3) Eastern lode (No. 3):	50 × (1 – 3.5) m	inconsistant,	0.72 to 1.17% Cu.

On the basis of the above, Pattanaik et al., (1998) computed the reserve and grade of the Adash deposit as follows:

(1) Block:	Tonnage/Tenor (0.8% Cu cut-off)	Tonnage/Tenor (0.4% Cu cut-off)
(2) School:	(Probable) 0.93 million tones/1.46%	1.81 million tonnes/1.04%
	(Possible) nil,	0.60 million tonnes/1.04%
(3) Rampali:	(Probable) 0.17 million tonnes/1.66%,	nil.

Graphite The major graphite occurrences of EGMB are in Orissa state, and are confined within migmatised quartz-garnet-sillimanite-graphite schists of the Khondalite Group. The major graphite belts, deposits/occurrences and mine workings in this region are given in detail in Table 3.2.6.

According to Mishra et al. (1998), the crystalline and flaky graphite bodies in khondalitic rocks can be classified into four types, viz, (i) Disseminated and streaky type; (ii) Vein, stringer and schlieren type (iii) Sheet type; and (iv) Pockety type.

The concentration and abundance of graphite appear to be related to the degree of migmatisation of the khondalite suite. The grade of graphite is variable, ranging from 1 – 2% 'fixed carbon' (F.C.) in disseminated types to 85% F.C. in rich graphite lodes/veins. In general, the F.C. content of graphite ores varies between 5 – 40%, and the ore samples respond well to physical beneficiation (Acharya and Rao, 1998). Bulk of the total of 1.35 million tonnes of recoverable reserves of graphite in the Orissa State is confined within the Eastern Ghats terrain. Acharya and Rao (1998) critically discuss the various theories put forward regarding the origin of the graphite deposits of the Eastern Ghats belt, viz, metamorphism of carbonaceous matter, reduction of oxides of carbon, granulite grade metamorphism in presence of CO_2 rich fluid under low fO_2 conditions, methanation of carbonate minerals etc. According to Acharya and Rao (1998), methanation of carbonate minerals followed by subsequent pyrolysis is responsible for graphitisation and remobilisation of graphite in the solid state. $^{12}C/^{13}C$ data, if present, would suggest whether the original material for the graphite could be a biomass or not.

Table 3.2.6 *Major graphite belts and deposits/occurrences within the confines of EGMB in Orissa (after Mishra et al., 1998)*

Graphite belts/districts	Area/ No. of occurrences	Major occurrences/deposits/mine quarries
Bargad, Bolangir and Kalahandi districts	More than 150 occurrences	Major occurrences are at Darrhamunda, Sapmuna, Dharukhaman, Rangali, Mohanilah, Raju–Nagfenda and Benimal areas
Titlagarh belt: Bolangir and Kalahandi districts	Over an area of 3500 Sq km ~ 120 occurrences	Prominent quarries are near Titlagarh, Boroni, Malisira, Singjharan and Loitora
Tumudlbandh belt: Phulbani, Rayagada and Gajapati districts	~ 9600 Sq km more than 50 occurrences	Main workings at Tumudibandha, Lakhajorna, Palur, Raisil, Ambaguda and Bandhamundi
Dandatopa belt: Dhenkanal and Anugul districts	1000 Sq km 22 occurrences	Prominent quarries at Dandatopa, Adeswar, Kamalpur, Tileswar, Akharkata and, Girida

Tungsten in the form of ferberite and scheelite occurs in quertzo-feldspathic pegmatites at a number of places in gneissic rocks, often graphite-bearing. The latter could initially have been khondalites. The aggregate strike lengths of the pegmatites could be several hundred metres and the width a few metres. The most well-known amongst these occurrences is near Burugubanda, East Godavari district, Andhra Pradesh. Small-scale mining took place on some of them.

Heavy Minerals Heavy mineral accumulation has taken place in sections of the EGMB, particularly at the Donkuru-Sanaikasingi (-Gopalpur)-Chhatrapur-Rushikulya-Palur section of the Ganjam coast of Orissa. Heavy mineral-content of the sands at the Chhatrapur area may be as high as 23% by weight. The heavy minerals comprise ilmenite, rutile, monazite, zircon and sillimanite. Rare Earth India Ltd., a Government of India undertaking, started a mineral separation plant at Chhatrapur. The source of the heavies remains the metamorphic-plutonic rock-complex that constitutes the EGMB.

4

Eastern India

4.1 Geology and Crustal Evolution

Introduction

The eastern segment of the Indian Precambrian shield preserves records of geological events spanning over Early Archean to Late Proterozoic time. There is a debate about the definition of the Eastern Indian craton. Should it comprise only the Archean basement rocks of South Singhbhum and some adjacent parts of Orissa and the banded iron formation (BIF)-bearing belts of supracrustal rocks flanking them? Or should it also include the physically contiguous and geologically related Proterozoic mobile belt of North Singhbhum and the Chhotanagpur Gneissic Complex (CGC), lying to the north of this belt. We, in the present context, define the area in its conventional sense, viz, from the Chhotanagpur Gneissic Complex in the north, upto the margin of Eastern Ghat granulite terrain in the south, and the eastern margin of the Mahanadi graben in the west to the last exposure of Precambrian rocks in the east. Considered in the context of crustal evolution of the region, these limitations should satisfy the definition of a craton (cf. Goodwin, 1996; Hoffman, 1988). However, the Archean granitic basement and the ancient supracrustal rocks of South Singhbhum–North Orissa may be considered to constitute the Archean cratonic nucleus (Fig.4.1.1).

Distinctly recognisable provinces within the Eastern Indian craton are:

1. The Archean cratonic nucleus (3.55–2.7 Ga), represented by the Singhbhum Granite batholithic massif (SBG), with enclaves of the Older Metamorphic Group (OMG) and Older Metamorphic Tonalite Gneiss (OMTG). Late Archean Tamperkola Granite (2.8 Ga), Rengali–Raimal Granulite (2.7–2.8 Ga) and Mayurbhanj Granite (3.0–2.8 Ga) occur along fringes of the terrain. Gabbro–anorthosite–ultramafic intrusives characterise the eastern domain from north to south. The SBG is further characterised by the presence of intersecting sets of 'Newer Dolerite' dyke swarm (oldest date: 2.6 Ga).

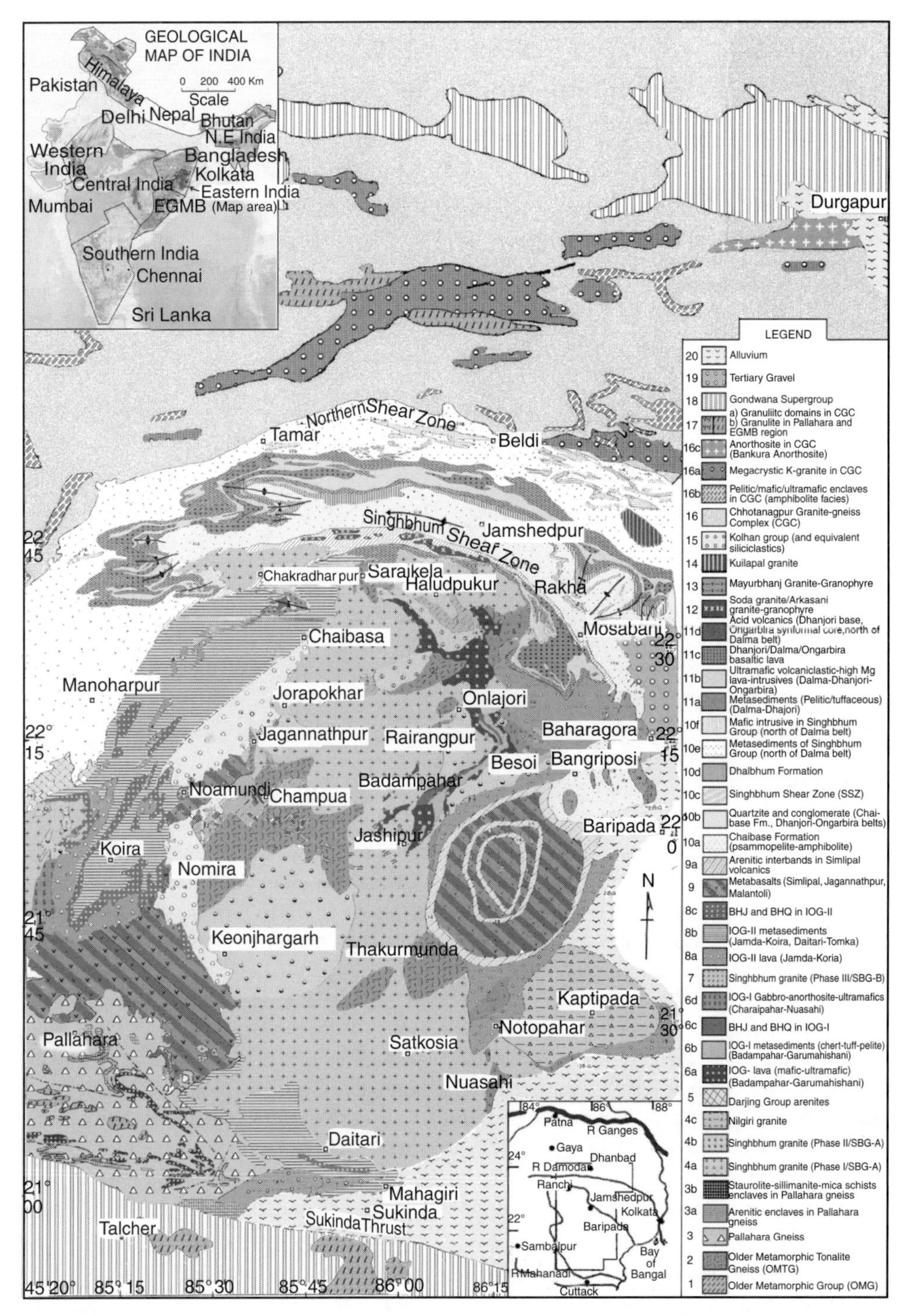

Fig. 4.1.1 Geological map of a part of the Precambrian terrain of Eastern India (Largely drawn from Dunn, 1929, 1942, Jones, 1934 Saha 1994; Sarkar and Saha, 1962; Gupta et al., 1980, 1981, 1985; Basu, 1985; Basu, 1993; Chattopadhyay and Ray, 1997; Blackburn and Srivastava, 1994; Ghosh and Bhattacharya (1998); Mazumder, 1996, 1988; Jena and Behera, 2000)

2. Iron Ore Provinces (volcanisedimentary sequences ± BIF,commonly described as Iron Ore group, or IOG) occur along the eastern, western and southern peripheries of the cratonic nucleus (Badampahar-Gorumahisani belt, Jamda-Koira-Noamundi belt and Tomka–Daitari belt). There are also other sedimentary, volcanisedimentary and volcanic sequences of varied ages in South Singhbhum.
3. The North Singhbhum Proterozoic mobile belt (Singhbhum Group: Chaibasa and Dhalbhum Formations., Dhanjori Group, Dalma Group, Ongarbira Group and their equivalents).
4. The Chhotanagpur Granite–Gneiss Complex with metasedimentary, mafic-ultramafic enclaves and granulite slivers.
5. Northern parts of Raimal–Rengali sector of Eastern Ghats Mobile Belt (EGMB).

Archean Cratonic Nucleus

The Singhbhum–Orissa Archean cratonic nucleus is a sub-rectangular or ovoid body (150 × 50 km) with its longer axis aligned to north-south. The ovoid massif, called Singhbhum Granite (SBG-A and SBG-B), is a composite granitic batholith, having innumerable large and small enclaves of Older Metamorphics (OMG), trondhjemite grading into tonalites (OMTG) and several unclassified rafts and inclusions, comparable to the components of typical Archean greenstones. Along the eastern and northeastern edge of the massif, there is a prominent linear greenstone belt (Badampahar–Gorumahisani iron ore belt) having tongues extending far into the Singhbhum batholithic massif and profusely intruded by the last phase of Singhbhum Granite (SBG-B). Surrounding the main Singhbhum Granite massif, there are quite a few detached granitic bodies, named as Bonai Granite, Chakradharpur Granite–Gneiss, Nilgiri (Kaptipada) Granite etc., which are considered to be equivalents of the main batholith.

Older Metamorphic Group (OMG) and Older Metamorphic Tonalite-Gneiss (OMTG) The Older Metamorphic Group (OMG), the oldest supracrustals recognised so far in eastern India, comprises medium grade (amphibolite facies) pelitic schists, arenites, calc-magnesian metasediments, ortho- and para-amphibolites etc., which are synkinematically intruded and partially digested by a suite of biotite (hornblende) tonalite gneiss grading into trondhjemite/ granodiorite (OMTG) (Sarkar and Saha, 1977, 1983; Saha, 1994).

The pelitic suite is the dominant constituent of OMG in the type area near Champua (22°04′: 85°40′), the important variants being muscovite-biotite (-sillimanite) schist, quartz sericite schist grading into quartzite, and quartz-magnetite-cummingtonite schist. The calc-magnesian metamorphites include banded calc-gneiss. Massive and schistose amphibolites are considered as ortho-amphibolites whereas the banded amphibolite is regarded as para-amphibolite. The ortho-amphiboites show tholeiitic character (Ray et al., 1987). Saha (op. cit.) was inclined to assign a komatiitic affinity to some of the ortho-amphibolites on the basis of CaO-MgO-Al_2O_3 plots but this postulation does not stand a closer scrutiny of the petrochemical data. Ultramafic komatiitic members are altogether absent in this suite, which are otherwise so very common in the oldest Archean supracrustals. The OMG supracrustals display intense deformation and are estimated to be metamorphosed at 5–5.5 kb and 600–630° C, which is indicative of substantial thickening of the crust at that time (Saha, op. cit.).

Besides the type area of OMG and OMTG in the western parts of the cratonic nucleus, there are numerous enclaves of varied size and shape of older supracrustals throughout the batholithic massif, which were partly grouped as OMG and OMTG by Saha (1994). However, the litho-packages and metamorphic grade of none of the OMG enclaves noted by Saha (mostly as amphibolites) in the south of Holudpukur, west of Bahalda near Rairangpur and Onlajori, and around Thakurmunda conform to those of Champua area. In the Holudpukur–Kalikapur stretch there are extensive exposures of older supracrustals comprising pillow basalt, gabbro, ultramafics, ferruginous arenites and tuffaceous rocks of acid-intermediate composition. Similarly, several enclaves of the OMTG were also marked in the regional map by Saha (1994) in the south of Saraikela, near Rairangpur, Asana–Manda–Besoi, south of Kalikapur, and around Thakurmunda. Most of the available petrochemical and other details on the OMG and OMTG rocks are from the type areas, and many of the newly identified enclaves outside the type areas are yet to be studied and precisely classified.

Majority of OMTG falls in the trondhjemite field while only a few plot in tonalite and grandiorite fields. Geochemically, OMTG shows remarkable similarity with the early Archean tonalite-trondhjemites from different parts of the world except for having conspicuously low Ti, relatively low V, Y and Zr, enriched Sr and Mn, and impoverished Ba and Rb.

The major element, REE and isotopic data are consistent with the derivation of the OMTG from partial melting of OMG amphibolites or equivalent rocks at the amphibolites–garnet field at depths. An initial ϵ_{Nd} (t) value of $+ 0.9 \pm 0.7$ for the amphibolites indicates the presence of a slightly depleted mantle source at 3.3 Ga with $^{147}Sm/^{144}Nd$ between 0.20 and 0.22 according to Sharma et al. (1994).

An age of 3700 – 3850 Ma (Sm-Nd) obtained for OMTG in the early eighties of the last century (Basu et al., 1981), created much interest in these and their associated rocks. Moorbath and Taylor (1988), however, obtained a much younger age of 3350 – 3400 Ma (Pb-Pb, Rb-Sr) for the same rock and Misra et al. (1999) reported a date of 3440 Ma (Pb-Pb) for OMTG. Sharma et al. (1994) obtained 3300 Ma (Sm-Nd) age for the OMG rocks and concluded that the age of OMTG could not be older than that. However, the dates for OMG rocks by Goswami et al. (1995) was 3550 Ma (Pb-Pb) and those by Misra et al. (1999) was 3550 – 3600 Ma (Pb-Pb). Recently two samples from the OMTG outcrop, located by Saha (1994) between west of Runkini temple and Potka (Fig. 4.1.1), yielded U-Pb (zircon) dates of 3527 ± 17 Ma and 3448 ± 19 Ma (Acharyya, Gupta and Orihashi, 2010). Nelson et al. (2007) have reported igneous crystallisation age of 3380 ± 11 Ma (Pb-Pb, zircon SHRIMP) age for OMTG exposed near Jagannathpur.

Singhbhum Granite (SBG) The Singhbhum Granite batholithic complex comprises at least twelve granitic units (plutons), which are considered to have been emplaced in three successive but closely related phases: Phase I and II dated at c 3.3 Ga and Phase III at c 3.2 – 3.1 Ga (Saha et al., 1986; Saha, 1994 and references therein). The Pb-Pb (WR) isochron ages of SBG-A (Phase-I and II) were, however, reported as 3442 ± 26 Ma and 3298 ± 63 Ma by Ghosh et al.(1996).

Each of these phases is reported to be compositionally distinct in terms of both major and trace element distribution. The twelve units of Singhbhum Granite, having distinctive shapes and foliation pattern, as described by Saha et al., and Saha (op. cit.), are enlisted below.

1. Saraikela–Jorapokhar–Tiring unit (737 sq km)
2. Haludpukur–Chapra unit (168 sq km)
3. Rajnagar–Kuyali unit (82 sq km)
4. Kalikapur–Matku unit (61 sq km)
5. Dalima unit (107 sq km)
6. Gorumahisani unit (33 sq km)
7. Rairangpur–Onlajhori unit (74 sq km)
8. Gamaria–Khorband–Karanjia unit (1792 sq km)
9. Hatgamaria unit (61 sq km)
10. Bara Nanda unit (5 sq km)
11. Keonjhargarh-Bhaunra unit (704 sq km)
12. Manda–Asana–Besoi (MAB) unit (200 sq km)

Based on critical field relations and composition, the twelve units were considered by Saha and his co-workers (op. cit.) to be the products of three successive phases and were grouped as follows:

1. Phase I: Rajnagar–Kuyali (3) and Dalima (5) units.
2. Phase II: Rairangpur–Onlajhori (7), Hatgamaria (9), Keonjhar–Bhaunra (11), and Besoi (12) units.
3. Phase III: Saraikela–Jorapokhar–Tiring (1), Haludpukhur–Chapra (2), Kalikapur–Matku (4), Gorumahisani (6), Gamaria–Khorband–Karanjia (8) and Bara Nanda (10) units.

The Phase I units are K_2O-poor granodiorite to tonalite, Phase II rocks are mostly granodiorite and the Phase III rocks are generally granite with highest average K_2O and SiO_2 among all the three.

The overlapping character of the different phases of Singhbhum Granite in respect of alkali contents is revealed in the Na_2O-K_2O diagram (Fig.4.1.2,a), while the distinction between Phase-I &-II and Phase-III is seen in the (Ga/Al) vs FeO/MgO diagram (Fig.4.1.2,b).

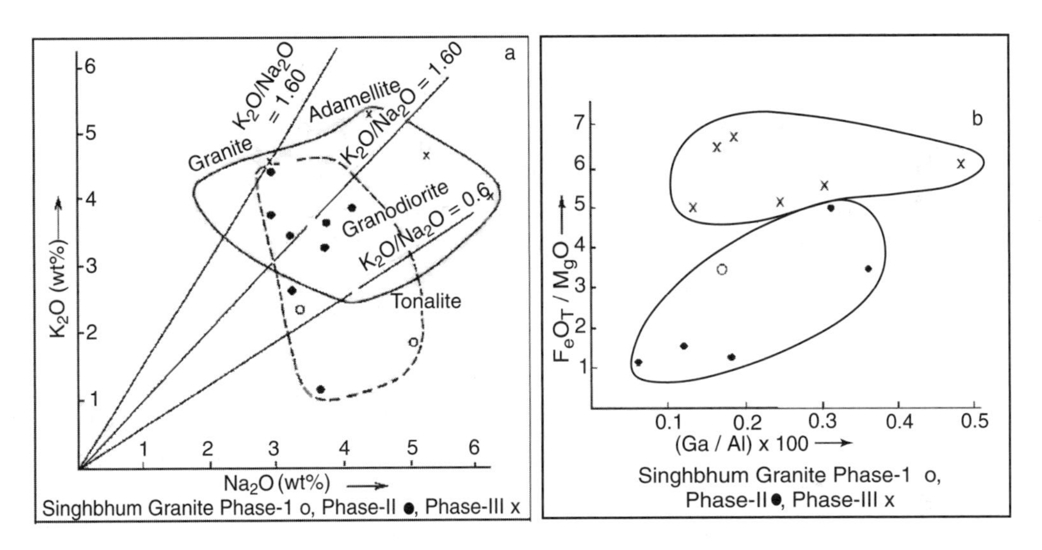

Fig. 4.1.2 (a) Na_2O-K_2O diagram showing overlap of the Singhbhum Granite phases (b) Ga/Al vs FeO/MgO diagram showing distinction between Singhbhum Granite Phase I and Phase II with the Phase III (slightly modified after Saha, 1994).

On the basis of REE-content, the Phase I and Phase II are further grouped together as SBG-A; and Phase III as SBG-B. The SBG-A is characterised by gently sloping REE-pattern with slightly depleted HREE and very weak or no Eu-anomaly. The SBG-B shows LREE enriched fractionated pattern with moderate to strong negative Eu-anomaly (Fig.4.1.3) (Saha and Ray, 1984; Saha, 1994). The SBG-A shows considerable compositional variation compared to OMTG, indicating pronounced fractionation of the former. The REE-pattern of the SBG-A is similar to those of the OMTG, except the following differences:

1. Overall REE abundance in SBG-A is 20% higher, and the slope is steeper than those in OMTG.
2. The average La/Yb ratio in OMTG is 26.7 as against 32.8 for SBG-A.

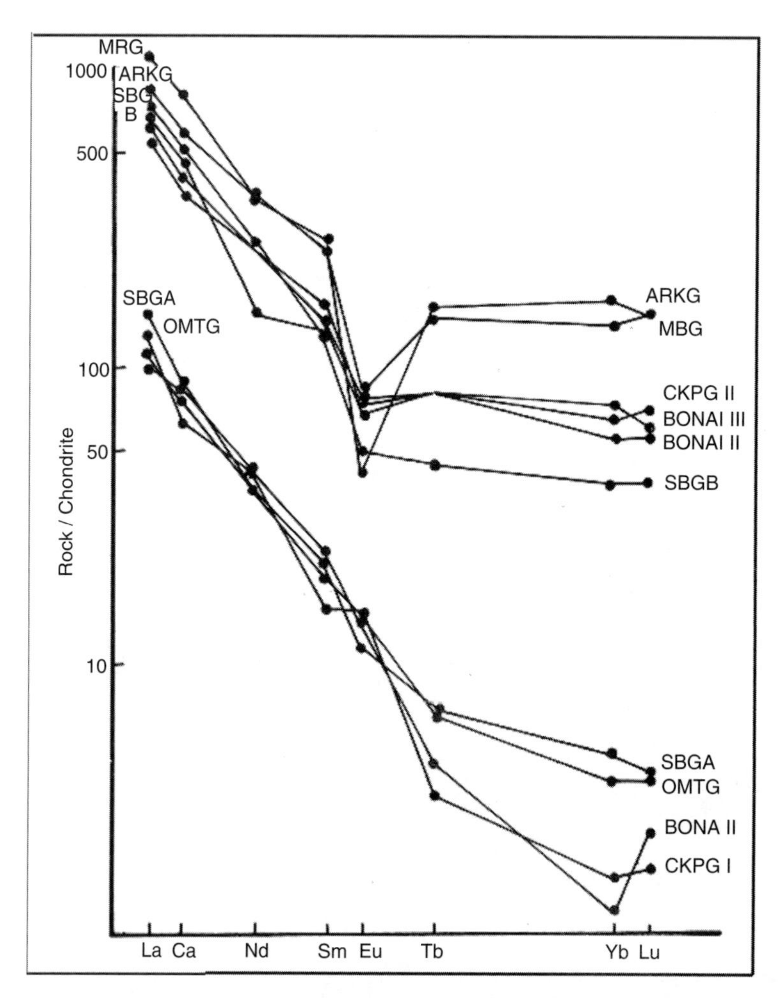

Fig. 4.1.3 Chondrite normalised REE plots of SBG-A and SBG-B showing distinctive patterns and their similarity or contrast with other granitic phases of the region (modified after Sengupta et al., 1991).OMTG: Older Metamorphic Tonalite Gneiss; Bonai I, II and III: Bonai older and younger phases; CKPG I and II: older and younger phases of Chakradharpur Granite; MBG: Mayurbhanj Granite; ARKG: Arkasani Granite

Saha (op. cit.) mentions that enclaves of OMG amphibolites and the OMTG are widely noted within SBG-A, while those of IOG rocks are quite common in SBG-B but are never recorded within SBG-A. The observations of Jena and Behera (2000), however, contradict the above. They mapped trails of numerous IOG- enclaves within SBG-B as well as in the OMTG (mapped by Saha, op.cit.), extending westward from the southern parts of the Badampahar-Gorumahisani (IOG) belt to the Champua area (Fig. 4.1.1). In the Rairangpur–Besoi sector (22° 0′–22° 25′: 86°5′–86°30′), Dey (1991) recorded banded and dark-coloured tonalitic rafts floating in the Singhbhum Granite. The older rafts do not have any inclusion of the IOG components whereas all the phases of Singhbhum Granite profusely intrude them. He also observed that the conglomerates within IOG sequence of the Badampahar–Gorumahisani belt are intraformational, and do not represent basal unconformity overlying any phase of Singhbhum Granite. The above observations, which are indicative of an older status of the eastern belt of the IOG with respect to Singhbhum Granite (SBG-A and B), are also shared by the present authors. The tonalitic rafts noted by Dey (op. cit.) might represent the vestiges of ancient sialic crust in the region.

The geochronological data obtained so far for the granitic rocks from the Singhbhum Granite terrain also lead us to debatable inferences due to significant overlaps. It may be noted that bulk of the age data available so far for OMTG range from 3.55–3.35 Ga, while those for SBG-A are 3.4–3.3 Ga. It may not, therefore, be unreasonable to consider that the OMTG and SBG-A, which are geochemically similar, were also near contemporaneous in their emplacement. In view of the blurring of real distinction between the OMTG and the early phases of SBG with the inflow of more and more data on petrochemical, geochronological and field criteria, the possibility of their continuous evolution as batholithic components between ~ 3.55–3.3 Ga cannot be ruled out. Older supracrustals (OMG and vestiges of other unclassified suites) of appropriate composition could be the protoliths for generation of granitic magma.

Though the chronology of OMG, OMTG, SBG-A and IOG are yet to be adequately understood, it may be reasonable to conclude that in Eastern India the major crust-forming events during Archean time closed at around 3.1 Ga with the emplacement of SBG-B, but for some later reactivation (2.8–2.7 Ga) in isolated sectors, discussed later.

Other major granitic bodies in this terrain also deserve a brief discussion here.

Bonai Granite (BG) Bonai Granite occupies about 700 sq km of a roughly triangular area in the west of the main cratonic massif of Singhbhum Granite, separated by the Iron Ore Group (IOG) rocks of Jamda–Koira–Noamundi basin. The eastern boundary of the body is in contact with the western limb of the IOG-'horse shoe' synclinorium of Jones, while to its west and north west occur the rocks equivalent to IOG (without BIF), Darjing Group (Mahalik, 1987) and Tamparkola Granite.

Based on field criteria and chemical composition, the granitic members of Bonai batolith are divided into three types (Saha, 1994):

1. Small patches of migmatitic tonalite (Unit-I)
2. Medium-grained, equigranular, two-mica trodhjemite with moderately developed foliation, also not making large bodies (Unit-II)
3. Porphyritic granite-granodiorite (Unit-III).

The bulk of Bonai Granite consists of a porphyritic granite/ granodiorite. The fairly common xenolithic bodies within the porphyritic granite are composed of migmatised tonalite (Unit-I) and equigranular trondhjemite (Unit-II). The porphyritic granite intersects and also has large enclaves of amphibolite, quartzite and pelitic metasediments, and meta-lavas, which are described as IOG-rocks by Saha (1994). The trondhjemitic xenoliths yielded an age of 3380 Ma (U-Pb, Pb-Pb) for magmatic zircon and a minimum age of 3448 Ma for xenocrystic zircon grains (Sengupta et al., 1994; Sengupta et al., 1996). The tonalite–trondhjemitic components may be reasonably correlated with OMTG based on major element geochemistry, REE pattern and age data. The dominant unit of Bonai Granite, dated at 3163 ± 126 Ma (Pb-Pb) (Sengupta et al., 1991), is comparable with the major unit of Singhbhum Granite (SBG-B) but differs in having higher K_2O and SiO_2 and lower MgO and CaO contents. The REE contents and patterns of the older phase of Bonai Granite (Unit-I) are more or less comparable with those of OMTG, SBG-A, CKPG-I, except for having much lower Yb content (Fig.4.1.3).

Chakradharpur Granite-Gneiss (CKPG) This granite–gneiss complex occupies about 200 sq km of an E–W trending lensoid shaped area in the western part of the North Singhbhum Proterozoic belt, separated from the main Singhbhum Granite massif by a sequence of low-grade metasediments and the Ongarbira volcanics. The northern limb of the forked Singhbhum shear zone (SSZ) in this region traverses along its northern margin and the southern limb along its southern margin (Fig. 4.1.1). The CKPG body was earlier considered as a late Proterozoic intrusive, equivalent to Arkasani Granite and Soda granite occurring along SSZ (Sarkar and Saha, 1962). Though Saha

(1994) maintained the same view by grouping it under Proterozoic granitoids, the characteristics of CKPG suggest it to be equivalent of the main Singhbhum Granite massif. Dependable age data on it, however, are still lacking.

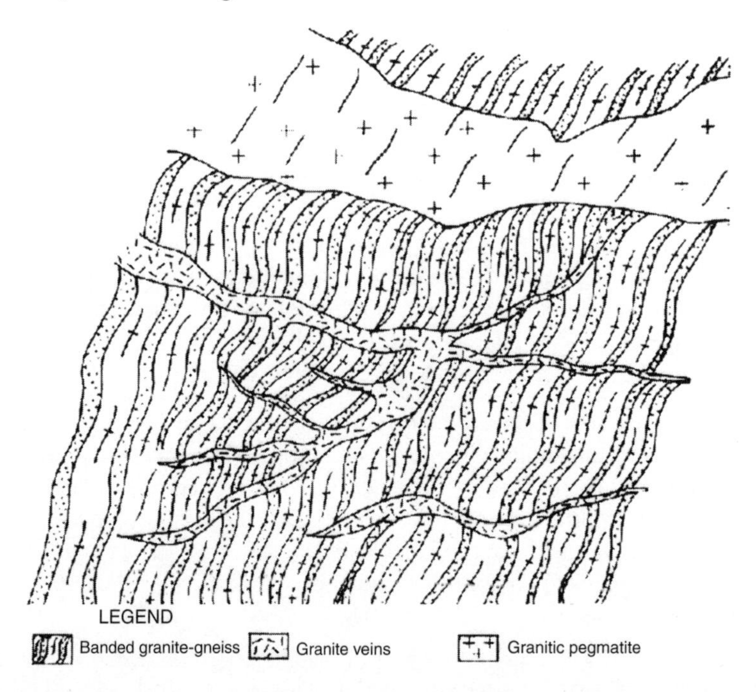

Fig. 4.1.4 Sketch showing field relationship of the different phases of CKPG Complex, 1 km north of Chakradharpur town limit (not to exact scale) (after Sarkar, 1984)

The CKPG complex consists of banded gneissic rocks, patches of metasediments comprising phyllite, quartzite, and conglomerate, variously transformed mafic and ultramafic rocks, late granitoid bands and stockworks, and finally a pegmatitic phase. A field sketch (Fig. 4.1.4) of CKPG exposure is indicative of the interrelationship of the various granitic phases in this complex. Newer Dolerite dikes traverse this complex. Available information on this complex (Dunn, 1929; Bandyopadhyay, 1981; Sengupta et al., 1983; Bhaumik and Basu, 1984; Sarkar, 1984) is outlined below.

The banded gneiss is tonalite–trondhjemite, grading upto diorite in composition. There are steeply inclined to reclined folds with steeply plunging northerly axes and E–W trending axial planes, a feature common in the entire shear zone. The late granitoid member is also granodiorite-trondhjemite in composition.

Alike many other Early Precambrian gneissic rocks of similar composition, these rocks have low REE, large LREE/ HREE ratios and positive Eu anomaly (Fig. 4.1.3) and an amphibolite, garnet amphibolite or eclogite as their source rock for the origin of these rocks on melting. However, the Chakradharpur gneisses differ from similar rocks in proven Archean terrains as these rocks occur in a low metamorphic (greenschist) environment, whereas most others occur in environments of amphibolite–granulite facies of metamorphism. Indeed, the mafic and ultramafic enclaves within the Chakradharpur Gneissic Complex can be likened to the Older Metamorphic Group (OMG) and tonalite- trondhjemite enclaves of the Older Metamorphic Tonalite Gneiss (OMTG) of South Singhbhum (Ghose et al., 1984).

Nilgiri Granite (Kaptipada Granite) The Nilgiri Granite, spanning over an area of about 150 sq km, is the southeastern arm of SBG and hardly differs from the latter. However, Saha (1994 and earlier publications) named this massif as Nilgiri Granite and separated it from Singhbhum Granite massif through the intervention of a tongue of Mayurbhanj Granite and N–S trending septum of IOG. Compositionally, it is a tonalite–granodiorite–granite massif and plots in the VAG field in a (Y+Nb)–Rb diagram. Saha (op. cit.) divided this massif into five discrete magmatic units, viz, (i) Kaptipada Grandiorite–Tonalite (central and SW part of the batholith), (ii) Korakad–Salchua Granite–Granodiorite, (iii) and (iv) Poradiha and Salchua Tonalite and (v) Sarat–Salchua Tonalite–Granodiorite. Vohra et al. (1991), naming the whole massif as Kaptipada Granite, reported a nine-point Rb-Sr isochron date of 3275 ± 81 Ma (MSWD = 2.38) for the tonalite–granodiorite suite of Kaptipada area.

Mayurbhanj Granite (MBG) The elongate Mayurbhanj Granite (MBG) pluton, occurring in the NE part of the Singhbhum–North Orissa craton, extends from near Butagora (22°33′: 86°20′) in the north to Kuliana (22°05′: 86°40′) in the south. The crescent shaped batholith skirts the southern boundary of Dhanjori basin in the north, tapering down towards southwest along the western boundary of the Simlipal basin. The small pluton of Romapahari in the east and the elongate stretch of granitoids occurring between Nilgiri (Kaptipada) Granite and Simlipal are also correlated with MBG.

Saha and his co-workers (Saha et al., 1977; Saha, 1994) divided the Mayurbhanj Granite into the following units:

1. A fine-grained near homophanous granophyric biotite-hornblende–alkali feldspar granite
2. A coarse grained near homophanous to well foliated ferrohastingsite–biotite granite, locally with ferrohedenburgite
3. A biotite aplogranite.

These rocks have $K_2O/Na_2O \geq 1$, in contrast to the SBG where it is <1. These are believed to be anorogenic (Saha, 1994). According to the author, (2) is intrusive into (1) and (3) is intrusive into (2). He further reported that Mayurbhanj Granite was intrusive into Singhbhum Granite, associated gabbro (-anorthosite) and even the metasediments of the Singhbhum Group, the Chaibasa Formation in particular. The REE pattern of MBG with prominent nagative Eu-anomaly is comparable with that of Arkasani Granite and the total content-wise it may be grouped with the later phases of the granitoids in this terrain (Fig.4.1.3).

The MBG was dated to be ~ 2200 Ma (Rb-Sr, WR) by Iyenger et al. (1981) and 1895 ± 46 Ma (Rb-Sr, WR) by Vohra et al. (1991). Misra et al. (1999) reported a Pb-Pb zircon age of 3080 ± 8 and 3092 ± 5 Ma from two groups of zircon grains from these rocks. This suggests that emplacement of Mayurbhanj Granite was coeval with the last major phase of Singhbhum Granite (SBG-B). This creates as many problems as it solves. This no doubt, supports the contention of Dunn and Dey (1942) that the Mayurbhanj Granite is a continuation of the Singhbhum Granite massif itself. It also indirectly confirms another early view that the granophyre of Gorumahisani are gradational into the nearby gabbro-anorthosite, the southern representative of which at Baula–Nuasahi has recently been radiometrically analysed for a similar Pb-Pb zircon age of 3122 ± 5 Ma (Auge et al., 2003). On the other hand, it becomes difficult to explain the so-called intrusive relation of the Mayurbhanj Granite with the volcanisedimentary rocks of the Dhanjori Group and the Chaibasa Formation sediments. The paper by Misra et al. (op.cit.) does not include a sample-map or at least the mention of the locations of the samples they collected for this study, nor the petrochemistry of the rocks sampled. This concern becomes all the more important in view of the fact that the Mayurbhanj Granite obviously contains numerous enclaves of earlier granitoid rocks, particularly in the areas close to the SBG.

In course of a recent study, a fine grained granophyric MBG from near Pitamahali (22°21′: 86°26′) yielded 2827 ± 200 Ma U-Pb zircon age, with a fairly strong overprinting of a possible thermal and/or acid magmatic event at 793 + 37 Ma (Acharyya et al., 2008). The diverse age data obtained so far from the MBG may be suggestive of the possibility of its being a multi-phased emplacement.

Pala Lahara (Pallahara) Gneiss and Rengali-Raimal Granulite A vast expanse of granite gneisses occupies the area between southwestern limit of Singhbhum Granite massif in the north and the Eastern Ghat granulite belt (EGMB) in the south. These gneissic rocks formally called 'unclassified gneiss' and later renamed as Pala Lahara Gneiss (Sarkar et al., 1990), have noticeable difference with other granitoid massifs of the region, besides being gneissic. The Pala Lahara Gneiss plots in the 'granite field', while others plot more in the tonalite–granodiorite–trondhjemite field. No radiometric age data of this rock is available. But these are believed to underlie the rocks of the Malayagiri–Sundermundi and Deogarh iron ore (IOG) basins (Saha, 1994). The geology of this gneissic terrain is further complicated by the occurrence of E–W trending sub-parallel domains of 'charnockite' (granulite), IOG and Kolhan-type sediments in its southern part. The Pala Lahara Gneisses are considered by Saha (op. cit.) to be a metamorphosed and thermally activated product of an ancient andesite-rhyodacite lava complex, which represents the Archean basement in this region.

There are charnockite bodies trending almost E–W, in the west of Pala Lahara in the Rangali–Riamal area. These granulitic rocks have been dated at 2.74 Ga (Rb-Sr, WR) (Sarkar et al., 2000). These are some of the oldest granulite rocks of the country, next only to the products of a suspected 3000 Ma granulite metamorphic event in South India (cf. Meen et al., 1992; Jayananda and Peucat, 1996). Their relationship with the EGMB is worthy of investigation.

Tamperkola Granite and Acid Volcanics Prasada Rao et al. (1964) recorded the occurrence these rocks, but it was Mahalik (1987) who studied these rocks in some details and named the granitoid member as the Tamperkola Granite. This body occurs to the west of Bonai Granite (Fig.4.1.11). It has yielded insitu Pb-Pb zircon age of 2809 ± 12 Ma (Bandyopadhyay et al., 2001). The granite is high in $(K_2O + Na_2O)/SiO_2$ and K_2O / Na_2O ratios and satisfy most criteria of an anorogenic granite. The associated acid volcanic rocks closely resemble the granitoid rocks in composition and fall in the rhyolite, rather trachyte, field in a TAS diagram.

Gabbro-anorthosite and Ultramafic Suites Besides the granitic rocks described above, the Archean cratonic nucleus of Eastern India is characterised by large bodies of gabbro-anorthosite intrusives and ultramafic suites.

Several gabbro-anorthosite bodies occur along the eastern margin of the Singhbhum Granite massif, extending from Butgora (22°35′: 86°20′) in the north to Nuasahi (21°16′: 86°20′) in the south. Locally, they are associated with ultramafic rocks such as peridotite, pyroxenite and lherzolite. Their field relationships with other major rock units have been controversial. These rocks, together with those in Charaipahar–Dublabera areas, are often associated with vanadiferous and titaniferous magnetite bands and aggregates.

Dunn and Dey (1942) believed the gabbro-anorthosite suites to be pre-Singhbhum Granite in age. Chatterjee (1945) on the other hand, presented evidence that the relationship was just the opposite. Sarkar and Saha (1962, 1963) shared the view of Chatterjee and even concluded that the gabbro-anorthosite rocks and the granophyres of Mayurbhanj were possibly co-magmatic. Chakraborty et al. (1981) also concluded that the gabbro–anorthosite–granophyres–leucogranite in the Gorumahisani area constituted a co-magmatic suite. Years later, on a revision of their earlier conclusion Sarkar and his co-workers (cf. Saha et al., 1977) declared that the gabbro-anorthosite suite and the granophyre were not co-magmatic and that the former was obviously older than the granophyre.

At Baula- Nuasahi down south along the belt occurs a basic-ultrabasic complex, consisting of dunite–peridolite–pyroxenite and it hosts a gabbro-anorthosite massif, the whole complex being intrusives into the Iron Ore Group (IOG) rocks. Recently, Auge et al. (2003) seperated fresh zircon from a gabbro member (Bangur Gabbro) and the gabbro matrix of the breccia zone. The Pb-Pb age obtained from these zircon grains ranged from 3119 ± 6 to 3123 ± 7 Ma. If this gabbro was emplaced nearly at the same time as those occurring further north in the belt, then those gabbro-anorthosite suites are not Proterozoic in age, as some have suggested (Saha, 1994), but were emplaced in the late Middle Archean. It at the same time proves that the IOG in the eastern belt (Badampahar–Gorumahisani) is older in age than the above. If a similar age obtained for the Mayurbhanj Granite by Misra et al. (1999) is truly representative, then this date would support a contention that the granophyre (Mayurbhanj Granite) and the gabbro-anorthosites are contemporaneous, if not consanguinous. In that case, the problem of explaining the 'invasion' of Mayurbhanj Granite into the basal Dhanjori and also the Chaibasa Formation belonging to the North Singhbhum Proterozoic belt would remain an enigma, at least for the time being.

A body of ultramafic rocks occurs in the Sukinda valley in the southwest of Nuasahi, within the basalts bounded by BIF of Daitari and quartzite of Mahagiri. Major members constituting this body are dunite, bronzitite, websterite and harzburgite. The rocks are intensely deformed. More details on both Nuasahi and Sukinda suites are provided in Section 4.2 in the context of PGM and chromite mineralisation. Many isolated gabbro and ultramafic intrusions are also noted within the SBG massif. Partially layered cumulate ultramafics associated with non-cumulate mafic-ultramafic differentiates are found intruded into the phyllitic rocks near Jojohatu–Roro, in the Jamda–Koira valley.

Newer Dolerite The Singhbhum cratonic nucleus is characterised by a spectacular cross-cutting basic dyke swarm, with dominantly NNE–SSW to NE–SW and NW–SE trends. These dykes were named 'Newer Dolerite' by Dunn and Dey (1942) and were described as dominantly quartz dolerite with minor noritic and porphyritic variants. The dykes range in length from a couple of metres to over 20 km and in width from less than a meter to nearly 1 km. The spacing of the dykes ranges from one to four per kilometer. The larger dykes form prominent rectilinear hillocks in the flat undulating granitic country.

Saha and his coworkers (Saha, 1994) described a number of additional lithologic types within Newer Dolerite suite (viz, ultramafic-, syenodioritic-, troctolitic- micropegmatitic-, xenolithic-, lamprophyre-dykes etc.), but almost 90% of the dykes are doleritic in composition conforming to the description of Dunn (1929) and Dunn and Dey (op. cit.). Though the Newer Dolerites are most spectacular within the Singhbhum Granite massif, they also traverse the Bonai Granite, Chakradharpur Granite–Gneiss and Nilgiri (Kaptipada) Granite with lesser degree of frequency and persistence. The dyke system uninterruptedly cuts across the OMG, OMTG and IOG rocks of the Badampahar Gorumahisani belt but never crosses the margins of these granitic massifs. It is noteworthy that the Mayurbhanj Granite country is devoid of these dykes. Besides, neither the Proterozoic supracrustals of North and East Singhbhum, nor the IOG rocks of Jamda–Koira–Noamundi belt are penetrated by the Newer Dolerites. The same holds true for the iron ore provinces of Daitari–Tomka and Pala Lahara region. It is, however, intriguing that dismembered dolerite dikes similar to those in CKPG body are also seen within the Arkasani Granophyre, which occurs well within the SSZ.

Verma and Prasad (1974) on the basis of paleomagnetic studies identified at least three episodes of dolerite intrusion during Late Precambrian at wide time intervals.

The field relations and geochemical characters indicate the existence of two generations of dykes in the Newer Dolerite swarm, viz, Dolerite-I with higher MgO, Cr, Ni and Ca/Na ratio and Dolerite-II with higher Al_2O_3, Sr, Cu and Co/Ni ratio (Saha, 1994). The effects of metamorphism are distinct in parts of the dyke system. Sarkar et al. (1969, 1977) presented age data (K-Ar/WR) ranging from 950–1960 Ma for the Newer Dolerite, mostly from the eastern sector. Mallik and Sarkar (1994) obtained K-Ar (WR) ages of c 1250 and 2100 Ma for two sets of younger and older dykes from the western parts of the Singhbhum Granite massif. So far, the oldest date (Rb-Sr, WR) of ca 2.6 Ga has been obtained for an ultramafic Newer Dolerite dyke in the western part of the SBG terrain (Roy, 1998).

Dunn and Dey (1942) and many later workers placed the Newer Dolerite at the top of the stratigraphic succession proposed by them for the Precambrians of Singhbhum region. The spatial distribution, field relationship and the available age data of these dykes do not subscribe to this long held contention. It is obvious that these dykes are not as new as were considered earlier, but the appellation 'Newer' is still in use.

The Newer Dolerite dykes were emplaced mainly along two fracture systems produced by brittle deformation, the latter divided into two end members, i.e. tensile fracturing and shear fracturing with combinations thereof. The overall geometry of fracture distribution patterns that guided the intrusion of Newer Dolerites in the Singhbhum Granite terrain resembles conjugate shear fractures and appears to have developed in response to an approximately N–S acting horizontal stress (Mandal et al., 2006; Sarkar, 1984; Saha, 1994; Gupta and Basu, 2000). An interesting finding of Mandal et al. (2006) is that the size distributions of these dolerite dykes are non-fractal in nature and show strong departures from power law distributions.

Iron Ore Provinces and Other Sedimentary/Volcanisedimentary Belts in South Singhbhum-North Orissa

Banded Iron Formations (BIF) associated with metasedimentary and metavolcanic rocks, commonly referred to as Iron Ore Group (IOG), skirt the Singhbhum Granite massif as well as occur as linear strips or small enclaves within the latter. These iron ore bearing provinces are of varied dimensions and extents, the principal ones of which are:

1. Badampahar–Gorumahisani belt in the east (IOG-I)
2. Jamda–Koira–Noamundi belt ('Horse-shoe' synclinorium of Jones) in the west (IOG-II) (Fig.4.1.6 a–b)
3. Tomka–Daitari and Malayagiri–Sundarmundi belts in the south and southwest (IOG-III).

In addition, there is the Gandhamardan iron ore deposit, occurring isolated, further south of (2), the E–W trending Deogarh belt in the west of Pala Lahara and another one west of Sundarmundi.

In general, the IOG comprises clastic sediments, basic volcanic rocks and Banded Iron Formations (BIFs). BIF in the (1) and (3) are associated with a thick sequence of quartz arenite, fuchsite quartzite, basic flows, tuff, chert and conglomerate. Sahu and Mukherjee (2001) reported spinifex-textured komatiite from several locations in (1). On the other hand, BIF in (2) is overlain by shale (and volcanics) and underlain by shale and basic volcanics with subordinate acid- intermediate members, local dolostone, sandstone and conglomerate. Sedimentation was generally shallow water type. The IOG rocks are metamorphosed to low greenschist facies.

A major point of controversy in Singhbhum geology pivots on whether all the BIFs in this region belong to one group/age or not. Dunn and Dey (1942), Sarkar and Saha (1962, 1977, 1983), Saha (1988, 1994) and several other workers considered all of them to be broadly of the same age (Iron Ore Group, IOG) and older than SBG or at least older than SBG-B (3.1 Ga). On the other hand Prasada Rao et al., (1964), Banerji (1974), Iyengar and Banerjee(1964) and Acharya (1984) recognised at least two different BIF sequences: an older one of Gorumahisani-Badampahar (IOG-I) and a younger one of Jamda-Koira-Noamundi areas (IOG-II), both being younger than SBG. Many later workers paid credence to the view that there may be more than one BIF sequence in the belt. The BIF of Gorumahisani-Badampahar greenstones are welded into the SBG craton with ample evidence of profuse intrusion of SBG into the sequence. Besides, there are quite a few narrow tongues and detached enclaves of Badampahar-Gorumahisani greenstone sequence extending well into the Singhbhum Granite massif, displaying the status of 'older enclaves' rather than that of 'infolded younger cover sequence' (cf. geological map of Dunn and Dey, 1942; Iyengar and Murthy, 1982; Jena and Behra, 2000; Dey, 1991). On the other hand, the BIF of Jamda–Koira valley is hosted by a thick pelitic sequence interbanded with tuffs, cherts, dolostone, conglomerates

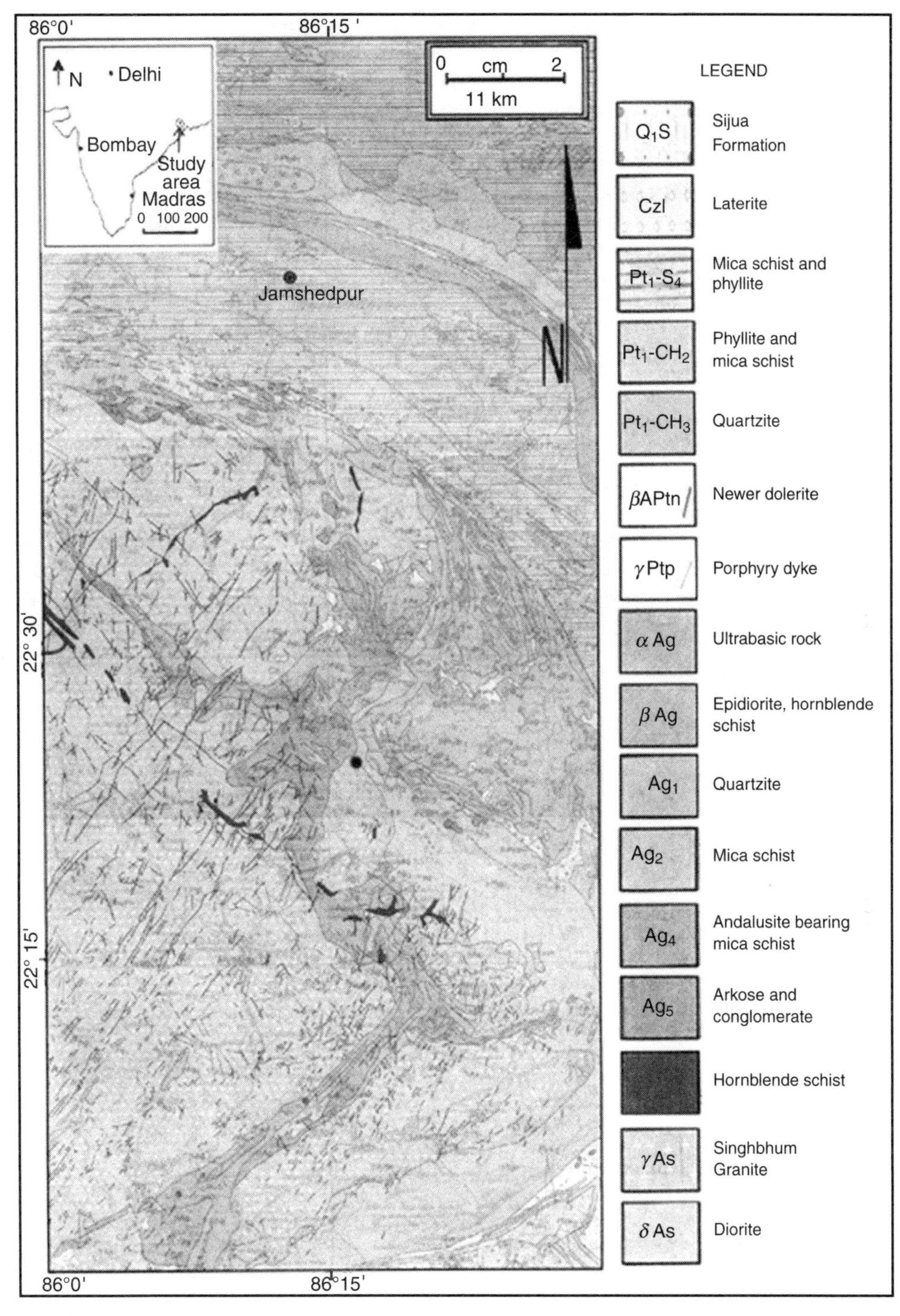

Fig. 4.1.5 Geological map of a part of Singhbhum-North Orissa cratonic nucleus showing Newer Dolerite dyke swarms (dark green) in the Singhbhum Granite country (pink) (from GSI map of Quadrangle sheet 73 J, reproduced by Mandal et al., 2006)

and minor lava, which flank the western margin of the SBG craton. Except in Deo river section, there is no reported crosscutting relation of SBG with the BIF and associated formations of the Jamda–Koira valley. Along Deo river section there are detached rafts of BIF (BHQ/BHJ) within an epidotised granitic suite. This granite has yielded an errorchron age of 3145 ± 282 Ma (Paul et al., 1991) and is petrographically correlated to Phase III (SBG-B) of Singhbhum Granite. Though a group of workers consider this as the most convincing evidence in favour of placing the BIF of Jamda–Koira Valley (and Iron Ore Group in general) as older than 3.1 Ga, the possibility of the BIF enclave in Deo river belonging to an older cycle compared to the BIF of Jamda-Koira valley cannot be ruled out at this stage. A look at the original map of Jones (1934) or any later map would clearly indicate that the BIF enclaves exposed in Deo river section cannot be physically correlated with the BIF of the Jamda–Koira Valley, which passes far through the west, with a thick package of 'Lower phyllite' intervening the SBG margin and the BIF of IOG-II. Moreover, the associated rock types with the BIF enclaves at Deo river section are hornblende schist, mica schist and tuffaceous rocks, which are akin to the older 'Dharwar' rocks or Jones's OMG suite. It may also be kept in view that within the OMTG, near Champua, there are several magnetite and cummingtonite-grunerite bearing banded quartzites. BIF in association with the arenites is also not uncommon within the greenstones of Badampahar-Gorumahisani affinity, which are scattered as enclaves within SBG craton.

Besides the two iron ore provinces described above, there is another extensive iron ore province girdling the southern margin of the SBG craton from Tomka-Daitari in the east to Pala Lahara and beyond in the west. The axial planes of major folds change from NE–SW near Nuasahi to ENE–WSW near Tomka–Daitari to NW–SE near Pala Lahara and finally to NNE–SSW in Jamda–Koira valley. The curvature suggests that the folds in the supracrustals were moulded around the rigid basement block. In contrast to this structural pattern, the greenstones hosting BIF in Badampahar-Gorumahisani belt display variable structural trends cutting across the SBG massif. The same holds true for the OMG rocks. Recently, Mukhopadhyay et al. (2008) separated zircons from a dacitic lava from a poorly metamorphosed low-strain IOG greenstone succession of Daitari. The mineral yielded a U-Pb SHRIMP age of 3506.8 ± 2.3 Ma.

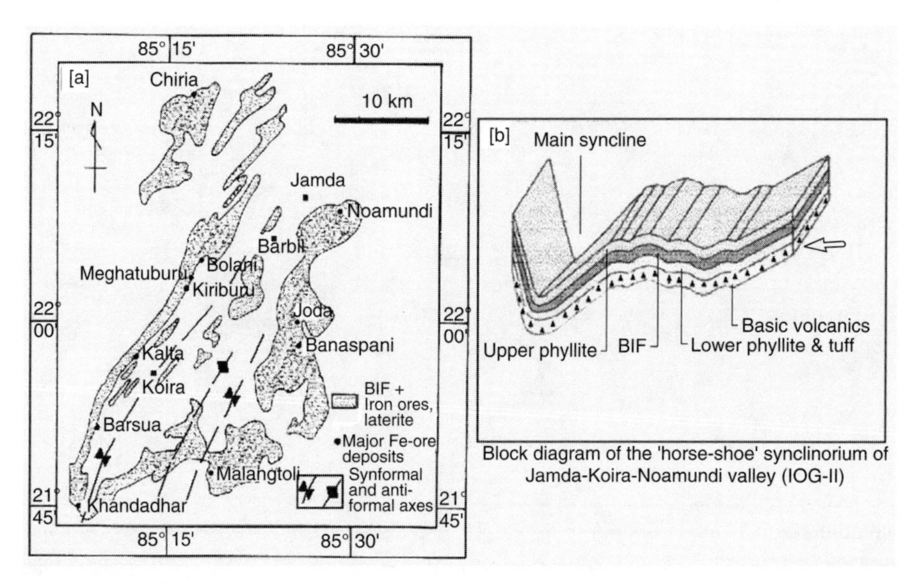

Fig.4.1.6 (a) Sketch map of the Jones (1934) 'horse-shoe' synclinorium of Jamda–Koira–Noamundi (IOG-II) belt (b) block diagram of the synclinorium (modified after Ghosh and Mukhopadhyay, 2007)

Summing up, as of today, IOG rocks of the eastern belt have been found to be older than at least the last phase of Singhbhum Granite (SBG-B). There is little doubt at the same time that these belts developed in the background of a sialic crust. This contention is supported by the following observations: (1) There are arkose and quartzwacke in the lower part of the sequence in tangible proportions in most of the places, (2) some of the rocks contain primary zircon grains encapsulated in clastic quartz grains and there is profusion of clastic monazite grains in the matrix (Sarkar, 2002), (3) presence of uranium (+ Th + Au + REE) at the base of IOG sequences in both the eastern and western basins. If one has to guess about the protolith of these materials, an alkali feldspar granite will be preferred to a typical Archean TTG. Trace element contents of some Daitari IOG rocks suggest their emplacement in an arc environment (Mukhopadhyay et al., 2008)

Banded Iron Formations (BIFs) The definition of banded iron-formations (BIFs) accepted in this work has been briefly discussed in Chapter 1. Banded Iron Formations (BIFs) are important not only as the source of iron in most of the mineable iron ore deposits in the world, but also as a valuable source of information on the nature of atmosphere and the ocean and their evolution during the early Precambrians (Middle–Late Archean to the Paleoproterozoic). The source of iron and silica in the BIF, their collection, transport and the geological-geochemical environment of deposition and the subsequent transformation have been the subjects of keenly contested debates in sedimentology/ore geology for the last one century or so. May be some of these questions do not have an unique answer.

Besides BIF, which are constituted of chemical sediments, there is another type of iron formation called 'Granular iron formation' (GIF), which is deposited as chemical sands analogous to calc-arenites (Clout and Simpson, 2005). GIF normally has thicker and more discontinuous bedding layers compared to those in BIF. Majority of GIF belongs to the oxide and silicate mineral facies and consists of (1) a framework of clasts, (2) finer grained matrix, and (3) authigenic cementing material. Many granules and ooids contain small seplarian-style cracks formed by post-depositional shrinkage suggesting they originally consisted of amorphous, gelatinous material. Han (1978, 1988), by a series of textural studies, demonstrated that most of the crystalline magnetite in GIF and BIF are diagenetic in origin rather that a direct precipitation of detritals. Thus diagenesis alone can give rise to low-grade syngenetic ore deposits called taconite. However, the quartz grains in a GIF are much coarser. Cross-stratification is the dominant depositional structure in most GIF, which has not undergone very pronounced diagenetic alteration. GIF thicker than a few metres is rare, whereas BIF could be much more extensive, both laterally and vertically. In nature a mixture of BIF and GIF are more common than pure GIF.

Before going further, let us briefly discuss the nature of the BIFs in the region under discussion.

The BIFs are composed mainly of alternate iron-rich and silica-rich bands, although shales may locally take up the place of the latter. The bands in the BIFs of this region vary from fractions of a mm to a meter or more. Acharya (2000) reported bands as thick as 30 m from Gandhamardan. The common range, however, is in terms of mm and less commonly in cms. The bands are laminar and parallel in a general look, but often show diagenetic modifications on close inspection. Local occurrence of such features as cross laminations, ripple marks, scour and fill, desiccation cracks are reported from these iron formations (Majumdar and Chakraborty, 1977; Rai et al., 1980). These features suggest a shallow depth (< 100m) of precipitation of the BIFs involved. However, it cannot be generalised as much more common parallel micro-banding and the lack of evidence of wave activity would suggest water depths of 200 m or more. The BIFs have undergone deformations of all the phases that the rock sequence, as a whole, went through (Acharya 2000; Sarkar, 2000). The deformation was dominantly ductile. But local brittle deformation at different times is indicated by the development of brecciated ore, presence of clasts of banded hematite quartzite/ feldspar in massive hematite ore, BHQ/BHJ breccia and faults of different scales affecting the BIF.

The BIFs of this area are principally oxidic. Sulphide facies BIF is conspicuous by its absence. The Ca-, Mg-, Fe-carbonates, which are quite common in the BIFs of some other regions are also very rare in Eastern India. The iron rich bands are composed of hematite, magnetite, martite, maghemite and goethite in different proportions. The silica-rich bands in the eastern deposits of Gorumahisani–Badampahar region consist mainly of quartz with tangible proportions of siderite and grunerite (± stilplomelane) and cummingtonite at many places. Mukhopadhyay and Chanda (1972) studied the petrography of the chert and jasper in the BIF from the Koira valley, particularly their diagenetic transformations. The jasper bands contained fine granules of hematite and magnetite dispersed in cryptocrystalline-crystalline quartz. Carbonate, a minor phase, occurs as rhombs and stringers. Traces of zeolite and clay have also been identified in some specimens.

The ore bands in the BIF generally consist of magnetite and martite in varying proportions. The position of martite may be taken up by bladed hematite at places upto about 30% locally; siderite present with magnetite is also oxidised. The intervening quartzose bands may contain blades of hematite or equivalent grains of magnetite-martite as dissemination. Magnetite-martite pair has a replacement relationship. Bladed hematite grains are reformed or neo-formed.

The fabric of quartz in quartzose cherty bands in the BIF varies widely in grain size and shape. Microcrystalline (< 1–30μ) to macrocrystalline (~ 30 μ to ≥ 1 mm) in size, the grains may be xenocrystic to nearly polygonised. Numerous transverse veins of quartz/quartz + Fe-oxide cut across the layering. They may be penecontemporaneous with sedimentation or diagenetic (compactional). Grain size is generally larger in the veinlets. Coarser grains, rather cluster of coarse grains, may be strain-induced or fluid-induced, or both. Some of the carbonate grains within the quartzose layer are replaced by quartz.

Banded hematite (-magnetite) jasper/quartzite locally grades into banded hematite (-magnetite) shale (Fig. 4.2.40 a). Intricately folded thin interbands of BIF and blue dust are also not uncommon (Fig. 4.2.46). Locally BIFs are underlain by massive 'cherty' bodies.

It is difficult to take a generalised view on the chemical composition of BIFs of such a vast region without adequately large number of representative analytical data. In the absence of such a database, limited numbers of randomly selected samples of BIFs from different areas may at the best provide an approximate idea of the composition of this rock formation in a region (Tables 4.1.1 and 4.1.2). It is worth noting in Table 4.1.1 that the contents of Al_2O_3, CaO, MgO, MnO and particularly alkalies are lower in the Indian BIFs than in others (see also Bhattacharya et al., 2007).

Table 4.1.1 *Major element composition (wt%) of some BIF samples from Eastern India and the world (Sarkar and Gupta, 2005)*

	1	2	3	4	5	6	7	8	9	10	11	12	13	14
SiO_2	41.55	57.50	60.28	56.85	51.83	56.75	57.61	51.48	46.20	51.77	53.2	47.02	47.3	46.90
TiO_2	0.00	0.00	0.01	0.00	0.00	0.03	0.01	0.01	0.01	0.00	0.007	-	-	
Fe_2O_3	56.61	41.98	39.45	43.16	47.73	43.07	40.73	48.74	53.26	46.77	46.20	50.00	45.80	43.78
Al_2O_3	0.26	0.21	0.13	0.10	0.42	0.16	0.08	0.30	0.21	0.08	0.20	0.07	1.25	1.15
CaO	0.06	0.09	0.13	0.10	0.09	0.11	0.10	0.10	0.10	0.10	0.10	0.17	2.84	4.43
MgO	0.07	0.02	0.02	0.02	0.02	0.02	0.02	0.02	0.02	0.02	0.02	0.13	3.66	4.41
MnO	0.16	0.01	0.00	0.01	0.01	0.01	0.01	0.01	0.03	0.00	0.03	0.06	0.59	0.59
Na_2O	-	-	-	-	-	-	-	-	-	-	-	0.12	0.22	0.29
K_2O	-	-	-	-	-	-	-	-	-	-	-	0.14	0.09	0.48
P_2O_5	0.01	0.04	0.02	0.02	0.02	0.05	0.09	0.02	0.05	0.07	0.04	0.07	0.22	0.14
SO_3	<0.01	<0.01	<0.01	<0.01	<0.01	<0.01	<0.01	<0.01	<0.01	20.01	<0.01	-	-	-

Note: Fe_2O_3 – Total iron oxide

Samples: 1. Barsua, 2.Barsua, 3.Gorumahisani (Foot wall), 4.Thakurani, 5.Gandhamardan, 6.Gandhamardan, 7. Royda, 8. Noamundi, 9. Bolani, 9, 10. Kalta (1 – 10: Analyses at the University of Bonn. Na_2O and K_2O undetectable), 11. Average of 1 – 10, 12. Average of ten samples reported by Chakraborty and Majumdar (2002), 13. Average Archean BIF, 14. Average Proterozoic BIF (13 and 14 from Gole and Klein, 1981)

Table 4.1.2 *Trace element contents (ppm) in some BIFs from Eastern India (Sarkar and Gupta, 2005)*

	1	2	3	4	5	6	7	8	9	10
V	3	9	16	3	3	3	3	3	7	3
Cr	218	331	302	774	850	706	557	472	385	296
Co	2	44	43	51	47	37	48	48	39	47
Ni	27	28	28	26	36	40	21	27	26	17
Cu	16	3	3	3	3	3	3	3	3	3
Zn	24	11	11	40	3	28	26	6	21	35
Ga	8	8	2	2	2	2	2	2	2	2
As	18	4	4	4	4	5	4	4	4	4
Nb	3	2	2	2	2	3	6	2	2	2
Ta	2	2	2	2	2	2	4	6	2	2
Mo	5	2	2	2	2	6	6	7	2	2
Pb	3	3	3	6	3	11	3	19	7	22
U	5	5	14	4	3	7	3	13	8	4
Th	7	2	30	2	2	2	43	37	9	2
Ba	16	64	55	32	87	28	62	81	45	18
Rb	5	9	5	4	6	2	5	6	11	4
Sr	6	13	12	8	14	14	13	13	10	15

The sample analysed for trace element contents here are the same as 1 – 10 in Table 4.1.1.

Some details on the modes of occurrence, important characteristics and genetic considerations of the BIFs have already been presented in Chapter 1. In the context of Eastern India, the major opinions on the origin of this rock on the basis of available information on the geochemistry and geological environment are briefly discussed here. Table 4.1.1 presents analysis of ten samples of BIF randomly chosen from different mining areas in the region. Their average is then placed against the average of another ten samples from the region analysed by previous workers (Majumdar et al., 1982) and also compared with the average composition of the Archean and Proterozoic BIFs as calculated by Gole and Klein (1981). H_2O-content is not shown in Table 4.1.1, but Majumdar et al. (op. cit.) shows an average content of ~2% (wt) H_2O. Bulk of it should come from goethite and clay mineral(s), if present. Nature of the BIF being what it is, no body should expect very rigid average values, particularly of SiO_2 and Fe_2O_3. But what is noticeable is that the BIFs of this region are particularly poor in Al_2O_3, CaO, MgO, MnO and also P_2O_5 compared to the average Archean and Proterozoic BIFs. Amongst the trace elements contents of Cr, Co, Ni, Zn, and to some extent Th are high (Table 4.1.2). Available chondrite–normalised data and the plottings of REE are shown

Table 4.1.3 *REE contents (ppm) in Fe-oxide phases of BIFs from the Jharkhand–Orissa area (from Bhattacharya et al., 2007)*

	Noamundi					Gandhamardan					Gorumahisani (Oxide facies)					Gorumahisani (Silicate facies)				
Sample No	**1**	**2**	**3**	**4**	**5**	**6**	**7**	**8**	**9**	**10**	**11**	**12**	**23**	**14**	**15**	**16**	**17**	**18**	**29**	**30**
La	1.09	0.78	0.83	0.98	0.86	2.16	2.20	2.18	2.16	2.20	0.08	0.07	0.08	0.08	0.08	0.99	1.18	1.17	0.99	1.04
Ce	3.37	2.61	3.01	3.12	2.99	4.50	4.09	4.12	4.43	4.30	0.05	0.05	0.05	0.05	0.05	1.75	2.02	1.83	1.78	1.99
Pr	0.43	0.39	0.41	0.41	0.41	0.59	0.59	0.59	0.59	0.59	0.01	0.01	0.01	0.01	0.01	0.20	0.23	0.23	0.23	0.23
Nd	2.05	1.84	1.89	1.88	2.03	2.65	2.67	2.66	2.65	2.67	0.09	0.10	0.11	0.10	0.09	0.86	1.06	0.88	0.87	1.06
Sm	0.81	0.83	0.82	0.82	0.81	0.64	0.64	0.64	0.64	0.64	0.00	0.04	0.04	0.03	0.03	0.28	0.35	0.29	0.31	0.28
Eu	0.37	0.42	0.39	0.41	0.38	0.36	0.35	0.36	0.36	0.36	0.03	0.03	0.03	0.01	0.02	0.18	0.23	0.20	0.20	0.20
Gd	0.97	0.99	0.97	0.99	0.97	0.66	0.62	0.62	0.64	0.65	0.09	0.09	0.07	0.09	0.08	0.28	0.39	0.31	0.29	0.31
Tb	0.17	0.19	0.18	0.18	0.17	0.12	0.12	0.12	0.12	0.12	0.01	0.01	0.01	0.01	0.01	0.05	0.08	0.05	0.08	0.06
Dy	1.13	1.09	1.12	1.10	1.10	0.88	0.77	0.77	0.77	0.82	0.07	0.06	0.07	0.08	0.06	0.25	0.49	0.48	0.26	0.39
Ho	0.24	0.22	0.23	0.22	0.23	0.18	0.16	0.17	0.17	0.17	0.02	0.02	0.02	0.02	0.02	0.05	0.09	0.07	0.08	0.05
Er	0.66	0.64	0.65	0.64	0.64	0.49	0.41	0.68	0.46	0.42	0.00	0.00	0.00	0.00	0.00	0.14	0.23	0.20	0.19	0.17
Tm	0.09	0.09	0.09	0.09	0.09	0.07	0.06	0.07	0.06	0.07	0.00	0.01	0.01	0.00	0.01	0.02	0.03	0.02	0.03	0.02
Yb	0.57	0.54	0.56	0.57	0.56	0.43	0.41	0.42	0.42	0.43	0.09	0.08	0.08	0.08	0.10	0.10	0.18	0.16	0.14	0.18
Lu	0.08	0.07	0.08	0.07	0.08	0.06	0.06	0.06	0.06	0.06	0.01	0.01	0.01	0.01	0.01	0.02	0.03	0.03	0.02	0.02
LREE	7.77	6.45	6.97	7.21	7.11	10.54	10.2	10.10	10.47	10.4	0.23	0.27	0.29	0.27	0.26	4.08	4.84	4.40	4.18	4.60
HREE	3.90	3.82	3.86	3.85	3.84	2.89	2.61	2.70	2.70	2.75	0.29	0.28	0.27	0.29	0.29	0,91	1.52	1.32	1.09	1.20
Σ REE	12.0	10.1	11.2	11.48	11.3	13.79	13.1	13.26	13.53	13.5	0.55	0.58	0.59	0.57	0.57	5.17	6.59	5.92	5.47	6.00
Eu/Eu*	1.95	2.16	2.04	2.12	2.00	2.58	2.59	2.67	2.62	2.60	0.00	2.33	2.64	0.90	1.90	3.00	2.90	3.11	3.11	3.17

respectively in Table 4.1.3 and Fig. 4.1.7. It is obvious from these that $(Eu/Sm)_{CN}$ of the analysed BIF samples are > 1. In the context of the discussions in earlier paragraphs, it is suggestive of volcanogenic hydrothermal origin of the BIF (Sarkar, 2002; cf. Bhattacharya et al., 2007). Acharya (2000) and Banerji (1984) came to a similar conclusion on the basis of intimate association of the basic volcanic rocks and the BIFs in the IOG sequences of this region. However, the opposite view that the BIFs of the region are non-volcanogenic also exists (Majumdar et al., 1982; Chakraborty and Majumdar, 2002). This problem may be seriously persued in future.

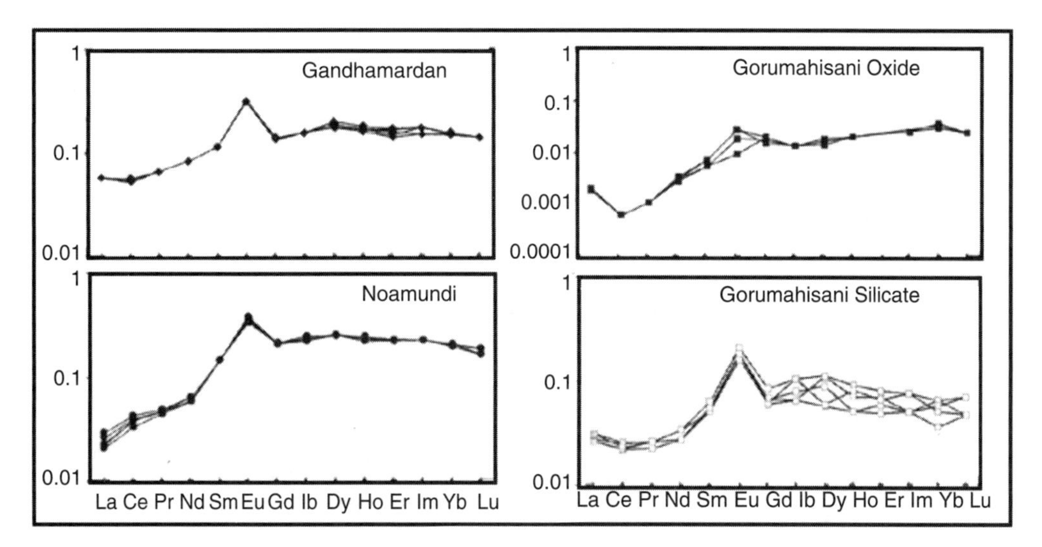

Fig. 4.1.7 PASS normalised REE plots of oxide phases of the BIFs of Eastern India (Bhattacharya et al., 2007)

The BIF-bearing belts in the region occur as curvilinear belts discontinuously surrounding the Singhbhum Granite massif. The BIF-iron ore association of Jamda-Koira-Noamundi belt may largely be assigned to the Lake Superior type, but in a liberal sense in the context of our discussions on their 'classification' in Chapter 1. With the same liberal sense, the iron ores of Badampahar–Gorumahisani could perhaps be designated as Algoma type.

As mentioned earlier, the BIFs and associated iron ores of Badampahar-Gorumahisani belt are considered by a group of workers as the components of an Archean greenstone complex, which is profusely invaded by at least the later phase of Singhbhum Granite (SBG-B) and the Newer Dolerite dyke swarm. Trails of these rocks and BIFs have been traced all across the Singhbhum–Orissa craton from east to the western margin. The BIFs and iron ores of Jamda–Koira valley are regarded by this school as to represent a younger sediment-dominated greenstone cycle. The other group of workers considers all the iron formations of this region to represent rift-controlled greenstone belts of sorts, belonging to a more or less single cycle.

Sarkar (2000) pointed out that the volcanics of the Badampahar–Gorumahisani belt were more primitive compared to those in the Jamda–Koira valley. That the IOG rocks of the western belt are localised on a sialic basement is suggested by the presence of tangible proportions of sandstones and quartzwackes in the lower part of the sequence. Some of these rocks contain primary zircon grains encapsulated in clastic quartz grains and a fair number of monazite grains in the matrix (Sarkar, 2002).

Non-amenability of the BIFs to the age determinations by a robust geochronological method remains a great impediment in direct estimation of the age(s) of the BIFs of this region. The mafic-ultramafic rock complex of Boula–Nuasahi, Orissa, is intrusive into the Iron Ore Group (IOG) rocks of the area.We have already mentioned that zircon separated from the Bangur gabbro, a member of the complex, yielded $^{207}Pb - ^{206}Pb$ ages of 3119 ± 6 to 3123 ± 7 Ma (Auge et al., 2003). They obtained a comparable Sm-Nd date of 3205 ± 280 Ma from 8 gabbro samples of the complex. Sm-Nd whole rock isochron age of Singhbhum Granite Phase II and Phase III, that reportedly intruded the IOG rocks of Badampahar–Gorumahisani, yielded an age of 3120 ± 100 Ma (Majumdar in Saha, 1994). Such data are not available for the BIFs of Jamda–Koira valley, except for the 3145 ± 282 Ma (Rb-Sr, WR–errorchron) date obtained by Paul et al.(1991) for an epidotised phase of Singhbhum Granite containing some rafts of BIF in Deo river section. Apart from the fact that the obtained age of the intrusive granite is an errochron, there is uncertainty about the correlativity of the BIF here with the BIFs of the Jamda–Koira valley.

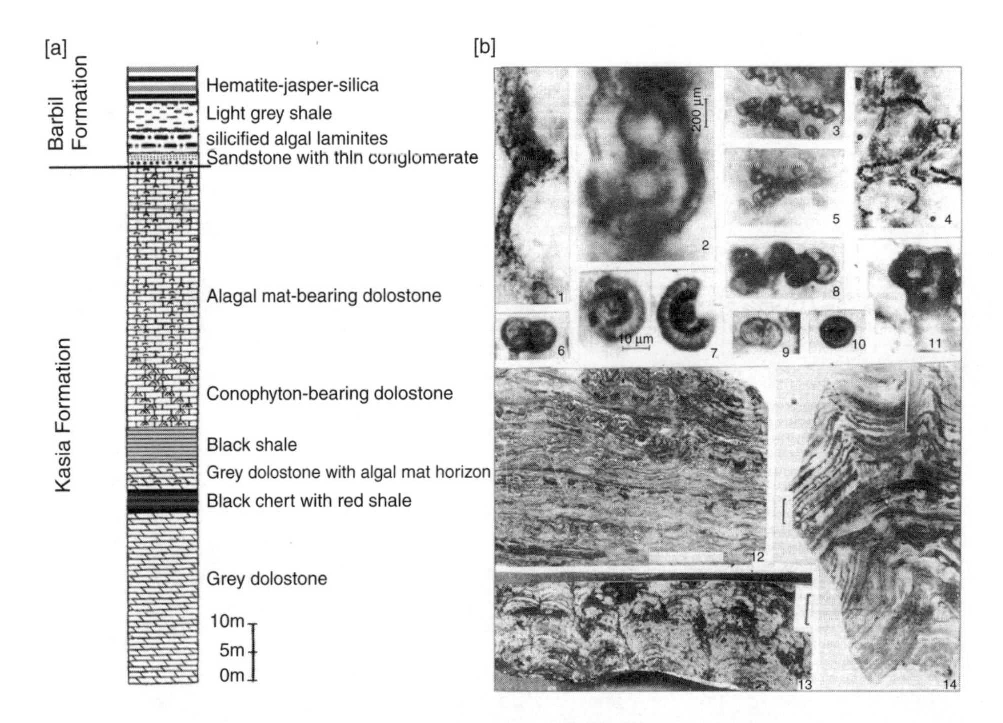

Fig. 4.1.8 (a) Profile of the iron ore bearing formations in Kasia mines, Jamda–Koira valley, Orissa. (b) Microfossils from dolostone in Kasia mine section (Maithy et al., 2000)

Rich remains of cyanophycean microfossils have been recorded from the chert occurring within the stromatolitic dolomite unit of the IOG succession exposed in the Kasia mine working (now abandoned), located ~6 km southwest of Barabil, Orissa, in the Jamda–Koira valley (Maithy et al., 2000 (Fig. 4.1.8 a–b). The fossils that can be divided into (a) spheroidal cells and (b) filamentous forms, are embedded within the chert matrix. The spheroidal forms, much more abundant than the filamentous, have features that show cyanobacterial chroococcalaean affinities. Like all cyanobacteria, chroococcales are capable of oxygenic photosynthesis. In fact, oxygen-producing photo-autotrophy is characteristic of this group. Therefore, its local contribution of oxygen to the development of BIF may be surmised, but no generalisation can be made in the absence of information from other parts of the region. In such a situation, abiogenic development of oxygen for BIF formation remains an eligible mechanism. However, the mounting paleontological and isotopic evidence for photosynthetic oxygenation (Schoff and Packer, 1987; Buick, 1992) may suggest that the oxygenic photosynthesis was an important natural process since about 3.5 Ga. Only their remains may be ill-preserved, because of the primitive nature of the fossils and the thermodynamic processes they were later subjected to.

If this fossil is really chroococcalean as suggested, then it is a precursor of the form that became well developed in the Proterozoic. Oxygen producing photoautotrophy is characteristic of this group. The presence of stromatolites, microbial photoautotrops and BIF development are geochemically compatible (Maithy et al., op. cit.).

Mafic (–Ultramafic) Volcanic Rocks in the IOG Mafic (–ultramafic) volcanic rocks, as already mentioned, constitute an important member of IOG package in this region. In the eastern belt, comprising Gorumahisani-Badampahar and Baula-Nuasahi, they occur as linear bodies hosted by Singhbhum Granite. Compositional banding in them is steep dipping. In contrast, the much more extensive volcanic rocks in the western belt of Jamda–Koira–Noamundi are characterised mostly by low dipping primary banding.

On the basis of major element composition, these volcanic rocks can be divided into tholeiitic basalt, calc-alkaline basalt and high-Mg tholeiitic basalt/basaltic komatiite (Fig. 4.1.9).

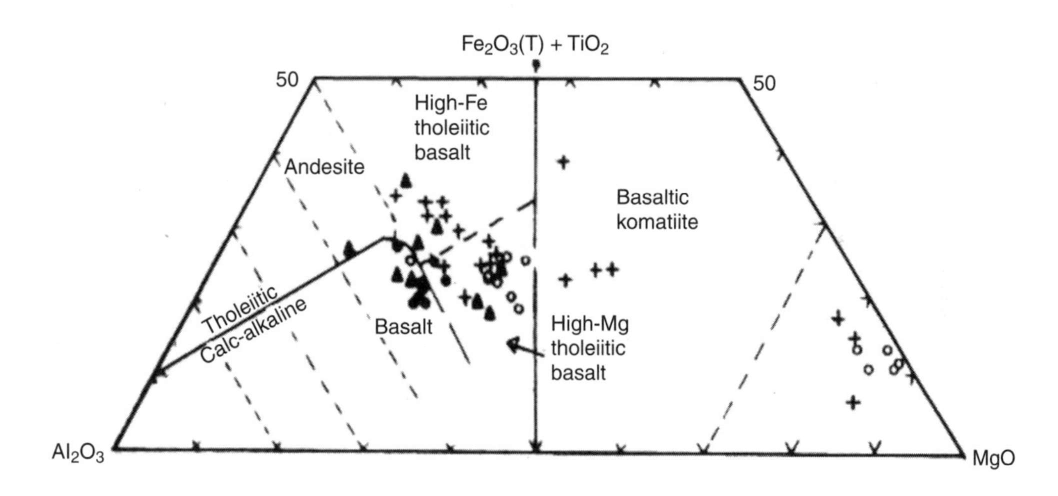

Fig. 4.1.9 Al_2O_3 -MgO-(Fe_2O_3 + TiO_2) diagram showing the plots of mafic–ultramafic volcanics of different IOG provinces, Singhbhum–North Orissa (after Sengupta et al., 1997)

Tholeiite basalt with minor peridotitic komatiite bodies is common in the eastern belt, whereas the basalt in the western belt is dominantly calc-alkaline. Rare ultramafic cumulates are present in both the belts (Sengupta et al., 1997). On the basis of geochemical characteristic particularly, Zr/Nb ratios and REE-contents, the non-komatiitic basalts of the IOG are seperated into three groups: the tholeiites, calc-alkaline basalts and a group intermediate between these two (Fig. 4.1.10 a–c).

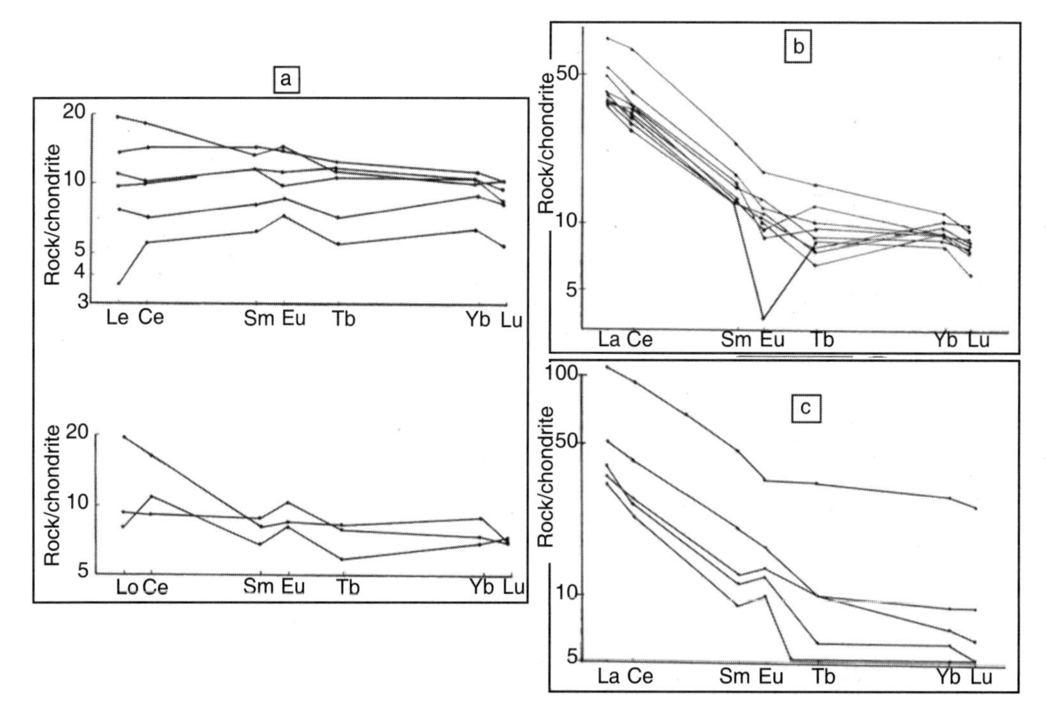

Fig.4.1.10 (a–c) Chondrite–normalised REE plots for samples of IOG volcanics (a) Group-I (tholeiitic basalt), (b) Group-II (calc-alkaline basalt), and (c) Group-III (intermediate type) (after Sengupta et al., 1997)

The tholeiitic basalt is suggested to have generated from a source characterised by chondritic REE contents, whereas the calc-alkaline variety formed from a source that was LREE-enriched. Sengupta et al. (1997) further suggested that the tholeiitic basalt formed rather early from a source of chronditic composition. Down-buckled, this material added LREE-enriched melts to the source, thereby modifying its character. This modified source on melting produced the calc-alkaline basalt. An obvious fall out of this model is that the basalts (and hence the IOG as a whole) in the eastern belt is older than those in the west. But at the present moment we have no data to numerically estimate this time difference.

Malangtoli and Jagannathpur Lavas There are lavas south of the Jamda–Koira valley and to the east of Noamundi (Fig. 4.1.1), which were previously called Nuakot volcanics (Dunn, 1940; Banerjee, 1982) and Dangoaposi lava (Iynger and Murthy, 1982) and are later called Malangtoli and Jagannathpur lavas respectively (Saha, 1994). Saha distinguished these lavas from the IOG lavas in the neighbourhood and thought the former to be younger, i.e. Proterozoic in age. However, Sengupta et al. (1997) observed no difference between these lavas and those which are a part of the IOG. According to them these can be mapped continuously and are also petrochemically indistinguishable. Bose (2000) appears to be somewhat in favour of these rocks being younger to the IOG volcanics of the region. Gupta and Basu (1991), based on field criteria considered the Jagannathpur Lava to be younger than Malantoli Lava. Nelson et al. (2007) have reported Pb-Pb date (zircon, SHRIMP) of 2806 $\pm$ 6 Ma for a dacitic tuff occurring in the basal part of Malantoli lava near Golbhand village.

Jagannathpur lava, best exposed around Dangoaposi Railway Station is massive, fine grained and composed of actinolite, albitised plagioclase and relict augite with chlorite, clinozoisite, quartz and calcite as accessories. A four point Pb-Pb isochron age of the volcanics is 2250 ± 81 Ma (MSWD = 5.8), which is considered to be the emplacement date of these volcanics as these are almost unmetamorphosed (Misra and Johnson, 2005).

Darjing Group A sequence of sedimentary rocks occurs between the Gangpur Group of rocks to the north and the IOG and the Bonai Granite to the south (Fig. 4.1.11). It occurs as a 20 km wide and 100 km long belt, extending from Monoharpur in the east to Bamra (on the South Eastern Railway) in the west, through Bisra, Rourkela and Rajgangpur. For many years these rocks did not draw as much attention from the researchers as the rocks occuring to the north and south did. This is reflected in scanty, if at all, characterisation of these rocks and wide spectrum of opinions on the stratigraphic position of this unit that varied from as old as the pre-IOG to the Dhalbhum Formation (Krishnan, 1937; Sarkar et al., 1969; Iyengar and Murthy, 1980). The first detailed reporting on these rocks, however, was by Mahalik (1987).

Mahalik thought that these rocks qualified for the lithostratigraphic status of a 'Group' and assigned the name of Darjing Group to this package, after the name of a village where its basal Formation was well-developed. These rocks overlie the IOG rocks and the Bonai Granite and are in turn overlain by the Gangapur Group of rocks. This package of clastic sediments were divided by Mahalik into three formations: the lower one (Birtola Formation) consisted of polymictic conglomerate overlain by feldspathic quartzite-arenite, the middle one (Kumakela Formation) made dominantly of carbonaceous sediments and the upper one (Jalda Formation) consisting of matapelites with a subordinate proportion of calc-schist and calc-gneiss. They have been multiply deformed with the dominant schistosity trending almost E–W. The maximum metamorphism registered by these rocks is of 'staurolite grade' (Mahalik, op. cit.). A major breakthrough into the problem of anitquity of these rocks came from a work of Bandyopadhyay and his group (Bandyopadhyay et al., 2001), when they showed that the Tamperkola Granite, the age of emplacement of which is well-constrained at 2800 Ma, is intrusive into the Darjing Group sediments.

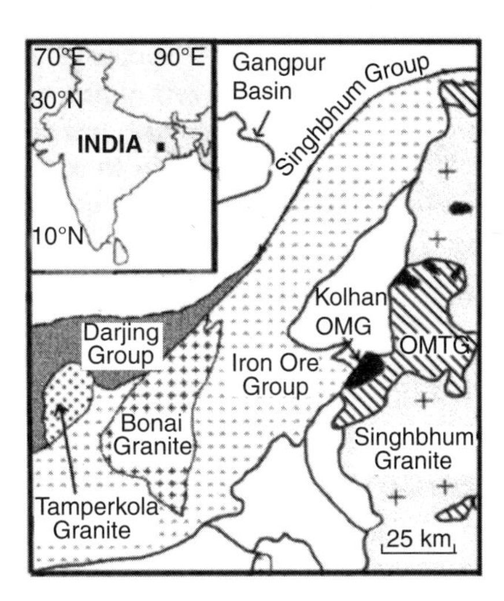

Fig. 4.1.11 Generalised geological map of the western parts of Eastern Indian craton showing the locations of Darging Group and Tamperkola Granite (modified after Bandyopadhyay et al., 2001)

In a still recent publication (Saha et al., 2004), we have valuable information on the geochemistry and petrology of some rocks (sandstones belonging to the Birtola Formation of Mahalik) and the plausible interpretation of this information in terms of the nature, age and tectonic environment of the development of the protolith(s), which was the source of these sediments. The sandstone chosen for this study consisted of monocrystalline quartz ~87%, polycrystalline quartz ~2.5%, microcrystalline quartz/chert < 2%, K-feldspar ~10% of the frame work grains and plagioclase < 3%, lithic fragments ~1%. Accessory minerals consisted of well-rounded zircons, rutile and opaques. Most of the pseudomatrix (1–25%) that consisted of sericite and chlorite formed in situ by the alteration of clay minerals and lithic grains. Chondrite-normalised REE patterns of this sandstone show LREE-enrichment with no Eu- anomaly, while the HREE show a flattish disposition. Primitive mantle-normalised compatible and incompatible trace element plots of this sandstone show similarity with global Archean-Proterozoic sandstones including strong Nb-Ta negative anomalies. From the REE abundances and the $f_{Sm/Nd}$ vs $E_{Nd}(O)$ plot, authors conclude that the sandstones represent a bimodal mechanical mixture of OMG and OMTG. Strong Nb-Ta depletion relative to the primitive mantle suggests that the sandstones were derived from subduction-related magmatic arc sources. Low Nb/Ta and high Zr/Sm ratios of these sandstones, similar to the Archean tonalites-trondhjemites, further support this suggestion. The Nd model ages of these sandstones range, from 3600–4000 Ma. The OMG and OMTG, the suspected source rocks of the sandstones, formed in a subduction-related arc setting.The basement rocks upon which this arc was constructed had to be older. The Nd model ages of the sandstone suggest the presence of continental crust in the Eastern Indian craton older than heretofore estimated 3300 Ma, possibly even as old as 4000 Ma, according to these authors.

Kunjar Group Prasada Rao et al. (1964), while presenting the stratigraphy of the BIFs and associated rocks in Orissa, identified a litho-package in the west of Tamparkola Granite, comprising, from bottom to top, conglomerate–sandstone, mafic lava–tuff, shale–phyllite–calcareous schist and siliceous limestone and named it as ‘sixth sequence’. The rocks of Darjing Group, named as ‘fifth sequence’ were placed below the aforesaid litho-packge with an unconformity in between. Iyengar and Murthy (1982) first named the basin occupied by this younger group of rocks as Kunjar basin and correlated it with the Kolhan and Gangpur basins. It was suggested that all these basins developed after the “emplacement” of Bonai Granite and culminated with the intrusion of Tamperkola Granite. Behra et al. (2005), however, concluded that the rocks of Kunjar basin conformably overlie the Darjing Group.

Kolhan Group The Kolhan Group of rocks, constituting sandstone-carbonate-shale sequence, were described first from the Chaibasa–Noamundi basin along the western periphery of the Singhbhum Granite country (the erstwhile Kolhan State) in the thirtees of the last century by Jones (1933) and then by Dunn (1940). Besides the type area of Kolhan Group described by the pioneers, Saha (1994) regarded a few other siliciclastic-dominant sedimentary sequences as Kolhan. These occur along the westen periphery of the Singhbhum Granite country to the south of Noamundi (Champakpur–Keonjhargarh basin), southern boundary of Malantoli volcanics (Mankarchua basin) and also further down south along Saraipalli–Kamakhyanagar–Mahagiri area. Of these, the type area is better studied (Ghosh and Chatterjee, 1990; Singh, 1998; Bandyopadhyay and Sengupta, 2004; Chakraborty et al., 2005; Mukhopadhyay et al., 2006).

Chaibasa–Noamundi basin The rocks of the Kolhan Group exposed in the Chaibasa–Noamundi basin extend in strike for about 60 km in NNE–SSW direction with an average width of 10–12 km (Fig. 4.1.12 a). The poorly deformed sedimentary sequence displays low westerly dip with broad warps and dome-basin structures and unconformably overlies the Singhbhum Granite, Jagannathpur Lava and the IOG rocks of the Jamda–Koira sector. Along the eastern margin of the basin the unconformable contact of the Kolhan sediments with the older rocks is very distinct (with several small outliers), but the same is not clear along its western margin, where the formations have much steeper dips and are faulted against the IOG rocks (Saha, 1948; Roy et al., 1999). The western boundary of the basin, which was tentatively marked by Dunn (1940), still remains somewhat uncertain due to the lack of detailed mapping along this tract. Singh (1998) described the Kolhan succession as follows.

Kolhan Group	Jetia Shale (up to > 1000 m)
	Jhinkpani Limestone (maximum – 80 m)
	Mungra Sandstone (maximum – 25 m)
-------------------- Unconformity ----------------------	
Basement	

Saha (1994), however, estimated the maximum thickness of the impersistent argillaceous limestone overlying the basal sandstone-conglomerate as 20 m and that of the topmost phyllitic shale as exceeding 200 m. The distribution of different formations of Kolhan Group in parts of Chaibasa-Noamundi basin is depicted in the Fig. 4.1.12 (b).

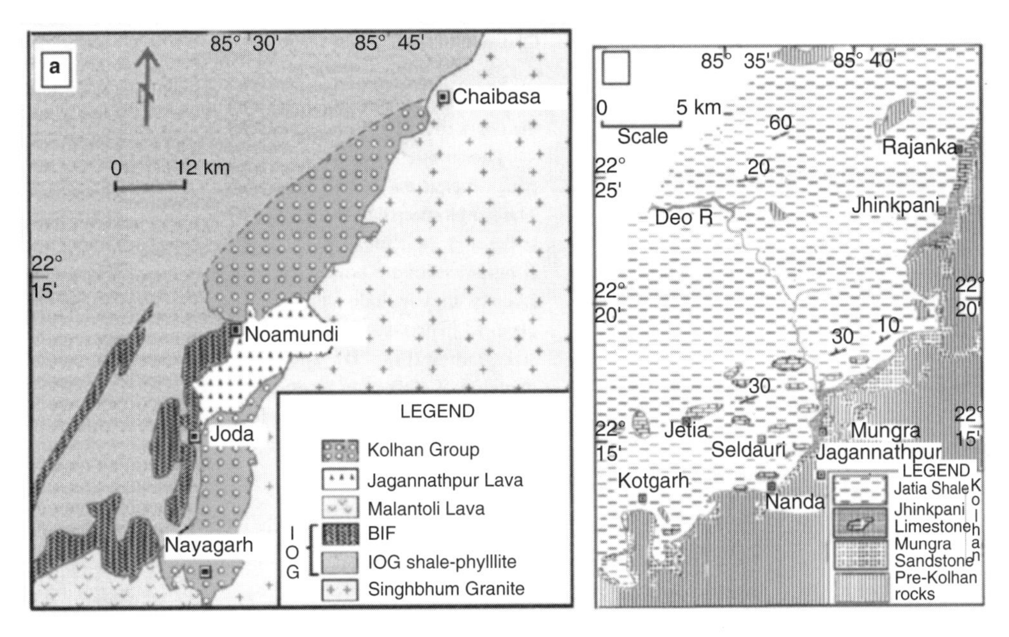

Fig. 4.1.12 (a) Generalised geological map of the western parts of Singhbhum–North Orissa craton showing disposition of Kolhan Group of rocks in the type area and part of Chamakpur–Keonjhargarh basin, (b) Geological map of a part of the type area of Kolhan (Chaibasa–Noamundi) basin (modified after Singh 1998).

The Mungra Sandstone contains polymictic conglomerate at the base, overlain by arkosic and sub-arkosic sandstone. The massive and crudely stratified conglomerate is poorly sorted and matrix supported. The conglomerates may interlaminate with cross-stratified coarse-grained arkose. The pebbles are composed of chert, jasper, phyllite, quartzite, vein quartz, BIF and granite. The occurrence of massive iron ore pebbles at places in Kolhan conglomerate has important bearing on the genesis and time of the formation of the iron ores in this sector (discussed in Chapter 4.2). The arkose and sub-arkose occur either as wavy-planar beds or hummocky cross stratifications.

The Jhinkpani Limestone, comprising impersistent limestone bands at the bottom and calcareous shale towards top, directly overlies the Mungra Sandstone with a sharp contact. The limestone bands, divisible into lower grey and upper purple limestone become siliceous laterally. The limestone is generally composed of 10–30 cm thick micritic limestone separated by 1–2 cm thick shale laminae. The argillaceous content increases upwards. Manganese nodules occur within the limestone near Chaibasa. A cement factory at Jhinkpani is based on this limestone.

The Jhinkpani Limestone horizon grades upward into a thick sequence of thinly bedded shale called the Jetia Shale. The Jetia Shale comprises three facies viz, purple shale, calcareous shale and limestone, with gradational boundaries. Towards the west a steep dipping fracture cleavage system largely obliterates the bedding (Saha, 1994).

Mukhopadhyay et al. (2006) considered the Kolhan succession in the type area to represent a major transgression and relative sealevel rise. The basal alluvial fan-braid plane deposits and the wave reworked conglomerates are likely to represent the late lowstand to early Transgressive Systems Tract (TST) (Fig. 4.1.13). The arkosic rocks interbedded with hummocky cross-laminations were laid in the storm-dominated shore zone and marks the rapid transgression that reworked the earlier deposits. The limestones were deposited below the wave base, particularly when the siliciclastic depositional system retrograded towards the craton interior during a major sealevel rise facilitating carbonate deposition behind. Jetia Shale was deposited during continued transgression at the maximum relative sealevel of the 'Kolhan Sea'.

Champakpur–Keonjhargarh basin This basin, comprising a sequence of low westerly dipping grey orthoquartzite with minor arkose (-conglomerate) and several bands of greyish yellow shale, extends in N–S direction for about 50 km from Champakpur in the north to south of Keonjhargarh with an average width of 6–8 km. The sequence unconformably overlies the Singhbhum Granite in the east. West of Keonjhargarh, a 10 m thick pyrophyllite band occuring below orthoquartzite cover is being mined, but the bottom is not exposed. Saha (1994) considered this sequence to be overlying the Malantoli Lava and to have faulted contact with the Gandhamardhan BIF as well as the main IOG rocks of Jamda–Koira valley. Acharya (1984) considered the sequence to underlie the IOG rocks in the west, thereby contesting its equivalence with the Kolhan Group. The occurrence of arkose and conglomerates, containing BIF and massive hematite pebbles at the basal parts of the sequence in Chamakpur–Nayagarh sector is very similar to such occurrences in the Mungra Sandstone in the Baljori–Merelgara area (NW of Noamundi) in the type Kolhan basin (Mukhopadyay et al., 2007). Ghosh and Chatterjee (1990) had earlier detailed the the lithologic sequence of this basin and had correlated it with the Kolhan Group.

Mankarchua basin Saha (1994) has correlated the highly dissected narrow tract of thin (~50 m thickness) quartzitic sandstone beds that overlie the southern periphery of the Malantoli Lava with the Kolhan Group. These rocks occupy about 100 sq km area to the north of Pala Lahara (21°26′:

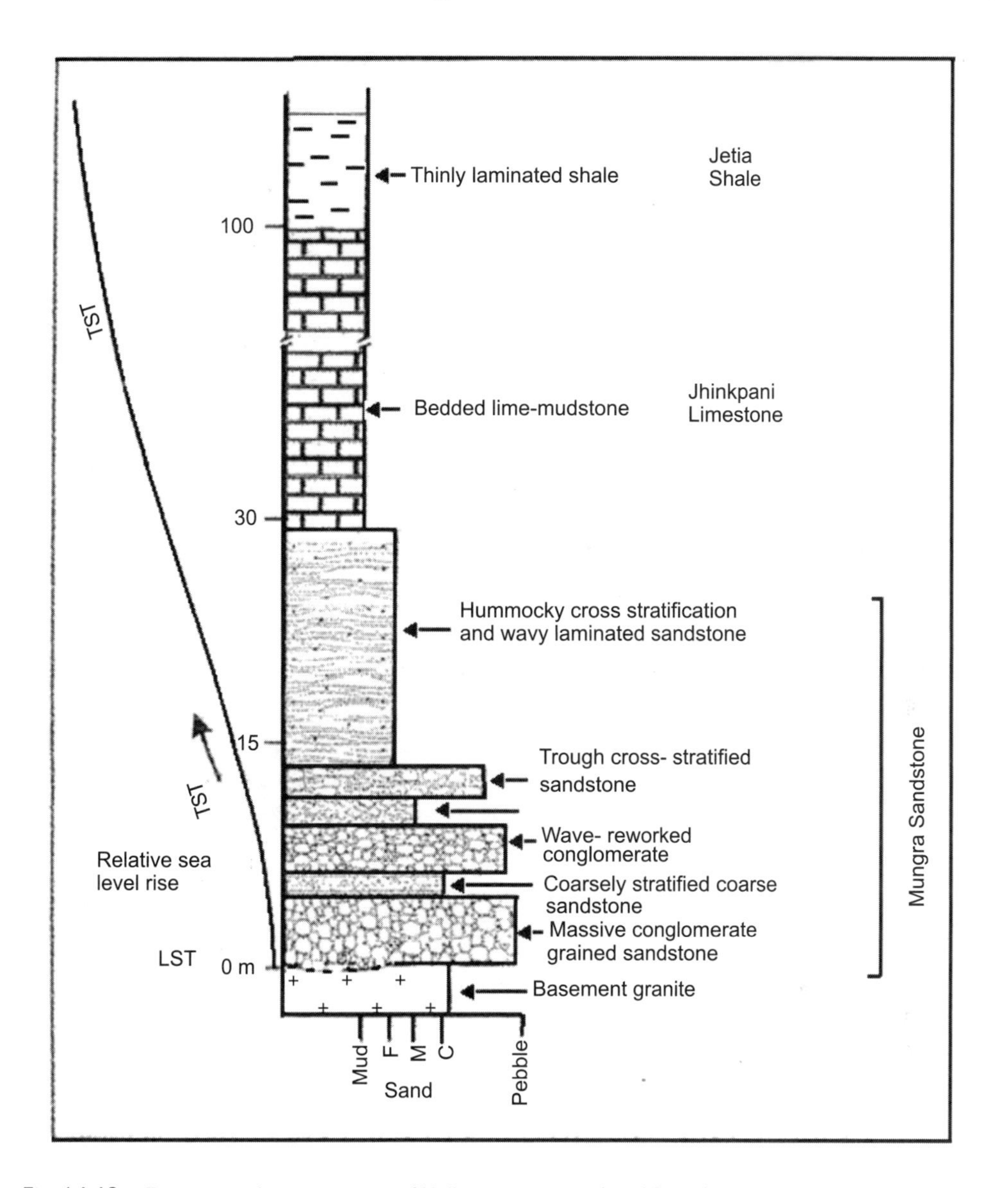

Fig. 4.1.13 Depositional system tracts of Kolhan succession (modified after Mukhopadhyay et al., 2006)

85°11′), displaying faulted contact with the Pala Lahara gneisses in its south. It is suggested that these outcrops represent remnants of a much wider basin that extended over the entire southern flank of Malantoli lava.

Saraipalli-Kamakhyanagar-Mahagiri area The roughly E–W trending arcuate quartzite bands extending for about 100 km from Saraipalli (21°10′: 84°30′) in the west to Kamakhyanagar (20°58′:85°35′) and further east upto Mahagiri have also been correlated with Kolhan Group by Saha (1994).

Alternate current-bedded orthoquartzite and cherty quartzite display steep southerly dip along this zone, which more or less demarcates the Singhbhum- North Orissa craton from the Eastern Ghat granulite belt. Near Kamakhyanagar the NW–SE trending folded ferruginous IOG sequence abut against the quartzite horizon with a fault running in between. The suggested correlation of this horizon with Kolhan is yet to be generally accepted

Age of the Kolhan Group The poorly metamorphosed sedimentary package of Kolhan Group is not amenable to any robust system of geochronological determination. However, two samples of weakly metamorphosed phyllitic shale of Kolhan Group from near Noamundi yielded K-Ar (WR) dates of 988 and 1531 Ma (Sarkar and Saha, 1977). The Kolhans are reported to overlie the Jagannathpur Lava, which was dated at 1629 ± 39 Ma (K-Ar) (Sarkar and Saha, 1977) and at 2250 ± 81 Ma (Pb-Pb) (Misra and Johnson, 2005). Saha (1994) considered the older1531 Ma date of Kolhan shale as the approximate age of metamorphism of these rocks, the age of actual deposition being around 2000 Ma. To the north of Hatgamaria, the Kolhan shales are intruded by Newer Dolerite dykes, the emplacement of which is radiometrically constrained over the age range of 2.1 – 1.2 Ga (Rb-Sr) by Mallick and Sarkar (1989), and as 2.6 Ga (Rb-Sr) by Roy (1998). Earlier, Sarkar, et al. (1969) and Sarkar and Saha (1977) reported K-Ar (WR) dates for Newer Dolerite in the range of 1069 Ma to1960 ± 40 Ma. The above data renders Kolhan Group to be even older than the early phases of Newer dolerite but younger than Jagannathpur and Malantoli Lava. Thus, according to the above quoted data sets, the age of Kolhan Group could be at least 2.0 – 2.2 Ga.

According to Saha (op. cit.) Kolhan basins represented several intracratonic marine basins developed within the Singhbhum–North Orissa craton, having varied basin-fill sediments depending on the provenance and depositional environment. The Chaibasa-Noamundi basin received the material both from eastern Singhbhum Granite terrain and the IOG of Jamda–Koira–Noamundi belt in the southwest, as obvious from high content of Fe in the sediments. The source area for the Champakpur–Keonjhargarh basin was mainly the Singhbhum Granite area while the Pala Lahara gneisses mainly supplied material to the Mankarchua basin. The southern Saripalli–Kamakhyanagar belt is remnant of a much wider basin, which derived mixed granite and IOG material from the northern craton and also witnessed chemical siliceous precipitation. Mahadevan (2002) postulated that the formation of the Kolhan basins could have initiated during large withdrwal of magma for emplacement of extensive Newer Dolerite dyke swarm and thermal subsidence that followed.

Mukhopadhyay et al. (2006) consider Kolhan deposition as a part of a pan-southern Indian shield event (encompassing Dharwar, Bastar and Singhbhum cratons) of large scale subsidence leading to the formation of a vast deep water intracratonic basin during Mesoproterozoic–Neoproterozoic transition time (~1.0 Ga), possibly related to the fragmentation of the Rodinia. However, it was the time of Rodinia assembly rather than its breaking (vide Chapter 8). The model appears interesting but mere similarity of lithologs of the Meso–Neoproterozoic intra-continental basins of peninsular India may not be sufficient to prove that a vast cratonic deep water basin existed covering most of the Precambrian shield. How such a basin developed is, however, a different question, which needs be thoroughly addressed. The contention also does not take into account the earliest metamorphic age of Kolhan shale (1.5 Ga) and the relation of the Kolhan sediments with the adjacent rocks, specially the intrusive Newer Dolerite dykes.

North Singhbhum Mobile Belt

The region between the terrain occupied by the Singhbhum Granite massif and the associated rocks in the south and that by the Chhotanagpur Gneissic Complex (CGC) in the north is commonly referred to as the North Singhbhum mobile belt (NSMB). It is nearly co-extensive with the CGC in the east but falls far short of the latter in its western boundary. It is difficult to pinpoint where it ends in the west. But it may be reasonable to terminate it with the western margin of the Dalma suite of rocks at that end. Based on the difference of geology and hence convenience of discussion the North Singhbhum mobile belt is sub-divided into five longitudinal segments or domains (Sarkar et al, 1992; Gupta and Basu, 2000). Counted from the south to north, these are:

1. Domain I – The matasedimentary belt in the south of Singhbhum shear zone (SSZ) and the Dhanjori–Ongarbira–Simlipal vocanisedimentary basins
2. Domain II – The Singhbhum shear zone (SSZ)
3. Domain III – The area between the Singhbhum shear zone and Dalma belt
4. Domain IV – The Dalma volcanic belt
5. Domain V – The area between the Dalma volcanic belt and CGC.

Domain I: The Matasedimentary belt and Dhanjori–Ongarbira volcanisedimentary basins in the south of Singhbhum Shear Zone (SSZ)

The domain comprises the earliest Proterozoic cover sediments of NSMB, having prominent polymictic conglomerate–arkose–wacke along or close to the contact with the basement, overlain by grey and purple low-grade metapelites, banded magnetite quartzite, variable contents of tuffaceous rocks and mafic/ultrmafic intrusives. There are also intra-formational conglomerates of limited strike extension at different stratigraphic levels. The basal conglomerate, extending from Runkini temple (22°38′: 86°20′) in the east to Saraikela (22°41′:85°46′) and beyond (Raj Kharsawan–Barabambu) in the west, is constituted of pebbles of quartzite, chert, BHJ, granite, phyllites and green fuchsite quartzite similar to older components within the Singhbhum Granite country (Fig.4.1.15 a–h). The most prominent basal conglomerate horizon is developed near Bisrampur (22°41′: 86°10′) and to its west near Jaikan (Fig.4.1.14). Arkosic sandstone is often closely associated with the basal conglomerate. The Bisrampur conglomerate is described as submarine debri flow deposited at the base of a rifted basin (Bose and Ghosh, 1996). This domain is further characterised by intracratonic sub-basins containing volcanisedimentary suites of Dhanjori and Ongarbira in the eastern and western extremities of NSMB. The Simlipal volcanics occurring in the east of the Singhbhum craton is also considered to be time equivalent of the Dhanjori volcanics.

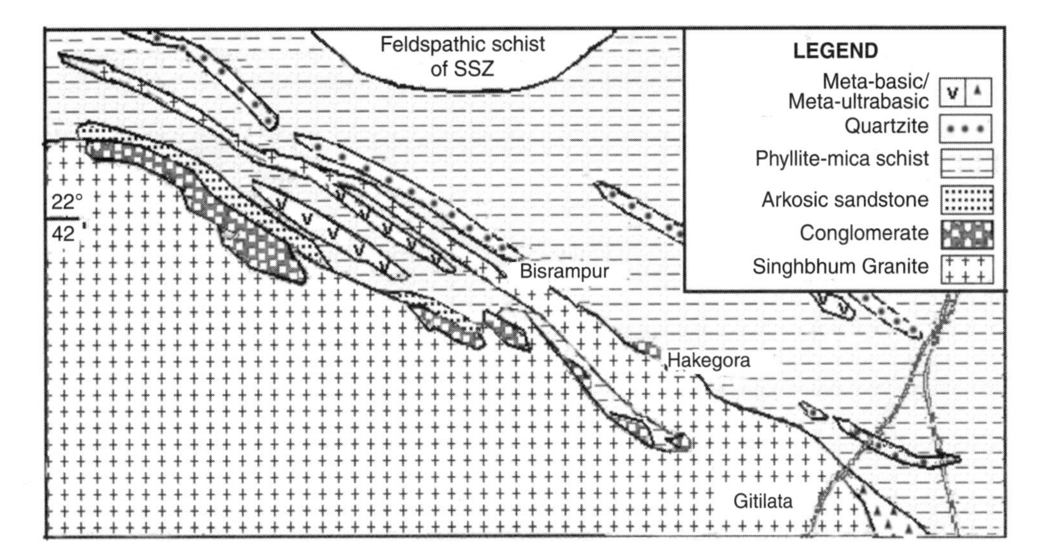

Fig. 4.1.14 Geological map of Bisrampur and adjacent area showing the disposition of conglomerate and arkosic sandstone beds along the contact of Singhbhum Granite (modified after Rath et al., 2005)

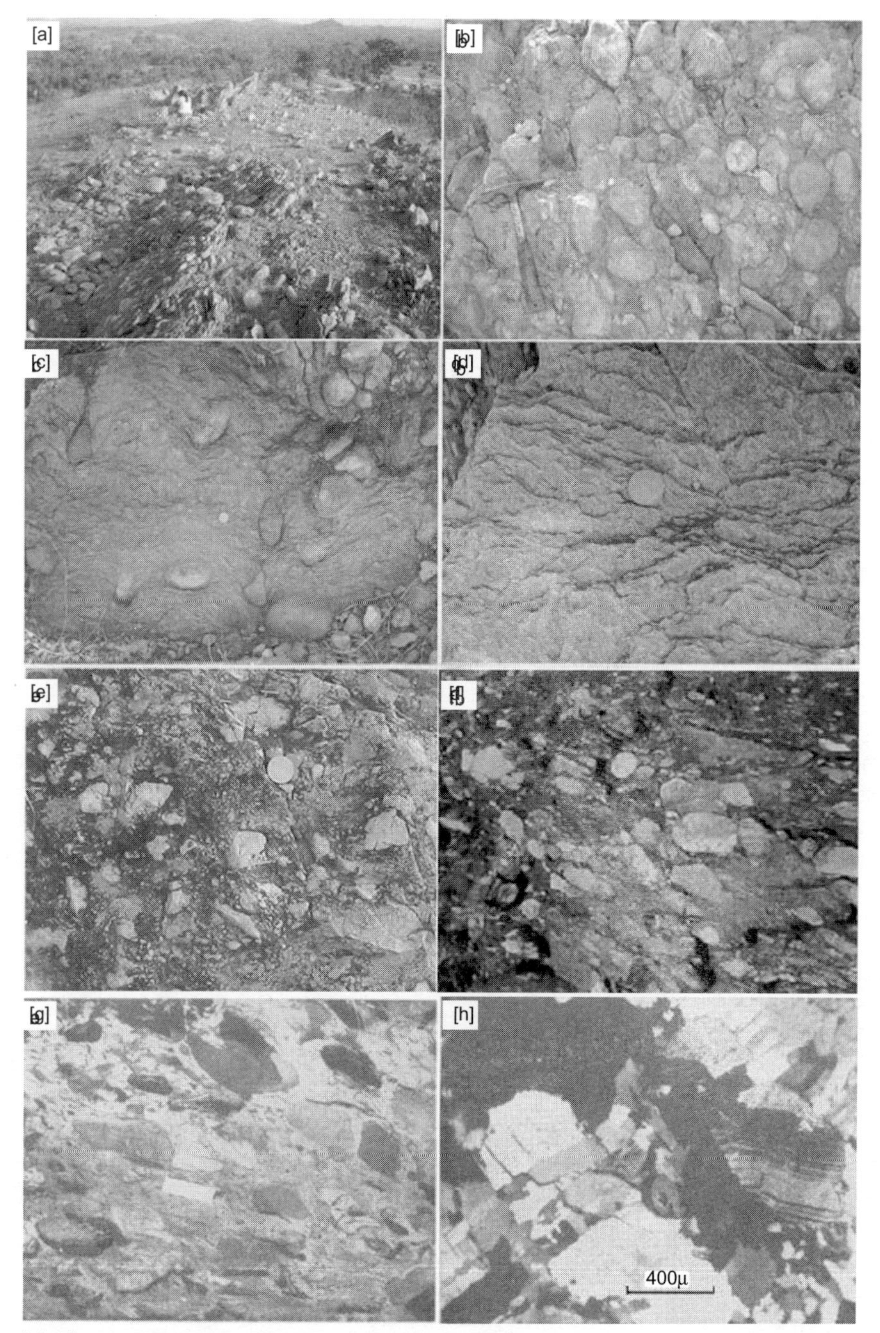

Fig. 4.1.15 (a) Exposure of basal polymictic conglomerate at Bishrampur. (b) Close-up view of polymictic conglomerate showing assorted pebbles, Bishrampur.(c) View of 'drop-stones' in arkosic matrix, Bishrampur.(d) 'Sand wash' bed associated with conglomerate horizon, Bishrampur. (e) Polymictic conglomerate with angular fragments and arkosic matrix overlying Singhbhum Granite, near Runkini temple, (f) Polymictic conglomerate, Hakegora. (g) Polymictic conglomerate at Udalkham village, south of Rajkharsawan Railway Station. (h) Photomicrograph of a granitic pebble in basal conglomerate, south of Narwapahar; (a–e: photographs by the authors and P. Sengupta; f: from Rath et al., 2005; g and h from Sarkar, 1984).

The deformative features in the Domain I, south of SSZ, are dominated by D_1 (F_1)-structures on large and small scales with limited effects of the D_2 (F_2)-structures. In the Dhanjori basin the regional F_1-synclinorium with closure in the SE comprises several inclined to reclined synforms and antiforms, having moderate plunge towards NE (Gupta et al., 1985, 1996; Gupta and Basu, 1990, 2000; Basu, 1985). Similarly, the Ongarbira suite displays a regional southerly overturned F_1-synform with low westerly plunge (Gupta et al., 1981; Blackburn and Srivastava, 1994; Gupta and Basu, op. cit.). The metasediments of this domain show excellent preservation of mesoscopic F_1-folds on bedding (best seen in BMQ bands), which are of inclined to reclined geometry in the central and south-eastern parts and low E-W plunging upright/ overturned folds in the western parts. The

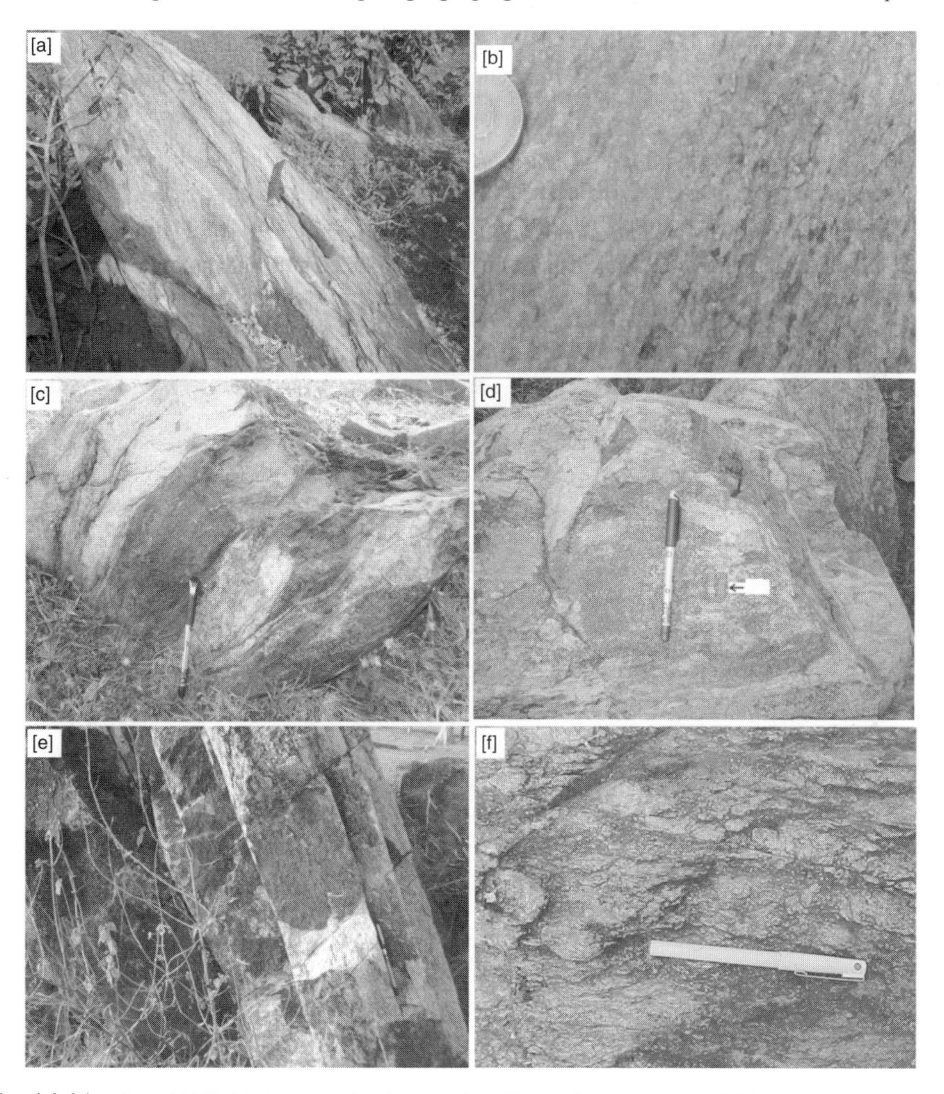

Fig. 4.1.16 (a and b) Field photographs showng the effects of shearing along SSZ far to its south in the rocks of Domain I and basement granite; (a) Down-dip stretcheng lineation in SBG, north of Gitilata; (b) Close-up view of (a) showing the elongate hornblende laths (dark), besides quartz and felspar,defining a strong lineation, (c) and (d) partially digested ultramafic enclave in SBG, north of Gitilata; (d) shows asbestos (As) vein parallel to the pen; (e) Sheeted and sheared basement granite, Saraikela; (f) Sheared basal polymictic conglomerate, west of Runkini temple (photographs by the authors).

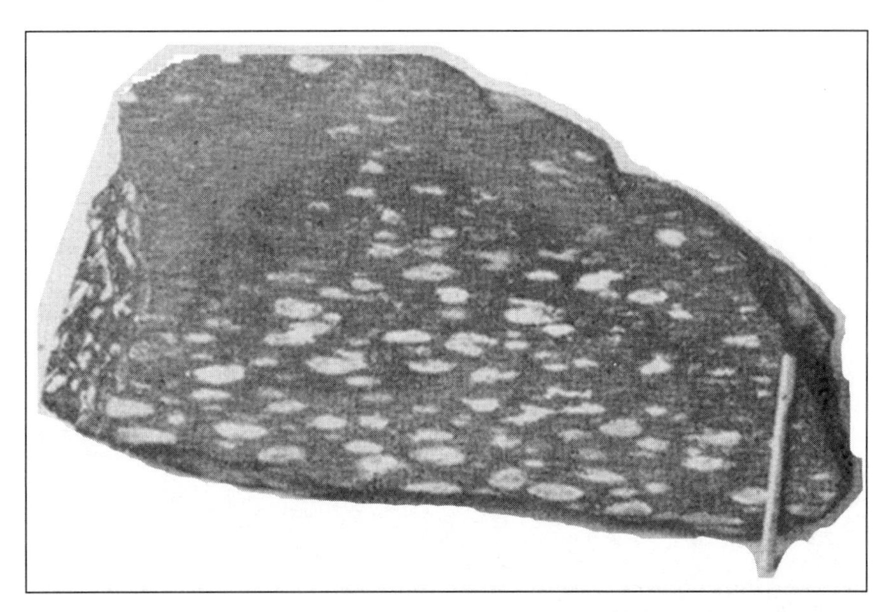

Fig.4.1.17 Stretched amygdules in Dhanjori basalt, Kulamara village SSZ (Sarkar, 1984)

imprints of later folding (F2, F3) and multiple mylonitic fabric become prominent and pervasive towards the north with the approach of the shear zone (SSZ) and the F_1-folds gradually become vestigial, specially in the south-eastern sector. Throughout this domain the mineral schistosity (S_1), which is axial planar to F_1 folds, is the most pervasive planar structure. The F_2-crenulations and S_2 -cleavages are feeble and non-pervasive. However, strong shear fabric similar to those in SSZ is developed along some discrete zones within the pile and the basement granite, close to their contact. There are slices of basement granite at places within the basal conglomerate-arkose horizon, as seen in the Bahar Dhari–Bisrampur area (Fig. 4.1.14). To the south east of Kudada and north of Gitilata, there is an asbestos quarry in a serpentinite body at the contact of the Domain I with the basement SBG. Here, the basement SBG displays strong down-dip lineation and also partially digested enclaves of the asbestos-veined ultramafic rock (Fig. 4.1.16 a–f). Prominent downdip stretching of amygdules in the Dhanjori Basalt is noted well inside Domain-I, up to quite some distance from the SSZ proper (Fig. 4.1.17).

Dhanjori volcanics The type Dhanjori basin extends from Narwapahar in the northwest to Singpura in the southeast. A map of the main part of the basin is presented in Fig. 4.1.18. The volcanisedimentary sequence, that occupies this basin starts with phyllite, gritty shale (acid tuff), rhyolite porphyry, Quartz pebble conglomerate (QPC) interlayered with mafic-ultramafic bodies, followed upwards by quartzite and polymictic conglomerate. The Upper Dhanjori sequence is predominantly volcanic in character comprising high-Mg volcaniclastics, alkali olivine basalt, basaltic and peridotitic komatiite, topped by low-K tholeiites. Komatiitic peridotite with olivine micro-spinifex was first reported from the Kulamara section of Dhanjori basin by Vishwanathan (1978) (Fig. 4.1.19). The Dhanjori Group was divided into two Formations, Upper Dhanjori and the Lower Dhanjori by Gupta et al. (1985) and the same classification was adopted later by Sarkar et al. (1992) (Table 4.1.4).

Table 4.1.4 *Sub-divisions of the Dhanjori Group (Gupta et al., 1985)*

Upper Dhanjori Formation	Tholeiitic basalt (locally pillowed) mafic– ultramafic tuffs, mafic–ultramafic intrusives, tuffaceous mixed sediments
Lower Dhanjori Formation	Phyllite, gritty and pebbly shale with quartzite, ultramafics and mafics Arkose, grit and conglomerate
---Basement---	

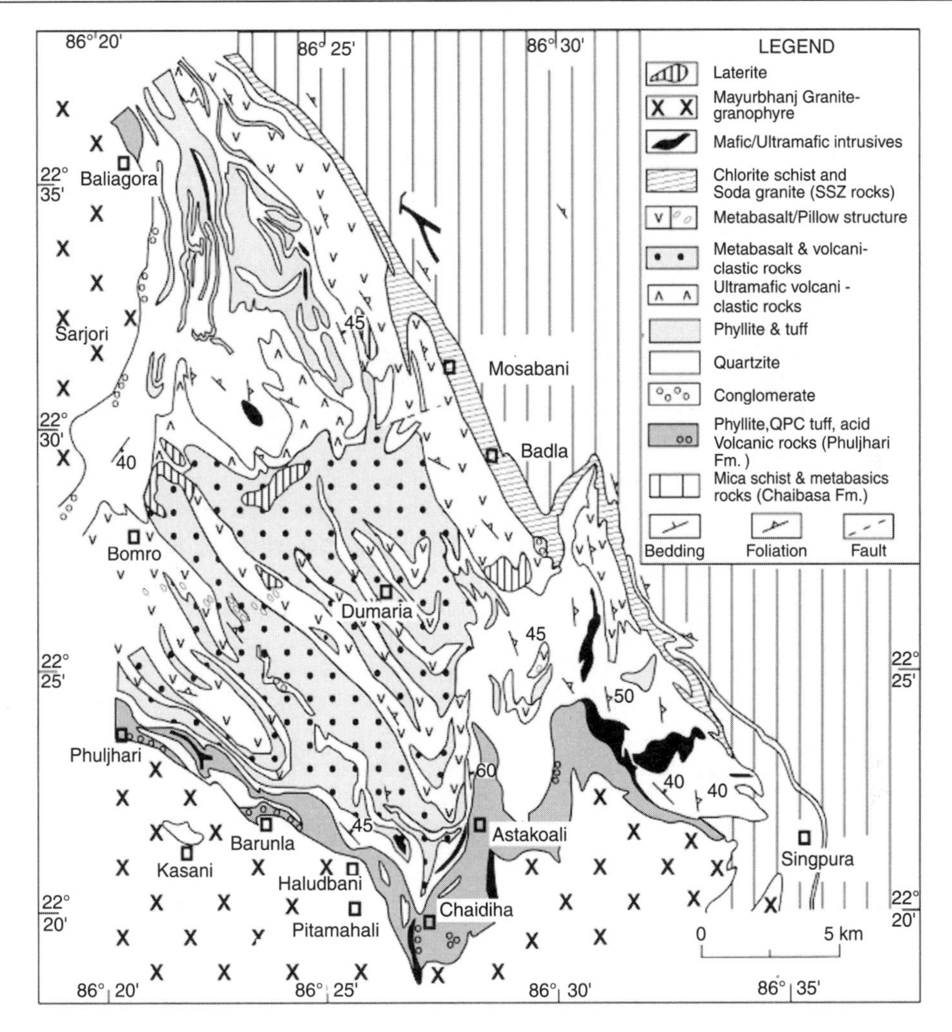

Fig.4.1.18 Geological map of Dhanjori Basin and adjacent rocks, Singhbhum (East) district, Jharkhand (Gupta et al., 1985–legend slightly modified after Acharyya et al., 2008)

Acid volcanics (rhyolitic porphyry and acidic crystal tuff) and quartz pebble conglomerates (QPC) were subsequently recognised as important components of the basal Dhanjori sequence (cf. Gupta and Basu, 2000), the latter attracting special attention of GSI and AMD due to the incidences of uranium and gold (discussed under Section 4.2). Acharyya et al. (2008) have recently designated the bottom most part of Lower Dhanjori Formation as Phuljhari Formation. According to this stratigraphic scheme, the polymictic conglomerate–quartzite horizon that underlies the volcanic dominated assemblage of Upper Dhanjori Formation and defines the basin configuration, represents the base of the Proterozoic Dhanjori Group, while the Phuljhari Formation, comprising QPC, acid volcanics and tuff, is an older suite of end Archean age.

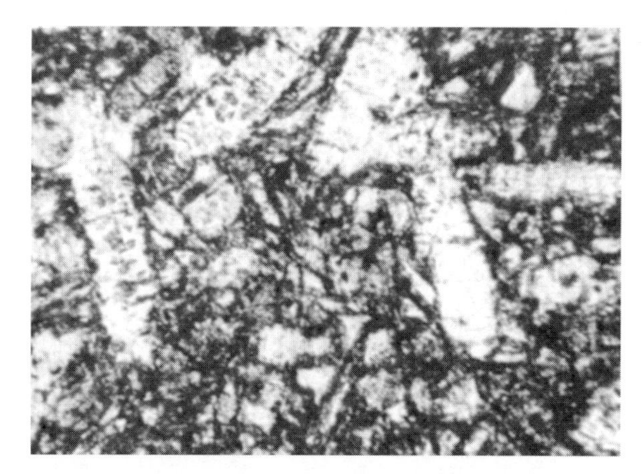

Fig.4.1.19 Photomicrograph showing micro-spinifex texture displayed by elongate olivine and skeletal pyroxene in partially serpentinised peridotite (5.5 mm across field of view; crossed nicol) (Gupta et al., 1980)

Between Singpura and Jaduguda–Runkini temple, the basal Dhanjori formations (renamed as Phuljhari Formation by Acharyya et al. (op. cit.) are intruded by the Mayurbhanj Granite–Granophyre (dated 2.8 Ga), thereby obscuring the contact relation of the former with Singhbhum Granite or the IOG rocks of the eastern belt.

Further to the west, the basal polymictic conglomerate, coarse grained arkosic sandstone and phyllite rest directly on the Singhbhum Granite basement and could represent distal fringe of an alluvial fan complex (Majumdar and Sarkar, 2004).

Table 4.1.5 *Chemical composition of Dhanjori volcanic rocks (Sarkar et al., 1992)*

Major element oxides (wt%)	Ultramafic volcaniclastic rocks (1) $\bar{x}$ (SD)	Peridotitic komatiite (6) $\bar{x}$ (SD)	Basaltic Komatiites (5) $\bar{x}$ (SD)	Upper Dhanjori basalt (13) $\bar{x}$ (SD)	Lower Dhanjori basalt (3) $\bar{x}$ (SD)
SiO_2	46.38	48.35 (2.74)	50.31 (2.97)	51.94 (2.50)	49.10 (5.00)
$Al_2 O_3$	3.57	3.68 (1.05)	6.84 (2.12)	12.60 (2.92)	15.31 (1.98)
Fe_2O_3	6.40	5.33 (4.37)	2.33 (1.41)	3.31 (0.90)	7.06 (6.70)
FeO	7.20	8.13 (2.72)	9.68 (0.63)	9.55 (1.61)	7.80 (2.14)
MnO	0.15	0.24 (0.04)	0.19 (0.13)	0.22 (0.12)	0.16 (0.02)
MgO	25.00	20.20 (1.85)	13.53 (1.99)	7.97 (5.32)	3.56 (3.04)
CaO	4.08	8.38 (4.74)	10.30 (1.80)	8.73 (2.23)	8.63 (3.03)
Na_2O	0.27	0.21 (0.13)	1.88 (0.21)	2.52 (1.14)	2.54 (1.29
K_2O	0.03	0.05 (0.03)	0.23 (0.14)	0.60 (0.51)	1.01 (0.55)
TiO_2	0.30	0.41 (0.23)	0.81 (0.52)	1.09 (0.64)	1.54 (1.26)
M-Value	78.00	74.33	50.60	50.84	42.00

Table 4.1.6 *REE-composition (ppm) of some mafic-ultramafic rocks from the Dhanjori basin (Roy et al., 1997)*

Sample No.	Cr	Co	Ni	La	Ce	Nd	Sm	Eu	Tb	Yb	Lu	(Ce/ Yb)	(∑LREE/ ∑HREE)
1	1132	68.8	252	6.39	13.3	6.4	1.72	0.59	0.30	1.46	0.20	2.36	2.12
2	438	69.6	332	15.9	32.8	17.8	4.49	1.37	0.78	2.08	0.25	4.09	3.40
3	1934	78.7	459	7.17	14.8	5.7	1.50	0.37	0.28	0.89	0.16	5.99	3.24
4	1298	81.4	308	12.2	24.0	11.4	2.89	1.02	0.54	1.36	0.18	4.57	3.63

In the Dhanjori Group of rocks, F_1 folds with inclined to reclined geometry are the map-scale structures (Gupta et al., 1985, 1996), discussed later in more details. Beyond the limits of the type area (Singpura in the south-east to Narwapahar in the north-west), the sediment-rich lower Dhanjori Group extends westward and merges with the infra-Ongarbira suite near the western end of the mobile belt. At the Dhoba–Ukampahar region the ultramafic bodies and the tuffaceous rocks are almost similar to what we get at the type area of the Dhanjoris. The upper Dhanjori basic - ultrabasic rocks have been dated as 2072 + 106Ma (Sm-Nd) (Roy et al., 2002b). ^{40}Ar- ^{39}Ar dating of Dhanjori lava revealed emplacement date of 2.4–2.3 Ga, with the imprints of two prominent metamorphic events at 1.8–1.7 Ga and ~1.2 Ga (unpublished work by S.K.Acharyya and A. Gupta).

The petrochemistry of the Dhanjori and Ongarbira volcanic suites are discussed later in this chapter along with that of the Dalma volcanics.

Ongarbira volcanics The Ongarbira volcanic suite, occurring in the western NSMB between Singhbhum Granite (SBG) and Chakradharpur Granite–Gneiss (CKPG), comprises K-poor tholeiite, high-Mg basalt, gabbro-pyroxenite intrusives, basic/andesitic tuff and acid volcanics (Gupta et al., 1981; Blackburn and Srivastava, 1994; Chattopadhyay and Ray, 1997). The underlying metasediments, consisting of slate–phyllite, psammopelite, quartzite and dolomitic limestone pass into the IOG-II province of Jamda–Koira–Noamundi belt towards the south (Fig.4.1.20). The structures in the Ongarbira suite and IOG are rather incongruous, a fact that is obvious from the regional geological map itself. The folds affecting the Ongarbira suite, together with the rocks of Roro–Jojuhatu area to the further south, have east–west striking axial planes and variably plunging axes (F_1 according to Blackburn and Srivastava), and make high angles with F_1 in the main IOG syclinorium, the latter having NNE-striking axial planes and northerly plunging axes (Fig 4.1.1). Blackburn and Srivastava (1994) felt that the Ongarbira and Roro–Jojohatu rocks belong to the NSMB rather than to the IOG as suggested by Sarkar and Saha, (1977, 1983).

In the west of the Ongarbira synformal core there is a large expanse occupied by felsic pyroclasts (ignimbrite) and ash beds (Chattopadhyay and Ray, op. cit.), interlayered with low grade metapelites and hybrid tuffaceous metasediments. This litho-assemblage overlies the Ongarbira mafic volcanic suite and extends westward to appear as infra-Dalma in Sonua area. Arkose and conglomerate occur both along the margin of the main Singhbhum Granite terrain in the southeast and along the contact of CKP Granite in the north and northwest (Fig. 4.1.21). The southern fork of the SSZ passes through north of Ongarbira synform. Fold and shear structures are prominent in the quartzite bands occurring along the shear zone (Fig. 4.1.22 a–b).

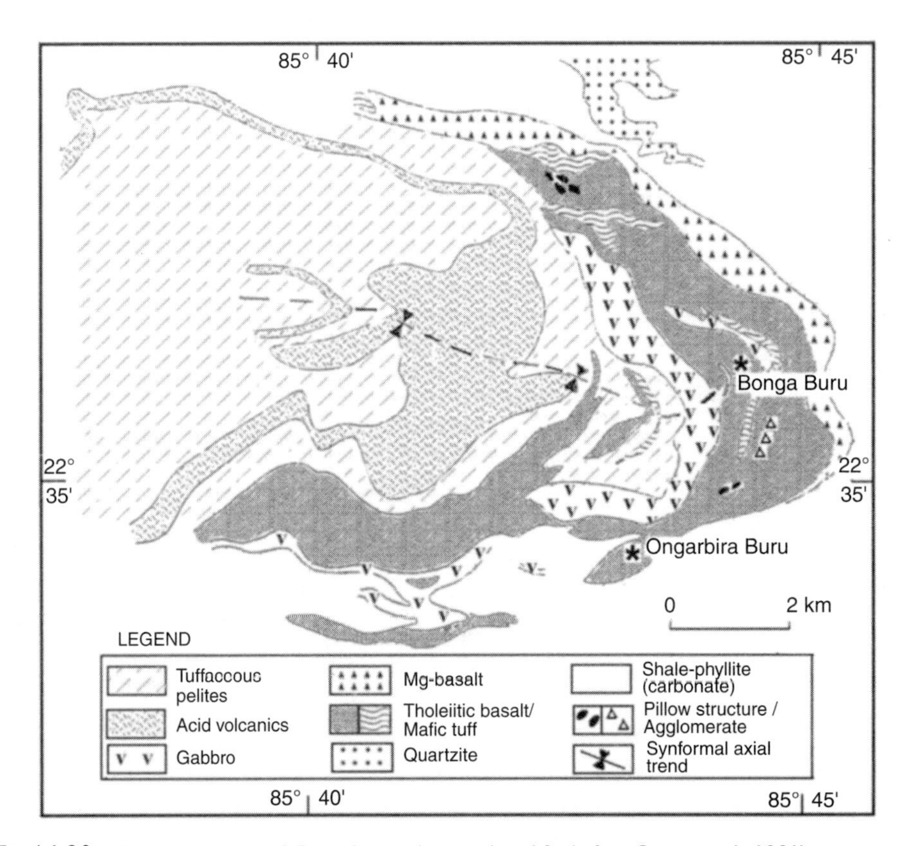

Fig.4.1.20 Geological map of Ongarbira volcanics (modified after Gupta et al., 1981)

Fig. 4.1.21 Polymictic conglomerate with arkosic matrix along southern margin of CKPG (Gupta and Basu, 1992)

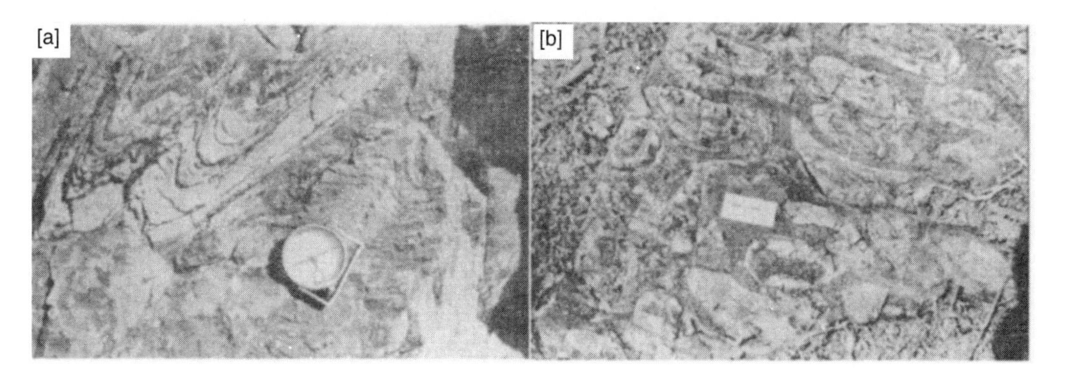

Fig. 4.1.22 Structures in quartzite occurring along SSZ, north of Ongarbira (a) southerly overturned F_2 Folds and (b) closely packed sheath folds with steep plunge in quartzite (Gupta and Basu, 1992).

Blackburn and Srivastava (1994) reported tholeiitic and sub-alkaline character for the Ongarbira volcanics and co-magmatic intrusives as reflected by FeO/MgO vs SiO_2, alkali vs SiO_2 and alkali-FeO-MgO diagrams (Fig. 4.1.23). Three Groups (I, II and III) are recognised in the package from bottom to top based on trace element and REE contents. The Group I and Group II indicate REE depletion, while the upper most Group III shows more evolved character by having LREE enrichment (Fig. 4.1.24). It is suggested that these volcanics were derived from a single LREE depleted source and evolved upwards through their stratigraphy. The authors further suggested that these volcanics were generated in an extensional environment.

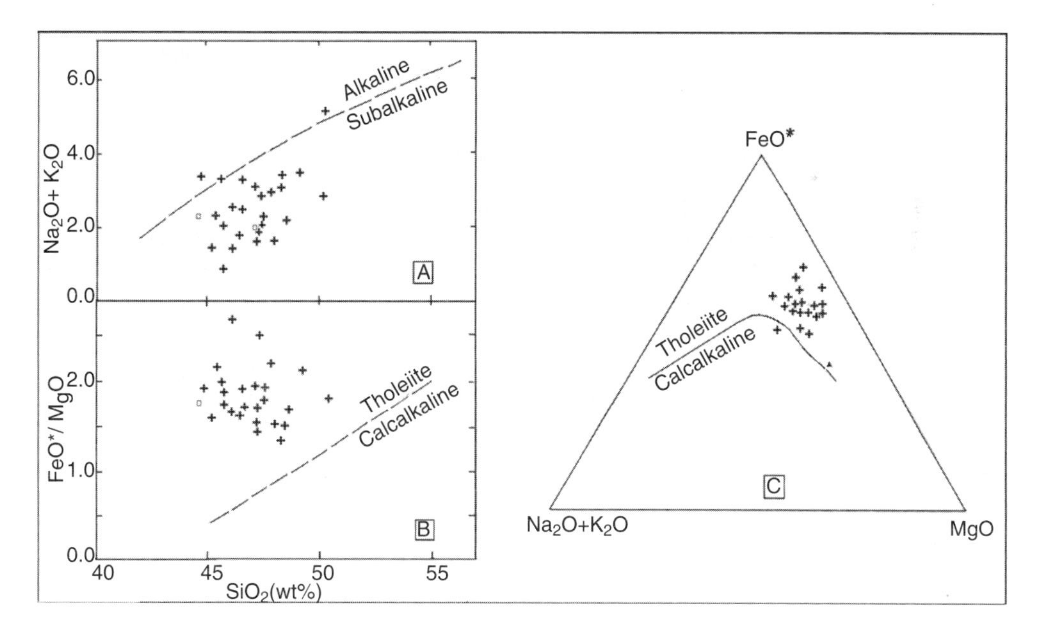

Fig. 4.1.23 (a) Alkali-silica diagram, (b) FeO*/MgO – SiO_2 diagram, and (c) Alkali-FeO*-MgO diagram showing sub-alkaline tholeiitic character of Ongarbira basalts (after Blackburn and Srivastava, 1994).

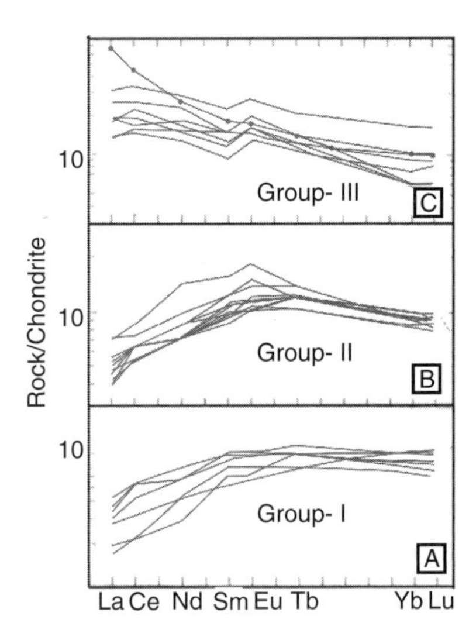

Fig.4.1.24 Chondrite-normalised REE distribution in Ongarbira volcanics of Group I, Group II and Group III (after Blackburn and Srivastava, 1994)

^{39}Ar-^{40}Ar dating experiments with the gabbro in Ongarbira sequence yeilded emplacement age of 2350 Ma and imprints of a metamorphic event at 1400–1300 Ma (Y. Takigami, 2005 – personal communication to A. Gupta).

Simlipal volcanics The Simlipal volcanic complex, occurring close to the eastern margin of the cratonic area, outside the limits of NSMB, is striking for its circular outcrop pattern (Fig. 4.1.1). The suite is reported to consist of spilitic lava of ocean-floor affinity with co-magmatic mafic–ultramafic intrusives and some felsic volcanics, alternating with units of clastic sediments (Iyenger et al., 1981; Basu et al., 1989). In reality, there is a considerable volume of acid volcanics in the Simlipal complex and the whole sequence unconformably overlies the Singhbhum Granite. The rocks in the complex are folded around a NE–SW trending axis, complying well with the fold-patterns in the Dhanjori basin. The field relation between Dhanjori and Simlipal volcanics is obscured due to the invasion of Mayurbhanj Granite in the intervening region, but many workers generally suggest time equivalence between the two. However, geological information available till date on the Simplipal region is meagre as a whole.

Domain II: The Singhbhum Shear Zone This narrow zone girdling the cratonic boundary for about 120 km along the north of Domain I mainly comprises chloritic schists of basic volcanic and heterogenous parentage (possibly a tuff-graywacke assemblage), 'Soda granite'/feldspathic schists, Arkasani Granite–Granophyre, quartzite (kyanite bearing at places), tourmalinite, conglomerates (both oligomictic and polymictic), sericitic and biotitic schists and amphibolites. Soda granite, quartzite/quartz schists and vein quartz have been rendered mylonitic (Fig. 4.1.25 a), while quartz–kyanite rocks and the basic rocks have been converted to phyllonites (Fig. 4.1.25 b) along discrete zones in this belt.

The Singhbhum shear zone (SSZ) was described by Dunn and Dey (1942) as a 'great shear or thrust zone, which has formed along the overfolded southern limb of the geo-anticline'. The absence of any reversal of stratigraphy and lack of clear cut break in structure and metamorphism, together with profuse development of mylonites, phyllonites and L–S fabric in the rocks of the zone, led many later workers to call it a ductile shear zone rather than a major thrust in the classical sense of the term.

Fig. 4.1.25 (a) Thinly laminated and striated mylonitised quartzite, Rakha Mines; (b) Phyllonite with down-dip stretching, Rakha Mines; (c) Boudines and stretched/fragmented quartz veins in sheared chlorite-quartz schist (granular rock), Rakha Mines (a – from Sarkar, 1984; b–c Photographs by the authors).

In the eastern parts of the SSZ, from Narwapahar (22°42′: 86°16′) to Singpura (22°21′: 86°22′), the zone is quite narrow (< 1 km in width). To further southeast it is traced through Baharagora (22°16′: 86°43′) upto Jamsola–Kesarpur in Mayurbhanj district of Orissa, outside the limits of NSMB. In the west, from Ramchandra Pahar (22°43′: 86°13′) onwards, the SSZ becomes much wider (> 3 km) and passes through Turamdih along south of Tatanagar (Jamshedpur) and gets bifurcated near Raj Kharsawan (22°44′: 85°49′), with the Chakradharpur Granite coming in between. The northern limb of the SSZ-fork extends westward upto Duarpuram (22°48′: 85°34′) while the southern limb is traced upto Lotapahar (22°37′: 85°34′). Beyond this point, the possible continuity of the shear zone with a southwesterly trend has been suggested by some workers but it has remained controversial. Similarly, the continuity of the shear zone is also suggested in the south of Kesarpur upto the Sukinda thrust, based on the interpretations of satellite imagery (Das, 1998). The SSZ grades into the rocks of Chaibasa Formation in the hanging wall and the Dhanjori Group rocks in the footwall.

Within the confines of the SSZ, mesoscopic reclined F_1-folds on compositional banding (S_0) are well preserved (Fig.4.1.26 a), especially in the central and western sectors where the shear zone is much wider. Diverse plunge directions of these folds are noted from west to the east (WNW in Lotapahar–Chakradharpur sector, NE in central sector and ENE in Kanyaluka–Khejurdari sector). There are also profusion of reclined folds defined by mylonite bands and mylonitic foliation (sub-parallel to S_0–S_1), best seen in the quartzites and feldspathic rocks. The mesoscopic F_2-folds are asymmetric in style (having shorter and steeper southern limbs and longer and gentler northern limbs) and are southerly overturned with low plunges towards E or W in the western and central parts and low NW or SE plunges in the southeastern sector. Ghosh and Sengupta (1987) described these folds as 'last set of subhorizontal folds' in SSZ with 'arcuation of hinge lines'. The F_2-folds and crenulations are defined by the flexturing of S_1-schistosity, with varied development of axial planar (S_2) surface in form of pervasive schistosity, discrete cleavages or lamellae with S–C fabric (/sygmoids) or mylonitic foliation (Fig. 4.1.26 b–c). The S_2 has steeper northerly or northeasterly dip than S_1 wherever the former is not fully transposed on all earlier planar structures. In the profiles, the asymmetricity of F_2- folds and the sygmoidal patterns in the shear lamellae are indicative of a sinistral shift when viewed from east and relative up-dip (thrust) sense of movement of northern block. In the central and western sector, the F_2-folds on plan show sinistral shift indicative of anticlockwise movement, while the early reclined folds (F_1) display both S and Z shapes in profile and plan (Mukhopadhyay and Deb, 1995).

The characteristic down-dip lineations in the SSZ are defined by elongate minerals, elongate mineral aggregates, stretched pebbles and amygdules, pressure shadows, striae on the shear surfaces in rigid rocks such as quartzite and Soda granite.

A set of broad cross-warps and local crenulations (F_3) with axial planes transverse to the general trend and prominent planar fabric of the belt are developed with steep to moderate down dip plunge. Mukhopadhyay and Deb (op. cit.) considered these transverse warps as products of a distinct later episode of longitudinal compression. Basu et al. (1979) have described a set of low northerly dipping cleavages (Riedel shear) from Turamdih mines, which cut across all earlier structures and are seen to have remobilised sulphides to form richer ore shoots. Down dip lineations (fold axes, mineral- and stretching lineations) are very prominent throughout the SSZ, which may be related to the multiple phases of folding and shear movement along this high strain zone, but cannot always be chronologically discriminated.

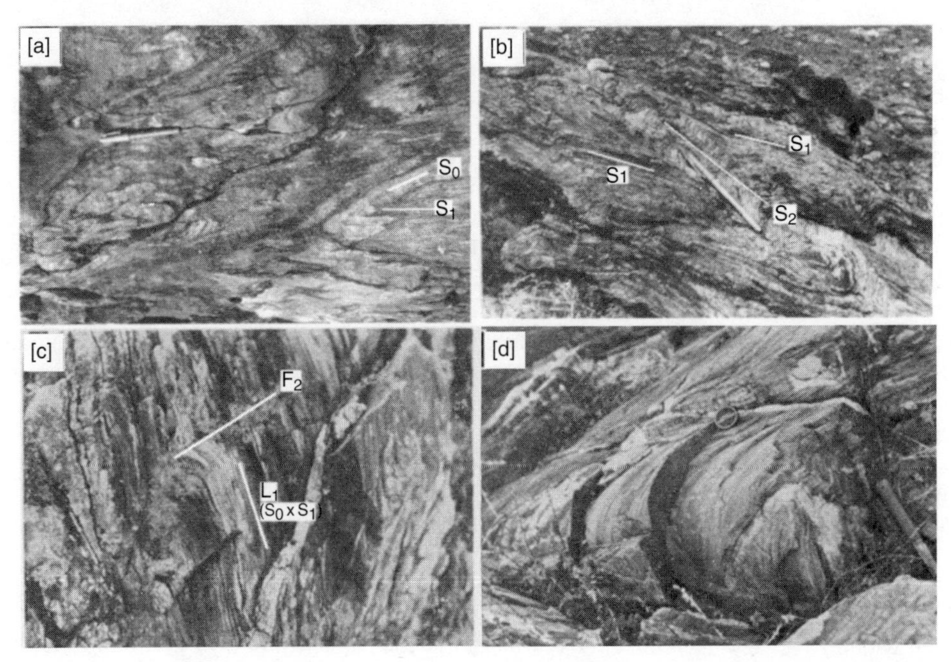

Fig. 4.1.26 (a) Plan view of reclined folds on compositional bands (S_0) and strong axial planar schistosity (S_1), Quartzo-felspathic schist, Turamdih; (b) Profile section view, looking from east to west, of southerly vergent asymmetric F_2 folds on schistosity (S_1) with axial planar cleavage (S_2), Chlorite quartz schist, Nandup hillock; (c) F_2-Folds on schistosity (S_1) with subhorizontal axis, showing flexturing of down-dip (S_0 x S_1) intersection lineation, Quartzose schist, Nandup; (d) Fold on lineated Banded ferrugenous quartzite, Turamdih (a–b by the authors; c–d Sarkar, 1984).

The deformation features displayed by the minerals and the mylonitic fabric in the SSZ rocks (Fig. 4.1.27, a–d) further testify ductile deformation along this zone, with an up-dip sense of movement (Mukhopadhyay and Deb, op. cit.). There are successive sets of mylonitic foliation, folding and dissection by later sets of axial planar schistosity and or mylonitic foliation. The most prominent mylonitisation in SSZ represents an early L–S deformation prior to the earliest folding. Where the mylonitic fabric is not developed, the schistosity, axial planar to a set of reclined folds on compositional banding, are the earliest secondary planar structure. These planes are parallel to the mylonitic foliation. Strong down-dip stretching lineation developed both on mylonitic foliation and schistosity planes represent the direction of tectonic transport during progressive shearing along SSZ. The oblique to near parallel S- and C- planes representing the mylonitic foliation and sygmoidal curvature of earlier foliations are conspicuous all along the zone and serve as indicators of the stage and intensity of shear deformation.

Mylonitisation though common, is not uniform. Lenses of less deformed rocks may be jacketed by zones of intense shear. The mylonitic foliation may be planar or deformed into folds of different generations, varying in style and orientation. The earliest folds in the shear zone are the reclined to inclined in nature, affected by two sets of later folds: one with sub-horizontal axis and overturned towards the south and the other one (younger) is transverse upright fold. Petrofabric analysis of quartz c-axis from sections cut normal to the down dip lineations in quartzose rocks in the shear zone show 'bc' girdles (Sarkar, 1984 and references therein). The occurrence of sheath folds and U-shaped deformed lineations indicate that the reclined folds were produced by rotation of fold-hinges through large angles (Sengupta and Ghosh, 1997). Moderately deformed rocks of the shear zone, particularly the Soda granite/feldspathic schist and the quartz schist gave rise to S–C fabric, while intensely deformed variety gave rise to ultramylonite (Sarkar, 1984; Mukhopadhyay and Deb, 1995).

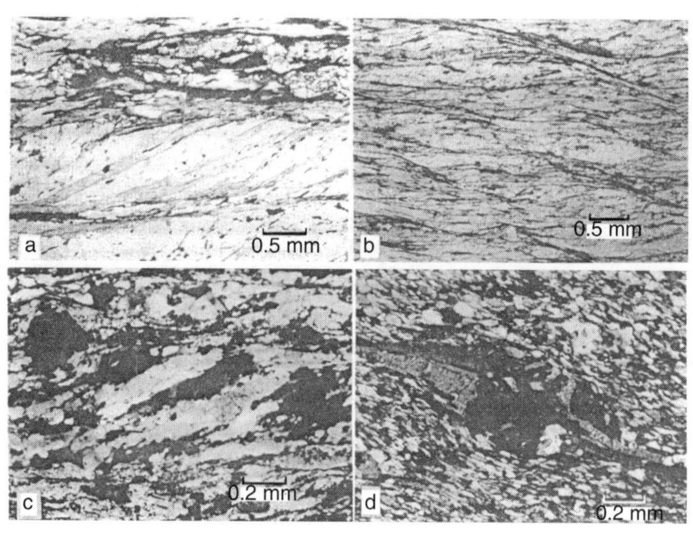

Fig. 4.1.27 (a) Earlier mylonitic foliation showing sigmoidal curvatures is cut across by later micaceous laminae indicating sense of movement along the later planes; (b) Shear bands causing sigmoidal curvatures to earlier planes; (c) S–C mylonite: elongate quartz ribbons (S) oblique to C-planes; (d) σ-structure defined by pressure fringes made up of chlorite flakes at the edges of a tourmaline–quartz knot (Mukhopadhyay and Deb, 1995).

The mylonitic foliation in the shear zone is subparallel with the axial plane cleavage of the dominant fold set outside the domain. According to Sengupta and Ghosh (op. cit.) even the reclined first generation folds in the shear zone developed on mylonitic foliation. It leaves an old question about the genetic and temporal relationship of the deformation within the shear zone and outside of it unanswered. If the reclined folds in the zone developed by rotation of fold hinges by large angles, then the original fold hinges may reasonably be suggested to have been sub-horizontal. Such folds can develop during ductile shearing when the homogenous shear stress is perturbed due to the development of phyllosilicates, such as biotite and sericite in discrete surfaces in the zone (Mandal et al., 2004). The development of the intense shear stress is the culmination of the grossly unequal compression of rocks in the zone. This may be penecontemporaneous with the major folds north of the zone. Worth noting in this context is the fact that there are few discontinuous zones in the mica schist country to the north where most of the shear zone features have developed.

The rocks of the SSZ structurally overlie the Dhanjori rocks in the eastern sector, the Ongarbira rocks in the western parts and the metasediments with mafic–ultramafic suites of Dhoba Pahar–Ukham Pahar stretch of Domain I in the central parts. The interface is often marked by one or more conglomeratic horizon(s), with highly flattened and stretched pebbles at places (Fig. 4.1.28 a–d) (Sengupta, 1977; Sarkar, 1984; Srivastava, 1985; Gupta and Basu, 2000).

Near Mainajharia, pebbles of Dhanjori basalt are abundant in the conglomerate of this level (Gupta et al., 1985). Some of the Cu-lode zones in the Rakha–Roam sector are hosted by oligomictic quartz conglomerate. The Chaibasa rocks occur along the hanging wall of the shear

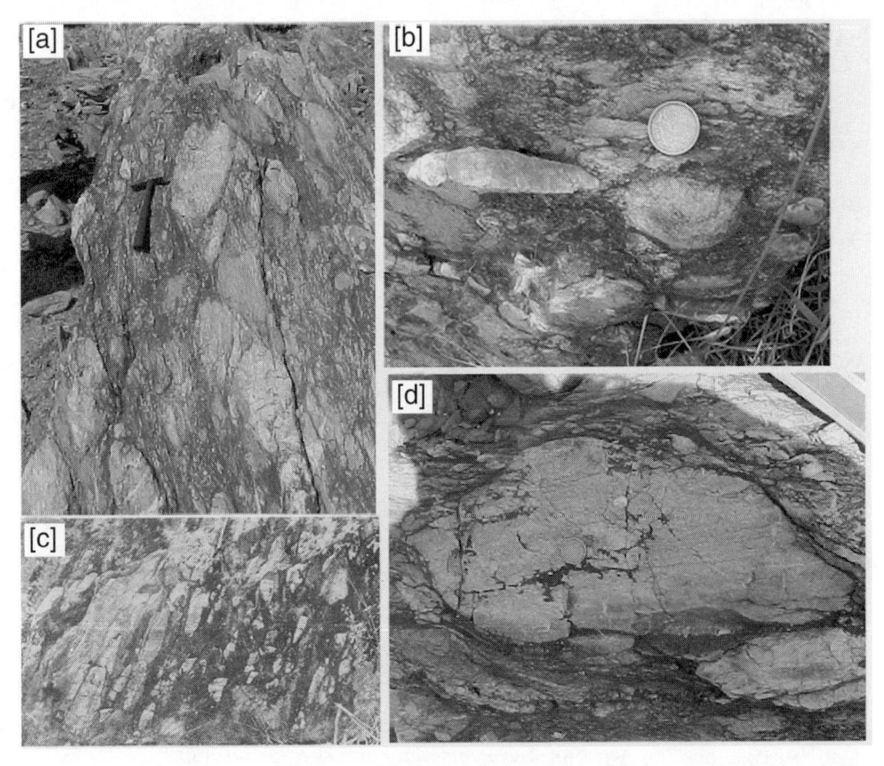

Fig. 4.1.28 (a– b) Flattened pebbles in oligomictic conglomerate at the contact of SSZ and the low-grade metapelies of Domain I, near Lailam; (c) Stretched pebbles in conglomerate along the footwall of Cu (-U) ore zone, Rakha Mines; (d) Bedding laminates in an elongate pebble in sheared conglomerate (photographs a, b, d: P. Sengupta and S.C.Sarkar; c: from Sarkar, 1984).

zone. For an appreciable stretch from Bhatin to Surda the Cu-U-apatite mineralised rocks of SSZ are found sandwiched between Dhanjori quartzite along footwall and a fairly continuous horizon of kyanitiferous quartzite along the hanging wall. The effect of shearing locally transgresses both into the rocks lying in the south and north of the main SSZ.

The rocks occurring within the SSZ are briefly described here.

Chlorite–quartz schist/Quartz–chlorite schist/Biotite–quartz schist Chlorite-quartz schist, commonly referred to as the 'granular rock' by the people associated with the exploration and mining industry in the belt, is best developed in the Surda-Bhatin sector and is the host to most of the ore deposits located in this zone. Impersistent patches of conglomerate mark its interface with the biotite–quartz schist at the footwall. Quartz and chlorite are the principal constituents of the rock. Biotite, magnetite, tourmaline, sodic feldspar, apatite and epidote–zoisite occur in subordinate, but varying proportions. West of Bhatin, this rock gradually changes its composition where chlorite becomes the more dominant phase and continues as such through Rajdah and Bayanbil to Turamdih and beyond. This rock, which should better be called quartz–chlorite schist, hosts the sulphide mineralisation in this sector. With increased proportion of magnetite and apatite, some horizons of the chlorite schist host the apatite ore bodies. Sericite quartz schist occurs at places interleaved with the chlorite-rich schists. In the southeastern sector, another rock dominantly made up of biotite with chlorite, sericite and quartz and minor proportions of apatite, tourmaline and magnetite occur at the Pathargarah–Sankh Nala section as patches within the Soda granite–feldspathic schists or along their northern margins. Fe-, Mg-, Ti- and V-contents of biotite in this rock are greater than in biotite in average metapelities of low to moderate grade. Biotite–quartz schist occurring south of the kyanite–quartz unit in the Surda–Jaduguda sector is mylonitic at places and laterally grades into biotite schists. Chlorite-bearing schist constitutes a common rock type along the northern and eastern fringe of the Dhanjori metabasites. In several borehole cores at the Rakha mines area, the rock showed all gradations to amphibolitic rocks.

Chloritoid is a common mineral in the above described rocks of the shear zone and adjacent mica schists of Chaibasa Formation, particularly in the eastern part. The common chloritoid-bearing assemblages are:

1. Chloritoid + quartz + muscovite + biotite + chlorite + magnetite
2. Garnet + chloritoid + quartz + muscovite + biotite + magnetite + ilmenite
3. Kyanite + quartz + chloritoid + muscovite + magnetite.

Chloritoid in these associations are Fe^{2+} rich and so also is the co-existing garnet (Bandyopadhyay, 2002). Grain size varies widely. Elongate grains participating in downdip lineations are characteristic of the zone.

The different assemblages may be found at short distances apart, suggesting variation in the bulk chemistry rather than in P-T. Garnet stabilisation is increased by the increase in the concentration of CaO and MnO. (CaO + MnO)-poor and Mg- rich bulk composition and high μH_2O stabilise the chloritoid-bearing assemblages.

'Soda granite', feldspathic gneisses and schists 'Soda granite' and the associated feldspathic gneisses and schists constitute an important rock suite in the shear zone. Described first by Dunn and Dey (1942), the suite drew attention of some later workers also (Sarkar, 1984; Dasgupta et al., 1993), in view of its unique mode of occurrence, uncommon petrography and close association with the copper mineralisation at Mosabani–Badia sector.

It is noteworthy that except in Mosabani–Badia sector, there is no spatial correlativity between Cu-mineralisation and the feldspathic rocks.

The massive variety of Soda granite occurs discontinuously along the shear zone from near Kukudungri (22° 47′: 86° 01′) in the west to Khejurdari (22°24′: 86° 34′) in the east. However, the schistose feldspathic rocks extend far beyond Kukudungri in the west. They occur generally as sheet-like bodies and share their internal schistosity with the country rocks (Fig. 4.1.29).

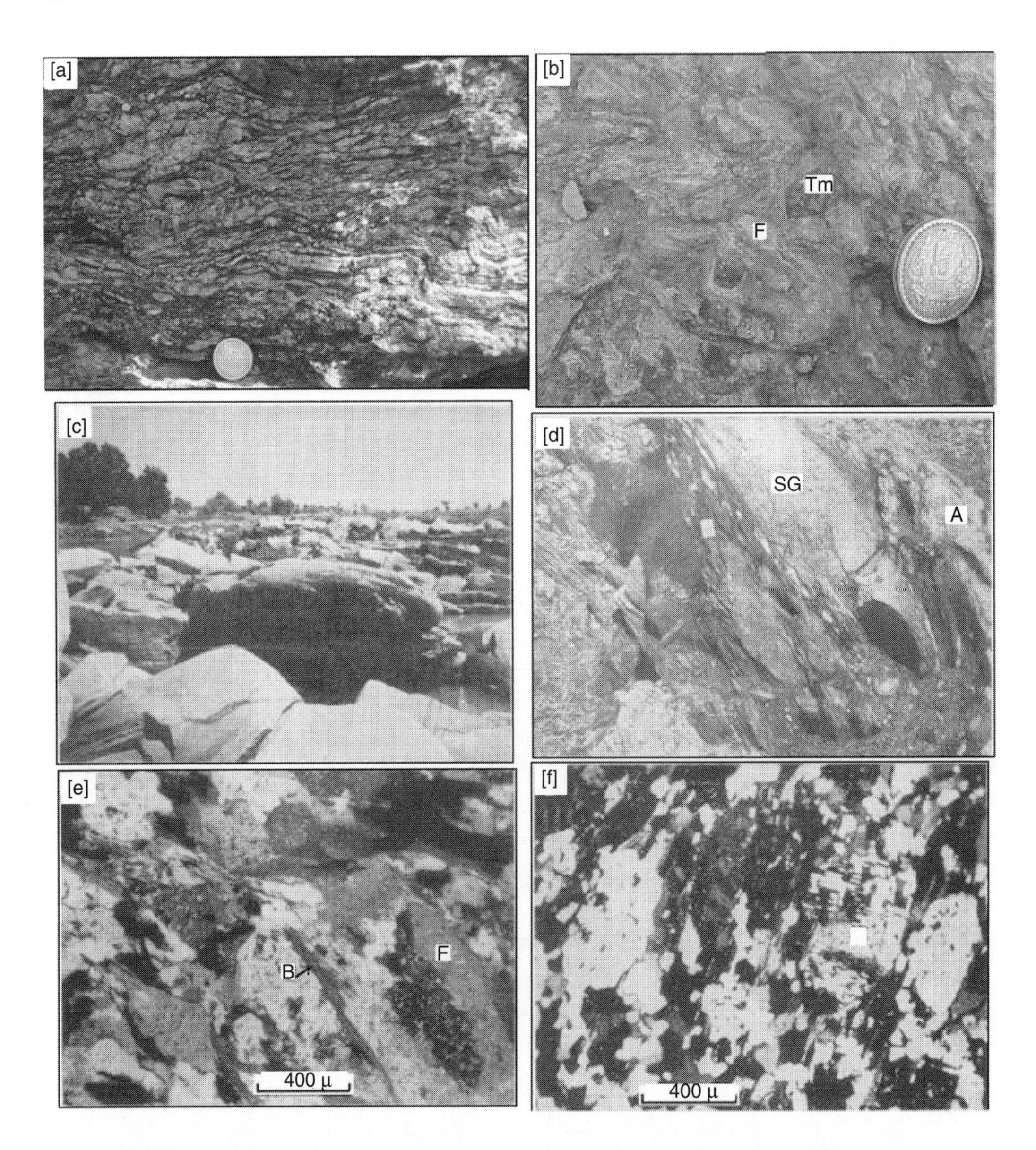

Fig. 4.1.29 (a) Feldspathic schist (Soda granite), north of Tamadungri; (b) Deformed feldspathic schist with porphyroblasts of albite (F) and tourmaline (Tm), Pathargorah; (c) Exposure of Soda granite/feldspathic schist in Sankh River section, south of Badia mines; (d) Soda granite (SG) and apatite (A) lenses in chlorite quartz schist, Dhantuppa; (e) Segregation of feldspar (F) in massive variety of Soda granite; B-biotite/chlorite, Mosabani; (f) Feldspar porphyroblasts participated in overall schistose fabric, Soda granite, Mosabani (a–b: photographs by authors; c–f : from Sarkar, 1984).

Smaller bodies and tongues occur lit-per-lit in the host rocks. In the field these rocks vary from feldspathic biotite-chlorite-quartz schists, through migmatites and augen gneisses (metatexite) to massive varieties (diatexite). The massive variety of Soda granite consists principally of feldspar and quartz. Minor phases include biotite, chlorite, muscovite/ sericite, tourmaline and apatite. In less massive varieties, allanite, epidote, zoisite and sphene may be present. The plagioclase feldspar is dominantly albitic ($An_{5\text{-}6}$) and rarely oligoclasic ($\sim An_{15}$). K_2O-content is represented more by muscovite than by a K-feldspar. There is variation in the SiO_2 content of these rocks. But the variation is most conspicuous in the K_2O/Na_2O ratios. Normative corundum is present in some samples studied (Sarkar, 1984; Dasgupta et al., 1993). In these rocks, contents of Cu, Ni, Co, Cr and locally V, are higher and those of Zr and Li are lower compared to the low-Ca granites (Sarkar, 1964; Mukherjee, 1968). Sr_i ranges between 0.7314 and 0.7449 (Sarkar et al., 1986). It is difficult to smugly fit the Singhbhum Soda granite in the classification of igneous rocks recommended by the IUGS sub-commission on the systematics of igneous rocks (Streckeisen, 1976).

It, however, closely approaches the 'alkali feldspar granite' that has no restriction on the K-feldspar/ Na-feldspar ratio. It falls on the low-Or domain of the Ab-Or-Q diagram (Fig. 4.1.30). It is too Ca- poor to be called a tonalite or granodiorite. Massive variety of the Soda granite does not show typical granitoid texture (Fig. 4.1.28, c) and its bulk chemistry (generally $Na_2O > K_2O$) does not tally with the Rb/ Sr ratios (~6) (Das Gupta et al., 1993).

Dunn and Dey (1942), Sarkar (1964), Das Gupta et al., (1993) concluded that the suite is of magmatic origin, although they differed in details. Later, on a closer look at many occurrences of this suite of rocks along the shear zone, Sarkar (1984) concurred with the view of Banerji and his coworkers (Banerji and Talapatra, 1966) that these rocks were essentially migmatites.

Observations in scales varying from macroscopic to mesoscopic, suggest that the migmatising fluid was hydrothermal in nature, rather than a silicate melt. Pollard (2000) proferred an interesting explanation for the development of albitic rocks. Fluids composed essentially of H_2O-CO_2-salts, unmix during decompression and cooling, sequestering CO_2 to the vapour phase and salts to the liquid (Bowers and Helgeson, 1983). He concludes that albitisation results from further unmixing of CO_2, as the resulting hypersaline fluid will have $Na/(Na+K) >$ equilibrium value at the given condition and will lead to albitisation. Further cooling, particularly in Cl-bearing fluids, will lead to potassic alteration (Orville, 1963). This is a meritorious proposition but does not explain the absence of carbonate minerals along the Singhbhum belt or the profuse precipitation of quartz, more or less along with albite. The results of Burnham's now classical experiment (Burham, 1967) involving granite + water may also be recalled here. It showed that at high pressure and temperature (10 kb/650°C) the solute content in aqueous solution was ~ 9%, containing Si, K and Na directly in proportion to that what characterises the 'minimum melt' composition in the Q-Ab-Or system. With the fall of pressure and temperature, the fluid composition will move towards the quartz peak. At high a H_2O, muscovite rather than a K-feldspar, will be stable. This model apparently better explains the development of the Soda granite + albitisation in this belt. Profuse B-metasomatism and hydration along the belt is also better explained by this model.

Quartzites and Kyanite–Quartz Rocks A prominent, though discontinuous zone, comprising quartzite bands occur almost as a marker horizon along the interface of the shear zone and the garnetiferous mica schist country (Chaibasa Fm.) of the north. The bedding and subparallel early schistosity are well-preserved, though mylonitic foliation and mylonite bands often obscure other features. Excellent F_1 and F_2 folds are preserved at many places in the quartzites.

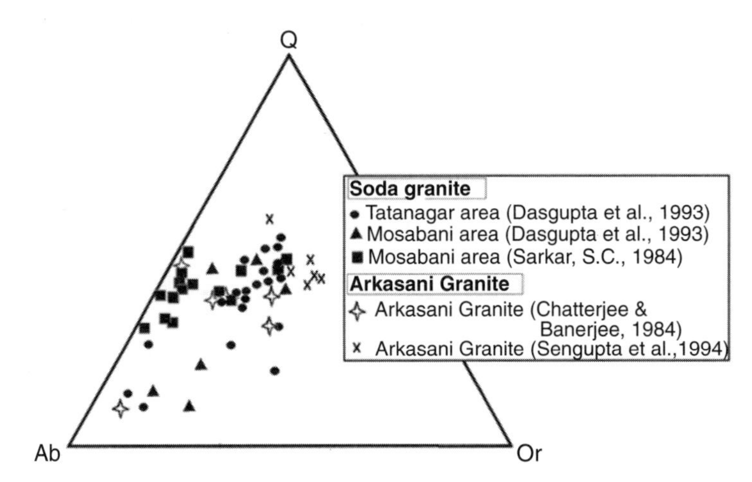

Fig. 4.1.30 Ab-Or-Q diagram based on composition of Soda granite-feldspathic schist (Sarkar, 1984; Dasgupta et al., 1993) and Arkasani Granite (Chatterjee and Banerji, 1984; Sengupta et al., 1994)

The quartzites are characterised by the association of kyanite–quartz laminates, kyanite–muscovite defining the prominent schistosity and also as lenses, pockets and nodules of kyanite–quartz rock. Through much of the shear zone from south of Surda to the west of Bhatin, the kyanite-bearing quartzite and muscovite–quartz schist (kyanite–quartz rock) constitute a continuous unit. Here, the rock is composed of equant grains of quartz, subordinate plagioclase and orthoclase, aggregates of sericite/muscovite, minor magnetite and ilmenite, and fine needles or blades of kyanite (named by Dunn and Dey, 1942, as 'kyanite-quartz-granulite'). The kyanite rich pockets within this body have much coarse-grained quartz and lack the orientation of the kyanite blades and the polygonisation of quartz. Locally it may be even pegmatoid. Muscovite is commonly present, but generally as replacer of kyanite. Locally present are chloritoid, dumortierite and rutile. In thinly banded kyanite–quartz rock, both kyanite and quartz show varying effects of deformation, with or without recrystallisation. At places kyanite–quartz veins traverse the quartzite.

On either side of this stretch, the rock is more muscovitic and schistose and occurs mostly as small discontinuous bands, forming low ridges, spread over a wider zone, many a time well within the mica schist country, and can be traced upto Baharagora in the southeast and Lapsa Buru (22°48′; 85°44′) in the west. At both these places primary layering of alternate kyanite- and quartz-rich bands (± tourmalinite bands) is conspicuous (Fig.4.1.31 a–d).

Although it may be possible to hydrolyse granitic/pelitic schist material and produce alumino-silicates directly at elevated temperatures and pressures (Andreasson and Dallmeyer, 1995; Kerrich, 1990), it is surmised in view of the presence of finely laminated kyanite–quartz rocks at places that the protolith in this case was a sedimentary rythmite. The initially developed kyanite rock was later hydrothermally modified in most places. The oxygen isotope ($\delta^{18}O$) signature reported from this rock (Bandyopadhyay, 2003) could be imposed at that stage. Lazulite development is also related to this phase. Owens and Pasek (2007) explained the kyanite–quartzites of Piedmont, Virginia, USA, as the product of a high sulphidation system, followed by metamorphism. Their conclusion is based on intense Ga-depletion and LREE-enrichment with depletion in Gd through Ho.

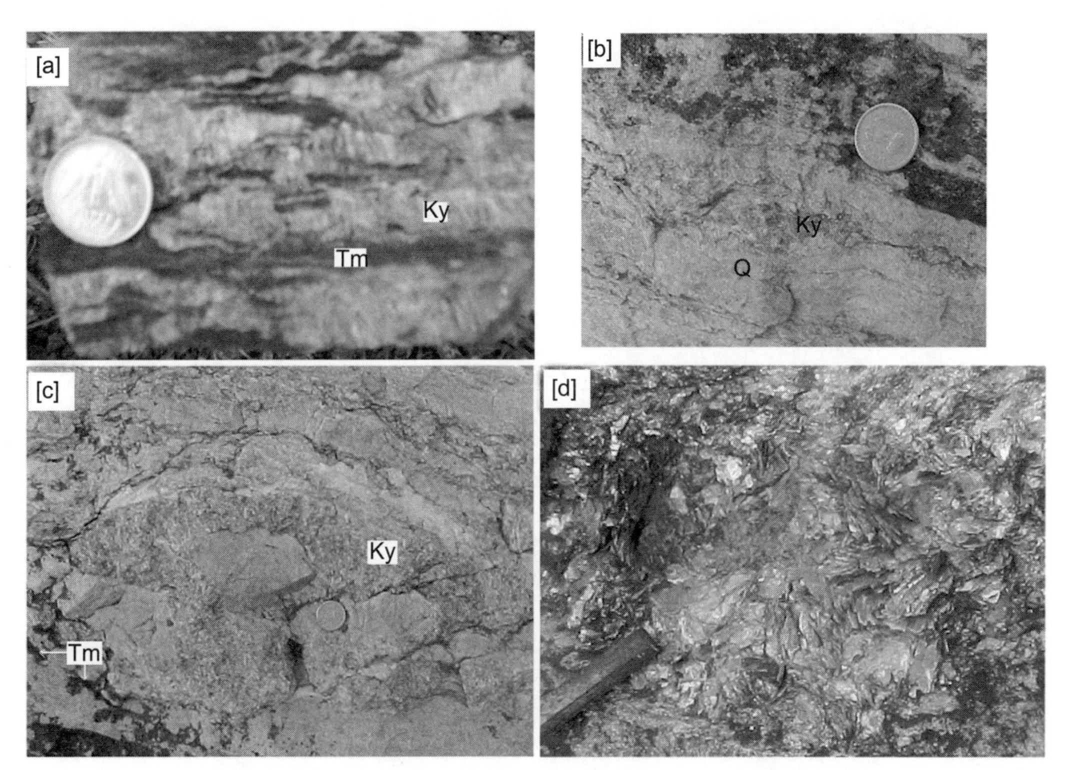

Fig. 4.1.31 (a) Kyanite (Ky)–Tourmaline (Tm) rhythmite, (b) Kyanite (Ky)-bearing bands in quartzite (Q), (c) Kyanite (Ky) in lensoid pocket within quartz-kyanite rock and tourmaline (Tm) as clots, (d) Post-tectonic kyanite blades and books of muscovite replacing kyanite (photographs by authors and P. Sengupta)

Dumortierite and pyrophyllite bearing rocks Dumortierite-bearing rocks, as already mentioned, occur in the SSZ from Bhatin westward up to some distance to the west of Turamdih. Dumortierite occurs as aggregates of individual grains or along grain boundaries and cleavages/ fractures in kyanite in the quartz-kyanite rocks or kyanite–quartz schists. The most prominent occurrence of the mineral is seen near Ujanpur (22°45′38″: 86°04′18″) on the western bank of the Kharkai river (near Sanjay river confluence), where a hillock made up of quartz–kyanite rock along the northern periphery of the shear zone exposes spectacular purple coloured bands (> 2–3 m across) and pockets of dumortierite-bearing rock. Coarse-grained tourmaline, in lenticular bands and pockets, is also closely associated with this rock (Fig. 4.1.32 a–e).

The textural relationship between kyanite and dumortierite suggests that the latter has developed from the former by the action of boron-bearing hydrothermal solution. Some of the suggested reactions involving kyanite that produced the accompanying minerals are:

Kyanite + Fe^{2+} (magnetite) + H_2O → Chloritoid + O_2 (Ganguly, 1969)
Kyanite + B^+ + H_2O → Dumortierite + quartz +H^+
Kyanite + Si^{4+} (quartz) + K + H_2O → Muscovite + H^+ (cf. Sarkar, 1984)
[Lazulite-bearing veins are locally developed in kyanite–quartz rocks.]

Fig. 4.1.32 (a) Dumortierite hillock at Ujanpur, west of Kharkai river; (b) Hand specimen showing dumortierite (Dm) and kyanite; (c) Schistosity parallel laths and boudins of dumortierite (Dm) in kyanite–quartz schist (Ky + Q); (d) Tourmaline pocket (Tm) surrounded by dumortierite (Dm), vein quartz (Q) in patches; (e) Kyanite (Ky) rimmed by dumortierite (Dm) (photographs by the authors).

Dumortierite, geologically the most important ternary phase of the ABSH system according to Werding and Schreyer (1996), is a minor but ubiquitous phase, that commonly replaces kyanite. Although the mineral is stable up to granulite facies, its lower P-T limits were not constrained. In an attempt to do so, Verding and Schreyer (op. cit.) investigated mixtures of kaolinite + diaspore + H_3BO_3 seeded with synthetic dumortierite. Growth of dumortierite took place at 3 kb/380°C and 5 kb/360°C. This seems reasonable even in the context of the observations along the Singhbhum belt.

In the same strike continuity of the dumortierite-hill, at about 1 km west of Rangamatia (22°45′28″: 86°01′30″), there is a fairly large pocket of pyrophyllite in the kyanite-bearing quartzite ridge, which is being quarried at present. Alteration of kyanite into pyrophyllite might have followed the reaction,

$$2\ \text{Kyanite} + 2\ H_2O + 6\ SiO_2 + 3\ O_2 = 2\ \text{Pyrophyllite}$$

$$(2\ Al_2O_3.SiO_2) \qquad \{2\ Al_2O_3.Si_4O_8(OH)_2\}$$

Tourmaline-bearing rocks and tourmalinite Tourmaline is the most common boron-bearing mineral in nature. Dunn and Dey (1942) described a rich concentration of the mineral in the form of tourmaline-quartz rock making a hillock to the southwest of Surda. Later, Sarkar (1984) reported concentration of the mineral at several other places along the belt. Since then there have been several interesting publications on it (Bhattacharya et al., 1992; Deb and Mukhopadhyay, 1991; Sengupta et al., 2005). The tourmaline-bearing rocks occur as patches and streaks in the shear zone, particularly at Rakha mines and further southeast. It also occurs well within the mica schists of Chaibasa Formation as at Tetuldanga in the east and Lapsa Buru in the west. Commonly, tourmaline-bearing rocks and tourmalinite in the belt are in form of, or associated with the following:

1. Tourmaline-quartz rock, banded or massive (Fig. 4.1.33 a)
2. Soda granite-feldspathic gneiss/ schist, particularly where deformed (Fig. 4.1.33 d)
3. Biotite-chlorite-quartz schist ('granular rock') (Fig. 4.1.33 b)
4. Mica schists (Fig. 4.1.35 b)
5. Apatite and magnetite are present in many of these rocks (Fig. 4.1.33 c).

The tourmaline concentration in the Surda hillock varies from being massive to a thinly laminated rock having alternate tourmaline rich and quartz-rich laminae (1 mm to > 1 cm thick), locally fragmented. Overall texture is granular with perceptible orientation in the fabric, revealed better under the microscope. Patches, veins and disseminations of apatite (± magnetite) are present (4.1.33 c). Some structures, locally resembling cross-laminations and slumps, were interpreted to have sedimentary origin by Deb and Mukhopadhyay (1991) and Bhattacharya et al. (1992).These authors even believed the boron to have come as an exhalate related to Dhanjori volcanics, where the host rock for tourmaline is quartz chlorite schist.

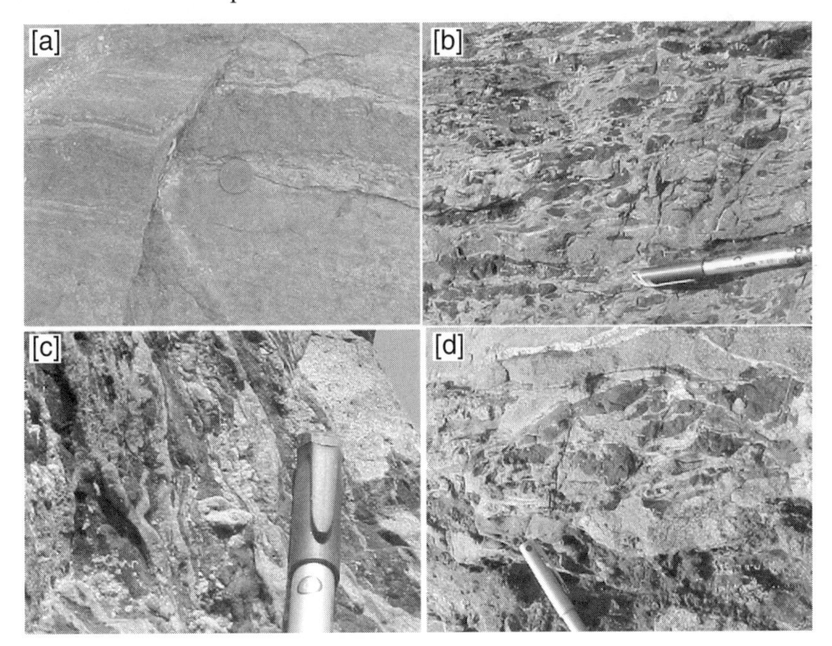

Fig. 4.1.33 (a–d) Tourmaline mineralisation in SSZ; (a) Thinly banded tourmalinite, Surda; (b) 'Granular rock' with stringers and pockets of tourmaline; (c) Pinch-and-swell tourmalinite bands showing a cross cutting lense of apatite contained in tourmalinite; (d) Fragments of tourmalinite in lensoid body of deformed Soda granite within granular rock (photographs by S.C. Sarkar and P. Sengupta).

Before commenting on this particular case, let us see how it is generally understood. An exhalate, rather exhalative sedimentary association, consists principally of interbedded volcaniclastic and chemical sediments. The volcaniclastic member typically consists of volcaniclastic sediment, siltstone and shale and the chemical component may include chert, Fe-Mn sediment (including coticules), sulphides, tourmalinites and barite-rich rock. On metamorphism, the meta-exhalites may contain several other minerals such as gahnite, Zn-staurolite, etc. (Spry et al., 2000). In the present case, most of the features are absent. While thinking about the origin of this tourmalinite, we have to also consider the tourmaline-rich excellantly layered pelitic metasediments at Tetuldanga, participating in the metamorphic fabric. Moreover, boron behaves as an incompatible element in igneous systems and its content in mafic igneous rocks is usually < 10 ppm (Spivack and Edmond, 1987; Chaussidon and Jambon, 1994). We should therefore, normally not expect the mafic (-ultramafic) Dhanjori volcanics to produce boron-rich exhalates.

Tourmaline concentration in deformed Soda granite is observed at many localities, but more copiously at the Pathargorah area. Tourmaline clasts and clusters, besides disseminations, are common in the granular rock (Deb, 1971; Sarkar, 1984). Many of the constituent grains in them have no finite orientation. Tourmaline replaces albite and quartz in Soda granite. Bandyopadhyay (2003) reported tourmaline bands occurring along the axial cleavage of folds on schistosity. These and the replacemant of kyanite by dumortierite in kyanite–quartz rocks in the shear zone, as recorded by us in the foregoing pages and observed by earlier workers (Deb, 1971; Sarkar, 1984 and Bandyopadhyay, 2003), testify to the metasomatic introduction of boron in a later stage. But what is the relationship between the late-introduced boron and the early tourmalinites? Even what was the source of boron in the latter is not clear. It is possible that the late diagenetic-metamorphic fluid produced the tourmalinites by replacement (cf. Steven and Moore, 1995; Slack, 1996; Dutrow et al., 1999). Boron- contents of pelitic protoliths are in range of 100–1000 ppm (London et al., 1996). A part of the boron from these early tourmalinites could be mobilised/remobilised during a late phase of shear movement(s) along the belt, accompanied by intense aqueous fluid activity. Tourmaline, stable in acidic to neutral aqueous fluid, becomes unstable in alkaline fluid (Morgan and London, 1989). Similar chemistries and boron isotopic values for tourmalines in scheelite-bearing exhalites and later tungsten-bearing quartz veins at Eastern Sirra Pampeanes, Argentina, led Tourn et al. (2004) to conclude that the boron in tourmaline in quartz veins was remobilised from the tourmaline-bearing exhalites. In the northern parts of NSMB, a situation comparable with the above mentioned area does exist. Here, in the tungsten field of Chhendapathar, tourmaline occurs both in tungsten bearing quartz veins as well as in form of thinly bedded tourmalinite in the adjacent metapelites (discussed in Section 4.2).

Boron is easily soluble in hydrothermal solutions and hence this element can be used as a 'tracer' for fluid activity in natural rock systems. Chemically, the stability field of tourmaline is located in the μB+-μK+ space (Slack, 1996; Dutrow et al., 1999; London, 1999; Bandyopadhyay, 2003). It may be noted that tourmaline is the principal sink of boron in nature and the mineral may form in conditions ranging from diagenesis to granulite facies conditions. But in terms of P-T, it is stable in a diagenetic environment compared to that of the granulites; the latter must of course develop in a low a_{H_2O} situation.

Sengupta et al. (2005) took up the potential of tourmaline as an independent recorder of fluid regime in a polymetamorphic terrain as the Singhbhum copper-uranium belt and its neighbourhood. The boron introduction in the belt took place at least three times during the latter's evolution. The first one appears to be pre-M_1 (their progressive dynamo-thermal metamorphic cycle), the second one during M_1. The last major event of boron introduction was post M_2 (their retrogressive metamorphic cycle), when individual nodular aggregates of undeformed tourmaline and dumortierite grains at the expense of kyanite developed. Granitic magma is the likely source of the later boron (vide discussions on Soda granite in the previous pages).[1]

Basic intrusives Similar to the innumerable sill-like bodies of amphibolites and hornblende schists occurring with the mica schists of Chaibasa Formation in the north, such rocks are present in the shear zone also, with the exception that these are more schistose with profuse development of chlorite and biotite. Amphibole is actinolite ± hornblende. Tourmaline and sulphide minerals are minor constituents in some.

A set of fresh doleritic intrusives that are holocrystalline at the core and tachylitic at the border, is also reported from the shear zone (Sarkar, 1969).Ray and Biswas (1951) reported a lamprophyric rock from Mosabani mine.

Ultramafic intrusives Ultramafic intrusives now represented by talc + tremolite ± Mg-chlorite ± magnetite ± chromite ± epidote ± plagioclase ± carbonate occur along the shear zone, concordant with the other lithounits (Dunn and Dey, 1942; Sarkar, 1984). One such body is traceable from the west of Surda upto Jaduguda and another major body occurs at Kudada–Pathar Pahar–Chhadbi Pahar region southwest of Tatanagar. Smaller bodies occur at several other places in the zone. Several ultramafic intrusives occur in the mica schist to the north and Dhanjori group rocks in the south. Their petrography varies within a small range. At Kudada, a gabbroic rock is associated with an ultramafic body.

Arkasani Granite–granophyre Granitic rocks with granophyric texture at places, and occurring as disconnected lenses along the northern limb of the forked Singhbhum shear zone (Fig. 4.1.1), particularly at Arkasani (Akarsani, the actual name of the place) hills (22°46': 86°52') and further east have been called Arkasani Granophyre by early workers (Dunn, 1929; Dunn and Dey, 1942) (Fig. 4.1.34 a–b).

The lenticular bodies of this rock are usually massive at the core and gneissose in the outer parts. Smaller bodies may be foliated all throughout. The non-massive variety, containing enclaves of mica schists and hornblende schists at places, could be migmatitic (metatextite). Many of the structures present in the country rocks are present in the migmatites also. Na_2O/K_2O in these rocks widely vary (0.74–7.8, n=6). Some of these rocks contain both plagioclase (albite–oligoclase) and K-feldspar, while others contain only sodic plagioclase, such that some plot in the granite field and others in the trondhjemite field in the An-Ab-Or diagram (cf. Fig. 3.41, Sarkar, 1984).

Opinions on the origin of these rocks range between two extremes: magmatic and metasomatic (Sarkar, 1984 and references therein). There are some similarities in petrochemistry and mode of occurrence between the Arkasani Granophyre and the Soda granite and the related rocks. Moreover, the Singhbhum shear zone apparently controls both. However, the bulk of the rock is characterised by a granitoid petrography in contrast to that in the Soda granite/feldspathic schist.

Reliable emplacement date of the Arkasani Granophyre is still lacking. However, a thermal event at around 1.0 Ga is recorded by Rb-Sr (WR) isochron data from this rock (Sengupta et al., 1994). This cannot be the emplacement age of these rocks as in that case it would be difficult to explain the presence of dolerite dikes in it and their absence in the Soda granite / feldspathic schists and other rocks in the neighbourhood.

[1] Pal et al. recently (2010) published some interesting boron-isotope data ($\delta^{11}B$ ranging between – 6.8 and + 17.2 per mil) from the Jaduguda mines area. They utilised the data to suggesst a 'marine evaporite or basinal brine' for the fluids involved. We shall wait till more data on this aspect, reasonably integrated with the regional geology, are available.

Fig. 4.1.34 (a) A view of the Akarsani Pahar located near Kharsawan, from where the name 'Arkasani' Granophyre is derived (photograph by the authors); (b) A close view of an outcrop of Arkasani Granophyre showing porphyritic texture; (c) Photomicrograph of Arkasani Granophyre showing graphic intergrowth of quartz and feldspar (photographs by the authors).

Domain III: The area between Singhbhum shear zone and Dalma volcanics The main metasedimentary belt (Ghatsila–Tatanagar–Kharsawan–Sonapet) lying to the north of Singhbhum shear zone essentially comprises a thick sequence of metapelites (± meta-psammitic interbeds) with numerous bands of amphibolite, hornblende schist and quartzite. Along the northern periphery of this sequence and immediately to the south of Dalma range passes a magnetitic phyllite horizon with prominent quartzite bands, which extends from the eastern extremity of the NSMB to the Sonapet sector in the west. The carbonaceous and ferruginous phyllites and tuffaceous sediments, constituting the basal sequence of the Dalma volcanisedimentary pile (Domain IV) occur along its north (Fig. 4.1.1). Dunn and Dey (1942) assigned the former to their Chaibasa Stage and the latter (along with the basal Dalma sediments) to the Iron Ore Stage, together constituting their Iron Ore Series. Sarkar and Saha (1962) renamed Dunn and Dey's Iron Ore Series as the Singhbhum Group and the Chaibasa Stage and the Iron Ore Stage rocks respectively as the Chaibasa Formation and Dalbhum Formation (Table 4.1.7).

Table 4.1.7 *Singhbhum Group (Sarkar and Saha, 1962)*

Formations	Characteristic lithology
Dalbhum Formation (> 4 km)	Ferruginous chloritic schists/phyllite, sericite phyllite, minor carbon phyllite, hematite phyllite, quartzite and hematite quartzite.
Chaibasa Formation (> 8 km)	Mica schists with a number of quartzite bands and numerous isofacially metamorphosed basic intrusives.

Gupta et al., 1980 and Sarkar et al. (1992), while presenting a detailed account of their Dalma Group, discriminated the rocks of Dhalbhum Formation from those belonging to Lower Dalma Formation, the latter characterised by prominence of carbonaceous and ferruginous phyllites–chert with profuse tuffaceous components(discussed later in this chapter). The two quartzite bands extending for several tens of kilometers along the southern base of Dalma hill range (included earlier in Dhalbhum Formation by Sarkar and Saha, op. cit.) were considered as part of Lower Dalma Formation by Gupta et al. (op. cit.).

The sedimentation of the psammopelites in this domain have been variously interpreted as deposition under neretic to epineretic condition (Sarkar and Saha, 1962), flyschoid sequence developed in an unstable geosynclinal trough (Naha and Ghosh, 1960; Gaal, 1964), shallow water shelf deposits with tide-storm interaction (Bhattacharya, 1991; Bose et al., 1997; Bhattacharya and Bandyopadhyay, 1998), submarine mid-fan deposition (Das, 1997; Biswas, 1998). The current beddings, convolute laminations (Fig. 4.1.35 a,c,d,e), slump sheets, wave ripples and desiccation structures, best preserved in the psammitic layers are indicative of an overall shallow water depositional environment. The limited studies on paleocurrents suggest southerly and south-westerly located provenance for these sediments (Sarkar and Saha, op. cit.; Das, op. cit.).

The Chaibasa rocks are commonly garnetiferous muscovite–quartz schists. Chloritoid, biotite, garnet, kyanite, staurolite and sillimanite occur progressively in space and apparently in time towards the median axis of the domain throughout its length. The metamorphism was generally assigned to Barrovian type (Naha, 1965; Roy, 1966; Lal and Singh, 1978). The peak of the first metamorphism event M_1 is commonly considered to be syn- to late- tectonic with respect to the first deformation D_1, which formed F_1 folds and S_1 schistosity. Sarkar et al.(1993), however, reported from the eastern part of the belt a pronounced early pre-tectonic metamorphic cycle, M_0, characterised by the development of andalusite, chloritoid, biotite, some garnet and 'perhaps some kyanite'. The dynamothernal event of metamorphism started with M_1, coeval with F_1 (D_1) and marked by the crystallisation of syn-to post-tectonic biotite, garnet, staurolite, kyanite and rare chloritoid. The peak of metamorphism that again produced some of these minerals coincided with F_2 (D_2) deformation. The estimated range of metamorphism was 3 – 6 kb and 415° – 600°C.

A subordinate but important rock type in this domain is kyanite–quartz rock. This rock type occurs as rather thin units towards the southern parts of this domain. At Lapsa Buru (22°48′: 85°44′), one of the largest repositories of commercial kyanite occurs well within the Chaibasa Formation. Rhythmic layers (S_0) of ultrafine massive kyanite, alternating with tourmaline, occur within the mica schist with local pockets of large bladed post-tectonic kyanite (Fig. 4.1.36 a). Kyanite–quartz rhythmites are found at places between Kanyaluka and Baharagora. Composition of the protolith obviously played an important role in their formation. The Dhalbhum phyllites contain large porphyroblasts of andalusite and staurolite.

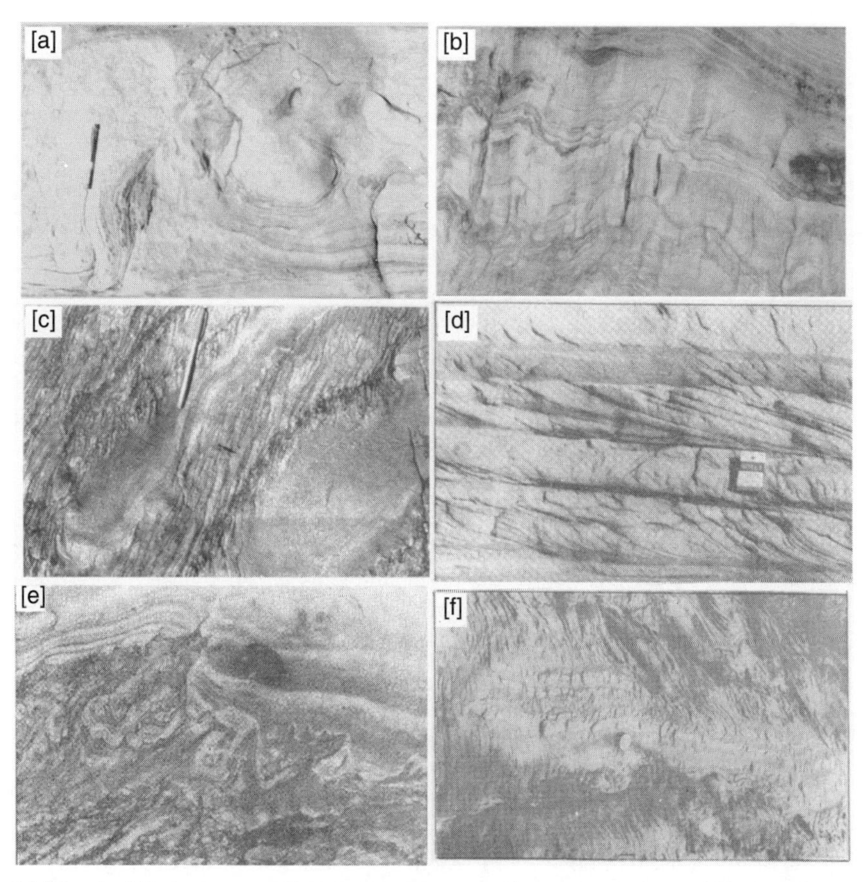

Fig. 4.1.35 (a–f) Sedimentary and tectonic structures in the psammopelites of Chaibasa Formation, Harindukri–Tetuldanga area, Ghatsila; (a) Bedding (S_0) and soft-sediment convolute structure in meta-psammite; (b) Puckered $S_0 \| S_1$, with strong axial planar cleavage (S_2); Undeformed tourmaline bearing laminates (dark) in top right; (c) Flattened convolutes in psammitic bands (S_0) and crenulation cleavage (S_2) in pelitic layers; (d) Cross bedding in meta-psammite; (e) Soft sediment deformation structures in psammopelites; (f) Refraction of crenulation cleavage (S_2) across psammitic layer.(a-c: photographs by the authors; d–f: from Gupta and Basu, 1992).

At Lapsa Buru, where kyanite was mined till late last century, the thin kynite bands and the best grade of ore (float ore–'elephant's egg') are made up of extremely fine-grained stout prisms of kyanite without having any preferred orientation (Fig. 4.1.36 b). Similar is the texture of tourmalinite bands. However, in the open-cast mine at the East hill of Lapsa Buru, large bladed kyanite occurs in coarse grained quartzose (almost pegmatoid) veins and pockets in the quartzite. The occurrence of topaz is also reported from similar rocks in the vicinity (Dunn, 1929).

The structural patterns in the Domain III (Chaibasa Formation – between SSZ and Dalma belt) are dominated by low to moderate E–W plunging antiforms and synforms belonging to D_2–F_2 deformation. Along the middle part of the basin these folds are upright asymmetric with axial planes (S_2) dipping to the north, while they become progressively overturned to the south with the approach of the SSZ. The incidences of F_1-folds in this domain are limited to discontinuous root-less hinges of inclined to reclined folds on bedding (S_0), mostly preserved in the psammopelitic and psammitic metasediments. The early mineral schistosity (S_1) is, however, ubiquitously developed as the most pervasive planar structure in Chaibasa rocks (Basu et al., 1979; Gupta and Basu, 1990, 2000; Gupta et al.,1996).

Fig. 4.1.36 (a) Thinly laminated kyanite-tourmaline rythmite, Lapsa Buru; (b) Float ore of massive kyanite, Lapsa Buru, Saraikela-Kharsawan district, Jharkhand (photographs by the authors).

In the eastern part of the domain, in the Galudih–Ghatsila area, the large-scale antiforms and synforms defined by the quartzites and psammopelitic bands within the mica schists country are now identified as F_2-structures (Basu and Gupta, 1998; Gupta and Basu, 2000; Mukhopadhyay et al., 1998; Ghosh et al., 2006) in variance with the long held contention of the previous workers (Sarkar and Saha, 1962; Naha, 1965) that these structures represented first generation folds in the belt. The pelites in the vicinity of the major F_2-folds on quartzite record coaxial F_2-crenulations with the development of prominent crenulation cleavages (S_2) (Fig. 4.1.35 b, c, f), which are often differentiated. S_1 in limited subdomains, specially in the phyllites of Dhalbhum Formation in the north, is almost totally transposed into S_2.

In the eastern sector of Domain III all earlier structures are overwhelmed by roughly N–S trending F_3-flextures on regional scale with the development of a series of depressions and culminations (Bankati depression–Deoli culmination–Paharpur depression etc.) (cf. Sarkar and Saha, 1962). To the east of Nischintapur there is an axial planar dislocation in a large scale F_3-hinge zone. The form surface of these large scale F_3-folds is mostly defined by transposed S_2 planes. The geological map of a part of this sector and the corresponding FCC prepared from satellite imagery are presented in Fig. 4.1.37 a, b.

According to Mukhopadhyay et al.(1998) and Ghosh et al.(2006), the quartzite bands in the north of Ghatsila and west of Dhalbhumgarh define steeply plunging U-shaped F_2-fold closures facing in opposite direction and form a steeply plunging sheath fold like structure with acute hairpin curvature of the fold axis, defining a culmination (Deoli culmination). These workers have also described mylonitic foliation parallel to S_2 in Dhalbhumgarh quartzite, thereby implying strong shear movements witnessed by the Chaibasa rocks during D_2 deformation.

Domain IV: The Dalma Volcanic Belt The Dalma belt, composed principally of volcanisedimentary rocks, constitutes an outstanding geomorphological and geological feature in the region. It is an arcuate ridge occurring almost along the middle of the North Singhbhum mobile belt, showing regional fold closures around E–W axial traces in the western and central sectors and around an NE-SW axis in the eastern part (Fig. 4.1.1). The belt is composed of a greenstone assemblage, which consists of shale/ phyllite, carbonaceous phyllite/ tuff with interlayered basic volcanics, overlain by a thick pile of high-Mg volcaniclastic rocks and co-magmatic flows of komatiitic composition.

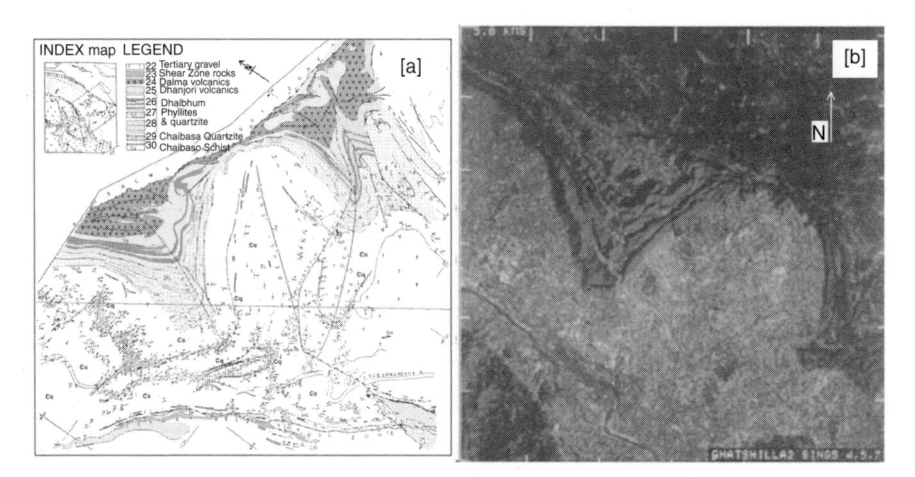

Fig. 4.1.37 (a) Geological map of Ghatsila–Galudih sector (modified after Sarkar and Saha, 1962), (b) FCC prepared from imagery for a part of the area shown in (a) (courtesy–P. Purkaet).

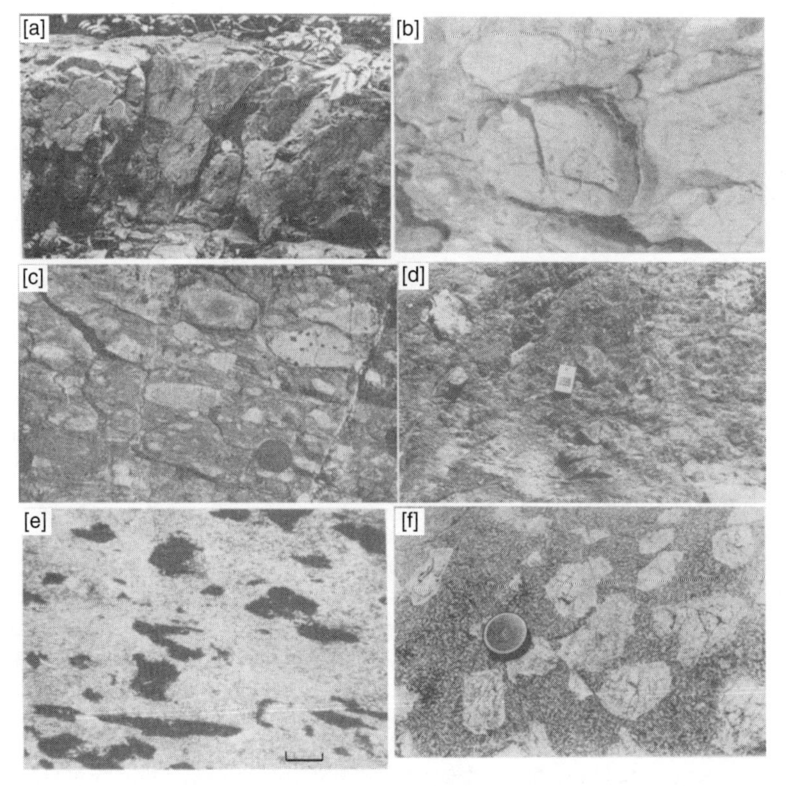

Fig. 4.1.38 (a, b) Pillow structures in Upper Dalma Basalt, north of Sonapet Valley; (c) Epiclastic conglomerate with flattened pebbles of basalt in Lower Dalma, north of Hesadih; (d) Outcrop of Lower Dalma volcaniclastics with angular blocks of ultramafics in tuffaceous matrix, Dimna Sector; (e) Schistosity aligned spindles and shards of partially devitrified glass (dark) in high-Mg vitric tuff represented by felted tremolite (white), Scale Bar 0.5 mm; (f) Feldspathoidal phenocrysts (light coloured) in Lower Dalma Basalt, Mailpir, Sonapet (photographs from Gupta and Basu, 1979, 1992 ; Gupta et al., 1980,1982).

This is followed upward by pillowed low- K, high-Mg tholeiites of ocean floor affinity. No felsic volcanics or sediments are reported to be overlying this basalt (Gupta and Basu, 1977, 1979, 1991; Gupta et al., 1977, 1980, 1982). The lower part of the sequence locally contains analcite basalt. Gabbro-pyroxenite bodies also intrude the sequence (Fig. 4.1.38, a–f). The stratigraphic succession of the Dalma volcanisedimentary suite, after Gupta et al. (1980), is furnished in Table 4.1.8 and detailed maps of some segments of the belt are presented in Fig. 4.1.39.

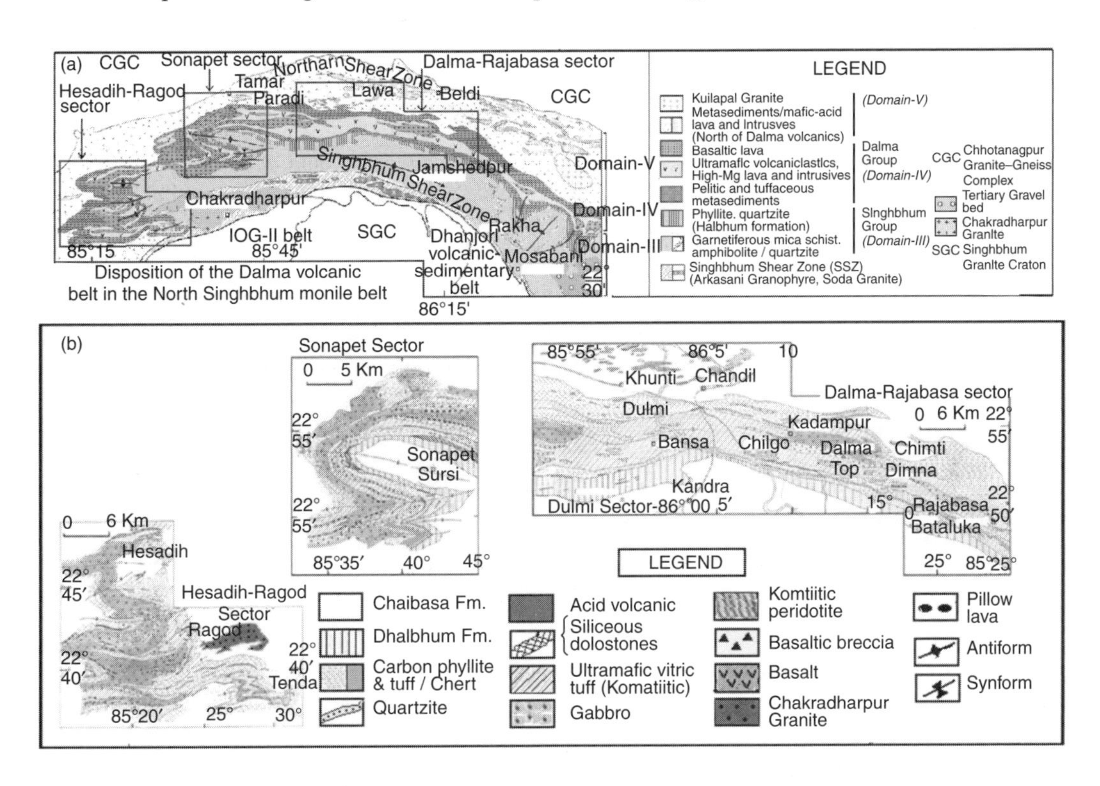

Fig. 4.1.39 Map showing disposition of Dalma volcanic belt in the NSMB (excerpt from Fig.4.1.1) and detailed geological map of some sectors of the belt (modified after Gupta et al., 1980)

Table 4.1.8 *Stratigraphic succession of Dalma volcanisedimentary suite (after Gupta et al., 1980)*

Group	Formation	Rock types
	Upper Dalma Formation	Tholeiitic basalt
Dalma Group	Lower Dalma Formation	Tuffaceous tremolitic schist (komatiitic vitric tuff) and chloritic schist with coarse pyroclasts, gabbro, peridotite and minor basaltic flows Phyllite-shale, carbonaceous phyllite-tuff, quartzite etc., with mafic-ultramafic intrusives

Age data on the Dalma volcanic rocks is extremely meagre. Sarkar et al. (1969) assigned a K-Ar age of 1547 ± 20 Ma to the volcanic rocks they studied, which was obviously the metamorphic age. Roy, Abhijit et al. (2002) obtained a Rb-Sr (WR) age of 1619 ± 38 Ma from the gabbro-pyroxenite rocks at Kuchia (22°46′: 86°26′) and interpreted it to be the emplacement date. It is also likely to be the metamorphic age because pyroxenite is schistose along its margins and occur within highly schistose tuffaceous volcanics of Lower Dalma sequence. Recently, ^{40}Ar-^{39}Ar dating of a pyroxenite sample from the same location yielded 2.2–2.3 Ga age range, while a basalt sample from Dalma Hill top showed the most prominent plateau at 2.0 Ga (unpublished work by S.K.Acharyya and A.Gupta).

The upper tholeiite in the Dalma sequence occurs over a limited area along the core of the main synclinal structure framed by the lower Dalma rocks. That the volcanism was at times violent is suggested by the profuse development of agglomerates and breccia at many places in the range.

The post-depositional deformational structures in the Dalma volcanisedimentary pile are similar to those found elsewhere in the mobile belt. The central sector of the belt displays a synclinal structure, defined by structural repetition of the Lower Dalma rocks around Upper Dalma rocks, with the axial surface trending E–W. The northern limb of the syncline is overturned to the south. Dunn (1929) and Dunn and Dey (1942) described a thrusted contact of the Dalma volcanics with the northern metasediments along the northern overturned limb of the syncline. The western sector of the belt (Sonapet–Hesadih–Tebo) displays spectacular regional folds (mostly F_2) with E–W axes and sinistral shifts (Gupta et al., 1980; Sarkar et al., 1992), while in the east it is folded (F_3) on large scale around N–S axes. The reclined F_1-folds are prominent at the two extremities of the belt near Sonua–Tenda in the west (Gupta et al., 1980; Bhaumik and Basu, 1984) and near Simulpal in the east (Mukhopadhyay and Sengupta, 1971). The regional folds displayed by the Dalma volcano-sedimentary sequence are also manifested by the adjacent marker horizons within Dhalbhum and Chaibasa Formations.

Petrochemistry of Volcanic Rocks and Their Significance The petrochemistry of the Dalma and Dhanjori mafic–ultramafic volcanics and the co-magmatic intrusives has been studied in some detail (Gupta et al., 1980, 1985; Gupta and Basu, 1991; Sarkar and Deb, 1971; Sarkar et al., 1992; Bose and Chakraborti, 1981; Chakraborti and Bose, 1985; Bose et al., 1989; Roy et al., 2002a, b).

Composition on the Dalma and Dhanjori lavas, plotted on a MCA diagram (Fig. 4.1.40) show close correspondence with komatiites and tholeiites of the known Archean greenstones of South Africa and Canada. The discriminant diagrams of Al-Mg-(Fe + Ti) (Fig. 4.1.41 a, b) and Al_2O_3 vs FeO/(FeO + MgO) and TiO_2 vs M-value also bear out the bi-modal (mafic–ultramafic) character of the magmatism in both the Dalma and the Dhanjoris (Fig. 4.1.42 a, b). In F_2–F_3 diagram of Pearce and Cann (1973), most of the analyses fall in the island arc type LKT field (Fig.4.1.43 a, b).

The trace element diagrams (Ti/100 – Zr – Y x 3; Ti/100 – Zr – Sr/ 2; Ti– Zr and Zr–Zr/ Y) suggest an OFB/ MORB affinity for the bulk of the upper Dalma basalt (cf. Chakraborti and Bose, 1985). Again, the high Th/Ta ratios (~10) obtained in the analysis of the Dalma basalt (Bose et al., 1989) are characteristic of subduction-associated island arcs. A perusal of the above diagrammatic interpretations suggests that none of them should be considered final. The final interpretation should be holistic. For example, the Dhanjori Group of rocks lie directly on the marginal part of the continent. The chemistry of the igneous components of the sequence, therefore, could not avoid the influence of the continental wedge, through which the magma penetrated.

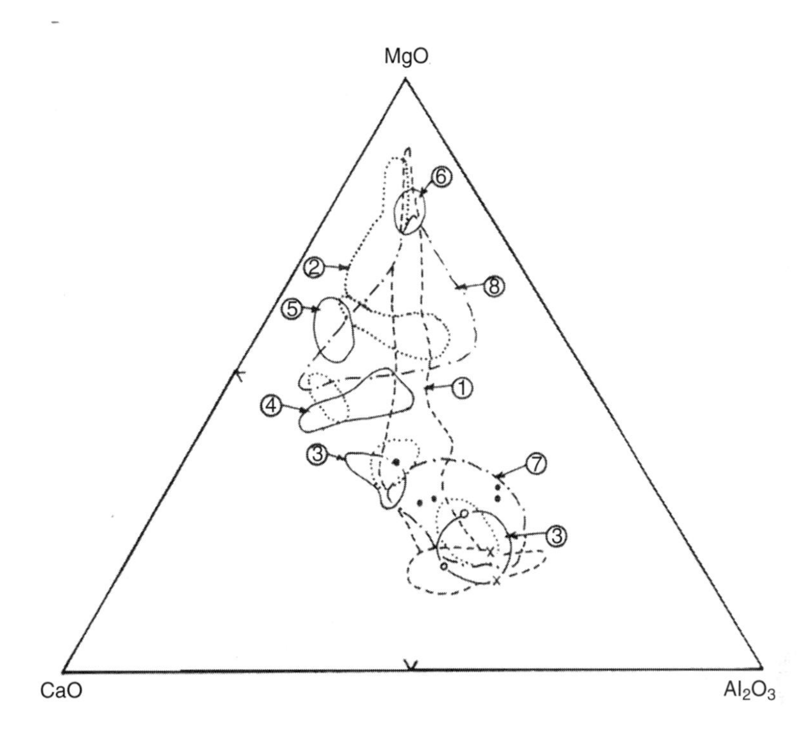

Fig. 4.1.40 MgO-CaO-Al_2O_3 diagram depicting fields of different volcanic units of North Singhbhum mobile belt and other areas. 1 – Munro Township (Pyke et al., 1973; Arndt et al., 1977); 2 – Barberton (Viljoen and Viljoen, 1969 a, b); 3 – Upper Dhanjori basalt; 4 – Dhanjori basaltic komatiite; 5 – Dhanjori komatiitic peridotite–pyroxenite; 6 – Dhanjori ultramafic volcaniclastics; 7 – Dalma basalt; 8 – Dalma ultramafic volcaniclastics; Solid circles – Ongarbira volcanics; hollow Circles – Jagannathpur Lava; cross – Malantoli Lava (after Gupta and Basu, 1991; Sarkar et al, 1992).

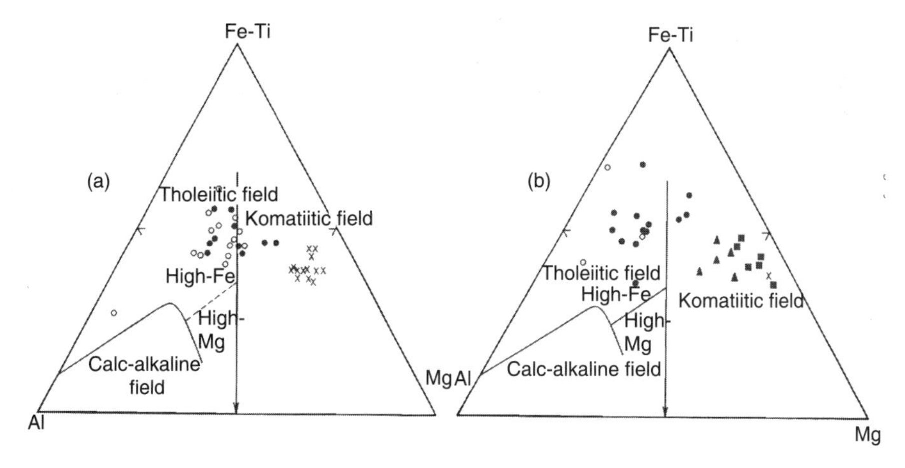

Fig. 4.1.41 Al-(Fe+Ti)-Mg diagram (after Jensen, 1976) for (a) Dalma volcanics and (b) for Dhanjori volcanics. Solid circle – upper basalt; hollow Circle – lower basalt; cross – ultramafic volcaniclastics; solid triangle – basalti komatiitic; Solid square – ultramafic komatiite (from Gupta and Basu, 1991; Sarkar et al., 1992).

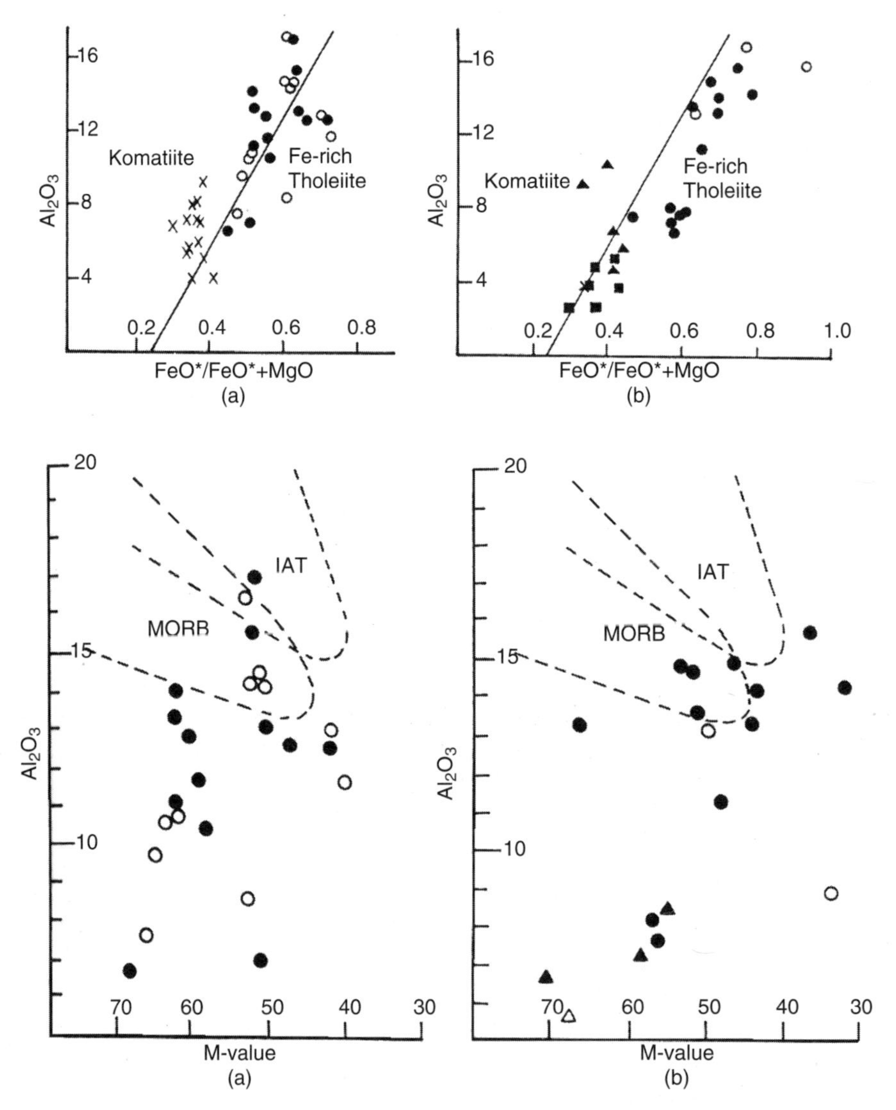

Fig. 4.1.42 Al_2O_3 vs FeO/FeO + MgO and Al_2O_3 vs M-value diagrams (after Gill, 1979) for (a) Dalma and (b) Dhanjori volcanics. Solid circle – Upper basalt; hollow circle – lower basalt; cross – ultramafic volcaniclastics; solid triangle – basaltic komatiitic; solid square – ultramafic komatiite (from Sarkar et al., 1992).

Chondrite-normalised REE plots of the mafic–ultramafic intrusives of the Dalma (Fig.4.1.44a), show depleted LREE and flat HREE patterns that are strikingly similar to the komatiitic and tholeiitic lavas from the belt. In primitive manle normalised plot (Fig. 4.1.44b), the intrusives show general depletion in HFSE and LREE but flat pattern in most compatible elements like Y, Yb and Lu. It is also notable that the plots of mafic–ultramafic rocks of Dalma in Nb/Y vs Zr/Y diagram fall within plume array (Fig. 4.1.44c). Nd-isotopic data with mean $f_{Sm/Nd} \sim + 0.2704$ and high ϵ_{Nd} (mean + 7.8)

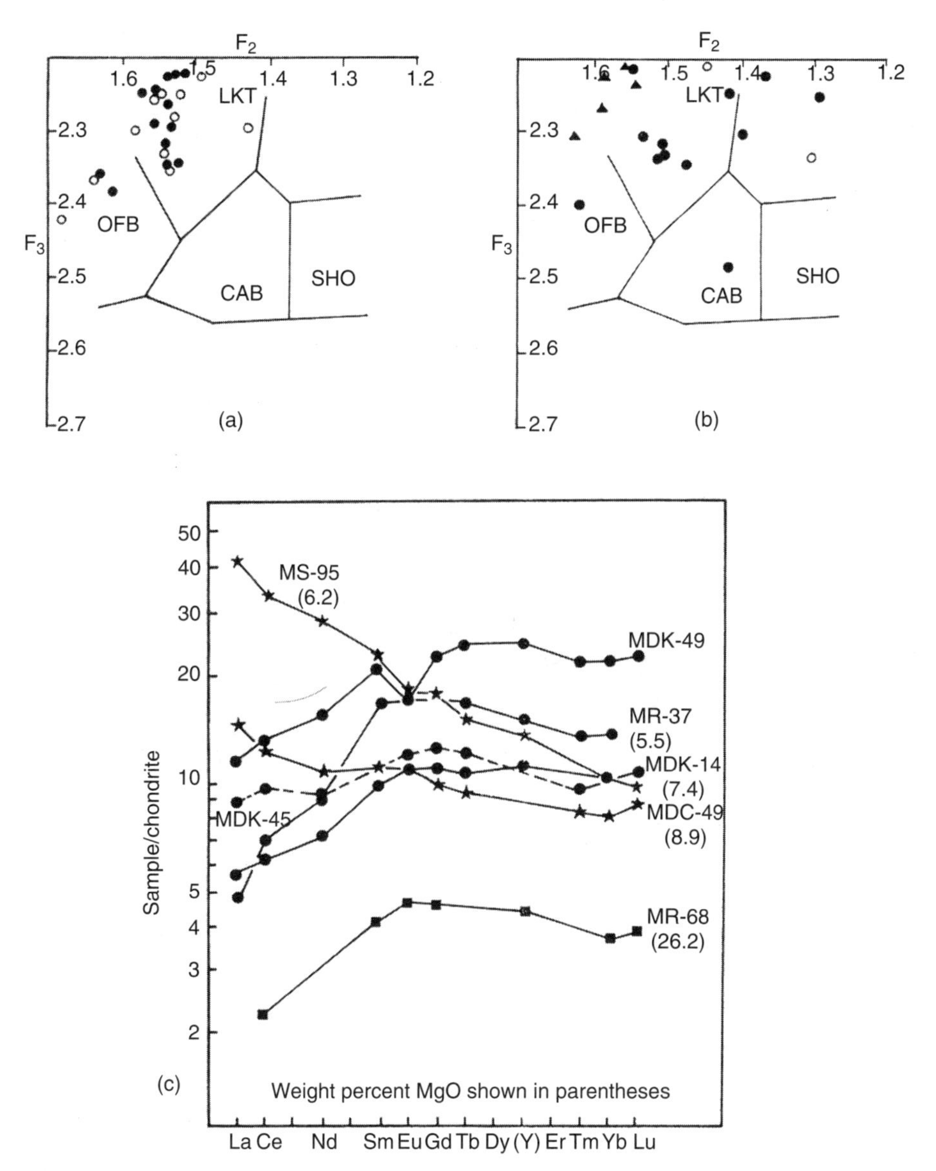

Fig. 4.1.43 (a, b) F_2–F_3 discriminant diagrams for (a) Dalma and (b) Dhanjori volcanics (after Pearce and Cann, 1973. Solid circle – Upper basalt; hollow circle – Lower basalt; cross – Ultramafic volcaniclastics; solid triangle – Basaltic komatiitic; solid square – Ultramafic komatiite (from Sarkar et al., 1992); (c) Chondrite-normalised REE contents of Dalma volcanics (after Bose and Chakraborty, 1989).

values indicate that the source of these rocks was depleted in LREE for a long time. Plotted on the global ϵ_{Nd} evolutionary path for upper mantle, the Dalma intrusives fall exactly around the depleted MORB-type mantle at 1600 Ma (Roy, Abhijit et al., 2002a).

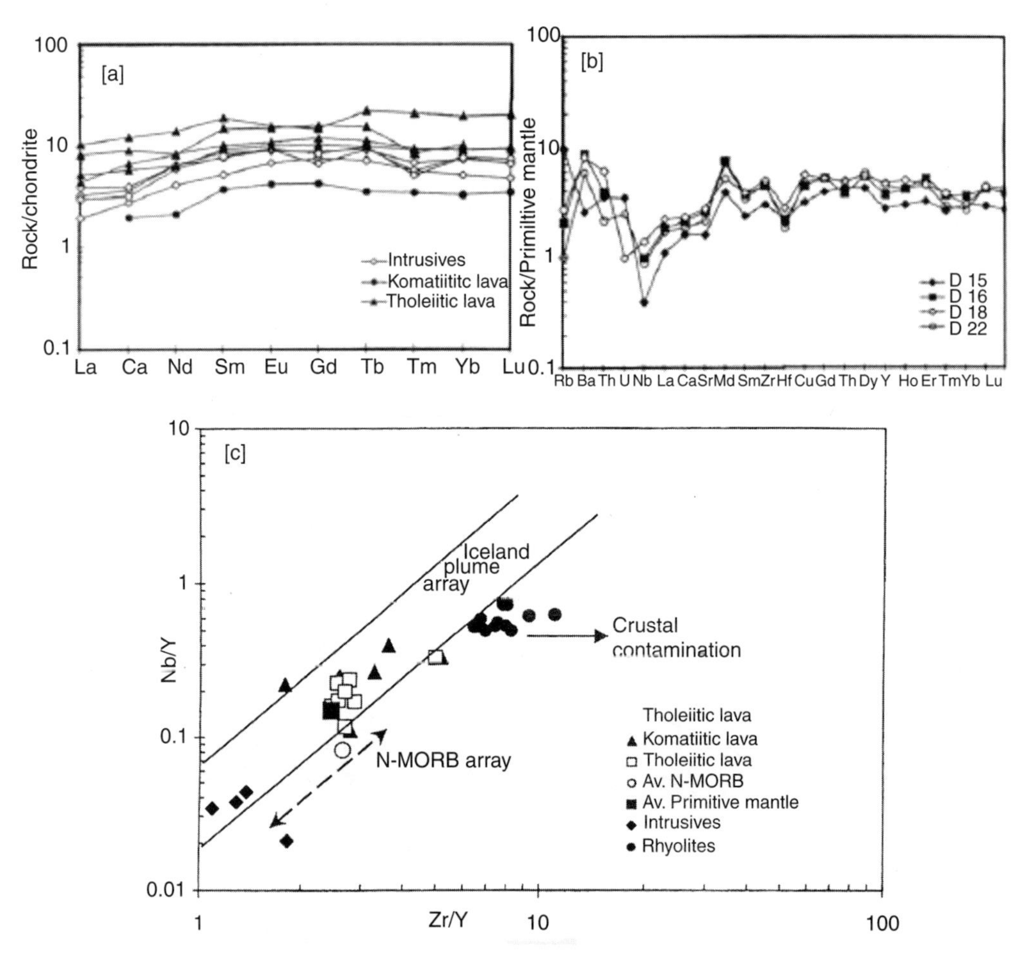

Fig. 4.1.44 (a) Chondrite normalised REE plots for Dalma mafic-ultramafic intrusives, (b) Primitive mantle normalised trace element patterns for Dalma mafic–ultramafic intrusives, (c) Nb/Y–Zr/Y discriminant diagram with plots of the volcanics and intrusives from Dalma sequence. Notable that the plots of mafic–ultramafic rocks fall within plume array (after Roy, Abhijit et al., 2002a).

Chondrite normalised REE plots of the Dhanjori mafic-ultramafic rocks (Fig.4.1.45a) show LREE enrichment. Primitive mantle normalised plots of the trace elements (Fig. 4.1.44b) also show shallow fractional pattern. There are depletions of Nb, Ba, Zr and enrichment of the LILE like, Rb, Th and U: a situation also noticed in the N–MORB rocks. Nd isotopic data suggest that an enriched ($\in_{Nd}$ = ~2.4) mantle existed below the Dhanjori basin during ~ 2400 Ma (Roy, Abhijit et al., 2002 b). According to these authors, the enrichment was possibly caused by continuous recycling of the earlier crust into the mantle, wherein the subducted slab-derived fluid modified surrounding mantle, affecting even the Rb-Sr systematics (Sr_i = 0.702–0.717). The enriched mantle material, part of a thermal plume, penetrated along a continental margin fracture(s) and erupted onto the Dhanjori basin. This has been taken to mean that a deep plume was fed by a recycled oceanic crust, via globally extensive subduction process, already initiated by the end-Archean period (Roy, Abhijit et al., 2002 b).

Absence of calc-alkaline suite in this belt and dominance of mafic-felsic bi-modal suite are more favourable for a continental and extensional tectonic environment. It is possible that the Dalma volcanics erupted in an extensional rift basin with an attenuated continental crust around (Singh, 1997).

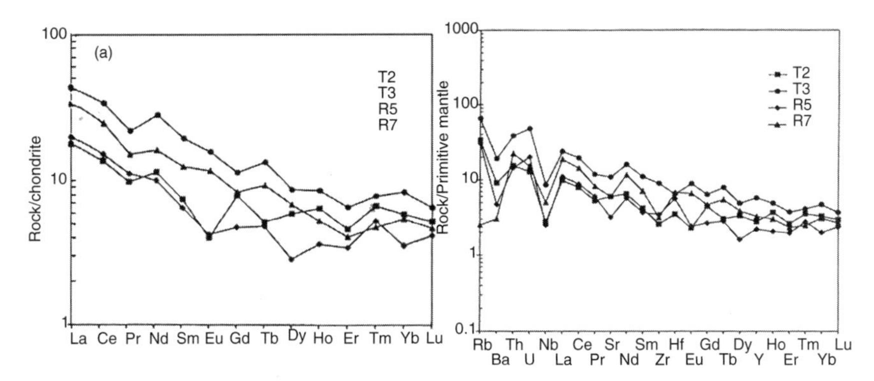

Fig. 4.1.45 (a) Chondrite normalised REE plots and (b) Primitive mantle normalised trace element patterns for Dhanjori mafic-ultramafic rocks (after Roy, Abhijit et al., 2002 b).

Domain V (Area Between the Dalma Belt and the CGC) This domain essentially comprises pelitic and tuffaceous metasediments but differs from the litho-assemblages of Domain III in having abundant chert and black shale–chert rhythmites, larger mafic–ultramafic bodies in the east, abundant acid volcanics and some syenitic rocks and scarcity of siliciclastics. Impure limestone/dolostone occurs along the southern edge close to the Dalma belt. Besides sporadic basemetal sulphide occurrences, a prominent phosphorite-bearing carbonatite body characterises the northern contact of this domain with the CGC, which is marked by a brittle-ductile shear zone (discussed in Section 4.2 in more details). Tungsten-bearing crustal S-type granites intrude the metamorphites near Kuilapal and Chhendapathar in the eastern part of the domain.

Though Dunn and Dey (1942) included these rocks in their 'Chaibasa Stage' with infoldings of 'Iron Ore Stage', the northern belt was recognised by them to be quite different in its lithological content and metamorphic grade compared to the one lying to the south of Dalma volcanics. Sarkar and Saha (1962) also grouped these rocks under their Singhbhum Group with tacit suggestion of its being structural repetition of the southern sequence across Dalma synform. Gupta et al. (1980) noted that besides the dissimilarities in the lithological character, there is striking absence of the rocks of Dhalbhum Formation all along the northern margin of the Dalma belt. There is a strong possibility that these rocks had developed in a different basin and brought against Dalma volcanics during inversion.

From Chandil (22°57′: 86°04′) westwards, the pelites comprising tuffaceous material and normal sediments are least metamorphosed while in the eastern extension of the same belt garnet and staurolite bearing mica schists are developed. In the eastern and central parts of the belt, the grade of metamorphism gradually rises northwards from low greenschist facies near Dalma to greenschist–amphibolite transition facies around Lawa–Maysera (23°01′: 86°00′). Gupta (1975) described well preserved thin cross laminations and flaser-wavy-lenticular bedding in the psammopelitic rhythmites around Lawa, indicating alternation of current and wave action in slack water deposition, typical of subtidal and intertidal zones. The thick quartzite bands in the sequence are also thinly bedded and do not display cross-beddings, suggesting non-detrital deposition.

Silicic volcanic rocks, generally rhyolite in composition, have been reported from a number of places in this terrain (Ray et al., 1996). Rb-concentration with respect to Y+Nb values together with low concentrations of Ti, Nb and Y and high Zr/Nb ratio favour volcanic arc environment.

Downstream of the dam on the Subarnarekha river near Chandil, a dark and compact fine-grained rhyolitic rock with primary layering and lapilli size fragments is exposed on the south bank. This rock was recently dated at 1631 ± 6 Ma (zircon/ Pb-Pb SHRIMP) (Nelson et al., 2007). At another location near Ankro village (22°55′: 86°34′) a dark grey colored acid tuff yielded 1487 ± 34 Ma (Rb-Sr /WR) date with Sr_i = 0.70825 ± 0.001 (Ghosh Roy and Gangopadhyay, 1998). Subsequently, Sengupta et al. (2000) reported (Rb-Sr/ WR) age of 1484 ± 44 Ma (Sr_i – 0.708545 ± 0.001) for the acid volcanics from Chandil area. All the above workers have regarded the respective age data as the emplacement dates for the acid vocanics in this belt, though they occur within a highly deformed and metamorphosed sequence. It is also suggested that these volcanics could be produced by differentiation of a mantle derived basic magma with some crustal contamination (Sengupta et al., op. cit.).

Dunn and Dey (1942) described an intrusive rock, 800 m long, within mica schists and phyllite in Sushnia hill (22°57′: 86°37′) as albitite. This was later identified as alkali syenite with nepheline and sodalite developed along thin bands (Ray et al., 1996).

Mafic–ultramafic rocks, represented by chlorite schists, talc schist, amygdaloidal basaltic flows and ultramafic intrusives are more common in the eastern sector, where they form long strike ridges and innumerable small concordant bodies within the mica schist country (Dunn and Dey, op. cit.).

An ellipsoidal area (12 × 6.5 km) of granite-gneiss, known as Kuilapal Granite, occur in the eastern extremity of the domain, ranging in composition from trondhjemite through dominant granodiorite to adamellite to granite proper (Dunn and Dey, op. cit.; Saha, 1994). Dunn and Dey (1942) originally described it as a gneissic body with evidence of injection of gneissic material within the surrounding mica schist along the schistosity. According to Ghosh (1963), the gneissic rocks were formed synchronously with later phase of deformation of surrounding metapelites but peak of metamorphism leading to development of staurolite and sillimanite was post-kinematic, outdating the granitisation. Later workers have more or less subscribed to this observation but have characterised Kuilapal body (having wolframite bearing veins) as a S-type crustal granite. Sengupta and Gangopadhyay (1994) described andalusite bearing mica schist from Satnala area (22°44′: 86°45′), located in the southeast of Kuilapal Granite body. These workers regard the andalusite porphyroblasts to be syn-to post-tectonic with respect to dominant schistosity.

According to Saha (1994), the K-Ar age of 1163 Ma for the Kuilapal Granite (Sarkar et al., 1969) probably denotes minimum estimate of time when K-Ar isotopic system closed at 300°C. An age of 1638 + 38 Ma (Rb-Sr WR) with a very high Sr_i (0.72173 ± 0.00156) is reported from a sample of this granite by Sengupta et al. (1994).

The rocks in Domain V display an overall conformity with the structural architecture of the NSMB. Like Chaibasa–Dhalbhum rocks to the south of Dalma, low-plunging and E–W trending F_2-folds on large and small scales dominate this domain. The only contrast is in the opposite sense of asymmetricity of thcsc folds (steeper northern limb and shallower southern limb) and opposite vergence of axial plane schistosity (southerly dip). The preservation of early F_1-folds having reclined geometry are rare but S_1 planes at low angle with bedding are ubiquitous and these together define the form surface of F_2-folds. In Lawa–Maysera area the southerly dipping S_2 planes are most prominent and pervasive, which are in turn crenulated (co-axial with F_2) and dissected by locally pervasive northerly dipping crenulation cleavages (S_3). The mineral schistosity and the transposed shear planes along the Tamar–Porapahar shear zone are characterised by steep northerly dip.

The grade of metamorphism of the rocks belongs to greenschist facies, except in the east where metamorphism attained amphibolite facies.

Structural framework is more or less the same as in most other domains of the North Singhbhum mobile belt. E–W trending folds (F_2) are most conspicuous and the F_1 hinges are generally ill preserved. Overprinting of the earlier structures by F_3- folds are distinct at places (Gupta and Basu, 1985).

Tectonic Architecture of NSMB In their classical work, Dunn and Dey (1942) postulated a southerly overturned 'geo-anticline' with a prominent southward thrust (shear zone) along its southern limb to explain the regional structural framework of the belt. The 'geo-anticline" is followed to the north by a narrow (Dalma) syncline, which is also overturned to the south. Sarkar and Saha (1962) more or less accepted this structural model, though differing in some details.

The NSMB essentially records three sets of folds (F_1, F_2 and F_3) and related planar and linear structures, which are variously interpreted as products of a single progressive deformation or three distinct deformative episodes (D_1, D_2, and D_3). However, it is generally accepted now that the Singhbhum shear zone (SSZ) represents a zone of intense ductile shearing recording repeated cycles of mylonitisation, folding and rotation/stretching of the pre-existing structures along the direction of tectonic transport over a prolonged period of time.

Similar to many other mobile belts, the different domains of NSMB record selective imprints of the deformative episodes with varying intensity depending on their location and disposition with respect to the stable cratonic block, thickness and rheology of the layered sedimentary-volcanic pile and the basement configuration.

Most of the workers have invoked an overall north-south compression, roughly orthogonal to the arcuate cratonic boundary, as responsible for the deformation in NSMB. The shearing along SSZ was considered by early workers as a late tectonic process, which, however, was later recognised to have initiated much earlier in the deformative history with subsequent repetitions over a prolonged time. The cross-folds (F_3-folds) were considered by some wokers as products of a distinct deformative phase with longitudinal compression (Sarkar and Saha, 1962), while these were regarded by others as accommodation structures in progressive continuity of earlier folding (Naha, 1965).

Gupta and Basu (1990) suggested transtensile and transpressive movements in the extensional and compressive regimes as responsible mechanisms for the formation of the basins/sub-basins and subsequent deformation. The sinistral pattern of large scale folds (F_2) in the western sector and mesoscopic reflections of the same throughout the belt are suggestive of an anticlockwise rotation of the tectogene against the southern stable block during compressional regime. The reversal in sense of vergence of the asymmetric F_2 folds across Dalma belt may also be important to note in this context. The marked fanning of the S_2 schistosity (axial planar to the F_2 folds) resembles 'flower' or 'palm tree' structure, which are produced by oblique overthrusts in transpressional domains

Mukhopadhyay (1984) observed the tectonic similarity of the southern part of NSMB with foreland thrust belts of younger orogenes and similar postulations were presented by a few others (Coulomb wedge model – Gupta et al., 1996; listric thrust system model – Joy and Saha, 1998). It may be worthwhile to discuss the theoretical aspects of fold–thrust belt model and to examine the possible relevance and applicability of the model to the evolution of the southern segment of NSMB.

Evolution of Southern NSMB as a Fold–Thrust Belt The fold–thrust belts (FTBs) are linear or curvilinear belts of intense folding and shearing along the margins of many orogenic belts. The accretionary prisms in the active convergent plate boundaries provide the present day analogues of the FTBs, revealing at least some clues in regard to their developmental history and the kinematics, though the details are yet to be fully understood (Mitra, 1997). In general, the kinematics of these belts (FTBs and present day accretionary wedges) are modeled after the 'critically tapered Coulomb wedge' (Chapple, 1978; Davis et al., 1983; Willett, 1992; Mitra, op. cit.). When a pile of non-cohesive sediments on a sloping continental shelf is pushed from behind (hinterland) against the basal slope and the stable continental block (foreland), the material shortens by foreland vergence internal thrusting and folding, leading to appreciable thickening of the rear part. This imparts a wedge like geometry to the sedimentary pile with forward surface slope and backward basal slope angles, and the progressive push from behind increases the taper angle, until it reaches a critical value. At this stage the material within the wedge is at the verge of failure while the wedge as a whole slides on its base. In classical Coulomb wedge model strain partitioning takes place along a foreland verging linear zone (with accumulation of shear strain) developed between the foreland and hinterland parts of the wedge. The accretion of the front of the wedge at its toe and internal failure (partly gravity failure) in the hinterland reduces the wedge taper below critical value. With continued squeezing, however, the taper angle again goes on increasing and internal deformation is renewed by out-of- sequence thrusting and folding of the earlier thrust planes within the wedge, resulting in a complex fold-thrust geometry in the belt. However, many variations are possible with changes in the governing parameters. In case of an increase in basal friction or slope the wedge may be overthickened and attain a supercritical stage leading to a more complex wedge dynamics. Though ideally the wedge material should be non-cohesive (Davis, et al., op. cit.), yet the wedge tectonics can be operative in cohesive material also, having significant shear strain or rigid plastic rheology (Chappel, op. cit.; Willett, op. cit.). Chattopadhyay and Mandal, (2002) have experimentally demonstrated the validity of wedge dynamics in case of even non-linear viscous material, very similar to the classical Coulomb wedge. Hatcher and Hooper (1992) showed that deeper sections of the crust comprising sedimentary package and crystalline rock mass can approximate wedge tectonic behaviour as a hybrid (partly Coulomb and partly non-Coulomb) wedge.

From the above discussions, it is apparent that the repeated folding and thrusting events in such orogenic belts should be the result of continuous deformation rather than time separated discrete deformative events. The kinematics and the resulting geometry of the structures will depend on the location of a segment with respect to the stable block, wedge morphology and thickness, besides the rheology of the wedge material.

In the context of the above discussions, let us examine if the southern part of the NSMB may be compared with FTB as postulated by Gupta et al. (1996) and Gupta and Basu (2000). Taking a closer view of the eastern part of the belt for this purpose, the Dhanjori volcanisedimentary pile may be considered as the foreland part of the wedge constituted of Proterozoic cover sequence which was compressed against the Archean cratonic block represented by the Singhbhum Granite massif (Fig. 4.1.46a). In the early phase of compression, the Dhanjori pile lying at the foreland largely remained passive except for possible development of mild curvatures on the bedding (S_0) and formation of fracture cleavages in restricted domains. The Chaibasa Formation, lying in the hinterland away from the cratonic margin, witnessed pronounced deformation at this stage resulting in the development of first generation buckle folds (F_1) on bedding (S_0) and axial planar regional schistosity (S_1). At this stage a foreland vergent zone of anisotropy (SSZ) accompanied by strain accumulation developed between the foreland and hinterland, which witnessed the first phase of ductile shearing resulting in mylonitic fabric(Fig. 4.1.46b). The earliest buckle folds developed along this zone involved both bedding (S_0) and mylonitic foliation. With the progression of the compressive regime under unchanged kinematic parameters, second generation folding (F_2) started in the hinterland block involving both S_0 and S_1 planes with the development of axial planar crenulation cleavage (S_2). The earlier folds (F_1) were refolded (F_2) with progressive overturning towards south. It was around this stage that the F_1-folds were rotated to near reclined geometry with the continuation of ductile shearing along the high strain domain (SSZ). The major bulk of Dhanjori volcanisedimentary sequence constituting the foreland part of the wedge was buckled at this stage to attain the final forms of F_1 folds with the development of pervasive axial planar S_1 schistosity (Fig. 4.1.46 b). However, in some isolated domains of foreland, where incipient S_1 planes were formed in response to initial stage of compression, the progressive compression buckled S_0 and S_1 together to form F_2 folds, nearly coaxial to the F_1 folds at places. The earlier structures were also refolded

(F_2) along the SSZ, with continued shearing, now parallel to the S_2 surfaces. The hinterland part of the wedge with increased thickness witnessed repeated in-sequence and out-of-sequence southward thrusting resulting in arcuation of F_2-fold axes and formation of large scale sheath folds in Ghatsila–Dhalbhumgarh area, as described by Mukhopadhyay et al. (1998) and Ghosh et al. (2006).

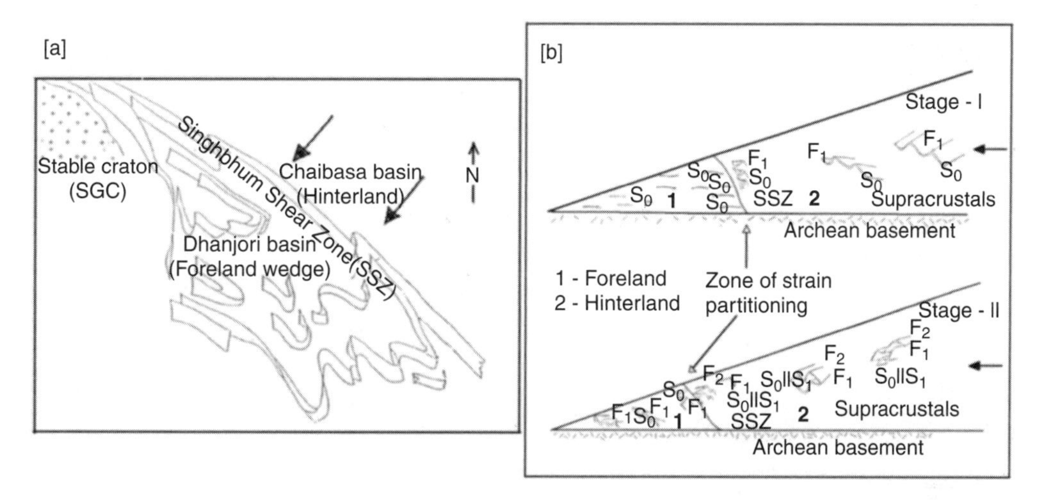

Fig. 4.1.46 (a) Sketch diagram showing the structural pattern of rocks in Dhanjori basin (b)Schematic diagram depicting progressive deformation under compression in the foreland and hinterland at two stages invoking Coulomb wedge model on broad terms (after Gupta et al., 1996).

At the final stage of compressive regime increasing lateral extensional component acting on the thickened part of the wedge resulted in the development of cross-folds (F_3) as accommodation structures. These structures (F_3-folds) are most prominent in Chaibasa Formation (hinterland) followed by the SSZ and least prominent in the Dhanjori basin (foreland). This heterogeneity was perhaps caused by the basement topography and difference in the extent of available space for lateral extension in different parts of the wedge under compression. The cross-folds in the SSZ are often manifested by a set of down-dip puckers, broad warps and also by the rotation of F_2-fold axes to some extent. The large-scale interference patterns seen in the Turamdih sector (vide Section 4.2) are the effects of broad F_3-flexture on the F_2-folds. The strong down-dip lineations along the SSZ (and down-dip orientation of the sulphide ore shoots, particularly in the southeastern part) may, however, be attributed to the formation/alignment of fabric parallel to the direction of tectonic transport during repeated thrusting/shearing along this zone since the initial partitioning of strain between the foreland and hinterland.

The continuous pervasive deformation model proposed above adequately explains overall dynamics and kinematic history of the southern stretch of NSMB, where the folding and shearing episodes are reasonably correlated with the progressive development and modification of strain domains in space and time across the supracrustal wedge under compression. Viewed in this context, the history of deformation and the entailing structural features as also their chronology are not expected to be uniform across the belt. It may be expected that the large-scale bedding folds (F_1), most prominent in the foreland, developed much later in the compressive regime, when the hinterland had already witnessed refolding (F_2). Ductile shearing and mylonitisation along SSZ commenced before the first buckle-deformation. The history of deformation and evolution of structures along this zone, when examined in detail, are not comparable with those in the adjacent fold-belts on either side. For that matter, all the different segments of the wedge are likely to respond differently to the compression and record varied succession of events and related features. The P–T environments for metamorphism in the FTBs are also not expected to be uniform in time and space and would deserve scrutiny accordingly. Juxtaposition of tectonolithons having varied metamorphic mineral assemblages due to stacking of the thrust sheets is common in FTBs.

Metamorphism Generally, the rocks of Domain I and Domain II are metamorphosed to greenschist facies, the grade somewhat increasing from south to north. Within Domain III the grade increases towards north reaching up to amphibolite facies along more or less the median part and thereafter gradually falls down towards north to reach low greenschist facies in the Domain IV (Dalma tract). In the Domain V (north of Dalma) the metamorphic grade again increases northwards from low-greenschist facies to greenschist–amphibolite transition facies. Around Kuilapal, in the eastern parts of this Domain, the grade reaches up to high-amphibolite facies. Close-spaced variations in the development of metamorphic minerals controlled by the original composition of the sediments are, however, noted throughout the NSMB (Gupta and Basu, 2000).

Polymetamorphic character is conspicuous in the belt, but temporal relation between the deformational and the metamorphic phases and a detailed assessment of P–T–t paths based on metamorphic reactions are yet to be established for major part of the mobile belt. However, the studies conducted in selected parts of the belt indicate that the most prominent metamorphic trend is that of intermediate-pressure medium-temperature (Barrovian) type (Naha, 1965; Lal and Singh, 1978; Roy, 1966), though there are distinct imprints of an earlier (pre-orogenic) low-pressure medium-temperature metamorphism (Buchan type) in the Dhalbhum Formation (all along the southern contact of Dalma volcanics) as well as in eastern and western extremities of Chaibasa Formation. The early growth of andalusite in the pelites of appropriate composition is considered to be due to a steeper thermal gradient attributed to maximum crustal thinning and mantle up-warp along Dalma volcanic tract and to the rising temperature in tensile regime of low pressure just before the onset of the compressional regime (Dunn and Dey, 1942; Naha, 1965; Sarkar et al., 1992; Sarkar et al., 1993; Mukhopadhyay et al., 1998). Besides widely reported andalusite, pre-tectonic growth of chloritoid, garnet and kyanite are also reported from Dhalbhum phyllite of eastern sector by Sarkar et al.(1993).All these minerals have, however, developed more widely during the syn- to post-tectonic regional metamorphism(s).

Though differing in details, many workers have described two successive episodes of progressive metamorphism (M_1 and M_2) coeval with two deformations (D_1 and D_2) and a third phase of diapthoresis (M_3) related to D_3 deformation. Peak metamorphism was reached during M_1 (syn- to post-F_1-S_1) in some parts of the belt, while M_2 (syn- to post- F_2–S_2) marks the highest grade in others. The growth of a particular mineral (biotite, garnet, staurolite etc.) is often related to both M_1 and M_2 phases, ranging from syn-tectonic to post-tectonic stages with respect to D_1 or D_2. This indicates an overlap of P–T regime during D_1 and D_2 deformations. The effects of M_3 are prominent in some zones in form of retrogression of the pre-existing assemblages or neomineralisation of low-grade minerals along axial planes of F_3 folds and puckers.

In both eastern Ghatsila–Galudih and western Sonapet–Hesadih sectors the major F_2/F_3 surfaces show distinct parallelism with M_1 isograds or isoreaction grads while M_2 isograds are seen to cut across F_2 structures in the Hesadih sector (Sarkar and Bhattacharya, 1978; Sarkar et al., 1993). In Kandra–Gamaria sector, the highest grade assemblage encloses an elongate E–W trending area (Roy, 1966), which may be the reflection of a doubly plunging F_2 structure. To the north of Dalma, in Chandil–Lawa sector, where metamorphism reached highest grade during D_2–M_2 phase, the M_2 isoreaction grads do not show any regard to the major F_2 structural surfaces (Gupta, 1975; Sarkar et al., 1992; Gupta and Basu, 2000).

There is a recent publication on the metamorphism of the rocks in Dhalbhumgarh area by Ghosh et al. (2006), in which the metamorphic history of the rocks is also divided into three episodes: M_1, M_2 and M_3. In M_1 episode, andalusite porphyroblasts developed in a matrix of quartz, chlorite and muscovite in Al-rich pelites. Andalusite could develop by any of the following reactions:

Pyrophyllite → Andalusite + quartz + H_2O
Paragonite + quartz → Andalusite + muscovite + albite +H_2O
(Holland, 1979; Okuyama–Kusunose, 1994)

M_1 is considered to be pre-orogenic. Heat for the metamorphism is provided by the asthenospheric upwelling that led to the extension and basin formation, as has been suggested in similar type of metamorphism in the Pyrenees (Wickham and Oxburg, 1987) and Welsh basin (Robinson et al., 1999). The main phase of regional metamorphism was M_2 that produced chloritoid, kyanite, garnet, and staurolite porphyroblasts. The peak of M_2, however, postdated D_2 deformation. The mica schists are distinguished into two types: (i) High-alumina pelites (HAP), and (ii) Low-alumina pelites (LAP). The HAP are represented by (i) kyanite–chloritoid zone and (ii) staurolite–garnet–chlorite zone. The corresponding zones of LAP are garnet–biotite–chlorite zone and staurolite–garnet–biotite zone. Presence of chlorite and magnetite in chloritoid suggests that chloritoid could have formed by one or more of the following reactions:

Fe-chlorite + pyrophyllite → Fe-chloritoid + quartz + H_2O
(Miyashiro, 1973; Spear, 1993)
Pyrophyllite + magnetite → Chloritoid + quartz + H_2O + O_2
Fe-chlorite + hematite → Chloritoid + magnetite + quartz + H_2O
(Thompson and Norton, 1968)

Staurolite formed by reaction involving chloritoid (Ganguly, 1969; Spear, 1993), or involving chlorite (Hoschek, 1969; Carmichael, 1970). The pressure and temperature during M_2 has been computed to be 5.5 kb and 550° ± 50° C, respectively.

M_3 was characterised by retrogressive stabilisation of chlorite at the expense of M_2 staurolite and garnet. The temperature of such reactions was not far too low (380–500° C) compared to those of M_2. It was difficult to constrain P at this stage. The P–T–t path was clockwise, supporting the tectonic model of initial rifting followed by compression (Gupta et al., 1980; Gupta and Basu, 2000; Sarkar, 1984; Mukhopadhyay, 1984). Another recent interesting study on the structures and petrology of a NW-SE trending strip of western NSMB obtained P–T values of 7 ± 1kb and ~ 620 ± 20°C in a stout band of rocks near the northern end of the strip and the P–T values of 10 ± 1.5 kb and 6.25 ± 50°C in another thick zone near the southern end of the strip (Mahato et al., 2008) This extra-high pressure needs be suitably explained.

It is interesting to note that Bandyopadhyay (2003) on the basis of her detailed study on the metamorphism of the rocks in the southeastern sector of the SSZ concluded that the variation in the assemblages of muscovite, biotite, chloritoid and kyanite-bearing schists in close contact can be better explained by variation in µ K+ , µ O_2 and µH_2O in isothermal isobaric condition. It renders the model of Barrovian metamorphism here still weak.

Gangpur Group

The Gangpur Group (Series in Krishnan, 1937) of rocks occurs as a Proterozoic intracratonic basin-fill, appearing nearly rectangular and trending almost E–W. In the south it virtually starts from where the Darjing Group ends in the north. According to Krishnan (1937), who was the first to study these rocks in some details, the Gangpur rocks formed an anticlinorium and are stratigraphically overlain by the rocks which in the grouping of Dunn and Dey (1942) came to be known as the Iron

Ore Series. Later workers (Kanungoo and Mahalik, 1975; Banerjee, 1967) differed with Krishnan in that they believed that the stratigraphic position between the Gangpur group of rocks and the Iron Ore Supergroup were just the opposite of what Krishnan had said. Kanungo–Mahalik and Banerjee established this on the basis of field evidence and Sarkar on radiometric (K-Ar) age data. On the suggested regional structure also the latter workers differed with Krishnan and even amongst themselves. Krishnan, Kanungo-Mahalik and Banerjee proposed different stratigraphic successions. After scrutiny of the logical basis of their proposals, the authors preferred the one suggested by Kanungo and Mahalik (1975) vide (Table 4.1.9).

The sediments have been metamorphosed to middle amphibolite facies, followed by granitic intrusion (Sarkar, 1974). Sarkar (1968) suggested a depositional age of 1700–2000 Ma for the Gangpur sediments.

Table 4.1.9 *Stratigrahic succession of the Gangpur Group (Kanungo and Mahalik, 1975)*

	Granites, metamorphosed basic intrusives.	
Gangpur Group	Ghoriajhor Stage	Staurolite and garnetiferous mica schists calc-epidote mica schists, quartzites and conglomerates
	Kumarmunda Stage	Carboniferous metapelites, overlying carbon bearing banded quartzites
	Birmitrapur Stage	Metamorphosed limestones and dolostones overlain by metamorphosed carbonate (dolomitic) arenaceous and pelitic sediments
	Lainger Stage	Metamorphosed pelitic sediments with thinbedded quartzite rocks
	Raghunathpalli Stage	Conglomerates, quartzites and slates
	---------- Unconformity ----------	
	Iron ore Supergroup rocks	

Note: According to the presently accepted code of stratigraphic nomenclature, the 'Stage' in above succession will be 'Formation'.

Chhotanagpur Granite–Gneiss Complex

The Chhotanagpur Granite–Gneiss Complex (CGC), covering an area of about 500 × 200 km, occurs with an approximately east–west trend, occupying parts of Uttar Pradesh, Madhya Pradesh, Bihar, Jharkhand and West Bengal. A large part of the complex is composed of the members of the granitic clan and granitic gneisses, with numerous enclaves of pelitic schists of varying composition and metamorphic grade, calc-silicates, marbles, quartzites, amphibolites, gabbro-norites, alkaline rocks, ultramafics, anorthosite, charnockite, pyroxene granulite and skarn rocks.

The granitic rocks vary in composition from granite through grannodiorite to tonalite (Ghose, 1983) though majority of them plot in the granite-adamellite field in a CaO-Na_2O-K_2O diagram (Fig. 4.1.47). Porphyritic monzogranite and non-porphyritic syenogranite occur in the Bihar mica belt (now in Jharkhand). The mica pegmatites are believed to be related to the late phase of high-level granite emplacement. Migmatites and gneisses are syn- to late- kinematic with respect to the peak of deformation and metamorphisn and have been suggested to be products of partial melting and anataxis (Ghose, op. cit.; Sarkar, 1988). The metasediments and most of the associated basic rocks are metamorphosed in the low amphibolite facies. However, locally they attained higher grades, reaching upto the granulite facies, as at places of Palamou and Dumka in Jharkhand and Bankura–Purulia in West Bengal.

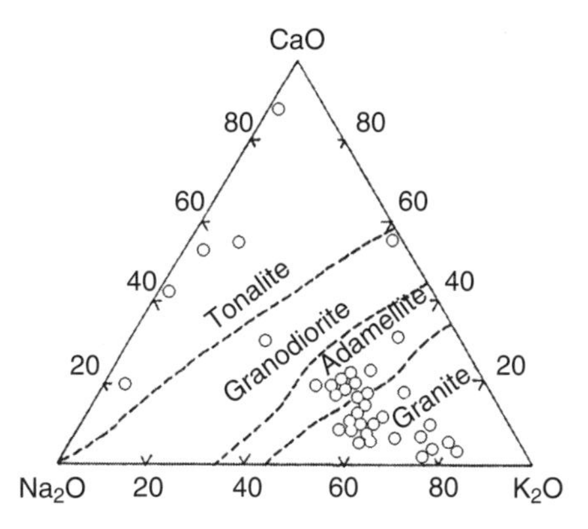

Fig. 4.1.47 CaO-Na_2O-K_2O diagram showing the composition variation of granitic rocks in CGC terrain (after Ramesh Babu, 1999)

The southern marginal part of the CGC is marked by brittle-ductile shearing at many places, from some of which the occurrence of carbonatite and alkaline rocks has been reported (Ghosh Roy, 1989). Besides, satellite imageries suggested very long lineaments, some of which were later identified as zones of intense shearing (Mazumder, 1988).

Sarkar (1980) proposed three sets of K-Ar age data for orogeny and the resulting metamorphism and petrogenesis.These are: 1416–1246 Ma for the north eastern part of the CGC, 1086–850 Ma for the Ranchi–Muri and the mica-belt regions and a very young, 420–358 Ma for rocks in the Mungher area. The last orogeny has been called the Indian Ocean Cycle by Sarkar (1968) and the 'Mongher Orogeny' by Banerji (1991). This age range, however, tallies well with that of the last phase of the Pan-African Orogeny, and is now reported from several places in India and Sri Lanka. Its significance is yet to be fully appreciated. Mazumder (1998) reported a Rb-Sr (WR) age of 2300 Ma from Dudhi, Uttar Pradesh. The other age data (Rb-Sr, WR), obtained mostly from the eastern part, range from 1600 Ma to 859 Ma (Pandey, et al., 1986; Roy, 1998).

In recent times, there have been attempts at classification of the CGC rocks in some chronostratigraphic/ litho-stratigraphic order (Ghose, 1983; Banerji, 1991; Singh, 1998). These attempts are appreciated. But as the rocks in this region apparently underwent multiple phases of deformation, metamorphism, magmatism and metasomatism and well-constrained radiometric age-data are still inadequate, it would be better to wait for more time before we can reasonably conclude on the sequence of geological events in the region.

Sanyal et al. (2007), working in the area around Deoghar, Jharkhand, in the eastern part of the belt, have demonstrated Ultra High Temperature metamorphism (c. 930°–950°C) at mid-crustal depth (5–6 kbar) corresponding to a shallow geothermal gradient of c. 50°C/km for the metapelitic granulite within CGC.The authors felt that an advective heat transport was required for this UHT metamorphism along the construed shallow thermal gradient. Three clusters of dates, the oldest being 1400–1500 Ma, were obtained from the rocks of CGC in this sector by chemical dating of monazite grains. The UHT metamorphism is linked to the oldest date (Sanyal et al., op. cit.)

For more details on the CGC, the reader is referred to the full text of the papers mentioned here and an excellent recent review by Mahadevan (2002).

The Bengal Anorthosite in the Bankura district, West Bengal, is an oval body, being massive at the core and banded at the periphery. The body (32×7 km^2) is structurally accordant with the surrounding rocks that comprise varieties of granulites and granitic gneisses. The anorthosite body, which ranges in composition from anorthosite to gabbronorite, through leucogabbro, shares the deformation and metamorphism of the ambient rocks. The country rocks belong to the Chhotanagpur

Gneissic Complex near its eastern margin. The time of emplacement of the massif has been estimated as 1550 ± 12 Ma by U-Pb dating of zircon from the anorthosite and U-Th-Pb chemical dating of monazite in the adjacent metapelitic granulite (Chatterjee et al., 2008)

Stratigraphic/Geochronologic Sequences in the Eastern India and their Regional Correlation

Trying to understand the stratigraphic/geochronologic sequences in a segment of the earth's crust and their correlation on a regional basis has been a practice as old as the modern phase of geological studies. It has been there to appreciate the generality, or otherwise, of the geological process in space and time. The work of Ball of the Geological Survey of India (Ball, 1881) needs be specially mentioned in this context. However, the most vigorous geological investigations in the region took place in the interlude between the last two World Wars. As a result, we are presented with excellent works of Krishnan on south Singhbhum (in Pascoe, 1928) and on Gangpur (Krishnan, 1937), Jones (1934) in south Singhbhum, Keonjhar and Bonai and Dunn in north and south Singhbhum (1929, 1940). Their detailed discussion is beyond the scope of the present work. It can only be said that virtually there was no difference between the stratigraphic sequences suggested by Krishnan and Jones for the south Singhbhum and the adjacent areas. Dunn differed with them by not recognising the 'Older Dharwar' rocks and adding a platformal sedimentary sequence he called the Kolhan Series after the name of the then Kolhan State in the area.

According to Dunn and his co-worker (Dunn and Dey, 1942) rocks in the north Singhbhum and south Singhbhum and the adjacent areas more or less shared the same geological history, but only that these two domains are separated by a 200 km long 'overthrust', in which the northern part overrode the south. This according to them explained the exposure of the 'oldest' Chaibasa Stage rocks only to the north of the thrust zone. This stratigraphic correlation by Dunn and Dey (Table 4.1.10) was generally accepted until Sarkar and Saha (1962) proposed a new scheme of stratigraphic correlation (Table 4.1.11), in which the rocks in the north and south broadly belonged to two contrasted age groups and represented different geological histories. They based their conclusions on the results of renewed field investigation by their teams and a set of radiometric age data on the rocks of the region. Over the time new information was collected by the same workers, and they finally presented (Saha et al., 1988; Saha, 1994) a revised geochronological succession (Tabel 4.1.12). One will notice in this table that although they accommodated some new information in their latest suggestion, still it is not free from their old bias. This particularly pertains to the age of the OMG and the OMTG, as well as the relative age of the Singhbhum Group and the Dhanjori Group. The latest status of the geochronology of the area, based on the to-date available radiometric age data supported by field observations is presented in Table 4.1.13. It should be mentioned at the same time that the controversy is not yet resolved and a lot of field investigations and many more properly constrained radiometric analyses of the rocks need be done before the geologic history of the region becomes clearer.

Table 4.1.10 *Regional correlation of the Precambrians of Singhbhum-North Orissa region (Dunn and Dey, 1942)*

North and East Singhbhum (North of shear zone)	South Singhbhum and South Dhalbhum (South of shear zone)
	Kolhan Series
	———*Unconformity*———
	Newer Dolerite

(Continued)

(Continued)

<table>
<tr><td colspan="3">North and East Singhbhum (North of shear zone)</td><td colspan="2">South Singhbhum and South Dhalbhum
(South of shear zone)</td></tr>
<tr><td colspan="3">Soda granite and granophyre</td><td colspan="2">Soda Granite, granophyre and biotite granite</td></tr>
<tr><td colspan="3"></td><td colspan="2">Singhbhum Granite (may be pre-Dhanjori), Diorite</td></tr>
<tr><td colspan="3">Dalma Lavas</td><td>Dhanjori Group</td><td>Lavas and thin phyllites
Quartzite-conglomerate</td></tr>
<tr><td colspan="3">-------------- Overlap -------------</td><td colspan="2">------------ Unconformity ---------</td></tr>
<tr><td rowspan="4">Iron Ore Series</td><td rowspan="3">Iron Ore Stage</td><td>Phyllite,</td><td rowspan="3">Iron Ore Stage</td><td>Phyllite and tuff, lenticular conglomerate and quartzite.</td></tr>
<tr><td>Quartzite often hematitic (impersistent)</td><td>Banded Hematite Quartzite (BHQ) phyllites, tuff and basic igneous rocks</td></tr>
<tr><td>Phyllite, calcareous rocks with tuffs and basic igneous rocks.</td><td>Phyllite and basic igneous rocks</td></tr>
<tr><td>Chaibasa Stage</td><td>Mica schist, hornblendeschist quartz granilite, quartz schist, tuffs (where less metamorphosed)</td><td></td><td></td></tr>
</table>

Table 4.1.11 *Revised correlation of the Precambrians of Singhbhum–North Orissa (Sarkar and Saha, 1962)*

<table>
<tr><td colspan="2">South of Copper Belt thrust zone</td><td colspan="2">North of Copper Belt thrust zone</td></tr>
<tr><td colspan="2">Newer Dolerite</td><td colspan="2">Newer Dolerite</td></tr>
<tr><td colspan="2">Granophyre, biotite granite
Ultrabasic of Jojohatu =? Ultrabasic bodies associated with Singhbhum granite.</td><td colspan="2">Ultrabasic sills and dykes
Soda granite, granophyre, Kuilapal granite, Chakradharpur granite gneiss.</td></tr>
<tr><td colspan="2">---------Singhbhum Orogeny-------------</td><td colspan="2">------Singhbhum Orogeny(905–937 my)-----</td></tr>
<tr><td>Dhanjori Group</td><td>Dhanjori lava
Quartzite-conglomerate</td><td colspan="2">Dalma Lava
-------------- Overlap ---------</td></tr>
<tr><td colspan="2">? Kolhan Series (> 1584 m)</td><td>Singhbhum Series</td><td>Dhalbhum Stage(= Iron Ore Stage of Dunn)
Chaibasa Stage</td></tr>
<tr><td colspan="2">---------- Unconformity ----------</td><td colspan="2" rowspan="10"></td></tr>
<tr><td colspan="2">Singhbhum granite</td></tr>
<tr><td colspan="2">Iron Ore Orogeny (2038 ma)</td></tr>
<tr><td colspan="2">Gabro- anorthosite, epidiorite</td></tr>
<tr><td rowspan="3">Iron Ore Series</td><td>Upper shales with sandstone and volacanics</td></tr>
<tr><td>Banded Hematilte Jasper (BHJ)</td></tr>
<tr><td>Lower Shales
Lavas
Sandstones and conglomerates</td></tr>
<tr><td colspan="2">--------------Unconformty----------------</td></tr>
<tr><td colspan="2">Older Metamorphics: Mica, sillimanite, hornblende, chlorite and quartz schist</td></tr>
</table>

Table 4.1.12 *A generalised chronostratigraphic succession of the Singhbhum–Orissa Iron Ore craton (Saha et al., 1988)*

Newer Dolerite dykes sills		c.1600–950 Ma
Mayurbhanj Granite		c. 2100 Ma
Gabbro–anorthosite–ultramafics		
Kolhan Group		c. 2100–2200 Ma
------------------------------ *Unconformity* ------------------------------		
Jagannathpur Lava	Dhanjori, Simlipal Lavas (C. 2300 Ma) Quartzite-conglomerate	
Malangtoli Lava		Dhanjori Group
Pelitic and arenaceous metasediments with mafic sills (c. 2300–2400 Ma)		Singhbhum Group
------------------------------ *Unconformity* ------------------------------		
Singhbhum Granite (SBG- B) (Phase III)		c. 3100 Ma
Mafic lava, tuff, acid volcanic tuffaceous shale, banded hematite quartzite with iron ores, ferruginous chert, local dolomite and quartzite sandstone		Iron Ore Group
Singhbhum Granite (SBG-A) (Phase I and II) c. 3300 Ma		Nilgiri Granite Bonai Granite
Folding and metamorphism of OMG and OMTG		c. (3400–3500 Ma)
Older Metamorphic Tonalite Gneiss (OMTG)		c. 3775 Ma
Older Metamorphic Group (OMG): pelitic schist, quartzite, para-amphibolite, ortho- amphibolite		c. 4000 Ma

Table 4.1.13 *Chronostratigraphy of rock formations in the Eastern Indian craton (This work)*

Chronostratigraphic units	Radiometric age	References
Main bulk of CGC	1.6–0.85 Ga (Rb- Sr) (metamorphic events)	Pandey et al., 1986; Mallik, 1998; Roy, 1998
Felsic volcanic rocks (north of Dalma belt)	1.5–1.45 Ga (Rb-Sr) 1.6 Ga (SHRIMP Pb-Pb, zircon)	Ghosh-Roy and Gangopadhyay 1998; Sengupta et al., 2000 Nelson et al., 2007
Arkasani Granophyre	1.0 Ga (Rb-Sr) (metamorphic event)	Sengupta et al., 1994
Chaibasa metasediments (overlying Dhanjori Group)	< 2.1 Ga (metamorphic event at ~1.6–1.5 Ga)	
Soda granite	1.63–1.66 (Rb-Sr) (metamorphic event), 2.2 Ga (Pb- Pb)	Sarkar, et al., 1986
Early member of CGC	2.3 Ga (Rb-Sr)	Mazumdar, 1998
Malantoli basaltic lava (underlain by felsic volcanics of 2.8 Ga*)	< 2.8 Ga	*Nelson et al., 2007
Jagannathpur basaltic lava	2.25 Ga (Pb-Pb)	Misra and Johnson, 2005

(Continued)

Chronostratigraphic units	Radiometric age	References
Dalma mafic–ultramafics Dalma volcanics (Upper) (Basalt-Dalma hill top) Dalma volcanics (Lower) (Pyroxenite- Kuchia)	1.6 Ga (K-Ar) 1.6 Ga (Rb-Sr) (metamorphic event) 2.0 (Ar-Ar) 2.3–2.2Ga	Sarkar et al., 1969; Roy, 1998 Takigami (personal communication to A. Gupta, 2007)
Dhanjori volcanics (Upper) Simlipal volcanics Ongarbira volcanic suit (gabbro), Dhanjori volcanics (Lower)	2.1 Ga (Sm-Nd) 2.4–2.3 Ga (Rb-Sr) 2.35 (Ar-Ar) 2.4–2.3 (Ar-Ar)	Roy et al, 2002b Iyenger et al., 1981; Y. Takigami in Acharyya et al.,2008; Y. Takigami in Acharyya et al.,2008
Newer Dolerite dyke system	2.2–1.2 Ga (Rb-Sr); 2.6 Ga (Rb-Sr)	Mallik and Sarkar, A., 1989; Roy Abhijit, 1998
Felsic volcanics (Base of Dhanjori, base of Malantoli lava)	2.8–2.7 Ga (U-Pb, zircon), 2.8 (SHRIMP Pb-Pb,zircon)	Acharyya et al., 2008; Nelson et al., 2007
Tamperkola Granite (TG)	2.8 Ga (Pb-Pb)	Bandyopadhyay et al., 2001
Rengali – Raimal granulites	2.8–2.7 Ga (Rb-Sr)	Sarkar, 2000
Darjing Group	> 2.8 Ga (TG–intruded)	
Mayurbhanj 'Granite'	3.1 Ga (Pb-Pb), 2.37–1.90 Ga (Rb-Sr), 2.8–0.8 Ga (U-Pb)	Misra et al., 1999; Vohra et al., 1991; Acharyya et al., 2008
Singhbhum Granite (SBG-B)	3.1 Ga (Pb-Pb), 3.28 Ga (SHRIMP Pb-Pb, zircon)	Saha, 1994; Nelson et al., 2007
Ultramafic-gabbro- anorthosite (UGA) complexes, Baula-Nuasahi.	3.12 Ga (Pb-Pb)	Auge et al., 2003
Iron Ore Group (IOG-I) – Badampahar-Gorumahisani belt (No data for the western IOG-II belt)	> 3.12 Ga (intruded by UGA)	
Bonai Granite (Porphyritic)	3.1 Ga (Pb-Pb)	Sengupta et al., 1991
Bonai Granite (Tonalitic)	3.3 Ga (Pb-Pb)	Sengupta et al., 1991
Nilgiri (Kaptipada) Granite	3.3 Ga (Rb-Sr), (Sm-Nd)	Vohra et al., 1991; Saha, 1994
Singhbhum Granite (SBG –A)	3.3 Ga (Pb-Pb), 3.38 Ga (SHRIMP Pb-Pb, zircon)	Moorbath et al., 1986; Saha, 1994; Misra et al., 1999; Nelson et al., 2007
Older Metamorphic Tonalite Gneiss (OMTG)	3.4–3.35 Ga (Pb-Pb, Rb-Sr); 3.44 Ga (Pb-Pb); 3.4 Ga (Sm-Nd, Ar-Ar); 3.5–3.4 Ga (U-Pb, zircon)	Moorbath and Taylor, 1988; Misra et al., 1999; Baksi et al., 1987; Acharyya et al., 2008
Older Metamorphic Group (OMG) IOG–III (Daitari–Tomka)	3.3 Ga (Sm-Nd); 3.6–3.5 Ga (Pb-Pb); 3.6–3.55 Ga (Pb-Pb); 3.51 Ga (SHRIMP Pb-Pb, zircon)	Sarma et al., 1994; Goswami et al., 1995; Misra et al., 1999 Mukhopadhyay et al., 2008

A Summary of the Crustal Evolution

If we try to reconstruct the sequence of geological events that led to the evolution of the segment of the earth's crust under review, we have to go back to the OMG rocks that have been dated 3550–3600 Ma (Goswami et al., 1995; Misra et al., 1999). Depleted mantle Nd model-ages for the Darjing Sandstone vary from 3600 Ma to 4000 Ma, suggesting that there could be an older continental crust in the area (Saha Aniki et al., 2004). This suggestion has the support of the composition of the OMG represented by such rocks as muscovite–biotite schist; quartz–sericite schist and quartzite; quartz–magnetite–cummingtonite schist and calc-magnesian metamorphites, a fair part of which must have been contributed by a neighbouring crust, dominantly sialic in composition. As for the TTG in OMTG, however, slab melting in a subduction model seems reasonable in the context of experimental results and observation in Phanerzoic orogens (Foley et al., 2002; Martin and Moyen, 2002). Contents of Sr, Ni, Cr, (Na_2O + CaO) and Mg number in the OMTG (Saha, 1994) are suggestive of shallow melting of this slab as was common during the Early Precambrian time. Regarding the tectonic setting during the emplacement of Singhbhum Granite and Nilgiri Granite we have no clue except that in a Ta-Yb diagram they plot mostly in the VAG (Volcanic arc granite) field for whatever it is worth. The Mayurbhanj Granite on the other hand plots in the WPG (Within plate granite) field. Saha (1994) called the latter anorogenic or A-type granite on the petrochemical basis.

The eastern volcanisedimentary belt (IOG-I – Badampahar–Gorumahisani belt), as already indicated, has a number of attributes similar to those of many Archean greenstone belts, where the volcanisedimentary rocks are older to the adjacent granitic rocks. The situation is visibly different in the western part i.e. in the IOG-II of Jamda-Koira–Noamundi belt or the 'horse-shoe' region. Here the sediments dominate over the volcanics and the petrochemistry of the volcanics tangibly differs from that of the volcanics in the eastern belt. According to one geochemical model, the volcanics in the western belt (by extension, the whole volcanisedimentary package) are younger to those in the east (Sengupta et al., 1997). The map produced by Misra et al. (1997) shows a volcanisedimentary pile overlying the Bonai Granite (3.3 – 3.1 Ga) with an unconformity and containing mappable units of banded iron formation. Being contiguous to the northern part of the western limb of the Jamda–Koira– Noamundi synclinoruim, it seems possible that the volcanisedimentary pile under reference belongs to IOG. This again suggests that the IOG in the western basin is younger to those in the east. The recently reported date of 2.8 Ga date for the felsic volcanics at the base of Malantoli volcanics by Nelson et al. (2007) further suggests a younger age for the IOG of the western belt, which overlies the former. The oldest date of 3.51 Ga has recently been reported (Mukhopadhyay, et al., 2008) from the IOG-III sequence of Tomka–Daitari belt. A gravity field measurement across the western belt suggests the continuation of a 12.6 km thick sialic crust below the supracrustals. How did the IOG basins form is another moot question. These could be riftogenic basins in an OMG– OMTG / SBG-A country. Worthy of notice is the fact that the BIF-bearing basins peripheral to Archean cratons are not unique to Eastern India. The Ungava craton in Canada is a spectacular example of this feature. Kulik and Korznew (1997) suggested that the BIF-rich Krivoy Rog basin of Ukraine was an intracratonic riftogenic trough. Another example of the same could be Carajas basin, Brazil (Machado et al., 1991).

The Kolhan sediments in the type area of Chaibasa–Jagannathpur are platformal, overlying the Singhbhum Granite at its northwestern margin. If the sediments of Champakpur–Keonjhargarh basin further south also belong to the Kolhan Group, then they overlie the Singhbhum Granite and the Malangtoli lava, which is commonly equated with Dhanjori lava. The felsic volcanics (Phuljhari Formation) below Dhanjori sequence dated recently as 2.8 Ga by Acharyya et al. (2008) substantiates its possible time-correlation with the felsic volcanics at the base of Malantoli lava, and as a corollary between the latter and Dhanjoris.

In a publication Erikson et al. (1999) stated that the '2.7 – 2.0 Ga volcanisedimentary record in Singhbhum begins with the Late Archean' and that 'the presence of inferred glacigenic rocks, evidence of higher free-board type continental sedimentation and plume related sub-aereal volcanism in the Dhanjori Formation suggest possible correlation with global ca 2.4–2.2 Ga glaciation event'. There are a few recent published age data which indeed support that the felsic volcanics, the precursors to the main Dhanjori volcanism, were emplaced at ~ 2.8–2.7 Ga followed by the Dhanjori mafic–ultramafic volcanism between 2.4–2.1 Ga (Acharyya et al., 2008; Roy et al., 2002b). Misra and Johnson (2005) have reported 2.8–2.7 Ga age for Dhanjori volcanics based on Pb-Pb and Sm-Nd (WR) isochron data. There is, however, no evidence of substantial sub-aerial volcanism in the Dhanjoris.

If we do not disagree with the premise that the protolith for quartz-kyanite rocks was siliceous kaolinitic clay then the ubiquitous presence of quartz-kyanite near the base of Chaibasa Formation and some isolated occurrences of the same further up, including the famous Lapsa Buru Kyanite deposit, are indicative of an intense weathering under hot humid conditions in the extended free board in the south i.e the Singhbhum Granite terrain. This is further supported by the bulk composition of the sediments of the Chaibasa Formation. Even if it is allowed that this pronounced weathering could effect draw down of atmospheric CO_2, the question remains if this local lowering of the atmospheric CO_2 content was good enough to cause local/regional glaciation, even in the context of reduced luminosity of the Paleoproterozoic Sun. The geologic evidence commonly sought for such a glaciation comprises the presence of diamictites, dropstones and the bulk chemistry (absence of Al_2O_3-enrichment) of the sediments. South Singhbhum and the adjacent areas of Orissa constitute a 'granite-greenstone terrain' of sorts, but as will be obvious from our discussion, the whole of Eastern Indian craton should better not be declared as a typical Archean tonalite–trondhjemite–granodiorite (TTG)–greenstone terrain in view of its geological characteristics already discussed.

If all of what is grouped under the Mayurbhanj Granite (Saha, 1994) was emplaced at 3100 Ma or thereabout (Misra et al., 1999), then the whole of south Singhbhum and the adjacent parts of Orissa were cratonised before the close of the Archean Era. The possibility of multiphase emplacement of the Mayurbhanj Granite from 3.00 Ga to 2.8 Ga, as suggested by Acharyya et al. (2008), helps us to correlate its later phase with the Tamperkola Granite (2.8 Ga). This, added with the exhumation of granulites in Raimal area at around 2.8–2.7 Ga, perhaps marked the final cratonisation of the Singhbhum–North Orissa region in the Late Archean time.

The North Singhbhum Mobile Belt and the Chhotanagpur Gneissic Complex further north are the Proterozoic developments. The Dhanjori Group of rocks occurring at the southern margin of this belt (and northern periphery of the Singhbhum Granite massif) constitutes the oldest deposits in the belt, starting their deposition some times in the Paleoproterozoic and closing not long after 2100 Ma. The metasediments of the Chaibasa Formation are younger to the Dhanjori Group. The rocks of Dhalbhum Formation appear younger to the Chaibasa Formation. When sedimentation was well advanced in North Singhbhum basin, a rift zone developed almost along the middle of this basin which was filled up by some sediments at the bottom, followed upward by ultramafic and mafic volcanics respectively to form the Dalma volcanisedimentary belt. As already mentioned, the obtained age for these rocks such as 1547 ± 20 Ma (K-Ar) by Sarkar et al. (1969) is obviously a metamorphic age. Roy et al. (2002a) reported 1619 ± 38 Ma (Rb-Sr) date from a pyroxenite body at Kuchia in the eastern part of the belt and inferred it to be its emplacement date. There were several contradictions at this point. Firstly, the pyroxenite sill dated by Roy et al. (op. cit.) occurs within a highly schistose tuffaceous sequence of Lower Dalma Formation and is itself schistose along the margins. Hence, it is unlikely that the pyroxenite band had escaped the effects of the overwhelming metamorphic event at ~1.6 Ga recorded by all other rocks in the NSMB. Secondly, there is a general consensus that the Dalmas developed in an extension regime of local crust. But the inferred emplacement date of Dalma according to Roy et al. (op. cit.) nearly coincides with the generally believed most important compressive regime and the development of most of the deformational structures and the penecontemporaneous metamorphism that localy reached upto the amphibolite facies. The situation was further complicated by the fact that the Dalma volcanics are seen affected by all the phases of deformation in the region (Gupta et al., 1980; Sarkar et al., 1992). This problem is largely resolved now with the availability of new set of age data.

The $^{40}Ar/^{39}Ar$ experimental step heating data of the pyroxenite sample collected from the same spot as Roy et al., is indicative of a much older age range of 2200–2300 Ma (Y.Takigami, 2005 – personal communication to A.Gupta). Mishra and Johnson (2005) have reported a five point Rb-Sr (WR) isochron age of 2396 ± 110 Ma (MSWD = 5.3) for Dalma volcanics. However, they inferred it as a metamorphic age.

Several models have been proposed to explain the tectonolithological developments in this region (NSMB). These briefly stated are:

1. Intraplate subduction model (Fig. 4.1.48)
2. Microcontinental collision model (Fig. 4.1.49)
3. Back-arc marginal basin model
4. Intracratonic extension, rifting and ensialic orogenesis model (Fig. 4.1.50).

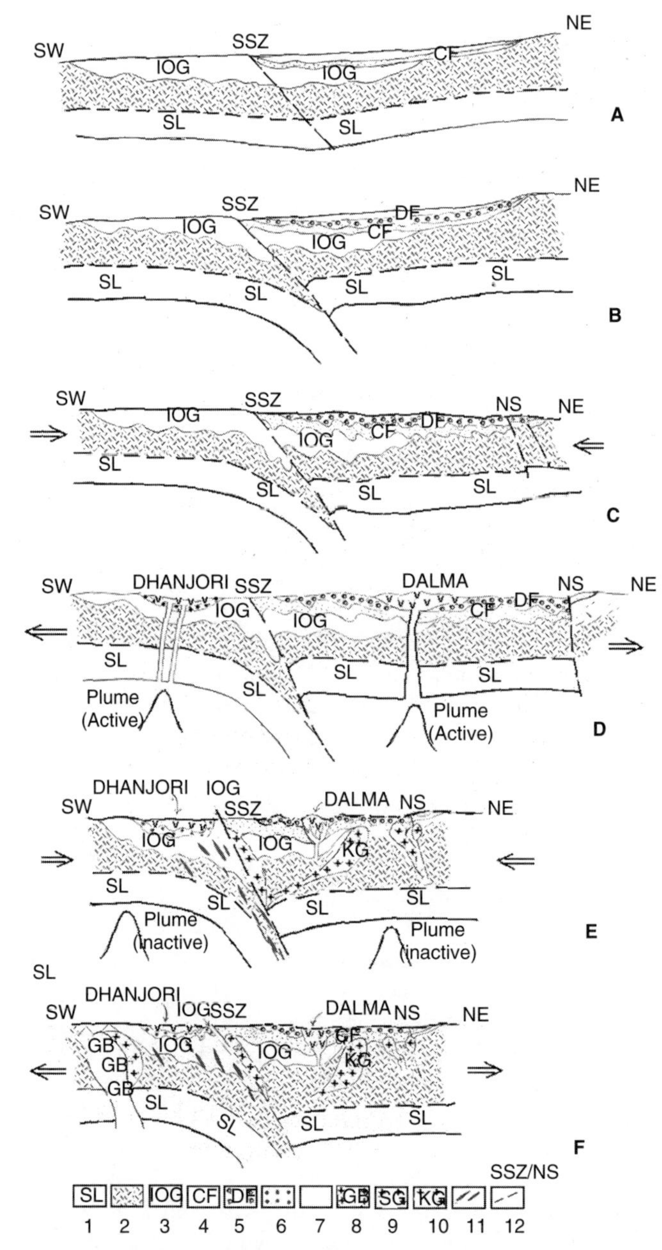

Fig. 4.1.48 Intra-plate subduction model proposed by Sarkar and Saha (1977) for the NSMB 1. Subcrustal lithosphere; 2. Continental crust; 3. Iron Ore Group; 4. Chaibasa Formation; 5. Dhalbhum Formation; 6. Dhanjori Sandstone and Conglomerate; 7. Dhanjori Lava, Dalma Lava; 8. Gabbro; 9. Granites of Singhbhum orogenic cycle; 10. Kuilapal Granite, Ultramafic injections; 11. Singhbhum shear zone and Northern shear zone (after Saha, 1994)

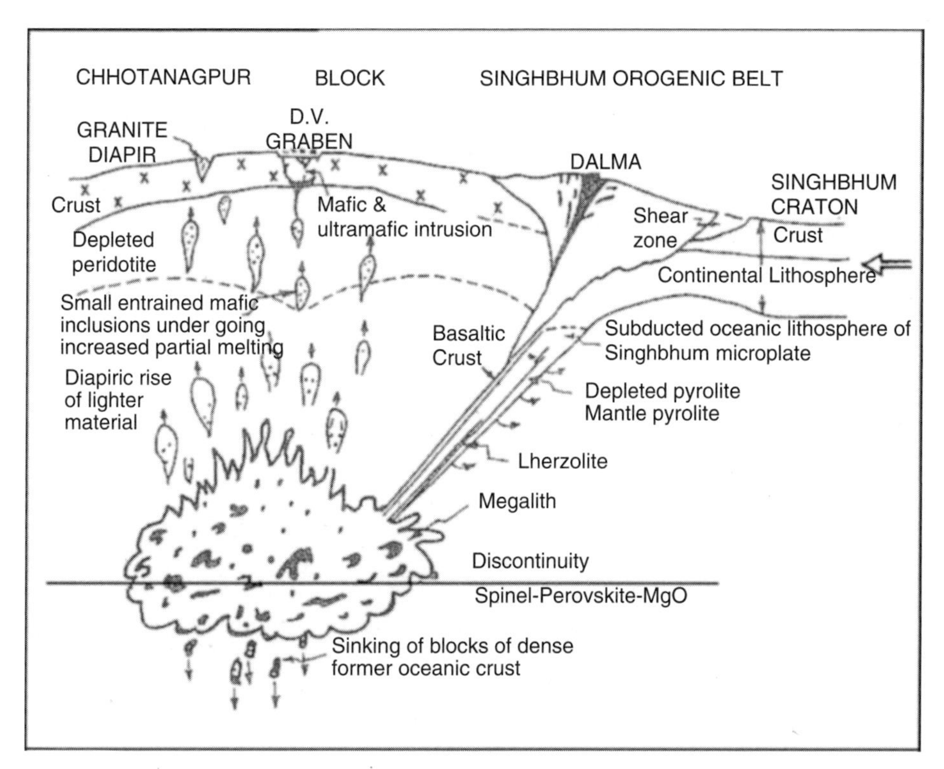

Fig. 4.1.49 Model for the tectono-magmatic evolution of the crustal segment of North Singhbhum–Chhotanagpur by Sarkar(1982)

The intraplate subduction model proposed by Sarkar and Saha (1977) suggested northward subduction of the Singhbhum granitic continent along the southern margin of the mobile belt, following what is now delineated as the Singhbhum shear zone (SSZ). Extensive outpouring of lavas of Dalma, Dhanjori and Simlipal marked extension regime between successive subduction episodes.

The microcontinental collison model proposed northward subduction of the Singhbhum microplate below the 'Chhotanagpur microplate' along the northern margin of the former (Sarkar, A.N. 1982). The Dalma mafic–ultramafic dominant complex is viewed as an obducted 'ophiolitic' suite in this model.

The back-arc marginal basin model suggested that the entire width of the mobile belt is a back arc domain, adjacent to the stable cratonic area of Singhbhum Granite. The Chhotanagpur Gneissic Complex, according to the authors of this model (Bose and Chakraborty, 1981; Bose et al., 1989), is the root zone of the volcanic arc. The zone of the southerly subduction is located further north.

The intracratonic extension (rifting and ensialic orogenesis) model (Gupta et al., 1980; Mukhopadhyay, 1984; Sarkar, 1984; Sarkar et al., 1992) is based on the perception of an attenuated crust due to thermal plume–generated extension of the continental crust leading to basin formation. Dislocation along the southern margin at an early stage produced the Dhanjori sub-basin (s) and ultimately the basin fill. The continued distension ultimately gave rise to rift zone along the spinal zone of the belt and filled it with some sediments and much volcanics.

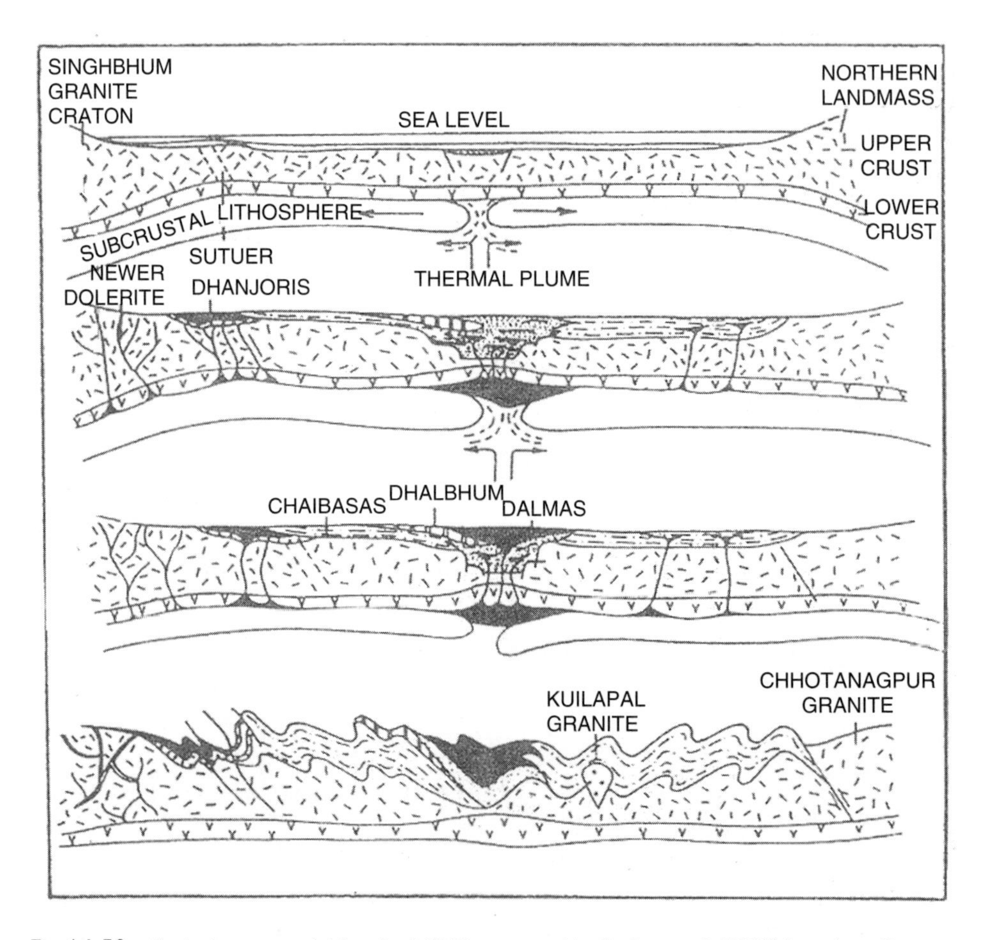

Fig.4.1.50 Evolutionary model for the NSMB proposed by Sarkar et al. (1992) based on Gupta et al. (1980) and Sarkar (1984)

Needless to say that none of the above models are adequately constrained. The last one, however, has some relative merit. The geophysical observation that the north Singbhum Mobile Belt is floored by continental crust (Qureshy et al. 1972; Verma et al., 1984) assigns it to a Highly Extended Terrain (HET) (cf. Olsen and Morgan, 1995) and supports the last model in general terms. However, no age data older than 2300 Ma is reported from the rocks on the north side of the NSMB to date, i.e. the CGC. In fact, we have little or no data to reasonably constrain the development and evolution of this terrain. The rocks of this belt range in age from 2300 Ma to 850 Ma, with a cluster around 1000 Ma, a probable impresses of Sausar (Satpura) orogeny.

Whatever may be the tectonic frame within which the region evolved geogically, there is little doubt that the interface between the southern Archean terrain and the northern mobile belt (NSMB) broadly coincides with a zone of high shear strain, commonly referred to as the Singhbhum thrust belt or Singhbhum shear zone (SSZ). The available evidences suggest that the zone alternated between brittle and ductile shearing for a long but unestimated period of time. It may be pointed out that Bandyopadhyay (2002) estimated the maximum pressure and temperature of progressive metmorphism (M_1) as ~ 500°C and 6.5 Kb from an area close to the SSZ. This pressure value translates into a crustal overburden of 20 (+) km. The calculated P–T plotted above the steady state geotherms of the Early Precambrian time and this thermal perturbation, citing model of Ganguly et al. (1995), she prefers to explain as thickening due to thrusting. An average of several pressure determinations from metamorphic tectonites collected from an area in SW NSMB, is 10 ± 1.5 kb (Mahato et al, 2008). This suggests a greater crustal thickening. This means, Singhbhum shear zone remains the Singhbhum thrust belt by its own right; only the case may not be as straight forward as Dunn and Dey (1942) thought of it.

4.2 Metallogeny

Introduction

Eastern India is as well rich in mineral deposits as it is interesting in its geological buildup. It is well-known for containing copper and uranium mineralisation along the Singhbhum copper uranium belt, BIF and the related iron ores, chromite and manganese ores in Orissa and the adjacent Jharkhand areas. The copper mines along the Singhbhum belt were responsible for the bulk of the copper producton in the country before their closure towards the end of the last century. More than 60% of

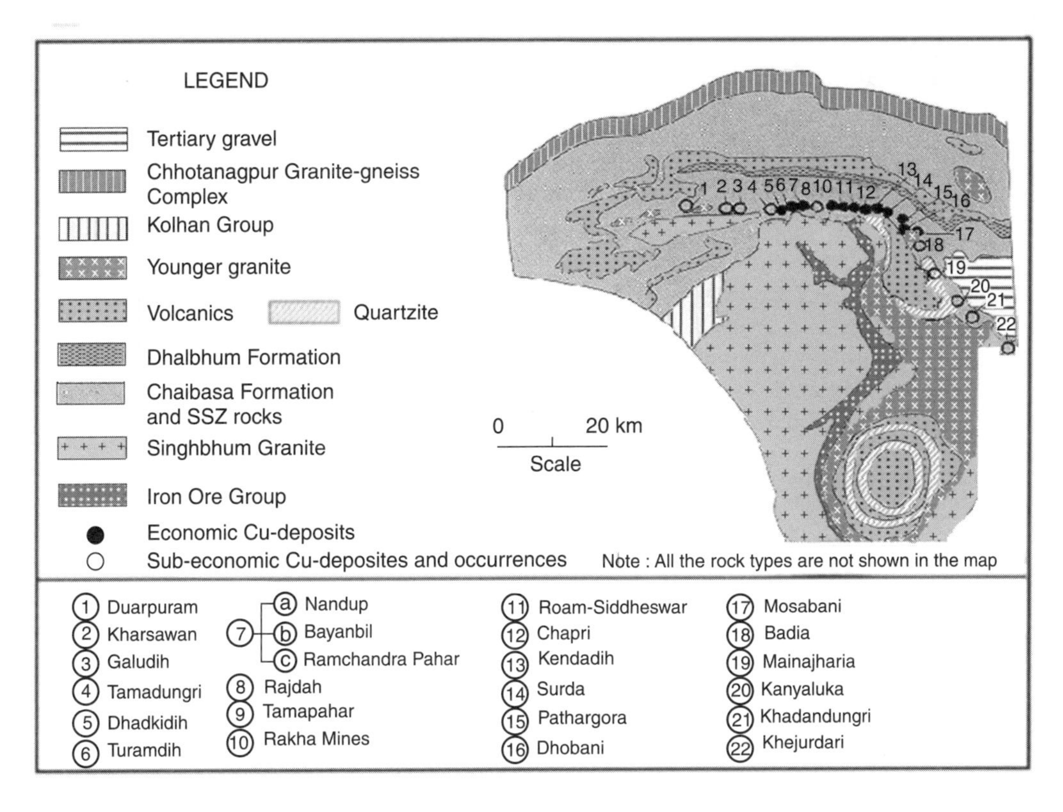

Fig. 4.2.1 Copper sulphide deposits and their occurrences along Singhbhum shear zone (modified after GSI, 2000)

the country's hematitic iron ores and > 90% of chromite ores occur in this part of the country. All the operative uranium mines of the country occur along the Singhbhum Cu-U belt. Quite a few deposits of rare-earth and rare metals occur in the northern part of this region, much of which is already mined out. Phosphate mineralisation has also taken place along the Singhbhum copper uranium belt and the Northern shear zone. Gold occurs at a number of places, amongst which Parasi in the Ranchi district holds out promise for the future and the old mine of Kunderkocha in East Singhbhum resumed mining in a small way. PGE mineralisation in the Nuasahi chromite mine, Keonjhar district, is a new find. Commercially important bauxite deposits occur in western Jharkhand. The aforesaid and a few other small deposits/occurrences of ores will be discussed in this subchapter.

Copper Sulphide Mineralisation along Singhbhum Shear Zone

Copper mineralisation along at least some parts of what is now known as the Singhbhum shear zone (SSZ) from tectonic point of view, or Singhbhum copper uranium belt (Fig. 4.2.1) from the metallogenic point of view, must have been known to the ancient people in the distant past. Several kinds of anthropological remains suggest that the earliest mining in the belt could be as old as 2000 BC and the latest phase of early (pre-modern) mining must have taken place sometimes during 300–600 AD. Local people appear to have had forgotten about these deposits during the rest of the Christian era, until Capt. J. C. Hughton rediscovered the mineralisation at Tama Dungri (22°45′: 86°02′) near Narainpur in Seraikela in 1854 (Dunn, 1937). It is also possible that the ancient people who knew the mining of copper ores and also their smelting, as evidenced by slag-heaps at many places on the Subarnarekha bank, were driven out of the land by later invaders (may be the 'Ho' tribes), for whom these activities could have been unnecessary sophistries.

Copper mineralisation is noted along the entire SSZ over 120 km length from Duarpuram in the west to Baharagora in the southeast. Out of this, about 70 km length in the eastern half of the belt comprises the best mineralised sections of economic value, which include Baharagora, Badia–Mosabani, Dhobani–Chirudih–Samaidih, Pathargora, Surda, Kendadih–Chapri, Roam–Rakha Mines–Tamapahar, Ramchandra Pahar (Ramchandrapahar)–Bayanbil–Nandup and Turamdih–Dhadkidih–Chhota Jamjora (Figs. 4.2.2, 4.2.3 and 4.2.4).

Modern copper mining in the Badia–Mosabani area started in 1928 by ICC (Indian Copper Company – a British company) and was later taken over and continued until 1998 by HCL (Hindusthan Copper Ltd. – a Public Sector undertaking). The Table 4.2.1 reflects the life span of all the mines in this belt, the closure of which were due to the economic non-viability under the then prevailing low metal price and high cost of production rather than non-availability of mineable ore.

Table 4.2.1 *Status of copper mines, Singhbhum copper-uranium belt*

Years	Badia-Mosabani	Dhobani	Pathargora	Surda	Kendadih	Roam- Rakha Mines
Year of opening	1928	1934	1957	1956	1973	1976
Year of closing	1998	1953	2000	2003*	2000	2001
Residual reserves (Mt)	19.80# (1.70%Cu)	1.4 (2.4%)	3.58 (1.32%)	26.09 (1.2%)	15.00 (1.73%)	85.36 (1.32%)

* A joint venture by HCL and M/s Monarch Gold, an Australian company has resumed mining in late 2007 and produced 397.569t ores (0.98% Cu) in 2010-2011.

Reserve inclusive of Dhobani. Some of these mines (Parthargora, Kendadih and Roam-Rakha) may reopen in near future.

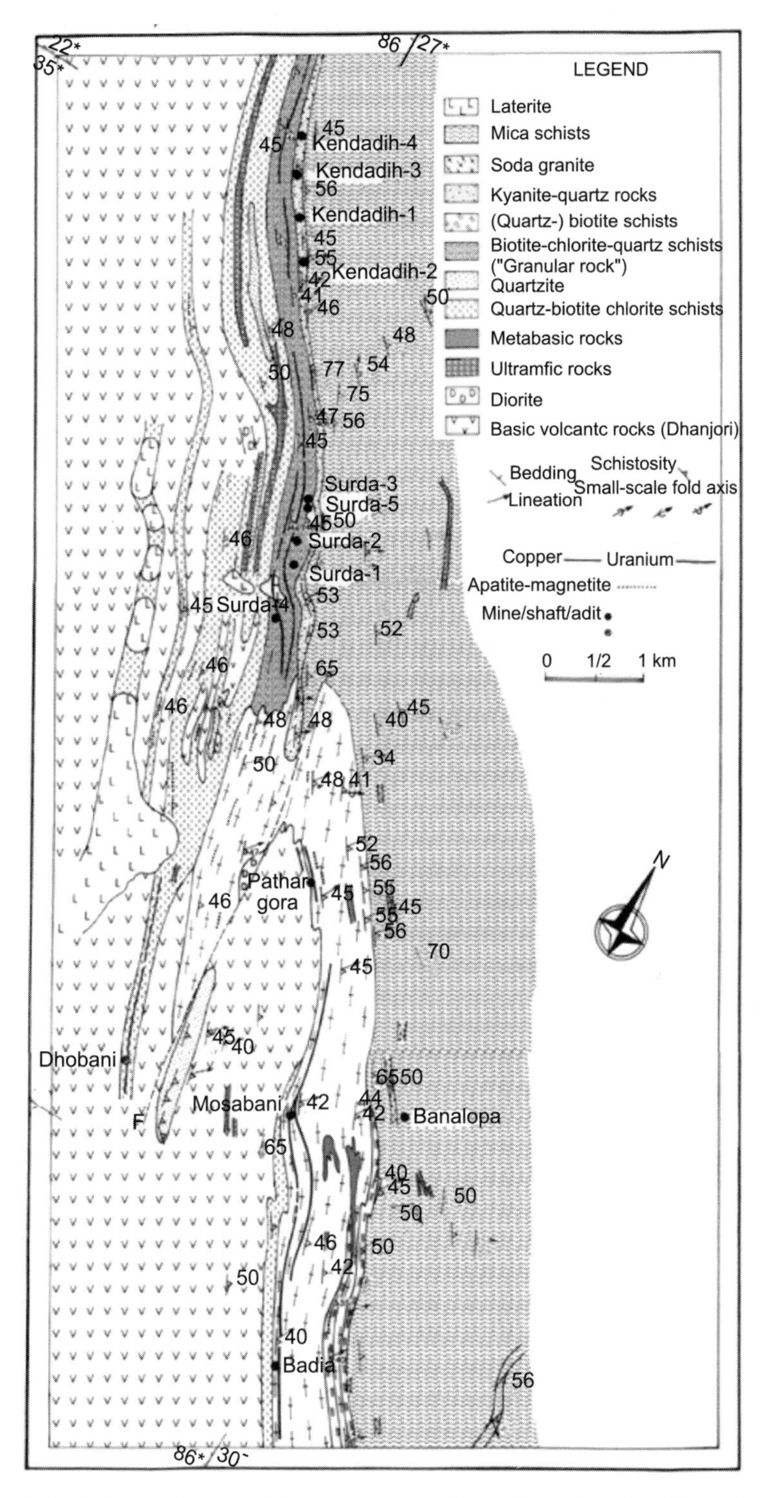

Fig. 4.2.2 Geological map of the Badia–Mosabani–Surda–Kendadih section, Singhbhum Cu-U belt, East Singhbhum district, Jharkhand (after Sarkar, 1984)

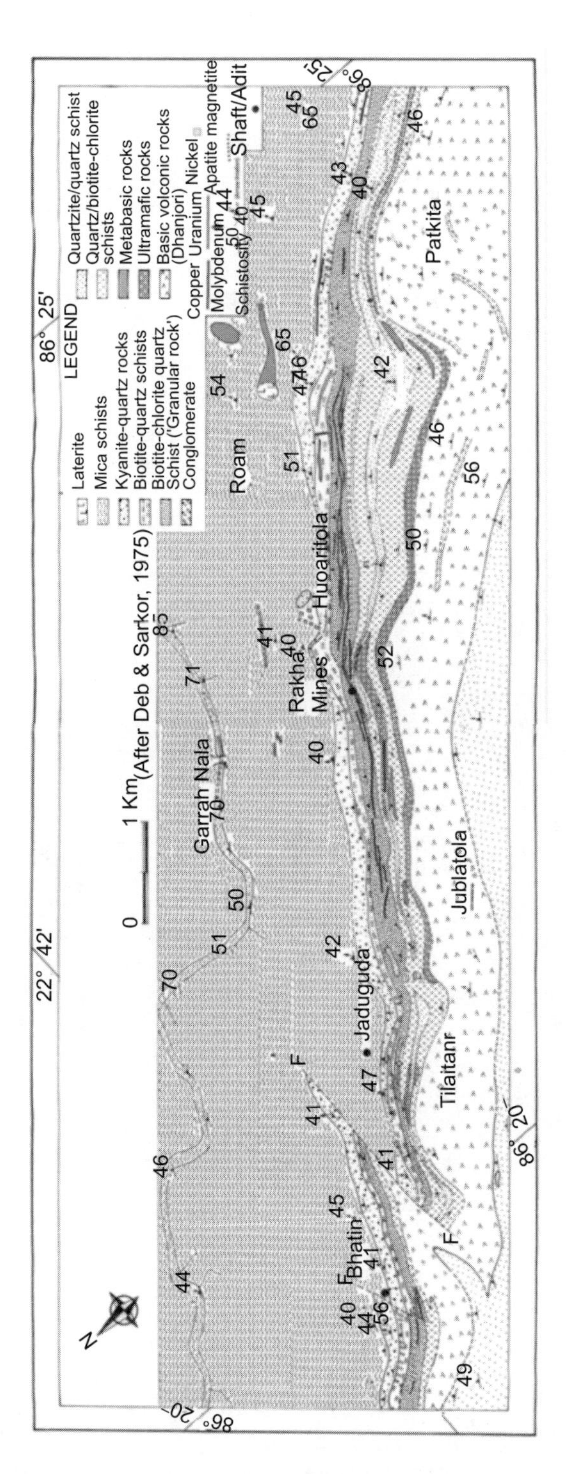

Fig. 4.2.3 Geological map of the Roam–Rakha Mines–Tamapahar–Jaduguda–Bhatin sector, Singhbhum Cu-U belt, East Singhbhum district, Jharkhand (after Deb and Sarkar, 1975)

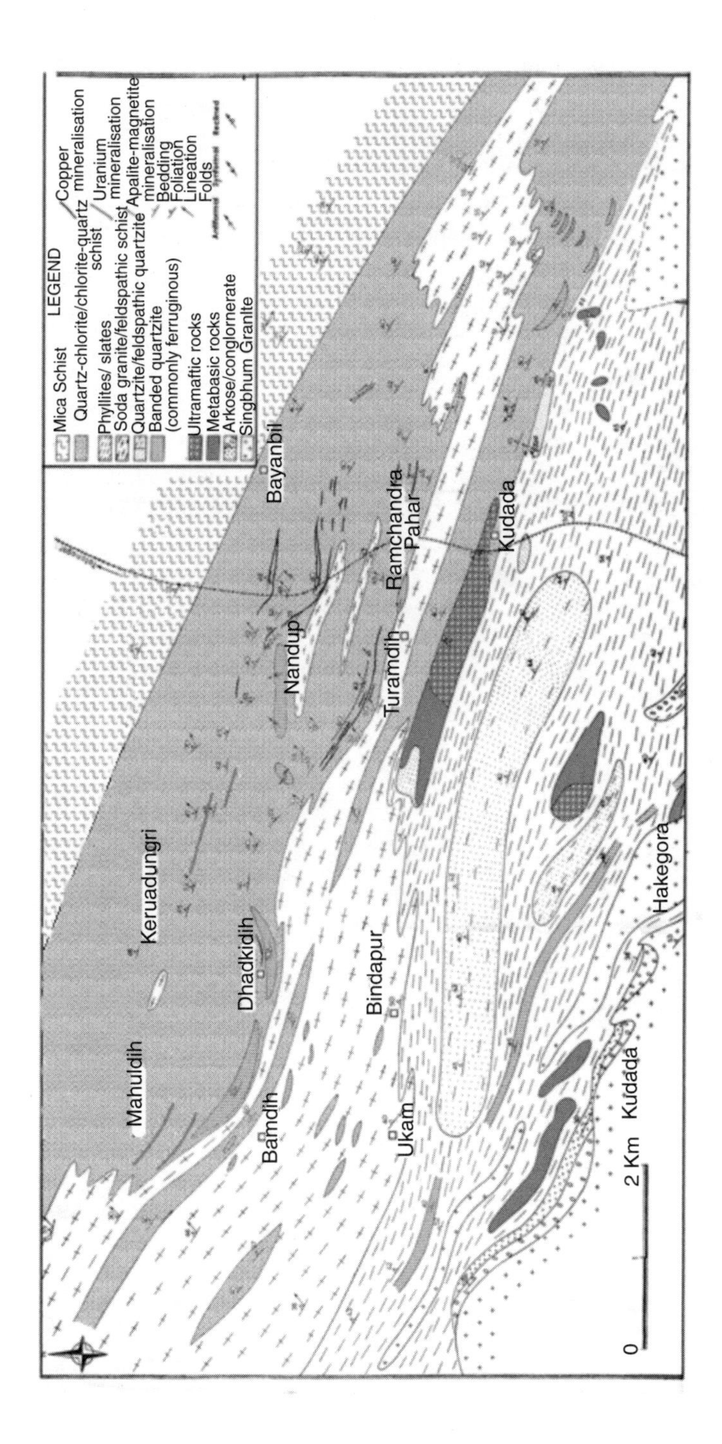

Fig. 4.2.4 Geological map of the Ramchandra Pahar–Nandup–Bayanbil–Turamdih Dhadkidih–Mahuldih sector, Singhbhum Cu-U belt, East Singhbhum district, Jharkhand (after Sarkar, 1984)

Approximately 40 Mt of copper ore have been mined out from this belt since 1928 until the mines closed down. About half of it had an average grade of 2% Cu and above while the rest had 1.6% Cu or slightly below. Besides producing copper, the Singhbhum mines before closure also produced substantial quantities of nickel sulphate, selenium, gold, silver and tellurium as by-products. Even 1–2 ppm of platinum and 180–200 ppm of palladium were obtained on fire assaying of anode slime from the Maubhandar Plant (Chowdhury et al., 1983).

Total residual reserve in these deposits is in the order of 150 Mt with average grade of 1.4% Cu. Besides, the explored but unexploited deposits of Turamdih, Bayanbil, Nandup, Dhadkidih, Ramchandra Pahar and Baharagora contain another 46.5 million tonnes (1.4% Cu) ores.

Baharagora Cu-sulphide mineralisation in Baharagora area is hosted by sheared basic (hornblende) schists, biotite quartz schist or quartz biotite schist, and granitic rocks. Out of the three blocks explored, the Mundadevta–Darkhuli block (1000 m strike length) is best mineralised with four subparallel orebodies, hosted by basic and biotitic schists and disposed as an asymmetric synform, plunging 45° due N 28°E (Fig. 4.2.5). The width of the orebodies varies from less than a meter to17 m and grade ranges up to 12% Cu. In other two blocks, Jharia and Charakmara, the mineralisation is in granitic rocks, where orebodies are delineated for about 500 m length in each. The total ore reserve of Baharagora deposit is estimated at 3.88 Mt with 0.9% Cu.

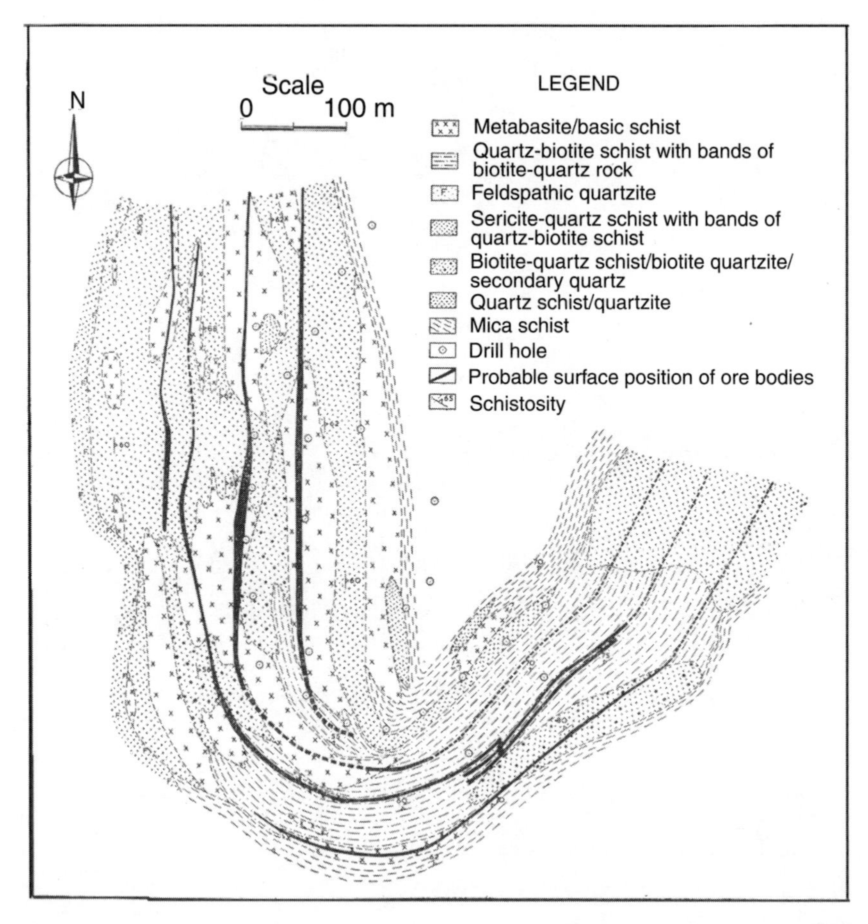

Fig. 4.2.5 Geological map of Mundadevta–Dharkuli block of Baharagora deposit (from Majumdar, 1979 in Sarkar, 1984)

Badia–Mosabani–Dhobani The Badia–Mosabani section has the richest Cu-mineralisation in the entire length of the Singhbhum shear zone (SSZ). Mineralisation is more intense in the Mosabani sector where mining developed more than 35 levels down to the depth of about 1300 m. Mineralisation has taken place in the form of two subparallel orebodies, of which the footwall orebody is better developed (Fig. 4.2.2). A cross section across Banalopa shaft gives an idea of the downdip extent of the copper lodes and the underground workings at Mosabani (Fig. 4.2.6). The 7 km long ore zone is generally, but not strictly, restricted to sheared Soda granite close to its contact with the underlying Dhanjori basic volcanics. Ore-shoots delineated on the assay plans are elongate and their long axes are subparallel to the downdip lineations that are so common in the SSZ rocks and discussed in the previous section. At Badia, mineralisation becomes poor below the 10th level. The ore–host rock relationship is better seen in smaller scales inside the mines (Fig. 4.2.7 a, b, c). The ores often occur as sharp veins and at places small spheroidal bodies of Soda granite are seen caught up in massive ore. Typical wall rock alteration in Soda granite is generally weak to absent. Late dolerite and lamprophyre intrusives cut through the deposits. The ore reserve of Mosabani-Badia stood at about 20 Mt x 1.7% Cu.

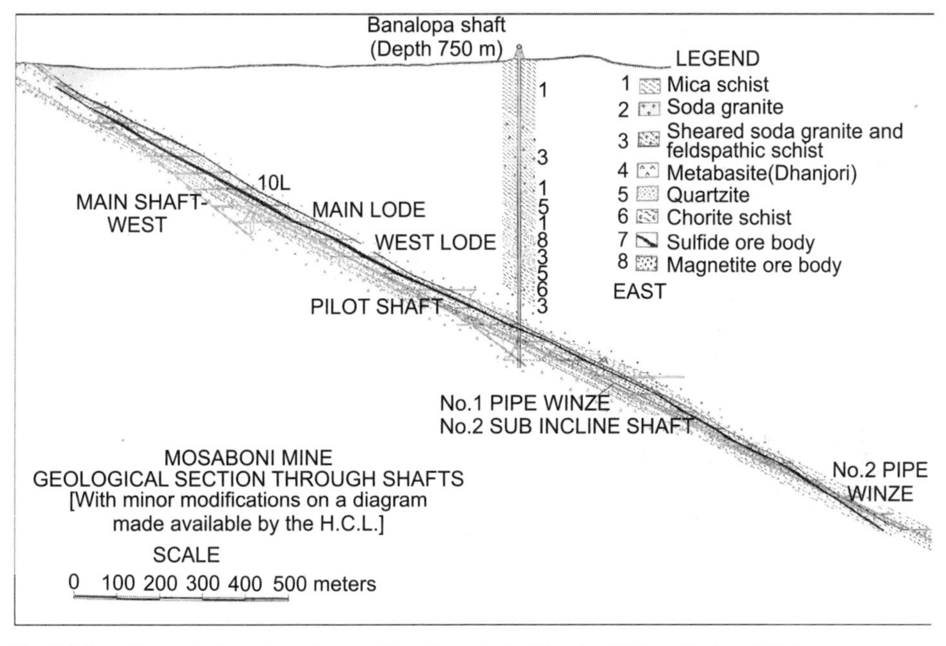

Fig. 4.2.6 Geological section through Banalopa shaft, Mosabani Mine (Sarkar, 1984)

About 1.4 km west of Mosabani there is an ore zone on the western flank of Dhobani Pahar within Dhanjori basic volcanic rocks that are now partially altered to biotite chlorite (± hornblende/actinolite) schist. Mining started here in 1934 and continued for two decades before it was closed. But the ores were not exhausted. A strike length of 3.2 km falling outside the mine area was later investigated by GSI by putting 19 shallow and eight deeper boreholes. Eight correlatable lodes with several ore shoots were delineated with ore reserve estimates of 1.4 Mt × 2.4% Cu under 0.5% cut-off (including 0.8 Mt × 3.13% Cu at 1.0% cut-off). The deposit remains unattended.

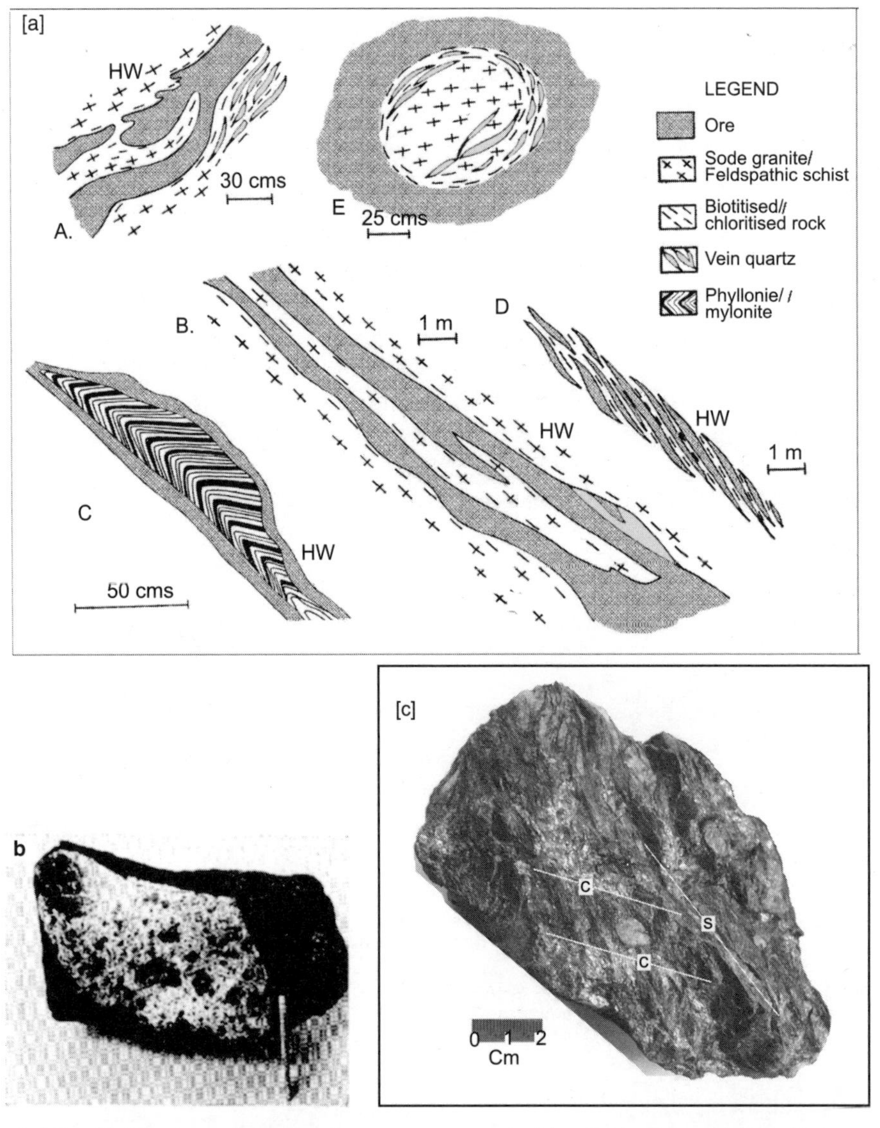

Fig. 4.2.7 (a) Sketches of small scale ore structures in Mosabani mines; (b) 'Durchbewegung' or kneading of vein-quartz and host rock (dark grey) in a sulphidic matrix (white to light grey) in Mosabani mines; (c) Chalcopyrite dominant sulphide mineralisation in chlorite-quartz schist, displaying S-C structure control (a and b from Sarkar, 1984; c by the authors)

Another close-by prospect is at Tamajhuri in the north-western extension of Dhobani area. Here also the host of copper mineralisation is in magnetitic chlorite schist of the Dhanjori Group. The ore zone with limited strike extension is up to 2.25 m in thickness. Down to 100 vertical depth the ore reserve estimate was 0.27 Mt × 1.70% Cu.

Pathargora The Pathargora mine is located to the immediate northwest of Mosabani deposit and the two are connected at the 5th level. This mine was developed for about 1 km strike length at the 7th to 9th levels. There are two principal ore bodies, of which the hanging wall(HW) lode in feldspathic schist sheared Soda granite, is continuous while the footwall lode in Dhanjori volcanics is prominent only in northern parts. The lode in the southern part of the deposit is richer and wider. The down-dip plunges of the ore shoots are similar to the Mosabani–Badia section, the grade of ore being slightly lower.

Surda Further to the northwest of Pathargora, the 2 km long mineralised stretch at Surda consists of several sub-parallel ore bodies packed in a 30–35m thick zone. The host rock of mineralisation is generally chlorite-biotite–bearing quartz schist ('granular rocks' of the local exploration geologists), which locally contains appreciable proportions of albite, apatite, magnetite, tourmaline, allanite etc., and rare hornblende. The orebodies are conformable to the regional schistosity/shear plane and formational boundaries and dip at about 40° towards NE. Three sub-parallel lodes in the southern section of Surda mine was stoped together. The mine was developed up to the 13th level. The residual reserve of this deposit (26 Mt × 1.2% Cu) is considerable.

Chirudih–Samaidih This mineralised zone hosted by the Dhanjori volcanics and associated variants is in the northern extension of Dhobani–Tamajhuri trend, passing through the west of Surda mines. The sulphide mineralisation is intermittently traceable for about 4.6 km. The lodes are less than a meter to 1.6 m in thickness and ranges in grade from 1–5% Cu. The ore reserve stands at 1.5 Mt with 1.4% Cu.

Kendadih–Chapri To the north west of Surda, there is another mineralised stretch of 600 m where five sub-parallel ore bodies with partial overlaps are hosted by biotite–chlorite–quartz schist (often massive, resembling quartzite). One of the lodes in the hanging wall is hosted by mica schist. Warps with sub-horizontal axes are prominent along the dipping surface of the lodes.

The orebodies are thin but rich in sulphides, with an average grade of ~ 1.7% Cu. Mining was started in this block by developing two adits of old Cape Copper Company. The northern part of this block continues into Chapri block, which in turn has continuity with the Roam-Sidheswar area. In Chapri block the two orebodies are thicker, one hosted by conglomeratic rock. The ores remaining in the Kendadih–Chapri sector are still economically mineable.

Roam–Rakha Mines–Tamapahar These deposits have a total strike length of 5 km, with Roam-Sidheswar block in the southeast, Rakha Mines block in the middle and Tamapahar block in the northwest (Fig. 4.2.3). The mineralised shear zone is quite narrow (300–500 m) in this stretch, occurring between two prominent quartzite horizons passing along the footwall and hanging wall. Nine major lodes are delineated over the strike length of 3.5 km in the Rakha and Roam blocks, disposed in en-echelon fashion with sinistral shift. In Tamapahar block there are two prominent ore bodies with leaner mineralisation. Most of the lodes are hosted by granular biotite–chlorite–quartz schist /rock ('granular rock'), except a few hosted by biotite– rich schist and conglomeratic rock. In the Rakha–Roam sector there are many shafts and adits with considerable underground development (nine levels in Rakha Mines) done by the Cape Copper Company, which discontinued activities in 1922 due to obscure reasons. These deposits were explored during sixties by IBM and GSI and were subsequently taken up by Hindusthan Copper Ltd., a PSU, for exploitation.

The ore bodies in this part of the belt are parallel to the pervasive schistosity / shear planes and the litho-boundaries and display a pinch and swell character which are best reflected in the assay plans, underground maps, sections and of course in exposures on the surface and underground workings. Fig. 4.2.8 depicts the disposition of very rich ore shoots in the assay plan of old Cape Copper Mines in the Main Shaft and No 4 Shaft workings of Rakha Mines area. The plunges of the ore shoots are parallel to the down-dip lineation similar to Mosabani and Surda. The lode continuity along strike and depth are excellently displayed by the longitudinal and transverse sections drawn on the basis of borehole intersections (Fig.4.2.9, a - b).

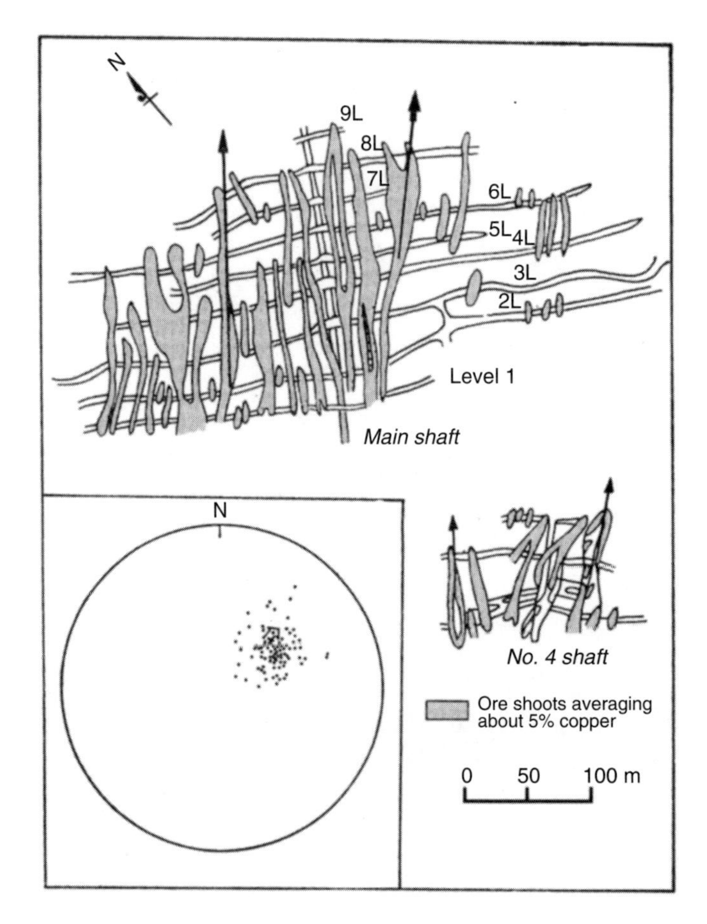

Fig. 4.2.8 Assay plan of Main Shaft and No.4 Shaft workings of Cape Copper Company at Rakha Mines displaying disposition of the ore shoots. The plunges of individual ore shoots plotted on a sterionet (from Sarkar, 1984; original published by Dunn, 1937)

Ramchandrapahar–Bayanbil–Nandup In the northwest of Tamapahar block, there is gap of about 15 km before another copper-rich zone is encountered at Ramchandrapahar. This gap, however, comprises a series of economic uranium deposits under active mining from Jadugoda to Narwapahar (Narwa Pahar). Copper is subordinate in this stretch, though Mo is a recoverable accessory. From Ramchandrapahar onwards in the west there are several discontinuous ore zones along the SSZ, which is much wider (up to 3 km) in this sector (Fig. 4.2.4).

The ore mineralisation in this sector is confined to chlorite–quartz/quartz–chlorite schists. The chloritic schist horizon often bifurcates and coalesces along the strike. The ore zones are silicified at places but silicification is also noted in adjacent sericite schists which are devoid of mineralisation. Soda granite/feldspathic schists in the neighbourhood are also devoid of mineralisation. The mineralised chloritic schists are locally sericitised but sulphide concentration bears an antipathic relationship with sericitisation (Sarkar, 1984). The ore bodies are parallel to the pervasive schistosity/ shear planes and are folded along with the enclosing rocks.

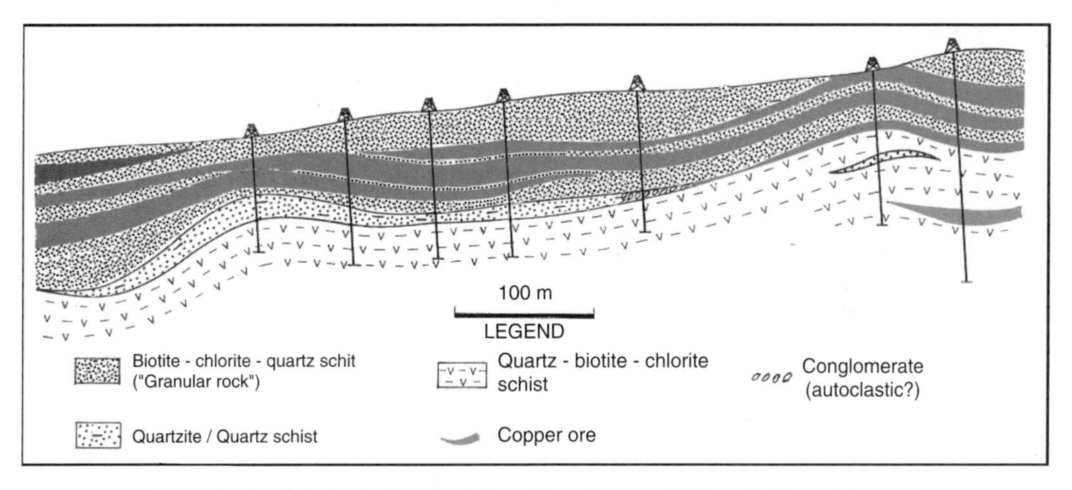

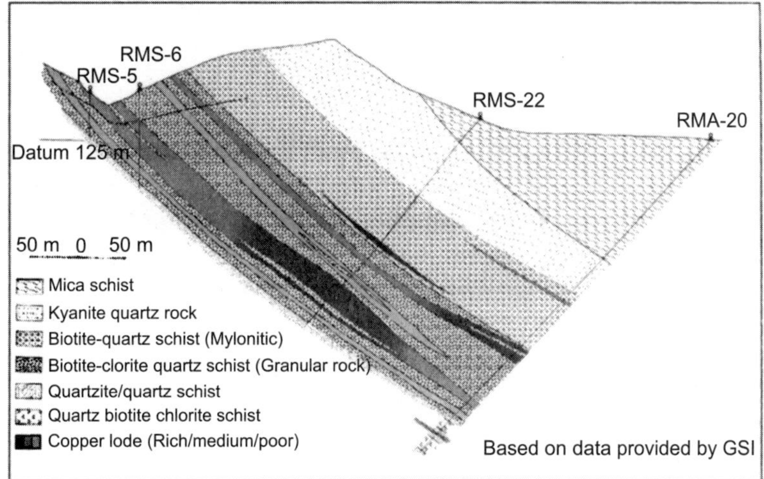

Fig. 4.2.9 (a) Longitudinal vertical section of a part of Rakha Mines deposit showing strike wise lode disposition and (b) Transverse geological section along boreholes in Roam–Sidheswar block showing subsurface lode correlation along the dip (after Sarkar, 1984).

At Ramchandrapahar, there are several ore bodies with limited strike and depth extension. The reserve estimate of this deposit is 1.7 Mt with average grade of 1.5% Cu down to a depth of 130 m.

The mineralised zone at Bayanbil has a strike extension of 1200 m. GSI estimated ore reserve of 1.36 Mt × 1.74% Cu (at 1.0% cut-off) for about 100 m depth. This was later upgraded by MECL to 9.82 Mt × 1.49% Cu (under 1.0% cut-off) by deeper drilling in the deposit.

Nandup is the western most deposit in this stretch where mineralisation was traced for nearly 1000 m. Old working is seen here in form of an incline. Reserve estimated by GSI was 4.0 Mt × 1.29% Cu.

Turamdih–Dhadkidih–Chhota Jamjora This is a very significant mineralised sector where both copper and uranium concentrations are noted. Apatite-magnetite mineralisation also occurs here. The sulphides and apatite–magnetite are hosted by chlorite quartz schists while the uranium mineralisation is in the chlorite–sericite–quartz schists and feldspathic schists. There are several ancient mine pits and inclines in this area.

Turamdih In Turamdih area there are two distinct horizons of chlorite–quartz schist forming strike ridges with subordinate sericite and minor biotite, tourmaline and epidote. The northern horizon extending from Nandup in the east to Keruadungri in the west is characterised by predominant apatite–magnetite mineralisation, while the southern horizon passing through Turamdih and Dhadkidih is marked by significant concentration of copper sulphides (Fig. 4.2.10).

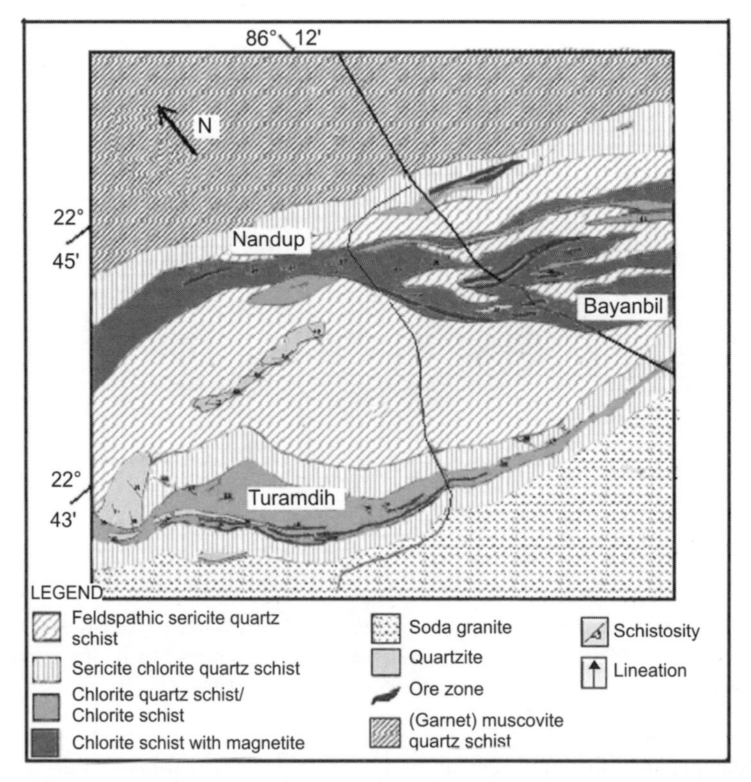

Fig. 4.2.10 Geology of Turamdih–Nandup–Bayanbil area showing Cu-sulphide ore zones (modified after Basu et al., 1979)

There is an ancient incline at the top southern flank of Turamdih ridge. The intervening area lying between the two chloritic schists comprises a thick sequence of banded feldspathic sericite–chlorite–quartz schists with quartzite and mylonitised quartzo-feldspathic bands. The ore bodies along with the enclosing rocks are conspicuously deformed by two sets of folds with E–W and N–S axes, resulting into interference patterns with dome and basin structures. The surface and subsurface structural patterns displayed by the ore body are excellently brought out in this deposit by integrating large number of drillhole data (Basu et al., 1979). The near surface dip of the ore body is 45–50° due north but there is a gradual shallowing of dip in depth with ultimate indication of a reversal of dip direction, depicting an open synform with roughly E–W axial trend. Due to the absence of any large scale reversal of dip direction on surface Basu et al. (op. cit.) inferred the phenomenon as a cumulative effect of small scale southerly vergent asymmetric folds, causing significant shallowing of the ore body at depth (Fig. 4.2.11 a, b).

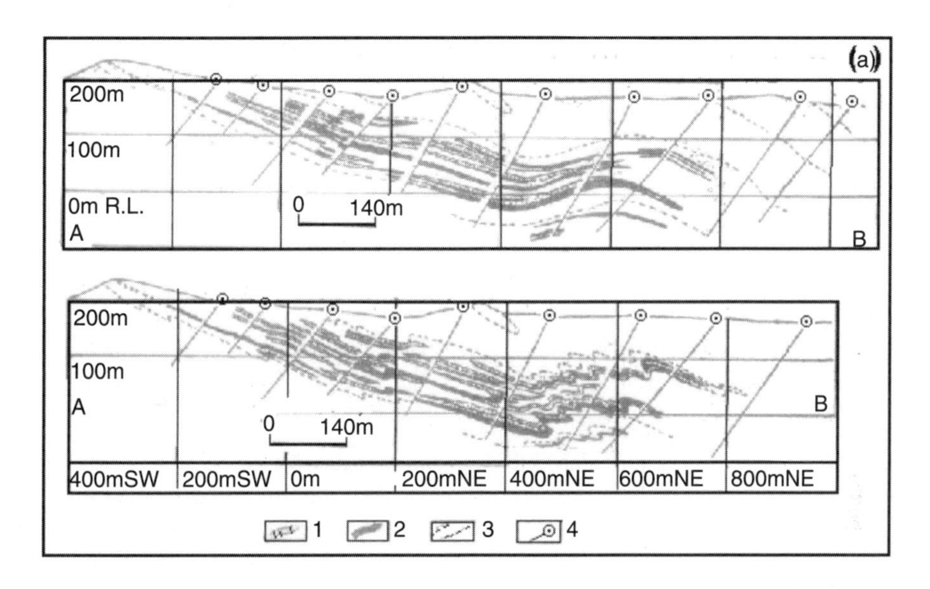

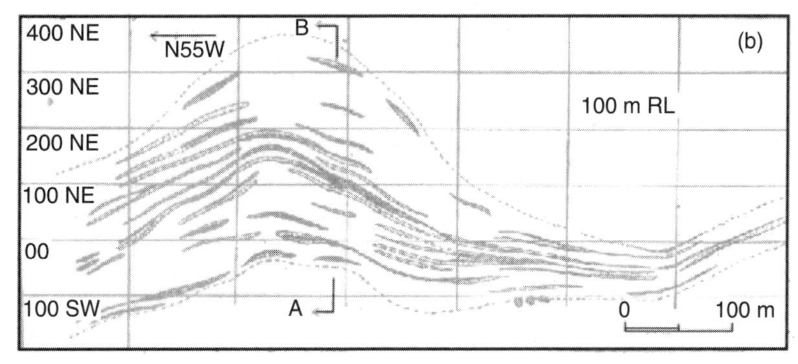

Fig. 4.2.11 (a) Transverse borehole section along A–B in (b), depicting depth-wise lode correlation, Turamdih copper deposit: 1. Cu-lode < 1%, 2. Cu-lode > 1%, 3. Enveloping surface of lode zone, 4. Borehole. (b) Lode correlation plan at 100 m RL, Turamdih copper deposit. Index same as in (a) (modified after Basu et al., 1979).

The strike length of the ore zone is 1300 m with 50 m width, in which there are three sub-parallel lodes. The central lode extends for 1200 m along the dip direction in the middle and upto 500 m in the two ends. The ore reserve is 17.85 Mt with average grade of 1.57% Cu. The orebodies are parallel to the regional schistosity, within which the richer shoots are controlled by a set of late shear with shallower dip (Lahiri et al., 1973).

At one point of time, the deposit was considered for open cast mining for winning both copper and uranium (in the hanging wall of copper lode). However, the Hindusthan Copper Ltd. (HCL) relinquished the leasehold of this deposit after remaining dormant for long years and ultimately the Uranium Corporation of India Ltd. (UCIL) took over the deposit and developed it for uranium ore mining without handling the copper ore in the footwall.

Dhadkidih The Dhadkidih deposit is located to the west of Turamdih deposit with a gap of about 1 km. The copper sulphide dominant mineralisation in chloritic schists is traced for 1500 m, in which five lodes varying in width from 1.2–10 m (with 1.06–2.98% Cu) have been delineated. The reserve estimate stands at 3.18 Mt with an average grade of 1.42% Cu (Chakraborty et al., 1984).

Chhota Jamjora This is the western-most prospect in this sector, which drew recent attention of GSI mainly because of old underground mining history pertaining to this area. The prospective area lies in the west of Dudra village and constitutes a flat soil covered ground with small and detached exposures of chloritic schists and haphazardly strewn old slag mounds. Four boreholes drilled at 200 m strike spacing encountered 2.25–5.43 m thick mineralised zone with 1.0–1.5% Cu. The ore mineral assemblage consists of pyrite, chalcopyrite, covellite and arsenopyrite in decreasing order of abundance. The host rock is sheared chlorite schist. The prospect and its extensions were not explored in detail.

Other Occurrences Besides the deposits and prospects described above, there are many more occurrences all over the belt both in the southeast of Mosabani–Badia and in the west of Turamdih–Chhota Jamjora sectors. In the western sector some of the explored blocks are Tamadungri, Kharsawan, Galudih, Bara–Chakri, Itihasa and in the southern sector Mainajhari, Gohala–Kanyaluka and Khadandungri–Andharika–Khejurdhari. The ore zones are either fairly rich but discontinuous, or more or less persistent but poorly mineralised. GSI in the past decade has established an ore zone of 900 m length at Khadandungri with a small reserve of 0.8 Mt x 1.0% Cu (Kishore and Kumar, 1998).

General Characteristics of the Orebodies and Mineralogy of Copper Ores Ore bodies of copper in the Singhbhum belt generally occur in the form of sheets, conformable with the dominant planar structures, including both the litho-boundaries and schistosity. Within the orebody, where the mineralisation is rather poor, a reticulate structure may form by the occurrence of ore, both in the S- planes and C- planes. As already mentioned, the delineable ore shoots are elongate, with the long axis sub-parallel with the down dip lineation common in the belt.

The sulphidic ores in this belt are massive, braided, stringery, streaky or disseminated. Quartz, chlorite and biotite dominate the gangue minerals. In the Mosabani sector, where the host is Soda granite, the ore minerals are closely associated with quartz and sodic feldspar.

The ore minerals occur as massive patches or disseminations in Soda granitic rock. The mineralisation hosted by schistose rocks in all other sectors is mostly in form of stringers, streaks or elongate blebs parallel to the schistosity and /or shear cleavages. In such rocks in which the schistosity is not pronounced, the ore minerals occur as non-oriented grains and clots. In the conglomeratic rocks the ore minerals are concentrated in the matrix. Veins and veinlets of quartz have rich concentration of ore minerals within an orebody irrespective of the composition of the host.

Wall rock alterations do not display a typical pattern in most places and similar alterations are noted with or without ore mineralisation (Fig. 4.2.12 a–e). At Mosabani the ore zone is characterised by the development of phyllosilicates and vein quartz, though there are copper ore-bearing veins with little or no development of hydrous minerals.

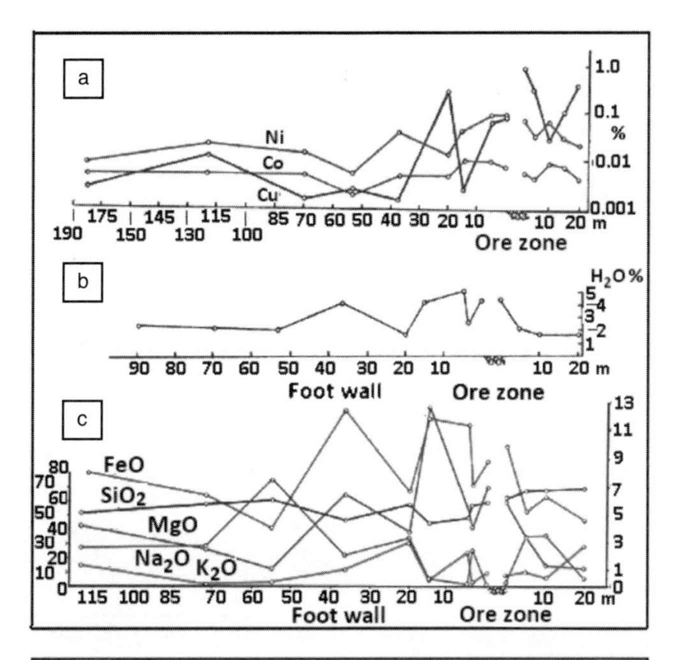

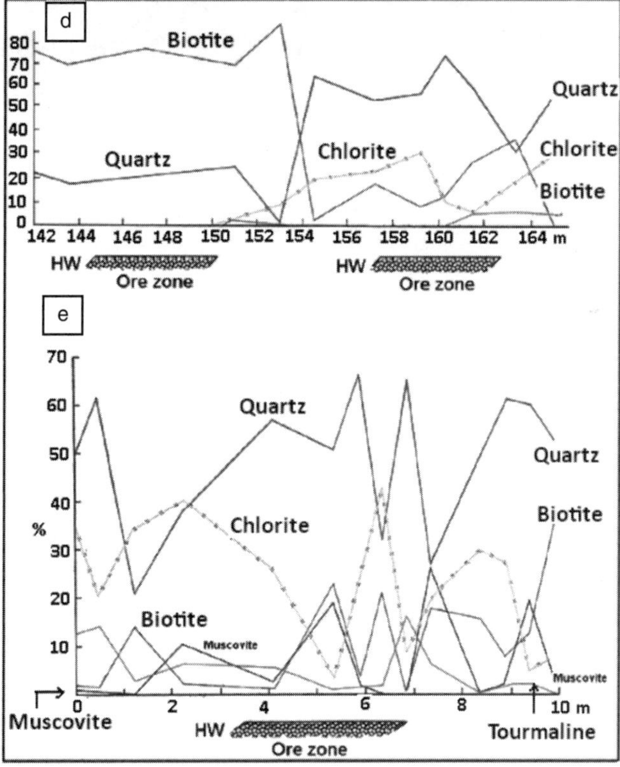

Fig. 4.2.12 Wall rock alteration in immediate vicinity of ore zones in (a–c) Mosabani–Badia and (d–e) Roam–Rakha Mines sectors of SSZ, displaying non-specific modal variation patterns unrelated to ore zones; (d) Results of core samples from RMS-27, Roam Block; (e) Cross-cut samples from Shaft No. 4, Rakha Mines (modified after Sarkar, 1984).

Throughout the belt the sulphides and associated ore minerals are deformed and metamorphosed together with the silicate hosts. Folds, puckers and kinks in ore bands are common and plastic flowage of ore minerals in late cleavages is also not rare. Single phase aggregates show polygonisation and development of straight boundaries triple points. Preferred orientation of elongate sulphide blebs parallel to schistosity is not uncommon. Mutual inclusion and mutual embayment of different phases are also indicative of metamorphic recrystallisation. Besides, there are prominent deformation twins in the sulphide minerals. At many places textural features suggest that deformation outlasted recrystallisation (Fig. 4.2.13 a–d).

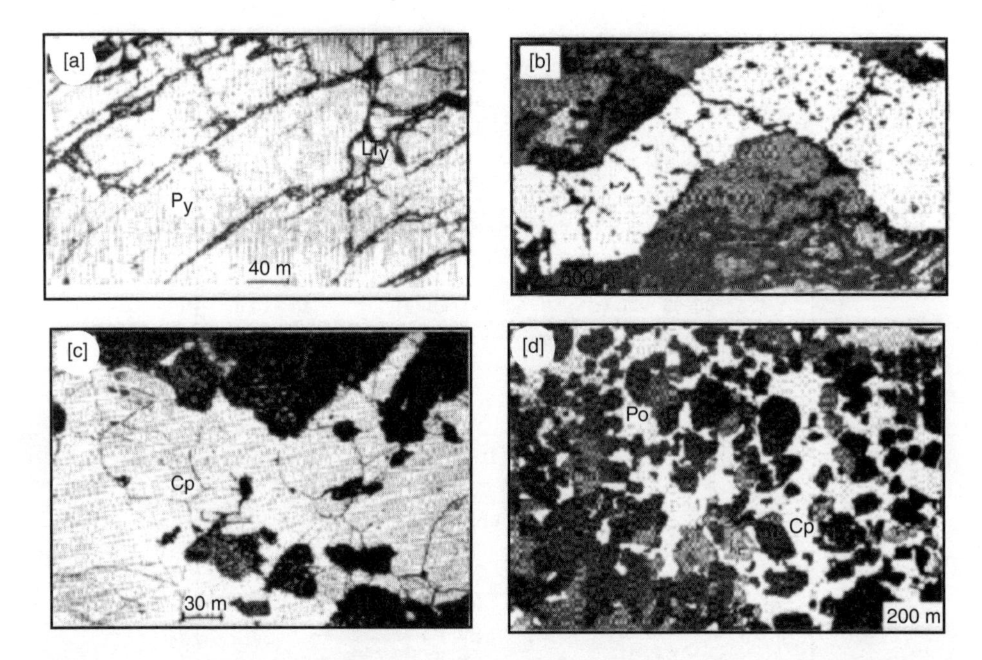

Fig. 4.2.13 Textures of recrystallised and deformed sulphidic ores in SSZ deposits. (a) Fractures in deformed pyrite (Py), cemented by chalcopyrite (Cp), Mosabani mines; (b) Folded pyrite (Py) grain with orthogonal fractures, Rakha Mines; (c) Polygonised chalcopyrite (Cp), Pathargora; (d) Granoblastic texture displayed by chalcopyrite(Cp), pyrrhotite (Po), magnetite (Mt), subhedral quartz (dark) and other gangue minerals, Rakha Mines (from Sarkar, 1984).

Very wide range of ore minerals (sulphides, oxides, tellurides, selenides, carbonates and native elements) have been identified in this belt by various workers (Dunn, 1937; Sarkar, 1984 and references therein). The principal sulphides are pyrite, pyrrhotite and chalcopyrite. The other minor sulphide phases include pentlandite, violarite, millerite, skutterudite, molybdenite, cubanite, sphalerite, bismuthinite, mackinawite, valeriite, galena, parkerite, wittichenite, gersdorffite, arsenopyrite, bornite and bravoite. In near surface ores chalcocite, covellite, marcasite, smythite are noted. The oxide minerals include magnetite, hematite, ilmenite, rutile and anatase, tenorite, cuprite, delafossite and uraninite and pitchblende. The tellurides are wehrlite, tetradymite and volynskite, and torbernite, malachite and azurite are the common phosphates and carbonates noted in surface exposures. Besides, in the oxidised zones there are native copper, gold, silver, bismuth and tellurium. Supergene alteration is generally not intense. $\delta^{34}S$ of the sulphides range between + 3.3 and + 7.2 (Fig. 4.2.14 a–d and Fig. 4.2.15 a–d).

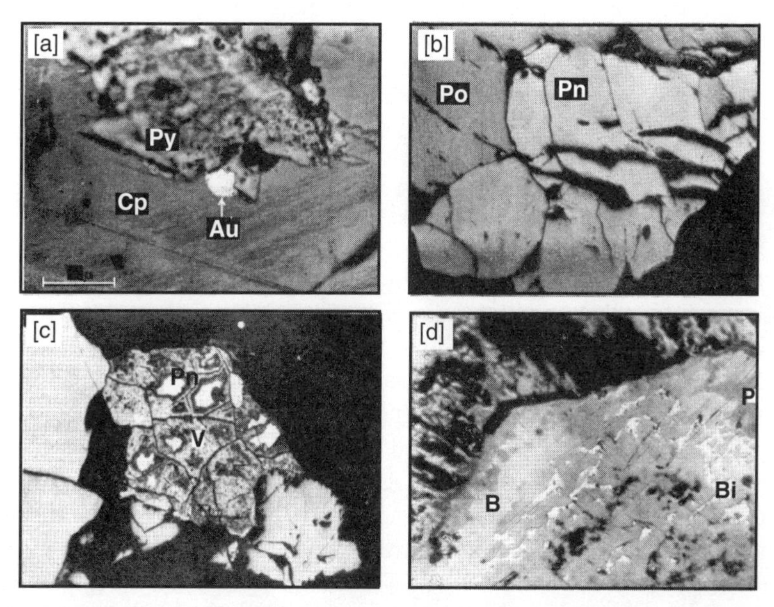

Fig. 4.2.14 Photomicrographs showing minor phases associated with Cu-sulphide ore along SSZ; (a) A grain of native gold (Au) in chalcopyrite(Cp)-pyrite(Py) ore, Mosabani Mines, (b)Stumpy pentlandite (Pn) grains with pyrrhotite(Po), Rakha Mines, (c) Pentlandite (Pn) relict left on replacement by violarite,(V), Mosabani Mines, (d) Perkerite(P) intergrown with bismuthinite(B) and natve bismuth (Bi), Rakha Mines (from Sarkar, 1984).

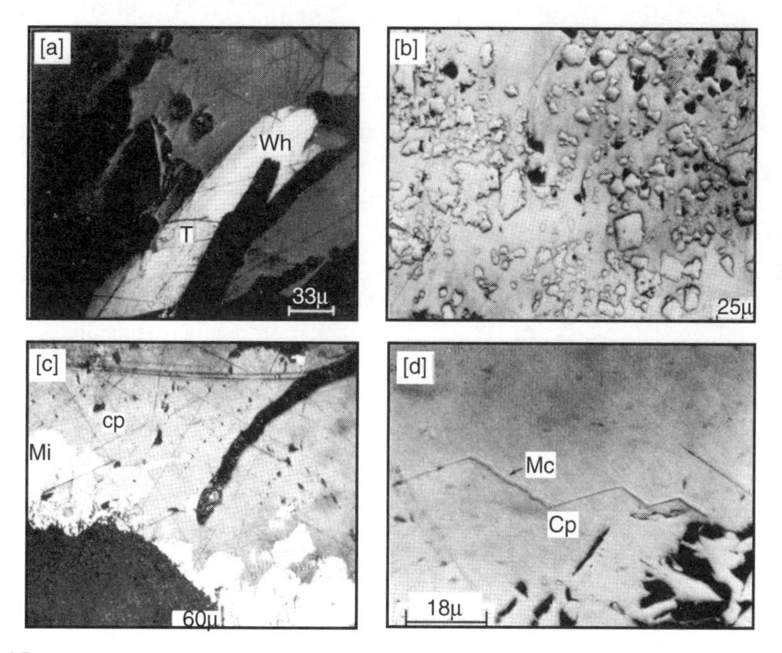

Fig. 4.2.15 Photomicrographs showing minor phases associated with Cu-sulphide ore along SSZ; (a) Tetradymite (T) and wehrlite (Wh), Mosabani Mines; (b) Subhedral to euhedral grains of skutterudite in chalcopyrite, Rakha Mines, (c) Millerite(Mi) replaced chalcopyrite (Cp) along the grain boundary, Mosabani Mines; (d) Mackinawite(Mc) in chalcopyrite(Cp), Rakha Mines(from Sarkar, 1984).

Recently, Pal et al. (2009) published an interesting paper on the pyrite textures and compositions in the Turamdih U-Cu (-Fe) deposit and commented in this context on the ore mineralisation along the entire belt. They sequenced various phases of mineralisation as follows:

1. magnetitic Fe (-Ti-Cr) oxide and Fe-Cu(-Ni) sulphide minerals inferred to be magmatic in origin
2. deposition of hydrothermal uranium, Fe-oxide and Fe-Cu(-Ni) sulphide minerals, predating most/ all ductile deformations.

Pyrites of more than one generation occur in the above associations. The oldest of the pyrites (Pyrite-A) occurring as globular-semiglobular composite inclusions of pyrite + chalcopyrite in magnetite is suspected to be of magmatic origin. This pyrite contains up to 3.07% Ni and low Co/Ni ratios (0.01–0.61). Ni-bearing pyrite is reported from Jaduguda also (Sarkar, 1984). The pyrites of second generation (Pyrite-B) are elongate, with Co-rich cores (up to 2.36% Co) and high Co/Ni ratios (7.2–140.9) have been interpreted to be hydrothermal in origin predating ductile deformation. The third or the last generation of pyrite (Pyrite-C) postdates ductile deformation and contain low Ni- and Co-contents. The study indeed added some valuable information to the ore mineralogy of the belt. However, the fact that the Ti-poor magnetite here contains all the high-Ni pyrite (Pyrite-A), quartz and chlorite, suggests that none of the phases are of magmatic origin. The content of Ni and Co in pyrite mentioned above can occur in hydrothermal pyrite in Fe-Ni-Co-S system. Chromite and Ti-rich magnetite grains could be accessory phases from the mafic protolith. Turamdih is virtually a spot in the long mineral belt and more such studies are warranted in other localities as well before drawing a generalised conclusion for the entire belt.

Uranium Mineralisation along the Singhbhum Shear Zone, Basal Dhanjori and the IOG

As in the case of copper, the Singhbhum shear zone (SSZ) is the largest repository of uranium ores in the country. Discovery of mineable ores in the belt dates back to early fifties of the last century, when India became an active member of the movement, ‘Atom for peace’, and embarked upon an ambitious programme of nationwide search for uranium. In that programme the Singhbhum shear zone came in handy. It was an already known metallotect and moreover there were already some indications of the presence of uranium in the belt. Fermor noticed incrustations of ‘uranium mica’ on the apatite-magnetite ores at Sungri (in Hayden, 1919, p.14). Pascoe (1930) reported some uranium-bearing incrustations, i.e. of secondary minerals, on apatite-magnetite ores from nearly the same area. Dunn (1937) identified a mineral as torbernite that was collected from the belt and handed over to him by E. O. Murray, a freelance prospector. In the same report, Dunn recorded that Murray collected a mineral from the apatite deposits of Kanyaluka which ‘resembled xenotime’. This preliminary information led to intense surface radiometric surveys, followed by drilling and assaying of drill-cores from selected areas. These intense exploration activities by the Department of Atomic Energy, Govt. of India, led to the discovery of numerous deposits of uranium ores in the belt (Bhola, 1965; Sarkar, 1984; Mahadevan, 1992). Important amongst these finds were, from east to west, Khadandungri (Dhantuppa), Purandungri, Bhalki–Kanyaluka, Bagjata–Moinajharia, Badia–Mosabani, Surda –Rakha Mines, Jaduguda –Bhatin, Narwapahar, Turamdih–Keruadungri and Mohuldih (Fig. 4.2.16). Geological reserve of uranium (U_3O_8) in Singhbhum belt is about 45,000 tonnes.

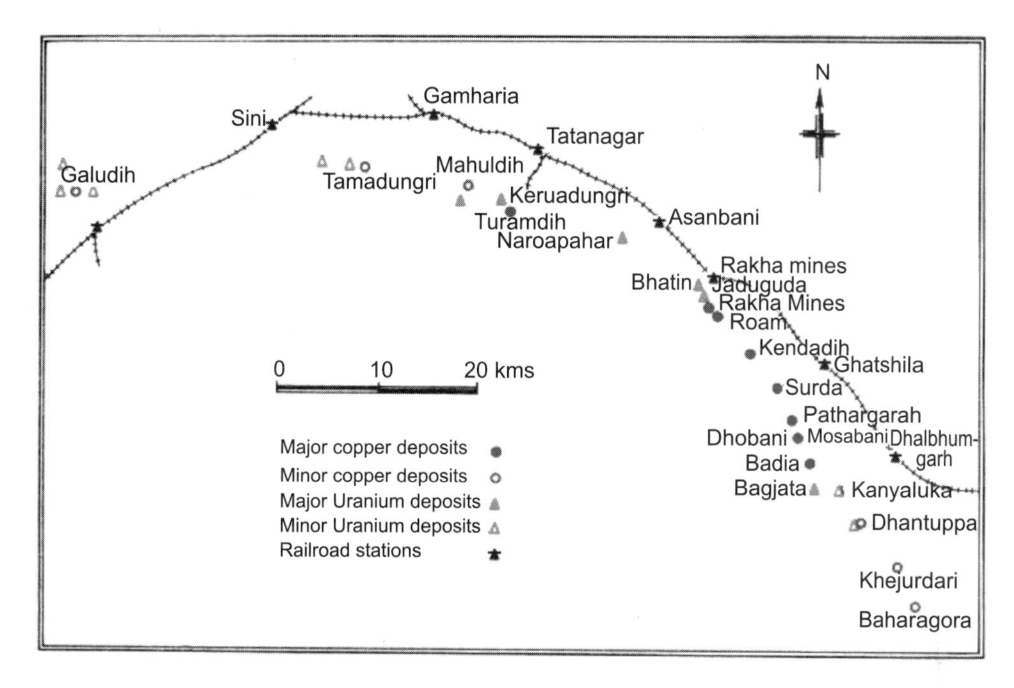

Fig. 4.2.16 Distribution of copper and uranium deposits along Singhbhum shear zone

At present mining is going on at Jaduguda, Bhatin, Narwapahar, Turamdih and Bandhuhurang–Keruadungri and development work is being carried out at Mohuldih in the west and Bagjata in the east. After outlining the geology of most important deposits from SE to NW, we will briefly discuss the characteristic features of this type of mineralisation in the belt, followed by comments on the genesis and evolution of the same.

Major deposits

Bagjata deposit At Bagjata, about 10 km south of the former Mosabani mining township, two uranium ore bodies, more than 600 m in strike lengths and occurring in quartz chlorite–biotite schists, have been proven upto a depth of 600 m. Uranium Corporation of India Ltd (UCIL) has undertaken a plan to construct an underground mine there with an inclined and a vertical shaft upto a depth of 300m. The ores produced from this mine will be treated at the Jaduguda Mine-Mill Complex, about 25 km away by road towards the north west and the deslimed mill tailings will be used for the mine-fills. The project has been commissioned in 2008.

Mosabani–Rakha Mines sector Overlapping of copper and uranium mineralisations is not rare in nature, though this relationship can not be generalised. It has been stated that in the USA alone, about 600,000 tonnes of U_3O_8 were recoverable as by-product from the copper ores. In the Singhbhum belt, tailings from the treatment of copper ores from Badia and Mosabani mines contained an average of 0.012% U_3O_8 (Bhola, 1965). At places, bulk samples assayed U_3O_8 as high as 0.09%. In the copper-rich sector of Surda to Rakha Mines, small lenses of uranium ores assaying upto 0.1% U_3O_8, were found close to the copper lodes. Three plants to treat the copper ore tailings for their uranium content were set up by the Department of Atomic Energy at Mosaboni, Surda and Rakha mines. But with the closure of copper mines these plants had to be closed down.

Jaduguda deposit Jaduguda is the best uranium ore deposit in the belt in terms of the total reserve of the ore, if not for the percentage of the U_3O_8 -content. The host rocks for mineralisation comprise apatite–magnetite–biotite–chlorite bearing quartz schist (the 'granular rock'), brecciated quartzite, conglomerate and biotite–chlorite schist. Two orebodies, about 60 m apart across the width occur at Jaduguda. They are referred to as the Footwall lode or Central Jaduguda lode, and the Hanging wall lode or Eastern Jaduguda lode. The Footwall lode which is 5–6 m thick in the upper levels, attains a thickness of over 20 m at about 300 m below the surface and then gradually becomes thin again (Fig. 4.2.17). The Hanging wall lode, which is generally 2–5m in thickness does not continue below a vertical depth of 600m, while the Footwall lode continues down to a vertical depth of 800 m. The dip of the ore bodies is about 40° (northerly) which steepens to about 60° at about 300 m depth and then flattens again to about 40° below about 400 m depth. The average ore grade at Jaduguda is 0.06% U_3O_8. Present production of the mine is about 700 tonnes per day. The mine started towards the end of the sixties (1967) and at the present rate of production is likely to continue for another 10–12 years.

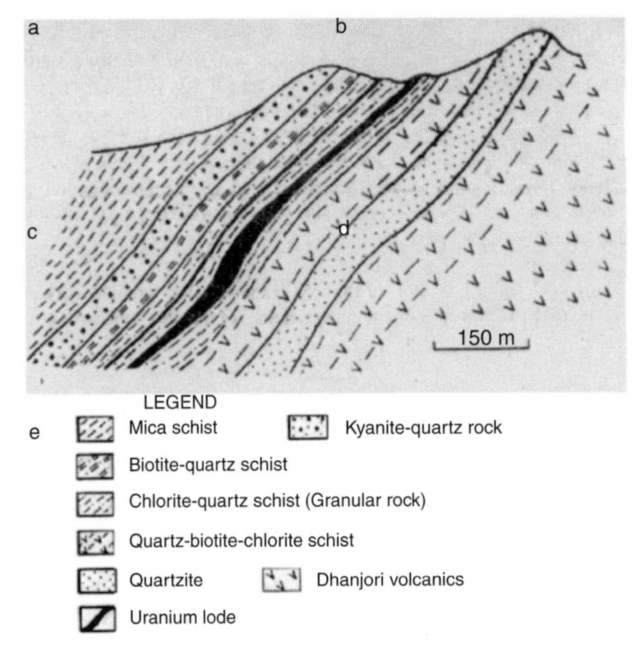

Fig. 4.2.17 Transverse geological section across Jaduguda deposit showing disposition of the two parallel U-lodes (from Sarkar, 1995)

At about 200 m west of the footwall of the Jaduguda ore zone, mineralisation of uranium together with associated minerals occurs at Jublitola. The country rock, which is Dhanjori basic volcanic in general, is transformed with the conspicuous development of such phyllosilicates as chlorite and biotite and introduction of quartz. Uranium mineralisation is mainly located in the chlorite–biotite and quartz–biotite schists. Metallic minerals other than uraninite (± pitchblende) include magnetite, ilmenite, rutile, pyrite, chalcopyrite and molybdenite (>100 ppm) locally apatite and tourmaline may be conspicuously present. The mineralisation is in general disseminated, with local fracture filling. The temporal relationship of uraninite with the associated sulphides is not unequivocal. As far as the authors are aware, the deposit is not in the priority list for mining in foreseeable future.

Bhatin deposit Bhatin is the next mining area for uranium at about 3 km to the northwest of Jaduguda. The two are separated by an oblique fault (Tirukocha fault) that has shifted the ore zone towards the hanging wall side by more than 500 m. Bhatin shares many features with Jaduguda but the deposit is small. There are two uranium ore bodies here in brecciated quartzite and granular rock. The mineralised zone has been traced over a strike length of 1300 m, but less than half of it is better mineralised. Sheet like ore bodies dip northerly. The ores are processed at Jaduguda. The mine was commissioned is 1986.

Narwapahar deposit The Narwapahar deposit is located at about 12 km to the north-west of Jaduguda. The underground mine here was commissioned in 1995. Although the average grade of the ore is somewhat lower than that of Jaduguda (0.04–0.05% U_3O_8), it is supposed to be the largest repository of uranium in the belt, if not in the country today.

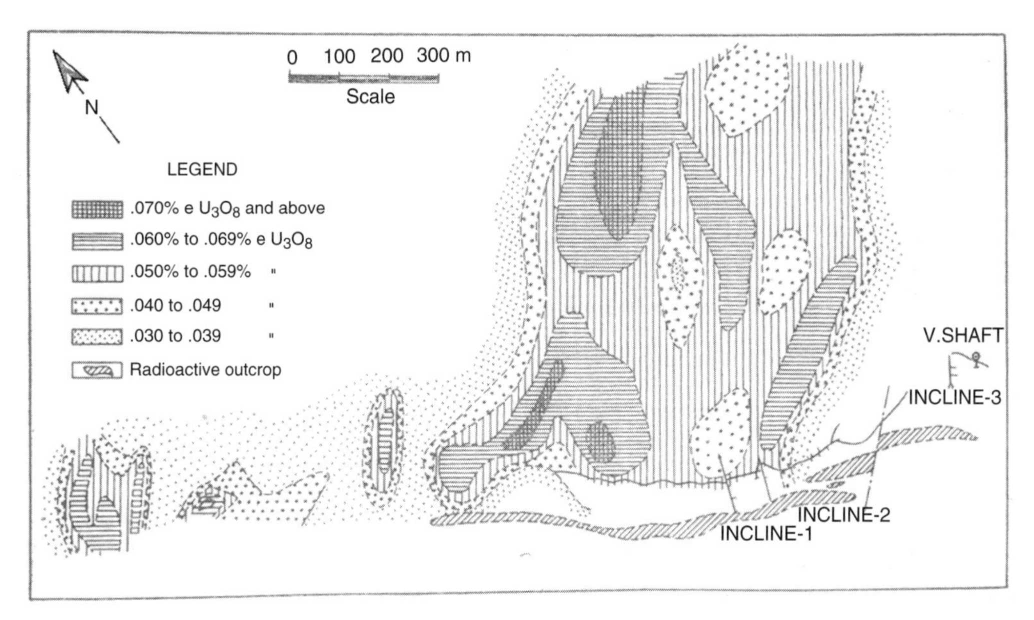

Fig. 4.2.18 Plan projection of the isogrades of uranium ore body on outcrop map showing near down-dip disposition of the ore shoots, Narwapahar (Sarkar, 1984)

The host rock is sericite chlorite schist with abundant albitic feldspar at places. Veinlets of fluorite and carbonates are not very rare in the ore zone. There are four sub-parallel ore zones at Narwapahar of which the footwall lodes are mineable. The mineralisation has been confirmed to a depth of about 500 m from the surface. The footwall-most orebody, averaging 5 m in thickness, is more persistent and is generally referred to as the Main lode. The isogrades delineated on assay plans (Fig. 4.2.18) is interesting and may be considered as generally representative of the whole belt. The mineralisation at Banadungri is likely to be the western extension of the deposit. Veinlets of scheelite have been noted in some drill-cores.

Turamdih–Bandhuhurang–Keruadungri sector Turamdih–Bandhuhurang–Keruadungri area, located at 5–6 km south of the Tatanagar Railway Station, is another area of interest in this belt with respect to uranium mineralisation. There are several ore bodies in the area (Fig. 4.2.19). The host rock is sericitised quartz–chlorite schist, often containing lenses of quartz and augens of albite.

The ore zone is about 800 m long, but the ore grade is low (~ 0.04% U_3O_8). Underground mining has started here in 2003. Banduhurang deposit is in fact the western extension of orebody at the Turamdih mine. The deposit has a large volume but low tenor. However, being located close to the surface it will be amenable to open cast mining and the mine has been commissioned here in June, 2007.

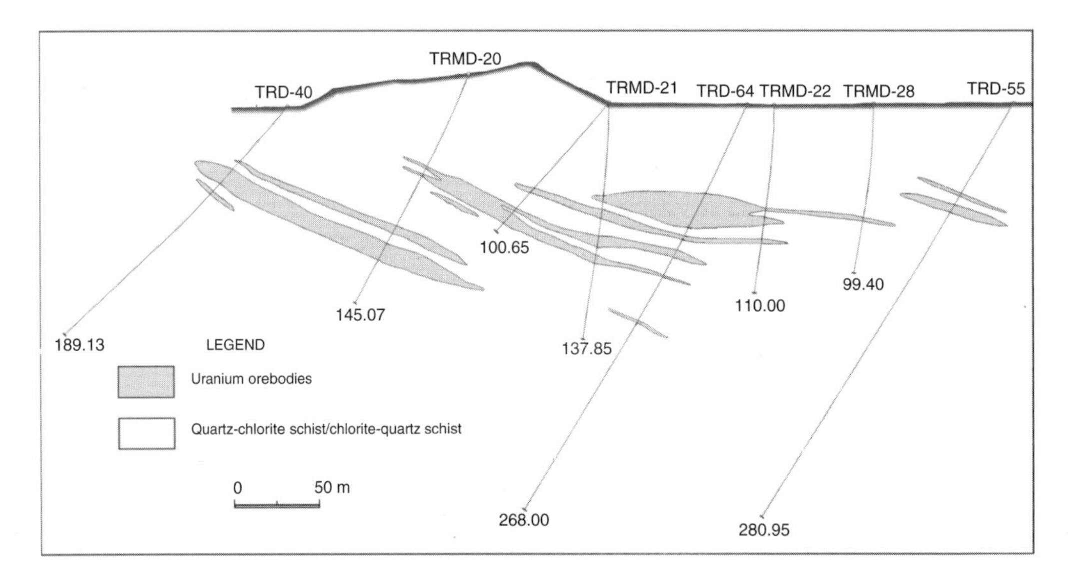

Fig. 4.2.19 Cross-section along drilled boreholes showing disposition of uranium ore bodies, Turamdih–Bandhuhurang–Keruadungri area (courtesy, UCIL)

Mahuldih Deposit Mahuldih deposit is located about 3 km west of Turamdih mine. Two prominent ore bodies in this deposit are hosted by tourmaline bearing quartzite and quartz schists passing through the south (footwall) of the copper sulphide bearing chlorite schist horizon of Turamdih. The mineralised zone is about 500 m long. The UCIL started opening up an underground mine here and it is expected to be productive in 2011. The ores produced from this mine along with those produced from Turamdih and Banduhurang mines will be treated in the processing plant constructed at Turamdih.

General Characteristics of the Uranium Ores and Ore Mineralogy Mineralisation of radioactive elements in the SSZ is not lithology-controlled, a fact borne out by the discussion in the preceding paragraphs. The ore bodies, however, are invariably sheet-like, which are parallel with the schistosity in the country rocks and the interface of the lithounits, the last two being generally parallel/sub-parallel. Ore bodies may have sharp to gradational (assay) contacts, generally the latter. Delineable ore shoots trend nearly down-dip (except having much shallower pitch in Turamdih area) as do the shoots of the copper ores in the belt. No definite wall rock alteration pattern is discerned around these deposits. At Jaduguda, Bhatin and Narwapahar mines quartz and albitic grains reddened with dusts of hematite are locally seen as disseminations or veinlets. But it is difficult to correlate the intensity of hematitisation with the intensity of uranium mineralisation.

The radioactive minerals containing uranium (and thorium) reported from the Singhbhum belt (Sarkar, 1984 and the references therein) comprise:

1. uraninite $(U^{4}_{1-x}U^{6+}_{x})O_{2+x}$
2. shooty pitchblende $(U^{4}_{1-x}U^{6+}_{x})O_{2+x}$
3. complex U-mineral U-Ti oxide ('pronto ore') $(UO_{2+x} + TiO_2)$
4. allanite $(Ca,Ce,U,Th,Y)_2(Al,Fe,Mn, Mg)_3 (SiO_4)_3(OH)$
5. xenotime $(Y,U,Th)PO_4(Ce,La,Th)PO_4$
6. monazite $(Ce,La,Th,Nd)PO_4$
7. davidite $(Fe^{2+},U,Ce,La)_2(Ti,F^{3+},Cr)_5O_{12}$
8. pyrochlore-microlite $(Na,Ca,U,Ce)_2(Nb,Ta)_2 O_6 (O,OH,F)$
9. clarkeite $(Na,Ca,Pb,Th)_2U_2(O,OH)_7$
10. autunite $Ca\{(UO_2)_2PO_4\} nH_2O$ (meta-autunite)
11. torbernite, $Cu(UO_2)_2(PO_4)_2 nH_2O$ (meta-torbernite)
12. schoepite $UO_3 2H_2O$
13. uranophane $Ca(UO_2)_2 Si_2O_7 \cdot 6H_2O$

Of these, uraninite is the principal source of uranium in the belt. Rao et al. (1999) have reported the presence of gold in sub-ppm level at the Jaduguda and Narwapahar mines.

Uraninite in the Singhbhum ores is usually disseminated. About 75% of the grains of uranium occur in the interstices of the gangue silicates and about 25% within these minerals (Fig. 4.2.20). Uraninite grains are found replaced by quartz and sulphides, when the latter occur in association. Galena veinlets in uraninite were possibly formed by the sulphidation of radiogenic lead. Singhbhum uraninite is low-thorium, high-lead and has low to moderate content of REE (Rao and Rao, 1980a) (Table 4.2.2).

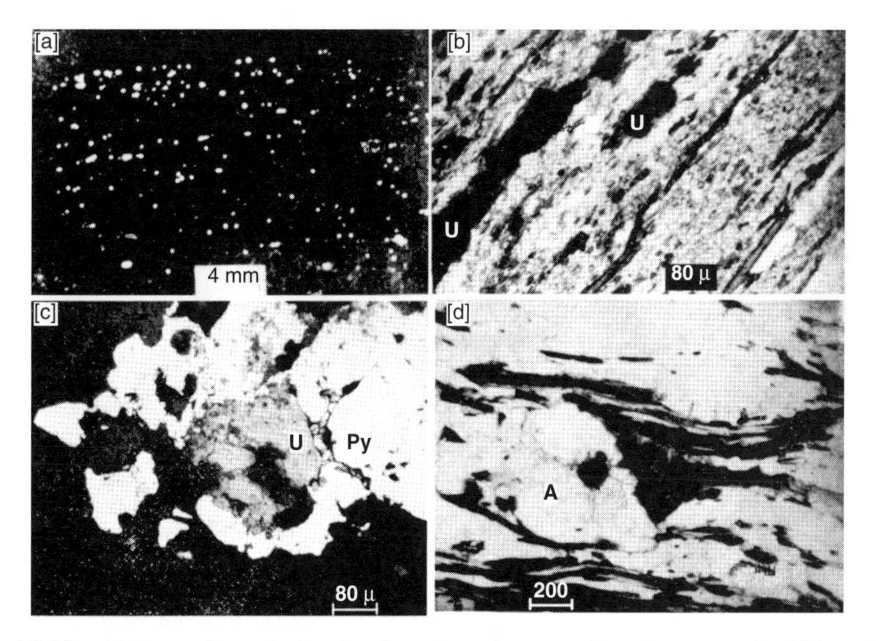

Fig. 4.2.20 (a) Autoradiograph showing disseminated nature of uranium ore in Singhbhum belt; (b) Trails of haloed uraninite grains (U) in mylonitic chlorite quartz schist, Jaduguda; (c) Uraninite (U) surrounded by pyrite (Py); and (d) Haloed uranium (U) in apatite, Jaduguda (Sarkar, 1984)

Table 4.2.2 *Composition of uraninite from the Singhbhum belt (Rao and Rao, 1980,a)*

Oxides (%)	Bhatin	Narwapahar-A	Narwapahar-B	Keruadungri
UO_2	55.06	54.08	49.51	43.10
UO_3	24.69	25.87	29.44	33.57
ThO_2	0.19	0.77	0.60	0.30
ZrO_2	0.45	0.17	0.14	0.20
$(REE)_2O_3$	5.07	4.32	4.46	8.16
CaO	0.92	0.89	1.15	1.13
PbO	13.62	13.90	14.70	13.54

Shankaran et al.(1970) analysed the Zr and REE contents of Jaduguda uraninite to be as follows: Zr 0.17%, Y-earth 3.59% (Gd 0.91%, Tb 0.26%, Dy 0.29%, Er 0.20%, Yb 0.48%, Y 1.45%) and Ce-earth 2.19% (Ce 0.88%, La 0.42%, Pr 0.2%, Nd 0.49%, Sm 0.2%).

Xenotime occurs at many places of the belt. But its maximum concentration is at the Khadandungri–Kanyaluka–Bhalki area. At Kanyaluka, xenotime crystals (Fig. 4.2.21), as large as 1 cm or more, may be found together with apatite and magnetite in the biotite schist (Sarkar, 1970). Rao (1977) reported an interesting observation in which tiny crystals of xenotime formed bands in quartzite, the individual grains being oriented parallel to lithological banding. A Kanyaluka xenotime analysed to Y-group REE 60%, Ce- group REE 0.5%, P_2O_5 36.4%, U_3O_8 0.14%, ThO_2 0.29%, ZrO_2 0.5%, SiO_2 0.89%, Fe_2O_3 0.69%, CaO 0.057% (RaO, 1977). Notable, the U_3O_8 /ThO_2 ratio here is reverse of what is normally found in xenotime.

Fig.4.2.21 Xenotime from Kanyaluka. Length of James clip–1.5 cm (authors' collection and photograph)

Nickel–Molybdenum (–cobalt) Mineralisation at Jaduguda and Bhatin In the Jaduguda Mine, close to the 'Foot-wall lode' and also at the Bhatin mine, nickel occurs as millerite, ferroan heazlewoodite, melonite, Ni-bearing pyrite, biotite and chlorite. Locally the total nickel content may be as high as 3% or more. Cobalt occurs in gersdorffite. At both Jaduguda and Bhatin, molybdenite is found thinly disseminated. More importantly, a 15 cm thick shear zone has been mineralised with molybdenite at central Jaduguda within the uranium-rich zone (Fig. 4.2.22). Around the latter there are patches which contain $\geq$ 85% allanite (volume), the rest being made of quartz, albite, chlorite, sericite and magnetite.

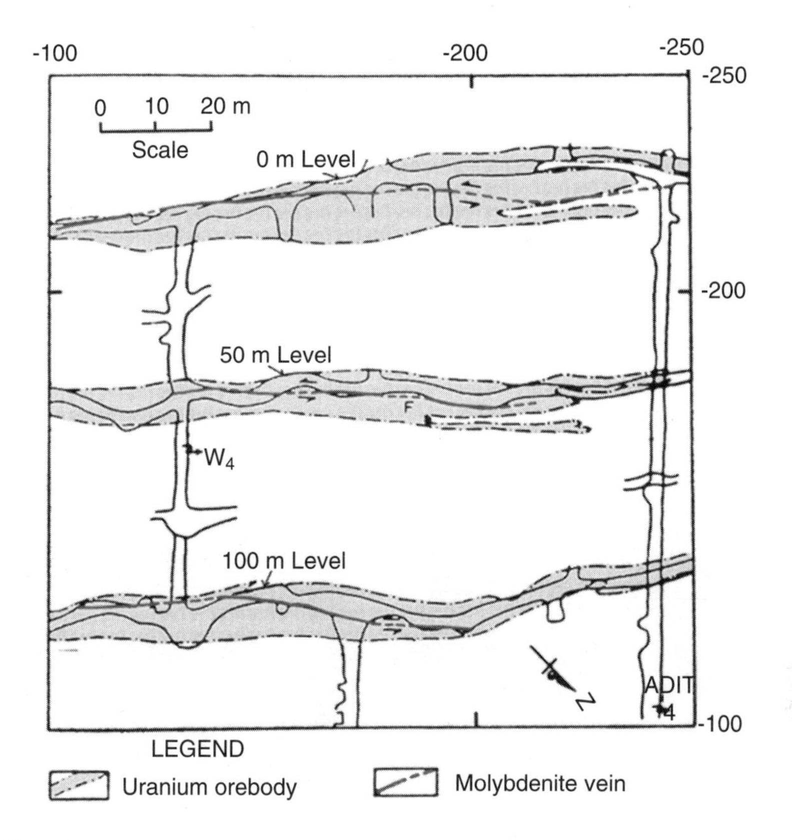

Fig. 4.2.22 Molybdenite vein within footwall uranium lode exposed at three levels of Jaduguda mines (after UCIL Mine Plan)

Age of Uranium Mineralisation Along the SSZ The radiometric age data obtained from the uranium ores in the Singhbhum belt are not convergent. A sample of apatite from Mosabani Mine and another magnetite sample containing uraninite from the Jaduguda Mine gave an age of 1950 Ma ($^{207}Pb/^{206}Pb$). Another sample of apatite gave an age of 1600 Ma on being analysed by both the Pb-U and Pb-Pb methods (Vinogradov et al., 1964; Tugarinov and Voitkevitch, 1970). Uraninite samples drawn from Narwapahar, Bhatin, Surda and Rakha Mines and analysed by both Pb-U and Pb-Pb methods produced an age range of 1500–1600 Ma (Rao et al., 1979). This broadly coincides with the first major metallogenic era of uranium on a global scale.

Uranium Mineralisation in the Basal Dhanjori and IOG Sediments Quartz pebble conglomerate (QPC) type uranium mineralisation has been reported from a number of places within the basal Dhanjori sediments (Das et al., 1988) and also in basal Iron Ore Group (IOG) sequence at Sayamba–Taldih area, Sundargarh district bordering the Bonai Granite (Misra et al., 1997). But the intensity of mineralisation at none of these places is high enough to qualify for exploitation. The age of mineralisation at the basal Dhanjori sediments may reasonably be considered as Late Archean to Early Paleoproterozoic. But for the Syamba–Taldih area, one has to be sure that the sequence is really IOG, before any age is assigned to the mineralisation. Could it be Darjing Group or its equivalent?

Phosphate (Apatite-Magnetite) Mineralisation in Singhbhum Shear Zone

Phosphate mineralisation occurs discontinuously along the Singhbhum shear zone, the first description of which was by Fermor in1908, followed amongst others by Dunn (1937). The easily accessible deposits along the Singhbhum shear zone are mined out. The present day exploitation of the mineral in the Eastern India is confined to the Purulia area only, across the Singhbhum mobile belt.

Distribution Phosphate (apatite-magnetite) mineralisation along the Singhbhum belt is concentrated along the following sectors:

1. Khejurdari–Dhantuppa–Gohala
2. Badia–Pathargora–Samaidih
3. Patharchakri–Kudada–Dudra

In the Khejurdari–Dhantuppa–Gohala sector phosphate mineralisation in the form of lensoid bodies of apatite–magnetite occurs associated with different rocks (Fig. 4.2.23). At Khejurdari the mineralisation is hosted by biotite schist. Elsewhere in the sector it is associated either with the Soda granite and/or chlorite–biotite schists, the petrology of which is briefly discussed in an earlier section.

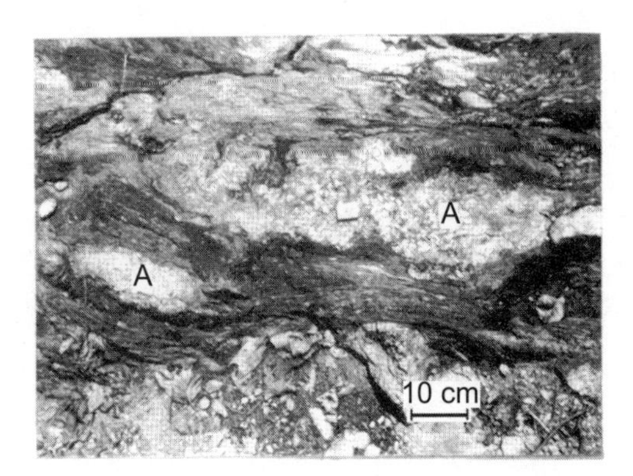

Fig. 4.2.23 (a) Lenses of apatite–magnetite (A) in the ore body of Dhantuppa; (b) An abandoned open cast apatite mine in Dantuppa area, indicative of the lensoid shape of the excavated ore (from Sarkar, 1984)

In the mine dumps at Dhantuppa, coarse apatite-magnetite ores contain serpentine/asbestos along grain boundaries or as veinlets. In others, orthoamphibole and phlogopite take the place of serpentine/ asbestos. Nests of apatite–magnetite occur in mica schist at Sungri. Another uncommon assemblage of quartz–kyanite–magnetite–apatite is observed at Gohala.

Biotite schists, east of Badia and Mosabani, contain pockets and lenses of apatite with varying proportions of magnetite. The mineralised zone is at the hanging wall side of the Soda granite/feldspathic schist. Similar is the situation in which this mineralisation has taken place at Rangamatia, east of the Pathargora mine shaft. The Pathargora–Samaidih part in this sector is much richer in this ore. At Pathargora and the neighbouring Ruhildih and Lutidhiri area apatite magnetite mineralisation took place along the interface of the feldspathic schists and the Dhanjori metabasites or their derivatives. At Dhobani, Tamajhuri, Kasidih and Samaidih the mineralisations are confined within Dhanjori metabasite. Apatite-magnetite mineralisation, though weak, has been noted in the mica schists of Chaibasa Formation at Barjudi, about 0.5 km away from the boundary of metabasite and mica schist (Sarkar, 1984).

There is another rich zone of apatite–magnetite mineralisation in the Patharchakri–Kudada–Dudra sector, located broadly to the south of Tatanagar.

The important localities of mineralisation in this sector are Patharchakri, Ramchandrapahar, Kudada and Dudra. The rocks hosting these deposits are:

Quartz–chlorite schists ± Soda granite/feldspathic schists

Ore Bodies and the Ores Wrapping around of pods or lenses of apatite-magnetite ores by the schistosity of host rocks is a common feature. Very rarely veins / veinlets of apatite-magnetite cut across the schistosity. Apatite–magnetite lenses seldom exceeded 100 m in strike, 10 m in thickness and 40 m in depth continuation. At places the apatite-magnetite ore bodies share the tectonic structures in the host rocks (Sarkar, 1984). Quartz veins and veinlets cut through the ore bodies at different angles. So does some chlorite-biotite rich veins and veinlets. Core of an apatite-magnetite vein is usually made of coarser grains.Small scale structures and textures commonly noted in apatite ore are presented in Fig. 4.2.24.

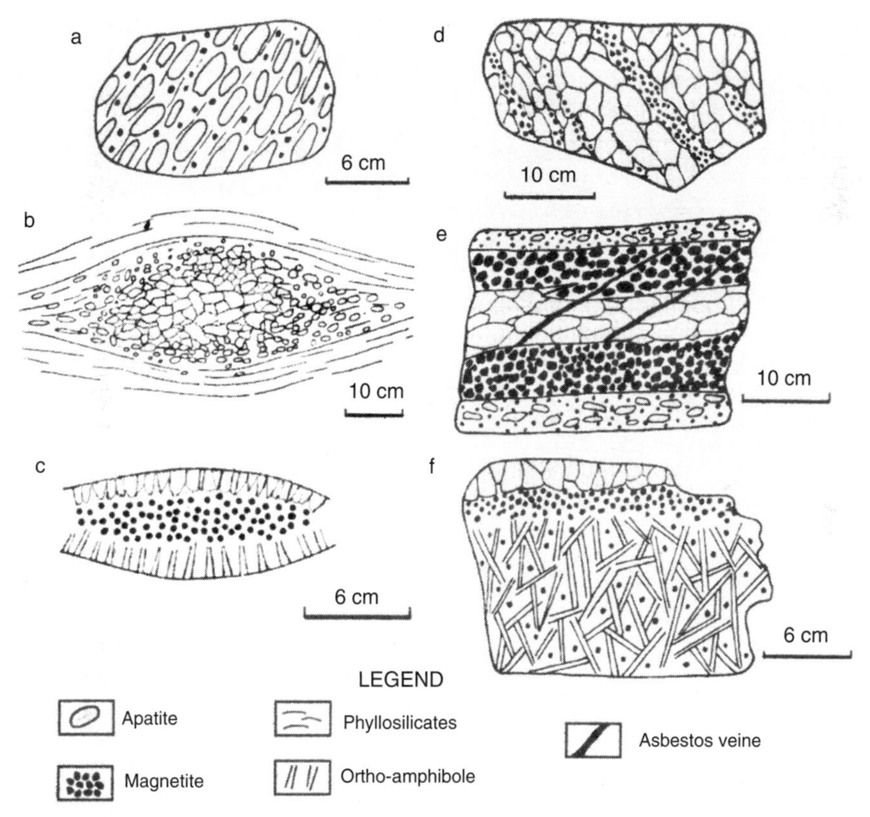

Fig. 4.2.24 Sketches of small-scale structures in apatite ore: (a) Grain orientation in dissiminated ore; (b) Apatite–magnetite lensoid with coarser grains in the core; (c) Apatite–magnetite lensoid with magnetite in the core and apatite as needles orthogonal to vein boundary; (d) Compositional banding in apatite–magnetite ore; (e) Monominerallic bands of magnetite and apatite are made up of coarser grains compared to mixed bands; (f) banded apatite–magnetite ore and a band with decussate orthoamphibole needles (after Sarkar, 1984).

Generally both apatite and magnetite are present in the ores, although in different proportions, which is reflected in their bulk composition (Table 4.2.3). In rare cases, one may be present to the exclusion of the other. Recently, in the mine dump of the now abandoned Badia Mine, massive magnetite with a little apatite has been noticed. The minerals other than apatite and magnetite present in these ores are biotite, chlorite, and quartz. Tourmaline, ilmenite, xenotime allanite and rutile may be present and as mentioned earlier and rarely serpentine, chrysotile, orthoamphibole may be present. Disseminated ores may show a sort of orientation of the apatite grains (Fig 4.2.24a). Sengupta (1972) in an interesting work showed that the optic axis orientation of apatite in ores from Pathargora showed a 'strong maximum parallel to the down dip lineation'. This appears generally true for the entire belt. Dunn's contention that the Singhbhum apatite is fluor-apatite has been confirmed by later workers also. Only a part of the F-site is occupied by (OH) and in some cases a still smaller part by Cl (Sarkar op. cit.). In some samples apatite grains are inclusion ridden. Inclusions identified in coarse grains are quartz, magnetite, allanite, xenotime and orthoamhibole. Fluid inclusions are apparently present but not studied. Chrysotile and serpentine where present, replace apatite. The length/ breadth ratio of the apatite grains from some of the deposits under review varies in the range of 1.2–2.5 (Sarkar, 1984), which is expected when the apatite crystals co-exist with liquid or vapour in the system, $CaO–CaF_2–P_2O_5–H_2O–CO_2$ (Willie et al., 1962). During deformation, individual apatite grains commonly yielded by grain rotation and fracturing. Locally, strain induced recrystallisation is also noticed, aided possibly by fluid focusing.

Table 4.2.3 *Composition (partial, in wt%) of representative apatite–magnetite ores from a few deposits along the Singhbhum belt (Sarkar, 1966b)*

Sample No	P_2O_5	Fe_2O_3	FeO	CaO	F	H_2O
1	13.49	9.42	4.02	17.51	1.08	1.01
2	27.74	3.58	4.88	35.60	1.50	0.22
3	17.82	30.12	1.99	26.61	1.40	0.40
4	38.88	1.97	0.72	50.86	1.89	0.21
5	33.74	3.18	0.58	44.86	1.59	0.61

Deposit locations: 1. Pathargora (close to copper mine shaft), 2. Dhobani, 3. Moinajharia, 4. Ruhildih, 5. Lutidhiri.

Rare Earth Elements (REE) are common minor/trace constituents of apatite. These may occur in cation (Ca) sites, adsorbed on the surface of the mineral grains, associated with crystal defects and may even occur as inclusions of REE-rich minerals. Fleischer and Altschuler (1969) concluded that apatite from mafic, ultramafic and alkaline igneous rocks were characterised by a higher content of Ce-group elements. Apatite from granitic rocks in contrast, has widely varying REE composition. Negative Eu-anomaly is generally accepted as an indicator of low fO_2, provided there was no dearth of Eu in the system. The apatite crystallised as early sedimentary apatite, i.e. phosphorite is generally low in REE content ($< 0.1\%$). Singhbhum apatite is high in Ce-content and has a district negative Eu–anomaly. Dy and Yb contents also are high (Table 4.2.4 and Fig. 4.2.25). $(REE)_2O_3$ in the Singhbhum apatites ranges between 0.167–0.39 wt% (n = 6), in some samples (n = 2), however, it is below detection limit. Magnetite is very poor in TiO_2, V_2O_3 and MgO. These are always < 0.5 wt% except in the apatite-poor specimens from Dhantuppa where these could be 1–2 wt% (Sarkar, 1984).

Table 4.2.4 *REE contents (ppm) of Singhbhum apatite (Rao and Rao, 1980b)*

Elements	1	2	3	4	5	6	7	9
La	12.2	100.0	146.0	292.0	571.0	95.7	42.2	100.0
Ce	324.0	799.0	885.0	1082.0	1582.0	721.0	343.0	599.0
Sm	12.8	81.3	184.0	102.0	271.0	89.0	43.6	114.0
Eu	1.05	6.4	13.2	8.2	26.4	8.4	2.8	9.0
Tb	4.6	16.1	25.7	11.2	22.9	18.6	7.1	14.0
Dy	86.5	163.0	191.0	93.2	72.0	170.0	89.5	101.0
Yb	126.0	97.0	57.9	79.5	26.0	107.0	101.0	65.2

Deposit localities: 1. Purandungri, 2.Kanyaluka, 3.Dhantuppa, 4.Jaduguda, 5. Dhadkidih, 6. Surda, 7. Bhatin, 8.Narwapahar.

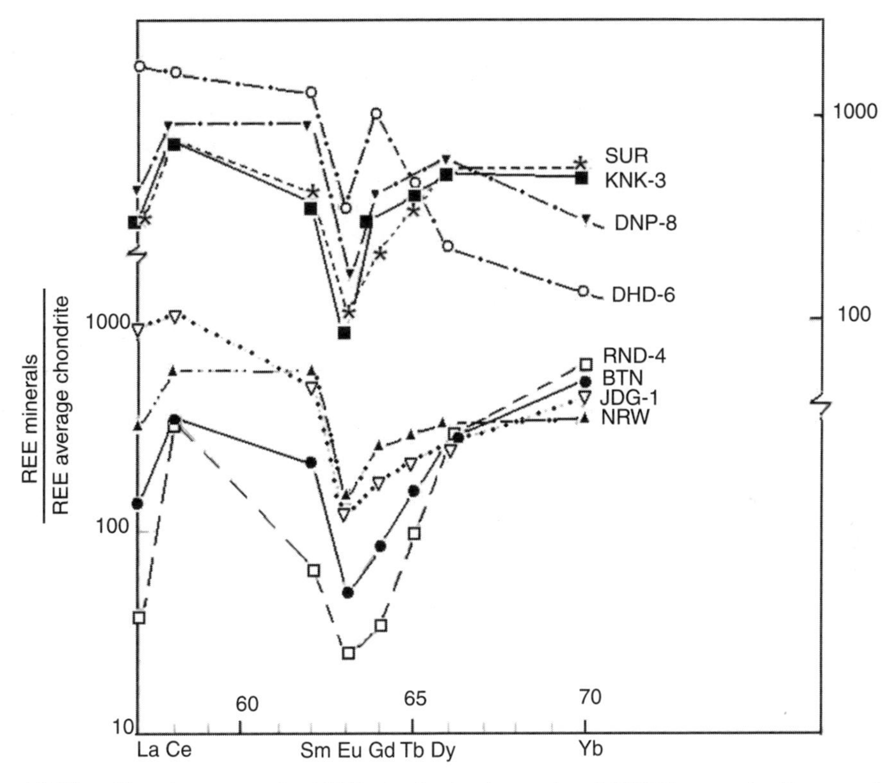

Fig. 4.2.25 Chondrite-normalised REE distribution in apatite of SSZ. The negative Eu-anomaly is conspicuous. SUR – Surda, KNK – Kanyaluka; DNP – Dhantuppa; DHD – Dhadkidih; BTN – Bhatin; JDG – Jaduguda; NRW – Narwapahar (after Rao and Rao, 1980 b in Sarkar, 1984).

Genesis and Evolution of Ores Along the Singhbhum Copper–Uranium Belt

The copper, uranium and phosphate mineralisations along this belt are described separately so far, for the sake of clarity and brevity. Now while discussing the ore genesis, it is thought desirable to take up all the mineralisations along the belt together, for a better appreciation of their metallogenic evolution and point out interrelation if any.

Copper and the Related Mineralisation The first published opinions on the copper mineralisation in the belt are more than 100 years old now (Ball, 1870; Stoehr, 1870). Since then the problem has remained alive and the hypotheses about the genesis of the ores along the belt have been swinging between two extremes: lithogenetic to allogenic, the former meaning the formation by the collection of the ore constituents from host/country rocks by geological process(es), the latter meaning subsequent introduction from an external source. In Ball's idea, the ore material was mechanically or chemically deposited along with the containing sediments and this ore material segregated along cracks and openings during later deformation and metamorphism. Stoehr, a Balls' contemporary, was less clear on this point. When Fermor took interest in this problem, the hypothesis of magma-related hydrothermal ore deposition was enjoying a position of primacy. So he suggested the mineralisation as hydrothermal, the hydrothermal solutions having contributed by the Singhbhum Granite, occurring to the south of the zone (Fermor, 1921). This was followed by a sustained programme of geological investigations along the belt under the leadership of J.A. Dunn. Results of these investigations (Dunn, 1929; 1937; Dunn and Dey, 1942) were substantial additions to the knowledge of geology and ore mineralisation in this part of the country. Dunn's final view was that the association of the copper lodes and apatite-magnetite bodies (uranium and the associated mineralisation along the belt was little known at that time) with the Soda granite was more important than the 'thrust' (shear) zone that more or less coincides with the ore zone in localising the deposits. He envisaged that the sulphidic melt developed by liquation from the residual Soda granite magma. The liquated melt separated into two, the upper apatite-magnetite liquid and the lower sulphidic liquid, which initially a melt, later changed to an aqueous fluid.

The next phase of study on ore mineralisation along the Singhbhum belt was conducted by two groups of workers one led by Banerji and the other by Sarkar. Banerji (1962) and Talapatra (1968) suggested dual sources for the ore constituents: part derived from a deeper region extraneous to the host volcanisedimentary pile, and part from the immediate host rocks. They later modified their view and suggested that whole of the ore material was collected from the country rocks (Banerji et al., 1972). Ghosh (1972 b) almost shared this view. When the model of origin of the Soda granite from a magmatic melt could no longer be sustained and no other felsic plutonic rocks that could be the source of ore fluid was discovered, the workers fell back upon the concept of lateral secretion by way of metamorphism or metasomatism of the country rocks. But the available metal distributions in the wall rocks (Fig. 4.2.12 a–e, this work) and in the feldpathic schists–Soda granite (Banerji et al., 1972) do not directly support this contention. Moreover, there is no direct relationship of the copper mineralisation with migmatisation i.e. development of feldpathic schist–Soda granite. The latter is associated with Cu-mineralisation in the Badia–Mosaboni section. Beyond this, i.e. from Surda onward, this relationship is non-existent. Again in the foot-wallside of the main ore zone in the Mosaboni area, Cu-Fe sulphide mineralisation (0.5 – 1.0%) occurs in chloritised Dhanjori basalt at a number of horizons over a 70 m zone (Das et al., 2008).

Sarkar (1964) from his initial study of the ore mineralisation, particularly at the Badia–Mosaboni area where the Soda granite locally had sort of a granitoid fabric, concluded that the mineralisation could be magmatic hydrothermal. But after extensive study at different parts of the belt Sarkar and his co-workers came to the conclusion that the mineralisation picture was not clear, but the Volcanogenic Massive Sulphide (VMS) model with subsequent modifications during deformation and metamorphism-metasomatism answered most of the questions related to the genetic problem better, but not all (Sarkar and Deb, 1974; Sarkar, 1984).

In order to understand the efficacy of drawing the VMS model by Sarkar and his co-workers for explaining copper mineralisation along the Singhbhum belt, let us quickly scan through some important characteristics of such deposits.

VMS deposits are generally classified into: (i) Cu-deposits, (ii) Zn-Cu deposits, and (iii) Zn-Pb-Cu deposits depending on the dominance of the ore metal(s), or into 'Cyprus type', 'Bessi type' or 'Hokuroku (Kuroko) type' depending on the type area. These classifications are handy but not comprehensive. Barrie and Hannington (1999) proposed a five-fold classification of VMS deposits based on a large volume of data collected from both ancient and modern occurrences of VMS-style mineralisations. The classification is based on the pre-alteration composition of the penecontemporaneous (within 3–4 Ma) volcanic host rocks. Rocks upto ~ 3 km into the stratigraphic footwall, ~1 km into the stratigraphic hanging wall and upto 5 km along the strike have been considered. Arranged from the most primitive to the most evolved, in a petro-chemical sense, these are

1. Mafic
2. Bimodal–mafic
3. Mafic–siliciclastic
4. Bimodal–felsic
5. Bimodal–siliciclastic

We may have to return to the VMS hypothesis in a final section of this discussion. In the meantime it may be conceded that if at all the VMS hypothesis is found most suitable to explain Cu (and the related) mineralisation along the Singhbhum copper–uranium belt, then it should relate to the 'Mafic' sub-type in the classification scheme of Barrie and Hannington (1999). The observations that would favour such a conclusion are (i) Closeness of the Dhanjori volcanics (inclusive of their derivatives such as magnetite-quartz–bearing chlorite schists) and the ore zone in general and occurrence of some deposits such as the Dhobani and several others in the same alignment; (ii) Similar deformation of the ore bodies and the host amphibolites at Baharagora (Fig. 4.2.5), extension of the ore shoots along prominent lineation (down dip in the SE) in the shear zone and the overall recrystallised nature of the ores (except where affected by late stage brittle deformation); (iii) Low and consistent $\delta\ ^{34}S$ values of the sulphides. This model presumes the development of the regional dislocation, now identified as the Singhbhum shear zone, penecontemporaneous with volcanism. This suggestion is supported *inter alia* by the emplacement of several ultramafic bodies along the zone that are thermo-tectonically modified along with other rocks. But the model fails as such to explain the high salinity in the fluid inclusions of the ore zone minerals, in the absence of undoubted acid plutonic rocks in the vicinity. We will return to it later.

Uranium and Related Mineralisation Like the sulphide mineralisation, the origin and evolution of uranium mineralisation along the Singhbhum Cu-U belt is also not very clear. Bhola and his co-workers (Bhola et al., 1966) and Gangadharan et al., (1963) concluded that the mineralisation was hydrothermal and the hydrotherms were produced by the cooling Soda granite. But Banerji and his colleagues (Banerji, 1962 and Banerji et al, 1972) thought that a part of uranium was derived from the country rocks and part came along with the migmatising material that constituted the Soda granite.

Rao (1977) and Mookherjee (1975) on the other hand suggested that the uranium mineralisation in the belt was the end product of a prolonged history of geological activities. They envisaged that the uranium released from the Singhbhum Granite in the south accumulated in the Chaibasa Group of sediments and later mobilised along the shear zone through metamorphic–metasomatic processes. Sarkar (1982 and 1984) generally agreed with this model and suggested that bulk of uranium released from the Singhbhum Granite before oxyatmoversion must have accumulated in the basal Dhanjori Group rather than in the Chaibasa Group of sediments. Part of uranium could of course, easily move on to the Chaibasa sediments, particularly after the oxyatmoversion. Development of uranium bearing metallotects during the earliest major Proterozoic orogeny (1600–1800 Ma ago) was a global phenomenon and the Singhbhum mineralisation was a part of it. This is generally an accepted model. A pertinent question pertaining to this model, however, still remains unanswered. To mobilise uranium from its distribution in sediments into orebodies during dynamothermal activities in the orogen, uranium will need to be re-oxidised to be labile again. The problem dissolves if such oxidising material as red bed etc. occurs in the right hydrologic position in the sequence (Rich et al., 1977). But that is not the situation, in most of the cases. It needs be mentioned here that Rao et al. (1999) commented that most of uraninite and brannerite grains from Jaduguda, Bhatin and Narwapahar are detrital. This opinion remains unconvincing in the context of the chemically labile nature of uranium and the apparent mobilisation of uranium in the belt. Neither do the petrographic observations support it (Sarkar, 1984).

There is a partial resemblance of uranium mineralisation in the Singhbhum belt with the 'albitite uranium deposits'. The resemblance is limited to the fact that both uranium mineralisation and albitisation have taken place along this highly tectonised belt. But on a closer look it becomes obvious that in the best mineralised section (Jaduguda–Narwapahar) there is little or no temporal or spatial relationship between the two. Moreover, the mineralogy of the ore assemblages from this belt and from those of type areas of albitite uranium mineralisation (Abou-Zied and Kerns 1980; Oesterlen and Vetter, 1986; Laznicka, 1993) obviously differs. In the present state of knowledge, the mineralisation may be likened to *sub-unconformity–epimetamorphic* type, using Dahlkamp's nomenclature (Dahlkamp, 1993).

Nickel–cobalt (–silver–bismuth) mineralisations, usually not very rich, have been noted in many vein deposits of uranium in Canada, Central and Eastern Europe, and Australia. They usually occur in metamorphosed volcanisedimentary rocks that included basic members. In contrast, uranium mineralisation in granitic rocks, such as those in Portugal, Massif Central of France or the Rossing deposit of Namibia, this subsidiary mineralisation has not been noted. It may, therefore, be reasonable to guess that the source of the ore elements in these subsidiary mineralisations is the basic members in the host rock associations.

At Jaduguda–Bhatin sector nickel (and cobalt) is present in the ore zone several times larger in concentration than elsewhere in the belt. A plausible explanation of this increased concentration of these elements is deposition from hydrothermal solutions that leached these elements from the mafic and ultramafic rocks that occur a little below the ore zone, particularly at Jaduguda.

Phosphate and Related Mineralisation For a better understanding of the phosphate deposits with or without Fe-Ti minerals we may divide them into the following principal modes of occurrence:

1. Phosphates in sedimentary rock associations
2. Phosphates associated with intrusive rock associations
3. Phosphates associated with volcanic/volcanisedimentary rock associations.

The first types of deposits are generally stratiform and extensive over several kilometers to tens, or even hundreds of kilometers along sedimentary basin margins. The rock types commonly associated with them are black shales/siltstones with chert and carbonate members, an association in which the phosphate rocks pass basin-ward into carbonates devoid of phosphorus. At places there are evidence suggesting an initial deposition of a phosphatic mud or phosphatic replacement of a carbonate substrate is followed by continuous fragmentation, transportation and deposition elsewhere in the shelf (Howard, 1972). Phosphorus is present here principally as collophane or cryptocrystalline carbonate–fluorapatite. Iron may also be present in these deposits, but in the form of siderite, or limonite (or even as pyrite, chlorite or chamosite as in some Australian deposits). REE-content is low (< 0.1 wt%). The preceding discussion on the phosphate mineralisation along the Singhbhum shear zone finds little, if any, similarity with this type of deposits.

Rocks rich in apatite and Fe-Ti oxides leading to economic deposits are found at many places in the world associated mainly with alkali-rich intermediate and ultramafic rocks and carbonatites, such as at Khibny and Kovdor, Kola Peninsula, Russia, Palabora Complex in South Africa, Jacupiranga Complex in Brazil, Fanshan Complex in Northern China. Rocks consisting almost wholly of apatite and Fe-Ti oxides, commonly called nelsonite, are also found associated with anorthosite in North America and elsewhere. Park, Jr. (1972) reported apatite - magnetite - amphibole rocks from a number of places in Chile. There is a sharp division of opinions on the genesis of these deposits. Some hold that these are orthomagmatic in origin (Park et al., 1972; Nashlund et al., 2002; Henriquez et al., 2003; Neng Jiang et al., 2004), while others believe them to be hydrothermal replacement deposits (Rhodes et al., 1999; Sillitoe and Burrows, 2002). Again the magmatists differ on details. Philpotts (1967) based on his experimental studies concluded that these ores formed by liquid immiscibility from a parent magma of appropriate composition. Clark and Kontak (2004) suggested that nelsonitic Fe-Ti-P oxide melt generated by liquid immiscibility from an andesitic magma formed by mixing of acid and basic magmas. Others, based on the field observations, would believe them to be magmatic cumulates (Dymek and Owens, 2001; Neng Jiang et al., 2004). The composition of the oxide-phosphate ores in reality is not restricted to the strict formulation (i.e. 2 parts magnetite + 1 part apatite) of Philpotts (op. cit.). A Ti-mineral (generally ilmenite) is commonly present and the mineral proportions can and often do vary. The deposits are generally LREE enriched, some with a negative Eu anomaly. The Singhbhum shear zone apatite (–magnetite) mineralisation has little in common with the type of deposits discussed above.

The best example of the third type of deposits are iron oxide (with apatite in varying proportion) deposits in North Sweden, spread over an area of about 200 × 200 sq km in clusters, popularly known as Kiruna deposits from the locality name of Kirunavaara, where the mineralisation is better developed. They are associated with Late Paleoproterozoic alkaline volcanic, subvolcanic and intrusive rocks. The deposits may be banded, stratiform, concordant to discordant, massive, brecciated and even in the form of veins. Some of the deposits are sheared, refolded and shoot-like plunging bodies. A few of the deposits are very large. The Kirunavaara deposit, which is approximately 5 km (strike) × 2 km (depth) × 100 m (thickness), contains an ore reserve of > 2 Bt having 60–62% Fe. Host rock alteration associated with the deposits is generally prominent. This in case of magnetite–hematite–apatite–actinolite deposits is characterised by actinolite and biotite (Kirunavaara, Luossovaara, Svappovaara, Maleaberger) and sericite-biotite–chlorite and carbonate alteration with barite, fluorite, tourmaline, scapolite and apatite around magnetite and hematite rich deposits. Red rock alteration (hematite dusting) is locally seen around iron ore bodies. The mineralisation is essentially magnetitic (with subordinate hematite) with apatite, tremolite, actinolite, biotite, chlorite and minor calcite. The Luossovaara deposit is relatively rich in phosphorus. TiO_2 -content is low (< 1%). Apatite is LREE- rich. Orebodies are commonly massive but can be deformed and recrystallised as at Malmberget (950 Mt, 60–62% Fe). Such contrasted features have been reported from these deposits as (i) cross-bedding in banded apatite-magnetite ores at Rektorn (Parak, 1975); (ii) locally vesicle-like voids in the ores. Kirunavaara deposit is almost indistinguishable from similar structures at Laco magnetite deposit, Chile, which is considered magmatic (Nystrom and Henriquez, 1994; Naslund et al., 2002). To these the omnipresent hydrothermal alteration is to be added. The mineralisation, obviously, is not controlled by any tectonic structure. No sulphide mineralisation is closely associated with this mineralisation except at Viscaria deposit, which occurs at a distance of ~ 1.5 km towards the footwall side of the Kiruna deposit. The volcanic-associated mineralisation at Viscaria deposit contains a reserve of 40 Mt with 1.5% Cu and 0.7% Zn. No wonder that such a large spectrum of variation in features led to a variety of genetic models for these deposits that include (i) magmatic segregation and injection, (ii) extrusion, (iii) crystal tuff settling,

(iv) exhalative-sedimentary, (v) hydrothermal deposition including metasomatic replacement (Geiger and Odman, 1974; Parak, 1975; Frietsch, 1978; Gilmour, 1985; Fleischer, 1983; Romer et al., 1994; Nystrom and Henriquez, 1994). Indeed no unique geological process explains all the reported features of these deposits. Rather they suggest operation of more than one geological process, with possible overprinting. A comparison of the apatite-magnetite deposits of the Singhbhum belt and those of Kiruna will show that the dissimilarity beteen them is more prominent than the similarity.

Dunn (1937) preferred a hydrothermal origin for the magnetite-apatite deposits related to the Soda granite magma. But the model becomes unsupported when we do not find convincing proof of the Soda granite being magmatic. Basic volcanic rocks are rich in both Fe and P. It has been shown by Froelich et al. (1977) that seawater–basalt interaction produces both iron and phosphorus. The two elements may form soluble complexes like Fe $(HPO_4)^-$ and Fe $(HPO_4)_2^-$ in alkaline media. Basaltic rocks have an average fluorine content of ~ 400 ppm. So the Dhanjori mafic volcanic rocks with which the Singhbhum deposits are so intimately associated could be a viable reservoir for the constituents of these deposits. But when were they scavenged into these deposits? Micro- to macro-scale mobilisation of apatite during metamorphism is reported in literature (Vernon, 1975; Parfenov et al., 1979). There is no tell-tale evidence in support of this proposition either. But three things are obvious in case of the apatite–magnetite mineralisation along the Singhbhum belt: its shear zone control, occurrence within or at the periphery of the mafic volcanic rocks or their derivatives, and generally intense hypothermal alteration accompanying the ore bodies. The mafic volcanic rocks could be the source of Fe, P and F but could not be the source of water. This water could be the water released during dynamothermal metamorphism of the clastic sediments in the sequence and on its way out along the pre-existing/penecontemporaneous shear zones could leach out Fe, P and F from the mafic volcanic rocks and precipitated in the form of ore deposits. Metasomatic development of Soda granite could locally cause its redistribution. A hydrothermal introduction of Fe-P into the belt is corroborated by the local presence of apatite-magnetite clusters within the Chaibasa mica schists at Sungri and Barjudi and the same in the quartz- kyanite rock at Gohala, as already mentioned.

The IOCG Hypothesis and Mineralisation Along the Singhbhum Copper-uranium Belt

Iron Oxide–Copper–Gold (IOCG) is a relatively new grouping of mineral deposits with shared characteristics. In less than two decades time, a large volume of literature has developed on this group of deposits (Barton and Johnson, 1996, 2000; Hitzman et al., 1992; Hitzman, 2000; Oliver et al., 2004; Pollard, 2000; Sillitoe, 2003; Skirrow and Walshe, 2002; Williams et al., 2005; Groves et al., 2010). They are briefly reviewed below, first on characteristics and then on genetic interpretations.

Salient Features of IOCG Deposits

1. Cu (± Au) is the principal economic metal. In > 60% Cu-Au deposits, the Cu-grade lies between 0.5% and 4.0%, averaging about 1%. In the rest it is of the order of 0.1%. Chalcopyrite is the principal Cu-mineral, in cases followed by bornite.Au usually < 1 ppm, in a few >1 ppm and in rare cases reaching upto 3.8 ppm as in the Starra deposit, Cloncurry district, North Australia (Rotherham et al., 1998).
2. Abundant Fe-oxide minerals (magnetite and/or hematite) are present in many such deposits, but not in all. Fe-oxides have low-Ti, compared to those in most igneous rocks.
3. Magnetite mineralisation at places is accompanied by abundant phosphate (apatite).
4. Pyrite-content is low in the sulfide assemblage.

5. Minor elements other than gold, in which the Cu-rich IOCG deposits are tangibly enriched, are P, Co, Ni, As, Mo, Ag, Ba and U. They are conspicuously poor in Pb and Zn. Variability is higher in Ba-, F-, and U-contents.
6. Almost all IOCG deposits are enriched in LREE, which is commonly of the order of 0.5%, but may be as high as 10%, rarely though.
7. Many of this type of deposits are associated with plutonic-hypabyssal intrusives that are variously composed of alkalic granitoids to intermediate (arc-related) rocks. In some cases, however, such intrusive rocks are not obvious.
8. Generally, structurally controlled pervasive alkali (mostly Na $\pm$ K $\pm$ Ca) metasomatism characterises most such ore fields, which may be upto several hundred square kilometres in extent. These alteration zones may be depleted in Fe, Cu, and Au. The alteration mineralogy within individual deposits depends on host lithology and depth of formation. There is a general trend from sodic alteration at depth to potassic alteration at intermediate to shallow levels, to sericitic (hydrolytic) alteration and silicification at very shallow levels. At places the early sodic/sodi-calcie alteration is overprinted by potassic (biotite $\pm$ K-feldspar) alteration. IOCG mineralisation has at places been found to lag behind regional alkali metasomatism by 10–20 Ma.
9. IOCG deposits are not time specific. They range in time from Late Archean to Cenozoic. Early Precambrian deposits, however, cluster around 1900 Ma.
10. Many of the Proterozoic deposits appear intra-cratonic, characterised by orogenic basin collapse or anorogenic magmatism while many important Phanerozoic deposits are located at continent margin arcs, above a subduction zone. All of these environments have significantly voluminous igneous activity, high heat flow and oxidised fluid flow.
11 IOCG deposits are generally localised along faults and shear zones, which are actually splays off major crustal-scale dislocations. They occur at different depth zones: brittle to ductile.
12. Hydrothermal veins are generally breccia-filling and/or replacement in style of mineralisation in structure-specific sites. Breccias may be either of hydraulic or tectonic origin.
13. Where both the Fe-oxide $\pm$ apatite and Cu-Au mineralisations are present, the latter is in most cases found to have overprinted the former.
14. The Fe-oxide $\pm$ apatite and Fe-oxide-Cu-Au deposits particularly the pre- and syngenetic ores, generally show co-existing, high temperature (> 250°C), hypersaline and CO_2-rich fluid inclusions.
15. Sulphur isotope data ($\delta^{34}S$) are typically, but not universally close to 0 ± 5 per mil. $\delta\ ^{18}O$ data, particularly in cases of late mineralisations in shallow crustal depths, show evidence of surface water contribution. Otherwise, the magmatic signature is much more prominent.
16. Data on radiogenic isotopes are relatively few. From the available Sm-Nd isotope data on the Olympic Dam deposit, Australia, Johnson and McCulloch (1995) inferred that a portion of Cu and REE was sourced from mafic or ultramafic rocks, or magmas that are similar to those emplaced penecontemporaneously with the mineralisation. Sm-Nd data from the Tenant Creek deposit are also suggestive of contribution of REE, and by implication ore metals, from a primitive source.
17. Evaporite beds may or may not be present in the sequence of host/country rocks.
18. The deposits are characterised by low vein quartz content, compared to many established hydrothermal ore deposits.

Genetic Interpretations From the above list of observations on this group of deposits it is obvious that certain features are more common than others and as Williams et al. (2005) have pointed out, there are gaps in our knowledge about them on certain counts. The most common features amongst these deposits are as follows:

1. Localisation along disjunctive structures, i.e. faults, ductile shear zones.
2. Occurrence as hydrothermal veins and breccia filling and/or replacement modes.
3. The chemistry and mineralogy of the ores.
4. Cu (± Au) is the principal economic metal although it always does not make the grade. REE-content is generally high. Abundant Fe-oxide(s) ± apatite in majority of the cases. Magnetite is Ti-poor.
5. Hot (>250°–500°C), hypersaline and CO_2-rich ore fluids.
6. $\delta^{34}S$ clustering around 0 ± 5 per mil.
7. Common association of alkali metasomatism (Na ± Ca–K) at and around the ore zone.

The above features prima facie suggest that these mineralisations are hypogene hydrothermal. But what were the source(s) of ore constituents and the hot aqueous solutions, i.e. hydrotherms that carried them? In cases where all the above situations are satisfied the source of ore-bearing hydrotherms are generally believed to be granitic magmas without /with limited interference by the host/country rocks. However, as pointed out by Williams et al. (2005), there is only limited direct evidence for, or against, fractionation of either F- and P-rich melts/aqueous fluids and Cu (-Au) rich hydrothermal solutions from magmas constraining the IOCG system. In some belts contemporaneous granitoid bodies are absent, or at least not detected. Limited radiogenic isotope data from these ores (Point 15 above) suggest some link to the mantle. More data are awaited, on these. High CO_2-content in the magmas of the associated intrusive bodies and the hydrothermal carbonates in the ores also warrant a suitable explanation.

Another aspect, which needs more studies, is about the alkali metasomatism, principally albitisation. It is not yet clear if it is a prerequisite for mineralisation, a co-product of mineralisation, or an unrelated phenomenon. It can only be recapitulated here that it is present in most fields of IOCG mineralisation but not in all and wherever present, the intensity of mineralisation is not necessarily directly proportional to the scale of alkali metasomatism. What intrigues many researchers is the mechanism of its development. A phase of albitisation apparently preceeded ore mineralisation.

The interesting model of albitisation by Pollard (2000) discussed in an earlier section of this Chapter is recalled in this context, but need not be repeated. Oliver et al. (2004) have written the following reactions (unbalanced) while modelling the role of sodic alteration in the genesis of iron-oxide-copper gold deposits, Eastern Mount Isa block, Australia.

1. Biotite + plagioclase ± K-feldspar ± magnetite ± ilmenite + NaCl ± $CaCl_2$ ± CO_2 ⟶ albite + titanite ± scapolite ± calcite + KCl (in intrusive rocks)
2. Biotite + calcite + quartz + NaCl ⟶ actinolite + albite + titanite + KCl + CO_2 ± microcline (in impure marbles)
3. Muscovite + biotite + plagioclase (or calcite) + KCl ⟶ tremolite/actinolite + albite + titanite + KCl (in weakly calcareous metapelites)

Not all workers on IOCG deposits believe in the magmatic model of genesis of these deposits and the associated petrochemical changes in the host rocks, not in all cases at least. Their views are divisible into two groups: Those which relate the hydrothermal fluids involved with surface waters or brines in shallow basins, and those which relate the fluids to dehydration during lower and midcrustal metamorphism. In the former, igneous intrusions (not necessarily of granite magma) are involved but that is mainly as thermal source for convection. Ore metals are provided by the wall-rocks. Hydrothermal alteration in the host rocks in the first type is characterised by K, H+ ± Na (Ca) in upwelling zones and Na (Ca) ± K in recharge zones. Fluid salinity may be derived by the action of circulating water on pre-existing evaporite deposits (Barton and Johnson, 1996, 2000) or from the evaporated surface water in warm and arid situations (Williams et al., 2005). Metamorphic models do not need magmatogenous heat source as a necessary pre-requisite. However, coeval magmatic intrusions may be present and they may contribute both the thermal energy and the ore-constituents (Williams, 1994; De Jong et al., 1998; Hitzman, 2000).

The majority opinion today favours the idea that most of the IOCG deposits formed from structure (fault and/or shear zone) controlled distal magmatic systems, with varying but subordinate contribution by ground water and/or country rocks. The so-called 'Non-magmatic' models similarly admit of contributions by penecontemporaneous magmatism in varying measures. The hydrothermal source is imagined to be distal (in contrast to proximal in case of porphyry Cu-deposits, skarn deposits) as in majority of cases such intrusives are not exposed. Moreover, these should desirably be distal so that the hydrothermal fluids produced by them become regionally circulative. It is possible and even likely that these magmas developed at the lower crust – upper mantle region and have some petrochemical specialisation such as enrichment in CO_2 and REE, primitive ε_{ND} signature etc. Mineralisation took place by one or more of the following: (i) cooling, (ii) fluid-mixing, (iii) increase in pH, and (iv) reaction with the country rocks. Vide Groves et al (2010) for a more detailed discussion.

Mineralisations along the Singhbhum Cu-U belt, no doubt, have some resemblance to those of the IOCG group, as enumerated below.

1. The mineralisation is principally of copper, uranium, apatite-magnetite with Au, Ag, Mo, Te, Se, Ni, Co as trace /minor elements.
2. The mineralisation is controlled by a regional disjunctive zone/ductile shear zone.
3. The same zone controlled alkali (Na ± K) metasomatism.
4. Hypersaline nature of fluid inclusions in gangue quartz.
5. Presence of low Ti-Fe-oxides.
6. Absence of suitable igneous intrusions.

But the features which are generally considered not very characteristic of the IOCG group and observed in the ores of the Singhbhum Cu-U belt include:

1. The metallotect as it is now, shows features of ductite-brittle-ductile shear zones and the orebodies deformed (and reformed) and the ores are dynamothermally recrystallised.
2. The tectonic setting, as can be interpreted now, is not 'anorogenic' or 'intracratonic', affected by 'far-field stress related to distant orogenesis'.
3. Both pyrite and pyrrhotite are present in subequal proportion in the belt scale.
4. The Singhbhum ores cannot be designated as LREE-enriched. Apatite there, is indeed LREE-enriched but ΣREO < 0.5%. Instead, xenotime, essentially a phosphate of Y and HREE, is ubiquious in the belt, being locally exploitable.

5. Ore metals suggest both mafic (Cu, Fe, Au, Ag, Ni, Co) and felsic (U, B) sources for them.
6. SiO_2-depletion in the ore zone is not established (Sarkar, 1984).

Conclusions on the Genesis and Evolution of the ore Deposits Along the Singhbhum Cu–U belt While considering the genesis of Cu–Fe–U–P mineralisations (and associated phases) along the SSZ, it appears that the Cu–Fe sulphides were deposited earliest in the metallogenic history of the belt before deformation (D_1) and metamorphism (M_1) of the tectonite under greenschist to amphibolite facies. The host rock of this mineralisation is dominantly the chloritic schists of tholeiitic derivation, except for Mosabani–Badia sector (Main lode in Soda granite) and Baharagora (mineralisations in muscovite schists and amphibolite of Chaibasa Formation). It may, however, be noted that even at Mosabani–Badia, where the Soda granite broadly forms the host for Cu–Fe sulphide ore bodies, the mineralisation is associated with streaks, stringers and patches of chlorite-quartz schist enclosed in Soda granite. Albite never occurs as a prominent gangue mineral in the vicinity of the ore mineral assemblages.

Throughout the belt copper and iron sulphides are seen to be mainly disposed along the pervasive S_1 schistosity and shear planes, sub-parallel to the lithological boundaries and the ore bands also display small scale folds along with the enclosing host. The sulphide assemblage records wide spread deformation and recrystallisation textures and are often seen remobilised along the locally pervasive later developed cleavages intersecting the S_1 planes (steeper crenulation cleavages S_2 at Nandup and shallower Riedel shears at Turamdih). That the disposition of the sulphide ore bodies is structurally controlled is amply evident from the parallelism of the ore-shoots with the most prominent linear structures such as fold axes, intersection lineation, stretching lineation etc. From Mosabani to Narwapahar the ore shoots have down-dip disposition showing parallelism with the strong down-dip orientation of all linear features. Further west, in the Turamdih sector the ore shoots display fold interference pattern with low westerly or easterly pitch, as displayed by the transverse sections, level correlation plans and stratum contours. The richer ore-shoots in this sector, as seen in the underground mines, are disposed parallel to the very low dipping Riedel shears, contained within the enveloping surface of the steeper dipping orebodies bounded by S_1 planes.

The grade of metamorphism is generally of greenschist facies along the SSZ, as discerned from the silicate mineral assemblages. Though diapthoretic effects are locally present, the metamorphism in general records a prograde event marked by the development of chloritoid at its peak. The metamorphic mineral assemblage of the adjacent Chaibasa rocks is characterised by the development of muscovite–garnet, garnet–staurolite–andalusite etc. denoting amphibolite facies (550° C/6 kb). The sulphide assemblage displays recrystallisation twins, mutual boundary, mutual embayment, preferred dimensional orientation of grains and annealed textures with triple points. These features indicate metamorphic reconstitution of the sulphides along with the gangue minerals under dynamothermal condition followed by post-tectonic static recrystallisation, with the waning of pressure. The peak metamorphism in most of NSMB is dated at 1.6 Ga. Except at Baharagora, where the mineralisation might have witnessed a higher P-T regime, the metamorphic condition for rest of the SSZ corresponded to that of greenschist facies.

The apatite–magnetite mineralisation along SSZ shows more or less the same host rock affinity as the sulphides but for its preferred location along the contact of feldspathic schists/Soda granite at places. In Turamdih sector, the apatite–magnetite bearing chlorite schist forms a mappable band, which is spatially separated from the sulphide-rich chlorite schist horizon by intervening feldspathic sericite quartz schists. Preferred orientation of apatite grains parallel to the schistosity and swerving of schistosity around lensoid apatite–magnetite aggregates, rhythmic banding of apatite and magnetite are the features indicative of the mineralisation being pre- to syn-tectonic. Post-tectonic remobilisation of these ore minerals is displayed by the pegmatoid apatite–magnetite veins noted at places (Dhantuppa). It may therefore be reasonably presumed that the oxide phases of P and Fe were also emplaced during or before the D_1–M_1 event and not far separated from the sulphide precipitation in respect of both time and space.

The uranium-oxide mineralisation along SSZ, like the sulphides and other oxides, occur within the chloritic schists in the best mineralised section of Jaduguda to Narwapahar. The uranium lodes are disposed towards the hanging wall of the low-grade Cu lodes/ore zones, wherever developed. In the Rakha Mines–Mosabani sector too uranium is present within the chloritic schists or in chloritic patches included in Soda granite as at Mosabani, that hosts copper sulphides as well as apatite-magnetite mineralisation. In the Turamdih-Bandhuhurang sector, however, it is concentrated in the feldspathic chlorite sericite quartz schist occurring in the hanging wall of the chlorite schist hosted copper sulphide lodes. At Keruadungri, uranium concentration is in the magnetite bearing chlorite schist. Further west in the Mahuldih deposit the mineralisation is in tourmaline bearing feldspathic sericite quartz schist (HW-lode) and magnetite bearing quartzite (FW-lode), which occurs along the footwall of the copper bearing chlorite schist horizon of Turamdih. Molybdenite concentration is conspicuous in some of the uranium lodes in contrast to the Cu-sulphide lodes. On the other hand, both Cu-Fe sulphides and U- oxide mineralisations are associated with traces of Au. Haloed uraninite occurs as dissemination in the host silicates as well as in form of specks within the Cu-/Fe-sulphides, apatite and other ore minerals. Boron metasomatism, giving rise to at least a phase of tourmaline and dumortierite mineralisation along the belt, is very prominent.

The chronological order of formation of ore minerals, i.e. the paragenesis of different ore mineral phases in the polymetallic ensemble of the SSZ cannot be determined by following the conventional textural criteria, as because the assemblage is thoroughly deformed and metamorphosed leading to simultaneous recrystallisation and (re)mobilisation of the ore minerals at multiple stages.

Let us now examine the possible modes and order of ore deposition along this belt. Though there are ample evidences of primary sedimentary deposition of Fe-Cu sulphides in form of thin bedding parallel laminates in the carbonaceous rocks of NSMB, such incidences are totally absent along SSZ. Moreover, the hosts for the mineralisation along this zone are mostly variously transformed rocks of basic volcanic parentage. As such, a volcanogenic origin for Cu-Fe (± Au) -P mineralisation along this zone becomes a possible choice while considering its genesis. Under the said premises, the VMS model and volcanogenic hydrothermal processes may be considered as a possible genetic option for the sulphides in the SSZ at the waning stage of Dhanjori volcanic outpouring. Theoretically, the ore forming elements like Cu, Fe, P and Au might have been provided to the volcanic exhalatives/hydrotherms by the basic volcanic members emplaced along the zone as well as may be directly derived from the magmatic hearth. These metals could get into the solution as hydroxyls, chlorides etc. and volcanogenic sulphur may be in form of H_2S or sulphates in the solution. The Cu-Fe sulphides may be precipitated from the circulating hydrotherms along formational interfaces, fractures or any other low pressure domains in the volcanic pile by one or more of fluid mixing, cooling and reaction with wall rocks under low fO_2 condition. With subsequent increase in oxygen fugacity after the deposition of sulphides, the same circulating hydrotherms may precipitate the minerals, specially Fe and P in form of oxides at the same locales or new adjacent receptacles within the volcanic strata. Ca for apatite could easily be derived from the host basic volcanics.

It is obvious that it would be extremely difficult or altogether impossible to accommodate uranium, molybdenum and also the influx of alkali and boron in the above genetic scheme as because the basic volcanics or their parent magma cannot provide these elements in such proportions. Therefore, one would be obliged to invoke hydrotherms from a preferably felsic magmatic source, which may bring in appreciable quantities of above elements and also explain high salinity of fluids entrapped in the gangue quartz and high $\delta^{18}O$ in some shear zone rocks. For uranium, one may consider scavenging of the constituents from a pre-existing fertile granite or some uranium bearing formation by the circulating hydrotherms. In the case of SSZ, no adjacent granite intrusion is exposed which could be directly related to these mineralisations. The Soda granite and feldspathic schists are metasomatic in origin (as discussed in Section 4.1), representing a phase or phases of albitisation along the disjunctive zone. The Rb-Sr date available for Soda granite is 2.3 Ga. The feldspathic schists are seen to be strongly mylonitised and folded by F_1 folds. At Mosabani, the reported 'Durchbewegung' structure involving Soda granite clasts in copper ore matrix is indicative of kneading of the sulphides with the clamts of associated quartzo-feldspathic rocks during deformation. These would suggest that the process of feldspathisation (albitsation) along the SSZ got initiated rather early, before or concomitant with the first ductile shearing and certainly prior to the first folding. The albitised zones do not show any close spatial or temporal relation with the Cu-Fe sulphides or apatite–magnetite mineralisation except a spatial coincidence in Mosabani–Badia Section. Similarly, the uranium mineralisation also does not show any spatial relation with the feldspathised zones except in the Turamdih sector. Therefore, the event(s) of albitisation cannot possibly be genetically linked except very generally with the matallogeny along SSZ. But the source of such extensive alkali influx must be magmatic (not related to the envisaged basic volcanism) and the same or some other concealed granitic source could provide hydrotherms responsible for bringing in U-Mo in the system directly from the source or concentrated from the basement Singhbhum Granite. The QPC-hosted placer uraninite and gold at the base of Dhanjori sequence could also be considered as a supplier of the constituents to the circulating hydrothermal system. In that case we presume that U-mineralisation in general and Mo-concentration locally, as at Jaduguda–Bhatin sector, were preceded by the apatite – magnetite and Cu-Fe sulphides, in that order. The first phase of albitisation may be placed very close to or just after the emplacement of the Cu-Fe sulphides and apatite-magnetite ores. Thus the age of these mineralisations may also be close to 2.3 Ga. The bulk of uranium mineralisation on the other hand shows an age range of 1.5–1.6 Ga, which matches well with the age of peak metamorphism in the belt.

Ore mineralisation in the Singhbhum Cu-U belt has a complex history, consisting of several events, one tangibly different from the other. However, we need more information to define them better. Those events can be at the moment viewed in terms of classical models of ore genesis, or perceived in the context of IOCG deposits. We hope, with more work in future on Singhbhum geology and ore mineralisation, some of the ambiguous observations will be better understood and we will be able to arrive at more reasonable conclusions on the ore mineralisation along the Singhbhum Cu-U belt.

Cu-Pb Sulphide Mineralisation Along Northern Shear Zone

As mentioned earlier a brittle (-ductile) shear also passes through the northern contact of the NSMB with the CGC and is named variously as Northern shear, Tamar–Porapahar shear or South Purulia Shear. The shear zone trends WNW–ESE for more than 100 km, extending from Tamar in the west in Ranchi district, Jharkhand to Porapahar in the east in Bankura district, West Bengal, passing through Singhbhum, Saraikela–Kharsawan and Purulia districts. A number of basemetal sulphide occurrences and apatite bearing carbonatites are reported from this belt. The important sulphide occurrences are located at Jhimri and Hesalong in western sector, at Beldih and Biramdih in central sector and at Thakurdongri, Tamakhun and Kutni–Dandodih in the eastern sector.

The country rock comprises a meta-psammopelitic assemblage with mafic and acid volcanic rocks, which are accompanied by extensive silicification, brecciation, ferruginisation and alkaline carbonatite intrusions along the shear zone. Sporadic Cu-Pb sulphide mineralisation is noted along this zone with evidence of old mining activity at places. The most prominent in these is the old copper mine at Tamakhun and excavations for lead at Beldih. Besides basemetal, a number of apatite deposits, associated with alkaline carbonatite rocks, are located along the belt.

The sulphide mineralisation in the area may be broadly categorised into three types as stated below.

1. Cu-sulphides in the silicified rocks of Tamakhun and Beldih: Chalcopyrite is the dominant ore mineral at Tamakhun, which occurs within silicified and carbonate rich rocks that are sheared and brecciated. Significant incidences of Y (250–350 ppm), Zr (400 ppm), Sr (500 ppm), Ba (600 ppm) have been recorded in the ore zone. Presence of upto 530 ppb Au was also recorded. The ore reserve estimate of Tamakhun was 0.11 Mt with 1.85% Cu. In Beldih there is 150 m long and 5.25 m wide zone with poor disseminations of chalcopyrite.
2. Cu-Pb sulphide in alkaline carbonatite complex: The alkaline carbonatite complex at Beldih has lean Cu-Pb mineralisation in form of disseminated chalcopyrite and galena in thin lenticular bands (0.6% Cu and 0.9% Pb) with 5.2 ppm Ag, 750 ppb Au. Pb-sulphides are also reported from the apatite deposit at Kutni–Dandodih, associoated with > 1000 ppm of Cu, Pb and Zn. The metallogenic environment is somewhat comparable to the Palabora copper-apatite deposit of South Africa (Basu, 1993).
3. Cu-Au in ferruginous and hematitic rocks along with volcanics: The ferruginous rocks in Thakurdongri and Jhimri–Hesalong contain upto 1000 ppm Cu and 500 ppb Au in surface samples. Though the available information is scanty, the situation appears to be similar to that of Malati Pahar and its extensions described below.

Phosphate Deposits in the Northern Shear Zone

Phosphatic ore deposition has been located at a number of places along the Northern shear zone, south of Purulia town, particularly at Beldih, Mednitanr, Kutni and Churingora, in close association with carbonatite in a metamorphosed volcanisedimentary rocks. The ore grade is generally low. At the Beldih area a reserve of 4.56 million tonnes of ores with an average P_2O_5 -content of 13 wt% has been estimated (Kumar et al., 1985). The ores here occur in three modes: (i) banded, (ii) brecciated, and (iii) as veins. The apatite is fluorine-bearing and the other minerals present include quartz, Mn-oxides, magnetite and pyrite[2]. Singh et al. (1977) reported Nb_2O_3 and Ta_2O_3 contents from the Beldih ores in the range of 0.06–3.6 wt% and 0.2–0.35 wt%. The genesis of these ores is not clear. Bhattacharyya and Bhattacharya (1989) suggested a volcanisedimentary origin for these ores. But there should be more studies on this aspect.

Apatite mineralisation in the Chhotanagpur Granite Gneiss is reported from Panrkidih in north Purulia district. It is estimated to contain a reserve of 1,20,900 tonnes of ore having an average grade of 20 wt% P_2O_5 (Baidya, 1982). The geology of these ores is different from that of the Beldih ores. But the literature on this deposit is scant or inaccessible

[2] Under study by P. Sengupta and others, Jadavpur University.

Iron Oxide Breccia Hosted Cu–Au–U Mineralisation in NSMB (Between Dalma Volcanics and CGC)

Evidences of this type of metallisation are recorded from the following two areas in the north of Chandil and possibly also at the localities mentioned in the preceding paragraph. These occurrences have not been explored in depth and available information is sketchy.

1. Iron-oxide breccia type polymetallic mineralisation around Chandil, West Singhbhum district, Jharkhand: Promising uranium mineralisation (Fe-U-Cu-Au-REE) associated with iron oxide breccia is reported (Mishra, 1998) from the neighbourhood of Chandil (22°57′: 86°04′) over 3.5 sq km area. The radioactive brecciated carbon phyllite and ferruginised quartzite along an E–W trending regional shear/fault zone recorded 90 to 770 ppm U_3O_8 between Kantaldih and Bandhdih. The ferruginised breccia zone is also associated with Cu (191–498 ppm), Ag (1.2–4.29 ppm), Au (0.25–1.6 ppm) with high content of Fe_2O_3 (10–60%).
2. The iron-oxide breccia along regional E–W trending shear zone passing through the north of Lawa-Maysera gold prospects, Singhbhum district: In the metasedimentary belt lying to the north of Dalma volcanic belt there is a prominent E–W trending shear/fault zone passing between the Lawa–Mysera tract of gold prospects and the Northern shear zone. The shear/fault zone under reference is characteristically marked by linear ferruginised and silicified breccia resembling lateritic capping over the country rock, dominantly consisting of mica schist, carbon phyllite and cherty rhythmite. Strikewise, this zone with local forks and flays is intermittently traceable from Tamar (Ranchi district, Jharkhand) in the west to south of Beldih, Purulia district, West Bengal in the east. The zone passes through Malati Pahar on the Chandil–Purulia road, between Raghunathpur and Balarampur, where the country rocks are involved in shearing and extensive alteration to form a clay (ochre) deposit.
   The ferruginised breccia is made up of angular fragments of hematite, quartzite, chert and some phyllite, cemented by ferruginous material (hematite, goethite, limonite) and silica. There is intricate network of silica veinlets in the breccia, which show specs and disseminations of pyrite and chalcopyrite. At some places limonitic box work is also seen. The breccia zone is also reported to be significantly radioactive. The individual stretches of breccia outcrops, forming low strike ridges in the rolling country, are from a few hundred meters to more than a kilometre in length, the width generally not exceeding 50 m. The Atomic Minerals Directorate conducted drilling in some of the target areas along this zone, identified through radiometric surveys.

Metallogeny in the Gangpur Basin

The Gangpur basin contains ore deposits of base metals and manganese. But they are not very large. The base metal deposit, predominantly of lead, occurs at the Sargipalli (22°03′N : 83°55′E) area and the manganese deposit at Ghoriajhor.

Sargipalli Lead (– Zinc–Copper) Ore Deposit Ball (1877) first reported some roundish pebbles consisting of a mixture of oxide and carbonate of lead in a section of the Ib river near Talpatia (21°57′N : 84°04′E). Apparently led by this clue, Krishnan discovered the site of the ore mineralisation close to the village Sargipalli in 1937. Detailed exploration started towards late sixties, proving it to be a deposit having an ore reserve of ~ 6 million tonnes with an average metal content of 5.75%. Underground mining on this deposit started towards mid-seventies and closed down in late nineties.

Productivity aside, Sargipalli is an interesting base metal deposit from geological point of view. The mineralisation, which is generally stratiform (Fig. 4.2.26 and Fig. 4.2.27), apparently occurs within the Ghoriajhor Stage in the stratigraphic succession proposed by Kanungo and Mahalik (1975). There are two parallel orebodies in the deposit, located almost at the interface of the banded calc-silicate rocks and the overlying pelitic schists. Of these, the hangwall orebody is larger, more than 2 km along the strike and 5–12 m in thickness. Interestingly the orebodies are not only stratiform, but also stratified except where affected by later deformation and metamorphism. 'Wall rock alternations', whatsoever are absent. The main orebody has been cannibalised by later granitic intrusives.

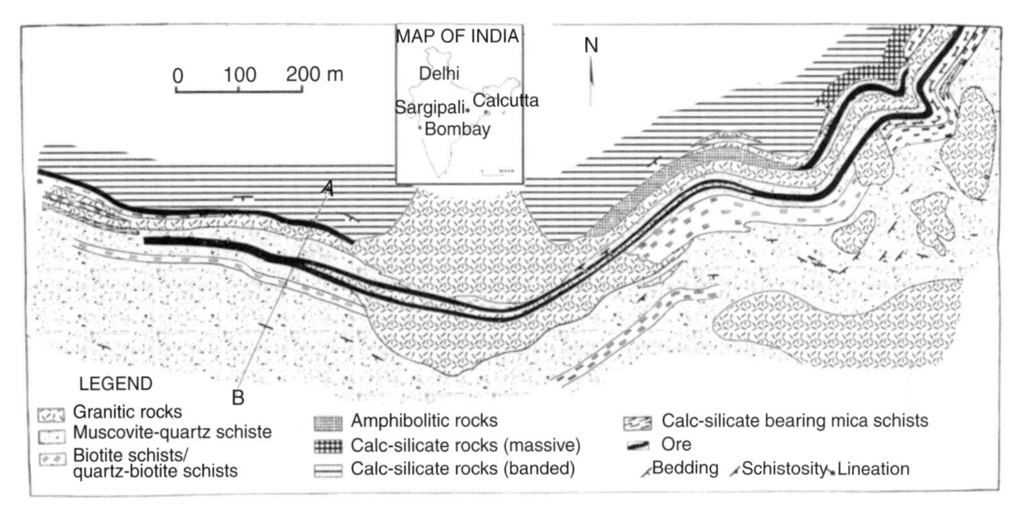

Fig. 4.2.26 Geological map of the Sargipalli lead (-zinc-copper) deposit (after Sarkar, 1974)

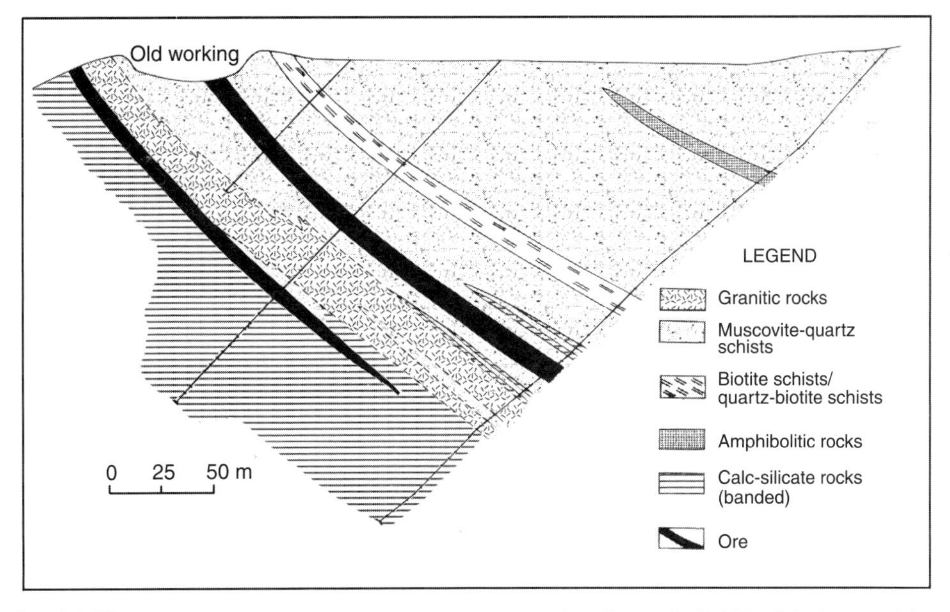

Fig. 4.2.27 Transverse geological section along boreholes, Sargipalli Pb(-Zn-Cu) deposit (after Sarkar, 1974)

The country rocks are garnetiferous mica schists in the hangwall side and calc-silicates, composed of quartz + calcite + plagioclase ± hornblende ± diopside ± clinozoisite, in the footwall side. The silicate assemblages correspond to upper green schist - low amphibolite facies.

The principal ore mineral is galena. The abundant phases comprise sphalerite, chalcopyrite, pyrrhotite and pyrite. Arsenopyrite, tennantite, tetrahedrite and loellingite are the minor phases. The gangue minerals are same as those, which constitute the immediate host rocks. The ores show metamorphic fabric. From an invariant assemblage of pyrite + arsenopyrite + pyrrhotite (+ S-As liquid + vapour, now absent) a temperature of equilibration of ~ 500°C (494° ± 12° C), with a positive pressure correction of 18°C/kb, according to Clark, 1960 and Barton, 1969. These observations prove the isofacial metamorphism of the ores and their host rocks (Sarkar, 1974). $\delta^{34}S$ in galena ranged between – 2.9 and + 4.51 per mil (Rai and Changkakoti, 1995). Pb-isotope composition of the Sargipalli galena shows very small variation in isotopic ratios and a high μ- value ($^{238}U/^{204}Pb$) of 11.74 (Viswakarma and Ulabhaje, 1991).

Regarding the origin of this deposit opinions vary. Kar Ray (1972) and Rajarajan (1978) proposed a hydrothermal origin of this deposit, whereas Sarkar (1974), Rai and Changkakoti(1995) and Vishwakarma (1996) preferred a sedimentary diagenetic origin of the deposit. The principal arguments on which the latter view is based are (i) stratiform and stratified nature of the ore bodies, (ii) lack of wall rock alteration (iii) isofacial metamorphism of the ores and the host rocks, (iv) homogeneity of $^{206}Pb/^{204}Pb$ ratio throughout the sedimentary sequence. The high value of μ undoubtedly points to the fact that the main source of lead for the deposit was crustal rock of granitic composition. The obtained model-Pb age for the Sargipalli galena ores is 1665 Ma (Viswakarma and Ulabhaje, op.cit.). This does not contradict a time range of 1700 – 2000 Ma preferred by Sarkar, S. N. (1980) for the deposition of the Gangpur sediments.

Manganese Mineralisation in Gangpur Fermor (1909) first reported the presence of manganese ore bodies associated with gonditic rocks (metamorphic rocks rich in Mn-silicates, such as spessertine garnet, rhodonitic pyroxene, etc.) and metamorphosed pelitic sediments in the then Gangpur State, now in the Sundargarh district of Orissa. Krishnan (1937), in course of his detailed study of the geology of Gangpur also studied the manganese ores and manganese-bearing rocks in fair details. Manganese-bearing rocks and manganese ores occur at a number of places but their best development took place in the Ghoriajhor–Monomundra area. According to Krishnan (op. cit.) this manganese bearing rocks and ores occur nearly at the bottom of the Gangpur 'Series' (Group) in his suggested stratigraphic succession. The very basis of his proposed succession was his interpretation of the regional structure as an anticlinorium. Applying the presently accepted top-bottom criteria, Kanungo and Mahalik (op. cit.) interpreted the regional structure to be rather a synclinorium. That reverses the whole sequence. We accept the latter interpretation and think that the manganese bearing Ghoriajhor Formation is located near the top of the stratigraphic succession of the Gangpur Group, rather than near the bottom, as suggested by Krishnan (op. cit).

The Mn-ore minerals reported from the ores comprise braunite, bixbyite, hollandite, jacobsite, hausmannite, vredenburgite, manganite, pyrolusite and cryptomelane (Deb and Majumdar, 1963). Petrography of these ores agrees with other gonditic ores from elsewhere in the country (Roy, 1966). However, Nicholson et al (1997) suggested a mixed source for Mn.

Gold in Singhbhum and Adjoining Areas

Gold mineralisation in Singhbhum region can be broadly grouped under three major geological set-ups (Gupta, 2010) (Fig.4.2.28):

1. in the volcanisedimentary supracrustals (Badampahar–Gorumahisani belt, IOG-I) occurring along the eastern edge of the Archean cratonic nucleus (SGC)

2. in the south-western flank of the SGC, along the base of IOG-II of Jamda–Koira–Noamundi valley
3. in the North Singhbhum (Proterozoic) Mobile Belt (NSMB).

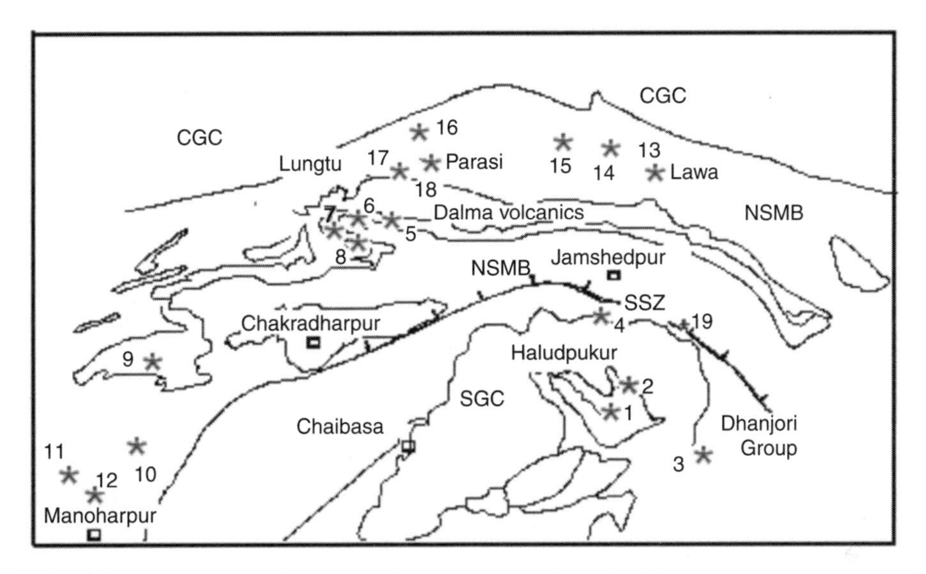

Fig. 4.2.28 Map showing locations of gold prospects in Singhbhum region, Eastern India: 1.Kundarkocha, 2. Porojharna, 3. Barunia, 4. Bhitardari, 5. Chatuhasa, 6.Taramba, 7. Siadih, 8. Rugudih, 9. Sausel, 10.Ankua, 11. Pahardia, 12. Rungikocha, 13. Lawa, 14. Maysera, 15. Ichagarh, 16. Babaikundi, 17. Parasi, 18. Sinduari, 19.Associated with Cu-U ore in SSZ.

Gold in Singhbhum-Orissa Archean Cratonic Nucleus (SGC) Gold mineralisation in this set-up is mainly restricted to the supracrustals of the eastern IOG-I belt, which is profusely intruded by 3.1 Ga old SBG-B. The most notable concentration is located in Kundarkocha area in the East Singhbhum (e) district.

Kundarkocha Gold Deposit The gold mineralisation at Kundarkocha (22° 28′: 86° 15′), occurring within the NNE–SSW trending Iron Ore Group (IOG) rocks of the Badampahar–Gorumahisani belt passing through the eastern edge of the craton, belongs to the first category. Here, the IOG rocks crop out in an extensive fork shaped disposition with one of its arms passing from Kundarkocha to Rajnagar and the other extending from Digharsai to Jaikan. The litho-assemblage is represented by a basal cherty quartz-arenite (± fuchsite quartzite), overlain by two distinct horizons of BIF separated by a 15–20 m thick banded quartzite. The banded quartzite has numerous intercalations of black chert, carbonaceous (graphitic) phyllite and hybrid tuffaceous rocks. The sedimentary package is intimately associated with mafic-ultramafic volcanic rocks, including typical komatiites.

The Kundarkocha deposit, which was mined in the past and recently reopened, is the best prospect known so far in the Eastern Indian shield. It is situated at 24 km SSE of Haludpukhar on the Tatanagar–Badampahar section of South Eastern Railway. Dunn (1937, 1941) first described the deposit as extraordinarily rich in visible gold, sometimes assaying several hundred ounces per ton.

Gold mineralisation is in smoky and bluish quartz veins (often with fine carbon dust) traversing the black chert, gaphitic phyllite and the talc chlorite schist. It occurs as sparsely disseminated fine specks. Sulphides, mostly pyrite, are associated with some of the gold veins. Better concentration of gold is noted at Porajharna, Kariam, Bodadungri, Rangara and Suraikhadan spread over 2 sq km area. Unlike these localities, the mineralisation at Digarsai is in white quartz veins. There are several old workings in the area of which the one at Porajharna had two shafts and one adit. A rich ore shoot of 600 m length with fairly steep pitch was delineated during later exploration, having records of upto 1300 gm/t Au in some underground samples (Ziauddin and Narayanaswami, 1974; Banerjee and Thiagarajan, 1969).

MECL, during 1984–88, conducted limited exploratory mining in the Porajharna block of Kundarkocha and estimated an ore reserve of about 4500 tonnes with 19.5 gm/t Au, for a rich ore shoot (Singh et al., 2005). The two-level mine development by MECL has brought out a bonanza ore zone corresponding to main lode striking E–W and another feebly mineralised zone striking NNW–SSE. The main lode was explored between the two levels by a raise and below second level by a winze. The strike and grade of ore does not show continuity even in close proximity. The borehole intersections have revealed presence of parallel lodes. The property is presently held under mining lease by a private company.

Gold Occurrences in the South-western Flank of SGC Minor occurrences of gold are reported from the south-western margin of the cratonic area from the metavolcanics the siliciclastic dominant metasediments bordering the south of the western IOG -II belt of Jamda–Koira–Noamundi valley and flanking the south-western fringe of SGC. Along this zone several old workings/ancient pits for gold were located in the Keonjhar district, Orissa, extending over 60 km along a NNW-SSE trending tract between Nuakot (21°46′: 85°12′45″) in the north and Madra (21°20′42″: 85°20′24″) in the south.

The gold mineralisation in this area is hosted by a litho-package comprising basalt, conglomerate–arenite–BIF, mafic–ultramafic intrusives, sub-volcanic acidic rocks and quartz veins. In Gopur–Banspal prospect (21°23′10″: 85°20′24″), auriferous quartz veins traverse the volcanisedimentary sequence along schistosity planes. Drilling by the Directorate of Mining and Geology (DMG), Orissa, revealed three sub-parallel lode zones characterised by Au-bearing smoky quartz veins,having cumulative strike length of 1.2 km, width ranging from < 50 cm to maximum of 4 m. The middle lode is most persistent with grade ranging upto 20 g/t Au. The auriferous quartz veins contain pyrite, pyrrhotite and ilmenite. In another nearby locality around Telkoi, the Geological Survey of India (GSI) identified an auriferous zone spread over 15 sq km area, based on surface primary gold dispersion pattern in the volcanisedimentary assemblage, comprising phyllite, sericite schist, conglomerate and metabasics. In the same area U-Au-bearing QPCs were also identified by the Atomic Minerals Directorate (AMD). The gold content is generally 1–2 g/t. DMG, Orissa, has also delineated another U-Au bearing QPC horizon in the vicinity.

Besides the above two areas, preliminary geochemical surveys were conducted in several other blocks like Raiguda, Sukdera, Gajpur, Chulia, Chuahuli, etc., in the Nuakot–Madra stretch, but none of those seemed to be potential, except for being indicative of the existence of a Late Archean to Paleoproterozoic auriferous environment in the region.

Gold in North Singhbhum Mobile Belt (NSMB) The river Subarnarekha (meaning golden streak), which is the major river flowing through NSMB is known for extensive gold panning since time immemorial. Gold mineralisation, mostly of moderate concentration, is distributed over wide tracts of NSMB (Fig. 4.2.27).

In the Domain I, i.e. the metasedimentary belt in the south of SSZ, there are sporadic occurrences of vein quartz hosted gold-sulphide mineralisation in the low-grade phyllites. At Bhitardari and adjacent area, south of Tatanagar, there are several auriferous quartz veins of limited extension. Some of these are exposed in old pits excavated by ancient miners. Gold mineralisation as detrital grains in the matrix of the Quartz pebble conglomerate (QPC) of the Lower Dhanjori Formation is noted at Haludbani, Barunia and Sanudor. The matrix is composed of quartz, fine-grained phyllosilicates, chlorite and pyrite. Other heavy minerals are ilmenite, magnetite, rutile and zircon. Besides gold, the association of uranium in the matrix of Dhanjori QPC has already been described. Three bulk samples collected from QPC bands exposed in Sanudor area yielded average grades of 0.68 gm/t, 0.51 gm/t and 0.41 gm/t Au.

The Domain II, i.e. the Singhbhum shear zone (SSZ) or Singhbhum Cu-U belt is known for the association of gold with the copper as well as uranium mineralisation. By-product gold to the tune of 500–600 kg has been produced per annum from this belt.

Gold content in run-off-mine (ROM) and mill-concentrate samples of some of the copper and uranium mines in Singhbhum Cu-U belt are furnished in Tables 4.2.6.

Table 4.2.6 *Gold content in ROM and Mill concentrate samples of some Cu / U mines, Jharkhand (Rao et al., 1999)*

Ore	Nature of the sample	Au range (ppm)
Mosabani copper ore	ROM ore	0.25-0.41
	Cu-concentrate	3.0
Rakha copper ore	ROM ore	0.07
	Cu-concentrate	9.6
Surda copper ore	Gravity concentrate from floatation tailings	7.0
Jaduguda uranium ore	ROM ore	0.11
Narwapahar uranium ore	ROM ore	0.07-0.11

Gold in Mosabani and Jaduguda ores occurs in native state with fineness > 900. Rao et al. (1999) suspected that it could originally be detrital at Jaduguda. Gold was recovered as by-product from the Singhbhum ores during 1985–98. During 1997–98, 612 kg of gold was extracted in the Maubhandar plant.

In Domain III and IV, i.e. the Chaibasa, Dhalbhum and Dalma belt, the gold mineralisation is mainly located in the Sonapet valley (antiformal) area and Manoharpur area. In Sonapet the mineralisation is in quartz reefs and veins penetrating the Chaibasa, Dhalbhum and Lower Dalma Formations. The important localities in this area where gold occurrences have been prospected are Chatuhasa, Taramba, Siadih, Bandih, Dalbhanga and Rugudih. In the western folded extremity of the Dalma belt there is another occurrence of gold mineralisation in Lower Dalma sequence at Sausel, where six old pits aligned roughly in ENE–WSW direction exposes auriferous quartz veins along sheared carbonated ultramafic volcanic suite.

Along the northern limb of the Sonapet antiform, auriferous blue and dark grey quartz veins of about 1 m thickness occur in thinly laminated magnetite-sericite-biotite schist and hornblende schist occurring between Chatuhasa and Dalbhanga. The western reef at Chatuhasa is reported to have yielded 63 gm/t Au, the middle reef 36 gm/t Au while at Dalbhanga 6 gm/t Au (Ziauddin and Narayanaswami, 1974). The auriferous vein at Siadih analysed average 1.3 gm/t Au. The river draining the Sonapet valley carries placer gold in the recent alluvial sand, the source of which are the auriferous zones in the surrounding hills as well as a thick gravel bed dissected by the river.

Pahardia Gold Deposit In Manoharpur area the most important prospect is located at Pahardia, where the mineralisation is traced over a strike length of 600 m, width varying from 5–10 m. Interbanded quartz–chlorite schist, tremolite–actinolite schist and sericite-quartz schist/sericite quartzite are intruded by thin gold-sulphide bearing quartz veins and quartz–carbonate veins. The mineralised zone trends NE–SW with 50–70° northerly dip. Pyrite is the most common sulphide in the auriferous zone, occurring with quartz-carbonate gangue as isolated idiomorphic grains, sometimes in perfect cubic shape. Chalcopyrite occurs in association with pyrite as sub-idiomorphic grains. Magnetite is fairly abundant as discrete grains while ilmenite occurs as laths and needles aligned parallel to the schistosity. Gold occurs mostly as independent grains and flakes. At places, gold grains of 30-40 microns size are seen to occur as mesh within pyrite

There are two old vertical shafts testifying serious prospecting ventures of the past. Samples from one of the shafts analysed 7.5 and 5.5 gm/t Au over widths of 3.6 m each for two zones. The other shaft yielded 13.2 and 8.7 gm/t over widths of 0.9 and 1.0 m respectively. Two dump samples analysed 18.5 and 8.3 gm/t Au (Ziauddin and Narayanaswami, op. cit.). Exploration by drilling (GSI) established resource of 0.25 Mt with 3.85 gm/t Au upto 65 m vertical depth. The total resource includes ore reserve of 22,000 tonnes of 13.11 gm/t Au and 27,000 tonnes of 7.81 gm/t Au (Singh et al., 2005).

In Domain V, i.e. the metasedimentary belt lying to the north of Dalma volcanic belt, there are some of the most prominent gold deposits and occurrences found so far in the NSMB. From east to west these are located at Lawa, Maysera, Ichhagarh, Burudih, Babaikundi and Parasi over an E–W stretch of about 80 km.

The gold deposits are located along the middle of the western half of the belt having comparatively lower metamorphic grade.

Lawa Gold Deposit The deposit is located near Lawa village in the north-west of Chandil (22°57′: 86°04′) at a road distance of about 20 km. Rhythmic alternations of pelitic, psammopelitic and psammitic layers with occasional carbonaceous beds, amphibolites and acid volcanic rocks constitute the regional litho-assemblage. The deposit area exposes a large-scale upright asymmetric antiformal F_2-fold, plunging moderately towards WNW, recorded by a thick quartzite band interlayered with pelites and psammopelites (Fig.4.2.29). Gupta (1975, 1986) provided a detailed description of this deposit and various aspects of metallisation.

There are three sub-parallel discontinuous ore zones within a strike length of 1000 m in the thick southerly dipping quartzite horizon (southern limb of the fold) that forms the Bhalukkhad–Tamapahar section of the southern ridge (Fig. 4.2.30). The footwall-most northern ore body, which is characterised by a pyrrhotite-pyrite rich sulphide assemblage, has a strike persistence of 500 m, width varying from 10–20 m. The centrally occurring ore zone consists of a few discontinuous orebodies characterised by chalcopyrite rich sulphide-gold assemblage which were worked in the past for gold in Bhalukkhad (East and West) and Tamapahar mines. The individual orebodies of Bhalukkhad and Tamapahar are about 100 m in length separated by 250 m of barren ground. The width of these lensoid bodies vary from 3–10 m. There is another small mineralised lens in the same strike towards the west of Tamapahar having a gap of 400 m. Drilling has proved that these ore bodies are of lenticular shape without much strike or depth extension beyond the limits of old

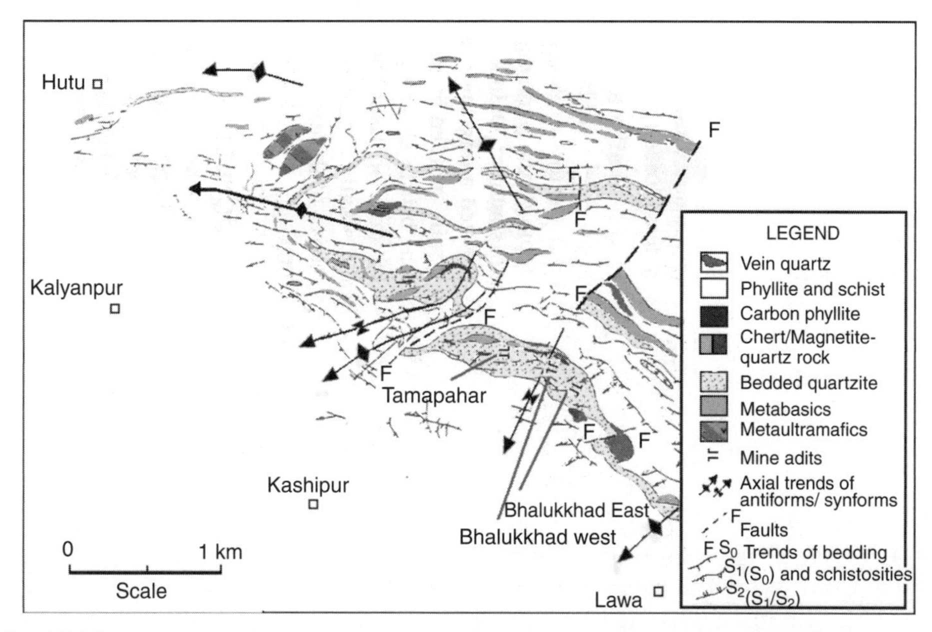

Fig. 4.2.29 Geological and structural map of Lawa area showing locations of the old gold mines at Bhalukkhad East and West and Tamapahar (Gupta, 1975, 1986)

workings. The hangingwall (southern) ore zone comprises a few detached lenticular bodies with maximum length of 200 m, occurring close to the southern contact of the quartzite horizon. These orebodies also have chalcopyrite rich sulphide-gold assemblage.

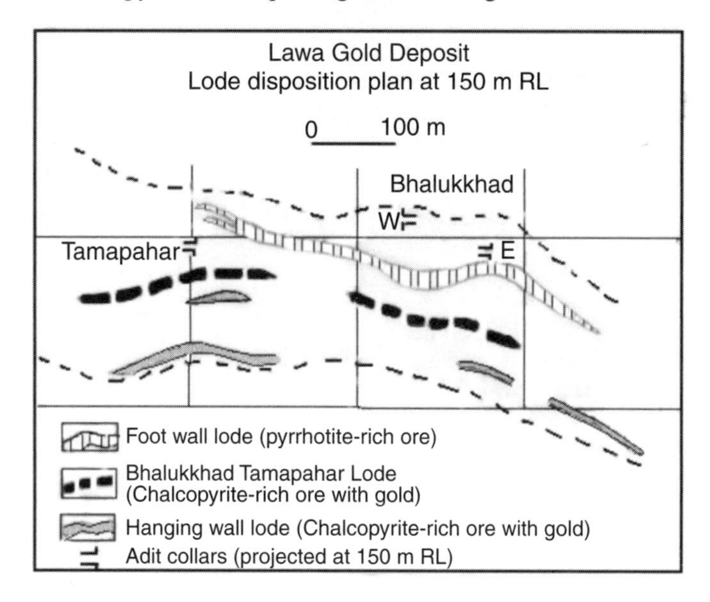

Fig. 4.2.30 Lode disposition at Lawa gold deposit (Gupta, 1975)

All the three old workings at Lawa are located at ridge tops, above the water table, having strike adit entrances. The Bhalukkhad orebody has been worked from two sides in the East and West mines by separate adit entrance and two levels developed over a total strike length of 80 m up to a depth of about 30 m. These two old workings are not connected in the underground. At the western part of the deposit Tamapahar working itself covers a similar strike length with one adit entrance and two levels. Underground channel samples drawn from the adit levels (Arogyaswami and Dutta, 1948) of Bhalukkhad workings showed auriferous bodies with 11.85 gm/t Au over 22.5 m length x 2.44 m width in Bhalukkhad East mine and 11.08 gm/t Au over 37.95 m length x 1.76 m width in Bhalukkhad West mine. About 20 years later the exploration was resumed in this belt by GSI and the underground mines were sampled upto deeper levels (revealing grades upto > 20 gm/t), followed by drilling. However, no strike or depth continuity of the ore shoots mined in the past or existence of new blind shoots could be proved within the limits of drilling down to about 150 m vertical depth from the surface exposures.

The two sulphide assemblages described above are also distinctive in their mode of occurrence. The pyrrhotite rich ore occurs mostly in the form of thin layers parallel to bedding stratification or as minute elongate grains disposed parallel to the schistosity. The chalcopyrite (+ Au) mineralisation occurs more often as patches and clots occupying the fractures and interstices of quartz veins and lenticles. There is no finite wall rock alteration pattern associated with either of the ore types, though neoformed minerals like sericite and chlorite are prominent in their proximity. However, sericitisation and chloritisation are also seen away from ore zones and the same is true with silicification and tourmalinisation also.

The ore minerals in all the lodes are mainly sulphides though there are also other phases like oxides, diarsenides, native elements etc., found intimately associated with them. Besides the principal sulphides, both the ore types have minor proportions of sphalerite, cubanite, mackinawite, skutterudite, safflorite, löllingite, bismuthinite, magnetite and ilmenite. Covellite, chalcocite, pyrite, malachite, smythite, native copper and hydrous iron oxides are important supergene minerals. The ore assemblages at Lawa record ample evidence of deformation and metamorphism along with the host rocks in form of recrystallisation in presence of stress, static annealing as well as low temperature deformation (Fig.4.2.31.). The pyrrhotite rich mineralisation appears to be syngenetic while the chalcopyrite-gold mineralisation is related to a later hydrothermal phase of ore emplacement (Gupta op. cit.).

Two levels of drilling in the deposit revealed several interesting aspects of metallogeny in this belt. The two types of sulphide assemblages, which have separate spatial distribution in the area, are distinctive in many respects. The pyrrhotite rich ore occurs in a black phyllite-carbonate-chert sequence having mafic tuffaceous bands in proximity. The other assemblage, which is rich in chalcopyrite, occupies highly sheared segments of the quartzite where there has been profuse impregnation of silica in form of veins, veinlets, lenses and stringers. Chemically, the pyrrhotite rich ore has comparatively higher Mn (400–800 ppm), higher Pb (upto 200 ppm) and no Ag or Au. The chalcopyrite rich ore has higher concentration of Co (upto 1000 ppm), Mo (100–250 ppm), Zn (> 1000 ppm), As (upto 1000 ppm) and average content of 50 ppm Ag and 0.75 ppm Au. Besides the much richer concentration of gold encountered in this ore type in the old mines, the drill intersection of a small chalcopyrite rich mineralised body to the west of Tamapahar revealed two copper lodes (1.0 m × 2.8% Cu and 0.8 m × 1.25% Cu) with a 3.0 m parting with gold content of 4 gm/t.

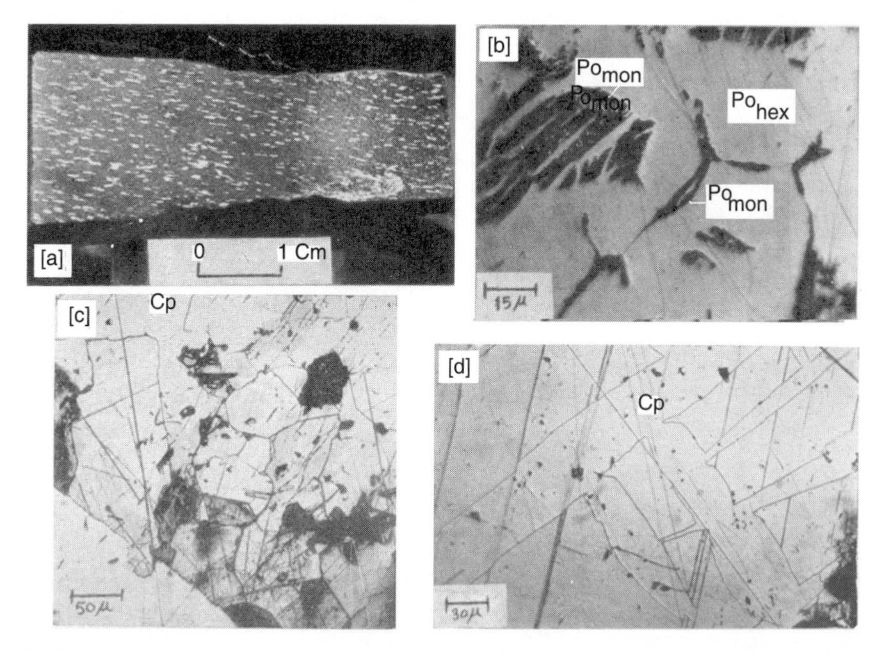

Fig. 4.2.31 Lawa ores: (a) Drill core specimen showing 'sulphide schistosity' displayed by pyrrhotite grains in carbonaceous phyllite; (b) Photomicrograph of etched pyrrhotite showing triple point, development of dark coloured monoclinic pyrrhotite Po_{mon} within and along straight boundaries and triple points of light coloured hexagonal pyrrhotite and Po_{Hex}; (c) Photomicrograph of etched chalcopyrite showing annealed texture with straight boundaries and triple points; (d) Photomicrograph showing two sets of spindle shaped deformation twins in chalcopyrite (Gupta, 1986).

Mayasera Gold Deposit At about 25 km north-west of Lawa, auriferous quartz reefs occur within metapelitic country rock comprising phyllite and sericite-quartz schist. The E–W trending quartz reef, ranging in width from 1–2 m, is intermittently exposed over a strike length of about 600 m and is marked by a chain of old mining pits and excavations. The quartz reef and adjacent quartz veins dip to the north parallel to a set of crenulation cleavages (S_3), which dissect the southerly dipping regional schistosity.

Lungtu–Parasi–Sinduari Gold Prospects This is a recent find by GSI where exploration is still continuing. The area lies in the western part of Domain V, in the north-west of Sonapet antiform. This shear controlled mineralised zone occurs close to the northern contact of Dalma belt and appears to be hosted by the metapelites–volcanics of northern belt rather than lower Dalma volcanisedimentary package. The latter represented by carbonaceous phyllites is exposed along the southern part of the area. The rock types encountered in the deposit area (Western, Central and Eastern blocks) are magnetite–biotite–quartz–sericite schist (most extensive), phyllite with quartz schist (mineralised horizon) ferruginous quartzite, tuffaceous quartzo-feldspathic rock (acid tuff), rhyolite and acid agglomerates, ultramafic tuffaceous actinolite-tremolite schists, metabasics and siliceous carbonates (Fig. 4.2.32).

Sheared quartzose schists, mylonitic at places, and traversed by quartz and quartz–calcite veins and phyllites with quartz ribbons are the major hosts for gold mineralisation (upto 10 gm/t). The ore zone extends for more than 700 m, trending NE–SW with steep dip on either side. The width of the orebody varies from less than a meter to > 7 m. (GSI, ER- News 2004–2005; Singh et al., 2005). A photomicrograph showing dissemination of native gold and arsenopyrite grains in the orebody of Parasi is presented in Fig. 4.2.33.

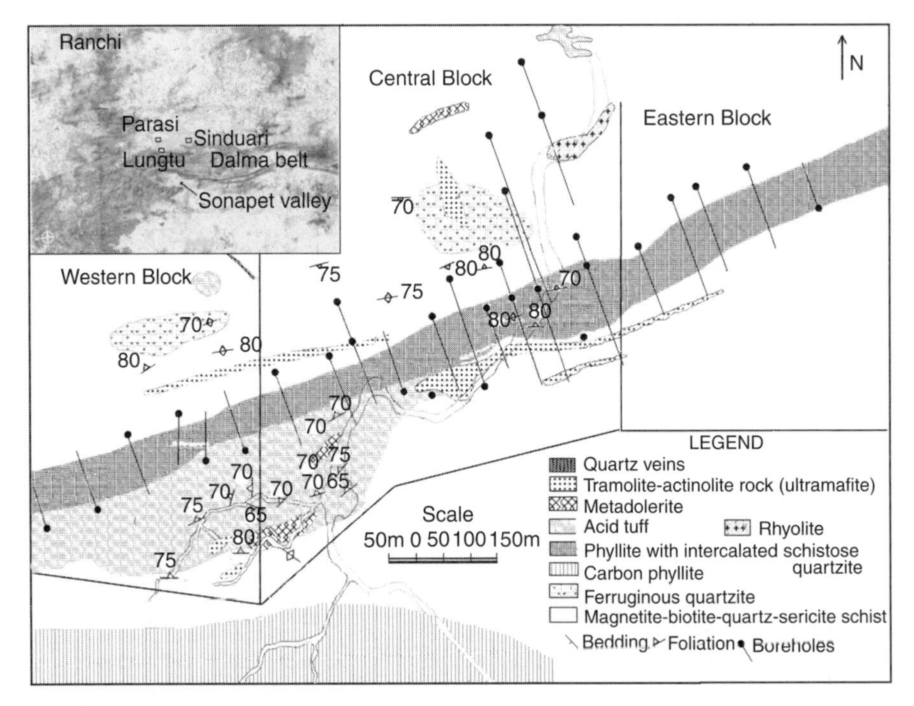

Fig. 4.2.32 Geological map of Parasi gold prospect, Ranchi district, Jharkhand (modified after GSI – courtesy Dr. R.N.Singh)

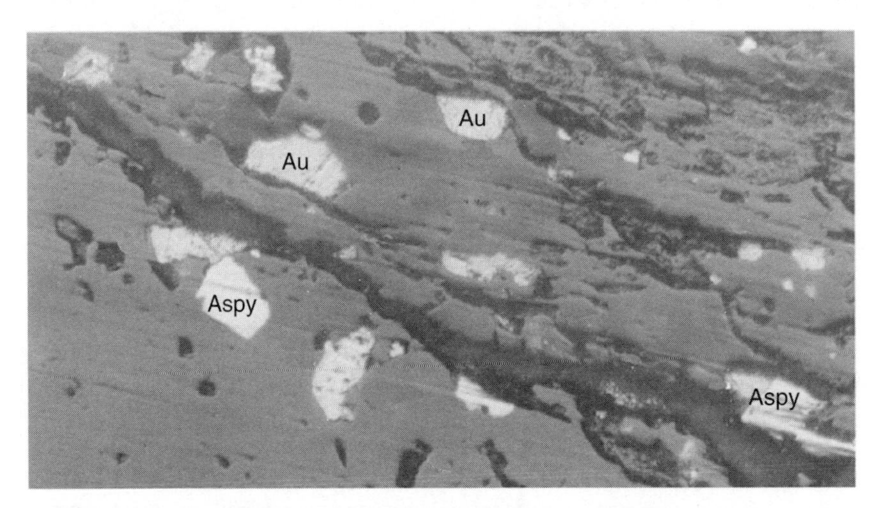

Fig. 4.2.33 Photomicrograph of disseminated gold (Au) and arsenopyrite (Aspy) in the ore body (reflected light), Parasi prospect, Jharkhand (courtesy: Dr. R.N.Singh, GSI)

The reserve estimates of 'Indicated' category worked out so far at different cut-off grades of 0.5 ppm, 1.0 ppm and 3.0 ppm are 1.19 Mt with 1.26 gm/t , 0.54 Mt with 2.44 gm/t and 0.14 Mt with 6 gm/t average grades respectively. Adding up the above with the estimates under 'Inferred category' the total reserve estimates stand at 2.38 Mt (1.25 g/t) under 0.5 ppm cut-off, including 1.0 Mt (2.52 g/t) at 1 ppm cut-off and 0.26 Mt (6.48 g/t) at 3 ppm cut off.

Tungsten Mineralisation in West Bengal

In Eastern India there are two known nearby prospects of tungsten ore around Chhendapathar (22° 45′: 86° 45′) and Porapahar (22° 57′ 30″: 86° 49′), in the Bankura district of West Bengal (Fig.4.2.34).

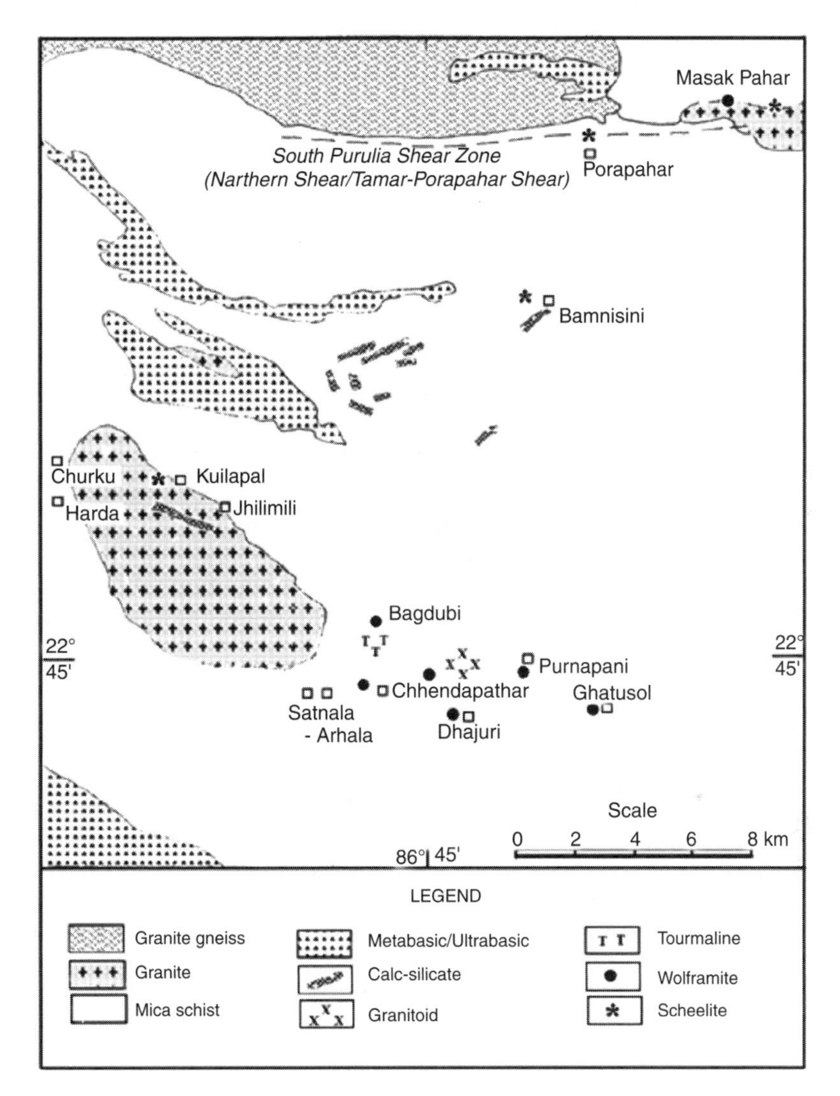

Fig. 4.2.34 Generalised geological map of northeastern segment of NSMB showing locations of tungsten mineralisation in Bankura district, West Bengal (modified after Sengupta and Gangopadhyay, 1994)

The Chendapathar area for some time now is known in Indian geology for economic concentration and mining of tungsten ore. M/s Tata Sons and Co. Ltd. reportedly attempted to mine tungsten ores from this area in early twenties of the last century. M/s Gouripur Industries Pvt. Ltd. mined the ore during 1952–1970, with occasional breaks. The area outside private leasehold (Porapahar and surroundings) was taken up for detailed exploration by the Geological Survey of

India (Kar et al., 1975; Ghosh and Ghosh, 1979). The Directorate of Geology and Mines, West Bengal, and the universities also took up studies, mainly around Chhendapathar area (Singh and Verma, 1975; Misra et al., 1999; Baidya and Dasgupta, 2002).

The country rocks in Chhendapathar area is the metamorphosed sedimentary–volcanic rocks of the Singhbhum Group (Domain V of NSMB, discussed in Section 4.1), and a few granitic stocks, collectively known as the Chhendapathar Granite. The Porapahar tungsten occurrences lie close to the northern contact of the Domain V with the CGC, which defines the Tamar–Porapahar (Northern shear) zone. The metasedimentary rocks in the area are represented mainly by mica schists that contain besides quartz and muscovite such minerals as almandine garnet, staurolite and kyanite. Locally present tourmaline, "mainly controlled by sedimentary layering (So) and S_1-schistosity" (Baidya and Dasgupta, 2002), is an interesting accessory mineral in these rocks. These pelitic metasediments are interbanded with felsic and mafic volcanics. The Chhendapathar granitic rocks are specified by Baidya and Dasgupta (op. cit.) to be granite–granodiorite and assigned to the 'volcanic arc granite' and 'syn-collisional granite', without presenting any discriminant diagram in support of their contention. A larger and better studied granitic pluton of Kuilpal is located at 6–7 km in the northwest from Chhendapathar. The Chhendapathar Granite is not dated, but as already mentioned, the Kuilapal Granite is dated to be 1600 Ma (Rb-Sr / WR).

The nature of the mineralised zones and assessed tonnage and grade of the individual deposits are described below.

Chhendapathar Area In Chhendapathar area, there are two prominent sets of mineralised veins, viz, (i) the NNW–SSE trending low westerly dipping Thanpahar–Cheradungri veins and (ii) the E–W trending Purnapani group of veins with northerly dip (Fig. 4.2.35). The wolfram bearing minerals occur as distinct bladed crystals, stringers, lenses and pockets, which are irregularly distributed in the quartz veins. The main ore mineral is ferberite. It is associated with scheelite, hydrotungstite, tungstite, magnetite, hematite (specularite), martite, goethite, manganese oxide, chalcopyrite, pyrite, siderite, etc. Goethite, limonite and a yellowish bloom of hydrotungstite and tungstite characterise the weathered and oxidised products of the wolfram bearing ore zones.

Thanpahar–Cheradungri The Thanpahar vein is a single bodied lens with an outcrop length of about 90 m and average width of 25 m. The vein strikes at NNW with about 15°dip towards west. The vein is weakly mineralised with spotty and sporadic pockets, lenses, stringers and disseminations of wolframite. Limonitised zones within the vein have WO3 ranging normally between 0.03% and 0.7%. Significant mineralisation is restricted to about 30 m length of the vein.

The Cheradungri vein is a composite flat lying vein with an outcrop length of about 80 m and breadth of 30 m. The trend is NNW–SSE. The vein is widely mineralised with pockets, lenses, stringers and grains of wolframite, occurring within limonitised zones characterised by WO3 content ranging from 0.05% to 0.17%. The mineralisation is occasionally associated with scheelite.

Purnapani The Purnapani group comprises a number of en-echelon mineralised veins trending WNW–ESE to EW (Fig.4.2.35). The total srike length of the veins in the Purnapani Group is 500 m and the average width is nearly 2 m. The nature of mineralisation and also the grade of ore within the limonitised zone are similar to that in the Thanpahar and Cheradungri veins.

Other Mineralised Veins in the Surrounding The Dhajuri veins comprise two NE–SW trending sub-parallel veins of quartz located in the vicinity of Dhajuri village. The distance between the two

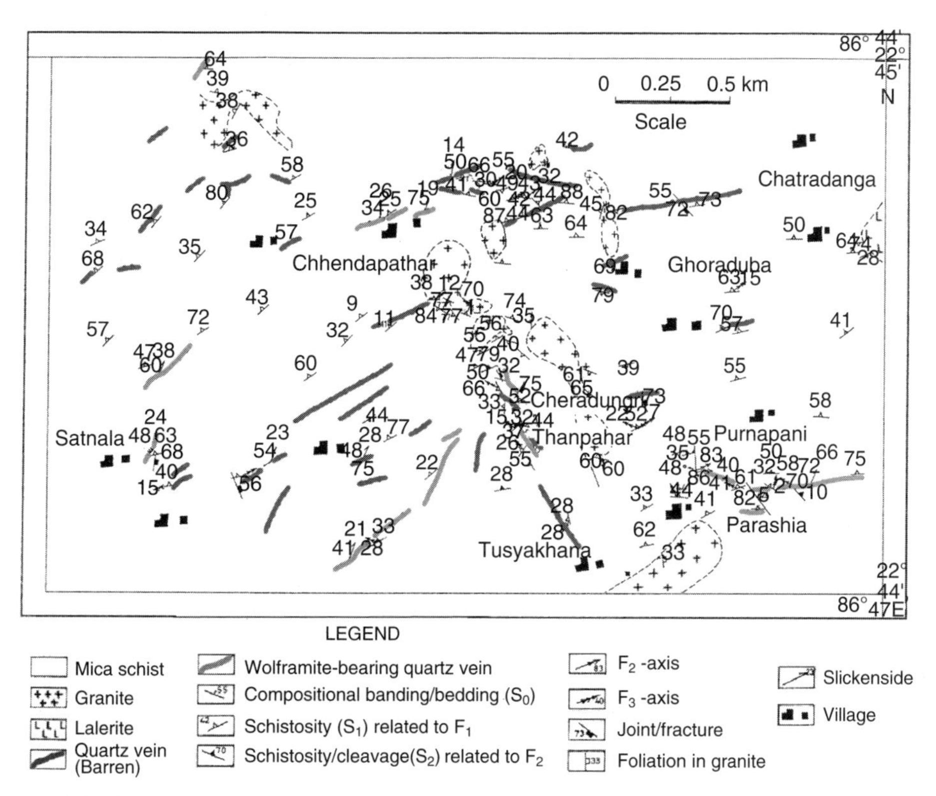

Fig. 4.2.35 Geological map of Chhendapathar area showing the distribution of W-bearing and barren quartz veins (modified after Baidya and Dasgupta, 2002)

veins is 120 m, the dip of both the veins varying from 45°–55° northwesterly. The veins are fissure filling type emplaced in oblique shear fractures. Wolframite occurs within the oxidised zones in the veins defined by oxides of iron and manganese. A number of shoots, pockets and lenses, composed essentially of varying proportion of magnetite and wolframite with some sulphides, such as pyrite and chalcopyrite, occur within the cavities in quartz. In the oxidised zone WO_3 content ranges from 0.05% to 0.44%.

The Ghatusol vein system consists of a number of closely connected en echelon veins of quartz, traversing a highly puckered staurolite–garnet–quartz–muscovite schist, occasionally containing graphite. The strike of the vein varies in trend from N15°E in the northern end to N10°W in the southern end and the dip varies from 40° to 50° towards east. The vein is intensely oxidised. Relicts of wolframite are characterised by the presence of altered siderite, magnetite and manganese ores.

At Satnala-Arhala area, lying close to the southeastern margin of Kuilapal Granite, ferberite-bearing quartz veins are disposed in en-echelon pattern along axial planar S_2 schistosity within the garnetiferous andalusite- bearing mica schist. The mineralised zone, traced over a strike length of 500 m, is impersistent and oxidised with pockety occurrence of the tungsten bearing minerals. The individual quartz veins, however, do not exceed 35 m in length and 2.5 m in width. Stratified tourmalinite composed of millimetre thick rhythmic alternations of tourmaline and quartz-rich layers and layer parallel or cross-cutting tourmaline-quartz veins (largest 100 m long and 30 m wide) are reported from the vicinity of the mineralised zone near Bagdubi. The tourmaline-rich layers and veins are composed of dravite, quartz and ilmenite with minor sulphides. The tourmalinite bands display excellent preservation of sedimentary structures and also the bedding folds (Sengupta and Gangopadhyay, 1994; Gangopadhyay and Sengupta, 2004).

Subordinate mineralisations are also reported from Parasia, Tusya-Khana, Satnala, Bheduasol, Ratanpur, Ghoraduba, Chatradanga. The Marupahar vein has been intensely oxidised and only very low-grade wolfram mineralisation has been encountered.

Ore Reserve The GSI estimated about 2, 14,380 tonnes of ore reserves (average grade 0.06% WO_3) in the three blocks of Chhendapathar area with the assumption of 20 m of depth continuity of the ore bodies. In Thanpahar, the mineralised veins considered for reserve estimate of 56,700 tonnes was 35 m in length and 30 m in width. The estimates for Chheradungri veins for 70 m length and 28 m width was 1,05,840 tonnes. In Purnapani, estimated reserve was 51,840 tonnes for a mineralised vein of about 500 m length and nearly 2 m thickness (Ghosh and Ghosh, 1979). Susequently, the Directorate of Mines and Minerals, based on further suface sampling and drilling in the aforesaid area, re-estimated the total ore reserves as 1,73,063 tonnes with 0.26% WO_3 , down to 20 m vertical depth (Baidya and Dasgupta, 2002).

Copper is another important metal that is associated with tungsten ores, but it never makes the grade. Along with chalcopyrite and pyrite, other sulphides present either as minor or trace phases are galena, sphalerite, pyrrhotite, molybdenite, arsenopyrite and bismuthinite. At Dhajuri, a probable reserve of 30,375 tonnes of ore has been estimated for the two mineralised veins.

Porapahar Area The Porapahar hill, which is situated at about 3 km WSW of Khatra, hosts wolfram mineralisation at about 200 m NNW of the highest peak of the hill. A number of quartzite bands, trending E–W to WNW–ESE with steep southerly dip, are interbedded with schists. A mylonitised quartzite band occurring towards the north of the main peak of the Porapahar hill is irregularly fractured and traversed by numerous criss-cross veinlets of quartz. The veinlets are mineralised with tungsten ore in varying degrees of concentration and by their anastomsing character form a stockwork in the sheared quartzite. The extension of the stock work has been established over a length of 750 meters along the strike. The mineralisation is in the form of disseminated oxide ores of tungsten with value of W ranging from 0.002 to 0.50%. The oxidised ores have been identified to be hydrotungstite, tungstite and tungsten bearing limonite/goethite. The wolframite occurring in pockets is found to be oxidised into iron and manganese minerals along the margins.

The estimated probable reserves of the ore zone (length: 105 m; width: 6 m; expected vertical continuity: 40 m) in the Porapahar stockwork are 68,364 tonnes. The possible reserve, inclusive of probable reserve, calculated for a 245 m long ore zone with 6 m width came to 1,58,760 tonnes.

Mineralisation Tungsten, mainly in the form of wolframite, occurs in quartz veins traversing the mica schists and is to-date unreported from the granitic rocks of the area. Other than wolframite, tungsten occurs in form of scheelite, cupro-tungstite, ferritungstite and hydrotungstite (Baidya and Dasgupta, 2002). The main phase of tungsten bearing mineral, however, is reported as ferberite by all other workers referred to above.

Baidya and Dasgupta report the presence of tungsten in the metapelitic and metavolcanic rocks, besides quartz-veins, but do not indicate the mode of its occurrence in former two rocks (Table 4.2.7). As we know, tungsten is unlikely to go into early silicate structures. Instead it could occur as minute/small quartz-wolframite inclusions, or as wolframite dissemination. If that is the case, it may not be qualitatively much different from the quartz-wolframite veins. According to Sengupta and Gangopadhyay (1994), the wolframite mineralisation is entirely within the quartz veins while scheelite is noted within the country rocks.

Table 4.2.7 *Tungsten contents (WO_3%) in different rocks in the Chhendapathar area, Bankura, West Bengal (Baidya and Dasgupta, 2002)*

Location	Vein quartz	Metapelitic rocks	Metavolcanic rocks
Thanpahar	0.066–2.244	0.008–0.042	0.021
Cheradungri	0.008–0.180	0.01–0.185	0.008
Purnapani (including Parasia)	0.021–0.403	0.012–0.31	-
Chatradanga	0.016	0.276	-
Ratanpur	0.022	0.012	-
Satnala	0.062	0.010	-
Sankhagara	0.049	0.015–0.017	-
Birkanr	-	0.009–0.012	-

Note: The number of samples analysed for each type from each area is not mentioned in the publication.

All former investigators, including most of those referred to in an earlier paragraph, shared the view that the Chhendapathar mineralisation was an example of typical hydrothermal mineralisation related to granitic intrusives. Baidya and Dasgupta (2002) tried to detail this aspect. According to them, sedimentation and concomitant acid volcanism in an island arc type environment resulted in syngenetic concentration of ore metals in the sediments and volcanics. Anatexis at the culminating stage of the first regional deformation and metamorphism (D_1–M_1) of the sedimentary–volcanic pile generated the Chhendapathar Granite. Wolframite–quartz veins, formed from the hydrotherms given out by the anatectic magma, were located along the schistosity (S_1) planes. The second event of deformation and metamorphism (D_2 –M_2) followed. Formation of ore veins along S_2 and remobilisation of the earlier veins into other disjunctive structures characterised this stage. Still later, anastomosing ore veins and ore-stockworks took place 'during and after the events of faulting and shearing'.

Normally, a granitic magma, anatectic or otherwise, becomes fractionated to be enriched in LILE and HFS elements (eg., W, Mo, Nb). After partial cooling and /or depressurisation, the magma becomes H_2O-saturated to produce hydrothermal solutions carrying incompatible/excess elements in solution. Ore mineralisation takes place at the endo- and exo-contact zones of the intrusive body at its upper parts. Tungsten mineralisation in such a situation will be ferberite-dominant, with minor scheelite, molybdenite, bismuth, bismuthimite, chalcopyrite etc. Mineralogically, the tungsten mineralisation at Chhendapathar resembles the granite-related type. The stratabound tungsten deposits, mainly as scheelite, occur in specific lithologies: metabasites, meta-carbonate and calc-silicate rocks and rarely in tourmalinite. Syngenetic/syndiagenetic origin of many such deposits have been discussed by a number of workers (Plimer 1987; Palmer and Slack, 1989; Raith and Prochaska, 1995). None of these criteria are satisfied by observations on the ore-field under review, except in that tourmalinite and calc-silicates are present. Gangopadhyay and Sengupta (2004) described tourmalinites interlaminated with pelitic sediments and also the tourmaline-quartz veins from this area. The former is considered by them to have volcanic exhalative origin while the latter is attributed to remobilisation during deformation and metamorphism and the tungsten mineralisation is also tacitly correlated with these processes. However, a clear spatial or temporal relation between tungsten and tourmaline or the calc-silicates is lacking, except for their general association at places. At this moment the evidences are far from definitive to suggest an overall SEDEX type metallogenic environment as suggested by Gangopadhyay and Sengupta (op. cit.), though such a possibility may be kept open. On the other hand if the tungsten could have been concentrated in the protolith from hydrothermal solution derived from a granitic melt, it remains inconceivable at the moment, why and how the tungsten bearing quartz veins and the granitic bodies of the area remained spatially unrelated all over.

The P–T of Chhendapathar tungsten mineralisation as determined from fluid inclusions in vein quartz ranged between 2.3kb/385°C and 1.63 kb/364°C, probably with an initial temperature of more than 450°C (Mishra et al., 1999). The fluid was not only generally highly saline, its carbonic composition included CH_4 besides CO_2, suggesting reduced nature of the granitic magma, at least at the time of separation of the ore fluid. At a late stage, less saline fluid that is recorded in the fluid inclusions could have a large meteoric contribution.

BIF Associated Hematitic Iron Ores of Eastern India

The banded iron formations (BIF), along with associated hematitic iron ores were discovered by P.N.Bose in the wee years of the last century on the Gorumahisani hills of Mayurbhanj, then a small princely state and now a district of Eastern Orissa. Investigations resulting in discovery of more or less similar mineralisation took place at the Sulaipat and Badampahar region further south. Investigations in the neighbouring regions of Keonjhar (Kendujhar) and Bonai led to the discovery of large deposits of BIF and hematitic iron ores in the Jamda–Koira Valley, particularly along the limbs of the 'horse-shoe' structure of Jones (1934) (Fig. 4.2.36). Some other important reports made on these deposits until about the middle of the last century are by Dunn (1937, 1941, and 1953), Spencer and Percival (1952).

The second phase of the study of the BIFs of this region was initiated by Sarkar and Saha (1962). But this time it was on the stratigraphic position of this important rock formation in the geological sequence of the region. The latest stratigraphic position of this rock formation has already been discussed in the Section 4.1. In Eastern India, the banded iron formations (BIFs) are often designated as banded hematite jasper (BHJ) or banded hematite quartzite (BHQ) by the exploring and mining agencies, depending on the nature of the silica-rich bands in the BIF. These nomenclatures have been retained in the maps, sections and other figures hereunder, as per the usage by the different agencies. Petrologic, structural and chemical studies followed which though not adequate by themselves, yet produced some results that are interesting for whatever they are worth. An account of the iron ore resources of this region and the mode of occurrence of the ores in the important deposits, their physical and chemical properties, as also their genesis, are discussed here.

Major Deposits and Resources The Indian reserve of hematitic ores is 12.2 billion tonnes, which consists of 6.8 billion tonnes of 'proved' category, 2.12 billion tonnes of 'probable' category and 3.40 billion tonnes of 'possible' category. Besides, there is a reserve of greenstone-associated banded magnetite ore, amounting to > 6 billion tonnes. Out of the above, the reserves of hematitic ore in the Eastern India stood at 7458 Mt (Orissa – 4177 Mt and Jharkhand – 3281 Mt) as on 1st April, 2000, accounting for more than 60% of the national iron ore resources (Indian Mineral Yearbook, IBM, 2003)[3]. The Eastern India in the recent years shared about 45% of the country's annual production of (> 200 Mt) of hematitic ores (2009–2010).

More recent data being inaccessible to us, the grade-wise reserves of hematitic ores in Eastern India, as reported by Chatterjee, 1993, are presented in Table 4.2.8. The data, though somewhat old, might none the less reflect the overall proportion of the various types and grades of iron ore available in the region. As the country is presently passing through a boom in the iron ore mining, exports and related industries, there is also a fast growth in the fields of resource augmentation as well as production. As such, the above mentioned data are rather indicative and may not be accurate for future references.

[3] Indian Bureau of Mines (IBM) declared India's hematitie iron ore reserve to be 14.6 Bt and magnetitic iron ore reserve to be 10.62 Bt (yearbook, 2005). These are geological estimations.

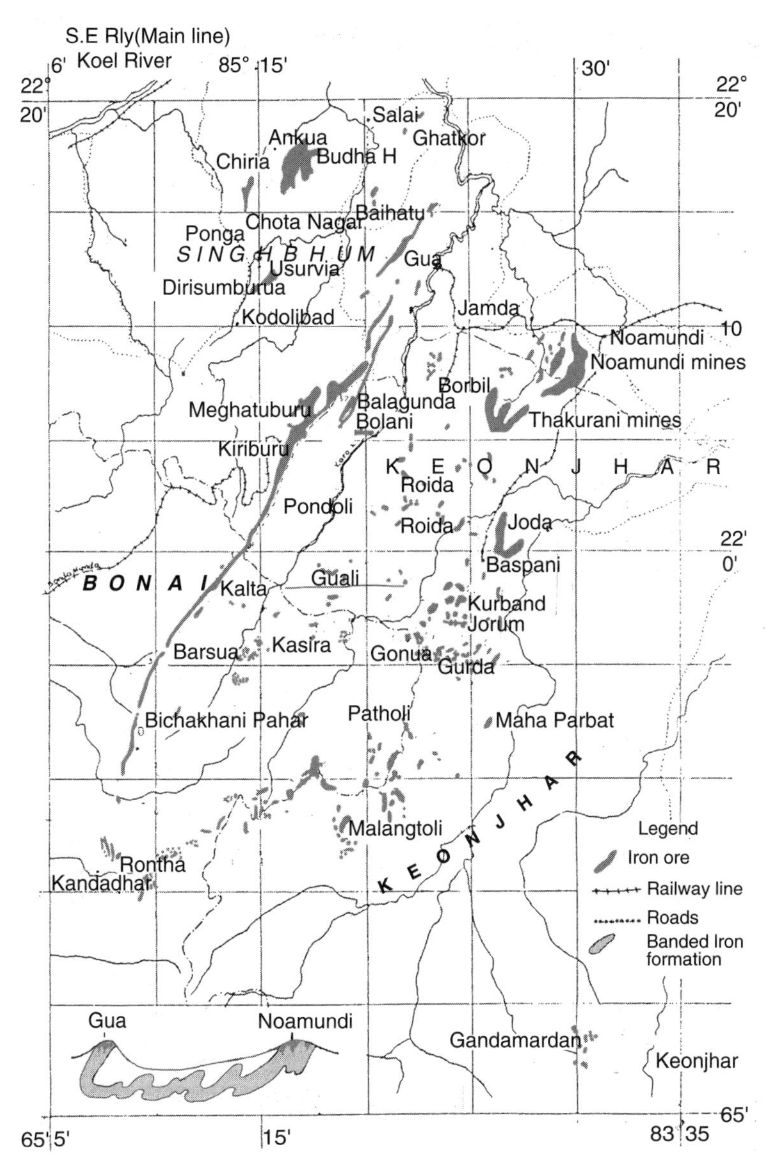

Fig. 4.2.36 Map showing distribution of iron ores along the 'horse shoe', Jamda-Koira valley, Singhbhum–Orissa region (after Jones, 1934)

Table 4.2.8 *Grade-wise reserves (Mt) of hematitic ores in Eastern India (Chatterjee, 1993)*

Region	High Grade + 65% Fe	Medium Grade 62–65% Fe	Low Grade < 62% Fe	Unclassified	Blue dust/ Black Iron ore*	Total
Jharkhand	-	1660.43	111.96	310.86	-	3083.25
Orissa	82.39	1440.54	728.80	342.01	8.68	2602.34
Total	82.39	3100.94	1840.76	652.87	8.68	5685.59

* Black iron ores: iron ores containing ≥ 10% Mn

During the years since 1993 the region produced 300–400 Mt of iron ores of high and medium grades. Total reserves should therefore diminish by that amount. But even otherwise these values are proximate. The present authors through their personal queries with the mines in this region have come to know, that still there should be at least 80–100 Mt of ores having + 65% Fe in the Jharkhand area. If the claim of Kiriburu mine is justified then one may add another 180 Mt of ores to this category. Enquiry with the authorities of the operating mines shows that the reserve of the medium grade (62–65% Fe) in the Jharkhand deposits should be around 1970 Mt. Further, there is no data on the blue dust in the above table for Jharkhand. But we know from personal experience that all the deposits contain blue dusts and in some deposits in substantial quantities. These are now used in sinters and pellets. It may be pointed out that the estimate of the iron ore reserves in the country is less well-constrained, compared to many other mineral commodities.

The Eastern Indian deposits are located in the southern part of the western Singhbhum district of Jharkhand (Chiria, Gua, Noamundi, Kiriburu, Meghahatuburu deposits), Keonjhar–Sundergarh–Cuttack districts of South-Central Orissa (Thakurani, Joda, Banspani, Malantoli, Bolani, Kalta, Barsua, Gandhamardan, Daitari–Tomka deposits) and the Mayurbhanj district of Eastern Orissa (Gorumahisani, Badampahar, Sulaipat deposits). The iron and steel plants in India that are run with the ores from this region are Tata Steel (erstwhile TISCO), Rourkela Steel Plant (RSP), Durgapur Steel Plant (DSP), Bokaro Steel Plant (BSP) and Indian Iron and Steel Company Ltd (IISCO). Each of these plants has its captive mines (Noamundi, Joda and Khondbond for Tata Steel; Barsua and Kalta for RSP; Bolani for DSP and Gua and Monoharpur (Chiria) for IISCO. The Chiria deposit is the largest single deposit of iron ore (> 2000 Mt) in the country, but only marginally exploited so far (Fig. 4.2.37 a–c). Gua deposit, located at about 40 km north of Chiria, sustained extensive mining (Fig. 4.2.38).

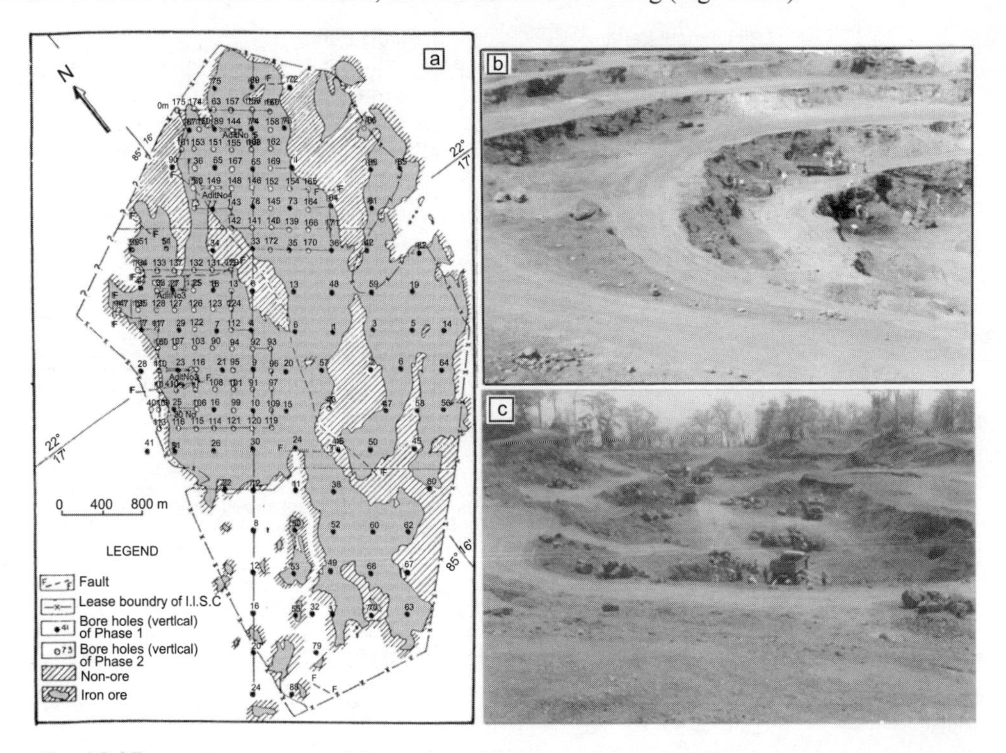

Fig. 4.2.37 (a) Deposit map of Chiria, West Singhbhum district, Jharkhand, showing extension of high grade iron ore (reserve > 2 Bt) delineated by systematic drilling (courtesy: Mine authorities); (b) and (c) Open-cast mines in the southern parts of Chiria deposit, showing shallow excavations done till 2005 (photographs by the authors).

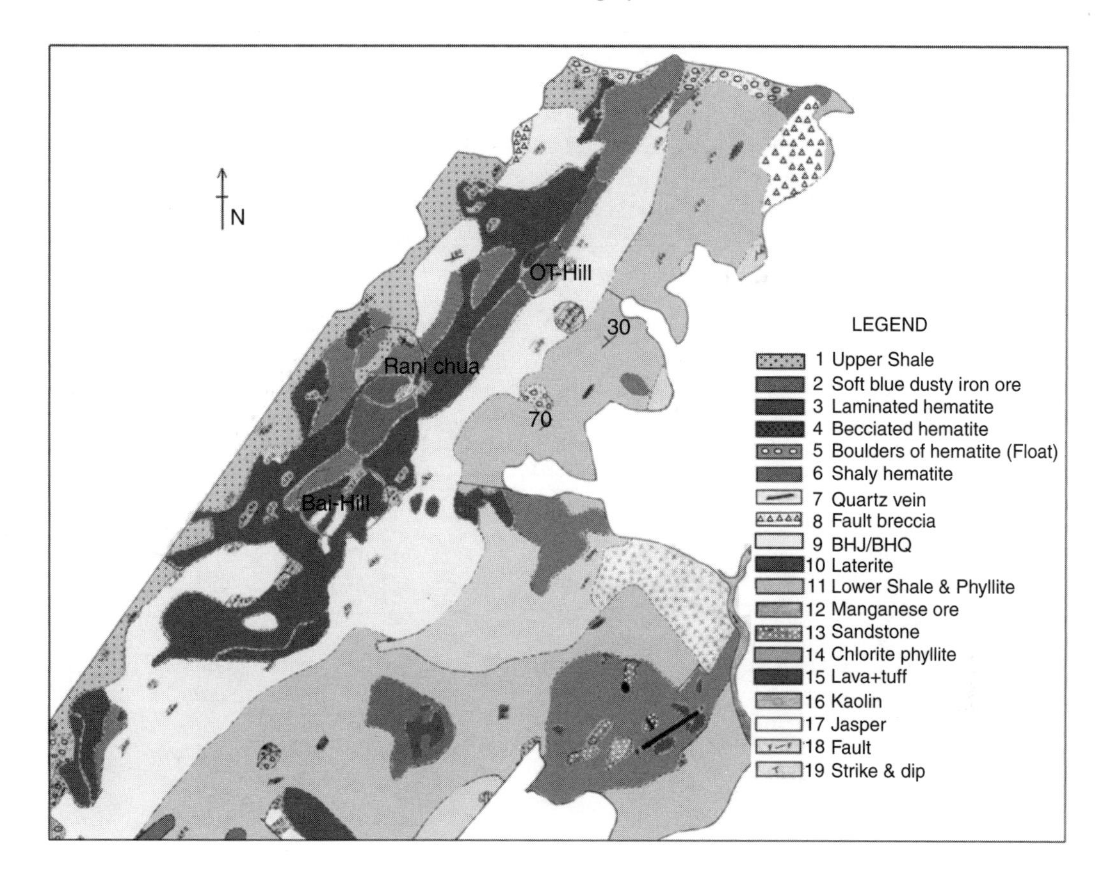

Fig. 4.2.38 Geological map of Gua iron ore deposit, West Singhbhum district, Jharkhand, showing ore body overlying BIF (BHJ/BHQ), which together are sandwiched between two shale horizons. The Lower Shale is associated with basic volcanics, represented by chloritic schists. Interfingering of ore types is notable (courtesy: mines authorities).

There are also several major iron ore deposits in the Noamundi–Thakurani Pahar–Joda areas of Jharkhand and Orissa, which are being mined since long. Panoramic views of a few mines of this area are presented (Fig. 4.2.39).

Classification of Ore-types The nomenclature of the various ore-types found in the iron ore deposits of Eastern India vary to some extent depending on the usage by the mining companies. However, based mainly on their physical properties, the ores are generally categorised as Hard Ore (HO), Friable, Flaky or Platy Ore (FO), Blue Dust (BD), Lateritic or Goethitic Ore (LO) and canga.. The terms 'biscuity' and 'laminated' are also in use for the flaky ores. The principal characteristics of the major ore-types, including their physical property, Fe-content and mineral compositions are tabulated in Table 4.2.9.

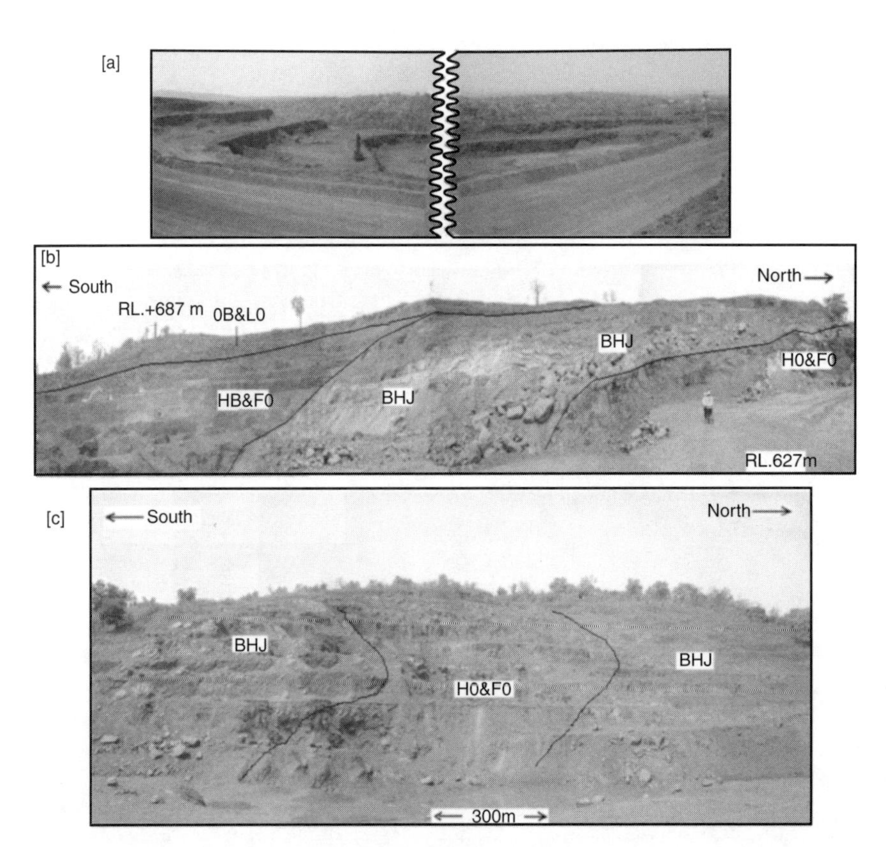

Fig. 4.2.39 (a) View of Noamundi East Mine (Tata Steel Ltd.), Jharkhand, a view from east to west; (b) Mine face at Khondbond 'Q' deposit (Tata Steel Ltd.), Orissa, showing ore zone both above and below BHJ horizon; (c) View of an OMC mine between Joda and Khondbond, Orissa, showing nearly 100 m of excavation by the development of 12 benches, each with 8 m height. Note flat ridge profile parallel to the strike in all the deposit areas (photographs by the authors).

Table 4.2.9 *Physical and chemical properties and mineral composition of the East Indian iron ores*

Ore-types	Important physical properties	Fe-content	Ore minerals	Gangue minerals (Minor–trace phases)
Hard or massive ore (HO)	Hard, massive to blocky;often microlaminated; steel-grey	65–68%	Hematite>> martite ± goethite	Quartz, fengite, kaolinite, talc, Al-phosphate
Friable, flaky or laminated ore (FO) (including biscuity/platy varieties)	Cohesive; flaky, laminated; commonly fragmented	62–66 %	Hematite, martite	Kaolinite
Blue dust (BD)	Fine (fractions of a mm), cohesion-less granular; deep bluish grey/ black	65–68%	Hematite. with minor martite	Quartz
Lateritic, goethitic ore (LO)	Highly porous; brown/ yellow	56–58%	Goethite, limonite	Kaolinite, rutile
Canga	Pieces of BIF/hard ore and other rocks cemented by Fe-oxide (hydroxide)	58–60%	Hematite, goethite	Cryptocrystalline silica, rock-fragments

In any iron ore deposit, all the above ore-types are not always present, but most of them are. They may interfinger as seen in the geological map of Gua deposit (Fig. 4.2.37) and in the transverse sections across Noamundi deposits (Fig. 4.2.39 a). Generally BIF underlies the ore zone, but in cases the ores may be directly overlying the shales (Fig. 4.2.39b). BIF overlying the ore zone is also not uncommon (Fig. 4.2.38b; 4.2.41 c, d, f and g).

The Indian iron ores are generally characterised by a high Al_2O_3-content and low SiO_2 -content, which result in the Al_2O_3/SiO_2 ratio being > 2, while the market demand of this ratio in the world today is <1. This is so in spite of the fact that the Indian BIFs are generally poorer in aluminium. The ores of the Jharkhand–Orissa region are more aluminous. Today the maximum tolerance limit of ($Al_2O_3 \pm SiO_2$) in the ore-feed for direct reduction route is 2 and that of blast furnace route is 5. The Al_2O_3-content is generally higher in the 'fines' obtained from the washing plants. The mines of the Tata Steel Ltd., ensure maintaining the Al_2O_3 content of sized and fine iron ores to 1.4% and 2.0% respectively, by upgrading the run-of-mine (ROM) ores by dry or wet processes.

In the rich hard ores of Thakurani mines (Block-B), the authors detected (SEM-EDX) minor proportions of phengitic mica, kaolinite, talc and a Fe-bearing alumino-phosphate. The first three phases contribute Si. Goethite–limonite aggregates often do contain some alumina. Another possible source of alumina and silica in the ores could be decomposed micro/meso S-bands originally present in the BIF. A conclusion that the quality of iron ores, in particular respect of ($Al_2O_3 \pm SiO_2$)-contents, improves with depth (Mahapatra, 1993), if confirmed, should be a useful guide in exploitation. Distribution of phosphorus in these deposits is not uniform. It is somewhat higher in the western limb (0.05–0.08%). In Noamundi it is 0.05% on the average, while being much lower (0.02–0.025%) at Khondbond.

It will be interesting to mention why the steel industry throughout the world is so particular today about the quality of raw materials used. Sustained research has shown the following: (i) 1.5% reduction in SiO_2-content in the ores leads to a reduction of 65 kg of slag, i.e. 6.5% volume/ tonne of pig iron. It reduces the coke consumption, and hence reduces the cost. (ii) Al_2O_3 contributes to improved slag-fluidity, but if it exceeds the optimum level, it consumes more energy. (iii) Some experiments in the former USSR had shown that 1% increase in Fe in the furnace feed raises productivity by 2% and reduces coke consumption by 3% (iv) Alkalies are very harmful to the blast furnace as they react with the furnace lining and damage it. They also choke off the flow of gases and make the coke friable. (v) Phosphorous raises the fuel consumption of blast furnaces. Conversion of P-rich pig iron to steel is expensive. (vi) Reduced lump sizes are more fuel efficient. (vii) Sinters are preferable because of fuel efficiency, and also because the ore is partially pre-reduced. Reduction of Al_2O_3 -content to the desired level may be possible by mixing with Al_2O_3-poor ores. Jigging and improved washing methods do reduce the Al_2O_3-content.

Disposition of Orebodies and the Ore Characters The iron ore deposits along the eastern margin of the Jamda–Koria Valley, i.e. along the eastern limb of the 'horse shoe' synclinorium of Jones (1934) together contain the bulk of the iron ores of this region. Ghosh (1993) believes that the deposits along the eastern limb (normal) and western limbs (overturned) of the regional synformal structure differ considerably in terms of iron concentration, thickness of the ore bodies, ore types and the nature of overburden. In the eastern limb the ores are more often richer, hard and brecciated. Thickness of the ore bodies usually varies between 20–50 meters. Lateritic zone is thin to absent. Association with manganese ores is more common. In the western limb, in contrast, the iron ores are generally somewhat poor in grade (58–62% Fe), ores are mostly friable and laminated. Thickness of the ore bodies can be as high as 200 m. The 'hard cap' (lateritic cover) is also thick.

An almost similar case may be cited from the Lake Superior region. There in the Cuyuma Range and in the Marquette Range, where the dips of the BIF are steep, ore developed to depths of 1000–1100 meters, the best however, occupying the top 300 meters or so. In the Mesabi Range, which produced about 70% of the Lake Superior ores, the BIFs dip at shallow angles and the depth of mineralisation is generally < 60 meters. A plausible explanation is that in the steeper parts of the iron formations there is greater penetrability of descending meteoric water.

In the Noamundi–Joda sector, the BIF, shale and the ore bodies are generally low dipping, 20°–25° towards west. There is no evidence whatsoever of overall steep dip (about 60° due west) of the ore zone and the adjacent lithounits at Noamundi or any other deposit in this region, as depicted for Noamundi by Beukes et al. (2002). Locally, the lithounits and the ore zones assume steeper dips due to the folding but they do not affect the general low-dipping disposition of orebodies. The cross sections of this deposit, as authenticated by the company sources (Fig 4.2.40a) represent the actual ground situation. Similar is the overall pattern of ore body disposition (low westerly or easterly dipping) in all other deposits located along the eastern limb of the horse-shoe synclinorium, some examples of which are presented here (Fig. 4.2.40 b–c)

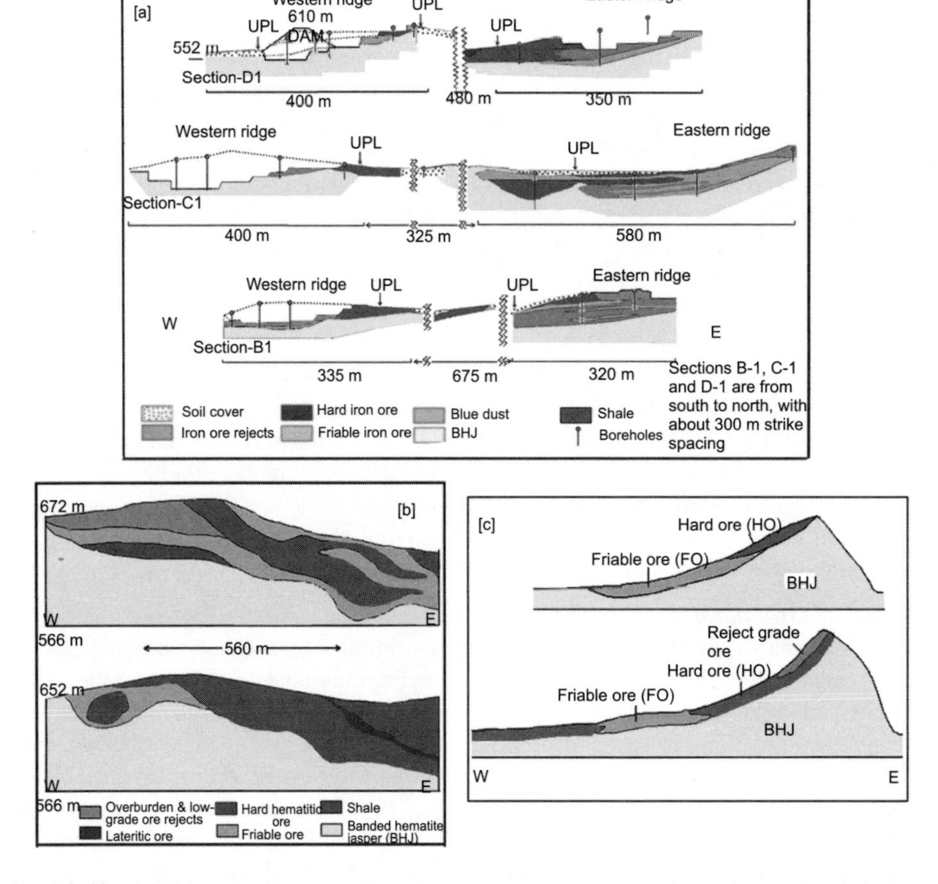

Fig. 4.2.40 (a) Three profile sections across Noamundi iron ore mines, showing low westerly dip of the ore zones and interfingering of different ore types. Borehole cross-sections of (b) Two profile sections across Khondbond deposit, Orissa, showing low easterly dip of the ore zone and (c) Two profile sections across Banspani deposit, Orissa, showing low westerly dip of the ore zone. (Data source: Tata Steel authorities).

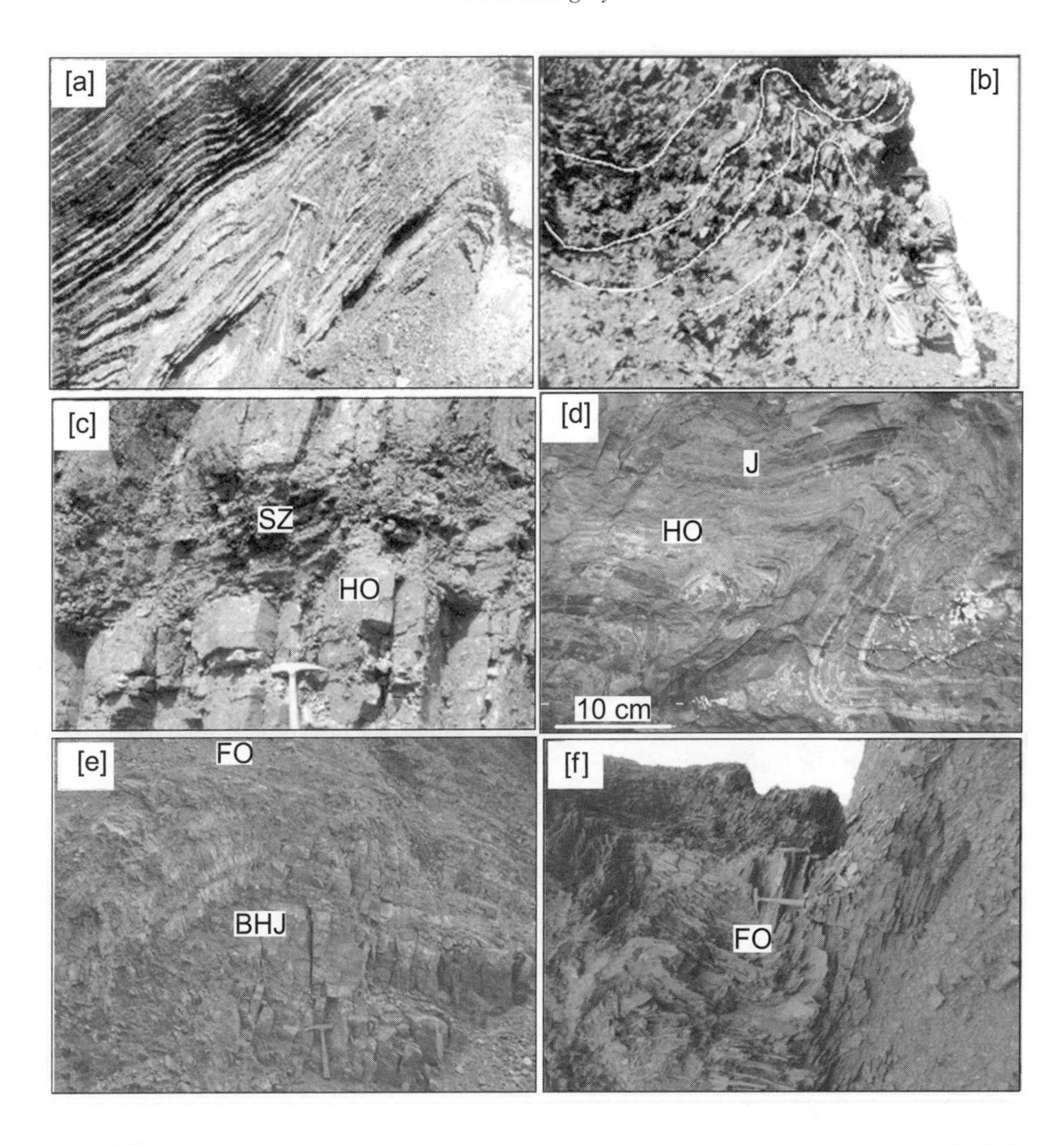

Fig.4.2.41 Field photographs showing (a) Chevron folded BHQ (top left) changing over to Banded hematite shale, Gua mines, Singhbhum; (b) Flexure in high-grade iron ore bands with well-developed axial plane schistosity visible on close inspection, Gua mines, Singhbhum; (c) Hard and blocky high-grade iron ore (HO) with well-developed shears planes (SZ) and fractures of tectonic origin, Thakurani Pahar, Orissa; (d) Folded BHJ with alternate hard ore (dark grey-HO) and jasper (brown-J) bands in Joda mines; (e) Axial planar cleavages and joints developed in a folded BHJ band occurring within friable iron ore (FO), Hill 6, Noamundi East mine; (f) Highly folded friable iron ore (FO), Hill 4, Noamundi East mine. (Photographs by authors).

Tectonic deformations in form of folds and shears are recorded on all scales in the BIF, shale and the various types of iron ore, though folds are more prominently developed in the BIF (Fig. 4.2.41 a–f). In the iron ore deposits located in the eastern limb of the horse-shoe synclinorium of the Jamda–Koira valley, these are mostly upright N–S trending folds developed on the sub-horizontal or low dipping bedding surfaces. At places pervasive sub-vertical or steep dipping axial planar cleavages and joints are developed in the hinge zone of folded BHJ, with local fanning and refraction (Fig. 4.2.41e). The N–S trending sub-vertical/steep dipping joints in the hard massive ore Fig. 4.2.41c) may represent the cleaving under the same compressive stress that folded the adjacent rocks, specially the BIF.

Hard Massive Ore As described earlier in Chapter 4.1, the rhythmic alternation of silica rich or sometimes shaly bands with iron-rich bands in the BIF widely varies in thickness from less than a cm to several metres. The Fe-content of the iron rich bands in the BIF is often very high and such bands are also at places compact and hard similar to the economic hard ore variety. The hard and massive iron ores of considerable width and economic grade (Fig. 4.2.42a), which constitute distinct entities in the sedimentary sequence, occur at many places above the BIF, but it is not necessarily so everywhere. In many deposits the hard ore is overlain by BIF as observed in parts of Joda mines (Fig. 4.2.42c), Roida (Fig. 4.2.42f), Gandhamardan (Fig. 4.2.42g), Kondbond mines (Fig. 4.2.39b) and Katamati deposits. It also occurs above or below shale band or a friable ore zone without showing any direct contact with BIF in the Noamundi–Joda sector. BIF occurring as thin layer-parallel partings of a few meters or lenticular bodies of appreciable dimension within friable ore or blue dust is also not uncommon (Fig. 4.2.42 d, e).

The hard hematitic ore is commonly characterised by thin laminations, often less than a mm in thickness, extending for several meters without interruption. Such laminations, as seen in the microscope, are mainly defined by varying grain size of the ore minerals besides differing proportions of microplaty hematite and martite. Very fine streaks of clay minerals and other silicates are found disposed parallel to the laminae (Fig. 4.2.43 a–d). The fine laminations also display warps, kinks and contortions in geometric harmony with the mesoscopic and larger folds recorded in the rocks.

The banding in hard massive ore is widely believed to be 'relict bedding' showing self-preservation of the primary structure, even after complete transformation of BIF into rich iron ore by supergene or hypogene processes of alteration. However, the morphology of the thin laminations in the hard and compact ores of the present area suggest it to be a primary feature, which is not disturbed by any later process, hypogene or supergene, except to a minor extent evidenced by the presence of silicate minerals in traces. Such ores (with ~ 68% Fe) are almost entirely made up of Fe-oxide without any silica-rich bands. Megascopically, the Fe-oxide lamellae have variable lusture and display textural variation under the microscope. With greater magnification the interlocking microplaty hematite prisms are seen to be the dominant phase along with martite in patches (Fig. 4.2.44).

At places, however, features of replacement and redeposition are seen along discrete bands parallel to the thinly laminated hard ore (Fig. 4.2.45).

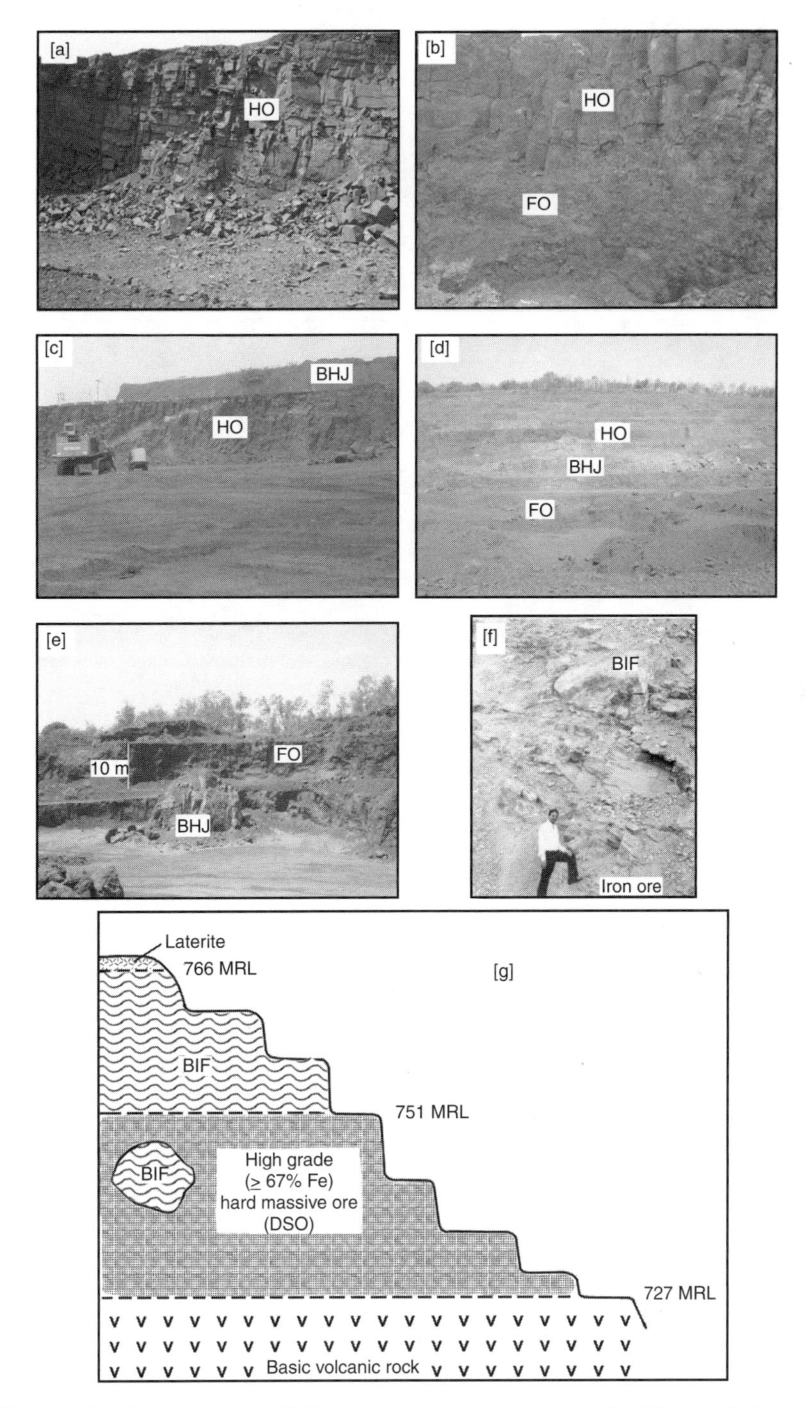

Fig. 4.2.42 (a) Hard blocky iron ore (HO) with a prominent low dipping (bedding parallel) joint and two sets of steep dipping joints, Katamati mines, Orissa; (b) Friable ore (FO) underlying hard ore (HO), Joda mines, Orissa; (c) Bench face in Joda mines showing BHJ overlying hard ore (HO); (d) A lenticular BHJ body within iron ore exposed in the mine face, Sarda (Jindal) mines, Soabil, Orissa; (e) BHJ remnant within iron ore exposed at the floor of Noamundi north-west mine, Jharkhand; (f) BHJ overlying massive iron ore in Roida deposit; (g) High grade iron ore overlain by BIF, Gandhamardan (photographs and sketch by authors).

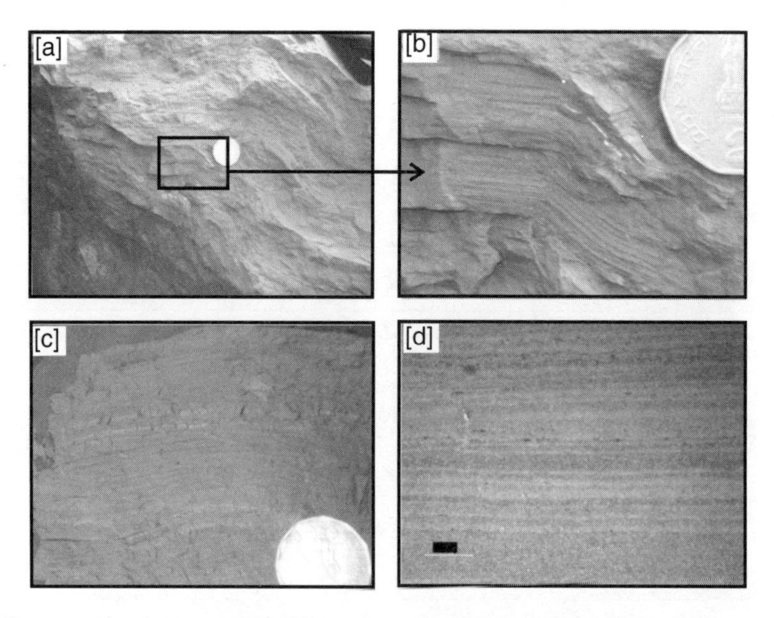

Fig.4.2.43 (a) Hard iron ore showing warped laminations; (b) Enlarged view of a part of (a) showing kinked laminations, Joda mine; (c) Very thin laminations in hard ore, Khonbond Q mine; (d) Photomicrograph of a polished section of hard ore showing micro-bandings, Noamundi (phographs by the authors).

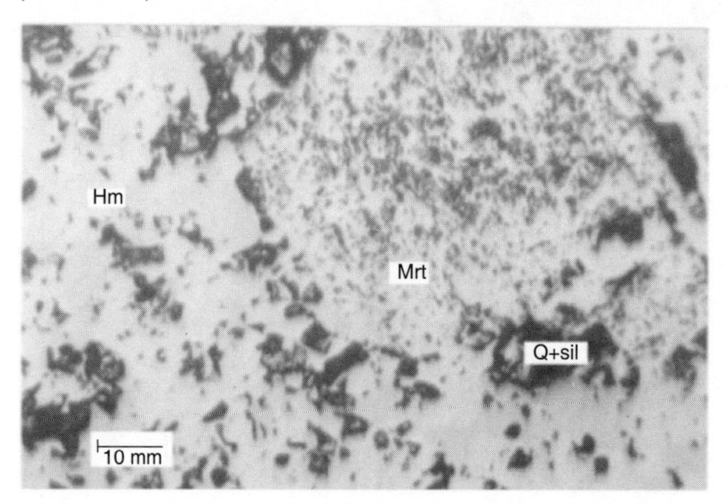

Fig. 4.2.44 Photomicrograph of hard massive ore, Thakurani Pahar, Orissa, showing microplaty hematite (Hm) and martite (Mrt) with quartz and silicates (Q+Sil) as gangue minerals (photograph by authors)

In contrast to the undisturbed finely laminated hard ores or those with replacement lamellae in small-scale (vide Fig. 4.2.43 and 4.2.44), there are ample instances where such ores are affected by wide spread secondary alteration by solution activity, giving rise to vugs and cavities with secondary deposition of hematite, goethite, limonite, clay etc. (Fig. 4.2.46 a–f). Near surface limonitisation and formation of 'canga' are indicative of percolating surface water, while the presence of vein quartz in such zones would suggest invasion of hydrotherms from below. However, such secondary alterations have rather degraded the hard massive ore than enriched them.

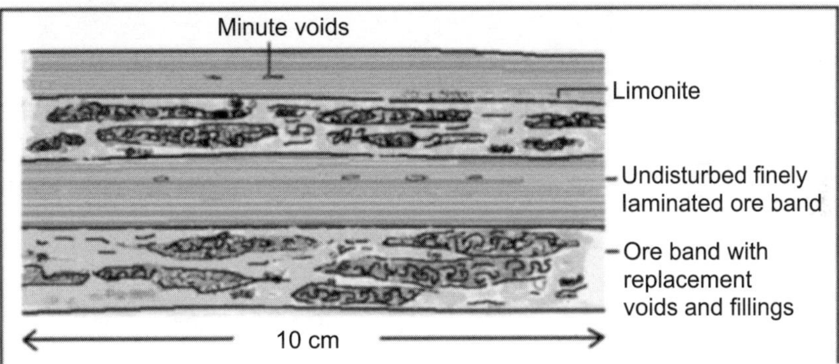

Fig.4.2.45 Photograph and sketch showing hard hematitic iron ore (> 66% Fe) with fine undisturbed laminations alternating with non-laminated scoraceous bands. Sporadic limonite streaks/specs and minute cavities are noted in the laminated bands, Katamati mine, southern extension of Noamundi deposit.

Soft friable/Flaky ores In case of soft friable ore or blue dust a gradual transition of these from BIF is noted at many places showing distinct evidence of the leaching out of silica and enrichment of Fe-oxide within short span (Fig. 4.2.47 a, b, d). The original laminations of BIF (and fold-contortions, wherever present) are maintained in most cases even after its transformation to powdery blue dust through intermediate stage of friable platy ore. At places thin blue dust layers alternate with silica-rich laminates instead of platy hematite (Fig. 4.2.48).

There are instances in Noamundi mines, where a BIF (with 52% SiO_2) band is seen to be enriched into a friable ore (with 66% Fe) within a distance of less than a meter. However, it is observed that the compositional transformation has taken place strictly along the bedding planes and not across (Fig. 4.2.47 d).

The friable ore consists of microplates of hematite ± martite, displaying end to end enmeshing but not as well interlocked as seen in case of rich hard ores. The contacts of the friable ore/ blue dust zones with BIF or other rocks are controlled either by the low dipping bedding planes or the steeply dipping joints, or both. All these features are suggestive of solution activity leading to the removal of silica and the enrichment of BIF into the ore types described above. The proto-botroidal structure on the hematite plates in friable ore (Fig. 4.2.47c) weighs more on the possibility of supergene enrichment in these zones.

Fig. 4.2.46 (a) Near-surface limonitic ore overlying friable ore (FO) and blue dust (BD), (b) Boulders of 'Canga' derived from the surface, (c) Friable and hard ore with solution cavities parallel to the bedding surfaces, location of (a–c): Noamundi west mine, (d) Hard laminated ore alternating with contorted ore bands with cavities and solution breccia, (e) Highly scoraceous hard massive ore with goethitic patches, (f) Quartz vein in Low-grade friable ore with limonitic layers, location (d–f): Joda mines (photographs by authors).

In the Noamundi–Joda sector, the hard massive ore does not show a gradational contact with BIF (BHJ), but its transition to FO and BD are common (Fig. 4.2.49). There is a complete absence of direct contact between HO and BHJ in the mine excavations of this area. The same is revealed in the lithologs of boreholes drilled in these deposits, some examples of which are presented in Fig. 4.2.50. The lithologs also reveal that ores of economic grade do occur below BHJ. It is further noted that the litho-contacts between BHJ and ores are very sharp, marked by abrupt increase in Fe-content from about 41–45% in the former to more than 64% in the latter within a few centimeters.

Fig. 4.2.47 (a–b) Field photograph and representative sketch of intertwined friable ore (FO) and blue dust (BD) showing remnants of primary laminates in form of hematite plates; (c) Close-up view of a hematite plate from location of (a), showing proto-botryoid texture on the plate surface, Katamati mines, Orissa; (d) sketch showing transition of BHJ into blue dust through friable platy ore within less than a meter; note layer parallel compositional gradation along band 1 and 3, and no layer-across change to the unaltered band 2, Hill 4, Noamundi East mine (Photographs and sketches by the authors).

Fig. 4.2.48 Intricately folded BIF with quartz laminates (light coloured, high relief) interlayered with blue dust (dark grey coloured, low relief), gua mines (from Sarkar and Gupta, 2005).

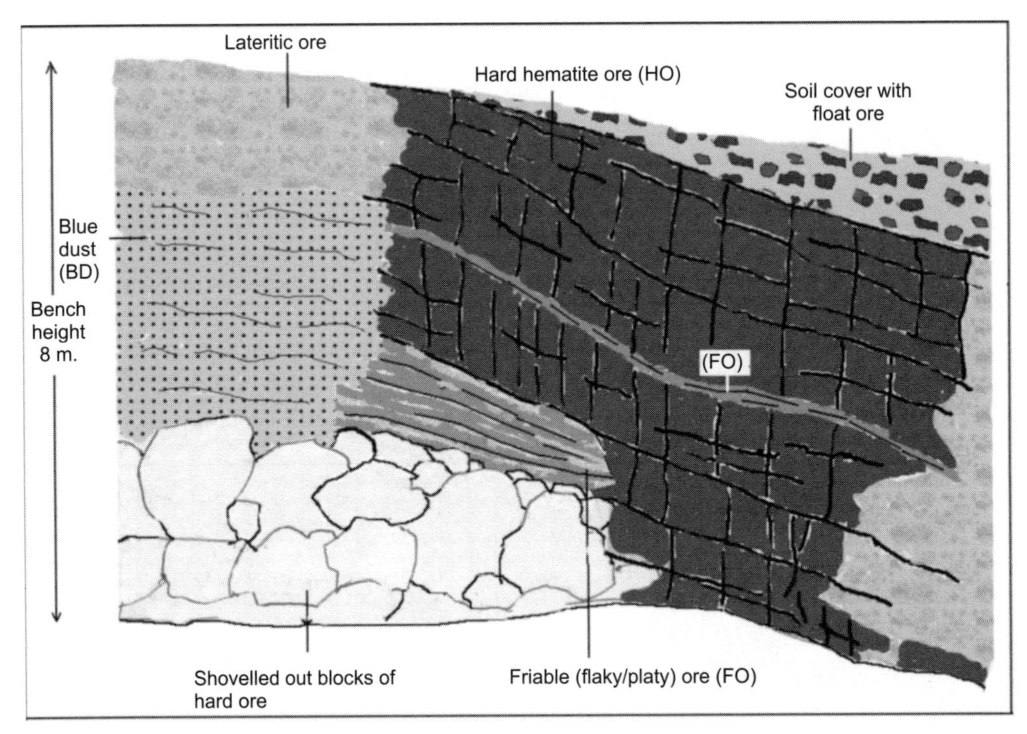

Fig. 4.2.49 Sketch of a bench face in Katamati mine showing the disposition and interrelationship between hard ore (HO) and other ore types (FO, BD, lateritic and float ore)

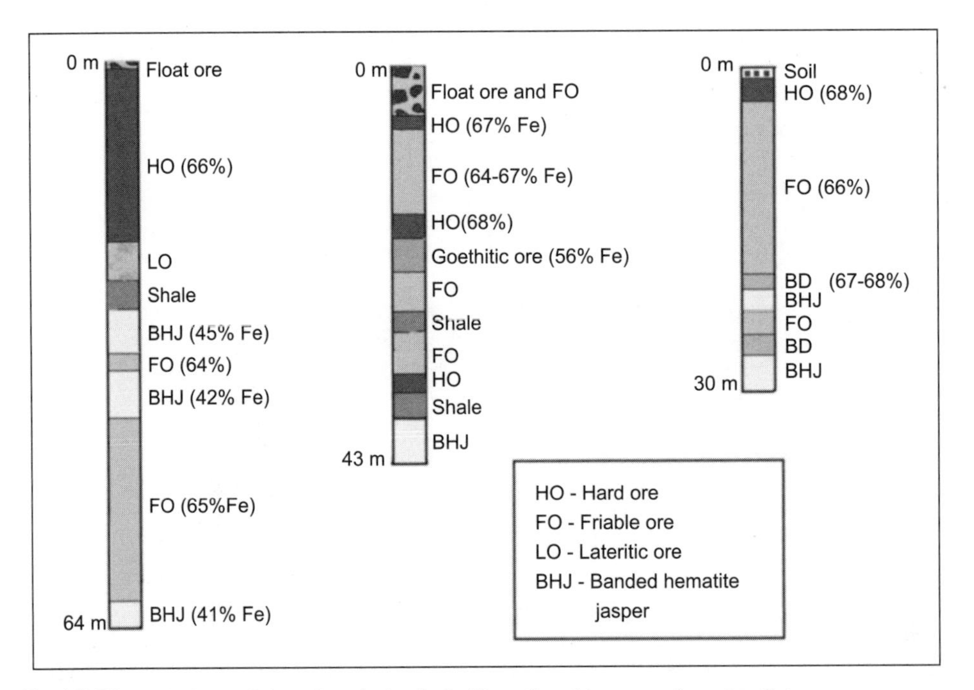

Fig.4.2.50 Lithologs of three boreholes, Joda-Khondbond iron ore deposits, Orissa

Some earlier workers (Beukes et al., 2002;Gutzmar et al., 2006) showed that rich hard ore was predominant and almost entirely occupied the lower levels of the Noamundi west mine.In fact, the ore is dominantly of friable type in Noamundi deposit and the ore zone is bottomed by low dipping BHJ, as shown in Fig. 4.2.40a.

On the Genesis of Iron Ores with Special Reference to the Eastern Indian Deposits The problem of development of rich iron ore in Banded Iron Formations is already briefly discussed. It is also seen that the issue has been a subject of animated debate for sometime now. Inflowing informations from different parts of the world, no doubt, have reduced the area of differences, but their final conclusion differs in details. The overall weightage at present is on the hypogene origin of the ores, though the role played by the supergene processes in concentrating the ores in some cases is also not denied. However, some of the more accepted genetic models tangibly vary (vide Morris, 1985; Powell et al., 1999; Taylor et al., 2001). Further, some generalisations suffer from the lack of universal support. For example, Dalstra and Guedes (2004) concluded that the development of magnetite-carbonate-amphibole, magnetite-carbonate and hematite-carbonate, the precursors of hydrothermal origin were essential in the history of hydrothermal origin of rich hard iron ores in the BIF. These Ca-Fe-Mg metasomatites later became the protore, which on supergene alteration produced the hard rich ores. In some such cases the carbonates are thought to be the constituents of the original sedimentary sequence, while in still others carbonates and the microplaty hematite appear to have formed more or less simultaneously. Such evidences are conspicuous by their absence in the iron ore deposits of Eastern India as well as in Central India. Upto 30% porosity has been described from hard massive ore from Mount Tom Price deposit of Hamersley Province, Western Australia and this has been generalised as a prime evidence of the selective replacement of silica giving rise to the enriched hard ores (Taylor et al., op. cit.). This criterion is also not applicable to the rich hard ores of the deposits under reference, specially the hard laminated ores. Marked reduction of the formational thickness, wherever rich ore body has developed in the BIF, compared to barren BIF in strike continuity is another evidence cited from parts of Hamersley Province to account for mass removal of silica from BIF by hypogene process leading to the formation of hard massive ore. Such situations are not observed in the Indian deposits under reference.

Indeed there is more uncertainty about the origin of rich friable ores. In many deposits including the giant deposits of Karajas and Quadrilatero Ferrifero, bulk of the important ore is soft high grade ore or friable ore in which rich hard orebodies ‘float’. The situation is similar in case of Noamundi and many other deposits in the vicinity, where the rich soft ore is far more dominant in total volume compared to the hard ore. Gutzmer et al. (2006) concluded that such ores developed through geologically recent supergene enrichment of carbonate-metasomatised iron formation. We do not have any evidence of carbonate metasomatism in the Eastern Indian deposits. Though the features of the soft ores described above (vide Fig. 4.2.42, 4.2.47, 4.2.48) are suggestive of their origin by alteration of BIF and the hard ores, it may not be largely attributed to a recent supergene process. However, near-surface replacement features in hard ore and development of goethite-limonite bearing soft ore are certainly indicative of weathering and supergene alteration (Fig. 4.2.46a). The botryoidal growth (Fig. 4.2.47c) on platy hematite in soft ore also supports secondary solution activity.

Blue dust appears to be more common in the Indian deposits and may be so in Australian deposits. Their origin is another riddle. Harmsworth et al. (1990) and Morris (2002) believe that simple supergene leaching of silicates in BIF is locally responsible for the development of blue dust in high rainfall areas. But how does that explain alternating bands of blue dust with both hematite (Fig. 4.2.47a,b) and quartz (Fig.4.2.48) laminates, which crumble down as fine powdery dust even when lightly touched. Gradual layer-parallel gradation of BIF into blue dust within a very short distance of < 1 m, with an intermediate stage of platy friable ore (Fig.4.2.47d), is noted in the Noamundi and Joda deposits.

Regarding BIF associated iron ores of Eastern India, the pioneering workers Percival and Jones made some interesting conclusions many decades ago, even when many of the major deposits of the region were yet to be discovered. They concluded that such ores formed by leaching of silica in BIF either in the 'Archean' (meaning Pre-Phanerozoic) or in the recent geological time. Dunn was in favour of the former possibility. On the question whether Fe-enrichment in the BIF took place by descending meteoric or 'ascending hot magmatic water' (the issue constituted a hot debate at that time, particularly with respect to Lake Superior ores), Dunn (1937) concluded: "It seems unnecessary to subscribe to such arguments for it is apparent here that concentration has arisen under both sets of conditions. Any circulating waters seem capable of performing the work." He did not elaborate further. His contention that at least part of these ores formed during the 'Archaean' was later vindicated when he discovered iron-ore pebbles in the basal Kolhan conglomerate overlying the IOG (Dunn, 1941). Recently, two such occurrences, one near Champakpur village, Keonjhar district and the other at the Baljori–Melalgara villages, north of Noamundi, West Singhbhum, have been studied in some details by Singh (2005) and Mukhopadhyay et al. (2007).

Chatterjee and Mukherjee (1981, 1985) observed structural domes and basins in the Malantoli iron ore deposit, which apperently developed due to the interference of folds. They reported that the ore concentration took place in the basinal parts as opposed to the domes. This observation was inferred as an evidence of ore concentration by supergene process. The authors feel that ore concentration in dilated loci in a folded sequence may take place by both descending and ascending fluids. Sarkar and Gupta (2005), while reviewing the BIFs and BIF related iron ores of the eastern India, opined that the ore concentration could be either by hypogene or supergene processes or by a combination of the two, varying from deposit to deposit. They, however, appreciated that the well preserved and perfectly parallel very thin (< 1mm) ore laminations are enigmatic and as such did not altogether preclude the possibility of their sedimentary-diagenetic origin, at least at places.

An opinion regarding the genesis of the hard massive iron ores in association of BIF in the Eastern India, specially the thinly laminated ore-type, has been in favour of syngenetic sedimentary-diagenetic process (Rai et al., 1980; Chakraborty and Majumder, 1992; Acharya, 1982, 2000, and references therein). According to this view, both BIF and the associated massive iron ores in Jamda–Koira–Noamundi (IOG-II) and Tomka–Daitari (IOG-III) belts were deposited primarily under sedimentary conditions and were subsequently subjected to coeval and or later effects of diagenesis and low-grade burial metamorphism. In Noamundi area the total thickness of 350–400 m of BIF, sandwiched between Lower and Upper Shale Formations, consists of banded hematite jasper (BHJ) rock and a few abnormally thick interstratified iron ore beds of 20–60 m (Rai et al., op. cit.). As per our observation, generally, the width of rich hard ore (with + 65% Fe) macrobands in the area ranges between 10–30 m. Such ores are often characterised by internal meso- and microbands. Rai et al., (op. cit) recorded numerous syn-depositional and diagenetic structures and distinctive biogenic forms preserved both in the BHJ and iron-ore lithosomes are comparable to other well recognised chemical sediments. Besides, features like cross-bedding, ripple marks, graded bedding, scour-and-fill structures etc., are reported from near shore basinal margins from both BHJ and iron ore lithosomes, which are described these days as granular iron formations (GIF) (Trendall, 2002). The occurrence of both BIF and GIF within the same stratigraphically defined horizon and their transition into each other, as described from this belt (Rai et al., op. cit.), have also been reported from many deposits of Circum–Ungava , Lake Superior Provinces of USA and Canada (James and Sims, 1973; Morey, 1983; Simpson, 1987), Transvaal region (Beukes and Klein, 1990). These iron formations are fundamentally dissimilar to the Hamersly BIFs (Trendal, 2002).

However, the thinly laminated hard ores in the Noamundi–Joda sector are comparable with the similar ore-types of the Hope Downs deposit of Hamersley Province, which are attributed by Lascelles (2006) to the syngenetic development of 'cherty BIF' with 'chert-free BIFs', the protore formed by diagenetic removal of silica before lithification.

In summary we may draw the following conclusions:

1. Iron formations in the eastern belt of Badampahar–Gorumahisani (IOG-I) are closer to the Algoma type and rest i.e. those in the western Jamda–Koira–Noamundi valley 'horse-shoe' belt (IOG-II) and in the southern Tomka–Daitari belt (IOG-III) are closer to Superior type of Gross (1973, 1980).
2. The iron ores of the region belong to both hypogene and supergene types and many of them a combination thereof. The syngenetic sedimentary–diagenetic model proposed by some workers for the hard laminated ores in IOG-II and IOG-III belts are yet to be well constrained in respect of the possible depositional environment and source of such large quantity of iron, though a volcanogenic source same as for common BIF remains a distinct possibility.

As per present day knowledge, the hypogene origin refers particularly to the hard high-grade ores that are characterised by the following: (i) micro-platy hematite–martite mineralogy with little or no goethite, (ii) such ores participating in early diastropic structures, (iii) rich hard ores locally overlain by BIF, (iv) depletion in ^{18}O in some samples of Noamundi. Future studies should concentrate on fluid inclusions, paleomagnetism, oxygen isotopes and anything else that would appear relevant.

That the recent supergene processes have been operative are proven by (i) presence of goethite in tangible proportions near the present-day erosional surface i.e. below the hard cap, (ii) development of ores along dislocation structures connecting to the surface, (iii) some ^{18}O values that are better explained by hematite–meteoric water equilibration, other factors remaining the same (Gutzmer, 2006).

The origin of friable ores here, as anywhere else, is yet to be fully understood. However, formation of this ore-type by gradual alteration of BIF, leading to removal of silica and enrichment of Fe is evidenced in many deposits of Jamda–Koira–Noamundi belt.

Chromite Mineralisation

Eastern Indian craton, particularly the region south of the Singhbhum Cu-U belt, is very rich in chromite ores. More than 90% of India's insitu reserve of about 179 million tonnes (Ambesh, 2006) of economic grade chromium ores occurs in this part of the country. The mineralisations hosted by mafic-ultramafic complexes occur in the following regions:

1. Jojohatu–Hatgamaria area (West Singhbhum district), Jharkhand
2. Sukinda (Jajpur–Dhenkanal distrits)–Baula Nuasahi (Keonjhar district), Orissa.

Besides the above, chromite-bearing ultramafic sills are also located at Amjhori in the Simlipal volcanic complex. Deposits in the Jojohatu–Hatgamaria area were small and already mined out. Sukinda and Nuasahi deposits are large and hold the bulk of chromite reserves of the country. Between the two of them, Sukinda is larger.

Out of 179 million tonnes of all India insitu resource of chromite, about 90 million tonnes are categorised as recoverable. More than 97% of the recoverable reserves are located in Orissa (Sukinda and Baula–Nuasahi deposits). The production of chromite stood at about 4.1 million tonnes during 2006–07, the contribution of Orissa being 99%. Four principal producers (Tata Steel, OMC, FACOR and IDC Ltd) operating 10 mines, accounted for 92% of the total annual production (Gupta, 2009).

Jojohatu Deposit The chromite bearing ultramafic bodies in Jojohatu occur in three adjoining hills, viz, Roroburu–Chitamburu, Kimsiburu and Kittaburu, as intrusives into the IOG rocks of Jamda–Koira valley. The ultramafic bodies, composed of rhythmically layered dunite–chromitite, clot peridotite (poikilitic harzburgite) and pyroxenite, are co-folded with the country rocks forming an east-west trending synform at Roroburu hill. Mineral graded layers, < 1 cm–30 cm in thickness, are composed of dunite, chromite–dunite, olivine–chromitite and chromitite. The chromite-rich layers also exhibit magmatic sedimentation structures similar to those described from Nuasahi complex (Haldar et al., 2005)

In spite of wide spread alteration of the ultramafics to serpentinite, the original grain boundaries and primary textures of the rock forming minerals are largely preserved and the chromite grains mostly remain unaffected. Depending on the ratio of olivine to chromite content the texture grades from a network of tiny chromite grains sandwitched between larger olivines as seen in chromite-dunite, through chain structure in olivine-chromitite, to occluded silicate textures where isolated olivine grains occur in a mosaic of chromite cumulates. Dunite and pyroxenite show adcumulus to mesocumulus textures, while the clot-peridotite represents the ideal heteradcumulates where smaller grains of olivine are studded within larger oikocrysts of orthopyroxene (Haldar et al., op. cit.).

The range of major element composition (wt%) of Jojohatu chromite (n = 9) is restricted to Cr_2O_3: 52.67–55.85, Al_2O_3: 6.89–10.20, FeO: 18.54 –22.85, MgO: 11.00–12.10, TiO_2: 0.58–0.83. All the analyses lie within the field of Al-chromite compositional range of chromite (plotted in Cr_2O_3–Al_2O_3–Fe_2O_3 diagram), compares with the podiform chromitites and chromites from the stratiform Stillwater Complex. The Cr/Fe ratio varies from 2.04 to 2.61 with the average being 2.32 and Mg″/Fe″ varies only from 1.32 to 1.46 with the average value of 1.33, while Cr^{+3}/Al^{+3} ratio varying considerably from 6.80 to 9.95 with the average value being 8.55. The unit cell dimension varies form 8.30 to 8.32 A° (Haldar et al., op. cit)

Some smaller chromite-bearing ultramafic bodies are located near Hatgamaria and Kusumita–Bichaburu in West Singhbhum district, which display features similar to those in Jojohatu area (cf. Haldar and Mukhopadhyay, 1991).

Sukinda and Baula–Nuasahi Deposits

Sukinda The Sukinda mafic–ultramafic complex is the largest of its kind in the Singhbhum–Orissa craton, extending for about 25 km in strike and ~ 400 m in plan width (Fig. 4.2.51). The rock exposures are scanty in the area because of the development of a thick laterite and soil cover. Basu et al. (1997), while interpreting the geology of the area based on surface as well as extensive drill core data, reported the Sukinda body as a package of ultramafic intrusives and basaltic effusives, occurring as a folded sequence within the Iron ore Province of Tomka–Daitari.

The folded mafic-ultramafic complex forms a WSW–SW plunging synform with moderately (40°–50°) dipping northern limb and subvertical southern limb. Serpentinised dunite and orthopyroxenite (locally transformed into talc-tremolite schist) are the dominant ultramafic rocks in the sequence. The orthopyroxenites are relatively less altered than the dunites and display sharp contact with the latter. Basu et al. (op. cit.) has described two main horizons of metabasalts, one along the base of the ultramafics and the other at the core of the synform. The contact of the lower metabasalt with the ultramafics is highly fractured and brecciated. Secondary chert (birbirite) is also common in the sequence. Granophyric granite and dolerite intrude the mafic-ultramafic complex. Sarkar et al. (2001) reported a breccia zone from Katpal in the western extremity of the belt, which is similar to the PGM-bearing breccia zone in Baula–Nuasahi area.

The chromitite seams at Sukinda occur as disconnected bands and lenses within the serpentinised dunite. The tabular and massive ore bodies, 6 in number and concordant with the folded host rocks, extend in individual strike lengths from 200 m to as much as 3 km with thickness ranging upto 20 m. The ore bodies, being mostly contained in the southern limb of the synform, have sub-vertical dip (Fig. 4.2.52). The total ore reserve (ores mined out + ore remaining) down to 250 m depth is > 112 million tonnes with a grade of > 42% Cr_2O_3 or > 32 Mt Cr metal. Sukinda is the largest deposit of chromite in the country, but it is much smaller compared to the super-large deposit of Bushveld, and even smaller than the chromite deposits in the Great Dyke of Zimbabwe.

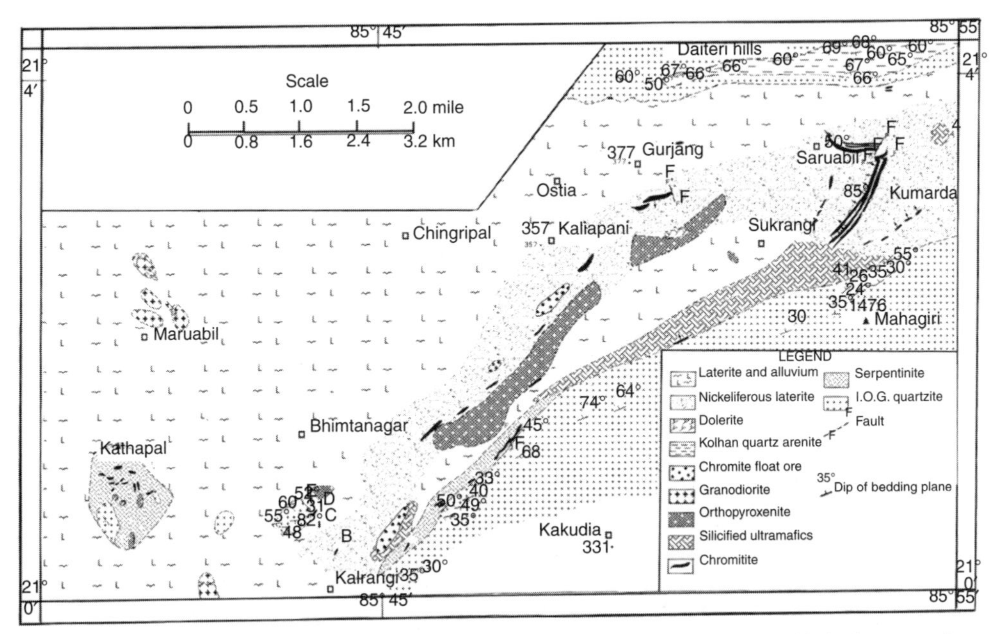

Fig. 4.2.51 Regional geological map of Sukinda area, Orissa (modified after Chakraborty and Chakraborty, 1984)

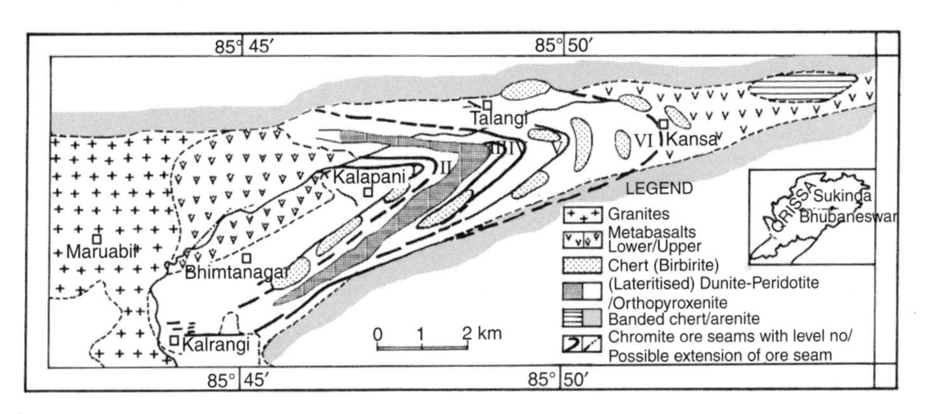

Fig. 4.2.52 Inferred geological map of the synformal zone of Sukinda belt showing the disposition of chromite orebodies (modified after Basu et al., 1997)

Baula–Nuasahi The mafic-ultramafic complex at Baula–Nuasahi occupies nearly 10 sq km area in the southeastern part of the craton. The rocks are much better exposed and better preserved in this area compared to Sukinda. A nearly N–S trending ultramafic massif of 4 km length and stratigraphic thickness of 800 m, consisting of pyroxenite (enstatitite) and dunite-peridotite is apparently intruded by a gabbro-anorthositic complex. Chromite ore bodies occurring within the dunite-peridotite unit are commonly referred to as the Durga Lode, the Laximi Lode and the Ganga–Shankar Lode (Fig. 4.2.58). The steeply east-dipping orebodies, have a sharp footwall and gradational hanging wall. Compositional banding defined by varying chromite/ silicate ratios, is generally conspicuous within the orebodies. Magma-sedimentary and clot structures have been reported from these ores (Mukherjee, 1969; Chakraborty, 1972; Mukherjee and Halder, 1975; Sahoo, 1998). The former structure shows right side up. The eastern margin of the ultramafic body at Baula–Nuasahi is brecciated. The ore reserve of this deposit is much lower than that at Sukinda.

The range of major element oxides (wt%; n =16) of Nuasahi chromite reported as: Cr_2O_3: 54.77–62.61, Al_2O_3: 5.72–12.43, FeO*: 13.23–20.13, MgO: 12.38–16.71 and TiO_2: trace–0.25. The chromites lie in the field of Al-chromite composition in Cr_2O_3-Al_2O_3-Fe_2O_3 diagram. The Cr/Fe ratio varies from 2.46 to 4.02 with an average being 3.47. The variation of both Mg″/Fe″ and Cr/Al is limited within 1.63 to 2.6 (av. 2.15) and 3.75 to 4.15 (av. 3.89) respectively.. The cell parameters range from 8.24 Å to 8.33 Å (Haldar et al., 2005).

Characteristics of Chromite Ores In both Nuasahi and Sukinda, chromite seams display magmatic-sedimentation structures such as rhythmic layering, pseudo-current bedding etc. and various gravity-controlled features such as slumping, overturned slump folds and convolute structures (Fig. 4.2.53). In Sukinda the massive chromitite bands are often thinly laminated with alternate thin laminae of granular chromite and serpentinised olivine. (Haldar et al.,2005; Mondal et al.,2006).

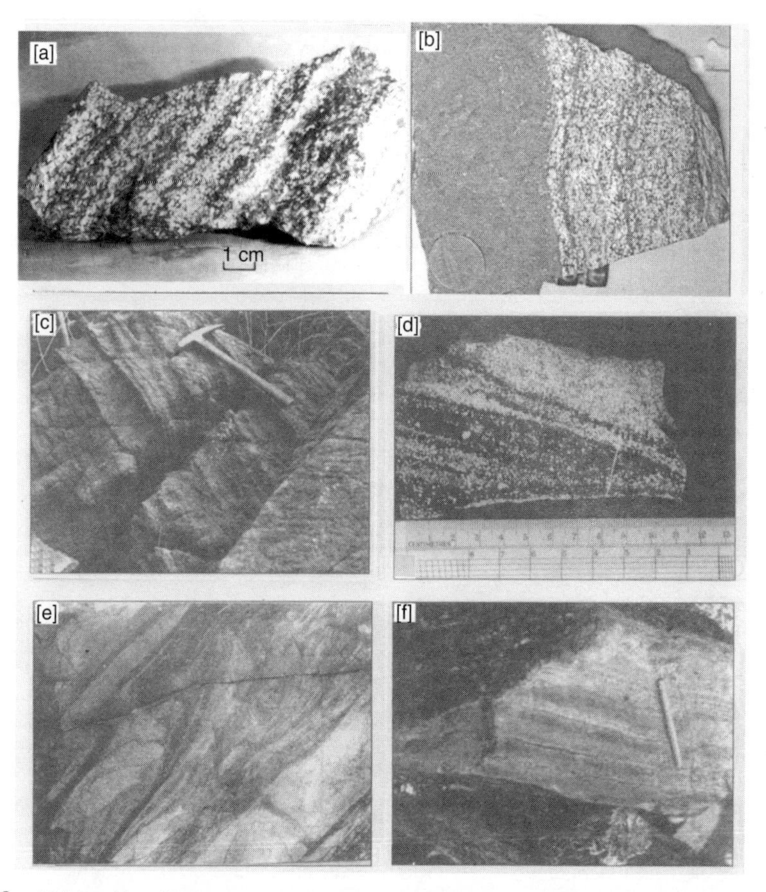

Fig. 4.2.53 Field and hand specimen photographs of chromite ore: (a) Crudely banded chromite ore, Sukinda (from Chakraborty and Chakraborty, 1984); (b) Chromitite- and olivine-rich schlieren bands in massive ore, Nuasahi (from Mondal et al.,2006); (c) Outcrop of mineral graded rhythmic layering in chromite bearing ultramafics, Jojohatu; (d) Angular discordance in alternate chromite- and olivine-rich bands simulating cross-stratification, Nuasahi; (e) Psudo-sedimentary penecontemporaneous structures with straight banded chromitite, Nuasahi; (f) Rhythmic layering in dunite-peridotite-chromitite, Jojohatu (from Haldar et al., 2005).

The ores are fine to coarse grained and occur disseminated, layered or massive. The interstices are occupied by the silicates. The larger grains of chromite display polygonal or sub-rounded character. There are some interesting textures in the ores of both the deposits. They range from initial or primary growth, through post-cumulus modifications by sintering, to deformation with or without recrystallisation (Mukherjee, 1969; Mitra, 1972). Chromitite seams in both Nuasahi and Sukinda deposits share the common features like schlieren and spotted layers, clot-textured bands, net-textured massive bands, nodule bearing bands, besides rhythmic layering (Mondal et al., 2006). The nodular chromitite is characterised by fine-grained olivine-chromitite nodules embedded in massive coarse-grained chromitite. Some textures and microstructures of chromite ores, Eastern India are presented in Fig. 4.2.54.

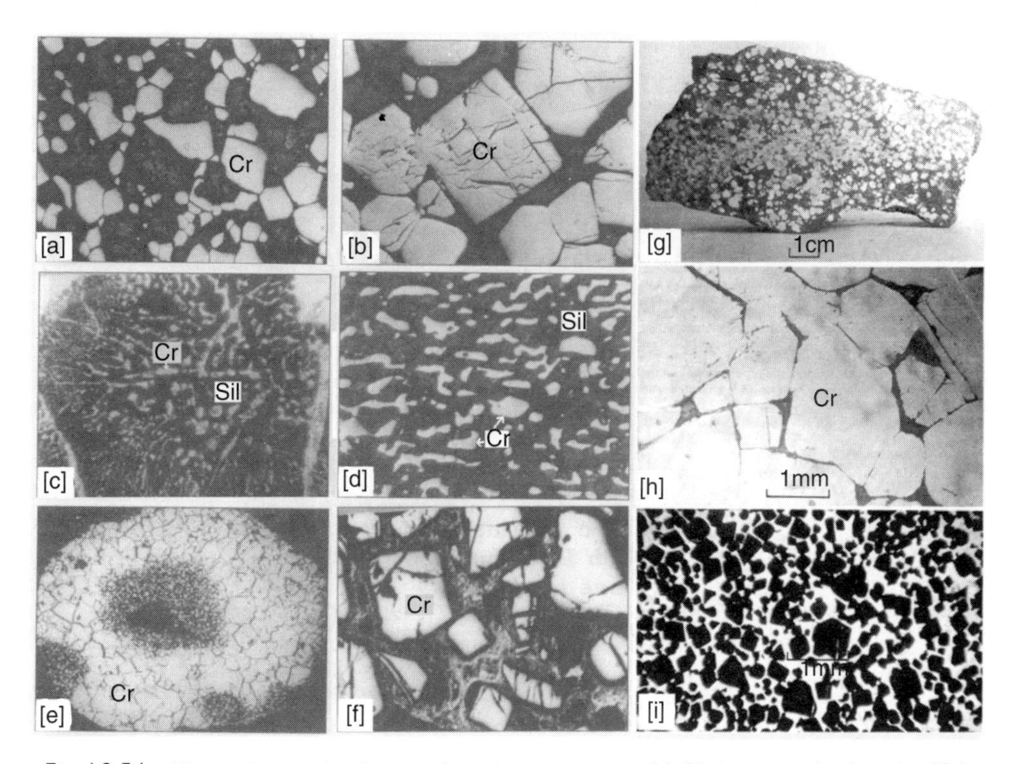

Fig. 4.2.54 Photomicrographs showing chromite ore textures: (a) Chain texture in chromite (Cr)-olivine cumulate (reflected light; 80 X), Nuasahi; (b) Chromite cumulate with euhedral chromite (Cr) grains; intercumulate space occupied by serpentinised olivine (reflected light, 50 X), Nuasahi; (c) Unusual graphic intergrowth of chromite (Cr-light) and olivine (Sil-dark) in harzburgite (reflected light; 200 X), Nuasahi; (d) Enlarged view of (c), chromite (Cr – white), silicate (Sil – dark) (reflected light, 400X); (e) Occluded silicate texture in chromite cumulate, showing markedly fine grained chromite in the silicate compared to much corser grains outside (reflected light, 50 X), Nuasahi; (f) Unaltered chromite grains (Cr) and altered silicates (grey and black) in lateritised ore (reflected light; 100 X), Sukinda (a–f: from Haldar et al., 2005); (g) Photograph of hand specimen showing antiorbicular chromite ore (white rounded spots of serpentine in ground mass of chromite), Sukinda; (h) Granular mossaic of chromite in chromitite showing straight boundary and triple points suggesting sintering, Sukinda; (i) Cumulus texture of chromite (dark) with interstitial serpentine (white) in thin section, Sukinda (g–i: from Chakraborty and Chakraborty, 1984).

In both Sukinda and Nuasahi, the chromite in massive chromitite layers is characterised by high Cr-number [Cr/ (Cr+Al), molar] and high Mg-number [Mg/(Fe^{2+}+Mg), molar] as listed here (Mondal et al., 2006).

1. Nuasahi massive ore deposit – Cr-number: 0.78–0.87, Mg-number: 0.66–0.82
2. Sukinda massive ore deposit – Cr-number: 0.75–0.81, Mg-number: 0.62–0.73

This characteristic is absent in accessary chromite veins in dunite, orthopyroxenite and net-textured chromitite layers, where these values show considerable variations. This variation, obviously the result of sub-solidus chromite–silicate re-equilibration, may be observed even in thin section scale. However, chromite in harzburgite shows a restricted variation. Chromite grains in serpentinite are relatively enriched in TiO_2 (Mondal et al., 2006). Locally, the NiO content in chromite may be as high as 6.27% (Pal et al., 1994). Other mineralogical characteristic is the high Fo content (Fo_{92-95}) for olivine in dunite and high En-content (En_{89-94}) in orthopyroxene. There is substance in the claim of Mondal, et al. (2006) that the primitive composition of minerals have been retained in the respective monomineralic rocks.

Ore Genetic Considerations How well do these deposits fit with the standard varieties commonly known as the Bushveld and Alpine types? Mitra (1960) assigned the Sukinda deposit to the Bushveld type and Banerjee (1972) to the Ophiolite or Alpine type. Mukherjee and his co-workers noticed features intermediate between those of Bushveld and the Ophiolite types (Mukherjee, 1969; Mukherjee and Halder, 1975). Haldar et al. (2005) opined that these Precambrian deposits represent a distinct clan, not belonging either to the stratified Bushveld type or the podiform Alpine(Ophiolite) type. Page et al. (1985) on the basis of a not-too-well constrained geochemical model with respect to PGE, suggested the two complexes to belong to the Ophiolitic type. Before we record our opinion on this aspect, we may check with the state-of-the-art approach in distinguishing these two varieties of chromite mineralisation.

Stratiform Chromite Deposits (Bushveld type) Bushveld type stratiform chromite deposits occur in layered intrusions. The deposits themselves are internally layered in which the chromite ore or the chromitite layers range in thickness from a millimetre or less to several metres and may be traceable for tens of kilometres. Thickness is generally uniform and the contacts of the chromitite layers with the silicate host rocks are sharp and planar. These deposits may occur within a nearly massive ultramafic rock or may occur close to the base of the cycles. Such deposits are generally believed to have been emplaced in extensional settings.

Podiform Chromite Deposits (Ophiolite type) The ophiolitic chromite deposits occur in ophiolitic sequences in island arc or back arc or even in retro-arc environments. The ophiolitic sequences, as are well known, consist of an upper mantle zone of harzburgite and dunite, overlain by a lower crustal sequence of layered ultramafic-mafic cumulates, passing upwards into massive gabbro, sheeted dykes, pillow basalt, radiolarian chert etc. The upper part of the mantle zone is highly tectonised. A transitional dunite unit is present along the mantle-crust boundary.

Stratiform chromitite layers, commonly deformed, occur in the dunites, overlying the lower cumulates and the transition zone. Podiform chromite deposits, on the other hand, occur in the underlying mantle tectonites. Chromitite orebodies in ophiolite type deposits are intensely folded and faulted, disrupting them into apparent pods and lenses. Compared to the Bushveld type, these deposits are generally smaller, hardly exceeding 1200 m in length and 150 m in thickness, with ore reserves usually within 2 million tonnes. Pods consist of centimetre sized nodules of chromitite in dunitic matrix. Less commonly, decimetre sized dunitic spheres or lenses, often with a chromititic core, is formed within chromitite. These structures are exclusive to the podiform chromite. Chromitite bodies in ophiolite type deposits are generally located close to the petrologic Moho. The chromitite bodies themselves are deformed resulting in extreme situations in the development of dislocated pods and lenses, attesting to an unstable tectonic setting. In outcrop scales many podiform chromite bodies show discordant relationship with their host rocks, while some are elongate (elongated?) parallel to the linear tectonic fabric of the enclosing rocks (Lago et al, 1982; Stowe, 1994; Zhou and Robinson, 1997; Cawthorn et al., 2005).

Composition Fields of Different Types of Chromite Deposits Let us see if the composition fields of Bushveld type layered chromitite and those of ophiolitic chromitite, particularly podiform variety, widely differed. These fields indeed show significant variance, with areas of overlap, in the various binary and ternary chemical diagrams. Barnes and Roeder (2001) compiled and plotted the available data in Fe^{3+}-Al-Cr and Cr / (Cr+Al)–Fe^{2+}/(Fe^{2+} + Mg) diagrams (Fig. 4.2.55 a–b). These data plotted in three fields of layered intrusives, ophiolite and boninites show large overlaps. In Bushveld type deposits, Cr-number often decreases upwards. This number, varying widely in the ophiolitic chromite, is generally higher for the mantle zone podiform variety compared to that in the cumulate zone.

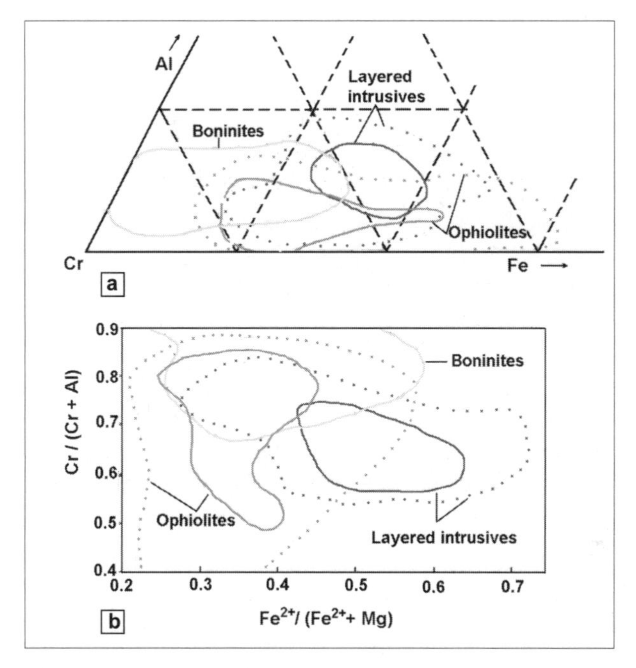

Fig. 4.2.55 (a–b) Chromite composition from different intrusive types. The fields denoted by solid and dotted lines represent 50 and 100 percentiles, respectively (after Barnes and Roeder, 2001).

Plots of 14 cleaned chromite samples from Sukinda in the Cr/(Cr+Al) vs Fe^{2+}/(Mg + Fe^{2+}) diagram (Chakraborty and Chakraborty, 1984) show a restricted compositional field, partly falling in the area of overlap between the stratiform and Alpine-type deposits (Fig. 4.2.56 a–b). Similarly, 16 and 9 cleaned chromite samples from Nuasahi and Jojohatu also display very restricted compositional fields overlapping with the tails of the field of stratiform and ophiolitic podiform deposits towards Cr-end in the Al_2O_3-Cr_2O_3-Fe_2O_3 ternary diagram of Thayer, 1964 (Haldar et al., 2005).

Neoproterozoic and Phanerozoic ophiolites are found either at nascent spreading centres, or at island arcs (supra-subduction zone). Precise identification of an ophiolitic assemblage in the Archean–Paleoproterozoic sequence is usually fraught with a measure of uncertainty. Suggesting it with the help of some inadequately constrained geochemical models (Page et al., 1985) is neither more dependable.

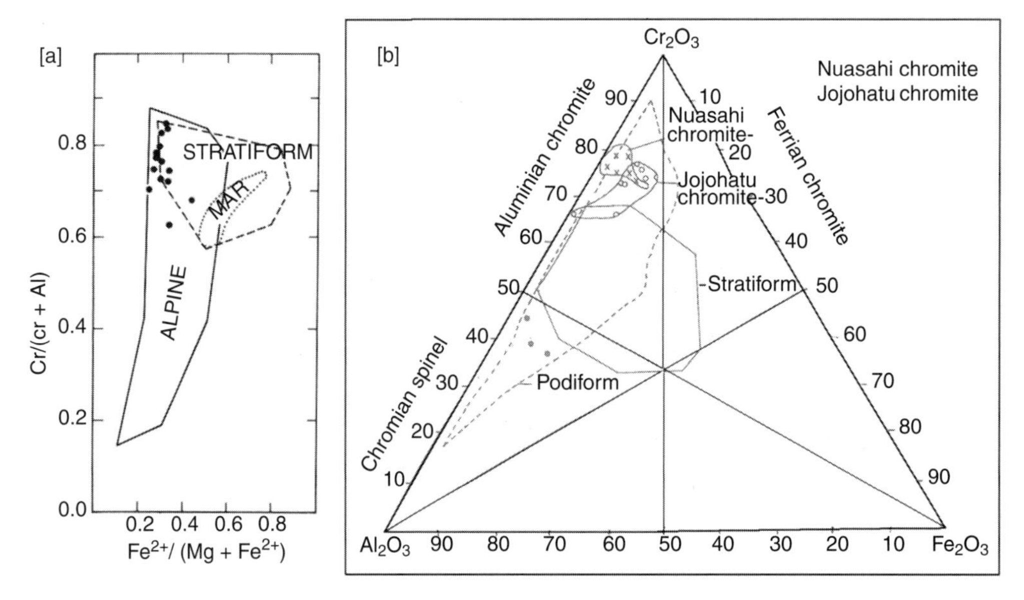

Fig. 4.2.56 (a) Chemical composition of chromite from Sukinda, compared with the composition fields (Greenbaum, 1977) of Alpine, stratiform and Mid-Atlantic Ridge chromitite complexes (after Chakraborty and Chakraborty, 1984); (b) Plots of Nuasahi and Jojohatu chromite on Cr_2O_3–Al_2O_3–Fe_2O_3 diagram of Thayer (1964), showing the fields of stratiform and podiform chromites and those of Nuasahi and Jojohatu (modified after Haldar et al., 2005).

We are often confronted with a question: Is the chromite composition an indicator of the tectonic setting of the deposit? The reasonable answer is, 'generally yes' (Dick and Bullen, 1984; Stowe, 1994; Jahn et al., 1985). But for a better result, the mineral paragenesis should be simultaneously worked out (Sarkar, 2000). The element partitioning in chromite, particularly [Cr/(Cr + Al)] and [Mg/(Mg + Fe^{2+}] etc may be modified during post-cumulus re-equilibration and later hydrothermal alteration.

Magmatic Crystallisation of Chromite and Olivine Magmatic origin of the chromite orebodies (chromitites), particularly the Bushveld type, can not be doubted on the basis of generally accepted field and petrographic criteria, although the effects of late stage hydrothermal activities have been superimposed locally. But the crux of the problem lies elsewhere. Why does a magma that was crystallising olivine and chromite in co-tectic relationship, stop precipitating olivine, making chromite the only liquidus phase? After a lapse of time olivine and chromite again start precipitating until the residual liquid reacts with olivine to produce enstatite. As Lipin (1993) has pointed out, cessation and resumption of the Mg-rich olivine precipitation is not the normal course of fractional crystallisation of a basic magma. Moreover, Lambert et al. (1989) showed that there is geochemical diversity in the chromite seams within the cyclic units. Their data on the Re/Os and Sm/Nd systematics indicated that different chromitite seams crystallised from isotopically distinct magmas. Several explanations have been proferred in interpretation of the situation. These, *inter alia*, are as follows:

1. Liquid immiscibility (McDonald, 1965)
2. Increase of f_{O_2} in the system (Ulmer, 1969; Murck and Campbell, 1986)
3. Change of total pressure in the system (Cameron, 1980; Lipin, 1993)

4. Crustal contamination or silica addition (Irvine, 1975; Rollinson, 1997; Marques and Feraira Filho, 2003)
5. Mixing of a fractionated magma with a primitive magma (Irvine, 1977; Campbell and Murck, 1993)
6. Mixing of an ultrabasic parental magma (U-type) with an evolved (?) magma (A-type) where plagioclase formed first (Sharpe and Irvine, 1983).

The presence of a chromite rich immiscible liquid was suggested by McDonald on the basis of an interpretation of silicate inclusions in chromite grains and development of chromium rich immiscible liquid in chrome smelters. But these situations are unlikely to be the same in magma chambers. Ulmer (1969) suggested that the development of chromitite layers in the critical zone of Bushveld Complex was initiated by the increase of f_{O_2} in the magma, the most probable cause of the increase of oxygen fugacity being surmised as the assimilation of wall rocks. Murck and Campbell (1986) elaborated on this model. Cameron's idea in this context was the rise of tectonically induced total pressure of the system, including the magma chamber, should be conducive to the precipitation of chromitite layers. Lipin subscribed to this view, though with some reservations. Contamination of the parental magma by the siliceous country rocks, pushing the crystallisation along the olivine-chromite co-tectic to the chromite-alone field is now believed by several workers. The remaining two models deal with magma mixing, i.e. mixing of a fractionated magma with a primitive magma, or mixing of an ultrabasic parental magma with an feldspar-rich magma. Some differences in details notwithstanding, most workers on chromite deposits now believe that the repeated development of chromitite seams in a deposit is not the product of simple differentiation of the parent magma, but that of contamination by country rocks or by mixing with other magmas such that the bulk composition of the system is shifted to that of the chromite field in the quartz-olivine-chromite ternary (Fig. 4.2.57). It may be pointed out that many of the above hypotheses are meritorious but as Cawthorn et al.(2005) concluded, no single hypothesis satisfactorily explains most features of stratiform chromite deposits.

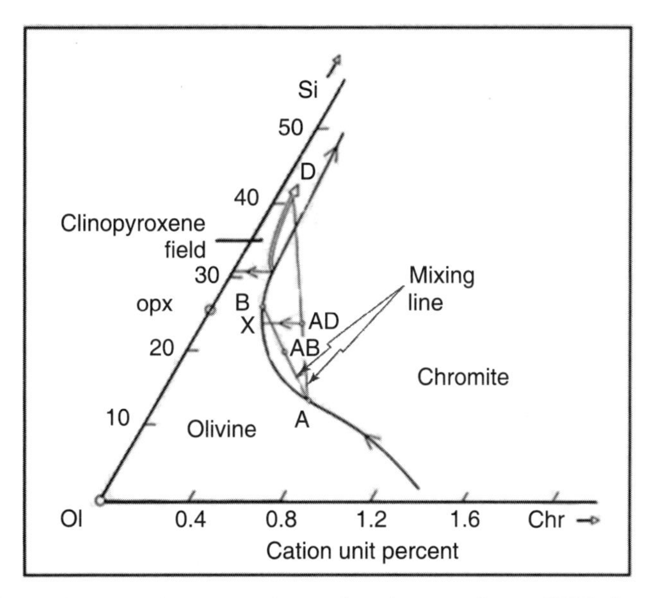

Fig.4.2.57 Phase relations in the system olivine-silica-chromite (Irvine, 1977), illustrating results of mixing primitive magma (A) with well fractionated (D) and slightly fractionated (B) variants of the same primitive magma

The models proposed to explain the origin of Bushveld type layered chromitite deposits with their strengths and weaknesses should also be relevant in explaining the origin of the chromitites in the lower crustal cumulates of an ophiolite sequence, except possibly the crustal contamination, as silica rich rocks are less likely to be available in an ophiolite setting. But the main problem is faced in explaining the podiform chromite. Various suggestions made to explain these structures include (i) sinking of the crustal cumulates into the underlying mantle and production of these structures by tight folding and attenuation (Dickey, 1975); (ii) multistage melting and segregation of boninitic and tholeiitic magma and their continuous reaction to produce chromitite segregates (Paktune,1990); (iii) early formed chromite in the melt accumulated in the steeply dipping magma conduits in the tectonites (Lago et al.,1982; Stowe, 1994). These models, interesting as they may be, did not satisfactorily answer some of the debated questions, particularly the ore textures. Matveev and Ballhaus (2002) suggested a new model, in which equilibrium between an olivine-chromite saturated magma and a H_2O-rich fluid at high P-T is involved. The experiment produced nodular and orbicular textures, but these were very small in size. In the plate margin setting, partial melts from the mantle will have saturated H_2O content through the dehydration of the subducting oceanic crust.

Genesis of Sukinda–Nuasahi Deposits The relevance of the discussed genetic models with respect to Sukinda and Nuasahi chromite ores is examined here. A large change in the magma oxidation is expected to alter the Mg-numbers of both the chromite and olivine, at least along the footwall and hanging wall of the chromitite bodies. This is absent in case of Sukinda and Nuasahi deposits. Moreover, these chromites show reduced character (Mondal et al., 2006), as is common in Al-depleted komatiitic or ophiolitic chromite (Barnes and Roeder, 2001). Nuasahi chromite has a mantle-like $\delta^{18}O$ values. These should be higher in case the parent magma assimilated country rocks that could contain sediments also. Chromite-saturated melts at different temperatures on mixing will induce chromite supersaturation (Murck and Campbell, 1986). Magmas of different temperatures are expected to have different compositions also. Evidence of such mixing is absent in Nuasahi and Sukinda magmatic complexes (Mondal et al., 2006). Their genetic model may be outlined as follows.

Plots of liquidus composition and calculations of the composition of parental magma suggest that the parent magma of chromitites and associated ultramafic rocks in Sukinda and Nuasahi complexes was a low-Ti and high-Mg siliceous magma that corresponds to boninite. This parental magma may have been generated by the interaction of subjacent depleted mantle with a fluid enriched melt derived by roasting of the subducting oceanic crust below. Precipitation of monomineralic chromitites still remains for detailed investigation. Alternatively, elevated H_2O content in the parent magma may suppress the crystallisation of the silicates by lowering their liquidus temperature and advancing the crystallisation of chromite (Nicholson and Mathez, 1991)

According to Mondal et al. (2006), the chromitites of Nuasahi and Sukinda massif were crystallised from a mantle-derived boninitic magma, which was produced due to second stage melting of depleted upper mantle in presence of fluids generated from a subducting slab in supra-subduction zone setting. Further, the stable isotopic content of the Nuasahi complex is indicative of the overprinting of a hydrothermal activity on the primary igneous process. The $\delta^{18}O$ and δD values suggest that the hydrothermal fluid, responsible for alteration and isotopic exchange might have been the evolved seawater and the $\delta^{34}S$ values of the sulphides from the breccia zone show a narrow range of 1.2 to 2.3 per mil, consistent with magma-derived sulphur.

Os-isotopic character of chromites from Nuasahi and Sukinda defines the age of initial melt extraction from the mantle as 3.7 Ga (Mondal et al., 2007). The isotope values of sub-chondritic component of Os are suggestive of its derivation from the sub-continental lithospheric mantle beneath the Singhbhum-North Orissa craton.

Lateritic Nickel

Nickel together with cobalt has been concentrated in the regolith 5–75m thick (average 22 m) in a well-developed lateritic profile over an area of about 40 sq km on the top-part of the ultramafic complex at Sukinda. Estimated secondary nickel-ore reserve is 1.2 Mt with 1% Ni, including 72 thousand tonnes of ore with 0.06% Co (Ziauddin et al., 1979).

Chakraborty and Chakraborty (1976) explained the concentration of nickel in the lateritic profile of Sukinda through the release of Ni from the primary minerals like olivine and chromite and its subsequent precipitation in the lateritic profile. Recently Som and Joshi (2002) made a detailed mineralogical and chemical study of the weathering profile at Sukinda. The essential mineralogical features of the weathering profile reported by them are as follows. They have divided the profile into four divisions: Transported laterite (0–10m, where present), in-situ laterite (0–4.5m), saprolite (1 to > 50m and partially altered serpentine > 1m). Serpentine chrysotile and lizardite with traces of antigorite with a Mg/Ni ratio of upto 7, is the principal phase. With more substitution of Mg by Ni, lizardite gives way to nepouite $[(NiMg)_3\ Si_2O_5\ (OH)_4]$ with a Mg/Ni ratio of about 1. Fe and Cr content in nepouite is higher than in lizardite. A phase close in structure to imogolite $[Al_2SiO_3\ (OH)_4]$ is also reported. The minor phases (all secondary) comprise fine-grained chlorite, kerolite $[Mg_3Si_4O_{10}(OH)_2]$, pimallite $[Ni_3Si_4(OH_2)]$, smeatite and sepiolite $[Mg_4Si_6O_{15}\ (OH)_2,\ 6H_2O]$ Cryptocrystalline silica is common. Nepoutite and talc are the principal phases in the partially altered serpentinite zone, whereas smeatite, goethite (thematite) and cryptocrystalline silica are the major phases in the upper levels, i.e saprolite and laterite zones.

The results of chemical analyses of the samples from the profile show that most of the elements are highly variable, except of course the SiO_2 content in the partially altered serpentine. It is further notable that the mean content (limited to the number of samples studied), the content of not only Ni and Co but also of SiO_2, P, Ba, Zr, Pb are much larger in all the sections of the profile compared to their Clarke values. An extraordinary high content of Ba, Zr and Pb, even in the partially altered serpentine, may be suggestive of a possibility that the ultramafic magma interacted with the gneissic basement rocks. Ni content shows some positive correlation with Fe in the profile, but the relationship is not linear. It is likely that a part of the nickel in the system is accommodated in the goethite structure.

Platinum Group Element (PGE) Mineralisation at Baula–Nuasahi, Orissa

The PGE mineralisation at Baula–Nuasahi in the Keonjhar district of Orissa occurs in a basic–ultrabasic complex that consists of an ultramafic body consisting of dunite–peridotite and pyroxenite and its host, a gabbro anorthosite massif. The complex is intrusive into the Iron Ore Group (IOG) rocks. The dunite–peridotite rocks structurally overlying the pyroxenite are hosts to Durga and Laxmi chromitite bodies (Fig. 4.2.58). The ultramafic complex contains another chromitite body, called the 'Ganga–Shankar lode', at the upper part close to the hanging wall gabbro. This orebody is broken into pieces through shearing and this shear zone has apparently been the site of introduction of a younger gabbro, named 'Bangur Gabbro' (Auge et al., 2003).

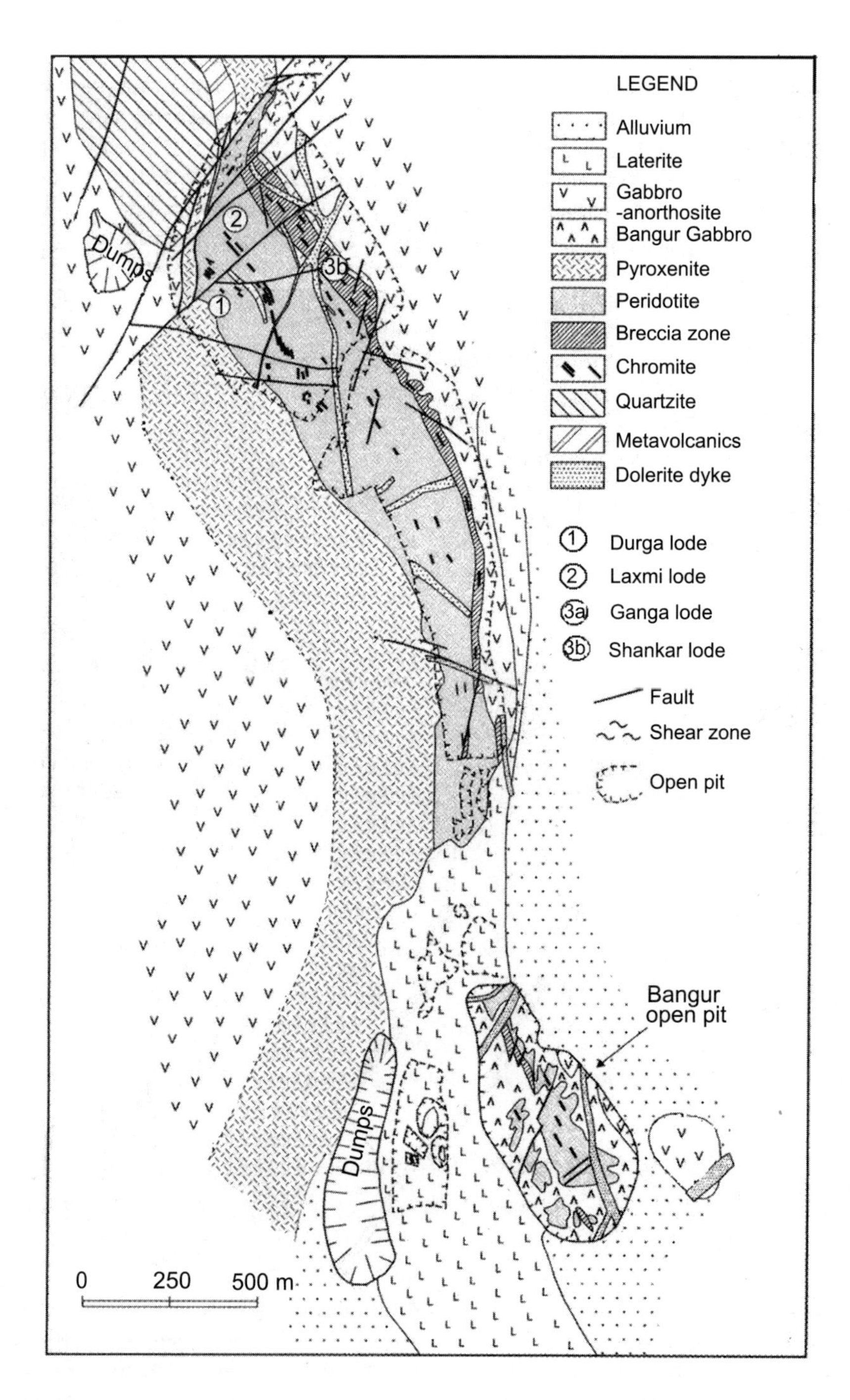

Fig. 4.2.58 Geological map of Baula-Nuasahi complex (after Auge et al., 2003 with limited modifications)

The Bangur Gabbro is a coarse grained gabbro–pyroxenite that contains xenoliths of dunite-peridotite and chromitite. Fragments of chromitite of varying sizes are incorporated in the gabbroic matrix based on the field relationship and geochemical characteristics, the gabbroic matrix may be assigned to the Bangur Gabbro only (Auge et al., 2003).

PGE mineralisation occurs in a 1 km × (2–40) m zone, closely associated with Ganga Shankar lode (Fig. 4.2.59).

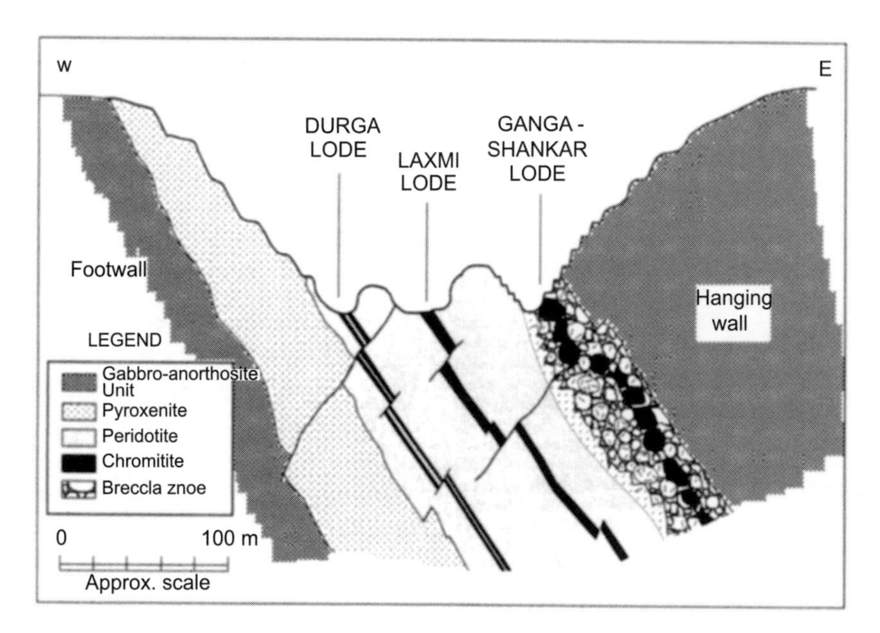

Fig. 4.2.59 Geological section across Baula mine showing disposition of the chromite lodes. PGM is associated with the Ganga–Shankar lode system hosted by breccia zone (after Auge et al., 2002)

PGE besides occurring as native elements, and natural alloys are present as complex compounds of Sb, Te, As and Bi. Auge and his co-workers in a relatively recent publication (Auge et al., 2002) grouped the PGE mineralisation of the area into two broad divisions: magmatic and hydrothermal (Table 4.2.10).

Symbol size represents abundance of the mineral phase present only as an inclusion in chromite. The unnamed gabbroic matrix in the breccia zone is generally in association with the base metal sulphides and ferrian chromite (cf. Nanda et al., 1996; Patra and Mukherjee, 1996; Mondal and Baidya 1997; Auge et al., 2002).

Auge et al., (2003) concluded that the source of PGE in the Type 1 mineralisation was the magma that was also the progenitor of Bangur Gabbro. Both the sub-types under the Type 2 mineralisation are restricted to the hydrothermally altered matrix of the breccia zone (Fig. 4.2.60). The composition of the oxygen and hydrogen isotopes in the minerals suggested that the hydrothermal fluid was magmatogenous (deuteric).

Mondal et al. (2006), believing in a somewhat different model, suggested that the whole mineralisation was magmatic, produced by a hybrid magma that resulted from incorporation of serpentinised chromite –bearing ultramafic rocks by a younger gabbroic magma.

Table 4.2.10 *PGM assemblages, Baula–Nuasahi, Orissa and their distribution (after Auge et al., 2002)*

Mineralisation type		I	2A	2B
Ferrian chromite		■	■	■
Chromite		■		
Sulphide			■	
Pt		■	■	■
Pd		■	■	■
Au			■	
Pt/Pd		8–8	0.5	2–3
PGM				
Isoferroplatinum	Pt_3Fe	■		
Braggite	(Pt,Pd,Ni)S	■		■
Sperrylite	$PtAS_2$	■	■	■
Geversite	$Pt(Sb,Bi)_2$			■
Malanite	$Cu(Pt,Ir)_2S_4$	■		
Ni-dominant analogue of Cuprorodsite	$Ni(Rh,Pt)_2S_4$	■		
Moncherite	$(Pt,Pd)(Te,Bi)_2$			■
Mertieite	$Pd_8(Sb,As)_3$		■	■
Sudburyte	(Pd,Ni)Sb		■	
Merenskyite	$(Pd,Pt)(Te,Bi)_2$		■	
Laurite	RuS_2	■	■	■
Osmium	Os,Ir			
Hollingworthite	(Rh,Pt,Pd)AsS	■	■	■
Un 1	Pd(Sb,Te,Bi)		■	
Un 2	$(Pd,Ni)_3(Sb,Te)_4$		■	
Un 3	Pd,Ag,Sb,Te,Bi		■	
Stiboan merenskyite un4	$Pd_2(Te,Sb,Bi)_3$		■	
Bismuthoan merenskyite un5	$Pd(Te,Bi,Sb)_2$		■	
Bismuthoan sudburyite un6	Pd(Sb,Bi,Te)		■	
Potarite	PdHg	■		
Argentian gold	Au,Ag		■	

Note:
Type 1 – Magmatic: In Bangur gabbro, chromite and ferrian chromite nodules
Type 2A – Hydrothermal: With basemetal sulphides and ferrian chromite
Type 2B – Hydrothermal: With ferroan chromite. No basemetal sulphide

Symbol size represents abundance of the mineral phase present only as an inclusion in chromite. Un-unnamed. Gabbroic matrix in the breccia zone, generally in association with the basemetal sulphides and ferrian chromite (cf. Nanda et al., 1996; Patra and Mukherjee, 1996; Mondal and Baidya 1997; Auge et al., 2002)

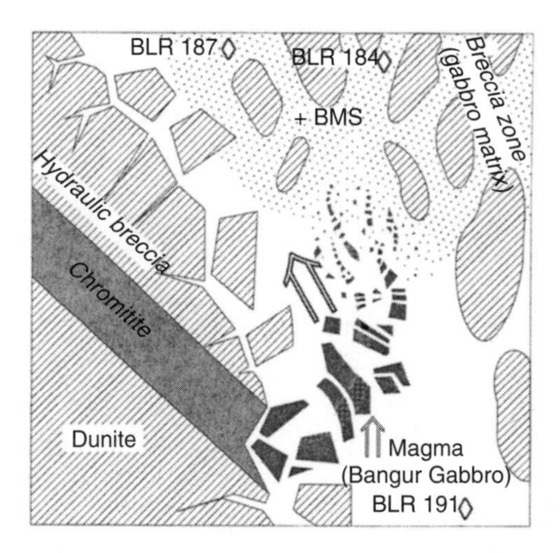

Fig. 4.2.60 Schematic diagram depicting the intrusion of Bangur Gabbro and hydraulic brecciation of dunite, Baula–Nuasahi complex + PGM and + BMS show the PGM- and base metals- enriched domains in the breccia zone; BLR – Sample Nos (after Auge et al., 2002). PGM-PGE minerals

As mentioned in an earlier section, Auge et al. (2003) determined the radiometric age of some Baula rocks. The Bangur Gabbro and the matrix of the breccia zone yielded Sm-Nd isochron age of 3205 ± 280 Ma (Fig. 4.2.61a). They also successfully separated fresh zircon from the above mentioned rocks. The Pb-Pb ages obtained from these zircon grains ranged from 3119 ± 6 to 3123 ± 7 Ma (Fig. 4.2. 61b).

The ultramafic rocks including the enclosed chromitite defied any means of direct dating. But since they occur as enclaves within the Bangur Gabbro, they should be older than the later. Neither the gabbro–anorthosite rocks could be dated. But it may be a permissible speculation that these rocks and the Bangur Gabbro, which intruded them, may not be very far from each other in time of emplacement. These suggest that all the members in the Baula basic-ultrabasic complex to be close in age, which is about 3120 Ma. This is also possible, and even probable, that they are genetically related. The view of Saha (1994) stands out in contrast to this conclusion. He believed these rocks and the enclosed chromitite bodies were Proterozoic – a view that has little support in the context of the above findings.

The PGE and the PGM assemblages of the Baula-Nuasahi area may be taken to suggest an environment, intermediate between the Bushveld and Ophiolite types. In the Bushveld type the PGE are dominated by Pd and Pt and they may be hosted by chrome-spinel or sulphides, where as the ophiolitic chromite is strongly enriched in Os, Ir and Ru. But that will be too simplistic an explanation, as there are instances where this distinction can hardly be made (Bacuta et al., 1988). The relationship may even be reversed as in the MLA chromitites, Coolac serpentine belt, SE Australia (Franklin et al., 1992). Spectral variation in PGE distribution in ores is mainly due to late-stage hydrothermal activity modifying the pristine distribution and superimposing a new signature. Such a conclusion is based upon a commonly developed hydrothermal alteration in the ore zone, results of experimental studies on the solubility and transport of PGE (Mountain and Wood, 1987) and fluid inclusions in PGM and PGE-bearing sulphides and the paragenetically related silicates (Ferrow et al., 1994). This must have happened here also.

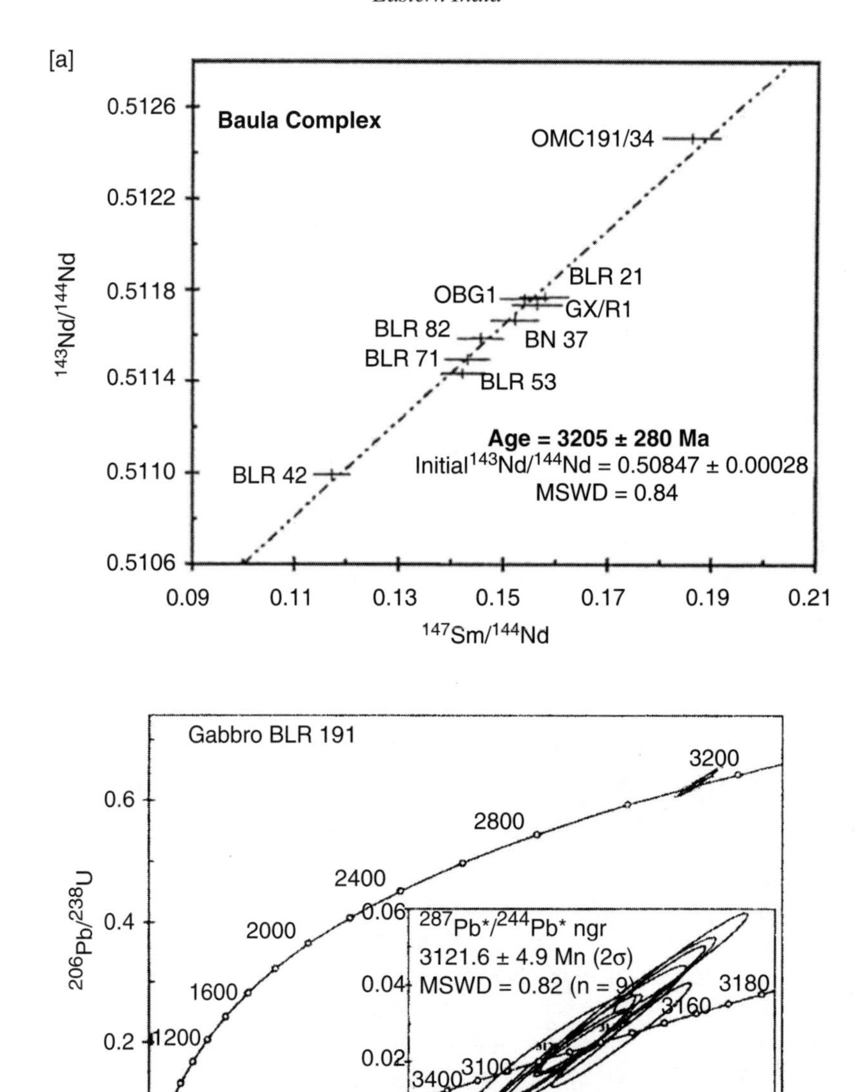

Fig. 4.2.61 (a) Nine point Sm-Nd isochron age (3205 ± 280 Ma; MSWD = 0.84) of Bangur Gabbro and matrix of breccia zone, Baula–Nuasahi complex, Orissa (from Auge et al., 2003);
(b) Concordia diagram of nine samples of zircon from Bangur Gabbro, depicting mean ^{207}Pb-^{206}Pb age of 3122 ± 5 Ma; MSWD = 0.82 (from Auge et al., 2003).

Primary partitioning of (Pt +Pd) vis-a-vis (Os + Ir + Ru) is thought to be controlled by the presence of sulphides in the system, being an order higher therein (Naldrett and Duke, 1980; Sattari et al., 2002). It can be many times increased by hydrothermal reworking (Rowell and Edger, 1986). These observations appear valid in case of the Baula–Nuasahi mineralisation.

The deposit may turn out to be economic on final evaluation.

Manganese Mineralistion in Singhbhum–North Orissa Region

The manganese mineralistion in this region is mainly concentrated in the iron ore province (IOG) of Jamda–Koira valley in Singhbhum (West) district of Jharkand and Kendujhar (Keonjhar) district of Orissa. Besides, weaker mineralistions are also reported from Lower Dalma sediments, both from the eastern and western parts of Dalma volcanisedimentary belt in the NSMB.

Mn-Mineralistion in Jamda–Koira Valley The mineralistion occurs within the 'horse-shoe' synclinorium of Jones (1934). About 200 small-scale mines are reportedly in operation in this area. There was some uncertainty about the Mn-ore bearing shale of this area. Recently, Patel et al. (2005) have stratigraphically delineated the manganiferous horizon comprising all the known deposits. This horizon, about 100 km in length with < 1 to > 5 km in width, has been traced from Gamaria (Jharkhand) in the western limb of the 'horse shoe' through Nalda–Pacheri–Panduliposi–Kusumdihi–Bhutula (Orissa). In the east it can be traced from Gitilip and Surjabas (both in Jharkand), through Soyabali–Joda–Bamebari –Gurda–Dobuna (Orissa). They support the contention of Mahapatra et al. (1991) that the shale unit, which hosts the Mn-deposits, forms a part of the 'Upper shales' of the earlier workers. The stratigraphic sequence proposed by them is significant in the context of exploration geology (Table. 4.2.11).

Table 4.2.11 *Stratigraphic sequence of the IOG rocks in Jamda–Koira Valley, showing the position of manganiferous formation (after Patel et al., 2005)*

Formations	Rock units
Upper volcanics	Mafic lava and tuff
Ferruginous Shale Formation	Shales of various description, some apparently tuffaceous; narrow BIF bands with derivative iron ores
Manganiferous Formation	Cherty Member: Massive chert grading to dolomite Shale Member: Manganiferous shale with Mn-ore
Banded Iron Formation	Finely and coarsely laminated jaspillite (BHJ) and BHQ with Fe-ores
Volcanic Formation	Upper unit: Tuffaceous shale Lower unit: Mafic lava
Basal Formation	Sandstone–grit–conglomerate

The Mn-ore mineralisation in this region may be divided into two principal types:

1. Stratabound, even stratiform
2. Lateritoid.

The stratabound ores occur interbanded with shales. The ore bands are usually concordant with planar structures in the host rocks. Locally the ore bands also display folded structures. Lateritoid ores in general occur as tabular bodies and have limited depth extension compared to the other type. These are also lower in grade than the stratabound ores. At places lateritisation of stratabound ore is also seen. A manganese mine near Thakurani Pahar exposes lateritoid ore and an overlying bed of kaolinite (Fig. 4.2.62). Locally Mn-ores have accumulated in brecciated chert, but these hardly make the grade.

Fig. 4.2.62 Lateritoid manganese ore overlain by kaolinite layer (white) exposed in a manganese quarry near Thakurani Pahar, Orissa (photograph by authors)

Cryptomelane, pyrolusite, lithiophorite are the principal ore minerals, while romanechite, manganite, and chalcophanite are locally present. Other minerals present are: hematite, goethite, lepidochrocite. Kaolinite is ubiquitous in the stratabound ores (Mahapatra et al., 1989).

A study of REE distribution in the (ferro-) manganese oxide ores of the western parts of the area throws some interesting light on the possible origin of these ores. There is a conspicuous enrichment of REE and some other trace elements (Sc, Sr, Y, Cs, Hf, Co, and Cu) in the stratabound ore bodies in contrast to the depletion of REE in the lateritoid ores. The REE patterns in stratabound ores display pronounced negative anomaly of Ce, which is interpreted to be due to the hydrothermal nature of the REE-bearing fluid (Elderfield and Greaves, 1981). It is suggested that the stratabound ores formed by the leaching of Mn-protores or disseminated Mn (oxide) in the shale by diagenetic fluids along the bedding/bedding parallel cleavages. This is supported by the depletion of REE in the surrounding phyllite (Mahapatra et al., 1996).

The Mn-deposits of the area being small and scattered and being mostly in the hands of private operators, there is no dependable reserve estimation of these deposits.

Mn-mineralisation in the NSMB Though Mn-mineralisation is little known in this belt, Dunn reported a few such occurrences way back in 1937. These occurrences are reported from an area between Miritanr and Basadera, located at about 10 km to the north of Ghatsila. Mining started there in 1927, but had to be given up by 1931, as the venture turned out to be non-viable. Similar mineralisation was also traced at Jhatijharna further to the east. Patches of Mn-mineralisations reportedly occur along the north of Dalma volcanics. Even lateritic low-grade ores are reported from Dalma hills, occurring on the basic volcanic rocks (Dunn, 1937). In the phyllites, underlying the volcanics, secondary Mn-oxides are found accumulated along cleavages and fractures. That the source of manganese for these occurrences was the host phyllites is supported by the presence of Mn-ottrelite and piedmontite in the latter. No mineable manganese deposit has been discovered in this area by later exploration.

Occurrences of manganese mineralisation, similar to the above, were also reported by Gupta and Basu (1976–77) while mapping the western parts of Dalma belt. Secondary Mn-oxides occur at various locations in the Hesadih sector as fracture fillings and clots within the ferruginous phyllite and interlayered iron stones, which along with carbonaceous phyllites and tuffaceous sediments constitute the basal parts of Lower Dalma Formation. The source of manganese may be the pyroclstics, part of which now appears as phyllite. Similar occurrences of Mn-oxides are also noted in the central part of the belt in Dimna–Patamda area.

Copper (Gold) Mineralisation in Chhotanagpur Granite -Gneiss Complex (CGC)

There are numerous occurrences of polymetallic mineralisation in the Chhotanagpur Granite–Gneiss Complex (CGC), mostly hosted by the older supracrustals. As mentioned in the previous section, the older supracrustals in CGC terrain include pelitic schists of greenschist to highest amphibolite facies, calc-silicate, marble, quartzite, amphibolite, skarn rocks, pyroxene–granulites, anorthosite and charnockite. The metamorphic grade is generally of almandine–amphibolite facies, reaching at places the high amphibolite facies with local transition to anatectic assemblage or granulite facies. The migmatisation is syn- to late- kinematic with respect to peak metamorphism. Polyphase fold-deformation is widely manifested in the supracrustals and granite-gneiss. The available age data range from 1.6 to 0.85 Ga.

The most prominently mineralised belts in the CGC are the copper dominant mineralisation in Pareshnath–Baraganda–Bagodar belt and the polymetallic mineralisation along Hesathu–Belbathan belt (Fig. 4.2.63).

Baraganda Copper Deposit The Baraganda copper deposit (24°05′: 86°03′), now in Giridih district of Jharkhand, is located at about 23 km from the nearest railway station Pareshnath on the Grand Chord Railway line, and can be approached by a 9 km long forest road from Dhawatanr on the Dumri–Giridih Trunk Road.

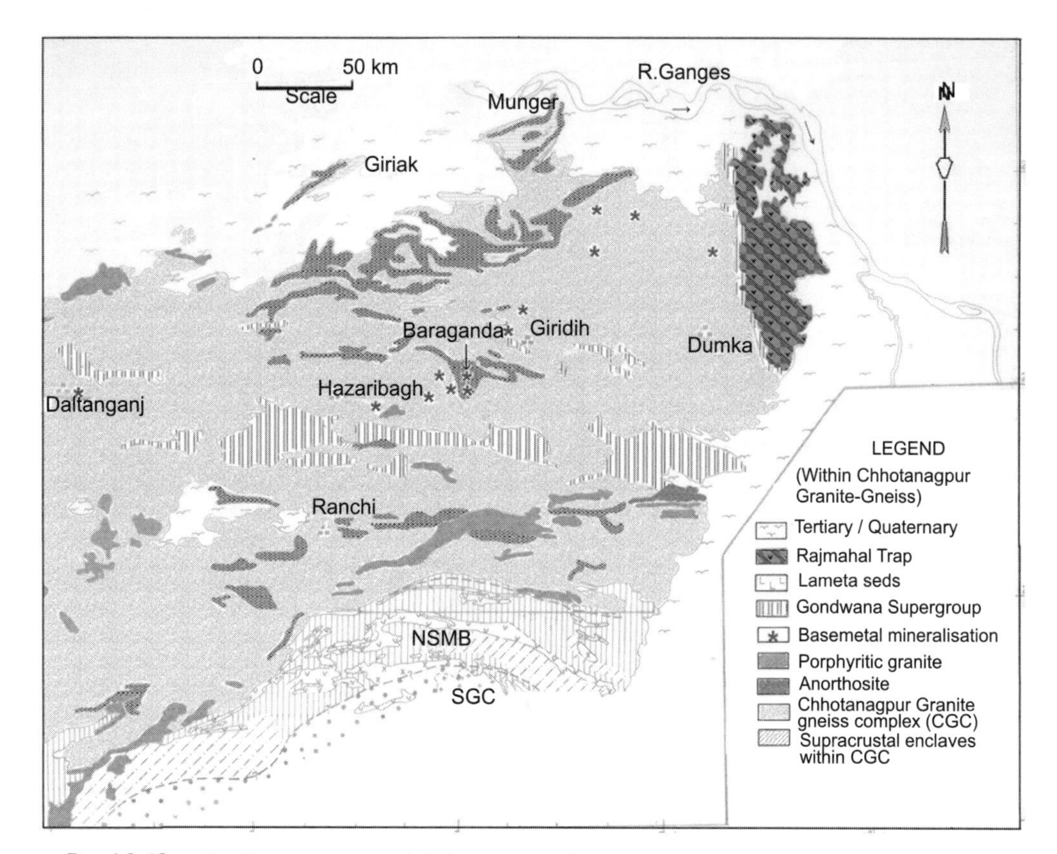

Fig. 4.2.63 Northeastern parts of Chhotanagpur Granite–Gneiss Complex (CGC) showing the locations of basemetal sulphide prospects and deposits (compiled from Fig. 4.1.1 and Mazumdar, 1988)

It was a site of intense mining activity in the ancient time when the deposit was worked by open excavation extending down to about 30 m. Later, between 1882 and 1891, the lodes were worked by underground mine development comprising five shafts and six levels upto a depth of 100 m. The western shaft was deepest reaching down to 330 ft. The production of copper ingots per month was in the tune of 40–65 tonnes. In 1888, about 220 tonnes of refined copper was produced from the smelter at Giridih. It is reported that mining difficulties and lack of funds caused the ultimate closure of the mine.

Geologically, the Baraganda area is located within an extensive WNW–ESE trending supracrustal belt (120 km long and 5–20 km wide) in the CGC country, passing through north of Hazaribagh and south of Giridih. In the deposit area, feldspathic (biotite) muscovite-quartz schist is the most dominant rock type with subordinate quartz–muscovite schist, quartzite, amphibolite, chlorite–biotite quartz schist, feldspathic chlorite–biotite–quartz schist and talc–tremolite–actinolite schist. The general trend of the rock formations and the most prominent schistosity is N 70° W–S 70° E (with local swings upto 50°) and the dip is sub-vertical with tilts on either side. Regionally the rocks are deformed into isoclinal folds, as depicted by a number of large amphibolite bodies (one at the eastern end of the deposit). There are cross puckers on the schistosity plane with local development of two sets of crenulation cleavages.

The sulphide mineralisation is mostly confined to silicified, brecciated and sheared feldspathic muscovite–chlorite–biotite–quartz schist. Chalcopyrite is the dominant sulphide with subordinate pyrite, galena, sphalerite, pyrrhotite, marcasite and arsenopyrite.

The total strike length of the deposit is 1020 m, out of which 715 m is best mineralised. There are four discontinuous lodes, concordant to the schistosity, with en-echelon disposition within the said strike length, which are numbered from north to south. The northern most Lode-1 is hosted by muscovite quartz schist and extends for 600 m, of which about 200 m contains more than 1.0% Cu. Lode-2, occurring within feldspathic chlorite-biotite quartz schist, is the richest (2.28% Cu) and thickest (2.84 m) and extends for 260 m. Lode-3 is located along the central part of the same host rock as Lode-2 and have three richer shoots. The southern most Lode-4 (and Lode - 4A) are lean orebodies of 60m lengths with barren ground in between. The total reserve of all the lodes come to 0.6 Mt with average grade of 2.28% Cu, down to 80 m vertical depth.

Hesathu–Belbathan Belt In the north of Pareshnath–Baraganda–Bagodar supracrustal belt and south of Bihar mica belt (now in Jharkhand) there are numerous small enclaves of mineralised tremolite–actinolite schist, associated with calc-silicate rocks, quartzite, bedded chert and amphibolite within the granite–gneiss country. The 30 km wide zone roughly trends WSW–ENE to SW–NE and extends for nearly 120 km through parts of Giridih, Deoghar, Jamui, Banka, Bhagalpur, Munger and Godda districts along northern Jharkhand and southern Bihar. Similar mineralised enclaves are also noted in the south west of Baraganda in Hesathu (23°59′20″: 87°00′50″) area near Hazaribagh. In our opinion, the nomenclature 'Hesathu–Belbathan Belt' is a misnomer, which is imagined to have joined Hesathu in the southwest near Hazaribagh with Belbathan (24°43′30″: 87°18′30″) in the extreme northeast in Godda district through an unconfirmed fault zone across the rock formation. The available geological informations are sketchy (Mahadevan, 2002) and do not permit a reliable correlation. However, the individual mineralised zones are marked by residual gravity-highs of 3.5–4.0 m Gal. and these occurrences fall within a ENE-WSW trending 50 km wide zone passing through the south of Bihar mica belt.

The 100–200 m long actinolite-tremolite enclaves in the area defined above are characterised by sporadic polymetallic mineralisation (Zn + Cu + Pb) and are marked at many places by the evidences of ancient mining activity. In Pindara block, Banka district, a small ore resource of 0.43 Mt with average grade of 3.93% of Zn, Cu and Pb (including 0.22 Mt × 1.13% Cu) was located. The ore minerals are sphalerite, chalcopyrite, galena, bornite, pyrrhotite and pyrite with minor bismuthinite. In the area around Dhawa (24°45′15″: 86°38′5″) and Biharbari (24°38′10″: 86°33′15″) in Banka district, very small ore resources of 0.13 Mt × 3.4% (Cu,Pb,Zn) and 0.13 Mt × 9.54% (Cu, Pb, Zn) were established in actinolite-tremolite schist (Singh et al., 1998). Low concentration of gold 100–150 ppb) in soil and bedrock is reported from some sectors of the belt like Charkipahari–Tulsitanr, Deoghar and Banka districts.

Gold Mineralisation in Sono Area In Sono–Karmatiya area, Jamui district, gold is recovered from the alluvium and laterite gravels. There has been intense prospecting by both GSI and the State Government for gold in this area since last several decades without much avail. The primary gold mineralisation in this area is in bedded chert, grunerite quartzite and schistose amphibolite bands occurring within CGC. Average gold content is below 0.2 gm/t, ranging upto 2 gm/t at places. No ore zone of significance could so far be located in the area in spite of much effort.

Rare Metal (RM) and Rare Earth Elements (REE) Mineralisation in Jharkhand ('Bihar') Mica Belt, Chhotanagpur Granite–Gneiss Complex

The well known 'Bihar mica-belt', now in Jharkhand, is located along the north eastern part of CGC. The belt derives its name from its being the repository of high quality mica for industrial uses in numerous quartzo–feldspathic pegmatites. Such pegmatites, besides containing quartz, feldspar and mica(s), may contain other economic minerals as beryl, columbite–tantalite, lepidolite and rarely cassiterite in exploitable concentrations. Accessory minerals, sometimes in economic proportions, include tourmaline, biotite, ilmenite, magnetite, zircon and xenotime. Monazite, uraninite, pitchblende, fergusonite, bismuthinite, arsenopyrite, cryolite, fluorite, apatite, allanite, pyrochlore–microlite, triplite, frondelite, betafite, tapiolite, euxenite, amblygonite, wolframite and samarskite are the rare minerals in some of these pegmatites (Sinha, 1999) (Fig. 4.2.64).

The pegmatites of the region were previously exploited for muscovite books, with beryl, lepidolite, and tourmaline and in rare cases, Nb-Ta minerals as byproducts. But it was due to a vigorous exploration programme taken up by the Atomic Minerals Division (now Directorate) (AMD), Govenrment of India, during the sixties and seventies of the last century that new reserves of these minerals were established. As a result more than 200 tonnes of columbite–tantalite analysing 10–50% Ta_2O_5 and 40–65% Nb_2O_5 and several hundred of tonnes of beryl have been produced from this belt during the last few decades. The ore minerals are available from the pegmatites as well as from the gravels (alluvium and deluvium) that surround them. Obviously these were derived from the surface decomposition of parts of the neighbouring pegmatite(s).

Quartzo–feldspathic pegmatites of the area under review are usually divided into the following types, depending on their internal/external structures and mineral composition:

1. Sill-like, unzoned and mineralogically simple type
2. Dyke-like, lenticular and stock-like zoned pegmatites containing commercial mica and beryl
3. Sill-, dyke- or stock-like weakly zoned bodies with complex mineral composition and containing RM and REE minerals.

Type-1 pegmatites occur within both mica schists and granite/granite gneisses. Type-2 is, in general, a preserve of the mica schists. Type-3 pegmatites occur in all types of rocks, such as mica schists, granitic rocks and amphibolites. There are, however, some pegmatites such as of Manbagawa, Bhanakhap and Bishanpur, which contain both columbite–tantalite in fair quantities and also commercial muscovite. Beryl, common in mica-pegmatites, may be found in considerable quantities in some Nb-Ta pegmatites as at Dhab–Dhajua area. Cassiterite–lepidolite pegmatites are numerous in a NE–SW trending zone from Kherdotoli to Charki. A wolframite-bearing pegmatite

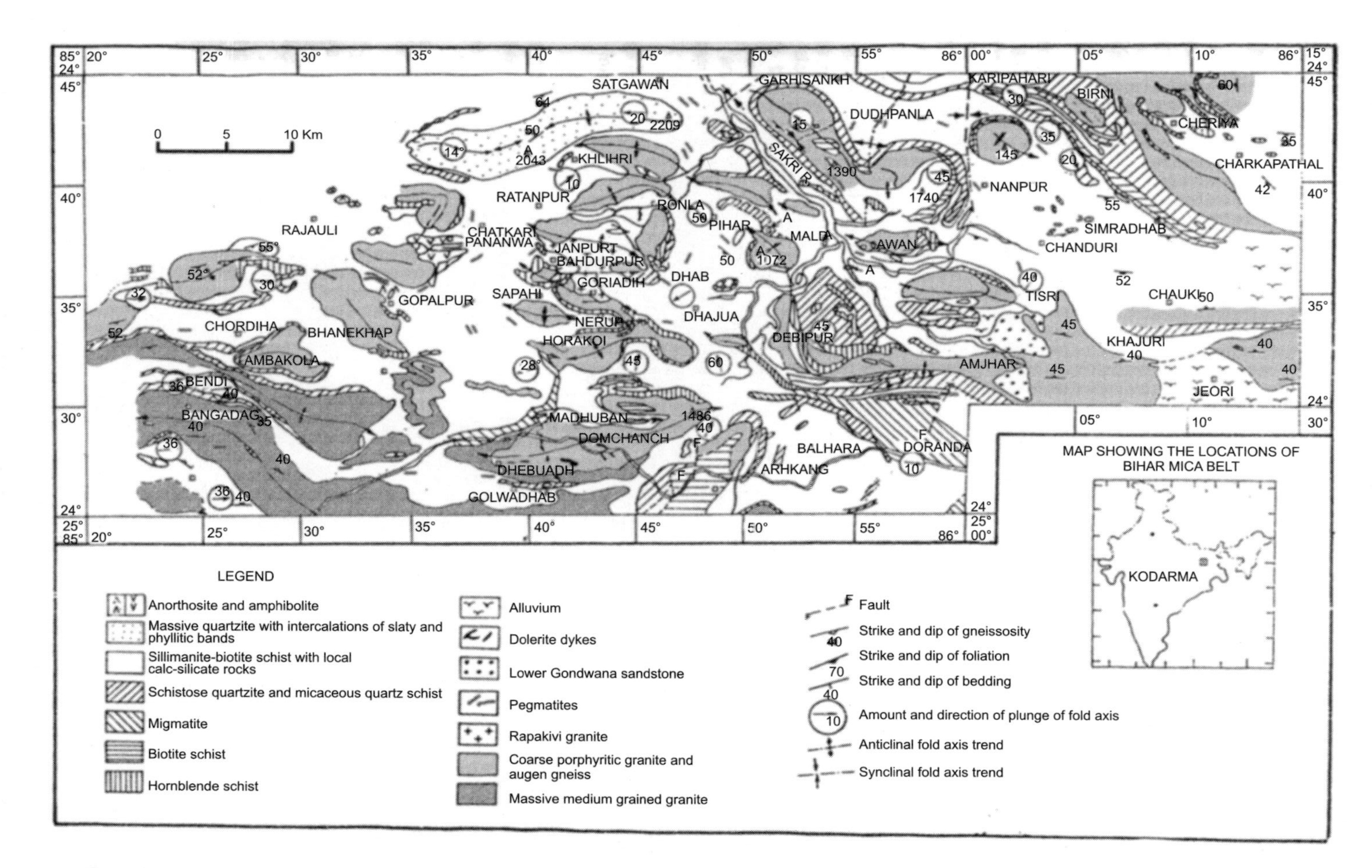

Fig. 4.2.64 Regional geological map of Bihar (Jharkhand) mica belt (after Nateswara et al., 1981, in Sinha, 1999)

is also reported, but is a rare case (Sinha, 1999). There have been attempts to distinguish the Ta-Nb bearing pegmatites from other varieties on several criteria. The most dependable amongst these appear to be the high contents of a Ta and Nb in muscovite mica in the potential pegmatites (Ramachandran and Sinha, 1992).

The Ta–Nb bearing pegmatites in the Jharkhand mica belt are mainly localised in the following four sectors: (i) Bhanakhop–Pichli–Ghortuppa–Bendi in the west, (ii) Pihara–Satgawan–Ghatkari–Manbhagwa in the mid-west, (iii) Khairadih–Bekabar in south-west, and (iv) Bijaia–Batra–Asarhwa–Amarjharna in the east.

By way of an example of rare metal pegmatite of this region, the Khairidih pegmatite is discussed here. Khairidih pegmatite is 188 m long and 60 m wide with prominent quartz-core. Beryl, tantalite–columbite and monazite occur here in mineable quantities. Beryl occurs at the margin of the quartz-core together with quartz and muscovite. Tantalite–columbite (40.8% Ta_2O_5 and 36.50% Nb_2O_5 and 18% U_3O_8) occurs between quartz core and adjoining muscovite or the intergrowth zone. More than 2.5 tones of tantalite–columbite have been produced from this mine. Monazite occurs in massive to crystalline (granular) forms and contains 6.2% ThO_2 and traces of U_3O_8. More than 25 kg of monazite has been obtained from this pegmatite (Sinha, 1999).

Regarding the origin of these pegmatites it has been suggested that these were probably derived from the fractionation of granitic melt generated from the Chhotanagpur Gneissic Complex around 1600 Ma at a depth of 30–35 km and 750°–800°C. This age is discrepant with the pegmatite age of 960 ± 50 Ma (Pb-isochron) suggested by Vinogradov et al. (in Sarkar, 1968).

Mineralisations in Purulia District, West Bengal The district is located at the southeastern part of Chhotanagpur plateau and hence geologically belongs to the CGC. The rocks are granite–gneisses, granitoids and several types of highly metamorphosed rocks as meta-norite, meta-gabbro, pyroxene granulite, ortho-amphibolite and metamorphosed anorthosite.

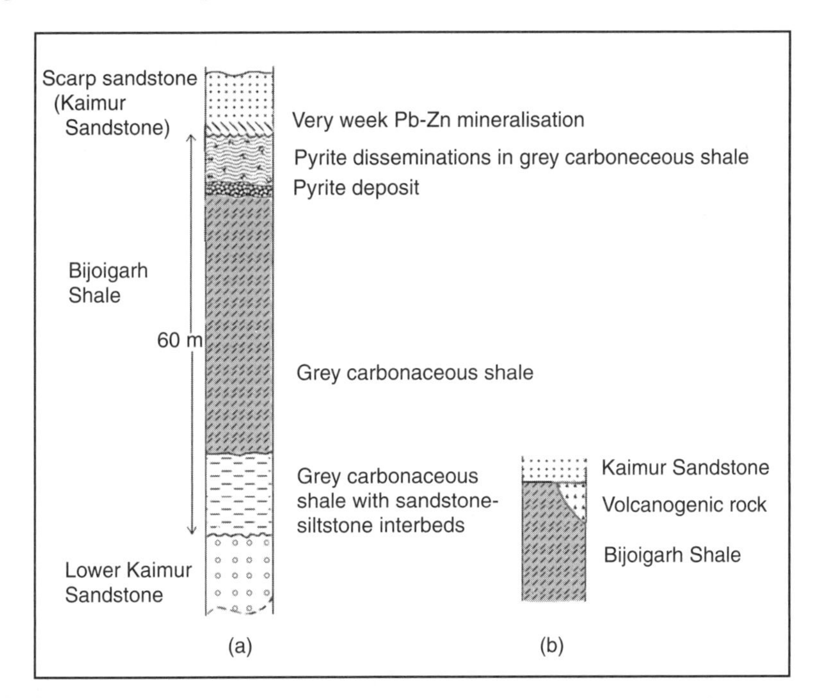

Fig. 4.2.65 Stratigraphic column showing pyrite mineralisation in the Kaimur Group at Amjhore, Rohtas district, Bihar (based on Banerjee and Om Prakash, 1975 and Chakraborty et al., 1996)

Broadly three types of RM and REE mineralisations are reported from this region (Basu and Ghosh, 1989): (i) La, Ce and Nb mineralisations along with carbonatite rocks emplaced along the South Purulia Shear Zone (SPSZ), also referred to as 'Northern shear zone' with reference to Singhbhum shear zone, (ii) Apatite–magnetite–barite–rare earth (La, Ce) mineralisation along North Purulia Shear Zone (NPSZ), (iii) Rare metal (Li, Cs, Rb) mineralisation along a megalineament within CGC. In addition to these, monazite occurs as inland placers at places in Purulia and neighbouring areas of Ranchi.

Apatite (-magnetite) mineralisation of commercial interest is noted at a number of places along the SPSZ, including Chirugora, Kutni, Medinitanr and Beldih, associated with an alkaline-carbonatite complex consisting of alkali syenite, carbonatite, alkaline ultramafite and phologopite amphibolite, not necessarily all the rocks occurring at every point. The phosphate ores of this belt are enriched (>1000 ppm) in Nb, Zr, and Ba. Beldih ores analysed 0.06–3.6% of Nb_2O_5. At the Beldih area a reserve of 4.56 million tonnes of ore with an average content of 13.1% P_2O_5 has been estimated (Kumar et al., 1985).

An interesting pegmatitic mineralisation of REE is located near the Panrkidih village (Basu et al., 1996). The apatite–magnetite bearing pegmatite trends E–W and has a strike length of 600 m. The REE contents of the ores have been found to be as follows – La: 500–1754 ppm, Ce: 1053–3137 ppm, Nd: 346–1809 ppm, Sm: 125–340 ppm, Gd: 24–146 ppm, and Y: 68–208 ppm. The P_2O_5-contents in the ores ranged between 8.14% and 22.15%.

An allanite–bastnaesite–barite mineralization in pegmatite (NYF type of Cerney, 1991) with REE_2O_3-contents upto 35.23% is reported from Nowahatu in NPSZ. Uranium and thorium contents of these ores are high (U_3O_8: 2.2–3.0%; ThO_2: 0.10 – 0.15%) (Talapatra et al., 1995). Interestingly a few tonnes of highly radioactive allanite–bastnaesite ore occur at the surface.

Further north of the NPSZ, along an ENE–WSW trending megalineament Li, Cs and Rb mineralisation in pegmatites have been noted at a few places such as Beku, Jabor, Belamu and Porgu (Talapatra et al., 1995). The pegmatites have anlysed 0.04–0.11% Li, 0.05–11.0% Cs, 0.43–0.49% Rb and resemble the LCT type of Cerney, 1991 (Talapatra et al., 1995). Alkali feldspar has been found to contain 0.55–0.94% Li, 0.03–15.70% Cs and 0.24–0.49% Rb.

The Beku Pegmatites in the light of available information may be called pollucite–lepidolite granite pegmatite and spodumene–pollucite granite pegmatite. The Beku Granite, from which the Beku Pegmatite is thought to have formed by fractionation, is peraluminous, belonging to the S-type and is a constituent of the Chhotonagpur Gneissic Complex at its eastern extremity (Som et al., 2002). An important occurrence of pollucite (ore mineral of caesium) bearing pegmatite/aplite has recently been investigated and assessed by GSI near Beku in Purulia district, West Bengal, having a resource of 0.09 Mt ore with 0.6 to 1.27% Cs (+ 0.1% Rb and Li).

Pyrite Deposit at Amjhore, Rhotas District, Bihar

The pyrite deposit at Amjhore, Rhotas district (formerly a part of Sahabad district), Bihar, occurs within the Bijaygarh Shale, a Formation within the Kaimur Group, which in turn is a part of the Vindhyan Supergroup in the Sone Valley. The location is almost at the eastern fringe of the Vindhyan basin.

The pyrite deposit at Amjhore is a subhorizontal stratiform body nearly 1 m in average thickness and extended over an area of about 2 sq km. The host rock, the Bijaygarh Shale, is an almost unmetamorphosed argillite, containing 3–5% non-carbonate carbon. The presence of the latter has rendered the rock generally gray. The unit is about 60 m thick, of course with the interbeds of fine-grained sandstone-silt. The pyrite orebody occurs near the top of this unit, i.e. Bijaygarh Shale (Fig. 4.2.65). Pyrite disseminations are observed in thin zones within this unit, not commonly though. Research workers on the mineralisation at Amjhore (Muktinath, 1960; Guha, 1971; Pandalai et al.,

1983) did not report the presence of volcanic rocks in the immediate vicinity of the deposit, or even within the Kaimur Group itself. But a group of workers that studied the geological build-up of the area reported the presence of volcaniclastic rocks that overlie the Bijaygarh Shale below the Kaimur Sandstone (Chakraborty et. al., 1996). The volcaniclastic rock unit, according to them is felsic in composition and varies in thickness from few mm to about 50 m.

The ore mineral in the deposit is pyrite. A minute phase is locally seen that resembles tetragonal FeS or mackinawite. The ores may be broadly divided into two types: the massive ore and the richly disseminated ore. There are obvious textural variations within the massive ores. Some contain spheroid and ovoid structures of pyrite that initially could have been framboids. In some ores of the massive variety, relatively coarse grains of pyrite occur in a cryptocrystalline groundmass of pyrite. Disseminated individual grains or clusters of grains of pyrite occur in a black shale matrix. Pandalai et al. (1983) studied the trace element contents in the monomineralic fraction of Amjhore pyrite as well as in the wall rocks. They found the Co/Ni ratios in pyrite to average 2.3. Guha (1971) analysed a number of (n = 36) pyrite samples from Amjhore for the determination of the sulphur isotope composition. He obtained a $\delta^{34}S$-range of + 19.96 to + 4.52 per mil, with an average of + 9.33 per mil. Some of the samples are isotopically heavier than the sulphur in the sea water sulphate at that time. This could happen by the bacterial reduction of sulphate in a comparatively restricted part of the basin where sulphate gradually became heavier through evaporation. Pandalai et al. (op. cit.) on the other hand, though agreeing with Guha (op. cit.) that the Amjhore pyritic ores were sedimentary-diagenetic/diagenetic, believed that the ore-forming elements were probably largely contributed by submarine exhalative activity. Their conclusion is based mainly on the trace element contents. Well, ore constituents supplied by submarine volcanic activity could, under limiting conditions be deposited with the help of bacterial activity. Moreover, the volcanic rock fabric in nature, overlie the pyrite deposit separated by black shale.

Banerjee and Om Prakash (1975) reported a weak galena mineralisation 6–10 m above the pyrite deposit, in Pb + Zn (Pb >> Zn) content is about 0.4% and the ore horizon is thin (< 1m).

Bauxite Deposits of Eastern India

The bauxite belt of Eastern India forms the eastern part of the 400 km x 50 km belt, extending from Jharkhand in the ENE to Chhattisgarh and Madhya Pradesh in the WSW. It comprises chains of bauxitised mesas, including those of Bagru Hills (Lohardaga district), Richiguda (Ranchi district), Shrendag and Netarhat (Gumla district) in Jharkhand. The Chhotanagpur Granite–Gneiss Complex represents the country rock along this belt. Before the discovery of Eastern Coast Bauxite, this region of Jharkhand together with its western extension was the largest repository of bauxitic ores in the country.

Jharkhand Deposits Roy Chowdhury (1965) reported 80 bauxite deposits/occurrences from the region, which has recently been curved out of Bihar and named the state of Jharkhand. But it seems some of them still deserve detailed exploration. The Bagru deposits amongst them are being mined from the first half of the last century. However, published geological information on them is generally meagre.

Bagru Deposits The Bagru deposits, comprising the lease-holds at Bagru, Bhusar, Hisri and Hisri New, were being exploited by the Indian Aluminium Company Ltd (INDALCO), until the property changed hands recently and became a part of a larger company, Hindusthan Aluminium Company Ltd. (Hindalco Ltd.), which reportedly adopted quite a few prospective areas of the region for further development. The Bagru deposits are located between 23°28′20′′ and 23°29′50′′ N and 84°35′05′′ and 84°36′40′′ E latitudes and longitudes (Figs. 4.2.66 and 4.2.67).

Fig. 4.2.66 (a) Panaromic view of Bagru hills and (b) a view of the flat top of the hill ('Bagdu Pat') About 1000 m MSL, Lohardaga district, Jharkhand (from Gangopadhyay, 1998)

Fig. 4.2.67 (a, b) Views of the Bagru mines showing bauxite horizon below soil and laterite capping (Gangopadhyay, 1998)

The four leaseholds mentioned above are located on a plateau, which has an elevation of 1025–1055 m above MSL, and is known locally as 'Bagdu Pat'. The mine-head is connected to the district town of Lohardaga by an all-weather motorable road, about 16 km long. The Bagdu Pat is also called Dudmatia (milky soil) Pat because of the exposure of the clay unit underlying the bauxite unit on scarp faces of the plateau.

The weathering profile of the Bagru hills is characterised by a 5–14 m thick soil + murram zone underlain by a lateritic zone of 1–9 m thick. Below the laterite developed a 5–17 m thick bauxite zone. The bauxite zone gives way to the kaolinitic clay zone intervened by a transitional or lithomerge zone of 1–4 m thick. The interface between the clay zone and the country rock below is hardly revealed. A schematic outline of this profile is presented in Fig 4.2.68.

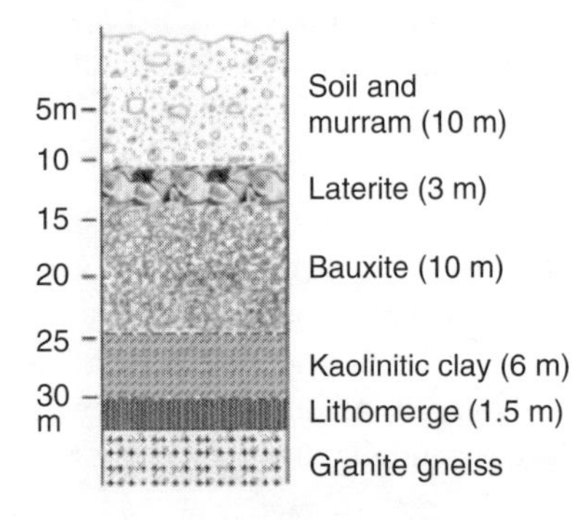

Fig. 4.2.68 Representative weathering profile of Bagru hill

The ores at the Bagru deposits are characterised by the presence of gibbsite, as the major phase with Fe-oxide/hydroxide (goethite) as subordinate and Ti- and Fe-Ti-oixdes as minor/trace phases. The ores are physically divisible into the following textual varieties:

1. Hard and pisolitic (Fig. 4.2.69 a–b)
2. Hard and massive with a few pisoliths
3. Granular and compact, but finely porous.

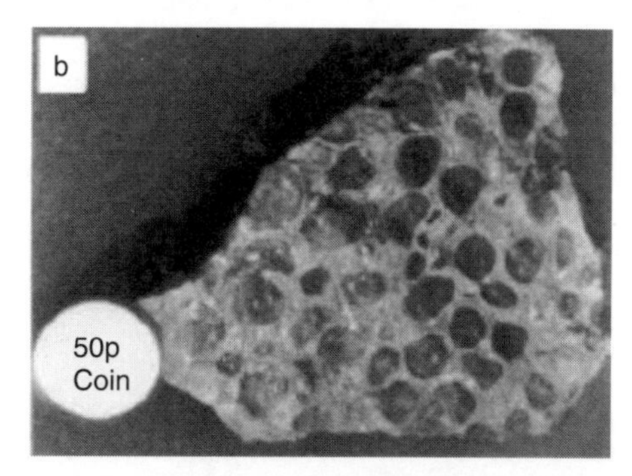

Fig. 4.2.69 (a) Hard pisolitic bauxite (b) Pisolites in bauxite replaced by iron oxide (Gangopadhyay, 1998).

Pieces of earlier formed bauxite cemented by a structure-less matrix (Fig. 4.2.70a) suggests polygenetic history of the ores, at least locally. Gibbsite growth in pisolitic bauxite makes interesting patterns (Fig. 4.2.70 b–c), while the texture in the granular variety is simple.

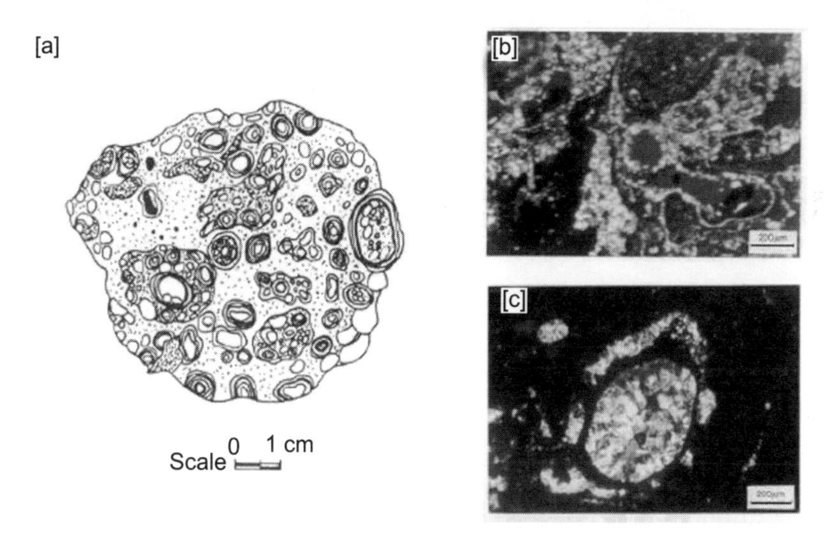

Fig. 4.2.70 (a) Sketch of composite bauxite with pisolitic fragments and pisoliths embedded in structure-less fine grained bauxite. Photomicrographs of pisolitic bauxite (cross-nichols), showing (b) annular and (c) radial growth of gibbsite (Gangopadhyay, 1998).

The principal mineral phase in the clay-zone is kaolinite, with quartz, illite and muscovite as minor phases. Tourmaline, zircon, rutile, kyanite and sillimanite are in trace phases.Possibly no other bauxite deposit in India is as much debated as the Bagru hill deposits with respect to the ore genesis. One model suggested kaolinitisation of the granite gneiss by hydrothermal water, transformation of the clay into bauxite by desilication of kaoline and subsequent lateritisation of bauxite through the deposition of Fe-hydroxides/oxides from the iron dissolved in meteoric waters (Fox, 1923). Valeton (1972), on the otherhand, suspected the basalt to be the protolith for the bauxite deposits in the Bagru area.

The nature and composition of the minor and trace minerals (kyanite, sillimanite, tourmaline, zircon) in the clay zone material (unpublished observations, S.C. Sarkar) do not allow assumption of a basaltic protolith for the clay and if clay (kaolinite) gave rise to bauxite by desilication, the same would hold also for the bauxite. Neither there is any field or petrographic proof in support of clayey Gondwana sediments to be the protolith of bauxite. The present authors support derivation of the clay and ultimately bauxite, from the gneissic rocks with the help of meteoric water, rather than by hydrothermal water. The blanket nature of the clay and the bauxite zones, relict gneissic fabric in the lithomerge and the nature of the accessory/trace mineral phases in the clay zone lend strong support to this conclusion. The chemical compositions of the basement rocks and the bauxite developed on the Bagru hills are presented in Table 4.2.12 and 4.2.13.

Table 4.2.12 *Chemical composition (wt%) of the basement rocks below the Bagru hill deposits, Lohardaga (Gangopadhyay, 1998)*

Rock type	SiO_2	TiO_2	Fe_2O_3	Al_2O_3	CaO	Na_2O	K_2O	LOM	Total
Granite gneiss	65.80	0.65	8.04	11.50	2.65	3.13	7.49	0.27	99.33
Granite gneiss	66.80	0.58	7.34	11.40	2.10	2.95	7.94	0.45	99.56
Granitoid rock	70.80	0.24	2.50	10.90	1.24	2.52	10.50	0.94	99.64
Granite gneiss	64.00	1.07	8.21	12.80	4.25	2.88	4.92	1.21	99.34
Granitoid rock	70.90	Tr	1.20	11.50	0.60	2.10	13.00	0.27	99.57
Granite gneiss	72.2	0.17	2.28	11.40	0.89	2.27	9.55	0.88	99.64

Note: Tr – trace. MgO, MnO and P_2O_5 were not analysed. Obviously they may be present only in insignificant proportions.

Table 4.2.13 *Chemical composition (wt%) of some samples of bauxite from the Bhusar East, Bagru NE and Bagru Central blocks, Bagru hills (Gangopadhyay, 1998)*

Sample No.	SiO_2	TiO_2	Fe_2O_3	Al_2O_3	LOM
1	0.53	12.8	3.35	58.7	24.4
2	1.20	18.3	4.42	50.6	25.3
3	3.04	9.46	3.90	56.4	27.0
4	0.84	12.3	3.00	57.8	25.7
5	7.96	10.9	3.34	53.1	24.5
6	0.46	13.6	3.55	58.2	24.0
7	1.09	10.9	4.52	56.6	26.6
8	1.19	5.73	31.80	41.4	19.3
9	1.10	10.1	3.62	56.1	28.7
10	0.90	5.92	32.90	39.4	20.3

Shrendag and Netarhat Deposits Shrendag (23°27′: 84°25′) area contains a cluster of bauxite deposits. A total reserve of ~19 million tonnes of ore has been reported for this area (GSI Report, 1994). The Netarhat plateau is covered by laterite–bauxite capping. Important deposits have been located at Amtipam, Chirodih and Gurdari. Bauxite occurs as bands, lenses and impersistent sheets or blankets. They vary in thickness and grade within short distances. At Kunjam deposit the bauxite sheet is very persistent but for the variation in thickness (Fig. 4.2.71).

The generalised weathering profile of the area, as discerned from escarpments, pits and drill holes, is as follows:

1. Top soil and murrum
2. Lateritic bauxite (pisolitic)
3. Bauxite (impersistent lenses/boulders in laterite)
4. Aluminous laterite
5. Lithomerge/variegated clay
6. Altered basement gneiss.

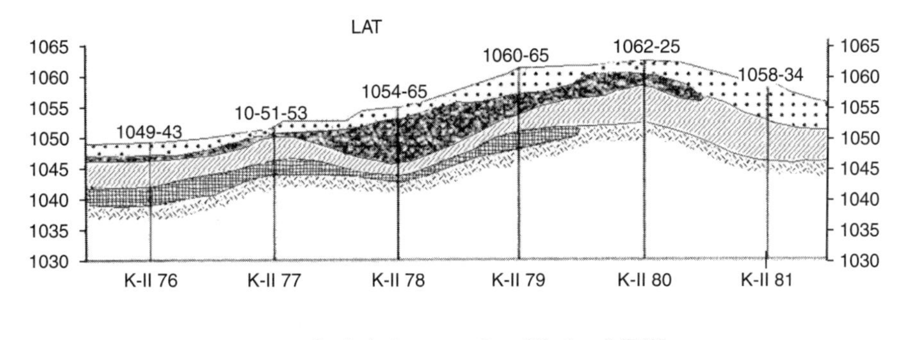

Geological cross-section of Kunjam-II (E-W)

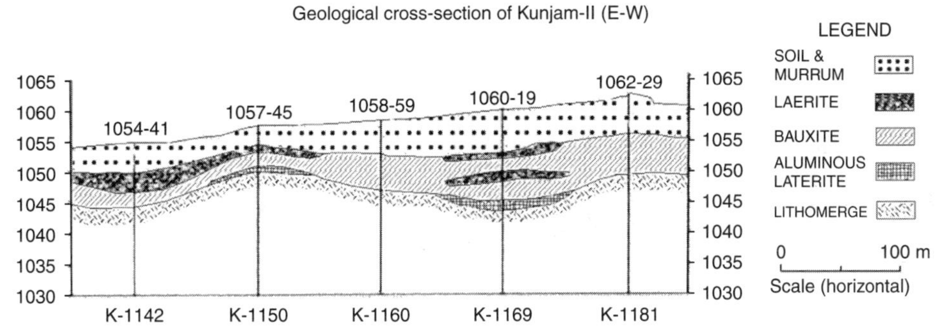

Geological cross-section of Kunjam-II (N-S)

Fig. 4.2.71 Geological cross-sections of Kunjam-II deposit along E–W and N–S profiles prepared from drill hole data (courtesy: Dr. H.Banerjee, Hindalco Ltd.)

The principal bauxite mineral in the ore is generally boehmite, a rather uncommon development. The ores are light grey to grey in colour, but may locally be pink or pinkish grey. They are hard and compact, being concretionary to pisolitic.The ores have apparently formed from the gneissic rocks by supergene alteration. Gibbsite formed either by desilication of kaolinite, or directly by the breakdown of feldspar, or both. But the moot question is about the prolific development of boehmite. As discussed in an earlier section, boehmite is commonly formed by the dehydration of gibbsite. The accompanying Fe-mineral should be hematite, rather than goethite.

5

North-East India

5.1 Geology and Crustal Evolution

Introduction

North-Eastern India, comprising seven states of the Indian Union, viz, Assam, Arunachal, Meghalaya, Nagaland, Mizoram, Manipur and Tripura, geologically represents a collage of different tectonic blocks with distinctive geological history (Fig. 5.1.1).

The central part of the terrain constitutes the Shillong–Mikir Precambrian massif (Meghalaya plateau and Mikir Hills of Assam), representing the north-eastern continuation of the Chhotanagpur Gneissic Complex (CGC) across the Bengal Basin (Ganges–Brahmaputra valley). The Dauki Fault demarcates the southern boundary of the plateau, while the northern and eastern edges are covered by alluvials of the Brahmaputra river valley in the Assam plains. Several inselbergs of the basement jut out in the Brahmaputra alluvial plains, of which those at Goalpara and Dhubri are the most prominent. The eastern most segment of the Himalaya including the 'Eastern Himalayan Syntaxis' (occupying Arunachal Pradesh) and the Indo-Burman Range (IBR) passing through Nagaland–Manipur, binds the region along its north and east. Along the west of the IBR, there are N–S to NE–SW trending Neogene molasse sediments of shelf facies, the southern parts of which make up the low hill ranges of Tripura–Mizoram. The Bengal Basin (Rajmahal–Garo Hills gap) intervenes between the Indian Peninsular shield and the North-Eastern region, though with uninterrupted continuation of the Himalayan Range along the northern territory.

The geology and metallogeny of the Himalaya is discussed in a succeeding chapter and hence not elaborated here. However, it may be very briefly mentioned that while the Himalaya is regarded as a classical example of continent–continent collision between Gondwanic Indian and Eurasian blocks, the Indo-Burman Range (IBR) is considered to be the on-land continuation of the Andaman and Nicobar Island Arc–a product of continued accretion since late Mesozoic time along the proto-

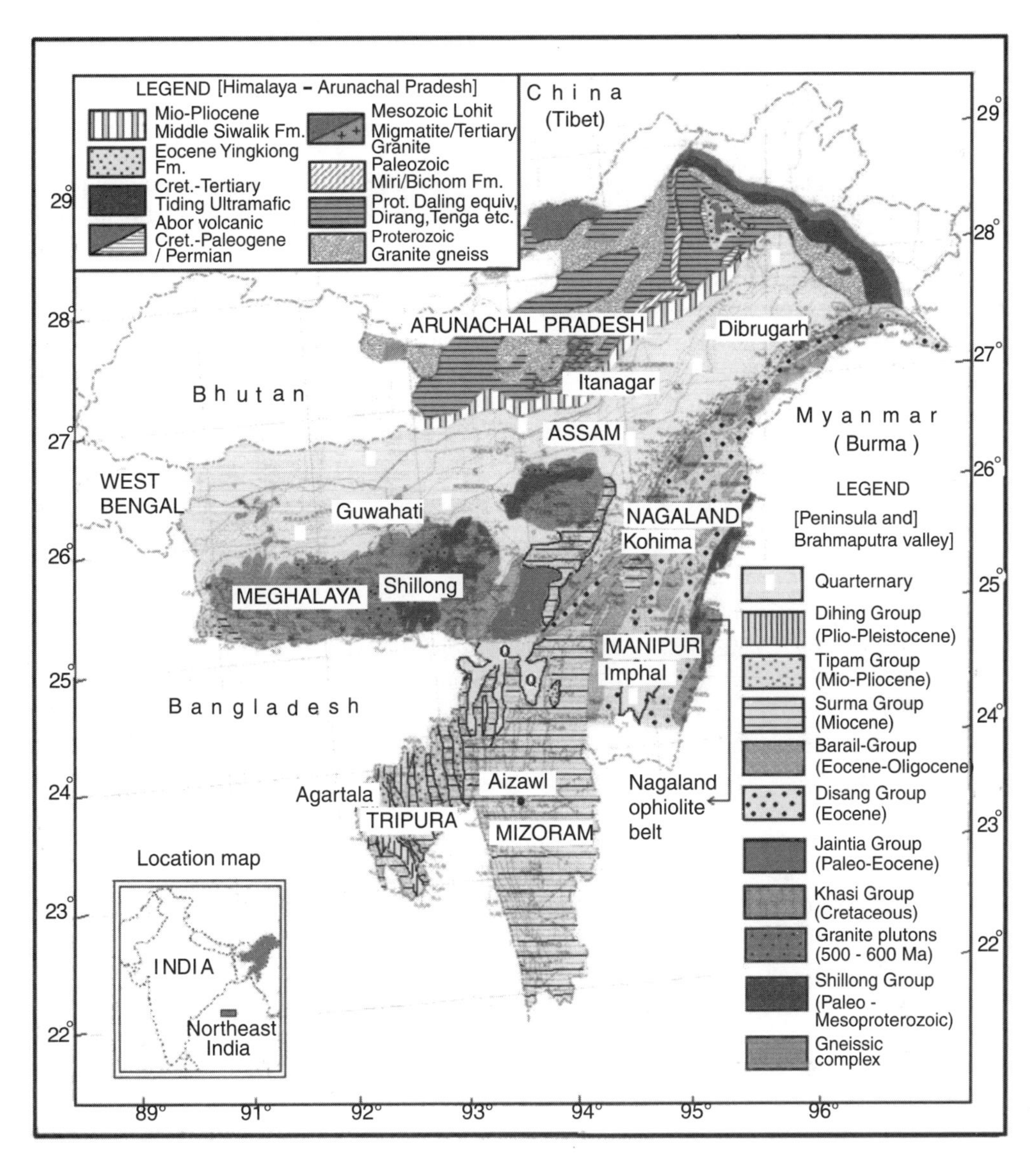

Fig. 5.1.1 Generalised geological map of the Northeastern India (partially modified from GSI's map on 1:5 million scale)

Andaman–Java subduction zone (Mukhopadhyay and Dasgupta, 1988; Dasgupta and Biswas, 2000; Currey, 2005). According to some researchers, the Himalaya and IBR are considered to have developed through two stages: (i) early phase of rifting of the Gondwana Supercontinent and (ii) collision and accretion of continental fragments with Eurasia from Permo–Carboniferous time onwards (Mitchel, 1981; Metcalfe, 1988; Acharyya, 1998, 2001). Acharyya (2005) proposed that the IBR and Andaman–Nicobar Island Arc evolved through the process of collision of the micro-continents followed by active subduction continuing since Neogene time.

Some geological details of the Shillong–Mikir massif, the granitic plutons and the Proterozoic cover sediments of Shillong Group and the Cretaceous (Mahadek and Langpar Formations)–Eocene sequence (Sylhet Trap and Jaintia Group) overlying the basement rocks in the Shillong–Mikir Hills plateau are given here, followed by a brief account of IBR rocks of of Nagaland–Manipur. However, the main focus is on the cratonic segment(s) of this region.

Shillong–Mikir Massif (Meghalaya and Assam)

The Shillong (Meghalaya) plateau comprises the Garo, Khasi and Jaintia Hills from east to west. Not very far to its northeast are the detached Mikir Hills, with intervening alluvial plains of the Naogaon district, Assam (Fig. 5.1.1, 5.1.2).

The gneissic complex, of the Shillong–Mikir plateaus, is exposed on an elevated block (~1500 m MSL), probably 'popped up' between the Dauki Fault (exposed reverse fault along south) and the Oldham Fault (buried reverse fault along north, marking the southern edge of Brahmaputra trough) during Pliocene (Johnson and Alam, 1991), triggered by the collision of Indo-Burman Block with the Shillong massif (Acharyya, 1998; 2005). The lithology, structural frame, metamorphism of dominantly amphibolite grade with local granulitic assemblages and prominent Proterozoic tectonothermal imprint on the massif are closely comparable with the CGC of Eastern India. Some workers are even inclined to consider the terrain as an extension of the CITZ of Central India (Acharyya and Roy, 2000).

Gneissic Complex The gneissic complex largely comprises migmatites, banded gneiss, granulites etc., with metasedimentary and amphibolitic enclaves and several granitic plutons of varied dimension. The general strike of the foliation is NE–SW, except in the Garo Hills where the trend is E–W. The migmatites with enclaves of high grade sillimanite–corundum (sapphirine) bearing rocks in Sonapahar, located in the central part of the plateau, have yielded a Rb–Sr (WR) age of 1700 Ma, whereas the migmatite in Nongpoh area located in the north-eastern parts was dated at 1150 Ma by the same method (Ghosh et al., 1994, 2005). The older component of the gneissic complex is grey coloured, banded, and is a composite biotite granite gneiss. The granitic constituents are fine to medium grained, sometimes porphyritic or aplitic, and are characterised by abundance of microcline. Locally, hornblende biotite gneisses and biotite cordierite gneisses are present (Mazumder, 1986). Magnetite and minor apatite are almost ubiquitous. The older gneisses are invaded by a later phase of pink coloured gneisses of granitic to granodioritic composition (Mitra and Mitra, 2001).

The gneissic complex exhibits intense deformation. The pervasive gneissic foliation and compositional banding are axial planar to a set of tightly appressed isoclinal folds, preserved mostly as rootless hinges. These earlier structures have in turn gone through multiple phases of later folding.

All the components of the gneissic complex and the late granite of the Shillong plateau are also observed in the Mikir Hills. Here, the general strike of the gneissic foliation is E–W in contrast to the NE–SW trend in the Khasi and Jaintia Hills. Referring to an old report of Smith (1898), Pascoe (1950) described the occurrence of foliated charnockite and hypersthene-bearing gneiss interbanded with ordinary gneiss at the foothills of Mikir Hills near Miji.

Granite Plutons There are several granitic plutons in this terrain, some of the larger ones in the central and eastern part of the plateau are Rongjen pluton (~50 sq km), Sinduli pluton (200 sq km), Mawdoh pluton (100 sq km.), Kylang pluton (80 sq km), South Khasi batholith (600 sq km), Mylliem pluton (100 sq km), Kyrdem pluton (200 sq km) and Nongpoh pluton (500 sq km). These plutons and many more similar intrusive bodies are mostly characterised by coarsely crystalline flesh colored granites (± grey and rose pink medium grained variants), without any structural fabric. Pegmatites, aplites and ramifying quartz veins related to the late granite are common. The South Khasi batholith, extending E–W between Mawphlang and Nongstoin, intrudes the basement gneisses and Shillong Group and is overlain by the Cretaceous–Tertiary rocks along south. The South Khasi Granite is

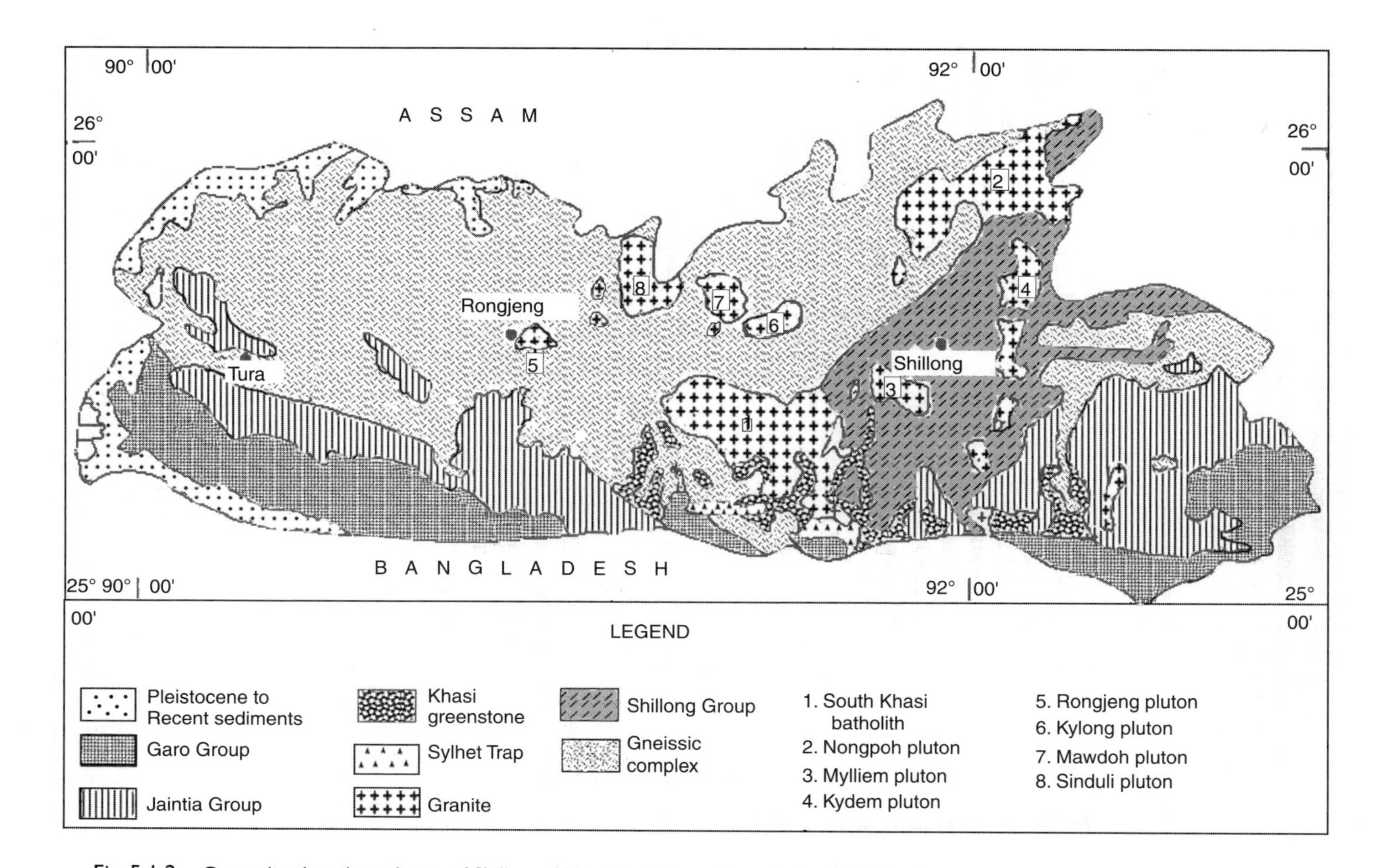

Fig. 5.1.2 Generalised geological map of Shillong–Mikir (Meghalaya– Assam) plateau, NE India

dominantly very coarse grained and porphyritic, corundum normative and per-aluminous (Das and Mahapatro, 2005), with a later emplaced variant of fine grained homophanous pink granite. The porphyritic variety falls in granite/quartz monzonite/quartz syenite in Q–A–P diagram (Ghosh et al., 1991). The Mylliem Granite, intruding the Shillong Group, is also meta- to per-aluminous in nature. The Nogpoh pluton, located midway between Shillong and Guwahati, is dominantly characterised by coarse porphyritic grey granite with crude preferred orientation of the magacrysts at places. It falls in the granite/quartz monzonite/quartz syenite field in the Q–A–P diagram (Ghosh et al., op. cit.) The elliptical Kyrdem pluton occurring in the north of Sung valley is emplaced within the Shillong Group of metasediments and dominantly comprises porphyritic hornblende–biotite–quartz diorite with minor components of quartz monzonite, quartz syenite and biotite aplite (Das and Mahapatro, 2005). Kumar (1990) has described a post-kinematic pink granite body from Songsak, East Garo Hills, which has high K_2O (5.72%), low CaO (0.88%) and of per-aluminous nature.

The granitic plutons in the Khasi Hills, East Garo Hills and Goalpara (Assam) have so far yielded the following Rb–Sr (WR) dates:

1. Kyrdem	479 ± 26 Ma (Sr_i – 0.71482 ± 72)
2. Nongpoh	550 ±15 Ma (Sr_i – 0.70948 ± 47)
3. Mylliem	607±13 Ma (Sr_i – 0.71187 ± 47)
4. South Khasi	690 ±19 Ma (Sr_i – 0.71074 ± 29)
4a. – do –	757 ± 60 Ma (Sr_i – 0.71069 ± 0.00092)
5. Songsak, East Garo Hills	500 ± 40 Ma —
6. Goalpara, Assam	647 ±122 Ma (Sr_i – 0.711 ± 6)

[S. No. 1, 2 and 4 from Ghosh et al., 1991, 1994 ; 4a Selvam et al., 1995; 3 from Chimote et al.,1988 ; 5 from Van Breemen et al., 1989 ; 6 from Kumar, 1990]

Of the available dates, the most well-constrained age of 550 ± 15 Ma for the Nongpoh Granite is considered to denote more precisely the time of the partial melting of the rocks of basic composition to give rise to this pluton, while the wide scatter of age data from other plutons might represent incomplete homogenisation of strontium isotope due to greater incorporation of crustal material at higher levels (Bhattacharya and Ray Barman, 2000). This granitic activity is correlated with the Pan-African thermal event marked elsewhere in the Indian Peninsula.

Metamorphic History of Basement Rocks The available geological account of the crystallines and the older enclaves of this region testify multiple phases of metamorphism, ranging from granulite facies to medium or high amphibolite facies. However, the data are still inadequate for reconstructing the evolutionary history of the basement gneisses with a reasonable confidence. Moreover, the meager age data referred to above is merely suggestive of a major crust forming process during Mid- to Late Proterozoic, without any certainty about the antiquity of the process. Lal et al. (1978) estimated P–T of 5 kb and 750°C for some reaction equilibria pertaining to the sillimanite–cordierite–sapphirine bearing rocks of Sonapahar. While reinterpreting the available isotopic and petrographic data (Ghosh et al., op. cit. and Gogoi, 1973), Bhattacharya and Ray Barman (op. cit.) pointed out that the formation of sapphirine at the expense of spinel (and cordierite) suggested greater than 850°C temperature at a high-pressure zone. It is also inferred that magnetite, a common accessory in the gneisses, exolved from associated dark green hercynite with the fall of temperature below 850°C. The petrographic similarity of the hypersthene–plagioclase–quartz–biotite rocks of Mikir Hills with Madras charnockite also indicates antecedence of a high P–T metamorphic regime in this crystalline massif. The coexistence of scapolite and plagioclase in Goalpara inlier, in the eastern extremity, also suggests minimum temperature of 780°C temperature for metamorphism. These evidences, coupled with mathematical modelling, prompted Bhattacharya and Ray Barman (op. cit.) to suggest that there was a garnet-forming early stage of metamorphism (M_1), followed by garnet-consuming M_2 decompression path, culminating into large scale dehydration (M_3 – 550 Ma).

Shillong Group The siliciclastic and pelitic metasediments (± BIF) of Shillong Group unconformably overlie the basement in the eastern part of the plateau. These rocks display a NE–SW trend and extend for 240 km from the Jadukata river section in southern Meghalaya to the north of Mikir Hills in central Assam. Sills and dykes, known as 'Khasi greenstone', penetrate the rocks of Shillong Group and predate the deformation and metamorphism of the latter and the emplacement of the late granite. The Pb–Pb age of 1530–1550 Ma from recrystallised galena in lower Shillong Group metasediments may be indicative of the metamorphic age of the host (Mitra and Mitra, 2001).

The Shillong Group comprises thick sequences of well-bedded quartzite and conglomerate/grit, interlayered with phyllite, mica schist, chlorite schist, hornblende schist and locally carbonaceous shales and slates. The metapelitic rocks dominate the lower part of the sequence while the upper part is more arenaceous. Phyllites and mica schists are by far the most dominant constituents. Tyrsad–Barapani shear passes close to the western boundary of the Shillong Group, hosting minor and sporadic sulphide mineralisation (Fig. 5.1.3).

In partial modification of the stratigraphy of the Precambrian rocks of Khasi Hills, Meghalaya, established by Mazumder (1986), Mitra (2005) has proposed the following stratigraphic succession for the Shillong Group in consideration of lithology and structural evidences (Table 5.1.1).

Table 5.1.1 *Stratigraphic succession of Shillong Group (after Mitra, 2005)*

		Intrusive granites (South Khasi, Mylliem etc.)
		Intrusive contact
Shillong Group	Khasi greenstone	Basic dykes and sills
	Upper Formation	Cream coloured quartzite
		Conglomerate/gritty quartzite
	Lower Formation	Slate/carbonaceous shale
		Phyllite with bands of massive quartzite and quartz sericite schist
		Quartzite
		Basal conglomerate/gritty quartzite
		Mica schist
		Unconformity
		Gneissic complex

The regional strike of dominant foliation in the Shillong Group of rocks is NE–SW with a steep southeasterly dip. The rocks are involved in four generations of folding. The bedding planes (S_0) are folded into isoclinal (F_1) folds with the development of the NE–SW trending pervasive axial planar schistosity (S_1). These in turn are coaxially refolded into open upright F_2 folds with an axial plane striking NNE, exhibiting local development of S_2 cleavages. Two more fold deformations have affected the rocks in the form of NE-plunging recumbent folds (F_3) and NW trending open cross folds (F_4) (Mitra, 1998, 2005).

Cretaceous–Tertiary Sedimentary Sequences and Intrusives in Meghalaya Plateau

Cretaceous Sequence In the Garo, Khasi and Jaintia Hills, the basement gneisses and the Proterozoic cover rocks of the Shillong Group (wherever developed) are overlain by gently dipping Cretaceous marine sediments of the Khasi Group, which are dominantly arenaceous with subordinate shales and carbonaceous (coaly) layers along the southern edge of the plateau. The basal rocks in Khasi and Jaintia Hills are conglomeratic. This is overlain by hard gritty and massive sandstone (with sandy shale towards top) of Mahadek Formation, up to 250 m in thickness, followed upwards by shale- and limestone-dominant Langpar Formation (up to 100 m thickness). The Mahadek

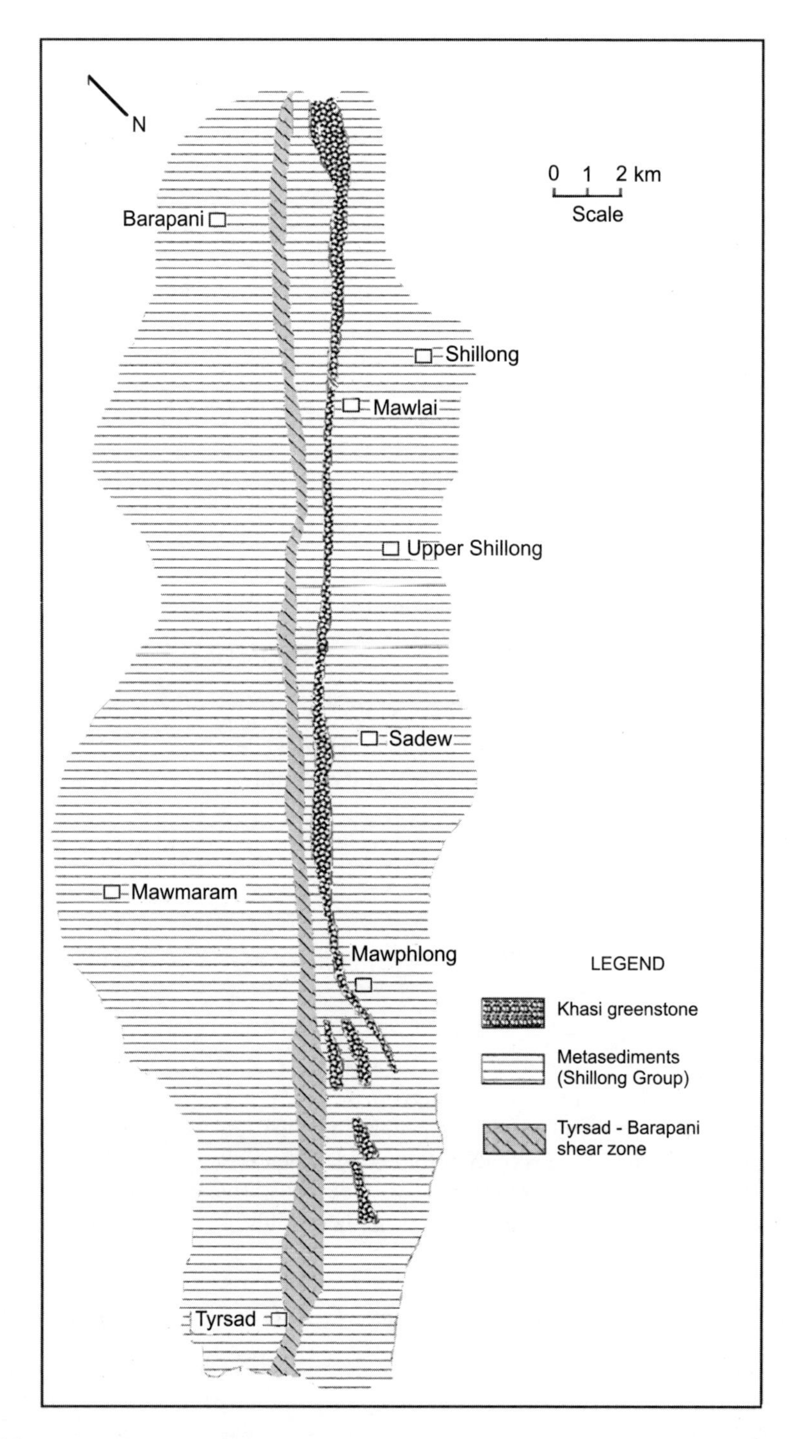

Fig. 5.1.3 Geological map of Tyrsad–Barapani shear zone, Meghalaya (after Mitra and Mitra, 2001)

Formation is the repository of very significant uranium deposits, which are described in some detail later in the chapter.

The basaltic rocks known as the Sylhet Trap, overlying at places the Cretaceous sediments, are in general correlated with the Rajmahal Trap of Eastern India and assigned an Upper Cretaceous age.

Tertiary sequence In southern and eastern parts of the Meghalaya plateau, the Tertiaries are represented by the a well-developed limestone–sandstone sequence (> 1000 m thick) of the Jaintia Group, which is subdivided from bottom to top into the Therria Formation, Sylhet Limestone Formation and Kopili Formation, spanning in age from Lower Paleocene to Upper Eocene. The Sylhet Formation in Jaintia and Khasi Hills enshrines a very large resource of cement- and high-grade limestone, characterised by the presence of abundant foraminifera. Some other formations also consist of limestone deposits of economic importance in this sector. However, the limestones in general become siliceous and of lower grade both in the eastern and western extensions of the Garo Hills.

It may be mentioned that Tertiary rocks ranging from Paleocene to Pliocene of varied litho-facies are developed in wide territories of this region. The Disang Group (Eocene), well-developed in the east of the Haflong–Disang Fault in Upper Assam and in the Naga hills–Manipur region, is broadly equivalent to the Jaintia Group but is represented by a different facies comprising unfossiliferous dark splintery shale and fine grained sandstone. The Barail Group in Central and Lower Assam and Meghalaya and the Tikak Parbat and Baragoloi Groups in Upper Assam represent the Oligocene sequences. These are essentially shale–sandstone sequences having prominent coal seams and oil resources in Upper Assam. The Miocene is represented by sandstone–shale with local lignite belonging to the Surma and Tipam Groups, while Mio–Pliocene sequences of clay–shale–sandstone are variously named as Dupi Tila and Namsang Groups. The youngest Tertiary rocks of Plio-Pleistocene age, comprising pebble beds–sandstone–clay, are named as the Dihing Group.

Carbonatites and Alkali Intrusives of Sung Valley The alkali ultramafic–carbonatite complex of the Cretaceous age intrudes the Shillong Group Proterozoic metasediments in the Sung valley, East Khasi and Jaintia Hills at the intersection of two prominent lineaments trending NNE–SSW and E–W (Fig. 5.1.4). Serpentinised peridotite core rimmed by pyroxenite comprises 80–90% of the oval shaped exposure (~ 26 sq km) of the complex. This is intruded by medium and small bodies of uncompahgrite, ijolite, syenite and carbonatite of varied shapes (Chattopadhyay and Hashimi, 1984; Krishnamurthy, 1985). The three phases of emplacement recognised in the complex are, (i) Early stage peridotite (olivine–diopside–augite ± enstatite) and pyroxenite (diopside–augite ± aegirine–augite), (ii) Intermediate stage uncomprahgrite (mellilite–diopside ± olivine monticellite ± melanite garnet), ijolite–meltigite (aegirine–augite + nepheline urtite) and nepheline syenite (K-feldspar–albite–nepheline ± alkali amphibole) and (iii) Late stage carbonatite (calcite–dolomite ± apatite ± magnetite). The carbonatite emplacement is in the form of small dykes, lenses, veins and stocks close to the ijolite bodies. Fenitisation of the country rock is noted around the periphery of the complex. Apatite–magnetite bands and veins are commonly associated with the complex, which are considered to be genetically related to the carbonatite emplacement.

Jaireth et al. (1991), described two types of fluid inclusions, i.e. solidified S–type and fluid F-type, co-existing in single apatite grains. The S-type inclusions homogenised into relatively viscous melt-like fluid at 740–806°C, while F-type inclusions homogenised into a liquid phase at 178–488°C. Compositionally, the solid phases of the former are carbonates and silicates while those of the latter are carbonates and halides. According to these authors the coexistence of such contrasting inclusions (apparently syngenetic) with wide gap between their homogenisation temperatures might be suggestive of immiscibility between a hydrous carbonatitic melt and a saline aqueous fluid at some stage.

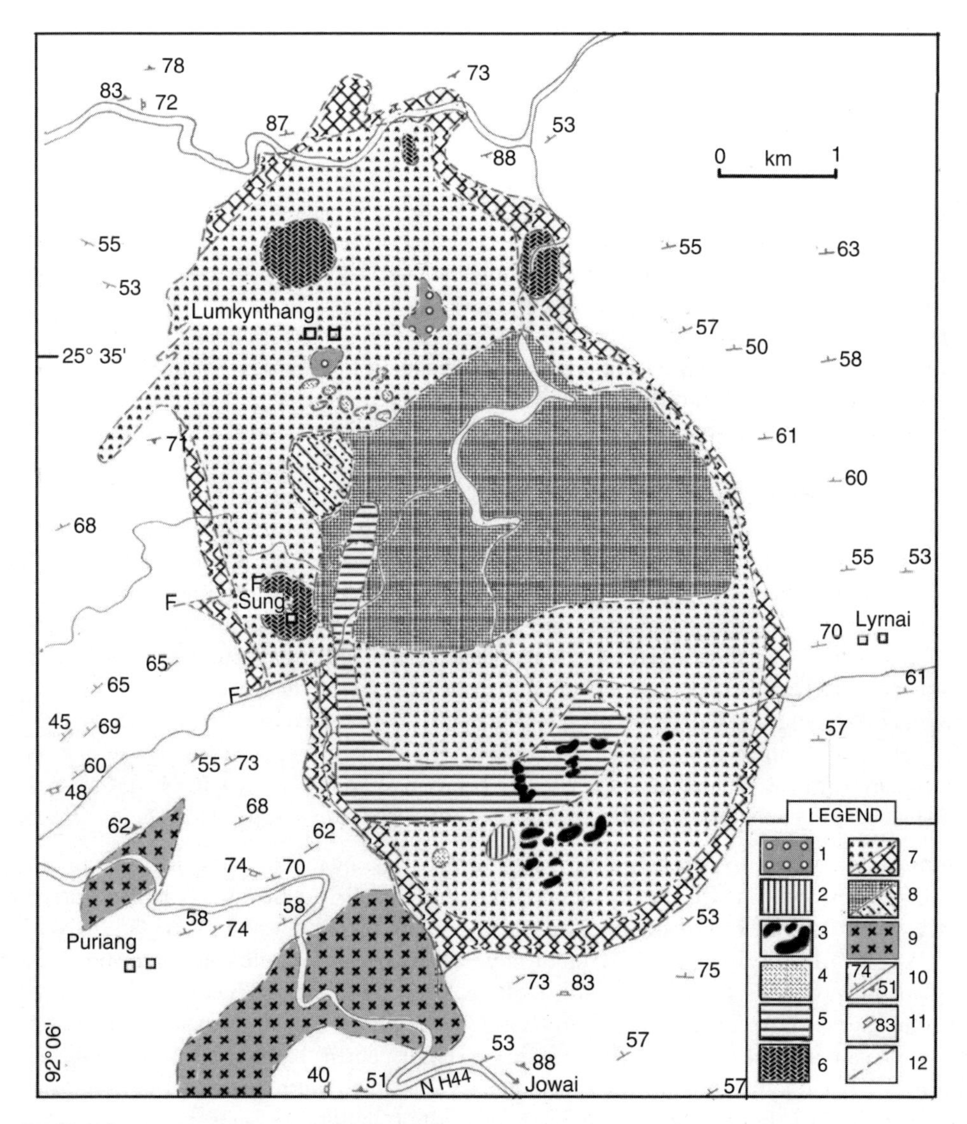

Fig. 5.1.4 Geological map of the Sung Valley and surrounding area, Khasi and Jaintia Hill districts, Meghalaya (modified after Jaireth et al., 1991)
Legend: 1 – Laterite cover, 2 – Feldspathic breccia, 3 – Carbonatite, 4 – Nepheline syenite, 5 – Ijolite group of rocks, 6 – Uncompahgrite, 7 – Pyroxenite/feldspathic pyroxenite, 8 – Peridotite/perovskite–magnetite–peridotite, 9 – Amphibolite, 10 – Bedding/foliation, 11 – Joint, 12 – Fault. Blank space – Rocks of Shillong Group

Indo-Burman Range (IBR)

The Naga–Manipur Hills comprise the northern parts of the Indo-Burman Range (IBR). The easterly dipping and dominantly pelitic rocks of Disang and the overlying psammitic sequences of lower Barail constitute the western flank of Naga–Manipur Hills and the 'schuppen belt' along Haflong–Disang Fault in its west (Fig. 5.1.5).

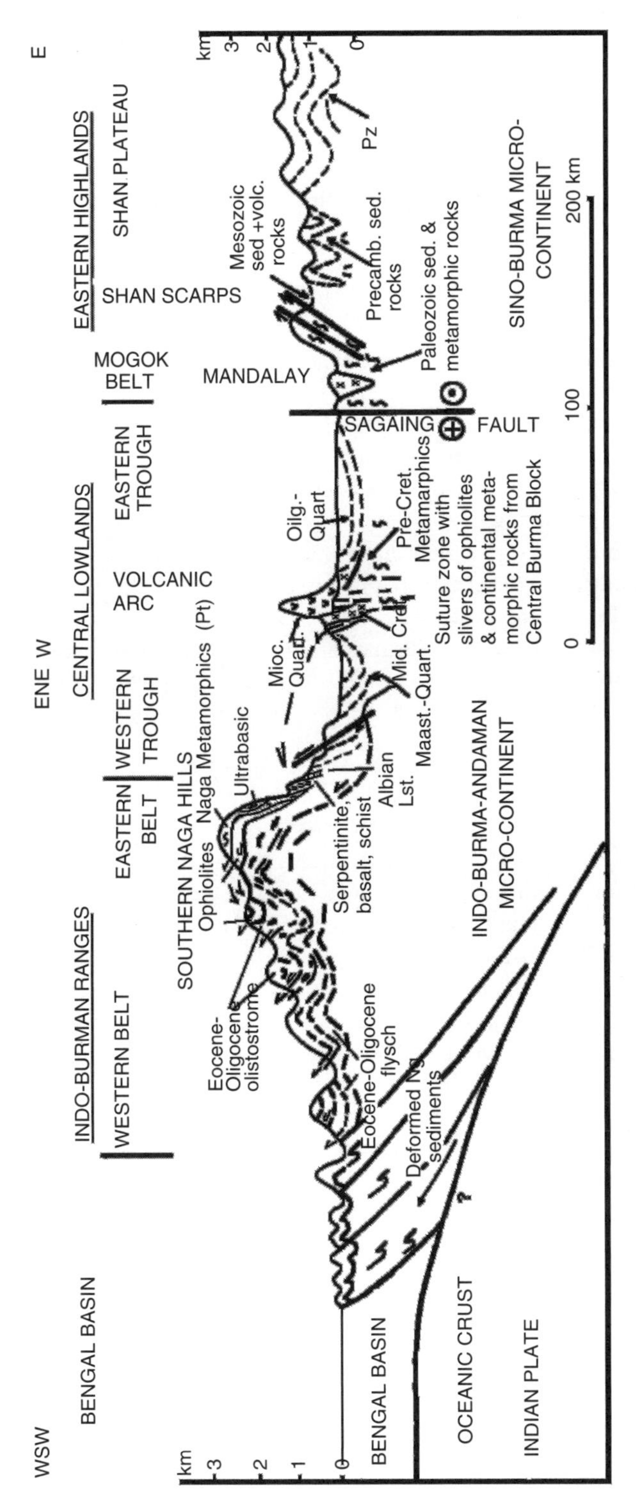

Fig. 5.1.5 Schematic section across Indo-Burma Range and Central Burma basin (after Acharyya, 2005)

The central zone of the range is characterised by olistostromal Paleogene flysch (with olisoliths of ophiolites and fossiliferous limestone – prominent in Manipur Hills) overlying the Disang pelites. The eastern flank exposes dismembered ophiolites, ophiolite-derived Eocene cover sediments and thrust sheets of metamorphic rocks overriding the rocks of the central zone. In the Naga Hills, the ophiolites are in turn overthrusted from east by the Naga Metamorphics (quartz mica schists ± garnet, quartzite and granite gneisses). Mid-Cretaceous arkose and limestone (Nimi limestone) unconformably overlie the tectonic contact of the Naga Metamorphics with the ophiolites. Along the eastern margin of the IBR, the easterly dipping Naga Metamorphics are overlain by the Miocene molasses of the Central Burma basin (Acharyya, 2005 and references therein).

The ophiolite belt is about 200 m wide in the Naga Hills, while it branches into multiple thinner isolated bodies in the south. The ophiolites and associated sedimentary piles record isoclinal F_1 folds on mesoscopic scale and westerly vergent upright asymmetric F_2 folds on both small and large-scales, often defining the map pattern (Acharyya, 1986, 2005; Acharyya, et al., 1990). The subduction related ophiolite accretion along this zone is considered to be just preceding the mid-Eocene age on the basis of fossil ages determined from the overlying and underlying sediments (Acharyya, 2005). The floor of the ophiolite belt is inferred as a continental crust by some workers (Acharyya, op. cit. and references therein) implying the IBR to be a part of the IBA (Indo-Burmese Arc) micro-continent lying between the Burmese micro-continent and Indian continental blocks. The ultramafic rocks of the ophiolite sequence host several small podiform chromite deposits, some of which were locally mined out to shallow depths in the past.

It is postulated that the folded belt (with N–S axes) of Surma and Tipam Groups of Miocene age in the Mizoram–Tripura hills were possibly developed over a subducting ocean floor which flanked the western margin of the IBA (Acharyya, op. cit.).

5.2 Metallogeny

Introduction

Metalliferous deposits of economic significance are yet to be located in any of the varied geological milieu of the NE India, except those of uranium in the Cretaceous (Mahadek) formation of Meghalaya. However, the region is well-known for its oil resources in the Tertiaries (Barail) of Upper Assam. The first oil well was drilled in Assam in 1866; only seven years after Drake Well in USA ushered in modern petroleum era. Besides oil, there are moderate resources of both Tertiary and Gondwana coal in Assam, Meghalaya and Arunachal Pradesh, superior quality sillimanite at Sonpahar, Maghalaya, extensive resource of cement- and high-grade Tertiary (Sylhet Formation) limestone in Meghalaya and Assam, dolomite in the Himalayan foothills of Arunachal Pradesh, and several industrial minerals. Very small but high-grade chromite occurrences in the ophiolite belt of Manipur were exploited in the past. There are some occurrences of basemetal sulphides in the Khasi Hills, which do not have much economic value but are otherwise quite interesting.

Zinc–Copper–Lead Sulphide Deposit, Umpirtha, Meghalaya

Polymetallic sulphides of Zn, Cu and Pb, mainly hosted by cummingtonite–anthophyllite–cordierite rock occurring as enclaves within the biotite gneisses of the basement complex, is located at Umpirtha, East Khasi Hills, Meghalaya. The gneissic complex in this area is also characterised by the enclaves of actinolite–tremolite schist, some carbonates and quartz sillimanite schist, of which the former two are also mineralised to varied extent.

The mineralised rocks are lensoid in shape, disposed along the generally ENE–WSW trending but highly contorted foliation of the surrounding gneisses. About 20 mineralised lenses were located within 1200 m × 200 m stretch of main block and 1000 m strike length of the extension block by means of drilling and geophysical investigation. The dimension of the individual ore lenses were found as 100–140 m in length and 20–30 m in width, with average 3.6 m thickness. The most prominent mineralised body in the main block is of 110 m in strike extension, with inferred reserve of 118,000 tonnes with average 4.58% metal content (Zn 2.83%, Cu 1.35%, Pb 0.40%). The total ore reserve of the deposit was about 0.8 Mt (Gupta et al., 2002).

Lead Sulphide Mineralisation, Mawmaram, Meghalaya

Galena dominant sulphide mineralisation in white quartzite unit of lower Shillong Group (Proterozoic) metasediments was located by GSI near Mawmaram (25°31′05″ : 91°41′55″) in the east of Mawphlang sector of the Barapani–Tyrsad shear zone (Fig. 5.1.3). Though the occurrence does not hold any economic promise, it is unique on many counts. The mineralised layers and lenses are disposed along two zones parallel to the primary compositional banding of the host quartzite, which records three sets of fold-deformations and related secondary planar structures (Mitra, 1998). Both the mineralised zones (0.30–1.5 m thick) are characterised by the association of ferruginised layers within the quartzite. Besides galena, the ore mineral phases include arsenopyrite, löllingite, gersdorffite, safflorite and varying amounts of chalcopyrite, pyrite, pyrrhotite, cubanite, covellite and minor sulphosalts of Ag (Mitra, 2005). Pyrite is also disseminated along cross cutting sets of late fracture. The bedding–parallel mineral assemblages record effects of metamorphism in form of recrystallisation and polygonisation. Ore samples from the mineralised zone-I and II analysed 3.76–29.20% Pb, maximum 1.26% Zn and 0.11% Cu, 1.00–2700 ppm Ag, 25–260 ppb Au, 5–80 ppm Mo, 10–350 ppm Bi, 5–150 ppm Sn and 1.28–9.56% Fe. The zone-I is richer in Ag while Au is more in the zone-II. EPMA results clearly depict partitioning of Ag in galena and that of Au in arsenopyrite and diarsenide mineral phases (Mitra op. cit.). The wide separation of $\delta^{34}S$ values of two ore samples as + 11.9 and + 6.7 per mil is interpreted (Mitra op. cit.) as the result of sulphur derivation due to sulphate reduction within a limited reservoir in a sedimentary domain, bearing no mantle signature. The author also advocates syngenetic/synsedimentary origin for this ‘sandstone – Pb (+Ag)’ deposit. The Pb–Pb model age range of 1530–1550 Ma determined for the Pb (Ag) mineralisation at Mawmaram is considered to be the age of the mineralisation.

Uranium Mineralisation in Meghalaya

Uranium mineralisation is noted in the sandstone of lower Mahadek Formation of the Khasi Group (Upper Cretaceous), which overlies the basement gneisses and Shillong Group of rocks in south-eastern Meghalaya. Domiasiat and Gomaghat are the two blocks of the uranium deposit delineated in this area by the Atomic Minerals Directorate (AMD) (Fig. 5.2.1). The deposits are located respectively at 110 km and 130 km SSW of Shillong in an extremely high rainfall zone and are almost inaccessible for half of the year. With total uranium metal content of 7819 tonnes and average ore grade of 0.085% U_3O_8, the two deposits are considered to be of high grade and medium tonnage. The UCIL plans to develop two large open-cast mines in the two blocks of the deposit. The mill is planned for treating about 1300 tonnes of ore per day, aiming at annual production of 160–200 tonnes of uranium per year (Murthy et al., 2006).

The host rock of uranium mineralisation is coarse grained feldspathic quartz arenite with 70–90% clasts and rest matrix. The framework constituents comprise quartz, subordinate K-feldspar (3–15%), fragmental coal and other lithic fragments. Apatite, garnet, zircon, monazite, sphene,

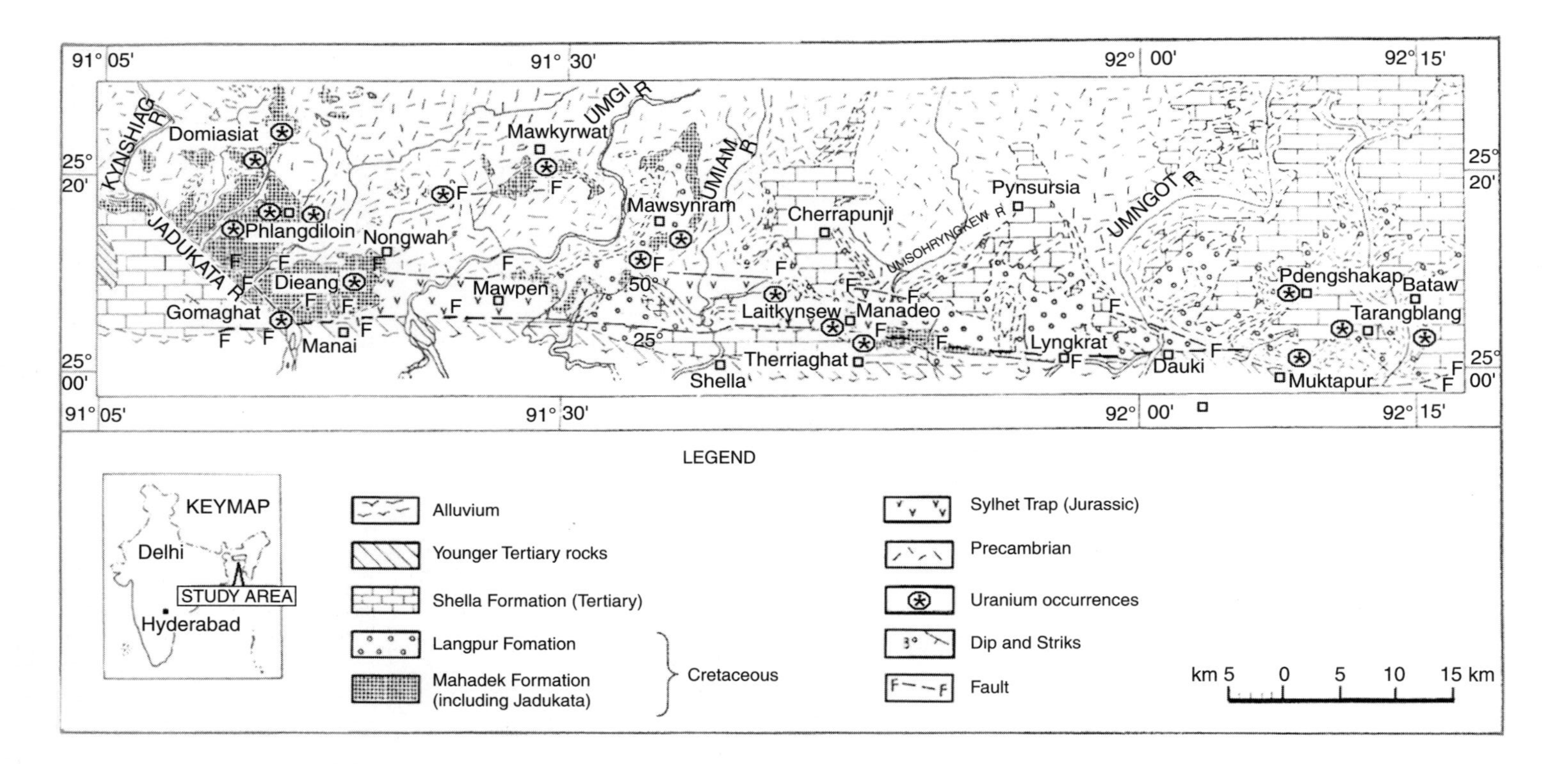

Fig. 5.2.1 Geological map of southern Meghalaya showing uranium occurrences and associated rocks (modified after Dhana Raju et al., 1989)

sphalerite and oxidised ilmenite and magnetite are present as trace phases. Muscovite and chlorite are abundant in the matrix while the cementing material is clayey (± micro-cryptocrystalline silica, calcite, pyrite–marcasite) and bituminous. The uranium content in the ore zones varies between 0.04 to 0.60% e U_3O_8 (Dhanaraju et al., 1989; Krishna Rao et al., 1995). There is a positive correlation of uranium mineralisation with Se, As, Mo and Pb and a negative correlation with Cu (Kaul, et al., 1991).

Besides pitchblende and coffinite, uranium is also associated with both varieties of cementing material. Petrography of the ores is interesting. Uranium occurs mainly as ultrafine pitchblende and adsorbed in the organic matter (OM), the later occurring in several forms such as lumps, layers, veins, fracture fillings, coatings, dispersions and impregnations (Krisna Rao et al., op. cit.). The proportion of coffinite is always minor. The OM is type II-kerogens with macerals mainly belonging to the liptinite group. Some of the primary structures apparently controlled the deposition of uraninite and other authigenic minerals like Fe-disulphides, sphalerite and apatite. Development of a secondary granular structure and blades of secondary OM are features related to maturation according to Krishna Rao et al. (op. cit.). The stage of maturity of the OM, as indicated by its reflectivity, varies from high volatile bituminous to nearly semi-anthracite. It is another 'sandstone type' uranium mineralisation.

The source of uranium for these deposits can reasonably be located at the 'fertile' granite and gneiss(es) in the South Khasi batholith, which form the main provenance of the Mahadek Sandstone. Saraswat et al. (1977) reported U-content as high as 40 ppm from the South Khasi Granite batholith. Uranium might have been leached from the fertile source rock by the atmosphere-oxidised meteoric water and/or released from the arkosic sandstone during syn-diagenetic transformation of the latter. Shivananda et al.(1992), based on the results of uranium extraction experiments conducted on these ores, suggested polygenic character of the migration and fixation of uranium in the Mahadek arenites, including biogenic processes.

Apatite and Rock Phosphate Mineralisation in Meghalaya

Apatite, Sung Valley In the Sung valley, East Khasi and Jaintia Hills, an oval shaped zoned alkaline ultramafic–carbonatite complex (~ 26 sq km area) intrude the metasediments of Shillong Group (25°31′ to 25°37′ and 92°05′ to 92°10′). The ultramafic complex comprises pyroxenite, which is closely associated with melilite–pyroxene rock, ijolite, magnetite rock, syenite and carbonatite. The pyroxenite bodies, closely associated with carbonatite, host apatite–magnetite lenses, plugs and veins varying in length from 100 to 225 m and width from 30 to 35 m. A total ore reserve of 4.56 Mt with 11% P_2O_5 was calculated for three major and three minor bodies at 5% P_2O_5 cut-off. In addition to this another 4.87 Mt of probable and possible reserves with 10–11% P_2O_5 were estimated in the area (Gupta et al., 2002; Mathur, 2005).

Phosphatic Nodule, Garo Hills In Siju area of Garo Hills (25°19′ to 25°21′30″ and 90°38′55″ to 90°41′50″), a phosphatic nodule-bearing shale layer,1–6 m thick, is intermittently exposed along the base of Kopili Formation for about 200 km . A total reserve of 119,240 tonnes of ore with 19–29% P_2O_5 has been estimated (GSI Report, 2002).

In Bandarigiri area of Garo Hills (25°23′ : 90°19′30″), the sediments of Jaintia Group (Tertiary) unconformably overlie the basement gneissic complex. A fairly thick horizon (up to 30 m) containing phosphatic nodules are recorded from the dark grey shales at the basal parts of Kopili Formation, conformably overlying the highly fossiliferous Siji limestone. Near shore platformal deposition is represented by the sediments underlying and overlying the phosphatic horizon. The average P_2O_5 content of the nodules is 10% (Gupta et al., 2002).

Chromite Mineralisation in Manipur–Nagaland

Manipur Chromite in the Ukhrul and Chandel districts of Manipur occurs as small pockets, lenses and pods associated with ultramafic bodies in the ophiolite suite exposed along the eastern margin of the IBR, falling within Indian Territory. Chromitè is of massive, granular, nodular, banded or podiform types, hosted by harzburgite, dunite and sepentinite. Disseminations are also noted in the dunites and peridotite. The important occurrences are located in Sirohi–Gamnon areas of Ukhrul and Moreh area of Chandel districts. No large deposit has been found so far. The occurrences are of a few tens of meter in extension, but characterised by high Cr_2O_3 content (44–59%), low TiO_2 (trace), Cr/(Cr +Al) = 0.59–0, 88 and Mg/(Mg +Fe^{2+}) = 0.46–0.74. The overall physical and chemical attributes are similar to the 'Alpine type chromite'. Low SiO_2 and Fe_2O_3 in most places make the ores suitable for refractory purpose. From metallurgical point of view the ores are of grade I and II (Gupta et al., 2002; Mathur, 2005).

Sirohi Area, Ukhrul District The area is located at 19 km to the east of Ukhrul. A small chromite lens (11 × 8 m) occurring in the north of Sirohi peak has massive, lumpy and compact variety of ore hosted by sepentinite. The chromite from this spot analysed 47.68 to 56.59% Cr_2O_3 and 13.91 to 15.21% Fe_2O_3. There is also a nodular variety composed of rounded to ellipsoidal chromite grains/ aggregates of 1–3 cm across, enclosed by thin films of serpentinite.

The chromite bodies of Sirohi were being mined by M/s Orissa Industries Ltd. by selective quarrying. Most of the pits are of small dimension, ranging from 5m × 5 m to 12 m × 12 m, except the one of larger dimension (20 m × 15 m) in the north of Ranshokhong.

Gamnom Area, Ukhrul District There are nine small chromite pockets located over 0.4 sq km area, at about 10 km southeast of Ukhrul. The chromite bodies are in the form of 5–20 m long and 1–5 m wide lenses, disposed in en–echelon pattern. The host rock is serpentinised dunite–harzburgite. The exposed chromitites are mostly massive, though disseminated and nodular varieties are also noted, having transitional relation at places with the former type. Among the three types of chromite of this area, the massive variety analysed 44–49% Cr_2O_3, 18.4–20.5% Al_2O_3, 15.6–17.6% Fe_2O_3, 2.6–3.9% SiO_2, the nodular variety analysed 45.6–46.4% Cr_2O_3, 16.3–20.0% Al_2O_3, 16.0–16.4% Fe_2O_3, 3.8–4.3% SiO_2, and the disseminated type analysed 35.75% Cr_2O_3, 10.2% Al_2O_3, 9% SiO_2.

One small lensoid pocket of 20 m × 10 m dimension is located in Harbui–Khajui area comprising coarse grained disseminated chromite in serpentinite host.

Moreh area, Chandel District Small pockets of chromite occur at 5 km north of Moreh near Khundang Thabi, which were worked in the past by M/s Orissa Industries Ltd. There are also float ores of chromite of different varieties over an area of 0.5 sq km near Minou villege. Chemically these ores contain 40–51% Cr_2O_3, very low Al_2O_3 1.32–1.47%, 13–14% Fe_2O_3, and higher SiO_2 content of 13–16%.

Nagaland The chromite occurrences in the ultramafites of the ophiolite belt of Nagaland are even smaller and sporadic compared to those in the Manipur sector. Minor occurrences are located in Reguri and Washello areas of Phek district and Pang, Pokhur and Wui areas of Tuensang district.

Magnetite Deposit in Nagaland

A magnetite deposit occurs at 2.5 km east of Pokhur village in Tuensang district. About 1 m thick tabular ore body extends for nearly a kilometre in NNE–SSW direction with 30–40 dip towards WNW. The deposit received attention because of appreciable content of nickel and cobalt in association with magnetite.

The GSI and the State DGM estimated a total of 4.43 Mt ore reserves, covering the North and South blocks of the deposit, with average 65.25% Fe_2O_3, 4.45% Cr_2O_3, 0.63% Ni and 0.1% Co. Ni and Co apparently occur in the magnetite structure. The technical feasibility of utilisation of the ore in ferro-alloy industry and the extraction of the associated elements as by-products are being examined (Gupta et al., 2002; Mathur, 2004).

6

Western India

6.1 Geology and Crustal Evolution

Introduction

'Western India' in the present context, encompasses a region now mostly covered by the states of Gujarat, Rajasthan, Haryana and Delhi (Fig. 6.1.1). Geologically the region includes the Aravalli–Delhi orogenic belt and the lithostratigraphic units adjacent thereto.

Investigations into the geology of parts of this region started in the nineteenth century itself (Hacket, 1881). With the beginning of the twentieth century, the Geological Survey of India undertook an elaborate programme of geological investigation in some regions of the country, which included Rajasthan and its adjacent areas. Some outstanding results of these investigations are contained in the publications of Heron (1917 a, b; 1936 and 1953), Gupta (1934), Gupta and Mukherjee (1938). Modern concepts of structure–tectonics, petrology and geochronology hardly, if at all, had developed by that time. The earlier concepts radically changed since about the middle of the last century. As a fallout now, we have also collected a fair amount of new data related to those aspects, and new interpretations of the geology of the region based thereon have been offered. However, as we will see later in this section, some of the basic tenets of early work remain more or less valid. This particularly refers to the regional geological maps prepared and regional stratigraphy suggested.

The basic conclusion of early work was that the basement rocks in that region were a Banded Gneissic Complex (BGC) and a granitoid massif, called Berach Granite, which was correlated with the Bundelkhand Gneiss. This was overlain unconformably by a 'System' (Supergroup) of supracrustal rocks, namely the Aravalli System. This in turn was succeeded by another 'System' of supracrustal rocks, called the Delhi System. In between these two 'Systems' of supracrustal rocks, they placed a carbonate-rich formation that was called Raialo 'Series' (Group). The Delhi 'System'

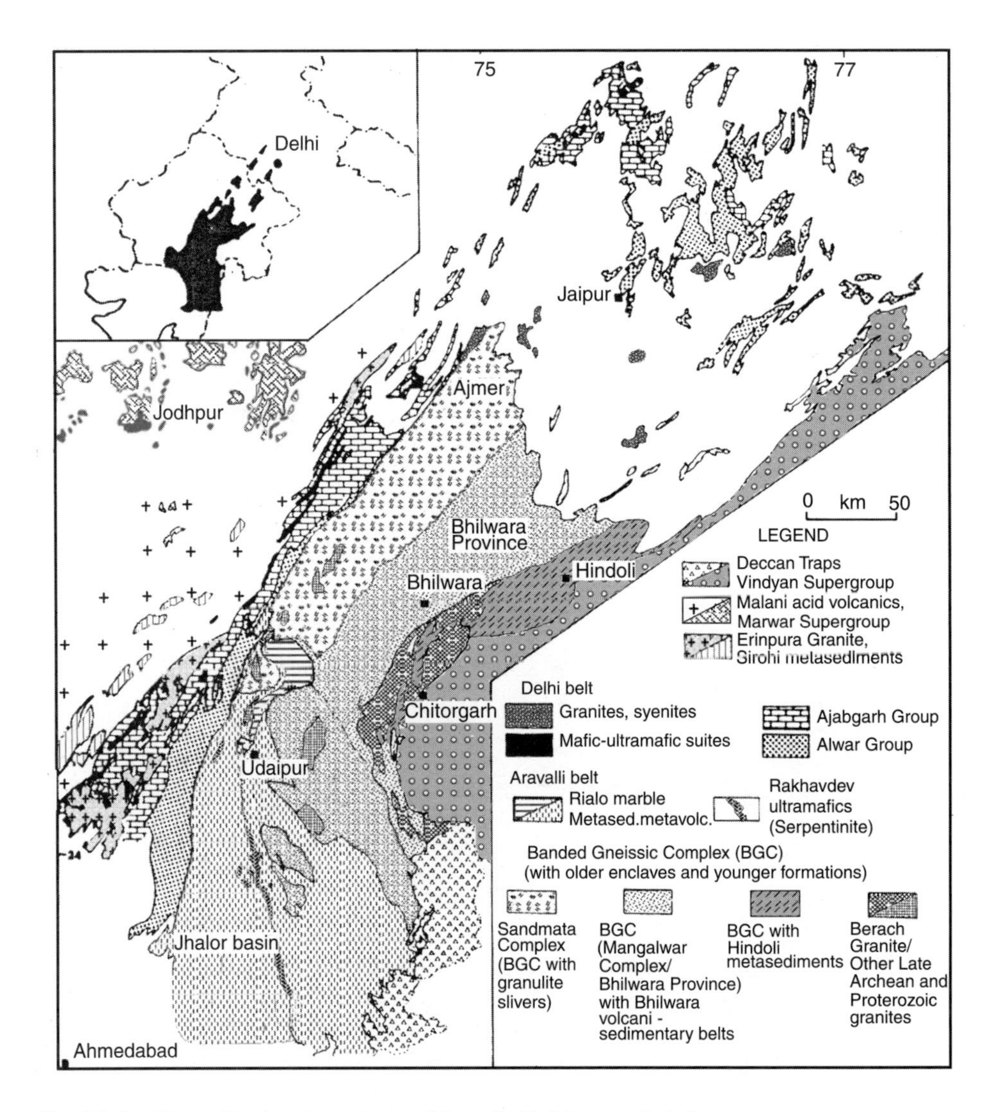

Fig. 6.1.1 Generalised geological map of Aravalli–Delhi orogenic belt

in turn was divided into two 'Series': the Alwar Series (lower) and the Ajabgarh Series (upper). A suite of granitoid rocks, which were called Erinpura Granite (Table 6.1.1), intruded the supracrustal rocks.

A challenge to the stratigraphic sequence suggested by Gupta (1934) and Heron (1953) came from Crookshank (1948), who suggested a continuous sequence of the BGC, the Arvalli System (Supergroup) and the 'Raialo Series' and concluded that the BGC was the syn-kinematically migmatised part of the Aravallis. Naha and Halybarton (1974) initially supported the idea, mainly based on structural similarity between the BCG and the associated Aravalli rocks. However, later work by Naha and Roy (1983) and Roy et al. (1981) re-established that the BGC was really the basement to the Aravalli supracrustals and that the structural similarity between the supracrustals and the basement rocks was due to mobilisation of the basement during deformation of the supracrustals.

Table 6.1.1 *Pre-Vindhyan stratigraphy of Rajasthan and Gujarat (after Heron, 1953)*

	Malani Series	Rhyolite, tuffs	Granite, ultrabasic rocks, Erinpura Granite, pegmatite aplite, epidiorites and hornblende schists
Delhi System	Ajabgarh Series	Upper phyllites limestones, biotitic limestones and calc–gneisses	
		Calc–schists, phyllites, biotite–schists and composite gneiss	
	Alwar Series	Quartzites, arkose, grits, and conglomerates	
Raialo Series		Garnetiferous biotite schists, limestone (marble), local basal grit.	
		Unconformity	
Aravalli System		Impure limestones, quartzites, phyllites, biotite schists, composite gneiss.	Aplogranites, epidiorites and hornblende–schists, ultrabasics
		Quartzites, grits and local soda–syenites, conglomerates	
		Local amygdaloids and tuffs	
		Unconformity	
Banded Gneissic Complex (BGC)		Schists, gneisses and composite gneiss	Pegmatites, granites, aplites and basic rocks
		Quartzites	
Berach Granite (≅ Bundelkhand Gneiss)		Granitoid to granite gneiss	

Modifications to the Heron's stratigraphic sequence of Rajasthan and the adjacent area, proposed by Raja Rao et al. (1971) and Gupta et al. (1980) were nonetheless drastic. They introduced a new stratigraphic unit, the 'Bhilwara Supergroup' of Archean age, in which they included the BGC, the Aravalli Supergroup of the Bhilwara area, Raialo Group of the Jahazpura area and part of the Gwalior facies rocks of earlier workers. They proposed the name Mangalwar Group for the lower division of the Bhilwara Supergroup. Mangalwar Group included BGC and other supracrustal rocks. The upper division they called Hindoli Group.

The Delhi fold belt rocks have been suggested to be tangibly different in the northern and southern parts and hence the entire Delhi belt is divided into North Delhi Fold Belt (NDFB) and South Delhi Fold Belt (SDFB) (Sinha Roy, 1984). Even it has been suggested that the geological sequence in the Delhi Supergroup is diachronous. In fact, people have different views on the nature and evolution of the Proterozoic Delhi–Aravalli mobile belt. The proposed tectonic models vary widely from one of intra-continental rifts to subduction of plates.

In the following pages, the status of these and some other issues relating to the geology of western India are briefly reviewed.

Basement Complex

By 'basement complex', we refer to the rock complex Heron called 'Banded Gneissic Complex', abbreviated to BGC. Although granite–gneiss is a major rock type in this complex, the granitoid rocks intrusive into the gneissic member, or occurring as inliers in the younger supracrustals, occur in considerable proportions. Besides, there are enclaves of amphibolites and metasediments within this complex. As already briefly discussed, a doubt was raised about the basement status of this complex with respect to the associated supracrustal rocks. Hopefully, it has been removed by (i) re-establishing that the conglomerates occurring at the interface of this complex and the associated supracrustals were sedimentary rather than autoclastic; (ii) identification of paleo-weathering surface above the basement rocks and below the overlying supracrustal rocks at places (Roy and Paliwal, 1981; Roy et al., 1981; Banerjee,1996); (iii) obtaining well-constrained radiometric age data from the representative members of this complex (Table 6.1.2), as well as the supracrustal rocks wherever possible.

Table 6.1.2 *Geochronology of the members of the basement complex of Rajasthan area*

Age (Ma)	Method used	Rock analysed	References
2610 ± 50	Single zircon U–Pb	Berach Granite	Widenbeck et al. (1996)
2505 ± 3	Single zircon U–Pb	Untala Pink Granite	Widenbeck et al. (1996)
2506 ± 4	Single zircon U–Pb	Vali River Granite,	Widenbeck et al. (1996)
2532 ± 5	Single zircon U–Pb	Jhamarkotra Granite	Widenbeck et al. (1996)
2562 ± 6	Single zircon U–Pb	Ahar River Granite	Widenbeck et al. (1996)
2620 ± 5	Single zircon evaporation	Gingla Granite	Roy and Kroner (1996)
2658 ± 5	Single zircon evaporation	Jagat Granitoids	Roy and Kroner (1996)
2666 ± 6	Single zircon evaporation	Untala Trondhjemite Gneiss (enclave within pink granite)	Roy and Kroner (1996)
2828 ± 46	Sm–Nd isochron age	Mafic dikes of Mavli	Gopalan et al. (1990)
2830	Sm–Nd isochron age	Nandwara Granitoid	Tobisch et al. (1994)
2887±5	Single zircon evaporation	Banded TTG gneiss of Jagat	Roy and Kroner (1996)
3230	Single zircon evaporation	Trondhjemite intrusive	Roy and Kroner (1996)
3281 ± 3	Single zircon evaporation	Igneous protolith, banded TTG gneisses, Jhamarkotra	Widenbeck and Goswami (1994)
3307 ± 65	Sm–Nd isochron age	Igneous protolith, banded TTG gneisses, Jhamarkotra	Gopalan et al. (1990)

Gneissic Rocks

The gneissic rocks are usually compositionally banded, varying in composition from tonalite–trondhjemite–granodiorite (TTG) to granite, with enclaves of generally isofacially metamorphosed mafic and sedimentary rocks (Fig. 6.1.2 a–d).

The rocks have been multiply deformed and finally metamorphosed to the amphibolite–granulite facies. Sen (1980) suggested that the banded gneissic terrain might be divided into two blocks: north and south. The north block underwent a relatively higher grade of metamorphism (amphibolite to granulite) than the southern block where metamorphism did not exceed amphibolite facies.

Fig. 6.1.2 Field photographs showing (a) Remnants of dark, foliated biotite gneiss within pink Untala Granite; (b) Transposition of deformed early foliation in basement gneiss; (c) Flattened sheath folds developed in basement gneiss; (d) Coaxial refolding of isoclinal fold in Banded Gneiss from Kankroli. (a–c after Roy and Jakhar, 2002).

Early determination of radiometric ages of some samples from this gneissic complex (Crawford, 1970; Chaudhary et al., 1984) were informative. But single zircon ion–microprobe U–Pb age, single zircon evaporation age, Sm–Nd whole rock isochron ages, relatively recently obtained (Table 6.1.2), are on the other hand better constrained and hence geologically more significant.

The older members of the complex occur mostly in the southern part of the BGC terrain, known as Mewar. This led Roy to rename the gneissic rocks of this region as Mewar Gneiss and those of the northern part as the Sandmata Complex (Roy, 1988; Roy and Jakhar, 2002). Proliferation of new names in geology may desirably be kept in check in the interest of the readers, if not for the subject itself. The present authors would prefer to retain the nomenclature of Gupta (1934), BGC–I and BGC–II, or at the most call them BGC (south), BGC (north) if that is more acceptable (cf. Bose, 2000). The oldest rock so far directly dated in Rajasthan is a 3.3 Ga banded TTG gneiss from Jhamarkotra area. This is supported by the ca 3230 Ma and ca 3281 Ma single zircon ages reported by Roy and Kroner (1996) and Widenbeck and Goswami (1994). This renders an earlier Pb-isochron age of 3.5 Ga (Vinogradov et al., 1964), obtained from a detrital zircon in Aravalli metasediments close to the BGC border near Udaipur, more significant. This means that the rocks older than 3.3 Ga may also be found one day on detailed search. Within the limits of the available data, the ages of TTG gneisses in Rajasthan varied from 3.3 to 2.66 Ga (Table 6.1.2).

The gneissic rocks of the Mewar region vary considerably in the contents of SiO_2, Al_2O_3, CaO, Na_2O, K_2O and Na_2O/K_2O ratios (Table 6.1.3) but in a normative Ab–An–Or diagram, they plot in the tonalite–trondhjemite and granodiorite fields. Gneissic rocks of TTG composition are common in Archean terrains. Chondrite normalised REE distribution pattern in these rocks show LREE enrichment, a feature also common in the Archean (TTG) gneisses.

Table 6.1.3 *Chemical analyses of major (wt.%), trace and rare earth elements (ppm) of some banded gneissic rocks of the Mewar region (Roy and Jakhar, 2002; Upadhyay et al., 1992)*

	1	2	3	4	5	6	7	8	9	10 (n = 15)	11 (n = 9)
SiO_2	75.81	59.63	76.00	71.36	74.84	59.03	57.53	66.82	63.58	68.43	65.00
TiO_2	0.39	0.93	0.54	0.40	0.61	0.47	2.18	0.58	0.39	0.38	0.56
Al_2O_3	10.47	13.59	12.56	15.28	12.02	20.46	13.78	14.51	19.68	16.31	17.44
FeO(T)	4.57	7.44	4.11	2.55	5.19	4.35	12.69	5.34	3.26	297	5.46
MgO	1.25	0.07	0.90	1.35	1.87	1.82	2.55	2.57	1.07	1.11	2.08
CaO	1.66	2.46	2.06	1.63	1.70	4.38	4.87	5.61	5.67	2.88	3.01
Na_2O	2.8	2.06	4.96	5.61	3.16	6.78	3.04	3.38	7.25	4.91	3.98
K_2O	1.35	4.31	1.21	2.09	2.44	2.05	2.14	0.99	0.07	1.55	2.11
P_2O_5	0.05	2.67	0.15	0.07	0.19	0.22	0.99	0.17	0.22	0.29	0.23
Total	98.35	93.16	102.5	100.3	102.0	99.56	99.72	99.97	101.2	99.43	99.87
Rb	44	73	39	41	60	38	58	19	1	51	98
Sr	190	279	239	302	141	492	256	217	572	401	415
Ba	342	400	–	700	–	–	–	–	60	1742	1519
Ni	3	50	1	7	–	–	–	–	–	37	48
Cr	16	40	4	9	4	18	13	–	13	41	43
Co	9	20	31	25	9	18	30	–	13	–	–
Sm	–	–	9	6	7	6	18	6	4	–	–
Nd	–	–	39	49	35	40	87	30	16	–	–

Subordinate Rock Types Subordinate rocks within the BGC comprise metasediments (pelites, arenites and carbonates) and amphibolites and a few enclaves of ultramafic rocks, the latter represented mainly by talc–antigorite–tremolite–actinolite schists. Quartzites and carbonate bodies occur in mappable dimensions. Quartzites contain ≥ 90% quartz. Cr-bearing greenish muscovite is present in the matrix similar to such rocks in many other Archean terrains. Cross-bedding present in these rocks proves that they are not recrystallised cherty rocks. The metamorphosed pelitic rocks contain quartz, muscovite, biotite, garnet with/without staurolite and sillimanite. Carbonate rocks are crystalline, low-Mg type marbles with a little or no impurities.

Amphibolite bodies, usually not too large, occur as enclaves within the gneisses. In composition most of these amphibolites correspond to tholeiitic composition, but some could stray into the high-Mg field. Trace element composition also supports their derivation from igneous source. The Sm/Nd isochron age of these amphibolites is 2828 ± 6 Ma (Gopalan et al., 1990).

Late Archean Granitoid Intrusives It has already been noted (Table 6.1.2) that there are a number of late Archean granitoid bodies within the basement complex, or occurring as inliers surrounded by the rocks of the Aravalli Supergroup. Of these, the Untala Granite and the Gingla Granite occur within the gneissic terrain of Mewar. Two other important ones, the Berach Granite and Ahar River Granite occur respectively near Chittaurgarh and Udaipur.

Untala Granite It is a lenticular body, about 30 km x 10 km in surface extension. The granite is composite in nature and consists of grey diorite–tonalite–trondhjemite (locally gneissose) and late pink/grey granite (Chattopadhyay et al., 1987). The average Rb- and Sr-contents in the grey phase are 50 ppm and 175 ppm respectively, as contrasted to 150 ppm and 50 ppm in the pink variety. The pink granitoid variety is the dominant phase in the massif. The single zircon U–Pb age of the pink variety has been determined to be 2505 ± 3 Ma (Widenbeck et al., 1996).

Gingla Granite Gingla Granite occurs at the Jai Samand Lake area, about 100 km southeast of the Udaipur city. A fair part of it is believed to lie under the water body of the lake. It is partly granitoid and partly migmatitic. In a normative Ab–An–Or diagram, the migmatites of the Gingla massif plot in the tonalite–trondhjemite field, while the granitoids, apparently consolidates of partial melts, plot in the fields of granite–adamellite. In SiO_2 vs FeO/(FeO + MgO) diagram, both the granitoid and migmatitic members of Gingla Granite plot in the 'orogenic' field (Bose et al., 1996).

Berach Granite The Berach Granite in the Chittaurgarh area occupies an area of about 300 sq km in the form of an elongate body, trending NNE in the southern part and NE in the northern.

The Berach Granite also consists of two phases: a foliated, dark grey or pink coloured older granodiorite and a younger granitic phase, which occupies mostly the eastern part. K-rich variety dominates over the Na-rich. The trace-element contents of Berach Granite are as follows: Sn ~10 ppm, Be ~5 ppm, Li 7–30 ppm. Locally Ba content is 1770–3000 ppm, V 70–100 ppm, Y 5 ppm and Mn 1000–1500 ppm (Prasad, 1987). Felsic volcanic rocks of rhyodacitic composition are reported from the Berach Granite massif, particularly in its southeastern part. That the associated supracrustals overlie the Berach Granite with an erosional unconformity is not doubted (Gupta, 1934; Heron, 1953; Bose, et al., 1996).

Interestingly, for a long time the Berach Granite enjoyed a very 'old' status, Heron (1953) put it stratigraphically below the gneissic complex (BGC) (Table 6.1.1). He and Pascoe (1950) even considered the Berach Granite to be the oldest rock in India. However, the relatively recent determinations of the radiometric age of this massif are as follows: Rb/Sr whole rock age – 2555 Ma (Crawford, 1970), U–Pb/zircon age – 2440 ± 8 Ma, (Srinivasan and Odam, 1982), Rb/Sr whole rock isochron age, 2533 ma (Choudhary et al., 1984) and the single zircon U–Pb age, 2610 ± 50 Ma (Widenbeck et al., 1996). The ages obtained by utilising different systematics are remarkably close to each other.

Ahar River Granite Ahar River Granite is in fact a granitoid body named as such by Crawford (1970). The body contains enclaves of metamorphosed mafic and sedimentary rocks. In fact, most of the sedimentary rock bodies are outliers of Aravalli Supergroup and the massif suffered severe post-crystalline deformation during the deformation of the Aravalli cover sediments. Prominent ductile shear zones developed along the margin of the body and elsewhere (Roy and Jakhar, 2002).

Petrologically, the Ahar River Granite varies from alkali granite to tonalite through granodiorite. It is rather mica-poor, presumably due to water deficiency in the crystallising magma. Widenbeck et al. (1996) obtained a single zircon U–Pb age of 2562 ± 6 Ma for this rock.

Nandwara Granitoids These granitoid rocks were named as such by Heron (1953) with reference to their type area of occurrence. They occur in two NE–SW trending linear bodies. The western amongst the two occurs along the interface between the BGC and the Delhi Supergroup. Besides, there are patches and screens of highly deformed rocks of the same within the metasediments of the area. The Nandwara granitoids are K_2O-rich, and have moderate Al_2O_3- and moderate to high SiO_2-contents. In a SiO_2 vs FeO/(FeO + MgO) diagram, these rocks plot in the field of orogenic granite.

Tobisch et al. (1994) obtained a Sm–Nd isochron age for these rocks, which is 2830 Ma. Expectedly, these rocks are more radiogenic than the ~3.3 Ga tonalitic and granodioritic gneisses, reported by Gopalan et al. (1990) from southwestern Rajasthan.

Proterozoic Magmatism within the Banded Gneissic Complex (BGC) The thermo-tectonic perturbations that led to the magmatism of the basement complex during the Archean time did not completely cease to be operative during the Proterozoic. Locally there has been granitoid-emplacement as at Anjana near Devgarh in Central Rajasthan, carbonatite intrusive within the Untala Granite near Newania and emplacement of alkaline syenite at Kishengarh.

Anjana Granite A nearly dome shaped porphyritic granite body that intruded the augen gneiss of BGC near Devgarh in Central Rajasthan (Heron, 1953).

Two varieties of granitic rocks constitute the pluton. The earlier one is foliated, while the latter one is massive. The younger one intruded the granulitic rocks occurring in the neighbourhood. The composition of the members of the Anjana Granite is not well-constrained and limited data suggest them to vary between granite to granodiorite–tonalite in composition (cf. Gangopadhyay and Lahiri, 1988).

No radiometrically determined age is available for the Anjana Granite. However, its Proterozoic age is suggested by its intrusion into granulitic rocks within the BGC.

Newania Carbonatite Newania Carbonatite bodies occur within the Untala Granite (–migmatite) near Newania in Udaipur district. Though known for a long time as a carbonate-rich rock, it was identified as a carbonatite (banded sovite) for the first time by Phadkey and Jhingran (1968). Chattopadhyay et al. (1988) divided the Newania Carbonatites into two varieties: (i) a medium to coarse grained crystalline variety, made of ferroan dolomite (ankerite) or calcite with the accessory phases that included alkali-bearing mafic phases, apatite and one or more opaque phases, (ii) a medium to fine grained variety, made principally of carbonate and hematite (hence reddish in colour). The (i) and (ii) are generally interlaminates. Smaller bodies of carbonatite occur elsewhere in the region.

Deans and Powell (1968) dated the Newania Carbonatite at 955 ± 24 Ma. Carbonatite dikes have intruded into Vindhyan Supergroup rocks at Nayaphala, Chittaurgarh district, attesting to the minimum age of the Vindhyan rocks at the place.

Kishengarh Nepheline Syenite The Kishengarh Syenite occurs as a 25 km x 1–5 km wide lenticular body in Kishengarh–Patan–Hamara area, located between gneiss–metasediment–metavolcanic rocks (BGC) in the east and Delhi Supergroup rocks in the west. The alkaline rock is mainly nepheline syenite in composition. Neogi (1965) conducted preliminary studies on the structure and petrology of the rocks. In a recent work, involving detailed field mapping. Roy and Dutt (1995) showed that the nepheline syenite is an elongate flattened dome, refolded into a hook shape. Compositionally, the syenite contains sub-equal proportions of nepheline and andesinic plagioclase with subordinate amounts of magnesio–riebeckite, biotite and sodic pyroxene. The massif contains enclaves of theralite and dikes of camptonite near Mandoria. Gabbroid rocks occur in association.

Srivastava (1988) shares the view of Heron (1924) that the syenite body is an igneous intrusive. Neogi (1965) on the other hand, believes that the rocks formed by alkaline metasomatism of gabbroid rocks. Crawford (1970) dated (Rb–Sr) the rock to be 1490 ± 150 Ma old.

Deformation and Metamorphism of the BGC – an Outline An Early Precambrian basement complex has generally a complex history of deformation and metamorphism. The BGC is no exception to this generality.

The rocks of the Banded Gneissic Complex bear evidence of 3–4 phases of deformation, depending on the area (Naha, 1983; Srivastava et al., 1995). The structures and the sequence of their development in the basement rocks and the supracrustal sequences have much in common, a fact that led some earlier workers to strip the basement complex of its basement status. However, locally deformational structures (folds and the related planar and linear structures) have been identified that appear older than those that are common in the basement rocks and the supracrustals (Mohanty and Naha, 1986; Roy,1988). There was mobilisation in the basement rocks during deformation that caused obliteration of earlier structures in many places leaving behind the imprints of later structures only. Foliation in the gneissic rocks in the BGC, even if primary to start with, is now mostly modified. For details of the structures studied in different segments of the area, references may be made to Naha and Mohanty (1990), Roy and Rathore (1999), Sharma (1977), Sharma and Upadhyay (1975), Mukhopadhyay and Das Gupta (1978), Pyne and Bandyopadhyay (1985), Mohanty and Guha (1995), besides what have been mentioned earlier.

The grade of metamorphism in the gneissic schistose–amphibolitic rocks of the BGC generally varies from lower middle to upper amphibolite facies, locally passing into granulite facies as at Bandanwara in Ajmer district (Pyne and Bandyopadhyay, 1985). The geology of the Sandmata Complex carrying granulites is discussed separately.

Sandmata Complex It is a 200 km x 50 km body consisting of an ensemble of such rocks as migmatites, varieties of gneisses, mica schists, garnet–sillimanite schists, garnet–staurolite–sillimanite schist, cordierite–garnet gneisses, charnockites, enderbite, two–pyroxene granulite, leptynites, amphibolites, quartzites, conglomerates and dolomitic marbles. Hence, it is a 'complex' in the sense of the term as used in geology. It derives its proper name from the Sandmata Temple that is located near Kekri. It is bordered by the BGC (II) in the east and in the west it is supposedly overlain by the rocks of the Delhi Supergroup that we shall discuss in a following section. The granulite facies rocks occur in a milieu of gneiss, amphibolite and metamorphosed pelitic rocks. The granulites, particularly charnockite, enderbite and mafic granulites, are generally massive except along the marginal parts where they are commonly foliated. A profusion of migmatites developed around the granulite bodies. The granulite composition varies in space. While the pelitic and psammopelitic granulites are the major members of the Sandmata Hill area, charnockite, enderbite and mafic granulites are common in the Bhinai–Bandanwara–Nasirabad region in the north. Quartzo–feldspathic granulites vary in composition from that of granite to tonalite (Gyani and Omar, 1999). The pelitic granulites are foliated and resemble khondalites of the Eastern Ghat mobile belt to some extent (Roy and Jakhar, 2002). The calc–silicate granulites are rather rare.

Opinions on the origin, evolution and emplacement of the granulites of the Sandmata Complex vary, at least on some aspects. Sinha Roy et al. (1992, 1995) suggested that the granulitic pelitic rocks of Sandmata are 'tectonically interleaved' bodies with the supracrustal rocks, metamorphosed in the amphibolite facies. According to Sarkar et al. (1989), the charnockites–granodiorite suite in this tract was emplaced at 1723+14/–7 Ma (U–Pb/zircon). In their interpretation the granulite facies metamorphism probably resulted from the high temperature regime associated with underplating of basic magma in a continental arc setting. Two contrasting events were diagnosed by Guha and Bhattacharya (1995) in the metamorphic history of these rocks. The first high grade metamorphic event (M_1), which was older than 1900 Ma, occurred before the onset of the major tectonic event and during the development of penetrative gneissosity (S_1). The M_1 metamorphism was coeval with the a melt (partial)-forming event leading to the development of stromatic migmatite and granodioritic plutons. This was followed by the overthrusting of high-grade (Sandmata Complex) and underthrusting of the low to middle grade BGC rocks during orogeny in the neighbouring Proterozoic fold belt.

Sharma (1995) suggested thickening of the local continental crust through undersoling by basalt, created through partial melting of the asthenoshpere. Granulites were later shifted to a shallow level (~20 km depth) along major shear zones. Sharma also believed that the 1725 Ma U–Pb zircon age of charnockite indicated reworking (reheating) of the granulites during final emplacement.

Dasgupta et al. (1997) confirmed the earlier view that the granulites of the Sandmata Complex are of deep crustal origin and were subsequently tectonically emplaced at higher levels along ductile shear zones. These authors also subscribe to the view that the peak metamorphism had taken place before emplacement of the charnockite–enderbite pluton at ca 1725 Ma. However, the granulites were again reworked during the next upliftment at ca 1000 Ma.

Aravalli Belt

The Aravalli belt is an N–S trending belt, widest south of Jharol and tapering gradually northward up to about Katar. The rocks of this belt comprise what is commonly known as the Aravalli Supergroup. The supracrustal rocks present in the Bhilwara belt are also believed to belong to this Supergroup by some workers, while others prefer to give them an older (Archean) status. We will return to this in a later section while discussing the geology of the Bhilwara belt separately. The Aravalli belt is often divided into two sectors: the northern, Udaipur sector and the southern, Lunavada sector. The Aravalli Supergroup is best exposed in the Udaipur region. Structurally the Lunavada sector is more complex. The Aravalli Supergroup in the Udaipur area contains the marbles and metapelitic rocks of the Raialo Group. The relationship of the Aravalli Supergroup rocks with the basement complex is no doubt locally blurred, but as we have already pointed out, there are ample convincing evidences to establish their true temporal relationship.

Rock Types and Their Stratigraphic Sequence There have been several attempts to classify further the Aravalli Supergroup. Banerjee (1971 a, b) classified the Aravalli rocks (his 'Group') into three formations: the Debari Formation (oldest), the Matoon Formation and the Udaipur Formation. Gupta et al. (1980, 1997) also divided the Aravalli rocks of the type area into several groups, such as Debari Group, Udaipur Group, Bari Lake Group (= Kankroli Group), Jharol Group (= Dovada Group, Nathdwara Group), Lunavada Group and Champaner Group. Each of the Groups in turn is subdivided into several Formations. This Group-division is based on lithostratigraphy, structure, metamorphism, tectonic set-up and depositional environment. Accordingly, all the Group divisions are not sequential, some being temporally equivalent to the others. For example, the Kankroli Group is equivalent to the Bari Lake Group while the Dovda and Nathdwara Groups are temporally correlated with the Jharol Group (Gupta et al., 1997). The stratigraphic classifications for the Aravallis proposed by different workers are presented in Table 6.1.4.

Roy and his co-workers came out with a different scheme of stratigraphy for the Aravalli Supergroup with a three-fold classification into Lower, Middle and Upper Aravalli Groups, with further sub-divisions into Formations under each (Roy and Kataria, 1999). In contrast to the views of Gupta et al., 1997, they recognised Delwara Formation in the Lower Aravallis as the basal formation and placed the Debari Formation at much higher stratigraphic level at the base of their Middle Aravalli Group. The stratigraphic classification proposed earlier by Sinha Roy et al. (1993), was also three fold with Delwara Formation placed below Debari Formation (Table 6.1.4). Roy and Jakhar (2002) proposed a modified scheme for the Aravallis (Table 6.1.5, Fig. 6.1.3), in which the Jharol Formation, representing deep-water facies, was placed separately from rest of the sequence representing shelf facies, as temporal equivalent of Lakhwali Phyllite of the Upper Aravallis of shelf

Table 6.1.4 *Stratigraphic schemes proposed by different workers for the Aravalli sequence of Rajasthan and Gujarat*

	Heron (1953)		Gupta et al. (1997)			Roy and Kataria (1999)		Sinha Roy et al. (1993)
	Delhi System		Delhi Supergroup			Delhi Supergroup		Delhi Supergroup
-----------------------Unconformity/Tectonic contact ---------------								
Rialo Series	Garnet–biotite schist Limestone (marble) Local grit		Champaner Group					
			Lunavada Group					
	Aplogranite, epidiorite		Darwal Granite					
	Hornblende schist, ultramafics		Rakhabdev ultramafics					
Aravalli System	Impure limestone, quartzite, phyllite, biotite schist and composite gneiss	Aravalli Supergroup	Jharol Group (= Dovda = Nathdwara Groups)	Aravalli Supergroup	Upper	Serpentinite	Aravalli Supergroup	Jharol Group (Upper Aravalli)
						Lakhwali Phyllite		
						Jharol Formation		
						Kabita Dolomite		
						Debari Formation		
					--Unconformity--			--Unconformity--
	Quartzite, grit, local soda syenites, conglomerate, amygdaloids and tuffs		Bari Lake Group (Kankroli Group		Middle	Tidi Formation		Leucogranite Debari Group (Middle Aravalli)
						Bowa Formation		
			Udaipur Group			Mochia Formation		
			Debari Group			Udaipur Formation		
					--Unconformity--			--Unconformity--
					Lower	Jhamarkotra Formation Delwara Formation		Pink granite Delwara Group (Lower Aravalli)

facies. Difference of this scheme with others is obvious. The shelf facies is represented principally by coarse clastics and carbonate rocks, while the deep-water facies further west (Jharol Formation) is composed of pelitic sediments now transformed to phyllites and mica schists with thin beds of quartzites. Ultramafic rocks, apparently intrusive and now fully altered, occur along the margin between the shelf and deep-water sequences and form a sort of lineament, commonly referred to as Rakhabdev lineament (Fig. 6.1.1). In the south, the Deccan Trap and Tertiary sediments cover the Aravalli Supergroup.

Table 6.1.5 *Stratigraphic succession of the Aravalli Supergroup (Roy and Jakhar, 2002)*

Shelf sequences			Deep water sequences	
Upper Aravalli Group	Serpentinites (intrusions)			
	Lakhawali phyllite		Jharol Formation	Mica schists with thin quartzite beds, serpentinites
	Kabita Dolomite			
	Debari Formation	Quartzite, arkose, conglomerate		
------------------------------- *Unconformity* ----------------------------------				
Middle Aravalli Group	Tidi Formation	Slate, phyllite with thin bands of dolomite and quartzites		
	Bowa Formation (Machhia Mogra Formation, Roy et al., 1988)	Quartzite and quartzose phyllite		
	Mochia Formation (≃ Zawar Formation, Roy et al., 1988)	Dolomite, carbonaceous phyllite, greywacke/ phyllite and conglomerate (Hosts the Zn–Pb deposits of Zawar belt)		
	Udaipur Formation (Kallalia Formation of Zawar, Straczek and Srikantan, 1966)	Greywacke/phyllite, Conglomerate		
-------------------------------- *Unconformity* -------------------------				
Lower Aravalli Group	Jhamarkotra Formation (= Mandli Formation of Zawar, Straczeck and Srikantan, 1966)	Dolomite, quartzite carbon phyllite, phyllite, thin local bands of stromatolytic phosphorite (≅ Raialo marble of Iswal and Rajsamand) (contains weak U-mineralisations of Umra–Udaisagar and Pavati–Selpur belts		
	Delwara Formation	Metabasalts with thin bands of dolomite / quartzite, veins of barite		
------------------------------------- Archean–Proterozoic unconformity ----------------------				
Mewar Gneissic Complex, Granitoids (Archean)				Pre-Aravalli gneisses, amphibolites, granitoids and metasediments

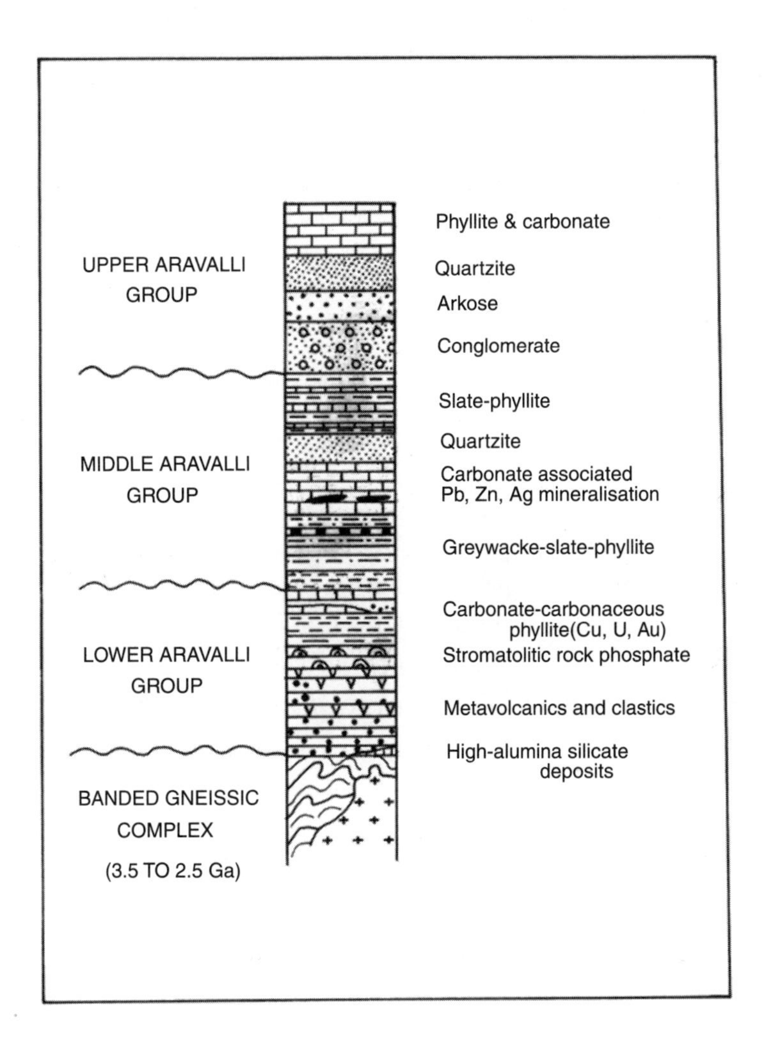

Fig. 6.1.3 Suggested stratigraphic column of Aravalli sequence in Udaipur area (after Roy, 2000)

A horizon of metavolcanics (mostly mafic with rare thin bands of felsic meta-tuff) with quartzite and conglomerate ± carbonate lenses, overlie the gneissic complex at a number of places and this Roy and his co-workers named Delwara Formation and placed as the lowest Formation of the Lower Aravalli Group (Table 6.1.5). The Delwara Formation passes upward into a carbonate (dolomitic)-rich horizon that also contains phyllite/mica schist, with/without carbonaceous rocks and stromatolytic phosphorite in the Jhamarkotra area. This unit locally lies directly over the BGC. This unit was previously called Mandli Formation by Straczeck and Srikantan (1966) and now designated as Jhamarkotra Formation by Roy and co-workers (op. cit.). This Formation includes the Raialo metasediments which were included in a 'Series' between the Aravalli 'System' (Supergroup) and the Delhi 'System' (Supergroup) by Heron (1953). Many workers already denied a separate stratigraphic status to the Raialos.

The Jhamarkotra Formation is unconformably overlain by a thick sequence of greywacke–shale (phyllite)–conglomerate that is called the Udaipur Formation (Kathalia Formation of Straczeck and Srikantan, 1966). This is the lowest Formation of the Middle Aravalli Group of Roy et al. (op. cit.). The Udaipur Formation is overlain by Mochia or Zawar Formation of Roy et al. (1988). This is a dolomite-dominant assemblage, which contains, besides massive dolostones, such rocks as carbonaceous and dolomite-bearing phyllite, greywacke rather lith–arenite etc. This Formation contains the well-known Pb–Zn deposit of the Zawar belt. The overlying Bowa Formation is quartzitic and the top most Formation of the Middle Aravalli Group in the classification scheme of Roy et al. (op. cit.) is the Tidi Formation, which is composed mainly of slate–phyllite with thin bands of dolostone and quartzite.

The Upper Aravalli Group consists of coarse clastic–dominant Debari Formation, the dolomitic Kabita Formation and the phyllitic Lakhawali Formation, arranged from the lowest upward (Table 6.1.5). It may be noted here that there is a strongly discordant view with respect to the stratigraphic position of the Debari Formation. Banerjee (1971 a, b), Gupta et al. (1980), Sinha Roy et al. (1998) are of the view that it should, on the other hand, be the basal Formation of the Aravalli Supergroup. This dispute should be resolved sooner than later.

The fine-grained metasediments (phyllites/mica schists ± thin beds of quartzite) are apparently correlatable with the similar metasediments of the upper Aravalli Group of the shelf facies. The ultramafic rocks occurring along with the Jharol sediments are suggested to be flows by Roy and Jakhar (2002) and intrusives by Deb and Thorpe (2004).

Deformation and Metamorphism of the Rocks of the Aravalli belt

Structures Geological structures of the rocks in the Aravalli–Jharol belt, particularly in the type area of Udaipur and the neighbourhood, have been studied by a number of workers and they are too numerous to be referred to individually (vide Roy and Jakhar, 2002 and references therein for some details). The salient aspects of the results of these studies are as follows. The geometry and disposition of the tectonic structures in the rocks of this belt are the products of polyphase deformations, D_1, D_2 and D_3, particularly interaction of the related fold-phases, F_1, F_2 and F_3. The F_1 folds are appressed isoclinal and commonly reclined. Axial plane cleavage/schistosity is generally developed. Lineations related to F_1 are both stretching and intersection (bedding–cleavage) types. In the areas of high strain, the axial plane cleavage becomes mylonitic and the general fabric of the rocks changes to L–S type. The F_2 folds are generally upright with sub-vertical axial plane cleavage. The F_2 folds involved both the bedding and the planar and linear structures related to F_1. The superimposition of F_2 on F_1 has given rise to complex structures even on map scales (Banerjee et al., 1998, Mukhopadhyay and Sengupta, 1979; Roy and Bejarniya, 1990). In cases, F_1 has undergone co-axial refolding before being superimposed by F_2 (Naha and Mohanty, 1990). F_3 folds refolded both F_1 and F_2 folds. Unlike the earlier folds, these are more open and have sub-vertical axial surfaces, generally with E–W or WNW–ESE strikes. F_3 folds vary in scale from crenulations to megascopic. An example of the latter, together with related small-scale structures may be found in the Zawar area (Fig. 6.1.4). Locally, a still younger set of folds (F_4), are found with subhorizontal axial cleavage (S_4). Differential movement along the zones of ductile shearing around relatively low strain domains has resulted in the development of complex outcrop patterns, according to Drury (1990). The development of these ductile shear zones of regional extent and the related differential movements along them, took place during the major crustal shortening accompanied by F_2 folding (Roy, 1995). However, the expected low plunging linear structures supporting the suggested regional shear movement are poorly developed, if at all.

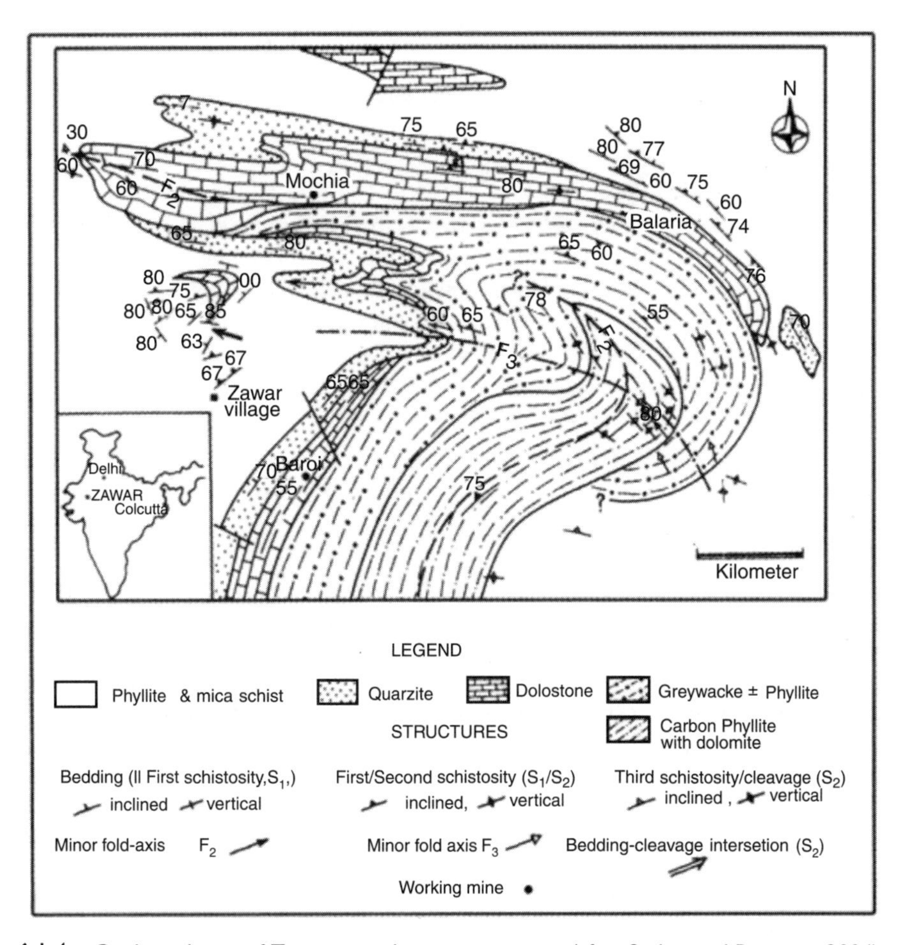

Fig. 6.1.4 Geological map of Zawar area showing structures (after Sarkar and Banerjee, 2004)

The structures in Lunavada sector have apparently been studied less. However, the geological maps of this region produced by some workers (Gupta and Mukherjee, 1938; Mamtani et al., 1999 a, b) suggest that the deformation here has not been less complex. However, Roy (2000) notes that at least some of the fold phases that affected the rocks of the Lunavada sector cannot be related to the Aravalli and Delhi orogenic cycles. If true, this observation demands a satisfactory explanation. Rakhabdev lineament has been a zone of regional dislocation (shear) in the region.

Metamorphism The intensity of metamorphism of the rocks of the Aravalli–Jharol belt varies. Garnet–kyanite–staurolite-bearing mica schists have developed in the metapelites of the Raialo Series of Heron (1953) in the hammer head syncline area. A progressive increase of metamorphic grade from chlorite zone to kyanite zone has been reported by Ghosh (1983) from Batia to Fatehpur.

In the Udaipur–Zawar area the grade of metamorphism, however, is low. Here the metamorphosed pelitic rocks and greywacke now consist of quartz, muscovite, biotite and alkali feldspar in different proportions. In the SiO_2-bearing dolomitic rocks, tremolite has not developed. The tremolite-forming reaction, is constrained by X_{CO_2} in the fluid present and P_{total} of the system. The obtained $X^{Fl}_{CO_2}$ = 0.3 from fluid inclusions, corrected for P = 2.9 kb, following Skippen's (1974) and Greenwood's (1976) diagrams, yielded the upper stability limit of the quartz + dolomite pair to be 442°C. So if the fluid in the fluid inclusions were cognate to the system at large, then the inability to develop tremolite suggests that the temperature of metamorphism here was < 440° C (Banerjee and Sarkar, 1998). The prograde metamorphism is essentially coeval with F_1 deformation (Roy 1991; Banerjee and Sarkar, op. cit.).

Dolomite + quartz → tremolite + calcite

Age of the Aravalli Rocks There are a few granitic bodies within the Aravalli–Jharol belt, such as the Ahar River Granite or the Udaipur Granite, Udaisagar Granite and the Darwal Granite. The temporal relationship of the first two with the enclosing sediments is controversial. Some believe they are pre-Aravalli (Roy, 1988). The post-Aravalli status of the alkaline Darwal Granite is virtually uncontested. Chaudhary et al. (1984) determined an age of 1900 ± 80 Ma (Rb–Sr) for this granite. This puts the minimum age of the sediments at 1900 Ma. Recently galena was separated from thin barite lenses in metavolcanic rocks at Negaria, belonging to the Delwara Group, the basal Formation of the Aravalli Supergroup. A Pb–Pb model age of 2040 Ma for the galena in the barite vein was obtained (Deb and Thorpe, 2004). This supports the above conclusion.

Bhilwara Province and the Status of the Bhilwara Supergroup

The Bhilwara belt (named here as Bhilwara Province to distinguish the region as a whole from the narrow supracrustal belts within it) in Rajasthan geology means an elongate area trending NE–SW and is bound by the central Rajasthan part of the BGC (BGC II of Gupta, 1934, or the Sandmata Complex of Roy, 1988) in the west and the Great Boundary Fault in the east, marking the western margin of the Vindhyan basin. This province in turn is divisible longitudinally into two domains: the high-grade western domain (Sandmata) and the low-grade eastern domain (Mangalwar and Hindoli). The interface between the two domains bears evidence of being a zone of dislocation, along which the western block has been uplifted, ultimately exposing relatively higher-grade rocks. In the north, its boundary with the rocks of Delhi Supergroup is covered. In the south, it is lost in the gneissic complex of Mewar or the moderately metamorphosed Aravalli rocks in the Nathdwara area, or covered by the Deccan Traps. The Bhilwara rocks have for long been believed by a group of workers to belong to the Aravalli Supergroup (formerly 'System') (Gupta, 1934; Heron, 1953; Roy, 1988: Deb and Sarkar, 1990). Raja Rao (1970), Gupta et al., (1980) proposed a major change in the Precambrian stratigraphy of Rajasthan. They introduced a new stratigraphic nomenclature, the 'Bhilwara Supergroup', that included the entire litho-package from Hindaun to Gurli, comprising the BGC, the Aravalli of the Bhilwara area, the Raialo Group rocks of Heron (1953) particularly of the Jahazpur area and the volcanisedimentary rocks of 'Gwalior facies' of Heron, (op. cit.). The sequence was believed to be Archean in age. The more metamorphosed part of it was given the name, Mangalwar Group and the weakly metamorphosed rocks of the eastern most domain – the Hindoli Group. This is a major difference of view and both cannot be correct at the same time.

Bhilwara Supergroup (Archean)	Hindoli Group	Same as the 'Gwalior facies' of the Aravalli System, Heron, 1953
	Mangalwar Group	Includes the BGC–II, Aravallis of the Bhilwara region, and Raialos of the Jahazpur area

In the following paragraphs, the situation in the context of the available information shall be critically reviewed.

Classification of the Bhilwara Supergroup, as proposed by GSI (2001), in partial modification of Gupta et al., 1997, is reflected in Table 6.1.6. It may be seen that in this version of Bhilwara stratigraphy, there are both Archean and Lower Proterozoic components. This modification departs from the original views of Raja Rao (op. cit.) and Gupta et al. (op. cit.), which regarded their Bhilwara Supergroup as entirely Archean in age. The recognition of the carbonate bearing volcanisedimentary sequences exposed in the isolated elongate belts of Rajpura–Dariba, Pur–Banera, Sawar, Jahazpur etc., in the Bhilwara Province as of Proterozoic age appears to be quite acceptable in the light of the recently available evidences. However, at the same time, as per the code of stratigraphic nomenclature, it demolishes the concept of 'Bhilwara Supergroup', as it encompasses together the rocks of both Archean and Proterozoic. Even with the partial convergence of the views of different workers about the Proterozoic status of the sequences of Bhilwara belt under reference, the answer to the obvious question of their correlatability or otherwise with the Aravallis remains largely speculative.

Table 6.1.6 *Classification of Bhilwara Supergroup (GSI, 2001)*

Lower Proterozoic	Unclassified granites and basic rocks
	Ranthambhor Group
	Rajpura–Dariba = Pur–Banera = Sawar = Jahazpur Groups (from west to east)
Archean	Intrusives: Berach Granite & Gneiss (2585 Ma*) Untala & Gingla Granites (2860 Ma*), mafic–ultramafic intrusives
	Sandmata Complex = Mangalwar Complex = Hindoli Group (from west to east)

*The latest available single zircon U–Pb ages of Berach Granite and Untala Granite are 2450 ± 8 Ma and 2505 ± 3 Ma respectively (Widenbeck et al., 1996)

The geology of the BGC in central Rajasthan (BGC–II) is already briefly reviewed. Let us now discuss the geology of the high-grade domain of the Bhilwara belt, in brief. Before entering into that discussion, it may not be out of place to note that Sinha Roy once (1985) called BGC a reworked granite–greenstone terrain. Some other workers have since reported greenstone enclaves in the BGC (Upadhyay et al., 1992; Mohanty and Guha, 1995). Somewhat liberalisation of the definition of the term 'granite–greenstone belts' in recent years notwithstanding, it may be premature to call the BGC a 'reworked granite–greenstone terrain' on the basis of limited studies on a few stray occurrences of 'greenstones' here and there.

The major rock type in the high-grade domain in the Bhilwara province is mica schist that grades into migmatitic paragneisses. Several large and elongate outcrops of metamorphosed shale–carbonate–sandstone association with black shales ± ironstones are present at a number places such a Rampura–Agucha, Jahazpur, Pur–Banera–Gangapur, Bethummi–Rajpura–Dariba and Bhinder. Most of the supracrustal sequences at these areas contain sulphide mineralisation, dominated by lead and zinc (discussed in a following section). Banded garnet–biotite–sillimanite schist and gneiss ± graphite are the major members in the rock association containing the Rampura–Agucha deposit. Calc–silicate rocks dominate over the marbles in contrast to the Aravalli type area. Amphibolites present at some of these areas have been suggested to be meta-igneous by some workers (Deb, 1992) and para-amphibolite by others (Roy and Jakhar, 2002). Graphite in these rocks is interpreted to be dominated by biogenic carbon.

The low-grade rocks along the eastern part of the Bhilwara province constitute a volcanisedimentary sequence. Here quartz–wacke and pelitic sediments (turbidites) are interlayered with flows and tuffaceous material of compositions varying from basalt through andesite to rhyodacite (Bose and Sharma, 1992). They constitute the Hindoli Group of Gupta et al. (op. cit.). Three lithofacies are identifiable in this sequence:

1. turbidites comprised of quartz–wackes and slate/phyllite, interlayered with volcaniclastic rocks
2. shelf facies of quartz arenite and carbonate (dolostone)
3. volcanic flows and tuffaceous material of mafic to felsic composition.

The volcaniclastic rocks of the sequence show LREE-enrichment with a sharp negative Eu-anomaly. The low-grade volcani-sedimentary sequence is correlated with the basal unit of the Aravalli succession (Delwara Formation) in the Udaipur type area. Again, the upper carbonate unit in the Aravalli succession that contains the Pb–Zn mineralisation in the Zawar belt is correlated with the nearly similar rock associations that contain base metal mineralisations in the Bhilwara province (Roy and Jakhar, 2002). However, these are not strong enough reasons in support of correlativity of the rock successions in the Aravalli belt and Bhilwara province. In the next two paragraphs, the other observations relevant in this context are discussed.

The gneissic rocks in the high-grade domain of the Bhilwara province are locally migmatised paragneisses. The TTG members, characteristic components of Archean basement complexes, are conspicuous here by their absence. The Bhilwara supracrustals (Jahazpur Group) overlie the Berach Granite, separated by a pronounced band of sedimentary conglomerate (Gupta, 1934). The Berach Granite is dated 2533 Ma (whole rock Rb–Sr) by Crawford (1970) and 2450 ± 8 Ma (single zircon ion–micropobe U–Pb age by Widenbeck et al., 1996). Results of dating zircon from the so-called Hindoli volcanic rocks are interesting. Five of six zircon samples collected from a felsic volcanic rock interlayered with sediments yielded ^{207}Pb/^{206}Pb ages between 1844 and 1888 Ma. The remaining sample yielded a higher age of 2480 Ma. One possible explanation of this finding is that some of the zircon grains could be inherited. Deb and Thorpe (2004) are inclined to believe that. The apparent local intrusion of the Berach Granite into the overlying supracrustals could be due to marginal remobilization of the former into the latter during later tectonic events (Bose and Sharma, 1992).

The model Pb – Pb age of the sulphide deposits in the Bhilwara supracrustals is ~1800 Ma (Deb et al., 1989; Deb and Thorpe, 2004). These are all sedimentary–diagenetic deposits (see a following discussion for details). However, as is known, the model Pb–Pb ages are generally proximate. A study of U–Pb ages of zircon from a tuff band on the hanging wall side of the Rajpura–Dariba deposit was taken for comparison. The zircon has the features of detrital zircon grains. That accepted, the youngest ^{207}Pb/ ^{206}Pb age of 1808 Ma (Deb and Thorpe, 2004) should be the maximum depositional age of the tuff. In that case, they have little difference with the depositional age of the stratiform ore deposits. This may be taken to mean that the model Pb-ages of the stratiform ore deposits is proximate estimates of the age of the sedimentary (–volcanic) sequence in which they occur. In the context of the above discussion it is difficult to sustain the proposal of the Bhilwara Supergroup put forward by Raja Rao (1970, 1976) and Raja Rao et al. (1971), supported by Gupta et al. (1980, 1997). Nearly the same Pb–Pb age of the ore deposits in the rocks of the Aravalli belt and the Bhilwara belts are suggestive of near contemporaneity of their developments. The obvious difference in the geology of the two belts can be explained by the initial facies difference, superimposed by high-grade metamorphism with/without anatexis, particularly in the western domain.

Delhi Fold Belt

The Delhi Fold Belt (DFB) in Indian geology is commonly understood as a Proterozoic mobile belt that is extensive over a distance of ~700 km from Delhi or beyond in the north and to the south of Idar in the south. The belt forms the backbone of the Aravalli Mountains in the northwestern part of the Indian Shield and is home to quite a few base metal ore deposits. Beyond these, much about it is debated. Some of the controversies that surround it are: (i) Is it a single crustal province[1], or composed of more than one? (ii) What is the nature of its boundaries – unconformity, paleosuture, modified rift margin, mélange zone or anything else? (iii) What is the proper stratigraphic sequence, belt-wise and segment or province-wise? (iv) Is it a Proterozoic continent-based riftogenic basin-fill, or is of a Phanerozoic type plate tectonics related origin? In the discussion that immediately follows, we shall briefly review the status of these and other relevant problems in the context of available information. For the ease of discussion, we geographically divide the DFB into the North Delhi fold belt (NDFB) and the South Delhi Fold Belt (SDFB), the dividing point being Ajmer. Geographically this division is nearly same as what was proposed by Sinha Roy in 1984. The South Delhi fold belt spatially coincides with the 'Main Delhi synclinorium' of Heron (1953). As the later workers are unable to confirm the 'synclinorium' status of this part of the belt, they prefer to refer to it as the "Main Delhi basin". Heron did not give much importance to the lithological difference in the Delhi fold belt in its northern part, situated in the Bayana–Alwar–Khetri region and the southern part occupying the Main Delhi fold belt and drew up a generalised stratigraphic sequence for the entire belt.

However, later workers (Singh 1988b; Gupta et al., 1995; Gupta et al., 1997; Sinha Roy et al., 1998) pointed out that a generalised stratigraphic sequence does not satisfactorily explain the reality on the ground and that it would be more reasonable to assume that rocks of the Delhi fold belt, commonly bracketed under the 'Delhi Supergroup', were deposited in multiple basins, or rather subbasins. These are:

1. Khetri–Saladipura basin
2. Bayana–Alwar (sub-) basins, North Delhi fold belt
3. Main Delhi basin, South Delhi fold belt.

North Delhi Fold Belt

Khetri (sub-) Basin Today's 'Khetri belt' was once known as the Khetri (sub-) basin. It is indeed a part of the Delhi fold belt. The Khetri belt is particularly well known in Indian geology and in the Indian mineral industry, because of having the 'Khetri copper belt' located along it. Besides, there is a large iron-sulphide deposit (still unexploited) at Saladipura. Further, the so-called 'Albite line' (Ray, 1990, 2004) in it is gaining importance for its being a potential zone for mineralisation of Li, Th and other metals.

The rocks constituting this 'belt' can also be referred to the QPC association (Condie, 1989) with subordinate volcanic rocks. Besides, there are a number of granitoid intrusives. Heron (1923, 1953) divided the rock-sequence here also into the lower Alwar 'Series' and the upper Ajabgarh 'Series', the former being arenite-rich and the latter, pelite-rich. Many later workers (Dasgupta, 1968; Sarkar and Dasgupta, 1980; Sarkar et al., 1980; Ray, 1974) followed this classification, admitting explicitly or implicitly that the lithologic development here and elsewhere in DFB are not the same. The rocks are multiply deformed and poly-metamorphosed. The geometry and disposition of deformational structures vary because of variation in the stress field, physical properties of the

[1] A **crustal province** is a segment of the continental crust that has a similar range of radiometric dates and shares a similar deformational history. Its limits are delineated by one or more of the following: (i) fault or shear zone, (ii) unconformity, (iii) sharp variation in metamorphic grade (iv) contacts with intrusive bodies.

rocks involved, and superimposition by later structures and interference by igneous intrusive bodies. However, several fold sets (F_1, F_2 and F_3) are generally distinguished, of which the first two (F_1 and F_2) are co-axial in many places and the axial trace of F_3 is at high angles with them (Naha et al., 1988; Sarkar and Dasgupta, 1980; Sarkar et al., 1980). The maximum grade of metamorphism reached is amphibolite facies, with the P/T gradient varying in space and time (Sarkar, 1973; Lal and Shukla, 1975; Sarkar et al., 1980; Sharma, 1988), causing variation in the mineral assemblages in comparable bulk lithologies, i.e. andalusite ± cordierite vis-a-vis kyanite ± staurolite.

There are a number of granitoid intrusive bodies in the Khetri belt of which the Dadikar Granite, Bairat Granite, Chapoli Granite, Gothra Granite and the Udaipurwati Granite are more important. They are generally granite–adamellite in composition and are generally granitoid in texture, although some are partially gneissose. There is controversy about whether at least some of these bodies are really intrusive into the supracrustals, or in fact constitute their basement. This particularly refers to the Dadikar Granite, Bairat Granite and the Udaipurwati Granite (Gangopadhyay, 1972; Chakrabarti and Gupta, 1990).

There are a few moderately well constrained geochronological data on the rocks of this belt. Zircon from the Manaksas felsic volcanics has been dated at 1832 ± 3 Ma (U–Pb) (Guerot, 1993). Granitic rocks from Dadikar, Bairat, Udaipurwati and Harsora have been dated in the range of 1700–1500 Ma (Rb–Sr/WR) (Crawford, 1970; Gopalan et al., 1979; Shastry, 1992). The U–Pb age of Anasagar Granite, Ajmer, is 1850 Ma (Mukhopadhyay et al., 2000). The model Pb–Pb age of syn-diagenetic Saladipura ores is 1780 Ma (Deb et al., 1989) and the same for Ghugra ores, Ajmer, also syn-diagenetic (see later) is 1787 Ma (Deb and Thorpe, 2004). However, the Rb–Sr (WR) isochron age of Jhunjhunu Granite Porphyry has been determined to be 805±18 Ma (Chaudhary et al., 1984).

Some people divide the Khetri belt into a North Khetri belt and a South Khetri belt with respect to the transverse Kantli fault that runs sub-parallel with the Kantli River. Recently, Gupta et al. (1998) and Gupta and Guha (1998) proposed some drastic changes in the geology of the Khetri belt. They think that the Khetri geology should not be discussed in the framework of the Delhi Supergroup alone. They choose to refer to it in terms of a cover sequence unconformably overlying another dominantly metasedimentary sequence. The cover sequence in the so-called north Khetri belt they call Khetri Group and that in the South Khetri belt, Shyamgarh Group. The Khetri Group and the Shyamgarh Group are thought Proterozoic (and belonging to the Delhi Supergroup?) and the basement rocks according to these authors, have striking similarity with the so-called Archean Mangalwar Complex down south. The main arguments in support of their conclusion are: (i) The first deformation of the cover rocks, in the northern block coincides with the second deformation in the local basement rocks (which have still two more deformational phases behind and the same in the southern block coincided with the third phase of deformation in the basement around, (ii) The basement metapelitic rocks have prismatic sillimanite whereas the cover rocks of the similar bulk composition have fibrolitic sillimanite, (iii) Apparent similarity of the basement rocks with the so-called Mangalwar Complex. However, the correlation of two deformational phases, except when regional and supported by radiometric age data, is generally fraught with uncertainty. Conclusion of a 'metamorphic hiatus' based on the crystal habits of sillimanite does not stand on a firm ground, as the development of stumpy or acicular grains of sillimanite, or for that matter any other mineral, would depend more on the fluid flow and fluid composition of the system and the substrate, rather than on pressure and temperature. The single zircon ($^{208}Pb/^{206}Pb$) age of Manaksas felsic volcanics fixes the minimum age of the sequence that contains it. For the basement rocks, we have no age data. However, simple physical similarity with the so-called Mangalwar Complex may not go far. The lower member of the Mangalwar Complex, commonly identified with the BGC II, has TTG as an important constituent. These rocks are not reported from the so-called basement rocks here.

The other members of the Mangalwar Complex are indeed sediment-rich, but no Archean age is ascertained from them (vide, discussion in an earlier part of this Chapter). May it be made clear here that the present authors have no bias against the presence of Pre-Delhi rocks in this region. It is possible, even probable, that they are there. However, their presence is to be established with convincing observations in the field and laboratory. We remember, Heron (1953) suggested the presence of Pre-Delhi, even Pre-Aravalli rocks here.

Bayana–Alwar (sub-) Basins These are located in northeastern Rajasthan and comprise three NE–SW trending adjacent subparallel riftogenic basins (subbasins). The basin fill is over 10 km thick and compositionally may be referred to the common Proterozoic association consisting of quartzite, pelites and carbonates (QPC of Condie, 1989). Heron (1917a), with the exception of some volcanic rocks near the bottom of the sequence, proposed a stratigraphic sequence for the rocks of this area as shown in Table 6.1.7a.

Table 6.1.7a *Stratigraphic sequence of the Delhi System (Supergroup) rocks of Northeastern Rajasthan (after Heron, 1917a)*

Ajabgarh Series (Several thousand feet thick)	Slates, phyllites, quartzitic sandstones and quartzites, impure limestones.
Hornstone breccia (Thickness variable)	
Kushalgarh limestone (1500 ft)	
Alwar Series (10,000–13,000 ft)	Quartzites, arkose, grits and conglomerates limestones and mica schists, volcanic rocks.
Raialo Series (2000 ft)	Limestone and quartzite
-------------------------------- *Unconformity* --------------------------------	
Pre-Delhi rocks	Mica schist, crystalline limestone, quartzite and schistose conglomerate.

Later workers have elaborated this scheme of classification, with small modifications here and there, which is presented in Table 6.1.7b. In this version, the Kushalgarh Limestone and Hornstone breccia are included in the Ajabgarh Group (neé Series). Roy (1988) prefers to call the Raialo Group rocks of this area as Rayanhalla Group to distinguish these rocks from the similarly called rocks, i.e. 'Raialo Series', elsewhere (Heron, 1953).

There is, however, some difference in lithology, structures and metamorphism between the Bayana (sub-) basin and Alwar (sub-) basin. The Bayana rocks are less deformed and less metamorphosed than the rocks of Alwar. The latter have been metamorphosed to amphibolite facies, whereas the former has been stuck up at the low green schist facies (Singh, 1988b). Detailed structural analyses in the parts of the Alwar (sub-) basin show that the rocks are multiply deformed and the F_1 and F_2 folds developed in more ductile regime than the F_3 (Roy and Patil, 1996; Das, 1988). Metamorphic history is also polyphase.

Granitoids in North Delhi Fold Belt There are numerous poorly exposed granitoid bodies throughout the area, some of which are recrystallised and foliated parallel to the most dominant schistosity in the country rocks, while others occur as small undeformed bodies showing post-tectonic nature with respect to the peak deformation fabric. Compositionally, the granitoids vary from composite bodies of grey-white tonalite and pinkish red syenite, to homogenous massive pinkish red syenite, and porphyritic grey tonalite (Knight et al., 2002).

Table 6.1.7 b *Stratigraphic sequence of rocks in the Delhi fold belt of northeastern Rajasthan (Banerjee, 1980; Singh, 1988 a, b; Roy, 1988)*

Group	Formation	Lithology
Ajabgarh Group	Acid intrusive rocks	Granite, pegmatite
	Basic intrusive rocks	Amphibolite and metadolerite
	Arauli–Mandhan Formation	Quartzite, staurolite garnet schist, carbon phyllite, slate
	Bharkol Formation	Quartz with interbedded phyllite, and carbon phyllite.
	Thana–Ghazi Formation	Carbon phyllite, tuffaceous phyllite, sericite schist, quartzite and marble.
	Seriska/ Weir Formation	Brecciated and ferruginous quartzite, chert and marble
	Kushalgarh Formation	Impure marble with lenses of phosphorite, basic flows, agglomerate tuff.
	-------------------------------- Unconformity	----------------------------------
Alwar Group	Pratapgarh Formation (= Bayana and Damdama Formations)	Quartzite, quartz–sericite schist and conglomerate.
	Kankwarhi Formation (=Badalgarh Formation)	Quartz–sericite schist, quartz schist and quartzite with thin lenses of marble and conglomerate.
	Rajgarh Formation (=Jogipura Formation	Quartzite, marble, gritty quartzite, conglomerate
	-------------------------------- Unconformity	----------------------------------
Rayanhalla Group (=Raialo Group of Heron)	Tehla Formation (=Jahaz–Govindpura volcanics)	Lava flow, pillow lava, agglomerate, pyroclastic breccia, conglomerate, quartzite, schist, phyllite, marble.
	Serrate Quartzite (=Nithar Formation)	Quartzite with wedge and lenses of oligomictic conglomerate.
	Dogeta Formation	Banded siliceous marble, quartzite, phyllite, and dolomitic marble with bands of conglomerate, quartzite, phyllite and schist.
	-------------------------------- Unconformity	----------------------------------
Pre-Delhi formations		

The major granitic plutons in the northern Khetri belt (north of Kantli River) are Gothra (Gotro) Granite and Jasrapura Granite, while those in the southern Khetri belt are Chapoli Granite, Udaipurwati Granite, Paniras Granite, Seoli Granite and Saladipura Granite. Several workers have attempted to discern the petrology and field relationship of some of the granitoids of Khetri as well as Alwar–Bayana belts with the country rocks of NDFB, and have correlated the emplacement phases with the various deformative episodes recorded therein (Heron, 1917, 1953; Chakraborti and Gupta, 1992; Gupta et al., 1998).

The pre- or syn-tectonic Gothra (Gotro) Granite, occurring near Khetri township, is characterised by strongly fractionated REE with strong enrichment in LREE, depletion in HREE, no Eu-anomaly, (Fig. 6.1.5a), enriched Th and HFSE, and high Na_2O content. Albitisation of primary K-feldspar is widely evidenced in this granite (Knight et al., op. cit.). Gupta et al. (1998) has classified it as mantle derived I-type granite.

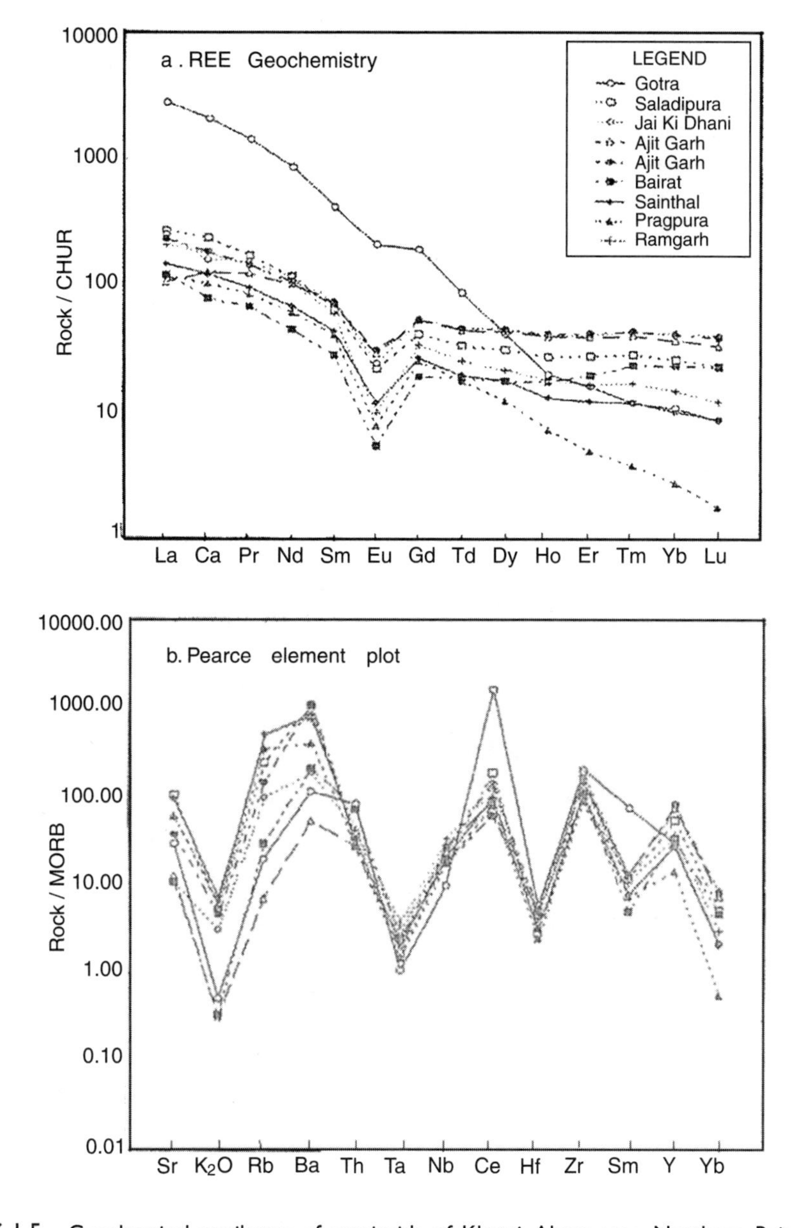

Fig. 6.1.5 Geochemical attributes of granitoids of Khetri–Alwar area, Northern Rajasthan: (a) Chondrite–normalised REE Pattern (b) MORB-normalised spider diagram (Knight et al., 2002)

The Jasrapura Granitoid (9 km x 1 km sheet like body) in the northern parts of KCB is strongly foliated and has been characterised as peraluminous, I-type, calc–alkaline granite with subordinate granodioritic component. There is a late phase of crosscutting equigranular leucogranite dykelets and veins (Kaur and Mehta, 2007). According to these workers, the calc–alkaline trend of evolution, high K_2O/Na_2O ratios, enriched incompatible elements (Rb, Ba, Th, U etc.), flat HREE profile, low

$(La/Yb)_N$ ratios and high Yb_N values, and the plots in various tectonic discriminant diagrams and multi-element spidergrams (Fig. 6.1.6 a–d) are suggestive of subduction related island arc setting for the Jasrapura Granitoid.

To the east and southeast of the Khetri belt proper, the entire terrain extending from Dhanuta–Chaukri to Alwar–Bayana–Jaipur there are exposures of several small granitic bodies. The Bairat Granite exposed in the SW of Alwar town is a composite body comprising an older syn-tectonic foliated variety and a younger post-tectonic variety of homophanous pink granite. The Dadikar Granite exposed in the NW of Alwar has intruded the Delhi metamorphites (GSI Mannual, 2004). Several other granitoid bodies occur in Alwar and Jaipur districts, the Ajitgarh Granite and Barodia Granite being important amongst them. The Ajitgarh pluton, according to Pandit et al. (1996), is an anorogenic trondhjemite–alkali granite body, which was emplaced in pulses.

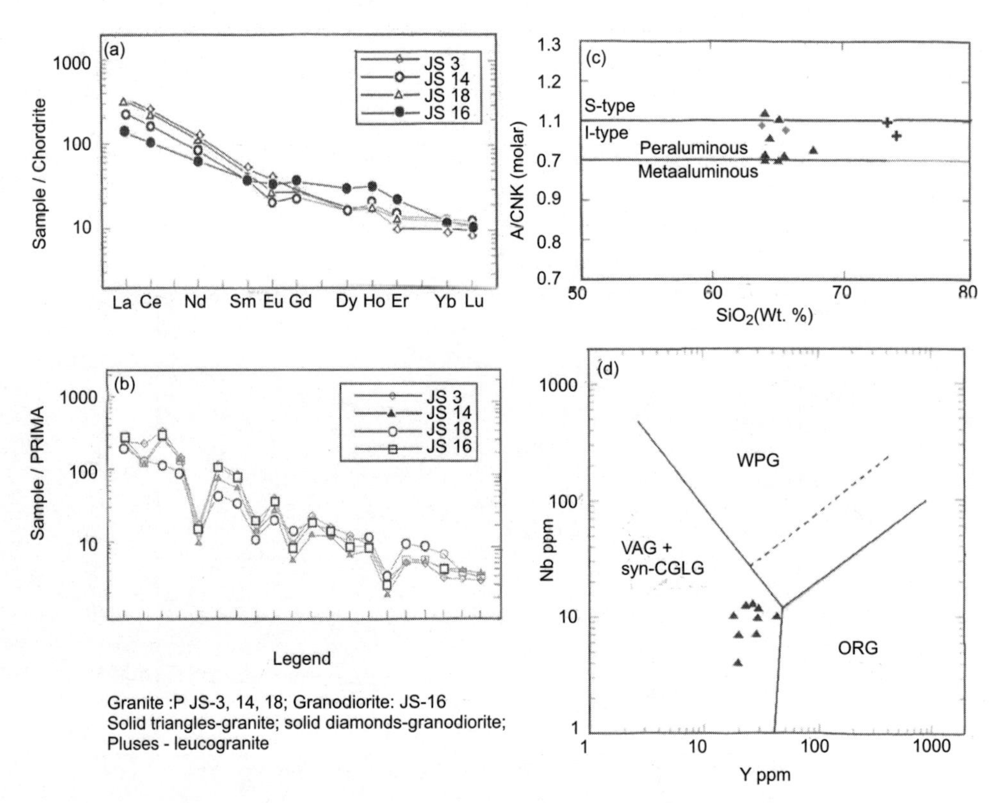

Fig. 6.1.6 (a) Chondrite normalised REE patterns of Jasrapura Granitoids, (b) Primitive mantle(PRIMA– normalised multi-element spider diagram of Jasrapura Granite and Granodiorite, (c) Molar A/CNK vs SiO_2 diagram(after Chappell and White, 1974), (d) Nb vs Y discriminant diagram (after Pearce et al., 1984) for Jasrapura Granitoids (from Kaur and Mehta, 2007)

Knight et al. (op. cit.) describes the post-tectonic granitoids in the North Delhi fold belt as alkali rich and meta-aluminous (A/CKN < 1.1) with a Na/K ratio ranging from < 1 to > 20. These are further characterised by uniform SiO_2 (69.5–77.4%), low Al and Ca, high Th (28.3–85.3 ppm) and moderate HFSE (Fig. 6.1.5b; Table 6.1.8). Moderately enriched LREE, moderately depleted HREE, pronounced negative Eu-anomaly and weakly fractionated REE pattern make this group distinctive from the pre-/syn-tectonic granitoids of this area. The mineralogical and geochemical features of the post-tectonic granitoids are suggestive of their affinity to A-type granitoids (Knight et al., op. cit.).

Table 6.1.8 *Chemical composition of the granitoids from the Khetri–Alwar region, northern Rajasthan (after Knight et al., 2002)*

	Gothra	Saladipura	Jat Ki Dhani	Ajitgarh (1)	Ajitgarh (2)	Bairat	Sainthal	Pragpura	Ramgarh
Major element oxides (%)									
SiO_2	70.80	69.52	76.58	77.43	76.15	76.58	73.80	73.80	72.51
TiO_2	0.28	0.58	0.50	0.45	0.45	0.07	0.28	0.27	0.50
Al_2O_3	13.51	13.09	10.58	11.94	12.07	12.01	14.19	13.15	11.75
Fe_2O_3	2.67	4.79	3.52	2.16	5.48	1.02	2.99	2.86	4.16
MnO	0.03	0.04	0.02	0.01	0.03	0.01	0.03	0.04	0.03
MgO	0.35	0.76	0.46	0.40	0.32	0.25	0.41	0.48	0.93
CaO	0.71	1.76	0.95	1.44	1.90	0.27	0.84	1.13	0.91
Na_2O	8.30	2.74	4.57	7.43	3.88	7.49	3.22	2.59	2.25
K_2O	0.52	5.64	3.18	0.31	4.78	0.34	7.37	6.10	5.96
P_2O_5	0.26	0.17	0.08	0.09	0.07	0.01	0.24	0.26	0.41
A/CKN	1.01	0.87	0.83	0.93	0.78	1.05	0.82	0.89	0.86
Na/K	14.33	0.43	1.28	21.19	0.73	19.86	0.39	0.38	0.34
Trace elements (ppm)									
Ba	117	1080	191	55	1110	213	837	405	763
Hf	6.4	3	5.8	5	5.8	5.2	3.4	2.6	4.4
Li	2.5	25	3	6.5	22.5	15	51	35	81.5
Nb	10.5	21	29.5	27	28	18.5	21.5	18.5	35.5
Rb	20.3	245	98.8	7.2	149	30	496	352	492
Sr	29.6	106	10.7	12.8	38.6	11.1	94.4	61	60.2
Ta	1.1	1.7	3.15	1.9	2.5	2.6	2.25	1.5	3.95
V	40	34	24	20	14	8	16	18	34
Y	32.7	59.8	80.5	82.4	91.6	36.3	30.4	15.9	41
Zr	219	110	199	160	175	116	114	94	155
U	8.8	2.35	6.35	3.85	6.2	17.5	16.3	4.65	10.5
Th	85.3	37.3	41	28.5	28.8	74	34.5	28.3	43.5
Metals and ore elements (ppm unless otherwise indicated)									
Au (ppb)	bld	bld	bld	bld	bld	bld	1	bld	bld
Ag	bld	bld	bld	bld	bld	bld	bld	bld	bld

(Continued)

(Continued)

	Gothra	Saladipura	Jat Ki Dhani	Ajitgarh (1)	Ajitgarh (2)	Bairat	Sainthal	Pragpura	Ramgarh
Bi	bld	bld	bld	0.1	0.8	bld	0.3	0.1	2.3
Cu	67	10	15	10	15	44	101	119	62
Mo	3	1.2	0.4	2.6	2.2	1.8	2	0.4	2.4
Ni	8	8	4	4	6	6	42	8	10
Pb	10	26	3	8	40	8	53	45	27
Pt(ppb)	bld	bld	bld	bld	bld	bld	bld	bld	bld
Pd(ppb)	bld	bld	bld	bld	bld	bld	bld	bld	bld
S	100	60	40	40	140	60	120	220	120
Sb	bld	bld	bld	bld	bld	bld	bld	bld	bld
W	1	1	bdl	0.5	1	1	1.5	1.5	3.5
Zn	13	47	11	22	23	11	66	55	36
Rare earth elements (ppb)									
La	870	86.1	79.9	33.5	74.1	38.4	46.3	37.9	66.1
Ce	1680	196	130	104	151	65.4	101	84.9	146
Pr	167	20	18	14.6	17.1	8.02	11.3	10.1	17.1
Nd	523	71.6	71	62.5	66.1	27.6	42	37.3	63.5
Sm	81.6	13	14.6	14	14.5	5.6	8.6	8.05	11.7
Eu	15.5	1.64	1.82	2.16	2.32	0.4	0.88	0.58	0.72
Gd	51.6	11.2	14.6	14.4	14.6	5.2	7.2	6.8	9.2
Tb	4.48	1.72	2.36	2.28	2.36	0.94	1	0.9	1.3
Dy	14.2	10.7	15.5	14.9	15.3	6	6.1	4.2	7.45
Ho	1.54	2.12	3.08	3.06	3.2	1.34	1	0.56	1.38
Er	3.65	6.3	9.3	8.9	9.6	4.45	2.75	1.1	3.75
Tm	0.38	0.92	1.38	1.3	1.4	0.76	0.38	0.12	0.54
Yb	2.4	5.85	9.45	8.25	9.2	5.25	2.25	0.6	3.3
Lu	0.3	0.8	1.4	1.16	1.34	0.78	0.3	0.06	0.42
(La/Lu)n	294.6	10.9	5.8	2.9	5.6	5.0	15.7	64.2	16.0
(La/Yb)n	239.4	9.7	5.6	2.7	5.3	4.8	13.6	41.7	13.2
Eu*	0.7	0.4	0.4	0.5	0.5	0.2	0.3	0.2	0.2
La/Sm	6.5	4.0	3.3	1.5	3.1	4.2	3.3	2.9	3.4
Gd/Yb	17.3	1.5	1.2	1.4	1.3	0.8	2.6	9.1	2.2
Total REE	3415.7	428.0	372.4	285.0	382.1	170.1	231.1	193.2	332.5

Analytical methods: Combination of ICP–MS and ICP–OES techniques. Bld = Below detection limit

The available age data of the granitoids from NDFB are as follows:

1. Gothra (Gotro): 1691± 4 Ma [SHRIMP U–Pb zircon] – Knight et al., 2002
   Granite: ~1700 Ma [^{208}Pb–^{206}Pb zircon] – BRGM, 1993 in Gupta et al., 1998
2. Jasrapura Granite: 1844 ± 7 Ma [Rb/Sr WR] – BRGM, 1993 in Gupta et al., 1998
3. Chapoli Granite: 1313 ± 270 Ma [Rb/Sr WR] – Chaudhary et al., 1984
4. Saladipura Granitoid: 1480 ± 40 Ma [Rb/Sr WR] – Gopalan et al., 1979
   Biotite separates: 700 Ma [Rb/Sr min.] – Gopalan et al., 1979
5. Seoli Granite: 1494 ± 206 Ma [Rb/Sr WR] – Chaudhary, et al., 1984
6. Udaipurwati Granite: 1780 Ma [Pb–Pb model] – Deb et al., 2001; 1480 ± 40 Ma [Rb/Sr WR] – Gopalan et al., (1979)
7. Bairat Granite: 1660 Ma [Rb/Sr WR model age] – Crawford, 1970
8. Ajitgarh Granite:
   Trondhjemite – 1725 Ma [EPMA zircon] – Biju Sekhar et al., 2002
   Alkali granite – 1719–1741 Ma [do]
9. Haton (SW of Kho–Dariba): 1876 Ma [U–Pb zircon] – Deb et al., 2001
10. Granite (West of Khetri): 820 ±18 [Rb/Sr WR] – Deb et al., 2001
11. Chula Granite, Alwar: 745 ±29 Ma [Rb/Sr WR] – Pandey et al., 1995.

The age data of the syn-sedimentary felsic volcanics of the North Delhi fold belt, which are considered as coeval to an early phase of granite emplacement in this belt (Gupta et al., 1998 and references therein), are as follows:

1. Hathori felsic tuff, Bayana, yielded an age range of 1878–1862 Ma (U–Pb/zircon) with a peak at 1876 Ma (Deb and Thorpe, 2004).
2. Felsic volcanics near Udaipurwati, southern KCB, yielded 1832 ± 3 Ma (Pb^{208}–Pb^{206}/single crystal zircon) age (Unpublished BRGM report, 1993 quoted by Gupta et al., 1998).
3. Manaksas felsic volcanics, South Khetri belt produced an age of 1832 ± 3 Ma (U–Pb zircon) (Guerot, 1993).

Knight et al. (2002) correlated the post-tectonic granitoids of NDFB with the Erinpura Granite and other younger granitoids of SDFB, which were dated at 0.75–0.85 Ga (U–Pb zircon) by Deb et al., 2001. A close scrutiny of the above data and those presented in Fig. 6.1.9 and Table 6.1.13, does not subscribe to such generalisation based on meagre data. However, the petrochemical and structural data available on the granitoids are suggestive of at least two generations of granitic emplacement in the region. Evidently, more geochronological data would be needed for age wise discrimination of the various petrogenetic types of granitoids studied so far in the NDFB.

Albitite Zone, Northern Rajasthan Extensive albitisation along the eastern parts of Khetri belt was recorded by Dasgupta (1968), followed by similar reports from different areas of NDFB and BGC by several later workers (Ray, 1990, 2004; Sinha et al., 2000). A major disjunctive tectonic structure associated with quartzo–feldspathic rocks, extending in NNE–SSW direction for about 170 km from Sikar district in northern Rajasthan to southwest of Jaipur, was identified as 'Albitite zone or Albitite line' by Ray (op. cit.). There are three sectors with well-exposed albitite rock in this zone from north to south, viz, The Maonda–Sior, Khandela–Guhala and Sakhun–Ladera sectors (Fig. 6.1.7). There is a distinct possibility of further extension of this zone in the NNW into Haryana and to the further south of Sakhun–Ladera area. Besides, there are reports of albitite bearing sheared rocks from nearby areas outside the Albitite Line, many of which are described as products of soda–metasomatism (Sinha et al., op. cit.) in variance to the intrusive character inferred by Ray (op. cit.) for the rock in the aforesaid sectors. This may warrant a closer look to ascertain the actual character of the rock.

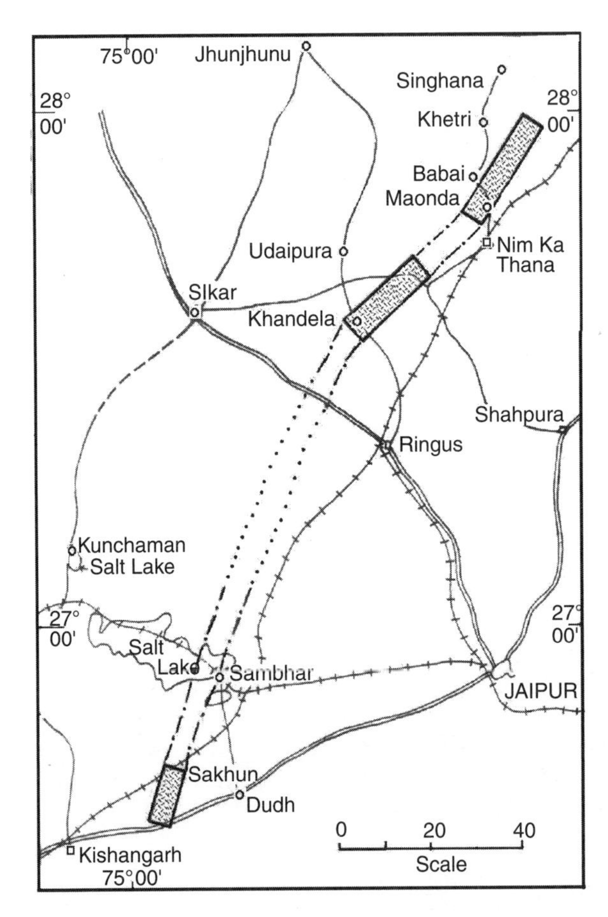

Fig. 6.1.7 Map showing the Albitite zone in northern Rajasthan (chain and dotted line): 1. Maonda–Sior sector 2. Khandela–Guhala sector 3. Sakun–Ladera sector (modified after Ray, 1990)

The principal mineral of albitite rock is fine-grained red coloured aventurine variety of albite. The rock is deep pink to brick red in colour with 80–90% modal and normative content of albite. The albitites are mainly composed of randomly oriented laths of aventurine albite of uniform grain size (0.75–1 mm), showing clouded character (due to minute iron oxide dust), complex undulose extinction and very rare twining. The accessory minerals include calcite, fluorite, quartz, apatite, chlorite, epidote, allunite, corundum and variable proportions of opaques. Sphene is abundant and widely distributed. Magnetite and ilmenite are also locally abundant and occur as discontinuous layers and bands. Coarse-grained mafic cumulates (pyroxenite), occurring as dismembered lenticular bodies are common constituents along the zone. The magnetite–ilmenite accumulations also occurs as large segregated pockets within the pyroxenite bodies (Ray, 1990). The albitite is variously associated with concentrations of calcite, magnetite, apatite, fluorite, uraninite, chalcopyrite, molybdenite, brannerite, ilmeno–magnetite and bismuthinite.

The model of deep fracture-controlled emplacement of the albitites in intra-cratonic rift environment (Ray op. cit.), alkaline affinity of the adjacent volcaniplutonic rocks and their association with fluorite–magnetite–ilmenite–Cu sulphide–uraninite mineralisation may be suggestive of a phase of IOCG type metallogeny in this terrain (Knight et al., 2002). This aspect is discussed in greater detail in the succeeding part of this chapter. It would be important to examine whether the wide spread soda–metasomatism reported from this terrain is related to the late granitic activity.

South Delhi Fold Belt The South Delhi Fold Belt (SDFB) is part of the Delhi fold belt that lies south of Ajmer. With it coincides Heron's 'Main Delhi synclinorium', which, however, the later works did not confirm. It is now also referred to as the Main Delhi basin. It may be mentioned at the beginning, that the basin and its evolution deserved a much greater attention than it received so far. In the pages that follow, we briefly review the status of the geology of the basin, rather the SDFB, at the present time.

Metamorphosed clastic sediments, both coarse (quartzite, arkose and conglomerate) and fine (pelites), are the principal constituents of the SDFB as we know it, except in its western part. Particularly remarkable is a conglomeratic (polymictic) horizon along the western margin of the basin, extending for about 40 km on either side of Barr (26°55′: 74°20′) and is commonly referred to as the 'Barr Conglomerate'. The pebbles are generally highly deformed, (meta-) arkose forms the matrix and more importantly, some of the pebbles are granite–granite gneiss in composition, suggesting pre-existence of a granite-dominant country to the west. Another important conglomerate horizon occurs, but this time along the eastern margin of the basin, north of Bithur. This of course is beyond the limits of the Main Delhi basin or the South Delhi fold belt as defined here.

The clastic sediments gradually pass upward to calcareous sediments, now represented by impure dolomitic marble and calc silicate rocks with the deepening of the basin, or due to a marine transgression. Undoubted volcanic rocks in the sequence are conspicuous by their absence in the eastern part.

An elongate body of gneissic rocks, apparently remnant of a basement horst, longitudinally divides the basin into two sub-basins (cf. Roy and Jakhar, 2002). There is another similar body just west of Ajmer. The eastern sub-basin is called Bhim–Shyamgarh basin and the western one, 'Senda–Barotia' or 'Birantiya' basin. The lithologic build-up of the two sub-basins is sharply contrasted. While metamorphosed sandstone–shale–carbonate assemblage represents the lithology of the eastern sub-basin, a dominant proportion of volcanic rocks characterises the Sendra–Barotia basin. The basic volcanic rocks dominate over the felsic volcanics (Bhattacharjee et al., 1988; Tiwary and Deb, 1997). Besides, low-K tholeiite, pillowed dacitic andesite, banded meta-gabbro and meta-pyroxenite, all variously metamorphosed (Bhattacharya and Mukherjee, 1984; Patwardhan and Oka, 1988), occur along a 100 km long narrow belt passing through Basantgarh (24°44 : 76°00′) Some authors, think that these rocks represent an ophiolite assemblage (Sinha Roy and Mohanty, 1988; Sugden et al., 1990).

As we have already mentioned, there are quite a few contrasted views on the geology of the South Delhi Fold Belt (SDFB). Heron (1953) had no difficulty in identifying the Alwar and Ajabgarh Groups in the SDB also. Gupta et al. (1997) accepted Heron's view in general terms and named the SDFB equivalent of the Alwar Group as the Gogunda Group and the SDFB equivalent of the Ajabgarh group as Kumbhalgar Group. Coulson (1933) had considered both the highly deformed and poorly deformed rocks of the Sirohi region to belong to the 'Aravalli Series'. Roy and Sharma (1999) in the recent time studied the geology of this region in some details and concluded that the rocks here belonged to two stratigraphic sequences separated by a distinct unconformity. They assigned the more deformed older unit to the Sirohi Group, which according to them was a post-Delhi Supergroup. The younger unmetamorphosed and the little deformed sequence was called Sindreth Group by them. These molasse-like shallow water sediments are also known as the Punagarh Group. Gupta et al. (1997) included the Sirohi Group and Punagarh (Sindreth) Group under the Delhi Supergroup. This is included in the GSI's classification of the Delhi Supergroup

Table 6.1.9 *Chemical composition of Barotiya and Sendra volcanics (Bose et al., 1990)*

	Barotiya volcanics									Sendra volcanics		
	Ultramafic			Mafic			Felsic					
SiO_2	43.57	44.50	38.00	47.06	48.09	50.00	77.00	76.12	80.12	58.44	55.35	51.73
Al_2O_3	10.39	11.00	20.00	16.10	15.50	14.50	12.00	11.08	11.01	13.00	12.86	16.32
TiO_2	1.38	1.20	2.10	1.30	1.20	1.20	0.40	0.20	0.20	0.96	0.69	2.63
Fe_2O_3	3.61	14.00	16.20	11.80	12.50	11.50	2.10	1.90	1.50	0.61	0.60	2.71
FeO	11.88	–	–	–	–	–	–	–	–	7.92	3.60	9.17
CaO	2.90	8.50	5.50	12.00	12.00	11.20	0.45	0.10	0.30	10.82	11.48	6.94
MgO	22.26	15.90	15.30	6.40	7.00	6.30	0.25	0.22	0.11	6.37	7.56	4.02
MnO	0.23	0.40	0.10	2.90	2.40	1.50	4.60	4.52	4.70	0.43	0.76	3.31
Na_2O	0.15	0.03	0.02	0.30	0.25	0.40	2.52	2.85	2.00	0.26	3.60	2.00
K_2O	0.13	ND	ND	ND	ND	ND	ND	ND	ND	0.28	0.19	0.45
P_2O_5	0.22	0.19	0.21	0.18	0.18	0.18	0.01	0.01	0.01	0.26	0.06	0.21
H_2O	3.37	3.21	2.44	ND	ND	ND	ND	ND	ND	ND	ND	0.56
Total	101.04	98.93	99.87	98.04	99.12	96.78	99.33	97.00	99.92	99.35	96.75	100.00
Ni	1000	220	250	80	40	40	30	10	10	20	10	20
Co	150	20	30	50	10	20	10	10	10	20	20	30
Cr	1000	500	400	80	120	200	10	10	10	20	20	30
Zr	30	30	30	30	20	120	300	350	200	–	–	–

proposed in 2001 (Table 6.1.10). Gupta et al. (1995) divided the Delhi Supergroup rocks in the SDFP particularly in the central Aravalli Ranges, into Barotiya, Sendra, Rajgarh and Bhim Groups. The Barotiya Group is characteristically composed of a thick mafic felsic volcanic sequence underlain by and interlayered with terrigenous clastic and carbonate sediments. The Sendra Group is composed of calc–gneisses, marble, pelitic schists, metamorphosed wacke, mafic rocks and conglomerate. The major rock type in the Rajgarh Group is quartz arenite, with subordinate proportions of pelitic schists and calc–gneisses. A basement made of granite–gneiss and high-grade metapelites underlies the Bhim Group, which represents a platformal association of metamorphosed quartzite–carbonate–shale. The Rajgarh Group in contrast is dominantly arenaceous. The latter is thought to be younger than the Bhim Group (Gupta et al., 1991). There is no direct way to establish the age relationship between the sequences on either side of the Pre-Delhi inlier.

Geochemistry of the mafic volcanic rocks of Basantgarh–Ajari and Deri–Ambaji (Deb and Sarkar, 1990) deserve a mention here. The Basantgarh–Ajari mafic volcanic rocks are low-K tholeiite to calc–alkaline basalt, with highly variable Fe ($Fe_2O_{3\,T}$ = 3.8–16.7 wt.%) and less variable Mg (MgO = 4.9–10.4 wt.%). Strong enrichment in some LIL elements (Rb, Ba, Th) compared with HFS elements (Nb, Zr, Ti) is noted. Low $\sum$REE content ($\leq$ 15 ppm), no or little LREE enrichment (Ce_N/Yb_N: 0.97–1.24) and distinct positive Eu-anomaly are other characteristics. The Basantgarh–Ajari basic volcanic rocks resemble basalts from modern island arcs in chemical composition.

Deri–Ambaji mafic volcanics are low-K tholeiite to picrite in bulk composition. They are further characterised by a medium abundance of ∑REE (57.3–97.9 ppm), poor REE fractionation and a faint Eu-anomaly. These chemical characters partly resemble those of mafic volcanic rocks from island arc environment and partly those of ocean floor basalts (back-arc environment?)

Table 6.1.10 *Classification of Delhi Supergroup (after GSI,2001)*

	South-western Rajasthan & North-eastern Gujarat		Ajmer Sector	North-eastern Rajasthan	
Post-Delhi intrusives	Malani Igneous Suite (volcanic & plutonic) Erinpura Granite Godhra Granite				
Delhi Supergroup	Bambolani Group	Sojat Punagarh Khambal SowaniaFms	Sindreth Group (Angor and Goyali Fms.)		
	Sirohi Group	Jiyapura Reodar Ambeshwar Khiwandi Fms.			
	Sendra–Ambaji Granite and Gneiss		Kishengarh Syenite	Dadikar, Bairath, Ajitgarh, Sikar and Chapoli Granites	
	Phulad Ophiolite Suite				
	Kumbhalgarh Group	Todgarh, Bewar, Katra Sendra, Ras, Barr, Basantgarh, Kalakot Fms.	Ajabgarh Group (Ajmer Fm.)	Ajabgarh Group	Kushalgarh, Sariska, Thanagazi, Bhakrol, Arauli Fms.
	Gonguda Group	Richer, Antalia, Kelwara Fms.	Alwar Group (Srinagar and Naulakha Fms.)	Alwar Group	Rajgarh, Kankarhi, Pratapgrah, Nithar, Badalgarhy, Biana Fms.
				Raialo Group	Dogeta and Tehla Fms.

Fm(s) = Formation(s)

Structures and Metamorphism of the SDFB Rocks South Delhi fold belt rocks, like many other Precambrian fold belts are multiply deformed and poly-metamorphosed.

Structures There are, as is common in the Proterozoic mobile belts, generally three sets of folds (rarely four) with/without well-developed axial plane cleavage/schistosity and lineations.

Gupta et al. (1995) have recognised four sets of folds and associated shears in the rocks of their Barotiya, Sendra and Rajgarh Groups. The earliest folds are rootless, flattened isoclinal and inclined to reclined in disposition. The second-generation folds are tight, isoclinal, and large in dimensions such that they control the map pattern. The third phase is open to tight, and is associated with, longitudinal shears. The fourth or the final phase of folding produced transverse warps with variable axial plunge. Tight major folds with variable plunge in the Barotiya Group have also been recognised as of second generation by Mukhopadhyay and Bhattacharyya (2000). Dextral shape of these folds, as well as shapes and dispositions of some linear structures such as pebbles within conglomerates were subscribed to subhorizontal dextral shear by these authors.

Mukhopadhyay and Bhattacharyya (op. cit.) also reported that the first folds in the Sendra Formation were isoclinal small-scale folds. Gangopadhyay and Mukhopadhyay (1987) mapped large-scale second-generation folds from Sendra area. They also reported syn-/post-tectonically emplaced diapiric granitoid intrusive bodies that affected the tectonic structures in their neighbourhood.

In the Bhim Group, however, there are three phases of co-axial deformation. The first generation structures are represented by isoclinal folds of both large and small scales. The second-generation folds (F_2) are also tight (and sub-vertical) with north to northeasterly axial trace. The fourth phase folds (F_4) in the Bhim Group are steeply plunging and have E–W to EWE–WSW trending axial planes (Mukhopadhyay and Bhatttacharyya, 2000).

It bears mention here that Gupta et al. (1995) and Gupta and Bose (2000) stated that the central Aravalli Range consisted of multiple tectonic units. Almost the same opinion i.e. the central Rajasthan is an ensemble of different tectono–stratigraphic units delimited by zones of intense movement, has been expressed by Mukhopadhyay and Bhattacharyya (op. cit.). These in fact confirm an earlier conclusion of Sen (1980, 1981, and 1983) that the SDFB was divisible into five lithologically, structurally and tectonically distinct units, 1–5 (Fig. 6.1.8). Zones 1,2, 3 and 4 of Sen correspond respectively to the classification into Barotia, Sendra, Rajgarh and Bhim Groups by Gupta et al. (1995). Zone 5, according to Sen, was constituted of pre-Delhi rocks.

Metamorphism The South Delhi fold belt is generally characterised by a moderately high-temperature intermediate-pressure type metamorphism (amphibolite facies), with local occurrence of green schist metamorphism and development of granulite facies metamorphism. The P–T regime changed character with time, such that the hornblende hornfels facies metamorphism could be superimposed on amphibolite facies metamorphism (cf. Desai et al., 1978; Deb, 1980). Locally, as at Balaram and Mawal in north Gujarat, granulites of various protolithic compositions (pelitic, calc–silicate, mafic/ultramafic) have developed. The granulite facies metamorphism took place at 900°–1000°C /7–8 kb (Sharma, 1988). However, at places the granulites have retrograded to the rocks of amphibolite facies (Desai et al., 1978).

The main metamorphism with fabric development is suggested to have taken place during the first deformation (D_1), the second deformation (D_2), or both (Sharma, 1988). This is understandable, since the build-up of stress in a rock and thermal peaking in it during dynamothermal metamorphism do not progress at the same rate. Deformation is a much faster process and volume-specific, compared to thermal perturbation.

High-pressure metamorphism, particularly in the Phulad belt, is reported by Sinha Roy (1988), but contested by Munshi (1993) and Sharma (1995). Such metamorphic assemblages are unlikely to survive in the area that has been the site of intense granitic activities, during or immediately after the orogeny. Moreover, P–T–t trajectories for blue schists suggest that there may be increase in temperature before uplift, resulting in the conversion of blue schist facies assemblages to amphibolite, or even greenschist facies assemblages.

The granulites occurring in Balaram and Mawal area, Gujarat, are associated with mafic intrusive rocks such as gabbro and norite. Biswal et al. (1998a) suggest that the mafic granulites, together with gabbro–norite are intrusive into the meta-sedimentary granulites and granite gneisses.

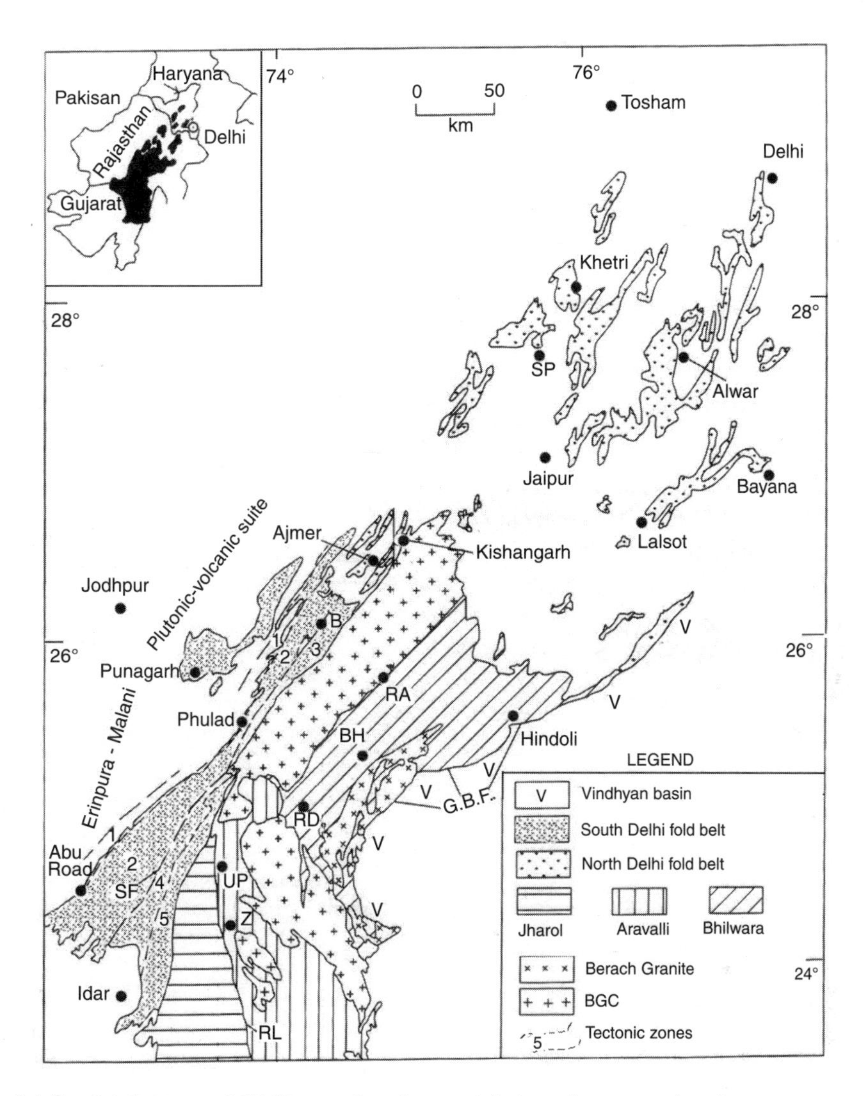

Fig. 6.1.8 Subdivisions of SDFB into five distinct lithological, structural and tectonic zones as proposed by Sen (1980).

Intrusive Rocks in the South Delhi Fold Belt Intrusive rocks in the SDFB vary in composition from felsic through intermediate (diorite) and mafic to ultramafic rocks. The felsic rocks are generally granitoids, varying in composition from granite to tonalite and fall both in 'I' and 'S' fields. The most important ones amongst them are discussed below.

Important granitoid bodies occur mainly in the Barotiya and Sendra Groups. Some such intrusive bodies are Sewariya Granite, Sendra Granite, Erinpura Granite and Degana Granite.

Sewariya Granite The Sewariya Granite is an elongate body, occurring close to the western margin of the belt. It is a massive two–feldspar granite with mica, tourmaline and opaques. W-bearing veins within this granite, occurring in the Pali district have raised interest in this granite body from exploration point of view. The average chemical composition of Sewariya Granite is presented in (Table 6.1.11).

Table 6.1.11 *Average chemical composition of the Balda and Sewariya Granites (Krishnamurthi and Prabhakar, 2000)*

Major and trace elements	Balda Granite	Sewariya Granite
SiO_2	73.44	71.60
TiO_2	0.20	0.25
Al_2O_3	15.80	13.70
Fe_2O_3	0.09	2.05
FeO	0.64	1.88
MnO	0.02	0.04
MgO	0.03	0.58
CaO	0.10	1.23
Na_2O	3.16	2.99
K_2O	5.14	4.94
P_2O_5	0.16	0.14
W	255	125
Li	267	140
Rb	411	278
Ba	41	245
Sn	40	15
B	130	50

Note: Major elements (oxides) in percent and trace elements in ppm.

Sendra Granites The Sendra Granites vary in composition from granite through granodiorite to tonalite (Table 6.1.12). Gangopadhyay and Mukhopadhyay (1987) believe that some of the Sendra Granite bodies were diapirically emplaced. Tobisch et al. (1994) on the strength of $^{87}Sr/^{86}Sr$ ratios, concluded that the Sendra Granites are neither juvenile in origin, nor are the products of crustal melting derived from the BGC by anatexis. Instead, the protoliths were low Rb/Sr Mesoproterozoic crust (Delhi Supergroup) with contributions from BGC. Another possibility is the mixing of crustal anatectic melts with mafic magmas.

Table 6.1.12 *Chemical composition of the granites, granodiorites and tonalites of the Sendra region (Agarwal and Srivastava, 1997)*

Major and trace elements	Granite (n = 20)	Granodiorite (n = 46)	Tonalite (n = 19)
SiO_2	73.18	75.27	74.49
TiO_2	0.13	0.12	0.14
Al_2O_3	14.42	13.49	13.95
Fe_2O_3	0.97	0.77	1.22
FeO	1.67	0.91	1.74
MnO	0.07	0.03	0.05
MgO	0.15	0.13	0.13

(Continued)

(Continued)

Major and trace elements	Granite (n = 20)	Granodiorite (n = 46)	Tonalite (n = 19)
CaO	1.12	0.99	2.35
Na_2O	2.61	4.18	4.19
K_2O	4.57	3.54	1.41
P_2O_5	0.07	0.04	0.06
Rb	245	159	45
Sr	47	25	120
Ba	627	497	412
Li	47	45	24
Ni	8	9	8
Cr	5	7	7
Co	6	6	7
Cu	6	5	6
Zn	69	68	71
K_2O/ Na_2O	1.76	0.85	0.34
ASI	1.28	1.08	1.11

Note: Major elements in wt.% oxide and trace elements in ppm. ASI: Alumina Saturation Index.

Erinpura Granites The term 'Erinpura Granites' is well-entrenched in the geology of Rajasthan since the days of Heron (1953) or even earlier. Heron described all the intrusive-looking granites occurring within the Delhi fold belt as Erinpura Granites and considered them to be generally post-Delhi in origin. As no radiometric age data on these rocks were available in those days, their contemporaneity or otherwise could not be ascertained. Results of the study of Chaudhary et al. (1984) threw some light on this aspect. It showed that the so-called Erinpura Granites were in fact divisible into two age groups: 1.7–1.5 Ga (Ajmer and north) and 0.85–0.75 Ga (South of Ajmer).

The name Erinpura Granite is now restricted to a large batholith, occupying Erinpura and the neighbouring areas in the district of Sirohi, Gujarat. Mount Abu is a part of it. It is a polyphase granitic–massif, which varies from porphyritic granites to porphyritic gneisses. The porphyritic variety is a two–feldspar granite with biotite and a little muscovite and the accessories like epidote, sphene, apatite, zircon and opaques. The porphyritic variety yielded a Rb/Sr (WR) age of 815 ± 30 (Chaudhary et al., 1984).

Balda Granite The inclusion of Balda Granite under the Erinpura Granite is now questioned. This W-mineralisation-bearing granitic pluton is now suggested as belonging to the Malani suite (Bhattacharjee et al., 1984; Chattopadhyay et al., 1982). Its relatively young Rb–Sr/WR age of 763 ± 22 Ma (Sarkar et al., 1992), lends support to this view. The average chemical composition of Balda Granite is presented Table 6.1.11.

Degana Granite Degana Granite, Nagaur district, Rajasthan is another Neoproterozoic (740Ma, Rb–Sr/WR) alkali granite intrusion that varies from medium to coarse-grained, porphyritic to aporphyritic types. They are strongly peraluminous, enriched in F, Rb and Li and depleted in Sr. W-mineralisation associated with the Degana Granite is discussed in a later section.

Tenability of a N–S Divide of the Delhi Fold Belt Sinha Roy (1984) first proposed a division of the rocks of the Delhi fold belt (DFB) into an older North Delhi fold belt (NDFB) and a younger South Delhi fold belt (SDFB) based on the limited geochronological data of Chaudhary et al. (1984). Some later workers also (Bose, 1989; Gupta et al., 1991) accepted this diachronous development of the Delhi fold belt. Diachronous development of the Delhi fold belt is suggested also by Deb et al. (2001), but with a difference. Roy (1988), Roy and Jakhar (2002), on the other hand believe in a single stage evolution of the Main Delhi basin. However, their Main Delhi basin also includes the Ajmer–Khetri belt in its northward extension and they suggest that the northern formations unconformably overlie the southern ones across the Bithur–Pisangan join.

We have collated the available geochronologic information of rocks available from both the northern and southern segments of the DFB in Table 6.1.13. The data pertain to felsic volcanic and plutonic rocks and syngenetic/diagenetic sulphide deposits. Herein we find that the two felsic volcanic rocks, one each from Bayana and South Khetri show U–Pb (zircon) ages of 1832 ± 3 Ma and 1876 Ma (peak) respectively. In the absence of any information on the inherited nature, if any, of the zircon grains, we assume them to be primary only. Anasagar Granite and Jasrapura Granite are both older than 1800 Ma. Pb–Pb model age obtained from the syngenetic/diagenetic sulphidic ores of Saladipura and Ghugra yielded ages ground 1800 Ma. These consistent age data from the rocks and ores of NDFB cannot be ignored. Neither can we ignore 1000–800 Ma age range of felsic volcanics and granitoids emplaced early-/late-/,even post-kinematically in the southwestern part of the Delhi fold belt. Some of these granitoids are granodiorite–tonalite in composition, though they are not entirely of juvenile origin. It will be more reasonable to suggest that the SW part of the SDFB, extending from Birantiya–Sendra in the north till its southern end and limited approximately by a zone sub-parallel with the Sabarmati Fault in the east and including the Sirohi Group and even the Punagarh Group in the west, define a younger orogen (cf. Deb et al., 2001). If on the other hand we believe that the history of origin and evolution of the Delhi fold belt is a single stage affair, then we have to imagine that the belt was dynamically alive without a break throughout a period of about one billion years. It will be a very extraordinary situation. As pointed out by Condie (1989), most orogenies lasted about 50 Ma during the last 3 Ga. It is possible, even probable that the rocks in zones 3, 4 and 5 of Sen (op. cit.), i.e. the Gogunda and Kumbhalgarh Groups are co-eval with the rocks in the Khetri–Ajmer belt in the north. Unfortunately we do not have any reliable age data on these rocks. We have no age data on the rocks of the Sirohi Group either. There are some young ages also for the North Delhi Fold Belt rocks, such as for the Jhunjhunu Granite Porphyry and the Tosham Granite and Rhyolite. It is possible that the Tosham Granite and Rhyolite are related to the Malani suite. Otherwise their emplacement has to be explained by tectonic re-activation during orogeny in the south. Summing up, the history of the origin and the evolution of the Delhi fold belt does not appear to be a single stage affair. We do not have at the moment enough data to suggest that the SDFB is made of only Neoproterozoic rocks. On the other hand, some geochronological data are available to suggest that the NFDB was reactivated in the Neoproterozoic time.

Table 6.1.13 *Radiometric age data from the rocks and ores of the Delhi fold belt*

Rock/mineral analysed	Systematics used	Age (Ma) obtained	Reference
Ajmer and North			
Manaksas felsic volcanics, South Khetri	U–Pb/zircon	1832±3	Guerot (1993)
Hathori felsic tuff, Bayana	U–Pb/zircon	1978–1862 (Peak 1876)	Deb and Thorpe (2004)

(Continued)

(Continued)

Rock/mineral analysed	Systematics used	Age (Ma) obtained	Reference
Galena from Saladipura deposit, Sikar	Model Pb–Pb	1780	Deb et al. (1989)
Jasrapura Granite, West of Khetri	Rb–Sr/WR	1844 ± 7	Gupta et al. (1998)
Anasagar Granite, Ajmer	U–Pb/zircon	1850	Mukhopadhyay et al. (2000)
Granites from Khetri –Saladipura belt (Udaipurwati, Harsan, Dadikar and Bairat)	Rb–Sr/WR	1700–1500	Crawford (1970), Gopalan et al. (1979), Shastry (1992)
Galena from Ghugra ores, Ajmer	Model Pb–Pb	1787	Deb and Thorpe (2004)
Gotro Granite, Khetri	$^{208}Pb/^{206}Pb$	1700	Gupta et al. (1998)
Chapoli Granite, Khetri	Rb–Sr/WR	1313 ± 270	Choudhary et al. (1984)
Pilwa–Chinawali magmatic charnockite	Single zircon evaporation	1434	Fareeduddin and Kronar (1998)
Chula Granite, Alwar	Rb–Sr/WR	745 ± 29	Pandey et al. (1995)
Tosham Granite	Ar–Ar/WR	818 ± 36	Murao et al. (2000)
Tosham Granite	Rb–Sr/WR	732 ± 41	Kocher et al. (1985)
Tosham Rhyolite	Ar–Ar/WR	793 ± 18	Murao et al. (2000)
Tosham Rhyolite	Rb–Sr/Wr	770 ± 20	Kocher et al. (1985)
Khanak Granite	Ar–Ar/ WR	803.6 ± 3.4	Murao et al. (2000)
South of Ajmer			
Deri Rhyolite	U–Pb / zircon	987±6.4	Deb et al. (2001)
Birantya Khurd Rhyolite	U–Pb/ zircon	986.3±2.4	Deb et al. (2001)
Sendra Granite	Rb–Sr/WR	~835	Chowdhary et al. (1984)
Pali Granite	Rb–Sr/WR	815±30	Chowdhary et al. (1984)
Jalore Granite	Rb–Sr/WR	723±6	Dhar et al. (1996)
Erinpura Granite	Rb–Sr/WR	815±30	Chaudhary et al. (1984)
Sadri–Ranakpur Granite	Rb–Sr/WR	~835	Chaudhary et al. (1984)
Sai Granite	Rb–Sr/WR	~835	Chaudhary et al. (1984)
Belka Pahar Granite	Rb–Sr/WR	825±25	Chakranarayan et al. (1986)
Balda Granite	Rb–Sr/WR	763±22	Sarkar et al. (1992)
Kui–Chitrasani Granite	Rb–Sr/ WR	740±11	Pandey et al. (1995)
Ranakpur Diorite	Sm–Nd/WR	1012±78	Volpe and Macdougall (1990)

At this stage it may be worthwhile to synthesise the geochronological data of the Rajasthan area, available until date and to plot them on a generalised geological map. The authors have tried to do so by compiling a few different maps of Deb and Thorpe (2004) and the data therein (Fig. 6.1.9, Table 6.1.14).

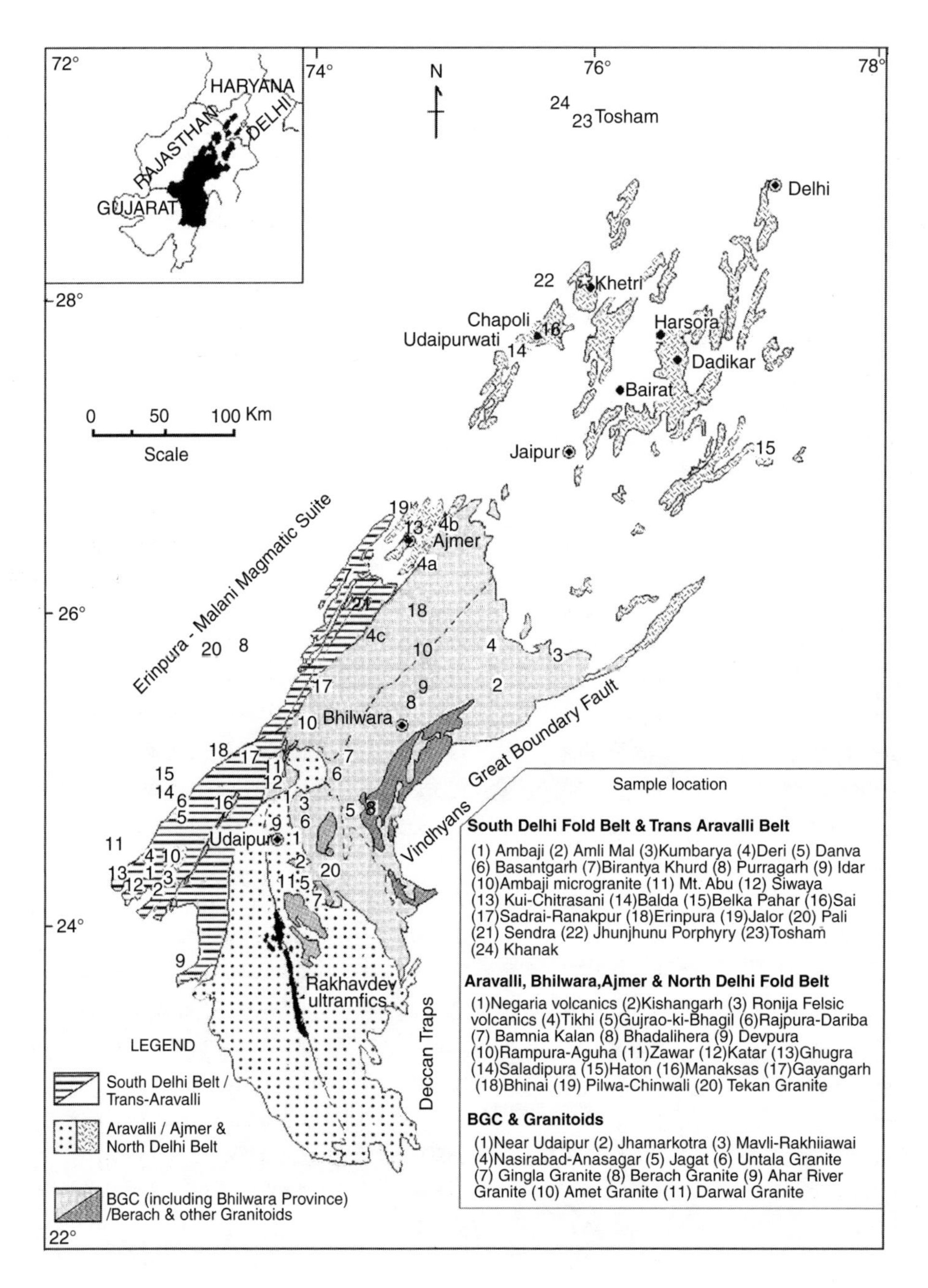

Fig. 6.1.9 Map showing locations of geochronological data available so far from different belts of Western India (Synthesised and Modified after Deb et al., 2001)

Data source: BGC and granitoids: (1) Macdougall et al., 1983; Vinogradov et al., 1964, (2) Gopalan et al., 1990; Wiedenbeck, and Goswami, 1994; Roy and Kroner,1996; Wiedenbeck et al., 1996, (3) Gopalan et al., 1990, (4) Tobisch et al., 1994, (5) Roy and Kroner ,1996, (6) Choudhary et al., 1984, (7) Roy and Kroner,1996, (8) Sivaraman, 1982; Choudhary et al., 1984; Wiedenbeck et al., 1996, (9) Guha and Garkhal, 1993; Goswami et al., 1994, (10) and (11) Choudhary et al.,1984.

Table 6.1.14 *Geochronological data from different belts of Rajasthan*

BGC & granitoids	Aravalli-Jhalor-Bhilwara, Ajmer -North Delhi fold belt	South Delhi fold belt & Trans-Aravalli rocks
11. ⬡ 1900 ± 80 Ma	20. ⬡ 1554 ± 50 Ma	24. ■ 817.1 ± 2.2 Ma
10. ⬡ 1870 ± 200 Ma	19. ⧅ 1128 ± 0.7 Ma 1434 ± 0.6 Ma(Ch.Gr)	23. ☆ 818 ± 3.6 Ma
9. ⧅ 2560 Ma (min) ⬡ 2026 ± 54 Ma	18. ⧅ 1621 ± 0.5 Ma. ⧅ 1641 ± 14 Ma ⧅ 1675 ± 0.6 Ma	22. 820 ± 18 Ma 21. ~ 835 Ma 20. 815 ± 30 Ma
✜ 2440 ± 8 Ma 8. ⬡ 2533 Ma □ 2610 ± 50 Ma	17. ■ 1723 + 14/–7 Ma ⧅ 1692 ± 0.5 Ma	19. 723 ± 6 Ma 18. ⬡ 825 ± 30 Ma 17. ~ 835 Ma
7. ⧅ 2620 ± 5 Ma	16. ■ 1832 ± 3 Ma	16. ~ 835 Ma
6. ✜ 2505 ± 3 Ma ⧅ 2669 ± 6 Ma ⬡ 2950 ±150 Ma	15 ■ ~ 1857 Ma 14. 1780 Ma	15. 825 ± 25 Ma 14. 763 ± 22 Ma 13. 740 ± 11 Ma
	13. 1787 Ma 12. 1800 Ma	12. ■ 836 +7/–5 Ma 11. ■ ~ 780 Ma
5. ⧅ 2658 ± 5 Ma 2887 ± 5 Ma	11. ▲ 1694-1812 Ma 10. 1804-1812 Ma	10. ⬡ 745 ± 50 Ma 9. ⬡ 721 ± 5 Ma
4c. △ 2.8 Ga 4b. △ 2.8 Ga □ 1849 ± 8 Ma 4a. 2.8 Ga △	9. 1802 Ma 8. 1813 Ma 7. 1792 Ma	8. ▲ ~ 940 Ma 7. ■ 986 Ma 6. 5. ▲ 990 Ma
3. △ 2828 46 Ma	6. ▲ 1799 Ma ■ 1808-1931 Ma	4. ■ 987 ± 6 Ma
✜ 2532 ± 5 Ma ✜ 3230 ± 5 Ma 2. ✜ 3281 ± 3 Ma △ 3307 ± 65 Ma	5. ▲ 1709 Ma 4. ▲ 2030 Ma 1735 (remobil.) 3. ■ 1854 Ma	3. 2. ▲ ~ 990 Ma 1.
1. ♦ □ ~ 3.5 Ga	2. ▲ 2024 Ma 1. ▲ 2075-2150 Ma	

Index of dating methods

♦ Nd model Age	▲ Pb-Pb model age	□ Discordant U-Pb zircon age
△ Sm-Nd isochron	■ U-Pb zircon age	✜ Ion microprobe zircon
⧅ Single zircon evaporation	☆ ^{39}Ar - ^{40}Ar	⬡ Rb- Sr Whole rock isochron

Data Source:

BGC & granitoids: (1) Macdougall et al., 1983; Vinogradov, et al., 1964 (2) Gopalan et al., 1990; Wiedenbeck, and Goswami, 1994; Roy and Kroner, 1996; Wiedenbeck et al., 1996; (3) Gopalan et al., 1990; (4) Tobisch et al., 1994 (5) Roy and Kroner, 1996; (6) Choudhary et al., 1984 (7) Roy and Kroner, 1996 (8) Sivaraman 1982; Choudhary et al., 1984; Wiedenbeck et al., 1996; (9) Guha and Garkhal, 1993; Goswami et al., 1994 (10) & (11) Choudhary et al., 1984.
Aravalli-Jhalor-Bhilwara belt: (1-15) Deb et al., 2001; (16) Guerot, 1993(17) Fareeduddin and Kroner, 1988 (18) Rathore, 2001; Wiedenbeck et al., 1996 (19) Fareeduddin and Kroner, 1988 (20) Sarkar et al., 1992.
Delhi fold belt and Trans Aravalli Belt: Deb et al., 2001.

Tectonic Models for the Origin and Evolution of Delhi–Aravalli Mobile Belt(s)

The most divided opinion on the geology of the Delhi–Aravalli mobile belt(s) centres on the choice of a suitable tectonic model to explain it. The first division is into Plate tectonists and non-Plate tectonists and then there are obvious differences of views on the details within each of these two groups: The salient aspects of the suggested models are outlined below.

Plate Tectonics Related Views Sen (1981, 1983) as a result of his prolonged study on the structure-tectonics of the Delhi (–Aravalli) fold belt, came to the conclusion that the DFB in central Rajasthan consisted of several longitudinal belts which have discordant contacts between them. He numbered them 1–5 (Fig. 6.1.8). Sen proposed that the initial development of his Zone-2 and Zone-3 took place in an intracratonic rift basin that evolved into an arc–trench system. Sedimentation according to him took place in an eastern marginal sea (Zone-5 being considered by him as pre-Delhi), in the arc–trench zone (Zone-3) and in a western shallow sea (Zone-1). Subduction directed towards the east led to the overriding of the island arc by the oceanic crust of Zone-2. Ultimately all the three (1–3) Zones were accreted against the continent in the east (Fig. 6.1.10).

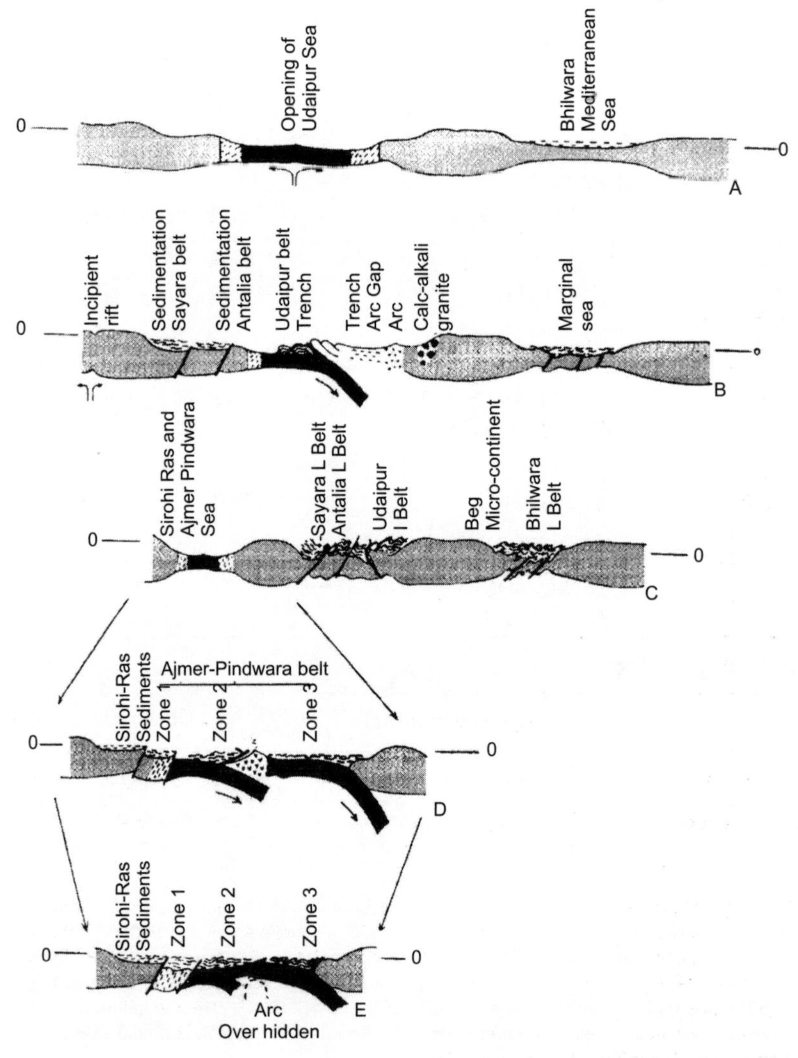

Fig. 6.1.10 Plate tectonic model of the Precambrian rocks of Rajasthan (after Sen, 1981)

Shychanthavang and Mehr (1984) also supported a plate tectonic model to explain the deformation, metamorphism and igneous activities in the belt. They suggested westerly subduction of the eastern plate not once but twice.

Sinha Roy (2004) suggested two stages of westerly subduction resulting in collision between an arc sequence in the west and a trench sequence in the east and linked the tectonic model to metallogeny (Fig. 6.1.11).

Radhakrishna (1989) not only thought that plate tectonics was involved in some form or other in the origin and the evolution of the Delhi–Aravalli belt, but also believed that it is a striking and picturesque example of the Wilson Cycle of separation and subsequent collision between two continental blocks – the Bundelkhand block to the east and the West Rajasthan (Marwar) block to the west. Qureshy and Iqbaluddin (1992) share almost the same opinion.

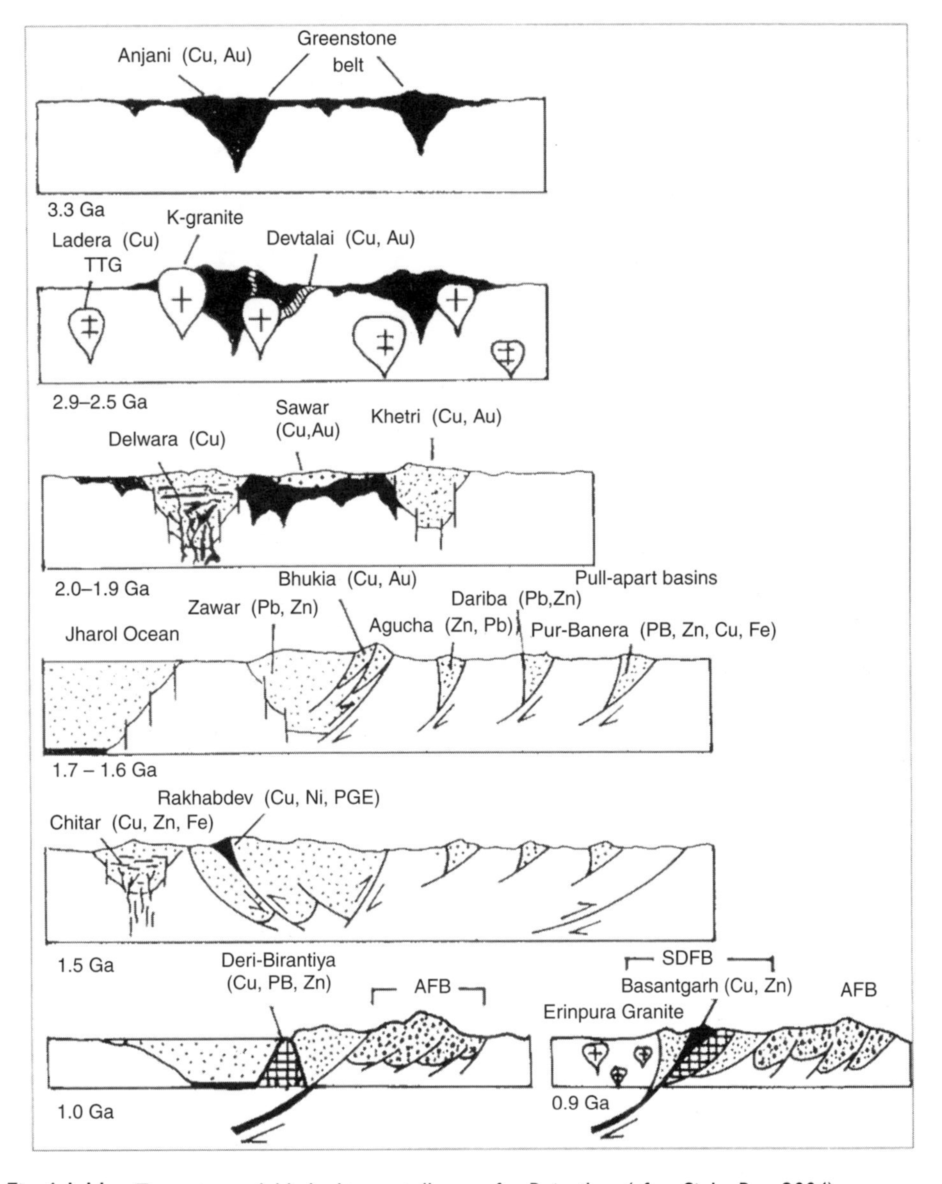

Fig. 6.1.11 Tectonic model linked to metallogeny for Rajasthan (after Sinha Roy, 2004)

The models suggested by Sugden et al. (1990) and Deb and Sarkar (1990) are nearly the same. Their model includes the origin of the Bhilwara belt in addition to that of the Delhi–Aravalli belt, as they think that these two belts are tectonically related. The Bhilwara belt was an intracratonic rift, rather a 'failed rift' or aulacogen. Another rifting-effect of the same thermal plume was the development of the Rakhavdev lineament. Aravalli–Jharol sediments were laid in a sub-basin on the passive margin close to the BGC country in the east, apparently through the period 2.0–1.7 Ga. A complete separation of the protocontinent took place along a line parallel to the present Rakhabdev lineament at around 1.8 Ga. Westward subduction of the oceanic crust around 1.7 underneath an attenuated block of the separated continent to the west, resulted in the formation of the Jharol accretionary prism and an incipient arc. Eventual closure of this ocean led to the collision of the Aravalli continental margin with this arc-microcontinent around 1.5 Ga and eastward obduction of the oceanic crust along the Rakhadew lineament, which also marked the shelf-rise boundary. Later, a marginal sea developed along the western periphery and the mafic crust of this marginal sea began subducting below the early arc and ultimately led to a terminal collision at ~1.0 Ga (Fig. 6.1.12).

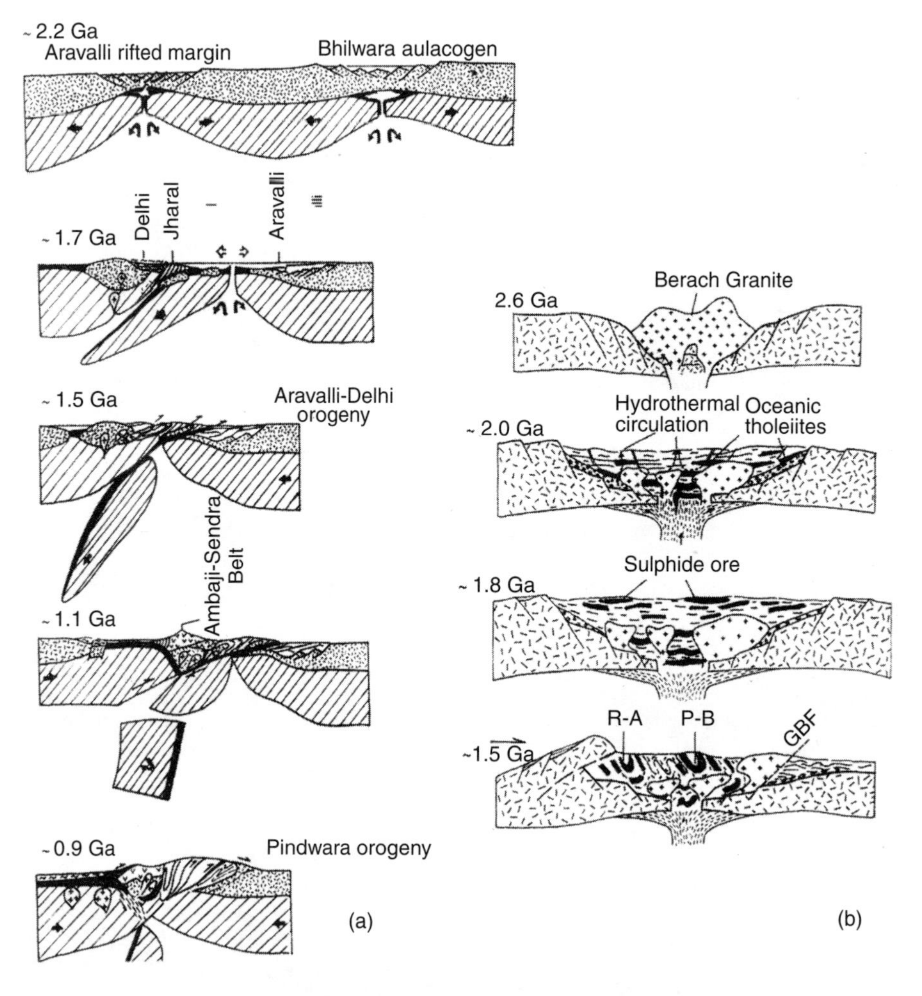

Fig. 6.1.12 Plate Tectonic model for (a) Aravalli–Delhi fold belt, Rajasthan (after Deb and Sarkar, 1990) (b) Bhilwara belt, Rajasthan (after Deb, 1993).

Volpe and Macdougall (1990) who studied the Phulad–Barr–Jetgarh (PBJ) and the Ranakpur–Desuri (RD) belts concluded that the chemical data of the rock complexes of these two subbelts suggested fragments of Proterozoic oceanic crust and island arc complexes respectively. It is relevant here to note that the so-called 'Phulad ophiolite suite' (Gupta et al., 1980) occurs discontinuously over a stretch of about 500 km, from near Ajmer to Khed Brahma and Kui in Gujarat. Some rock units, such as serpentinised ultramafic rocks, mafic volcanics and meta-wackes are no doubt present, but dismembered. Evidence for pillowed meta- basalts near Phulad is not unequivocal and sheeted dike or layered gabbroic complexes, the other characteristics of an ophiolite suite have not been identified (Volpe and Macdougall, 1990). The ultramafic rocks in an ophiolite suit are lherzolite and harzburgite. Their presence is not reported. As mentioned already, Munshi (1993) and Sharma (1995) have contested the presence of blue schist along this zone reported by Sinha Roy (1988). Indeed, it is less likely to survive in the given area of prolific late/ post- tectonic granitic activity even if it was ever there (cf. Deb and Sarkar, 1990).

Biswal et al. (1998b) based on the geochemistry of a plutonic complex of mafic intrusive rocks, consisting of gabbro, norite and basic granulites, in the Balaram–Mawal area in Gujarat, concluded that the complex signified subduction and continental collision in the region during upper Proterozoic.

Views Non-supportive of Plate Tectonics Roy (1988) and Roy and Jakhar (2002) suggested a single stage orogenic evolution of the entire belt. The nature of the sedimentary package, volcanicity and isotope signatures of intrusive bodies and the metallogenic character in the main Delhi basin attest to an intracratonic setting, according to these authors. The Main Delhi basin according to them was an aulacogen. However, they concede that the occurrence of low-K pillowed tholeiite and chemical characteristics of the basaltic rocks corresponding to trench or ocean floor type chemistry (Bhattacharyya and Mukherjee, 1984; Volpe and Macdougall, 1990), suggest the possibility of the limited opening of the ocean along the floor of the aulacogen that constituted the Main Delhi basin.

Sharma (1995) proposed ensialic orogenesis in his model (Fig. 6.1.13), which briefly stated, will be as follows: the continental crust of Rajasthan after stabilisation at ~2.58 Ga became stressed, apparently due to the introduction of a thermal plume at ~ 2000 Ma. Continued stretching led to the deepening of the basin, accumulation of a thick pile of sediments, particularly the Jharol Group. At one stage, the thin crust was torn apart to let in mafic/ultramafic lavas of tholeiite–komatiite composition.

The mantle lithosphere was disrupted and the eastern unit got decoupled and moved westward, i.e. A-subduction. This caused sagging in the superjacent crust; Delhi rift basins formed and filled up with volcanisedimentary piles (c 1800–1700 Ma). A-subduction allowed hot mantle material to underplate the overlying crust. This produced the granulite rocks as those of the Sand Mata Complex. Melts from an asthenospheric mantle source produced norite dykes and the granodiorite–charnockite rocks. Compression continued leading to the folding of not only the volcani-sedimentary piles in the Delhi basins, but also refolding of the Aravalli sediments and the older basement. At the final stage of this compressional event, ductile shearing, accompanied by upper level transfer of granulites now exposed within the BGC and in the Delhi Fold Belt, took place.

Malani Group

There is a lot of confusing nomenclature appearing in literature about this Group of rocks. Some of these are Malani Igneous Suite, Malani Igneous Complex, Malani Volcanic Suite, Malani Volcanic Series, and Malani Rhyolites. We prefer the name Malani Group and define it as a distinct litho–stratigraphic unit for an association of rocks that occur as an unconformity bounded sequence of

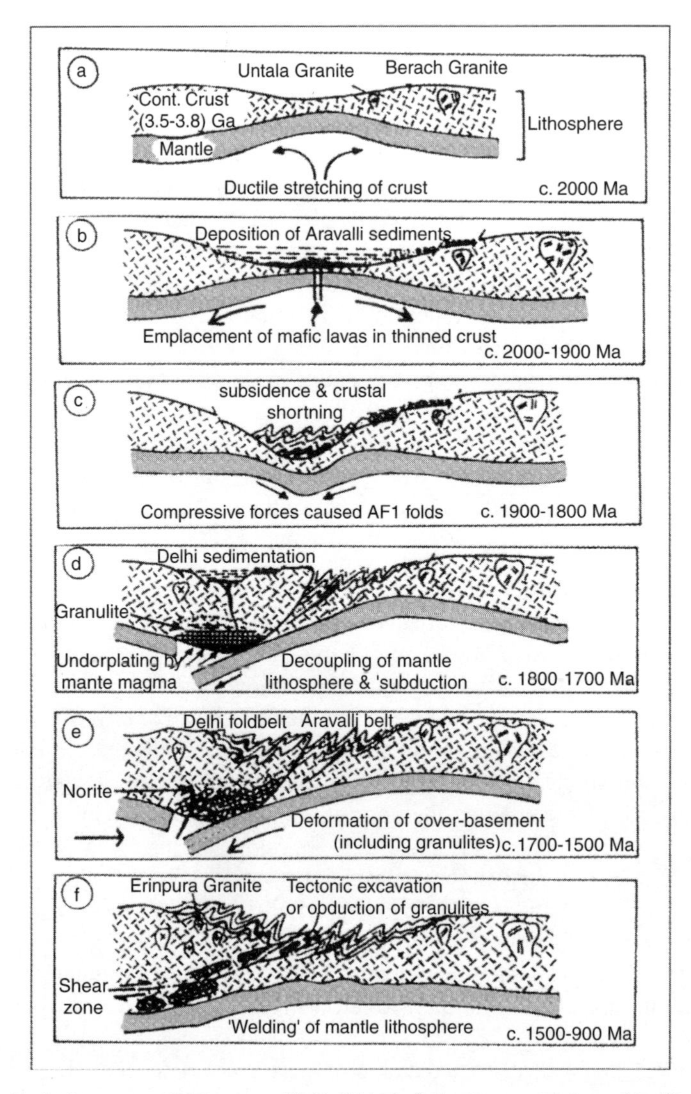

Fig. 6.1.13 Evolutionary model for Aravalli–Delhi belt, Rajasthan, as envisaged by Sharma (1995)

dominantly bimodal volcanics with minor interbedded sediments and peralkaline meta-aluminous to peraluminous granites. Most of the major outcrops of this Group occur in the Barmer, Jodhpur, Pali and Sirohi areas. These rocks also occur at places like Tosham (28°22′:75°35′) in Hariyana, and as per the reports of Basu (1982) and Srivastava (1988), in the Churu and Jhunjhunu districts of northern Rajasthan.

The Malani Group consists of the following lithological types:

1. Lava flows
2. Tuffs and pyroclastic materials
3. Granitoids
4. Sedimentary rocks.

The volcanogenic rocks, i.e. the lava-flows and the tuffs and pyroclastic materials together overwhelm the surficial exposures of the group. The lava flows are compositionally bi-modal: felsic (mostly rhyolitic) and mafic. Bhusan reported the presence of trachyandesite amongst the volcanic rocks. Felsic volcanics >> mafic volcanics. No imprint of metamorphism has been noticed so far. Basalt occurs either as basal lava flows, or interlayered with rhyolite. Tuffs and pyroclastics are generally associated with felsic lava flows. Bhusan (2000) reported 1km thick lapilli tuff sandwiched between two felsic flows near Jodhpur. The SiO_2-content in Malani Rhyolites may be up to 75.52%, iron (total) up to 7.11% and K_2O up to 9.38%. Commonly these are peraluminous, but locally they can be peralkaline also (Table 6.1.15). Both types of rhyolites show enrichment in LREE, a strong Eu anomaly and a flat HREE pattern (Bhusan, 1985, 2000). According to Srivastava (1988), some of the peralkaline rhyolites of the Malani Group are in fact comendite and pantellerite. He holds that the low CaO content and high Na_2O/K_2O and high (Na +K)/Ca ratios suggest these volcanics to be 'A-type'.

Table 6.1.15 *Average chemical composition of some Malani Group rocks (Bhusan, 2000)*

Major oxides (wt.%) and trace elements (ppm)	Basalt (n = 19)	Trachyte (n = 20)	Rhyolite (Peraluminous) (n = 13)	Rhyolite (Peralkaline) (n = 55)	Siwana Granite (n = 13)	Jalore Granite (n = 12)	Malani Granite (n = 3)	Rhyolitic tuff
SiO_2	47.96 (2.97)	62.83 (1.95)	71.52 (2.05)	71.70 (0.96)	73.06 (1.99)	72.70 (2.05)	72.31 (2.57)	71.75 (2.35)
TiO_2	2.93 (1.21)	0.89 (0.23)	0.70 (0.15)	0.56 (0.10)	0.38 (0.05)	0.22 (0.13)	0.50 (0.12)	0.38 (0.23)
Al_2O_3	15.05 (1.30)	13.03 (1.10)	10.07 (1.42)	9.54 (0.91)	10.23 (0.74)	13.54 (0.59)	10.02 (1.06)	13.76 (2.59)
FeOT	13.54 (2.14)	9.42 (1.28)	7.23 (1.33)	7.87 (0.67)	6.34 (0.76)	3.01 (1.11)	8.30 (1.19)	3.55 (2.81)
K_2O+Na_2O	5.70 (2.54)	9.64 (0.96)	6.14 (1.35)	7.14 (1.57)	8.57 (0.77)	7.82 (0.88)	5.72 (1.67)	6.84 (1.22)
MgO	5.52 (1.11)	0.42 (0.17)	1.26 (0.90)	0.78 (0.13)	0.51 (0.19)	0.27 (0.15)	0.13 (0.05)	0.83 (0.65)
CaO	7.29 (2.13)	1.24 (0.40)	0.33 (0.18)	0.65 (0.18)	0.48 (0.26)	1.22 (0.71)	2.01 (0.87)	1.18 (0.91)
La	18.5 (9.75)	214.66 (108)	182.69 (17.64)	497.00 (36.77)	230 (94.05)	75 (8)	NA	NA
Ce	45.75 (24.52)	484.33 (264.18)	182.66 (17.67)	1047.50 (53.03)	539.75 (217.53)	164 (25.4)	–	–
Nd	28.25 (13.72)	247.66 (150.10)	94.67 (9.80	544.50 (62.39)	110.88 (66.54)	90 (15.12)	–	–
Sm	7.9 (3.2)	49.00 (20.66)	25.33 (4.16)	136.50 (12.02)	63.94 (26.47)	70 (12.20)	–	–
Eu	2.63 (0.96)	6.20 (2.40)	3.90 (0.10)	11.22 (1.06)	6.61 (2.39)	1.70 (0.21)	–	–
Tb	1.23 (0.43)	8.73 (5.75)	3.63 (0.45)	29.00 (4.24)	11.72 (5.20)	1.70 (0.12)	–	–

(Continued)

(Continued)

Major oxides (wt.%) and trace elements (ppm)	Basalt (n = 19)	Trachyte (n = 20)	Rhyolite (Peral-uminous) (n = 13)	Rhyolite (Peral-kaline) (n = 55)	Siwana Granite (n = 13)	Jalore Granite (n = 12)	Malani Granite (n = 3)	Rhyolitic tuff
Yb	3.93 (1.54)	34 (23.06)	14.33 (2.52)	127.50 (13.44)	38.25 (16.07)	9.10 (1.37)	–	–
Lu	0.52 (0.23)	4.47 (3.26)	2.17 (0.35)	15.00 (2.83)	4.82 (1.98)	2.20 (0.18)	–	–
Th	2.5 (0.23)	34.33 (31.21)	9.20 (0.75)	401.00 (7.22)	29.73 (13.13	25.00 (7.15)	–	–
Sc	37.75 (9.7)	503 (3.26)	8.7 (1.01)	1.12 (0.97)	2.64 (3.69)	8.90 (1.69)	–	–
Co	45.03 (28.20)	1.32 (0.72)	2.23 (0.32)	1.03 (0.53)	1.06 (1.57)	3.50 (0.87)	–	–
Cr	25 (11.02)	10.67 (4.04)	11.03 (1.45)	28.00 (4.15)	4.67 (2.53)	2.60 (0.55)	–	–
Hf	6.10 (2.37)	60 (31.20)	29.00 (3.46)	315.00 (22.03)	72.31 (38.25)	12.00 (3.13)	–	–
Ta	0.81 (0.45)	93 (10.97)	21 (0.44)	33.16 (4.14)	9.45 (5.30)	1.40 (0.24)	–	–

Granitic members of the Malani Group include Siwana Granite, Jalor Granite, Degana Granite and Balda Granite. The Siwana Granite is a discontinuous elliptical ring-dyke covering about 300 sq km in area. Average compositions of Siwana Granites are as follows: SiO_2 69–73.87%, alkalies 7.54–11.35%, Al_2O_3 7.74–12.98%, total iron 6.48–9.81% and low in MgO (0.10%) and CaO (< 1.0%). Maheswari et al. (2001) described the Siwana Granites as essentially peralkaline, 'Within plate' and 'A-type' granites based on the results of detailed geochemical studies.

Jalor Granites also are characterised by high SiO_2, $Na_2O + K_2O$, Fe/Mg, Zr, Ga, Y, U and Th. They are commonly peraluminous. Srivastava et al. (1988) and Kochhar and Dhar, (1993) characterised these granites as 'Within plate' and 'A-type' granites.

The petrology and geochemistry of the ore-bearing granitoids of Degana, Tosham and Balda are briefly discussed in the later section.

The sequence of igneous activities involving the Malani Group is as follows (Bhusan, 2000):

Phase I: Eruption of mafic flows followed by the felsic flows, ending up with ash flows.
Phase II: Emplacement of granitic plutons, ring dykes etc.
Phase III: Emplacement of felsic to mafic dykes.

The Malani volcanics overlie the metamorphic tectonites of the Sirohi Group (cf. Sharma, 1996; Bhusan, 1985). Rb–Sr isochron ages obtained for the Malani volcanic–plutonic igneous rocks range from 680–750 Ma (Crawford and Compston, 1970; Dhar et al., 1996). Sediments in the Malani Group are subordinate to the igneous rocks in volume. However, their occurrence is none the less interesting from a geological point of view. At Sindreth, the volcanisedimentary sequence starts with a conglomerate horizon at the base. Well-bedded arkosic rocks overlie volcanic rocks at Sindreth and terrigenous sediments interlayered with lava flows are reported from a number of places. Sedimentary depositional structures are present in some of them (Bhusan, 2000).

Marwar Supergroup

The Marwar Supergroup is a platformal succession without metamorphism and tangible deformation, occurring in the Marwar region of western Rajasthan, to the west of the Aravalli Mountain Range. Its apparent resemblance to the Vindhyan Supergroup led some early workers (Heron, 1936; Pascoe, 1959) to call this sequence the Trans-Aravalli Vindhyans. The > 1 km thick Marwar Supergroup rests unconformably over the rocks of the Malani Group, the Sirohi Group and the Delhi Supergroup. Interestingly, recent studies on the sediments and fossils (stromatolites) from this Supergroup suggested its closer affinity, even connection, with the Tethyan sediments of the Salt Range, now in Pakistan (Awasthi and Prakash, 1981; Barman, 1987). The Marwar Supergroup occurs in two separate basins: the Nagaur basin and the Birmania basin. The Nagaur basin is also called the Main Marwar basin. The Marwar Supergroup is composed mainly of coarse clastic sediments (pebbles, cobbles, sandstone) and carbonates, including both limestone and dolostone. Fine sediments (shale, siltstone) are relatively less abundant. Pareek (1984) suggested the following stratigraphic sequence for the Marwar Supergroup in the Nagaur basin (Tables 6.1.16 and 6.1.17).

Table 6.1.16 *Stratigraphic succession of the Marwar Supergroup in the Nagaur basin (after Pareekh, 1984)*

Marwar Supergroup	Nagaur Group (75–500 m)	Tunklian Sandstone Nagaur Sandstone
	Bilara Group (100–300 m)	Pondolo Dolomite Gotan Limestone Dhanapa Dolomite
	Jodhpur Group (125–240 m)	Girbhakar Sandstone Sonia Sandstone Pokhran boulder bed
Basement rocks		

The Birmania basin, nearly oval in shape and covering an area of about 100 sq km, occurs at a distance of about 125 km SW of the Nagaur or the Main Marwar basin. Presence of a phosphorite bed in the sequence is of special interest. Muktinath (1969) suggested the following stratigraphic sequence in the Birmania basin:

Table 6.1.17 *Stratigraphic sequence of the Marwar Supergroup in the Birmania basin (Muktinath, 1969)*

Lathi Formation (Jurassic)		
-----------------------	--	
Marwar Supergroup	Birmania Formation	Dolomite to cherty limestone, shale and phosphorite
	Randha Formation	Sandstone and shale.

Basement		

The Henseran evaporiate group (= Bilara Group) outcropping in the northern part of the basin sub-conformably overlies the Jodhpur Group. It contains cyclic deposits of halite, i.e. K-minerals, alternating with clay, dolomite anhydrite and magnesite. K-minerals comprise polyhalite, sylvinite and sylvite with langbeinite and carnallite. Up to seven halite cycles have been identified. Evaporites in the same stratigraphic horizon (Bilara Group) contain gypsum/ anhydrite together with limestone, marl and chert. In fact, Nagaur gypsum is well-known in the mineral industry.

The Great Boundary Fault: Its Nature, Origin and Evolution

The Great Boundary Fault (GBF), an important feature in the geology of the northwestern part of the Indian Shield, broadly separates the Bhilwara Province of Rajasthan from the Vindhyan basin that occurs to its east (Fig. 6.1.1). It extends for more than 300 km from south of Chittaurgarh in the southwest to Machilpur in the northeast (Fig. 6.1.14). Recent geophysical studies suggest that it continues below the Gangetic basin-fill (Tiwari, 1995). The principal controversy centres on whether it is a post-depositional fault (Heron, 1953; Sinha Roy, 1985), or a rejuvenated pre-Vindhyan regional dislocation (Ramaswamy, 1995; Verma, 1996). In fact, Heron (1953) believed that the Vindhyans were much extensive beyond the GBF one day.

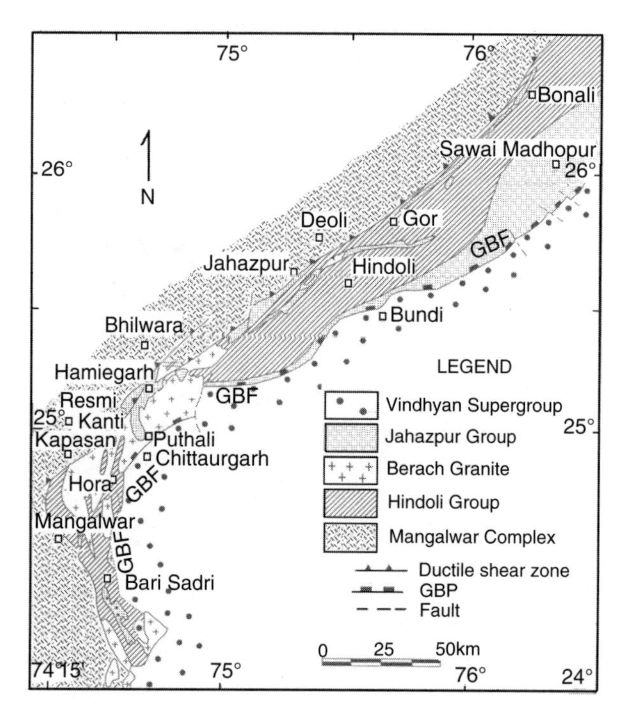

Fig. 6.1.14 Generalised geological map of Eastern Rajasthan showing the Great Boundary Fault (GBF) extending along the contact of the Bhilwara Province and Vindhyan Supergroup (modified after Sinha Roy et al., 1998)

Recent studies conducted by the GSI along and across this zone threw some valuable light on the GBF and related problems. Guha and Rai Choudhuri (1999) and Rai Choudhuri and Guha (2004), based on detailed mapping and structural studies on the GBF near Sawai Madhopur town in Rajasthan observed that the Vindhyan sequence unconformably overlies the Hindoli Group, comprising metavolcanic rocks, pelitic and semi-pelitic rocks, greywacke and arenite. According to them, the GBF is merely a post-Vindhyan thrust, developed in a brittle–ductile regime, rather than a major tectonic feature with significant antecedence. It is also not a boundary fault between the basement and the Vindhyan Supergroup rocks, as thought of earlier and hence, the term 'GBF' is a misnomer.

Other group of workers (Bhattacharya et al., 1999; Dutta et al., 1999), who studied larger stretch of the GBF in the Bundi–Chittaurgarh–Bari Sadri area did not subscribe to the above views and regarded the GBF as a zone of major dislocation between the cratonic terrain (BGC and Bhilwara Supergroup) and the Meso–Neoproterozoic cover sediments of the Vindhyan basin.

GBF is not necessarily a finite dislocation surface. At many places, it is a zone, being up to 1 km or more wide. The western i.e. Bhilwara– Hindoli side is the hanging wall one and the Vindhyan basin is in the footwall side. Striking fault scarps are characteristic morphotectonic feature along the GBF. Four sets of folds and some related structures are visible in the hanging wall side rocks and only two of its younger sets are observed in the footwall rocks. Majority opinion supports a westerly dip of GBF. Folds are tighter close to the GBF and they are overturned towards the east. The manifestations of tectonic activity along the GBF have been noticed in the form of truncation of beds, silicification, brecciation and ferrugenisation in the Vindhyan limestone, disharmonic folding along the GBF and steep dips in the Vindhyan rocks, near the GBF. Ramaswamy (1995) reported development of regional scale folds within the Vindhyans close to the GBF.

In proximity to the GBF, the Hindoli Group of rocks shows rotation of axial plane through large angles culminating in the development of near-isoclinal folds. The nature of movements related to the GBF is reverse with strike-slip component, the sense of movement being dextral. The tectonic nature of this contact is also evidenced by the development of GBF-parallel schistosity in the Berach Granite close to the contact, which is otherwise a non-schistose granite.

All these structural features suggest the GBF to be a northwesterly dipping reverse or thrust fault affecting both the basement complex and cover sediments. Deep seismic reflection profiling also support a northwesterly dip of the GBF. The previously mentioned geological observations in support of a prolonged tectonic history of the GBF, has no significant imprint on either gravity profile (Sinha Roy et al., 1995) or in magneto–telluric data-set (Gokaran et al., 1995). The latter suggested possible presence of GBF in deeper parts of the crust.

That the GBF merely reflects a post-Vindhyan shallow structural feature is unlikely to be the whole story of this prominent lineament demarcating the western boundary of the Vindhyan basin. The Vindhyan basin is an intra-continental fault-controlled depositional bowl, in which more than two thousand meter thick sediments could not be deposited without syn-depositional subsidence of the basin floor. This makes it more likely that the GBF had an earlier history that predates the post-Vindhyan thrust. In this context, it may not be out of place to mention that three models have been proposed for the origin of the Vindhyan Basin: (i) intracontinental rift, (ii) continental interior sag, and (iii) foreland flexture. The age-old idea of Oldham et al, (1901) that the Vindhyan basin could be a foreland basin in rifted crust is widely accepted even today. The development of the basin appears to be a cumulative effect of (i) loading of the tectonic masses causing flexture in the foreland region, (ii) thermal cooling and contraction of the crust causing sagging and (iii) rejuvenation of the pre-existing fault along GBF lineament due to thrust loading in the adjacent fold belt.

6.2 Metallogeny

Introduction

Rajasthan is rich in basemetal deposits, particularly in lead–zinc in Bhilwara and Aravalli belts and copper in the Delhi fold belt. Tungsten mineralisation associated with Neoproterozoic granitoids occur at several places, some of which were economically exploited in the past. Gold mineralisation at Bhukia–Jagpura, Banswara district, Rajasthan, is promising. The Aravalli belt contains phosphate deposits. Uranium, rare earth and rare metal mineralisations are also reported from a number of places. Bauxite deposits occur mostly along the Saurashtra coast of Gujarat.

Ancient Mining and Metallurgy in Western India

Like many other countries distinguished by old civilisations, India also bears testimony to ancient mining and metallurgical practices at many sites. The most numerous and well-studied ones occur in Western India, particularly in the present day Gujarat–Rajasthan–Haryana region (Fig. 6.2.1).

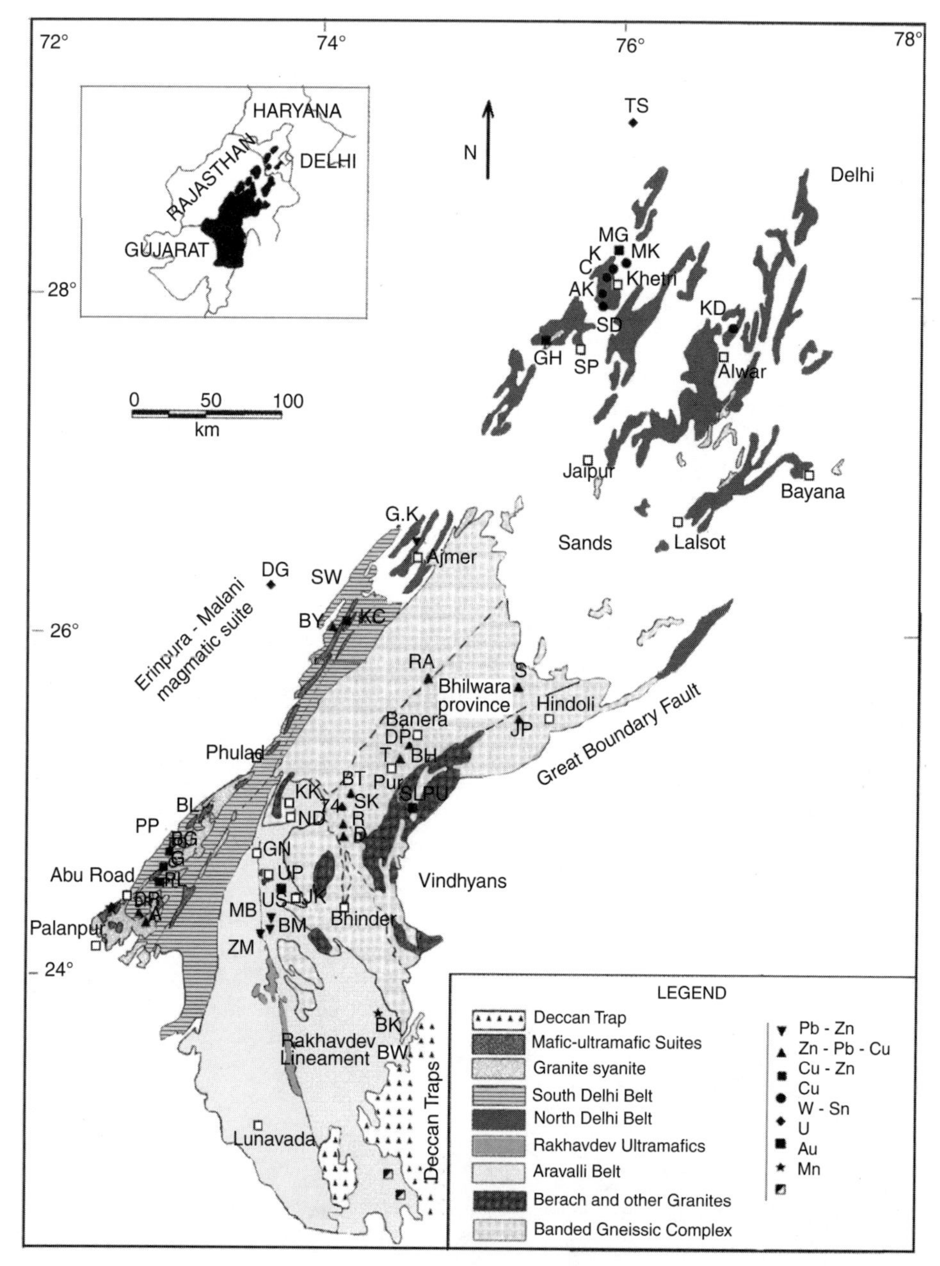

Fig. 6.2.1 Simplified regional geological map of Aravalli–Delhi orogenic belt (modified after Deb and Sarkar, 1990; Deb et al., 2001) showing locations of mineral occurrences and deposits: AK – Akwali; BG – Basantgarh; BH – Bhilwara; BK – Bhukia; BL – Balda; BM – Baroimagra; BT – Bethumni; BW – Banswara; BY – Birantiya; C – Chandmari; D – Dariba; DL – Diri; DG – Degana; DR – Deri; G – Golia; GH – Ghateswar; GK – Ghugra–Kayar; GN – Gogunda; GS – Gura Pratap Singh; JK – Jhamarkotra; JP – Jahazpur; K – Kolihan; K–C – Kalabar Chitar; KD – Kho Dariba; KK – Kankroli; MB – Mochiamagra Balaria; MG – Mewar Gujarwas; MK – Madan Kudan; ND –Nathdwara; PL – Pipela; PP – Phalwadi – Positara; R – Rajpura; RA – Rampura–Agucha; S – Sawar–Bajta; SD –Satkui–Dhanota; SL–PU – Salera–Putholi; SK – Sindesar–Kalan; SP – Saladipura; SW – Sewaria; T – Tiranga; TS – Tosham; UP – Udaipur; US – Umra–Udaisagar; ZM – Zawarmala

Although it has not been possible to trace the beginning of such activities in the region, discovery of evidence of chalcolithic culture at many places of western India and neighbouring Pakistan, suggests that it could have been pre-Harappan (c 3200–2500 BC) in age. Thanks to the interest shown and serious research done on various aspects of these ancient activities by a number of archaeologists, earth-scientists, mining engineers and metallurgists (Gandhi, 2000; Craddock et al., 1989, 1990; Hegde, 1991) that we now have such a clear view of the technological excellence shown by the Indians so long ago (Fig. 6.2.2). In fact, these are some of the milestones in the march of human civilisation, as we understand it today.

Fig. 6.2.2 Site of ancient zinc–lead smelting at Zawar (a) A battery of earthen retorts used for smelting, (b) An ancient temple not far from (a), showing cultural excellence of the mining community living around (photographs by the authors)

Keeping in mind that application of geochemical and geophysical methods in exploration of mineral deposits areTwentieth Century developments, one wonders how the early workers could detect the sites of ore mineralisation, particularly of the basemetals. Exposed parts of such ore deposits could hardly avoid weathering and decomposition into gossans, except in rare cases where erosion was faster than the weathering as in cases of some Himalayan deposits. There must not have been much information in the gossans evident in those days.

In fact, most of the ore deposits of the region have undergone mining at some point of time or the other. There has been opencast mining. More importantly, they also took to underground mining using inclines, adits, shafts, stopes, trenches and galleries. They tried to identify and delineate an ore shoot, if there was any, and tried to follow it in mining, a practice adopted even today. Pick and chisel were no doubt used for digging, but fire-setting was widely used to crack the rocks. Some of the galleries and stopes were very large as at Zawar and Dariba. Pillars were left at calculated distances to support the roof. Timbering was widely used, both for support and making ladder way. Carbon dating of some of these timbers along with the charcoal from fire-setting, helped in dating some of these activities (Table 6.2.1).

At Dariba, ancient mining went down to 265 m below the water table, which means, they had to tackle the ground water problem too. There is evidence to suggest that the early workers beneficiated the ores mined by crushing and then handpicking. Even probable gravity separation is indicated. Reasonably, this was all done at the mine head, to reduce labour.

Table 6.2.1 *Radiocarbon dates of different materials related to ancient mining at Zawar, Dariba and Agucha mines (Craddock et al., 1989)*

Location	Material	Age in years
Zawar		
Ash dump near Ramnath Temple	Charcoal	890 ± 130
Retort heap	Charcoal	500 ± 50
Large retort dump	Charcoal	350 ± 130
Lead smelting stag	Charcoal	1950 ± 60
Lead smelting stag	Charcoal	1930 ± 80
Zawarmala		
Zawarmala mines	Charcoal	2150 ± 110
Scaffold	Wood	2180 ± 35
	Wood	2350 ± 120
Launder	Wood	2140 ± 110
Mochia Pit Prop	Terminalia	2360 ± 50
Rajpura–Dariba		
South lode, 100 m depth	Timber	3040 ± 150
South lode, 100 m depth	Rope	2100 ± 280
Dariba underground	Timber	2245 ± 100
East lode, 263 m depth	Bamboo	1790 ± 120
East lode, wood riveting	Timber	2140 ± 100
Rampura–Agucha		
Mine gallery (30 m below surface)	Timber	2240 ± 130
Basket, footwall	Twigs	2380 ± 130
Cupel, debris area	Charcoal	2340 ± 40

Hegde (1991) discussed copper smelting in western India, taking into consideration the copper ingots at the chalcolithic sites and furnace remains in the Aravalli region. An approximately 35 cm high furnace with a 20 cm diameter at the rim and 10 cm diameter at the base was deployed. The roasted ore concentrate was mixed with crushed quartz (SiO_2), about equal to the weight of the ore, and crushed charcoal about twice the weight of the ore. The whole feed was charged into the furnace in the form of small hand-made balls or lumps. Quartz was used to flux the ore, and more probably to separate iron from copper in the form of fayalitic slag. All the chalcolithic copper ingots assayed > 98% Cu, very close to 'Blister copper' of today. This suggests that high quality copper metallurgy was known in the Indian sub-continent during the third millennium BC (Hegde, 1991). However, Iran is credited with the first development of copper-metallurgy in the world.

At Zawar, lead was smelted on a small scale around 500 BC, probably as a by-product of zinc mining (Craddock et al., 1989). Available evidence suggests that there was production of lead/silver at both Agucha and Dariba in the latter part of the first millennium BC (Gandhi, 2000). Dixon (Ball, 1881) described a lead-smelting furnace from the Ajmer area. Silver was probably separated from lead by cupellation.

Smelting of zinc in a big way has been a much later achievement in human history, although brass was apparently in use before the Christian Era. Suggestively, it was produced on an industrial scale by reacting calcined zinc ore with charcoal and finely divided copper in a closed crucible at 1000°C (Gandhi, 2000). There is reference to metallic zinc and its extraction from ores in Charaka Samhita, an outstanding Indian medical treatise of the late first millennium BC. Craddock et al. (1990) also suggest that at least part of the brass available during old times was made by alloying copper with metallic zinc.

Recent joint investigation by the British Museum, London, University of Baroda, Vadodara and the Hindusthan Zinc Ltd, Zawar, has unearthed intact ancient zinc distillation furnaces that were used for one of the most sophisticated pyrometallurgical operations before the 'Industrial Revolution'. The process is still considered basic to all high temperature distillation (Craddock et al., 1990).

Mineralisation in Basement Rocks

Ore mineralisation in the basement rocks of Western India is scanty and the known ones have so far been considered subeconomic. The copper mineralisation near Anjeri, Udaipur district, Rajasthan, seems most important and interesting amongst all these occurrences. Some details about this prospect are available from Ray et al. (1988) and are summed up below.

The mineralisation occurs in several chlorite-rich zones within 'granite', i.e. tonalitic in composition. These mineralised zones are considered ductile shear zones, as the wall rocks of the ore zones are schistose, the schistosity gradually disappears some distance away, and the rock is granitoid. The shear zone material is thoroughly recrystallised. Interestingly, some fold structures are noticed in these ore zones that may be comparable to those (AF_2 and AF_3) observed in the Aravalli cover sediments. The schistose metabasic rocks are almost free of mineralisation.

Mineralogically the ore consists of chalcopyrite, pyrite and pyrrhotite, with covellite and magnetite as minor phases. A signature of recrystallisation is present in the ore minerals. The authors in the context of above observations came to an obvious conclusion that the mineralisation took place from hydrothermal fluids guided by structures that developed in the Proterozoic. The source of ore fluids, however, remains a matter of speculation.

Ore Mineralisation in Bhilwara Province

In an earlier section, we have already discussed what is commonly understood by the Bhilwara Province ('Bhilwara belt'), its extent and geological framework. For convenience of discussion in the present context, the essential aspects are recapitulated below.

It may be mentioned here that in existing literature, the term 'belt' has been used for both the Bhilwara region as a whole, as well as for the linear ore bearing supracrustal tracts occurring within it. In order to make a clear distinction between the two, we prefer to name the former as 'province' and the latter as 'belt'. The Bhilwara Province is a broadly elongate terrain of about 200 km in length, about 100 km at its widest in the northern part, narrowing down to about 10 km in the south near Bhinder. It trends NW–SE in the south and NE–SW for its greater length. To its west lies the BGC (Heron, 1953) and the Vindhyan Supergroup rocks are located to the east (Fig. 6.2.3).

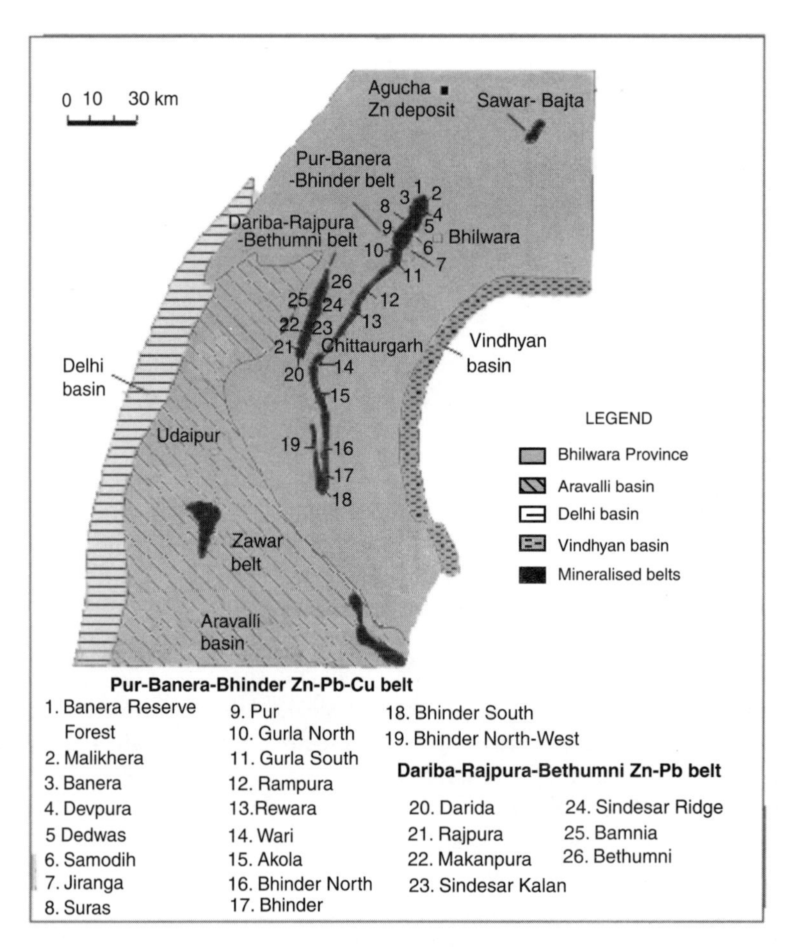

Fig.6.2.3 Locations of Pur–Banera–Bhinder, Dariba–Rajpura–Bethumni, Zawar, Sawar–Bajta mineralised belts and Agucha deposit in Southern Rajasthan

The Bhilwara Province consists of several sub-parallel, sub-linear belts of differently metamorphosed sedimentary (–volcanic) rocks separated from each other by members of BGC, or soil cover, and rarely by unconformity surfaces. The grade of metamorphism increases from the east to west. Several ore deposits, rather 'ore belts', occur within the Bhilwara Province, including the Rampura–Agucha deposit, Dariba–Rajpura–Sindheswar Kalan–Bethumni belt, Pur–Banera belt, Sawar belt and Jahazpur belt. The mineralisations are associated with supracrustal rocks that were deposited in the continental rifts (Sinha Roy, 1984; Deb and Sarkar, 1990).

Rampura–Agucha Deposit Initially a chance find, rather rediscovery, during the early seventies, the Rampura–Agucha deposit, on detailed exploration by the State Department of Geology and Mines, Rajasthan, Geological Survey of India and the Hindusthan Zinc Ltd., turned out to be the most important Zn–Pb (–Ag) deposit of India, producing ~ 9 x 10^5 tonnes of ore per annum. Total ore reserve is 63.7 Mt (Proved 39.2 Mt + probable 13.8 Mt + possible 10.7 Mt) with an average grade of 13.6% Zn, 1.9% Pb and 45 ppm Ag (Holter and Gandhi, 1995). As mentioned in an earlier section, Rampura–Agucha was one of the sites of intensive mining and metallurgical activities during the Pre-Christian Era.

Geological Setting Rock exposures at the Rampura–Agucha area cannot be deemed excellent. Field-mapping and projection of sub-surface geology suggest the following sequence from the east to the west: (i) Garnet–biotite–sillimanite–gneiss (GBSG), with bands of amphibolites and calc–silicate rocks, intruded by aplites and migmatites, (ii) graphite–mica–sillimanite schist containing the ores, (iii) GBSG with lenses of amphibolites, quartzo–feldpathic bands and calc–silicate rocks intruded by pegmatite/aplite veins, (iv) granite gneiss, (v) banded gneiss (Fig. 6.2.4).

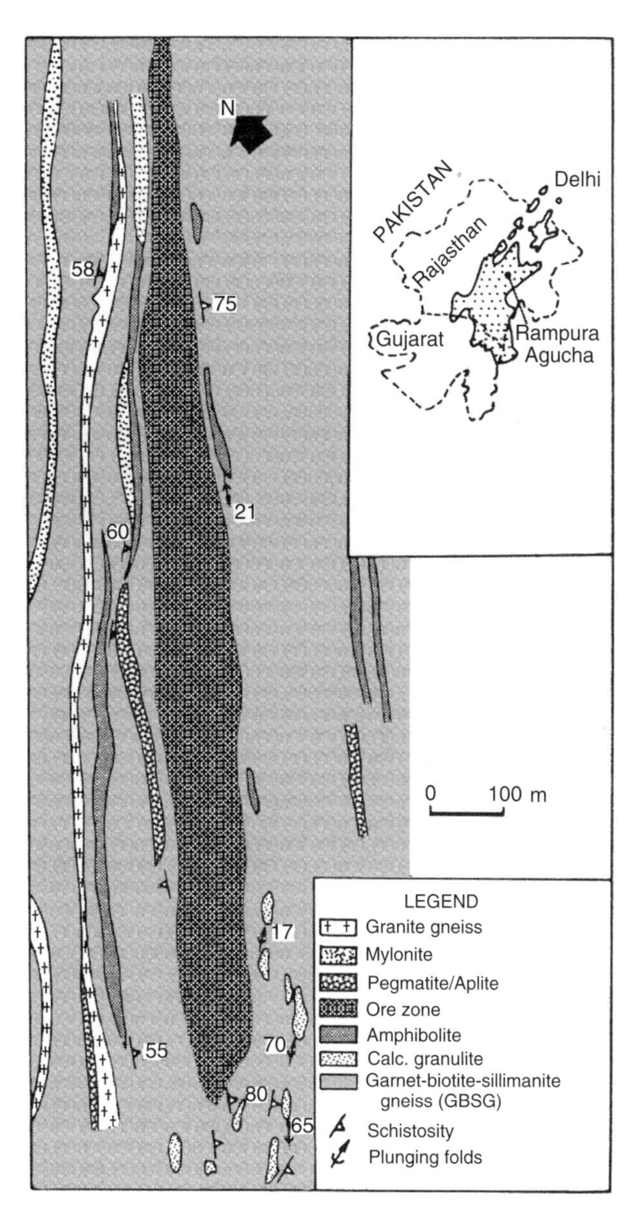

Fig. 6.2.4 Geological map of Rampura–Agucha deposit, Rajasthan (after Deb and Sehgal, 1997)

There are several conspicuous zones of ductile–brittle shearing in the footwall zone. A clearer idea about the protolith of the amphibolitic rocks of this area is important for a better appreciation of the tectono–thermal history of the area, as well as the metallogenesis. Various discriminant ratios of FeO, MgO, SiO_2, TiO_2 and Al_2O_3 together with Ti/Zr, Zr/Zr–Y, Ti/Zr/Sr, Ti/Zr/Y, $(Ba/La)_N$ and $(La/Sm)_N$ are consistent with the conclusion that the precursor of the amphibolites was a plume-generated basalt. The REE distribution patterns in the metasediments, however, suggest a continental (BGC?) derivation of the sediments (Deb and Sarkar, 1990; Deb, 1992). Although the interface between the granite gneiss and the associated supracrustal rocks are now highly tectonised, the available observations are in favour of the former being the basement rocks.

Structural history of the Rampura–Agucha area is more complex (Roy et al., 2004) than what was earlier concluded. Structures produced by ductile shearing at an elevated PT are now referred to as D_1 structures. The later deformation, related to the low-temperature shearing, took place subsequent to the phase of exhumation and 'unroofing' of the high grade metamorphic rocks and have been referred to as the D_2 structures. D_2 was heterogeneous in distribution and in intensity. Deb and Sehgal (1997) determined the T_{max} and P_{max} of metamorphism to be 650°C and 6 kb respectively, using garnet–biotite and garnet–hornblende geothermometer and garnet–plagioclase–sillimanite–quartz geobarometer.

Ore Mineralisation Gandhi et al. (1994) suggested the ore body to be a synformal structure. There is no convincing evidence in support of this conclusion though. It could be a simple ore lens, as is so common in layered rocks with some modification of the original form and concentration during deformation and metamorphism. A sectional view of the Rampura–Agucha deposit is furnished (Fig. 6.2.5) showing zonal distribution of the ore metals (Zn+Pb) and superficial alteration.

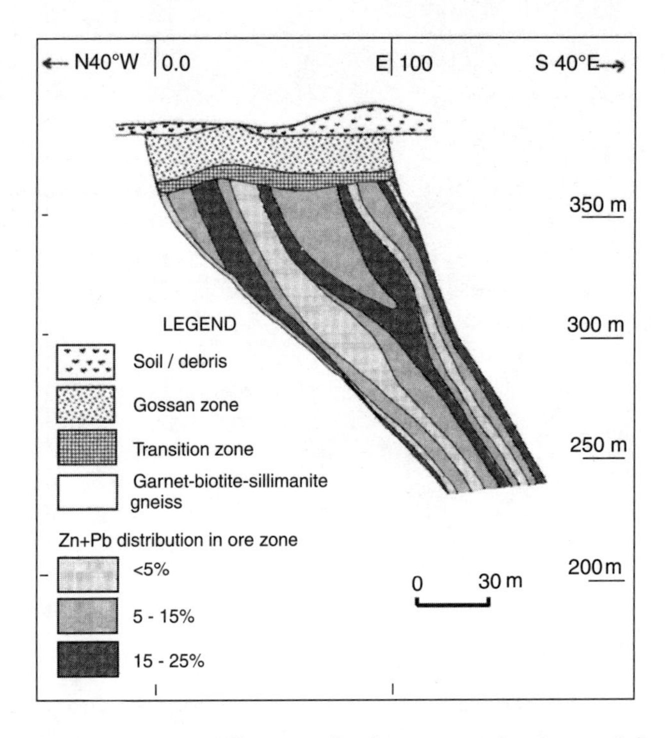

Fig. 6.2.5 Transverse section across Rampura–Agucha ore zone showing zonal distribution of ore metals (after Roy et al., 2004)

The major ore minerals are sphalerite, pyrrhotite, pyrite and galena. Minor phases comprise chalcopyrite, arsenopyrite, and tenantite–tetrahedrite. Trace phases include pyrargyrite, freibergite, gudmundite, breithauptite, ullmannite, greenockite, boulangerite, molybdenite and a thallium-bearing phase [(Cu, Ag, Tl)FeS_2], Ag–Pb–Sb sulphosalts, dyscrasite, stephanite, argentite, polybasite, a V–Cr–Fe–Mg–Mn–Zn–O and a V–Cr–Fe–O phase (Gandhi et al., 1984; Ranawat et al., 1988; Genkin and Schmidt, 1991; Höller and Stumpfl, 1995; Höller and Gandhi, 1995; Mukherjee et al., 1991). δ^{34} (sphalerite) varies from + 7 to +10.3 per mil and $\delta^{13}C$ in the associated graphite varies from –24 to –29 per mil (Deb, 1989; Deb, 1990a). The principal gangue minerals comprise quartz, feldspars, sillimanite, graphite and phyllosilicates (Fig. 6.2.6).

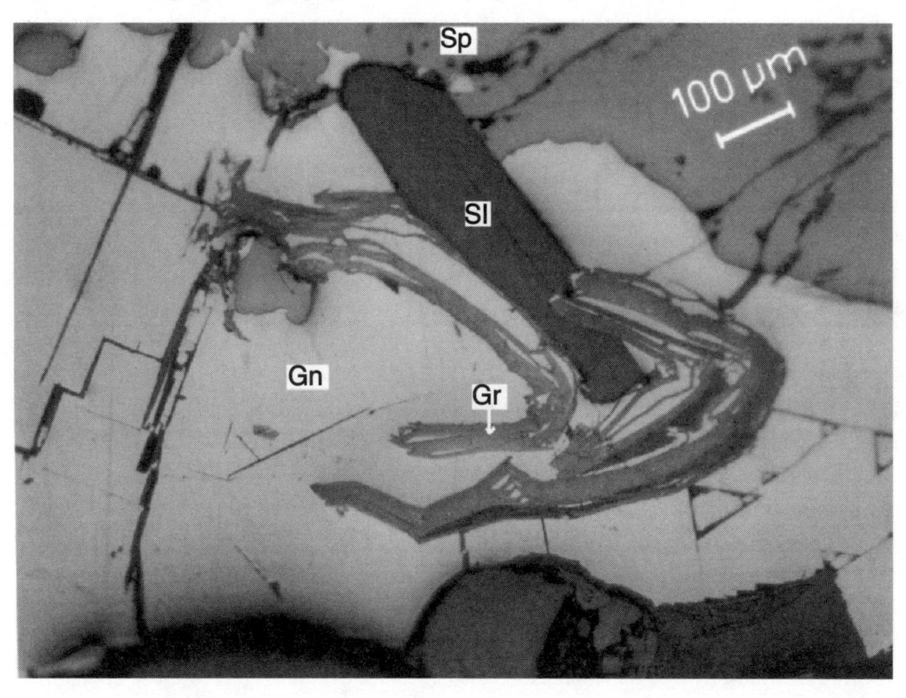

Fig. 6.2.6 Photomicrograph of an ore sample from Rampura–Agucha mine. Gn – galena, Sp – sphalerite, Gr – graphite, Sl – sillimanite. Graphite interleaved with galena shows post-crystallisation deformation (courtesy, M. Deb).

Overall fabric of the ores in hand specimen is schistose. Recrystallisation fabric in the ores is common except where deformation outlasted recrystallisation. P_{max} obtained from the sphalerite geobarometer tallies with the P_{max} obtained from the silicate geobarometer. The metamorphic fluid with which the ore equilibrated was H_2O-rich with a fair proportion of CO_2 (X_{CO_2}= 0.39). Minor gas components in the fluids are H_2S (X_{H_2S} = 0.043) and CH_4 (X_{CH_4}=0.025) (Deb and Sehgal, 1997). Holler et al. (1996) studying the fluid inclusions by microthermometry as well as Raman microspectrometry, distinguished three main types of fluid inclusions in quartz: (i) gaseous (CO_2, partially mixed with CH_4–N_2) (ii) low salinity (0–8 equivalent wt.% NaCl) aqueous inclusions and (iii) high salinity aqueous inclusions (NaCl+$MgCl_2$ – $CaCl_2$). Low density CO_2-rich and low salinity H_2O-rich inclusions are contemporaneous and occur together with CH_4–N_2 inclusions. The authors suggest that it implies immiscibility between the gaseous and aqueous phases and 'participation of these fluids in deposition or remobilisation of the ore' at P = 1220 – 200 bar and T = 450° – 250°C.

Raman spectra of graphite indicated upper greenschist facies of metamorphism. However, in view of the peak of the metamorphism being established at upper amphibolite facies (T = 650°C, P = 6 Kb), the lower values of T and P should relate to a phase of retrograde metamorphism. Another interesting report is the loss of 45 ± 25% of boron from the tourmaline during the upper amphibolite facies of regional metamorphism (Deb et al., 1997).

The Rampura–Agucha deposit belongs to the 'Sediment-hosted' type from a simple descriptive point of view and from the genetic point of view, it is difficult to assign it convincingly to any genetic model because of its intense post-depositional transformation. Some of the present features may be reconstructed into those of the SEDEX type, however. Absence of a 'coherent element dispersion halo' around the ore body at Rampura–Agucha led some workers to suggest that the deposit was allochthouous (Shah, 2004). But origin of such a huge deposit by an allochthonous mechanism is more difficult to explain by the state of the art knowledge of ore geology, than admitting that we do not have a satisfactory explanation for the absence of a coherent 'element dispersion halo' around the ore body.

Dariba–Rajpura–Bethumni Belt The Dariba–Rajpura–Bethumni ore belt, having a westerly convex crescent shape, extends for about 20 km from Dariba (24° 57′ E: 74° 08′N) in the south to Bethumni (village named as Bethumbi in Toposheet 45K/4) in the north, with probable extension up to Surawas. The maximum plan width of the belt across the central part is nearly 6 km, tapering on both ends. The belt, from south to north, comprises several economic and sub-economic zinc–lead deposits viz, Dariba (South and North), Rajpura (A, B, C), Mokanpura (North and South), Sindesar Ridge (North and South), Sindesar Kalan (East and West), Lathiyakheri (Latio ka khera), Jitawas and Bamnia deposits. The ore belt has been divided into 18 blocks by the GSI for the sake of systematic exploration (Fig. 6.2.7).

The mining township of Dariba–Rajpura is located at 16 km NNE of Fatehnagar railway station (Chittaurgarh–Udaipur section of Western Railways) and at about 85 km from Udaipur. Geologically, it is located within the Bhilwara basin of Paleoproterozoic age. This ore belt comprises one of the three clusters of major basemetal deposits of south central Rajasthan.

As it has been already discussed in an earlier section, there are impressive evidences of exploitation of Dariba–Rajpura deposit before the Christian Era. Remains of these ancient workings, such as heaps of mine debris and slags together with the presence of a spectacular gossan zone (Fig. 6.2.8), particularly in the southern part of Dariba made the geologists of the Geological Survey of India (GSI) aware of this ore zone as early as in 1934..

However, a programme of detailed exploration in this belt, particularly in the Rajpura–Dariba, Bethumni and a few other places by the GSI was conducted during 1962–70 only. This was followed by final exploration and exploitation in Dariba–Rajpura by the Hindusthan Zinc Ltd. Continued exploration by the GSI led to the discovery of several additional deposits/blocks in this belt (Table 6.2.2) and a total reserve of 310 Mt of ore with 3–8% (Zn+Pb) was established. The richer part of it is ~ 45 Mt with 6.5% Zn and 2.4% Pb (Haldar, 2004). Of the newly explored blocks, the North Sindesar Ridge is the most promising one, where mining has already started in the southern sub-block of 1400 m strike length. Here and in Lathiyakeri block, the ore bodies are totally concealed without any surface expression.

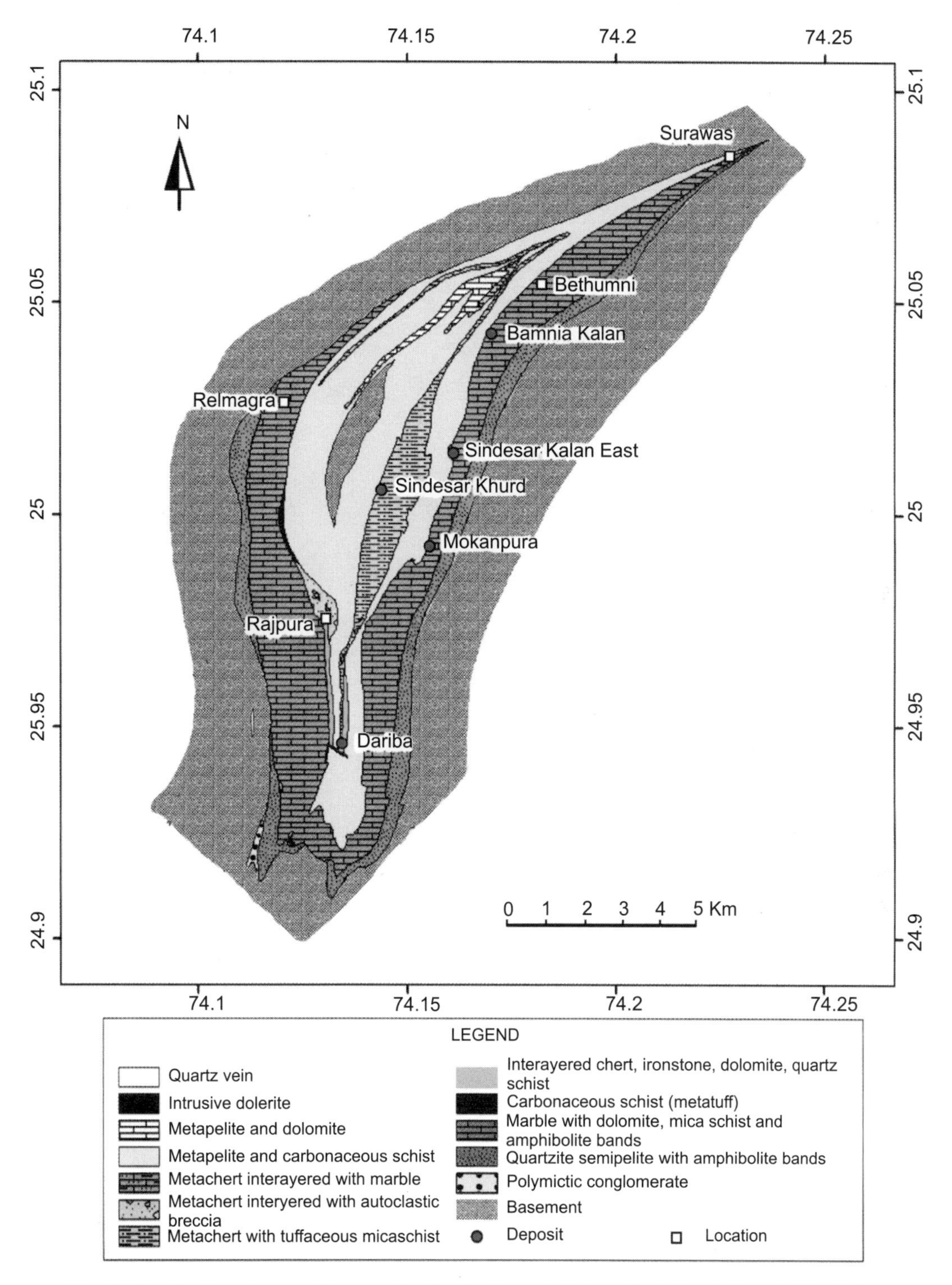

Fig. 6.2.7 Geological map of Rajpura–Dariba–Bethumni belt (modified after Deb and Pal, 2004)

6.2.8 Photograph of the gossan exposure at Dariba (courtesy: M. Deb)

Table 6.2.2 *Reserve and grade of Zn–Pb ore established in different deposits/blocks of Dariba–Rajpura–Bethumni ore belt (Data source: GSI)*

S.No.	Deposits/Blocks	Reserve (Mt)	Grade (Zn+Pb)%
1	Dariba South	13.01	8.05
2	Dariba North	7.87	6.16
3	Rajpura (A,B and C blocks)	25.91	4.21
4	Malikhera	–	poor
5	Sindhesar Kalan (West)	0.74	6.07
6	Dariba East Lode	8.67	4.34
7	Mokanpura South	–	poor
8	Mokanpura North	40.0	6.0
9	Sindesar Kalan East	70.0	2.65
10	Bamnia	56.12	4.54
11	North Sindesar Ridge	57.37	4.20
12	Lathiyakheri East	13.20	4.32
13	South Sindesar Ridge	3.8	2.5
14	Bethumni	0.36	3.53

Zinc–lead deposits of various sizes and grades occur in this ore-belt, the most important ones amongst which are located in the southern part, i.e. at Dariba and Rajpura. Exploration is being continued by GSI in this belt in several blocks like Jitawas, Sunariya Khera, Sindesar Khurd, Lathiyakheri West etc. Host rocks are both recrystallised dolostones and graphite-bearing mica schists. Mineralisation in the latter is usually large in volume, but low in grade. Copper and silver are obtained as by products. Cadmium, mercury and thallium, present as trace metals in the ores, are important not as much for their economic values as for their roles as probable environmental pollutants, if proper care is not taken during metallurgy.

Geological Setting The major rock types comprising the Dariba–Rajpura–Bethumni belt are the metamorphic equivalents of sedimentary (–volcanic) rocks, belonging to the orthoquartzite–carbonate–carbonaceous pelitic facies, interlayered with recrystallised chert and tuffaceous units. These tuffaceous rocks and occasionally found basic sills are now represented by amphibolites. Patches of barite and fluorite are also present. The carbonate rocks contain calcite and dolomite in sub-equal proportions, together with quartz, tremolite, diopside, phlogopite, biotite and scapolite in small but varying proportions. The pelitic rocks are now composed mainly of quartz, micas and graphite, with garnet, staurolite and kyanite suggesting a moderately high grade of metamorphism. The basement complex consisting of schists, gneisses and migmatites unconformably underlies this supracrustal sequence.

As mentioned above, the overall map-pattern of the belt is crescent shaped representing a regional synform with a closure situated to the south of Dariba, the axis plunging steeply (55°–60°) towards ENE. The rocks generally dip easterly (ESE) and bear evidence of multiple deformations. The first phase folds (F_1) are tightly appressed and symmetrical, rarely overturned. The F_2 folds with NNE to SSW axes are coaxial with F_1. They are asymmetrical with easterly dipping axial planes. The regional synformal structure of the area belongs to this phase. The F_3 folds are large and open with WNW to ESE axes, and caused the crescent shape of the regional structure.

Ore Mineralisation There are two orebodies at Rajpura–Dariba area, called Main South Lode and East Lode (Fig. 6.2.9). Of these, the Main Lode is larger and richer and presently the mining activity is confined to the Main Lode only. The Main Lode, extending over a strike length of 1700 m, is separated into two ore bodies, by a barren zone of about 300m length. The ore bodies are generally of the shapes of flattened lenses with pinches and swells.

The major ore bodies in the Dariba–Rajpura deposits are not only stratabound but also stratified at places into laminations of contrasted compositions (Fig. 6.2.10). The laminations vary in thickness amongst them and also within themselves. In rare cases, layers of fluorite were found interlaminated with sphalerite and quartz–carbonate association. The ore layers have been affected by phases of deformation. The close association of laminated sulphides with carbonaceous matter, their rhythmic alteration with black cherts, occurrence of spheroidal (framboidal) pyrites along undulating laminae of carbonaceous cherts that have close resemblance to cryptalgal laminae, are suggestive of primary deposition of the sulphides in a sedimentary – diagenetic environment. Later polyphase deformation and metamorphism, including recrystallisation and limited remobilisation modified the primary ores and converted them to essentially metamorphic tectonites. These features are common throughout the belt (Deb and Bhattacharya, 1980; Deb and Pal, 2004; Poddar, 1974).

In the North Lode, mineralisation is restricted to the calc–silicate dolostone with sharp contacts. In contrast to the other ore bodies in the deposit, the South Dariba contains stratabound but discordant coarse-grained sulphide bodies occurring as pods, lenses and veins along fold closures and fractures. Such ore-veins locally transect laminated ore. The East Lode is highly pyritic, hosted by graphite mica schist. Because of it's not being exploited until the date, information on it is very limited. Sindesar Khurd, Sindesar Kalan and Mohanpura deposits occur to the north–northeast of the Dariba deposit. At Sindesar Khurd, a massive ore body is located in the central part of the eastern limits of the Dariba–Bethumni regional fold, hosted by a calc–silicate bearing metamorphosed dolostone, close to the latter's boundary with carbonaceous metapelites. An ore zone from Sindeswar Kalan to Mokhanpura has been traced over a strike length of about 4 km in nearly the same geological

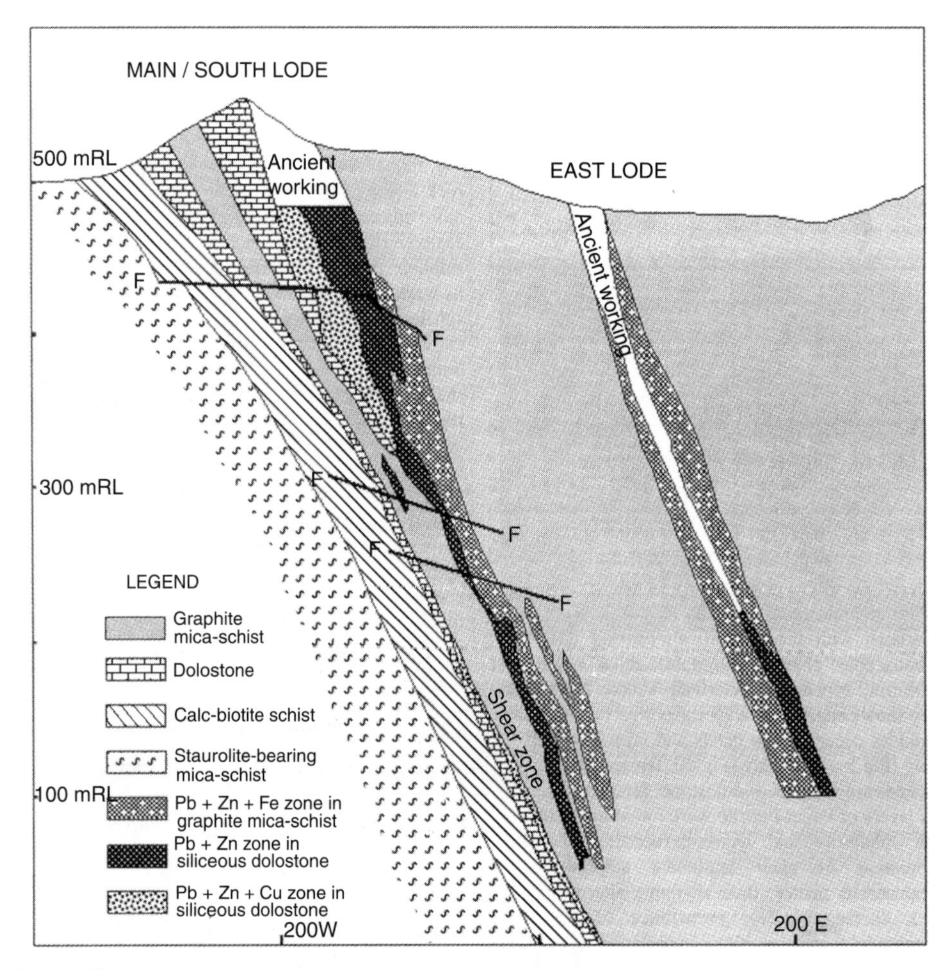

Fig. 6.2.9 Transverse geological section across Dariba–Rajpura ore zone. Each of the two lodes South (Main) and East shows large-scale zoning. The extent of ancient mining in the East lode down to 300 m from the surface is awesome (after Haldar and Deb, 2001).

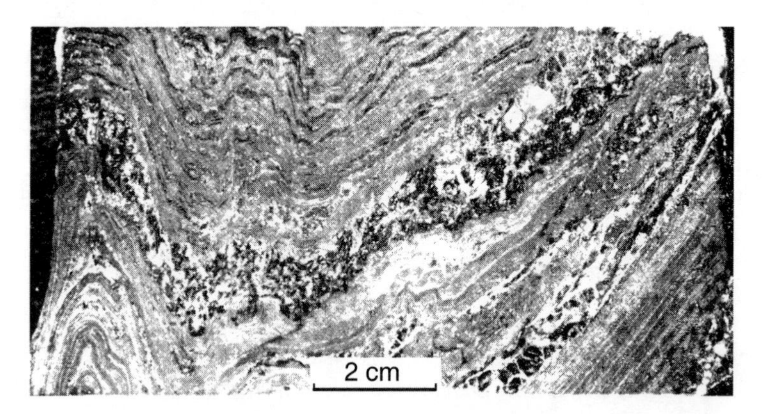

Fig. 6.2.10 Laminated sulphidic graphite mica schist from the hanging wall of Dariba Main lode (Poddar, 1974)

environment. Ore reserves at Sindesar Khurd, Sindesar Kalan East, Mokhanpura are 14 Mt (with 6.71% Zn and 3.2% Pb), 94 Mt (with 2.1% Zn + 0.6% Pb) and 63 Mt (with 2.2 Zn + 0.7% Pb). A similar deposit at Bamnia Kalan, further NNE, extends over a strike length of 1.2 km and has an ore reserve of 5 Mt (with 5.7% Zn + 2.5% Pb).

Ore from the different deposits of the belt have nearly the same mineral assemblages, differing only in their relative proportions. The common mineral assemblages in the Dariba–Rajpura, Sindesar Kalan and Sindesar Khurd deposits are as follows, arranged in order of their decreasing abundance (Deb and Pal, 2004):

1. Pyrite + sphalerite + galena
2. Sphalerite + galena + pyrite + fahlore (tetrahedrite – tenantite)
3. Chalcopyrite + sphalerite + galena + arsenopyrite + pyrrhotite
4. Chalcopyrite+sphalerite+galena+pyrrhotite+pyrite+ arsenopyrite +fahlore
5. Fahlore + galena + sphalerite + chalcopyrite

Pyrite in the Dariba–Rajpura ores occur as fine grained groundmass in laminated ores or as coarse porphyroblasts co-existing with pyrrhotite, sphalerite and galena. Pyrite spheroids, with carbonaceous matter at the core, are likely to be framboids. Sphalerites from stratiform ores of Dariba–Rajpura show variation in colour from lemon yellow, through light brown to dark brown, apparently due to variation in composition, but their significance in terms of genesis, has not been assessed. Mercury is not yet reported from Dariba–Rajpura sphalerite. However, high Hg-contents in the gossans of the area suggest that this aspect may be looked into for possible Hg-recovery as by product; more importantly, for control of Hg-pollution during smelting. Quite a few rare phases are reported from these ores. These are geocronite [Pb_5 (Sb, As)$_2$ S_8], pyrargyrite [Ag_3SbS_3], boumontite ($CuPbSbS_3$), owyheeite (Ag_2 Pb_5 Sb_6S_{18}), argentopyrite ($AgFe_2S_3$), pearcite/polybasite [(Ag,Cu)$_{16}$ (As <> Sb)$_2$ S_{11}], native silver (Ag), electrum (AgAu), enargite (Cu_3AsS_4), aurostibite ($AuSb_2$), argentite (Ag_2S), gudmundite (FeSbS), intermetallic compound ($Ag_{74.2}$ $Au_{16.4}$ $Hg_{9.4}$), ray ite[Pb_8(Ag,Tl)$_2Sb_8S_{21}$(Ag>Tl)], plumbian tetrahedrite [$(Me^{+}{}_{10}Me^{2+}{}_{2})_{12}$ X_4S_{12}($Pb_{1\text{–}2per\ cent}$)], thalcusite ($Tl_{1.94}Cu_3Fe_{1.05}S_4$), renierite [$(Cu_{10}Cu_{0.09}Zn_{0.71}Fe_{0.15})_{0.95}$ $Fe_4(Ge_{1.68}$ $V_{0.03}$ $As_{0.27})_{1.98}]_{16.93}$ $S_{16.08}$] (Poddar, 1965; Basu et al., 1981a,b, 1983, Deb, 1982; Deb and Bhattacharya, 1980; Mozgova et al., 1992). In the footwall quartzose dolostone of the South lode, copper is present and its content may go as high as 5%. Zinc is consistently high (up to 25%) in the dolostone, whereas, lead (up to 30%) is concentrated mainly in the quartzose parts. The stratigraphically, from bottom upwards, metal zoning is Cu → (Pb+Zn) → (Zn + Fe). The ores were metamorphosed at T_{max} = 550° C and P_{max} ≈ 5.4 Kb (Deb and Pal, 2004) (Fig. 6.2.11). The corresponding values obtained by Mishra (2000) are 525° C and 5 Kb.

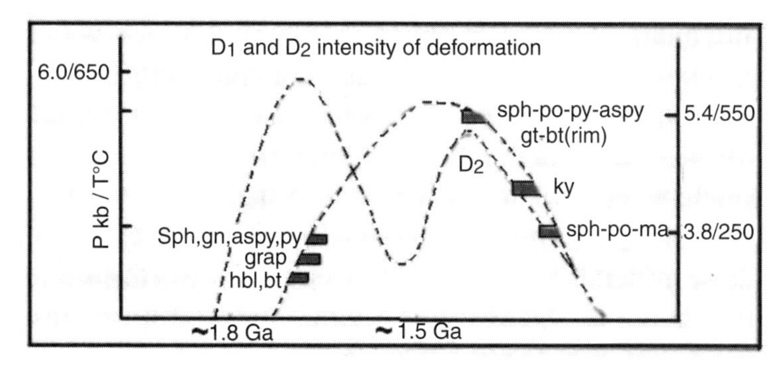

Fig. 6.2.11 P–T conditions of peak metamorphism in the Rajpura Dariba belt in relation to deformations D_1 and D_2 (Deb and Pal, 2004). Sph-sphalerite, gn-galena, asp-arsenopyrite py-pyrite, grap-graphite, hbl-hornblende, bt-biotite, po-phyrrhotite, gt-garnet, ky-kyanite.

It may be worthwhile to elaborate a little in this context another type of heterogeneity in the ores, parlty mentioned in an earlier section i.e. lenses and discordant coarse galena veins in coarse diopside-rich rocks. There are concentrations of Ag-and As- rich minerals such as geochronite and polybasite together with massive tetrahedrite–tennantite. Mishra and Mookherjee (1986) tried to characterise the ores, mineralogically and chemically, occurring in these two contrasted modes. While the banded ores contain pyrite, sphalerite, galena and chalcopyrite as the major phases and arsenopyrite, tetrahedrite as the minor ones, the coarse-grained variety consisted mainly of galena with geochronite, tetrahedrite and native As. They also contain almost all the rare minerals mentioned earlier. The second type could be products of metamorphic differentiation of the first type or introduced from an external source. Mishra (2000) has addressed this question, as reported below.

Temperatures obtained for Cd fractionation in co-existing sphalerite and galena in metamorphosed banded ore is 523° ± 37°C and that in vein ores, 430°± 30° C. The post-metamorphic growth of arsenopyrite in the banded ore, suggested by ore textures, is confirmed by the temperature range, 442°–483°C, calculated by arsenopyrite (+pyrite) thermometry. The arsenopyrite (+pyrrhotite) geothermometric values obtained from the vein ores, in contrast, yielded a temperature range of 341°–424°C. This, however, compares well with the temperature range of 358°–395°C, obtained using geothermometry based on Sb–As fractionation between co-existing tenantite–fahlore and bournonite in the vein ore. Primary fluid inclusions in quartz, in vein ores have also been studied. Compositionally they are divided into two types: (i) metastable halite-saturated and (ii) almost pure CH_4 fluids. They are supposed to be coeval. Inclusion thermobarometry, based on intersecting isochores of coeval aqueous and carbonic inclusions yielded three intersections in the P–T range of 1.7 kb/440° C to 0.97 kb/357° C. In this context, the author visualises a model in which metamorphogenic metal bearing aqueous carbonic fluid split into high saline aqueous and carbonic (CO_2 + CH_4) components due to decompression. This possibly was followed by cooling and mixing of the aqueous fluid with low-salt heated meteoric water, diluting metal-bearing chloride complexes, leading to the precipitation of the ore minerals in the veins. Regarding the source of the ore-constituents, black shales, now metamorphosed to graphitic schists, offer to be the best candidate, as they are known to contain generally Tl, As, Sb, Ag, and Hg more than many other rocks. In this model, the layered ores are in general metamorphosed, while the vein type is metamorphogenic. However, possible contribution by the pre-existing ore minerals on the way or at the site of precipitation cannot be ruled out. Moreover, one may still ask, why were the decompression-led metamorphogenic fluids restricted to the southern part of the Dariba ore body? Deb and Pal (2004) suggest that the vein type mineralisation under discussion could be the relict vein complex through which the hydrothermal fluids in the SEDEX model emanated. Their preliminary study on the fluid inclusions in the layered ores show that these are not much different from those described from the vein-type ores by Mishra (2000). Obviously, more studies are needed to resolve this problem.

The isotopic compositions of sulphur and carbon in the ores and host rocks of Dariba–Rajpura have been studied by Deb and co-workers (Deb, 1986, 1990a). Summary of these results are: (i) The sulphur isotopes in the ores have an unimodal distribution with $\delta^{34}S$ ranging between + 9.1 and – 6.7 per mil, (ii) There is a conspicuous trend of ^{34}S getting heavier stratigraphically upward, (iii) There is an apparent depletion of ^{34}S from south to the north, (iv) Metamorphism of the ores did not significantly modify the bulk isotopic compositions. Fractionation temperatures of sulphur isotopes between coexisting sulphides ranged between 190° C and 610° C. (v) That isotopic equilibration was not attained for all sulphides, not only in interphase distributions but also in intra-phase distribution, as is illustrated by $\Delta^{34}S$ = 10 amongst three successive layers of sphalerite in a single grain. (vi) Organic carbon (C_{org}) ranges from 0.5 to 9.3 wt.% in the host rocks. (vii) $\delta^{13}C_{org}$ values are similar in all rock types, being in the range of –21 to –31 per mil (mean –25.4 per mil), that indicates biogenic derivation of the carbon. $\delta^{13}C_{org}$ of ore zone dolostone was found to be –14.4 per mil.

Nine samples of galena from the layered ores and one sample from the vein ore from Dariba were analysed and they have yielded almost identical Pb-isotope ratios, giving a Pb–Pb model age of 1799 Ma (Deb et al., 1989). Similar ages have been obtained from the ores from Sindesar Khurd and Bamnia Kalan (Deb and Thorpe, 2004) (Fig. 6.2.12). The $^{207}Pb/^{204}Pb$ vs $^{206}Pb/^{204}Pb$ plot above the average curve for the upper crust (Zartman and Doe, 1981). A μ-value of 10.73 suggests that Pb was probably derived from the host sediments that presumably contained components of moderately old recycled crustal lead (Deb and Pal, 2004).

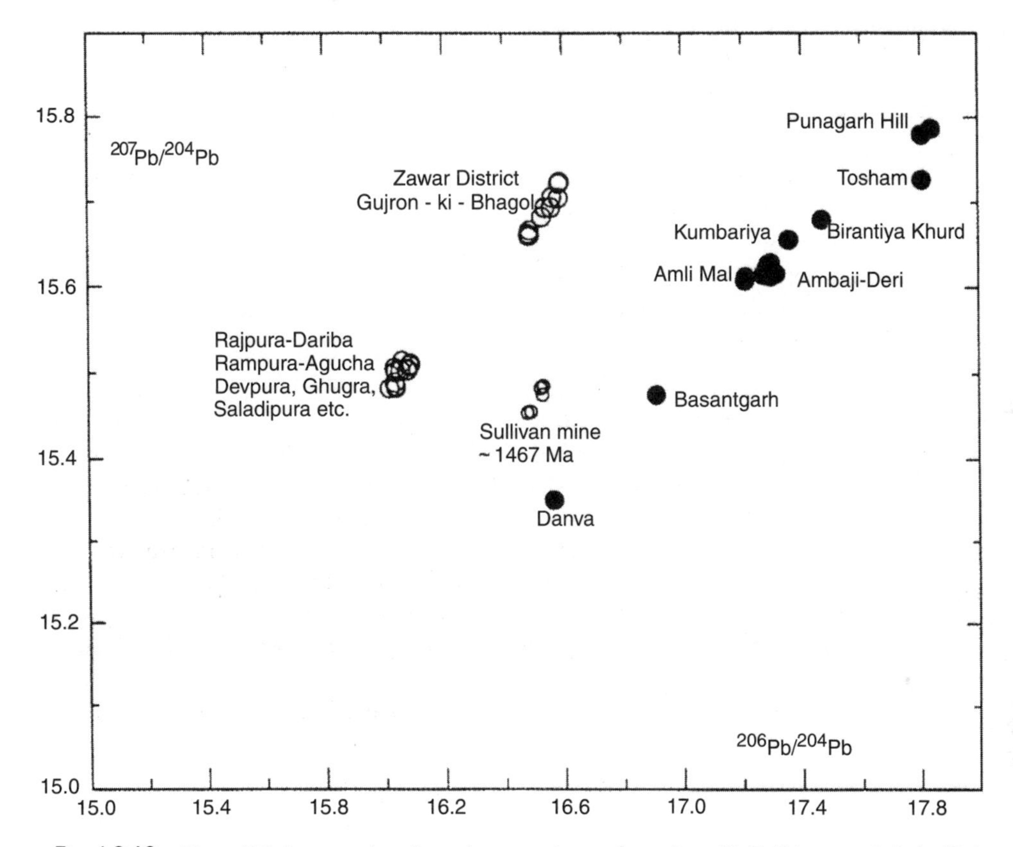

Fig. 6.2.12 Plot of Pb isotope data for galena specimens from Aravalli–Delhi orogenic belt (Deb and Thorpe, 2004)

If we do not overemphasise the importance of the veins, patches and pods of coarse grained galena-rich ores (with some concentrations of Ag, As, Tl, Hg and Au) in diopsidic rocks occurring in a small lenticular zone in the southern part of the Dariba ore body, the genesis and evolution of the ore deposits in the Dariba–Rajpura–Bethumni belt should not appear obscure in the context of the following general observations:

1. (Volcanic-) sedimentary geologic setting in an elongate zone or belt.
2. An overwhelming mass of the ores occurring in the stratiform or even stratified mode, congruently with the host rocks.
3. Participation of ore layers in all sets of tectonic structures.
4. Superposition of metamorphic fabric on the depositional- diagenetic textures/structures in the ores.

5. Profuse biogeneic activity in the ore zone, as suggested by the C_{org} content and $\delta^{13}C_{org}$ values.
6. Positive sulphur isotope values (mean $\delta^{34}S$ 1.9 per mil) that became heavier stratigraphically upwards.
7. Distinctly radiogenic lead in the ores that plot above the average upper crustal curve.

All the above features are suggestive of the stratiform ores of the Dariba–Rajpura–Bethumni belt to be sedimentary-diagenetic in origin in a rift basin (Deb et al., 1978; Deb and Pal, 2004; Sarkar, 2000) through a sedimentary-exhalative (SEDEX) mechanism. That the exhalating hydrotherms were produced by deep convective circulation is indicated by the nature of the S-isotopes, more particularly by the Pb-isotope composition.

Pur–Banera Belt Another belt (~35 km long) of basemetal deposits occurs in the central part of the Bhilwara province, extending from Pur in the southwest to Banera in the northeast (Fig. 6.2.3). The mineralised belt, though without much economic significance, discontinuously extends further to the south-west and south comprising Wari–Akola–Bhinder sector.

The rock association hosting the mineralisation comprises argillaceous, arenaceous and calcareous metasediments with mafic metavolcanics (Basu, 1971). The ore belt is about 5 km wide and contains low-grade Zn–Pb and copper mineralisations. Several prospects have been delineated in this belt, of which three deposits are important. Two of these at South Dedwas and Devpura are zinc–lead type and the other at Banera is essentially a copper deposit.

South Dedwas Zinc–Lead Deposit Muscovite–quartzite, quartz–biotite–muscovite–garnet schist, calc–schist and gneisses are the principal rock types in the area. Mineralisation occurs preferably along the interface of the mica schist and the underlying calcareous rocks. Generally sub-vertical, the rocks trend N30°E in conformity with the regional structural trend.

The mineralisation is generally stratabound. Surface indications of the mineralisation are gossan and old workings in magnetite-bearing mica schists close to its contact with the calc–schist. In the ores, sphalerite dominates over galena. Pyrrhotite and magnetite occur in considerable proportions. The gangue minerals comprise of quartz, cummingtonite, grunerite, anthophyllite, actinolite – tremolite, biotite, garnet, calcite and muscovite.

Estimated ore reserve is 8 Mt, with 2.24% Zn, 1.25% Pb and 37g/t Ag at 2% (Zn+Pb) cut off (Sahai et al., 1997).

Devpura Zinc–Lead Deposit The Devpura zinc–lead deposit has been traced over a strike length of 1 km. The major rock types of the area are calc–gneiss, overlain successively by quartz–magnetite, amphibole–biotite gneiss, and quartz– biotite–muscovite–garnet schist. Quartz–magnetite–amphibole–biotite gneiss is the principal host rock.

The ore bodies are generally stratiform and have been folded and otherwise deformed along with the host rocks. Ore concentration is noticeable along fold hinges. The ore is composed of the ore minerals like pyrrhotite, sphalerite, magnetite, galena, arsenopyrite, pyrite, chalcopyrite and gangue minerals are quartz, biotite, garnet, amphibole, magnetite, calcite, sericite and graphite.

The estimated ore reserve is 6 Mt with 2.45% Zn and 0.65% Pb at 2% (Zn+Pb) cut off (Sahai et al., 1997).

Banera Copper Deposit Following the gossan zones and old workings, detailed exploration proved an ore zone of about 450m length along the interface between the muscovite–quartzite and the calc–gneiss. Dominant sulphide minerals in the main ore zone are chalcopyrite and pyrrhotite with cubanite. Magnetite is the principal oxide phase. Disseminations of sphalerite galena and chalcopyrite are noticed in the adjacent, mainly footwall rocks.

The estimated ore reserve is only 1.5 Mt with 1.33% Cu at 0.8% cut off (Sahai et al., 1997).

The ores of Pur–Banera are also sedimentary–diagenetic and are characterised by low sulphur supply during deposition (including diagenesis–metamorphism), except locally.

Sawar–Bajta Belt Several isolated patches of metasediments, comprising calcareous and calc–magnesian silicate rocks with small proportions of carbonaceous mica schists, quartzite and iron formations occur in the Bhilwara province, one of which is exposed at the Sawar–Bajta area, located at a distance of about 140 km south Jaipur (Fig. 6.2.13).

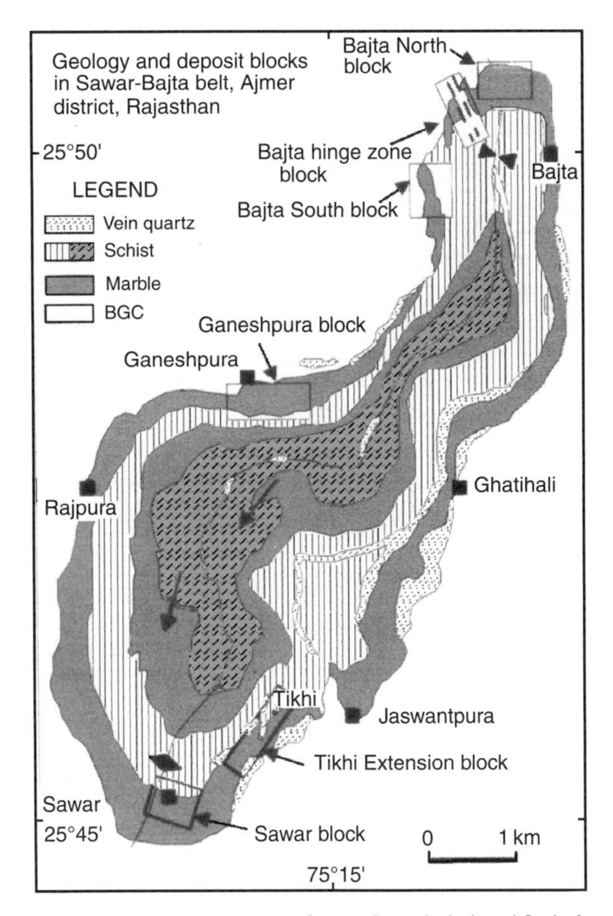

Fig. 6.2.13 Geology and ore deposit blocks in Sawar–Bajta belt (modified after Ray, 1988)

The N–S trending deformed elliptical belt of Sawar–Bajta extends for about 12 km, with a synformal hinge at Bajta in the north and antiformal hinge at Sawar in the south. Two levels of dolomitic marble, interlayered with schists and quartzite constitute the lithological sequence. There are several low tonnage Zn–Pb, Cu and Pb–Zn deposits in this area.

Zinc–lead mineralisation is concentrated in the hinge-zone of the southerly plunging antiform near Sawar. The mineralisation is confined mainly to the lower dolomitic marble unit. Estimated combined ore reserve at the Sawar Fort block and Tikhi block is 2–4 Mt with 3.99 – 4.99% of (Zn + Pb) (Sahai et al., 1997). Bhatnagar and Mathur (1989) reported appreciable quantity of cadmium and silver from these ores. In Bajta area, marble hosted stratiform Zn–Pb mineralisation occurs in the Central block [1.37 Mt with 3.0–6.4% (Zn + Pb)], Ganeshpura block [0.766 Mt with 5.75% (Zn+Pb)] and Cu mineralisation in structurally controlled locales of the North block (1.06 Mt with 1.15% Cu) and Hinge Zone blocks (3.07 Mt with 0.5% Cu). A prominent zone of Pb–Ag rich mineralisation, hosted by impure siliceous marble and quartzite has been located by the GSI in the Tikhi Extension block (southeast of Sawar hinge) over a strike length of 500 m, having ore reserve estimates of 2.58 Mt with 5.54% Pb and 85 ppm Ag.

Jahazpur Belt An 80 km long ore belt runs through Jahazpur and extends from Nainwa in Tonk district to Bigod in Bhilwara district. The rock types are dolomitic carbonate, chert, banded ferruginous chert and carbonaceous phyllite. The BIF and the carbonaceous phyllites have been the target lithologies for the exploration of base metals. Old workings in the form of trenches, pits and inclined shafts are present at several places. Bedrock samples of BIF and dolomite on analysis showed up to 0.4% Cu.

BIF contains several zones rich in ore metals but never rich enough to be of economic interest (Cu 350–4000 ppm, Pb 225–5000 ppm and Ni and Co between 150–1500 ppm). Interestingly slags collected from the region and analysed give contents of the above metals almost in the same order. Mo is present in the range of 150–1500 ppm. However, its content shows an inverse relationship with those of base metals (Malhotra and Pandit, 2000).

Recently, Ranawat et al. (2005) reported lumps of native antimony with minute grains of Pb–Te, as floats from this area. Its significance is yet to be fully understood.

The salient features of the sulphide deposits of Bhilwara Province are presented in Table 6.2.3 for comparison.

Table 6.2.3 *Sediment-hosted sulphide deposits in Bhilwara Province, Rajasthan*

Deposits/ore belts	Host and associated rocks	Thermo- tectonic transformation of the host and associated rocks	Ore bodies and their relation with the host rocks	Ore reserve, ore metals etc.
Rampura–Agucha deposit	Ore body hosted by graphite–muscovite–sillimanite schist. Garnet– biotite–sillimanite gneiss with amphibolite and calc–silicate rocks on two sides.	Several phases of deformation. Regional metamorphism: upper amphibolite facies. No sharp consistent compositional banding present.	A lensoid body, one-half of which is eroded. Ore body is concordant with the schistosity of the host rocks.	63.7 Mt; Zn, Pb, Ag. Zn/ Pb ≈ 7

(Continued)

(Continued)

Deposits/ore belts	Host and associated rocks	Thermo- tectonic transformation of the host and associated rocks	Ore bodies and their relation with the host rocks	Ore reserve, ore metals etc.
Dariba –Rajpura–Bethumni belt	Recrystallised dolostone and graphite-bearing mica- schists. Calcareous biotite schist occurs on the two sides. Subordinate chert and tuff are also present.	Several phases of folding and metamorphism (up to middle amphibolite facies). Sharp and persistent compositional banding well-preserved in some rocks.	Sheet-like ore bodies structurally congruent with the host rocks. Internally stratified at many places. Ore layers participated in all phases of deformation. Ores isofacially metamorphosed with the host rocks	29.5 Mt; Zn, Pb, Cu, Ag. Zn/Pb ≈ 3
South Dedwas deposit (Pur–Banera belt)	Mica schist, calc–schist and micaceous quartzite. Mineralisation along the contact of mica schist and calc–schist.	Complexly folded, and metamorphosed to amphibolite facies. Sharp and persistent compositional layering preserved.	Ore bodies stratiform and even stratified. Participated in all phases of fold movements. Ores metamorphosed	8 Mt; Zn, Pb, Ag. Zn/Pb≈ 2
Devpura deposit (Pur–Banera belt)	Ore hosted by quartz–magnetite–amphibole– biotite gneiss. Other associated rocks: calc–gneiss and garnetiferous mica schist.	Complexly folded and metamorphosed to amphibolite facies. Compositional banding preserved.	Ore bodies stratiform and even stratified. Participated in all phases of fold movements. Ores metamorphosed and contain graphite.	6 Mt; Zn, Pb. Zn/Pb ≈ 5
Banera deposit (Pur–Banera belt)	Ore mineralisation at the interface of micaceous quartzite and calc–gneiss.	Complexly deformed and metamorphosed to amphibolite facies.	Broadly, stratiform ores metamorphosed and contains graphite.	1.5Mt; Cu–1.35%
Sawar–Bajta belt	Mineralisation hosted mainly by dolomitic marble. Also present are calc–silicate rocks, carbonaceous mica schists, quartzite and iron formations.	Multiply folded and metamorphosed to amphibolite facies.	Stratiform ore bodies. Ores metamorphosed.	2–4 Mt at Sawar Fort and Tikhi blocks. Zn, Pb, Ag, Cd.

Sulphide Mineralisations at Zawar, Aravalli Belt

The ore mineralisation in the Aravalli belt, principally of zinc and lead, is confined to the Zawar belt, Udaipur district (Fig. 6.2.14). This is a nearly 20 km long mineralised belt, which is both geologically interesting and economically important. The Zn–Pb ores in the belt are mined at Balaria, Mochia Magra, Baroi Magra and Zawarmala. The intensity of mineralisation can be estimated from the ore reserves (including the tonnage already mined out) and the corresponding ore grade or the metal content (Table 6.2.4). Of these deposits, the Mochia Magra has been in operation for more than 50 years (leaving aside the ancient mining, of course!) and the Zawarmala is being mined for about two decades. A rich but small deposit occurs at Hameta Magra, 3–4 km away towards NNW from the Mochia (west) mine. Ag and Cd are valuable accessory metals in the Zawar ores and are extracted as by-products.

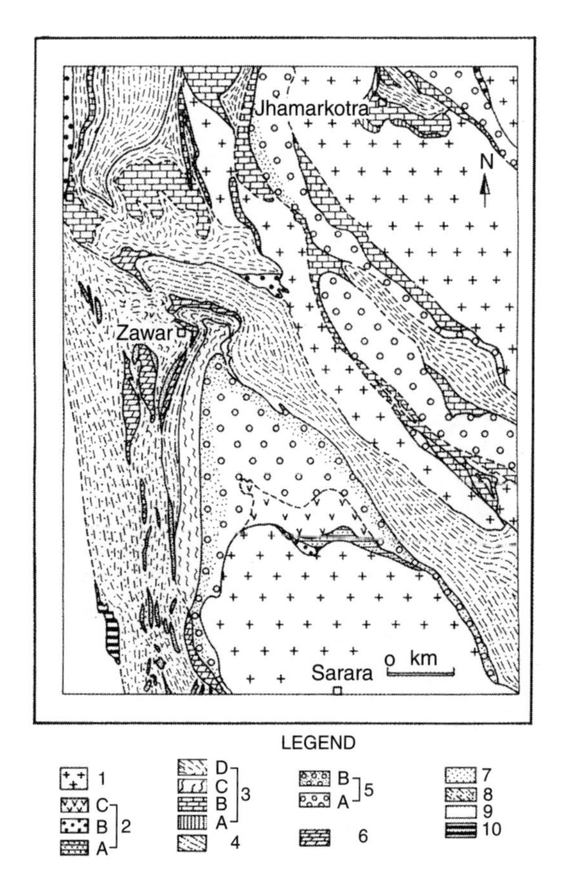

Fig. 6.2.14 Geological map of Jhamarkotra–Zawar–Sarara area, Aravalli belt, Rajasthan (after Roy et al., 1988)

Table 6.2.4 *Ore reserves and principal ore metal contents of Zawar deposits (Sarkar and Banerjee, 2004)*

Deposits	Ore reserves (Mt)	Average grade	Ore metals (x 10^5 t)
Balaria	16.0	5.66% Zn, 1.44% Pb	9.06 Zn + 2.30 Pb
Mochia Magra	19.3	3.8% Zn , 1.7% Pb	9.3 Zn + 3.28 Pb
Baroi Magra	11.0	1.33% Zn , 4.22% Pb	4.46 Zn + 4.72 Pb
Zawarmala	16.0	3.72% Zn , 2.16% Pb	5.95 Zn +3.46 Pb

Geologic Setting The rock association in the Zawar belt consists of phyllite–quartzite–carbonate (dolostone) with greywacke (Fig. 6.2.15). Besides, there is an uncommon rock unit, carbonaceous and dolomitic phyllite. The generally accepted stratigraphic sequence in the area is as follows:

1. Phyllite and mica schists
2. Quartzite
3. Dolostone
4. Greywacke
5. Carbonaceous and dolomitic phyllite.

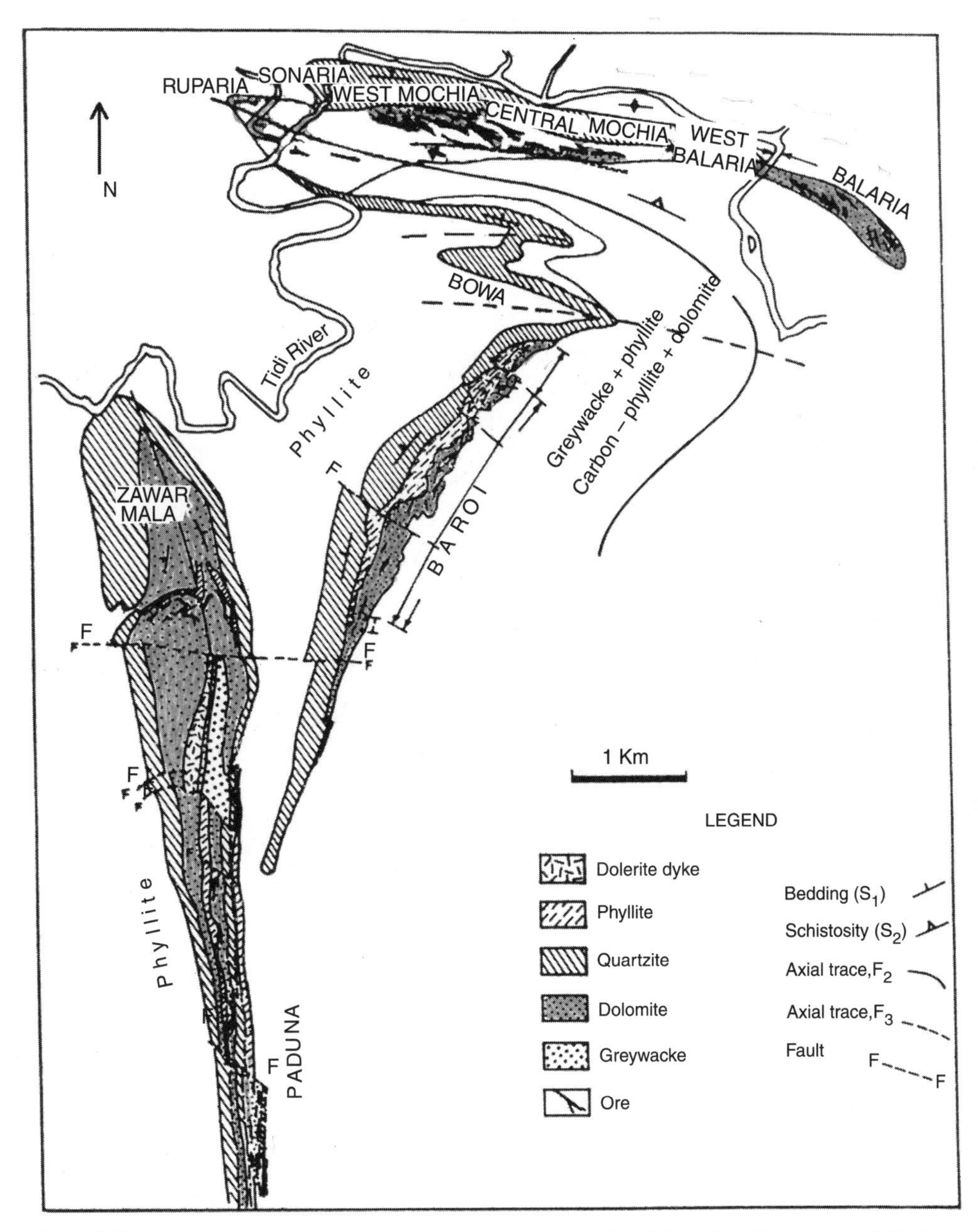

Fig. 6.2.15 Generalised geological map of Zawar ore belt (modified after Bhatnagar and Mathur, 1989)

The nearest exposure of the Precambrian basement rocks is at Sarara, 1–2 km from Zawar belt in the S–SE direction. The above sequence is separated from the basement rocks by the local development of conglomerate and basic volcanic rocks. The basal members are represented elsewhere by dolomitic carbonate, stromatolytic phosphorite, quartzite, and phyllite. The mineralisation occurs in the Middle Aravalli Group in a 3–tier classification of the Aravalli Group of rocks in the Udaipur area already discussed. It may be mentioned here that the depositional breaks reported from there are not always obvious in the Zawar area.

Host Rocks The carbonaceous and dolomitic phyllite consists essentially of muscovite + chlorite + ferroan dolomite (FeO/MgO up to 3.7, n = 7) + microcrystalline carbon (C_{org}/C_{total} up to 0.92, n = 7). The maximum Zn- content and Pb-content in the rock are 380 ppm and 45 ppm (n = 6) respectively. The greywacke is composed of quartz (45%–55%), alkali feldspar (25%–40%), rock fragments (up to 10%), mica (+ chlorite) (up to 10%) (Banerjee and Sarkar, 1998)). In high strain domain the rock is conspicuously schistose (Fig. 6.2.16 a).

The principal mineral in the carbonate rock is dolomite. However, with the increase of other minerals, the rock locally grades into quartzose or feldspathic dolostone (Fig. 6.2.16b). There are also interbeds of phyllitic dolomite and dolomitic phyllite. Laminations in the carbonate rock are present but not ubiquitous. Locally, rocks resembling debris flow are present. Sachan (1993) suggested, based on low REE-content and $\delta^{13}C$, that the dolomitisation process must have been initiated by seawater through tidal pumping in the tidal flat environment. Xenomorphic texture, low Na and Sr contents and light oxygen composition suggest that it subsequently underwent burial dolomitization. Many depositional–diagenetic features in dolomite, such as micrite, peloids, catagraphs, pisolites, and ooliths are in general absent in the dolostone.

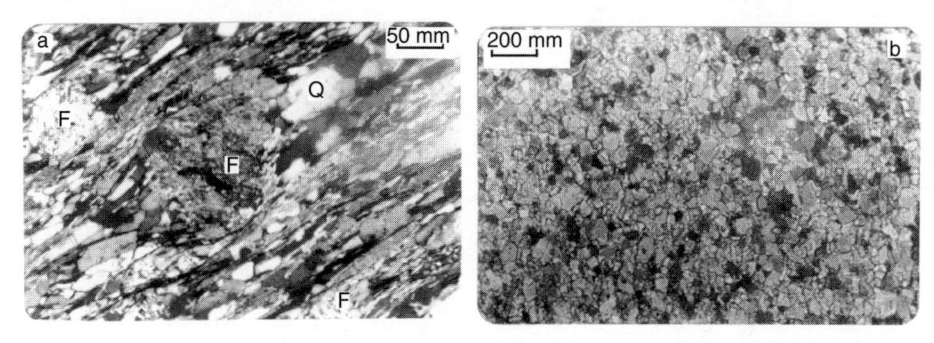

Fig. 6.2.16 Host rocks of Zawar Zn–Pb deposit (a) schistose greywake, (b) recrystallised impure dolostone (Banerjee and Sarkar, 1998). F-feldspar, Q-quartz

There are no highly sensitive T and P sensors amongst the metamorphic minerals or mineral assemblages in the rocks. However, well-crystallised coarse biotite (K-feldspar/phengite + chlorite → biotite) in greywacke and stoichiometric muscovite in more than one rock type suggests that the metamorphic temperature for these rocks was ≥ 400°C (cf. Barker, 1990). The tremolite forming reaction, dolomite + quartz → tremolite + calcite, is constrained by X_{CO_2} in the fluid present and P_{total} of the system. Fluid inclusion study of the siliceous dolomite in the ore zone showed X_{CO_2} to be around 0.3 (vide discussion on fluid inclusions in a following section). Allowed the assumption that the fluid in the ore zone is cognate to the system at large, the inability to develop tremolite in the given rock may mean that the metamorphic temperature was < 440°C (Banerjee and Sarkar, 1998).

Structures and Tectonics The area is located along the eastern fringe of the Proterozoic Aravalli basin, which, as we have already discussed at an earlier stage, was constrained by vertical tectonics, i.e. rifting, producing a number of horsts and grabens. In the inversion stage the sediment-dominant pile was deformed through several events, designated as D_1 to D_3, by a number of workers. The F_1 folds on bedding (S_0) are generally compressed and produced S_1 cleavage/schistosity during D_1. F_1 is generally small in scale, but a F_1 fold in quartzite, just west of Zawarmala is megascopic. The fold structure that controls the map-pattern in the Baleria–Mochia Magra–Bawa Magra sector and also in Zawarmala is F_2, developed during D_2 deformation on both S_0 and S_1 and produced S_2 cleavage at the same time. The third phase of deformation D_3 produced both megascopic and mesoscopic folds and a cleavage, S_3. There is yet another relatively weak phase of deformation D_4 with sub-horizontal axes, conspicuously developed on mesoscopic scale in the phyllites.

Mineralisation

Ore Bodies and Their Structures All the deposits in the Zawar belt are stratabound be they at Balaria, Mochia Magra, Baroi Magra or Zawarmala. The confining rock unit is dolomite, belonging to the middle Aravalli Group. In details they, however, vary. Mineable ore bodies in the Balaria and Mochia Magra sectors vary in length, thickness and disposition. They are discordant to the boundaries of the host dolostone unit in general (Fig. 6.2.17 and Fig. 6.2.18).

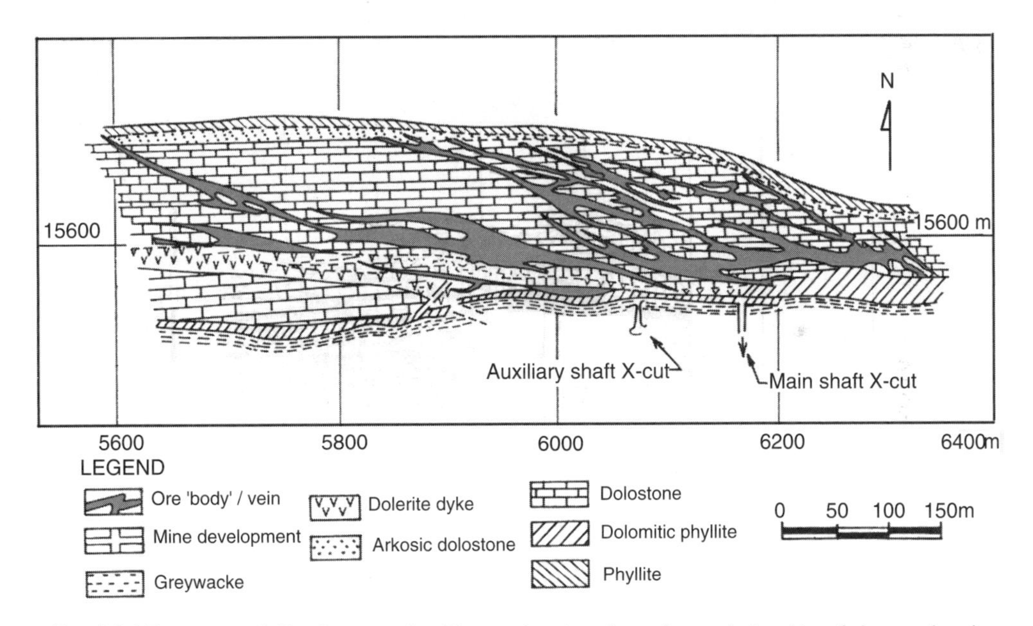

Fig. 6.2.17 Mochia 240 mL mine plan, Zawar, showing discordant relationship of the ore bands with the enclosing host dolostone (after HZL mine plans)

Within the dolostone unit, they cut across the arkosic dolostone interbeds, but generally avoid the phyllitic interbeds. In small scales also the ores are veinitic (Fig. 6.2.19). In some of the phyllitic interbeds, apparently laminated ores are noticed at places, which bear features that are commonly accepted as diagenetic to tectonic. Volumetrically, these laminated ores are insignificant compared to the vein-type ore bodies.

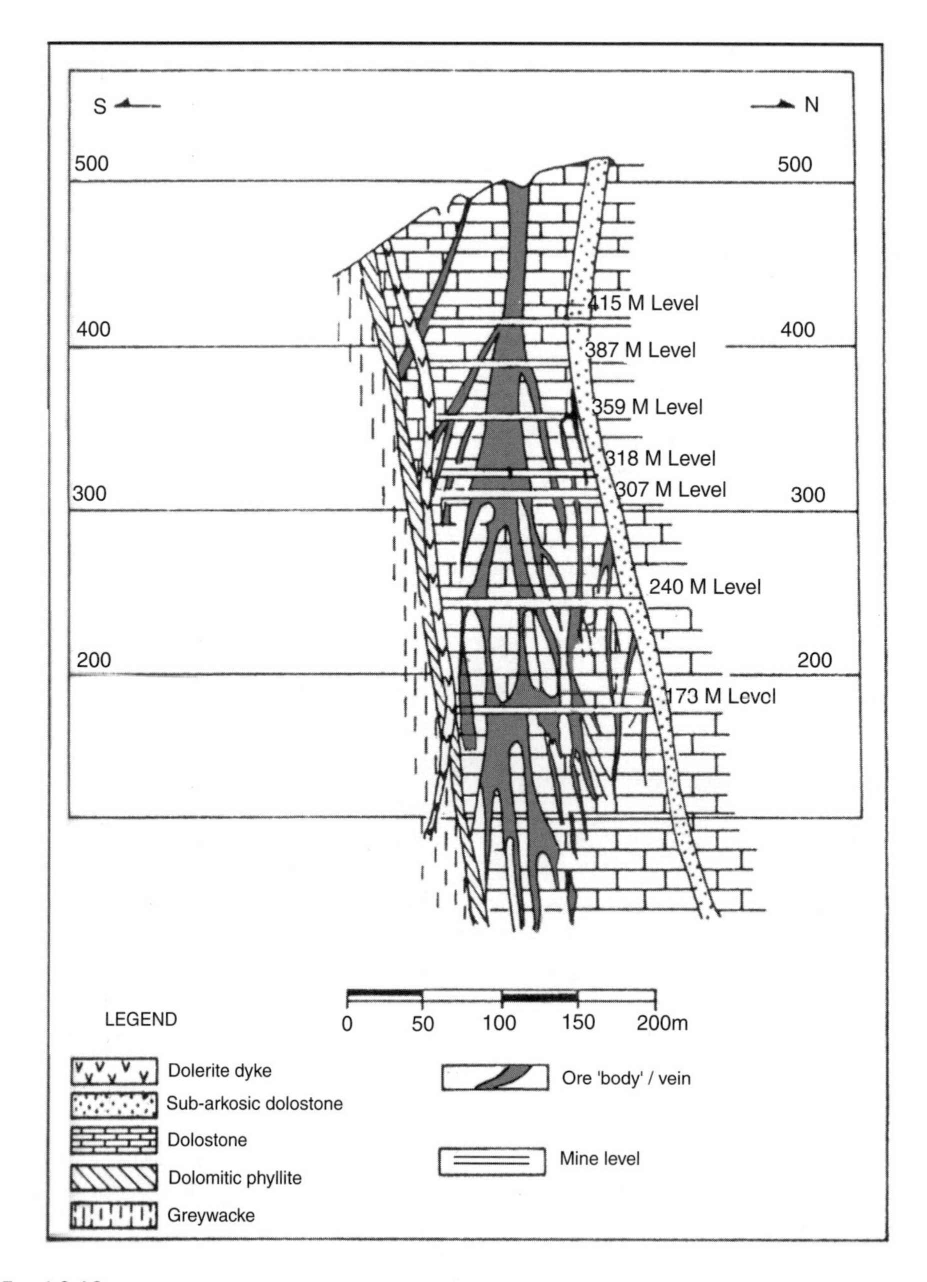

Fig. 6.2.18 Transverse section across ore body in Mochia Mines, Zawar, displaying the pattern of ore body disposition (after HZL Mine office)

At Baroi Magra, mineralisation occurs as a few impersistent sub-parallel sheets. At Zawarmala, the mineralised zone may be construed to have been a thick lens, which has been later flattened folded and broken into blocks varying in ore concentration (Fig. 6.2.20). Here the ores vary from laminated to vein type in details, the latter being more common (cf. Talluri et al., 2000).

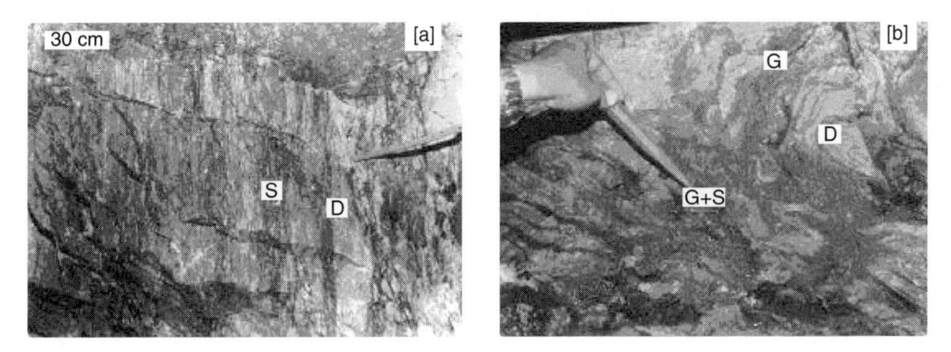

Fig.6.2.19 (a, b) Small scale structures displayed by sphalerite (S)–galena (G) ores in dolomite (D) in the underground mines (photographs by authors)

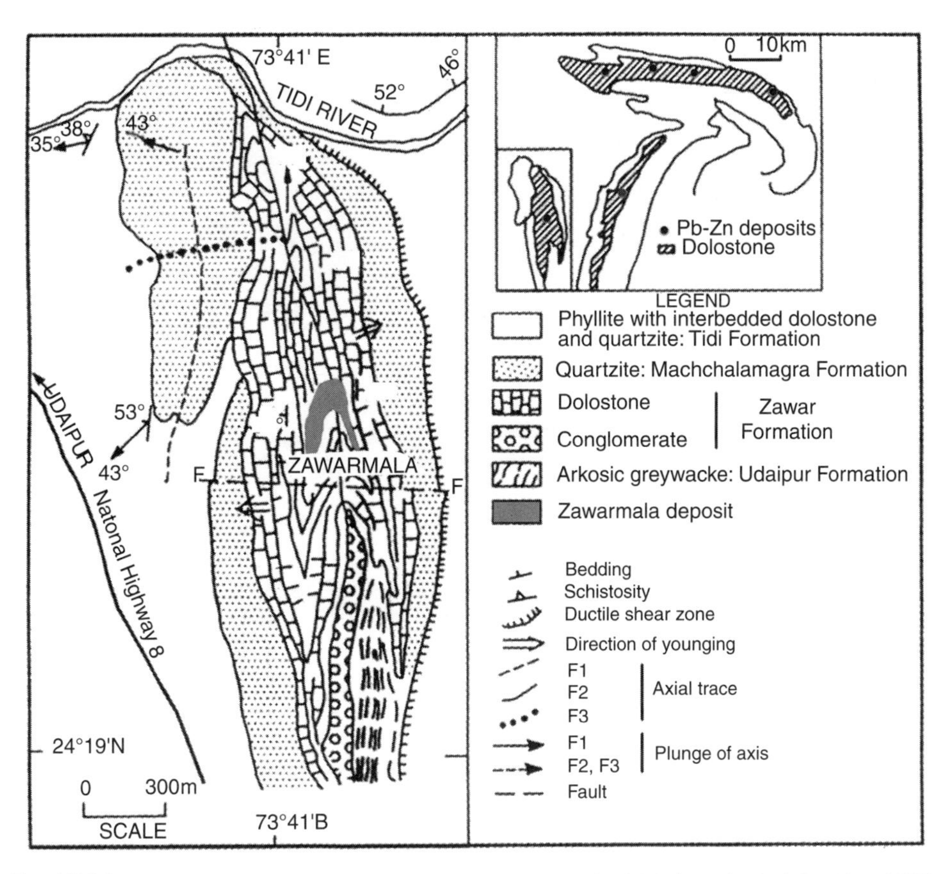

Fig. 6.2.20 Geological map of Zawarmala deposit showing the folded ore body (after Roy, 1995)

Ore Petrography The principal ore minerals in the Zawar deposits are sphalerite, galena and pyrite, occurring in different proportions. Arsenopyrite is a minor phase, but ubiquitous. Pyrrhotite, chalcopyrite, argyrodite and native silver occur as trace phases. Cadmium and silver, as already mentioned, are by-products from these ores. Cadmium occurs entirely within the sphalerite structure, while silver does so in galena, only partially (Table 6.2.5). A fair part of the silver occurs as native silver and as argyrodite. In hand specimen most ores appear massive, veined or crudely layered. Late sphalerite, small in proportion, is generally honey-brown, in contrast to the greyish nature of the dominant variety.

Pyritic ores occur as massive aggregates or crudely banded masses, with impersistent sub-parallel patches of sphalerite within. Sphalerite–galena ores occur as massive aggregates varying in grain sizes (>1 cm < 0.1 mm). Brecciated and crudely banded ores are locally present.

Table 6.2.5 *Composition (wt%) of some sphalerite samples from the Zawar belt (Sarkar, and Banerjee, 2004)*

Elements	WM/200 (n = 4)	M–9 /91 (n = 5)	M9(A)/91 n=2	M/100 (n = 4)	BL/18 (n = 5)	BL/19 (n = 6)
Zn	60.64	60.46	64.18	64.31	64.48	61.10
Fe	6.60	6.17	2.72	3.24	2.91	5.32
Cd	0.27	0.18	0.25	0.28	0.07	0.33
Mn	0.34	0.03	–	0.21	0.03	–
Pb	0.77	1.18	0.51	0.73	–	–
Ag	0.25	–	–	–	–	–
Co	0.48	0.002	–	0.001	0.01	0.01
Cu	0.26	0.004	0.03	0.175	0.001	0.02
S	31.82	31.06	31.35	31.72	32.25	32.30

Note: WM and M refer to Mochia mine, BL to Balaria mine

Fluid inclusions in light-coloured sphalerite and co-existing quartz grains were found to be composed of brine, undersaturated in NaCl. Other dissolved salts are KCl, $CaCl_2$ (+$MgCl_2$). CO_2 is present both as a gas and as a liquid. CH_4 is also present.

Microstructures and texture in the Zawar ores range from diagenetic through deformational to recrystallisational varieties. The primary, i.e. diagenetic microstructures, better preserved in the low-strain areas are represented by pyrite framboids, zoned pyrite grains and pyrite–sphalerite and pyrite–galena zonal intergrowths. Cataclastic fabric is best shown by the tensional fracturing of pyrite aggregates (pyritite), tension and compression related fracturing of pyrite and sphalerite, cataclastic flow in sphalerite and dislocation creep in galena. Recrystallisation in the ores is achieved in varying degrees, through the development of crenulation cleavage in galena to polygonisation in sphalerite and galena, annealing twins in sphalerite and martensite in galena, where recrystallisation outlasted deformation (Sarkar and Banerjee, 2004) (Fig. 6.2.21).

Gangue minerals consist predominantly of dolomite and quartzite, with alkali feldspar and muscovite (sericite) as sub-ordinate, but important phases. Tourmaline is generally a trace phase, except very locally. Grain-size increase in varying degrees is noticeable in the ore veins. Even patches of pegmatoid ores are found, rarely though. There is small but perceptible increase of Fe-content in dolostone in the ore zone.

Stratabound Native Sulphur and Patches of Gypsum A stratabound, even stratiform, native sulphur 'deposit' is reported from the eastern part of Mochia Magra (Basu, 1976). The body is 1.5 m thick and consists of alternating layers of native sulphur and carbonaceous matter. The native sulphur-bearing unit is underlain successively by carbonaceous shale, gypsiferous shale and ferruginous shale, and is overlain successively by carbonaceous shale, gypsiferous shale and ferruginous argillites. The principal ore-bearing carbonate horizon overlies the native sulphur-bearing horizon by several meters. The inescapable conclusion from this observation now is that the native sulphur was produced by organic reduction of the evaporitic sulphate in the system.

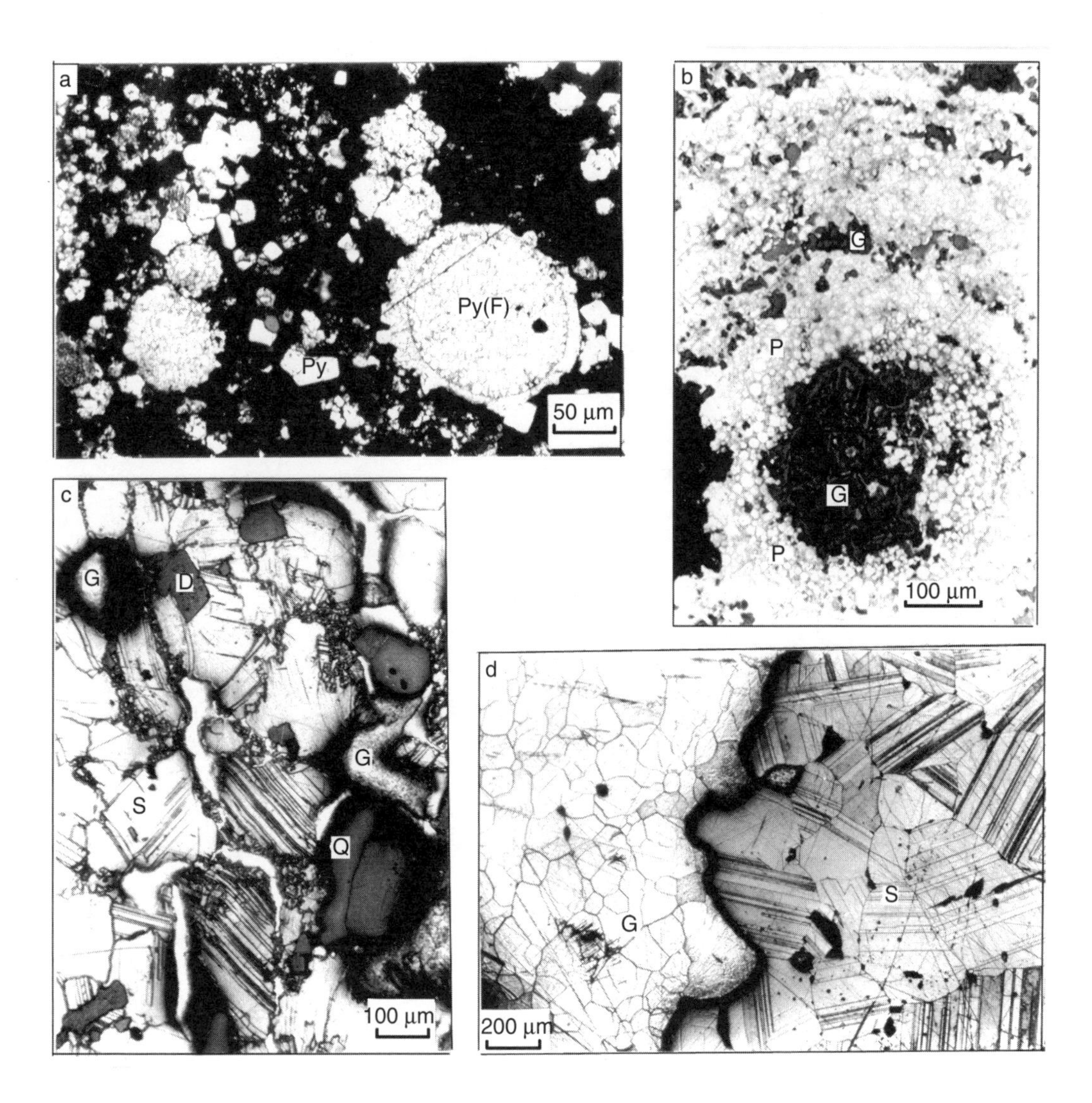

Fig. 6.2.21 Photomicrographs showing textures in the Zawar ore: (a) Framboidal pyrite displays low temperature (diagenetic) Growth; (b) Zonal intergrowth of galena (G – blackened by chemical etching) and pyrite (P); (c) Deformation twin and cataclastic flow of sphalerite, and dislocation creep in galena; (d) Recrystallisation fabric in sphalerite–galena .(G – galena, S – sphalerite, Py/P – pyrite, Py(F) – framboidal pyrite) (unpublished work, Sarkar and Banerjee).

Isotope Data Limited data on sulphur, carbon, oxygen and lead isotopes from the ore and associated minerals are available from the belt. The $\delta^{34}S$ data obtained by Deb (1990a) and Sarkar and Banerjee (2004) are presented in Table 6.2.6 and Table 6.2.7 and Fig. 6.2.22 respectively.

Table 6.2.6 *Sulphur isotope data for Zawar ores (Deb, 1990 a)*

Deposit	Sample No.	Remarks (Host rock, sulphide minerals)	$\delta^{34}S$ (per mil)		
			Galena	Sphalerite	Pyrite
Mochia Magra	ZMC 2	Dolomite, galena	2.8		
	ZMC 6	Dolomite, galena–sphalerite	5.6	7.8	
	ZMC 7	Dolomite, galena	5.6		
	ZMC 9	Dolomite, galena	7.1		
	ZMC 17A	Dolomite, galena	5.4		
	ZMC 18	Dolomite, sphalerite	4.8	8.4	
	ZMC 20	Dolomite, galena–sphalerite	6.0	7.7	
			X = 5.3	Z = 80	
Balaria	BL2	Dolomite; sphalerite–pyrite		3.5	
	BL4	Dolomite; sphalerite–galena	0.9	1.3	4.1
	BL5	Dolomite; sphalerite–pyrite		0.9	
	BL6	Dolomite; sphalerite–pyrite		0.1	1.7
	BL7	Dolomite; sphalerite–pyrite		7.1	10.6
	BL8	Dolomite; sphalerite–pyrite		1.3	
	BL9	Dolomite; sphalerite–galena		1.7	
	BL10	Dolomite; sphalerite–galena		1.3	
	BL11	Dolomite; sphalerite–galena		1.3	
				X = 2.1	X = 5.4
Zawarmala	ZM 4	Dolomite, sphalerite–pyrite	6.5		
	ZM5	Dolomite, sphalerite–galena–pyrite	9.3		
	ZM12	Dolomite, sphalerite–pyrite			
	ZM22	Dolomite, sphalerite–pyrite– galena			
	ZM25	Dolomite, sphalerite–pyrite			
	ZM26	Dolomite sphalerite			
	ZM27	Dolomite, sphalerite–galena	6.7		
			X = 8.0	X = 10.5	X = 10.3
Baroi Magra	BM 3	Quartzite, sphalerite		2.0	
	BM 6	Quartzite, sphalerite		3.8	
	BM 7	Quartzite, galena banded	4.2	X = 2.9	

Table 6.2.7 *Sulphur isotope composition of Zawar sulphides (Sarkar and Banerjee, 2004)*

Sample No	$\delta^{34}S$ (per mil) of co-existing sulphide pairs	
	Sphalerite	Galena
BL/250	5.8	4.4
BL/21	20.7	20.4
BL/7	5.8	2.3
BL/120	1.4	1.2
BL/8	1.7	-0.1
BL/9	0.7	-2.3
M/52	6.1	5.6
M/13 (3W)	5.1	7.1
M/15	7.8	6.0
B-5	10.7	7.6
B-13	8.5	5.9
Zm/443	16.2	16.2
Zm/443/90	18.8	18.6
	sphalerite	pyrite
BL/11	3.4	2.3
M/4	4.6	5.2

Note: BL – Balaria mine, M – Mochia mine, Zm – Zawarmala

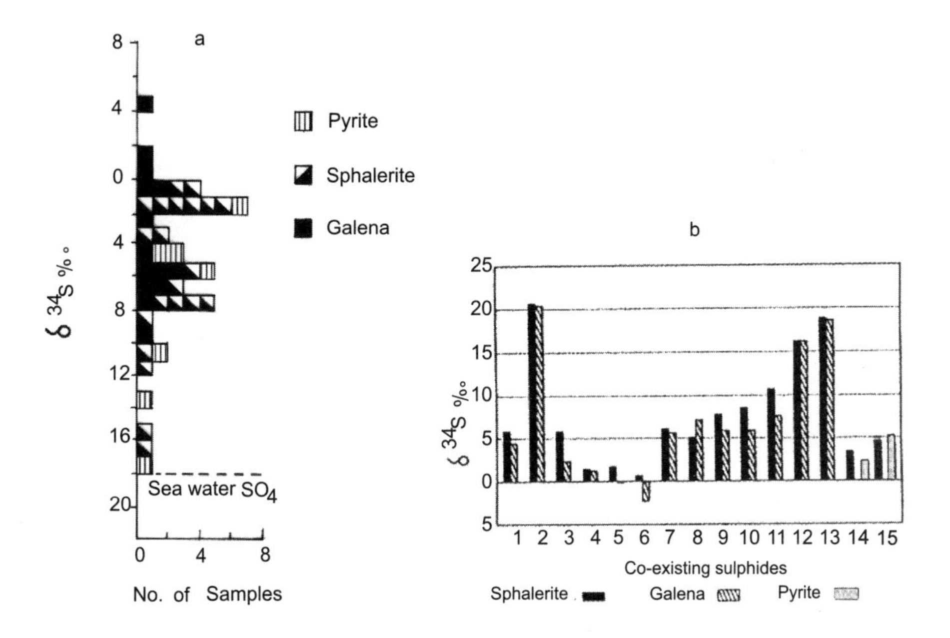

Fig. 6.2.22 (a, b) Sulphur isotope ($\delta^{34}S$) data of sphalerite, galena and pyrite from Zawar ores (Sarkar and Banerjee, 2004)

If the theoretical predictions of sulphur isotope fractionation based on bond-strength (Bachinski, 1969; Rye and Ohmoto, 1974) are correct, then the $\delta^{34}S$ enrichment as a function of temperature should be in the order, pyrite > sphalerite > chalcopyrite > galena. This is not found to have been strictly followed in all the samples from Zawar, suggesting non-attainment of equilibrium in some cases. Alternatively, the large spread of $\delta^{34}S$ in the Zawar belt sulphides suggests at least partial biogenic fractionation of the sulphur isotopes. Equilibrium partitioning of sulphur isotopes between the co-existing sulphide pairs has been only partial.

Carbon and oxygen isotope results from the ore zone carbonates (dolomite) are given in Table 6.2.8 and Fig. 6.2.23. The original carbon-isotope signatures may be overprinted during diagenesis and metamorphism but not erased (Schidlowski and Ahron, 1992). Decarbonation reactions during metamorphism may take place in impure carbonate rocks and this may shift the $\delta^{13}C$ towards negative value up to 2.5 per mil (PDB). This did not happen in this case (Sarkar and Banerjee, 2004). The mean positive $\delta^{13}C$ values of the analysed carbonate rocks from Balaria – Mochia and the adjacent areas vary from +1.4 to +4.32 per mil (PDB). These values are consistent with a marine environment. Another cluster characterised by negative values indicates biogenic process, which has already been suggested by sulphur isotope compositions. The $\delta^{18}O$ values of the same carbonate, barring some very low values (-24.41, -31.84 and -40.63 per mil), also indicate a marine origin (Hoefs, 1973)

Table 6.2.8 *Carbon and oxygen isotope composition of ore zone carbonate, Zawar belt (Sarkar and Banerjee, 2004)*

Sample No.	$\delta^{13}C$ (per mil)	$\delta^{18}O$ (per mil)
BL/9	-7.98	-2.81
	-9.65	+0.36
	+1.68	+4.49
	-1.32	-0.72
	-4.82	+1.24
	-18.88	-24.41
BL/14	-0.23	-3.74
	+2.50	-2.96
	+5.64	+1.15
	-1.70	-4.96
M-9	+2.081	-31.84
	+0.237	-40.69
	+4.14	-2.30
Zm-7	+3.07	-1.33
	+2.08	-2.10
	+1.28	-4.44
	+2.58	-3.07
	+3.57	-0.96
Zm -3B/2	+3.34	+0.48
	+2.20	-1.13
Zm-3B-1	+5.07	-3.00
	+3.57	-3.29
Zm-3B-2	+3.34	-0.48
	+2.20	-1.13

Note: BL – Balaria, M – Mochia Magra, Zm – Zawarmala

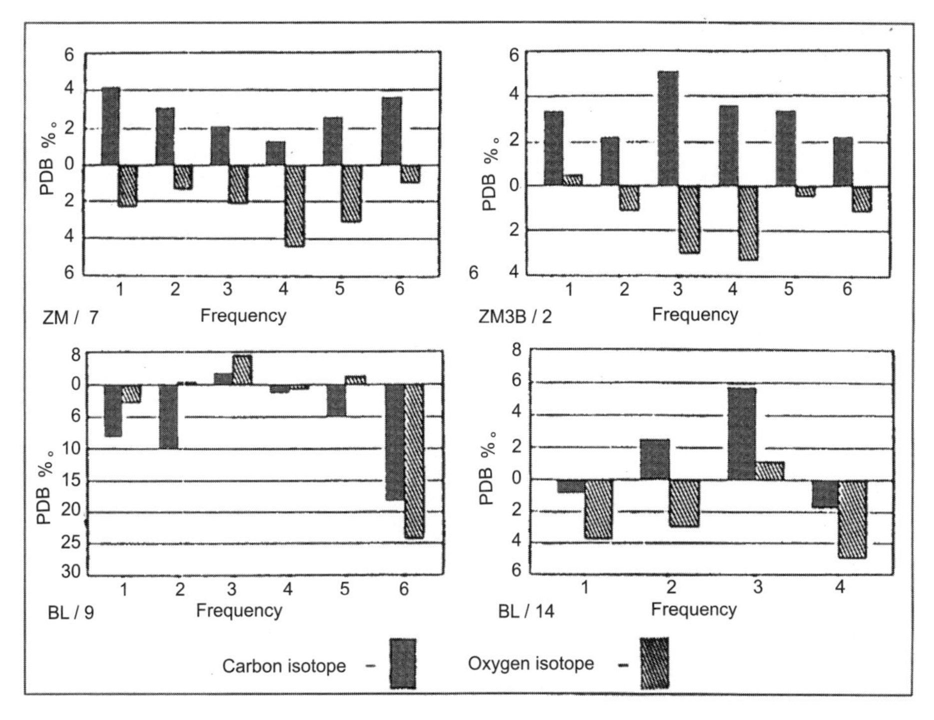

Fig. 6.2.23 Carbon and oxygen isotope data of Zawar ore (Sarkar and Banerjee, 2004)

Pb- isotopes from the Zawar belt have distinct characteristics (Deb et al., 1989). This lead is not only inhomogenous, but bears evidence of having been derived from a source that had a high μ-value (calculated $^{238}U/^{204}Pb$) for a considerable period of geological time (Fig. 6.2.24). The Zawar ores are very rich in ^{207}Pb and the data plot above the growth curves of Stacey and Kramers (1975). The Pb shows an appreciable compositional variation and plot along a steep linear trend with a steep slope of ~ 0.51. The determined (model) age of Zawar Pb is ~1700 Ma (Table 6.2.9).

Table 6.2.9 *Composition of Zawar lead and model age(s) of mineralisation (Deb et al., 1989)*

Deposit	N	$^{206}Pb/^{204}Pb$	$^{207}Pb/^{204}Pb$	$^{208}Pb/^{204}Pb$	'The Proterozoic model'	
					Source μ	Model age (Ma)
Mochia Magra	3	16.489	15.888	36.383	11.244	1710
Balaria	3	16.488	15.661	36.354	11.205	1702
Zawarmala (ZM5)	1	16.587	15.722	36.527	11.446	1712
Zawarmala (ZM22)	1	16.561	15.691	36.449	11.314	1691
Zawarmala (ZM27)	1	16.540	15.694	36.480	11.337	1708

Origin and Evolution of Zawar Ores The characteristics of the ore deposits of the Zawar belt are presented in Table 6.2.10. To understand and explain the features of the Zawar ores, as in case of most other deposits, we need to understand the source of the ore forming elements, their collection and transport to the site of deposition, deposition of the ores and their post-depositional transformation, remobilisation/translocation, if any.

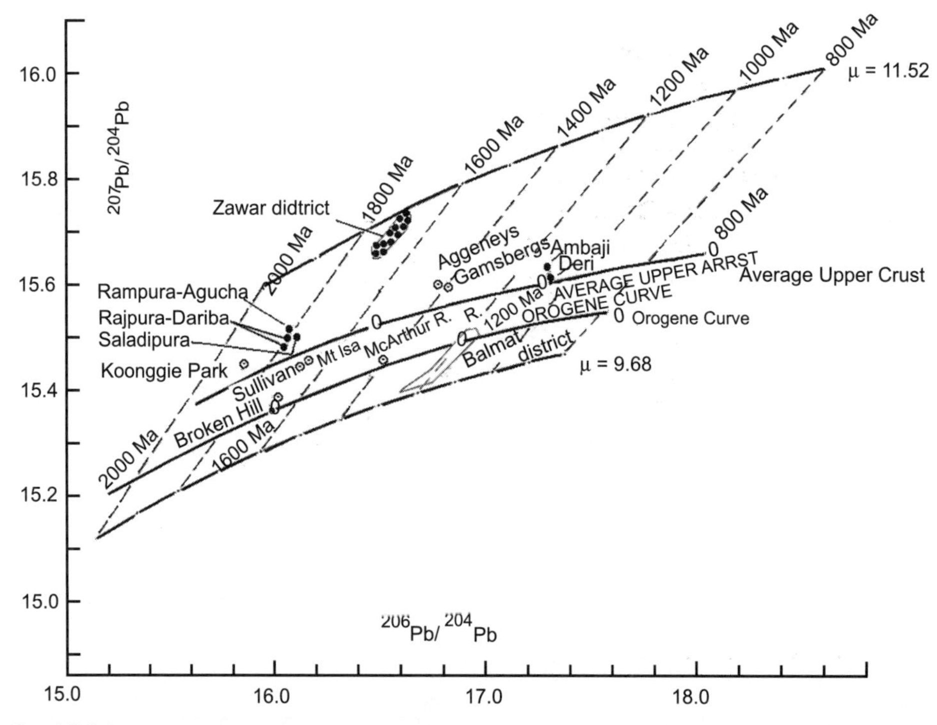

Fig. 6.2.24 Lead isotope data for sediment-hosted Zn–Pb deposits in Aravalli–Delhi belts, Rajasthan (adapted and modified from Deb et al., 1989)

Table 6.2.10 *Characteristics of sediment (carbonate)-hosted Pb–Zn ore deposits in Zawar belt, Rajasthan*

Deposits	Balaria–Mochia Magra (BMM)	Baroi Magra (BM)	Zawarmala (ZM)
Ore reserves, ore metals and metal ratio	Balaria 16 Mt; Zn, Pb, Ag, Cd; Zn/(Zn+Pb) = ~ 0.8 Mochia Magra 19.3 Mt; Zn, Pb, Cd, Ag; Zn/(Zn+Pb) = ~0.69	11Mt; Pb, Zn, Ag, Cd; Zn/(Zn+Pb) = ~0.24	16Mt; Zn, Pb, Cd, Ag; Zn/(Zn+Pb) = ~0.63
Host and the associated rocks	Dolomite, with quartzite in the hanging wall side and greywacke/litharenite in the footwall side	Same as BMM	Same as at BMM and BM
Ore bodies and their relation with the host rocks	Stratabound, but discordant to the host rock. Host and the associated rocks dip northward and ore bodies towards south; both dip steeply	Stratabound and broadly stratiform	Originally peneconcordant, discordant second order structures are of tectonic origin
Wall rock alterations	Recrystallisation and limited silicification, sericitization. Fe-increase in carbonates, local development of alkali–feldspar, tourmaline	Same as BMM, but less intense	Same as at BMM, but less intense

(Continued)

(Continued)

Deposits	Balaria–Mochia Magra (BMM)	Baroi Magra (BM)	Zawarmala (ZM)
Ore mineralogy	Sphalerite, galena and pyrite are the major phases. Arsenopyrite and argyrodite minor phases. Cd in sphalerite. Ag also in sphalerite and galena. Growth, deformation and recrystallisation fabrics	Same as BMM, except that galena > sphalerite and Ag-content	Same as at BMM
Sulphur and lead composition	$\delta^{34}S$ widely variable. Pb predominantly radiogenic, having high (above crustal) μ- values	Same as BMM	Same as at BMM
Deformation and metamorphism.	Obvious. Low temperature growth textures and micro- structures are preserved at low strain areas. Middle greenschist facies	Deformation less than at BMM. Ore fabric more or less as at BMM	Nearly same as at BMM
Tectonic regime of mineralisation	Emplacement in the present forms and disposition during orogeny	Could be pre-/ early orogenic	Pre-/early orogenic, but intensely modified during orogeny
Origin	SEDEX–epigenetic (latter overwhelmingly dominant)	Emplaced during diagenesis/early orogeny	SEDEX–epigenetic

Ghosh (1957) and Mookherjee (1964) proposed a classical magmatogenous hydrothermal replacement model for the ores of the Balaria–Mochia Magra sector (Zawarmala was not explored at that time). They apparently based their conclusion on the following observations: (i) An overall structural control of mineralisation, (ii) high temperature of mineralisation (~500°C according to Mookherjee op. cit.), (iii) presence of wall rock alteration, and (iv) mineral zoning. Another group of workers, including Smith (1964), Poddar (1965), Chakraborty (1965) suggested a sedimentary–diagenetic origin, followed by later remobilisation of sulphides. Their conclusions were mainly based on (i) the overall lithological (carbonate) control of mineralisation (ii) local presence of compositional layering at places and the absence of wall rock alteration that would be expected in a high-middle temperature hydrothermal mineralisation in a carbonate rock. Sarkar (2000) and Sarkar and Banerjee (2004) generally supported the second model. However, whether it should be assigned to the typical SEDEX type (Bhattacharya, 2004; Sarkar, 2000), or a SEDEX–MVT mix type (cf. Sarkar and Banerjee, 2004) is a matter of consideration. However, before proceeding further on this issue, let us briefly review the similarity and contrast between these two so-called different types of sediment-hosted zinc–lead deposits.

Obviously, SEDEX is a genetic term. When applied to an ore deposit it means deposition of the ores on the ocean floor ('bedded ore facies' of Goodfellow, 2004), or replacement of a suitable rock by ore-forming hydrothermal solution ('vent facies' of Goodfellow op. cit.). The sedimentary basin containing a SEDEX deposit is a part of an intra-continental rift system, the deposits generally occurring in a second order (lateral extent > 10 Km) or a third order (< 10 Km to several hundred meters) basin (Large, 1983). It has been even suggested that such a rift-system is a forerunner of an ultimate break in supercontinent (Lydon, 2004). Most SEDEX deposits occur in rift sag or rift cover stage of sedimentation characterised by sedimentary shallow water facies deposited during the thermal subsidence, covering both the burial site of the rift and its adjacent platform shoulders (Lydon, 2004). Sullivan is a glaring exception to the above generalisation. Here the mineralisation took place in the rift-fill sequence of coarse clastics where a basic sill occupied a sizeable volume. SEDEX deposits are spatially associated with syn-sedimentary faults, associated with the second and third order basins and these faults are generally the 'feeders'. Deposition and preservation of the SEDEX ores required an anoxic condition. Whether that was obtained locally or globally (Goodfellow, 2004; Lydon, 2004) or both, is a subject for investigation.

The most common hydrothermal minerals in a SEDEX deposit are pyrite, sphalerite, galena, pyrrhotite (rather rare except at Sullivan), chert, carbonates and barite (more common in the Phanerozoic deposits). Majority opinion holds that the ore fluids for the SEDEX deposits are formational waters of the sedimentary basin that became unusually hot, saline and metalliferous in the rift zone. Some authors (Russell, 1978; Samson and Russell, 1998; Boyce et al., 1983a) even suggested that high heat flow in such a situation would drive convective sea-water even through a depth of ~ 15 km and would leach lead from the quartzo –feldspathic basement rocks, before it undertakes a return journey upward, finally unloading the ore minerals and associated elements to form the deposits. This explanation, proffered mainly to explain the variable lead-isotope composition of the Irish zinc–lead deposits, has been contradicted by some later works, contending that there was not enough incontrovertible evidence to support it (Hitzman and Beaty, 1996).

In contrast to the genetic basis of the term SEDEX, the Mississippi valley type (M/T) nomenclature primarily connotes the type area of occurrence of this group of deposits. As a result, the genetic aspect of this group of deposits remains open for interpretation. But with the availability of more and more information about these deposits, the basic difference between these two 'types' or groups have been reduced to a minimum (Sangster, 1990; Hutchinson, 1996; Goodfellow, 2004; Sarkar and Banerjee, 2004). According to Hutchinson (op. cit.) the major difference between the two 'types' lies in their depositional setting or mode of emplacement. MVT deposits are epigenetic, emplaced in older strata and in secondary porosity of differing nature, whereas carbonate-hosted massive sulphide deposits, i.e. SEDEX type, are syn-depositional. It may, however, be mentioned that in some cases the field relationships are not clear enough to assign them to one or the other type. Goodfellow (2004) chooses the main difference between MVT and the SEDEX deposits in their depositional setting i.e. open space precipitation (within carbonate platformal sequences, versus sedimentation on the seafloor (SEDEX). Indeed, as Sangster (1993) pointed out, intermediate deposits possessing some characteristics of both do exit.

There are well-known carbonate-hosted zinc–lead ore deposits where the interpretation of the same set of observations swayed from epigenetic to syn-sedimentary/syn-diagenetic to epigenetic again. The Irish Zn–Pb (–Ba) ore field is under references in this context. Boast et al. (1981) highlighted the epigenetic features of these ores. However, the publications during 1970s and 1980s emphasised the syn-sedimentary/ syn-diagenetic characters of these ores (summarised by Andrew 1993). These interpretations are contradicted (Hitzman and Beaty, 1996), inter alia, on the following grounds.

1. Although the ore bodies are strata-bound, they are not strictly stratiform. Local discordance is common and the deposits like Tynag are of replacement origin. The really bedded sulphides are in fact geopetal cavity-fillings. A similar case in literature is the San Giovani Formation of Sardinia, where stratiform structure in the ores is inherited from the sediments replaced.
2. Crosscutting relationship in some deposit shows that the host rocks went through a substantial history of modification prior to mineralisation, including burial dolomitisation and their hydrothermal brecciation. Sulphides in the deposits like Navan and Tynagh cut diagenetic cements, indicating post-burial mineralisation.
3. Sulphide, barite and iron oxide textures show that majority of these minerals replace carbonate.

The authors conclude that the Irish deposits are 'true hybrids'. They were formed in tensional tectonic regime as in the case of SEDEX deposits but otherwise resemble the MVT deposits (except that they contain more Cu, Ag and Fe due to higher temperature). In the Irish ore field, according to Hitzman and Beaty (op. cit.), the hydrothermal ore flux encountered reactive and permeable rocks (prior to reaching the sea floor) and unloaded the ore-content. Almost a similar model was proposed for the genesis of the Silesea–Cracow zinc–lead district (Leach et al., 1996).

Now returning to the Zawar belt, only an insignificant portion of the ores show sedimentary diagenetic feature. Bulk is structurally controlled and is epigenetic. This epigenetic nature is suggested (but not proven) also by the difference in model Pb–Pb age of the galena in the basic volcanic rock occurring near the base of the Aravalli Supergroup and that of galena in the Zawar deposit. It would perhaps be better to call it a hybrid or intermediate between SEDEX and MVT types, as they are understood by many (cf. Sarkar and Banerjee, 2004). Lead for the ores could be derived from the feldspar bearing clastic sediments in the sequence or even the granitic basement, or both, by a density driven convective system (Fig. 6.2.25). One source for zinc and sulphur for the ores is the carbonate- bearing carbonaceous phyllite that underlies the ore-bearing carbonate horizon. The highly faulted nature of the metallogen is supportive of this model.

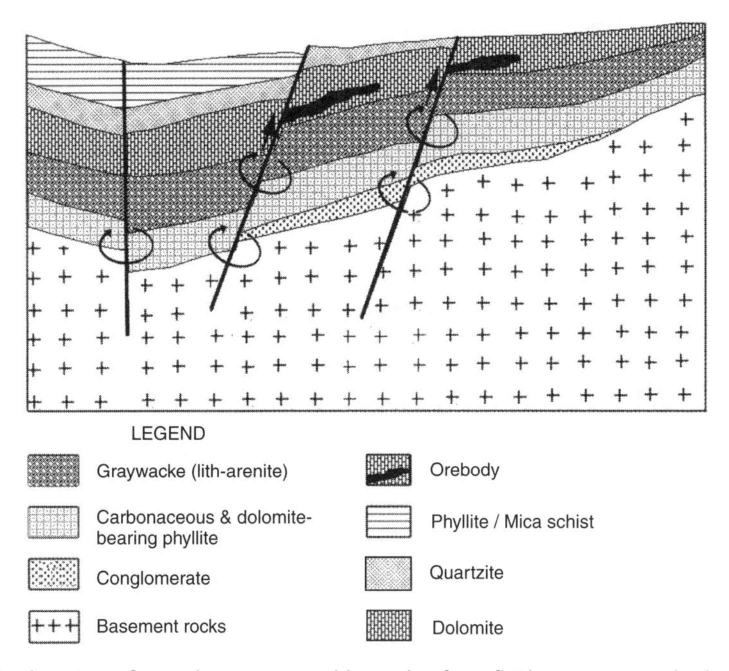

Fig. 6.2.25 A cartoon figure showing a possible mode of ore fluid movement and subsequent ore deposition in Zawar area (Sarkar and Banerjee, 2004)

There is a debate whether carbonate unit that principally contains the ore was deposited in a shallow marine (shelf facies) (Roy, 1988; Sachan, 1993) or deep marine environment (Bhattacharya, 2004). Bhattacharya cites the following in support of his conclusion: (i) no algal structure is reported from the Zawar carbonate unit, (ii) presence of mass flow features in sediments, (iii) alternation of clastics with carbonates, (iv) accumulation of thick black shale. Most of these arguments are meritorious but not unexceptionable. For example, absence of stromatolite should not necessarily mean a deep-water environment. Mass flows may operate in shallow waters, and even on land. Greywacke is not necessarily a primary rock. It is now believed to be the alteration (diagenetic) product of primary sediment, commonly lith–arenite. Therefore, it is not invariably a deepwater deposit, though often it is. Black shales need not always be deposited at depth, particularly if they contain enough evaporite in the form of sulphates that later underwent bacterial reduction (discussed). Presence of small patches of gypsum in the carbonate rock in textural continuity with carbonate exposed in the mine working does neither support a deep marine environment. The Sarara inlier (basement) is too close to the Zawar region to render its sedimentary environment 'distal' in the usual sense of the term. The size of the large clasts in the basal sediments may also be explained by steepness of the marginal basin slope. Interestingly, in a recent publication of Bhattacharya and Bhattacharya (2005), the clastic sedimentary unit just overlying the carbonate unit at Zawar is reported to contain storm event deposits. The reconstruction of the sedimentary environment of a tectono–thermally highly modified Proterozoic basin fill is often not easy. Added to it is the protracted tectonic instability of a riftogenic continental basin. The Zawar basin is another case in example of the situation.

Shear Controlled Ore Bodies/Ore Veins at Balaria–Mochia Magra Section As we have already mentioned most of the ore bodies/ ore veins are structurally controlled in the Balaria–Mochia Magra section and these structures developed late in the tectonic evolution of the belt (Banerjee et al., 1998; Roy 2001). Their probable mechanism of origin merits a brief discussion here. It has been suggested (Banerjee et al., op. cit.) that during the formation of the major F_3 folds, a dextral shear component started acting along the surfaces parallel to S_2 (the form surface) due to body rotation. As a result, different sets of fractures, both shear (Riedel P- and D-shears) and extensional, developed. The fractures being zones of increased porosity and permeability attracted increased fluid movement, and ultimately mineralisation in the form of ore bodies/ ore veins. One can imagine the following possible mechanisms to explain the development of these ore bodies/ ore veins: (i) the ore bodies formed directly from the convective fluids, already discussed; (ii) they formed by mobilisation of disseminated ore matter during metamorphism, prograde or retrograde; (iii) they formed by remobilisation from several small accumulations during a tectonic phase; (iv) they formed by remobilisation from one large ore accumulation during a dominantly brittle deformation phase. Of these four propositions, (i) appears to be the least meritorious, as the emplacement of these discordant ore bodies/veins were controlled by structures related to F_2 and F_3 folds (Banerjee et al., 1998; Roy, 2001). By that time, the phase of convective circulation of ore fluids must have been over. Regarding (ii), it is known that the development of F_3 was post peak metamorphism and there is no evidence that their development was accompanied by an intense wet metamorphism, i.e. retrogression (although, as we have already discussed, these ore bodies/ore veins are accompanied by a weak development of sericitisation and silicification). In the two remaining models, the translocation of an ore body could take place by liquid-/solid-/mixed -state transfer.

The liquid state transfer should involve dissolution of earlier ore minerals by chemically aggressive hypothermal fluids, the spatial transfer of the dissolved ore material following pressure gradients, their focusing at suitable locales and reprecipitation. Generally speaking, these fluids could be any of (i) deeply circulating meteoric/sea water; (ii) low temperature aqueous fluids accumulating as result of diagenesis; (iii) water available from metamorphic reactions; (iv) hydrothermal fluids given out by a contemporaneous igneous intrusion; (v) any combination of a–d (Marshall et al., 2000 and some references therein). The probable mechanisms of hydrothermal fluid transfer in this process are fluid flow around grain boundaries and fluid movement through dynamic micro-fracture systems, or channelised fluid flow along larger fractures/ small-scale shear zones. For the liquid state transfer to be viable, the ore components should be redissolved in situ, the time-integrated fluid flux should be sufficient to generate a genetically significant volume of secondary ore fluid and this fluid be focused, and the ore is to be again precipitated in a suitable locale. These neo-formed ores are expected to have some characteristics (chemical, mineralogical, textural) of their own. However, the main problem in the model is with the availability of a large volume of water (even if its supply is time integrated) and its access to the pre-existing massive ores for interaction. In most cases, the water molecules available in gangue minerals and (OH)-bearing minerals in the immediate country rocks will be inadequate for the purpose. Access of water to the massive pre-existing ores should be along tectonic fractures aided by original compositional inhomogeneity. Reaction induced permeability, thought important in oil geology, is unlikely to be important in massive sulphides. For the above reasons, in the present state of our knowledge, the liquid state transfer does not appeal to us as a major process in metamorphic remobilisation. The host rock, in the present case, is a dolomite. Water in such a rock, at the initiation of metamorphism, must have been be very little in volume and located physically along grain boundaries. This small amount of water, if still available for liberation during the middle greenschist facies metamorphism to which the rocks in the area were subjected, will be of little use for the purpose. Other rock types in the sequence, i.e., the pelitic sediments, if present, could be more useful, however. Relevant in this context is the fact that these ore bodies show little or no evidence of replacement and / or intense wall rock alteration at the contact with the host rocks. Hydrous minerals, such as micas, have developed but not copiously in the ore zone as a whole. Veinlets, patches, small lenses containing ore minerals commonly show some grain enlargement.

The solid-state mass transfer in case of sulphidic ores may take place by cataclastic flow, dislocation flow and diffusive mass transfer, or a combination thereof. Examples of incontrovertibly solid-state external remobilisation are few, possibly non-existent (Marshal and Gilligan, 1993). However, a small volume of water present in the system may induce dislocation flow through crystalloplastic process. Fluid assisted diffusive mass transfer may not be the cause of extensive remobilisation unless there is recharging of the fluid. However, locally there may be wet state diffusion, subordinate to dislocation flow in a mixed-state transfer of sulphidic ores, when affected in their post depositional history in an intense phase of thermotectonic regime. A similar mechanism for the translocation of the ore bodies under discussion from earlier concentrations is suggested (Sarkar and Banerjee, 2004). One may surmise if originally the ores were concentrated into one or more ore bodies. Difficult to answer, but remobilisation of ores from a number of pre-existent bodies would be easier than from one or two large bodies.

Gold Mineralisation in the Aravalli Sequence, Southeastern Rajasthan

Gold in Bhukia–Jagpura, Banswara District and Hinglaz Mata–Bharkundi, Dungarpur District, SE Rajasthan

A new and distinct type of gold mineralisation has been located in association with sulphides in amphibole bearing dolomitic marble, calc silicate, mica schist, quartzite and amphibolite of the Paleoproterozoic Delwara Group (Grover and Verma, 1993; Sinha Roy, 1996; Garia et al., 1998, 2001), belonging to the lower Aravalli sequence, in the south-eastern Rajasthan (Fig. 6.2.26 and Fig. 6.2.27).

The Delwara Group (vide Table 6.1.4), defined by Sinha Roy (1996) was regarded as the lower part of Debari Group by Gupta et al. (1997) and was subsequently included into Jhamarkotra Formation by Roy and Jakhar (2002). As the sequence exposed in the Bhukia–Jagpura area distinctly underlies the Jhamarkotra Formation exposed in type section at Udaipur, the Aravalli sequence in this part of Rajasthan is described as either Delwara Group or Jagpura Formation of Debari Group.

According to the classification of Sinha Roy (op. cit.), the Delwara Group, underlying Debari Group, is subdivided into lower Jakham Formation and upper Bhukia Formation, the mineralisation hosted by the latter. The stratigrafic column of Bhukia area is presented in Fig. 6.2.28.

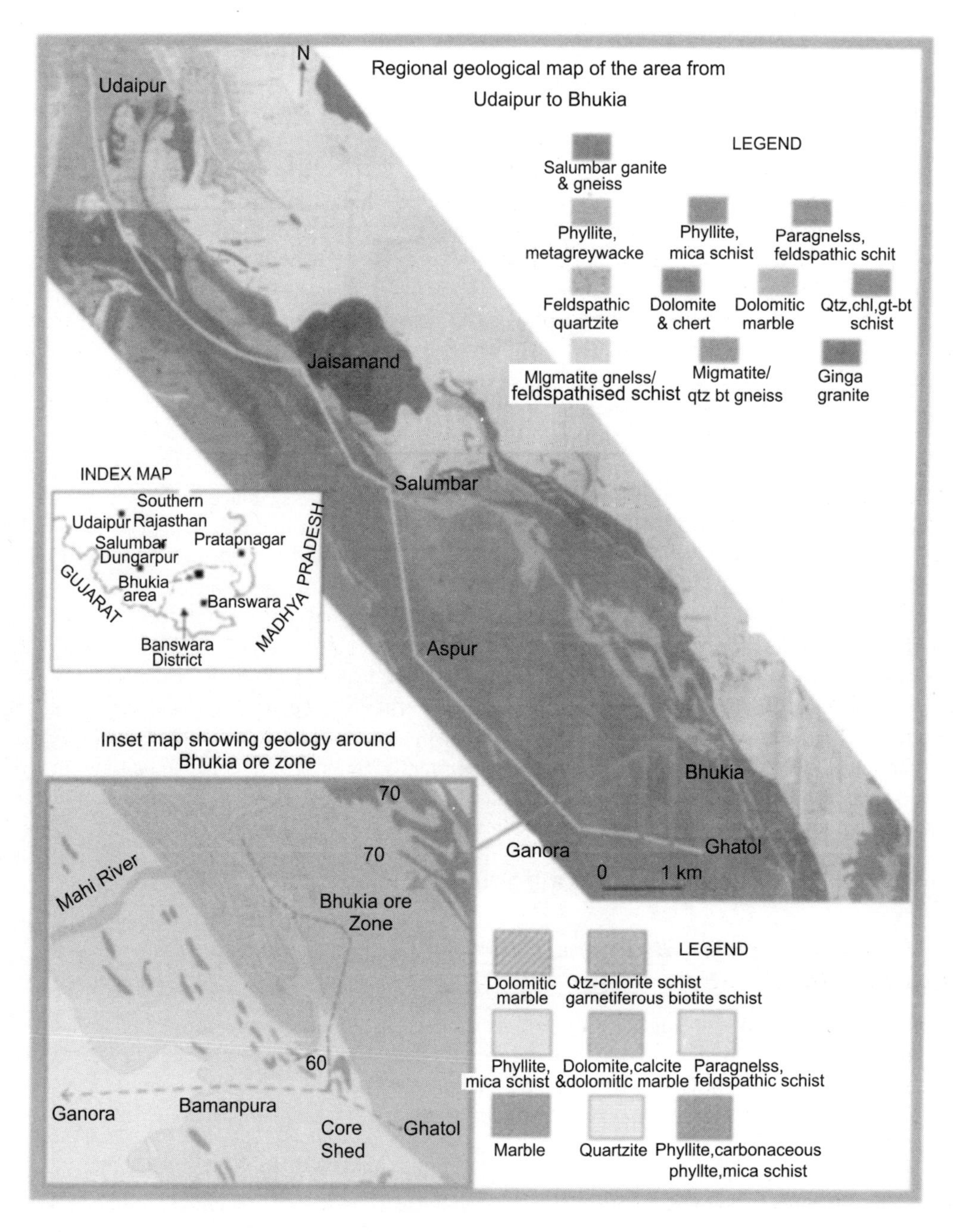

Fig. 6.2.26 Regional geological map of Udaipur–Bhukia area with inset map showing generalised geology around Bhukia Gold Prospect (modified after the GSI map)

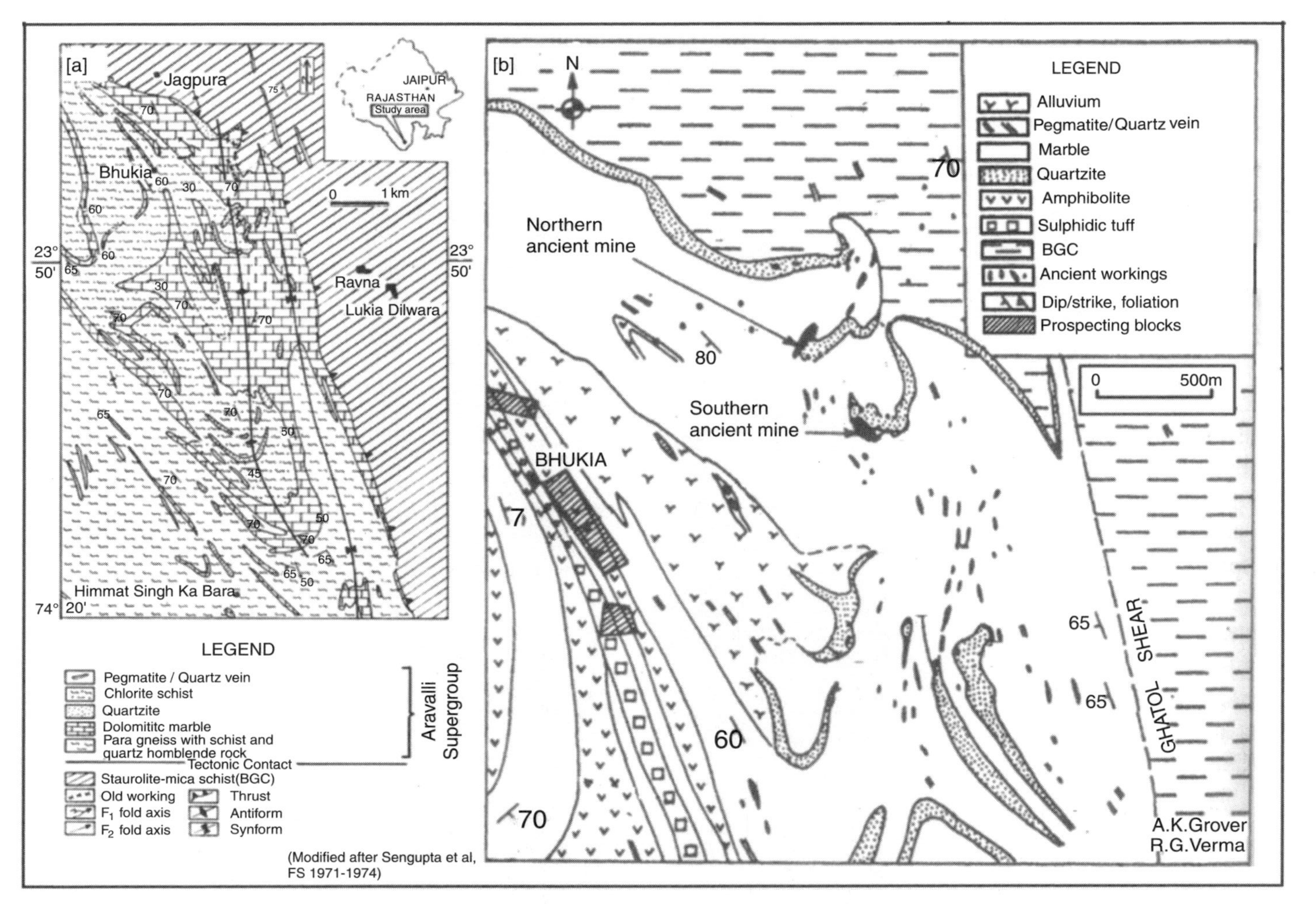

Fig. 6.2.27 (a) Regional geological map of parts of Banswara district showing Bhukia–Jagpura gold belt (after Garia et al., 2001); (b) Geological map of Bhukia area showing the ancient mine workings for gold and areas demarcated for prospecting by GSI/HZL (Grover and Verma, 1993).

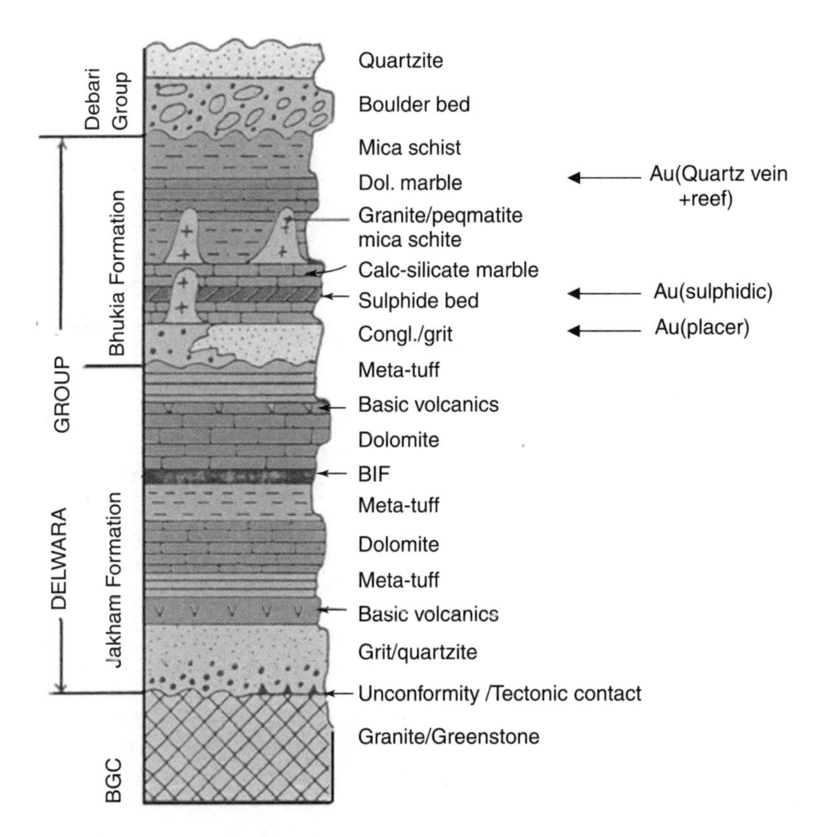

Fig. 6.2.28 Lithostratigraphic section of Bhukia area (Sinha Roy, 1996)

The sulphide minerals in order of decreasing abundance are pyrrhotite, pyrite, arsenopyrite and chalcopyrite. There are hundreds of open pits and shafts in the area and its neighbourhood, which were earlier believed to be ancient workings for copper. However, later prospecting efforts guided by abnormally high arsenic content (up to 2000 ppm) in the rocks and the gold related name of a local deity (Hiranyabeejwa Ma) led to the discovery of visible gold in many of the gossans in the area (Radhakrishna and Curtis, 1999). The largest ancient mine working is 250 m in length and 30 m in width (Grover and Verma, 1993). Gold incidences in the mine dumps also confirmed a past history of flourishing gold mining industry around Bhukia–Jagpura and Hinglaz Mata–Bharkundi areas, spread over a strike length of about 15 km.

In the recent past, the Geological Survey of India conducted preliminary prospecting followed by detailing in some blocks. Subsequently the Hindusthan Zinc Ltd. Undertook exploration in parts of Bhukia and conducted pilot scale heap leaching for the treatment of the ore, without much success. Of late, an Australian company M/s Indo Gold under joint venture with an Indian company M/s Metal Mining India carried out systematic soil sampling in the area followed by detailed exploration by drilling. An overlay of the soil sample anomalies on the geological map is presented in Fig 6.2.29.

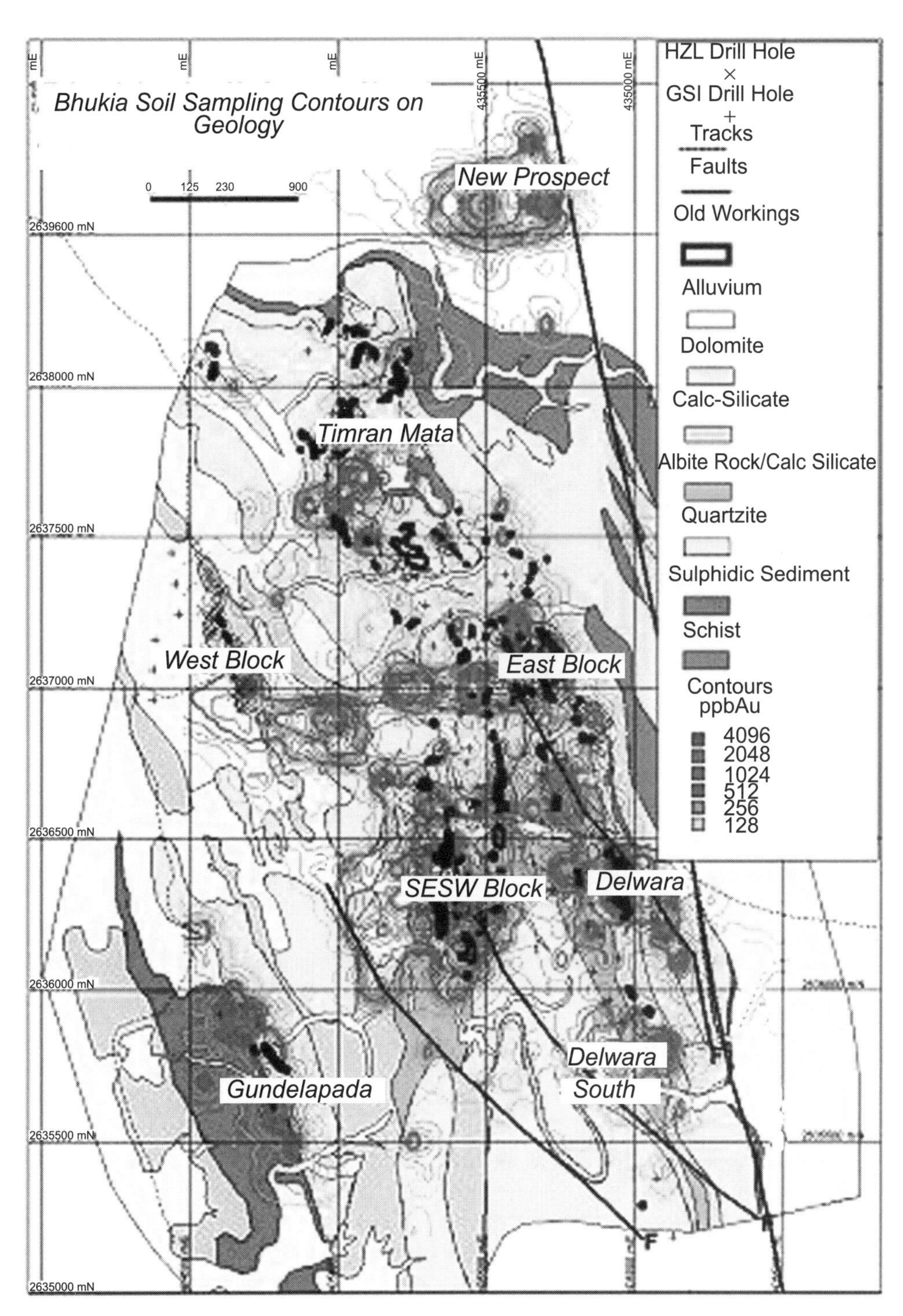

Fig. 6.2.29 Soil sample anomaly contours superposed on the geological map Bhukia–Jagpura prospect (prepared by M/s Indo Gold; www.indogold.com.au)

The highlights of the results of exploration venture by M/s Indo Gold in Bhukia are:

1. plus 100 ppb gold soil anomaly confirmed over at least 4.0 km strike, open to both the north and south
2. mineralisation now identified over at least 5.5 km strike length
3. open-ended 3.5 km × 1.1 km plus 100 ppb gold soil anomaly over the main Bhukia Prospect
4. at least 12 discrete plus 500 ppb Au anomalies within this broad anomalous zone, with assays up to 7932 ppb gold (7.93g/tAu), eight of these have no significant known mineralisation and little or no drilling
5. parallel open-ended 1.3 km x 0.2 km plus 100 ppb gold soil anomaly at Gundelapada, west of Bhukia, with assays up to 3057ppb gold (3.06g/tAu)
6. new prospect defined by 1.0 km × 0.6 km plus 100 ppb gold soil anomaly identified northeast of Bhukia as follow up of plus 300ppb gold stream sediment anomaly, assays up to 3366 ppb gold (3.37g/tAu).

In addition to the above, several other virgin prospects with no significant known mineralisation and limited or no drilling were identified including the following.

1. Timran Mata area
2. Northwest of Timran Mata
3. North of East Block
4. West of East Block
5. Northern extensions of SESW Block (including East Central Zone)
6. East of Delwara
7. South of Delwara
8. West of SESW Block.

Majority of the ancient mine workings are located within the marble and calc–silicates and a few in other lithounits along the contact of the formers. The rocks in this area have undergone three phases of folding and strong brittle–ductile shearing along the axial planes of F_2 folds. The sulphide bodies are deformed in conformity with the host rocks and the varying plunge of the major F_2 folds often control the disposition of the ore bodies and ore shoots Fig. 6.2.30 (a–b). At places sulphide bearing quartz–calcite veinlets occur along the F_2 axial plane shears.

The rocks are metamorphosed to greenschist to lower amphibolite facies. Wall rock alterations include chloritisation, biotitisation and sericitisation. At places gold and sulphide are also hosted by Quartzo–Feldspathic Rock (QFR) and tourmalinite in addition to the dolomitic marbles (Golani, 1996; Golani et al., 1999). Gold mineralisation averaging about 10 ppm and 5 ppm over varying widths of 5.65 m and 6 m have been encountered in some boreholes drilled in the East block of Bhukia. A transverse section across lode zones along boreholes drilled by M/s Indo Gold, corresponding to the largest old working in the vicinity are presented in Fig. 6.2.31 a–b. Pyrrhotite and arsenopyrite followed in abundance by chalcopyrite, are the principal sulphides along with free native gold. Gold occurs generally in close association with arsenopyrite. Size of the gold grains varies from 1–35 μm, average range being 10–20 μm (Grover and Verma, 1997; Verma and Golani, 1999).

In Hinglaz Mata–Bharkundi area lying in the northwest of Bhukia–Jagpura, disseminated copper–gold mineralisation is noted in the dolomitic marble. The mineralisation is associated with the process of jasperisation of dolomite along favourable structures. Sulphides are represented here by pyrite, chalcopyrite and cobaltite, and gold (0.8–20 ppm) is mostly associated with pyrite.

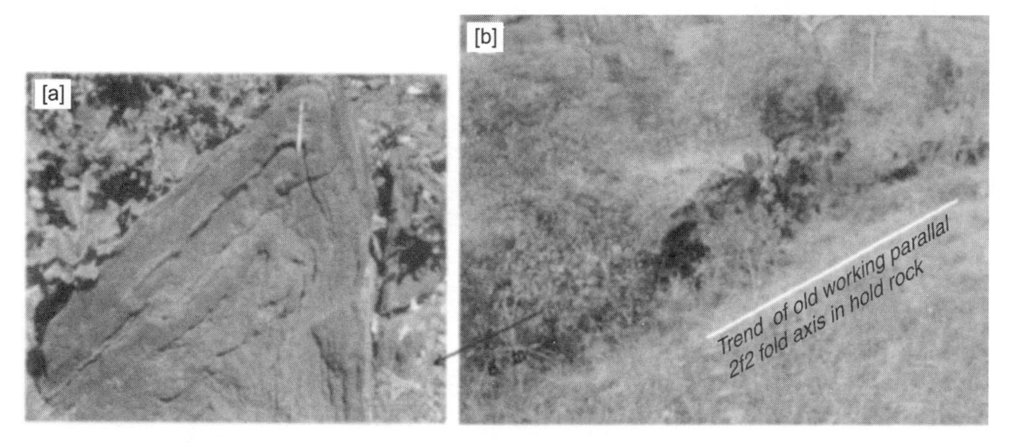

Fig. 6.2.30 Some field features of Bhukia deposit (a) F_2 fold in calc–silicate rock, exposed at about 6 m towards arrow direction from the spot in (b); (b) Old workings aligned parallel to axial plane of F_2 folds, auriferous quartz veins mobilised along the plane during F_2 folding, now mostly excavated in the quarry [photograph (a) Deb, (b) by authors]

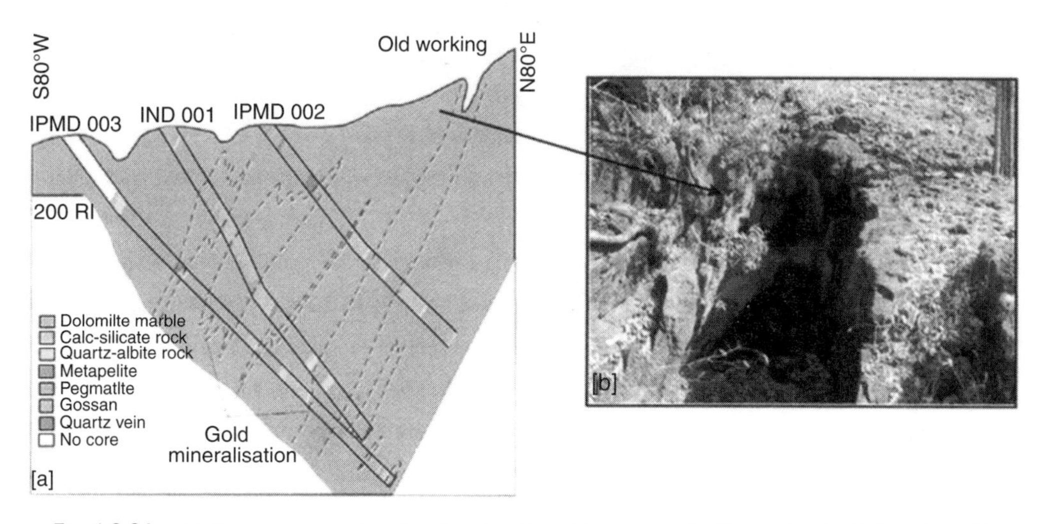

Fig. 6.2.31 (a) Transverse section showing the disposition of ore bodies at depth, (b) A major old working corresponding to the section Bhukia deposit (after Deol et al., 2008)

Deol et al. (2008) reported the mineralisation to be intimately associated quartz–albite rocks (Fig. 6.2.31) with or without tourmaline, with bedded tourmalinites and with calc–silicate-bearing impure carbonate rocks in the ore zone. The sulphide, silicate and carbonate assemblages were deformed and recrystallised during the tectono–thermal events, as evidenced by the textural features such as triple points, deformed lamellae and 'durchbewegung' texture (Vokes, 1973) in pyrrhotite(Fig. 6.2.32a). The temperature of the formation of ore minerals is deduced as 460°–480°C. The fluid inclusion studies indicate both high (~30–15 equivalent wt.% NaCl) and low (9.0–0.33 equivalent wt.% NaCl) salinity of the ore fluids, suggesting the possibility of fluid mixing. Based on the observed thin sulphide laminae parallel to S_0 and S_1 in metapelites and tourmalinite, it is envisaged that the sulphide mineralisation was pre-D_1 and probably of sedimentary–diagenetic origin, affected by subsequent deformation, metamorphism and attendant remobilisation. At a later stage during D_2 deformation epigenetic gold–sulphide–carbonate–quartz veins were emplaced concomitant with albitisation in the ore zone and granitic intrusion in the vicinity. The highly saline nature of some of the ore fluid inclusions is suggestive of a possible role of plutonism in the mineralisation at a particular stage (Deol et al., 2008). Golani (2008), on the other hand envisaged mantle flux into the gold-bearing carbonates of this area, based on the presence of light carbon, depleted $\delta^{18}O$, an unusually high content (5.8–70 ppm) of scandium, with relatively high Na, Ti, Ni, Co and Au in the carbonates.

Fig. 6.2.32 Gold-sulphide-bearing albite-rich and calc–silicate-rich bands in carbonate units of Bhukia Formation, exposed on a hillock with several old workings. Inset-1: closer view of albititic rock exposure, Inset-2: hand specimen of sheared albite–quartz rock (photographs by authors and M. Deb).

The Hindusthan Zinc Ltd took up a part of the Bhukia area in 1993, but the leasehold was relinquished in 2002. It is reported that 342,000 tonnes of bulk samples collected from gossan zones analysed average 1.23 gm/t Au in course of this project (Radhakrishna and Curtis, 1999). The deposit was considered as uneconomic under the then existing low price of gold in view of low tenor, limited tonnage and poor recovery of the metal.

The M/s Indo Gold under joint venture with an Indian company M/s Metal Mining India is reported to have established a gold deposit of economic significance (38.5 Mt of 'low grade' gold ore; average grade 1.4 g/t Au) in the area, with 2.33 M ounces or 66 tonnes of gold reserve (Times of India, New Delhi issue, dated 22nd February 2007; Chaku et al., 2008). It was further reported that the mine and the processing plants be ready for the annual production of 8 tonnes of gold per annum in four years time, from the date of clearance of the Project. The mining lease was yet to be granted until the end of 2008.

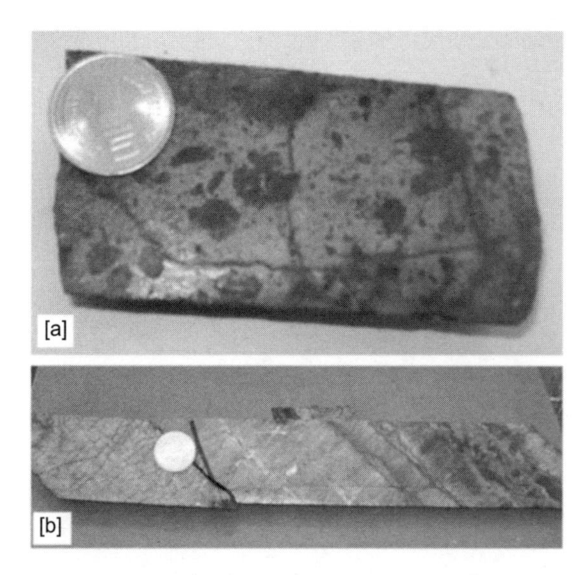

Fig. 6.2.33 Split core samples in Indo Gold core shed showing (a) 'Durchbewegung' structure displaying rounded gangue mineral aggregates in massive pyrrhotite, (b) Sulphide-bearing albite-rich veins in marble (photographs by authors)

Sulphide Mineralisation in Delhi Fold Belt

North Delhi Fold Belt In the North Delhi fold belt (NDFB), comprising the Khetri–Alwar–Bayana mineralised province (Fig. 6.2.34), the sulphide deposits of economic interest occur along the Khetri copper belt (KCB), at Kho–Dabriba in the Alwar district, and to the SW of KCB at Saladipura (27°38′N : 75°23′E). As the name suggests, the principal ore element in the Khetri deposits is copper. That is true also for the Kho–Dariba deposit. However, the Saladipura deposit is a massive sulphide deposit of pyrite–pyrrhotite with insignificant proportions of sphalerite.

Both the Khetri belt and Kho–Dariba bear evidence of ancient mining for copper ores. During modern period some small time mining is reported to have taken place at Madan Kudan (–Kolihan) area during the World War II. A vigorous exploration programme was, however, taken up after Independence by the concerned Government agencies. As a result of these investigations, a total reserve (mined + remaining tonnage) of 83 Mt of ores with a copper content of 0.88–1.5% was estimated at the KCB. Kho–Dariba is a small deposit. Hindusthan Copper Ltd. (HCL) produced about 0.5 Mt ores from this deposit having an average grade of 1.3% Cu.

Khetri Copper Belt

Geologic Setting The NDFB, as already discussed, is considered to consist of three sub-basins filled up by sediments $\pm$ volcanic rocks. From west to east, these are the Khetri (–Ajmer–Pindwara) sub-basin, the Alwar sub-basin and the Byana–Lalsot sub-basin (Singh, 1988a,b). The mineralisation in the Khetri belt is located in the northern part of the first sub-basin and Kho–Dariba in the Alwar sub-basin. The Saladipura deposit, though offset by a distance of about 15 km to the east of the Khetri copper belt, is supposed to occur within the Khetri sub-basin only.

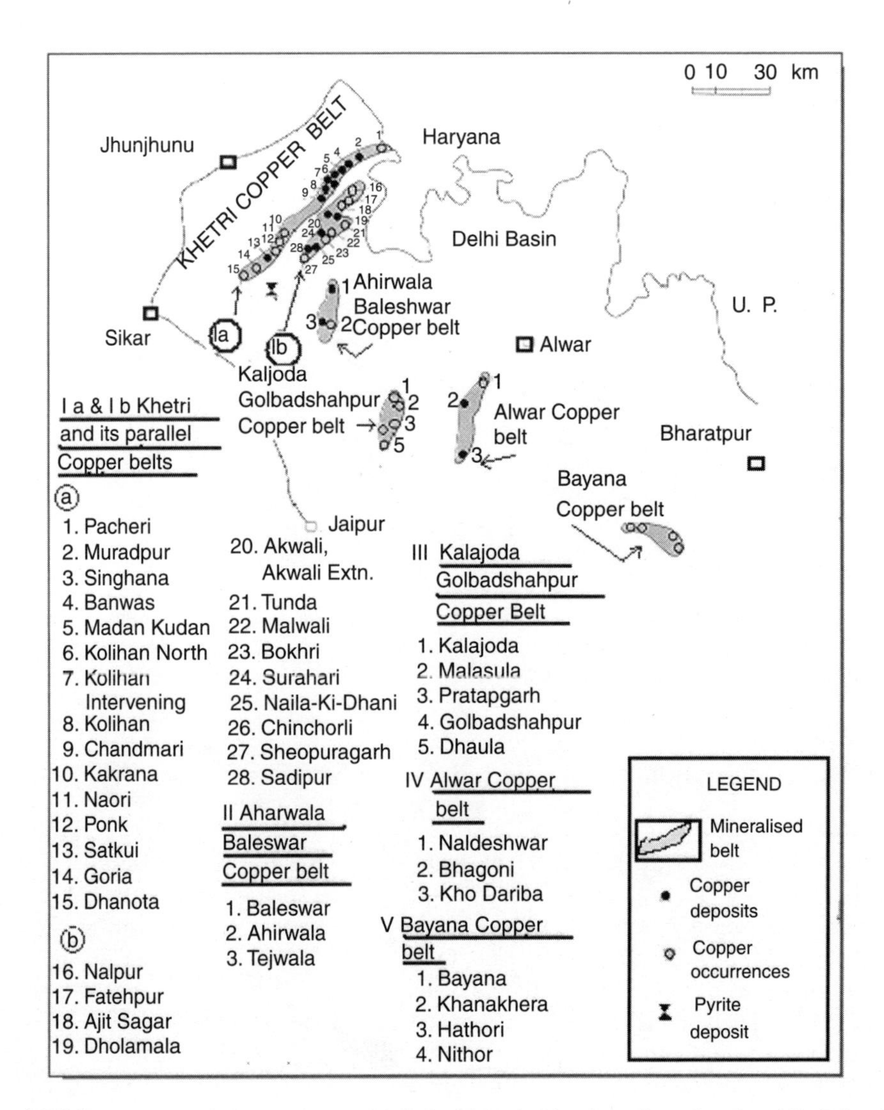

Fig. 6.2.34 Mineralised belts in the North Delhi fold belt, Northern Rajasthan (modified after GSI map, 1996)

The tectono–stratigraphic sequence of the major rock units of the Khetri area (Fig 6.2.35) is as follows:

1. Mica schists/phyllites
2. Quartzite
3. Mica schists
4. Garnetiferous chlorite schists
5. Banded amphibolite –quartzite
6. Feldspathic quartzite with magnetite (oldest)
7. Basement (not visible)

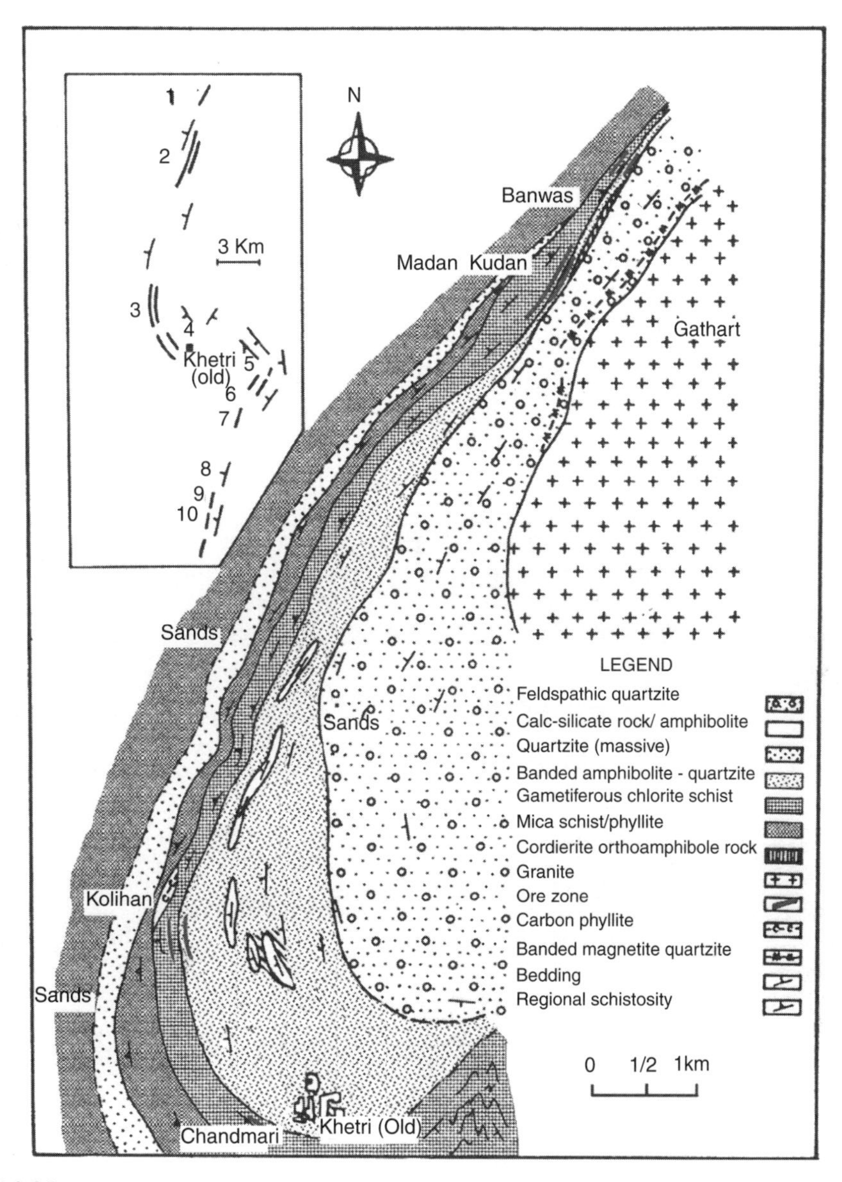

Fig. 6.2.35 Geological map of the Khetri copper belt showing the major deposits. The inset shows the folded nature of the ore belt and enclosing host rocks (after Sarkar and Dasgupta, 1980a).

A rather small cordierite–anthophyllite rock unit, occurring at the Madan Kudan–Banwas area at the interface of the feldspathic quartzite and banded amphibolite–quartzite unit, has been considered to be originally sedimentary in origin (Sarkar and Das Gupta, 1980a) and is therefore, assigned a stratigraphic position intermediate between these two. Volcanic rocks are not conspicuous in the KCB, except in the southern part, i.e. in the Shyamgarh Group of Gupta and Guha (1998). A zircon analysed from the acid volcanics yielded U–Pb age of 1832 Ma (Guerot, unpublished BRGM report, 1993). A late kinematic alkali granite massif occurs to the east of the Madan Kudan–Kolihan belt. It is intrusive into the sedimentary sequence.

Primary sedimentary and penecontemporaneous deformation structures are variously preserved in the metasediments of the KCB. These comprise normal, wavy and lenticular bedding, cross laminations, ripple marks, mud-cracks, load structures and convolute laminations. The way-up structures indicate normal sequence in the area. While the presence of load structures suggested a relatively rapid rate of sedimentation for the unit, rectilinear bifurcated ripple marks are believed to be the products of tidal activity in the shallow marine environment. Occasional development of evaporative conditions is suggested by the presence of such structures as mud cracks and scapolite-bearing rocks, whose development here is best explained by the metamorphism of evaporates-bearing protoliths. Presence of carbonaceous matter in the fine grained sediments in the upper part, e.g., mica schist, phyllites, indicate that a euxenic environment developed during their deposition.

Khetri rocks are multiply deformed. Dasgupta (1968) interpreted the structures of the Madan Kudan–Kolihan area to consist of a series of anticlines and synclines of regional dimensions. Absence of repetition of the key horizons and unimodal younging of the lithounits towards the west rather suggest that the rock units in the belt belong to the homoclinal sequence (except where locally complexly deformed, as in the Usri–Dholamala–Akwali sector) that consitituted one limb of a regional fold (Sarkar, 1973; Sarkar and Das Gupta, 1980a).

The intensity of regional metamorphism varies from the north to the south. While in the north the intensity of metamorphism is indicated by the attainment of amphibolite facies (T_{max} = 550° – 600°C; P_{max} = 5.5 kb), in the southern part it did not exceed middle greenschist facies. Empirically determined P–T–t has an anticlockwise trend (Sarkar and Das Gupta, 1980a; Sarkar et al., 1980).

Ore Deposits The Khetri copper belt is about 80 km long and comprises copper mineralisation at Madan Kudan (28°05′: 75°45′), Kolihan (28°02′:75°47′), Chandmari, Akwali (27°56′:75°48′), Satkui (27°49′:75°33′), Dhanota (27°45′:75°36′) and Charana, from north to south. Besides, there are several branching and parallel mineralised zones, viz, Deoru (28°03′:75°48′), Banwas (28°05′:75°48′), Dholamala (27°59′:75°48′), Rajotha (28°01′:75°49′), Usri (27°49′:75°48′), Ajitsagar (28°03′:75°52′) Taonda (27°58′: 75°53′), Malwali (27°57′: 75°50′), Chinchorli (27°48′:75°45′), Bokri, Naori (27°50′: 75°38′), Ponk (27°48′:75°38′) and Goria (27°45′:75°51′). Of these, the deposits at Madan Kudan, Kolihan and Chandmari are larger and have been/are being exploited for a little more than three decades. During 2010–11, Khetri and Kolihan mines produced 481966t (0.97% Cu) and 489654t (0.965% Cu) of ores respectively. Some details pertaining to a few important deposits are given below. Reserve estimates are mainly from GSI (2001).

Madan–Kudan deposit: The deposit is located in the northern part of the Khetri belt. The principal host for Cu-mineralisation is garnetiferous chlorite–quartz–amphibole schist and quartzite. Irregular network of impersistent stringers, disseminations, blebs and patches of chalcopyrite, pyrrhotite and pyrite characterise the mineralisation. The total drill indicated estimate of ore reserve (by GSI/IBM and MECL) is in the order of 92.8 Mt with 1.0% Cu (Proved – 46.2 Mt, Probable – 46.6 Mt), while demonstrated reserve proved so far by HCL is 43.92 Mt with 0.91% Cu at 0.5% cut off, mostly in the central block.

Kolihan–Chandmari deposit: This deposit is located at 6.5 km to the south of Madan–Kudan deposit and 3 km from the Khetri Township. For the sake of systematic exploration, the deposit was divided into seven blocks, of which the North Block, Central Block (main Kolihan deposit) and South Block (Chandmari deposit) have significant copper mineralisation with minor association of Ni, As and Ag. The reserve estimates of these blocks are 1.34 Mt with 1.0% Cu, 24.43 Mt with 1.4% Cu and 3.5 Mt with 0.88% Cu respectively.

Akwali deposit: This deposit, positioned at the middle of the Khetri copper belt, is located at about 6 km south of Khetri township. Out of 1000 m strike length of the mineralised zone, 700 m is more promising. The reserve estimate is about 1.65 Mt with 1.5% Cu.

Satkui deposit: This deposit forms a part of Goria–Satkui–Dhanota sector of the Khetri copper belt, located at about 20 km east of Neem–ka–Thana Railway station. The mineralisation is traced over a strike length of 1000 m along the contact of massive quartzite and phyllitic quartzite units of Ajabgarh Group. The drill-indicated reserve estimates are 3.88 Mt with 1.19% Cu (at 0.5% cutoff).

Banwas deposit: The deposit is located to the north of Madan Kudan in a largely alluvium covered stretch. The mineralisation occurs in the foliation and fractures of siliceous dolomite and schists and in the footwall quartzite. The reserve estimates by GSI over 800 m of the mineralised zones was 2.64 Mt with average grade of 0.9% Cu. Subsequent exploration by MECL by deeper drilling identified additional lodes in biotite quartzite and amphibole quartzite in the footwall and established a total reserve of 12 Mt with 2.0% Cu.

Dholamala deposit: The deposit with about 1.5 km strike length occurs in an eastern zone parallel to the main Khetri belt. The mineralisation, massive and disseminated, is restricted to a sheared and silicified zone occurring along the contact of garnetiferous chlorite schist and amphibolite. The estimated ore reserve is in the order of 1.5 Mt with 0.9% Cu.

Other small deposits where ore bodies of smaller dimensions have been established through limited exploratory in-puts are at Deoru, Kalapahar (a block of Kolihan deposit), Singhana–Muradpur–Pacheri, Surahari, Usri, Naori–Ponk, Dhanota, Tonda, Malwai, Chinchorli, Bokri etc. Besides, there are many reported occurrences of sulphide (mostly copper dominant) scattered over the three sub-basin areas of the North Delhi Fold Belt.

Mineralisation Location and disposition of the ore deposits along KCB have been controlled by the broad structures. Ore bodies are usually in the form of single or compound lenses, usually with trace to sub-economic mineralisation in the interspace. They are generally, but not exclusively, confined to single lithounits and are pene-conformable with the bounding surfaces of the unit containing them (Fig. 6.2.36 a, b).

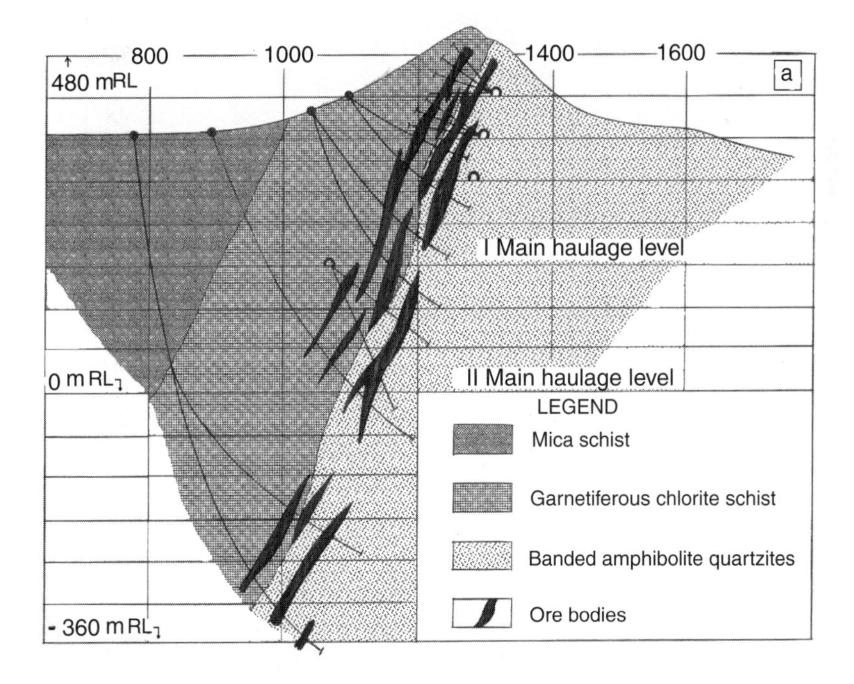

Fig. 6.2.36(a)

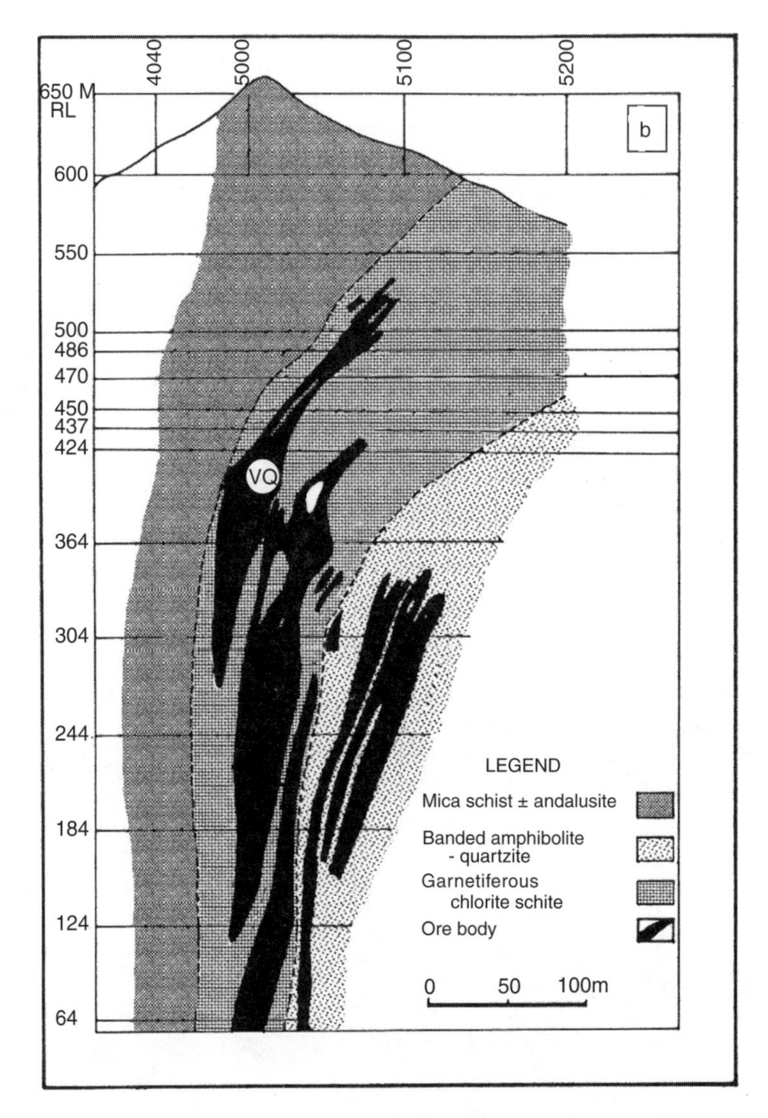

Fig. 6.2.36(b)

Fig. 6.2.36 Transverse sections across the orebodies at (a) Madan Kudan (b) Kolihan in Khetri copper belt, showing their conformity with the enclosing hosts of varied lithology (after Sarkar and Dasgupta, 1980a)

As mentioned earlier, the host rocks at Madan Kudan and Kolihan are both garnetiferous chlorite schist and banded amphibolite–quartzite, while at Chandmari it is garnetiferous chlorite schist alone. Carbonaceous phyllite hosts the mineralisation in the southern part of the belt. An earlier report of a regional shear zone controlling the mineralisation (Roy Chowdhury and Das Gupta, 1965) has not been corroborated by later workers. Neither have the reports of intense wall rocks alterations related to the ore mineralisation along the belt by some early workers (Roy Chowdhury and Das Gupta op. cit.) been corroborated by later studies. Cordierite– anthophyllite–cummingtonite rocks of Madan

Kudan area do not constitute immediate wall rocks of any ore body. Rather, they are the products of isochemical metamorphism of Mg-rich sediments. Other reported alterations are intangible, inconsistent or unspecific to the mineralised zone (Sarkar and Dasgupta, 1980a) (Fig. 6.2.37).

Copper has been the only metal extracted from the ores of KCB for many years. An unpublished recent statement of the Hindusthan Copper Ltd. (HCL) shows that they obtained about 10 kg of gold in the previous year on treating anode slime from Rajasthan. Discounting the small production of copper ores from Kho–Dariba, the HCL mines in the KCB are the likely sources of this gold.

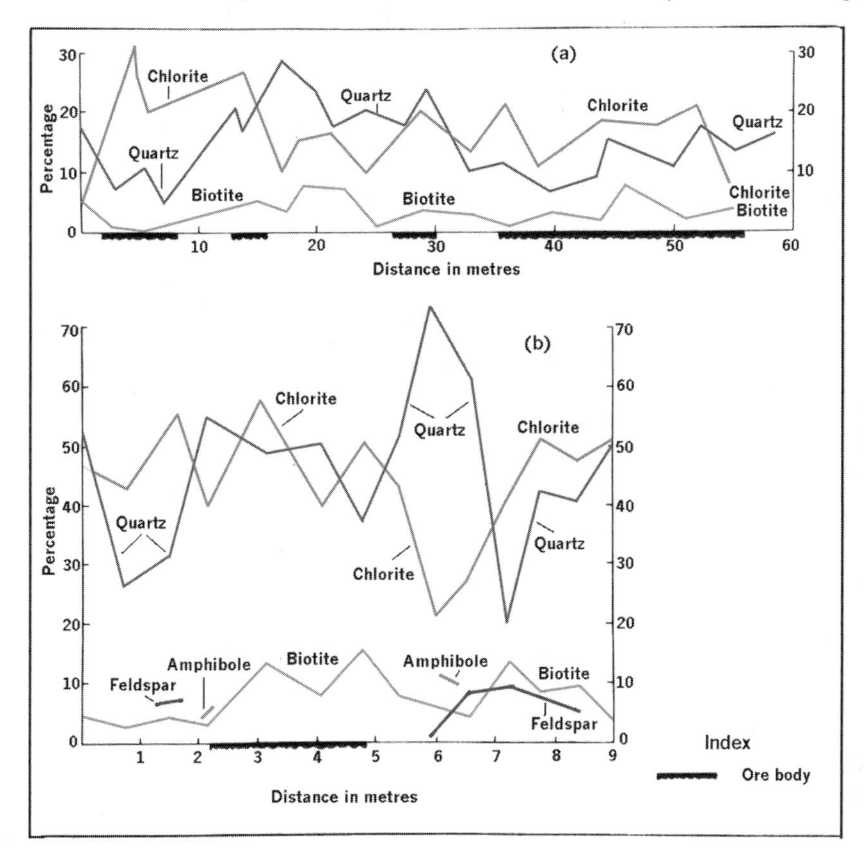

Fig. 6.2.37 Curves showing modal variation of the principal silicate minerals in the wall rocks of orebodies (a) Along 3425 grid, 364 m level, Kolihan mines; (b) Along 3125 grid across Central ore body, 350 m level, Madan Kudan.

Chalcopyrite, pyrite and pyrrhotite are the principal sulphide phases in the ores of Madan Kudan. In contrast, those in Kolihan are chalcopyrite, pyrrhotite and cubanite. Minor/trace phases comprise magnetite, sphalerite, ilmenite, arsenopyrite, molybdenite, cobaltite, pentlandite and mackinawite. The ores are metamorphosed, though relict primary banding of ores parallel to lithological strata is preserved at many places (Fig. 6.2.38 a–c). The difference in the mineralogy of major sulphide phases of the two deposits is explained by the difference in f_{S_2}, constraining the Cu–Fe–S system in the two fields during metamorphism. A peak metamorphic temperature of > 491°C at 1 atm (±18°C/kb) is suggested by the univariant assemblage, pyrrhotite + arsenopyrite (+As–S_l + V). This suggests an isofaciality between the metamorphism of the ores and their host rocks (Sarkar and Dasgupta 1980a). $\delta^{34}S$ (pyrite, pyrrhotite) ranges between + 7 to + 11 per mil (Deb, 1990a).

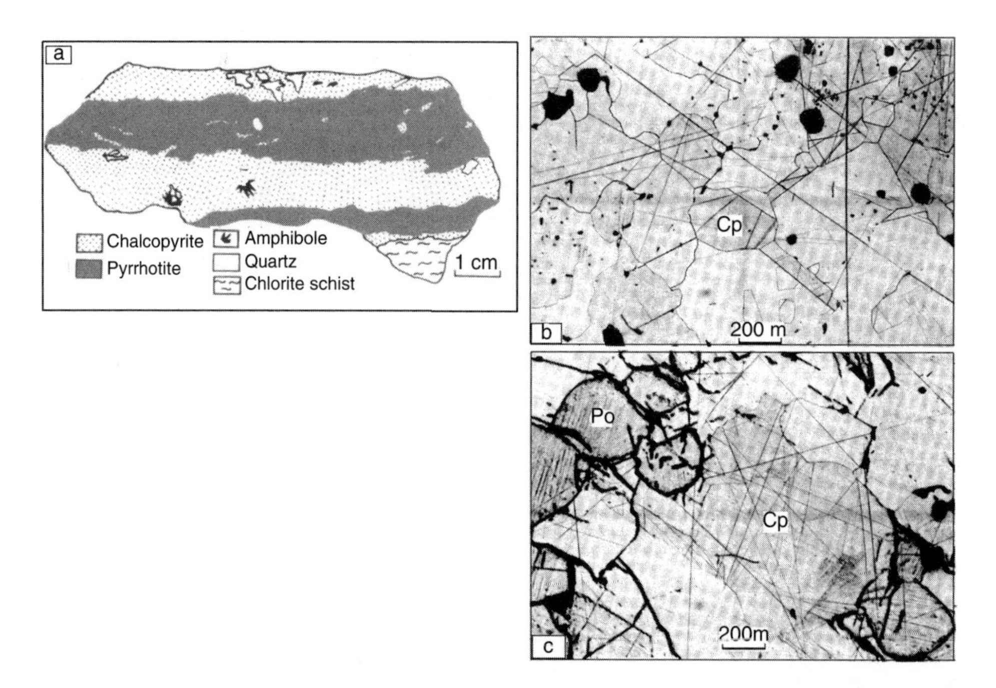

Fig. 6.2.38 (a) Primary banding in Khetri ores. Metamorphic textures of Khetri ores: (b) Polygonisation in chalcopyrite (Cp); (c) Polygonisation in pyrrhotite (Po) and chalcopyrite (Cp). (from Sarkar and Das Gupta, 1980a)

Locally, there are pods and lenses of pegmatoid ores, suggestively formed due to locally increased fluid activity.

Opinions on the origin of the copper ores of KCB vary. Roy Chowdhury and Das Gupta (1965) suggested the mineralisation to be epigenetic hydrothermal related to a nearby granitic source. However, Sarkar and Dasgupta (1980a), based on overall stratabound nature of the ore bodies, locally preserved ore layering, absence of appropriate wall rock alterations and metamorphosed nature of the ores suggested the ores to have initially developed through sedimentation and diagenesis. Close association of ore and carbonaceous matter and the isofacial metamorphism of the ores and the host rocks support a sedimentary–diagenetic origin of the ores. The locally developed variety of pegmatoid ores in the Madan Kudan and Kolihan mines formed by the activity of locally ponded H_2O, released during metamorphism.

A few dolerite dikes intersected the copper ores at both Madan Kudan and Kolihan (Fig. 6.2.39a). They do not show conspicuous thermal effects on the ores possibly because of being small bodies. On the other hand, remobilised veinlets containing copper minerals penetrated the intrusive bodies from the ore-bearing wall rocks (Fig. 6.2.39b). This is best explained by the dissolution of ores at the contact by the hydrothermal fluids given out by the cooling intrusives and their emplacement in the cooling cracks at the endo-contact of the intrusive bodies (Sarkar and Dasgupta, 1980b).

South and East of Khetri Belt In between the southern Khetri copper belt and the Alwar copper belt (closer to the former), there are several polymetallic sulphide deposits ranging in variety from the massive pyrite–pyrrhotite deposit of Saladipura to zinc-rich mineralisation at Manaksas (an exception in this part of NDFB). Before describing the Saladipura pyrite deposit in some detail, the base metal prospects in this belt at Naila–Ki–Dhani–Tonda–Rampura are described below, followed by a brief account on Manaksas and some other sulphide occurrences. The terrain to the south

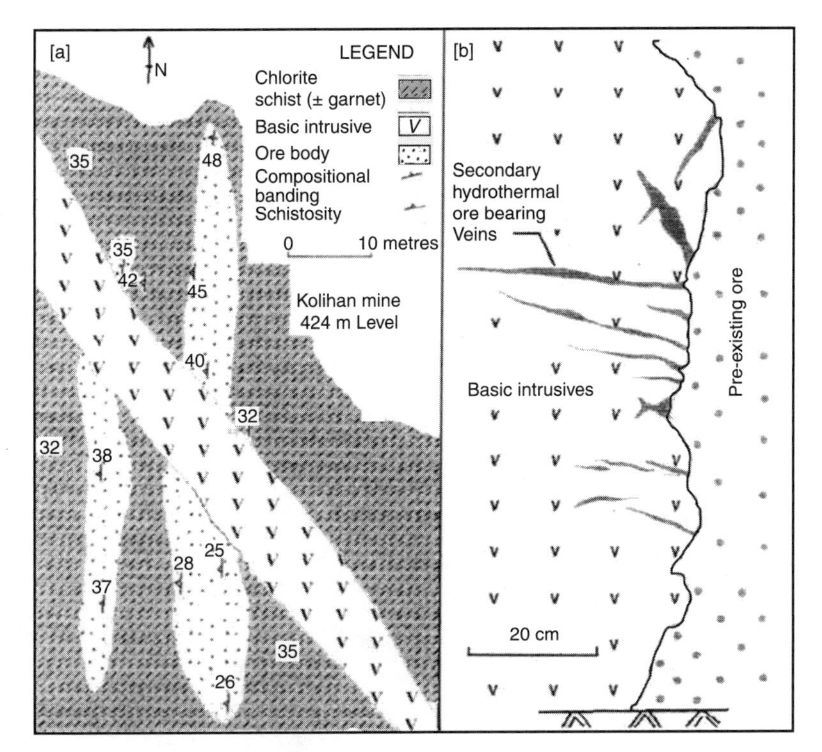

Fig. 6.2.39 Field sketches showing (a) plan disposition of basic intrusive cutting across ore bodies at 424 m level (Das Gupta, 1978), (b) secondary hydrothermal veins in the basic intrusive vis-a-vis the pre-existent ore in a subvertical mine face at 424 m level, Kolihan mines (Sarkar and Dasgupta, 1980b).

and east of KCB is also characterised by the occurrence of fluorite deposit at Chowkri–Chhapoli and some other localities, hosted mainly by feldspathic quatzite with albite–magnetite wall rock alteration. Chalcopyrite, hematite–magnetite and barite bearing quartz veins (< 1 m wide) along brittle fractures are seen around Alwar, which are enveloped by sericite–hematite–magnetite alteration zones and also have albite–hematite–magnetite alterations in the proximity. Specular hematite and magnetite veins and pods with lesser carbonates, quartz and apatite are common in the terrain (Knight et al., 2002).

Rampura–Tonda–Naila Ki Dhani Deposits, Sikar District The metapelites, quartzites (feldspathic) and garnetiferous amphibolites, belonging to Alwar and Ajabgarh Groups of the Proterozoic NDFB constitute the country rocks in Rampura–Tonda–Naila Ki Dhani area. The rocks are multiply deformed and metamorphosed to amphibolite facies. Garnetiferous (± kyanite) biotite schist represents the metapelites. There are carbonate and carbonaceous intercalations with the metapelites.

The polymetallic (Fe–Ti–Cu–As) mineralised zone in Rampura, marked by ancient workings and mine dumps, is parallel to the general NE–SW trend of the NDFB. The copper-rich mineralisation in Naila Ki Dhani (27°59′35″: 75°52′25″) and Tonda blocks are along an ill exposed shear/fault zone which runs NW–SE with steep dips towards southwest, oblique to the general NE–SW trend of the formations (Fig. 6.2.40).

The fault zone is brecciated and is characterised by extreme soda–metasomatism (albitisation). The fault zone is traced for 1.3 km and shows strong IP, SP and magnetic anomalies. Geochemical soil sampling at 50 × 25 m grid showed copper anomaly (121–635 ppm Cu) along this zone against the

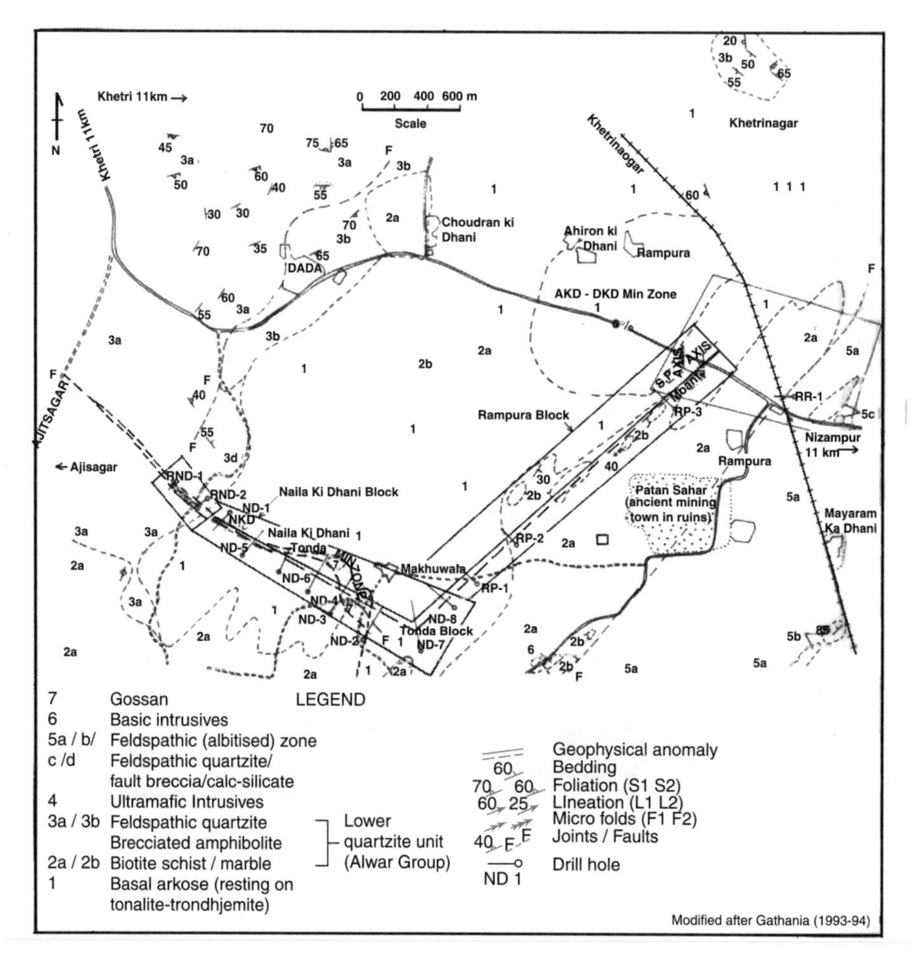

Fig. 6.2.40 Rampura–Tonda–Naila ki Dhani copper deposits, East Khetri belt, Rajasthan

background value of around 20 ppm. Shallow level drilling in this stretch intersected discontinuous mineralisation ranging from 6 m × 1.05% Cu to 5.65–10.65 m x 0.2–0.5% Cu (Sharma and Kumar, 1999). Most of the boreholes did not intersect any significant copper concentration. The sulphide mineralisation, represented by pyrite, pyrrhotite, chalcopyrite and arsenopyrite, occurs mostly as fracture fillings within albitised and magnetite bearing quartzite and garnetiferous schist.

A grain of native gold was seen in the polished section of a drill core sample. Channel samples in the old mine at Tonda also revealed anomalous gold values ranging from 0.10–1.38 ppm. A borehole drilled in the Rampura block near Patan Sahar mine dump intersected a 2.0 m thick zone at depth within ultramafic rocks having significant concentration of titaniferous magnetite and ilmenite (Sharma and Sekhawat, 1999).

Manaksas Zn–Pb Prospect, Jhunjhunu District This Zn-rich prospect is unique in otherwise Cu-dominant metallotect of Khetri. The prospect is located to the north of Saladipura and the lithounits represent the volcanisedimentary sequence of Ajabgarh Group (Delhi Supergroup), comprising graphitic tuffs, lapilli tuff, vesicular acid volcanics, andesite, chert, marble and greywacke. The mineralised zone is marked by a well-developed N–S trending gossan that extends for 500 m in strike with 20–40 m width. One of the boreholes drilled in the prospect revealed a highly sheared, 3.65 m wide, sphalerite and minor galena bearing mineralised zone analysing 9.24% Zn, 0.55% Pb,

222 ppb Au, 518 ppm Cd and 36 ppm Ag. A few other boreholes intersected wider zones (13–19 m) with much lower metal content (3% Zn), corresponding to much higher values obtained from the gossans (Sharma, 2000).

Adwana–Jahaz Block, Jhunjhunu District The Cu-bearing Adwana–Jahaz prospect (27°44′30″: 27°38′55″) is located at 2 km east of Zn prospect of Manaksas and 10 km NNE of the massive pyrite–pyrrhotite deposit of Saladipura. The host lithology is similar to that in Manaksas. A gossan zone extends for nearly 2 km with 10–20 m width. Channel sampling across the gossan (over seven 150–200 m spaced lines) yielded weighted average of 1200 ppm Cu, 292 ppm Ni and 193 ppm Co, over an outcrop width of 10–15 m. The shallow drill holes intersected a zone of very low tenor of copper (0.2–0.3%) corresponding to the gossan zone.

Baniwala Ki Dhani and Dokan Prospects, Sikar District Another copper prospect in the East Khetri belt has been located in the Baniwala Ki Dhani and Dokan area. The prospect area comprises Ajabgarh Group rocks represented by biotite–muscovite quartz schist/amphibole marble, (garnet) quatrz schist, kyanite–staurolite–quartz schist and amphibolite. The formations trend in NE–SW direction with steep dips on either side. Chalcopyrite, bornite and covellite are the main copper sulphides, associated with minor pyrite, arsenopyrite and galena. The mineralisation is hosted by the biotitic (± amphibole) schists and impure marble with quartz–calcite veins. Drilling has established a 1000 m long mineralised zone with several lodes varying in width from 1–9 m and assaying up to 2.0% Cu at 0.5% cut-off. The silver values range from 5 to 39 ppm. There are possibilities of further extension of this mineralised zone to its south in Kundla Ki Dhani block where surface geochemical sampling has yielded encouraging results (Geol.Surv.India, 2006).

Bhojgarh Block, South Khetri Belt, Jhunjhunu District A ferruginised breccia zone comprising altered tuffs with limonite–hematite veins and silicification at places represents the Satkui–Raghunathgarh fault, which is 10–50 m wide and extends for ~ 3 km in NE–SW direction. Drilling in this zone indicated very low tenor of Cu and associated Zn, Pb, Cd and Ag (Sharma and Maura, 2004). Locally, copper content in the breccia zone ranges up to around 1.0%. The occurrence at the present does not appear attractive.

Saladipura Deposit The Saladipura deposit in the Sikar district of Rajasthan is a large massive sulphide deposit of iron, with < 1% of Zn. It has a total reserve of 115 Mt of ore with average sulphur content of 22.5%. It is not located on the Khetri copper belt *sensu stricto*, but occurs about 15 km to its east. Its exploitation for sulphur alone is not economically viable at the present moment.

Geology: The Saladipura deposit occurs within the Ajabgarh Group of rocks belonging to the Delhi Supergroup (Fig. 6.2.35). The rocks here are multiply deformed, as elsewhere in the belt. The primary compositional layering is identifiable at many places. They have been deformed into compressed F_1 folds. The larger folds, which control the map pattern, however, belong to F_2 type and are nearly co-axial with F_1. F_3 folds, which occur at high angles to the F_1/ F_2 axes, are open (Roy, 1974; Sarkar et al., 1980). No fault or shear zone worth the mention is present in the area.

Quartzite, carbonaceous phyllite, chlorite quartz schists, amphibolites with intercalated metasediments such as mica schists, impure marble and calc–silicate rock, banded amphibolite–quartzite, are the major rock types in the area. The rocks evolved through a low- pressure intermediate facies series of metamorphism to a stage where the condition for the co-existence of the triple alumino–silicate polymorphs was closely approached at ~ 600°C and 5.5 Kb. The protolith of the amphibolitic rock associated with the ores still remains indeterminate. The granitoid rocks were emplaced at a late stage of the thermo–tectonic evolution of the region.

Mineralisation: There are five orebodies at Saladipura, four of which are parabolic on plan in conformity with the fold structures of the host rocks (Fig. 6.2.41). The largest amongst these ore bodies is ~7 km along the strike when straightened. The ore bodies having parabolic shapes on plan have been proved to be plunging folds on drilling. There are many interesting small-scale structures in the ores, some of which are 'soft sediment' type and the others are diastropic (Sarkar et al., 1980; Mukhopadhyay and Mookherjee, 1978). They provide with insight with regards to the origin and evolution of these ores (Fig. 6.2.42 a–d). It needs be mentioned that in the ore zone, the amphibolitic rocks are composed of cordierite + orthoamphiboles + cummingtonite ± scapolite and the pelitic rocks contain muscovite + biotite + quartz ± andalusite ± graphite.

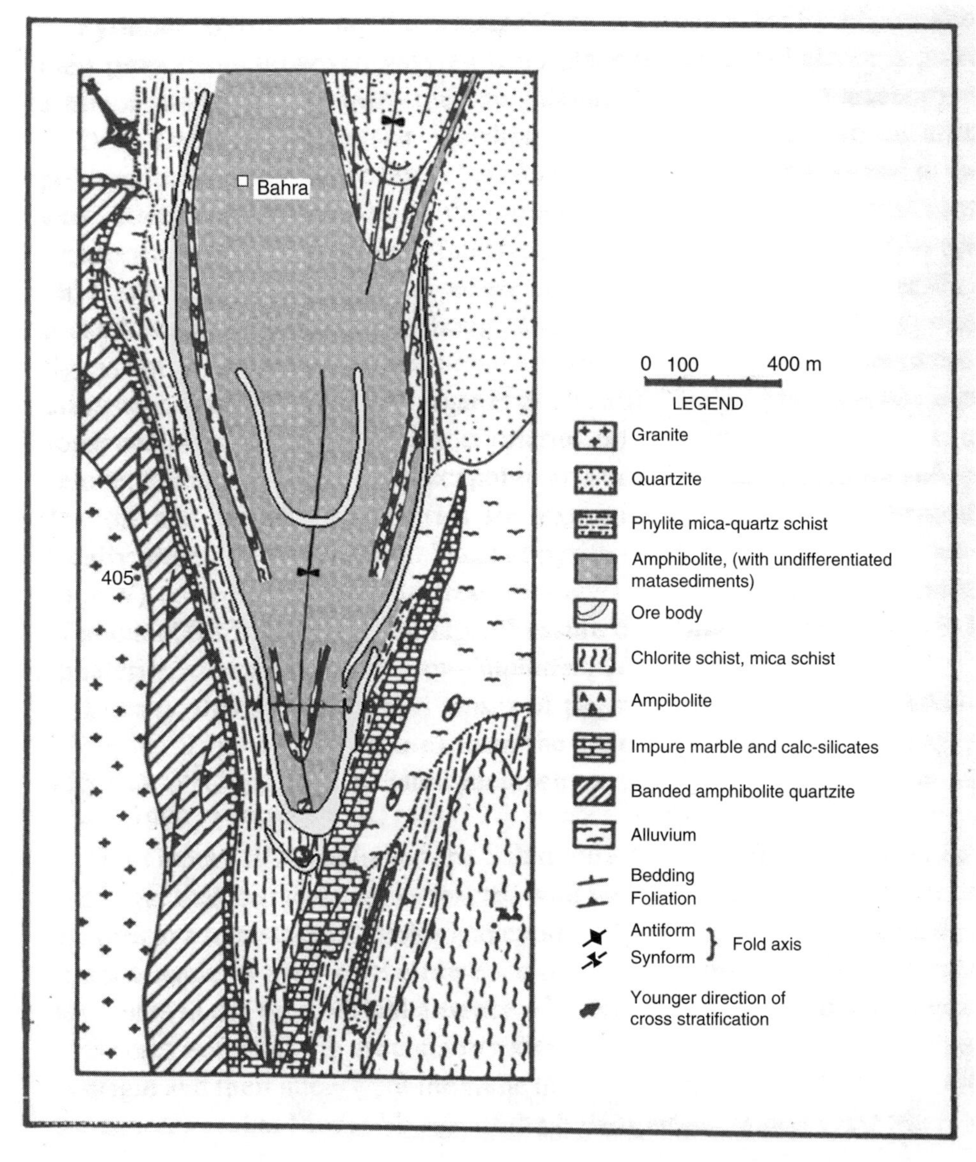

Fig. 6.2.41 Geological map of Saladipura area, Sikar district, Rajasthan (after Sarkar, et at. 1980)

Fig. 6.2.42 (a–d) Structures in Saladipura ores: (a) Slump fold in ore–silicate rhythmite, eastern limb, Main orebody; (b) Compositional layers flow around sulphidic concretions. Specimen length 5.2 cm; (c) Ore–silicate rhythmite, penecontemporaneous with sedimentation–diagenesis; (d) Sulphide–silicate interlaminates; a bundle of amphibolite needles cut across the lamination plane at top-right part of the specimen (from Sarkar et al., 1980).

Pyrite and pyrrhotite are the principal ore minerals in the Saladipura deposits. Their proportion varies from place to place. Sphalerite is a minor phase. Galena, chalcopyrite and arsenopyrite are present as trace phases. Ores are metamorphosed. Pyrrhotite aggregates in low strain areas are polygonised (Fig. 6.2.43 a, b). In folded ore layers, pyrite grains in sections cut normal to the fold axis often look elongate and their orientation is controlled by the axial cleavage. This feature could be an effect of fabric controlled growth i.e. growth modulated by oriented phyllosilicates/amphiboles, with or without the effects of stress. Pyrrhotite in general is not a breakdown product of pyrite, though very locally (particularly in sulphide-poor domains) petrographic evidence of such a relationship (Pyrite $\leftrightarrow$ pyrrhotite $+S_2$) does exist (Fig. 6.2.43b). The pair pyrite + pyrrhotite is buffered by ambient fs_2 at any given temperature and such a large Fe- sulfide deposit has an infinitely large reservoir of sulphur not to allow the reaction to proceed from the left to the right in such a situation. The gangue minerals comprise combinations of the phases cordierite, orthoamphiboles, phlogopite, quartz, calcite, tremolite, plagioclase, scapolite ($\pm$ rutile $\pm$ sphene). The ore and the gangue minerals are in textural equilibrium in general. The locally present invariant assemblage of pyrite + pyrrhotite + arsenopyrite (+As – S_L +V), corrected for possible pressure values, suggests a peak metamorphic temperature of ~ 600° C for the assemblage. Pressure determined from the mol% FeS in sphalerite, coexisting with iron sulphides, is 5.5 kb (Sarkar et al., 1980) (Fig. 6.2.44). $\delta^{34}S$ from the Saladipura ores (pyrite, pyrrhotite) vary widely (+ 3 to + 55 per mil). $\delta^{13}C$ in the associated metapelite varied but a little (-23 to -26 per mil) (Deb, 1989, 1990a).

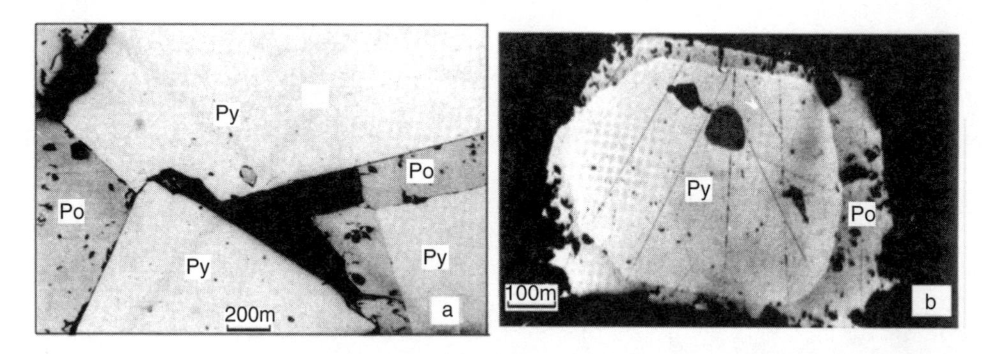

Fig. 6.2.43 (a) Co-existing pyrite (Py) and pyrrhotite (Po), a common relationship in the massive iron-sulphide ores, Saladipura; (b) A pyrrhotite (Po) rim around a pyrite (Py) grain in a disseminated ore, Saladlpura (from Sarkar et al., 1980).

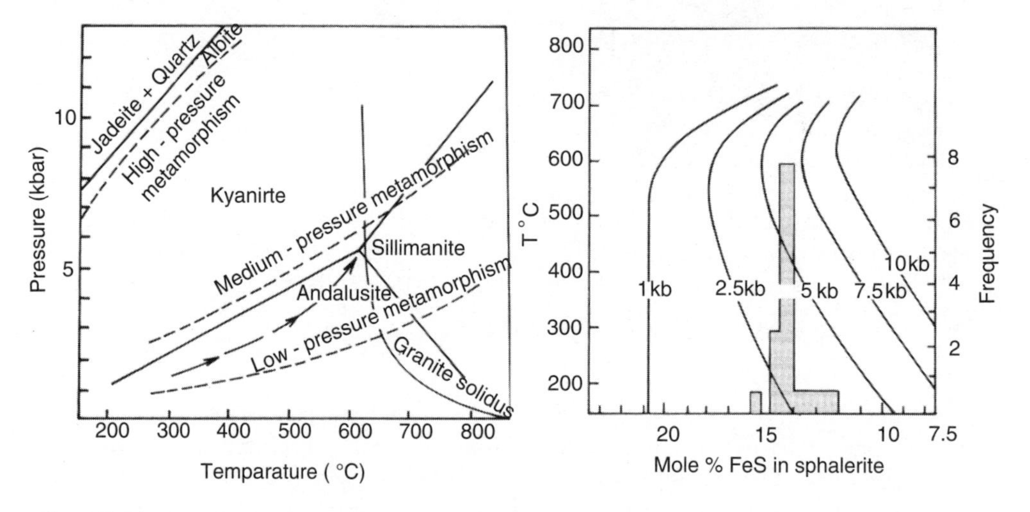

Fig. 6.2.44 P–T fields determined for sphalerite in Saladipura ore (after Sarkar et al., 1980)

Das Gupta (1970) proposed the Saladipura amphibolites to be meta- dolerites and suggested that the latter reacted with the introduced K,C and S to produce the ores as a late stage of geological evolution of the belt. However, this genetic model is difficult to sustain in view of the presence of diastropic and non-diastropic structures in the ores, virtual absence of copper and a metamorphism of the ores isofacial with the country rocks. The ores were initially sedimentary–diagnetic in origin and then underwent deformation and metamorphism along with the country rocks (Sarkar et al., 1980). The model Pb-age of the Saladipura ores is ~ 1700 Ma (Deb et al., 1989). It fits well with the zircon age obtained from the felsic rocks in the South Khetri belt mentioned earlier.

Alwar Copper Belt

Kho–Dariba Deposit The copper ore deposit at Kho–Dariba (27°10′N : 76°24′E) occurring to the south of Sariska Reserve Forest in Alwar district is a small deposit compared to Madan Kudan and Kolihan in KCB. Here also there is evidence of old mining, reportedly during the Moghul period (16th–19th centuries AD). After the Independence, exploration was taken up, followed by renewed exploitation. Hindusthan Copper Ltd. reportedly produced only about 0.5 Mt ores with average grade of 1.36% Cu.

The copper mineralisation in the area is essentially hosted by phyllites, which occur as minor intercalated bands within arkosic quartzites of Alwar Group. The leaner mineralisation in quartzite is mainly disseminated. Roy (1988) described the host rocks at Kho–Dariba as Kho–Dariba Formation of his Rayanhalla Group, which was earlier called Raialo Series by Heron (1935, 1953). The Rayanhalla Group here constitutes the lower most Group of the Delhi Supergroup, overlain by the Alwar Group (Roy and Patil, 1996).

The rocks of the Kho–Dariba area are multiply deformed, resulting in complex structures. The structural framework, according to Patil and Roy (2000), has been controlled by three factors: (i) stratigraphic overlapping of the Alwar conglomerate and arkose; (ii) DF_2 – folding, and (iii) faulting, initiated during the first stage of deformation (DF_1) and reactivated during the next (DF_2). The intensity of metamorphism is low, corresponding to that of greenschist facies.

The total strike length of Kho–Dariba deposit is 3.3 km, consisting of two main blocks: the Mine Block and Nala Block. The estimated reserves of both the blocks are about 0.56 Mt with an average grade of 2.46% Cu over a width of > 6 m (GSI, 2001). The Mine Block has been largely worked out. Varadhan et al. (1971) reported the occurrence of ore shoots along the 'axes of plunging synclinal structures in the ridge quartzite' in the Dariba Mine Block. In the Dariba Nala Block, ore occurs in the form of discrete lensoid bodies, oriented parallel to the dominant schistosity, S_2. Mineralisation is mainly confined to carbonaceous phyllite, but locally transgresses into the overlying quartzite.

The principal ore minerals are chalcopyrite, pyrite and pyrrhotite. Minor/trace phases are cubanite, sphalerite, pentlandite and mackinawite. The ores were subjected to regional metamorphism.

On the ore genesis, Patil and Roy (2000) surmised that the ore material was initially hosted by phyllites of Dogeta Formation. This was later remobilised to form structurally controlled ore bodies/ore shoots during deformation and metamorphism. They base their conclusion mainly on the 'congruous relationship' of the ore shoots with the S_2 planes and the plunge of DF_2 folds. This argument is not without merit, but needs sharpening.

Bhagoni Deposit The deposit, located at a distance of about one kilometer to the north of Bhagoni (27°17′: 76°21′), has profuse evidence of a thriving mining industry in the past in form of old shafts and scattered slag heaps. Exploration in the post-Independence period established mineralisation over a strike length of 1500 m, over the widths of 5–30 m (average 7 m). The mineralisation is restricted to a shear zone passing along the interface of biotite quartz schist and amphibolite of Ajabgarh Group. Chalcopyrite, pyrrhotite and pyrite are the principal sulphides occurring as disseminations and also in fracture and cleavages. The ore reserve estimates are in the order of 5.22 Mt of average 1.0% Cu, down to the depth of 250m (GSI, 2001).

Other smaller sub-economic deposits of Alwar copper belt are located at Baldeogarh (27°08′:76°25′), Kalighati (27°19′: 76°25′), Nalladeshwar (27°25′: 76°27′), Kalajoda (27°26′: 76°11′), Pratapgarh (27°15′: 76°09′).

Multimetal Iron-oxide Breccia Complex at Rohil, Sikar District, Rajasthan Multimetal iron-oxide–breccia type mineralisation is reported (Yadav et al., 2002) from a major NE–SW structural lineament in Mid-Proterozoic North Delhi fold belt near Rohil (27°34′: 75°29′), Sikar district, Rajasthan (Fig. 6.2.45 a, b).

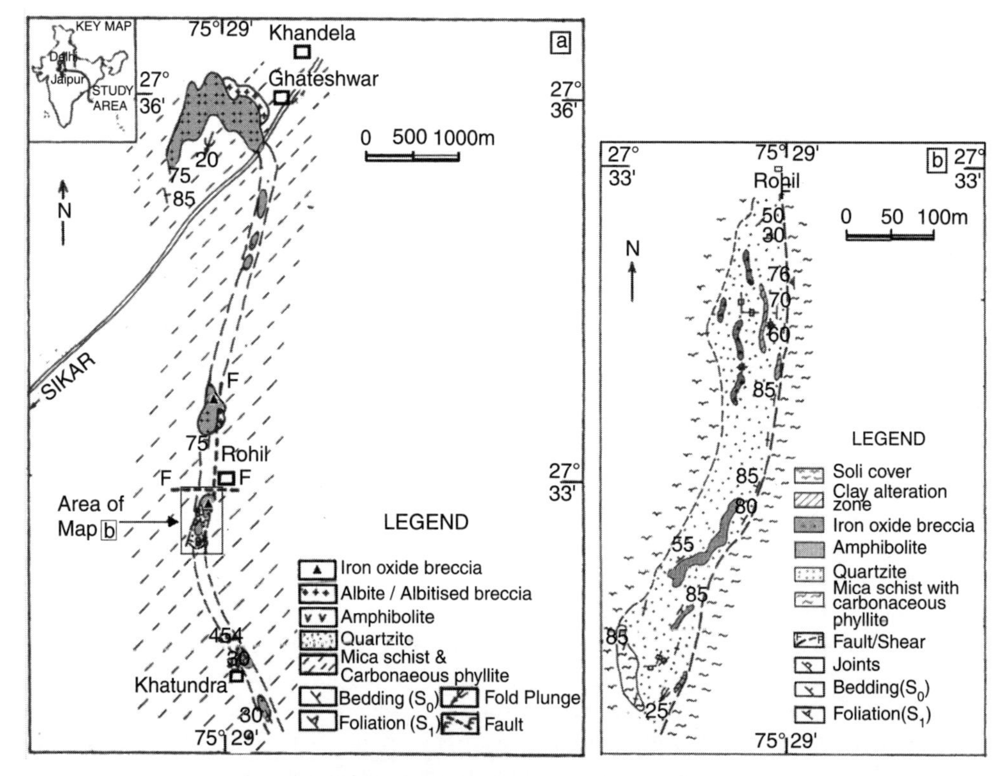

Fig. 6.2.45 (a) Regional geological map of Khandela–Ghateshwar–Rohil–Khatundra sector showing the zones of iron oxide breccia (b) Detailed map of Rohil area (after Yadav et al., 2002)

The breccia zone analysed Cu (average = 1824 ppm), Zn (average = 687 ppm), Ni (average = 214 ppm), U (average = 195 ppm) and Pb (average = 168 ppm). There are alternate horst and graben structures in the North Delhi fold belt, and the area under reference falls in the Khetri basin. An albitite zone occurs at deeper level along the periphery of the iron-oxide breccia, which is also enriched in Cu (average = 1082 ppm), U (average = 472 ppm) and Mo (average = 270 ppm).

The iron-oxide breccia occurs as fracture filling along the NE–SW shear/fault zone or as lump like masses at the intersection of the shear/fault with local EW fractures. The breccia zone is exposed intermittently in the quartzite over a length of 1.5 km and dips steeply. The individual bodies/veins range from 1–50 m in length and a few centimeters to 5 m in width. Hematitic and clayey alterations are prominent in the shallower sections while sodic alterations are extensive at deeper level. The breccia zone at Rohil coincides with a linear magnetic-low and high-chargeability/low-resistivity zone.

The iron-oxide breccia is composed of angular fragments of quartz, quartzite and massive to brecciated hematite, all components cemented by ferruginous matrix. Fe-metasomatism and epidotisation in the host quartzite and silicification in the breccia are prominent.

Uranium occurs in adsorbed state in the goethitic material in the iron-oxide breccia and as discrete grains of uraninite and minor brannerite in the albitite/albitised breccia. No thorium mineral could be identified even in the high-thorium zones. Pyrite, pyrrhotite, chalcopyrite and bornite occur as small specs and disseminations in the iron oxide breccia.

The major and trace element data pertaining to the iron-oxide breccia, albitised breccia, albitite, and quartzite of Rohil area are presented in the following Table 6.2.11.

Table 6.2.11 *Major and trace element data of iron-oxide breccia and other associated rocks from Rohil area, Sikar district, Rajasthan (Yadav, 2002)*

Oxides (wt%)/ Elements (ppm)	1	2	3	4	5	6	7	8	9
SiO_2	4.55	34.17	2.48	31.60	66.28	54.27	49.05	93.18	88.57
TiO_2	0.12	0.31	0.09	0.13	0.46	0.81	0.69	0.07	0.08
Al_2O_3	3.06	1.00	3.02	2.41	11.92	10.86	11.30	2.97	2.45
Fe_2O_3	78.38	53.54	81.00	56.17	6.62	15.63	18.87	NA	NA
FeO	0.30	NA	0.30	0.48	0.25	2.67	3.17	2.27*	4.81*
MnO	<0.01	<0.04	<0.01	<0.01	<0.01	0.01	<0.02	<0.01	0.01
MgO	0.05	0.06	0.06	0.04	0.17	1.81	3.27	0.20	0.28
CaO	<0.01	0.17	<0.01	<0.01	0.47	0.38	1.42	<0.01	0.15
Na_2O	0.29	0.16	0.59	0.19	5.70	5.94	5.67	<0.01	0.86
K_2O	0.09	0.01	0.12	0.18	0.72	1.56	1.56	<0.01	0.26
P_2O_5	1.40	1.06	1.46	0.80	0.08	0.19	0.16	0.06	0.09
LOI	11.58	8.40	11.42	8.17	6.55	5.33	4.64	NA	NA
Total	99.82	99.88	100.5	100.17	99.22	99.46	99.82	98.75	97.56
V	583	NA	252	311	132	327	413	17	34
Cr	138	NA	60	124	151	201	176	39	37
Cu	2424	1387	1818	1667	372	1457	707	56	102
Mo	NA	NA	NA	NA	NA	220	319	<5	<5
Ni	267	137	193	260	65	81	<50	30	25
Pb	186	227	51	209	97	<200	<200	26	20
Zn	862	747	554	585	<5	<20	<20	51	<5
Co	50	39	43	50	25	200	146	10	11
Y	25	NA	28	35	11	<10	45	22	8
La	NA	NA	NA	NA	NA	<50	846	NA	NA
Zr	43	NA	26	68	162	277	233	121	90
Rb	18	NA	18	18	19	<50	<50	37	27
Ba	172	NA	117	98	270	<100	44	<5	
Sr	<5	NA	<5	<5	223	30	60	23	32
U	153	263	212	153	14	374	569	6	<5
Th	396	<43	<43	<43	NA	<43	<43	<25	26

Note: * FeO (total); NA – Not analysed.

Sample Nos. 1,2,3 and 4: Iron oxide breccias; 5: Albitised breccia (n = 2); 6 and 7: Albitite (drill core samples); 8: Fresh quartzite; 9: Altered quartzite.

The major oxides of the iron-oxide breccia are seen to vary widely, e.g., SiO_2 2.48–34.17%; Fe_2O_3 53.54–81.00%; Al_2O_3 1.00–3.06% and so on. Yadav et al. (2002) are of the view that the overall geological set-up and nature of mineralisation at Rohil is very similar to those reported from Olympic Dam and Kiruna type deposits. Yes, it partly resembles Olympic Dam mineralisation.

Multimetal Mineralisation Associated with Albitite Zone, Northern Rajasthan The albitite zone delineated in northern Rajasthan has already been described in the preceding chapter. The ore mineral occurrences studied so far in the three sectors of the albitite zone are presented below in brief.

Maonda–Sior Sector At Sior and Maonda iron ore resources of 0.38 Mt (59–65% Fe) and 0.25 Mt (65–70% Fe) were established. The ore bodies have very close spatial relation with the albitites in this sector and these along with the ultramafic bands could be co-genetic. Spatially close to the albitites, there are massive bismuth mineral occurrences near Nim- Ka-Thana.

Fluorite occurrences in Kelapura–Salwari area are well known and have been locally mined. The fluorite or fluorite–calcite veins occur within the albitites as well as in the mafic–ultramafic intrusives. The metallic mineral occurrences include ilmeno–magnetite and massive nodules and pods of covellite–chacocite–bornite–cuprite–malachite aggregates. In a 50 m x 20 m outcrop coarse crystals of ilmeno–magnetite (21.83% Fe_2O_3, 26.10% FeO and 34.08% TiO_2) are seen interlaced with very large laths (several cm across) of hornblende (Ray, 2004). The metallic pods, containing upto 38.35% Cu, have a chalcocite–covellite core with an altered rim (48 52% Cu) and occur within the mafic–ultramafic rocks. It is suggested that these pods and nodules may owe their initial origin to immiscible sulphide liquid fractions segregated with heavy mafic–ultrmafic cumulates in the source magma (Ray op. cit.).

Khandela–Guhala Sector Besides the mineralisation reported from the Rohil area, there are reports of high U_3O_8 from different parts of this sector. Several molybdenite bearing (upto 0.22% Mo) and chalcopyrite bearing (0.33% Cu) zones in association with pyrite, pyrrhotite and sphalerite were encountered within the albitite, pyroxenite and albitite–pyroxenite breccia in shallow boreholes (upto 150 m depth) drilled in this sector, close to the Khetri copper belt (Ray, 1987). The radioactive zones were found to be more widely distributed ranging from the sulphide bearing albitites to other cogenetic rocks as well as in the metasomatically altered country rocks in the surrounding.

Sakhun–Ladera Sector Low concentration of Cu–Au mineralisation associated with pyrite and pyrrhotite is reported from this sector. In one borehole, 11 zones with 0.4–2.0 m thickness and 0.2–2.65% Cu were intersected. Some of these zones recorded 100–1000 ppm Mo and 100–2000 ppb Au. There are also fluorite–calcite veins in this sector. Magnetite is very prominently associated with albitite and therefore the rock is named here as 'magnetite albitite'.

Ore Genesis Let us have a look at the problem of ore genesis along the Khetri belt in the background of the world scenario. Observation at the Kupferschiefer in Europe, Zambian and Zairian copper belts in Central Africa, White Pine in the USA, Redstone in Canada, Udokan in Russia and many others, lead to the following generalisation about the characteristics of such deposits (Brown, 1997):

1. Widespread copper zones along preferred, relatively thin stratigraphic unit of major sedimentary basins.
2. Disseminated fine- grained Cu-sulphide minerals, concentrated along locally favourable strata.
3. Irregular but continuous mineralisation throughout well-defined cupriferous zones.

4. (a) Red bed below, (b) cupriferous zones located immediately adjacent to and on the reducing side of a distinctly variable redoxcline, (c) unmineralised grey beds (pyritic) above and also laterally beyond the mineralised zones.
5. Evaporites occur in close association.
6. A peneconformable configuration of the outer limit of the total cupriferous zones.
7. Ag and Co, if present, accompany minerals of Cu and other metals, e.g., Pb, Zn, Cd and Hg and form zones beyond the outer limits of copper mineralisation.
8. A low temperature mineral paragenesis trending from early syn-diagenetic (pre-ore) Fe-sulphide, generally pyrite (+sulfates) to subsequent post-sedimentary (ore stage) cupriferous sulphides, which generally have a sequence of formation, chalcopyrite → bornite → chalcocite. This paragenesis is likely to change on metamorphism.

Resuts of petrographic studies and some field features of ores in this type of deposits suggest that the copper mineralisation is diagenetic (Rentzsch, 1974; Chartrand and Brown, 1985; Hays and Einaudi, 1986; others) rather than syngenetic or otherwise. A late diagenesis is preferred to an early one (Jowett, 1986; Maynard, 1991). Fe–Mg minerals in the clastic sediments decompose to clay + hematite during red bed formation during early diagenesis. Cu and other ore metals released in the process are adsorbed in the Fe-oxides produced. The adsorbed metals should be redissolved in chloride-rich brine flowing through this red sandstone during late diagenesis. Moreover, the temperature obtained in the system during early diagenesis is inadequate to dissolve sufficient Cu.

The metallotect that hosts sediment-hosted Cu-ores is generally a rift basin, with some early volcanism. The volcanic flows, alkali basalt– tholeiite dominated, are commonly covered by volcanic derived clastics. Early diagenesis decomposes the Fe–Mg minerals, releasing Fe-oxide to coat the stable sand grains. It seems also possible that in situ destruction of mafic minerals in a nearby pediment formed in an arid climate could be the source of Fe- oxide in the red bed. Copper released from the basic rocks below, or from the underlying sediments derived from a granitoid/rhyolite/mixed provenance is adsorbed in the Fe- oxide, i.e. ferric oxyhydroxide(s). Upward moving brines, [(Na+K ± Ca ± Mg) – chloride ± sulphate]- bearing water according to Haynes and Blooms (1987) leach ore metals on the way as chloride complexes and precipitate then as sulphide minerals at suitable locales in the reduced zone above the redoxcline. Sulphur source need not be unique. It could be available from the breakdown of early diagenetic pyrite or evaporite sulphate in the reduced zone, or it could be introduced at least partially. Reduced sulphur activity controls which metal will precipitate as a sulphide. In one estimate $f_{H_2S} \leq 10^{-14}$ is to be maintained at pH = 7.6 (T = 25° C) for the precipitation of Cu and Ag, with no Pb, Zn and Co (Haynes and Bloom, 1987). Brines being heavier than normal water, their upward movement cannot be automatic. Suggested mechanism for this include highland recharge during basin-filling, renewed rifting during later crustal extension ('seismic pumping'), simple compaction, anomalous heating from below, due possibly to the rise of asthenosphere (Maynard, 1991; Brown, 1997), or any combination thereof.

How does the model of ore mineralisation outlined above relate to the Cu-mineralisation in the KCB? The lithologic build-up in the Khetri belt can be explained as rift filling, agreeing with the general suggestion by Singh (1988) for the North Delhi Supergroup. Ore bodies are generally conformant with their host-rocks and there is no epigenetic wall rock alteration around the ore bodies. The Cu-sulphide mineralisation is concentrated along thin zones of fine-grained sediments, containing carbonaceous matter. The host rocks beyond the ore zone locally contain chalcopyrite as an accessory mineral besides Fe-sulphides (pyrite ± pyrrhotite) Evaporite association of the ores is suggested by the presence of scapolite in the ore zone. Red bed as such has not been found out. Red bed could be represented by the now magnetite-bearing arkosic sandstone below the ore horizon. Another contender for the interpretation of KCB mineralisaton is the IOCG model (Knight et al.,

2002). However, the following observations, inter alia, do not support the application of the model here in its present form: 1) The ores show relict compositional banding and metamorphic fabric, 2) Pyrite is the principal Fe-sulphide in the Madan Kudan ores and also in some other places. Saladipura pyrite deposit in south Khetri still retains some deformed sedimentary structures, 3) The wall rocks are silica-enriched at places, 4) Neither sodi-calcic, nor late potassic alterations are important in KCB, 5) Zn-(Pb) mineralisation at Manaksas within the KCB. Direct determination of the age of the mineralisation by a robust method, if possible, will go a long way in solving the problem, aided by other valid observations.

Sulphide Mineralisation in Ajmer Area Large old workings, mine dumps and slag heaps at or close to the Ajmer city indicate a site of old mining and metallurgy. Even an ancient lead mine 'Sishakhan' occurs at the southwestern part of the city and 'Lohakhan', actually mined for lead and zinc, at the northeastern part. Conspicuous evidence of ore mineralisation notwithstanding, the area was not included in the priority list for detailed modern exploration for long. Later, the Geological Survey of India, followed by the Hindusthan Zinc Ltd. (HZL) took up the detailed exploration of the area, some results of which are included in the reports of Raghunandan (1994) and Sahai et al. (1997).

Preliminary exploration activity indicated significant mineralisation of lead and zinc at Ghugra and Kayar and zinc mineralisation at Madarpura, within a space of about 20 sq km. Principal rock types in the area are quartzite, migmatite, dolomite, calc–silicates and garnetiferous mica schists, belonging to the Ajabgarh Group of the Delhi Supergroup. General structural trend is NNE–SSW, in conformity with the Delhi–Aravalli tectonic grain (Fig. 6.2.46).

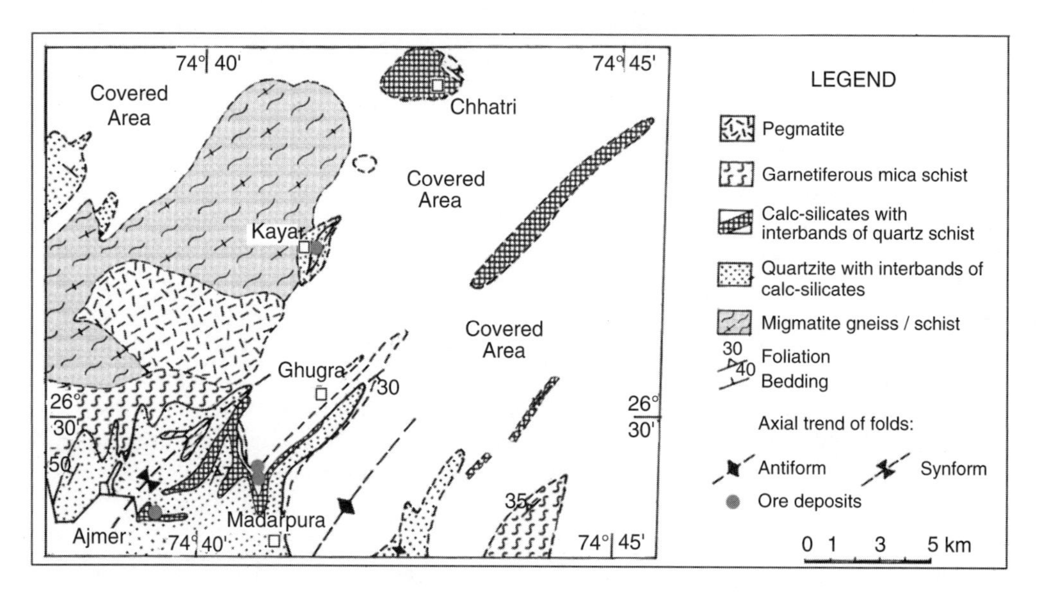

Fig. 6.2.46 Generalised geological map of Ajmer area showing locations of Zn–Pb ore deposits at Ghugra, Kayar, Madarpura and Lohakhan

At the *Ghugra* area mineralisation is hosted mainly by dolomite and calc–silicate rocks, spilling over to meta-pelites and quartzites. The orebody is located in the hinge zone of a synformal fold. The mineralisation has been proved over a strike length of a little over 1 km and down to the depth of 200 m. The estimated ore reserve is 9.0 Mt with an average grade of 5.78% (Zn 3.22% + Pb 2.56%)

at 4% cut-off. Principal ore minerals are sphalerite and galena. Pyrrhotite and chalcopyrite are minor phases. Co, Ag, Cd and Hg are present as trace metals.

At *Kayar* ore mineralisation occurs mainly within graphite-bearing mica schists and partly in quartzite and calc–silicate rocks. The ores occur along the foliation, cleavage and fractures. Besides the major ore minerals sphalerite and galena, chalcopyrite, pyrite, pyrrhotite, cubanite and pentlandite are present as minor to trace phases. In an ore zone of a little more than 1 km in extent, the estimated ore reserve is 9.2 Mt with 12% Zn, 1.2% Pb and minor copper. Ag, Cd, Hg, Sn and W are present as minor/trace metals.

Madarpura is located south of Ghugra in the dolomitic marble and calc–silicate rocks. Mineralisation reportedly occurs 'on an antiform complementary to the adjacent Ghugra synform' (Raghunandan, 1994). Sphalerite is the principal ore mineral. Chalcopyrite, galena, pyrite, pyrrhotite and molybdenite are minor to trace phase. Cd and Hg are reported as trace metals. A few million tonnes of low-grade (2–3per cent Zn) ores have been estimated.

The *Lohakhan* deposit enjoys the unique position of being located in the northeastern part of the Ajmer town itself. The host rocks are quartzite, biotite schists, carbonates and calc–schists striking NNE–SSW. The (Pb+Zn)- content varies from traces upto 25% and the mineralisation has been locally confirmed down to a depth of 300m. However, the necessary exploratory activity has been constrained by urban development. Detailed exploration of the *Taragarh* copper prospect, which also falls within the city limit, could not be taken up for the same reason.

The genetic problem of the Ajmer deposits will be taken up along with the other sediment-hosted zinc – lead deposits of Rajasthan.

South Delhi Fold Belt The polymetallic ore mineralisation (Zn–Cu–Pb ± Au) in the South Delhi fold belt (SDFB) occurs in three sectors along the Aravalli hill range. From northeast to southwest, these are Birantiya Khurd–Kalabar–Chitar, Goliya–Pipela–Basantgarh–Ajari–Danva–Pindwara–Watera and Deri–Ambaji sectors. Out of the above, the mineralisation at Ambaji, Deri and Birantiya Khurd belong to the Zn–Pb–Cu type, hosted by mafic–felsic bimodal volcanic rocks, while the rest belong to Zn–Cu/Cu type (Table 6.2.12). The Ambaji deposit is the largest amongst those discovered so far in SDFB. Ambaji and Deri deposits have been partly exploited by Gujarat Mineral Development Corporation and Rajasthan Mineral Development Corporation Ltd respectively. There is an exploratory mine at Basantgarh.

Table 6.2.12 *Reserves and grades of volcanic-associated sulphide deposits of the SDFB (Deb, 2000; GSI, 2001)*

Deposit	Reserve (million tonnes)	Zn (%)	Cu (%)	Pb (%)	Ag (ppm)	Au (ppm)
Kalabar	1.25	6.76	0.36			0.06–0.12
Chitar	0.19	0.28	0.66–1.5			
Birantiya Khurd (North block)	0.045	10.60	4.60	5.1		0.2–2.0 (Phulad – 1.55)
Basantgarh	3.58	1.27	1.74		100	
Pipela	1.5	0.5–0.7	0.5–2.7		5.8	0.04–0.09
Golia	1.0	Upto 2.5	0.7–1.4			
Ajari	0.65	0.57–3.30	0.2–1.9		20	0.58–1.37
Danva	0.31	6.42	1.40	0.53	10–110	0.18–2.57
Deri	1.0	7.32	1.98	5.4		
Ambaji	8.29	5.52	1.75	4.91	2–130	

Different aspects of these deposits and prospects have been discussed already by various workers (Deb, 1980, 1982; Patwardhan and Oka, 1988; Deb and Sarkar, 1990; Bhattacharjee et al., 1991; Mukherjee et al., 1992; Mukherjee and Bhattacharya, 1997 and Tiwary, 2000). We shall try to make a brief critical review of their findings in the following pages.

Geologic Setting The geologic formations that contain the above deposits were earlier thought to belong to the Ajabgarh Group of the Delhi Supergroup and the minimum age of the rocks was considered as 1565 $\pm$ 100 Ma (Choudhary et al., 1984).

As discussed in an earlier section, detailed determination U–Pb ages of zircon from felsic volcanic members of the host rock associations and Pb–Pb model ages of the stratiform ore deposits convincingly prove that the belt under discussion is a metallotect much younger than the typical Delhi–Aravalli ones (Deb et al., 1989; Deb et al., 2001; Sychanthavong, 1990) and may be broadly dated at $\sim$ 1 Ga.

Deri–Ambaji deposits, and may be, also the minor deposits of Danva and Birantiya–Khurd, belong to Type III (vide Chapter 4.2) in our classification and are hosted by felsic–mafic rocks (+ sediments) and the rest, belonging to Type II, are hosted by mafic rocks, all of course metamorphosed. The volcanic suite in the Deri–Ambaji (– Jharivav) area is in fact polymodal, comprising rhyolite, dacite, andesite, basaltic andesite and tholeiite (Tiwary and Deb, 1997). The host rocks at Deri are hydrothermally altered rhyolite and tholeiite, later metamorphosed. At Ambaji hydrothermally altered and metamorphosed rhyolite host the deposit. The altered mafic rocks are now represented by hornblende–biotite –plagioclase– quartz schists/hornfels with sharp variations in mineral proportions. The hydrothermally altered felsic volcanic rocks are instead represented by cordierite–anthophyllite–plagioclase–sericite–chlorite–quartz schists/hornfels, or phlogopitic–biotite–chlorite (–sericite– quartz) schists/hornfels. In the latter rocks also there are wide variations in the proportions of Mg-rich minerals. Zn-rich pleonaste is present in the first type. The large broadly ovoid body of rhyolite in the Ambaji mine-pit has some resemblance to rhyolite domes reported from some type areas. The volcanisedimentary sequence at Deri and Ambaji are intruded by later intermediate-acid plutonic rocks (Deb, 1980, 2000).

At the Birantiya–Khurd prospect, mafic volcanic rocks now represented by amphibolites are interlayered with the host felsic volcanic rocks. Concordant barite layers are intimately associated with the mineralisation. Late pegmatites intruded the ore zone. The Type-II deposits are hosted by altered pillow lavas, low- K tholeiites in primary composition. These tholeiitic rocks are associated with pyroxenite, layered gabbro (+ plagiogranite) and ferruginous chert in the Pindwara–Watera sector. The mafic rocks are now represented by varieties of amphibolites. The ore zone rocks of the Pindwara–Watera sector comprise different types of amphibolites and metapelites that contain andalusite and kyanite besides other minerals. Host rocks at Kalabar and Chitar deposits are also mafic volcanic rocks, now modified into amphibolites and chloritic schists. Late granitic plutons are present in all these areas. Deb and Sarkar (1990), based on the geochemistry of the mafic volcanic rocks, suggested that the tectonic setting in the Ambaji–Deri sector was that of a back-arc and that of the Basantgarh area showed an island-arc affinity. However, Golani et al. (1996) suggested that the former was rift-related and the latter represented a dismembered ophiolite emplaced in the proximity of a continental margin. Again the tectonic setting (s) of the Birantiya–Khurd and Kalabar–Chitar deposits with contrasted petro-mineralogical associations, occurring close to each other, is an enigma.

The deformation and metamorphism of the rocks of this belt have been studied by several workers (Roy, 1988; Deb, 1980, 1990b; Mukherjee et al., 1992; Sachan, 1993. Though varying in details the conclusion common in these works is that the rocks are multiply deformed and poly-metamorphosed. There is an apparent increase in the grade of metamorphism along the southwestern fringe, from amphibolite faces in the Ambaji–Deri and Basantgarh–Ajari areas to intermediate granulite facies at the Abu Road area (Deb, 2000).

Ore Mineralisation The ore bodies in all the deposits in the SDFB are lensoid to stratiform and peneconcordant with the ambient schistosity and compositional layering in the host rocks (Fig. 6.2.47 and Fig. 6.2.48). The original morphology of the ore bodies have conspicuously been modified at places.

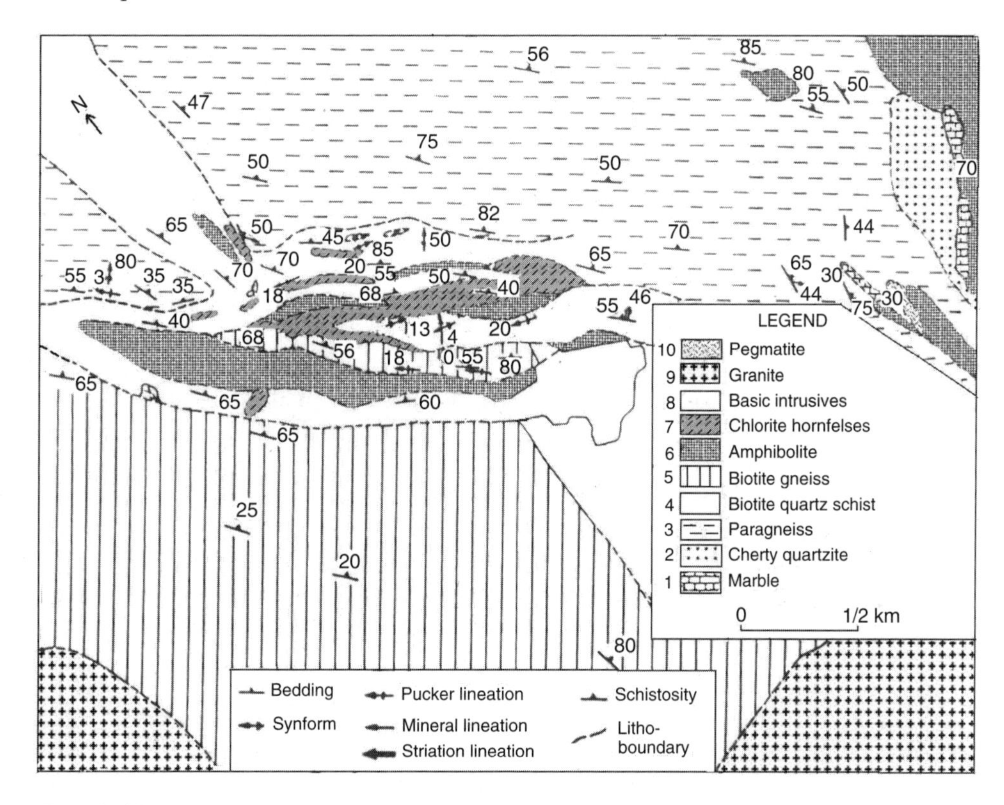

Fig. 6.2.47 Geological map of Ambaji ore zone (after Deb, 1980, 2000)

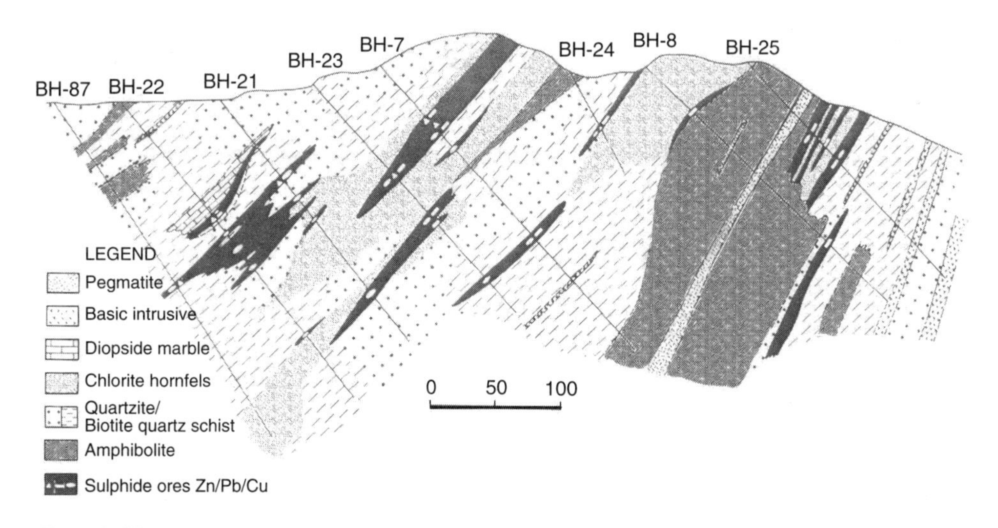

Fig. 6.2.48 Transverse section across the ore zones at Ambaji (after Deb, 1980, 2000)

Volcanic-associated basemetal sulphide deposits, as we have mentioned in an earlier section, are in general accompanied by wall rock alterations, commonly non-uniform in nature and intensity. The deposits/prospects of the SDFB also testify to this. Golani and Narayan (1990) studied the petrographic changes related to some of these deposits/occurrences. Tiwary and Deb (1997), however, made a detailed mass-balance calculation of alterations around the Deri deposit and showed Mg–Fe (–K) enrichment and Na-depletion in all the alteration facies. Kirmani and Fareeduddin (1998) studied the alterations at Danva. They showed that compared to the unaltered part of the host amphibolite, some altered rocks, now represented by quartz –biotite schist or quartz–sericite–biotite schists, FeO, CaO and Na_2O were considerably leached, while K_2O and H_2O were increased several folds. Tiwari (2000) also made a mass-balance calculation of alteration related to mineralisation at Kalabar–Chitar. She showed that the original host rocks of basaltic andesite to andesite composition gained in Mg, Fe, Si and H_2O and lost Na and Ca in hydrothermally altered rocks. Tiwary and Deb (op. cit.) also suggested that the extreme depletion of silica, enhanced negative Eu-anomaly and marked increase of LREE in felsic volcanic rocks at Deri, could be a signature of 'proximal' alteration. Used with discretion, all these alteration patterns could be useful tools in future exploration in the region.

Interlayering of sulphides and sulphides + silicate minerals is a common feature in both the two types of ores. Secondary structures such as folding, fracturing and cataclasis have been superimposed on them. In the Zn–Pb–Cu type deposits, massive sphalerite ores contain idiomorphic pyrite crystals. Galena and chalcopyrite rich ores are generally in form of stringers. Pyrite is ubiquitous and locally forms massive layers of pyrite. Pyrrhotite is present only at Deri. Locally magnetite is abundant. The Ambaji–Deri ores contain a large number of Bi-bearing sulphosalt minerals as minor phases. These are more common in chalcopyrite and chalcopyrite–galena rich ores. These comprise native bismuth, bismuthinite, pecoite, wittichenite and possibly bonchevite. Gold has been found in bismuthinite and silver in wittichenite and pecoite. The so-called tetrahedrite in Ambaji–Deri ores is in fact an Ag-bearing freibergite. Ores are metamorphosed (Deb, 1979, 2000; Tiwary et al., 1998). In the Cu–Zn type of ores in the Pindwara–Watera sector, pyrite is the major phase, followed by chalcopyrite and sphalerite. Pyrrhotite, galena, cubanite, mackinawite, marcasite and freibergite are present as minor phases. Pyrite is Co-rich (800–1000 ppm). Gold is reported from the Danva ores (Bhattacharya et al., 1991). There are interesting ore silicate relationships, such that trails of pyrite–pyrrhotite–sphalerite ± chalcopyrite ± ilmenite are present in garnet porphyroblasts, suggesting presence of these minerals at the site of garnet blasteisis.

Results of sulphur-isotope analysis of 50 monominerallic sulphide fractions from the Ambaji, Deri and Basantgarh deposits (Fig. 6.2.49) show that the sulphides of Amaji and Deri are isotopically similar with marked enrichment in ^{34}S, whereas those of Basantgarh are much less enriched (Deb, 1990b). $\delta^{34}S$ range from +15.7 to +20.0 per mil (mean + 17.2 per mil) in Ambaji sulphides and + 17.9 to +20.8 per mil (mean + 19.5 per mil) in the sulphides from the Deri deposit. In contrast, in the Basantgarh ores this value ranges from + 5.4 to +7.6 per mil (mean + 7.2 per mil). These observations are in agreement with the findings of Lydon (1984) that the sulphur isotopes in Zn – Pb – Cu sulphide ores are heavier than those in the Cu – Zn type volcanic associated ores. It may be mentioned here that the $\delta^{34}S$ in Ambaji – Deri ores is close to coeval seawater (~ + 20 per mil) at 1.0 Ga (Strauss, 1993).

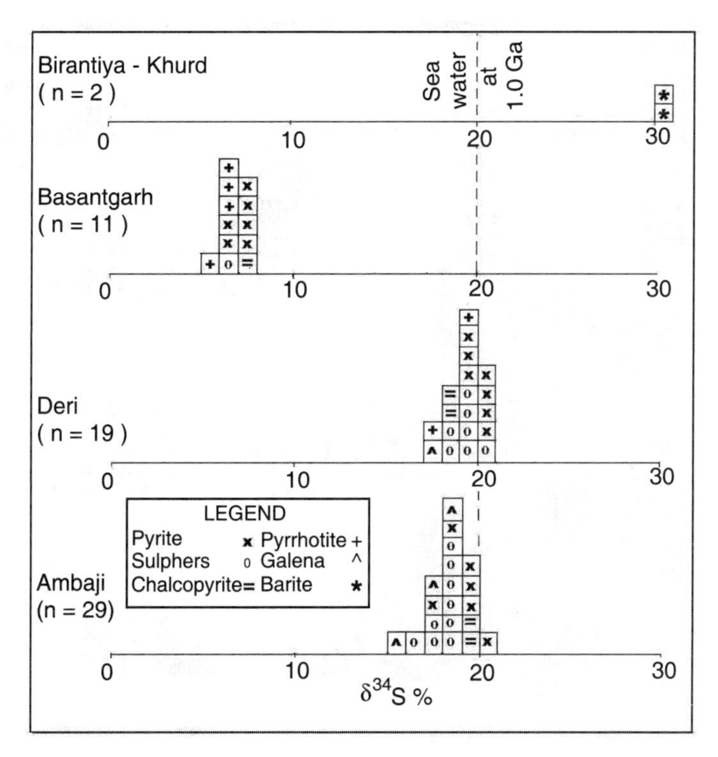

Fig. 6.2.49 Sulphur isotope data of ore forming dulphides from some deposits of SDFB

The order of enrichment in ^{34}S in the samples studied by Deb (1990a) is as follows: pyrite > pyrrhotite = sphalerite > chalcopyrite > galena, which suggests a close approach to equilibrium in the sulphide associations (Bachinski, 1969). The $\Delta^{34}S$ values between the physically coexisting pairs, such as pyrite–sphalerite and sphalerite – galena have been found to be relatively uniform (Deb, 1990a) suggesting isotopic equilibrium between the coexisting pairs in the temperature range of 423°–629°C for a majority of the co-existing pairs. These values are consistent with those obtained otherwise (Deb, 1990b). The source of sulphur in the ore depositing system is difficult to ascertain in the given geological context. There is no convincing evidence of any of the commonly conceived sources: bacterial reduction of seawater sulphate, reaction with Fe^{2+}-minerals in the hydrothermal reaction zone in or around the vent complex, or leaching during hydrothermal fluid flow.

We have already briefly discussed the nature of the Pb-isotopes from a number of these deposits/prospects (Table 6.2.13). Expectedly, there is conspicuous variation amongst these. Pb in the Birantiya-Khurd deposit is the most radiogenic. It is least so in the Danva ores. In other words, the Birantiya lead has the greatest crustal contribution, whereas the lead at Danva carries a strong juvenile, mantle signature. Pb-isotopes from Ambaji and Deri deposits are remarkably homogenous and fairly radiogenic. In fact, they are mixtures of lead from older supracrustal sources, including the basement rocks and the mafic – felsic members of the volcanic rocks in the belt.

Stratiform tourmalinite layers, commonly associated with metachert and Zn–Pb–Cu sulfides in the Deri deposit, are apparently volcanogenic. The relatively light $\delta^{11}B$ values of – 15.5 and – 16.4 per mil in these tourmalines suggest that the boron concentration of the hydrothermal fluid involved was much higher than the seawater (4.6 ppm). According to Deb et al. (1997), this could be the result of a low water/rock ratio within the hydrothermal system, or high boron concentration in the underlying felsic volcanic rocks and sediments that were involved in the formation of the hydrothermal fluids.

Table 6.2.13 *Pb-isotope data for VMS deposits in the Ambaji–Sendra belt, Western India (from Deb, 2000)*

Deposits	$^{206}Pb/^{204}Pb$	$^{207}Pb/^{204}Pb$	$^{208}Pb/^{204}Pb$
Danva	16.565	15.351	36.254
Basantgarh	16.911	15.476	36.707
Amli Mal	17.212	15.612	37.357
Ambaji	17.289	15.615	37.281
Deri	17.290	15.625	37.307
Birantiya—Khurd	17.467	15.680	37.556

There does not appear to be much controversy on some aspects of metallogenesis in the SDFB. These, *inter alia* comprise affiliation of the ore deposits/ occurrences with volcanic rocks, some compositional (including isotopic) correlation of the ores with the composition of the associated volcanic rocks, deformation and metamorphism of the ore deposits and the ores suggesting their early emplacement, wall rock alterations. These assign the deposits / occurrences to the Volcanic Associated Sulphide (VAS), or Volcanic Associated Massive sulphide (VMS) deposits (Fig. 6.2.50). Whether some of them belong to Kuroko sub-type or Bessi sub-type (Sarkar, 1992; Mukherjee and Bhattacharya, 1997) that is of secondary importance. However, the complex lithological makeup of the belt has kept the debate on the development of this metallotect still alive. Sugden et al. (1990) suggested that the sulphide ore deposits of the SDFB developed in a subduction related arc environment. Mukherjee et al. (1992) on the other hand, suggested a graben-type intra-cratonic basin structure particularly for the southern part of the Ajmer–Pindwara fold belt of Sen (1981).

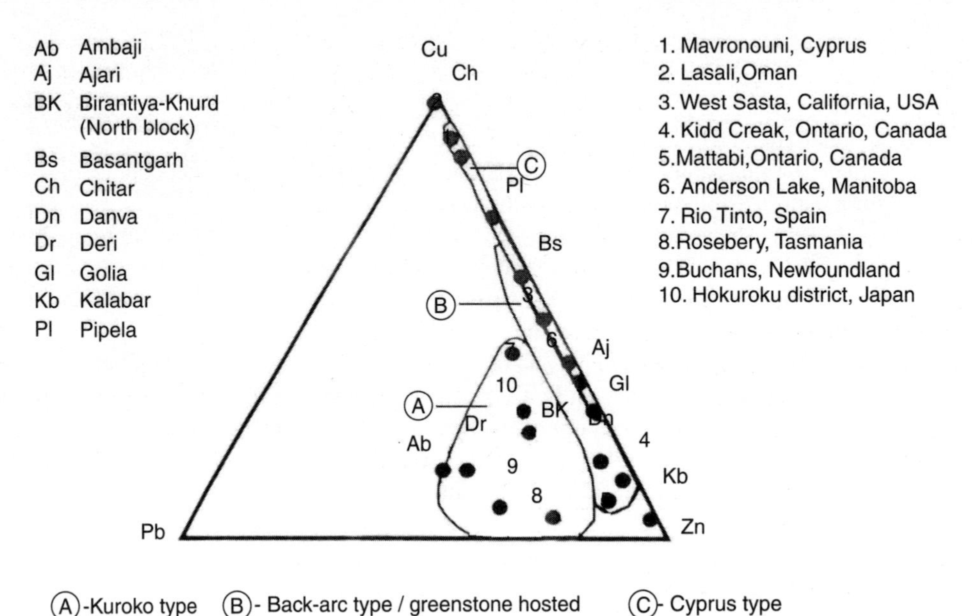

Fig. 6.2.50 A triangular diagram with plots of three major metals in VMS deposits, showing the different compositional fields (after Sawkins, 1990) and the positions of VMS deposits/prospects in Ambaji–Sendra belt

Mukherjee and Bhattacharya (1997) later modified the model and suggested that the low-K tholeiites of the Ajari–Basantgarh–Pipela zone were emplaced in a fault-controlled structure (graben) located in a fore-arc setting. Indeed, in a nothing very 'tell-tale' situation it seems reasonable to base the conclusion on whatever geochemical data is available on the mafic volcanic rocks of the belt.

Deb and Sarkar (1990) drew attention to the fact that the volcanic rocks of Basantgarh–Golia–Pipela zone, as well as those at Phulad represent low K–tholeiitic to calc–alkaline trend, characteristic of volcanic rocks occurring close to trenches in modern island arcs. The mafic rocks of the Ambaji–Deri zone show typical ocean floor geochemical characteristics, together with some geochemical signatures common in island arcs. These mafic volcanic rocks together with the associated felsic members can best be interpreted as representing a back arc environment, with possible attenuated segments of the continental crusts, a suggestion borne out by the Pb-isotope signatures of the ores. The lithological and Pb-isotopic similarities of this zone with the Birantiya—Khurd deposit suggest a similar tectonic setting for the latter. The Chitar ore zone occurs close to the northern extension of the so-called 'Phulad ophiolite zone'. That possibly explains the similarity of these deposits/prospects with those of the Basantgarh–Golia–Pipela zone.

Mineralised Hematite Breccia in Deldar–Kui–Chitrasani Fault Zone Hematite breccia, reported to occur extensively along Deldar–Kui–Chitrasani fault zone, marks the contact between Erinpura Granite and volcanisedimentary terrains of the east in Sirohi district, southern Rajasthan and adjacent parts of Gujarat. This fault zone with a length of 45 km and width of 50–80 m. traverses through magmatic rocks, both the mafic volcanic dominant southern parts of the SDFB and the adjacent Erinpura Granite pluton. The feature is described by Golani et al. (2001) as a brittle structure with anastomosing fractures and epithermal breccia filled with chalcedonic silica and products of argillic alteration. Pyrite, galena, chalcopyrite and minor sphalerite occur as disseminations and stringers. Srivastava et al. (1990) report pitchblende and secondary uranyl bearing siliceous breccia from the fault zone (with 0.02–0.35% U_3O_8 and < 0.005 ThO_2). The breccia zone analysed 50–200 ppb gold from widely spaced localities.

Uranium Metallogeny in Western Indian Craton

Uranium contents of 5–200 ppm, averaging 20 ppm, are reported from some granitoid members of the basement complex of this region (Singh, 2000) and yet there is no Quartz–pebble conglomerate (QPC) type uranium mineralisation in the region worth the mention. It may be pointed out at the same time that as yet no exploitable QPC type mineralisation is reported from India. However, as we have seen in earlier sections, some mineralisations of this type, lean though, are reported from the Dharwar region in the South India and the Singhbhum and adjacent areas in Eastern India. The thorium bearing polymictic conglomerates of Kharbar and Parsad, assigned to this type by Mahadevan (1986), do not really belong to the QPC type, as pointed out by Singh (2000).

Early Proterozoic Uranium Mineralisation in Metamorphic Tectonites Early Proterozoic or Paleoproterozoic uranium mineralisations in shelf facies shale–sandstone–carbonate dominant assemblages, variously metamorphosed and deformed, are reported from a number of places in the region, the most important of which are the Umra–Udaisagar–Kalamagra–Haldighati in the Udaipur district, followed by Putholi, Salera, Jojron ka Khera, Dadiya and Rundiyan in the Chittaurgarh district. Such mineralisations are also reported from a number of places in eastern and western Jahazpur (Table 6.2.14, Fig. 6.2.51). The host rocks are carbonaceous phyllites, calc–arenitic schists, metabasites and gneisses.

Table 6.2.14 *Uranium occurrences in Paleo–Mesoproterozoic rocks of Western Indian craton (modified after Singh, 2000)*

Locality (District)	Host rock	Mode of occurrence	Age	Grade and Extent
Umra, Udaisagar, Kalamagra, Denkli, Haldighati (Udaipur)	Carbon phyllite, metapelites and calc. arenite	Veinlets and disseminations along shear zones	Paleoproterozoic	5 4 1
Putholi, Salera, Jojron ka Khera, Dadiya, Rundiyan (Chittaurgarh)	Quartzite	Fracture- controlled veins and also as disseminations in quartzites overlying basement granites	Paleoproterozoic	4
Chainpura, Bhadurpura, Amalda, Devpura (Bhilwara)	Carbon phyllite	Along fractures and shears in the synclinal core	Paleoproterozoic	3
Antri–Beharipur (Mahendragarh)	Biotite schist	Disseminations	Mesoproterozoic	2
Kanthi (Sikar)	Carbon phyllite/ limestone	Disseminations along sheared contact between carbon phyllite and limestone.	Mesoproterozoic	2

Note: 1. Patchy, measuring a few tens of meters, U_3O_8% 0.02–0.03; 2. Extensive, U_3O_8% 0.02–0.03; 3. Locally measuring a few hundreds of meters, U_3O_8% 0.03–0.05; 4. Extensive/intermittent, U_3O_8% 0.03–0.30; 5. Extensive, U_3O_8 0.06–0.30.

Uranium mineralisation at Umra was discovered at an early stage of exploration for uranium in India (Bhola et al., 1958). The mineralisations at Umra and Udaisagar occur at the transition zone between the dolomitic carbonate rock and carbonaceous phyllite. Uranium occurs along schistosity and fractures mainly as pitchblende with some secondary uranium minerals (near the surface) and rare roundish uraninite (Srivastava et al., 1996). Coffinite is locally present. Detailed EPMA analysis revealed the presence of a U–Si–Ti phase, a U–Ti phase, a U–Si–Ti–Fe phase and a U–Ti–Ca phase (uranophane). Pitchblende and coffinite are rich in Pb, containing 9.27%–20.9% and upto 10.4% PbO respectively. Th-contents are negligible. Pyrite, chalcopyrite, pyrrhotite and galena are present as minor phases. $\delta^{34}S$ variation between – 13.5 to 21.3 per mil is suggestive of an important role played by microorganisms in sulphur isotope differentiation.

Dhar (1964) and Bhola (1965) suggested that the uranium ores of Umra and Udaisagar had formed from hydrothermal solutions given out by the cooling granitoid bodies now occurring in the neighbourhood. However, later work proved these granitoids to be older than the metasediments that contain the ores. It therefore, seems reasonable to believe that uranium at these places accumulated first during sedimentation–diagenesis and later reworked during metamorphism and deformation. The stratigraphic position of the mineralisation zone at Umra and certain alteration features such as silicification carbonatisation, feldspathisation, sericitisation and chloritisation may assign this mineralisation to the Proterozoic unconformity-related or unconformity–proximal type of deposit in a broad sense (Singh et al., 1995). The suggested possibility of occurrence of uranium mineralisation along Aravalli basal unconformity around Sarara and Kharbar Parsad, and Hindoli unconformity around Putholi and Salera (Singh et al., 1995), is worth detailed investigation.

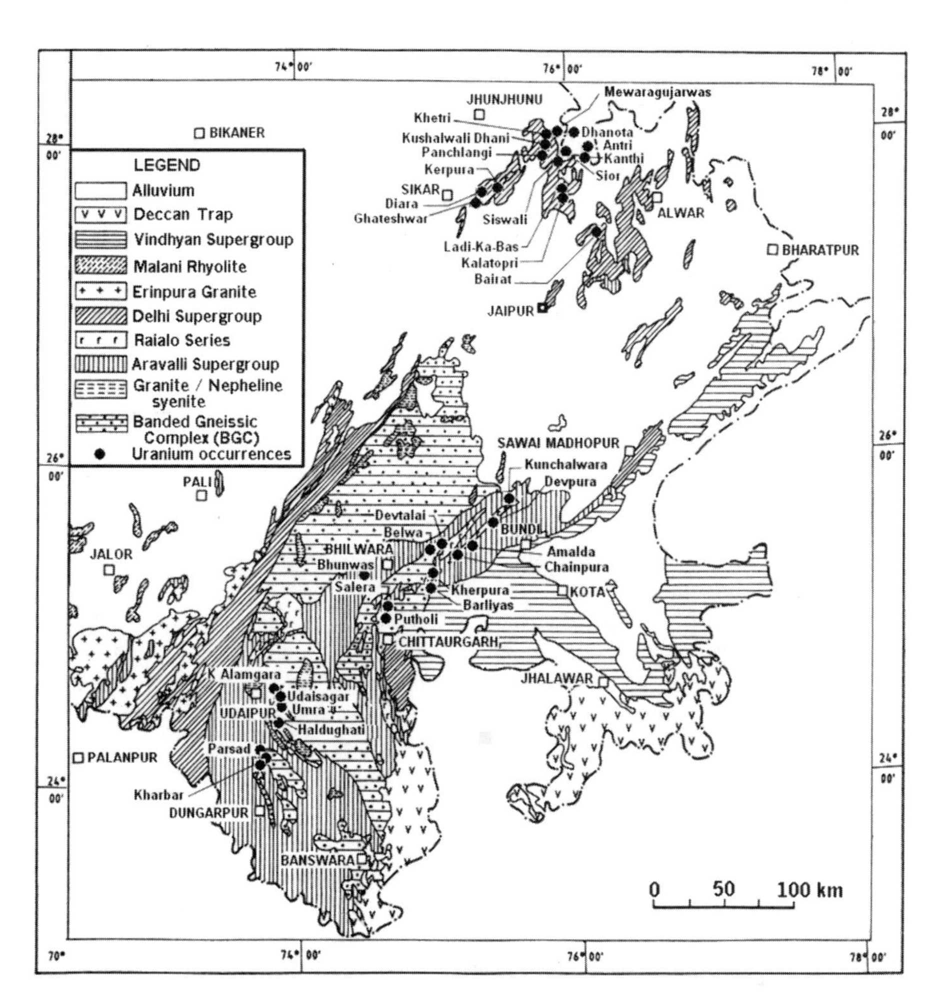

Fig. 6.2.51 Map showing distribution of U-mineralisation in Western India (from Singh, 2000)

Note: The authors have left the map unmodified although they do not agree with some parts of the map. The presence of Aravalli Supergroup rocks to the west of the Delhi fold belt is not confirmed by radiometric dating. These are rocks of a younger orogen, Ambaji–Sendra, as already discussed. Similarly, occurrence of Delhi Supergroup rocks to the east of Aravalli–Bhilwara rocks is also unproven.

Uranium mineralisation occurs along a zone of silicified and fractured cherty quartzite close to the unconformity at Putholi, Salera and adjacent areas over long strike distance. Secondary uranium minerals, such as uranophane and phurcalite, hydrous silicate and phosphates of Ca and U and grains of pyrite and other common sulphides are reported from this zone (Bhatt et al., 1995). Here also the Th-content in the uranium ores is negligible.

Uranium mineralisation is traced over a belt of about 70 km between Jawaj and Kunchalwara along the interface of the so-called Mangalwar Group granitoids and migmatites and the metasediments of the Jahazpur Group. Uranium and rare earth mineralisations are reported from Kunchalwara, Mayala, Polia and Kuradiya areas further north associated with migmatites (Singh, 2000).

Uranium Mineralisations Associated with Some Neoproterozoic Rocks of Magmatic Affiliation Uranium mineralisations of this type were located at places of Kolihan, Saladipura and Ghateswar areas in north Rajasthan mainly in the Sikar district. They occur within the Ajabgarh Group of rocks, but closely associated with quartzo–feldspathic rocks (Bhola, 1965; Narayan Das et al., 1980). These rocks have later been identified as albitites (Ray, 1990). Albite associated uranium mineralisation is a well-established type today and is reported from places in the erstwhile USSR, Brazil, Sweden, Canada and China (Sarkar, 1995 and references therein). Continued investigation has revealed the presence of a number of uranium and uranium–thorium mineralisations in a 50 x 10 km zone stretching in the NE–SW direction through a part of Sikar district in Rajasthan and Narnaul district of Haryana. Host rocks are one or more types of altered granites, albitites, pyroxenites, quartzites, metapelites including graphitic schists (Table 6.2.15). Uraninite, brannerite and davidite are the principal uranium bearing minerals. They may or may not contain thorium. Uranium shows a positive correlation with yttrium. Uranium also shows a preference for the altered sodic and peralkaline rocks (Singh et al., 1998). The common features of albitite associated uranium and other related mineralisations are Na–metasomatism and overall control by disjunctive tectonic structures of regional extents. However, the influx of the metasomatic fluids, including the ore-elements along such zones, still remains to be better understood, not only here, but also elsewhere.

Veins and disseminations of uraninite and pitchblende are reported from the metamorphic tectonites in south Rajasthan, associated with Godra Granite, Erinpura Granite and the Malani igneous suite. However, the role of acid magmatism in these mineralisations is not clear. Uraninite and U-bearing zircon occur as pockets in mica pegmatites of Bhunas, Bhilwara, and have been mined out. Sharma (1988) dated the emplacement of these pegmatites at ~0.95 ± 0.05 Ga.

Kui–Chitrasani fault zone along the border of the Erinpura Granite with the Ajabgarh Group of rocks in the southern part of the Sirohi district, Rajasthan, and neighbouring Banaskantha district of Gujarat contains pitchblende–bearing quartz veins at places (Srivastava et al., 1990). The reported extensive uranium mineralisation along a dislocation zone between Mawal and Awal within granite gneisses (Singh, 2000), await confirmation by detailed investigation. Some features of albitite-associated U-mineralisation of Western India are presented in Tables 6.2.15 and Table 6.2.16.

Minor uraninite-bearing veins in Champaner Group dolomitic limestone occur near Garumal in Panchmahal district, Gujarat.

Table 6.2.15 *Uranium mineralisation along with the albitite and associated rocks of Western Indian craton*

Locality (District)	Host rocks	Mode of occurrence	Age of the host rocks	Grade and extent
Maonda, Bandha ki Dhani, Banjaron ki Dhani, Sardarpura, Sagdu ki Dhani, Pachlangi, Karoth ki Dhaniyan, Diara, Kerpura, Ghateswar (Sikar)	Albitites/ pyroxenites/ magnetite – apatite rock	Disseminations in albitised granitoids, quartz–mica schists, carbon phyllites and pyroxenites along major fault and shear zones	Meso–Neoproterozoic	2
Ramsinghpura Mothaka (Sikar)	Albitites		Meso–Neoproterozoic	1
Dhhancholi, Dhanota (Mahendragarh)	Albitites			

Notes: 1. Local, measuring a few hundred meters; U_3O_8% 0.03–0.05
2. Extensive/intermittent: U_3O_8% 0.06–0.30

Table 6.2.16 *Some compositional characteristics of albitites and associated rocks, Sikar district, Rajasthan (Singh, 2000)*

Components (Elements in ppm/oxides in %)	1	2	3	4	5	6	7	8
U_3O_8%	1.60	0.24	0.055	0.007	0.43	0.13	0.03	0.12
ThO_2%	0.041	<0.005	0.015	0.035	< 0.005	< 0.005	0.013	0.012
Y	1349	117	56	21	148	135	96	191
Nb	825	32	220	<25	<10	116	192	27
Zr	132	507	1608	792	252	385	287	289
Cu	378	2165	28	33	>1000	18	21	28
Ni	36	78	4	11	180	57	31	48
Pb	327	178	49	24	621	105	34	157
La	16530	318	438	< 100	< 100	150	4248	1206
Ce	17868	Nd	Nd	Nd	Nd	Nd	2720	1408
Na_2O/K_2O	11.0	16.9	2.4	31.7	2.7	2.2	141	1.8

Absence of Quartz–pebble–conglomerate (QPC) type, and little or no development of Proterozoic unconformity proximal deposits like Athabasca and Thelon, Canada, appears due mainly to the lack of U-fertile granitoid rocks or the absence of their exposure in the distant past in view. The same argument should explain the absence of sandstone type deposits also. Uranium mineralisation associated with hematite breccias in granitoid rocks as at Olympic Dam, Australia, in a strict sense still remains unique in spite of a worldwide search

Tin–Polymetallic Mineralisation in Tosham Area, Haryana

The Tosham deposit and the Khanak quarry, situated at about 120 km NNW of Delhi in the Bhiwai district of Haryana, is known for the unique Cu (–Au) –Ag–Sn–W–Bi mineralisation (Kochhar, 1983; Seetharam, 1990; Khorana,1993, 2000; Murao et al., 2000; Khorana et al., 2004).

The mineralised area is located in an isolated dome shaped hillock, surrounded by a sandy country. The major rock types constituting this hillock are granite, rhyolite (partly andesite) and metasediments represented by andalusite bearing quartz–mica schist/ quartzite. The association is traversed by dykes of quartz–feldspar porphyry and pegmatite. Both the eastern and the western contacts of the rhyolite body at the centre are faulted. Fragments of granitic rocks are present within rhyolite near the latter's contact with the former (Fig. 6.2.52). At Khanak, aplite and pegmatite, which grades into each other, intrude the granite.

Mineralisation occurs mainly in and around this western fault as quartz–green biotite–sulphide veinlets and network of quartz–cassiterite veinlets. The principal ore minerals are cassiterite, wolframite and chalcopyrite. The other associated minerals/metals are gold, galena, pyrite, arsenopyrite, sphalerite, pyrrhotite, bismuthinite, skutterudite, rammelsbergite, cubanite and covellite (Khorana et al., 2004). Indium-bearing mineral roquesite and Pb–Bi sulphosalts has also been reported (Seetharam, 1990; Khorana et al., 2004). Greisenisation took place locally along the western fault zone. The greisen is composed of phlogopite, muscovite, hematite, quartz, chalcopyrite, pyrite and minor native sulphur. However, some pyrite mineralisations are also noticed at places in the metasediments north of the Tosham hill (Murao et al., 2000).

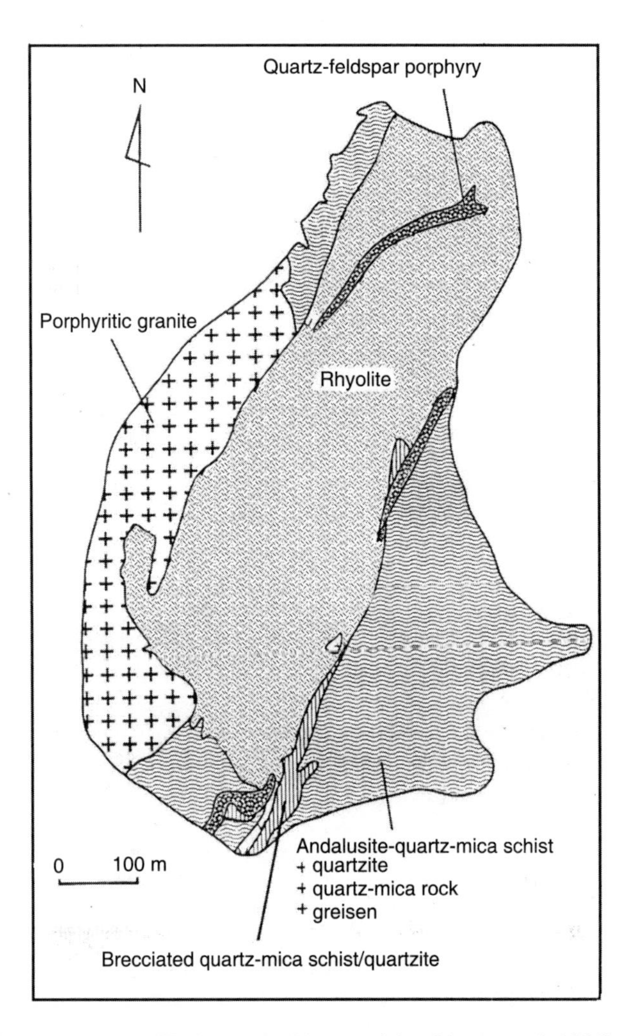

Fig. 6.2.52 Geological map of Tosham area, Haryana (after Murao et al., 2000)

The chemical composition of the igneous rocks from Tosham and Khanak are presented in Table 6.2.17. It is obvious from the table that the granites are characterised by high K/Na ratios and are peraluminous.

In the discrimination diagram of Pearce et al. (1984) the Tosham Granites plot in the field of 'syn-collision type granite'. According to Kochar (1983) the Tosham Granites belong to the A-type, whereas the data of Murao et al. (2000) plot them outside of this field i.e. in the field of I and S types of White and Chappel (1977) (Fig. 6.2.53). In contrast, rhyolite has the characteristics of A-type with high Ga/Al ratio between 4.1 and 5.6 and higher values of Zr, Y, Nb and Ce than granite. Interestingly, the rhyolites plot in the 'within plate granite' (WPG) in contrast to the 'syn-collision' type of the associated granites. These differences notwithstanding, the distribution pattern of REE in these rocks suggests that the granites and rhyolite of Tosham are consanguinous (Murao et al., 2000). The S-isotopes in the associated sulphides is a little too heavy for tin–tungsten mineralisations. One possible explanation of this fact is the assimilation of sulphate sulphur by the intrusive granitic magma.

Table 6.2.17 *Major element oxides (wt.%) and trace element (ppm/ppb) composition of Tosham Granite, Tosham Rhyolite and Khanak Granite (Murao et al., 2000)*

Sample No.	1	2	3	4	5	6	7	8	9	10	11	12
Rock type	Rhyolite	Rhyolite	Rhyolite	Rhyolite	Granite	Granite	Granite	Granite	Aplite	Granite	Granite	QFP
Locality	Tosham	Tosham	Tosham	Tosham	Tosham	Tosham	Tosham	Tosham	Khanak	Khanak	Khanak	Tosham
SiO_2	71.85	72.29	71.24	63.26	67.35	65.86	79.87	70.34	73.93	67.13	65.77	68.32
TiO_2	0.06	0.06	0.05	0.30	0.34	0.52	0.21	0.33	0.09	0.57	0.58	0.61
Al_2O_3	12.34	12.44	13.05	12.53	15.04	14.08	9.18	13.46	13.23	13.58	14.50	13.60
Fe_2O_3	10.68	8.91	6.14	16.43	4.02	8.08	2.85	4.99	1.76	4.93	4.95	5.10
MnO	0.08	0.07	0.22	0.21	0.03	0.15	0.05	0.15	0.025	0.08	0.09	0.16
MgO	0.14	0.12	0.08	0.32	0.45	0.87	0.50	0.26	0.13	0.55	0.67	0.68
CaO	0.16	0.40	0.90	0.06	0.18	0.25	0.11	0.18	0.93	1.96	2.23	1.60
Na_2O	0.23	0.22	0.22	0.25	0.24	0.19	0.14	0.10	3.44	1.98	2.46	2.16
K_2O	1.91	2.46	3.76	4.08	8.77	6.41	4.44	4.29	5.22	6.47	5.85	5.67
P_2O_5	0.01	0.01	0.01	0.06	0.19	0.20	0.11	0.17	0.05	0.18	0.18	0.19
LOI	2.33	2.50	2.79	1.15	2.31	1.90	1.79	4.85	1.08	2.09	1.91	1.02
Total	99.80	99.49	98.66	99.65	98.92	98.51	99.25	99.12	99.89	99.52	99.19	99.11
Rb	186	235	440	571	332	343	291	284	409	328	290	397
Sr	9.7	15	2.8	44	128	36	27	99	35	153	210	152
Zr	194	188	193	167	188	231	133	178	135	412	433	419
Y	159	148	218	204	12	25	10	13	107	68	61	67
Zn	118	286	81	169	35	92	72	231	12	28	63	122
Cu	44.8	47	22.6	23	9	10	0	14	53	2	1	32
Pb	83	91	51	133	128	104	57	123	21	14	26	100
Ba	218	31	17	466	923	580	240	478	209	910	1792	910
U	32	32	37	13	–	–	–	–	17	–	–	–
Th	128	111	97	17	–	–	–	–	67	–	–	–
Ce	66	30	148	95	74	70	52	70	133	280	250	280
La	9.9	21	61	31	–	–	–	–	60	–	–	–
Ga	36.3	32.1	28.3	28	17	19	11	16	22	18	18	19
Nb	108	102	–	–	24	24	10	14	28	24	22	22
Ta	12	12	–	–	< 2	< 2	< 2	< 2	6	2	< 2	2
Hf	8	8	–	–	6	6	2	4	4	12	12	10
$10^4 \times$ Ga/Al	5.6	4.9	4.1	4.2	2.1	2.5	2.2	2.2	3.1	2.5	2.3	2.6

Note: QFP – Quartz feldspar porphyry; Not analysed – ; LOI – Loss on ignition

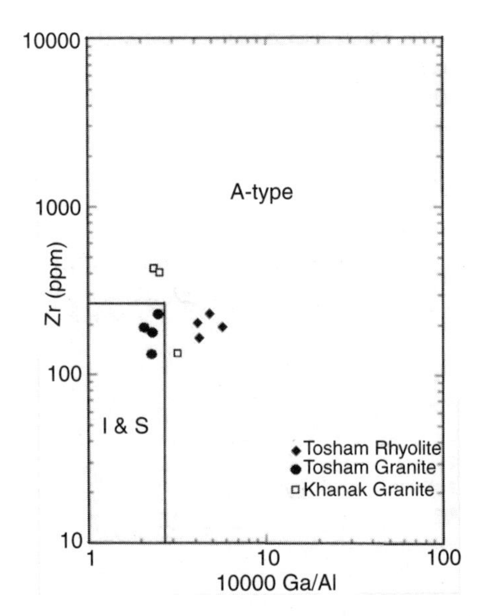

Fig. 6.2.53 Plots of Tosham Granite, Tosham Rhyolite and Khanak Granite in the discriminant diagram of White and Chappel (1977) (after Murao et al., 2000)

Khorana et al. (2004) described the composition of Khanak granitoid as varying from monzonite to granite/granite porphyry through granodiorite. They also assigned it a metaluminous character (A/CNK $\leq$ 1) in contrast to the peraluminous nature of Tosham granitoid. While Tosham granitoid is rich in normative corundum, it is absent or negligible in Khanak granitoid. Gold is reported to occur in association with quartz–sulphide–tungsten veins traversing the sheared and fractured metasediments and granite. The alteration zones bordering the mineralised veins and the fractures are often strongly hematitised (Khorana, 2004).

Determination of Ar–Ar plateau ages of rhyolite and biotite in granite from Tosham yielded values of 793 ± 18 Ma and 818.8 ± 3.6 Ma respectively and that of Khanak, 813 ± 3.4 Ma. K–Ar age determined from a hydrothermal phlogopite, collected from the ore-bearing greisen zone along the Western fault, is 704 ± 36 Ma (Murao et al., 2000). Kochar et al. (1985) obtained an Rb–Sr isochron age of 770 Ma from the Tosham Granite.

The mineralisation is obviously hydrothermal, although the temporal gap between the granite emplacement and hydrothermal mineralisation is ~100 Ma. Heavy S-isotope dominance in the associated sulphides and the homogeneity in the $\delta^{34}S$ per mil suggest sulphate assimilation by the upcoming granitic magma rather than modifying the sulphur isotope composition at the site of ore deposition. The change of the granite from the S (–I) type to A-type in the rhyolite, as obtained in the chemically discriminant diagrams, is not easy to explain in terms of tectonics or metallogenesis.

Tungsten Mineralisation at Degana, Balda and Sewariya, Rajasthan

Degana Tungsten mineralisation associated with Neoproterozoic granitic intrusives are reported from a number of places, located along the western fringe of the Delhi–Aravalli range, Rajasthan. The most important amongst these deposits is the one located at Degana, a few kilometers to the northwest of the Degana railway station. There was evidence of old working in the deposit. The Geological Survey of India and the Indian Bureau of Mines undertook detailed exploration of the deposit. Later, exploitation of the deposit was handed over to the state Directorate of Geology and Mines, Rajasthan.

The mineralisation occurs within the polyphase S-type Malani Granite, as well as in the adjacent Delhi/ Sirohi phyllites in the form of greisenised veins and disseminations. Chattopadhyay et al. (1982) correlated it with Jhalore Granite that was dated 735 Ma by Crawford (1975). The veins are short and hardly exceed 75 cm in length. Sulphides were introduced at a later phase of mineralisation. Ferberite is the principal tungsten mineral. Mineralisation occurs in three neighbouring but isolated hillocks, Rewat Hill, Tikli Hill and the Phyllite Hill (Fig. 6.2.54). Estimated ore reserve at Degana is 3440 tonnes of ore of all types, including a small portion of 'gravel ore'. This estimation is based on an average grade of 65% WO_3 concentrate (Sahai et al., 1997).

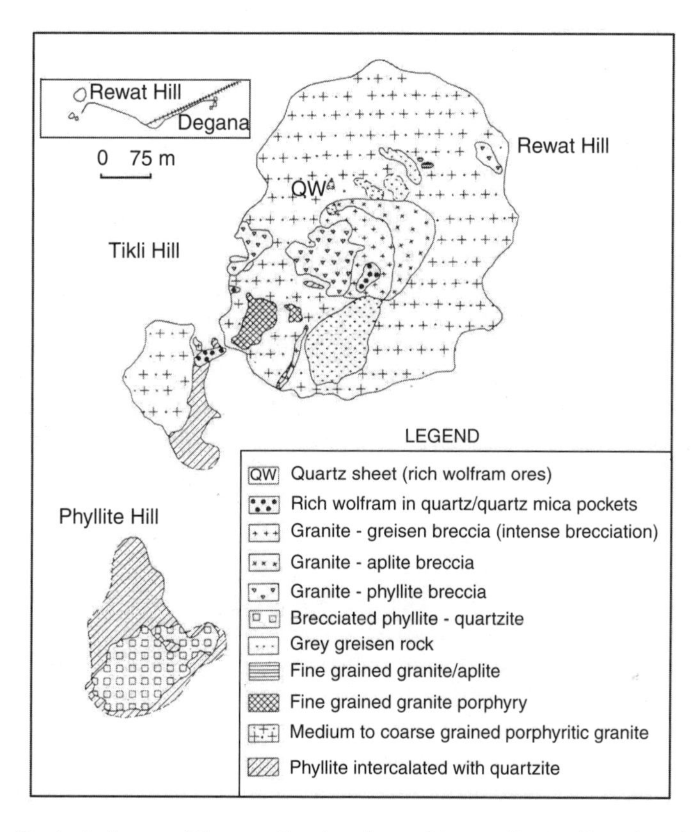

Fig. 6.2.54 Geological map of Degana Granite pluton, Nagaur district, Rajasthan (modified after Chattopahyay et al., 1982)

Balda The tungsten mineralisation at Balda, occurring at a distance of about 6 km from the Sirohi (24° 53′ : 72° 52′) was first discovered by A Mukhopadhyay (1980) of the Geological Survey of India. Our present day knowledge of this deposit is acquired from the results of follow up studies by several workers (Bhattacharjee et al., 1984; Krishnamurthy and Pravakar, 2000; Singh and Singh, 2002). At Balda, quartz–mica schists (/phyllite) of the Sirohi Group have been intruded by Neoproterozoic granite. Of the two phases of granitic intrusives in the area, the first one is coarse/ medium-grained porphyritic Erinpura Granite. The other one is the medium-grained leucogranite, i.e. the more fractionated Balda Granite with which is associated the tungsten mineralisation. The mineralisation has taken place along greisenised veins and silicified shear zones in both granite and mica schists (Fig. 6.2.55). The estimated ore reserve at Balda is 500 tonnes at 65% WO_3.

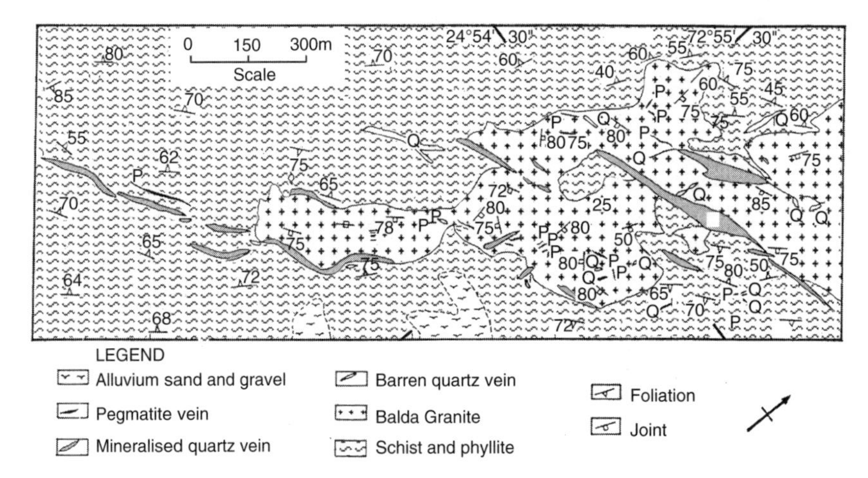

Fig. 6.2.55 Geological map of Balda area showing the disposition of tungsten-bearing quartz veins within granite and mica schist (after Singh and Singh, 2002)

Sewariya Sewariya tungsten mineralisations are relatively recent finds by the Geological Survey of India. The tungsten mineralisation occurs in quartz veins around Sewariya village (26°21′ : 75°15′) in Pali district of Rajasthan. The mineralisation is associated with Sewariya Granite, a polyphase granitoid massif consisting of major biotite-rich granite and one or more younger phases. The W (–Sn–Li) mineralisation occurs mainly as quartz veins, particularly with the set that is restricted to the N–S to NNE–SSW trending ductile shear zones, both within and outside the granitoid body (Fareeduddin and Bhattacharjee, 2000). The ore minerals comprise wolframite (principal phase), sicklerite, ferrisicklerite, triphylite and cassiterite. Host rocks show poor to intense alteration or greisenisation.

Mineralisation is proved in Pipaliya and Motiya areas. In Pipaliya, there are two mineralised zones, Pipaliya North and Pipaliya South. The ore body in Pipaliya North Block is 300 m long, 1.50 m wide and has a grade of 0.23% WO_3, whereas that in the South Block is 400 m long, 1.30 m thick and contains 0.18% WO_3. In Motiya, the ore zone extends for 200m with the average thickness of 0.74 m and average grade of 1.0% WO_3 (GSI, 2001).

Some common characteristics of all the tungsten mineralisations outlined above are as follows. They are all related to the younger (Neoproterozoic) granitoids and had therefore, better chance of avoiding erosion of the mineralised upper parts. Available geochemical data i.e. high SiO_2, Al_2O_3, K_2O and Rb/Sr ratios suggest that the granitoids were fractionated. The low Fe_2O_3/ FeO testify to the reduced state of the granitoid magma that produced the W-bearing ore solutions on vapour saturation. Although for all these deposits, as well as that for the Chhendapathar deposit in Eastern India, related granitoids are identified as of S-type, W-bearing ore fluids may be produced from both the S-type and I-type granitoids (Blevin and Chappel, 1995) provided the other conditions are satisfied. Pb-isotopic studies of the Early Precambrian tungsten deposits suggest that during the period 3000–2400 Ma, the primitive mantle was possibly the principal source of tungsten. The higher μ ($^{238}U/^{204}Pb$) and W ($^{232}Th/^{204}Pb$) values of the later tungsten accumulations suggest crustal evolution and the metal concentration at the suitable locales with time (Chiaradio, 2003).

RM and REE Mineralisations in Rajasthan–Gujarat Region

As is known and discussed already, a large part of the mineralisation of rare metals (RM) and a fair part of rare earth elements (REE) generally take place in pegmatites that are commonly, but not necessarily, mica-bearing. In India, the Rajasthan–Gujarat region was included in the national programme of exploration of rare metals and rare earth mineral deposits after the Independence. The Atomic Minerals Directorate (AMD) was the nodal agency chosen for this job. Serious search since about the middle of the last century, led to the discovery and later mining of a number of pegmatites in Rajasthan and Gujarat that contained substantial quantities of Be, Li, Nb and Ta (locally with little of uranium), besides commercial grades and quantities of mica. Important mines of such deposits are located at Basundui, Nasirabad, Rajgarh and Bhunas in the Ajmer and Bhilwara districts of Rajasthan. It is worth mentioning here that gadolinite was reported from Hussainpura in the Banaskantha district of Gujarat a century ago by Holland). The post-Independence exploration proved Nb–Ta mineralisation in pegmatites near Idar town, Sabarkantha district. Other places in the district, namely Sarangpur, Umednagar, Bhavangad were found to contain mineralisations of yttrotantalite, betafite, fersmanite, tantalite etc. Pegmatites occur in a cluster, extending for about 50 km from west of Jharol to Som in the south. In fact, the Rajasthan mica-belt is larger than the mica belts in Jharkhand (the former Bihar mica belt) and Andhra Pradesh (Nellore mica belt).

Characteristics and Varieties of the RM (–REE)-bearing Pegmatites These pegmatites are dominantly LCT in type, being enriched in Be, Li, Nb and Ta. Morphologically, they may be tabular, lensoid or poddy, including strings of pods, and irregular. They occur in both the gneissic basement (BGC) and the supracrustal rocks, i.e. Aravalli and Delhi Supergroups. While occurring within the deformed metasediments, the pegmatites are broadly conformant with the structure in the host rocks, including occupying fold hinges. In massive rocks, such as amphibolites, they may occupy tensional joints as at Dos–Datri, Deoria and Gujarwas in Nasirabad area of Ajmer district. The metamorphic facies of the host supracrustal rocks are upper greenschist–amphibolite. In size, these pegmatites vary from 10 m to 150 m in length and 1.5 m to 10 m in width. The pegmatites may be zoned or unzoned. The pegmatites in the basement rocks are generally less productive.

Ages of the pegmatites under review have not been taken up for serious radiometric determinations in the recent past. The few ages obtained decades back vary from 1200 Ma to 500 Ma. These may be constrained with the help of more precise systematics in use today.

Some important RM- and REE-bearing pegmatites of Rajasthan–Gujarat area are shown in Fig. 6.2.56 and described below

Beryl (–columbite–tantalite) Bearing Pegmatites The beryl-bearing pegmatites occur in the entire Aravalli mountain, starting from Narnaul area in Haryana and then passing through Alwar, Jhunjhunu and Ajmer in northern Rajasthan and Bhim–Bhilwara in Central Rajasthan upto Umedpur in the south. The mineralisation has been traced for a distance of about 500 km, but the richer part of the belt is about 250 km in length with a maximum width of about 150 km, covering the Jaipur, Ajmer, Bhilwara, Tonk and Udaipur districts (Maithani and Nagar, 1999).

Of more than 600 beryl bearing pegmatites that are known from this region, Bhola (1977) designated 6 (six) deposits (Basundui, Makera, Deoria, Bari Shikar Bari, Gujarwara and Kharwa) to be very 'important' and another 24–25 falling under his 'very good' category. Each of the former group produced several hundred to > 1000 tonnes of beryl, and each of the latter group, in terms of 100 tonnes or more. Beryl crystals in the first group are generally large. In rare cases, these have been found to be as large as 6m in length and 1.5 m in diameter. The large to very large beryl crystals occur in the intermediate zone of zoned potassic pegmatites at the contact with the quartzose core (Rao, 1977). So far the pegmatitic deposits of Rajasthan–Gujarat area produced several thousand tonnes of high-grade beryl ore, analysing>10% (ranging upto 14%) BeO (Maithani and Nagar, 1999).

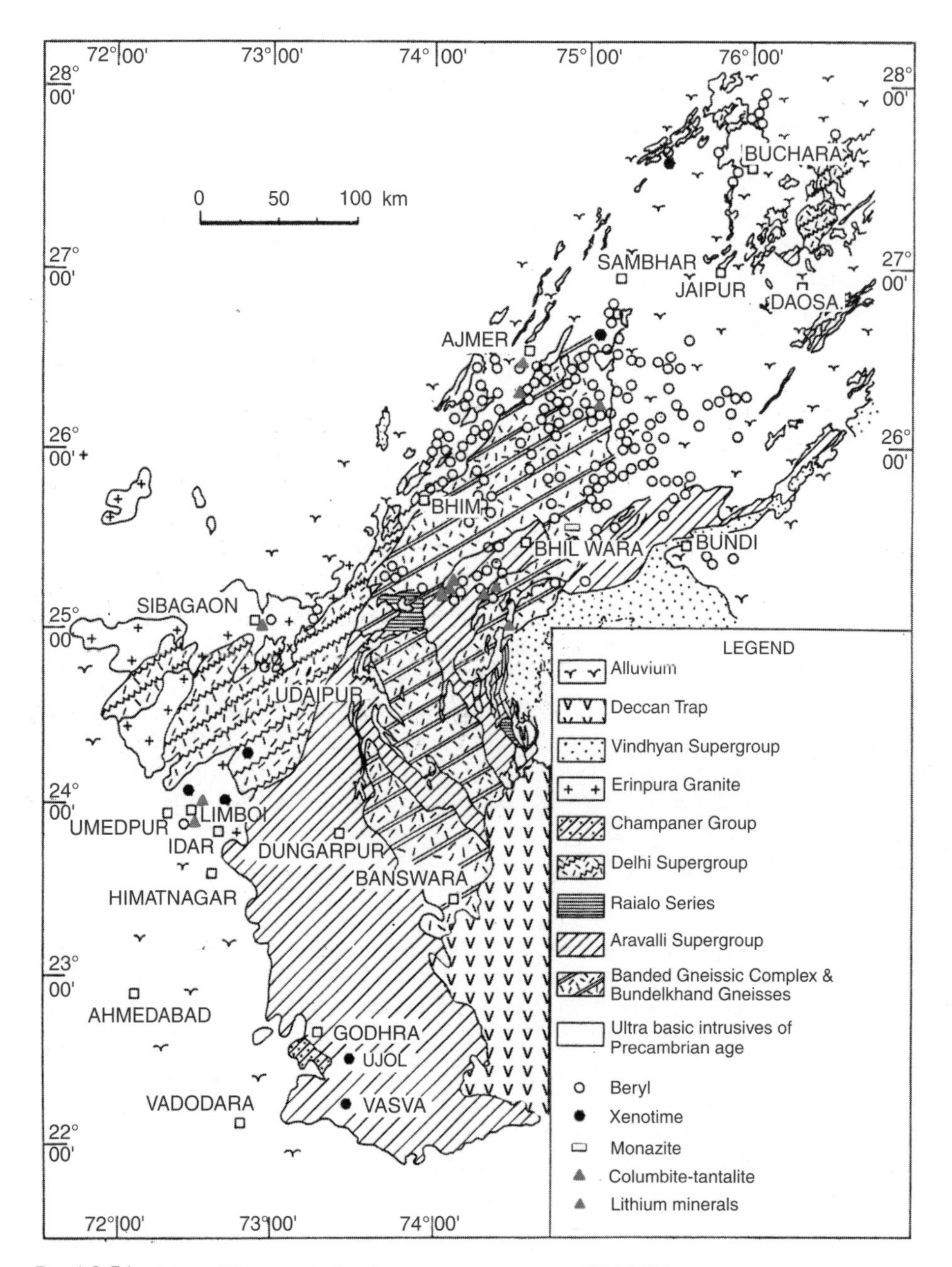

Fig. 6.2.56 Map of Western India showing occurrences of RM–REE pegmatites

Beryl and columbite–tantalite have been mined from the zoned pegmatites in the pelitic metasediments of the Aravalli belt, particularly in the Jharol–Gorimari sector.

Niobium–tantalum Pegmatites The columbite–tantalite rich pegmatites of Rajasthan are concentrated in the central part of the belt e.g., in the Borara–Deoria–Badri–Gulgadu area in Malpura sub-division, Kayar–Quazipura–Bir–Makarwali–Lohga–Narel–Bhudol–Alran area in Ajmer, Bandar, Sindri–Hatupura near Kishangarh and Sangwa–Lakhola–Soniana near Bhilwara.

The largest Sangwa pegmatite is 350 m x 180 m on surface and contained columbite–tantalite and beryl mineralisations in the intermediate perthite zone. About 10 tonnes of beryl were produced besides columbite–tantalite (Dubey, 1977). One of the Sangwa pegmatites produced 1500 kg of columbite–tantalite, analysing 49.24–59.88% Nb_2O_5 and 16.69–23.88% Ta_2O_5.

Foyasagar pegmatite, is a zoned pegmatite with a quartz core, rimmed by a zone of blocky feldspar. Columbite–tantalite occurs as pockets within the feldspar zone close to the quartz core. From this pegmatite, 850 kg of columbite–tantalite analysing 40.78–49.6% Nb_2O_5 and 23.88–31.5% Ta_2O_5 have been collected. Lakhola and Soniana pegmatites also produced columbite–tantalite together with beryl, but in smaller quantities (Dubey, 1977).

Danta–Bhunas Pegmatites, Bhilwara The zoned pegmatite of the Danta mine, Bhunas, Bhilwara district, occurs within the garnetiferous mica schists belonging to the Aravalli Supergroup. The core consists of isolated quartz segregations, surrounded by a zone of quartz–perthite–plagioclase–muscovite. Uraniferous columbite–tantalite from Danta not only contains 77% $(Nb, Ta)_2O_5$ but also 0.08% U_3O_8. Minor phases include uraninite, samarskite, fergusonite, brannerite, zircon, monazite, allanite, triplite and tschffkinite. The Danta pegmatites belong to the NYF type in Cerny's classification (Cerny, 1991a).

Pegmatites from Phalwadi–Positara area , Sirohi district had the uncommon ore assemblage of gadolinite (49% RE_2O_3), powellite, molybdenite and scheelite.

Limboi and Umedpur Pegmatites, Gujarat The Limboi pegmatite occurs within Idar Granite and is fairly large (350 m x 200 m). It is zoned with a quartz core, which is surrounded by a zone that consists of K-feldspar, muscovite, tourmaline with columbite–tantalite, fersmite, beryl and fluorite. Columbite–tantalite analysed 53.6–60.07% Nb_2O_5, 12.1–13.35% Ta_2O_5, 1.29–2.71% Y_2O_3, 1554 ppm Sn and 0.30% Ce_2O_3.

A radioactive Nb–Ta mineral together with topaz has been identified from pegmatite in the Umedpur area. The radioactive mineral analysed 41.49% Nb_2O_5, 8.75% Ta_2O_5, 2.77% SnO_2, 2.29% Y_2O_3, 1.43% Pb, 2244 ppm Zr, 1226 ppm Ce, 2766 ppm La and 3.93% Th. Some associated granites are abnormally high in the content of U, Nb, Ta, Sn, W, Ce, Y and Zr. The Limboi and Umedpur pegmatites closely approach the NYF or the Mixed Type pegmatites of Cerny (1991a).

Lithium-rich Pegmatites The development of lepidolite mica or spodumene or both express lithium enrichment in pegmatites. Bhola (1977) pointed out two locations where Li-bearing pegmatites were concentrated. These are Rajgarh in Ajmer district and Potlan in Bhilwara district of Rajasthan. Datta (1973) reported lepidolite from Rajgarh, analysing upto 5% LiO_2 and spodumene from Rajgarh and Potlan, analysing 9.42% and 6.6% LiO_2 respectively. Swarms of Li-bearing pegmatites have been reported from Sibagaon area Sirohi district, southern Rajasthan (Maithani and Nagar 1999).

Rock Phosphate Deposits of Rajasthan

The geologists of the Geological Survey of India first struck phosphorite mineralisation in the Aravalli Supergroup in late sixties of the last century in the Udaipur district. Before long, a number of such deposits were discovered in the Udaipur district and Banswara district of Rajasthan and also in the Jhabua district of Madhya Pradesh at nearly the southeastern tip of the Aravalli belt. Today the known deposits in the Udaipur district are Jhamarkotra, Maton, Kanpur, Kharwaria, Badgaon and Neemuch Mata, Dakankotra and Sisarama; Sallopat and Ram-Ka-Munna deposits in Banswara; and Khatama, Piploda, Kelkua and Amlamal deposits in Jhabua. Of these, the Jhamarkotra, Maton and Kanpur deposits in Udaipur and Khatama, Kelkua and Amlamal in Jhabua are better developed. A number of workers including Banerjee (1971a,b), Banerjee et al. (1980), Chauhan, 1979, Roy and Paliwal, 1981; and Nagori (1988) studied the deposits.

The estimated reserve (mined + remaining) of 85.40 million tonnes of these deposits (Table 6.2.18) and the production from the working mines in this region meet < 10% of the country's need. Moreover, most of these rock phorphates are low grade (≤ 20% P_2O_5). Rajasthan Mines and Minerals Ltd. are mining the Jhamarkotra deposit, and Hindusthan Zinc Ltd the Maton deposit.

Mode (s) of Occurrence and Characteristics of the Deposits The geology of the phosphorite deposits in the Udaipur region is relatively better studied. They occur in the shelf zone of the Aravalli basin. The phosphatic horizon is locally underlain by polymictic conglomerate with an arkosic matrix, or it lies directly on the BGC, the interface being an erosional unconformity. The phosphate deposit is associated with a stromatolitic dolomite, which is overlain by a carbonaceous slate. Obviously, the Udaipur Phosphorite is a part of the Lower Aravalli Supergroup in the area (Table 6.1.5).

Table 6.2.18 *Reserves of rock phosphates in Rajasthan (Source: IBM, GSI and PRCL)*

S. No.	Name of the deposit	Location	Reserves* (million tonnes)	Grade (% P_2O_5)
1	Jhamarkotra (High Gr+) (A, B, C, D, E and F Blocks)	Udaipur district	14.7	30
2	Jhamarkotra (Low Gr+) (A,B,C, D,E, F, G and A–Z Block)	Udaipur district	51.8	15–20
3	Jhamarkotra (H Block)	Udaipur district	1.8	22–25
4	Maton	Udaipur district	6.2	22–24
5	Kanpur Group	Udaipur district	8.0	15–25
6	Birmania		2.9	10–18
		Total	85.40	

Note: * Reserves of all categories; High grade ≥ 30% P_2O_5; Low grade ≤ 20% P_2O_5

The phosphorites under discussion have been divided into several types depending on their physical and chemical characteristics (Banerjee et al., 1980). However, the types of the phosphatic ores are not deposit-specific and are met with in all the deposits, albeit in differing proportions.

Type 1: Stromatolitic carbonate phosphorites (Both columnar and laminated)

Chemically characterised by relatively high CaO (> 52%), moderately high MgO (5–1.3%) moderate to high P_2O_5 (25–37%) and relatively low SiO_2 (3–5%) and Fe_2O_3 (< 1%) contents. The dominant mineralogy, compatible with the chemical composition is represented by calcite, dolomite, carbonate fluorapatite and detrital quartz.

Type 2: Massive bedded phosphorite

This type has a somewhat lesser content of CaO (46–50%), low to moderate P_2O_5 (20–30%), relatively high SiO_2 (2–10%) and moderate R_2O_3 (1–2.5%) contents. It consists of carbonate, fluorapatite, veinlets of silica, detrital quartz, detrital calcite and Fe-oxide coatings.

Type 3: Fragmental phosphorite

This type is divisible into several sub-types based on fragment shapes and clast–matrix relationships.

However, this division into subtypes is hardly sustainable by difference in chemical composition, that is characterised by low to moderate CaO (40–52%), Low MgO (~1%) high to very high P_2O_5 (32–37%), high SiO_2 (10–20%), moderately high Al_2O_3 (0.1–2%), low to moderate Fe_2O_3 (1–2%) and appreciable MnO (0.22–0.3%). Mineralogically, this type of the phosphate ore is composed of micritic carbonate detritus, calcite veinlets, chalcedonic silica, detrital quartz and fragments of microsphorite grains. The matrix is composed mainly of micrite, sericite, chlorite and a little Fe- and Mn-oxides (± hydroxides).

It may be mentioned here that the phosphatic ores briefly discussed above are generally microsphorite in nature that is microcrystalline carbonate fluorapatite, with or without a little Mg and/or Na. Some suggested structural formula for the mineral from these deposits (Banerjee et al., 1980) is as follows:

1. Jhamarkotra: $Ca_{8.89}\ Na_{0.08}\ Mg_{0.01}\ [(PO_4)_{5.88}\ (CO_3)_{0.2}]F_2$
2. Kanpur: $Ca_{9.56}Mg_{0.45}\ [(PO_4)_{5.4}\ (CO_3)_{0.6}]F_{1.1}$
3. Maton: $Ca_{9.77}\ Na_{0.18}\ MgO_{0.04}\ [(PO_4)_{5.7}\ (CO_3)_{0.30}]F_{1.42}$
4. Jhabua: $Ca_{9.85}\ Na_{0.14}\ MgO_{0.001}\ [(PO_4)_{5.82}\ (CO_3)_{0.18}]F_{1.5}$

Jhamarkotra Deposit Jhamarkotra is located at a distance of about 25 km from the Udaipur city. Of its total reserve of 65.5 million tonnes of rock phosphates, only 14.7 million tonnes qualify as of high-grade, i.e. ≥ 30% P_2O_5 and the rest is generally of low grade (15–20% P_2O_5). The high-grade ore is being mined at a rate of 600,000 tonnes per year and is utilised in making phosphate fertiliser.

The low-grade phosphate ore is beneficiated to concentrates having a composition of P_2O_5: 32–34%, CaO: 50.4%, SiO_2: 4.5%, Al_2O_3: 0.5%, Fe_2O_3: 0.35%, MgO: 1.4% and F: 3.2% and Cl: 100 ppm.

The low-grade ore from Jhamarkotra and the other places are directly usable in agriculture, particularly if the soil is slightly acidic (pH ≃ 6).

Origin Several workers (Muktinath, 1974; Chauhan, 1979; Banerjee et al., 1980) have studied origin of the rock phosphate deposits in the Aravalli basin. Muktinath concluded that the marine upwelling hypothesis explained best the deposition of the Aravalli phosphates. Chauhan thought that the metsomatic replacemet of carbonate rocks would better explain the origin of these deposits. Banerjee and his co-workers suggested that these phosphorites formed in protected shallow to intertidal water. Both stromatolitic and algal forms akin to such environs are associated with the phosphorite deposits. Water chemistry in various palaeodepressions where phosphorite accumulated, did not vary appreciably from place to place, thereby supporting a primary algal-induced biochemical origin of the phosphorites in these sub-basins. There is no convincing petrographic evidence in support of a replacement origin of these deposits. These authors (Banerjee et al., 1980) at the same time think that phosphorite genesis through the transformation of hypophosphites could be another possible mechanism, at least for some of these deposits.

Bauxite Deposits

In this part of the country, bauxite deposits occur only in the state of Gujarat.The state contains about 142 million tonnes of bauxite ores, distributed in three separate areas such as

1. Coastal areas of Bhavnagar, Amreli, Junagarh, Jamnagar
2. South and southwestern part of Kutch district
3. Ahmedabad region comprising parts of Sabarkantha, Kheda, and Surat districts.

Bauxite ores of some of these deposits (Jamnagar, Kutch, Sabarkantha and Kheda districts) are high in grade and are usable in refractory, chemical and abrasive industries.

Gujarat bauxites vary in origin. The suggested mechanisms comprise the following.

1. *In situ* alteration of Deccan Trap rocks
2. Reworking and deposition of the *in situ* deposits i.e., type-1
3. Lateritisation of supratrappean limestone.

Deposits of Bhavnagar, Amreli, Junagarh and Jamnagar Districts A number of small isolated deposits of bauxite occur in the laterite in a belt that fringes the Deccan Trap basalts all along the coastal region of Saurashtra, between Bhavnagar and Porbandar. Some of these are commercially exploitable.

Geological Survey of India estimated a total reserve (inferred) of 16.36 million tonnes of bauxite for the Porbandar–Veraval belt in Junagarh district. The bauxite pockets in this district are located at Aditiyana, Babda, Bakharia, Baradiya, Beran, Chotilibili, Gosa, Palakhada, Matadi, Simani and Una. The average ore grade is Al_2O_3 45%–54%, Fe_2O_3 1–15%, SiO_2 6–20%, TiO_2 1–4%, CaO 4.5%.

Bauxite deposits here occur along a narrow (1–6 km wide) belt of about 50 km length. The common lateritic profile of a bauxite–laterite zone (0.4–12 m thick) is underlain by clay that overlies Deccan Trap, the protolith. The profile includes an overburden of Gaj Series (Lower Miocene) and miliolitic limestone (Pleistocene–Recent). The Jamnagar bauxite is characterised by the chemical composition: 45%–58% Al_2O_3, 2%–4.5% SiO_2, 2.7%–5.7% TiO_2 and 3% Fe_2O_3. Deposits in the Kalyanpur area have an estimated reserve of about 20 Mt (GSI Report, 1994).

Deposits in Kutch District A large number of small deposits have been located in a narrow linear belt of ferruginous laterite overlying the Deccan Trap, about 200 km in length. Individual bauxite pockets cover 1–1.3 sq km area, thickness ranging upto 8 m. Important deposits comprise those in Abdasa, Lakhpat, Nakhtrana and Mandovi taluks. The deposits include both in situ and transported varieties. The average composition of bauxite is Al_2O_3 – 49.92%, SiO_2 – 5.96%, Fe_2O_3 – 11.63%, TiO_2 – 4.40%, CaO – 1.53%. A reserve of 29.55 million tonnes of bauxitic ores has been estimated for this area (GSI, 2001).

7

The Himalaya

7.1 Geology and Crustal Evolution

Introduction

'Hima' in Sanskrit – an old Indo-Aryan language–is chill/snow and 'Alaya' in the same is home. So Himalaya is the 'home to snow'. It is needless to say that the life and culture of the people of South Asia are beholden to the Himalaya in more than one way. It saves people of the Indian subcontinent from the freezing cold winds coming from the north during the winter. It, on the other hand, particularly in its eastern part, faces the monsoon coming from the Bay of Bengal during June–August; and is a major cause of heavy rains in the Ganga (Ganges)–Brahmaputra valleys. The Indus–Ganga–Brahmaputra river-system, originating from the Himalaya, is the lifeline of a majority of population in the corresponding valleys. To the Hindus, the Himalaya has been the abode of gods. Not infrequently, there have been major earth-shakes or earthquakes in the Himalaya causing huge damage to life and property, particularly in the region close to its southern border. The simple folk living there have accepted such disasters as divine retribution. The attitude is, however, changing.

Geological investigation of the Himalaya started with the setting up of the Geological Survey of India in 1851. Since then the scientists of the Geological Survey of India, other government agencies, academia and research institutes (for example, the Wadia Institute of Himalayan Geology at Dehradun) are continuing their investigations into the geology of this mountain belt. At the same time it must be mentioned that the contribution of the international community of earth scientists to the Himalayan geology has been great. In the pages that follow, the 'state of the art' knowledge about it is briefly discussed.

It hardly needs be pointed out to an informed reader that the importance of the Himalayan geology does not lie merely in the fact that it is the largest (2500 km x 200–250 km) or the loftiest (containing eight of the ten highest peaks on Earth) mountain chain of the world, but in the fact that it has a complex and protracted geological past and it is yet so vigorously dynamic. It has an orogenic style of its own, known in Plate Tectonic parlance as a 'collisional orogen', where two continental blocks dock against each other in their search for dynamic equilibrium.

The approximately 2500 km long mountain range stretched between two syntaxes at Nanga Parbat in the West and Namcha Barwa in the east, respectively, is commonly taken as the Himalaya. Laterally, it is divided into the Western, Central and Eastern Himalaya, mainly for the ease of reference. From the geological point of view, it consists of several distinct litho-tectonic units. Recorded from the south to north, these are the Sub-Himalaya (Foothill Himalaya), Lesser or the Lower Himalaya, the Higher or Central Himalaya and the Tethyan Himalaya or the Tethyan zone as in some literature. The Indus–Tsangpo suture zone (ITSZ) separates the Tethyan zone from the Kohistan–Ladakh–Lhasa terrain in the north. The Higher Himalaya is separated from the Tethyan zone by a thrust (± unconformity) and in the south it is separated from the Lesser Himalaya by a major thrust zone called the Main Central Thrust (MCT). Between the Lesser Himalaya and the Sub-Himalaya lies the Main Boundary Thrust (MBT). Southern limit of the Himalaya is defined by the Main Frontal Thrust (MFT), beyond which lies the Foreland basin(s) of the Indus–Ganga–Brahmaputra system (Fig 7.1.1 a–b). The lateral disposition of the aforesaid litho-tectonic units from north to south is presented in Table 7.1.1.

Table 7.1.1 *Disposition of the litho-tectonic zones of the Himalaya*

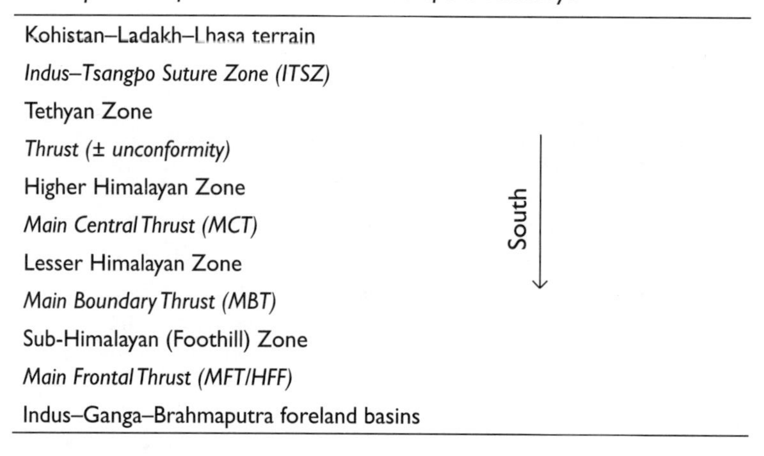

Kohistan–Ladakh–Lhasa terrain
Indus–Tsangpo Suture Zone (ITSZ)
Tethyan Zone
Thrust (± unconformity)
Higher Himalayan Zone
Main Central Thrust (MCT)
Lesser Himalayan Zone
Main Boundary Thrust (MBT)
Sub-Himalayan (Foothill) Zone
Main Frontal Thrust (MFT/HFF)
Indus–Ganga–Brahmaputra foreland basins

The Litho-Tectonic Zones

The Sub-Himalaya (Foothill Zone)

The Sub-Himalaya is comprised of late orogenic (Neogene–Quarternary) clastic sediments forming the southern marginal part or zone of the Himalaya, and as already mentioned, is delimited by the Main Frontal Thrust (MFT) in the south and the Main Boundary Thrust (MBT) in the north. The sediments constituting the unit are essentially molasse in nature and were deposited in a long narrow fore-deep in front of the rapidly rising and fast eroding Himalayan mountain. The fore-deep basin extends from Jammu-Kashmir in the west to Arunachal in the east. These sediments first started depositing in the western part, presumably after the first major uprise of the Himalaya, and are variously known as Murree–Dharamsala Group/Dagshai–Kasauli Formation. The Siwalik Group that overlie the above and extend upto Arunachal, is divided into three formations, viz, Lower, Middle and the Upper. The Siwalik Group started depositing in the Middle Miocene and continued

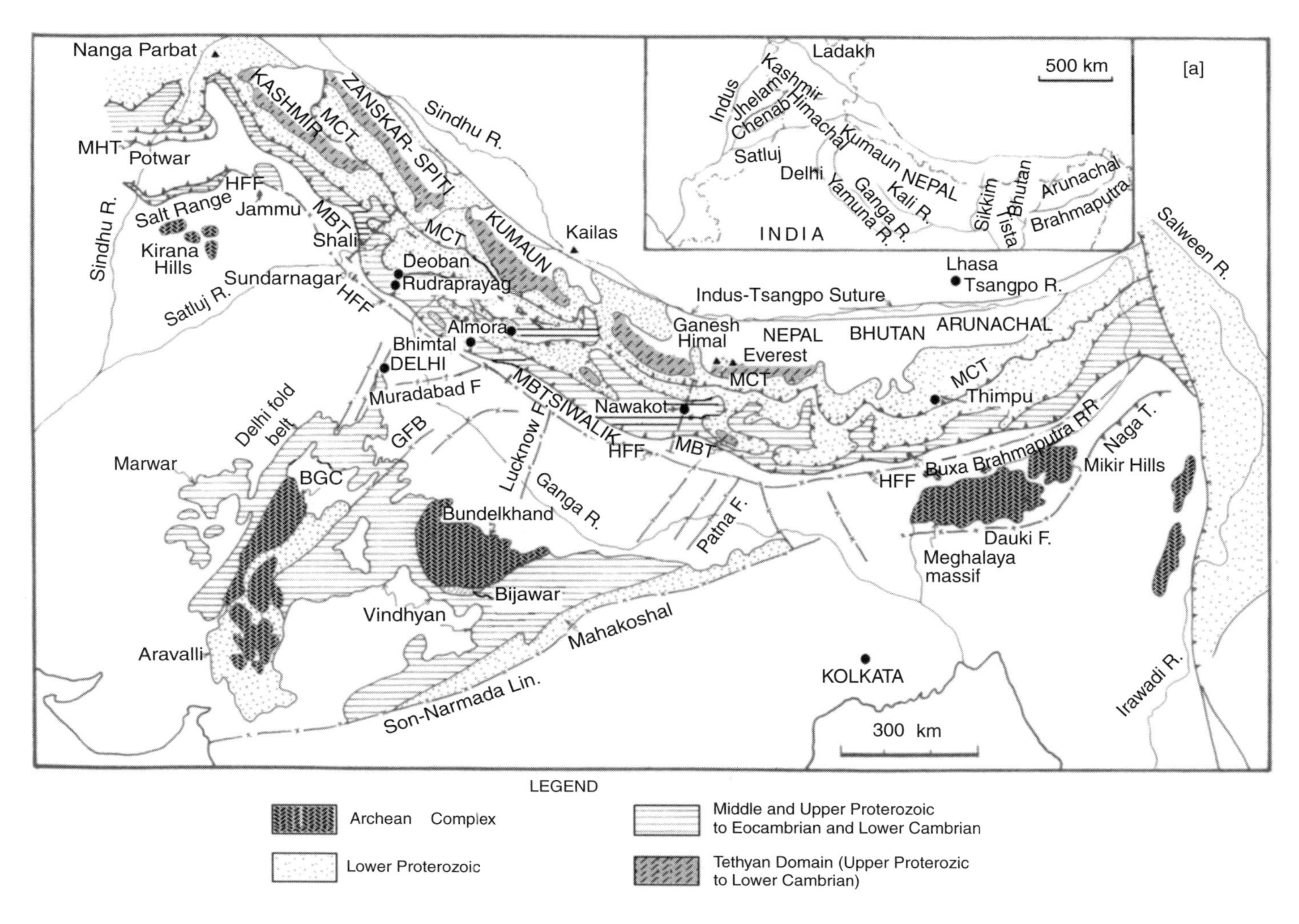

[a]
Ladakh
500 km
Indus
Jhelam
Chenab
Kashmir
Himachal
Satluj
Delhi
Kumaun
NEPAL
Yamuna R.
Ganga R.
Kali R.
Sikkim
Tista
Bhutan
Arunachal
Brahmaputra
INDIA
Nanga Parbat
KASHMIR
MCT
ZANSKAR- SPITI
Sindhu R.
MHT
Potwar
HFF
Salt Range
Jammu
MBT
Shali
Sindhu R.
Kirana
Hills
Sundarnagar
Satluj R.
HFF
MCT
KUMAUN
Kailas
Deoban
Rudraprayag
Almora
Bhimtal
DELHI
Muradabad F
Lhasa
Tsangpo R.
Salween R.
Indus-Tsangpo Suture
Ganesh
Himal
NEPAL
Everest
BHUTAN
ARUNACHAL
MCT
MCT
Thimpu
Nawakot
MBT
SIWALIK
HFF
MBT
Delhi fold
belt
GFB
Lucknow F
Marwar
BGC
Ganga R.
Bundelkhand
Patna F.
Buxa
Brahmaputra R.
Naga T.
Mikir Hills
HFF
Dauki F.
Meghalaya
massif
Bijawar
Vindhyan
Mahakoshal
Aravalli
KOLKATA
Irawadi R.
Son-Narmada Lin.
300 km
LEGEND
Archean Complex
Lower Proterozoic
Middle and Upper Proterozoic
to Eocambrian and Lower Cambrian
Tethyan Domain (Upper Proterozic
to Lower Cambrian)

(Continued)

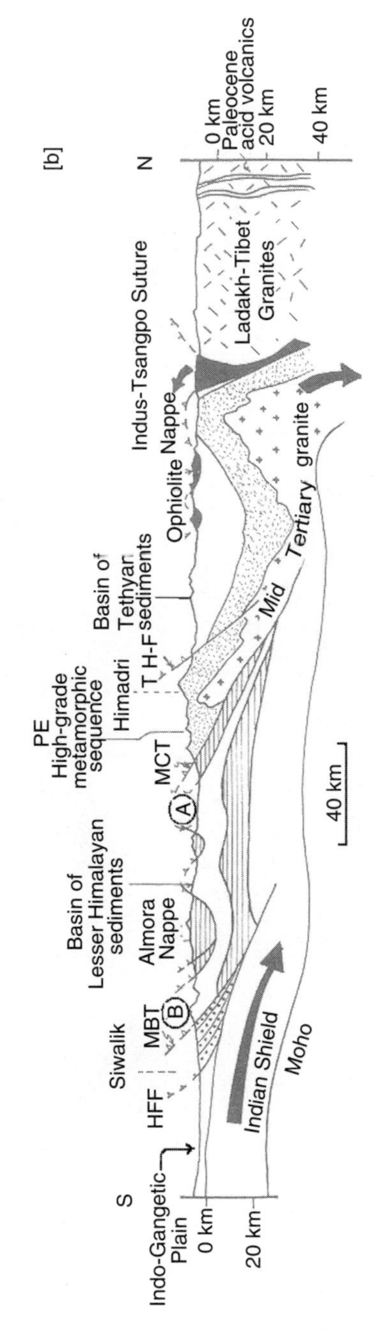

Fig. 7.1.1 (a) Generalised geological map of the Himalaya and Northern part of Indian Peninsula and the intervening Indus–Ganga–Brahmaputra alluvial valley (modified after Valdiya, 1995); (b) Schematic cross-section of the Himalaya (after Gansser, 1964).

till the Early Pleistocene. There is a pertinent suggestion that the Lower and Middle Siwaliks were deposited under a fluvial regime, whereas during the deposition of the Upper Siwalik Formation, the situation changed to that of fluvio–glacial (Yadav et al., 1996).

Another important feature of Sub-Himalaya is its neo-tectonics, expressed mostly as active faults and extraordinary seismicity of the recent past (Nakata, 1975, 1989; Valdiya et al., 1992; Guha et al., 2007). These active faults, defined as faults with repeated Quaternary slips, are rather common in the frontal Himalaya. Displacement during this time, north-over-south, along imbricate planar structures (thrusts) along the MBF are observed mostly along the eastern and the western Himalaya, and is a tenable explanation of the rapid rise of the Himalaya. However, in the Nepal Himalaya the situation is somewhat different. There the downthrow related to the active faults, is towards the north. In the active faults associated with MFT, or what is also known as the Himalayan Front Fault (HFF) (Nakata, 1975), dip-slip is the general observation where the fault trends E–W, the strike-slip is right-lateral (along NW–SE trending faults) or left-lateral (along NE–SW trending faults). Vertical displacement is towards the north except locally in the Bhutan Himalaya.

The average rate of vertical faulting is 3 – 4 m/1000 years along the Himalayan front and that of strike-slip about 1.2 m/1000 years (Nakata, 1989). The most conspicuous field features of active faults are the vertical displacements of the fluvial terraces and mountain slopes, in addition to offsets of ridges and stream courses. Guha et al. (2007) based on their studies of eastern Himalaya MFT, conclude that these structures are still in their early stage of growth and the later linkage between the active fault strands are expected in future. Based on the results of radiocarbon dating of organic-rich clay involved in the Gorubathan thrust and Chalsa thrust, they concluded that the development of these structures are post-33,875 ± 550 ybp and 22,030 ± 130 ybp, respectively.

However late-formed, the Sub-Himalaya could not escape the development of major structural features present in other parts of the Himalaya, i.e. folds and shears, the latter expressed as faults and thrusts. Petro-mineralogical transformation was just initiated. Intrusion by igneous rocks is not reported. The rocks are, however, relatively more modified close to the MBT.

The Lesser Himalaya The Lesser Himalaya, as already defined, is the litho-tectonic zone lying between the MBT in the south and the MCT in the north. It is the most complex of the divisions of the Himalaya, so far as the lithologic build up and tectonic developments are concerned. The tectono-stratigraphic units of the zone are commonly divided into the following:

1. Para-autochthonous unmetamorphosed to weakly metamorphosed sedimentary and volcanisedimentary sequences, predominantly of Precambrian (Proterozoic) age, covered by younger sedimentary rocks at places
2. Elongate and discontinuous masses of crystalline rocks, metamorphosed in the greenschist to amphibolite facies, and generally structurally overlying the rocks in (1).

To these may be added granite–gneisses and granitoid bodies that either form the basement or are intrusive into the above supracrustals.

Late Paleoproterozoic–Mesoproterozoic Developments The quartzose sediments, representing the earliest members of the sedimentary sequence, and interlayerd with rift-related mafic volcanics of Late Paleoproterozoic age (1800 Ma: Miller et al., 1999), is known as Kishtwar Group in Jammu-Kashmir, Sundernagar Group, Naraul Group and Rampur Group, in different parts of Himachal Pradesh and Garhwal, and Berinag Formation in Kumaon. This sedimentary-volcanic association is overlain by an orthoquartzite–carbonate sequence of Mesoproterozoic age, which is referred to as Riasi, Shali, Largi and Deoban Groups in different parts of the region (Bhat et al., 1998; Bhargava, 2000).

The equivalents of Lesser Himalayan Proterozoic volcanisedimentary formations of the western parts in the Lesser Himalayan domain of the Nepal sector, is known as the Kuncha Group and the Nawakot Group; the former generally being considered older, although the stratigraphic significance of their interface is not clear. Feldspathic augen gneisses interleaf with the Kuncha metasediments. It bears mention that the Kuncha Group is in general clastic in nature, while at least the upper part of the Nawakot Group is carbonatic.

The Late Paleoproterozoic–Mesoproterozoic sedimentary/volcanisedimentary formations continue into the Eastern Himalaya in the Sikkim–Darjeeling–Bhutan–Arunachal region, though in a somewhat reduced manner. The stratigraphic descriptions of the rocks of this region is highly confusing, particularly with respect to the Dalings and the Buxas, as will be obvious from Table 7.1.2. Different workers have proposed different stratigraphic sequences, utilising the same unit names. It is not known as to whether these differences have been resolved or reduced by now.

In the Bhutan sector, the Phuntsolling Formation is apparently overlain by the Buxa Formation. In Arunachal, the stratigraphic sequence is as follows:

Siang Group (oldest) → Miri Group → Abor Traps → Gondwanas

Table 7.1.2 *Stratigraphy of the Lesser Himalaya rocks in the Darjeeling–Sikkim sector, according to different authors*

Mallet (1875)	Le Fort (1975)	Ray (1971)		Acharya (1974)	
Gondwana	Gondwana	Gondwana ---Tectonic contact---		Gondwana ---Tectonic contact ---	
	Buxa (mainly dolomitic)				
Daling Series	Jayanti (quartzite) Sinchula (sandstone + greywacke)	Daling Group	Subgroup-II (Buxa) dolomite + quartzite + slates) Subgroup-I (Rayang) (Slate +Quartzite)	Buxa Group	Jayanti Formatotion (Dolostone + cherty quartzite + slate) Sinchu La Formation (Quartzite + slate + phyllite)
Buxa Series	Daling (Phyllite + schists)		Group-I (Gorubathan) (Slates + phyllites + schists)	Daling Formation (slate + phyllite)	

Neoproterozoic Formations Neoproterozoic formations are better developed in the Central sector of Lesser Himalaya in the Himachal–Uttaranchal region, in the Simla and Jaunsar belts. The sedimentary packages called the Simla and Jaunsar Groups are developed in a shallow marine stable shelf environment, but with little carbonates (Ravi Shanker et al., 1989). Penecontemporaneous Neoproterozoic deposits are yet to be established in the eastern part, starting from Nepal, through Darjeeling, Sikkim and Bhutan to Arunachal Pradesh.

Some Younger Formations Still younger sediments (Late Neoproterozoic–Early Phanerozoic) are present in the Lesser Himalaya, but at a few places only.

Late Neoproterozoic (Vendian) to Cambrian sediments are present as the Blaini–Infra Krol–Krol–Tal Formations of the Krol belt and these overlie the Simla and Jaunsar Groups with an unconformity. The Blainis are coarse clastics with dimictites at the base, while the Krols consist of fine clastics (black shale, siltstones) and carbonates with occasional phosphatic partings. The overlying Tal sediments are broadly similar to those of the Krols (Ravi Shanker et al., 1989). After the deposition of the Blaini–Infra Krol–Krol–Tal Formation, there was a long hiatus that extended upto the Cretaceous.

A pronounced linear belt of Permo–Carboniferous sediments (= Gondwana) developed along the stretch extending from Nepal to Arunachal Pradesh. Continental fluvial sediments (shale–sandstone) of central and eastern Nepal contain plant fossils that are of Permo–Carboniferous age. Trachyte–keratophyre type volcanic rocks are present in the lower part of Permo–Carboniferous sequence in eastern Nepal. In the Sikkim–Darjeeling area, the coarse sediments of the Rangit Formation is overlain by the continental Damuda Group. In Bhutan it is obviously coal-bearing. An almost isochronous sequence is obtained in Arunachal Pradesh, where it is divided into Miri, Bichom, Bhareli and Abor–volcanic Formations (from bottom upward) (Acharyya et al., 1975; Kumar, 1997).

It is interesting to note that even younger formations are locally present in the Lesser Himalaya. This is exemplified by the Palaeocene–Eocene Subathu Formation (and equivalent) in the Western Himalaya and Yinkiong Formation in Arunachal Pradesh. These were deposited during marine incursion at that time.

Granites and Granite Gneisses of Lesser Himalaya The granites and granite gneisses of the Lesser Himalaya have been studied in varying intensity. The Jeori–Wangtu Gneissic Complex (~ 2100 Ma) in Himachal Pradesh is fairly well studied (Bhargava, 2000; Ghose, 2000a, b). It does not show any intrusive relationship with the associated metasediments. The Wangtu Gneiss has a Rb-Sr age of 2068 ± 10 Ma, and has been intruded by granite, aplite, and pegmatite at 1860 ± 5 Ma. Parrish and Hodges (1996) reported an age of 2670 Ma for a detrital zircon obtained from Lesser Himalayan sedimentary rocks, suggesting presence of Archean rocks at the provenance.

The granitoid rocks in the Lesser Himalaya of Nepal are cordierite granite with muscovite, biotite and tourmaline (Le Fort et al., 1986). Available dates for these rocks range between 477 ± 7 Ma and 493 ± 11 Ma (UN, 1993: in Ghose, 2006).

Lingtse Gneiss is in fact a migmatite, where large augens of quartz + feldspar form neosomes. K-feldspar constitutes 2–30% and albite 10–40%. Mylonitisation parallel to gneissic foliation is common.

In Arunachal Pradesh biotite-bearing granite gneiss in the Siang Valley, also referred to as Zero Gneiss, yielded Rb-Sr (WR) ages, 1644 ± 40 and 1676 ± 122 Ma (Bhalla et al., 1989). The augen gneiss of Bomdila is somewhat older, with a Rb-Sr isochron age of 1914 ± 23 Ma. There are younger varieties also. The high-Ca granitoid intrusive at Salari yielded a Rb-Sr isochron age of 1536 ± 60 Ma (Dikshitulu et al., 1995). The intrusive granite at Deed, however, is much too younger (500 ± 19 Ma, Rb-Sr/WR).

Barrovian Metamorphism of Pelitic Rocks in Lesser Himalaya and the MCT Zone in Sikkim Himalaya The rising grades of Barrovian metamorphism in the pelitic rocks are noted from the Lesser Himalaya through MCT zone to HHC in the Sikkim–Darjeeling region. These, following Dasgupta et al. (2004), may be outlined, from south to north, as follows:

1. Lesser Himalayan Domain (LHD)
   Chlorite and part of biotite zone
2. Main Central Thrust Zone (MCT)
   Biotite zone (similar to that in LHD)
   Garnet zone
   Staurolite–kyanite zone (in high Al-protolith)

Staurolite zone
Kyanite zone

3. High Himalayan crystallines (HHC)
Sillimanite + K-feldspar zone

A compositional effect of these rocks on the mineral assemblages is conspicuous (Dasgupta et al., 2004, 2009). In the biotite zone, increasing Mg suppresses biotite in favour of chlorite and an increase of Mn facilitates the appearance of garnet. Development of chloritoid, rather than garnet, is favoured in the garnet zone with the increase of Fe at the expense of Al. In the staurolite zone the increase of Al favours appearance of kyanite, and increase of Ca suppresses the development of staurolite, even if there is increase of Fe and Zn. Appearance of staurolite may be hindered by the lowering of a_{H_2O} in any bulk composition within the sillimanite grade. In the kyanite zone, the garnet may be absent if the sediment was Al-rich. At any given grade there is a large difference in the composition of plagioclase between the garnet-bearing and garnet-free rocks. White mica is tangibly enriched in celadonite and Fe-celadonite components. At the sillimanite–K-feldspar isograd there is an abrupt increase in titanium content of biotite.

Dasgupta et al. (2009) carefully determined the peak metamorphic PT to range from 4.8 kb/490° C (garnet grade) to 8.4 kb/715° C (sillimanite–K-feldspar grade) in east Sikkim and 5 kb/ 525° C (garnet grade) and 7.9 kb/ 740° C (sillimanite–muscovite grade) in north Sikkim. The average metamorphic field gradient obtained by them in east and north Sikkim are + 60° C/kb and + 70° C/ kb respectively.

Inverted Metamorphic Sequence (IMS) in Darjeeling–Sikkim Himalaya Inverted sequence of Barrovian metamorphism in the pelitic rocks of Darjeeling–Sikkim Himalaya in the MCT footwall rocks was first reported by Santosh Ray (1947). In addition to this area being studied on these aspects by later workers (Lal et al., 1981; Dasgupta et al., 2004, 2009; others) this interesting feature was noticed in other parts of Himalaya also (Hubbard, 1996; Stephenson et al., 2000). Moreover, it is not exclusive to the Himalayan geology alone. Let us briefly review here the problem with unavoidable emphasis on eastern Himalaya as more data were available from this sector only.

Dasgupta et al. (2009) report on metamorphic field gradients along two separate traverses, one in east Sikkim and the other in north Sikkim. The first one is 60°C per kb and the other is 70°C per kb. The gradient is 'positive' as both the temperature and pressure increase upwards continuously. The models that have been proffered to explain this situation may be broadly divided into two groups: 1. 'Hot iron' type models, in which the thrusting of hotter deeply buried crust is thought to have taken place in one or more steps, 2. Models that assume post-/syn-metamorphic thrusting to be the principal cause that produce such results. Increase of both pressure and temperature upwards is not explained by the type-1 models, i.e. higher temperatures may be explained and not higher pressures. The type-2 model includes variants, some of which envisage situations in which pieces of 'right side down' metamorphic sequences (in which both P and T increase downwards) would be up-thrusted unit by unit, to ultimately invert the whole sequence (Hubbard, 1996; Grujic et al., 1996; Jain and Manickavasagam, 1993). This becomes an idealistic situation not normally expected in nature. On the other hand, out of sequence (in terms of metamorphism in particular) thrusting is common in nature including many other sectors, even in the Himalaya. Dominance of annealed fabric in metapelites does not support such a contention either. Dasgupta et al. (2009) contend that in the Sikkim Himalaya a right side down metamorphic sequence was formed first in the usual way, which was later inverted and exhumed as thrust blocks, virtually without being internally disturbed.

Dasgupta et al. (2009) find a plausible explanation in the numerical calculations of Faccenda et al. (2006, 2008), that modelised the thermal and dynamic effects of heat generated in the pieces of crust undergoing collision deformation. The Himalayan pelites contain an unusually high content of heat generating elements like uranium, thorium and potassium. They believe that the combined thermal effect of thrusting, shear heating and heat generation as an effect of burial of such radiogenic element bearing sediments, was capable of producing heat enough to cause metamorphism up to partial melting of pelites. In the model of Faccenda et al. (2008), the reduction of density and viscosity of a buried layer at the initiation of partial melting generates a driving force for uplift and exhumation. In a large number of simulations covering a range of parameters these experiments made a consistent conclusion that the Barrovian metamorphic sequence is inverted during such exhumation but in each case of exhumation the unmolten part of the metamorphic sequences occurred as a coherent block. Results of the above study suggest that the inverted sequence may be a rule rather than an exception in the metamorphic tectonites once the melting starts. The results of petrological studies of Dasgupta et al. (2009) and the suggestion of triggering effect of partial melting, entirely on structural grounds in Bhutan Himalaya (Davidson et al., 1997; Daniel et al., 2003) provide strong support to the applicability of this model in explaining the inverse metamorphic sequence in eastern Himalaya.

The Higher Himalaya The Higher Himalaya or the Higher Himalayan Zone, as it is commonly referred to in literature, is defined by the MCT in the south and the Tethyan sequence in the north. It is a continuous belt of metamorphosed sedimentary and igneous rocks, including granitoids, recorded from Nanga Parbat in the west to Namche Barwa in the east. Hence there are other names for this Zone, such as the Central Crystallines (CC) or the Higher Himalayan Crystalline Complex (HHC). These are represented by undifferentiated crystallines in Nanga Parbat, Suru Crystallines in Jammu-Kashmir, Vaikrita Group and Rohtang Gneiss in Himachal Pradesh, Central Crystallines in Uttaranchal, Himal Group in Nepal, Kanchenjunga Gneiss and Chungthang Gneiss in Sikkim, Darjeeling Gneiss in Darjeeling and Sikkim, Thimpu Group in Bhutan and Sela Group in Arunachal Pradesh. The Salkhala Group of Jammu-Kashmir is thought to be a relatively low grade representative of this zone. The best exposed sections of this zone are in the Kumaon Himalaya and Nepal Himalaya.

The Nanga Parbat–Haramosh massif occurring at the core of the Himalayan syntaxis in the west (Wadia, 1932; Misch, 1949, 1964) is made up of Precambrian gneisses, dated 1850 Ma (U-Pb zircon age) and intruded by Cambrian granitoids (Chamberlain et al., 1989).

The Central Crystallines in the Himachal Pradesh are represented by metapelitic rocks that show metamorphic gradations from the greenschist to amphibolite facies arranged in an inverted disposition, together with migmatites and anatectic granites. The most prolific development of the latter is noticed at different places in the Zanskar area, as well as in the Chamba–Dalhousi and Mandi areas in Himachal Pradesh. The granitic rocks of Zanskar have been dated to be 500 Ma (Le Fort et al., 1986). The same from the Mandi–Dalhousi area have age range of 456 ± 50 to 545 ± 12 Ma (Bhanot et al., 1975; Mehta, 1977).

The tectono-stratigraphic sequence across the Central Crystallines in Central Nepal is suggested to be as follows (Le Fort, 1994; Hodges, 2000; Ghose, 2006):

1. Formation III: Almost homogenous augen gneiss with some east–west trending enclaves of metasediments. The granitic fraction produced Rb-Sr ages corresponding to Cambrian–Ordovician (Ferreara et al., 1983; Projnnate et al., 1990). Some U-Pb ages correspond to Neogene time (Hodges et al., 1996).

2. Formation II: Calcareous rocks metamorphosed to amphibolite facies and now represented by banded calc-silicate gneiss (dominant), marble, calc-schist, psammitic schist, orthoquartzite and para-amphibolite (2–4 km thick), exposed mainly in Central Nepal.
3. Formation I: Consists of granitic gneisses, migmatites and sediments metamorphosed to the amphibolite facies. Compositional layering dip moderately towards the north.

The Higher Himalayan leucogranites, characterised by 70–75% SiO_2 and > 13% Al_2O_3, merit a commentary at this point, as the understanding of their nature and origin is essential for a satisfactory understanding of the evolution of the Himalaya. These are commonly divided into three groups:

1. Muscovite–and biotite-bearing granites with little or no tourmaline
2. Tourmaline- and muscovite-bearing granites
3. Muscovite-, biotite-, and tourmaline-bearing granites.

At one time there was a common belief that the Central Himalayan granites, particularly of the Western and Central Himalaya, formed by fluid-saturated melting of Formation I rocks (Vidal et al., 1982; Le Fort et al., 1987; others). However, results of the trace elements analysis (particularly Rb, Sr and Ba) suggested a fluid-undersaturated (dehydration) melting at high temperatures (≥ 750°C) on the other hand (Harris and Massey, 1994). This model is supported by the results of melting and crystallisation experiments conducted on the probable protoliths (Patino Douce and Harris, 1998; Scaillet et al., 1995). U-Th-Pb geochronology of accessory minerals from these rocks yielded an age range of 22–23 Ma to 12–13 Ma. A very young date of < 4 Ma was obtained from the Nanga Parvat area (Hodges, 2000).

HHC in the Sikkim–Darjeeling area is represented by Kanchanjungha Gneiss, Chungthang Gneiss and Darjeeling Gneiss, Darjeeling Gneiss of the Darjeeling–Sikkim area is composed essentially of quartz and alkali feldspars with subordinate proportions of perthite, biotite muscovite, garnet and sillimanite. Sillimanite content increases upward. Gansser (1964) even reported hornblende diopside and oligoclase/andesine from the Darjeeling hill section. Na_2O/K_2O ratio is variable; quartz also ranges between 30–50% (modal). Textural features range from deformational to crystallisational. All structures other than F_1 are recorded in these gneissic rocks.

In Bhutan the Central Crystallines, extending eastwards from Darjeeling–Sikkim area, are bifurcated into southern and northern belts, enclosing the metasediments of Late Precambrian (Maokhola Group: Hare Chu and Tirkhola Formations) to Paleozoic rocks (Black Mountain Group: Nake Chu, Mane Ting and Wachila Formations) of Tethyan affinity (Fig. 7.1.2). The Central Crystallines here are named as Thimpu Group, (earlier called Thimpu–Chekha Group by Jangpangi, 1978) comprising the gneissic complex of Thimpu Formation and high grade metasediments of Paro Formation (calc-silicate, marble, staurolite–kyanite–sillimanite schist, graphite schist). The rocks of Black Mountain Group (Ordovician–Devonian) are overlain in the north by the Tong Chu Formation (Devonian–Permo-Carboniferous). A Mesozoic sequence (Middle Jurassic to Cretaceous) of Lingshi Group occurs in the north western part of Bhutan (Gokul, 1983; Chaturvedi et al., 1983; GSI map, 1984).

The Thimpu Formation in Bhutan continues into Arunachal Pradesh, where it is called Se La Group (Jangpangi, 1978). Lum La Formation reported from northern Arunachal Pradesh (Tripathy et al., 1981) is also a pelitic–psammitic assemblage and could be equivalent to the Maokhola Group (earlier called Chekha) of Bhutan.

The MCT was for a long time believed to be an early Miocene structure but now its complex polyphase deformation history is established with well-constrained data. In Sikkim, the MCT was active at 22–20 Ma, 15–14 Ma and 12–10 Ma. Monazite from recrystallised MCT shear zone rocks in Central Nepal yielded ages as young as 3 Ma (Catlos et al., 2004). The 2–10 km thick

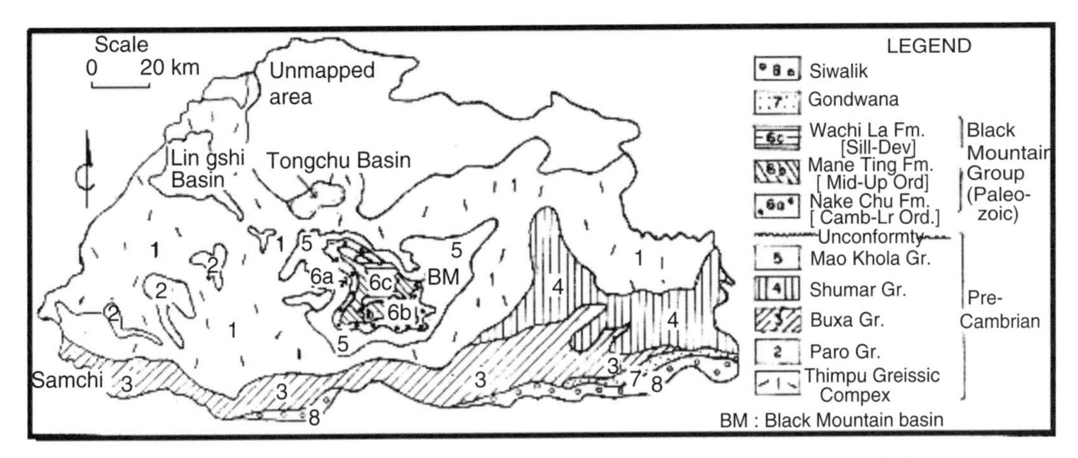

Fig. 7.1.2 Simplified geological map of Bhutan Himalaya (simplified from GSI map, 1984)

MCT system could have undergone a cumulative slip of as much as 250 km (Molnar, 1984). The MBT, in contrast, is a thin zone ($\leq$ 100 m) of cataclastic rocks with northerly dips. Based on the results of studies of sediments in the Sub-Himalaya, Meigs (1995) suggested an age of 11–9 Ma to the MBT. Translation along it could be several tens of kilometres (Srivastava and Mitra, 1994). This means that that the MCT involved a much larger zone of rocks, subjecting the latter to more intense thermo-tectonic modifications in a polyphase history, in contrast to the MBT. Mukul (2000) suggests that the region between the MBT and the MCT is the locus of 4.5–6.8 (Mb) magnitude earthquakes, as the records since 1960 show. Does that suggest that the MCT shear zone is presently tectonically active, at least locally? The MFT, though much less well-defined/definable, developed during Pliocene–Holocene time. Guha et al. (2007) dated, with the help of radiocarbons in organic-rich clay involved, the foreland-propagating blind thrusts of the FTB in the eastern Sub-Himalaya to be ~ 3.3×10^4 and ~ 2.2×10^4 ybp.

In an interesting study of the metapelites and migmatitic rocks of HHC in Sikkim, Neogi et al. (1998) determined several metamorphic episodes of the rocks, in which the first episode (M_1) represented pre-Himalayan metamorphism and decompression of the HHC. Later, collisional event led to the renewed burial of the HHC and M_2 metamorphism. M_2 metamorphism at P = 10–12 kb and T = 800°C–850°C, was characterised by dehydration melting of muscovite and biotite leading to the formation of granitic melts, which either migrated from the source accumulating into large granitic bodies or remained in-situ to form leucosomes. M_3 took place during decompression to ~ 5 kb heralding breakdown of porphyroblastic garnet in all HHC lithologies.

The Tethyan Zone The Tethyan zone of the Himalaya lies to the north of the Central Crystallines or the HHC and in the south of and along the Indus–Tsangpo Suture Zone (ITSZ). It consists mainly of a pack of fossiliferous sediments that were laid with occasional breaks, over a long period of time, ranging from Late Proterozoic to Early Tertiary. The base of the Tethyan sequence is the HHC, the interface having evidence of both depositional break (unconformity) and tectonic transport (thrust) (Gansser, 1964; Valdiya, 1987; Hodges, 2000).

The Tethyan sedimentation is best developed in several sub-basins that are now distributed in Kashmir, Zanskar–Spiti, Kumaon, Nepal and Bhutan. The sediments are generally platform type and their similarity over the entire zone is surprising (Gansser, 1981). However, the carbonate precipitation was more prolific in the Kashmir basin. In contrast, the Devonian Muth Quartzite and the Late Jurassic–Early Cretaceous pelagic sediments are uniform along the entire Himalaya (Gansser, 1981).

Gaetani and Garzanti (1981) divided the Tethyan sediments of the Zanskar region, Himachal Pradesh into Pre–Neo–Tethyan (Late Proterozoic–Early Permian) and Neo–Tethyan (Late Permian –Early Eocene). They are further divided and sub-divided (Gaetani and Garzanti, 1991; Ravi Shanker et al., 1996). The Tethyan sediments were deposited in shallow marine epi-continental environments, deep marine facies are present rarely though.

The Indus–Tsangpo Suture Zone The well-known Indus–Tsangpo Suture zone (ITS) marks the collision zone between the Indian, rather the Indo–Australian plate and northern Eurasian plate. It has been identified and traced from Kabul Spur in the west to the Namcha–Barwa in the east for a distance of about 2500 km, i.e. the length of the Himalaya proper. The zone, however, is discontinuous. It is possible, even likely, that the Main Mantle Thrust (MMT) separating the Nanga Parbat–Hazara block from the Kohistan belt in NW Himalaya is a continuation of the ITS (Ghose, 2006). The zone is represented by oceanic sedimentary rocks, ophiolite, ophiolitic melanges and basic volcanic rocks. Normal ophiolitic sequences are, however, rare. High pressure metamorphism is reported from the west, particularly in the Nanga Parbat Cross High region (Tahirkhali et al., 1979). Mesozoic flysch sediments occurring to the south of the ophiolite zone contains blocks (olistostromal?) of rocks of Permian–Triassic age. Many workers, including Dewey and Bird (1970), Gansser (1964, 1981) and Searle (1983), have contributed to the geology of this region.

The Trans-Himalayan Region Here the Trans-Himalayan region means the region that lies immediately to the north of ITS. The geology of this region is far from simple and in spite of some good work done on it, quite a few questions remain to be better understood and answered. This region includes Ladakh–Lasha belt, bound by the ITS in the south and Shyok–Bangong–Nujiang suture in the north. The Kohistan–Ladakh belt, originally an island arc complex, is taken to represent the last Pre-Himalayan accretion event in the northern (Eurasian) block in the Trans-Himalayan region (Hodges, 2000 and references therein). The rocks of the Kohistan block consist of Cretaceous to early Tertiary calc-alkaline plutons, basic plutonic/volcanic rocks and sediments. The Ladakh belt is dominated by tonalitic massifs with some basic and late granitic plutons. The ages of Kohistan and Ladakh plutons are 100 Ma and younger. The Karakoram represents an important morpho–tectonic belt further north having lithologic developments of the Late Paleozoic.

Himalayan Evolution – The Likely Story

There is a curiosity shared by many people about the origin and evolution of the Himalaya. Some of their questions are:

1. How old is this mountain range?
2. Did it form in a similar way, most other mountain ranges in the continents formed?
3. Is it true that it is still rising in perceptible rate(s)?
4. Why are earthquakes so common along this mountain range and that too along finite zones?
5. Large and lofty as it is, does it contain a proportionately large volume of mineral deposits? If not, why?

These are also some of the principal questions that earth scientists have asked themselves since they initiated studies of this extraordinary mountain range more than a century ago. Let it be mentioned here that definite answers to some of these questions are still elusive. Although it is now generally agreed that the Himalaya has a complex and protracted geological past, it is still geologically active. It is cited in geological literature as a typical example of an orogenic style, known in modern tectonic (Plate Tectonics) parlance as *collisional orogeny*. In collisional orogeny two crustal (mainly continental) blocks ('plates') lock horns and tectono-thermally affect the rocks occurring on either side of the zone of contact (suture). In case of the Himalaya, the Indo-Australian plate started moving NNE-ward sometimes in the Cretaceous until it stumbled against the Eurasian plate in the late Cretaceous–Early Eocene. As we have already seen, bulk of the rocks that constitute the Himalaya are old (Precambrian), but the entire making of the Himalaya as you see it, took place during the Cenozoic time. That explains how it earned the sobriquet *Enfant Terrible*.

In the brief discussion on Himalayan geology in the previous pages, we have seen that most of the rocks in the Himalaya, particularly in the Lesser Himalaya and Higher Himalaya, are Proterozoic (Early–Neoproterozoic) in age. In the Western Himalaya volcanisedimentary formations (1800 Ma) are known as Kistawar Group in Jammu-Kashmir, Sundernagar Group, Narnaul Group and Rampur Group in Himachal Pradesh and Garhwal and Berinag Formation in Kumaon. These Mesoproterozoic volcanisedimentary piles appear to be early rift-fills and the overlying quartzite–carbonate sediments (Riasi, Shali, Largi and Deoban Groups) represent platform facies. Some confusion about the details of the stratigraphy of Eastern Lesser Himalaya notwithstanding, there also clastic sediments are structurally overlain by carbonates.

The radiometric age obtained from Jeori–Wangtu gneissic complex, as we have already seen, yielded radiometric ages, which cluster around 2100 Ma (Rb-Sr). These intrude no metasediments, but are intruded by granitic rocks dated 1860 Ma. Granitic massif at the Nanga Parbat–Haramosh have also been dated 1850 Ma (U-Pb/zircon). The granite gneisses from the Siang Valley, Arunachal, yielded Rb-Sr (WR) ages of ~1600 Ma. The gneissic rocks of Bomdila are, however, older by about 300 Ma. Much younger (~500 Ma) granitic rocks are present in the Lesser Himalaya of Nepal and Arunachal.

In Nepal Himalaya highly metamorphosed (amphibolite facies) calcareous metasediments overlie gneisses and migmatites and are intruded by granitic intrusives of Cambrian–Ordovician to Neogene ages. In the Higher Himalayan zone of Eastern Himalaya (Sikkim–Darjeeling–Bhutan–Arunachal) the Paro Formation (= Chekha Group) is carbonate–dominant and overlies the pelitic Thimpu Group.

U-Pb and Sm-Nd isotopic studies of rocks from Langtang area, Central Nepal, showed important difference between rock packages on either side of MCT (Parrish and Hodges, 1996). The Greater Himalayan sequence had a sedimentary provenance that included a major source of 0.8 – 1.0 Ga zircons, implying a Late Proterozoic age of deposition. These determinations are broadly supported by the model Pb-Pb age of the sediment-hosted and metamorphosed lead-ores of the Ganesh Himal deposit of Nepal and the nearly similar ores of Genekha deposit, Bhutan, both belonging to the Higher Himalaya. The Precambrian part of the Lesser Himalayan sediments contained 2.60 – 1.87 Ga zircons, suggesting a source rock age of Late Archean to Middle Proterozoic or so. The Lesser Himalayan sediment-hosted metamorphosed deposits of Bageswar (Western Himalaya) and Rangpo–Gorubathan (Eastern Himalaya) yielded a model Pb-Pb ages of ~1800 Ma (vide next section for a somewhat detailed discussion on this). There was an Early Neoproterozoic cycle of sedimentation in places of the Lesser Himalayan belt, now represented by the Simla and Jaunsar Groups in the Himachal Predesh and Uttaranchal, and Novakot/Naukot Group in Nepal. Their tectonic setting is, however, contentious (Srikantia and Bhargava, 1982; Ravi Shanker et al., 1989). Towards the end of the Precambrian, the Tethys Himalayan Zone was a passive continental margin, characterised by shallow marine terrigenous sedimentation and facing an ocean, details of which are least known. This environment continued until the middle of Cambrian when a deepwater facies represented by shales and turbidites took over. This was in turn followed by the deposition of molassic sediments and widespread granitic intrusions. Gaetani and Garzanti (1991) believed that the association indicates a mild orogenic phase corresponding to the late Pan-African movements.

The final assembly of the Gondwana Supercontinent in the Ordovician was followed by a prolonged period of intermittent shallow marine sedimentation. Coastal deposition of quartz–arenitic sediment during the Silurian–Devonian (Muth Formation) was followed by brachiopod-bearing limestones and evaporites in early Carboniferous (Lipak Formation). These thin and commonly incomplete successions were deposited during periodic incursions of shallow epicontinental seas, which were probably linked to the Paleo–Tethys beyond the realms of the present-day Himalaya (Gaetani and Garzanti, 1991).

As the Paleozoic was nearing its conclusion, a major rifting event detached the Peri-Gondwanian micro-plate fringe (Cimmerian microcontinent, inclusive of Karakoram, Lhasa and Qiangtang blocks) and the opening of the Neo-Tethys was initiated north of the then Indian sub-continent (Dewy et al., 1988; Sengor, 1990). This was soon followed by the development of a series of rifts in the passive continental margin facing the neo-Tethys. This apparently progressively separated the Indian plate from the rest of Gondwana. Geological observations, however, suggest that the (paleo-) Peninsular India was possibly never far distant from the southern fringe of the Eurasian continent (Gansser, 1981; Ganging et al., 2003). The Neo-Tethys, as it seems, was a complicated zone of island arcs, internal basins and irregular slices of continental rocks (Gaitani and Garzanti, 1991).

Lifespan of the Neo-Tethys broadly coincided with the span of the Mesozoic period. According to Gaetani and Garzanti (op. cit.), the Mesozoic evolution of the passive continental margin, north of the then Indian Sub-continent is characterised by the operation of two rift/drift events documented by deposition of two sequences. The first of the two was operative during the Late Permian to the Jurassic and the second is reckoned from Early Cretaceous to the Late Cretaceous–Early Eocene. Early volcanism in the first phase was followed by sediment deposition dominated by carbonates. Suturing shifted from north to south with time with the development of the island arc type Bangong–Nujiang Suture (BNS) in the north of Lhasa block during Upper Jurassic–Lower Cretaceous time and the Andean type Indus–Tsangpo (IT) suture in the upper Cretaceous–Early Eocene period (Allegre et al., 1984). Commencement of the second neo-Tethyan mega-sequence deposition started in the Early Cretaceous corresponding to the final disintegration of the Gondwanaland and the opening of the Indian Ocean. Break up of the Indian plate from the Gondwana land and its northerly migration coincided with the gradual opening of the Indian Ocean and continued closure of the Neo-Tethys until the latter collided with the Eurasian block in the north sometimes in the Late Cretaceous–Paleocene time. (See also Chapter 8)

Phases of Himalayan Orogeny Hodges (2000) made a succinct review of the known facts related to Himalayan orogeny and divided the latter into three phases:

1. Neohimalayan
2. Eohimalayan
3. Protohimalayan.

The *Protohimalayan* Phase began in Late Cretaceous and extended to Early Eocene and comprised deformations that preceded India–Eurasia collision along the Trans-Himalayan Indus–Tsangpo (IT) suture zone and the Tibetan Zone. South–southwest verging fold and thrust structures of this age are common in the Trans-Himalayan region. The Kohistan terrane was also thrusted southward over the North Indian margin during this time. The Spontang ophiolite in the Tibetan zone in Zanskar range of Western Himalaya was obducted during the Protohimalayan Phase only.

The *Eohimalayan* Phase refers to the Middle–Late Eocene period when the main India–Eurasia collision took place generating the resultant structures. Beginning of the collision is read from the transition of marine to non-marine sedimentation and these processes were confined to the time-slot, 54–50 Ma (Rowley, 1986). The northward convergence of India into Eurasia during the last 50 Ma not only tectonically affected (and still affecting) the Eurasian plate, but also caused the Tibetan Plateau to develop. Lithospheric structure beneath Tibetan Plateau has been explained by complete or partial under-thrusting of the India (Indo-Australian) plate (Powell and Conaghan, 1973; De Celles et al., 2003). Recently Zhou and Murphy (2005), analysing a high resolution global tomographic model, suggested subduction of the Indian lithospheric mantle (and a low velocity zone above it – an asthenospheric wedge) sub-horizontally towards NNE. The high velocity Indian slab starts to dip towards the Tibetan Plateau hundreds of kilometres south of the Himalaya (Fig. 7.1.3).

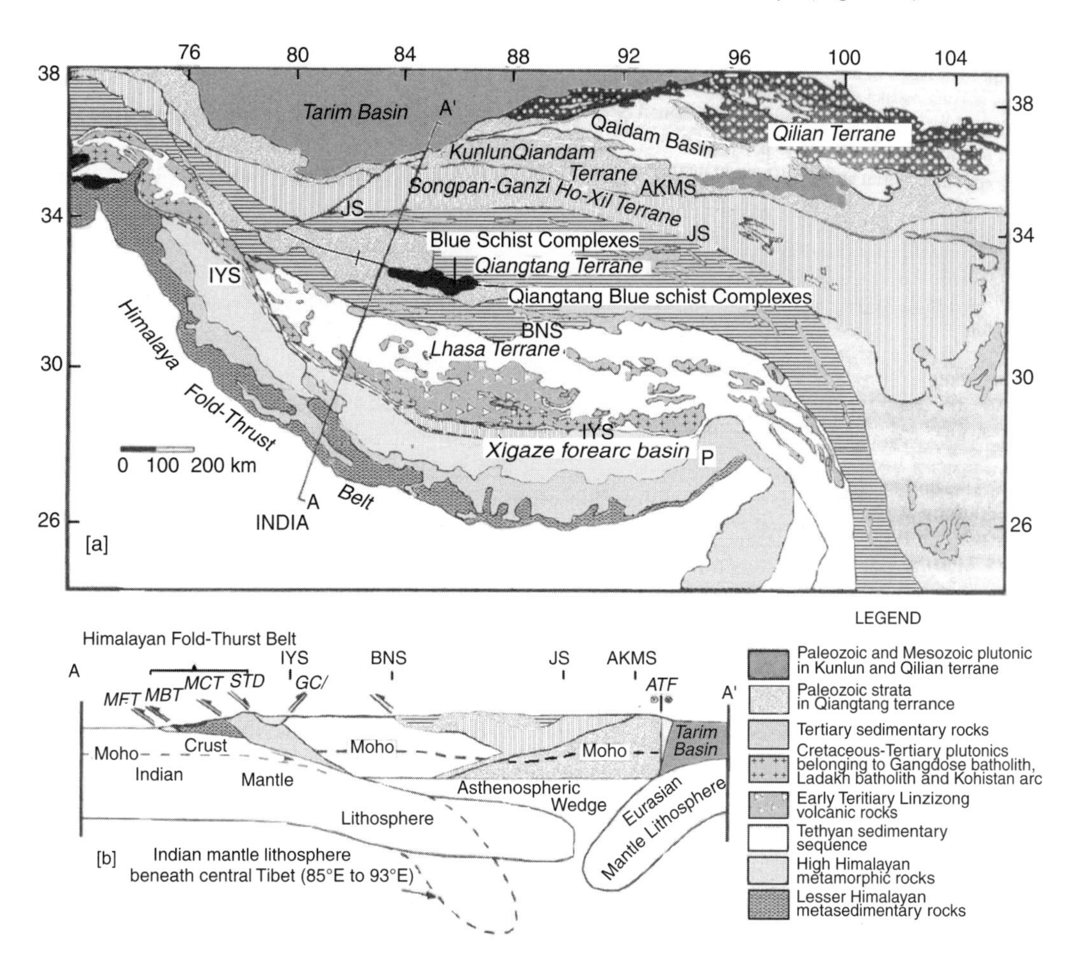

Fig. 7.1.3 (a) Tectonic map of the Tibet–Himalaya collision zone; (b) Schematic cross-section (A–A′) across Tibet–Himalaya orogen based on the interpretation of geological and tomographic data (after Zhou and Murphy, 2005).

The *Neohimalayan* structures are present throughout the tectono-stratigraphic zones and the architecture of the orogen is mainly a consequence of their making. These include the major thrust structures defining the litho-tectonic zones discussed in the previous pages, as well as smaller thrusts and other shear zones contained in these zones. A foreland basin is generally an unavoidable product of collisional orogeny. In the region under discussion, the alluvium-covered Indo-Gangetic basin and the frontal Himalaya (Sub-Himalaya) that borders its northern margin is represented by the Murees and the Siwaliks once occupied the Himalayan foreland basin. It initiates as a result of depression of the continental foreland on entering the subduction zone following closure of the remnant basin and later, as a result of isostatic adjustment in response to the rising thrust belt in the foreland (Mitchell and Garson, 1981). In the sub-Himalayan region, the sediments constituting the Murees were derived from the foreland to the south. But the major part, now represented by the Siwaliks, came from the north where the rocks in the rising mountains were already affected by weathering and erosion, including mass wasting. The lower part of the succession is correlatable with the main episode of orogeny as the foreland thrust belt and the upper part, commonly younger.

The south-dipping reverse faults and kilometres-long folds overturned towards the north, particularly in the Tethys zone, possibly took place during Pliocene–Pleistocene age, as they affected the molasses in the Zanskar–Ladakh area (Searle et al., 1999).

It has already been mentioned that the peak metamorphic conditions in the Sikkim Himalaya has been estimated to be 10.4 kb/800°C (Neogi et al., 1998). The observed retrograde breakdown, of garnet into spinel + cordierite, requires near-isothermal change and hence an extremely rapid (~15 mm/year) exhumation upto a depth of ~ 15 km. Numerical modelling suggests that the initial rapid exhumation must have been followed by a much slower process (~ 2mn/year) upto about 5 km depth, leading to the development of the observed compositional zoning in garnet. The dramatic change in the exhumation velocity might reflect a process of tectonic thinning, followed by erosion and/or horizontal flow at shallow depth, as suggested by Duchéne et al. (1997) from the depth-time -path analysis of eclogites. It is estimated that some High Himalayan Crystalline (HHC) complex rocks studied exhumed from a depth of ~ 34 km to ~8 km. A part of the change in the exhumation rate might have been accommodated by extension throughout the HHC (Ganguly et al., 2000). Mukul (2000) suggests that the region between the MCT and MBT is the locus of 4.5–6.8 (Mb) earthquakes since 1960 to 1998, which can be further extended to mean that the Sikkim MCT shear zone could be presently active (Catlos et al., 2004).

Obviously the Himalaya is not an example of the 'basin and range' model where a (volcanic-) sedimentary pile accumulates in a sedimentary basin which, during inversion, i.e. orogeny, is raised to a mountain range. It is essentially composed of Precambrian rocks, except the Sub-Himalayan zone in the south and the Tethys zone in the north. These rocks have litho-facial similarity as well as isochroneity with some rock formations of northern India. Some rock formations like the Gondwanas, can be straight away identified in the Himalaya. But how about the structures and metamorphism?

Regarding the HHC of Sikkim Himalaya, Neogi et al. (1998) and Dasgupta et al. (2004) determined three phases of metamorphism, M_1, M_2 and M_3. Of these, M_1 is pre-Himalaya according to them. Gansser however, found evidence of two pre-Himalayan metamorphic events – one that occurred sometimes during 1800–1500 Ma and the other during Late Precambrian. It seems highly plausible, even probable.

Examination of the published geological maps of the Himalaya reveals that although the major thrusts in the Himalaya have a general E–W trend parallel to the mountain front, in detail, they however show a sinuous pattern. The thrust traces are cuspate–lobate, with the broad lobes convex towards the foreland, joined by sharp re-entrants, which may be referred to as cusps and point to the hinterland. Larger lobes are often sub-divisible into smaller ones. Ray (2006) analysed the probability of the formation of curved thrust traces by several means, such as (i) the effect of topography, (ii) post-thrusting folding, (iii) differential transport and (iv) fault growth along laterally curved surfaces and confirmed that his explanation of these structures in the Shumar allochthon, eastern Bhutan (Ray, 1995), is generally valid for the entire range. The essence of it is that these trust surfaces are curved due to growth along laterally curved trajectories that may ultimately coalesce to form large thrust surfaces.

Continuation of the Peninsular Geology into the Himalaya

Many Himalayan geologists believe that the Precambrian rock formations in the Himalaya, south of the Indus–Tsangpo suture zone, are nothing but the continuation of the formations that now occur at the central and northern parts of peninsular India, including the Delhi–Aravalli Supergroups, the Bundelkhand Gneissic Complex, the Vindhyan Supergroup and the rocks of Chhotanagpur Gneissic Complex and even those of Meghalaya (Krishnan and Swaminath, 1959; Gansser, 1981; Srikantia and Bhargava, 1982; Bhargava, 2000; Ghosh, 2000). This conclusion is based mainly on the similarity of the lithology and age of the rock formations in these two regions. But many other details are lost because of the nature and intensity of the deformation that the Himalayan rocks were subjected to. Valdiya (1976) thought that the apparent parallelism between the transverse faults and folds in the Himalaya and the surface structures in north India is very significant in the context of correlation. He cited the subsurface transverse structures across the Gangetic basin, such as the 'Delhi–Hardwar ridge', 'Bundelkhand–Faizabad ridge' and the 'Mungher–Saharsa ridge' in this context (Valdiya, 1970). However, in Rao's (1973) interpretation of the subsurface geophysical data (ONGC) none of these ridges continue upto the Himalaya. Add to this the fact that the transverse structures in the Himalaya are not exceptional. They are common in all mountain belts, particularly where the translation across the trend has been considerable.

7.2 Metallogeny

Introduction

The Lesser Himalaya is relatively better mineralised. Lead–zinc–copper and uranium mineralisations occur at a number places, some of which are subeconomic to marginally economic. Economic deposits of magnesite occur in Kumaon and Garhwal Himalayas and of phosphate at Mussoori. Higher Himalaya contains some base metal deposits in Nepal. The type Tethys zone is still poorer in mineralisation. However, the Tethyan sediments in Bhutan host a copper deposit. Pb-isotopes from sulphide deposits threw some important light on the Himalayan geology and metallogeny.

The Alps and the Himalaya have disappointed us by their disproportionately small content of mineral deposits compared to their dimensions. We will try to find out a suitable explanation of this fact in a later section of this chapter. In the meanwhile, we briefly discuss the known ones, zonewise, in the same way we briefly discussed the geology in the previous section. The general distribution of base metal and uranium deposits/occurrences and major deposits of some other metals in the Himalaya are presented in Fig. 7.2.1.

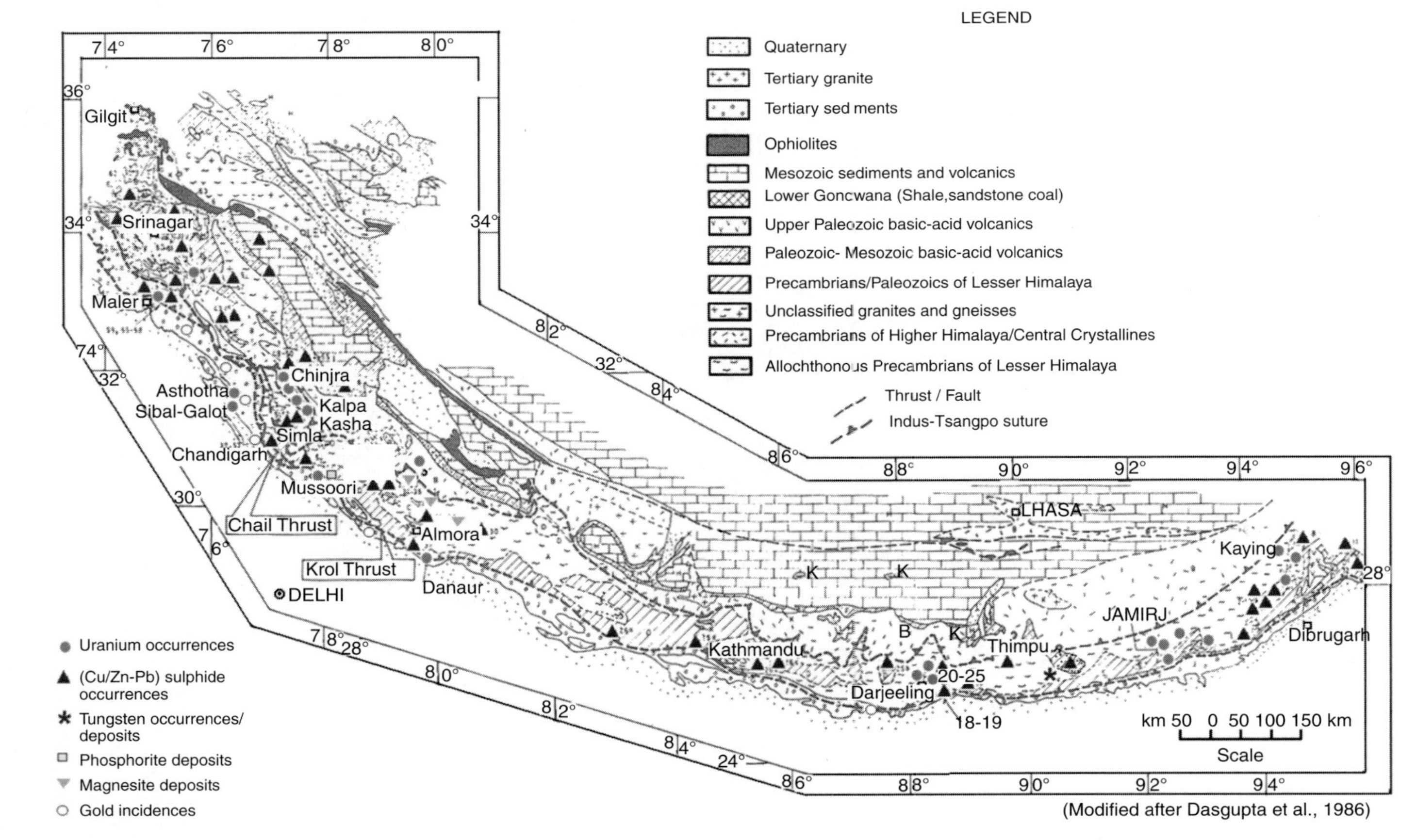

Fig. 7.2.1 Distribution of basemetal, uranium, tungsten, phosphorite, magnesite and gold mineralisation in the Himalaya (largely modified after Mahadevan, 1992)

The Sub-Himalayan Zone

This zone is very poor in mineralisation. Whatever is known about it refers to uranium and gold only. Uranium mineralisation is stratabound and its best development has been noted in the zone between the upper part of the Lower Siwalik and the transition between the Middle and the Upper Siwalik Formations, particularly in the Ranshahr-Kalka-Morni area near Chandigarh and also in the Pathankot–Jammu area. The U_3O_8-content (radiometric) in the samples ranged upto 0.06% (Mahadevan et al., 1986; Udas, 1986). The host rocks are current bedded feldspathic sandstones, medium–coarse in grain size. The ore zones contain carbonaceous vegetal matter, pyrite and clay (Dhanaraju et al., 1985). Laterally they may be extensive for upto 12 km (Yadav et al., 1996). Such deposits are also known to occur in central Nepal. Uranium occurs principally as uraninite, of course with sub-ordinate coffinite at places. The mineralisation may be assigned to 'sandstone' type.

The pre-eminently suitable explanation for such mineralisations is that the uranium was derived from the rising Himalaya in the north, by leaching of earlier mineralisations or fertile source rocks and deposited in suitably reduced environments. Uranium mineralisation in the Indian Siwalik continues into Pakistan within the corresponding stratigraphic position. Similar mineralisation also occurs in central Nepal.

Placer gold mineralisation is reported from the Middle-Upper Siwaliks of Western Himalaya and it locally (rarely though) may be upto 7–8 gm/tonne. Looking eastward, this becomes less obvious, though not totally absent (Kumar, 1997). Prospective zones with placer gold incidences in the Quaternary terrace and river sediments were located in north Bihar–eastern Nepal region, but none held much of economic promise. The source of gold in this context is, however, less clear. It could be the Lesser Himalayan rocks, or the Central Crystallines.

Lesser Himalaya

In a generally ore-poor Himalaya whatever significant mineralisation has been traced, they are mostly located in the Lesser Himalaya, whether in the para-autochthonous sedimentary and volcani-sedimentary rocks or crystalline allochthonous units (crystalline thrust sheets). We discuss these mineralisations from west to east. Commodity wise, these deposits are divided into: (i) uranium mineralisations, (ii) base metal sulphides.

(i) Uranium mineralisations related to para-autochthonous units are located in Khasa area, Simla district, Himachal Pradesh
(ii) Buddhakedar–Ghansyali–Chiratykhal in Tehri district, Uttaranchal.

Mineralisation is concentrated in the sheared rocks close to the thrust zone between granite gneisses and the Rampur Group (Lower–Middle Proterozoic). Uranium mineral is pitchblende/uraninite (± secondary phases) (Yadav et al., 1996). At Budhakedar and other places in the Tehri district, Uttaranchal, mineralisation occurs along gneiss–quartzite contact (Sharma and Nagar, 1996). The two determined ages of uranium-mineralisation are 1412 Ma and 313 Ma (Lall, 1985 in Sharma and Nagar, 1996). The mode of occurrence and high content of Sn, Ni, Cu and Pb in some of these occurrences suggest the mineralisation to be hydrothermal, irrespective of how the hydrothermal solutions were available.

Uranium mineralisation in schistose rocks is reported from a number of places in the Chamoli district. Particular mention may be made of Tungi–Bagarkhal–Dudham area, where U-content varies from traces upto 0.2% U_3O_8, with an average of 0.065% U_3O_8 along a good strike length (Udas, 1986). Here also the ores precipitated from hydrothermal solutions of obscure origin.

Besides the shear zone-controlled mineralisations mentioned above, there are also sediment-hosted uranium mineralisations in this part of the Lesser Himalaya. These are:

1. Sediment-hosted vein type uranium mineralisation, occurring at places in the Kistwar and Doda districts of Jammu and Kashmir; Kulu and Mahasu districts of Himachal Pradesh, Rudraprayag–Karnaprayag–Almora–Pithoragarh belt of Uttaranchal. Uraninite veins occurring along joints/fractures in quartzite are common.
2. Uranium occurs along with copper in sedimentary rocks in Chamoli district Uttaranchal.
3. Sediment-hosted Cu-Pb/ Pb mineralisations occur in the Kulu district, Himachal Pradesh.

Vein-type, sub-economic copper mineralisation has been traced along the Shallu valley and Tons valley, Himachal Pradesh.

Vein-type basemetal (Cu, Pb, Zu) sulphide mineralisations are reported from autochthonous to para-autochthouous Proterozoic rocks of the outer Lesser Himalaya around Golpakot, Khansiya and Amritpur, Nainital district of Uttaranchal. Occurrence of gold is also reported from here (Ghosh, 2006). Mineralisation is hosted by various metamorphic rocks including granite gneisses (Mangla Prasad et al., 1996). There are no tell-tale features in these ores that could suggest their origin.

Lead-zinc mineralisation occurs sporadically along a strike length of ~25 km within the Sirban Limestone Formation (Riphean age) in Riashi Tehsil, Jammu & Kashmir. The mineralisation is stratabound.

Sulphide mineralisation occurs in the Precambrian sericite-chlorite schists (Ascot Crystallines) at Barigaon and also at Gurji Gad within granite gneiss, near Askot, Kumaun, Uttaranchal. Pebble Creek Company, which staked a fortune on these deposits, announced an ore resource of 1.86 Mt of 2.62% Cu, 5.80% Zn, 3.83% Pb, 38g/t Ag, 0.48 g/t Au, based on 74 drill holes and 300m of drift (SEG News Letter 75, p.31, 2008). There is no tell-tale story about the origin of this deposit.

Bageshwar area in Almora district and Rain Agar/Bora Agar in Pithoragarh district are known for containing Pb (with Ag)- and Cu-mineralisation respectively. Both the mineralisations are located in silicified dolomitic carbonates of the Pithoragarh Formation of the Garhwal Group. The Pb-mineralisation may locally contain upto 14.3% Pb and be as thick as 6.75 m. The mineralisation is not persistent, so that it does not ultimately qualify for a viable mining venture today. The primary characteristics of the ores are hardly preserved, as they have been subjected to deformation and metamorphism. However, circumstantial evidence and some relict features suggest that these were probably sedimentary–diagenetic originally. In the Bageswar area Pb–Zn mineralisation occurs at Shukhani, Channa Panny and Bilamu. Pb–isotopes in galena from Shiskhani were analysed (Sarkar et al., 2000). One of the three analysed samples was enriched in ^{206}Pb by ca 1.2%, compared to the two others and has a lower model age of 1553 Ma. The other two samples, similar in ^{206}Pb content, showed ages older by 100 Ma. The μ-values ($^{238}U/^{204}Pb$) in these galena samples are high (10.68 – 10.84). This suggests that the principal supplier of Pb to these ores was crustal rocks of granitic composition.

In the Lesser Himalaya of Nepal, copper mineralisation occurs at Waps and some other places within the clastic metasedimentary rocks of the Kuncha Group. The mineralisation of Fe–Cu sulphides ± sphalerite ± galena occurs as cross-cutting veins. The estimated total reserve is 1.74 Mt with an average grade of 0.88% Cu. The richer parts have been subjected to mining. Nadrakhani deposit is estimated to have 2.6 Mt of ores with an average grade of 0.24% Cu. However, 0.87 Mt of this total reserve has an average composition of 0.87% Cu (UN, 1993).

The volcanisedimentary sequence constituting the Nawakot Group of Nepal shows Cu-occurrences at a number of places including Dhusa, Baisekhani, and Ningna. But the reserve is small (~1 Mt) and the grade is low (0.65% Cu). Sporadically minerals of Pb–Zn occur in the dolomitic rocks.

In the eastern Lesser Himalaya, base metal mineralisations have taken place at a number of centers. Cu-mineralisation is reported from a couple of places from the Phuntsholing Formation, Samchi district, Bhutan. Relatively important occurrences are at Bungthing, Chunpatany, Athaiskhola and Sundarikhola, but none of them are large enough to qualify for economic exploitation although some of them (Chunpatang) may contain as much as ~3.21% Cu. At its present state the mineralisation appears to be structurally controlled.

Mallet (1875) was the first to report the presence of a number of base metal deposits of Darjeeling–Sikkim Himalaya. At places the mineralisation was indicated by such secondary minerals as limonite and malachite etc. At some other places there were relatively fresh exposures of the mineralisations on the escarpments of streams and river sections, such as at Rangpo, Dikchu, Peshok, Pedong–Pachekhani and Gorubathan. Erosion (including mass wasting) faster than weathering characterise these sites. As a result, small-scale mining, not usually based on sound technical plans, took place at many of these places in the past. However, during the seventies and eighties of the last century modern exploration was carried out at most of these places by such government agencies as the Geological Survey of India (GSI), Indian Bureau of Mines (IBM), Mineral Exploration Corporation Ltd. (MECL) etc. In the following few pages the results of these investigations, as well as those of other research groups are briefly discussed.

Rangpo Deposit, Sikkim The Rangpo deposit is located at the Sikkim–Darjeeling border, but mostly within Sikkim itself. It stands partially exposed on the bank of the Tista River.

There are two ore bodies at Rangpo: the Bhotang Main Lode or the Lode-I and the smaller lode, called Lode II. The Bhotang Main Lode was traceable for about 250m and continued for a vertical depth of ~275 m. The ore bodies are, broadly speaking, accordant with the host rocks and folded and faulted along with the latter (Mukherjee and Dhruba Rao, 1974) (Fig.7.2.2). In smaller scales also the ore bands and laminations are folded and crenulated along with the non-ore bands (Sarkar and Banerjee, 1986; Banerjee, 1979). Nevertheless, local translocation through flowage of ores during post-depositional intense deformation and metamorphism is noticeable (Fig. 7.2.3 a–c). The immediate host-rocks are phyllites (locally carbonaceous), garnetiferous chlorite schists, chloritic quartzites (meta-wackes?) and a conformable band of amphibolite on the hanging wall side of the lode. Intense deformation rendered some of the rocks mylonitic and phyllonitic.

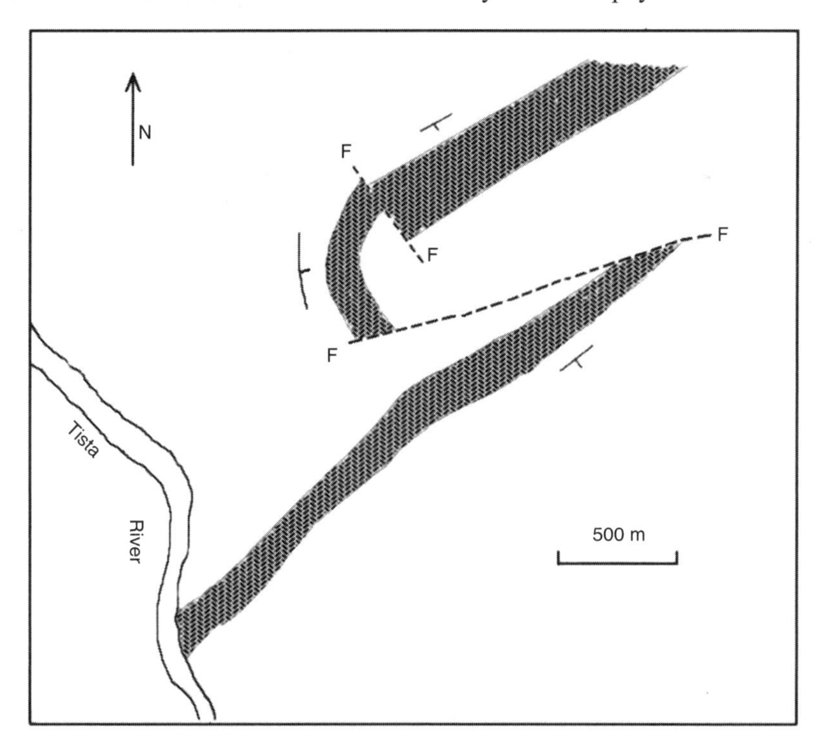

Fig. 7.2.2 Sketch map showing the ore zone folded and faulted with the associated country rocks, Rangpo, Sikkim (after Mukherjee and Dhruba Rao, 1974)

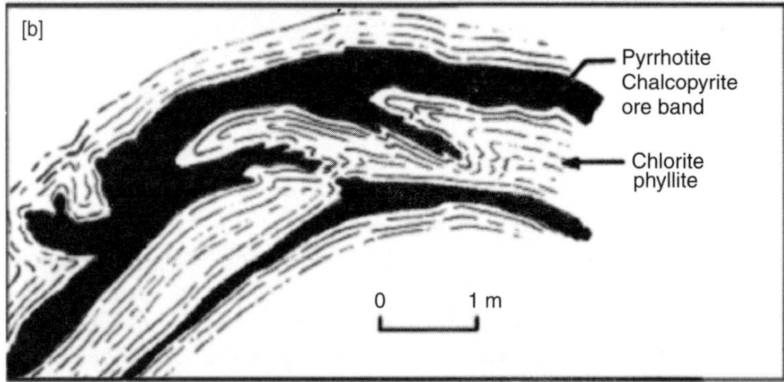

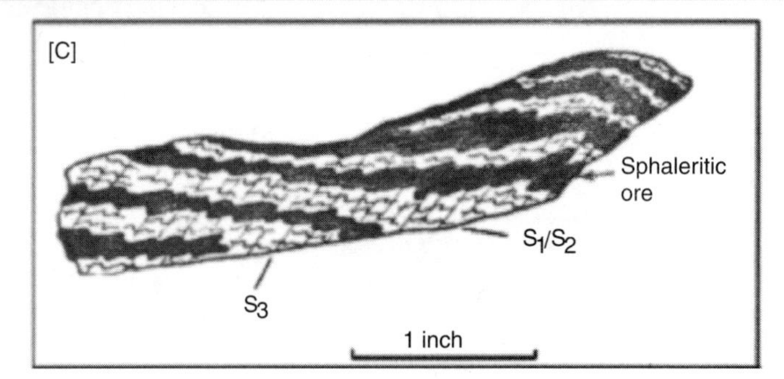

Fig. 7.2.3 Photograph and sketches showing ore structures on small-scale at Rangpo, Sikkim: (a) Primary-diagenetic banding in ore, (b) Tightly folded pyrrhotite + chalcopyrite-rich ore bands and associated chlorite phyllite, (c) crenulated sphalerite-rich ore laminations and associated sericite phyllite with the development of crenulation cleavage (from Sarkar and Banerjee, 1986)

Based on the dominance of the ore metals, Cu, Pb, Zn and Fe, the Main Lode is divisible into several zones. Cu- and Fe- sulphides are relatively concentrated in the north-eastern part, while the Pb–Zn ores are proportionately more in the south-eastern part. In the latter too, copper is more in the upper horizons. No wallrock alteration of any kind is noticeable. Taken the whole deposit together, the principal ore minerals at Rangpo are pyrite, pyrrhotite, chalcopyrite, sphalerite and galena. Magnetite is important at places where magnetite-rich layers may alternate with sulphide-dominant layers. However, magnetite hardly occurs to the exclusion of the sulphides. Subordinate phases include tetrahedrite, cubanite, arsenopyrite and glaucodot. Ores show annealing fabric except where superimposed by late low-temperature deformation (Sarkar and Banerjee, 1986; Banerjee, 1979).

Dikchu Deposit The Dikchu deposit, another Sikkimese deposit is also exposed on an escarpment face of the Dichu river.

Mukherjee and Jog (1963) were the first to report the presence of three ore bodies at the Dikchu deposit. Of these the Dikchu lode, traced over a distance of ~900m, is the most important. Its thickness varies from 20 cm to ≥ 1 m. The ore body is a tabular composite body, consisting of alternate bands/lenses of rich and lean ores. It is parallel to the strike and dip of the regional foliation (S_1/S_2) in the associated rocks. The host rocks of mineralisation are metamorphosed to the amphibolite facies. These consist of garnet + staurolite (± gahnite)-bearing mica schists, which locally may be graphitic. Granite gneiss and amphibolitic rocks also occur in association. The Dikchu ores consist of chalcopyrite + pyrrhotite + magnetite + sphalerite ± galena ± cobaltite ± tenantite–tetrahedrite ± energite + gundmundite ± arsenopyrite ± ilmenite (Banerjee, 1979). Some garnet grains contain 'rotated' trails of sulphides (chalcopyrite, pyrrhotite, and sphalerite). The average ore metal content in the massive ores of Dikchu is as follows: Cu ~ 3%, Zn ~ 1.5% and Pb ~ 0.9% (Mukherjee and Jog, 1963).

Gorubathan Deposit The Gorubathan deposit is also exposed along several stream sections. Two main lodes, Dalingkot lode-I and lode-II were detectable on the escarpment of the Dalingchu and Malkhola streams and finally confirmed by detailed drilling (Fig.7.2.4) (Ray, 1971, 1975, 1976). A separate ore body, exposed on the escarpment of the Kharkhola stream is called Kharkhola lode and occurs southeast of the Dalingkot ore bodies. The ore bodies are not only stratabound but also stratiform (Fig.7.2.4). Ore and silicate-rich laminae interleaf at many areas and have been co-deformed into F_1 folds. Effects of F_2 folds on the ore bands are more prominent.

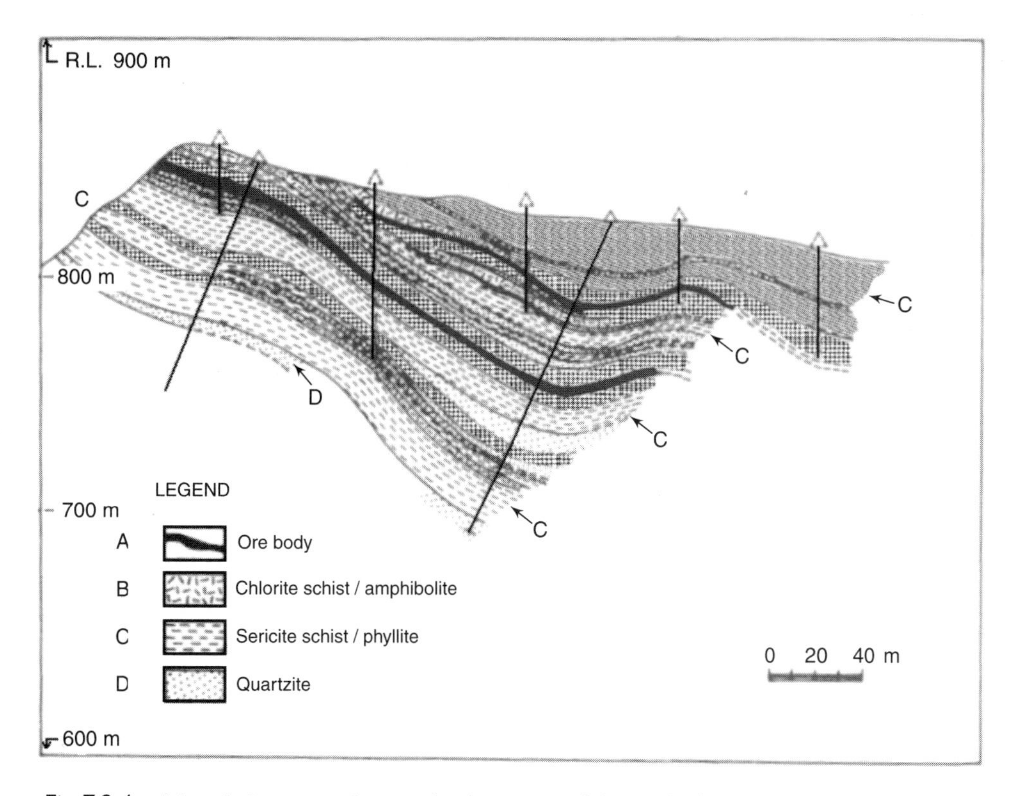

Fig. 7.2.4 A borehole section showing the disposition of the ore bodies at the Gorubathan deposit (after GSI in Sarkar et al., 2000)

Apart from the laminated ores, ores of other varieties such as disseminated, massive etc are also visible. The associated country rocks are sericite–chlorite–quartz schists, amphibolite (metamorphosed mafic tuffs?) garnetiferous chlorite schists, chloritic quartzite, carbon phyllite etc. Obviously, metamorphism did not exceed green schist facies.

Gorubathan ores have two major mineralogical compositions: (i) magnetite + galena + pyrite, (ii) magnetite + sphalerite + galena ± chalcopyrite ± pyrrhotite. The average contents of Zn and Pb in these ores are Zn –1.23% and Pb – 1%. It is obvious from the above that Cu is an unimportant metal in the Gorubathan ores in contrast to those of Rangpo and Dikchu.

Origin of the Ores in the Rangpo, Dikchu and Gorubathan Deposits The ore deposits of Rangpo, Dikchu and Gorubathan occur in the Daling Group of the Eastern Lesser Himalayan (Sikkim–Darjeeling) region and have many things in common. Dar (1968), Ghosh (1975) and Mukherjee and Dhruba Rao (1974) suggested an epigenetic hydrothermal origin for the Rangpo deposit. Ray (1976) suggested sedimentary–diagenetic origin for the Gorubathan deposit and Sarkar and Banerjee (1986) and Banerjee (1979) suggested an original sedimentary diagenetic (SEDEX) origin for all these three deposits, superimposed by later deformation and metamorphism. This is further discussed in the following section.

Isotopic Studies of Ore-lead and Their Significance Ten samples of galena (three from the Bageswar, four from the Rangpo and three from the Gorubathan deposits) were analysed by Sarkar et al. (2000), the results of which are presented in Table 7.2.1. It may, however, be pointed out that the results relating to the Bageswar ores were briefly mentioned in an earlier section.

Table 7.2.1 *Isolopic composition of Pb in galena from the ore deposits of Bageswar, Rangpo and Gorubathan, Lesser Himalaya (from Sarkar et al., 2000)*

S.No.	Ore deposits	Ore-characteristics	$^{206}Pb/^{204}Pb$	$^{207}Pb/^{204}Pb$	$^{208}Pb/^{204}Pb$	Stacey-Kramers model (1975)		Proterozoic model (1989)		
						t	μ_2	Th/U	t	$\mu(^{238}U/^{204}Pb)$
1	Bageswar	Carbonate-hosted	16.240	15.551	35.956	1683	10.67	3.88	1732	10.84
2	Bageswar	Pb-sulphide deposit	16.424	15.555	36.051	1553	10.48	3.80	1608	10.68
3	Bageswar		16.236	15.541	35.941	1672	10.61	3.87	1722	10.79
4	Rangpo	Pelitic-	15.818	15.400	35.439	1779	10.20	3.79	1839	10.46
5	Rangpo	psammopelitic host rocks; stratiform,	15.822	15.410	35.489	1791	10.26	3.83	1850	10.51
6	Rangpo	locally stratified sulphidic ore;	15.876	15.457	35.250	1819	10.51	3.58	1872	10.73
7	Rangpo	deformed and metamorphosed to greenschist facies	15.805	15.386	35.402	1767	10.12	3.77	1830	10.39
8	Gorubathan	Pelitic sediments	15.948	15.511	35.606	1843	10.79	3.85	1889	10.97
9	Gorubathan	metamorphosed to greenschist facies.	15.858	15.428	35.502	1790	10.33	3.82	1847	10.58
10	Gorubathan	Stratiform-stratified ores	15.830	15.420	35.509	1800	10.32	3.85	1857	10.56

In the table, model ages are listed following the model of Stacey and Kramers (1975) and the 'Proterozoic model' age used by Deb et al. (1989) in interpreting Pb-isotopes from the Proterozoic polymetallic deposits of Rajasthan. In the $^{206}Pb/^{204}Pb$– $^{207}Pb/^{204}Pb$ diagram (Fig. 7.2.5), the data are compared with the Pb-isotope growth curves of Stacey and Kramers (1975) and Zartman and Doe (1981). The isotopic composition of Pb in each of the deposits is somewhat inhomogeneous. The model Pb–Pb ages [modified by Stacey and Kramers (1974)] vary only between 1767 – 1819 and 1790 – 1843 Ma respectively in the Rangpo and Gorubathan deposits. The mean model ages for these deposits are 1789 ± 22 Ma (Rangpo) and 1811 ± 28 Ma (Gorubathan).

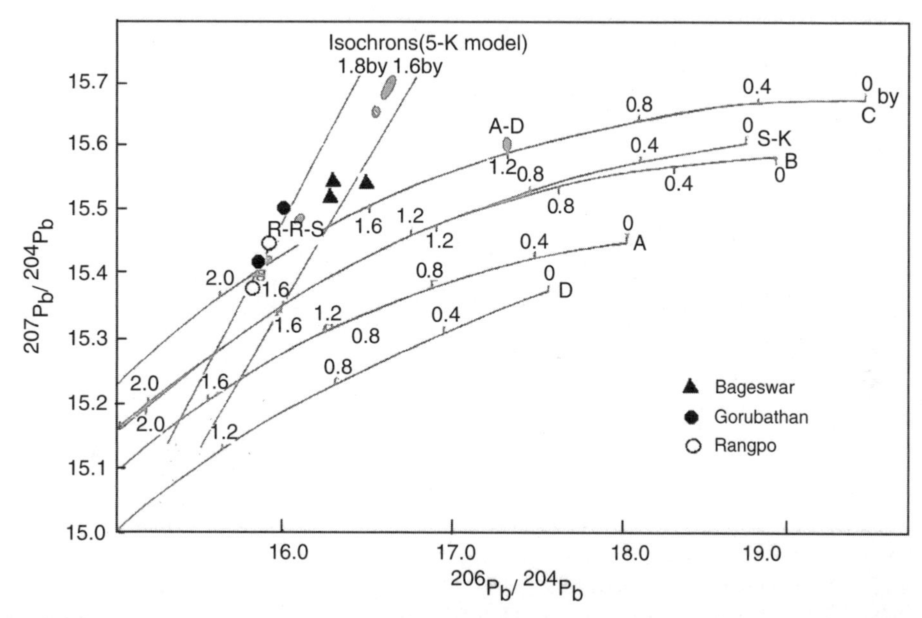

Fig. 7.2.5 Lead isotope evolution curves and isochron related to some Indian base metal deposits (cf. Deb et al., 1989). S–K: Evolution curve suggested by Stacey and Kramers (1975); A–D: Evolution curve suggested by Zartman and Doe (1981). A – Mantle, B – Orogen, C – Upper crust, D – Lower crust. Himalayan deposits shown by symbols. Z – Zawar group deposit, R–R–S – Rampura–Agucha, Rajpura–Dariba,and Saladipura deposits, A–D –Ambaji–Deri deposits (after Sarkar et al., 2000)

Indeed, the model Pb ages of ore deposits are not very exact, i.e. semi-quantitative. In case of Paleoproterozoic sediment-hosted deposits, they give a reasonably correlatable age of host rock deposition (Stacey and Kramers, 1975). A more accurate age of the host rocks would be obtained with the help of U–Pb systematics of authigenic zircon. This suggests that the ~1800 Ma age for the deposits of Rangpo and Gorubathan, should at least broadly correspond with the age of the Daling Group. The difference between the ages obtained with the models of Stacey–Kramers and the 'Proterozoic model' is small (ca 35 Ma). This suggests that the age of the Daling Group and the temporally equivalent rocks is much older than the 'Late Precambrian', conjectured by the great Himalayan geologist, Gansser (1993), in the absence of dependable radiometric data.

It is a known fact that the isotopic criterion of the source of lead in the crustal material is the value of μ ($^{238}U/^{204}Pb$). The obtained position of μ of all the deposits under discussion lies above the orogenic curve (Fig.7.2.5), which according to Zartman–Doe model represents evolution of lead in the reservoir with the mean composition of rocks in the Earth's crust. The value of μ for the given deposits lies in any model between 10 and 11, that may even partly include lead from deep magmatic source below the Proterozoic basin, which means the sources characterised by the value of μ=9.5 – 10. This could be basic volcanic rocks or even their direct derivatives, such as the greywacke. However, the high values of μ point to the fact that the principal supplier of lead to the strata–bound deposits was crustal rocks of granitic composition, a conclusion supported by the lithology of the host rocks.

The ore metals, Zn and Pb in particular, were probably leached from the basinal pile (sediments ± volcanics) by the basinal brine during burial and diagenesis, in cases extending upto the early stage of metamorphism. The source of Zn should be Fe–Mg minerals and that of Pb, K–feldspar(s). The hydrothermal fluid thus produced discharged into the basin bottom, or advected along stratification, guided by cross-stratal faults. It may be mentioned that some workers on similar deposits have suggested that ore metals in the basinal sedimentary-volcanic pile may be inadequate and that a part of the metals may be refluxed by a convective hydrothermal system in a tectonised (and hence rendered more permeable) granitoid–gneiss basement rocks (Russel, 1988; Russel et al., 1981; Strens et al., 1987). This possibly applies here also (Sarkar et al., 2000).

A corollary of the above observations is that since the basemetal mineralisations along the Lesser Himalaya are indeed Proterozoic, they should not be viewed in the context of Himalayan orogeny or any tectonics leading upto it. The Pb–Zn–Ag mineralisation in the underthrusting continent as shown in Fig. 11(H), Mitchell and Garson (1981), could only be inherited and hence incidental.

Pachekhani and Pedong–Peshok Deposits The ore mineralisations at Pachekhani and Pedong–Peshok areas are mainly of copper.

Pachekhani is located at about 18 km from Rangpo. The ore body is strike-short (150 m) but dip-long (300 m along the dip). Average ore metal content is 1.5% Cu. The ores consist of chalcopyrite, pyrite and pyrrhotite in the main. The host rocks are phyllites–chlorite schists belonging to the Daling Group (Raghunandan et al., 1981). The Sikkim Mining Corporation (SMC) at one time selectively mined the ores and blended these with the ores from the Rangpo Mine. However, presently both are closed.

Sulphide mineralisation has been traced for a strike continuity of 2.5 km in the Pedong area in the Darjeeling district and east Sikkim and for 600 m again in the Peshok area of the Darjeeling district. Mineralisations are confined to phyllite and chlorite schists of the Daling Group. The Pedong mineralisation has been traced for a strike length of ~2.5 km and the ore zone is very wide (70 – 114 m). The richer zones within the deposit have a grade of ~2.5 Cu (Raghunandan et al., 1981). At Peshok the copper content ranges between 0.5%–2.55% Cu (Raghundan et al., 1981).

Ore genesis of these deposits should not be much different from other deposits in the Daling Group rocks of Eastern Lesser Himalaya, already discussed.

Base Metal Mineralisations in Ranga Valley, Subansiri District, Arunachal Pradesh
Polymetallic base metal sulphide mineralisation occurs in the Potin Formation, a part of the Precambrian Bomdila Group in the Subansiri district, Arunachal Pradesh. According to Raghunandan et al. (1981) and Tripathy and Kaura (1982), the characteristics of the delineable mineralised zones are as follows:

1. The chalcopyrite–magnetite ore zone runs across the Ranga river for a strike length of 350 m or more. The ore zone is wide (7 – 18 m), but the average tenor is low, being in the range of 0.33 – 0.35% Cu. Some grab samples analysed upto 2.8% Cu, 0.5% Ni, 0.2% Zn and minor contents of Sn (0.05%), W (0.25%) and Ag (0.004%).

2. Cobaltiferous pyrite zone, 310m long and 1.5 – 7.4 m wide, contained Co in the range of 0.03 – 0.25%.
3. Chalcopyrite-bearing magnetite mineralisation in the form of an accordant body in the north bank of the Ranga river.
4. Ni-bearing pyrrhotitic mineralisation in the south bank of the Ranga river.

Some samples from the ore zone analysed > 1000 – 4000 ppm W and > 1000 ppm Sn. An estimated reserve of 1.44 Mt of ores in the Ranga valley analyses 0.33% Cu and 0.16 Mt Co-bearing pyrite (averaging 0.09% Co) (Raghunandan et al., 1981). Host rocks vary from chlorite schists, through garnetiferous biotite–chlorite schists to garnetiferous magnetite quartz schist. The mineralisation appears SEDEX type with some basic members in the litho-package.

(Iron–) Uranium Mineralisation in West Siang and Upper Subansiri Districts, Arunachal Pradesh The Fe–U mineralisation in the West Siang and upper Subansiri districts of Arunachal Pradesh occurs in association with ironstone (± sulphides), muscovite–magnetite–quartz schist, brecciated quartzite and quartzo–feldspathic veins (Bisht et al., 2005). Here iron formations are traceable for a strike length of over 30 km with an average width of 30 m. Uranium is locally concentrated along this zone (Fig. 7.2.6.). Uraniferous zones are relatively rich in LREE, Cu, Pb and Zn. Locally, the ironstones contain upto 100 ppb Au, 6 ppm Ag and 70 – 700 ppm Nb (Eshwara, 1991; Jain et al., 1996). Uranium mineralisation is also present at Ampuli and in the Ranga valley of the Subansiri district.

The middle Proterozoic Siang Group, that hosts the Fe–U mineralisation, is represented by a metamorphosed volcanisedimentary package and is correlatable with the Daling Group of Sikkim–Darjeeling Himalaya. The host rocks of the Fe–U mineralisation comprise quartzite, ironstone, pyrite–magnetite–quartz phyllite, garnetiferous mica schist, carbonaceous phyllite–graphitic schist, paragneiss and amphibolite. The maximum metamorphic temperature determined utilizing garnet–biotite geobarometer is 480°C (Bist, 1998 in Bisht et al., 2005).

The ironstone unit extensive over a strike length of 30 km and a thickness varying between 20 – 40 m merits a little discussion here. It is generally massive, constituted of magnetite, hematite and quartz in different proportions. Fe-oxides range between 44.03 and 63.26%. The contents of CaO, MgO, Al_2O_3, Na_2O, K_2O, TiO_2, V and P_2O_5 are low, adding upto < 2% The magnetite in this ironstone is characteristically low in Cr, Ti, Zr and Co/Ni values (Bisht et al., 2005). Sulphides (pyrite, chalcopyrite, molybdenite, glaucodot and galena) are apparently of two generations: syngenetic and epigenetic. This particularly applies to pyrite.

Uranium mineralisation occurs along fractures and foliation planes in different rock types of the area. However, it is commonly concentrated in Fe-rich rocks, such as the ironstones and garnet–magnetite–muscovite schist. U_3O_8 values in the Gamkak samples (n=126) range from 0.008 to 0.170% and some samples even assayed > 0.2%. The major U-minerals are uraninite, brannerite and davidite. This are very often accompanied by fluorite, tourmaline, green biotite and quartz (Bisht et al., 2005).

The uranium mineralisation is believed to be polyphase (Bisht et al., 2005). The first phase of mineralisation was sedimentary (–diagenetic), uranium suggestively was from the Sela Group granitoids that are thought to be the provenance of the Siang Group rocks. These granitoids are U-fertile (12 – 60 ppm).

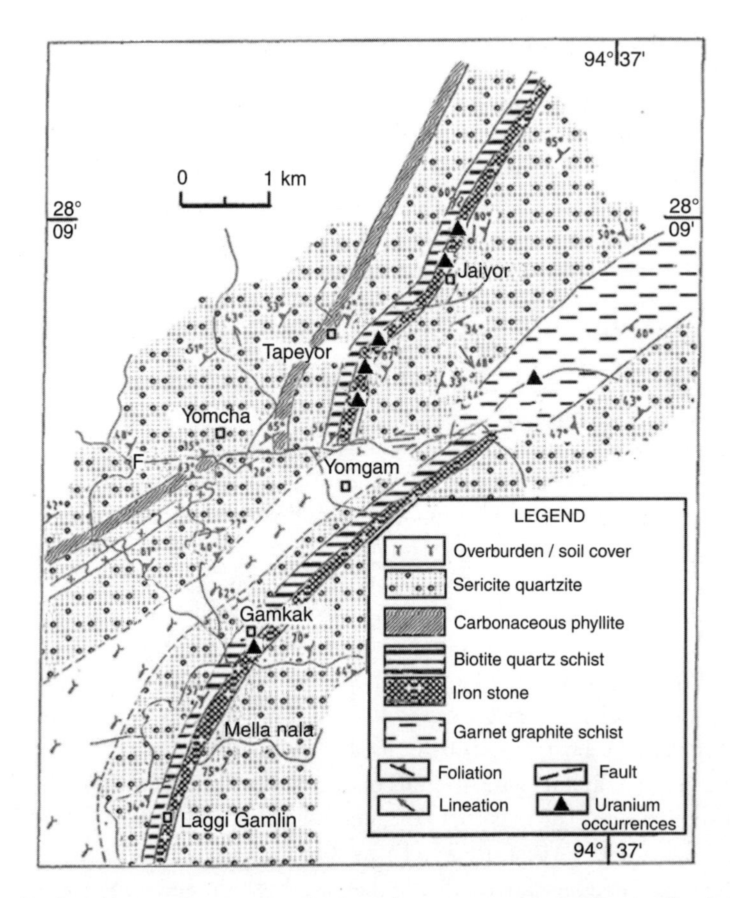

Fig. 7.2.6 Geological map showing distribution of uranium mineralisation in Gamkak–Yomgam –Jaiyor area, West Siang district, Arunachal Pradesh (after Bist et al., 2005)

Uraniferous Phosphorite Deposits in Mussoori Area, Uttaranchal Fairly large deposits of rock phosphate (phosphorite) occur in the upper part of the Lower Tal Formation, constituting the Mussoori synform. The phosphate is a constituent of chert–shale–phosphate assemblage, in which a thick sequence of black bedded chert, interbedded with light to dark grey shale, is overlain by the phosphate rock. Where the chert unit is absent, phosphate rock directly overlies Krol limestone and dolomite.

The phosphate of the Mussoori deposit is in fact collophane, a cryptocrystalline variety of apatite. It occurs as dark grey to black peloidal, stromatolitic and massive phosphorite interlayered with black chert and locally carbonaceous shale. The Mussoori phosphorite may be assigned to the basic varieties: grainstone type, pelletal variety and mudstone phosphorite. The grainstone variety includes the intraclastic phosphorite and also the stromatolitic–pyritic phosphorite. The mudstone phosphorite is in fact the massive variety. Overall bedded, rather layered feature is characteristic of these phosphate deposits (Banerjee, 1986; Patwardhan and Panchal, 1984).

The phosphorite bands vary in thickness from a few mm to ~10 m and the P_2O_5-content varies from about 15% to over 30%. Thick zones of better grade phosphate occur at Jalikhal in the northern limb and at Paritibba–Chamasari and Maldeota in the southern limb of the Mussoori synform. In the Durmala sector of Jalikhal the estimated reserve of the phosphatic ore is 2.40 Mt, with the P_2O_5 varying from 19 to 30%. At the Paritibba–Chamasari area near Mussoori, the estimated phosphorite reserve is 2.91 Mt with a P_2O_5 content of 11 – 31%. The largest reserve, however, is at the Maldeota area, which has been estimated to about 5.3 Mt with P_2O_5 content ranging between 15 and 31%. Total estimated reserve of Mussoori phosphate is ~18 Mt (Ghose, 2006).

Uranium occurs in association with phosphorite, particularly at the Mussoori–Sahasradhara area in the southern limb of the Mussoori syncline (Udas, 1986). U_3O_8-content in the phosphorite is as high as 0.03% for a stretch of 1.8 km over a thickness of about 1 m. The uraniferous phosphorite may contain upto 0.2% Ni and 0.16% Mo. Bulk of uranium in Mussoori phosphate apparently occurs in the Ca-position in apatite.This uranium-content, which is not good enough for consideration as an ore for uranium, is a bane for the phosphatic ore.

Banerjee and Mc Arthur (1989) analysed two samples of phosphatic rocks from Durmala and five samples from Maldeota. Results of this study are presented in Table 7.2.2 and 7.2.3.

Table 7.2.2 *Chemical composition (wt % except for Sr) of Mussoori Phosphate (Banerjee and McArthur, 1989)*

Major oxides	1	2	3	4	5	6	7
SiO_2	17.5	5.45	11.8	5.90	59.6	2.22	2.82
TiO_2	0.03	0.06	0.07	0.02	0.8	0.03	0.02
Al_2O_3	0.86	4.50	2.28	0.45	2.12	10.55	0.28
FeO	12.1	4.12	4.09	8.37	1.99	2.92	0.68
MgO	4.4	0.15	2.10	10.5	1.29	0.52	0.43
CaO	26.2	41.8	38.5	30.6	13.7	48.6	50.0
Na_2O	0.09	0.20	0.21	0.06	0.07	0.21	0.20
K_2O	0.20	0.29	0.73	0.18	0.70	0.13	0.05
P_2O_5	14.0	32.7	25.9	11.4	7.5	20.0	20.2
H_2O^+	1.4	3.5	1.6	0.95	1.4	1.1	0.74
H_2O^-	0.23	2.5	0.20	0.19	0.34	0.19	0.16
CO_2	11.8	2.02	6.41	24.5	7.30	18.5	22.2
S^{2-}	11.0	Nil	3.71	7.35	1.45	2.54	0.25
SO_4^-	0.48	0.40	0.38	0.46	0.34	0.38	0.35
F	1.49	3.21	2.89	1.18	0.79	2.22	2.20
C_{org}	0.65	0.15	0.25	0.22	0.42	0.42	0.17
Sr (ppm)	340	930	710	420	166	1040	818

Table 7.2.3 *Mineralogical composition (volume%) of Mussoori Phosphate (Banerjee and McArthur, 1989)*

Minerals	1	2	3	4	5	6	7
Quartz	20	5	10	0.5	60	Tr	Tr
Dolomite	20	0	10	50	5	0	0
Calcite	Tr.	0	0	0	10	40	50
Pyrite	30	0	10	20	5	5	Tr
Apatite	30	85	65	25	20	50	50

Compositionally Mussoori phosphorite is intermediate between francolite and fluorapatite, but is closer to the latter. The analytical data are consistent with the older age and tectonic setting of the deposit reflecting deep burial diagenesis and prolonged weathering (Banerjee and Mc Arthur, op. cit.).

Magnesite Deposits in Kumaon and Garhwal Himalaya An important mineral wealth of the Kumaon–Garhwal is magnesite occurring in a belt in the Almora, Pithoragarh and Chamoli districts of Uttaranchal. The magnesite belt is extensive over more than 100 kms (with interruptions) and contains a total reserve of 65 Mt of magnesite with ≥ 37% MgO (Table 7.2.4).

Table 7.2.4 *Distribution of magnesite in the Kumaon–Garhwal belt (after Kothyal et al., 2002; Ghose, 2006)*

Deposit	Locality	Geological Formation	Reserve and grade
Agar–Girechhina	Almora district	Pithoragarh/Tijam Formation	3.48 Mt with > 38 % MgO
Dewaldhar	Almora district	Pithoragarh/Tijam Formation	7.9 Mt with > 38 % MgO
Kanda–Masanli	Almora district	Pithoragarh/Tijam Formation	6.7 Mt with 39.67–44.50 % MgO
Dewalthal	Pithoragarh district	Pithoragarh/Tijam Formation	14.53 Mt with > 38% MgO
Bora Agar	Pithoragarh district	Pithoragarh/Tijam Formation	9.2 Mt with 37–44% MgO
Dwing–Tapoban	Chamoli district	Pipalkoti Formation	2.65 Mt with 38–41.61% MgO
Gulabkoti–Pagnao–Mamolta–Molta	Chamoli district	Gulabkoti Formation	10.56 Mt with > 38% MgO
Helang	Chamoli district	Gulabkoti Formation	1.38 Mt with 38–43.5% MgO
Palla–Jakhona–Kimana	Chamoli district	Gulabkoti Formation	1.54 Mt with 38–43.5% MgO
Mandra–Tarak–Tal	Chamoli district	Gulabkoti Formation	1.31 Mt with 43.48–45.96 % MgO
Ramini	Chamoli district	Gulabkoti Formation	5.45 Mt with 38–43.5% MgO

The Kumaon–Garhwal magnesite is associated with the dolomitic rocks of the Pithoragarh/Tejam, or Pipalkoti or Gulabkoti Formations. It is commonly coarse crystalline and occurs as bands/lenses of varying thickness. Magnesite is interlayered with carbonaceous mudstone/shale towards the bottom talc occurs locally.

There is some difference in the characteristics of the magnesite deposits of the area. The Garhwal magnesite occurs as lenticular to tabular bodies within dolomite and varies widely in colour and texture (Mangla Prasad et al., 1996). Magnesite of Kumaon are either grain-supported or 'mud-supported'. In the latter, the magnesite grains 'float' in black phosphatic and pyrite-bearing mudstone (Safaya, 1986). Magnesite showing algal lamination is higher in calcium and phosphorus content.

Origin of these magnesite deposits is not uniformly understood. Dubey and Dixit (1962) and Muktinath and Wakhaloo (1962) suggested that hydrothermal replacement of dolomite/dolomitic limestone by the fluids emanating from the basic intrusives, caused the origin of the magnesite deposits. Valdiya (1968) tried to explain the origin of these deposits by the penecontemporaneous replacement of the carbonate sediments in the sequence. Tiwary (1973) suggested that the primary mineral nesquehonite ($MgCO_3.3H_2O$) was the precursor mineral, which later dehydrated into magnesite. Whatever may be the ultimate origin of the Kumaon magnesite, it appears that the mineral was deposited in an overall tidal flat environment only.

The Higher Himalayan Zone

The Higher Himalayan zone or the Central Crystallines, discussed briefly in the previous section, consists principally of high grade metamorphic and plutonic rocks (granites, gneisses) of Proterozoic age. It lies between the MCT in the south and the Tethyan Zone metasediments in the north. Mitchell and Garson (1981) expressed the hope that the deposits like the Ordovician Pb–Zn deposits in carbonates of Burma and Malaysia (and less probably Kuroko–type Zn–Pb–Cu ores in the Cambrian volcanic rocks in Burma) could be expected in the Higher Himalaya. But that hope is yet to be realised.

Pb–Zn mineralisation, worth its name, is virtually absent in western Higher Himalaya. Whatever base metal mineralisations have been explored till the date within this zone are restricted to Central Himalaya (Nepal) and Eastern (Bhutan) Himalaya. In normal considerations they are not very attractive. However, there are some rare vein deposits of Sb, W, Au–Hg–Sb in this zone. Briefly mentioned, these deposits are:

1. Zn–Pb mineralisation: Ganesh Himal, Nepal
2. Pb–Zn mineralisation: Genekha, and Jemena, Bhutan
3. Tungsten mineralisation: Dholpani and Burkhola, Bhutan
4. Antimony mineralisation: Barashigri, Himachal.

Ganesh Himal Deposit, Central Nepal The location of the Ganesh Himal deposit of Central Nepal is controversial: Is it located within the Higher Himalaya or the Lesser Himalaya (Ghosh, 2005)? It occurs close to the MCT, which is well defined here. We have already mentioned the Himal Group under the Higher Himalaya in the preceding section and the same is maintained here (Fig.7.2.7 a–b).

The terrain is highly rugged and the mineralisation occurs at elevations of 4000 – 4900 m, which generally is not a favourable situation for exploitation, even if the deposit is moderately large and rich.

The sulphide mineralisation of Zn–Pb is hosted by a dolomite unit $< 1 - 29$ m thick, occurring within a repetitive sequence of garnetiferous mica schist, calc schist and quartzite. Amphibolitic rocks appear at various stratigraphic positions below the ore zone. According to Upreti (1999) the Ganesh Himal rocks belong to the upper Nawakot Group. The rocks have been metamorphosed to the amphibolite facies (Ghosh et al., 2005). Below the mineralised zone the carbonate and metapelitic rocks are carbonaceous, with the local development of graphitic schist the whole sequence is isoclinally folded, plunging at moderate angles towards north-east/east.

The mineralisation consists of several ore bodies, the best amongst which is known as Lari 1 ore body. The Lari 1 ore body contains 1.3 Mt ores with 13% Zn, and 2% Pb. Another deposit at Suple, < 2 km away from Lari 1 deposit, is estimated to contain ~ 1 Mt of ore with average 22% Zn and 4% Pb. The whole of the Ganesh Himal deposit is estimated to have metal content of $> 486 \times 10^3$ tonnes of Zn and 75×10^3 tonnes of Pb. Cd and Ag may be available as by products. The principal ore minerals are sphalerite and galena. Locally pyrite and magnetite, particularly pyrite is present in tangible proportions. The ores are co-deformed and metamorphosed in the amphibolite facies along with the host rocks.

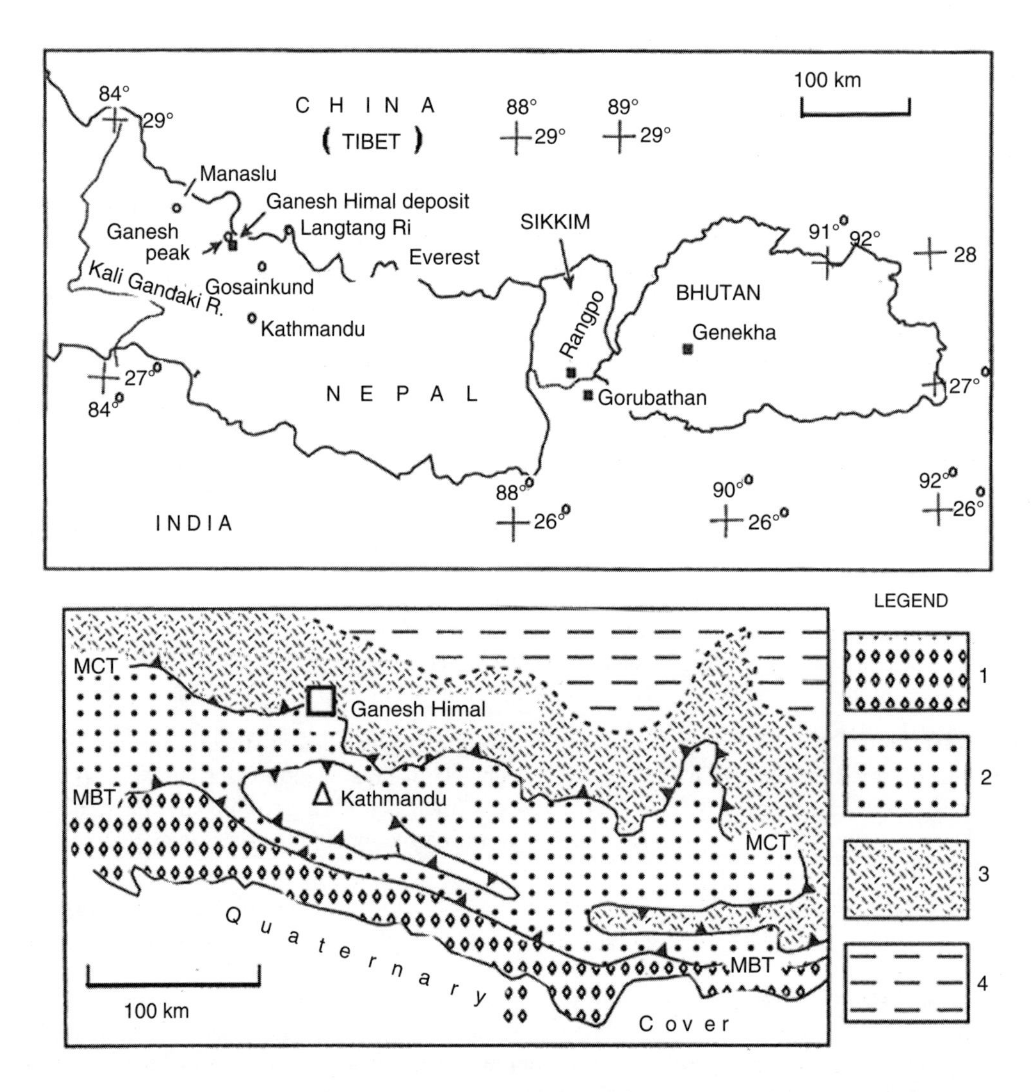

Fig. 7.2.7 (a) Location and (b) Geological environments of Ganesh Himal zinc–lead deposit, Western Nepal Himalaya. Legend: 1. Sub-Himalayan Tertiary Molasse and Flysch, 2. Lesser Himalayan metasediments, 3. Higher Himalayan Crystallines, 4. Tibetan Himalayan Tethyan Sediments, (after Ghosh et al., 2005).

Based on the Pb-isotope compositions, the model Pb age of Ganesh Hmial deposit has been determined to be in the range of 875 to 785 Ma (Ghosh et al., 2005). Interestingly, the metamorphosed carbonate-hosted Pb–Zn deposit of Genekha, Bhutan has a nearly same model Pb-age. The mineralisations apparently were sedimentary–diagenetic initially.

Genekha and Jemena Deposits, Bhutan The Chakula and Romegang Ri Pb–Zn deposits (altitude 3000 – 3700 m) near Genekha, Bhutan occur within the crystalline limestone, belonging to the Paro Formation. Metamorphosed carbonate rocks characterise the highly metamorphosed Paro Formation.

The ore bodies occur as several tabular/lenticular/pod-like bodies. The best one at Chakula (1100 m strike length) consists of 8 ore bodies with an estimated ore reserve of 3.12 Mt with and 6.33% Zn and 1.03% Pb. The other deposit at Romegang Ri (750 m strike length) has been estimated for a reserve of 0.514 Mt with 4.46% Zn and 3.74% Pb (GSI, 1986; UN, 1991).

The ore bodies at Chakula are almost entirely oxidised on surface or at depth, while those at Romegang Ri are partly fresh. The principal ore minerals are hemimorphite, hydrosphalerite, hydrozincite, smithsonite and cerussite. The completely oxidised ores, specially those from Chakula are not much amenable to beneficiation by floatation (GSI, 1986).

The Pb–Zn mineralisation also occurs at Jemena, Thimpu district, Bhutan, in coarse-grained carbonate rocks of the Paro Formation. The mineralisation occurs as stringers, pockets, veins and pods. The ores along with the host rocks are metamorphosed and even remobilised. Limited data show that the grade (11.31% Zn, 3.22% Pb) is good but the reserve is small (UN, 1991).

Dholpani and Bhurkhola Deposits, Bhutan Tungsten mineralisation is reported from the Dholpani and Bhurkhola area, Geylegphug district, Bhutan. The mineralisation in form of scheelite is located in the metamorphosed calcareous sediments occurring along the contact of the granite gneisses, feldspathic schists and garnetiferous mica schists belonging to the Thimpu Formation (GSI, 1984; UN, 1991).

The scheelite participates in the metamorphic fabric in small and larger scales, the scheelite-rich material forms bands sub-parallel to the schistosity. The ore zone at Dholpani, has a strike length of 450 m, with the width varying from 5 to 8 m. The estimated reserve is 0.2 Mt with an average grade of 0.25% WO_3, at 0.2% cut off (UN, 1991). The reserve will increase with lowering of the cut off grade.

The mineralisation is better at Bhurkhola. There the ore zone has been traced for a distance of 1.2 km and the width of the ore zone may go upto 50 m. A reserve of 1.38 Mt with an average grade of 0.45% WO_3 has been estimated (GSI, 1986; UN, 1991). Another estimate (Biswal and Sashidharan, 1992) puts the reserve at 3.6 Mt with an average grade of 0.22% WO_3, for a1500 x 15 m ore zone. Besides scheelite and the gangue minerals, some sulphides are also present in the ore zone.

Opinion on the ore genesis varies between a model of skarn-type mineralisation and a (meta-) sedimentary origin. Observed features weigh heavier in favour of the latter.

Barasigri Deposit, Himachal Pradesh Antimony mineralisation in sheeted quartz veins within coarse grained biotite gneiss (Rhotang Gneiss) is reported from Lahul–Spiti district, Himachal Pradesh. Antimony occurs as stibnite with Pb–Zn antimonates, pyrite, arsenopyrite, sphalerite, chalcopyrite and argentite (Dasgupta et al., 1986). The mineralisation is epigenetic hydrothermal. Mineralisation from hydrotherms given out by the cooling biotite granite is a reasonable guess. Estimated reserve: 10,568 tonnes (GSI, 1989).

Tethyan Zone

The Tethyan zone in the Himalaya, lying to the north of the Central Crystallines, or the Higher Himalaya, and south of the Indus–Tsangpo suture, is even poorer with respect to ore mineralisation, although the stratigraphic sequence represents a long period of time stretching from Late Proterozoic to Early Tertiary.

Western Himalayan Occurrences Veins of Cu-sulphides traverse the Late Proterozoic–Early Paleozoic sediments at several places of the Western Himalaya, such as Akche, Merling, King and Thalpe in Zanskar area (Srikantia et al., in Ghose, 2006). Mineralisations that hold any promise in this part of the Himalaya are those of Cu, W, As and Au occurring in the Late Proterozoic–Early Paleozoic successions in Malari–Barmatiya area, Chamoli district, Uttaranchal (Sharma et al., 1996). An outline of the geology of this area is presented in the Table 7.2.5.

Table 7.2.5 *Stratigraphic succession of Malari–Barmatiya area (after Sharma et al., 1996)*

Shiala Formation	Calcareous sandstone, shale and arenaceous limestone Calcareous shale with arenaceous bands
Garbyang Formation	Argillaceous dolomitic limestone, interleaved with shaly layers Dolomitic marl with gypsitic shale at the bottom
Rolam Formation	Massive quartzite with lenses of dolomite Purple quartzite with pebbly horizons
	---------- Unconformity (faulted/sheared)-------------
Martoli Group	Phyllites (some carbonaceous) and quartzites
	---------- Dar–Martoli Fault ----------------
Central Crystallines	

The mineralisation occurs in the form of disseminations and stringers along shear zones between the Martoli Group and Ralam Formation. Au, rarely upto 12 g/t and W upto 1000 ppm, are reported from the basal Rolam Formation. Locally lean Cu-mineralisation is noticed in the Garbyang Formation in Barmatiya block.The mineralization appears to have been syn-diagenetic originally, but then mobilised–remobilised during deformation and metamorphism.

Gongkhola Deposit, Bhutan The Gongkhola Cu-mineralisation is localized within the Mane Ting Formation, a fossiliferous Ordovician member of the Block Mountain Group. The main ore body is siderite hosted, occurring along the interface of hanging wall carbonaceous phyllite and rudaceous lithic wacke in the footwall. The latter represents a debris-flow deposit formed during basinal instability (Bandyopadhyay and Gupta, 1990). Besides, there are other parallel ore bodies in the wacke and quartzite with or without siderite as principal gangue. The ore zone extends parallel to the NE–SW trending formational boundaries for more than 5 km with local breaks (Fig. 7.2.8). The average width of the main ore body is 10 m, the subsidiary bands being 1 – 3 m wide. In spite of the overall conformable disposition of the ore bodies in respect of the lithological units, in detail there are also discordant veins and stringers. The principal sulphide assemblage in association with siderite and quartzitic host comprises chalcopyrite, pyrrhotite, pyrite arsenopyrite, galena and sphalerite. The siderite has appreciable proportion (average ~ 10%) of manganese. The carbon phyllite horizon displays very thin bedding parallel pyrrhotite–pyrite laminates.

Sulphide schistosity displayed by preferably oriented elongate blebs of pyrrhotite parallel to the dominant schistosity is very prominent within the wackes and quartzites. The polygonised sulphide texture at places in the siderite host is indicative of static recrystallisation, outdating the deformation (Gupta et al., 1983; Gupta and Bandyopadhyay, 2000, 2009).

Three blocks–Eastern, Central and the Western have been delineated in this deposit. The Eastern and the Central blocks together contain 2.24 Mt ores with an average grade of 1.52% Cu. An average of 0.20 g/t gold was analysed in the ores from the Eastern block (Gupta et al., 1983; UN, 1991).

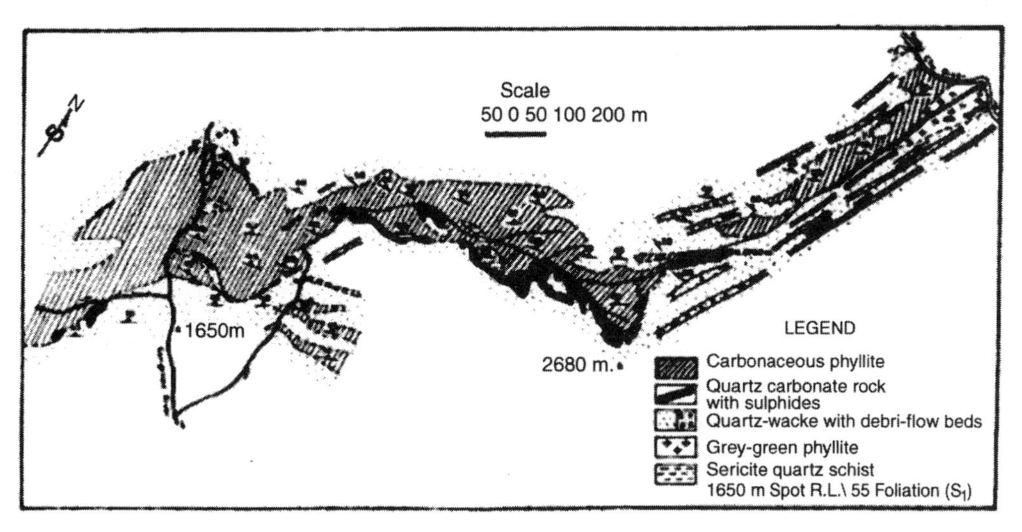

Fig. 7.2.8 Geological map of Gongkhola copper deposit, Bhutan (after Bandyopadhyay and Gupta, 1990)

Uranium Mineralisation Carbonaceous slates of the Haimanta Group occurring at the Kanu–Ropa–Thangi villages in Kinnaur district, Himachal Pradesh contain 0.03–0.33% U_3O_8 without any thorium. Trace metals associated with the uranium mineralisation comprise Ba, Cr, Ti, V, Y, Zr, Mo, Ni and Pb (Udas, 1986). Sedimentary–diagenetic origin in an euxenic situation is a reasonable explanation for it.

Indus–Tsangpo Suture Zone

The zone composed of ophiolites, ophiolitic melange and sedimentary rocks. The ophiolite assemblage is composed of ultramafic plutonic rocks (dunite and peridotite) at the bottom and basic volcanic rocks (Dras volcanics) at the top. No base metal (± noble metal) mineralisation has yet been discovered. However, Dras volcanics may still be searched for the purpose. Locally there are some spot values of Au, Pt and Pd in placers in the range of ppb (Wangdus et al., 1998).

Podiform chromite within the basal zone of Kyum Tso ophiolite body, exposed on the Hanle river section awaits detailed investigation.

Ladakh Belt The granitoids of the Ladakh belt are generally tonalitic in composition with low Sr_i. Most of the Porphyry-Cu deposits of the world are associated with such granitoids. But none has been found here till the date. Sporadic occurrence of gold and copper are reported from Sabut area, Leh district, Jammu and Kashmir (Singh, 1995)

Metallogenetic Analysis of the Himalaya

In the previous few pages, an attempt has been made to draw a picture of Himalayan metallogeny with the help of a broad brush. The picture, as it is, is not very attractive. None the less, it deserves an explanation, which is attempted below. Here again, our discussion will have a regional rather than a temporal basis in order to avoid any bias.

The Frontal Himalaya or the Sub-Himalayan zone is free of magmatism, as well as tangible metamorphism. Therefore, ore mineralisations that are commonly attributed to these two geological processes are excluded. In this domain the possible ore mineralisation can, therefore, be related to sedimentation and diagenesis only. These mineralisations can be both chemical and mechanical i.e. placer type. The most common chemical type deposition is that of uranium (± vanadium). Stratiform copper mineralisation in Oligocene succession is reported from Ebre basin, south of the Pyrenees as having formed in a post-Alpine stage (Caia, 1976). In Frontal Himalaya, particularly in the Western sector, 'Sandstone type' uranium mineralisation has taken place in the Middle Siwaliks of the Sulaiman Range of Pakistan (Moughal, 1974) and in the Indian sector, in the Middle–Upper Siwaliks of Chandigarh–Pathankot–Jammu area. A possible source for Sulaiman mineralisation is the U-rich carbonatites in the Indian plate, south of the collision zone in Pakistan (Mitchell and Garson, 1981). The occurrences in the Indian side, also 'Sandstone' type, are localised within sub-greywackes. U_3O_8 content ranges from 0.02 to 0.6%. Workers on these mineralisations have reasonably pointed to the granitic rocks of the rising Himalaya in the north (Udas and Mahadevan, 1974) as the source of uranium. Reduced metasediments rich in pre-existent uranium could act as other sources, particularly on exposure to weathering and erosion. Detrital gold reported from a still fewer places could have a similar provenance.

Lesser Himalaya is the principal repository of Himalayan ores. In the Western sector of the Lesser Himalayan zone, the ore mineralisations, none economically viable in any exploitation venture today, are mainly of uranium and base metals. Uranium mineralisations, most of which now show some epigenetic features, occur either within metasediments or along the interfaces of schists and gneisses. Base metal depositions (Pb–Zn ± Cu), locally occurring in platformal environments, are generally sub-economic in modern day mining parlance. The situation somewhat improves eastward from Nepal. There, copper mineralisation is associated with metasedimentary (clastic) rocks of Kuncha Group and volcanisedimentary rocks of Nawakot Group. Relatively better deposits occur in the Eastern Lesser Himalaya where base metal deposits occurring at Rangpo, Dikchu, Gorubathan, Pachekani and Pedong–Peshak were either mined for sometime in the relatively recent past, or await exploitation on small scales. The mineralisations in all these places appear to be rift-controlled SEDEX type. The phosphorite (uraniferous) deposits of Mussoori and the magnesite deposit of Almora were also laid in platformal environments. The tectonic environment of Fe–U mineralisation in the Middle Proterozoic Siang Group also does not appear much different.

Majority of the ore deposits in the Lesser Himalaya were metamorphosed up to amphibolite facies. Metamorphism is generally not expected to remove an ore deposit by dissolution or melting as such. Dewatering during metamorphism is not expected to produce enough focussed water and the necessary ligands to dissolve and remove a pre-existent ore body. That is why, we see ore deposits surviving metamorphism of all grades. Broken Hill deposit (amphibolite–granulite facies) of Australia is a glaring example of this fact. Deposits may, however, form from disseminations and older deposits partially remobilished over small distances during metamorphism with P or P–T gradients. These effects are visible in different degrees in the Lesser Himalaya.

The ore mineralisation picture in the Higher Himalayan zone is still dismal. Two base metal deposits of Pb–Zn sulphides, namely the Ganesh Himal deposit of Nepal and the Genekha deposit of Bhutan are worth the mention in this context. They also appear to have been laid in platformal environments. So also possibly was the environment of deposition of sediment-hosted bedded tungsten (as scheelite) mineralisation at Dholpani and Bhurkhola in Bhutan.

The antimony mineralisation within granite gneiss at Barasikri was hydrothermal in origin. However, the mineralisation is undated.

Bulk of the granite–granite gneiss masses in the Higher Himalayan zone, as well as in Lesser Himalaya is pre-Himalayan in age. A relatively small part developed during Himalayan orogeny. Limited available information suggests that again the bulk of these granitic masses developed by fluid-undersaturation or dehydration melting. In this context the absence of substantial Sn–W mineralisations in these granite–granite gneisses may be discussed.

Important tin mineralisations in the world occur as veins, breccias and replacement bodies in moderately shallow plutonic environments, although they may occur in volcanic and shallow sub-volcanic rocks (Heinrich, 1990 and references therein). These ilmenite type granitic rocks formed from reduced magmas that had f_{O_2} below Ni – NiO (NNO) and close to that of quartz–fayalite–magnetite (QFM) buffer (Ishihara, 1977; Burnham and Ohmoto, 1980). This is a handy explanation for many of the tin granites of the world falling under the S-type of Chappel and White (1974). The Al/(Na + K) ratio in the magma is important (Linnen et al.,1996). Fractionation of Sn in reduced peraluminous granites may be quite different from that in peralkaline granites at any redox state. Most of the Australian tin-bearing granites, show mixed S and I characteristics, but continue to be reduced ('ilmenite' type) (Solomon et al., 1991 in Heinrich, 1990). Thus when dissolved in fluids, it is commonly bi-valent (Sn-II). It precipitates in tetravalent state (Sn-IV) as cassiterite following such a reaction (Heinrich, 1990),

$$\underset{\text{(Aq)}}{Sn(II)Cl_x^{2-x}} + 2H_2O \rightarrow \underset{\text{(Cassiterite)}}{Sn\,(IV)\,O_2} + 2H^+ + x\,Cl^- + H_2^0$$

Tin granites are generally highly differentiated varieties. Granitic rocks associated with tungsten only are less well chemically constrained. They are generally type-I and may vary in composition from granodiorite to alkali–feldspar granites and contain magnetite rather than ilmenite as the characteristic phase belonging to Fe–Ti–O system (Kwak, 1987). A possible reaction like the following has been suggested to precipitate wolframite in the temperature range of 350°–300°C (Heinrich, 1990).

$$\underset{\text{(Aq)}}{H_2WO_4^0} + FeCl_2^0 \rightarrow \underset{\text{(Ferberite)}}{FeWO_4} + 2HCl^0$$

Depending on whether Fe or Ca will be available for reaction (commonly from wall rocks), wolframite or scheelite will form.

An anatectic granite to be the progenitor of one or more ore deposits must satisfy the conditions: (i) necessary ore metals and legands must be available in the system, (ii) there must be adequate aqueous-dominant fluid supply (less expected in fluid-undersaturated melting) in the system with an ultimate pressure-controlled focus, (iii) the redox-control of solution and precipitation, where involved, must be satisfied. To these may be added (iv) Mixing of the metamorphic fluid with meteoric water or increase of pH. If any of these conditions is not satisfied, the deposit will not be expected. Sn–W deposits are generally not emplaced near the top of the plutons. Even then they may be weathered and eroded away. Some deposits may even be covered by thrust-sheets. The above facts possibly explain the absence of tangible Sn–W (–Nb–Ta) mineralisation in the granitic massifs of the Himalayas.

The Tethyan sedimentation continued for a long time starting from Late Proterozoic to Tertiary. The environment was continental passive margin type during the sedimentation from the Late Proterozoic to the end of the Paleozoic. This was followed by the regime of Neo-Tethys, initiated north of the then Indian subcontinent. It was a collage of island arcs, internal basins and irregular bodies of continental rocks. Such regions are generally repositories of many ore deposits. But unfortunately we have none worth the mention except the Gongkhola Cu-deposit in Bhutan.

It is none very large either (vide discussion in an earlier section). Uranium mineralisation (with increased values of Ba, Cr, Ti, V, Y, Zr, Mo, Ni and Pb) in carbonaceous slates of the Haimanta Group represents passive margin mineralisation in a shallow restricted (sub-) basin.

Cu-rich base metal mineralisation occurs in ophiolitic association in a number of places including Troodos Mssif, Cyprus and Newfoundland, Canada. Podiform chromite deposits (± PGE ± Au) are also reported from Cyprus, Cuba, Greece and Philippines. The Indus–Tsangpo suture zone should be a prospective zone in this context. Although PGE and podiform chromite mineralisations are reported from this zone, the preliminary data are not very encouraging. The investigations may, however, be pursued. The tonalitic granitoids in the Ladakh belt are potential rocks for the discovery of Porphyry copper mineralisations if the ore forming conditions favour. But none has been discovered to date.

In conclusion it may be said that the poor ore mineralisation in the most magnificent mountain range of the world is not a freak of nature. It has to be understood in the context of its geology outlined above. So long we have tried to analyse the metallogeny in the Himalayan domain. But beyond it i.e. in Trans-Himalayan region, recently some unusual type of Porphyry-Cu mineralisations have been discovered and the researchers (Hou Zengqian et al., 2003) have assigned the mineralisation to an indirect effect of the Himalayan orogeny, or India–Eurasian collision. The mineralised zone, 300 km long and 15 – 30 km wide, is located in the Changdu continental block, particularly in the Qiangtang terrane of the Himalayan–Tibetan orogen. It is commonly referred to as the Yulong Porphyry copper belt. The belt hosts one giant deposit, two large deposits and two medium sized deposits, together with dozens of mineralised porphyry bodies. The largest member of the group, the Yulong deposit, contains 630 Mt ores with an average grade of 0.99% Cu and 0.028% Mo. The Yulong deposit consists of a steeply dipping pipe-like body (1000 m × 600 m × 500 m), hosted in a K-rich porphyritic granitoid intrusion, and extends into the overlying metasediments forming a stratabound/tabular ore zone at the top. The porphyritic intrusions and the associated volcanism took place at least in three phases, at around 52, 41 and 33 Ma, controlled by NS to NNW–SSE trending large-scale strike-slip faults, which are perpendicular to the collision zone between the Indian and the Eurasian continents. Although the Yulong belt Porphyry copper deposits developed in the intra-continental convergent environment and the ore-bearing granitoids are high-K (and enriched in Rb and Ba) in contrast to calc-alkaline granitoids in most other places, the mineralisation styles including wall-rock alterations, are comparable to common arc-located Porphyry copper systems. This discovery and the exploration and studies that followed are very significant contributions to ore geology, opening up new possible regions for the search and exploration for Porphyry Cu-mineralisation.

8

Crustal Evolution and Metallogeny in India: A Brief Review in the Context of the World Scenario

8.1 An Outline of the World Scenario

Introduction

The Earth's crust, as we see it, is the result of the geodynamic and geochemical activities that took place throughout the geological past in the crust–mantle system, added with the effects of evolution in the atmosphere, hydrosphere and biosphere. It is needless to mention that some of the crustal material must have been lost through erosion and tectonism (subduction). Moreover, the larger part of the crust is not accessible to direct observation. There the authors were obliged to take to indirect means, the results of which are fraught with a greater amount of uncertainty in many cases. In the present chapter, the global picture of crustal evolution in terms of the above is outlined, mentioning alongside the metallogeny which is now better understood in the above context. Projected on this background, a brief review of the salient aspects of crustal evolution and metallogeny in India is done, pointing out where these conform to the overall global picture and where they do not. This is likely to be of some use while making generalisations on the Earth's events, at the same time drawing attention of the potential researchers in the subject, particularly to the aspects, which deserve more attention in future.

Today the Earth's history is formally divided into the following broad divisions (IUGS Geologic Time Scale, 2002, 2003 and 2004):

PHANEROZOIC EON	Cenozoic Era (65 Ma – Present)
	Mesozoic Era (251 – 65.5 Ma)
	Paleozoic Era (542 – 251 Ma)

PROTEROZOIC EON	Neoproterozoic Era (1000 – 542 Ma)
	Mesoproterozoic Era (1600 – 1000 Ma)
	Paleoproterozoic Era (2500 – 1600 Ma)
ARCHEAN EON	Neoarchean Era (2800 – 2500 Ma)
	Mesoarchean Era (3200 – 2800 Ma)
	Paleoarchean Era (3600 – 3200 Ma)
	Eoarchean Era (3600 – Beginning not defined)
HADEAN EON	

We may fix the beginning of the Eoarchean at 4000 Ma, as the first 500–550 Ma of the Earth's history is convensionally called the Hadean Eon.

Probable History of the Early Earth

It is now agreed that the solar system was initiated by condensation of interstellar gas into planetesimals sweeping around the proto–Sun. Earth along with the other planets formed by accretion of these planetesimals. Their initial aggregate could be rather cold, but the impact of the increasing mass and gravitational attraction eventually raised the temperature to make it a molten mass. The primordial time of planet formation should have been complete by 4.55 Ga. Differentiation set in before long. Because of the lower melting point of the Fe–Ni alloys compared to Fe–Mg silicates, a Fe–Ni melt separated and gravitated to form the core of the Earth, overlain by a possibly semi-solid Mg–rich silicate mantle, in turn capped by a terrestrial magma mush ocean (Abbott et al., 1994; Condie, 2005; Ernst, 2007; Papanastassiou and Wasserberg, 1971).

By 4.4 Ga the magma ocean was essentially solidified and lithospheric platelets developed, at least locally. This conclusion is necessitated by the reports of 4.2–4.4 Ga zircons in the meta–conglomerates from Jack Hills, Yilgarn craton, Western Australia. The 4.4 zircon is zoned with respect to LREE and $\delta^{18}O$. The high LREE and high $\delta^{18}O$ in this and other old zircons (>4.0 Ga) are consistent with the growth in an evolved magma (granitic), that had interacted with supracrustal materials. Low temperature surficial processes, such as diagenesis, weathering and low temperature alteration took place before 4.0 Ga. The rock–water interaction needed to produce the obtained $\delta^{18}O$, suggests the presence of liquid H_2O, not ignoring an ocean at 4.4 Ga (Peck et al., 2001; Nutman et al., 2001; Wilde et al., 2001; Harrison et al., 2005). Hafnium isotope data from the Jack Hills zircons, however, suggest that both mafic and felsic crust were present in the Hadean and the early Archean (Hawkesworth et al., 2010). Recent discovery of metabasalts of Nuvvuagittuk, Canada, directly proves it. As very little Hadean rock has been discovered so far, it may not be an unreasonable presumption to assume that the crustal material of that age has been efficiently backed into the mantle (Amelin et al., 1999; Boyet and Carlson, 2005). The likely mechanism for this advection was caused by mantle overturns and rising plumes. As the Hadean Eon passed on to the Eoarchean, and then to the Paleoarchean, decompression fusion of hot up-welling asthenosphere and deep-seated large scale plumes generated great thickness of basaltic crust and through its rather shallow subduction, crustal suturing and partial fusion, produced granite–greenstone belts and tonalitic complexes. The gradual aggregation of these litho–tectonic units led to the amalgamation of island arcs and micro-continents to form protocontinents (Sleep and Windley, 1982; Ernst, 2007). With time the continental crust enlarged itself, such that it covered about 70 per cent of the present-day crust at the beginning of the Proterozoic (Veizer, 1976; Condie, 1998). The contention that the Plate Tectonics was operative through the Eo–Paleoarchean, is supported by the discovery of the Jamestown ophiolite (3.5 Ga) (deWit et al., 1987) and the vestige of an ophiolite (3.8 Ga) from southwest Greenland (Furnes et al., 2007). Even Archean komatiites could be the products of subduction (Parman et al., 2004). Modern-style subduction (deep subduction that involves a mantle wedge), however, does not show any evidence of having been operative before the Mesoarchean time.

The Eoarchean Crust and the First Record of Crustal Metallogeny

The oldest discovered crustal rock is a Hadean-age (4.28 Ga) ocean-floor basalt from Nuvvugittug, Qubec, Canada (R. Carlson and J. O'Neil, Carnegie Inst. Sci YB2008–2009, p. 48). Somewhat younger are the Acasta Gneisses from NW Canada. These early rocks are highly deformed and comprise such members as tonalite, amphibolite, ultramafic rocks, granites and locally even metasediments (calc-silicates, sillimanite-bearing metapelites and quartzite). Zircon from tonalite and amphibolite yielded ages in the range of 4.03–3.96 Ga (Bowring and Williams, 1999), with younger granite yielding an age of 3.6 Ga. The rock assemblages closely resemble a piece of an Archean terrain that contains greenstone–granite associations, set in a TTG–gneiss country. The Acasta body is small (a few hundred sq km) which may mean that as a primitive crustal nucleus it was small, or it is only a relic. No mineralisation is known from Acasta.

The next Early Archean rocks are reported from the Isaq Gneiss Complex in SW Greenland. One of the three terrains identified in this complex, namely the Akulleq terrain, the dominant member of the Amitosq Gneiss Complex, formed during 3.9–3.8 Ga. undergoing high grade metamorphism at 3.6 Ga (Nutman et al., 1996). These terrains collided at ~ 2.7 Ga.

The most interesting sector in this region is the Isua supracrustal belt, displaying sedimentary and volcanic stratigraphy and structures and the evidence of oldest mineralisation, sub-economic though. The sequence contains subrounded pebbles of granitoid rocks and chert and also pillow structures in the amphibolitic rocks. The oldest BIF, now represented by magnetite–chert rocks, is reported from here. Estimated iron ore reserve is ~2 billion tonnes with an average of 32 per cent Fe. Scheelite occurs in the amphibolite and calc-silicate rocks. Cu–Fe sulphides occur, locally though. The sequence was deposited at about 3.8–3.7 Ga (Rosing et al., 1996; Schidlowski et al., 1979). The sequence, including the contained mineralisation, resembles many Late Archean greenstone sequences, with the addition of remnants of an Early Archean (> 3.75 Ga) sea-floor hydrothermal system at Isua (Appel et al., 2001).

Increased $^{12}C/^{13}C$ ratios in some primitive organic matter suggest a prolific photosynthetic microbial life already in existence 3.8 Ga ago (Schidlowski, 1988; McKeegan et al., 2007). In fact life could have, or possibly did have appeared in the Hadean sea floor much earlier, promoted by hydrothermal activities. But the challenge to it is the commonly held belief that the early ocean was at least partially vaporised many times by meteorite impacts during the first 500–600 million years of Earth–history. But the partial vaporisation did not necessarily completely wipe out the earlier forms of life if present. It is interesting to note that hydrothermal activity on ocean floor is related to both base metal deposition and the origin and promotion of life. Fe, Zn, Ni, Co, Mo, Mn and S are typical constituents of many ore deposits. They are as well important constituents in enzymes, which are catalysts in many of the life processes (Russel et al., 2005). There is no reason to believe that there was dearth of hydrothermal systems on the floors of primitive oceans. Ore–microbiota relationship is noticed through much of the geologic time.

Archean Crust Development and Metallogeny Spread Over a Billion Years

At Warrawoona in the Pilbara craton, Western Australia, a well-preserved greenstone sequence (3.5–3.2 Ga) overlies an older greenstone–TTG basement. The Paleoarchean developments in the Barberton area, South Africa, have also been a subject of considerable study. Four greenstone–TTG terrains dated 3.55–3.2 Ga (Kroner et al., 1996) share similar stratigraphic sequences. The overlying Moodies Group includes organic sediments and appears to have been deposited after the four terrains united a little after 3.2 Ga (Condie, 1997). Thus the continental crust continued to grow through the Early Archean (3.6–3.0 Ga) as TTG–greenstone regions, but at much slower rates compared to Late Archean. A typical granitoid older than ~ 3.2 Ga was not known to the authors until a 3.58 Ga old granite was located in the Bastar craton, Central India (Rajesh et al., 2009). The Superior Province of eastern Canada is the largest Archean Province on Earth and has its Archean geological history limited to the time span 3.0–2.66 Ga. It consists of sub-provinces that are alternately dominated by greenstones and metasediments (metagraywacke). The TTG composition varied during the period 4.0–2.5 Ga due to the difference in the melting of the subducting slab and interaction of this melt with the overlying mantle wedge. Younger TTGs are richer in Mg, Ni, Cr and (CaO + Na_2O). According to Martin and Moyen (2002) it marks the progressive cooling of the Earth, as well as it proves the fact that Plate Tectonics were operative from 4.0 Ga, if not earlier.

A few mineral deposits developed during this period. BIF deposits of moderate sizes developed at Imataka (Venezuela), Hobei (China) and at a few places in the Pilbara and Yilgarn blocks in Western Australia. Some small (≤ 10 Mt) VMS type deposits developed in the Pilbara craton during 3.45–3.2 Ga. It was in contrast to somewhat younger deposits (also small) from the Yilgarn craton and the moderately large deposits from the Superior Province in Canada, the Pilbara ores contain Pb and Ba in tangible proportions, in addition to Zn and Cu. In this respect they resemble the Miocene Kuroko ores of Japan. The deposition of stratiform sulphates in these Archean oceans, suggest oxic conditions, even if local, during that time. The barite deposit as it is now, appears to be the replacement product of anhydrite (Buick and Dunlop, 1990)

The continental crust grew much faster during the Late Archean time. Volcanism, sedimentation, deformation, metamorphism and plutonism were all complete in < 100 Ma time or even in ~50 Ma, compared to similar history of up to 500 Ma in the Early Archean greenstone belts (Pilbara, Barberton) (Condie, 1997). There were, of course, more greenstone belts and these were complex in petrology and structures during Late Archean. At one end there were more komatiite rocks in the greenstone belts, at the other more of evolved (felsic) rocks. Greenstone belts have been divided into two types: platformal and basinal. The basinal type is generally larger and thicker and may have the platformal type at the base. At places the grade of metamorphism can be high (granulite facies). In bulk composition, some granulite protoliths appear to have been similar to the Phanerozoic shales from the cratonic basins. Red beds were still to appear.

From the metallogenic point of view, the Late Archean was most prolific in the history of the Earth. It was the most important period of deposition of orogenic gold ('lode gold'), Algoma type BIFs, komatiite-associated Cu–Ni (–PGE) ores, Quartz–pebble conglomerate (QPC)–associated Au–U ores, Cu–Zn/Zn–Cu type VMS ores (Superior Province, Canada) and Rare metal (Li ± Cs ±Ta ± Sn) pegmatites. Mn–mineralisation in the Archean is reported only from the ≥ 2700 Ma Rio das Vilas Series, Brazil. This of course excludes Indian occurrences already discussed. This aspect will be taken up in a later section. Besides, a supergiant BIF deposit, namely Hamersley of Western Australia, straddle along the Archean–Proterozoic boundary. In older greenstones (>3000 Ma), komatiites are found, but excepting a moderate deposit at Ruth Well, Pilbara, Western Australia, these rocks are characterised by the absence of Cu–Ni (PGE) mineralisation. Two outstanding ore deposits of the type are the Kambalda deposit in Western Australia and the Pikwe–Selebi deposit in Botsowana, both having developed in the Late Archean.

Another outstanding mineral deposit of the Meso–Neoarchean time (2900–2800 Ma), may be of all times, is the gold–uranium mineralisation in the coarse clastic sediments of the Witwatersrand basin, Kaapvaal craton, South Africa. It has already produced > 50,000 t gold and still there is a reserve of 35000–40,000 t, together composing nearly half of the total known gold (mined + yet to be mined) ores in the Earth's crust. Besides, 1.5 Mt U_3O_8 were produced from these ores during 1952–75. Pyrite and carbonaceous matter are conspicuous in the 'matrix'. Pyrite and uraninite grains are commonly roundish. Gold grains have two different morphologies: one typically detrital and the other irregularly shaped (Minter et al., 1993). There are two contesting views on the genesis of these deposits. These are: (i) modified paleoplacer model, (ii) hydrothermal model. In Model–1, rounded gold, uraninite and pyrite grains are thought detrital, which were partially modified during greenschist metamorphism to which they have been subjected (Frimmel, 1997; Frimmel et al., 2005). Model–2 advocates complete hydrothermal origin, interpreting the rounded mineral particles as pseudomorphic replacements and post–depositional dissolution–reprecipitation (Phillips and Law, 2000; Law and Phillips, 2005). Uranium, as per the hydrothermal model, was introduced into the basin through the leaching of the metal from the granite hinterland by meteoric water. While the 'modified paleoplacer model' assumes gold to have been derived from the older orogenic gold mineralisation (now eroded) by a dynamic fluvial system, followed by limited dissolution and precipitation, the hydrothermal model proposes gold to have been introduced in low salinity H_2O– CO_2 –H_2S fluids that were generated by devolatilisation of deeper sequences during green schist facies metamorphism and introduced along deep–seated thrust faults. The issue is not a mere ore genetic one, which of course, is none very unimportant. The two models have different views about the oxidation state of the then atmosphere. The first one suggests that the atmosphere at that time was anoxic or very poorly oxic, while the second view proposes that the atmosphere was oxic. However, the Re–Os ages of the Witwatersrand gold is older than the sedimentation (Kirk et al., 2002) and that this gold has considerably higher Os–concentration and near–mantle initial ^{187}Os /^{186}Os ratio (Frimmel et al., 2005). These findings along with the other features support the modified paleoplacer model only. Obviously the source was not too far away. But what made it so abnormally rich in gold? A suggestion that up to about 3000 Ma the extraction of Au (and Os) into the crust must have been higher than in the later times, indeed not improbable, but does not satisfactorily explain the accumulation of gold only at the source for the Witwatersrand or a similar deposit (Blind River, Canada). The whole problem would warrant clearer explanation.

The Crust Comes of Age during the Proterozoic with the Attendant Metallogeny

Condie (1989) succinctly outlined the geological development of the Proterozoic crust during nearly two billion years after the Archean. This refers to the Proterozoic developments in terms of the typical rock associations such as (i) Quartzite-pelite-carbonate (QPC)[1] association, (ii) Bimodal volcanics-arkose–conglomerate (BVAC) associations, (iii) greenstone associations. Of these the QPC is best developed, comprising > 60 per cent of the Proterozoic supracrustal rocks. Fluvial, tidal, and shallow marine deposits are most common in such an association. Interpreted in terms of the Phanerozoic QPC associations, these rocks are found in three tectonic settings:
(i) rifted continental margins, (ii) cratonic margins of the back arc basins, and (iii) intra-cratonic basins. Provenance of the QPC sediments was obviously dominated by granite–gneiss rocks.

In BVAC the proportion of volcanic rocks, which commonly consist of basalt and rhyolite, varies widely. Sediments, generally immature and dominated by arkose and quartzite, are influenced by the rapid uplift of the granite–gneiss dominant provenance. At places, however, pelites, mature massive quartzites, BIF and carbonates may be important constituents of the succession. Red beds first appeared 2.4 Ga ago, in both the QPC and BVAC. Modern BVAC associations are continental rift–fills. This could be a modern analogue of the Proterozoic situation. There were large rift systems in the Proterozoic.

Proterozoic greenstone belts are not much different from their Archean predecessors. Only these greenstone belts are smaller, contain less komatiite and the komatiite is less magnesian. The felsic volcanic rocks, of course, are richer in potassium and chert + BIF are less common.

[1] The acronym QPC these days is used in geology to represent two rock associations. For a relatively longer time it is being used to mean 'Quartz pebble conglomerate' association, as in Witwatersrand. Condie later adopted the same acronym to represent a major sedimentary rock association in the Proterozoic, as is used here.

Mafic dyke swarms became more common in the Proterozoic and the anorogenic granite–anorthosite suites, which bear geochemical signature of lower crustal origin, were not very uncommon either. Layered igneous complexes became more numerous and larger, compared to those in the Archean. Metamorphism varied from greenschist to granulite facies.

Proterozoic Eon, particularly its first half, witnessed the emplacement of many ore deposits or their protoliths. In fact, the overall trend in metallogeny was similar to that set by the Late Archean metallogenic epoch, with a few exceptions, of course. BIF deposition intensified during the period 2500–1800 Ma, mafic–ultramafic associated. Ni–Cu (+PGE) deposits continued to develop at suitable places. Some of these are Thompson Bay and Cape Smith deposits, (Canada), associated with komatiite, Pechanga–Monchegorsk (Russia) and Kabangu deposits (South Africa) associated with picrite. Origin of the Bushveld Complex (1900 Ma) is still debated. A lead view (Arndt et al., 2005) is that the original S–poor komatiite magma initially reacted with a large volume of crustal material precipitating olivine ± orthopyroxene. Only when did the Bushveld magma assimilate sedimentary rocks containing sulphur, did the PGE–rich sulphides precipitate due to S-saturation. The Sudbury deposit (1850 Ma) is believed to have a unique origin. It is an 'astroblem', where the ore filled structures, the igneous rock association and the ores were related to meteorite impacting around 1850 Ma ago (Dietz, 1964; Golightly, 1994; Naldrett, 2004). The Voisey's Bay deposit at Labrador, Canada, is associated with a troctolitic intrusion, such an association not being known earlier.

The ~1600 Ma hydrothermal Fe–oxide–Cu–Au (–REE) mineralisation (IOCG ore–type) within a granite–rhyolite complex at Olympic Dam, Roxby Down, South Australia, is another unique case of metallogenesis, that still awaits a clearer explanation.

Another interesting find in the metallogenic history of the Proterozoic is the unconformity related U (± Ni ± Au) mineralisation. Originally discovered at the Athabasca basin, Canada, and the Alligator river area in Australia, this type of mineralisation is now reported from other parts of the world.

The Proterozoic is not known for having many rich and large VMS deposits. Some deposits worth the mention are Mattagami and Matabi Cu–Zn deposits developed in the Canadian Shield during Paleoproterozoic. Some younger deposits developed at Flin Flon (Canada), Jerome and Bruce (USA) and Boliden and Skellefte (Sweden) during the period, 2000–1500 Ma. It may be pointed out that the period 1500–750 Ma is deficient in volcanic rocks and that reflects on the overall deficiency of VMS deposits during this period.

In contrast, there are many sediment-hosted base metal (Cu, Zn, and Pb) deposits in the Proterozoic without a conspicuous preference for time. Some of these are the 2000 Ma old Udokan Cu–deposit in Russia, 1700 Ma Broken Hill Pb–Zn deposit in Australia, 1650 Ma Zn–Pb ± Cu deposit at Mc Arthur, Australia, 1650 Ma Cu–Zn–Pb deposit at Mt Isa, Australia, 1430 Ma Pb–Zn deposit at Sullivan, Canada, 1320–840 Ma Cu–deposit at Mufulira, Zambia, 1050 Ma Cu deposit at White Pine, USA and the Neoproterozoic–Eocambrian Pb deposit at Laisvall, Sweden. Their host rock composition is not uniform but is broadly covered by the QPC type of Condie (1989) discussed earlier (see Gustafson and William, 1981, for details).

Sedimentary phosphate deposits were not common before 800 Ma. However, some important deposits of Paleoproterozoic (1900 Ma) do occur at some places, such is Rum Jungle and Broken Hill, (Australia) and Animikie–Gunflint region in Minnesota, USA (Cook and Mc Elhinny, 1979).

An Outline of Phanerozoic Geology and the Characteristic Mineralisations

The Phanerozoic geologic history is recorded in a number of well-known orogenic systems that include the Appalachian–Caledonian in eastern North America and West Europe, Hercynian and Uralian in parts of Europe, the Cordilleran system along the western margin of the Americas, the Alpo–Himalayan orogenic system in Europe–Asia. As pointed out by Condie (1989), the Phanerozoic orogenic belts have developed at convergent plate boundaries, as a result of complex plate interactions, involving collision. Expectedly, their records are generally better preserved and hence easier to interpret. The details of these systems are discussed in many texts and are unnecessary to repeat here. It may, however, be pointed out at this stage that this eon, however small, had a fair share of the global metallogeny. Particularly, the Mesozoic–Cenozoic eras within it had been most productive after the Late Archean.

The Phanerozoic Eon started with three supercontinents, namely the Eastern Gondwana, Western Gondwana and Laurentia (obviously synonymous with variously spelt Lauresia, Laurasia, Laurensia or Lauretia). The two Gondwanas first amalgamated into Gondwana at ~500 Ma and then this enlarged Gondwana and Laurentia united sometimes during 400–300 Ma to form Pangea, the largest supercontinent in Earth–history in today's estimation. The breaking and making associated with this supercontinent played their due roles in metallogeny.

During the Paleozoic, Caledonian–Appalachian orogenies that took place as an effect of the collision of Avalonia–Baltica, marking the initiation of Pangea, a number of large VMS deposits developed at many places, particularly in North America and Scandinavia. The Mississippi Valley type (MVT) Pb–Zn deposits of the southern USA are also considered related to the Appalachian orogeny. The collision of the Gondwana continent with the amalgamated Laurentia–Baltica–Avalonia is known as the Variscan orogeny. The Navan-type Irish Pb–Zn deposits, Rammelsberg and Meggen Pb–Zn deposits of Germany and the VMS deposits of the Iberian pyrite belt are related to the Variscan orogeny. Porphyry Cu (–Mo) deposits, repository of > 50 per cent of all types of Cu–ores in the world, occur mainly along the Mesozoic–Cenozoic convergent boundaries. Today they are dotted along Pacific sea–board. Phanerozoic is also known for Au-mineralisation. The super-large Mother Lode deposit of California is comparable to any large late Archean lode gold deposit in geology and ore reserve. However, such deposits are not many. There are, of course, Carlin type hydrothermal deposits in sediments. Epithermal Au (–Ag) deposits of 'World Class' occur along the circum–Pacific rejoin (Hishikari in Japan, Baguio and Lepanto in Phillipines etc). Mineralisation of Sn–W–U associated with S-type granitoids occurs in Central and Western Europe, including Cornwall. Sn–W–Ag–Bi mineralisations are related to Mesozoic granitoids of Bolivia and southern Peru and some places in Southeast Asia. The Central Tin Belt of Southeast Asia is associated with Upper Triassic granites. 'Sandstone type' uranium mineralisation, containing about 40 % of Western World's uranium resources, developed during the Phanerozoic. At Insizwa, South Africa, a small Phanerozoic Ni–Cu–sulphide deposit associated with mafic–ultramafic intrusives was known for long. But the relatively recently discovered Cu–Ni–sulphide deposit at Norilsk–Talnakh, Siberia (Russia) is uncommon in its geology and ore reserve. Here the ores are found at the base of 248 ± 4 Ma (Campbell et al., 1992) differentiated sills related to Siberian continental flood basalts. Assimilation by this magma of the underlying thick beds of anhydrite is believed to have saturated the magma with the metal sulphides (Naldrett et al., 1996; Lightfoot and Keays, 2005).

Supercontinents, Mantle Plumes – Plate Tectonics and Metallogeny

Rogers (1996) reports that the concept of assembly and dispersal of continental blocks, presumably throughout the geological time received a major stimulus in 1991, when a few regional geologists (Hoffman,1991; Dalziel,1991; Moores, 1991) separately proposed the existence of a supercontinent ~1 Ga ago. It was named Rodinia (apparently a corrupt rendition of the Russian word 'rodina', meaning motherland, or 'rodnia', meaning relative, or someone blood–related). The authors believed it to contain all the continental crust in existence at that time. Gradually there were many adherents of the core concept. At a fist glance it appears to be a modern day development of the earlier concept of 'Continental Drift'. It was, however, Rogers (op. cit.) who first endeavoured to outline a history of continents during the past three billion years. Other models followed.

Ur The oldest supercontinent in Rogers' scheme, receives its name from the ancient Sumerian city Ur that may be the World's oldest. This supercontinent contained the Kaapvaal, Western Dharwar, Bhandara (Bastar), Singhbhum and the Pilbara cratons. Figure 8.1 shows accretion to Ur along various marginal belts at ages upto 1.5 Ga. The areas included later on are the Zimbabwe craton (added to the Kaapvaal craton along the Limpopo belt), the Aravalli, Bundelkhand and a part of the Chhotanagpur area – all in North India and joined to South India along the Satpura (CITZ) and Singhbhum shear zones. The Yilgarn craton joined to the Pilbara craton along the Capricorn orogen and accretion of the Archean/Early Proterozoic Kimberly craton, Gawler craton and much of Eastern Australia took place.

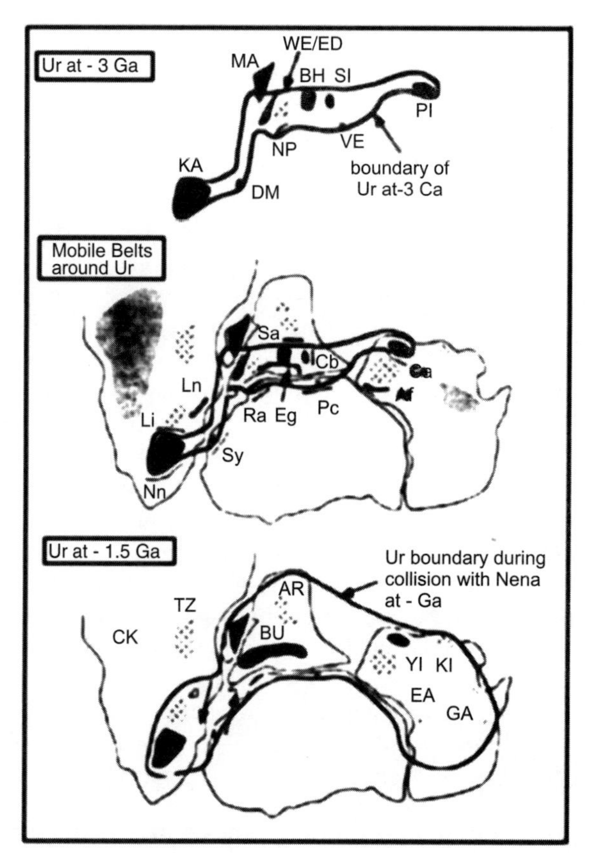

Fig. 8.1 The growth of Supercontinent Ur from ~3.0 Ga to ~1.5 Ga (from Rogers, 1996). Ur at ~ 3.0 Ga: KA –Kaapvaal, WD/ED – Western and Eastern Dharwar, BH – Bhandara, SI – Singhbhum, PL – Pilbara, DM – Western Dronning Maud Land, NP – Nappier, VE – Vestfold Hills. Mobile belts around Ur – Northern side: Li – Limpopo belt, Ln – Lurio and Namama belts, Sa – Satpura, Cb – Copper belt of India, Ca – Capricorn; Southern side: Nn –Namaqua – Natal, Sv – Sverdrupfiella, Eg – Eastern Ghats, Pc – Prince Charles and Vestfold, Af – Albany–Fraser belt in the south of Yilgarn craton, BU – Bundelkhand craton, YI – Yilgarn, KI – Kimberley, GA – Gawler, EA –early Proterozoic area in Eastern Australia. Arrows: Possible sinistral offset during consolidation of Gondwana.

Ur continued as a continent until about 1000 Ma, when it is thought to have given up its integrity to join other continents such as Nena and Atlantica to form Rodinia at ~1000 Ma, covering most continental blocks of the time, according to the early workers (see Rogers, 1996) (Fig. 8.2).

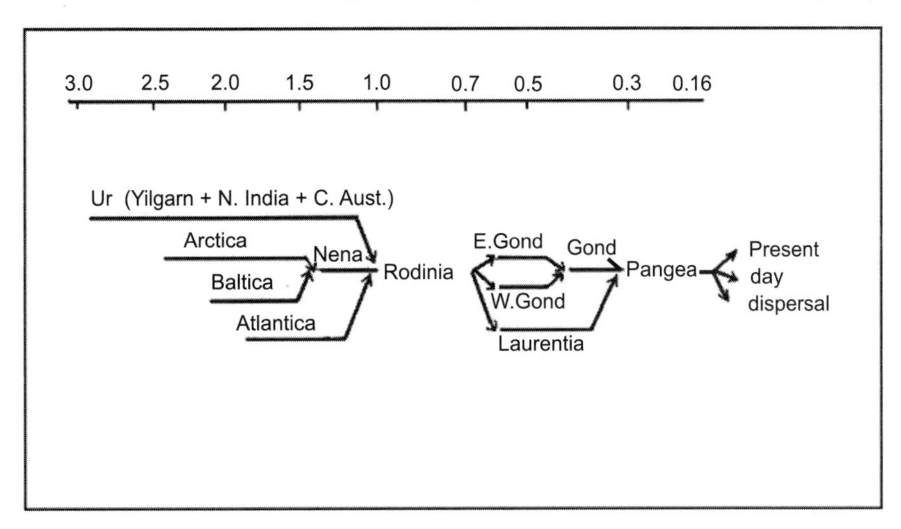

Fig. 8.2 Make up and break up of supercontinents (after Rogers, 1996)

Later Rogers and Santosh (2004) conceded that even if Ur did not constitute a coherent continent as old as 3.0 Ga, it might have formed the core of a continental block that became coherent before 1.8 Ga.

Columbia The first coherent supercontinent in Earth's history, according to Rogers and Santosh (2002), is a Paleoproterozoic assembly, named Columbia by them. These authors in a recent publication (Rogers and Santosh, 2009) suggested amalgamation into Columbia between 1.85 and 1.90Ga. The world-wide distribution of orogenic belts during the time may be significant in this context (cf. Condie, 2002; Zhao et al., 2002). Hou et al. (2008), reconstructing the Columbia supercontinent, based on a 1.85–1.75 Ga giant radiating dyke–swarm and Large igneous provinces (LIPs), suggested that the North China craton, Indian craton and Laurentia were united within Columbia before its extension and break up during 1.3–1.2 Ga due to plume activity.

Rodinia Few people doubt today that a large supercontinent, now called Rodinia, existed during the Neoproterozoic. But there are differences of opinion on details. A UNESCO–IGCP Project (No. 440/ 1999–2004) was, therefore, constituted to investigate the set up, amalgamation, configuration and break up of Rodinia and to construct a geodynamic map of Rodinia supercontinent. The results of investigations are recently published (Li et al., 2008).The conclusions made in the report are synopsis of the majority view based on the results of studies of geological correlations, paleomagnetism, orogenic histories, sedimentary provenances, developments of continental rifts and passive margins and mantle plume events. It is concluded that at 1100 Ma Laurentia, Siberia, North China, Cathaysia (part of the present day south China) and perhaps Rio de la Plata, were already together and the Yangtze had began its oblique collision with Laurentia. It was only by ca 900 Ma that all major continental blocks had aggregated to form the Rodinia Supercontinent (Fig. 8.3 a). During 990–900 Ma, high grade metamorphic event along the Eastern Ghat belt corresponded the Rayner Province in East Antarctica, incorporating both as parts of Rodinia. Rodinia dispersal became tangible by 700 Ma. Santosh et al., 2009 also included India in their reconstruction of Rodinea (Fig. 8.3b). These views are contrastable with one that says that possibly South African, Indian and South American cratons were not parts of Rodinia, *sensu stricto* (Kroner and Cordani, 2003; Piserevsky et al., 2003; cf Condie, 2005) (Fig. 8.4).

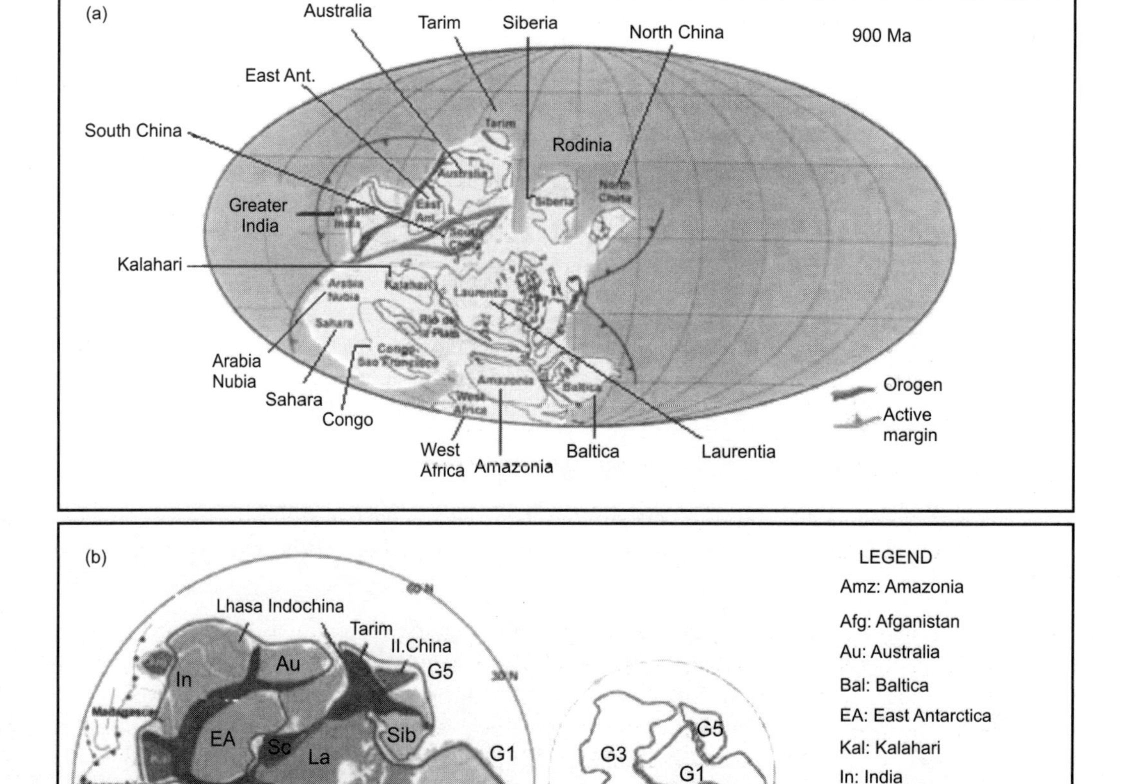

Fig. 8.3 Reconstruction of Rodinia: (a) modified after Li et al., 2008, (b) modified after Santosh et al., 2009

Gondwanaland Formation of the Gondwanaland (600–530 Ma) is another story of making and breaking of Supercontinents and continents. Hoffman (1991) was the first to suggest that the break–up of the Rodinia Supercontinent involved fragmentation around Laurasia (Laurentia) and clustering of some fragments into the Gondwanaland. By 550 Ma India had moved closer to its Gondwana position along the western margin of Australia (Fig. 8.5). Gondwanaland finally united around 540–530 Ma by the closure of the Mozambique ocean, causing Malagasy orogeny in the East African orogen and the final docking of India to the Australia – East Antarctica along the Pinjara orogen. Paleomagnetic data support the formation of Gondwanaland by ca 530 Ma. Gondwana was a southern hemispheric supercontinent, consisting of South America, Africa, Madagascar, India, Antarctica and Australia.

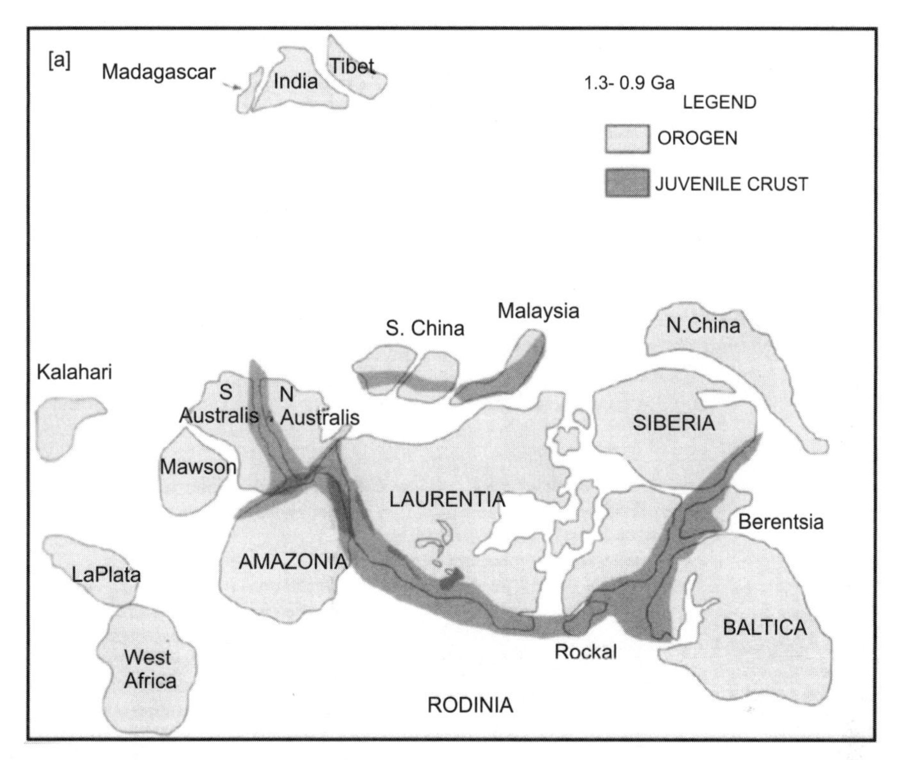

Fig. 8.4 Rodinia, a supercontinent formed between 1.3 and 0.9 Ga and fragmented into Gondwana and Laurentia at about 0.8–0.7 Ga (Condie, 2005)

There were three main episodes within the initial rifting stage, starting about 180 Ma ago (Hawkesworth et al., 1999). This led to the establishment of a seaway between South America and Africa in the west and Antarctica, Australia, India and New Zealand in the east and to the sea floor spreading in Somali, Mozambique sea basins. The second stage was in early Cretaceous (about 130 Ma) when South America separated from Africa and the Africa–India plate separated from Antarctica. Finally, at the end of Lower Cretaceous (90–100 Ma) the breakup of Gondwana was completed when Australia and New Zealand separated from Antarctica and other small continental blocks (Madagascar and Seychelles) separated from India, as the latter moved northward from Africa and Antarctica. In many cases these rifting events were accompanied by the development of Large igneous provinces (LIPs), such as the Deccan basalts during the Seychelles–India rifting and the Kerguelen–Rajmahal basalts during the Antarctica–India–Australia rifting.

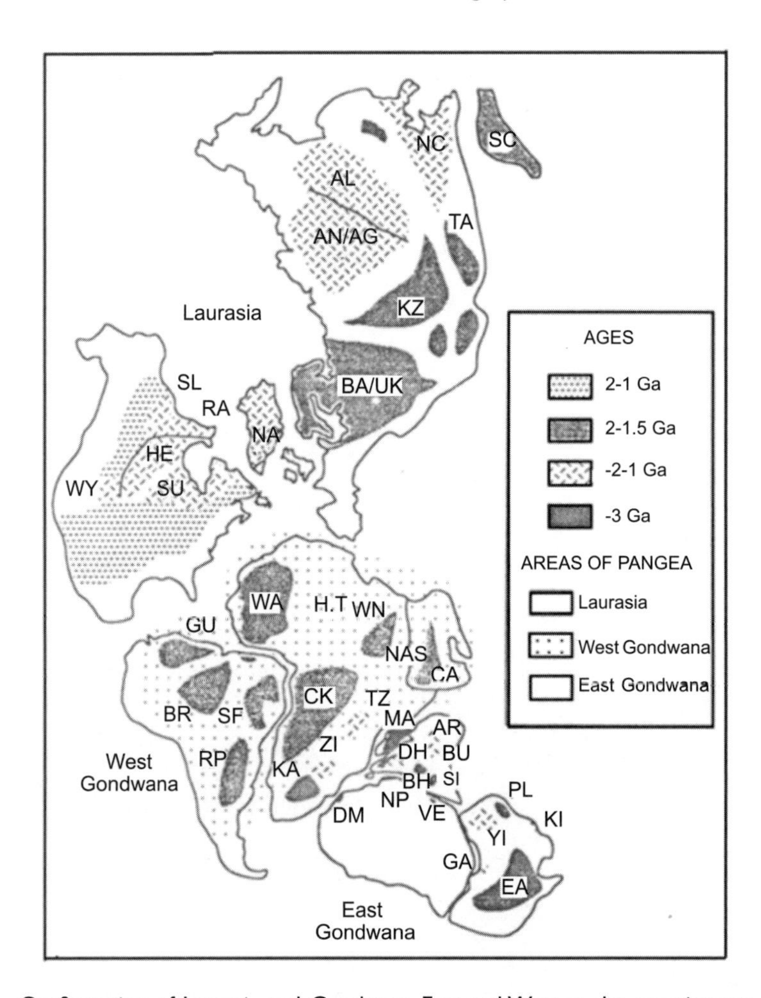

Fig. 8.5 Configuration of Laurasia and Gondwana East and West as the constituents of Pangea as conceived by Rogers (1996); AL – Aldan, AR – Aravalli, AN/AG – Anbar/Angara, BA/UK – Baltic/Ukraine, BH –Bhandara(Bastar), BR – Brazil(Guapore), BU – Bundelkhand, CA – Central Arabia, CK – Congo/Kasai, DH – Dharwar (West and East), DM – Western Dronning Maud Land, EA – an assembly of early Proterozoic areas in Eastern Australia, GA – Gawler, GU – Guyana, HE Hearne, H–T – Hogger and Tibesti, KA – Kaapvaal, KI –Kimberley, KZ – Kazakhstan, MA – Madagascar, NA – North Atlantic, NC – North China, NP – Napier, PI – Pilbara, RA – Rac, RP – Rio de la Plata, SC – South China(Yangtze), SF – Sao Fransisco (including Salvador), SI –Singhbhum, SL – Slave, SU – Superior, TA – Tarim, TZ – Tanzania, VE – Vestfold, WA – West Africa, WN – West Nile, YI – Yilgarn, ZI – Zimbabway.

The Gondwana supercontinent maintained its separate entity from the early Cambrian until its merger with Pangea in Mid.Carboniferous. It is after this event that the Gondwana facies sediments started precipitating with uneven depositional history. The Gondwana facies sediments, assigned a Supergroup status, are distributed along five discrete belts in peninsular India, such as: (i) Damodar belt, (ii) Satpura–Son belt, (iii) Mahanadi belt, (iv) Godavari belt, and (v) Rajmahal– Birbhum belt. In time they are distributed from Upper Carboniferous to the Lower Cretaceous. From the lowermost, i.e. Talchir Formation up to the Panchet Formation (Lower Triassic), the sequence is included under Lower Gondwana and the rest under the Upper Gondwana. The former contains > 99 per cent of India's bituminous coal reserve of ~250 billion tones and 564 billion cubic meters of coal-bed methane (Datta, 2009; Acharyya, 2000). The Gondwana basins under discussion are filled mainly with coarse to fine clastic sediments of continental origin. Only the early Gondwana basins had marine connections, i.e. at Umaria, Daltonganj, and its terminal stage is marked by the eruption of continental basalts as at Rajmahal in the present east and Panjal in the west. Besides the intra–continental belts mentioned above, the Gondwana facies rocks occur at a number of extra-peninsular areas along the east coast, Kutch in the west coast, Panjal in Jammu–Kashmir, Singrimari in Assam and Kemeng–Siang in Arunachal. The Gondwana basins are rifts (grabens and half-grabens) that developed unevenly within the stipulated time range. But how did these develop? Their development-ages scarcely correlate with the disintegration-age of the Gondwana Supercontinent. It is not that the development of the Gondwana basins under reference has not been Speculated upon. Tiwari and Casshyap (1994) state, "It is believed that the Early Permian reactivation of ancient Proterozoic lineaments initiated sedimentation and subsidence in the down–faulted linear troughs known as the Gondwana basins." This is possible but has to be established. Further, what caused the Early Permian reactivation?

Veevers (1993) looks into it somewhat deeply. He divides the tectonic history of the Gondwana rift basins into three stages:

1. 320–286 Ma – Marked by the non–development of structures and depositional lacunae as the pre-existing platform blocked the escape of heat from the interior.
2. 286–230 Ma – Period marked by the first Pangea-wide rifting that represents fast loss of subterranean heat along rifts.
3. 230–160 Ma – Represents the period of Pangea-wide rifting that represents fast loss of subterranean heat along rifts. Acharyya (2000) views the development of the Permian coal basins of Lower Gondwana as the failed rift-related intracratonic basins. The last two views are reconcilable.

Pangea Pangea was the latest supercontinent to form during 450–320 Ma. This new supercontinent is claimed to have covered most of the existing continents. Gondwana that started forming in Neoproterozoic, ultimately united with Pangea at ~300 Ma, sharing its history with the latter, as already briefly mentioned, until it disintegrated at ~160 Ma, leading to the present day configuration of the continents and oceans. The countries remaining together as Eastern Gondwana within Pangea instead of being randomly arranged, made Rogers (1993) suspect that the cluster is inherited from Ur. Hence the final disintegration of Ur he temporally drags beyond Pangea.

Other Suggestions Even amongst the believers of the concept of supercontinents, who today make a large number, there are variations in interpretations. Bleeker (2003) suggested that there were even three supercontinents, instead of one, during the late Archean. These are:

1. The supercontinent that contained the Slave province, Dharwar craton, Zimbabwe craton and the Wyoming craton, stabilised at ~2600 Ma and remained so, until it disintegrated during 2200–2000 Ma.
2. The supercontinent that contained Superior, Rae, Kola, Hearna and Volga.
3. Kaapvaal–Pilbara Supercontinent.

Aspler and Chiarenzelli (1998) suggested two supercontinents near the end of the Archean. These are: (i) Kenorland, containing Archean blocks in North America, the Baltics and Siberia; (ii) Southern continental block comprising present–day Zimbabwe, Kaapvaal and Pilbara cratons, São Francisco craton and some blocks in India.

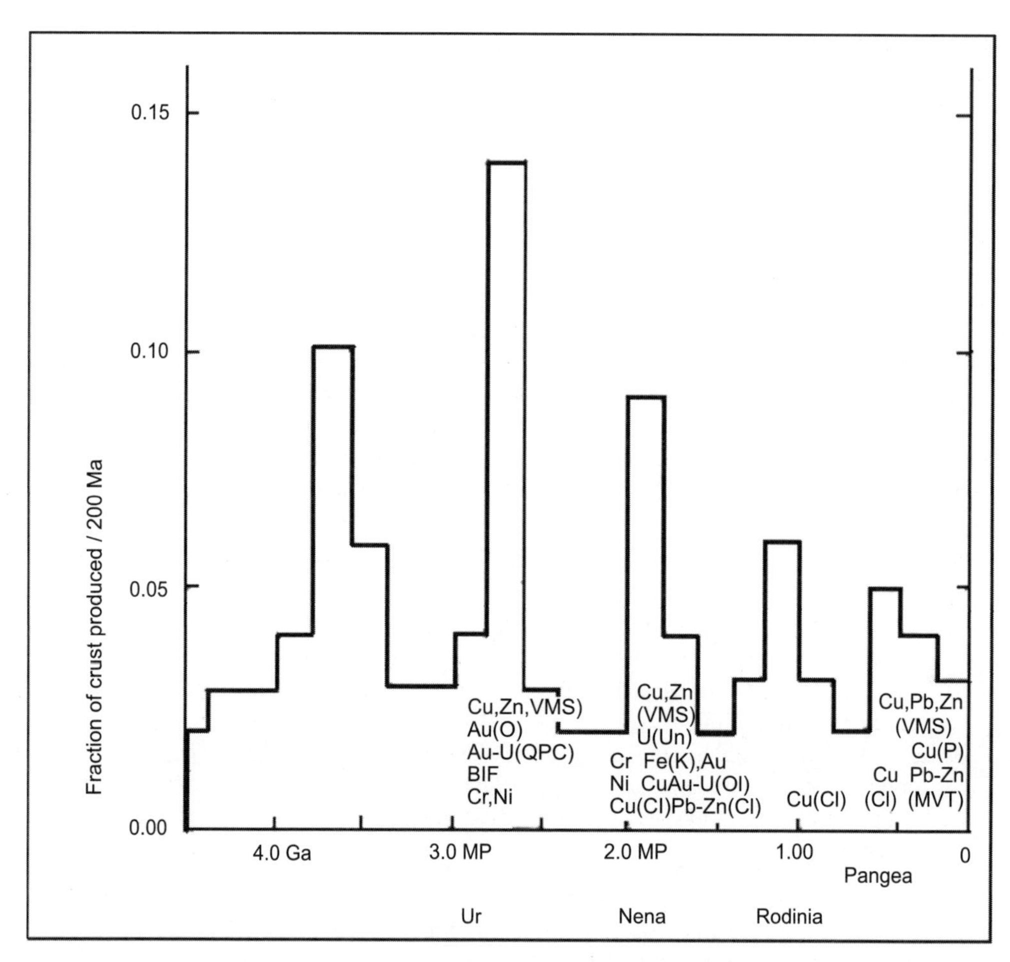

Fig. 8.6 Crustal growth per 200 Ma (Mc Culloch and Bennett, 1994) and the periods of supercontinent assembly and break-up and mega-plume (MP) activities (after Rogers, 1996) vis-a-vis metallogenic peaks in different periods of Earth's history: VMS – Volcanogenic massive sulphide, O – Orogenic, QPC – Quartz–pebble conglomerate, Un – Unconformity, K – Kiruna, Ol – Olympic Dam, P – porphyry, Cl – Clastic, MVT – Mississippi Valley Type

In both the supercontinents glacigenic deposits formed during 2400–2200 Ma. This is consistent with global cooling caused by CO_2-drawdown through intense weathering of emergent landmasses (buoyant over a mantle superplume?). Zegers et al. (1998) proposed yet another older supercontinent, they called Vaalbara. It possibly assembled at ~3600 Ma and fragmented at ~2100 Ma or more probably ~2700 Ma. There could as well be other supercontinents. While the concept seems valid, the parameters utilised, or more particularly their interpretations are always not well constrained and hence the variation in opinions.

Thermal Plumes Thermal plumes are believed to generate at the core–mantle boundary due to thermal disturbance. Another thermal boundary and hence plume depth may be at ~660 km. Plumes may even be generated at the chemical boundaries within the mantle. If a plume is less viscous compared to its surrounding mantle material, then the plume–head is roughly spherical, followed by a narrow tail. Head of a plume coming from the lowermost mantle may be several hundred to a thousand kilometres in diameter. Initially, plume material buoyantly uplifts the lithosphere and cause hotspot swell and ocean floor ridges. Distributions of komatiite, flood basalts, oceanic plateaus and layered mafic intrusions have been utilised to identity mantle plume events in the Precambrians (Isley and Abbott 1999, 2002). Based on this evidence, the mantle plume activity is thought to be episodic and Ernst and Buchan (2003) opined that there was no plume-free time gap greater than 200 Ma. Some major plume-events reportedly took place during 2750–2700, 2450–2400, 1800–1750, 1050 and 100–80 Ma (Isley and Abbott, 2002). Of these, as Condie has pointed out, the 2750–2700 and 1800–1750 Ma megaplume events correspond with the development of a considerable portion of the Earth's crust.

Metallogeny Late Archean (2770–2590 Ma) granitoid–greenstone belts are generally mineralised with Fe and Au and at places, also with Ni and Cu–Zn. The greenstone lavas mainly comprise basalts and komatiites. Chemistry of these lavas suggests that they were derived from deep mantle and brought to the surface by mantle plumes (Campbell and Hill, 1988; Xie et al., 1993). However, the development of rich Ni–Cu–sulphide ores in these rocks generally involved assimilation of sulphur–rich country rocks on the way.

Plumes reaching up to ocean bottom let out CO_2 to the ocean-atmosphere system. Increased CO_2 flux warms up the climate and increases chemical weathering and may cause extensive carbonate precipitation on extended platforms, made available through transgression. With the availability of metals these can be sites of Pb–Zn mineralisation. A plume breakout has been held responsible for the deposition of sulphidic black shale on a global scale, because of the anoxic atmosphere created. Arc extensions commonly lead to marginal basins, a favoured site for VMS deposition. Deposition of BIF type iron formations is favoured during marine transgressions. These transgressions take place during the emplacement of a large plume and the Fe and Si are expected to have been provided by episodic hydrothermal influxes related to the mantle–plume events (Barley et al., 1998; Isley and Abbott, 1999). A Paleoproterozoic (1900–1800 Ma) event of sedimentary phosphate deposition at different parts of the world has been related to contemporaneous mantle–plume event (Cook and McElhinny, 1970). Increased spreading rates, accompanied by terrain accretion, may even involve ridge (and oceanic plateau, if present) subduction. Such areas are often marked by lode gold deposition. This unusual type of subduction and the associated gold mineralisations are 'chaotic response' to plume breakout (Barley et al., 1998).

During a supercontinent making, the interface of the accreting blocks will be characterised by subduction and marginal basin development and collision accompanied by appropriate ore mineralisation. Its breaking will be initiated by rifting (Wilson Cycle) and the corresponding mineralisations. Repeated episodes of arc, marginal basin, plume magmatism and subduction in the convergent margins, wherever possible, will increase the ore potential in the region. Long term enrichment of the convergent margin lithosphere is believed to ultimately culminate in bonanza metallogenic events. Such a Late Archean tectonic setting is believed to have been at least obtained at the margins of a super–mature Archean ocean (Barley et al., 1998). Metallogenic peaks in Earth's history in the context of crustal growth, supercontinent assembly (– break up), megaplume activities, are outlined in a sketch–diagram (Fig. 8.6) on the previous page.

8.2 A Synoptic View of the Indian Situation and Its Comparison with the World

Barring some terrains covered by Phanerozoic rocks, such as the Upper Cretaceous Deccan Trap flood basalts of Central and Western India, terrigenous sediments of the Gondwana basins in Northern and Central India, sediments of a Jurassic incursion in Kutch and Cretaceous marine encroachments along the Tamilnadu coast, and of course the alluvial fill of the Indus–Ganga–Brahmaputra basin, the rest of peninsular India comprises rocks of different ages that spread from the Early to the Late Precambrian.

Within the Precambrian terrain of peninsular India occur a few pockets or 'nuclei' of very old rocks (≥ 3.0 Ga). These are: Western Dharwar (WDB) in present day South India, Bastar (Bhandara) (BC) in Central India, Singhbhum–North Orissa (SOC) in Eastern India, Bundelkhand (BDC) at the northern margin of Central India with Indo–Gangetic valley and Mewar and the adjacent areas (BGC) in Rajasthan. Not all rocks in these pockets are as old as > 3.0 Ga, however. They are commonly accompanied by rocks/rock formations that were emplaced during the period, <3.0 > 2.5 Ga. The only exception is the Shillong–Mikir Proterozoic massif of North Eastern India where no Archean rocks have so far been located.

Gneisses and granitoids dominate the older rocks. The geologically younger rocks have a large share of gneisses and granites, but variously deformed and metamorphosed volcanic (dominantly mafic–ultramafic) and sedimentary rocks in greenstone sequences are also important members.

The oldest rocks obtained from the Dharwar gneisses is 3.4 Ga. The maximum age obtained from detrital zircons is 3.6 Ga. Meen et al. (1992) would even extend the age of the crustal protolith of a rock they studied to beyond 3.8 Ga. Major episodes of crustal growth recorded in Western Dharwar block (WDB in Fig. 8.7) are estimated as 3.4 Ga, 3.0–2.9 Ga and 2.6 Ga. As in several other Early Precambrian terrains, there are two sets of greenstone belts in Western Dharwar. The older ones (3.3–3.0 Ga) called the Sargur type are smaller. 3352 ± 110 Ma (Sm–Nd) komatiites from Sargur showed some interesting features in a recent study (Jayananda, et al., 2008). The plume–generated komatiites were found coexisting with contemporaneous mafic–felsic volcanics and TTG accretion, suggesting a mixed plume–arc setting. The Nd-isotope composition of the studied komatiites, indicate depleted mantle reservoirs, which following Boyet and Carlson (2005), is conjectured to have evolved by early (> 4.53 Ga) global differentiation of the molten silicate Earth or extraction of continental crust during Early Archean.
The younger ones (2.7–2.6 Ga), called the Dharwar type, are larger and may be occupied by a volcanisedimentary sequence as thick as 8 km. Not all rocks of Dharwar share the same characters or evolutionary history. The Eastern Dharwar block (EDB in Fig. 8.7) is apparently separated from the Western Dharwar block by a brittle–ductile N–S trending shear zone passing along the eastern margin of the Chitradurga–Gadag belt. While the gneiss–granitoid rocks in Western Dharwar block are principally TTG in composition, those of Eastern Dharwar in general have a calc–alkaline trend. Age of these Eastern Dharwar rocks is restricted to the time-slot of 2.6–2.55 Ga. The greenstone belts of this region have been dated 2.75–2.6 Ga. The relationship of the Western Dharwar block with that of the Eastern is not clear. Chadwick et al. (1997, 2000) suggested that the Western Dharwar was the foreland to an accretionary arc in the east, represented by a late Archean batholith (their 'Dharwar batholith') with contained schist (greenstone) belts. Subducting the batholith is an uneasy proposition and the following two observations need to be explained: (i) transitional nature of the western part of eastern Dharwar, (ii) greenstone belts of same age in both the arc and the foreland. A simple proffered explanation could be as follows: when the ocean-crustal part of a plate subducts below a continental crust, the volcano–plutonic arc produced is normally located within the continent–margin. In the context of it Chadwick's model can be broadly accepted to explain the transitional nature of the western part of the Eastern Dharwar and a part of the calc–alkaline gneisses and granitoids could have been produced from anatectic melts from the basement TTGs. The isochronous development of greenstone belts in both the Western and Eastern Dharwar may be explained by the development of rift basins in the arc region, as well as in the cratonic margin due to extension caused by the rollback of the subducting block. Slab-melting by interacting with the

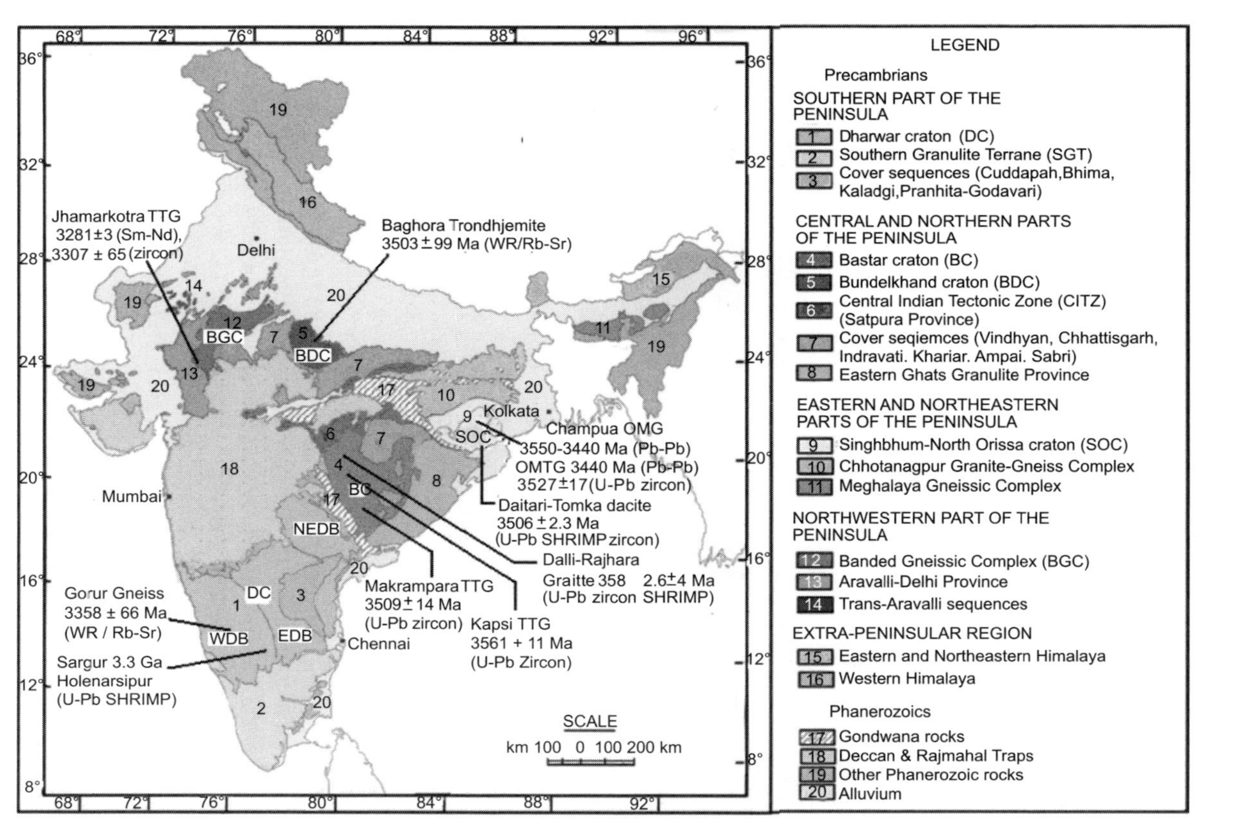

Fig. 8.7 Map showing geological provinces in India (Modified after GSI's map in Manual of Geology of India, Vol. I, Part 1, 2006) with the oldest dates from each craton: DC – Dharwar craton (WDB – Western Dharwar block, EDB – Eastern Dharwar block, NEDB – North Eastern Dharwar block), BC – Bastar craton, SOC – Singhbhum–Orissa craton, BDC – Bundelkhand craton, BGC – Banded Gneissic Complex

overlying lithosphere, might have produced different rocks now present. But the problem remains. How do we reconcile the individual greenstone belts explained by subduction tectonics, such as Kolar belt (Krogstadt et al., 1989), Sandur belt (Manikyamba et al., 1997), Ramagiri–Penkacherla belt (Manikyamba et al., 2004), etc., within the perimeter of the so-called Dharwar batholith?

Quite a number of granitoid massifs occupy a large area in the northeast Karnataka and Andhra Pradesh (NEDB in Fig. 8.7). The most important ones amongst them are Lepakshi, Kadiri, Tanakallu, Hampi, Hyderabad and Perur. In composition these 2.6–2.5 Ga granitoids range from granite, through quartz monzonite to granodiorite. However, the thermo–tectonic process involved in the generation of these granites needs be properly constrained.

In the Southern Granulite Terrain (SGT), particularly at the 'Northern block', 'Nilgiri block' and 'Madras block' an important event of crustal accretion and intense (granulite) metamorphism took place during 2.6–2.5 Ga. The above discussion shows that a large part of southern India was stabilised by 2.5 Ga. Let us see what happened at other places during this period.

At the Bastar craton (BC), the oldest member is trondhjemitic gneiss from Markampara (near Sukma) which has been dated ~3.5 Ga. A late granite from the neighbourhood is dated ~2.5 Ga, however. Recently, the basement TTG gneiss from Kotri belt and a granitoid from east of Dalli–Rajhara have been dated 3.56 Ga and 3.58 Ga respectively. The Sukma Group is overlain by a volcanisedimentary sequence with BIF (Bailadila Group). Indirect age dating suggest it to be Late Archean. Another Late Archean (2.5 Ga) accretion in this region is the emplacement of Malanjkhand Granite, located at the northern extension of the Dongargarh belt.

Trondhjemitic enclaves in the overall granite–gneiss massif of Bundelkhand (BDC) yielded Pb–Pb zircon ages of 3.5–3.0 Ga. The host granitoids range in age from ~ 2.5 to ~2.2 Ga.

In Eastern India, the Archean nucleus (SOC) consists of batholithic granitoids such as granodiorite, adamellite and older TTG and meta–sediments that range in age from ~3.6 to ~3.0 Ga. Besides, the contiguous Mayurbhanj Granite has recently been dated 3.0–2.8 Ga. Rather potassic Tamperkola Granite has been dated ~2.8–2.7 Ga. The BIF-bearing volcanisedimentary IOG rocks in the Gorumahisani–Badampahar belt in the east and Jamda–Koira–Noamundi belt in the west constitute greenstone belts of sorts. The eastern belt is older than the younger phase of Singhbhum Granite (SBG–A, > 3.1 Ga). The acid volcanics at the base of Malantoli Lava have recently been dated as 2.8 Ga. Interestingly, a dacitic lava from the southern IOG belt has recently been dated as 3.5 Ga, which is the oldest age determined for a greenstone member in India. That makes Singhbhum geology all the more interesting for further studies.

The oldest rocks in Western India, i.e. Rajasthan–Gujarat, the TTG gneisses with supracrustal enclaves, have been dated ~ 3.3 Ga. However, Vinogradov et al. (1964) dated the detrital zircon (Pb–isochron) from Aravalli sediments to be 3.5 Ga. There is no typical greenstone belt associated with these ancient gneissic rocks. Greenstone belts, *sensu stricto*, have not been established in this region. A couple of large granitoid intrusives, such as the Untala Granite (~2.6 Ga), Gingla Granite (~2.6 Ga), Berach Granite (2.5–2.45 Ga), Ahar River Granite (2.55 Ga) were emplaced during the Late Archean. Obviously the situation does not qualify the region for the development of major iron formations.

Summing up, there is yet no report of Eoarchean or Hadean rocks in India. But rocks whose dates of formation as determined by superior radiometric systematics are Early Archean, and are already there and recapitulated here. Continued investigations are expected to reveal in future more such or even older occurrences. In terms of Archean developments, India resembles in different measures many such regions of the world. However, the Archean history in India resembles more closely those in Pilbara craton of Australia and the Barberton craton of Africa.

A question may be asked about how the Indian protocontinents developed and also how the continents grew around them. Answers to these questions are not easy to find. An impressive concept at the moment is that the nuclei were initiated by the melting of the subducted wet oceanic crust, or the thickened roots of submarine plateaus. The TTG magma produced in the process (with or without reworking) produced the early cratonic nuclei in a vast oceanic crust. It is thought possible that the nuclei enlarged into continental blocks again through plate tectonics, the mechanism being the island arcs, submarine plateaus/oceanic ridges, colliding and accreting to the (proto-) continental margins (Condie, 1997, 2005). The authors are not aware of any India-specific discussion on this aspect. For the region in NE Karnataka and Andhra Pradesh where the extensive granitoid–gneiss terrain is composed of granite through quartz monzonite to granodiorite, a suggestion of island arc accretion may not be a unique solution. The rocks there could develop by the incompatible element influx caused by subduction-related processes involving a collisionally thickened crust (Condie, 1997).

The Proterozoic crustal developments in India are divisible into two broad types:

1. Mobile belts
2. Intracratonic basins.

Some of the better studied Proterozoic mobile belts in India comprise the North Singhbhum Mobile Belt (NSMB) girdling the Singhbhum–Orissa craton, the Delhi–Aravalli belt occurring along the western and southern limits of the BGC (–Bundelkhand) craton, the Chandenar–Tulsidongar (Bengpal) mobile belt, the Dongargarh–Kotri–Chilpi belt and the Sakoli belt of Central India, the 'Madurai block' or the 'Pandyan mobile belt' of Ramakrishnan (1994) in South India. The litho–tectonic developments in these belts have already been discussed in the context of available literature and the authors' personal experiences.

Proterozoic intra-cratonic basins in India, often referred to as the Purana basins, are quite a few in number. The most important and better studied amongst them are the Vindhyan basin, the Cuddapah basin and the Chhattisgarh basin. They developed, rather evolved, through hundreds of millions of years. They are poorly metamorphosed and deformed, except in case of the Cuddapah basin which has been highly tectonised along its eastern margin. The origin of these basins should be related to thermal perturbation in the mantle–crust system.

Coming to the picture of metallisation in the context of what we have so far discussed, we do not expect a mineralisation of the type and age of Isua in India. The earliest Archean deposits in India are reported from the Sargur type greenstone belts where a few ore deposits developed. Such mineralisations include the Cr–mineralisation at Byrapur in the Nuggihalli belt, Cu-mineralisation at Kalyadi and Alladahalli, V–Ti-bearing magnetite deposits in the Nuggihalli–Shimoga belts, stratiform barite at the Ghattihosahalli area and gold at a number of places, the most important of which was at Kempinkote in Nuggihalli belt. The greenstone-hosted small gold deposit at Kunderkocha in Eastern India is another such occurrence. The deposits are small, but many of them have been exploited in varying degrees. The Algoma type BIF deposition in the Badampahar–Gorumahisani belt (BG belt) in Singhbhum–North Orissa craton is a little younger. However, the bulk of the chromite mineralisations in India (90 per cent of India's total reserve of ~146 Mt) are located in Eastern India, particularly at Sukinda and Nuasahi. Their host rocks have been recently dated at 3.1 Ga (Auge et al., 2003). This age data relate them to the older greenstone belt of IOG–I. PGE are associated with Nuasahi chromite.

Late Archean has been a fertile epoch for ore mineralisation in India, as in many other countries sharing a Precambrian history. India's mineralisations are limited to iron (both Algoma and Superior types), orogenic (lode) gold, manganese, and an unusual deposit, the 2.5 Ga Malanjkhand Porphyry–style Cu deposit in the Balaghat district of Central India. No Archean VMS deposit other than a small Cu (–Zn) deposit (≤1 Mt.) from the Chitradurga belt (Ingaldhal), Karnataka, is known. The sub-economic (<1 Mt. with 0.4 per cent Cu, 3.5 per cent Zn, and 2 per cent Pb) deposit at Mamandur, in the Southern Granulite Province, Tamil Nadu, hosted in pelitic granulite and dated 2.6 Ga, possibly documents a fact that the Archean crust was not yet ready to form sediment–hosted economic deposits of Zn–Pb–Cu.

Bulk of the Indian iron formations (BIFs) is not as old as some people believed (Klein and Beukes, 1992). The Algoma type BIFs (35–40 per cent Fe) at and around Kudremukh with a huge reserve of nearly 10 billion tonnes were deposited at ~2.7 Ga and the Superior type (at Sandur) at ~2.6 Ga. The age of BIF deposition in the Bailadila belt of Central India is less well-constrained, but indirect evidence suggests that there also it must have taken place in the Late Archean. The Algoma type mineralisation in the Badampahar–Gorumahisani belt Eastern India is older (> 3.1 Ga). However, a recent report of zircon age (U–Pb) of 3.51 Ga from the felsic volcanic rock in the IOG occurrence at Daitari in the South is interesting. BIF lies stratigraphically below this rock. This confirms the contention that the BIF in the region is polychronous, extending over a considerable span of time. However, more data on this aspect is needed. In respect of environment of deposition, the greenstone associated Bababudan BIFs appear to have deposited from intra-basinal environment and the Superior type on continental shelves, in both the cases the intrabasinal magmatic hydrothermal activities played important roles. Not only the Indian deposits but also the Earth's thickest and most extensive BIFs were deposited during 2.7–2.45 Ga (Barley et al., 1998). It urges us to consider with open mind the oxygen-budget of the atmosphere through time. The remedy suggested by Casting (1993) in imagining 'oxygen oases' may not be enough. The ultimate development of hematitic Fe-ores from these iron formations (BIFs) has become a controversial issue once again. As we have discussed, in India, as in many other parts of the world, these ores formed through leaching of BIF by meteoric hydrothermal fluid not long after the deposition of the BIF, or through supergene leaching during the Cenozoic time, or a combination of the both. Syngenetic sedimentary–diagenetic concentration of GIF-related ores is noted at places. The possibility of local deposition of unusually thick BIF-related rich ore bands (beds) cannot also be ignored.

Manganese and iron have many chemical characters in common and may develop together in nature. However, Mn has a much higher solubility at low T–P. There are fewer and smaller Mn-ore deposits which themselves or whose protoliths formed during the Late Archean (2.9 to > 2.6 Ga). Of the five such deposits so far known in the world, three are located in India, viz, the Chitradurga–Tumkur and Sandur areas of Karnataka, Kodur in the Eastern Ghats and Singhbhum–Orissa (Jamda–Koira–Noamundi Valley) in Eastern India. The Jamda–Koira Mn–ore horizon is a part of the IOG–II in which they overlie the BIF unit. Precipitation of Mn alike that of Fe, is redox-controlled, i.e. the environment has to be oxic, but not necessary highly so. Roy (2006) felt that localised photosynthetic oxygen production in basin-margin euphotic zones (Casting, 1993) could play a vital role in initial Mn-precipitation in these Late Archean formations. It is not impossible that the metal (Mn) source for the Dharwar and Eastern Indian deposits was volcanogenic.

India's performance in the field of gold production is dismal. While it imports ~800 tonnes of gold per year (about a third of the world production), it produces a meagre 5–6 tonnes within the country. Again, the bulk of the latter is obtained by metallurgical treatment of the imported Cu–concentrates. India's most famous, gold deposit at Kolar produced ~800 tonnes of gold during a century of mining and is now abandoned. Ramagiri produced about 7 tonnes of the yellow metal before it last closed down. Hutti is still mining, with a record of production of > 70 tonnes in the past and a reserve of ~30 tonnes or more for the future. These and numerous other small deposits and occurrences of the metal in the Dharwar craton are located within the Late Archean greenstone belts. Serious efforts to discover workable gold deposits by the indigenous and foreign agencies for more than a decade, by and large remain fruitless, although the Dharwar craton prima facie appears to be a gold metallogen. In Eastern India there is a small gold deposit at Kunderkocha, in the Archean greenstone belt of Singhbhum, which is being exploited now.

Malanjkhand deposit (~2.5 Ga) is a unique case in the mineral inventory of the Late Archean (–Early Proterozoic) time. It has an ore reserve of 135 Mt with 1.32 per cent Cu and upto several hundred ppm of Mo and < 1 ppm Au. Discovery of this deposit has established that large granite-associated Cu-deposits may be discovered in Precambrian terrains, provided the deposit was saved from later erosion because of normal faulting involving the deposit. Porphyry Cu-deposits are generally located at the apical zone of shallow–level calc–alkaline granitoids and hence commonly eroded, if formed in the Precambrian.

Proper Proterozoic mineralisations in India include Mn-mineralisations in the Sausar Group in Central India and several other places, Cu–U mineralisation along the Singhbhum shear zone (SSZ), base metals (Zn, Pb and Cu) at several places in the Delhi–Aravalli belt, Gangpur Group and Cuddapah basin and some deposits along the Lesser Himalaya. Other mineralisations worth the mention in this context are the 'unconformity type' U–mineralisation and rare earth element (REE) and rare metal (RM) mineralisations at a number of places.

The bulk of Mn-mineralisation in India has taken place in the 200 km long Sausar belt which is the repository of 142 Mt of Mn-ores of all categories. Other locales are Goriajhor, Orissa (Gangpur Group), Jhabua, Madhya Pradesh (Aravalli Supergroup) and the Eastern Ghats. The present day ores may be metamorphosed sedimentary Mn-oxides, or lateritic ores formed by supergene alteration of rocks consisting of metamorphosed Mn-oxides, silicates and carbonates. Locally (Dongri Buzurg, Ukhuwa) in Sausar belt, hydrothermal remobilisation of Mn-oxides related to a syn–tectonic granitic activity is prominent. The Mn-oxide rich sediments were laid on shallow shelves, above the Mn^{4+}/Mn^{2+} redoxcline during the transgressive phase of the sea. This should hold good for all the above deposits (Dasgupta et al., 1992; Roy, 2006). Noteworthy, the Early Proterozoic (< 2.5–2.0 Ga) was the first major period of intense Mn-mineralisation in the world. This could be caused by perturbation in the mantle–crust system, promoting oceanic high stands and upwelling of the reduced Mn-bearing water on to the oxidised zone, the atmosphere and the upper zone of the seawater having already become oxidizing, though in a limited manner. Worth mentioning here are the results of experimental work of Anbar and Holland (1992) which established a high UV-dependence of manganese photo–reactions and its inhibition in the presence of iron. This may as well explain poor precipitation of manganese, compared to iron, in early Precambrian.

The mineralisation of copper, uranium and apatite–magnetite mineralisation along the Singhbhum shear zone have features that are not very common. Uncommon features of the copper mineralisation include (i) presence in the ores of such elements as Ni, Mo, Se, Te, Au and Ag in small but recoverable proportions, (ii) restriction of most of the deposits to basic volcanics (or their derivatives), (iii) prominent albitisation, (iv) deformation and metamorphism of ores, imparting parallelism of ore shoots with strong lineation in the host and (v) high-salinity of fluid inclusions. The U mineralisation, characterised by low-grade uraninite-disseminated deposition, has taken place at a number of places along the belt, while many rocks in the shear zone show much higher radiometric U-values compared to the background. All the working mines of uranium in the country are located along this belt. At Jaduguda, Mo–and Ni mineralisations took place in two later phases. Apatite–magnetite mineralisation along this belt has no parallel. The prolonged history of deformation, metamorphism and metasomatism in the belt has erased many primary features in the rocks and imposed newer ones. This has rendered the interpretation of the genesis of the ores more difficult. A visible tendency is to interpret the metallogeny in the belt in terms of the IOCG model. The model is interesting, but it is at the moment more descriptive than profoundly interpretative.

Unconformity proximal uranium mineralisation has taken place at a number of places in the Indian Peninsula, the most important amongst them being associated with the Papaghni group in the Cuddapah basin, the other located at Gogi–Ukinal in the Bhima basin. It is interesting to note that all these basins are set in 'fertile' granite background.

India has quite a few, but surely not many, sediment-hosted base metal deposits of Zn, Pb and Cu deposited during the Early–Middle Proterozoic. Majority of them, of course, are located in the Western India, particularly in the Rajasthan–Gujarat area. The Rajasthan deposits are strewn along three major belts: (i) the Khetri belt (Cu–dominant), (ii) the Bhilwara province (Zn >Pb > Cu) and (iii) the Zawar belt (Zn–Pb). Mineralisation took place within riftogenic basins or subbasins located within continents or at the continent margin (passive) and SEDEX in origin. It may be pointed out that the SEDEX deposits are not always the sedimentary equivalents of 'black smokers'. The ore material in the SEDEX deposits may be thrown up above the basinal floor along with the sediments when they form the syngenetic deposits. When such fluids come up along growth faults and advect along suitable horizons, then this mineralisation will be epigenetic and will have some similarity with MVT deposits, particularly if the host rock is carbonate. The ores under reference may be viewed in this context. Not only the bulk of the basin fills but that of the constituents were provided by the granitic crust already stabilised. However, it does not as easily explain the source of Cu for the Khetri belt as it does for the other base metals, i.e. Pb and Zn. In the south Delhi fold–belt Zn–Pb–Cu mineralisations have taken place in association with acid–basic volcanic rocks, including Ambaji, Deri and Birantia–Khurd. They are rather small reserve deposits, not amenable to largescale modern day mining. Most of the thermo–tectonic disturbances (basin formation and filling, followed by inversion, deformation, metamorphism and granitic intrusions) and mineralisations, including phosphate deposition at Jhamarkotra could be related to the 1800–1750 major plume event mentioned at an earlier section.

The Cuddapah basin has another distinction of hosting 95 per cent of country's barite resources, which translates into 25 per cent of the World's resource of this mineral. Here the barite mineralisation has taken place in two modes: (i) vein type, seen in most formations of the Cuddapah sequence and (ii) bedded type, restricted to Mangampeta in Rajampeta area. The type (ii) constitutes the main resource base. Two models may be invoked to explain bedded barite deposits: (i) hydrothermal or exhalative model, (ii) Biogenic model. Observations suggest that the two models can be compromised in the context of Mangampeta mineralisation. The thermo–tectonic perturbation, which is suspected to be at the root of the Cuddapah basin formation, might have played a direct/indirect role is sourcing the ore constituents.

Proterozoic gold in India is recorded in the Sakoli, Sonakhan and Mahakoshal Group rocks of Central India, in mainly the North Singhbhum Mobile Belt (NSMB) of Eastern India and in the Lower Aravalli sediments of southern Rajasthan in Western India. Besides, there are sporadic Au–U incidences in the Late Archean–Paleoproterozoic QPCs at the base of Dhanjori volcanics, Singhbhum district, Jharkhand, as well as along the south–western flank of the Singhbhum craton in Keojhar district, Orissa. Most of these occurrences are of sub-economic grade except a few prospects in the northern parts of NSMB. The Proterozoic gold, generally in association with base metal sulphides, is orogenic gold ('lode gold') in Proterozoic greenstones and in the 'turbidite type' and carbonatic metasediments. An acid magmatic phase is often, but not always, correlatable with the gold–bearing assemblages. The ores are mostly deformed and metamorphosed, except a late phase noted at places.

Table 8.1 *Geological developments and corresponding metallogeny through time in the various crustal segments of the Indian Peninsula and their age-wise correlation*

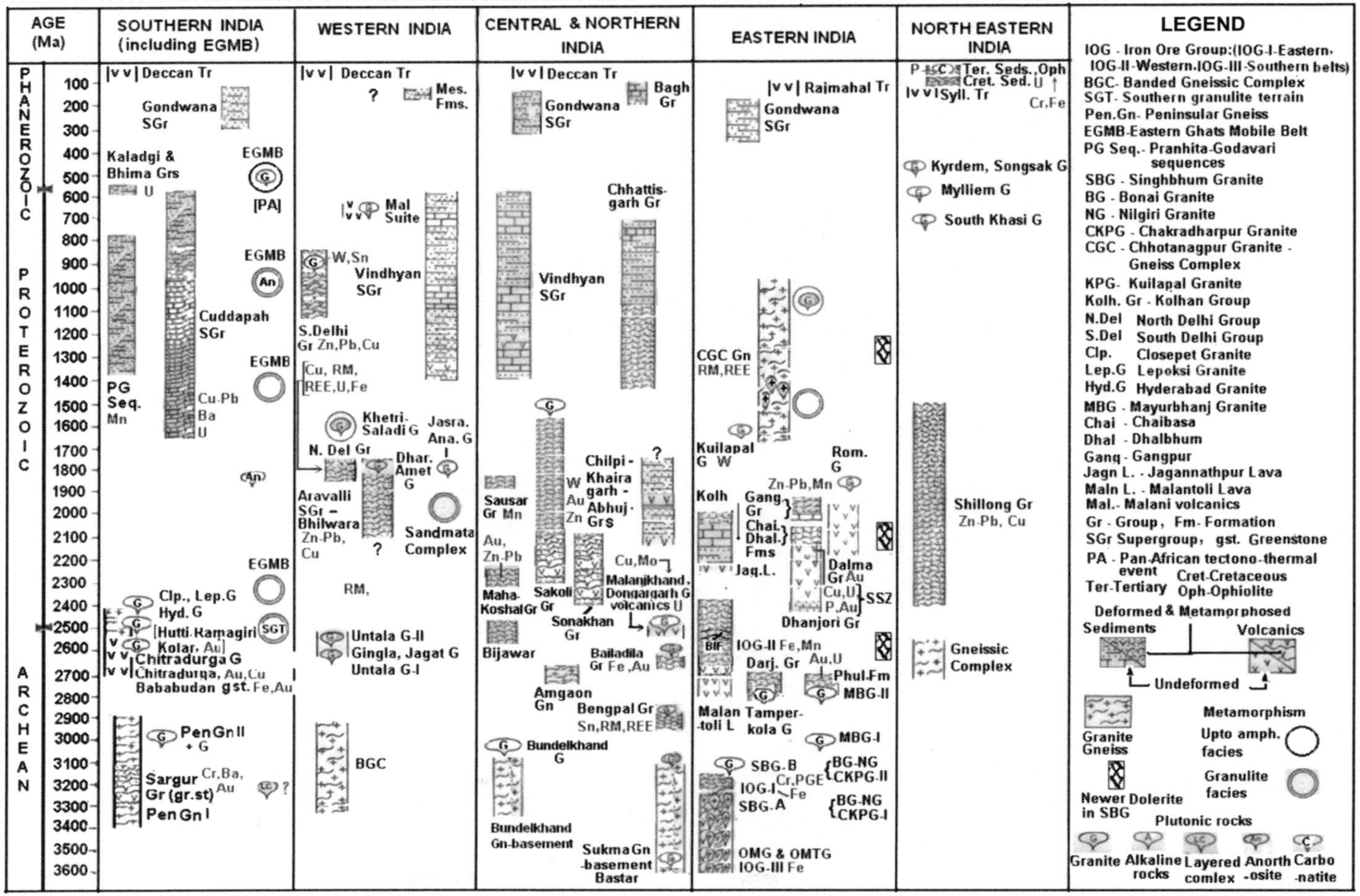

AGE (Ma)		SOUTHERN INDIA (including EGMB)	WESTERN INDIA	CENTRAL & NORTHERN INDIA	EASTERN INDIA	NORTH EASTERN INDIA
PHANEROZOIC	100	v v Deccan Tr	v v Deccan Tr; ? Mes. Fms.	v v Deccan Tr; Bagh Gr	v v Rajmahal Tr	P C Ter. Seds., Oph; Cret. Sed. U; v v Syll. Tr; Cr,Fe
	200	Gondwana SGr		Gondwana SGr	Gondwana SGr	
	300					
	400	Kaladgi & Bhima Grs; EGMB				Kyrdem, Songsak G
	500	U; (G) [PA]				Mylliem G
	600		Mal Suite			South Khasi G
PROTEROZOIC	700			Chhattis-garh Gr		
	800					
	900	EGMB	W,Sn			
	1000	(An)	Vindhyan SGr	Vindhyan SGr		
	1100	Cuddapah SGr				
	1200		S.Delhi Gr Zn,Pb,Cu			
	1300	EGMB	Cu, RM, REE,U,Fe		CGC Gn RM,REE	
	1400	PG Seq. Mn				
	1500	Cu-Pb, Ba, U	Khetri-Saladi G; Jasra. Ana. G	(G)		
	1600					
	1700		N. Del Gr; Dhar. Amet G	Chilpi-Khairagarh-Abhuj-Grs; ?	Kuilapal G W; Rom. G	
	1800	(An)		W, Au, Zn	Zn-Pb,Mn	
	1900		Aravalli SGr – Bhilwara Zn-Pb, Cu	Sausar Gr Mn	Kolh; Gang. Gr	Shillong Gr Zn-Pb, Cu
	2000		Sandmata Complex		Chai. Dhal. Fms	
	2100		?	Au, Zn-Pb		
	2200	EGMB		Cu,Mo	Jag.L.; Dalma Gr Au	
	2300	Clp., Lep.G	RM,	Malanjkhand, Dongargarh G volcanics U	Cu,U P,Au SSZ	
	2400	Hyd. G; (SGT)		Maha-Koshal Gr; Sakoli Gr		
ARCHEAN	2500	Hutti-Ramagiri	Untala G-II	Sonakhan Gr	Dhanjori Gr	
	2600	Kolar, Au; Chitradurga G	Gingla, Jagat G	Bijawar; Bailadila Gr Fe,Au	BIF; IOG-II Fe,Mn	Gneissic Complex
	2700	Chitradurga, Au,Cu; Bababudan gst. Fe,Au	Untala G-I		Darj. Gr Au,U; Phul-Fm	
	2800			Amgaon Gn	MBG-II	
	2900			Bengpal Gr Sn,RM,REE	Malan-toli L; Tamper-kola G	
	3000	PenGn II + G		Bundelkhand G	MBG-I	
	3100		BGC		SBG-B; BG-NG CKPG-II	
	3200	Sargur Gr (gr.st) Cr,Ba,Au; ?			IOG-I Cr,PGE Fe	
	3300	PenGn I			SBG-A; BG-NG CKPG-I	
	3400			Bundelkhand Gn-basement		
	3500			Sukma Gn-basement Bastar	OMG & OMTG	
	3600				IOG-III Fe	

LEGEND

IOG - Iron Ore Group:(IOG-I-Eastern, IOG-II-Western, IOG-III-Southern belts)
BGC- Banded Gneissic Complex
SGT- Southern granulite terrain
Pen.Gn- Peninsular Gneiss
EGMB-Eastern Ghats Mobile Belt
PG Seq.- Pranhita-Godavari sequences
SBG - Singhbhum Granite
BG - Bonai Granite
NG - Nilgiri Granite
CKPG - Chakradharpur Granite
CGC - Chhotanagpur Granite - Gneiss Complex
KPG- Kuilapal Granite
Kolh. Gr - Kolhan Group
N.Del North Delhi Group
S.Del South Delhi Group
Clp. Closepet Granite
Lep.G Lepoksi Granite
Hyd.G Hyderabad Granite
MBG - Mayurbhanj Granite
Chai - Chaibasa
Dhal - Dhalbhum
Ganq - Gangpur
Jagn L. - Jagannathpur Lava
Maln L. - Malantoli Lava
Mal.- Malani volcanics
Gr - Group, Fm- Formation
SGr Supergroup, gst. Greenstone
PA - Pan-African tectono-thermal event
Ter-Tertiary
Cret-Cretaceous
Oph-Ophiolite

Deformed & Metamorphosed: Sediments, Volcanics
Undeformed
Granite Gneiss
Newer Dolerite in SBG
Metamorphism: Upto amph. facies; Granulite facies
Plutonic rocks: Granite, Alkaline rocks, Layered comlex, Anorth-osite, Carbo-natite

REE and RM (rare metal)-bearing pegmatites occur aplenty in the Late Archeans of south India as well as in the Proterozoic rocks of Central India, Rajasthan, Bihar–Jharkand and West Bengal. In South India these occur in the Peninsular Gneiss, late granitoids and also within the greenstone belts. Such pegmatites are common in the Proterozoic granite–gneiss terrains of Bihar–Jharkhand, Chhattisgarh and Rajasthan. India is self-sufficient in these elements and is unlikely to be in short supply in near future.

Kimberlites and lamproitic intrusives carrying mantle derived (even crust derived) xenoliths are found at places in Central and Southern India (Majhgaon, Mainpur, Wajrakarur–Lattavaram, Narayanpet, Nallamalai, Krishna valley). Some of them are diamondiferous, some are not. Their mineral potential apparently has not been fully explored (cf. Viljoen, 2006). They are interesting even otherwise. Limited studies of ultramafic xenoliths from Wajrakarur Kimberlite showed that they formed at P = 70–32.5 kb and T = 1490º– 930º C and eclogitic xenoliths at P = 53–24 kb and T = 1240º – 800º C. The P–T estimates of these xenoliths suggested that the related lithosphere during the mid–late Proterozoic was at least 185 km thick and the peridotitic eclogitic xenoliths were collected from depths of about 100–180 km and 75–150 km respectively (Ganguli and Bhattacharya, 1987).

In the context of the foregoing discussions on the various crustal blocks of Indian shield, their evolutionary trends and the metallogenic characteristics, a comparative picture (Table 8.1) is presented for a quick reckoning of the time–space correlation of geologic and metallogenic events in the various sectors.

Before closing this discussion let us briefly touch upon two questions which disturb many Indian minds. First of these two is, why India does not have a Ni–Cu (–PGE) sulphide deposit associated with mafic (including troctolite) to ultramafic rocks while they exist in many other countries including Canada, South Africa, Australia, China and a couple of countries in Europe. Watson's (1980) suggestion that causing mineralisation of this type needed mantle inhomogenity below the region raised a hope in their mind, as India and some of these countries were possibly close neighbours during their journey through time. But today such a mineralisation is believed to be controlled mainly by the following: (i) abundance of ore metals in the magma, (ii) the state of sulphide saturation of the magma, (iii) capacity of the magma to react with the country rocks. The point (iii) above is very important if the magma is undersaturated in sulphide sulphur (which it often is) and sulphur in substantial quantity (in any form) is available to the magma in its upward journey. The point (i) is not very important in most cases as the ore metals are always there in such magmas unless incorporated in early Mg (–Fe) silicates (Arndt et al., 2005; Barnes and Lightfoot, 2005).

The other question is not much different. The frustration that there is no such mineralisation associated with the Deccan Trap while the Norilsk–Talnakh deposit in Siberia, Russia, is related to eruption of similar continental basalt. There, however, differentiated sills, derived from the flood basalt assimilated underlying thick beds of anhydrite. Not that such a situation is proved to be absent below the Deccan Traps all over. Available information shows that the Deccan Traps are S-undersaturated (Keays and Lightfoot, 2009). The study should cover more areas for a generalised view.

Notwithstanding the Himalaya being the loftiest and the largest mountain range of the world, known mineralisation within its limits is very poor. The likely explanation for this situation would be that the tectonics involved for its development, as we have already briefly discussed, is not known as an effective progenitor of much ores. Fyfe (1998), however, is pretty optimistic about such a situation.

India's sojourn to the tropics in the Tertiary had its quota of crustal evolution and mineralisation. It is because of this that large reserve of bauxitic ores and monazitic sands are available in the country. Where else should you expect such intense weathering and erosion? A part of the hematitic iron ores and lateritic manganese ores formed during this period.

References

Abbott, D., Burgess, L., Longhi, J., Smith,W.H.F.,1994. An empirical thermal history of the Earth's upper mantle. *Journal Geophys. Res.*, v. 99, pp. 13835–13850.

Abu–Hamattech, Z.S.H., Raza, M. and Ahmad, T., 1994. Geochemistry of Early Proterozoic mafic and ultramafic rocks of Jharol Group, Rajasthan, Northwestern India. *Jour.Geol. Soc. India*, v. 44; pp. 141–156.

Abu–Zied, S., and Kerns, G., 1980. Albitized uranium deposits: Six articles translated from *Russian literature. US Department of Energy,* Colorado, p.114.

Achar, K.K., Pandit, S.A., Natarajan, V., Kumar, M.K. and Dwivedy, K.K., 1997. Uranium mineralisation in the Neoproterozoic Bhima Basin, Karnataka, India. *Recent developments in uranium reserves, production and demand, IAEA,* Vienna, pp. 1–22.

Acharya, B.C. and Rao, D.S., 1998. Graphite in Eastern Ghat Complex of Orissa. *Geol. Survey of India, Spl. Pub.* No. 44, pp. 190–200.

Acharya, S., 1982. Diagenetic crystallization and migration in the Banded Iron Formations of Orissa, India. In G.C. Amstutz et al. (Eds.), Ore genesis–the state of the art. *Springer–Verlag,* Berlin, pp. 442–450.

Acharya, S., 1984. Stratigraphy and structural evolution of the rocks of the iron ore basin in Singhbhum–Orissa Iron Ore Province. *In Proc.Sem. on Crustal evolution of the Indian shield and its bearing on metallogery. Ind. Journal Earth Sci.*, pp. 19–28.

Acharya, S., 2000. Some observations in parts of banded iron formations of Eastern India. *Presidential address, 87th Indian Sci. Cong. (Earth System Science section).*

Acharyya, S.K., 1974. Stratigraphy and sedimentation of the Buxa Group, Eastern Himalaya. *Himalayan Geology*, v. 4, pp. 102–116.

Acharyya, S.K., 1975. Structure and stratigraphy of the Darjeeling frontal zone, Eastern Himalaya. *Geol. Survey India*, *Misc. Pub.*, v. 24(1), pp. 71–90.

Acharyya, S.K., 1986. Tectono–stratigraphic history of Naga Hills ophiolites. In *Geology of Nagaland Ophiolite, D.B. Ghosh Commemorative volume., Geol. Survey of India*, *Mem.* 119, pp. 94–103.

Acharyya, S.K., 1997. Evolutionary characters of Gondwanic Indian crust. *Indian Minerals*, v. 51, pp. 1–24.

Acharyya, S.K., 1998. Break-up of the Greater Indo–Australian continent and accretion of blocks framing South and East Asia. *Journal Geodynamics*, v. 26, pp. 149–170.

Acharyya, S.K., 2000. Coal and lignite resources of India: An overview. *Geological Society of India*, Bangalore, p. 50.

Acharyya, S.K., 2001a. Geodynamic setting of the Central Indian Tectonic Zone in central, eastern and north eastern India., *Geol. Survey of India Spl. Pub.*, v. 64, pp. 17–35.

Acharyya, S.K., 2001b. The role of Asia India collision in the amalgamation of Gondwana-derived blocks and deep-seated magmatism during Palaeogene at Himalayan foreland basin and around the Ghonga syntaxis in South China block. *Gondwana Research*, v. 4, pp. 61–74.

Acharyya, S.K., 2003. A plate tectonic model for Proterozoic crustal evolution of Central Indian Tectonic Zone. *Gondwana Geol. Mag*., v. 7, pp. 9–31.

Acharyya, S.K., 2005. NE India: A complex terrain of accreted continental blocks. In *Proc. Sem. Mineral and Energy resources of Eastern and Northeastern India. MGMI–Calcutta Branch,* Kolkata, India, pp. 29–54.

Acharyya, S.K., Gupta, A. and Orihashi, Y., 2010 a. New U–Pb zircon ages from Paleo–Mesoarchean TTG gneisses of Singhbhum Craton, India. *Geochem. Journal*, v. 44, pp. 81–88.

Acharyya, S.K., Gupta, A. and Orihashi, Y., 2010 b. Neoarchean-Paleoproterozoic stratigraphy of Dhanjori basin, Singhbhum craton, Eastern India: And recording of a few U-Pb zircon dates from its basal parts. *Jour. Of Asian Earth Sciences,* v.39, pp. 527–536.

Acharyya, S.K., Kayal, J.R., Roy, A. and Chaturvedi, R.K., 1998. Jabalpur Earthquake of May 22, 1997: constraint from aftershock study. *Jour. Geol. Soc. of India* , v. 51(3), pp. 295–304.

Acharyya, S.K., Ray, K.K. and Roy, D.K., 1989. Tectono–stratigraphy and emplacement history of ophiolite assemblage from Naga Hills and Andaman Island Arc, India. *Jour. Geol. Soc. of India*, v. 33(1), pp. 4–18.

Acharyya, S.K., Ray, K.K. and Sengupta, S., 1990. Tectonics of ophiolite belt from Naga Hills and Andaman Islands, India. In K. Naha, S.K. Ghosh and D. Mukhopadhyay (Eds.), *Structure and Tectonics: The Indian Scene. Nat. Acad. Sci., India*, (*Earth and Planet. Sci.*), v. 99, pp. 187–199.

Acharyya, S.K. and Roy, A., 2000. Tectonothermal history of the Central Indian Tectonic Zone and reactivation of major fault/shear zones. *Jour. Geol. Soc. of India*, v. 55 (3), pp.239–256.

Adamson, D.W. and Parslow, G.R., 1985. Granulite facies depletion of uranium: reality or not. In *Les Mechanismes due concentration de l'uraniun dans Les Environment Geologiques (Resumes),* Nancy, pp. 87–90.

Aerden, D.G.A.M., 1991. Foliation boudinage control on the formation of the Roseberry Pb–Zn Orebody, Tasmania. *Jour. Structurl Geology*, v. 13, pp. 759–775.

Aerden, D.G.A.M., 1994. Microstructural timing of the Rosebery massive sulphides, Tasmania: evidence for a metamorphic origin through mobilization of disseminated base metals. *Jour. Metamorphic Geol.*, v. 12, pp. 505–522.

Aftalion, M. , Bowes, D.R., Das, B. and Fallick, A.E., 1998. Pan–African thermal history of the Mid–Proterozoic Khariar Alkali Syenite in the Eastern Ghats, Orissa, India: a U–Pb and K–Ar isotopic study. *(Abst.) Internat. Sem. on Precambrian crust of Eastern and Central India (UNESCO–IUGS–IGCP–368)*, *Geol. Surv. India*, Bhubaneswar, pp. 10–12.

Agarwal, S. and Srivastava, R.K., 1997. Geochemistry of Late Proterozoic Sendra Granitoid Suite, Central Rajasthan, India: Role of magma mixing/hybridization process in their genesis. *Jour. Geol. Soc. of India.*, v. 50, pp. 607–618.

Ahmed, S., Manjunath, Khan, R. and Joshi, C., 1993. Geology of parts of Raichur, Manvi and Deodurg Taluk, Raichur dist., Karnataka. *Records, Geol. Surv. India*, 126(5). pp. 75–77.

Alexander, P.O., 1981. Age and duration of Deccan volcanism, K–Ar evidence. *Jour. Geol. Soc. of India, Mem.* 3, pp. 244–258.

Allegre, C.J. and others, 1984. Structure and evolution of the Himalaya–Tibet orogenic belt. *Nature* 307, pp. 17–22.

Ambesh, C.P., 2006. Indian non–coal mining sector and its future directions. In *Proc. First Asian Mining Congress*, *MGMI, Kolkata,* India, v. 1, pp. 83–94.

Amelin,Y., Lee, D.C., Hallyday, A.N., Pidgeon, R.T., 1999. Nature of the Earth's earliest crust from hafnium isotopes in single detrital zircons. *Nature*, v. 399, pp. 252–255.

Anantha Iyer, G.V. and Vasudev, V.N., 1979. Geochemistry of the Archean metavolcanic rocks of Kolar and Hutti goldfields, Karnataka, India. *Jour. Geol. Soc. of India*, v. 20, pp. 419–432.

Ananatha Iyer, G.V. and Vasudev, V.N., 1985. Copper metallogeny in the Jogimardi volcanics, Chitradurga greenstone belt, *Jour.Geol. Soc. India*, v. 26, pp. 580–598.

Anbar, A.D. and Holland, H.D., 1992. The photochemistry of manganese and the origin of the banded iron formation. *Geochim. Cosmochim. Acta*, v. 56, pp. 2595–2603.

Andreasson, P.G. and Dallmeyer, R.D., 1995. Tectenothermal evolution of high alumina rocks within Protogine Zone, Southern Sweden. *Jour. Metamorphic Geol.*, v. 13, pp. 461–474.

Andreoli, M.A.G., Smith, C.B., Watkeys, M., Moore, J.M., Ashwal, L.D. and Hart, R.J., 1994. The geology of the Steenkampskraal monazite deposit, South Africa: Implications for REE–Th–Cu mineralisation in charnockite granulite terranes. *Econ. Geol.*, v. 89, pp. 994–1016.

Andrew, C.J., 1993. Mineralisatin in the Irish Midlands: In Pattrick, R.A.D. and Poyla, D.A. (Eds.), *Mineralisation in the British Isles*, London Chapman Hall. pp. 208–269.

Anon, 1984. Explanatory Brochure on Geological and Mineral map of Bhutan (Scale 1: 500,000), First edition. *Geological Survey of India,* Calcutta.

Anon, 1984. Mineral Resources of Bhutan, *Geological Survey of India*, Bhutan Unit.

Anthony, E.Y. and Titley, S.R., 1988. Pre–mineralisation igneous process at the Sierrita porphyry copper deposit, Arizona. In: E. Zachrisson (Ed), *Proc. 7th Quad. IAGOD Symp., Lulea*. E. Schweiz., Stuttgart, pp. 536–546.

Appel, P.W.U, Rollinson, H.B. and Touret, J.L.R., 2001. Remnants of early Archean (> 3.75 Ga) sea-floor hydrothermal system in the Isua greenstone belt. *Precambrian Research.*, v. 112, pp. 27–49.

Arndt, N.T., Lesher, C.M. and Czamanske, G.K., 2005. Mantle derived magmas and magmatic Ni–Cu (–PGE) deposit. *Econ. Geol.*, v. *100th Anniversary vol.*, pp. 5–23.

Arndt, N.T., Naldrett, A.J. and Pyke, J.R., 1977. Komatiitic and iron rich tholeiitic lavas of Munro Township, northeast Ontario,. *Journal of Petrology*., v. 18, pp. 319–369.

Arogyaswamy, R.N.P., 1963. Gold occurrences in parts of Chotanagpur. Geological Society of India, Mem.1,

Arogyoswami, R.N.P. and Dutta, A.B. 1948. The Lowa mines, Singhbhum district, Bihar. *Records,Geol.Surv. India,* v. 91, pt 2.

Arora, M., Govil, P.K., Charan, S.N., Udairaj, B., Balaram, V., Manikyamba, C., Chatterjee, A.K. and Naqvi, S.M., 1995. Geochemistry and origin of Archean banded iron formation form the Bababudan Schist belt, India. *Econ. Geol.*, v. 90, pp. 2040–2057.

Aspler, L.B. and Chiarenzelli, J.R., 1998. Two Neoarchean Supercontinents ? Evidence from the Palaeoproterozoic. *Sed. Geol.*, v. 120, pp. 75–104.

Auden, J.B., 1933. Vindhyan Sedimentation in the Son Valley, Mirzapur district. *Geol. Surv. India, Mem.* 62, pp. 141–250.

Auge, T., Cocherie, A., Genna, A., Armstrong, R., Guerrot, C., Mukherjee, M.M. and Patra, R.N., 2003. Age of the Boula PGE mineralisation (Orissa, India) and its implications concerning the Singhbhum Archean nucleus. *Precamb. Research*, v. 121, pp. 85–101.

Auge, T., Salpateur, I, Bailly, L. Mukherjee, M.M. and Patra, R.N., 2002. Magmatic and hydrothermal platinum group minerals and base metal sulphides in Baula Complex, India. *Canadian Mineral*, v. 40, pp. 277–309.

Awasthi, A.K. and Prakash, B., 1981. Depositional environment of unfossilferous sediments from the Jodhpur Group, Western India. *Sed. Geol.*, v. 30, pp. 15–42.

Azam Ali, M., Banerjee, D.C., Chaturvedi, S.N., Paul, A.K., Saha, A., Gurjar, R., 1990. Uranium mineralization in the Proterozoic Abujhmar basin, Bastar District, Madhya Pradesh, India. *Exploration and Research for Atomic Minerals, AMD (India)*, v. 3, pp. 57–68.

Babruvahan Rao, K.,1989. The origin and evolution of the sand dune deposits of Ganjam coast, Orissa, India. *Exploration and Research for Atomic Minerals, AMD (India)*, v. 2, pp. 133– 146.

Babu, E.V.S.S.K., Nayak, S., Rai, A.K., Krishnamurthy, P.K.and Kak, S.N., 2002. Discrete selenide phases from the uraniferous Mahadek Sandstones of Domiasiat and Wahkyn South, West Khasi Hills districts Meghalaya. *Jour. Geol. Soc. of India*, v. 59, pp. 571–574.

Babu, H.P., Punuswami, M. and Krishamurthy, K.V., 1981. Shimoga belt. *Geol. Surv. India, Mem. 112*, pp. 199–217.

Babu, T.M., 1994. Tin in India. *Geological Society of India*, Bangalore, p. 217

Babu, V.R.R.M., 1970. Petrology of metamorphic rocks of almandine–amphibolite facies in Saidapuram–Podalakurau area, Nellore district, Andhra Pradesh, India. *Tsch. Mineral. Pet., Mitt.*, v. 14, pp. 171–194.

Bachinski, D.J., 1969. Bond Strength and Sulfur isotope fractionation in co-existing sulfides. *Econ.Geol*, v.64, pp.56–65.

Bacuta, G.C., Lipin, B.R., Gibbs, A.K. and Kay, R.W., 1988. Platinum-group element abundance in chromite deposits of the Akoje ophiolite block, Jambales ophiolite complex, Phillipines. In H.M. Prichard, P.J. Potts, J.F. Bowles and S.J. Cribb (Eds.), *Geo-Platinum Symp*., Elsevier, Amsterdam, pp. 381–382.

Baidya, T.K., 1982. Geological prospecting and exploratory drilling for apatite minerals in Pankridih area, Purulia district, West Bengal. *Report Directorate of Mines and Minerals, Govt. of West Bengal*, India.

Baidya, T.K. and Dasgupta, P., 2002. Syngenetic remobilized tungsten mineralisation in Chhendapathar, Bankura district, West Bengal, India: present status and prospect. *Indian Journal Geol*., v. 74, pp. 221–238.

Bailey, S.W. and Tyler, S.A., 1960. Clay minerals associated with Lake Superior iron ores. *Econ. Geol*., v. 55, pp. 150–175.

Baksi, A. K., 1999. Reevaluation of plate motion models based on hotspot tracks in the Atlantic and Indian Oceans. *Journal Geol*., v. 107, pp. 13–26.

Balakrishnan, S. and Rajamani, V., 1987. Geochemistry and petrogenesis of granitic gneisses around Kolar schist belt: constraints for the evolution of the crust in the Kolar area. *Journal Geol*., v. 75, pp. 219–240.

Balakrishna, S., Hanson, G.N. and Rajamani, V. 1990. Pb and Nd isotope constraints on the origin of high–Mg and theoretic amphibolites, Kolar schist belt, southern India. *Contrib. Min. Pet*., v. 107, pp. 272–292.

Balakrishna, S. Hanson, G.N. and Rajamani, V., 1999. U–Pb isotope study on zircons and sphenes from the Ramagiri area, southern India: evidence for accretionary origin of eastern Dharwar craton during late Archaean. *Journal Geol*., v. 107, pp. 69–89.

Balasubrahmanium, K.S., Surendra, M., and Ravi Kumar, 1987. Genesis of some bauxite profiles from India. *Chem. Geol*., v. 60, pp. 227–235.

Balasubrahmanyam, M.N., 1978. Geochronology and geochemistry of Archaean tonalitic gneisses and granites of South Kanara district, Karnataka State, India. In B.F. Windley and S.M. Naqvi (Eds.), *Archaean Geochemistry*, Elsevier, Amsterdam, pp. 59–77.

Ball, V., 1870. On the copper of Dhalbhum and Singhbhum. *Records, Geol. Surv. India*, v. 3, pp. 94–103.

Ball, V., 1877. On the geology of the Mahanadi Basin and its vicinity. *Records, Geol.Surv.India*, v.10 (4), pp. 167–186.

Ball, V., 1881. The geology of Manbhum and Singhbhum, *Records, Geol. Surv. India*, No.18(2).

Bandopadhyay, B.K., Bhoskar, K.G., Ramachandra, H.M., Roy, A., Khadse, V.K., Mohan, M., Sreeramchandra Rao, K., Roy Barman, T., Bishui, P.K. and Gupta, S.N., 1990. Recent geochronological studies in parts of the Precambrians of Central India. In: *Precambrian of Central India. Geol. Surv. India, Spl. Pub*., v. 28, pp. 199–211.

Bandyopadhyay, B.K. and Gupta, A., 1990. Submarine debri-flow deposits from Ordovician Mane Ting Formation in Tethyan Black Mountain basin, Central Bhutan. *Indian Journal Geol*, v. 36, pp. 277–289.

Bandyopadhyay. B.K., Roy, A., Shukla, R.S. and Guha Sarkar, T.K., 1992. Sedex type copper and zinc deposits in the Proterozoic Sakoli Group, Nagpur and Bhandara Districts, Central India. In S.C.Sarkar (Ed.), *Metallogeny related to tectonics of the Proterozoic mobile belts*. Oxford and IBH Co. Pvt. Ltd., New Delhi, India. pp. 53–102.

Bandyopadhyay, B.K., Roy, A. and Huin, A.K., 1995. Structure and tectonics of a part of the Central Indian shield. *Geological Society of India, Mem*. 31, pp. 433–467.

Bandyopadhyay, B.K., Sarkar, R.L., Dora, M.L. and Jeere, D.S., 2001. Metallogeny of Precambrian Central India. *Geol. Surv. India, Spl. Pub*. 64, v. 64, pp. 1–6.

Bandyopadhyay, D. and Hishikar, A.K., 1977. Stratigraphic sequence in the southern part of the Bailadila Range, district Bastar (MP). *Jour. Geol. Soc. India*, v. 18, pp. 240–245.

Bandyopadhyay, N., 2003. Metamorphic history of the rocks in the southeastern sector of the Proterozoic Singhbhum shear zone and its environs. *Unpub. Ph.D. thesis, University of Calcutta*, p. 256

Bandyopadhyay, P.C. and Sengupta, S., 2004. The Palaeoproterozoic supracrustal Kolhan Group in Singhbhum Craton, India and Indo–African Supercontinent. *Gondwana Research*, v. 7, pp. 1228–1235.

Bandyopadhyay, P.K., 1981. Chakradharpur Granite–gneiss–a composite batholith in Western part of Singhbhum Shear Zone, Bihar. *Indian Journal Earth Sciences*, v. 8, pp. 109–118.

Bandyopadhyay, P.K., Chakrabarti, A.K., Deomurari, M.P., Misra, S., 2001. 2.8 Ga old anorogenic granite-acid volcanic association from western margin of Singhbhum–Orissa craton, eastern India. *Gondwana Research*, v. 4, pp. 465–475.

Banerjee, A.K., 1980. Geology and mineral resources of Alwar district, Rajasthan. *Records, Geol. Surv. India*, v. 110, p. 137.

Banerjee, A.K. and Thiagarajan, T.A., 1969. Report on the investigation of gold at Kunderkocha, Singhbhum district, Bihar. *Geol. Surv. India, Unpub. Report*.

Banerjee, D.C., Krishna, K.V.G., Murthy, G.V.G.K., Srivastava, S.K. Sinha, R.P., 1994. Occurrence of spodumene in rare metal bearing pegmatites of Marlagallo–Allapatna area, Mandya district, Karnataka, India. *Jour. Geol. Soc. India* , v. 44, pp. 127–139.

Banerjee, D.M., 1971a. Aravallian stromatolites from Udaipur, Rajasthan. *Jour. Geol. Soc. India* , v. 12, pp. 349–355.

Banerjee, D.M., 1971b. Precambrian stromatolitic phosphorites of Udaipur, Rajasthan, India. *Bull. Geol. Soc. Amer.*, v. 82, pp. 2319–2380.

Banerjee, D.M., 1986. Proterozic Cambrian Phosphorites regional review. Indian Sub–continent. In P.J. Cook and J.H. Shergold (Eds), *Phosphorite deposits of the world*, V.1, Camb. Univ. Press, pp. 70–90.

Banerjee, D.M., 1996. A Lower Proterozoic palaeosol at BGC–Aravalli boundary in south–central Rajasthan, India. *Jour. Geol. Soc. India*, v. 48, pp. 277–288.

Banerjee, D.M. and Basu, P.C., 1979. Geology and structure of Precambrian Jhabua phosphorite deposit, M.P. *Indian Minerals*, v. 20, pp. 32–42.

Banerjee, D.M. and Bhattacharya, P., 1994. Petrology and geochemistry of graywackes from the Aravalli Supergroup, Rajasthan, India and the tectonic evolution of a Proterozoic sedimentary basin. *Precamb. Research.*, v. 67, pp. 11–35.

Banerjee, D.M., Basu, P.C. and Srivastava, N., 1980. Petrology, Mineralogy, geochemistry and origin of the Precambrian Aravalli phosphorite deposits of Udaipur and Jhabua, India. *Econ. Geol.*, v. 75, pp. 1181–1199.

Banerjee, D.M., Khan, M.W.Y., Srivastava, N. and Saigal, G.C., 1982. Precambrian phosphorites in the Bijawar rocks of Hirapur–Bassia area, Sagar district, Madhya Pradesh, India. *Mineral. Deposita*, v. 17(4), pp. 349–362.

Banerjee, H., 1979. A study of the geology and ore mineralisation at places of Darjeeling–Sikkim region, Eastern Himalayas. *Unpub. Ph.D. thesis, Jadavpur University*, p. 192.

Banerjee, P.K., 1967. Revision of the stratigraphy, structure and metamorphic history of the Gangapur series, Sundergarh distric, Orissa. *Records, Geol. Surv. India*, v. 92(2), pp. 327–346.

Banerjee, P.K., 1972. Geology and geochemistry of the Sukinda ultramafic field, Cuttack district, Orissa. *Geol. Surv. India, Mem.* 103, 171 p.

Banerjee, P.K., 1982. Stratigraphy, petrology and geochemistry of some Precambrian basic volcanic and associated rocks of Singhbhum district, Bihar and Mayurbhanj and Keonjhar districts, Orissa. *Geol. Surv. India, Mem.* 111, pp. 1–52.

Banerjee, P.K., 1990. On geology of Eastern Ghats of Orissa and Andhra Pradesh, India. In Naqvi, S.M. (Ed), *Precambrian continental crust and its mineral resources*, Elsevier, Amsterdam, pp. 391–407.

Banerjee, P.K. and Om Prakash, 1975. Galena mineralisation in the Vindhyan rocks of Amjhore, Sahabad district, Bihar, India. *Econ. Geol.*, v. 70, pp. 399–404.

Banerjee, S. and Sarkar, S.C., 1998. Petrological and Geochemical background of the zinc–lead mineralisation at Zawar, Rajasthan–a study. In B.S. Paliwal (Ed), *The Indian Precambrian*. Scientific Publishers (India), Jodhpur, pp. 312–326.

Banerjee, S., Mondal, N. and Sarkar, S.C., 1998. Geological structures of the northern part of the Zawar belt, Rajasthan, and the problem of location and translocation of the ores. *Indian Journal Geol.*, v. 70, pp. 171–183.

Banerji, A.K., 1962. Cross-folding, migmatisation and ore localization along parts of the Singhbhum Shear Zone, south of Tatanagar, Bihar, India. *Econ. Geol.*, v. 57, pp. 50–71.

Banerji, A.K., 1974. On the stratigraphiy and tectonic history of the iron ore bearing and associated rocks of Singhbhum and adjoining areas of Bihar and Orissa, *Jour. Geol. Soc. India*, v. 15(2), pp. 150–157.

Banerji, A.K., 1977. On the Precambrian banded iron formation and manganese ores of the Singhbhum region, Eastern India. *Econ. Geol.*, v. 72, pp. 90–98.

Banerji, A.K., 1991. Geology of the Chhotonagpur region. *Indian Journal Geol.*, v. 63, pp. 275–282.

Banerji, A.K. and Talapatra, A.K., 1966. Sodagranite from south of Tatanagar, Bihar, India. *Geol. Mag.*, v. 103, pp. 340–351.

Banerji, A.K. and Tatapatra, A., Shankaran, A.V. and Bhattacharya, T.K., 1972. Ore genetic significance of geochemical trends during progressive migmatisation within parts of the Singhbhum Shear Zone, Bihar. *Jour. Geol. Soc. India*, v. 13, pp. 39–50.

Bannister, V., Roeder, P. and Poustovetov, A., 1998. Chromite in the Pericutin lava flows (1943–1953). *Journal Volcano. Geotherm. Res.*, v. 87, pp. 151–171.

Bardossy, Gy., 1981. Palaeoenvironments of laterites and lateritic bauxites-effects of global tectonism on bauxite formation. *Lateritization processes. Oxford and IBH Publ. Co. Pvt. Ltd.,* New Delhi, pp. 287–294.

Bardossy, G. and Aleva, G.J.J., 1990. Lateritic bauxites. *Elsevier,* Amsterdam, p. 624.

Barker, A.J., 1990. Introduction to metamorphic textures and microstructures. Blackie & sons, Ltd. London.

Barker, F., 1979. Trondhjemite: definition, environment and hypothesis of origin. In F. Harker (Ed.), *Trondhjemites, dacites and related rocks. Elsevier*, Amsterdam, p. 1–12.

Barley, M.E., 1992. A review of Archean volcanic-hosted massive sulfide and sulfate mineralization in Western Australia. *Econ. Geol.*, v. 87, pp. 855–872.

Barley, M.E., Krapez, B., Groves, D.I. and Kerrich, R., 1998. Late Archean bonanza: metallogenic and environmental consequences of the interaction between mantle plumes , lithosperic tectonics and global cyclicity. *Precamb. Research*, v. 91, pp. 65–90.

Barley, M.E., Pickard, A.L. and Sylvester, P.J., 1977. Emplacement of a large igneous province as a possible case of Banded Iron Formation 2.45 billion years ago. *Nature*, v. 385 (2 Jan), pp. 55–58.

Barman, G., 1987. Stratigraphic position of Marwar Supergroup in the light of stromatolite studies. *Geol. Surv. India, Spl. Pub.*, v. 11, pp. 72–80.

Barnes, S.J. and Lightfoot, P.C., 2005. Formation of magmatic nickel sulphide ore deposits and processes affecting their copper and platinum group element contents. *Econ. Geol. 100th Anniversary vol.*, pp. 179–213.

Barnes, S.J. and Roeder, P.L., 2001. The range of spinel compositions in terrestrial mafic and ultramafic rocks. *Journal of Petrology.*, v. 42, pp. 2279–2302.

Barnicoat, A.C., Fare, R.J., Groves, D.I. and Mc Naughton, N.J., 1994. Symmetamorphic lode-gold deposits in high-grade Archean settings *Geology*, v. 19, pp. 921–924.

Barrie, C.T., Erendi, A. and Cathles, L. III, 2001. Palaeoseafloor volcanic associated massive sulphide mineralisation related to a cooling komatiite flow, Abitibi sub-province, Canada. *Econ. Geol.*, v. 96, pp. 1695–1700.

Barrie, C.T. and Hannington, M.D., 1999. Classification of volcanic associated massive sulphide deposits based on host-rock composition. In C.T. Barrie and M.D. Hannington (Eds.), Volcanic associated massive sulphide deposits: processes and examples in modern and ancient settings. *Review Econ. Geol.*, v. 8, pp. 1–11.

Bartlett, J.M., Harris, N.B.W., Hawkesnorth, C.J. and Santosh, M., 1995. New isotope constraints on the crustal evolution of southern India and Pan–African granulite metamorphism. *Geological Society of India, Mem. 34*, pp. 391–397.

Barton, M.D. and Jhonson, D.A., 1996. Evaporite source model for igneous related Fe–oxide–(REE–Cu–Au–U) mineralisation. *Geology*, v. 24, pp. 259–262.

Barton, M.D. and Jhonson, D.A., 2000. Alternative brine sources for Fe–oxide (–Cu–Au) systems: Implications for hydrothermal alterations and metals. In T.M. Porter (Ed.), Hydrothermal iron oxide copper–gold and related deposits: a global perspective. *Aust. Mineral Foundation Inc.*, pp. 43–59.

Barton, P. Jr., 1969. Thermochemical study of the system Fe–As–S. Geochim. Cosmochim. Acta, v. 33, p. 841–857. Basic volcanics from the Sakoli schist belt of Central India. *Jour. Geol. Soc. of India.* v. 50, pp. 209–221.

Basu, A., 1985. Structure and stratigraphy in and around SE part of Dhanjori Basin, Singhbhum, Bihar. *Records, Geol. Surv. India*, v. 113(3), pp. 59–67.

Basu, A. and Gupta, A., 1998. Structural analysis in eastern part of North Singhbhum Fold Belt around Ghatshila–Galudih : A re-appraisal (Abst.). *Nat. Sem. on Geoscience advances in Bihar in last decade, Geological Survey of India*, Patna, pp. 2–3.

Basu, A., Gupta, A. and Guha, P.K., 1979. Surface and subsurface structural framework of the area around Turamdih, Singhbhum district, Bihar. *Indian Journal Earth Sci.*, v. 6. pp. 52–66.

Basu, A., Maitra, M. and Roy, P.K., 1997. Petrology of mafic ultramafic complex of Sukinda valley, Orissa, India. *Indian Minerals*, v. 50, pp. 271–290.

Basu, A. K., 1986: Geology of the Bundelkhand Granite massif, Central India. *Records, Geol. Surv. India.*, v. 117 (2), pp. 61–124.

Basu, A.K., 2001. Some characteristics of Precambrian crust in the northern part of Central India. *Geol. Surv. India, Spl. Pub.*, v. 55, pp. 181–204.

Basu, A.K., Sen, A., Basumallick, S. and Mitra, A.S., 1989. Study of photogeological signatures to classify the various geotectonic domains in the eastern Indian craton. *Records, Geol. Surv. India*, v. 122 (3), pp. 51–53.

Basu, A.R., Ray, S.L., Saha, A.K. and Sarkar, S.N., 1981. Eastern Indian 3800 million years old crust and early mantle differentiation. *Science*, v. 212, pp. 1502–1505.

Basu (Sanyal), K., Bortnikov, N.S., Misra, B., Mookherjee, A., Mozgova, N.N. and Tsepin, A.L., 1984. Significance of transformation textures in fahlores from Rajpura–Dariba polymetallic deposit, Rajasthan, India. *Neues Jahrb. Miner. Abh.*, v. 149, pp. 143–161.

Basu, K., Bortnikov, N., Mookherjee, A., Mozgova, N.N., Tsepin, A.I. and Vyalsov, L.N., 1983. Rare metals from Rajpura–Dariba, Rajasthan, India. *Neues. Jahrb. Miner. Mh.*, v. 7, pp. 296–304.

Basu, K.K., 1971. Base metal mineralisation along the Pur–Banera belt, Bhilwara district, Rajasthan. *Geol. Surv. India, Misc. Publn.,* No. 16, pp. 153–159.

Basu, P.K., 1976. Geology of stratabound native sulfur deposits in Zawar lead-zinc belt, Udaipur district, Rajasthan. *Indian Minerals,* v. 30(2), pp. 30–35.

Basu, S.K., 1982. Phases of acid igneous activity in Malani suite of rocks around Jhunujhunu, Rajasthan. *Records, Geol. Surv. India*, v. 112 (7), pp. 89–94.

Basu, S.K., 1993. Alkaline carbonatite complex in the Precambrians of South Purulia Shear Zone, eastern India – its characteristics and mineral potential. *Indian Minerals*, v. 47(3), pp. 179–194.

Basu, S.K. and Ghosh, R.N., 1989. Rare earth elements and rare metal metallurgy in Precambrian terrains of Purulia district, West Bengal and its status of exploration. *Nat. Sem. Vol. on strategic rare and rare earth minerals and metals*, Hyderabad, pp. 120–140.

Basu, S.K., Mukhopadhyay, A. and Ghosh, T.K., 1996. Rare earth mineralisation in Precambrian rocks along South and North Purulia shear zones in Eastern India. (Abst.), *Workshop on geology and exploration of platinum group, rare metal and rare earth elements (GSI–JU–AMD),* Calcutta, pp. 58–59.

Bau, M. and Moller, P., 1993. Rare earth element systematics of the chemically precipitated component in Early Precambrian iron formation and the evolution of the terrestrial atmosphere–hydrosphere–lithosphere system. *Geochim. Cosmochim. Acta,* v. 57, pp. 2239–2249.

Beane, J.E., Turner, C.A., Hooper, P.R., Subbarao, K.V. and Walsh, J. N., 1986. Stratigraphy, composition and form of Deccan basalts, Western Ghats, India. *Bull. Volcanology*, v. 48, pp. 61–83.

Beckinsale, R.D., Drury, S.A. and Holt, R.W., 1980. 3360 m.y. Gneisses from south Indian craton. *Nature*, v. 283, pp. 469–470.

Behra, S.N., Jena, B.K. and Das, K.C., 2005. Younger BIF and associated iron ore deposits of Bonai–Kendujhar belt, North Orissa. In *Proc. Nat. Sem. on Mineral and Energy resources of Eastern and Northeastern India, MGMI–Calcutta Branch*, Kolkata, pp. 439–440.

Bekker, A., Slack, J.F., Planavsky, N., Krapež, B., Hofmann, A., Konhauser, K.O., and Rouxel, O.J., 2010, Iron formation: The sedimentary product of a complex interplay among mantle, tectonic, oceanic, and biospheric process. *Econ. Geol.*, v. 105, pp. 467–508

Bernard–Griffiths, J., Jahan, B.M. and Sen, S.K., 1987. Sn–Nd isotopes and REE geochemistry of Madras granulites, India: An introductory statement. *Precamb. Research*, v. 37, pp. 343–355.

Beukes, N.J., 1983. Palaeo–environmental setting of iron formations in the depositional basin of the Transvaal Supergroup, South Africa. In A.F. Trendall and R.C. Morris (Eds.), *Iron formations: Facts and problems. Elsevier,* Amsterdam, pp. 131–209.

Beukes, N.J. and Gutzmer, J. and Mukhopadhyay, J., 2002. The geology and genesis of high-grade hematite iron deposits. *Iron Ore Conf. vol., September 2002*, Perth, pp. 23–29.

Beukes, N.J.,and Klein, 1990. Geochemistry and sedimentology of a facies transition from microbanded to granular iron formation in early Proterozoic Transvaal Supergroup, South Africa. *Precamb. Research*, v. 47, pp. 99–139.

Beukes, N.J., Mukhopadhyay, J. and Gutzmer, H,, 2008. Genesis of high-grade iron ores of the Archean Iron Ore Group around Noamundi, India. *Econ. Geol.*, v. 103, pp. 365–386.

Bhairam, Ch., Raju, B.V.S.N., Koteswar Rao, M. and Duttanarayana, T.A., 2000. Delineation of uraniferous horizons in cataclasite zones using integrated techniques in Precambrian basement in the environs of Chhattisgarh basin, India. In *Proc. Internat. Sem. on Precambrian crust in Eastern and Central India (UNESCO–IUGS–IGCP 368), Geol. Surv. India, Spl. Pub.* 57, pp. 241–249.

Bhalla, J.K. and Bishui, P.K., 1989. Geochronology and geochemistry of granite emplacement and metamorphism in NE Himalaya. *Records, Geol. Surv. India*, v. 122, pp. 18–20.

Bhandari, A., Pant, N.C., Bhowmik, S.K., 2010. ~1.6 Ga Ultrahigh-temperature granulite metamorphism in the Central Indian Tectonic Zone: Insights from metamorphic phase relations and monazite chemical ages. In: Extreme metamorphism and Continental Dynamics. Santosh, M., Sajeev, K., Tsunogae, T. (Eds.), *Geological Journal*, v. x, (DOI: 10.1002/gj. 1221).

Bhanot, V.B., Goel, A.K. and Singh, V.P., 1975. Rb–Sr radiometric studies for Dalhousie and Rohtang areas. *Current Science*, v. 44, pp. 219–220.

Bhargava, M. and Pal, A..B. 1999. Anatomy of a porphyry copper deposit, Malanjkhand, Madhya Pradesh, *Jour. Geol. Soc. India*, v. 53, pp. 675–691.

Bhargava, M. and Pal, A.B., 2000. Cu–Mo–Au metallogeny associated with Proterozoic tectono–magmatism in Malanjkhand porphyry copper district, Madhya Pradesh. *Jour. Geol. Soc. India*, v. 56, pp. 395–413.

Bhargava, M., Pal, A.B. and Sindhupe, G.L., 1999. Molybdenum mineralisation and resources associated with copper at Malanjkhand deposit, Madhya Pradesh. *Proc. Nat. Seminar on Strategic, Rare and Rare Earth minerals and metals*, pp. 29–40.

Bhargava, O.N., 2000. The Precambrian sequences in the Western Himalaya. *Geol. Surv. India, Spl. Pub.* 55(1), pp. 84–96.

Bhaskar Rao, Y.J. and Naqvi, S.M., 1978. Geochemistry of metavolcanics form Bababudan schist belt: A late Archean/early Proterozoic volcanosedimetary pile from India. In B.F. Windly and S.M. Naqvi (Eds.), *Archean Geochemistry, Elsevier,* pp. 325–341.

Bhaskar Rao, Y.J., Naha, K., Srinivasan, R., Gopalan, K., 1991. Geology, geochemistry and geochronology of Peninsular Gneisses around Gorur, Hasan district, Karnataka. *Proc. Indian Acad. Sci* (*Earth Planet Sci*), v. 100, pp. 399–412.

Bhaskar Rao, Y.J., Sivaraman, T.V., Pantalu, C.V.C., Gopalan, K., and Naqvi, S.M., 1992. Ages of Late Archean metavolcanics and granites, Dharwar craton: Evidence for Early Proterozoic thermotectonic events. *Precamb. Reseasch*, v. 38, pp. 246–270.

Bhaskar Rao, Y.J., Vijaya Kumar, T., Dayal, A.M. and Janardhan, A.S., 2002. Significance of deep crustal shear zones in the Southern Granulite Terrain, South India : New Nd model ages of charnockites. (Absts.) *Goldschmidt Conference*, p. A 860.

Bhat, M.I., Claesson, S., Dubey, A.K. and Pande, K., 1998. Sm–Nd age of Garhwal–Bhowali volcanics, Western Himalayas: Vestiges of the Late Archean Rampur flood basalt province of the northern Indian craton. *Precamb. Research*, v. 87, pp. 217–231.

Bhat, P.G.K., Raju, D.P., Ahmed, S.A., Venkata–Subramanian, M., Anjanappa, B., Kanyanna and Mohiuddin, 1996. Specialised thematic mapping and geochemical surveys of Hungund–Kustagi–Hagari schist belt, Karnataka. *Records, Geol. Surv. India*, v. 129(5), pp. 71–77.

Bhatia, S.K., Bhoskar, K.G., Katti, S.Y., Babu Prem and Iyer, R.V., 2001. Status and potential of Konkan Bauxite. *Geol. Surv. India., Spl. Pub.* v. 64, pp. 399–408.

Bhatnagar, G.S., 1966. *Annual Report for Field Season 1965–66 (Unpubl. Report), Atomic Minerals Directorate,* Hyderabad.

Bhatnagar, S.N. and Mathur, S.B., 1989. Geology of lead–zinc resources. *HINDZINC TECH*, v. 1–2, pp. 15–28.

Bhatt, J.M. and Katti, G.H., 1996. Detailed exploration status of Gadag Gold Field, Dharwar district, Karnataka. *Workshop volume, Gold Resources of India,* Hyderabad, pp. 20–25.

Bhattacharjee, C.C., 1991. The ophiolites of northeast India – a subduction zone ophiolite complex of the Indo–Burman orogenic belt. *Tectonophysics*, v. 191, pp. 213–222.

Bhattacharjee, D.S., 1928–32. *Records, Geol. Surv. India.*, v. 65., pp. 107.

Bhattacharya, H.N., 2004. Analysis of the sedimentary succession hosting the Paleoproterozoic Zwar zinc-lead sulfide ore deposit, Rajasthan, India. In M.Deb and W.D.Goodfellow (Eds.), *Sediment-hosted lead-zinc sulfide deposits*. Narosa, New Delhi, pp. 350–361.

Bhattacharjee, J., Chattopadhyay, B., Mukhopadhyay, A., Chattopadhyay (Mrs.) S., Singhai, R.K. and More, M.K., 1984. Exploration for tungsten in Balda block, Sirohi district, Rajasthan. *Symp.MGMI. India, (8–9 March, 1984)*, Calcutta, pp. 116–132.

Bhattacharjee, J., Golani, P.R. and Reddy, A.B., 1998. Rift–related bi–modal volcanism and metallogeny in Delhi fold belt, Rajasthan and Gujarat. *Ind. Journal Earth Sci.*, v. 60, pp. 191–199.

Bhattacharjee, J., Ramji, Reddi, A.B., and Golani, P.R., 1991. Gold mineralisation associated with Cu–Zn massive sulphide in Danva area, Sirohi district, Rajasthan, *Indian Minerals*, v. 45(3). pp. 183–188.

Bhattacharjee, P.K. and Mukherjee, A.D., 1984. Petrochemistry of metamorphosed pillows and status of amphibolites (Proterozoic) from Sirohi district, Rajasthan, India. *Geol. Mag.*, v. 121, pp. 465–473.

Bhattacharji, S. and Singh, R.N., 1984. The mechanical structure of the southern part of the Indian shield and its relevance to Precambrian basin evolution,. *Tectonophysics*, v. 105, pp. 103–120.

Bhattacharya, A., 1996. (Ed.) Recent advances in Vindhyan Geology. *Geol. Soc. India, Mem. 36*, p. 331

Bhattacharya, A. and Bhattacharya, G., 2003. Petrotectonic study and evolution of Bilaspur–Raigarh Belt, Chhattisgarh. *Gondwana Geol. Mag.*, v. 7, pp. 89–100.

Bhattacharya, A.K., Gorikhan, R.A. Khazanchi, B.N., Fami, S., Singh, J. and Kaul, R., 1992. Uranium mineralisation hosted by migmatite mobilization and breccia zones in the north west Raihand valley, Sonbhadra district, Uttar Pradesh. *Indian Journal Geol.*, v. 64, pp. 259–275.

Bhattacharya, B.P. and Ray Baman, T., 2000. Precambrians of Meghalaya – A concept. In Proc. Sem.M.S. Krishnan Centenary (1998), *Geol. Surv. India, Spl. Pub.*, 55, pp. 95–100.

Bhattacharya, S., 1986. Mineral chemistry and petrology of the manganese silicate rocks of Vizianagram manganese belt, Andhra Pradesh. *Jour. Geol. Soc. India*, v. 27, pp. 169–184.

Bhattacharya, S., 1997. Evolution of Eastern Ghats granulite belt of India in a compressional tectonic regime and juxtaposition against Iron Ore Craton of Singhbhum by oblique collision–transpression. *Proc. Ind. Acad. Sci.* (*Earth and Planet. Sci.*), v. 106, pp. 65–75

Bhattacharya, S., Sen, S.K. and Acharya, A., 1994. The structural setting of the Chilka Lake granulite migmatite anorthosite suite with emphasis on the time relation of charnockites. *Precamb. Research*, v. 66, pp. 393–409.

Bhattacharya, S., Singh, R.J., Dutta, J.C. and Dutta, S., 1999. Detailed study along the Great Boundary Fault. *Records, Geol. Surv. India, (Extended Abstracts of W.Region FS–1995–96),* v. 130 (7), pp.1–5.

Bhattacharya, S.K., Behra, S.N., Acharya, B.K. and Hussain, A., 2000. Geology and mineralisation of the Eastern Ghats manganese deposits of Orissa. *Proc. Internat. Sem. on Precambrian crust of Eastern and Central India (UNESCO–IUGS–IGCP–368), 1998. Geol. Surv. India, Spl. Pub.* 57, pp. 260–270.

Bhattacharyya, H.N., 1991. A reappraisal of the depositional environment of Precambrian metasediments around Ghatsila–Galudih, Eastern Singhbhum. *Jour. Geol. Soc. India*, v. 37, pp. 47–54.

Bhattacharyya, H.N. and Bandyopadhyay, S., 1998. Seismites in a Proterozoic tidal succession, Singhbhum, Bihar, India. *Sed. Geol.*, v. 119, pp. 239–352.

Bhattacharyya, H.N. and Bhattacharya, A. 1989. Phosphorite deposits in Purulia and Singhbhum, Eastern India – a study in similarity and contrast. *Geological Society of India, Mem. 13*, pp. 83–95.

Bhattacharyya, H.N., Chatterjee, A., and Chaudhury, S., 1992. Tourmaline from Cu–U belt of Singhbum, Bihar, India. *Jour. Geol. Soc. India*, v. 39, pp. 191–195.

Bhattacharyya, H.N., Chakraborty, I., and Ghosh, K.K., 2007. Geochemistry of some Banded Iron Formations of the Archean supracrustals, Jharkhand–Orissa region, India. *Journal Earth System*, v. 116, pp. 245–259.

Bhattacharyya, H.N., Mukherjee, A.D. and Bhattacharyya P.K., 1993. Rock-water interaction and evolution of sulphide ores in the Late Archean Chitradurga greenstone belt of southern India. *Mineral. Deposita*, v. 28, pp. 303–309.

Bhattacharyya, P.K., Dasgupta, S. Fukuoka, M. and Roy, S., 1984. Geochemistry of braunite and associated phases in metamorphosed non–calcareous manganese ores India. *Contrib. Min. Pet.*, v. 87, pp. 65–71.

Bhaumik, N. and Basu, A., 1984. Tectonic and stratigraphic status of Chakradharpur Granite–gneiss, Singhbhum. *Ind. Journal Earth Sci (CEISM Seminar vol.)*, pp. 29–40.

Bhola, K.L., 1965. Radioactive deposits of India. In Y. N. Rama Rao (Ed.), *Symp vol. on uranuim prospecting and mining in India (7–9 October, 1964),* Jaduguda, Bihar, India, pp. 1–41.

Bhola, K.L., 1977. Atomic minerals of Rajasthan. In M.L. Roonwal (Ed), *the natural resources of Rajasthan*, v. 2, pp. 789–822.

Bhola, K.L. and Bhatnagar, G.S., 1969. Occurrence of beryl in Madhya Pradesh and adjoining Simdega subdivision of Bihar. *Quart. Journal Geol. Ming. Met. Inst*. India, v. XLI(1), pp. 37–44.

Bhola, K.L., Rama Rao, Y.N., Surishastry, C. and Mehta, N.R., 1966. Uranium mineralisation in the Singhbhum thrust belt, Bihar, India. *Econ. Geol.*, v. 61, pp. 162–173.

Bhola, K.L., Udas, G.R., Mehta, N.R. and Sahasrabudhe, G.H., 1958. Uranium ore deposits at Jaduguda in Bihar State, India. In Proc. 2nd Int. *Conf. on Peaceful Uses of Atomic Energy*, v. 2, pp. 704–708.

Bhoskar, K.G., 1983. Progress Report on the establishment of stratigraphy, structure and correlation of the Sakoli Group, *Geol. Surv. India, Unpub. Report*.

Bhoskar, K.G. and Saha, A.K., (2001). Polymetallic mineralisation in Sakoli fold belt and constraints of genetic modeling. *Geol. Surv. India*, *Spl. Pub.* No. 55, pp. 149–160.

Bhoskar, K.G., Saha, A.K. and Mahapatra, K.C. 2003. Targetting for platinum group metals in hyrothermal vein set–up: A case history from Sakoli fold belt, Maharashtra. *Gondwana. Geol. Mag.*, v. 7, pp. 421–429.

Bhowmik, S.K., 1997. Multiple episodes of tectonothermal processes in Eastern Ghats granulite belt. *Proc. Indian Acad. Sci* (*Earth Planet. Sci*). v. 106, pp. 131–146.

Bhowmik, S.K., 2006. Ultra high temperature metamorphism and its significance in the Central Indian Tectonic Zone. In: Extreme crustal metamorphism and related crust-mantle processes. Sajeev, K., Santosh, M. (Eds.) *Lithos*, v. 92, pp. 484–505.

Bhowmik, S.K. and Pal, T., 2000. Petrotectonic implication of the granulite suite north of the Sausar Mobile Belt in the overall tectonothermal evolution of the central Indian Mobile Belt. *Geol. Surv. India*, *Unpub. Report*, p. 82.

Bhowmik, S.K. and Roy, A., 2003. Garnetiferous metabasites from the Sausar Mobile Belt: Petrology, P-T pathe and implcations for the tectonothermal evolution of the Central Indian Tectonic Zone. *Journal of Petrology*, v. 44, pp. 387–420.

Bhowmik, S.K. and Spiering, B., 2004. Constraining the \prograde and retrograde P-T paths of granulites using decomoposition of initially zoned garnets: An example from the Central Indian Tectonic Zone. *Contrib. Min. Pet.*, v. 147, pp. 581–603.

Bhowmik, S., Pal, T., Roy, A. and Chatterjee, K.K., (1997). Penecontemporaneous deformation structures in relation to diagenesis of carbonate hosted manganese ores: An example from polydeformed and metamorphosed Sausar belt. *Indian Minerals*, v. 51(1 and 2), pp.149–164.

Bhowmik, S.K., Pal, T., Roy, A. and Pant, N.C., 1999. Evidence for Pre–Grenvillian high-pressure granulite metamorphism from the northern margin of the Sausar mobile belt in Central India. *Jour. Geol. Soc. India,* v. 53, pp. 385–399.

Bhowmik, S.K., Basu Sarbadhikari, A., Spiering, B. and Raith, M.M., 2005. Mesoproterozoic reworking of Palaeoproterozoic Ultrahigh Temperature Granulites in the Central Indian Tectonic Zone and its Implications. *Journal of Petrology*, v. 46, pp. 1085–1119.

Bhusan, S.K., 1985. The Malani volcanism in Western Rajasthan. *Indian Journal Earth Sci.*, v. 12(1), pp. 58–71.

Bhusan, S.K., 2000. Malani rhyolites – a review. *Gondwana Research*, v. 3, pp. 65–77.

Bierlein, F.P., Pisarevsky, S., 2008. Plume related oceanic plateaus as potential source of gold mineralisation. *Econ. Geol.*, v. 103, pp. 425–430.

Bierlein., F.P., Fuller, T., Stuwe, K. Arne, D.C. and Keays, R.R. 1998. Wall rock alterations associated with turbidite-hosted deposits – examples from Central Victoria, Australia. *Ore Geol. Rev.*, v. 13, pp. 345–380.

Binu–Lal, S.S., Swaki, T., Wada, H. and Santosh, M., 2003. Ore fluids associated with the Wynad gold mineralisation, Southern India and evidence from fluid–inclusion micro–thermomentry and gas analysis. *Journal Asian Earth Sciences*, v., pp. 1–17.

Bist, B.S., Ali, M.A., Pande, A.K. and Pavanagaru, R., 2005. Geological characteristics of the iron–uranium mineralisation within Lesser Himalaya region of Arunachal Pradesh. *Jour. Geol. Soc. India,* v. 66, pp. 185–202.

Biswal, T.K. and Jena, S.K., 1999. Large lateral ramp in the fold–thrust belts of Mesoproterozoic Eastern Ghats mobile belt, Eastern India. *Gondwana Research* (*Gondwana Newsletter section*), v. 2 (4), pp. 657–660.

Biswal, T.K. and Sashidharan, K., 1992. Scheelite mineralisation in Bhurkhola area, Sarbhang district of Bhutan: an example of syngenetic tungsten deposit. *Indian Minererals,* v. 46, pp. 47–52.

Biswal, T.K., Gyani, K.C., Parthasarathi, R. and Pant, D.R., 1998a. Implications of the geochemistry of the pelitic granulites of Delhi Supergroup, Aravalli Mountain Belt, North Western India. *Precamb. Research*, v. 87, pp. 75–85.

Biswal, T.K., Gyani, K.C., Parthasarathi, R. and Pant, D.R., 1998b. Tectonic implications of geochemistry of gabbro–norite–basic granulite suite in the Proterozoic Delhi Sugargroup, Rajasthan, India. *Jour. Geol. Soc. India,* v. 52, pp. 721–732.

Biswas, A., 1998. Preservation of sedimentary signatures in polyphased deformed Chaibasa quartzite, India, *(Abst.) v. XV, Conv. Ind. Assocn.Sedimentologists*, Gauhati University (Nov. 1998).

Biswas, S.K., 1990a. Gold mineralisation in Hutti–Maski greenstone belt, Karnataka, India. *Indian Minerals,* v. 44, pp. 1–14.

Biswas, S.K., 1990b. Gold mineralisation in Uti block, in Hutti–Maski supracrustal belt, Karnataka. *Jour. Geol. Soc. India,* v. 36, pp. 79–89.

Biswas, S.K., Prabhakara, K., and Rao, P.S., 1985. Preliminary exploration of auriferous lodes of Hutti–Maski schist belt, Karnataka, India. In *Proc. UN Interregional Sem. on gold exploration and development, Bangalore*, Section II, pp. 2.2.1–2.2.29.

Blackburn, W.H. and Srivastava, D.C., 1994. Geochemistry and tectonicsignificance of the Ongarbira metavolcanic rocks, Singhbhum district, India. *Precamb. Research,* v. 67, pp. 181–206.

Blanford, W.T., 1867. On the Traps and intertrappean beds of Western and Central India. *Geol. Surv. India, Mem.* 6(5), pp. 137–162.

Bleeker, W., 2003. The Late Archean record: A puzzle in ca 35 pieces. *Lithos* ,v. 71, pp. 99–134.

Blevin, P.L. and Chappel, B.W., 1995. Chemistry, origin and evolution of mineralized granites in the Lachland fold belt, Australia : the metallogeny of I- and S- type granites. *Econ. Geol.*, v. 90, pp. 1604–1619.

Boast, A.M., Coleman, M.L. and Italls, C., 1981. Textural and stable isotope evidence for the genesis of the Tynag basemetal deposit, Ireland. *Econ. Geol.*, v. 76, pp. 27–55.

Boger, S.D., Wilson, C.J.L. and Fanning, C.M., 2001. Early Paleozoic tectonism within East Antarctic craton: the final suture between east and west Gondwana. *Geology*, v. 29, pp. 463–466.

Bohlen, S.R., 1987. Pressure–temperature–time paths and a tectonic model for the evolution of granulites. *Journal Geol.*, v. 95, pp. 617–632.

Bohlen, S.R. and Mezger, K, 1989. Origin of granulite terranes and the formation of the lowermost continental crust. *Science*, v. 244, pp. 326–329.

Bose, M.K., 2000. Mafic ultramafic magmatism in the Eastern Indian craton – a review. *Geol. Surv. India, Spl. Pub.* 55, pp. 227–258.

Bose, M.K. and Chakraborti, M.K. and Saunders, A.D., 1989. Petrochemistry of the lavas from Proterozoic Dalma volcanic belt, Singhbhum, eastern India. *Geol. Rundsch.*, v. 78(2), pp. 633–648.

Bose, M.K. and Chakraborti, M.K., 1981. Fossil marginal basin from Indian shield: a model for the evolution of Singhbhum Precambrian belt, Eastern India. *Geol. Rundsch.*, v. 70, pp. 504–518.

Bose, M.K. and Chakraborty, M.K., 1989. Petrochemistry of the lavas from Proterozoic Dalma volcani belt, Singhbhum, Eastern India. *Geol. Rundsch.*, v. 78, pp. 633–648.

Bose, M.K. and Ghosh, A., 1996. Redeposited conglomerate on an Archean basement: debri flow in Proterozoic Singhbhum basin, Eastern India. *Indian Minerals,* v. 50(3), pp. 123–138.

Bose, M.K. and Goles, C.G., 1970. Chemical petrology of the ultramafic minor intrusions of Singhbhum, Bihar. *Symp. on Upper Mantle Project,* Hyderabad, pp. 305–325.

Bose, P.K., Majumdar, R., and Sarkar, S., 1997. Tidal sandwaves and related storm deposits in the transgressive Proterozoic Chaibasa Formation, India. *Precamb. Research*, v. 84, pp. 63–81.

Bose, P.N., 1887. Iron industry in the western portion of the district of Raipur. *Records, Geol. Surv. India*, v. 20, pp. 167–170.

Bose, P.N., 1898–99. *Unpublished Progress Report of Geol. Surv. India.*

Bose, P.N., 1899–1900. *Unpublished Progress Report of Geol. Surv. India.*

Bose, P.N. (cf.Griesbach, 1899, 1900). *Records, Geol. Surv. India*, v. 22, p. 23.

Bose, U., 1989. Correlation problems of the Proterozoic stratigraphy of Rajasthan. *Indian Minerals,* v. 43, pp. 183–193.

Bose, U., 2000. The nature, origin and evolution of the Banded Gneissic Complex beasement rocks of Rajasthan. In M. Deb. (Ed), *Crustal evolution and metallogeny in the northwestern Indian shield.* Narosa Publ. House, New Delhi, pp. 73–86.

Bose, U. and Sharma, A.K., 1992. The volcano-sedimentary association of the Precambrian Hindoli supracrustals in southeast Rajasthan. *Jour. Geol. Soc. India*, v. 40, pp. 359–369.

Bose, U., Fareeduddin and Reddy, M.S., 1990. Polymodal volcanism in parts of the south Delhi fold belt, Rajasthan. *Jour. Geol. Soc. India*, v. 36, pp. 263–276.

Bose, U., Mathur, A.K., Sahoo, K.C., Bhattacharyya, S., Dutt, K., Kunar, V.A., Sarkar, S.S., Chaudhury, S and Chaudhuri I., 1996. Event stratigraphy and physico–chemical characters of Banded Gneissic Complex and association supracrustals in the south Merwar plains of Rajasthan. *Jour. Geol. Soc. India*, v. 47, pp. 325–338.

Bouhallier, H., Choukrane, P., and Ballevre, M., 1993. Diapirism, bulk homogenous shortening, and transcurrent shearing in the Archean Dharwar craton: The Holenarsipur area, southern India. *Precamb. Research.*, v. 63, pp. 43–58.

Bowers, T.S. and Helgeson, H.C., 1983. Calculation of the thermodynamic and geochemical consequences of non–ideal mixing in the system H_2O–CO_2–NaCl on phase relations in the geologic system : Equation of state for H_2O–CO_2–NaCl fluids at high pressures and temperatures. *Geochim. Cosmochim. Acta,* v. 47, pp. 1247–1275.

Bowring, S.A. and Williams, I.S., 1999. Priscoan (4.00–4.03 Ga) orthogneisses from north–western Canada. *Contrib. Min. Pet.*, v. 134, pp. 3–16.

Boyce, A.J., Coleman, M.L. and Russel, M.J., 1983, Formation of fossil hydrothermal chimneys and mounds from silvermines, Ireland. *Nature*, v. 306, pp. 545–550.

Boyet, M. and Carlson, R.W.,2005. 142 Nd evidence for early (4.53 Ga) global differentiation of the silicate. *Earth. Science*, v. 309, pp. 576–581.

Braterman, P.S., Cairns–Smith, A.G. and Sloper, R., 1983. Photo–oxidation of hydrated Fe^{2+} – significance for banded iron formations. *Nature*, v. 303, pp. 163–164.

Brimhall, G.H. Jr. and Lewis, C.J. 1992. Bauxite and laterite soil ores. *Encyclop. Earth System Science*, v. 1, pp. 321–336.

Brimhall, G.H. Jr., 1980. Deep hypogene oxidation of Porphyry copper potassium–silicate protore at Butte, Montana : A theoretical evaluation of the copper remobilization hypothesis. *Econ. Geol.* v. 75, pp. 384–409.

Brown, A.C., 1997. World-class sediment-hosted stratiform copper deposits: characteristics, genetic concepts and metallotects. *Aust. Journal Earth Sci.*, v. 44, pp. 317–328.

Brown, A.S., 1969. Mineralisation in British Columbia and the copper and molybdenum deposits. *Canadian Mining and Metall. Bull.*, v. 62(681), pp. 26–40.

Buhl, D., 1987. U–Pb and Rb–Sr Altersbestimmungen und untersuchungen zum strontium isotopenaustausch und granuliten sudindiens. *Unpub. Ph.D. thesis, Nuremburg University.*

Buhl, D., Grauert, B., and Raith, M., 1983. U–Pb zircon dating of Archean rocks from the South Indian Craton: results from the amphibolite to granulite facies transition zone at Kabbal Quarry, southern Karnataka. *Fortsch. Miner.*, v. 61, pp. 43–45.

Buick, R., 1992. The antiquity of oxygenic photosynthesis: evidence from stromatolites in sulphate deficient lakes. *Science*, v. 225, pp. 74–77.

Buick, R. and Dunlop, J.S.R., 1990. Evaporatic sediments from Warrawoona Group, Western Australia. *Sedimentology*, v. 37, pp. 247–277.

Burnham, C.W., 1967. Hydrothermal fluids in the magmatic stage. In H.L. Barnes (Ed), *Geochemistry of hydrothermal ore deposits*. Holtz, Rinchart and Winston, pp. 34–76.

Burnham, C.W., 1979. Magmas and hydrothermal fluids. In H.L. Barnes (Ed.), *Geochemistry of hydrothermal ore deposits (Second Edition),* John Wiley and Sons, New York, pp. 71–136.

Burnham, C.W. and Ohmoto, H., 1980. Late stage processes of felsic magmatism. *Soc. Mining Geologists*, Japan, *Spl. Issue* 8, pp. 1–11.

Caia, J., 1976. Paleogeographical and sedimentological controls of copper, lead and zinc mineralizations. *Geology*, v. 71, pp. 409–422.

Cameron, E.N., 1980. Evolution of the Lower critical zone, central sector, eastern Bushveld Complex and its chromite deposits. *Econ. Geol.*, v. 75, pp. 845–871.

Campbell, I.H., Griffiths, R.W., 1990. Implications of mantle plume structure for the evolution of flood basalts. *Earth Planet. Sci. Lett.*, v. 99, pp. 79–93.

Campbell, I.H. and Hill, R.I., 1988. A two-stage model for the formation of granite–greenstone terranes of Kalgoorlie–Norsman area, Western Australia. *Earth Planet. Sci. Letters*, v. 90, pp. 11–25

Campbell, I.H. and Murck, B.W., 1993. The petrology of the G chromite zone in the Mountain View area of the Stillwater Complex. *Journal of Petrology*, v. 34, pp. 291–316.

Campbell, I.H., Czammanske, G.K., Fedorenko, V.A., Hill, R.J., Stepanov, V.A, and Kunilov, V.E., 1992. Synchronism of Siberian traps and Permian–Triassic boundary. *Science*, v. 258, pp. 1760–1763.

Candela, P.A. and Holland, H.D., 1986. A mass transfer model for copper and molybdenum in magmatic hydrothermal systems: the origin of porphyry type ore deposits. *Econ. Geol.*, v. 81, pp. 1–19.

Canfield, D.E., 1998. A new model for Proterozoic ocean chemistry. *Nature*, v. 396, pp. 450–453.

Cannell, J., Cooke, D,R,, Walshe, J.L. and Stein, H., 2005. Geology, mineralization, alteration and structural evolution of El Teniente Porphyry Cu-Mo deposit. *Econ. Geol.*, v. 100, pp. 979–1003.

Cannon, W.F., 1976. Hard iron ore of the Marguette range, Michigan. *Econ. Geol.*, v. 76, pp. 1012–1028.

Carmichael, D.M., 1970. Interesting isograds in Wheatstone Lake area, Ontario. *Journal of Petrology*, v. 11, pp. 147–181.

Casting, J.F., 1993. Earth's early atmosphere. *Science*, v. 259, pp. 920–926.

Castro, L.O., 1994. Genesis of Banded Iron Formations. *Econ. Geol.*, v. 89, pp. 1384–1397.

Catlos, E.J., Dubey, C.S., Harrison, T.M. and Edward, M.A., 2004. Late Miocene movement within the Himalayan Main Central Thrust Shear Zone, Sikkim, Northeast India. *Jour. Met. Geol.*, v. 22, pp. 207–226.

Cawthorn, R.G., Barnes, S.J., Ballhaus, C. and Malitch, K.N., 2005. Platinum group elements, chromium, and vanadium deposits in mafic and ultramafic rocks. *Econ. Geol*, 100th Anniversary vol., pp. 215–249.

Cerny, P., 1991. Rare–element granite pegmatite (Pt. I) – Anatomy and internal evolution of pegmatite deposits. *Geoscience Canada*, v. 18, pp. 49–67. (Pt. II) – Regional to global environments and petrogenesis. *Geoscience Canada*, v. 18, pp. 68–81.

Cerny, P., 1994. Fertile granites of Precambrian rare–element pegmatite fields: Is geochemistry controlled by tectonic setting, or source lithologies? *Precamb. Research,* v. 51, pp. 429–468.

Chadwick, B., Ramakrishnan, M. and Viswanatha, M.N., 1981. Structural and metamorphic relations between Sargur and Dharwar supracrustal rocks and Peninsular Gneiss in Central Karnataka. *Jour. Geol. Soc. India*, v. 22, pp. 557–569.

Chadwick, B., Ramakrishnan, M., and Viswanatha, M.N., 1985. Bababudan – A Late Archean intracratonic volcano-sedimentary basin, Karnataka, South India. Part I: Stratigraphy and basin development. *Jour. Geol. Soc. India*, v. 26, pp. 769–801.

Chadwick, B., Vasudev, V.N. and Ahmed, N., 1996. The Sandur schist belt and its adjacent plutonic rocks : Implications for Late Archean crustal evolution in Karnataka. *Jour. Geol. Soc. India*, v. 47, pp. 37–57.

Chadwick, B., Vasudev, V.N. and Hegde, G.V., 1997. The Dharwa Craton, Southern India, and its Late Archean plate tectonic setting: current interpretations and controversies. *Indian Acad. Sci (Earth and Planet Sci.)*, v. 106(4), pp. 1–10.

Chadwick, B., Vasudev, V.N. and Hedge, G.V., 2000. The Dharwar Craton, southern India, interpreted as the result of Late Archean oblique convergence. *Precamb. Research.*, v. 9, pp. 91–111.

Chako, T., Kumar, G.R.R. and Newton, R.C., 1987. Metamorphic P–T conditions of the Kerala (South India) khondalite belt: a granulitic facies supracrustal terrain. *Journal Geol.*, v. 95, pp. 343–358.

Chakrabarti, B., and Gupta, G.P., 1992. Stratigraphy and structure of the North Delhi basin. *Records, Geol. Surv*. India, v. 122(7), pp. 2–3.

Chakraborti, M.K. and Bose, M.K., 1985. Evolution of tectonic setting of Precambrian Dalma volcanic bellt, India, using major and trace element characters. *Precamb. Res*earch, v. 28, pp. 253–268.

Chakraborti, S., 1997. Elucidation of the sedimentary history of the Singhora Group of rocks, Chhattisgarh Supergroup, M.P. *Records, Geol. Surv*. India, v. 130(6), pp. 184–187.

Chakraborty, A.K., 1967. Genesis of lead–zinc deposits at Zawar, India. *Econ. Geol.*, v. 62, pp. 68–81.

Chakraborty, B.K. and Saha, A.K., 1974. Petrology and mode of emplacement of pegmatites in parts of Bihar Mica Belt. Proc. *All India Mica Convention* (Gudur), pp. 1–12.

Chakraborty, K.L., 1972. Some primary structures in the chromites of Orissa, India. *Mineral. Deposita*, v. 7, pp. 280–284.

Chakraborty, K.L. and Chakraborty, T.L., 1976, Source and fixation of the nickeliferous limonite profile of Sukinda valley, Orissa. *Jour. Geol. Soc. India*, v. 17, pp. 186–193.

Chakraborty, K.L. and Chakraborty, T.L., 1984. Geological features and the origin of chromite deposits of the Sukinda valley, Orissa, India. *Mineral. Deposita*, v. 19, pp. 256–265.

Chakraborty, K.L. and Majumder, T., 2002. Some important aspects of Banded Iron Formation (BIF) of Eastern Indian shield (Jharkhand and Orissa). Indian. *Journal Geol.*, v. 74, pp. 37–47.

Chakraborty, K.L. and Majumder, T., 1992. An unusual diagenetic structure in the Precambrian Banded Iron Formation (BIF) of Orissa, India and its interpretation. *Mineral. Deposita*, v. 27, pp. 55–57.

Chakraborty, K.L. and Majumder, T., 2000. Some important aspects of the banded iron formation (BIF) of Eastern Indian Shield (Jharkhand and Orissa). *Indian Journal Geol.*, v. 74, pp. 37–47.

Chakraborty, N., Mukherjee, K.K. and Chakraborty. S., 1984. Exploration for copper in the Singhbhum copper belt during the last decade. In *Significant discoveries of geology for mineral industries during the past decade*. Oxford and IBH, New Delhi, pp. 87–116.

Chakranarayan, A.B., Power, K.B., Pande, K. and Gopalan, K., 1986. Rb/Sr age of the Belka Pahar granite, Sirohi District, Rajasthan. *Jour. Geol. Soc. India*, v. 28, pp. 325–327.

Chaku, S.,Truelove, A.J., Singh, N.N., Sharma, S and Azad, H., 2008. Bhukia Gold Project, Banswara district, southern Rajasthan: An overview. *Int. Field Workshop on Gold Metallogeny in India*, Dec. 2008, pp. 114–116.

Chalapathi Rao, N.V., Burgess, R., Anand, M. and Mainkar, D., 2007. 40Ar–39Ar dating of Kodomali Pipe, Bastar craton. India: A Pan–African (491±11Ma) age of diamondiferous kimberlite emplacement. *Jour. Geol. Soc. India*, v. 69, pp. 539–546.

Chalapathi Rao, N.V., Gibson, S.A., Pyle, D.M. and Dickin, A.P., 1998. Contrasting isotopic mantle sources for Proterozoic lamproites and kinmshites from the Cuddapah basin and Eastern Dharwar Craton: implications for Proterozoic mantle heterogeneity beneath South India. *Jour. Geol. Soc. India*, v. 52, pp. 683–694.

Chamberlain, C.P., Zeitler, P.K. and Jain, M.Q., 1989. The dynamics of the suture between Kohistan island arc and Indian plate in the Himalaya of Pakistan. *Journal Met. Geol.*, v. 7, pp. 135–149.

Chappel, B.W. and White, A.J.R., 1974. Two contrasting granite types. *Pacific Geology*, v. 8, pp. 173–174.

Chapple, W.M., 1978. Mechanics of thin–skinned fold–and–thrust belts. *Geol.Soc.Amer. Bull.*, v. 89, pp. 1189–1198.

Chardon, D., Choukrane, P., and Jayananda, M., 1996. Strain patterns, decollement and incipient sagducted greenstone terrains in Archean Dharwar craton (South India). *Journal Struct. Geol.*, v. 18, pp. 991–1004.

Chartrand, F.M. and Brown, A.C., 1985. The diagenetic origin of stratiform mineralisation, Coates lake, Redstone Belt, NWT, Canada. *Econ. Geol.*, v. 80, pp. 325–343.

Chatterjee, A., 1964. Geology, mineralogy and genesis of iron ores of some deposits of the Bailadila Range, Bastar district, MP. Quart. *Journal Geol. Min. Met. Soc.* India, v. 36, pp. 12–57.

Chatterjee, A., 1970. Structure, tectonics and metamorphism in a part of South Bastar, MP. *Quart. Journal Geol. Min. Met. Soc. India*, v. 42, pp. 75–95.

Chatterjee, A.K., 1993. Indian iron ore resource base: an appraisal. In Proc. *Int. Sem. on Iron Ore 2000 and beyond*, SGAT, Bhubaneswar, India, pp. 80–90.

Chatterjee, A.K. and Mukherji, P., 1981. The structural set–up of a part of the Malantoli iron ore deposit, Orissa. *Jour. Geol. Soc. India*, v. 22, pp. 121–130.

Chatterjee, A.K. and Mukherji, P., 1985. Distribution of iron ore and ore types reflecting ore quality variations in Malantoli deposts, Orissa. *Earth resources for development, Geol. Surv*. India, p. 164

Chatterjee, A.K., Negi, M.S., Rao, K.S. and Lakbir Singh, 1987. Tungstean mineralisation in East Godavari district of Andhra Pradesh. In *Proc. Nat workshop on tungsten resources development*, Bhubaneswar, pp. 47–54.

Chatterjee, K.K., Gupta, Anupendu and Deshmukh, S.S., 2001. A theoretical approach to emplacement mechanism of Deccan flood basalt. *Geol. Surv. India, Spl. vol.* 64, pp. 529–541.

Chatterjee, N., Crawley, J.L. and Ghosh, N.C., 2008. Geochronology of the 1.55 Ga Bengal Anorthosite and Grenvillian Metamorphism in the Chhotanagpur Gneissic Complex, Eastern India. *Precambrian Research* , v. 161, pp. 303–316.

Chatterjee, S.C., 1945. The gabbro rocks found near Gorumahisani Pahar. *Proc. Nat. Inst. Sci.* India, v. 11, pp. 255–282.

Chatterjee, S.K., 1929–32. *General Reports section of Records, Geol. Surv. India.,* v. 65 p. 108.

Chattopadhyay, A, Bandyopadhyay, B.K. and Khan, A.S., 2001. Geology and structure of the Sausar Fold Belt: A retrospection and some new thoughts. *Geol. Surv. India, Spl. Pub.*, v. 64, pp. 251–263.

Chattopadhyay, A. and Mandal, N., 2002. Progressive changes in strain patterns and fold styles in a deforming ductile orogenic wedge: an experimental study. *Journal Geodynamics*, v. 33, pp. 353–376.

Chattopadhyay, A., Khasdeo, L., Holdsworth, R.E. and Smith, S.A.F., 2008. Fault reactivation and pseudotachylite generation in the semi-brittle and brittle regimes: examples from Gavilgarh–Tan Shear Zone, Central India. *Geol. Mag.*, v. 145(6), pp. 756–777.

Chattopadhyay, B. and Mukhopadhyay, A.K. and Singhal, R.K., 1982. Post-Erinpura acid magmatism in Sirohi, Rajasthan, and its bearing on tungsten mineralisation. *Proc. Symp. on Precambrian Metallogeny (IGCP Project 91)*, pp. 115–132.

Chattopadhyay, B., Chattopadhyay, S. and Bapna, V.S., 1987. The Untala Granite of Rajasthan – an overview. *Indian Minerals*, v. 41, pp. 9–18.

Chattopadhyay, B., Chattopadhyay, S. and Bapna, V.S., 1988. The Newania pluton, a Proterozoic carbonatite in an Archean envelope – a preliminary study. In A.B. Roy (Ed), Precambrian of the Aravalli Mountain, Rajasthan, India. *Geological Society of India, Mem.* 7, pp. 341–349.

Chattopadhyay, B., Mukhopadhyay, A.K., Singhai, R.K., Bhattacharjee, J. and Hore, M.K., 1982. Post-Erinpura magmatism in Sirohi, Rajasthan and its bearing on tungsten mineralisation. In *Proc. Symp. Metallogeny of the Precambrian* (IGCP–91), pp. 115–132.

Chattopadhyay, N, and Hashimi, S., 1984. The Sung valley alkaline ultramafic carbonatite complex, East Khasi and Jaintia Hills districts, Meghalaya. *Records.Geol. Surv. India, v.113* (4), pp. 24–33.

Chattopadhyay, P., 1995. High grade mtamorphism and the sulfidic ores at the Mamandur area, Arcot, Tamilnadu. *Unpub. Ph.D. Thesis, Jadavpur University, p. 176*

Chattopadhyay, S. and Ray, B., 1997. Evidence of felsic volcanism associated with Ongarbira volcanics, Singhbhum, Bihar. *Indian Minerals*, v. 51, pp. 239–246.

Chattopadhyay, S. and Saha, A.K., 1998. Lower Proterozoic gold–sulphide mineralisationin Pular–Parsori area, Maharashtra with a probable acid magmatic linkage.(Abst.). *Seminar on the Precambrian continental crust in Eastern and Central India, UNESCO–IUGS–IGCP–368, Geol. Surv. India*, Bhubaneswar, India, pp. 181–183.

Chaturvedi R.K., Mishra, S.N. and Mulay, V.V., 1983. On Tethyan sequence of Black Mountain Region, Central Bhutan. *Himalayan Geology*, v. 11, pp. 224–249.

Choudhary, A.K., Gopalan, K., and Anjaneya Sastri, C., 1984 Present status of geochronology of the Precambrian rocks of Rajasthan. *Tectonophysics*, v. 105, pp. 131–140.

Chaudhari, A.K., Harris, N.B.W., van Calsteren, P., and Hawkesworth, C.J., 1992. Pan–African charnockite formations in Kerala. *Geol. Mag.*, v. 129, pp. 257–264.

Chaudhuri, A., 1984. Origin and evolution of the ores and ore bearing formations in the Yarehalli–Jagalur sector, Chitradurga schist belt, Karnataka. *Unpub. Ph.D. thesis, Jadavpur University,* Kolkata, p. 235

Chaudhuri, A.K. 2003. Stratigraphy and palaeogeography of Godavari Supergroup in the South–central Pranhita–Godavari Valley, south India. *Journal Asian Earth Sci.*, v.21, pp. 595–611.

Chaudhuri, A.K. and Deb, 2004. Proterozoic Rifting in the Pranhita–Godavari valley: Implication on Indo–Antarctica Linkage. *Gondwana Research* , v. 7 (2), pp. 301–312

Chaudhuri, A.K., Dasgupta, S, Bandyopadhyay, G, Sarkar, S.,Bandyopadhyay, P.C. and Gopalan, K., 1989. Stratigraphy of the Penganga Group around Adilabad, Andhra Pradesh. *Jour. Geol. Soc. India*, v. 34, pp. 291–302.

Chaudhuri, A.K., Saha, D., Deb, G.K., Patranabis Deb, S., Mukherjee, M.K. and Ghosh, G., 2002. The Purana basins of southern cratonic province of India – a case for Mesoproterozoic fossil rift. *Gondwana Research*, v. 5, pp. 23–33.

Chaudhury, A., 1986. Geochemistry and petrology of Archaean gold-bearing volcanics in Ramagiri Goldfield, Andhra Pradesh, India. *Indian Minerals*, v. 40, pp. 13–30.

Chauhan, D.S., 1979. Phosphorite–bearing stromatolites of Precambrian phosphorite deposits of the Udaipur region, their environmental significance and genesis of phosphorite. *Precamb. Research*, v. 8., pp. 95–126.

Chaussidon, M. and Jambon, A., 1994. Boron-content and isotopic composition of oceanic basalts : Geochemical and cosmochemical implications. *Earth Planet Sci. Lett.*, v. 121, pp. 277–291.

Chellani, S. K., 2007. New find of diamonds in Basna Kimberlite field, district Mahasamund, Chhattisgarh. *Jour. Geol. Soc. India*, v. 69. pp. 619–624.

Chernyshev, I.V. and Safonov, Yu. G., 1988. The source and age of mineralisation. In F.V. Chukhrov et al (Eds.), *Zolorudnoe pole kolar (India)* (In Russian) Nauka Publication, Moscow, pp. 187.

Chetty, T.R.K., 1995. A correlation of Proterozoic shear zones between Eastern Ghats Belt , India and Enderby land , east Antarctica, based on LANDSAT imagery, *Geological Society of India, Mem.* 34, pp. 205–220.

Chetty, T.R.K., 2001. The Eastern Ghats Mobile Belt, India: a collage of juxtraposed terranes (?). *Gondwana Research*, v. 4, pp. 319–328.

Chetty, T.R.K. and Murthy, D.S.N., 1994. Collision tectonics in the Late Precambrian Eastern Ghats mobile belt: mesoscopic to satellite-scale structural observations. *Terra Nova*, v. 6, pp. 72–81.

Chetty, T.R.K. and Murthy, D.S.N., 1998. Regional tectonic framework of the Eastern Ghats Mobile Belt: A new interpretation. In *Proc. Workshop Eastern Ghats Mobile Belt, Geol. Surv. India*, Spl. Pub., v. 44, pp. 39–50.

Chiaradia, M., 2003. Evolution of tungsten sources in crustal mineralisation from Archean to Tertiary, inferred from lead isotopes. *Econ. Geol.* v. 98, pp. 1039–1045.

Chimote J.S., Pandey, B.K., Baggi, A.K., Basu, A.N., Gupta, J.N., and Saraswat, A.C., 1988. Rb–Sr whole rock isochron age for Mylliem Granite, East Khasi Hills, Meghalaya. *4th Nat. Symp. on Mass Spectrometry, Bangalore.*

Choukrane, P., Ludden, J.N., Chardon, D., Calvert, A.J. and Bouhallier, H., 1997. Archean crustal growth and tectonic processes : a comparison of the Superior Province, Canada, and the Dharwar craton, India. In J.P.Burgand M. Ford (Eds.), *Orogeny through time, Geol. Soc. Spl. Pub. No. 121*, pp. 63–98.

Chowdhury, A.N., Pal, J.C., Sen, B.N. and Das, K.M., 1983. Determination of platinum metal in rocks and minerals in parts per billion range by combined fire assay and spectrographic method. In M. Sankar Das (Ed.), *Trace analysis and technological development.* Wiley Eastern Ltd., New Delhi, pp. 243–247.

Christiansen, E.C. and Lee, D.E., 1986. Fluorine and chlorine in granitoids from Basin and Range–Province, Western United States. *Econ. Geol.* v. 81, pp. 1481–1494.

Clark, A.H., 1993. Are outsize porphyry copper deposits either anatomically or environmentally distinctive. *Soc. Econ. Geol., Spl. Pub.* 2, pp. 213–282.

Clark, A.H. and Kontak, D.J., 2004. Fe-Ti-P oxide melts generated through magma mixing in the Antanta subvolcanic centrer, Peru: Implications for the origin of nelsonite and iron oxide-dominated hydrothermal deposits. *Econ. Geol.*, v. 92, pp. 377–396.

Clark, L.A., 1960. The Fe–As–S system–phase relations and applications. *Econ. Geol.*, v. 55, pp. 1345–1381; 1631–1652.

Clark, S.H.B., Poole, S.G., Wang, Z., 2004. Comparison of some sediment-hosted stratiform barite deposits in China, the United States, and India. *Ore Geol. Rev*. v. 24, pp. 85–101.

Clayton, R.N., Oneil, J.R. and Mayeda, T., 1972. Oxygen isolope exchange between quartz and water. *Journal Geophys. Res.*, v. 77, pp. 3057–3067.

Clement, C.R. and Reid, A.M. 1989. The origin of kimberlite pipes: An interpretation based on synthesis of geological features displayed by southern African occurrences, *Geol. Soc. Austalia, Spl. Pub. 14*, pp. 632–646.

Cline, J.S. and Bodnar, R.J., 1991. Can economic porphyry copper mineralisation be generated by a typical calc–alkaline melt? *Journal Geophys. Res.*, v. 96(B5), pp. 8113–8126.

Cloud, P., 1972. A working model of the primitive earth. *Amer. Journal Sci.*, v. 272, pp. 537–548.

Cloud, P., 1973. Palaeontological significance of the banded iron formation. *Econ. Geol.*, v. 68, pp. 1135–1143.

Clout, J.M.F., and Simonson, B.M., 2005. *Precambrian iron formations and ironformation-hosted iron ore deposits: Econ. Geol., 100th Anniversary vol.*, pp.643–679

Coffin, M.F. and Eldholm, O., 1993. Scratching the surface: Estimating dimensions of large igneous provinces. *Geology*, v. 21, pp. 515–518.

Collins, W.J., Van Kranendonk, M.J., Teyssier, C., 1998. Partial convective orverturn of Archean crust in the east Pilbora Craton, Western Australia: driving mechanisms and tectonic implications. *Jour. Structural Geol.*, v. 200, pp. 1405–1424.

Condie, K.C., 1989. Plate tectonics and crustal evolution (Third Edn.), Pergamon, p. 475

Condie, K.C., 1997. Plate tectonics and crustal evolution (Fourth Edn.), Butterworth and Heinemann, Oxford, p. 282

Condie, K.,1998, Episodic continental growth and supercontinents : A mantle avalanche connection? *Earth and Planet Sci. Letts.*, v. 163, pp. 97–108.

Condie, K.C., 2002. Break up of a Paleoproterozoic supercontinent. *Gondwana Research*, v. 5, pp. 41–43.

Condie, K.C., 2005. Earth as an evolving planetary system, Elsevier, Amsterdam, 447 p.

Cook, P.J. and McElhinny, M.W., 1979. A reevaluation of the spatial and temporal distribution of sedimentary phophate deposits in the light of plate tectonics. *Econ. Geol.*, v. 74, pp. 315–330.

Cooke, D.R., Hollings, P. and Walshe, J.L., 2005. Giant Porphyry deposits: characteristics, distribution and tectonic controls. *Econ. Geol., 100th Anniversary vol* , pp. 801–818.

Cornell, D.H. and Schutte, S.S., 1995. A volcanic–exhalative origin for the World's largest (Kalahari) manganese field. *Mineral. Deposita*, v. 30, pp.146–151.

Cotterill, P., 1969. The chromite deposits of Selukwe, Rhodesia. *Econ. Geol. Monograph 4*, pp. 154–186.

Coulson, A.L., 1933. The geology of Sirdhi State, Rajputana. *Geol. Surv. India, Mem.* 63(I), p. 166

Craddock, P.T., Fresstone, I.C., Gurjar,Middleton, A.P. and Willies, L., 1989. The production of lead, silver and zinc in early India. In A. Hauptmann, E. Perrick and G.A.Wagner (Eds.), *Old world Archeometallurgy*, Bochun Bergbau Museum, pp. 51–70.

Craddock, P.T., Fresstone, I.C., Gurjar,Middleton, A.P. and Willies, L., 1990. Zinc in India. In P.T. Craddock (Ed.), *2000 years of zinc and brass. British Museum Occasional Paper 50*, pp. 29–72.

Crawford, A.R., 1969. Reconnaissance Rb–Sr dating of the Precambrian rocks in Southern peninsular India. *Jour. Geol. Soc. India*, v. 10, pp. 117–166.

Crawford, A.R., 1970. The Precambrian geochronology of Rajasthan and Bundelkhand, Northern India, *Canadian Jour. Earth Science.*, v. 125, pp. 91–110.

Crawford, A.R. and Compston, W., 1973. The age of the Cuddapah and Kurnool Systems, Southern India. *Journal Geol. Soc. Australia*, v. 19, pp. 453–463.

Crawford, A.R. and Compston, W., 1979: The age of the Vindhyan System of Peninsular India. *Quart. Journal Geol. Soc.* London, v. 125, pp. 351–372.

Crookshank, H., 1935. The Geology of northern slope of the Satpura between Morand and Sher river. *Geol. Surv. India, Mem.* 66, pp. 276–360.

Crookshank, H., 1938. The western margin of Eastern Ghats in South Jeypore. *Records, Geol. Surv. India,* v. 73.

Crookshank, H., 1948. Minerals of Rajputana pegmatites. *Mem. Geol. Met. Inst. India Trans.*, v. 42, pp. 105–189.

Crookshank, H., 1963. Geology of South Bastar and Jeypore from Boiladilla Range to Eastern Ghats. *Geol. Surv. India, Mem.*, v. 87, 149 p.

Crowe, W.A., Cosca, M.A. and Harris, L.B., 2001. $^{40}Ar/^{39}Ar$ geochronology and Neoproterozoic tectonics along the northern margin of the Eastern Ghats Belt in north Orissa. India. *Precamb. Research*, v. 108, pp. 237–266.

Cullers, R.L. and Graf. J.L., 1984. Rare earth elementars in igneous rocks of the continental crust: intermediate and silicic rocks ore petrogenesis. In P. Henderson (Ed.), *Rare earth element geochemistry. Elsevier,* Amsterdam, pp. 275–316.

Curray, J.R., 2005, Tectonic history of Andaman Sea region. *Journal Asian Earth Sci.*, v. 25, pp. 187–232.

Dahlkamp, F.J., 1993. Uranium ore deposits. *Springer Verlag,* Berlin, 460 p.

Dalstra, H. and Guedes, S., 2004. Giant hydrothermal hematite deposits with Mg–Fe metasomatism: A comparison of the Carajas, Hamersley and other iron ores. *Econ. Geol.*, v. 99, pp 1793–1800.

Dalton, J.A. and Presnall, C., 1998. The continuum of primary carbonatitic–kimberlitic melt compositions in equilibrium with lherzolite: Data from the system Cao–Mgo–Al_2O_3–SiO_2–CO_2 at 6 Gpa. *Journal of Petrology*, v. 39, pp. 1953–1964.

Dalziel, I.W.D., 1991. Pacific margins of Laurentia and East Antarctica as a conjugate rift pair: evidence and implications for an Eocambrian supercontinent. *Geology*, v. 19, pp. 598–601.

Daniel, C.G., Hollister, L.S., Parrish, R.R. and Grujic, D., 2003. Exhumation of the main Central Thrust From the lower crustal depths, eastern Bhutan Himalaya. *Jour. Met. Geol.*, v. 21, pp. 317–334.

Dar, K.K., 1968. Metallogeny in the Himalayas. 23rd Intemat. *Geol. Cong. Proc.* 7, pp. 35–42.

Das Gupta, S., Deb, G., Mukhopadhyay, D., Chadwick, B., Sen Gupta, S., and Van Deu Hul, H.J., 1993. A study of acid magmatism in the Eastern part of the Singhbhum shear zone, Bihar, India. *Proc. Nat. Acad. Sci. India*, v. 63(A) I, pp. 223–251.

Das, A.K., Awasti, A.B. and Sahoo, P., 1988. Quartz pebble conglomerate of Singhbhum Craton, Bihar and Orissa, *Geological Society of India*, *Mem. 9*, pp. 83–87.

Das, A.R., 1988. Geometry of the superposed deformation in the Delhi Supergroup of rocks, north of Jaipur, Rajasthan. In A.B. Roy (Ed.) Precambrian of the Aravalli Mountain, Rajasthan, India. *Geological Society of India*, *Mem. 7*, pp. 247–266.

Das, C.R., Jena, B.K., Behera, S.N. and Patel, S.N. 2005. Manganiferous horizon with syngenetic manganese ore association of the Eastern Ghats granulite belt, Orissa. *Proc. Seminar on mineral and energy resources of Eastern and Northeastern India, MGMI – Calcutta Branch,* Kolkata, India, pp. 189–202.

Das, D.P., 1998. Trace of Singhbhum Shear Zone east of Rakha Mines, as seen on satellite imagery. (Abst.) M.S.Krishnan Comm. *Nat. Sem. on Progress in Precambrian Geology of India, Geol. Surv. India,* Calcutta pp. 42–44.

Das, D.P., Kundu, A., Das, N., Dutta, D.R., Kumaran, A., Ramamurthy, S., Thanavelu, and Rajaiya, V., 1992, Lithostratigraphy and sedimentation of Chhattisgarh Basin. *Indian Minerals*, v. 46 (3 and 4), pp. 271–288.

Das, K.C. and Mahapatro, S., 2005. Dimension stone resources of Meghalaya. In *Proc. Nat. Sem. Mineral and Energy resources of Eastern and Northeastern India, MGMI – Calcutta Branch*, Kolkata, India, pp. 263–269.

Das, L.K. and Mall, R.P., 1995. Geophysical studies in the Son valley and the Gangetic Plains. *Geol. Surv. India, Spl. Pub.*, v. 10, pp. 260–284.

Das, L. K., Dasgupta, K. K. and De, M.K., 2008. Mineral potential of the Dhanjori metavolcanics, East Singhbhum, Jharkhand. *Indian Minerals,* v. 61 (3–4) & 62 (1–4), pp. 193–200.

Das, N., Dutta, D.R. and Das, D.P., 2001, Proterozoic cover sediments of southeastern Chhattisgarh state and adjoining parts of Orissa. *Geol. Surv. India, Spl. Pub.,* v. 55, pp. 237–262.

Das, N., Ganguly (Das) M. and Mishra, V.P., 1989. Geology of Bemetara–Kodwa–Bera area, Durg district, M.P., *Records, Geol.Surv.India*, v. 122(6), pp. 16–17.

Das, N., Roy Burman, K.J., Vatsa, U.S. and Mahurkar, Y.V., 1990. Sonakhan schistbelt: A Precambrian granite–greenstone complex. In: *Precambrians of Central India, Geol. Surv. India, Spl. Pub.* 28, pp. 118–132.

Das, S., 1997. Depositional framework of sandy mid–fan complex of the Proterozoic Chaibasa Formation, East Singhbhum, Bihar. *Jour. Geol Soc. India,* v. 50, pp. 541–557

Dasgupta, A.B. and Biswas, A.K., 2000. Geology of Assam. *Geolological Society of India, Text Book* Ser. 12, pp. 169

Dasgupta, S., 1978 a. A study of the geology and ore mineralisation of the Madan Kudan–Kolihan section, Khetri copper belt, Rajasthan. *Unpub. Ph.D thesis, Jadavpur University*, 215 p.

Dasgupta, S., 1978 b. Sedimentary structures in the Precambrian Delhi Supergroup rocks and their significance. *Indian Journal Geol.*, v. 5, pp. 177–182.

Dasgupta, S., 1993. Contrasting mineral parageneses in high-temperature calc–silicate granulites: examples from the Eastern Ghats, India. *Jour.Metamorphic Geol.*, v. 11, pp. 193–202.

Dasgupta, S., 1995. Pressure–Temperature evolutionary history of the Eastern Ghats granulite province: recent advances and some thoughts. *Geological Society of India, Mem.* 34, pp. 101–110.

Dasgupta, S., 1997. P–T–X relationships during metamorphism of manganese-rich sediments: Current status and future studies. In K. Nicholson, J.R. Hein, B.R. Buhn, and S.Dasgupta, (Eds.), Manganese mineralisation: Geochemistry and mineralogy of terrestrial and marine deposits. *Geol. Soc. London, Spl. Pub.* No. 119, pp. 327–337.

Dasgupta, S. and Ehl, J., 1993. Reaction textures in a spinel–sapphirine granulite from the Eastern Ghats, India, and their implications. *Europe. Journal Miner.*, v. 5, pp. 537–543.

Dasgupta, S. and Sengupta, P., 1998. Tectonothermal history of the Eastern Ghats Belt (Abst.) *M.S.Krishnan Comm.Nat. Seminar, Geol.Surv.India*, Calcutta, pp. 51–55.

Dasgupta, S. and Sengupta, P., 2000. Tectonothermal evolution of the Eastern Ghats Granulite Belt, India: A metamorphic perspective. *Geol. Surv. India, Spl. Pub.,v. 55*, pp. 259–274.

Dasgupta, S. and Sengupta, P., 2003. Indo–Antarctic correlation: a perspective from the Eastern Ghats Granulite Belt, India. In M. Yoshida, B. F.Windley and S. Dasgupta(Eds), Proterozoic East Gondwana: Supercontinent Assembly and Breakup. *Geol. Soc. London, Spl. Pub., v. 206*, pp. 131–143.

Dasgupta, S., Banerjee, H., Bhattacharya, P.K., Fukuoka, M., and Roy, S., 1990. Petrogenesis of metamorphosed manganese deposits and the nature of their precursor sediments. *Ore Geology Rev.*, v. 5, pp. 359–384.

Dasgupta, S. Banerjee, H. and Bandyopadhyay, G., 1992. Manganese deposition in the Proterozoic global perspective and Indian scenario. In S.C.Sarkar (Ed), *Metallogeny related to Tectonics of the Proterozoic mobile belt. Oxford and IBH Publ. Co. (Pvt.) Ltd.*, New Delhi, pp. 163–176.

Dasgupta, S., Chakraborty, S. and Neogi, S., 2009. Petrology of an inverted Barrovian sequence of metapelites in Sikkim Himalaya, India: constraints on the tectonics of inversion. *Amer. Jour. Sci.*, v. 309, pp. 43–84.

Dasgupta, S.,Ganguly, J. and Neogi, S., 2004. Inverted metamorphic sequence in the Sikkim Himalaya: Crystallization history, P–T gradient and implications. *Jour. Metamorphic Geol.*, v. 22, pp. 395–412.

Dasgupta, S., Hariya, Y. and Miura, H., 1993. Compositional limits of manganese carbonates and silicates in granulite facies metamorphosed deposits of Garbham, Eastern Ghats, India. *Resource Geology, Spl. Issue 17*, pp. 43–49.

Dasgupta, S., Sengupta, P., Bhattacharya, P.K. Mukherjee, M., Fukuoka, M., Banerjee, H. and Roy, S., 1989. Mineral reactions in manganese oxide rocks: P–T–X phase relations. *Econ. Geol.* v. 84, pp. 434–443.

Dasgupta, S., Sengupta, P., Fukuoka, M. and Bhattacharya, P.K., 1991. Mafic granulites from the Eastern Ghats, India: further evidence for extremely high temperature crustal metamorphism. *Journal. Geol.*, v. 99, pp. 124–133.

Dasgupta, S., Roy, S. and Fukuoka, M., 1992 a. Depositional models for manganese oxide and carbonate deposits of the Precambrian Sausar Group, *India. Econ. Geol.*, v. 87, pp. 1412–1418.

Dasgupta, S., Sengupta, P., Fukuoka, M., and Chakrabarti, S., 1992 b. Dehydration melting, fluid buffering and decompressional P–T path in a granulite complex from the Eastern Ghats, India. *Jour. Metamorphic Geol.*, v. 10, pp. 777–788.

Dasgupta, S., Sengupta, P., Mondal, A. and Fukuoka, M., 1994. Mineral chemistry and reaction textures in metabasites from Eastern Ghats belt, India and their implications. *Min. Mag.*, v. 57, pp. 113–120.

Dasgupta, S., Sengupta, P., Ehl, J., Raith, M. and Bardhan, S., 1995. Reaction textures in a suite of spinel granulite from Eastern Ghats Belt, India: evidence for polymetamorphism a partial petrogenetic grid in the system KFMASH and roles of ZnO and Fe_2O_3. *Journal of Petrology*, v. 36, pp. 435–461.

Dasgupta, S., Guha, D., Sengupta, P., Miura, H. and Ehl, J., 1997. Pressure–temperature–fluid evolutionary history of the polymetamorphic Sandmata Granulite Complex, North–Western India. *Precamb. Research*, v. 83, pp. 267–290.

Dasgupta, S.P., 1968. The structural history of the Khetri copper belt, Jhunujhunu and Sikar districts, Rajasthan. *Geol. Surv. India, Mem.* 98, p. 110

Dasgupta, S.P., 1970. Sulphide deposits of Saladipura, Khetri copper belt, Rajasthan. *Econ. Geol.*, v. 65, pp. 331–339.

Dasgupta, S.P., Sengupta, P.R., Poddar, B.C., Bhattacharyya, D.P., and Chakrabarty, B.K., 1986. Sulfide mineralisation in the Himalaya. *Geol. Surv. India, Misc. Pub.*, v. 44 (6), pp. 28-49.

Dash, C.R., Jena, B.K., Behra, S.N. and Patel, S.N., 2005. Manganiferous horizon with syngenetic manganese ore association of the Eastern Ghats Granulite Belt, Orissa. *Proc. Seminar on Mineral and energy resources of Eastern and Northeastern India, MGMI-Calcutta branch*, Kolkata pp. 189–202.

Datta, A.K., 1973. Internal structure, petrology and mineralogy of calc-alkaline pegmatites in parts of Rajasthan. *Geol. Surv. India, Mem.* v. 102, 112 p.

Datta, R.K., 2009. Indian coal scenario with focus on eastern and northeastern states. In Anupendu Gupta (Ed.), Scenario of Indian mineral resources and production. *Min. Geol. Met. Inst. India (MGMI), Calcutta Branch*, Kolkata, India, pp. 22–59.

Davidson, C., Grujic, D.E., Hollister, L.S. and Schmid, S.M., 1997. Metamorphic reactions related to decompression and synkinematic intrusion of leucogranite, High Himalayan Crystallines, Bhutan. *Jour. Met. Geol.* v. 15, pp. 593–612.

Davidson, C.F., 1956. The economic geology of thorium. *Mining Mag.*, v. 94, pp. 197–208.

Davidson, P., Kamenetsky, V., Cooke, D.R., Frikken, P., Hollings, P., Ryan, C., Achterberg, E.V., Mernagh, T., Skarmeta, J. Serrano, L. and Vargas, R., 2005. Magmatic precursors of hydrothermal fluids at the Rio Blanco Cu–Mo deposit, Chile: Links to silicate magmas and metal transport. *Econ. Geol.* v. 100, pp. 963–978.

Davis, D., Suppe, J and Dahlen, F.A.,1983. Mechanics of fold–and–thrust belts and accretionary wedges. *Journal Geophys. Res.*, v. 88, pp. 1153–1172.

Daziel, I.W.D., 1991. Pacific margins of Laurentia and East Antarctica–Australia as a conjugate rift pair: evidence and implications for an Eocambrian supercontinent. *Geology*, v. 19, pp. 559–601.

De Jong, G. Rotherham, J., Phillips, G.N. and Williams, P.J., 1998. Mobility of rare earth elements and copper during shear zone related retrograde metamorphism. *Geol. Mijnb.*, v. 76, pp. 311–319.

de Wit, M.J., 1998. On Archean granites, greenstones, cratons and tectonics: Does the evidence demand a verdict? *Precamb. Research*, v. 91, pp. 181–226

Deans, T. and Powell, J.L., 1968. Trace elements and strontium isotopes in carbonatites, fluorites and limestones from India and Pakistan. *Nature*, v. 218, pp. 750–752.

Deb, G. and Mukhopadhyay, D., 1991. An occurrence of tourmalinite along Singhbhum shear zone. *Indian Minerals*, v. 45, pp. 313–318.

Deb, M., 1971. Aspects of ore mineralisation in the central section of the Singhbhum copper belt, Bihar. Unpub. *Ph.D thesis, Jadavpur University*, Calcutta.

Deb, M., 1979. Polymetamorphism of ores in Precambrian stratiform sulphide deposits. *Mineral. Deposita*, v. 14, pp. 21–31.

Deb, M., 1980. Genesis and metamorphism of two stratiform massive sulphide deposits at Ambaji and Devi in the Precambrian of Western India. *Econ. Geol.*, v. 75, pp. 572–591.

Deb, M., 1982. Rare minerals in Rajpura–Dariba Ores—some further comments. *Jour. Geol. Soc. India*, v. 23, p.253–260.

Deb, M., 1986. Sulfur and carbon isotope compositions in the stratiform Zn–Pb–Cu sulphide deposits of the Rajpura–Dariba–belt, Rajasthan NW India : A model of ore genesis. *Mineral Deposita*, v. 21, pp. 313–321.

Deb, M., 1989. Isotopic composition of carbon in sulphide ore environments in Proterozoic Delhi–Aravalli orogenic belt, NW India. *Abst. 28th Internat. Geol. Cong.*, v. 1, pp. 379–380.

Deb, M., 1990a. Isotopic constitution of sulfur in the conformable base metal sulphide deposits in the Proterozoic Aravalli–Delhi orogenic belt, NW India. In S. M. Nagvi (Ed), *Developments in Precambrian Geology, No. 8, Elsevier,* Amsterdam, pp. 631–651.

Deb, M., 1990b. Regional metamorphism of sediment-hosted, conformable base metal sulphide deposits in the Aravalli–Delhi orogenic belt, NW India. In Spry, P. and Brindzia (Eds.), *Regional metamorphism of ore deposits*. VSP, Utrecht, pp. 117–140.

Deb, M., 1992. Lithogeochemistry of rocks around Rampura–Agucha massive zinc sulphide orebody, NW India – implications for the evolution of a Proterozoic aulacogen. In S.C. Sarkar (Ed), *Metallogeny related to tectonics of the Proterozoic mobile belts*. Oxford and IBH Publ. Co. (Pvt.) Ltd., New Delhi, pp. 1–35.

Deb, M., 1993. The Bhilwara belt in Rajasthan – a possible proterozoic anlacogen. In S.M. Kashyap (Ed.), *Rifted basins and anlacogens*. Gyanodaya Prakashan, Nainital. pp. 91–106.

Deb, M., 2000. VMS deposits: Geological characteristics, genetic models and a review of their metallogenesis in the Aravalli Range, NW India. In M.Deb (Ed.), *Crustal evolution and metallogeny in the north western Indian shield.* Narosa Publ. House, New Delhi, pp. 329–363.

Deb, M. and Bhattacharya, A.K., 1980. Geological setting and conditions of metamorphism of Rajpura–Dariba polymetallic ore deposit, Rajasthan, India. *Proc. 5th Quat. IAGOD Symp*, E. Schweiz. Verlag, Stuttgart, pp. 679–697.

Deb, M. and Joshi, A., 1984. Petrological studies on two East Coast bauxite deposits of India and implications on their genesis. *Sed. Geol.*, v. 39, pp. 121–139.

Deb, M. and Kumar, R. 1982. The volcano-sedimentary environment of Rajpura–Dariba polymetallic ore deposit, Udaipur district, Rajasthan, India. *Proc. Symp. on Metallogeny of the Precambrian (IGCP Project 91)*, pp. 1–17.

Deb, M. and Pal, T., 2004. Geology and genesis of basemetal sulphide deposits in the Dariba–Rajpura–Bethumni belt, Rajasthan, India, in the light of basin evolution. In M. Deb and W. D. Goodfellow (Eds.), *Sediment-hosted lead–zinc sulphide deposits*. Narosa Publ. House, New Delhi, pp. 304–327.

Deb, M. and Sarkar, S.C., 1975. Sulfide orebodies and their relation to structures at Roam-Rakha Mines–Tamapahar sections, Singhbhum copper belt, Bihar. In V.K. Verma (Ed.), *Recent researches in Geology*. v. 2, Hind. Pub. Corpn., Delhi, pp. 247–264.

Deb, M. and Sarkar, S.C., 1990. Proterozoic tectonic evolution and metallogenesis in the Aravalli–Delhi orgenic complex, Northwestern India. *Precamb. Research*, v. 46, pp. 115–137.

Deb, M. and Sehgal, U., 1997. Petrology, geothermobarometry and C–O–H–S fluid compositions in the environment of Rampura Agucha Zn (–Pb) ore deposit, Bhilwara, district, Rajasthan. Proc. *Indian Acad. Sci* (EPS), v. 106, pp. 343–356.

Deb, M. and Thorpe, R.I., 2004. Geochronological constraint in the Precambrian geology of Rajasthan and their metallogenic implications. In M. Deb and W. D. Good fellow (Eds.), *Sediment-hosted lead–zinc sulphide deposits*. Narosa Publishing House, New Delhi, pp. 246–263.

Deb, M., Banerjee, D.M. and Bhattacharya, A.K., 1978. Precambrian stromatolite and other structures in the Rajpura–Dariba polymetallic ore deposit, Rajasthan, India. *Mineral. Deposita*, v. 13, pp. 1–9.

Deb, M., Chattopadhyay, A and Deol, S., 2008. Conditions and timing of gold mineralisation in Gadag schist belt, constrained by structural and fluid inclusion data. In Pre–workshop (Extd. *Abst.) vol., Internat. Field Workshop on Gold Metallogeny in India, Dec. 2008,* NGRI – Hyderabad and Univ.of Delhi, p. 90–96.

Deb, M., Thorpe, R.I., Cumming, G.L. and Wagner, P.A., 1989. Age, source and stratigraphic implications of lead isotope data for conformable sediment hosted basemetal deposits in the Proterozoic Aravalli–Delhi orogenic belt, northwestern India. *Precamb. Research*, v. 43, pp. 1–22.

Deb, M., Thorpe, R.I., Krstic, D., Davis, D.W. and Corfu, F. 2001. Zircon Pb–U and galena Pb isotope evidence for an approximate 1.0 Ga terrane constituting the Western margin of the Aravalli–Delhi Orogenic belt, northwestern India. *Precamb. Research*, v. 108, pp. 195–213.

Deb, M., Tiwari, A. and Palmer, M.R., 1997. Tourmaline in Proterozoic massive sulphide deposits from Rajasthan, India. *Mineral. Deposita*, v. 32, pp. 94–99.

Deb, S. and Majumdar, P., 1963. Mineragraphic study of the manganese ores of Goriajhor, Sundergarh district, Orissa. *Proc. 50th Indian* Science. *Cong.*, p. 261.

DeCelles, P.G., Robinson, D.M., Zandt, G., 2003, Implications of shortening in the Himalayan fold–thrust felt for uplift of the Tibetan plateau. *Tectonics 21 doi: 10. 1029/2001 TC001322.*

Delian, F., Tiebing, L., and Jie, Y, 1992. The process of formation of manganese carbonate deposits hosted in black shale series. *Econ. Geol.*, v. 87, pp. 1419–1429.

Deol, S., Deb, M. and Chattopadhyay, A, 2008. Bhukia–Jagpura gold prospect, south Rajasthan: a preferred genetic model. In *Pre-workshop (Extd. Abst.) vol., Internat. Field Workshop on Gold Metallogeny in India, Dec. 2008, NGRI* – Hyderabad and University of Delhi, pp. 119–124.

Desai, S.J., Patel, M.P. and Mehr, S.S., 1978. Polymetamorphites of Balaram–Abu Road area, north Gujarat and southwestern Rajasthan. *Jour. Geol. Soc. India*, v. 19, pp. 383–391.

Deshmukh, S.S., Sano, T. and Nair, K.K.K., 1996. Geology and chemical stratigraphy of the Deccan basalts of Chikaldara and Behramghat sections from the eastern part of Deccan trap province, India. Proc. National Symp. on Deccan Flood Basalts, India, *Gondwana Geol. Mag., Spl. vol. 2*, pp. 1–22.

Deshpande, G.G., Mahoobey, N.K. and Deshpande, M.S. 1990. Petrography and tectonic setting of Dongargarh volcanics. *Geol. Surv. India, Spl. Pub. v. 28*, pp. 260–286.

Devrajan, M.K. and Hanuma Prasad, M., 1995–1996. Geological mapping in the western part of Mahakoshal belt, Jabalpur and Sahdohl districts, M.P., *Geol. Surv. India, Unpub. Report.*

Devarajan, M.K. and 20 others, 1997. Seismotectonic studies of Jabalpur earthquake of 22 May 1997, *Indian Miner*als, v. 50(4), pp. 377–396.

Devaraju, T.C., Halkoaho, T.A., Laajoki, K., Shreerama, M. and Subba Rao, G., 1996. Ag–Bi–tellurides from Kudremukh BIF, Karnataka. *Jour. Geol. Soc. India*, v. 47, pp. 263.

Devaraju, T.C.,Raith, M.M. and Spiering, B., 1999. Mineralogy of the Archean barite deposit of Ghattihosahalli, Karnataka, India. *Canadian Mineral.*, v. 37, pp. 603–617.

Devey, C.W. and Lightfoot, P.C., 1986. Volcanological and tectonic control of stratigraphy and structure in the western Deccan traps. *Bull. Volcanology*, v. 48, pp. 195–207.

Dewey, J.F. and Bird J.N. 1970: Mountain belts and the new global tectonics. *Journal Geophys. Res.*, v. 75, pp. 2625–2647

Dewey, J.F., Shackleton, R.M., Changfa, C. and Yiyin, S., 1988. The tectonic evolution of Tibetan Plateau. *Phil. Trans. Royal Soc. London*, v. A 327, pp. 379–413.

Dey, K.N., 1991. The older raft tonalite of Rairangpur and its bearing on the Precambrian stratigraphy of the Singhbhum craton. *Indian Journal Geol.*, v. 63(4), pp. 261–274.

Dhanaraju, R., Kumar, M.K., Babu, E.V.S.S.K and Pandit, S.A., 2002. Uranium mineralisation in the Neoproterozoic Bhima Basin at Gogoi and near Ukinal : An ore petrological study. *Jour.Geol.Soc. India*, v. 59, pp. 299–322.

Dhanaraju, R., Pasnnerselvan, A. and Virnave,S.N., 1989. Characterisation of Upper Cretaceous Lower Mahadek Sandstones and uranium mineralisation in Domiasiat–Gomaghat–Pdengshakap area, Meghalaya, India. *Expln. and Res. for Atomic Minerals, AMD Publ.*, v. 2, pp. 1–27.

Dhanaraju, R., Sinha, P.K., Umamaheswar, K., and Gorikhan A., 1985. Petrology of the uraniferous Siwalik Sandstones from near Loharkar, Himachal Pradesh, India, with a note on physico–chemical conditions of transformation and deposition of uranium. Journal Ind. Assocn. *Sedimentologists*, v. 5, pp. 12–22.

Dhar, S., Frei, R., Krammers, J.D., Nägler, T.F., Kochar, N., 1996. Sr, Pb and Nd istope studies and their bearing on the petrogenesis of the Jalor and Siwana complexes, Rajasthan, India. *Jour. Geol. Soc. India*, v. 48, pp. 151–160.

Dhoundial, D.P., Paul, D.K., Sarkar, A., Trivedi, J.R., Gopalan, K., and Potts, P., 1987. Geochronology and geochemistry of Precambrian granitic rocks of Goa, SW India. *Precamb. Research*, v. 36, pp. 287–302.

Dick, H.J.B. and Bullen, T., 1984. Chromian spinel as a petrogenetic indicator in abyssal and Alpine-type peridotites and spatially associated lavas. *Contrib. Min. Pet.*, v. 86, pp. 54–76.

Dickey, J.S., 1975. A hypothesis of origin for podiform chromite deposits. *Geochim. Cosmochim. Acta*, v. 39, pp. 1061–1074.

Dietz, V.V., 1964. Sudbury structure as an astroblem. *Journal Geol.*, v. 72, pp. 412–434.

Dikshitulu, G.R., Pandey, B.K., and Dhanaraju, R., 1995. Rb–Sr systematics of granitoids of central gneissic complex, Arunachal Himalaya : Implications on tectonics, stratigraphy, and source. *Jour. Geol. Soc. India*, v. 45, pp. 51–56.

Dimroth, E., 1975. Palaeo–environment of iron–rich sedimentary rocks. *Geol. Rundsch.*, v. 64, pp. 751–767.

Divakar Rao, V., Narayana, V. L., Rama Rao, P., Murthy, N. N., Subba Rao, M. V., Mallikarjuna Rao, J. and Reddy, G. L. N., 1996. Evolution of the Central Indian Craton, In *The Archaean and Proterozoic terrains in Southern India within East Gondwana, Mem.Gondwana Res.*, v. 3, pp. 317–325.

Divakar Rao, V., Narayana, V. L., Rama Rao, P., Murthy, N. N., Subba Rao, M. V.,Mallikarjuna Rao, J. and Reddy, G. L. N. 2000. Precambrian acid volcanism in Central India – geochemistry and origin. *Gondwana Res.*, v. 3, no. 2, pp. 215–226.

Dobmeier, C.J. and Raith, M., 2003. Crustal architecture and evolution of Eastern Ghats Belt and adjacent regions of India. In Yoshida, M., Windley, B.F. and Dasgupta, S. (Eds), Proterozoic East Gondwana: Supercontinent assembly and breakup. *Geol. Soc. London, Spl. Pub.*, v. 206, pp. 145–168

Dobmeier, C and Simmat, R., 2002. Post–Grenvillean transpression in the Chilka Lake area, Eastern Ghats Belt – implications for the geological evolution of peninsular India. *Precamb. Research*, v. 113, pp. 243–268.

Dostal, J. and Kapedri, S., 1978. Uranium in metamorphic rocks *Contrib. Mineral. Petrol.*, v. 66, pp. 409–414.

Doyle, M.G. and Allen, R.L., 2003. Sea-floor replacement in volcanic-hosted massive sulphide deposits. *Ore Geol. Rev.*, v. 23, pp. 183–222.

Drever, J.L., 1974. Geochemical model for the origin of Precambrian banded iron formations. *Geol. Soc. Amer. Bull.*, v. 85, pp. 1099–1106.

Drury, S.A., 1981. Geochemistry of Archaean metavolcanic rocks from the Kudremukh area, Karnataka. *Jour. Geol. Soc. India*, , v. 22, pp. 405 – 416.

Drury, S.A., 1983a. The petrogenesis and setting of Archaean metavolcanics from Karanataka state, South India. *Geochem. Cosmochem.* Acta, 47, pp. 317 – 329.

Drury, S.A., 1983b. A regional tectonic study of Archean Chitradurga greenstone belt, Karnataka, based on Landsat interpretations. *Jour. Geol. Soc. India*, v. 24, pp. 167–184.

Drury, S.A. 1984. A Proterozoic intracratonic basin, dyke swarms and thermal evolution of South India. *Jour. Geol. Soc. India*, v. 25, pp. 437–444.

Drury, S.A., 1990. SPOT image data as an aid to structural mapping in the southern Aravalli hills of Rajasthan. India. *Geol. Mag.*, v. 127, pp. 195–207.

Drury, S.A., Harris, N.B.W., Holt, R.W., Reeves–Smith, G.J., and Weightman, R.T., 1984. Precambrian tectonites and crustal evolution in South India. *Journal Geol.*, v. 92, pp. 3–20.

Drury, S.A., Holt, R.W., Van Calsteren, P.C., and Beckinsale, R.D., 1983. Sm–Nd and Rb–Sr ages for Archaean rocks in western Karnataka, South India. *Jour. Geol. Soc. India*, v. 24, pp. 454 – 459.

Dubey, V.S. and Dixit, P.C., 1962. A study of magnesite and its origin in Almora district, UP (Abst.). *49th Indian Science Cong.*, v. 3, pp. 140–191.

Duchene, S., Lardeaux, J.M. and Alberede, F., 1997. Exhumation of eclogites : Insights from depth–time path analysis. *Tectonophysics*, v. 280, pp. 125–140.

Duncan, R.A. and Pyle, D.G., 1988. Rapid eruption of Deccan flood basalts, Western India. In K. V. Subbarao (Ed), *Deccan Flood Basalts, Jour. Geol. Soc. India*, Bangalore, pp. 1–10.

Dunn, J.A., 1929. The geology of North Singhbhum including parts of Ranchi and Manbhum districts. *Geol. Surv. India, Mem.* 54.

Dunn, J.A., 1937. The mineral deposits of Eastern Singhbhum and surrounding areas. *Geol. Surv. India, Mem.* 69(1).

Dunn, J.A., 1940. The stratigraphy of South Singhbhum. *Geol. Surv. India*, Mem. 63(3).

Dunn, J.A., 1941a. Origin of the banded hematite ores in India. *Econ. Geol.*, v. 36, pp. 355–370.

Dunn, J.A., 1941b. The Economic geology and mineral resources of Bihar Province. *Geol. Surv. India.*, Mem. 78.

Dunn, J.A., 1953. Banded hematite quartzite. *Econ. Geol.*, v. 48, pp. 58–62.

Dunn, J.A. and Dey, A.K., 1942. The geology and petrology of eastern Singhbhum and surrounding areas. *Geol. Surv. India*, Mem. 69(2).

Dutrow, B.L., Foster, C.T. Jr. and Henry, D.J., 1999. Tourmaline rich pseudomorphs in sillimanite zone metapelites : Demarcation of an infiltration front. *Amer. Mineral*, v. 84, pp. 794–805.

Dutta, B., 1998. Stratigraphic and sedimetological evolution of Proterozoic siliciclastics in the southern parts of Chhattisgarh and Khariar, Central India. *Jour. Gol. Soc. India*, v. 51, pp. 345–360.

Dutta, D.R. and Dutta, N.K., 1990, Geology of Bilaigarg–Giraug area, Raipur district, M.P., *Records, Geol. Surv. India*,123(6), p. 46.

Dutta, J.C., Singh, R.J. and Dutta, S., 1999. Detailed study along Great Boundary Fault. *Records, Geol. Surv. India*, (Extended Abstracts of W.Region FS–1996–97), v. 131(7), pp. 5–8.

Dwivedi, G.N., Sharma, D.P., Mangla Prasad, Singh, V.P., Tripathi, A.K., Yadav, M.L., Mishra, R., Khan, M.A, and Absar, A., 2001. Gold exploration in Sona Pahari, Sonbhadra district, Uttar Pradesh – an introspect. *Geol. Surv. India, Spl. Pub.* 58, pp. 361–365.

Dwivedi, K.K., 1999. Mineral chemistry of pegmatitic minerals in parts of Bastar district, Madhya Pradesh, India. *Indian Journal Appld. Geochem.*, v.1(1), pp. 1–22.

Dwivedy, K. K.,1995. Economic aspects of the Cuddapah basin with special reference to uranium – an overview. *Sem. vol. on Cudapah basin (Tirupati'95), Geological Society of India*, pp. 116–138.

Dymek, R.F. and Owens, B.E., 2001. Petrogenesis of apatite-rich rocks (nelsonites and oxide–apatite gabbronorites) associated with massif anorthosites. *Econ. Geol.* v. 96, pp. 797–816.

Eales, H.V., de Klerk, W.J. and Teigler, B., 1990. Evidence for magma mixing processes within the critical and lower zones of the northwestern Bushveld Complex. *Chemical Geology*, v. 88., pp. 261–278.

Elderfield, H. and Greaves, M.J., 1981. Negative cerium anomalies in the rare earth element patterns of oceanic ferromanganese nodules. *Earth Planet Sci. Letters*, v. 55, pp. 163–170.

England, P. and Houseman, G., 1984. On the geodynamic setting of kimberlite genesis. *Earth Planet. Sci. Letters*. v. 67, pp. 109–122.

England, P.C. and Thompson, A., 1984. Pressure–temperature–time paths of regional metamorphism. I. Heat transfer during the evolution of regions of thickened continental crust. *Journal of Petrology*, v. 25, pp. 894–928.

Eriksson, P.G., Majumdar, R., Sarkar, S., Bose, P.K. and Altermann, W., 1999. The 2.7–2.0 Ga volcano-sedimentary record of Africa, India and Australia: evidence for global and local changes in sea level and continental free board. *Precamb. Research*, v. 97, pp. 269–302.

Ernst, R.E. and Buchan, K.L., 2003. Recognising mantle plumes in the geological record. *Earth Planet. Sci.*, v. 31, pp. 469–523.

Ernst,W.G., 2007. Speculations on evolution of terrestrial lithosphere–asthenosphere system——plumes and plates. *Gondwana Res.*, v. 11, pp. 38–49.

Eshwara, 1991. Preliminary investigation for sulphide mineralisation around Ragidoka, Tirbur, Tai Badak, West Siang district. *Records, Geol. Surv. India*, v. 124(4), pp. 22–25.

Faccenda, M., Gerya, T.V. and Chakraborty, S., 2006. Styles of incipient orogeny: insight from numerous modeling and observations. *EOS Trans. AGU*, 87 (52), *Suppl. Abst, T13C–0523.*

Faccenda, M., Gerya, T.V. and Chakraborty, S., 2008. Styles of post–subduction collisional orogeny: influence of convergence velocity, crustal rheology and radiogenic heat production. *Lithos*, v. 103, 257–287.

Fareeduddin and Bhattaacharjee, 2000. Tectonics and metallogeny along the Phulad lineatment zone, Rajasthan. In M. Deb (Ed.), *Crustal Evolution and metallogeny in the Northewestern Indian Shield.* Narosa, New Delhi, pp. 395–416.

Fareeduddin, and Kroner, A., 1998. Single zircon age constraints on the evolution of Rajasthan granulite. In B.S. Paliwal (Ed.), *The Indian Precambrian*, Scientific Publishers, Jodhpur, pp. 547–556.

Fareeduddin, Mishra, V.P. and Basu, S.K., 2007. Kimberlites, lamproites and lamprophyres of India: A petrographic atlas. *Jour. Gol. Soc. India*, v. 69, pp. 467–504.

Fermor, L.L., 1908. Apatite–magnetite rock from the Singhbhum district, Bihar. *Records, Geol. Surv. India*, v. 36, pp. 239–319.

Fermor, L.L., 1909. The manganese ore deposits of India. *Geol. Surv.India*, Mem. 37, pp. 1–4.

Fermor, L.L., 1921. The mineral resources of Bihar and Orissa. *Records, Geol. Surv. India*, v. 53, pp. 239–319.

Fermor, L.L., 1936 An attempt at the correlation of the ancient schistose formations of peninsular India. *Geol. Surv. India*, Mem. 70, Pts. 1 and 2.

Ferreara, G., Lombardo, B. and Tonarini, S., 1983. Rb/Sr geochronology of granites and gneisses from Mount Everest region, Nepal Himalaya. *Geol. Rundsch.*, v. 72, pp. 119–136.

Ferro, C.E.G., Watkinson, D.H. and Jones, P.C., 1994. Fluid inclusions in sulphides from North and South Range, Sudbury structure, Ontario. *Econ. Geol.*, v. 89, pp. 647–655.

Field, M. and Scott Smith. B.H., 1999. Contrasting geology and near surface emplacement of kimberlite pipes in South Africa and Canada. In Gurney, J.J. et. al (Eds), *Proc. 7th Internat. kimberlite conference*, Cape Town, pp. 214–237.

Fleischer, M., 1983. Distribution of lanthanides and yittrium in apatites and iron ores and its bearing on the genesis of the ores of Kiruna type. *Econ. Geol.*, v. 78, pp. 1007–1011.

Foley, S., Tiepolo, M. Vanancci, R., 2002. Growth of early continental crust controlled by the melting of amphibolite in subduction zones. *Nature* 417, pp. 837–840.

Force, E.R. and Cannon, W.F., 1988. Depositional model for shallow–marine manganese deposits around black shale basins. *Econ. Geol.*, v. 83, pp. 93–117.

Fournier, R.O., 1999. Hydrothermal processes related to movement of fluid from plastic to brittle rock in the magmatic–epithermal environment. *Econ. Geol.*, v. 94, pp. 1193–1212.

Fox, C.S., 1923. The bauxite and laterite occurrences of India. *Geol. Surv. India,* Mem. 49, p. 287.

Frakes, L.A. and Bolton, B.R., 1984. Origin of manganese giants: sea level change and anoxic–oxic history. *Geology*, v. 12, pp. 83–86.

Frakes, L.A. and Bolton, B.R., 1992. Effects of ocean chemistry, sealevel and climate on the formation of primary sedimentary manganese ore deposits. *Econ. Geol.*, v. 87, pp. 1207–1217.

Franklin, B.J., Marshall, B., Graham, I.T. and McAndrews, J., 1992. Remobilization of PGE in podiform chromite in the Coolac serpentine belt, Southeastern Australia. *Aust. Journal Earth Sci.*, v. 39, pp. 365–371.

Franklin, J.M., 1996. Volcanic–associated massive sulphide base metals. In O. R. Eckstrand, W. D., Sinclair and R. I. Thorpe (Eds.), Geology of Canadian mineral deposit types. *Geol. Surv. Canada*, pp. 158–183.

Franklin, J.M., Sangster, D.M. and Lydon, J.W., 1981. Volcanic associated massive sulphide deposits. *Econ. Geol.*, 75th Anniversary vol. pp. 485–627.

Freyssinet, Ph., Butt, C.R.M., Morris, R.C. and Piantone, P., 2005. Ore forming processes related to lateritic wealthering. *Econ. Geol.*, 100th Anniversary vol., pp. 681–722.

Friend, C.R.L. and Nutman, A.P., 1992. Response of U–Pb isotopes and whole rock geochemistry to CO_2 induced granulite mitamorphism, Kabbaldurga, Karnataka, South India. *Jour. Geol. Soc. India*, v. 38, pp. 357–368.

Friend, C.R.L. Nutman, A.P., 1991. SHRIMP U–Pb geochronology of the Closepet Granite and Peninsular Gneisses, Karnataka, South India. *Jour. Geol. Soc. India*, v. 38, pp. 357–368.

Frietsch, R., 1978. On the magmatic origin of iron ores of Kiruna type. *Econ. Geol.*, v. 73, pp. 478–485

Frimmel, H.E., 1997 Detrital origin of hydrothermal Witwatersrand gold – a review. *Terra Nova*, v. 9, pp. 192–197.

Frimmel, H.E., Groves, D.I., Kirk, J., Ruiz, J., Chesley, J., and Minter,W.E.L., 2005. The formation and preservation of the Witwatersrand gold–fields, the world's largest gold province. *Econ. Geol.*,100th Anniversary vol., pp. 769–797.

Froelich, P.N., Bender, M.L. and Health, G.R., 1977. Phosphorous accumulation rates in metalliferous sediments on the East Pacific Rise. *Earth Planet Sci. Letters*, v. 34, pp. 351–359.

Furnes, H., de Wit, M.J., Staudigel, H., Rosing, M. and Muehlenbachs, K.,2007. A vestige of Earth's oldest ophiolite. *Science*, v. 315, pp. 1704–1707.

Fyfe, W.S., 1978. The evolution of the Earth's crust: Modern plate tectonics to ancient hot spot tectonism. *Chem. Geol.*, v. 23, pp. 89–114.

Fyfe, W.S., 1998. Tectonics and chemical transport: The rules of the resource game. *Gondwana Res*. v.1(3/4), pp. 315–318

Gaal, G., 1964. Precambrian flysch and molasse–tectonics and sedimentation around Rakha Mines– Jaikan, Singhbhum dist., Bihar, India. *22nd Internat. Geol. Cong. Proc., Sec. 4*, pp. 331–356.

Gadadharan, A. and Jog, R.G., 1990. Stratabound zinc and tungsten mineralisation in Kolari–Bhaonri area, Maharashtra. *Geol. Surv. India., Spl. Pub. 28*, pp. 696–713.

Gaetani, M., and Garzanti, E., 1991. Multi-cyclic history of the Northern Indian continental margin (Northwestern Himalaya). *Amer. Asson. Pet. Geologists Bull.* 75, pp. 1427–1446.

Gandhi, S.M., 1983. Rampura Agucha lead–zinc deposit. *Mining Mag.*, November, pp. 315–321.

Gandhi, S.M., 2000. Ancient mining and metallurgy in Rajasthan. In M.Deb (Ed.), *Crustal evolution and metallogeny in the north western Indian shield*, Narosa Publ. House, New Delhi, pp. 29–48.

Gandhi, S.M., Paliwal, H.V. and Bhatnagar, S.N., 1984. Geology and ore reserve estimates of Rampura–Agucha lead–zinc deposit, Bhilwara district. *Jour. Gol. Soc. India*, v. 25, pp. 689–705.

Gangadharan, E.V., Rao, K.K. and Aswathanarayana, U., 1963. Distribution of radioactivity in Mosaboni copper mine, Bihar. *Econ. Geol.*, v. 58, pp. 506–514.

Ganging, J., Linda, E.S. and Christie–Blick, N., 2003. Proterozoic stratigraphic comparison of the Lesser Himalaya (India) and Youngte block (South China): Paleogeographic implications. *Geology*, v. 31, pp. 917–920.

Gangopadhyay, A., 1998. A preliminary study of the bauxitic ores and associated clay deposits at the Bagru Hills, Lohardaga, Bihar. *Unpub. M.Sc dissertation*, *Jadavpur University,* Kolkata, p. 42.

Gangopadhyay, K.K.and Sengupta, S., 2004. Tourmalinites and its significance as pathfinder in search of mineralisation in Singhbhum Group of rocks, Bankura district, West Bengal. *Geol. Surv. India, Spl. Pub.* 72, pp. 313–318.

Gangopadhyay, K.K.and Sengupta, S., 1998. Report of tourmalinite occurrence from Singhbhum Group of rocks in Bankura district, West Bengal and its significance. (Abst.) *M.S.Krishnan Comm. Nat. Sem. on progress in Precambrian geology of India. Geol. Surv. India*, Calcutta, pp. 62–63.

Gangopadhyay, P. and Roy, A. 1997. Petrotectonic implications of rhyolite–granite suite of Dongargarh Supergroup of Central India. *Indian Minerals,* v. 57, pp. 123–126.

Gangopadhyay, P.K., 1972. Structure and tectonics of Alwar region in Northeastern Rajasthan, India, with special reference to Precambrian stratigraphy. Proc. *24th Internat. Geol. Cong. Montreal*, Section 10, pp. 118–125.

Gangopadhyay, P.K. and Lahiri, A., 1988. Anjana Granite and associated rocks of Deogarh, Udaipur district, Rajasthan. In A.B. Roy (Ed.), Precambrians of the Aravalli Moventain Rajasthan, India. *Geological Society of India*, Mem. 7, pp. 307–316.

Gangopadhyay, P.K. and Mukhopadhyay, D., 1987. Structural geometry of the Delhi Supergroup near on tectonic structures. In A.K. Saha (Ed), *Geological evolution of Peninsular India – Petrological and structural aspects: Recent Researches in Geology*. Hindustan Publ. Corp., Delhi, pp. 45–60.

Ganguly, J., 1969. Chloritoid stability and related paragenesis: theory, experiments and applications. *Amer. Journal Sci.*, v. 267, pp. 910–944.

Ganguly, J. and Bhattacharya, P. K., 1987. Xenoliths in Proterozoic kimberlites from Southern India: Petrology and geophysical implication. In P. H. Nixon (Ed.), *Mantle Xenoliths*. John Wiley and Sons Ltd., pp. 249–265.

Ganguly, J. and Singh, R.N., 1988. Charnockite metamorphism: Evidence for thermal perturbation and potential role of CO_2. EOS *Amer. Geophys. Union Trans.*, v. 69, pp. 509.

Ganguly, J., Dasgupta, S., Cheng, W. and Neogi, S., 2000. Exhumation history of a section of the Sikkim Himalaya, India: records in the metamorphic mineral equilibria and compositional zoning of garnet. *Earth and Planet Sci. Letters*, v. 183, pp. 471–486.

Ganguly, J., Singh, R.N. and Ramana, D.V., 1995.Thermal perturbation during charnockitization and granulite facies metamorphism in the Southern India. *Journal Met. Geol.*, 13, pp. 419–430.

Gansser, A., 1964. Geology of Himalaya. *Interscience Publishers*, John Willey and Sons, London, p. 289.

Gansser, A., 1981. The geodynamic history of the Himalaya. In H.K. Gupta and F.M. Delany (Eds), Zagros–Hindukush–Himalaya geological evolution. *Geodynamic Series*, v.3, Amer. Geophys. Union. GSA., pp. 111–121.

Gansser, A., 1993. Facts and theories on the Himalayas. *Jour. Gol. Soc. India*, v. 41, pp. 487–508.

Gan Sengfei, 1992. To find gold deposits in Archean high grade terrains. *A print of the paper presented in 29th Dec,* Kyote, 19p.

Garia, S.S., Jat, R.L.. and Harpawat, C.L., 2001. Gold mineralisation and its controls in Bhukia prospect, southeastern Rajasthan. Proc. M.S.Krishnan Comm. Nat. Sem. on Progress in Precambrian Geology of India (1998), *Geol. Surv. India*, *Spl. Pub.* 55(II), pp. 129–136.

Garia, S.S., Rao, K.N. and Jat, R.L., 1998. Keratophyres in Lower ProterozoicsofAravalli fold belt, Rajasthan, India. (Abst.) *M.S.Krishnan Comm. Nat. Sem. on Progress in Precambrian Geology of India, Geol. Surv. India*, Calcutta, pp. 64–66.

Garrels, R.M. and Christ, C.L. 1965. Solutions, minerals and equilibria. Harper and Row, p. 450.

Garven, G.Ge.S., Person, M.A. and Sverjensky, D.A., 1993. Genesis of stratabound ore deposits in the mid-continent basins of North America: The role of regional groundwater flow. *Amer. Journal Sci.*, v. 293, pp. 497–568.

Gathania, R.C. and Golani, P.R., 1988. Felsic volcanics from the Khetri copper belt, Rajasthan. *Current Sci.*, v. 57, pp. 544–546.

Geiger, P. and Odman, O.H., 1974. The emplacement of the Kiruna ores and the related deposits. *Sverig. Geol. Undersok.* C 700, p. 48.

Genkin, A.D. and Schmidt, S.M., 1991. Preliminary data of a new thallium mineral from the lead–zinc ore deposit, Agucha, India. *Neues. Jahrb. Miner.Monash*, pp. 256–258.

Genkin, A.D., Safonov, Yu.G., Boronikhin, V.A., Krishna Rao, B. and Vasudev, V.N., 1988. Mineralogy and geochemistry of the Kolar Goldfield. In F.V.Chukhrov (Chief Editor), *Kolar Goldfield,* Nauka, Moscow, pp. 94–157 (In Russian).

Geological Survey of India (GSI), 1981. Brochure on geological and mineral map of Cuddapah basin, pp. 1–21.

Geological Survey of India (GSI), 1994. Detailed information on copper–lead–zinc ores in Madhya Pradesh–Maharashtra, India; Unpub. Report, GSI., pp. 1–17.

Geological survey of India (GSI) 1998. Geological map of the Eastern Ghat Mobile bell (Scale 1:1 million).

Geological Survey of India (GSI), 2000. Status of mineral resources in India (Part A) and Profiles of significant prospects (Part B), *GSI Professional document, Dec 2000.*

Geological Survey of India (GSI), 2001. Geology and mineral resources of Rajasthan *Misc. Pub.* 30(12), pp. 113

Geological Survey of India (GSI), 2004. Anupendu Gupta (Ed.), A Manual of the Geology of India, Vol. I, Part–IV: Northern and Northwestern Part of Peninsula,(Compiled by C. Chakrabarti., T. K. Pyne, P. Gupta, S. Basu Mallick, and D. Guha), *Geol. Surv. India, Spl. Pub.* 77, p. 257.

Geological Survey of India (GSI), 2004–05. *News Letters*, Eastern Region.

Geological Survey of India (GSI), 2006. A. Gupta and A. Chaudhuri (Eds.), A Manual of the Geology of India, Vol. I, Part–I: Southern Part of Peninsula. (Compiled by C. Chakrabarti, S. Basu Mallick, T. K. Pyne and D. Guha), *Geol. Surv. India, Spl. Pub.* 77, p. 572.

Geological Survey of India (GSI), 2006. Deccan Volcanics – A Field Guide book (Ed. K. S. Misra). *Geol. Surv. India*, Training Inst., Hyderabad.

Geological Survey of India (GSI),1996. Deccan Flood Basalt – A Pictorial Atlas. *Geol. Surv. India, Catalogue series*, p. 290.

George, T.S. 1995. Occurrence of gold in Bhima Basin, Gulbarga District, Karnataka, *Jour. Geol. Soc. India*, v.45, pp. 235–236.

Ghose, A., 2000a. Precambrian metamorphism, magmatism, and tectonic events in the Himalayan region. *Geol. Surv. India Spl. Publ.* 55, pp. 51–68.

Ghose, A., 2000b. Tectono–stratigraphic terrane elements and crustal evolution in the Himalayan and adjacent regions. *Indian Journal Geol.*, v. 72, pp. 233–255.

Ghose, Arabinda, 2006. Metallogenic characteristics in relation to tectonic framework of Himalaya. *Geol. Surv. India*, Mem. 132, 232 p.

Ghose, N.C., 1983. Geology, tectonics and evolution of Chhotanagpur granite–gneiss complex, Eastern India. *Recent Researches in Geology*, v. 10, pp. 211–247.

Ghose, N.C., Basu, A., Bhaumik, N. and Saran, A.K., 1984. Petrography and chemistry of the enclaves within the Chakradharpur granite and their significance in the crustal evolution of the Singhbhum craton, Eastern India. Indian Journal Earth Sciences, *(CEISM Seminar vol. I)*, pp. 50–63.

Ghosh, A., 1975. Abundance of Co. Ni, Cd, Ag and Hg in the polymetallic sulfide deposit of Rangpo, Sikkim. *Jour. Geol. Soc. India*, v. 16, pp. 99–102.

Ghosh, A. and Bhattacharya, A., 1998. Geological and tectonic significance of Adas–Rengali–Sukinda area, Orissa. (Abst.) M.S.Krishnan Comm. *Nat. Sem. on Progress in Precambrian Geology of India*, *Geol. Surv. India, Calcutta,* pp. 70–71.

Ghosh, A.K., 1972a. A preliminary evaluation of sulfur isotope studies of sulphide minerals from the copper ore deposits of the Singhbhum Shear Zone, Eastern India. *Econ. Geol.*, v. 67, pp. 818–820.

Ghosh, A.K., 1972b. Trace element geochemistry and genesis of copper ore deposits of the Singhbhum shear zone, Eastern India. *Mineral. Deposita,* v. 7, pp. 292–313.

Ghosh, A.K., 1995. Abundance of Co, Ni, Cd, Ag and Hg in polymetallic sulphide deposit of Rangpo, Sikkim *Journal Jour. Geol. Soc. India,* v. 16, pp. 99–102.

Ghosh, A.K., Thorpe, R.I.and Chakraborty, C.K., 2005. Interpretation of Pb and S isotope data for the Ganesh Himal metamorphosed stratiform Zn–Pb deposit, Central Nepal Himalaya. *Jour. Geol. Soc. India,* v. 65, pp. 725–737.

Ghosh, B., Parveen, M. N. and Srivastava, H.S., 2006. Gahnite chemistry from metamorphosed Zn–Pb–Cu sulphide occurrences of Betul belt, Central India. *Jour. Geol. Soc. India,* v. 67, pp. 17–20.

Ghosh, D., Das, J.N., Rao, A.K., Ray–Barman, T., Kolhapuri, V.K., and Sarkar, A., 1994. Fission–track and K–Ar dating of pegmatite and associated rocks of Nellore schishbelt, Andhra Pradesh: evidence of Middle to Late Proterozoic events. *Indian Minerals*, v. 48, pp. 95–102.

Ghosh, D.K., Sarkar, S.N., Saha, A.K., Ray, S.L., 1996. New insights on the early Archean crustal evolution in Eastern India: Re-evaluation of Pb-Pb, Sm-Nd and Rb-Sr geochemistry. *Indian Minerals*, v.50, pp. 175–188.

Ghosh, D.B., Sastry, B.B.K., Rao A.J. and Rahim, A.A., 1970. Ore environment and ore genesis in Ramagiri gold field, Andhra Pradesh, India. *Econ. Geol.*, v. 65, pp. 801–814.

Ghosh, G. and Mukhopadhyay, J., 2007. Reappraisal of the structure of the western Iron Ore Group, Singhbhum craton, eastern India: Implications for exploration of BIF-hosted iron ore deposits. *Gondwana Res.*, v. 12, pp. 525–532

Ghosh, G. and Saha, D., 2003. Deformation of the Proterozoic Somanpalli group, Pranhita–Godavari – implications for a Mesoproterozoic basin inversion. *Journal Asian Earth Sci.*, v. 21, pp. 579–594.

Ghosh, J.G., 2004. 3.56 Ga tonalite in the central part of the Bastar Craton,India: oldest Indian date. *Journal Asian Earth Sci.*, v. 23, pp. 359–364.

Ghosh, J.G. and Pillai, K.R., 1992. Tectono–magmatic evolution of the Kotri Lineament Zone – a study from Kondrunz–Mendra area, Bastar district, M. P. *Records, Geol. Surv. Ind*ia, 125(6), pp. 17–19.

Ghosh, J.G. and Pillai, K.R., 2002. Tectonostratigraphy of the Kotri linear belt and petrochemistry of Bijli rhyolite. *Geol. Surv. India*, *Unpub. Prof. Paper*.

Ghosh, K.P., 1983. Generalised structural and metamorphic map around Batia, Udaipur district, Rajasthan: A study. *Ind. Journal Earth Sci.*, v. 40, pp. 228–231.

Ghosh, K.P., 1993. Genesis of the Banded Iron Formation in the Jamda–Koira Valley – a tentative depositional model. *Ind. Journal Earth Sciences*, v. 20, pp. 163–172.

Ghosh, K.P. and Mc Farlane, M.J., 1984. Some aspects of the geology and geomorphology of the bauxite belt of Central India. In S. Sinha–Roy and S.K.Ghosh (Eds.), *Recent Researches in Geology*, v.11, Hindusthan Pub. Co., pp.118–135.

Ghosh, M., Mukhopadhyay, D. and Sengupta, P., 2006. Pressure–temperature–deformation history for a part of the Mesoproterozoic fold belt in North Singhbhum, Eastern India. *Journal Asian Earth Sci.*, v. 26, pp. 555–574.

Ghosh, P.K., 1941. The Charnockite Series of Bastar Series of Bastar State and western Jeypore. *Records, Geol. Surv. India*, v. 91, *Prof. paper* 15, p. 55.

Ghosh, P. K., Chandy, K.C., Bishui, P.K. and Prasad, R., 1986. Rb–Sr age of granite gneiss in Malanjkhand area, Balaghat district, Madhya Pradesh. *Indian Minerals*, v. 40, pp. 1–8.

Ghosh, S., 1957. Lead–zinc–silver mineralisation at Zawar, Rajasthan. *Quart. Jour. Geol. Min. Met. Soc. India*, v. 29, pp. 55–64.

Ghosh, S. and Ghosh, B.K., 1979. Some aspects of mineralisation in Chhendapather tungsten field Bankura, WB. In Mineralisation associated with acid magmatism, *Geol. Surv. India, Spl. Pub.*, v. 13, pp. 145–150.

Ghosh, S., Chakraborty, S., Bhalla, J.K., Paul, D.K., Sarkar, A., Bishui, P.K. and Gupta, S.N. ,1991. Geochronology and geochemistry of granite plutons from East Khasi Hills. *Jour. Geol. Soc. India,* v. 37(4), pp. 331–342.

Ghosh, S., Chakraborty, S., Paul, D.K., Bhalla, J.K., Bishui, P.K. and Gupta, S.N., 1994. New Rb–Sr isotopic ages and geochemistry of granitoids from Meghalaya and their significance in middle to late Proterozoic crustal evolution. *Indian Minerals,* v. 48(1and2), pp. 33–44.

Ghosh, S., Fallick, A.E., Paul, D.K., Potts, P.J., 2005. Geochemistry and origin of Neoproterozoic granitoids of Meghalaya, Northeast India: Implications for linkage with amalgamation of Gondwana Supercontinent. *Gondwana Res.*, v. 8, pp. 421–432.

Ghosh, S., Rajaya, V. and Ashiya, I.D., 1995. Rb–Sr dating of components from Sonakhan granite–greenstone belt, Raipur district, Madhya Pradesh. *Records, Geol. Surv. India*, v. 128(2), pp. 11–13.

Ghosh, S.K., 1963. Structural, metamorphic and mignatitic history of the area around Kuilopal, eastern India. *Quart. Jour. Geol. Min. Met. Soc. India*, v. 35, pp. 211–234.

Ghosh, S.K. and Chatterjee, B.K., 1990. Paleoenvironmental reconstruction of the early Proterozoic Kolhan siliciclastic rocks, Keonjhar district, Orissa, India. *Jour. Geol. Soc. India,* v. 35, pp. 273–286

Ghosh Roy, A.K., 1989. Investigation for apatite and other associated minerals in the Todra-Porapahas shear zone in Purulia district and the petrological studies of the asssociated carbonates with special reference to genetic aspects. *Records, Geol. Surv. India*. v. 122(3).

Ghosh Roy, A.K., and Gangopadhyay, P., 1998. Geology of the terrain north of the Dalma volcanic belt: some recent observations and their significance. (Abs.) *Nat. Sem. on Geosci. Advns. in Bihar*, India, pp. 7–8.

Gill, R.C.O., 1979. Comprehensive petrogenesis of Archean and modern low–K tholeiite – a critical review of some geochemical aspects. In L.H. Ahrens(Ed.), *Origin and Distribution of Elements*. Pergamon Oxford, pp. 431–447.

Gilmour, P., 1985. On the magmatic origin of iron ores of Kiruna type – A further discussion. *Econ. Geol.*, v. 80, pp. 1753–1757.

Godbole, S.M., Rana, R.S. and Natu S.R., 1996. Lava stratigraphy of Deccan basalts of western Maharashtra. Gondwana. Proc. National Symp. on Deccan Flood Basalts, India. *Geol. Mag.*, Spl. vol. 2, pp. 125–134.

Gogoi, K., 1973. Geology of Precambrian rocks in northwestern parts of the Khasi and Jaintia Hills, Meghalaya. *Geol. Surv. India, Misc. Pub.*, 23, pp. 37–48.

Gokaran, S.G., Rao, C.K. and Singh, B.P., 1995. Crustal structure in southeast Rajasthan using magneto–telluric technique. In S. Sinha Roy and K.R. Gupta (Eds), *Continental crust of northwestern and central India, Geol. Soc. India*, *Mem.* 31, pp. 373–383.

Gokul, A., 1983. Geological and mineral map of Bhutan (Scale 1:500,000), *Geol. Surv. India.*

Golani, P.R., 1996. Lithologic and structural controls on sulphide–gold mineralisation in Bhukia–Jagpura area, Banswara district, Rajasthan. (Abst.) *Workshop on gold resources of India, Geological Society of India*, NGRI, Hyderabad, p. 161.

Golani, P.R., Bandyopadhyay, B.K. and Gupta Anupendu, 2001. Gavilgarh–Tan shear: A prominent ductile shear zone in Central India with multiple reactivation history. *Geol. Surv. India*, *Spl. Pub. 64*, pp. 265–272.

Golani, P.R., Gathania, R.C., Grover, A.K. and Bhattacharjee, J., 1992. Felsic volcanics in South Khetri belt, Rajasthan and their metallogenic significance. *Journal Geological Society of India*, v. 40, pp. 79–87.

Golani, P.R., Rajawat, R.S., Pant, N.C. and Rao, M.S., 1999. Mineralogy of gold and associated alloys in the sulphides of Bhukia gold prospect in southeastern Rajasthan, Western India. *Jour. Geol. Soc. India*, v. 54, pp. 121–128.

Golani, P.R., Dora, M.L. and Bandopadhyay, B.K. 2006. Base metal mineralization associated with hydrothermal alteration in felsic volcanic rocks in Proterozoic Betul belt at Bhyari, Chhindwara distric, Madhya Pradesh. *Jour. Geol. Soc. India*, v. 68, pp. 797–808.

Golani, P.R., Reddi, A.B., Bhattacharjee, J. and Mathur, K.N., 1996. Crustal evolution and related contracting metallogeny in the Proterozoic South Delhi fold belt of Western India. *(Abst.) Nat. Sem. on Mineralisation in the Western Indian carton*, Univ. of Delhi, pp. 31.

Golani, P.R. and Narayan, S., 1990. Study of volcanics in Delhi supergroups of rocks in relation to base metal metallogeny in central and southern Rajasthan and morthern Gujarat. *Records, Geol. Surv. India* , v. 123(7), pp. 82–88.

Goldfarb R.J., Baker, T., Dube, B., Groves, D. I., Hart, C. J. R. and Gosselin, P.T., 2005. Distribution, character and genesis of gold deposits in metamorphic terranes. *Econ. Geol.*, 100th Anniversary vol., pp. 407–450

Goldfarb, R.J., Groves, D.I. and Gardoll, S., 2001. Orogenic gold and geologic time: A global synthesis. Ore *Geology Reviews*, v. 18, pp. 1–75.

Goldfarb, R.J., Hart, C., Davis, G. and Groves, D.I., 2007. East Asian gold deciphering the anomaly of Phanerozoic gold in Precambrian cratons. *Econ. Geol.*, v. 102, pp. 341–345.

Gole, M.J. and Klein, C., 1981. Banded iron formations through much of Precambrian time. *Journal Geol.*, v. 89, pp. 169–183.

Golightly, P.J., 1994. The Sudbury igneous complex as an impact melt: Evolution and ore genesis. *Ontario Geol. Surv.*, Spl. Pub. 5, pp. 105–117.

Goodfellow, W.D., 2004. Geology, genesis and exploration of SEDEX deposits, with emphasis on the selwin basin, Canada. In M. Deb and W.D. Goodfellow (Eds.), *Sediment-hosted lead-zinc sulfide deposits*. Narosa, New Delhi, pp. 24–99.

Goodwin, A.M., 1996. Principles of Precambrian Geology. *Acad Press,* London, 327p.

Gopala Reddy, T. and Viswanatha Rao, N., 1992. Study of granitic rocks in the vicinity of Raichur–Gadwal schist belt in Karnataka and Andhra Pradesh. *Records, Geol. Surv. India*, v. 125(5), pp. 247–248.

Gopalan, K. and Bhaskar Rao, Y.J., 1995. A critical appraisal of geochronological results on the Cuddapah and related rocks. *Seminar vol. on Cuddapah basin (Tirupati 95)*, pp.116–138.

Gopalan, K., MacDougall, J.D., Roy, A.B. and Murali, A.V., 1990. Sm–Nd evidence for 3.3 Ga old rock in Rajasthan, northwestern India. *Precamb. Research*, v. 48, pp. 287–297.

Gopalan, K., Trivedi, J.R., Balsubrahmanyan, M.N., Ray, S.K., and Sastry, C.A., 1979. Rb–Sr geochromology of the Khetri copper belt, Rajasthan. *Jour. Geol. Soc. India,* v. 20, pp. 450–456.

Goswami, J.N., Misra, S., Wiedenbeck, M., Ray, S.L., Saha, A.K., 1995. $^{207/206}$ Pb ages from the OMG, the oldest recognised rock unit from Singhbhum–Orissa Iron Ore craton, E. India. *Current Sci.*,v. 69, pp. 1008–1012.

Greenwood, H.J., 1976. Metamorphism at moderate temperatures and pressures. In D. K. Bailay and R. McDonald (Eds), *The evolution of crystalline rocks*. Acad. Press Inc., London, pp. 187–259.

Grew, E.S. and Manton, W.I., 1986. A new correlation of sapphirine granulite in Indo-Antarctic metamorphic terrane: Late Proterozoic dates from Eastern Ghats. *Precamb. Research,* v.33, pp.123-139.

Groenewald, P.B., 1993. Correlation of cratonic and orogenic provinces in south–eastern Africa and Dronning Maud Land, Antarctica. In Findley, R.H., Unrug, R., Banks, M.R. and Veevers, J.J. (Eds.), *Gondwana Eight: Assembly, evolution and dispersal*, A. A. Balkema, Rotterdam, pp. 111–122.

Gross, G.A., 1965. Geology of iron ore deposits of Canada. 1–General geology and valuation of iron deposits. *Geol. Surv. Canad. Econ. Geol. Rep. No. 22*, p. 181.

Gross, G.A., 1973. The depositional environment of the principal types of Precambrian iron formations. In Genesis of the Precambrian iron and manganese deposits. *UNESCO Earth Sci.*, v. 9, pp.15–21.

Gross, G.A., 1980. A classification of iron formations based on depositional environments. *Canadian Mineral.*, v. 18, pp. 215–222.

Grover, A.K. and Verma, R.G., 1993. Gold mineralisation in the Precambrians (Bhukia area) of southeastern Rajasthan – A new discovery. *Jour. Geol. Soc. India,* v. 42(3), pp. 181–188.

Grover, A.K. and Verma, R.G., 1997. Gold mineralisation in Hinglaz Mata–Bharkundi area, Dungarpur district – an extention of Bhukia–Kundli auriferous tract of Rajasthan. *Jour. Geol. Soc. India,* v. 49, pp. 443–447.

Groves, D.I. Bierlein, F.P., Meinert, L.D. and Hitzman, M.W., 2010. Iron oxide copper-gold (IOCG) deposits through Earth history: Implications for origin, lithospheric setting, and distinction from other epigenetic iron oxide deposits. *Econ. Geol.*, v. 105, pp. 641–654.

Groves, D.I., Goldfarb, R.J., Gehre–Marium, M., Hagemann, S.G. and Robert, F., 1998. Orogenic gold deposits: a proposed classification in the context of their crustal distribution and relation to other gold deposit types. *Ore Geol. Rev.*, v. 13, pp. 7–27.

Groves, D.I., Goldfarbs, R.J., Robert. F. and Hart, C.J.R., 2003. Gold deposits in metamorphic belts: Overview of current understanding, outstanding problems, future research, and exploration significance. *Econ. Geol.*, v. 98, pp. 1–30.

Grujic, D.E., Casey, M. Davidson, C., Hollister, L.S. Kusdig, R., Pavlis, T. and Schmid, S., 1996. Ductile extrusion of the High Hiamalayan Crystallines in Bhutan: evidence from quartz micro–fabrics. *Tectonophysics*, v. 260, pp. 21–43.

Grunner, J.W., 1930. Hydrothermal oxidation and leaching experiments, their bearing on the origin of Lake Superior hematite–limonite ores. *Econ. Geol.*, v. 25, pp. 697–719, 837–867.

Grunner, J.W., 1937. Hydrothermal leaching of iron ores of the Lake Superior type – a modified theory. *Econ. Geol.* v. 32, pp. 121–130.

Guerot, C., 1993, Geochronological results in the Khetri copper belt (Rajasthan, India). *App. 2–BRGM Report.*

Guha, D., Bardhan, S., Basir, S.R., De, A.K. and Sarkar, A., 2007. Imprints of Himalayan thrust tectonics on the Quaternary piedmont sediments of the Neora–Jaldhaka Valley, Darjeeling–Sikkim Sub-Himalaya, India. *Jour. Asian Earth Scs.*, v. 30, pp. 464–473.

Guha, D.B. and Rai Chowdhary, A., 1999. Detailed study along Great Boundary Fault in Sawai Madhopur sector, Rajasthan. *Rec. Geol. Surv. India, (Extended Abstracts of W. Region FS–1994–95)*, v. 129(7), pp. 5–8.

Guha, D.B. and Bhattacharya, A.K., 1995. Metamorphic evolution of high–grade reworking of the Sandmala Complex granulites. In S. Sinha Roy and K.R. Gupta (Eds.), *Continental crust of northwestern and Central India. Geological Society of India, Mem.* 31, pp. 163–18.

Guha, J., 1971. Sulfur isotope study of the pyrite deposit of Amjhore, Sahabad district, Bihar India. *Econ. Geol.*, v. 66, pp. 326–330.

Guild, P.W., 1972. Massive sulphides vs. porphyry deposits in their global tectonic settings. *Prof. MIJ–AIME Joint Meeting*, Tokyo, pp. 1–12.

Gupta, A., 1975. A study of geology and ore mineralisation at Lawa area, District Singhbhum, Bihar. *Unpub. Ph.D thesis, Jadavpur University*, Calcutta, p. 245

Gupta, A., 1986. Ore mineral assemblage at Lawa gold prospect, Singhbhum district, Bihar, India. *Indian Minerals*, v. 40(3), pp. 17–30.

Gupta, A., 2008. Gold mineralisation in the eastern segment of Indian Precambrian shield: A review. *Pre–workshop (Extd. Abst.) vol., Internat. Field Workshop on Gold Metallogeny in India, Dec. 2008,* NGRI – Hyderabad and Univ. of Delhi, pp. 138–143.

Gupta, A. (Ed.), 2009. Scenario of Indian mineral resources and production with focus on Eastern and Northeastern States and references to global scenario. *Mining Geol. Met. Inst. India* (MGMI) publication, Calcutta Branch, Kolkata, India, p. 92

Gupta, A., 2010. Gold mineralisation in the eastern segment of Indian Precambrian shield: A review: In M. Deb and R.J. Goldfarb (Eds), *Gold Metallogeny: India and beyond.* Narosa, New Delhi, pp. 256–280.

Gupta A., Kar, S.K. and Chakravarti, C., 2002. Report on mineral resources in the Northeastern Region. *Geol. Surv. India,Unpub. Report.*

Gupta, A. and Bandyopadhyay, B.K., 2000. Siderite hosted copper sulphide mineralisation in Palaeozoic sequence of east Himalaya, Bhutan. *(Abst.), IGC–2000.*

Gupta, A. and Bandyopadhyay, B.K., 2009. Palaeozoic metallogeny in the Tethyan Black Mountain Basin, Bhutan Himalaya and its regional implication. *Indian Journal Geosciences*, v. 63(1), pp. 97–106.

Gupta, A. and Basu, A., 1977. Report on the mapping of Dalma volcanic belt in the Tebo–Hesadih sector, Singhbhum district, Bihar. *Geol. Surv. India, Unpub. Report*. (FSP 1976–1977).

Gupta, A. and Basu, A., 1979. On occurrence of pillow lavas in Dalma volcanic suite, Singhbhum and Ranchi districts, Bihar. *Jour. Geol. Soc. India,* v. 20, pp. 42–44.

Gupta, A. and Basu, A., 1985. Structural evolution of the Precambrians in parts of North Singhbhum, Bihar. *Records, Geol. Surv. India*, v. 113(3), pp. 13–23.

Gupta, A. and Basu, A., 1990. Structural evolution of North Singhbhum mobile belt, East Indian shield. (Abst.) *Sem. on Evoln. of Precamb. Crust* (GSI, Calcutta), pp. 24–25.

Gupta, A. and Basu, A., 1991. Evolutionary trend of the mafic–ultramafic volcanism in the Proterozoic North Singhbhum mobile belt. *Indian Minerals*, v. 45(4), pp. 273–288.

Gupta, A. and Basu, A., 1992. Photo–illustration of some significant geological features in the North Singhbhum mobile belt. *Indian Minerals*, v. 46(2), pp. 177–184.

Gupta, A. and Basu, A., 2000. North Singhbhum Proterozoic mobile belt, Eastern India – a review. Proc. M.S. Krishnan Cent. Sem., Calcutta (1998), *Geol Surv. India, Spl. Pub.* 55, pp. 195–226.

Gupta, A., Augustine P.F. and Bandyopadhyay ,1983. Final report on Gongkhola copper investigation, Dist. Tongsa, Bhutan. *Geol. Surv. India, Unpubl. Report.*

Gupta, A., Basu, A. and Ghosh, P.K., 1980. The Proterozoic ultramafic and mafic lavas and tuff of the Dalma greenstone belt, Singhbhum, Eastern India. *Canadian Journal of Earth Sci.*, v. 17, pp. 210–231.

Gupta, A., Basu, A and Ghosh, P.K., 1982. Ultramafic volcaniclastics of the Precambrian Dalma volcanic belt, Singhbhum, Eastern India. *Geol. Mag.*, 119(5), pp. 505–510.

Gupta, A., Basu, A. and Singh, S.K., 1977. Occurrence of pyroclastic conglomerate in Dalma metavolcanics, Singhbhum district, Bihar. *Indian Journal Earth Sci.*, v. 4(2), pp. 160–168.

Gupta, A., Basu, and Singh, S.K., 1985. Stratigraphy and petrochemistry of Dhanjori greenstone belt, Eastern India. *Quart. J. Geol. Min. Met. Soc. India*, v. 57(4), pp. 248–263.

Gupta, A., Basu, A. and Singh, S.K., 1996. Structural analysis of the Dhanjori basin, Singhbhum, Bihar – a preliminary appraisal. In A.K.Saha (Ed.), *Recent Resarches in Geology, RRG*, v. 16, Hindustan Publn.Corp.(India), pp. 32–42.

Gupta, A., Basu, A. and Srivastava, D.C., 1981. Mafic and ultramafic volcanism of Ongarbira greenstone belt, Singhbhum, Bihar. *Jour. Geol. Soc. India,* v. 22, pp. 593–596.

Gupta, Abir, 1998a. Primordial storms: An overview of depositional environments in mid-late Proterozoic plateform of India. *Gondwana Res.*, v. 1(2), pp. 291–297.

Gupta, Abir, 1998b. Hummocky cross-stratification in Chhattisgarh basin., M.P. and its hydraulic and bathymetric impilcations. *Jour. Indian Assoc. Sed.* v. 17(2), pp. 213–224.

Gupta, B.C., 1934. The geology of central Mewar. *Geol. Surv. India*, *Mem.*65, pp. 107–168.

Gupta, B.C. and Mukherjee, P.N., 1938. Geology of Gujarat and southern Rajputana. *Records, Geol. Surv. India*, 73(2), pp. 103–208.

Gupta, P. and Guha, D.B.,1998. Stratigraphy, structure and basement–cover relationship in the Khetri copper belt and the emplacement mechanisms of the granite massifs, Rajasthan, India. *Jour Geol. Soc.India,* v.52, pp. 417–432.

Gupta, P., Fareeduddin, Reddy, M.S. and Mukhopadhyay, K., 1995. Stratigraphy and structure of Delhi Supergroup of rocks in central part of Aravalli Range. *Records, Geol. Surv. India*, v. 120(2–8), pp. 12–26.

Gupta, P., Guha, D.B. and Chattopadhyay, B., 1998. Basement–cover relationship in south Khetri belt and the emplacement mechanism of the granite massifs, Rajasthan, India. *Jour. Geol. Soc. India,* v. 52, pp. 417–432.

Gupta, P., Mukhopadhyay, K., Fareeduddin and Reddy, M.S., 1991. Tectono–stratigraphic framework and volcanic geology of the South Delhi fold belt in South–central Rajasthan. *Jour. Geol. Soc. India,* v. 37, pp. 431–441.

Gupta, R., Sarangi, A.K. and Bhattacharya, A., 2004. Uranium mining in Jharkhand – state-of-art new ventures. *In Sem. on Role of mining industry in economic and industrial development of Jharkhand: Problems and prospects*, pp. 70–80.

Gupta, S., Sengupta, P., Ehl, J., Raith, M. and Bardhan, S., 1998.The westward reach of the Eastern Ghats belt: evidence from the Dharamgarh granulites, Kalahandi district, Orissa. *(Abst.) Internat. Sem. on Precambrian crust of Eastern and Central India (UNESCO–IUGS–IGCP–368), Geol. Surv. India*, Bhubaneswar, pp. 21–24.

Gupta, S.N., Arora, S.N., Mathur, R.K., Iqbaluddin, Prasad, B., Sahai, T.N. and Sharma, S.B., 1980. Lithostratigraphic map of Aravalli region. *Geol. Surv. India*, Calcutta.

Gupta, S.N., Arora, Y.K., Mathur, Y.K., Iqbaluddin, Balmiki Prasad, Sahai, T.N. and Sharma, S.B., 1997. The Precambian geology of the Aravalli region, southern Rajasthan and northeastern Gujarat. *Geol. Surv. India*, Mem. 123, p. 263.

Gurney, J.J., Helmstaedt, H.H., le Roex, A.P., Nowicki, T.E., Richardson, S.H., and Westerhend, K.J., 2005. Diamonds: crustal distribution and formation processes in time and space and an integrated deposit model. *Econ. Geol. 100th Anniversary vol.*, pp. 143–178.

Gustafson, L.B. and Hunt, J.P., 1975. The porphyry copper deposits at El Salvador, Chile. *Econ. Geol.*, v. 70, pp. 857–912.

Gustafson, L.B. and Williams, N., 1981. Sediment hosted stratiform deposits of copper, lead and zinc. *Econ. Geol., 75th Anniversary vol.*, pp. 139–178.

Gustafson, L.B., Orquera, W., McWilliams, M., Castro, M., Olivares, O., Rojas Gonzalo, Malnenda, J. and Mendez, M., 2001. Multiple centers of mineralisation in the Indio Muerto District, El Salvador, Chile. *Econ. Geol.*, v. 96, pp. 325–350.

Gutscher, M.A., Maury, R., Essen, G.P. and Bourdon, E., 2000. Can slab melting be caused by flat subduction? *Geology*, v. 28, pp. 535–538.

Gutzmer, J. and Beukes, N.J., 1998. The manganese formation of the Neoproterozoic Penganga Group, India – revision of an enigma. *Econ. Geol.*, v. 93, pp. 1091–1102.

Gutzmer, J. and Beukes, N.J., 2000. The manganese formation of the Neoproterozoic Penganga Group, India – revision of an enigma – A reply (to Supriya Roy's discussion). *Econ. Geol.*, v. 95, pp. 239–240.

Gutzmer, J., Mukhopadhyay, J., Beukes, N.J., Pack, A. Hayashi, K. and Sharp, Z.D., 2006. Oxygen isotope composition of hematite and genesis of high–grade BIF–hosted iron ores. *Mem. Geol. Soc. Amer.*, v. 198, pp. 257–268.

Gyani, K.C., Omar, I.E.M., 1999. Charnockite from Central Rajasthan: geochemistry, thermo–barometry and petrogenesis. In N. G. K. Murthy and V. Ram Mohan (Eds.), *Charnockites and granulite facies rocks. Geol. Asson. Tamilnadu, Chennai,* India, pp. 187–206.

Hacket, C.A., 1881. On the geology of Aravalli region, central and eastern. *Records, Geol. Surv. India*, v. 14(4), pp. 279–303.

Hagemann, S.G. and Cassidy, K.F., 2000. Archean orogenic lode gold deposits. *Rev. Econ. Geol.*, v. 13., pp. 9–68.

Haggerty, SE., 2000. Diamond geology: A global view with glimpses of the Indian scene. In *Workshop volume on status, complexities and challenges of diamond exploration in India*, Raipur, pp. 1–19.

Haldar, D. and Ghosh, R.N., 2000. Eruption of Bijawar lava: an example of Precambrian volcanicity under stable cratonic condition. *Geol. Surv. India, Spl. Pub.*, v. 57, pp. 151–170.

Haldar, D., Mukhopadhyay, A.K. and Mukherjee, P., 2005. Petrotectonic environment of chromitites associated with the Precambrian ultramafic rocks of Jharkhand and Orissa – a reappraisal. *Proc. Sem. on Mineral and Energy Resources of Eastern and Northeastern India. MGMI, Calcutta,* Kolkata, India, pp. 203–225.

Haldar, D., Ghosh, R.N. and Ghosh, D.B., 1981. An Archean mafic–ultramafic suite in Bagrodha–Solda sector, M.P. and U.P. *Geol. Surv. India, Spl. Pub.*, v. 3, pp. 119–127.

Haldar, D. and Mutchopadhyay, A., 1991. Report on the comparative study of selected chromite ores of Singhbhum district, Bihar and the adjacent parts of Orissa. *Unpub. Rept., GS9*, 23p.

Haldar, S.K., 2004. Grade and tonnage relationships in sediment-hostsed lead-zinc sulfide deposits of Rajasthan, India. In M. Deb and W.D. Goodfellow (Eds.), *Sediment-hosted lead-zinc sulfide deposits*. Narosa, New Delhi, pp. 264–272.

Haldar, S.K. and Deb, M., 2001. Geology and mineralzation of Rajpura–Dariba lead–zinc belt, Rajasthan. In M.Deb and W.D.Goodfellow (Eds.), Sediment-hosted lead–zinc sulphide deposits in the north western Indian shield. *Pre-Seminar vol.*, pp. 177–187.

Halden, N.M., Bowes, D.R. and Dash, B., 1982. Structural evolution of migmatites in granulite facies terrane: Precambrian crystalline complex of Angul, Orissa, India. *Trans. Royal Soc. Edin. Earth Sci.*, v. 73, pp. 109–118.

Halter, W.E., Pettke, T. and Heinrich, C.A., 2002. The origin of Cu/Au ratios in the porphyry type ore deposits. *Science*, v. 296, pp. 1844–1846.

Hamilton, J.V. and Hodgson, C.J., 1986. Mineralisation and structure of Kolar gold field, India. In A.J. Macdonald (Ed.). *Proc. Gold 86, Toronto*, pp. 270–283.

Han, T.M., 1978. Microstructures of magnetite as guide to its origin in some Precambrian iron-formations. *Fortsch. Mineral.* v. 56, pp. 105–142.

Han, T.M., 1988. Origin of magnetite in Precambrian iron-formations of low metamorphic grade. *Proc. 7th Quad. IAGOD symp.*, Sch. Verlag., Stuttgart, pp. 641–656.

Hand, M., Reid, A., and Jagodzinski, L., 2007. Tectonic framework and evolution of the Gawler Craton, Southern Australia. *Econ. Geol.*, v. 102, pp. 1377–1395.

Hansen, E.C., Hickman, M.H., Grant, N.K. and Newton, R.C., 1985. Pan–African age of Peninsular Gneiss near Madurai, South India. *EOS 66*, pp. 419–420.

Hansen, E.C., Janardhan, A.S., Newton, R.C., Prame, W.K.B.N. and Rabindrakumar, G.R., 1987. Arrested charnockite formation in southern India and Srilanka. *Contrib. Min. Pet.*, v. 96, pp. 225–244.

Hansen, E.C., Newton, R.C. and Janardhan, A.S., 1984a. Pressure, temperature and metamorphic fluids across an unbroken amphibolite facies to granulite facies transition in solution Karnataka, India. In A. Kröner, G.N. Hansen and A.M. Goodwin (Eds.), *The origin and evolution of Archean continental crust*. Springer Verlag, New York, pp. 161–182.

Hansen, E.C., Newton, R.C. and Janardhan, A.S., 1984b. Fluid inclusions from amphibolite–granulite transition in Southern Karnataka, direct evidence concerning the fluids of granulite metamorphism. *Journal Met. Geol.*, v. 2, pp. 249–267.

Hanson, G.N., 1980. Rare earth elements in petrogenetic studies of igneous systems. *Ann. Rev., Earth Planet. Sci.*, v. 8, pp. 371–406.

Hanson, G.N., Krogstad, E.J. and Rajamani, V., 1988. Tectonic setting of the Kolar schist belt, Karnataka, India. *Jour. Geol. Soc. India,* v. 31, pp. 40–42.

Hansoti, S.K. and Sinha, D.K., 1995. Signatures of breccia complex/iron oxide-type U-REE mineralisation in the Khairagarh Basin with special reference to Dongargarh-Lohara area, Central India. *Exploration and Research for Atomic Minerals, AMD (India),* v. 8, pp. 155–172.

Hanuma Prasad, M., Krishna Rao, B., Vasdev, V.N., Srinivasan, R. and Balaram, V., 1997. Geochemistry of Archean bi–modal volcanic rocks of the Sandur supracrustal belt, Dharwar craton, Southern India. *Jour. Geol. Soc. India,* v. 49, pp. 307–322.

Hanuma Prasad, M., Hakim, A., Rao, B.K., 1999. Metavolcanic and metassedimetary inclusions in the Bundelkhand granite Complex in Tikamgarh district, M.P. *Jour. Geol. Soc. India*, v. 54, pp. 359–368.

Hari Narain, 1987. Geophysical constraints in the evolution of Purana basins of India, with special reference to Cuddapah, Godawari and Vindhyan basins. *Geological Society of India*, *Mem.* 6, pp. 5–32.

Harmsworth, R.A., Kneeshaw, M., Morris, R.C., Robinson, C.J. and Srivastava, P.K., 1990. BIF–derived iron ores of Hamersley Province, In Hugh, F.E. (Ed.), *Geology of mineral deposits of Australia and Papua New Guinea, v. 1: Australian Inst. of Mining and Metallurgy Monograph*, v. 14, pp. 617–642.

Harris, A.C., Golding, S.D. and White, N.C., 2005. Bajo de la Alumbrera copper-gold deposti: stable isotope evidence for a porphyry-related hydrothermal system dominated by magnitic aqueous fluids. *Econ. Geol.*, v. 100, pp. 863–886.

Harris, D.C., 1989. Mineralogy and geochemistry of the Hemlo gold deposit, Ontario. *Geol. Surv. Canada, Econ. Geol. Report No. 38*, p. 88.

Harris, N.B.W. and Massey, J.A., 1994. The decompression and anatexis of Himalayan metapelites. *Tectonics*, v. 13, pp. 1537–1546.

Harris, N.B.W., Holt, R.W. and Drury, S.A., 1982. Geobarometry–geothermometry and Late Archean geotherms from the granulite facies terrain of South India. *Journal Geol.*, v. 90, pp. 509–527.

Harris, N.B.W., Santosh, M. and Taylor, P.N., 1994. Crustal evolution in South India: constraints from Nd isotopes. *Journal Geol.*, v. 102, pp. 139–150.

Harrison,T.M., Blichert–Toft, J.,Muller, W., Alberade, F., Holden, P., Mojzsis, S.J., 2005. Heterogenous Hadean hafnium: evidence of continental crust at 4.4–4.5 Ga. *Science*, v. 310, pp. 1947–1950.

Haruna, M., Hanamuro, T., Uyeda, K., Fuzimaki, H. and Ohmoto, H., 2003. Chemical, isotopic and fluid inclusion evidence for the hydrothermal alteration of the footwall rocks of the BIF-hosted iron ore deposits in the Hamersley district, Western Australia. *Res. Geol.*, v. 53, pp. 75–88.

Hatcher, R.D.Jr. and Hooper, R.J., 1992. Evolution of crystalline thrust sheets in the internal parts of mountain chains. In K.R. McClay (Ed.), *Thrust Tectonics*, Chapman and Hall, London, pp. 217–233.

Hawkesworth, C.J., Kelly, S., Turner, S., Le Roex, A. And Storey, B., 1999. Mantle processes during Gondwana breakup and dispersal. *Journal African Earth Sci.*, v. 28(1), pp. 239–261.

Hawkerworth. C.J., Dhuime, B., Pietramick, A.B., Cawood, PA., Kemp, A.I.S. and Storey, C.D., 2010. The generation and evolution of the continental crush. *Jour. Geol. Soc. Lond.*, v. 167, pp. 229–248.

Hayden, H.H., 1919. General report of 1918. *Records, Geol. Surv. India*, v. 50.

Hayes, T.S. and Einaudi, M.T., 1986. Genesis of the Spar Lake stratabound copper-silver deposit, Montana Part 1. Controls inherited from sedimentation and pre-ore diagenesis. *Econ. Geol.*, v. 81, pp. 1899–1932.

Haynes, D.W. and Bloom, M.S., 1987. Stratiform copper deoposits hosted by low-energy sediments. pts. III and IV. *Econ. Geol.* v. 82, pp. 635–648, 875–893.

Hedenquist, J.W. and Lowenstern, J.B., 1994. The role of magma in the formation of hydrothermal ore deposits. *Nature*, v. 370, pp. 519–527.

Hedenquist, J.W. and Richards, J.P., 1998. The influence of geochemical techniques on the development of genetic models for Porphy copper deposits. *Rev. Econ. Geol. N. 10*, pp. 235–256.

Hegde, K.T.M., 1991. An introduction to ancient Indian metallurgy. *Geological Society of India,* 86p.

Heidrick, T.L. and Titley, S.R., 1982. Fracture and dike patterns in Laramide plutons and their structural and tectonic implications: American Southwest. In S.R.Titley (Ed.), *Advances in geology of porphyry copper deposits*, Tucoon, pp. 73–91.

Heinrich, C.A., 1990. The chemistry of hydrothermal tin (tungsten) ore deposition. *Econ. Geol.*, v. 85, pp. 457–481.

Heirer, K.S., 1979. The movement of uranium during higher grade metamorphic processes. *Phil. Trans. Royal. Soc. Lond.*, v. A291, pp. 413–421.

Henri, D.J. and Dutrow, B.L., 1996. Metamorphic tourmaline and its petrologic applications. In E.S. Grew and L.M. Anovitz (Eds), Reviews in mineralogy, v. 33, *Min. Soc. Amer*, pp. 503–557.

Henriquez, F., Nashlund, H.R., Nystrom, J.O., Vivallo, W., Aguirre, R., Dobbs, F.M. and Lledo, H., 2003. New field evidence bearing on the origin of the Laco magnetite deposit, Northern Chile a discussion. *Econ. Geol.*, v. 98, pp. 1497–1500.

Heron, A.M., 1917 a. Geology of Northeastern Rajputana and adjacent districts. *Geol. Surv. India*, *Mem.* 45, p. 128.

Heron, A.M., 1917 b. The Byana–Lalsot hills in eastern Rajputana. *Records, Geol. Surv. India*, v. 48, pp. 181–203.

Heron, A.M., 1923. The geology of Western Jaipur. *Records, Geol. Surv. India*, v. 54, pp. 345–397.

Heron, A.M., 1924. The Soda-bearing rocks of Rajputana. *Records, Geol. Surv. India*, v. 55, pp. 179–195.

Heron, A.M., 1935. Synopsis of Pre-Vindhyan geology of Rajputana. *Trans. Nat. Inst. Sci. India*, v. 1, pp. 17–33.

Heron, A.M., 1936. Geology of southeastern Mewar, Rajputana. *Geol. Surv. India, Mem. 68(1)*, p. 120.

Heron, A.M., 1949, Synopsis of Purana formation of Hyderabad. *Journal Hyderabad Geol. Surv.*, v. 5(2), 1–129.

Heron, A.M., 1953. The geology of Central Rajputana. *Geol. Surv. India, Mem. 79*, p. 385.

Hildreth, W., Moorbath, S., 1988. Crustal contributions to are magmatism in the Andes of Central Chile. *Contrib. Min. Pet.*, v. 98., pp. 455–489.

Hitzman, M.W., 2000. Iron oxide – Cu – Au deposits: What, where, when and why? In T.M.Porter (Ed.), *Hydrothermal iron oxide copper–gold and related deposits: a global perspective*. Aust. Mineral Foundation Inc., pp. 9–25.

Hitzman, M.W. and Beaty, D.W., 1996. The Irish Zn–Pb–(Ba) ore field. *Soc. Econ. Geol. Spec. Publn. No 4*, pp. 112–143.

Hitzman, M.W., Oreskes, N. and Einaudi, M.T., 1992. Geological characteristics and tectonic setting of Proterozoic iron oxide (Cu–U–Au–LREE) deposits. *Precamb. Research*, v. 58, pp. 241–287.

Hodges, K.V., 2000. Tectonics of the Himalaya and southern Tibet from two perspectives. *Geol. Soc. Amer. Bull.*, v. 112, pp. 324–350.

Hodges, K.V., Parrish, R.R. and Searle, M.P., 1996. Tectonic evolution of the central Annapurna Range, Nepalese Himalayas. *Tectonics*, v. 15, pp. 1264–1291.

Hoefs, J., 1973. Stable isotope geochemistry. *Springer-Vislag,* New York, 140p.

Hoering, T.C., 1989. The isotopic composition of bedded barite from the Archean southern India. *Jour. Geol. Soc. India*, v. 34, pp. 461–466.

Hoffman, P.F., 1988. United plates of America, the birth of a craton. *Ann. Rev. Earth Planet Sci.*, v. 16, pp. 543–603.

Hoffman, P.F., 1991. Did the breakup of Laurentia turn Gondwanaland inside out? *Science*, v. 252, pp. 1409–1412.

Holland, H. and Malinin, S.D., 1979. The solubility and occurrence of non–ore minerals. In H.L. Barnes (Ed), *Geochemistry of hydrothermal ore deposits* (2nd Edn.), John Wiley and Sons, New York, pp. 461–508.

Holland, H.D., 1973. The oceans: A possible source of iron in iron formations. *Econ. Geol.*, v. 68, pp. 1169–1172.

Holland, H.D., 1984. *The chemical evolution of the atmosphere and oceans*. Princeton Univ. Press, Princeton. N.J., 582p.

Holland, T.H., 1893. Petrology of Job Charnock's tombstone. *Journal Asiatic Society of Bengal*, v. 62, pp. 62–64.

Holland, T.H., 1900. The charanockite series: A group of Archean hypersthenic rocks in Peninsular India. *Geol. Surv. India, Mem. 2*, pp. 192–249.

Holland, T.H., 1907 Classification of the Indian strata. Presidential Address, *Transaction No.1, Mining and Geological Institute, Calcutta*, India.

Holland, T.J.B., 1979. Experimental determination of reaction, paragonite = jadeite + kyanite + H_2O, and internally consistant thermodynamic data for part of the system Na_2O–Al_2O_3–SiO_2–H_2O with application to eclogite and blue schist. *Contrib. Min. Pet.*, v. 68, pp. 292–301.

Holler, W. and Gandhi, S.M., 1995. Silver-bearing sulfosalts from metamorphosed Rampura Agucha Zn–Pb (–Ag) deposit, Rajasthan, India. *Canad. Mineral.*, v. 33, pp. 1047–1057.

Holler, W. and Stumpfl, E.F., 1995. Cr–V oxides from the Rampura Agucha Pb–Zn (–Ag) deposit, Rajasthan, India. *Canad. Mineral.*, v. 33, pp. 745–752.

Holler, W., Touret, J.L.R. and Stumpfl. E.F., 1996. Retrograde fluid evolution at Rampura Agucha Pb–Zn (–Ag) deposit, Rajasthan, India. *Mineral. Deposita*, v. 31, pp. 163–171.

Hoschek, G., 1969. The stability of staurolite and chloritoid and their significance in metamorphism of pelitic rocks. *Contrib. Min. Pet.*, v. 22, pp. 208–232.

Hou Zengquian, Ma Hongwen, Khin Zaw, Zhang Yuquan, Wang Mingjie,Wang Zeng, Pan Guitang and Tang Renli, 2003. The Himalayan Yulong porphyry copper belt: product of large-scale strike-slip faulting in Eastern Tibet. *Econ. Geol.*, v. 98, pp. 125–145.

Hou, G., Santosh, M.,Quian, X., Lister, G.S., Li, J., 2008. Configuration of the Late Paleoproterozoic supercontinent Columbia: Insight from radiating mafic dyke swarms.*Gondwana Res.*,v. 14, pp. 395–409.

Howard, P.F., 1972. Exploration for phosphorite in Austraila– a case history. *Econ. Geol.*, v. 67, pp. 1180–1192.

Howel, D.G., 1989. Tectonics of suspect terranes. Chapman and Hall, London, p. 232

Howie, R.A., 1955. The geochemistry of the charnockite series of Madras, India. *Trans. Royal Soc. Edin, v. 62*, pp. 162–164.

Hubbard, M.S., 1996. Ductile shear as a cause of inverted metamorphism: example from Nepal Himalaya *Jour. Geol.*, v. 104, pp. 493–499.

Huckriede, H. and Meischner, D., 1996. Origin and environment of manganese–rich sediments within blackshale basins. *Geochim. Cosmochim. Acta*, v. 60, pp. 1399–1413.

Hussain, S.M., Naqvi, S.M., Ramchandra, K.T., Sawkar, R.H. and Group, G.R., 1996. Geochemistry and genesis of Ajjanahalli gold diposit— a guide for exploration strategy in banded iron formations. *Proc. symp. Gold resources of India;, Hyderabad.* pp. 258–265.

Hutchinson, R.W., 1996. Regional metallogeny of carbonated-hosted ores by comparison of field relationships. In D.F. Sangster (Ed.), Carbonate-hosted lead-zinc deposites. *Soc. Econ. Geol. Spec. Publn.*, v. 4, pp. 8–17.

Indian Bureau of Mines, 1992. Indian Mineral Year Book.

Indian Bureau of Mines, 2003. Indian Mineral Year Book.

Inger, S. and Harris, N.BW. 1992. Tectonothermal evolution of the High Himalayan crystalline sequence, Longtang Valley, Northern Nepal, *Jour. Met Geol.* v. 10, pp. 439–452.

Irvine, T.N. 1995. Crystallization sequences in the Muscox intrusion and other layered intensions–II. Origin of chromite layers and similar deposits of other magmatic ores. *Geochim. Cosmochim. Acta*, v. 39, pp. 991–1020.

Irvine, T.N., 1977. Origin of chromite layers in the Muscox intrusion and other stratiform intrusions – a new interpretation. *Geology*, v. 5, pp. 273–277.

Ishihara, S., 1977. The Ilmenite–Series and Magnetite–Series-granitic rocks. *Mining Geology*. v. 27, pp. 293–305.

Ishihara, S., 1981. The granitoid series and mineralisation. *Econ. Geol. 75th Anniversary Vol.*, pp. 458–484.

Isley, A.E., 1994. Hydrothermal plumes and the delivery of iron to banded iron formation. *Journal Geol.*, v. 103., pp. 169–185.

Isley, A.E. and Abbot, D.H., 1999. Plume-related mafic volcanism and the deposition of banded iron formation. *Journal Geophys. Res.*, v. 104, pp. 15461–15477.

Isley, A.E. and Abbot, D.H., 2002. Implications for the temporal distribution of high–Mg magma for mantle plume volcanism through time. *Journal Geol.*, v. 110, pp. 141–158.

Iyenger, S.V.P. and Alwar, M.A., 1965. The Dhanjori eugeosyncline and its bearing on stratigraphy of Singhbhum, Keonjhar and Mayurbhanj districts. D.N. Wadia Comm. vol., *Min. Geol. Met. Inst. India* (MGMI), pp. 138–162.

Iyengar, S.V.P. and Banerjee, S., 1964. Magmatic phases associated with the Precambrian tectonics of Mayurbhanj district, Orissa. *Rep. 22nd Internat. Geol. Cong.*, 10, pp. 515–538.

Iyengar, S.V.P. and Murthy, Y.G.K., 1982: The evolution of Archaean –Proterozoic crust in parts of Bihar and Orissa, eastern India. *Records, Geol.Surv India*, v.112(3), pp. 1–6.

Iyenger, S.V.P., Kresten, P., Paul, D.K., Brunfelt, A.O., 1981. Geochemistry of Precambrian magmatic rocks of Mayurbhanj district, Orissa. *Jour. Geol. Soc. India,* v. 22, pp. 305–315.

Jacobsen, S.B. and Pimental–Klose, M.R., 1988. A Nd–isotopic study of the Hamersley and Michipicoten Banded Iron Formation: the source of REE and Fe in Archean oceans. *Earth Planet Sci. Lett*, v. 87, pp. 29–44.

Jagannatharao, J. and Krishnamurthy, C.V., 1981. Some observations on the mineralogy and geochemistry of Hazaridadar and Raktidadar plateaus, Amarkantak area, Madhya Pradesh,India. In *Lateritisation Processes* (Proc.Internat. Sem. on Lateritisation Processes), Oxford and IBH Pub. Co. Pvt. Ltd., New Delhi, pp. 89–108.

Jahn, M.P., Windley, B.P. and Khan, A.A., 1985. The Waziristan Ophiolite, Pakistan : General geology and chemistry of chromite and associated phases. *Econ. Geol.*, v. 89, pp. 249–306.

Jain, A.K. and Manickavasagam, R.M., 1993. Inverted metamorphsim in the intracontinental ductile shear zone during Himalayan collision tectonics. *Geology*, v. 21, pp. 407–410.

Jain, R.C., Bajpai, R.K. and Kumar, D., 1996. U–Cu mineralization in the Proterozoic metasediments of the Siyom Group around Kau, West Siang district, Arunachal Pradesh, India. *Jour. Atom. Mineral Sci.*, v. 4, pp. 33–36.

Jain, S.C., Nair, K.K.K. and Yedekar, D.B., 1995. Geology of the Son–Narmada–Tapti lineament zone in Central India. *Geol. Surv. India, Spl. Pub. v. 10*, pp. 1–154

Jain, S.C., Yedekar, D.B., and Nair, K.K.K., 1990. A review of the stratigraphic status of Bharweli–Ukwa Manganese Belt, Balaghat district, M. P. *Geol. Surv. India, Spl. Pub., v. 28*, pp. 323–353.

Jain, S.C., Yedekar, D.B., and Nair, K.K.K., 1991. Central Indian Shear Zone: A major Precambrian crustal boundary. *Journal Geological Society of India*, v. 37, pp. 521–548.

Jairam, M.S., Roop Kumar, D. and Srinivasan, K.N., 1998. Classification of greenstones and associated granitoids of Jonnagiri schist belt, Kurnool District, A.P. (Abst.) *M.S.Krishnan Comm. Nat. Sem. on Progress in Precambrian Geology of India, Geol. Surv. India*, Calcutta, pp. 98–99.

Jairam, M.S., Roop Kumar, D. and Srinivasan, K.N., 2001, Classification of greenstones and associated granitoids of Jonnagiri schist belt, Kurnool district, Andhra Pradesh. *Geol. Surv. India*, Spl. Pub. v. 55(2), pp. 59–66.

Jairath, S. and Sharma, M., 1986. Physico–chemical conditions of ore deposition in the Malanjkhand copper sulphide deposit. *Proc. Ind. Acad. Sci. Earth Planet Sci.*, v. 95, pp. 209–221.

Jaireth, S., Sen, A.K., and Varma, O.P., 1991. Fluid inclusion studies in apatite of the Sung Valley carbonatite complex, N.E.India: Evidence of melt–fluid immiscibility. *Journal Geological Society of India*, v. 37(6), pp. 547–559.

James, D., Sacks, I.S., 1999. Cenozoic formation of the central Andes: A geophysical perspective. *Soc. Econ. Geol. Spl. Pub. 7*, pp. 1–25.

James, H.L., 1954. Sedimentary facies of iron formation, *Econ. Geol.*, v. 49, pp. 235–293.

James, H.L. and Sims, P.K., 1973. (Eds) Precambrian iron–formations of the world. *Econ. Geol.*, v. 68, pp. 913–1179.

Janardhan, A.S. and Leake, B.E., 1975. The origin of metaanorthosite gabbros and garnetiferous granulites of the Sittampundi Complex, Madras, India. *Jour. Geol. Soc. India,* v. 16, pp. 391–408.

Janardhan, A.S. and Vidal, R.H., 1982. Rb–Sr dating of the Gundhupet Gneisses. *Jour. Geol. Soc. India,* v. 23, pp. 578–580.

Janardhan, A.S., Francis Anto, K. and Shivasubramanian, P., 1996. Evolution of the Proterozoic southern granulite terrain: its position in Eastern Gondwana. *Gondwana Research Group, Misc. Pub.* No. 4, pp. 38–39.

Janardhan, A.S., Newton, R.C. and Hansen, E.C., 1982. The transformation of amphibolite facies gneisses to charnockite in southern Karnataka and northern Tamil Nadu. *Contrib. Min. Pet.*, v. 79, pp. 139–149.

Janardhan, A.S., Shadakshara Swami, N., and Capdevila, R., 1990. Trace and REE geochemistry of pelites from Sargur high grade terrain, southern Karnataka. *Jour. Geol. Soc. India,* v. 36, pp. 27–35.

Janardhana Rao, L.H., Srinivasa Rao. C. and Ramakrishna, T.L., 1975. Reclassification of the rocks of Bhima basin, Gulbarga district, Karnataka State. *Geol. Surv. India, Misc. Pub.* v. 23(1), pp. 177–184.

Jangpangi, B.S., 1978. Stratigraphy and structure of Bhutan Himalaya. In P.S. Saklani (Ed), *Tectonic Geology of Himalaya*, pp. 221–242.

Jayananda, M. and Peucat, J.J., 1996. Geochronological framework of Southern India. In M. Santosh and Y. Yoshida (Eds.), Archean and Proterozoic terranes in Southern India within the East Gondwana. *Gondwana Res. Group, Mem. 3*, pp. 53–75.

Jayananda, M., Kano, T., Peucat, J.-J, Channabasappa, S. 2008. 3.35Ga Komatiite volcanism in the Western Dharwas Craton, Southern India: constraints from Nd isotopes and whole-rock geochemistry. *Precamb. Research*, v.162, pp.160–179.

Jayananda, M., Moyen, J.F., Martin, H., Pencat, J.J., Auvray, B., and Mahabaleswar, B, 2000. Late Archean (2550–2520 Ma) juvenile magmatism in the Eastern Dharwar craton, Southern India: Constraints from geochronology, Nd–Sr isotopes and whole rock geochemistry. *Precamb. Research*, v. 99, pp. 225–254.

Jayaprakash, A.V., 1999. Evolutionary history of Bhima Basin. (Absts.) Field–workshop on integrated evolution of the Kaladgi and Bhima basins, *Geological Society of India*, pp. 23–28.

Jayaprakash, A.V., Sundaram, V., Hans, S.K. and Misra, R.N., 1987. Geology of Kaladgi–Badami basin, Karnataka. *Geological Society of India, Mem. 6*, pp. 201–252.

Jena, B.K. and Behra, U.K., 2000. The oldest supracurtal belt from Singhbhum Craton and its possible correlation. Proc. *Internat. Sem. on Precambrian Crust in Eastern and Central India, (UNESCO–IUGS–IGCP 368), Geol. Surv. Ind., Spl. Pub. 57*, pp.106–121.

Johnson, J.P. and McCulloch, M.T., 1995. Sources of mineralizing fluids for the Olympic Dam deposit (South Australia): Sm-Nd isotopic constraints. *Chem. Geol.*, v. 121, pp. 177–199.

Johnson, K.S., 1982. Solubility of rhodochrosite ($MnCO_3$) in water and sea water. *Geochim Cosmochim. Acta*, v. 46, pp. 1805–1809.

Johnson, P.T., Dasgupta, D. And Smith, A.D., 1993. Pb–Pb systematics of copper sulphide mineralisation, Singhbhum area, Bihar. *Indian Journal Geol.*, v. 69, pp. 211–213.

Jones, H.C., 1934. The iron ore deposits of Bihar and Orissa. *Geol. Surv. India, Mem. 63*(2), pp. 167–302.

Joshi, A., Pant, N.C. and Bhatia, S.K., 2005 a. Petrological and geochemical studies on weathering profiles at three geomorphic levels: implications on regolith development and bauxite genesis in western Maharashtra. In: *16th Int. Symp. of ICSOBA on Status of Bauxite, Alumina, Downstream products and Future prospects. Tranvaux ICSOBA– Croatie*, v. 32(36), pp.133–155.

Joshi, A., Pant, N.C. and Bhatia, S.K., 2005b. On origin of nodular horizon in the coastal regolith profiles, Ratnagiri and Sindhudurg districts, Maharashtra. In *16th Int. Symp. on status of bauxite, alumina, downstream products and future prospects. Tranvaux ICSOBA– Croatie*, v. 32(36), pp. 201–212.

Jowett, E.C., 1986. Genesis of kupferschiefer Cu-Ag deposits by convective flow of Rotliegende brines during Triassic sifting. *Econ. Geol.*, v. 81, pp. 1823–1837.

Joy, S. and Saha, D., 1998. Comparable structures from the Dhanjori Group and Singhbhum shear zone – A linked thrust system model of deformation in the Singhbhum region. *Internat. Sem. on Precambrian crust in Eastern and Central India, (UNESCO–IUGS–IGCP 368), Bhubaneswar, India*, pp. 197–200.

Jwell, P.W., 2000. Bedded barite in the geologic record. In Ch. Glenn and L.J. Prevotclucas (Ed.), *Marine authigenesis: From global to microbial. SEPM Spl. Pub.*, v. 66, pp. 147–161.

Jwell, P.W. and Shallard, R.F., 1991. Geochemistry and paleographic setting of Central Nevada bedded barites. *Journal Geol.* v. 99, pp. 151–170.

Kaila, K.L. 1988: Mapping the thickness of Deccan Trap flows from DSS studies and inferences about a hidden Mesozoic basin in Narmada–Tapti region. *Geological Society of India, Mem. 10*, pp. 91–117.

Kaila, K. and Bhatia, S.C., 1981. Gravity study along the Kavali–Udipi deep seismic sounding profile in the Indian peninsular shield: some inferences about the origin of anorthosites and Eastern Ghats orogeny. *Tectonophysics*, v. 79, pp. 189–243.

Kaila, K.L., Roy Chowdhury, K., Reddy, P.R., Krishna, V.G., Hari Narayan, Subbotin, S.I., Sollogb, V.B., Chekunov, A.V., Kharetchko, G.E., Lazarenko, M.A. and Ilchenko, T.V., 1979. Crustal structure along Kavali–Udip, profile in the Indian peninsular shield from deep seismic soundings. *Jour. Geol. Soc. India,* v. 20, pp. 307–333.

Kak, S.N. and Hasan Mohammad, 1979. Genesis and characteristics of uranium mineralisation in Mahadek sandstones. *Indian Journal Earth Sci.*, v. 6, pp.152–161.

Kak, S.N. and Subrahmanyam, A.V., 2002. Depositioning environment and age of Mahadek Formation of Wahblei river section, West Khasi Hills, Meghalaya. *Jour. Geol. Soc. India,* v. 60, pp. 151 162.

Kale, V.S., 1991. Constraints on the evolution of the Purana basins of Peninsular India. *Jour. Geol. Soc. India,* v. 38, pp. 231–252.

Kale, V.S. and Phansalkar, 1991. Purana basins of Peninsular India: A review. *Basin Research*, v. 3, pp. 1–36.

Kamineni, D.C. and Rao, A.T., 1998. Sapphirine–bearing quartzite from the Eastern Ghats granulite terrain, Vizianagram, India. *Journal Geol.*, v. 96, pp. 209–220.

Kaneoka, I, Iwata, N., Nagao, K. and Deshmukh S.S. 1996. Period of volcanic activity of the Deccan plateau inferred from K–Ar and ^{40}Ar–^{39}Ar ages and problems relating to radiometric dating. Gondwana. Proc. National Symp. on Deccan Flood Basalts, India. *Geol. Mag., Spl. Vol. 2*, pp. 311–319.

Kanungo, D.N. and Mahalik, N.K., 1975,. A revision of stratigraphy and structure of Gangpur Series in Sundargarh district, Orissa and their tectonic history. *Geol. Surv. India, Misc. Pub. No. 23(2)*, pp. 129–138.

Kapdevila, R., Arnd, N., Letendre, J., Sauvage, J.F., 1999. Diamonds in volcaniclastic komatiite from French Guina. *Nature*, v. 399, pp. 456–458.

Kar Ray, M.K., 1972. Geology with special reference to the modes of occurrence and control of mineralisation of Sargipali lead deposit, Sundergarh district, Orissa. *Indian Minerals*, v. 26, pp. 37–47.

Kar, P., Datta, N.R., Roy, S., Bhowmick, N., Santra, D.K., and Chakravorty, K.K., 1975: Exploration of the tungsten mineralisation in Bankura district, West Bengal. *Indian Miner.*, v. 29(2), pp. 1–31.

Karmakar, S. and Fukuoka, M., 1998. Successive metamorphism from a suite of granulite facies rocks from Araku valley: implication for the P–T trajectory in the Eastern Ghats. *Geol. Surv. India, Spl. Pub. No.* 44, pp. 201–219.

Karunakaran, C., 1976. Sulfur isotope composition of barites and pyrites from Mangampeta, Cuddapah district, Andhra Pradesh. *Jour. Geol. Soc. India,* v. 17, pp. 187–195.

Kato, Y., Ohta, I., Tsunematsu, K.T., Watanabe, Y., Isozaki, Y., Maruyama, S., and Imai, N., 1998. Rare earth element variations in mid-Archean Banded Iron Formations: Implications for the chemistry of ocean and continent and plate tectonics. *Geochim. Cosmochim. Acta*, v. 62, pp. 3475–3497.

Kato, Y., Kano, T. and Kunugiza, K., 2002. Negative Ce anomaly in banded iron formations: evidence for the emergence of oxygenated deep sea at 2.9–2.7 Ga. *Resource Geol.*, v. 52, pp. 101–110.

Kaul, R., Singh, R., Shiv Kumar, K., Saran, R. and Sachan, A.S., 1991. Subsurface distribution pattern of uranium and some associated elements in the Upper Cretaceous Lower Mahadek sandstones of Gomaghat, West Khasi Hills district, Meghalaya, India. *Expln. Res. Atom. Minerals, v. 4*, pp. 61–67.

Kaur, G. and Mehta, P.K., 2007. Geochemistry and Petrogenesis of Jasrapura Granitoid, North Khetri copper belt, Rajasthan: Evidence for island arc magmatism. *Jour. Geol. Soc. India,* v. 69, pp. 319–330.

Kazansky, V. I., 1995. Evolution of ore–bearing Precambirian structures. *Oxford and IBH Pub. Co. (Pvt.) Ltd.*, New Delhi, p. 307.

Keays, R.R. and Dightfoot, P.C., 2009. Crustal sulfur is required to form magmatic Ni-Cu sulfide deposits: evidence from chalcophile element signatures of Siberian and Deccan Trap basalts. *Mineral. Deposita* v. 45, pp. 241–257.

Keays, R.R., and Scott, R.B., 1976. Precious metals in oceanic ridge basalts: Implication for basalts as source rocks for gold mineralization. *Econ. Geol.*, v. 71, pp. 705–720.

Kerrich, D.M.,1990. The Al_2O_5 polymorphs. *Review in Mineralogy*, v. 22, p. 406.

Khan, A.H., Raghvarayya and V.K. Gupta, 1998. Radiation and environmental safety in uranium mines and mill. Proc. 7th Nat. Symp. on Environment. Section: Environment *Management at Jaduguda and adjacent places*, pp. 5–11.

Khan, A.S., Huin, A.K. and Chattopadhyay, A., 1999. Specialised Thematic mapping in Sausar Fold Belt in Manegaon–Karwahi, Totaldph–Kirangi, Sarra and Susurdoh–Sitekasa area, Nagpur and Bhandara, Districts, Maharashtra for elucidation of stratigraphy, structure, metamorphic history and tectonics. *Records, Goel. Surv. India, v. 132*(6), pp. 31–35.

Khan, A.S., Huin, A.K. and Chattopadhyay, A. 2000. Specialised thematic mapping in Sausar Fold Belt in Ramtek–Mahuli–Chorbaoli area, Nagpur District, Maharashtra, for elucidation of stratigraphy, structure, metamorphic history and tectonics. *Records, Geol.Surv.India*, v. 133(6), pp. 25–29.

Khan, M.W.and Bhattacharya, T.K., 1993. A reappraisal of the stratigraphy of Bailadila Group, Bacheli, Bastar district, M. P. *Jour. Geol. Soc. Ind.*, v. 42(6), pp. 549–562.

Khan. M.W. and Mukherjee, A., 1993. Geochemical constraint on the genesis of phosphorites of Chhattisgarh Basin, Durg dist., M.P., *Jour. Geol. Soc. India,* v. 41, pp. 360–370.

Khan, R.M.K., Govil, P.K. and Naqvi, S.M., 1992. Geochemistry and genesis of banded iron formation from Kudremukh schist belt, Karnataka, India. *Jour. Geol. Soc. India,* v. 40, pp. 311–328.

Khan, R.M.K., Naqvi, S.M., 1996. Geology, geochemistry and genesis of BIF of Kustagi schist belt, Archean Dharwar craton, India. *Mineral. Deposita*, v. 31, pp. 123–133.

Khan, R.M.K., Nirmal Charan, S., Arora, M. and Naqvi, S.M., 1995. Mineral composition and its bearing on depositional history of Banded Iron Formation of Kudremukh schist belt, Karnataka. *Jour. Geol. Soc. India,* v. 46(6), pp. 603–610.

Khandali, S.D. and Devaraju, T.C., 1987. Laterites bauxites of Paduvari Plateau, South Kanara, Karnataka State. *Jour. Geol. Soc. India,* v. 30, pp. 255–266.

Khorana, R.K., 1993. Regional investigation for strategic minerals in the surrounding areas of Tosham hill, Bhilwara distric, Haryana. *Records, Geol. Surv. India,* v. 125 (8).

Khorana, R.K., 2000. Preliminary appraisal of gold mineralisation at Tosham hill, Bhiwani distric, Haryana. *Indian Minerals*, v. 54, pp/ 127–129.

Khorana, R.K., Dhir, N.K. and Srivastava, R.N., 2004. Petrochemical indicators for search of tin-tungaten mineralization associated with Malani igneous suite. *Geol. Surv. India, Spl. Pub. 72*, pp. 467–474.

Kimberly, M.M., 1978. Paleoenvironmental classification of iron formations. *Econ. Geol.*, v. 73, pp. 215–229.

King, W., 1872. The Cuddapah and Kurnool formations in Madras Presidency. *Geol. Surv. India, Mem. 8*(1), p. 346

King, W., 1885. A sketch of the progress of the geological work in Chattishgarh division of Central Provinces. *Rec. Geol. Surv. India*, v. 8, pp. 169–200.

King, W.,1881. Geology of Pranhita–Godavari Valley. *Geol. Surv. India, Mem.*18, pp. 151–311.

King, W., 1898–99. General Report for 1898–99. *Geol. Surv. India*, pp. 39–42.

Kirk, J., Ruiz, J., Chesley, J., Walshe, J. and England, G., 2002. A major Archean gold and crust forming event in Kaapvaal craton, South Africa. *Science*, v. 297, pp. 1856–1858.

Kirmani, I.R. and Farreduddin, 1998. Wall rock alteration in Cu-Zn-Au bearing volcanogenic massive sulfide deposit at Danva, distric Sirohi, Rajasthan. *Jour. Geol. Soc. India*, v. 52, pp. 391–402.

Kishore, M. and Kumar, R., 1998. Basemetal mineralisation in Khadandungri–Sudharika area, East Singhbhum district, Bihar. (Abst.) *Nat. Sem. on Geosci. Advns. in Bihar, India, in the Last Decade. Geol. Surv. India*, p. 70

Klein, C. and Beukes, N.J., 1989. Geochemistry and sedimentology of a facies transition from limestone to iron formation deposition in the Early Proterozoic Transvaal Supergroup, South Africa. *Econ. Geol.*, v. 84, pp. 1733–1774.

Klein, C and Beukes, N.J., 1992. Proterozoic iron formations. In K.C.Condie (Ed.), *Proterozoic crustal evolution*, Elsevier, Amsterdam, pp. 383–418.

Klein, C. and Ladeira, E.A., 2004. Geochemistry and mineralogy of Neoproterozoic banded iron formation and some selected siliceous manganese formations from Urukum district, Mato Grosso Dosul, Brazil. *Econ. Geol.*, v. 99, pp. 1233–1244.

Kloppenburg, A., White, S.H. and Zegers, T.E., 2001. Structural evolution of Warrawoona Greenstone belt and the adjoining granitoid complex, Pilbara craton, Australia: Implications for Archean tectonic processes. *Precamb. Research,* v. 12, pp. 107–147.

Knight, J., Lone, J., Joy, S., Gameron, J., Merrillee S.J., Nag, S., Shah, N., Dua, G., and Jhala, K., 2002. The Khetri copper belt, Rajasthan : Iron oxide copper–gold terrane in the Proterozoic of India. In T.M. Porter (Ed), Hydrothermal iron oxide copper gold and related deposits: a global perspective. *Aust. Mineral Foundation, Adelaide*, v. 2, pp. 321–341.

Kochar, N., 1983. Tusam ring coomplex, Bhiwani, India. *Proc. Indian Nat. Sci. Acad.*, v. 49, pp. 459–490.

Kochar, N., and Dhar, S., 1993. The association of hypersolvus–subsolvus granites : A study of Malani igneous suite, India. *Jour. Geol. Soc. India*, v. 42, pp. 449–476.

Kochar, N., Pande, K. and Gopalan, K., 1985. Rb-Srage of the Tusham ring complex, Bhiwani, India. *Jour. Geol. Soc. India*, v. 26, pp. 216–218.

Kolb, J, Rogers, A. Meyer, F.M. and Vennemann, T., 2004a.Development of fluid conduits in auriferous shear zones of Hutti gold mine, India: evidence for spatially and temporally fluid flow. *Tectonophysics*, v. 375, pp. 65–84.

Kolb, J., Hellmann, A., Rogers, A., Sindren, S., Vennemann, T., Bottcher, M.E. and Meyer, F.M., 2004b . The role of transcrustal shear zone in orogenic gold mineralisation at the Ajjanhalli Mine, Dharwar craton, South India. *Econ. Geol.*,v. 99, pp. 743–759.

Kollapuri, V.K., 1990. Petrological and ore mineragraphic studies in Manglur schist belt, Karnataka. *Records, Geol., Surv. India*, v. 123, pp.310–311.

Konhauser, K.O., Hamade, T., Raiswell, R., Morris, R.C. Ferris, F.G., Southern, G and Canfield, D.E., 2002. Could bacteria have formed the Precambrian banded iron formations ? *Geology*, v. 30, pp. 1079–1082.

Kothiyal, D.L., Srivastava, V.C. and Verma, R.N., 2002. Geology and mineral resources of the states of India : Uttar Pradesh and Uttaranchal. *Geol. Surv. India. Misc. Pub.*, v. 30(13).

Kovach, V.P., Salnikova, E.B., Kotov, A.B., Yakovleva, S.Z. and Rao, A.T., 1997. Pan– African U–Pb zircon age from apatite–magnetite veins of Eastern Ghats granulite belt, India. *Jour. Geol. Soc. India.*, v. 50, pp. 421–424.

Krause, O., Dobmeier, C., Raith. M.M. and Mezger, K., 2001. Age of emplacement of massif–type anorthosites in the Eastern Ghats belt, India : constraints from U–Pb zircon dating and structural studies. *Precamb. Res.*, v. 109, pp. 25–38.

Kretz, R., 1983. Symbols for rock-forming minerals. *Am. Mineralogist*, v. 68, pp. 277–279.

Krienitz, M.S., Trumbull, R.B., Hellmann, A., Kolb, J. Meyer, F.M. and Wiedenbeck, M., 2008. Hydrothermal gold mineralisation at Hira–Buddini gold mine , India: constraints on fluid evolution and fluid sources from boron isotopic compositions of tourmaline. *Mineral. Deposita* , v. 43, pp. 421–434.

Krishna, K.V.G. and Thirupathi, P.V., 1999. Rare metal and rare earth pegmatites of southern India. *Expln. and Research for Atomic Minerals, AMD*, pp. 133–167.

Krishna Rao, B., and Hanuma Prasad, M., 1995. Geochemistry of the phyllites of the Copper Mountain region, Sandur schist belt, Karnataka. *Jour. Geol. Soc. India,* v. 46, pp. 485–495.

Krishna Rao, B., Satis, P.N. and Sethumadhav, M.S., 1989. Syngenetic and epigenetic features and genesis of the bauxite–bearing laterite of Boknur–Navge Plateau, Belgao district, Karnataka. *Jour. Geol. Soc. India,* v. 34, pp. 46–60.

Krishna Rao, N., Sunil Kumar, T.S. and Narasimhan, D., 1995. Uraniferous organic matter in the sandstone–type uranium ore from Domiasat, Meghalaya, India. *Jour. Geol. Soc. India,* v. 45, pp. 407–425.

Krishna Rao, S.V.G., 2001. Geology of gold prospects in Karnataka. *Geol. Surv. India, Spl. Pub.* 58, pp. 59–68

Krishnamurthi, R., and Prabhakar, T., 2000. The nature, fluid characteristics and genesis of tungsten mineralisation in Rajasthan: A brief review. In: M. Deb (Ed.), *Crustal evolution and metallogeny in northeastern Indian shield.* Narosa Publ. House, New Delhi, pp. 417–429.

Krishnamurthy, P., 1985. Petrology of the carbonatite and associated rocks of Sung Valley, Jaintia Hill district, Meghalaya. *Jour. Geol. Soc. India*, v. 26, pp. 361–379.

Krishnamurthy, P. and Sarbajna, C., 1999. Late Archean granitoid with a ring structure around Chandra, Mandya district, Karnataka. *Jour. Geol. Soc. India*, v. 42. pp. 411–414.

Krishnamurthy, P., 1988. Carbonatites of India. *Expln. and Res. for Atomic Minerals. AMD, DAE, Govt.of India*, pp. 81–115.

Krishnamurthy, P., Chaki, A., Sinha, R.M., Singh, S.N., 1988. Geology, geochemistry and genesis of metabasalts, metarhyolites and associated uranium mineralisation at Bodal, Rajnandgaon district, M.P. and implications for uranium exploration in Central India. *Expln. and Res. for Atomic Minerals*, v. 1, pp. 13–19

Krishnamurthy, P., Sinha, D.K., Rai, A.K., Seth, D.K. and Singh, S.N., 1990. Magmatic rocks of the Dongargarh Supergroup, Central India – their petrological evolution and implications on metallogeny. *Geol. Surv. India, Spl. Pub.* 28, pp. 309–319.

Krishnan, M.S., 1937. The geology of Gangpur State, Eastern States, *Geol. Surv. India, Mem.* 71.

Krishnan, M.S., 1952. The iron ores of India. *Indian Minerals*, v. 6.

Krishnan, M.S., 1960. Geology of India and Burma. *Higginbothams (Pvt.) Ltd.*, Madras, 604 p.

Krishnan, M.S. and Swaminath, J., 1959. The great Indian basin of Northern India. *Jour. Geol. Soc. India*, v. 1, pp. 10–30.

Krogstad, E.J., Balakrishnan, S., Mukhopadhyay, D.K., Rajamani, V. and Hanson, G.N., 1989. Plate tectonics 2.5 billion years ago: evidence at Kolar, South India. *Science*, v. 243, pp. 1137–1340.

Krogstad, E.J., Hanson, G.N. and Rajamani, V., 1991. U–Pb ages of zircon and sphene for two gneiss terranes adjacent to the Kolar schist belt, South India: Evidence for separate crustal evolution histories. *Journal Geol.*, v. 99, pp. 801–816.

Krogstad, E.J., Hanson, G.N., and Rajamani, V., 1992. Archean sutures marked by schist belts, east Dharwar Craton, South India. *EOS Trans. Amer. Geophys. Union*, v. 73, p. 332.

Krogstad, E.J., Hanson, G.N. and Rajamani, V. 1995. Sources of continental magmatism adjacent to the late Archean Kolar suture zone, South India: distinct isotopic and elemental signatures of two Late Archean magmatic series. *Contrib. Min. Pet.*, v. 122, pp. 159–173.

Kröner, A., and Cordani, U., 2003. African, South Indian and South American cratons were not parts of the Rodinia supercontinent: evidence from field relationships and geochronology.*Tectonophysics*, v. 375, pp. 325– 352.

Kröner, A., Hagner, E., Wendt, J.L.,and Barely, G.R., 1996. The oldest part of Barberton granitoid–greenstone terrane, South Africa: Evidence for crust formation between 3.5 and 3.7 Ga. *Precamb. Research*, v. 78, pp. 105–124.

Kulik, D.A. and Korzhnev, M.N., 1997. Lithological and geochemical evidence of Fe and Mn pathways during depostion of lower Proterozoic banded iron formation in the Krivoy Rog basin (Ukraine). In K. Nicholson et al (Eds.), Manganese mineralisation geochemistry and mineralogy of terrestrial and marine deposits. *Spec. Publn. Geol. Soc. Lond.*, 119, pp. 43–80.

Kumar, B., Srivastava, R.K., Jha, D.K., Pant, N.C. and Bhandaru, B.K., 1990. A revised stratigraphy of the rocks of type area of the Bijawar Group in Central India. *Indian Minerals*, v. 44(4), pp. 303–314.

Kumar, G., 1997. Geology of Arunachal Pradesh. *Geological Society of India*, p. 217.

Kumar, M.N., Das, N. and Dasgupta, S., 1985. Geology and mineralisation along Northern shear zone, Purulia district, West Bengal – an upto-date appraisal. *Records, Geol. Surv. India*, v. 113, pp. 25–31.

Kumar, S., 1990. Petrochemistry and geochronology of pink granite from Songsak, East Garo Hills, Meghalaya. *Jour. Geol. Soc. India*, v. 35(1), pp. 39–45.

Kumar, V., 1993. Petrography and geochemistry of Jungel metavolcanics, Bijawar greenstone belt, Central India: a picritic basalt low K–andesite association. *Jour. Geol. Soc. India,* v. 41, pp. 9–19.

Kwak, T.A.P., 1987. W–Sn skarn deposits and related metamorphic skarns and granitoids. Elsevier, Amsterdam, 451p.

Lachenbruch A.H. and Saas J.H., 1977. Heat flow in United States and thermal regime of the crust. In J.G.Heacock(Ed.), *The earth's crust. AGU, Washington, Geophys, Mon. Ser.* 20, pp. 626–675.

Lago, B.L., Rabinowicz, M. and Nicholas, A., 1982. Podiform chromite ore bodies: A genetic model. *Journal of Petrology*, v. 23, pp. 103–125.

Lahiri, D., Gupta, A. and Kar, S.K.,1973. Report on detailed exploration of Turamdih copper deposit, Singhbhum district, Bihar. *Unpub. Report, GSI.*

Lakshminarayana, G., Bhattacharjee, S. and Ramanaidu, K.V., 2001. Sedimentation and stratigraphic framework in the Cuddapah basin. *Geol. Surv. India Spl. Pub. v. 55(2)*, pp. 31–58.

Lal, R.K. and Singh, J.B., 1978. Prograde polyphase regional metamorphism and metamorphic reactions in pelitic schists in Sini, district Singhbhum, *India. Neues. Jahrb. Miner. Abh.*, v. 131, pp. 304–333.

Lal, R.K. and Sukla, R.S., 1975. Low–pressure regional metamorphism in northern portion of the Khetri copper belt, Rajasthan, India. *Neues. Jahrb. Miner. Abh.*, v. 124, pp. 294–325.

Lal, R.K., Ackermand, D. and Upadhyay, H., 1987. P–T–X relationships deduced from corona textures in sapphirine–spinel–quartz assemblages from Paderu, South India. *Journal Pet.*, v. 28, pp. 1139–1168.

Lambert, D.D., Morgan, G.W., Walkar, R.J. Shiray, S.B., Carlson, R.W., Zientek, M.L. and Koski, M.S., 1989. Rhenium–osmium and samarium–neodymium isotopic systematics of the Stillwater complex. *Science*, v. 244, pp.1169–1173.

Lambert, I.H., Donelly, T.H., Dunlop, J.S.R. and Groves, D.I., 1978. Stable isotope composition of Early Archean sulfate deposits of probable evaporitic and volcanogenic origin. *Nature, v. 276*, pp. 808–811.

Lang, J.R., Baker, T., Hart, C.J.R. and Mortensen, J.K., 2000. An exploration model for instrusion-related gold systems *SEG News Letter*, No. 40, pp. 1,6–15.

Large, D.E., 1983. Sediment-hosted massive sulfide lead-zinc depposits: An empirical model. *Min. Assocn. Canad. Short course Handbook, v. 8*, pp. 1–30.

Lascelles, D.F., 2006. The genesis of Hope Downs iron ore deposit, Hamersley Province, Western Australia. *Econ. Geol.*, v. 101, pp. 1359–1376.

Law, J.D.M. and Phillips, G.N., 2005. Hydrothermal replacement model for Witwatersrand gold. *Econ. Geol., 100th Anniversary Vol.*, pp. 799–811.

Laznicka, P., 1993. *Precambrian empirical metallogeny, (Parts a and b)*, Elsevier, Amstardam, p. 1622.

Laznicka, P., 1999. Quantitative relationships amongst giant deposits of metals. *Econ. Geol.*, v. 94, pp. 455–474.

Leach. D.L., Viets, J.G., Kozlowski, A. and Kibitlweski, S., 1996. Geology, geochemistry and genesis of the Silesia-Gracow zinc-lead distric, southern Poland. *Soc. Econ. Geol. Spec. Publn 4*, pp. 144–170.

Le Fort, 1994. French Earth Sciences research in the Himalaya region, Kathmandu, Nepal. *Alliance Francaise*, 174p.

Le Fort, P., 1975. Himalayas: The collide range: Present knowledge of the continental arc. *Amer. Journal Sci.*, v. 275–A, pp. 1–44.

Le Fort, P., Guillot, S. and Pecher, A., 1987. HP metamorphic belt along the Indus Future Zone of NW Himalaya: New discoveries and significance. *Sciences de La Terra et des Planets*, v. 325, pp. 773–778.

Le Roux, A. P. Bell. D. R. and Davis, P. 2003. Petrogenesis of Group–I kimberlites from Kimberly, South Africa: Evidence from bulk rock geochemistry. *Journal Pet.*, v. 44, pp. 2261–2286.

Leelanandam, C., 1993. Almandine–rich garnets from fayalite–ferrosilite–quartz–bearing ferrosyenites of Andhra Pradesh, India. *Jour. Geol. Soc. India*, v. 41, pp. 437–443.

Leelanandam, C., and Ashwal, L.D., 2004. Precambrian ophiolite complex, schist belt, deformed alkaline rocks and suture zone in Andhra Pradesh, India. *(Abst.) Conf. on Plate tectonics, plumes and planetary lithospheres* (Kevin Burke Feshshrift), (Nov., 2001).

Leelanandam, C. and Vijay Kumar, K., 2007. Petrogenesis and tectonic setting of the chromitites and chromite bearing ultramafic cumulates of the Kondapalli Layered Complex. Eastern Ghats Belts, India. Evidences from the textural, mineral-chemical and whole rock geochemical studies. *IAGR Mem. 10*, pp. 89–107.

Lehmann, B., Storey, C., Mainkar, D., and Jeffries, T., 2007. *In-situ* U–Pb dating of titanite in the Tokapal–Bhejripadar Kimberlite system, central India. *Jour. Geol. Soc. India*, v. 69, pp. 553–556.

Leith, C.K., Lund, R.J., and Leith, A., 1935. Precambrian rocks of the Lake Superior region. *US Geol. Surv. Prof. Paper 84*, p. 34.

Li, Z.X., Bogdanova, S.V., Collins, A.S., Davidson, A, De Waele, B., Ernst, R.E. and others, 2008. Assembly, configuration, and break–up history of Rodinia: A synthesis. *Precamb. Research*, v.160, pp. 179–210.

Lightfoot, P.C. and Keays, P.R., 2005. Siderophile and chalcophile metal variations in flood basalts from Siberian Trap, Norilisk region: Implication for the origin of Ni–Cu–PGE sulphide ores. *Econ. Geol.*, 100th *Aniv. vol.*, pp. 439–462.

Lindsay, D.D., Zentilla, M., Rojas de la Rivera, J., 1995. Evolution of an active ductile to brittle shear system controlling mineralisation at the Chiquikamata porphyry copper deposit, northern Chile. *Internat. Geol. Rev.*, v. 37, pp. 945–958.

Linnen, R.L., Pichavant, M. and Holtz, F., 1996. The combined effect of fO2 and melt composition on SnO_2 solubility and tin diffusivity in haplogranitic melts. *Geochim. Cosmochim. Acta*, v. 60, pp. 4965–4976.

Lipin, B.R., 1993. Pressure increases in the formation of chromite seams and the development of the ultramafic serves in the Stillwater Complex, Montana. *Journal of Petrology*, v. 34, pp. 955–976.

Lippolt, H.J. and Hautmann, S., 1994: $^{40}Ar/^{39}Ar$ ages of Precambrian manganese ore minerals from Sweden, India and Morocco. *Mineral. Deposita*, v. 18, pp. 195–215.

Lisker, F. and Fachmann, S., 2001. Phanerozoic history of Mahanadi region. *Journal Geophys. Res.*, v. 106, B–10, pp. 22027–22050.

London, D., Morgan, G.B. IV, and Wolf, M.B., 1996. Boron in granitic rocks and their contact aureoles. In E.S. Grew and Anovitch, L.M. (Eds), Boron: Mineralogy, petrology and geochemistry. *Rev. Mineral.*, v. 33, pp. 299–330.

London, D., 1999. Stability of tourmaline in peraluminous granite systems: the boron cycle from anatexis to hydrothermal aureoles. *Eur. Jour. Mineral*, v. 11, pp. 253–262.

Longstaffe, F.J., Sarkar, G., Paul, D.K. and Potts, P.J., 2003. Oxygen isotope evidence for widespread hydrothermal water/rock interaction in Precambrian granitoid rocks from Bastar craton, India. *Gondwana Geol. Mag.*, v. 7, pp. 201–215.

Lorenz, V., 1985. Moars and diatremes of phreatomagmatic origin, a review. *Trans. Geol. Soc. South Africa.*, v. 88, pp. 459–470.

Lorenz, V., Zimanowski, B., Butner, R., Kurzlankis, S., 1999. Formation of kimberlite diatremes by explosive interaction of kimberlite magma with groundwater. Field and experimental aspects. In Gurney, J.J. et. al. (Eds), *Proc. 7th Internat. Kimberlite Conference*, Cape Town, pp. 522–528.

Loucks, R.R. and Mavrogenes, J.A.,1999. Gold solubility in supercritical hydrothermal brines measured in synthetic fluid inclusions. *Science*, v. 284, pp. 2159–2163.

Lowell, J.D. and Guilbert, J.M., 1970. Lateral and vertical alteration–mineralisation zoning in porphyry ore deposits. *Econ. Geol.*, v. 65, pp. 373–408.

Lydon, J.N., Goodfellow, W.D. and Jonasson, I.R., 1985. A genetic model for stratiform baritic deposits of the Selwin basin, Yakpu Territory and the district of Michigan. Curr. Res. Pl. 6A. *Geol. Surv. Canada. Paper*, v. 85–1A, pp. 651–660.

Lydon, J.W., 1984. Volcanogenic massive sulfide deposits. Pt 1: A descriptive model. *Geoscience Canada*, v. 15, pp. 195–202.

Lydon, J.W., 2004. Geology of the Bell-Purcell basin and the Sullivan deposit. In M. Deb and N.D. Goodfellow (Eds.), *Sediment-hosted lead-zinc sulfide deposits*. Narosa, New Delhi, pp. 100–148.

Machado, N., Lindenmayo, Z., Krogh, T.E. and Lindenmayor, D.C., 1991. U-Pb geochronlogy of Archean magmatism and basement reactivation in teh Karajas area, Amazon shield, Brazil. *Pricamb. Research*, v. 49, pp. 329.

Mahabaleswar, B., Jayananda, M., Peucat, J.J., Shadakshara Swami, N., 1995. Archean Gneiss Complex from Satnur–Halgur–Sivasamudram area : petrogenesis and crustal evolution. *Jour. Geol. Soc. India*, v. 45, pp. 33–49.

Mahadevan, C., Krishna Rao, J.S.R., 1956. Genesis of manganese ores of Visakhapatnam–Srikakulam districts (India). *Symp. on manganese, 20th Internat. Geol. Cong.*, pp. 134–138.

Mahadevan, T.M., 1986. Space-time controls in Precambrian uranium mineralisation in India. *Jour. Geol. Soc. India*, v. 27, pp. 47–62.

Mahadevan, T.M., 1992. Uranium metallogeny in the Proterozoic Mobile Belts in India in relation to tectonic developments. In S.C. Sarkar (Ed.), *Metallogeny related to tectonics of Proterozoic mobile belts*. Oxford and IBH Pub. Co. (Pvt.) Ltd., New Delhi, India, pp. 177–208.

Mahadevan, T.M., 2002. Geology of Bihar and Jharkhand. *Geological Society of India, Bangalore*, p. 563.

Mahadevan, T.M. and Dhanaraju, K., 1999. Rare metals and rare earth pegmatites of India. EARFAM Publ., Hyderabad, p. 171.

Mahadevan, T.M., Swarnakar, B.M. and Qidwai, H.A., 1986. Recognition and evaluation of uranium mineralisation in the Siwalik sedimentary basin of India. *Geol. Surv. India, Misc. Pub.*, v. 41 (IV), pp. 442–457.

Mahakud, S.P., 1993. Final report on the detailed exploration for polymetallic mineralisation in Bhanwara–Tekra block, Kherli Bazar–Baragaon area, Multai tehsil, Betul, M.P. *Geol. Surv. India*, Unpub. Report.

Mahakud S.P., Raut P. K., Hansda C., Ramteke P.F., Chakraborty U., Praveen M.N. and Sisodiya D.S., 2001. Sulphide mineralisation in the central part of Betul Belt around Ghisi–Mauriya–Koparpani area, Betul district, Madhya Pradesh. *Geol. Surv. India., Spl. Pub. 64*, pp. 377–385.

Mahalik, N.K. 1987. Geology of rocks lying between Gangpur Group and Iron Ore Group of the Horse–shoe Syncline in North Orissa. *Ind. Journal Earth Sci.*, v. 14, pp. 73–83.

Mahapatra, B.K., 1998. Manganese mineralisation in Precambrian rocks of Orissa and its optimum utilization. (Abst.) *Internat. Sem. on Precambrian crust in Eastern and Central India, UNESCO–IUGS–IGCP 368, Geol.Surv.India*, Bhubaneswar, pp.165–167.

Mahapatra, B.K., Paul, A.K. and Sahoo, R.K., 1989. Characterisation of manganese ores of a part of western Koira valley, Keonjhar district, Orissa. *Jour. Geol. Soc. India*, v. 34, pp. 643–646.

Mahapatra, B.K., Paul, A.K. and Sahoo, R.K., 1991. Characteristics of shale units from Iron Ore Group of rocks, Orissa. *Indian Journal Geol.*, v. 64, pp. 220–228.

Mahapatra, B.K., Paul, D., and Sahoo, R.K., 1996. REE distribution in ferromanganese, oxide ores from Iron Ore Group, Western Koira Valley, Orissa. India. *Journal Min. Pet. Econ. Geol.*, v. 91, pp. 266–274.

Mahapatra, K.C., Saha, A.K., Bhoskar, K.G. and Gupta, Anupendu, 2001. Platinum incidences from auriferous quartz veins of Bhimsain Killa Pahar area, Sakoli Fold Belt, Bhandara district, Maharashtra. *Geol. Surv. India, Spl. Pub.* 58, pp. 457–463.

Mahapatra, K.S., 1993. Development of India's iron ore resources. *Proc. Internat Sem. on Iron ore 2000 and beyond*, Bhubaneswar, pp. 10–37.

Mahendra, A.R., 1975. Apatite–magnetite veins of Kasipatnam hill, Visakhapatnam district, Andhra Pradesh. *Journal Geol.Soc.India.*, v. 16, No.2, pp.157–164.

Maheshwari, A., Sial, A.N., Coltorts, M., Chittora, V.K. and Cruz, M.J.M., 2001. Geochemistry and petrogenesis of Siwana peralkaline granite, west of Barmer, Rajasthan, India. *Gondwana Res.*, v. 4(1), pp. 87–96.

Mahoney, J. J., Sheth, H. C., Chandrasekharam, D. and Peng, Z. X., 2000. Geochemistry of flood basalts of the Toranmal Section, Northern Deccan Traps, India: Implications for regional Deccan stratigraphy. *Journal of Petrology*, v. 41(7), pp. 1099–1120.

Mahato, S., Goon, S., Bhattacharya, A, Misra, B., and B., and Bernhardt, H.-J., 2008. Therno-tectonic evolution of the North Singhbhum Mobile Belt (eastern India): A view from the western part of the belt. *Precamb. Research*, v.162, pp.102–127.

Mainkar, D. and Lehmann, B., 2007. The diamondiferous Behradih Kimberlite, Mainpur Kimberlite field, Chhattisgarh. *Jour. Geol. Soc. India*, v. 69, pp. 547–552.

Maithani, P.B. and Nagar, R.K., 1999. Rare metal and rare earth negmatites of western India. *EARFM, v. 12*, pp. 101–131.

Maithy, P.K., Kumar, S. and Rupendra Babu, 2000. Biological remains and organo–sedimentary structures from Iron Ore Group (Archean) Barbil area, Singhbhum–Orissa. *Proc. Internat. Sem. on Precamb. crust in Eastern and Central India, (UNESCO–IUGS–IGCP 368), (1998), Bhubaneswar, Geol. Surv. India, Spl. Pub.* 57, pp. 98–105.

Majumdar, P., 1979. A study of the geology and ore mineralisation at Baharagora area, Singhbhum district, Bihar. *Unpub. Ph.D thesis, Jadavpur University*, Calcutta.

Majumdar, R. and Sarkar, S., 2004. Sedimentation history of the Paleoproterozoic Dhanjori Formation, Singhbhum, Eastern India. *Precamb. Research*, v.130. pp. 267-287.

Majumdar, T.K. and Chakraborty, K.L. 1977. Primary sedimentary structures in Banded Iron Formations of Orissa, India. *Sed. Geol.*, v. 19. pp. 287–300.

Majumdar, T.K., Chakraborty, K.L. and Bhattacharyya, A., 1982. Geochemistry of Banded Iron Formation of Orissa, India. *Mineral. Deposita*. v. 17, pp. 107–118.

Malhotra, G. and Pandit, M.K., 2000. Geology and mineralisation of the Jahazpur belt, southeastern Rajasthan. In M. Deb (Ed), *Crustal evolution and metallogeny in Northwestern Indian shield.* Narosa Publ. House, New Delhi, pp. 115–125.

Mallet, F. R., 1869. On the Vindhyan Series as exhibited in the Northwestern and Central Provinces of India. *Geol. Surv. India*, Mem.7(1), pp. l–129.

Mallet, F.R., 1875. On the geology and mineral resources of the Darjeeling district and western Duars. *Geol. Surv. India*, Mem. 11, p. 50.

Mallik, A.K., 1998. Isotope studies of granite plutons of Bihar mica belt, Kodarma. *Abst. vol., M.S. Krishnan Cent. Symp., Geol. Surv. India*, pp. 115–116.

Mallik, A.K. and Sarkar, A., 1994. Geochronology and geochemistry of mafic dykes from the Precambrians of Keonjhar, Orissa. *Indian Mineals*, 48(1and2), pp. 13–24.

Mallik, T.K., Vasudevan, V., Aby Vergese, P. and Machado, T., 1987. The black sand placer deposits of Kerala beach, south west India. *Marine Geology*, v. 77, pp. 129–150.

Mamtani, M.A., Greiling, R.O., Karanth, R.V. and Mehr, S.S. 1999 a. Orogenic deformation and its relationship to AMS fabric – an example from the southern margin of the Aravalli Mountain belt, India. In T. Radhakrishna and J.D.A. Piper (Eds.), *Geological Society of India, Mem.* 44, pp. 25–32.

Mamtani, M.A., Karanth, R.V., Mehr, S.S. and Greiling, R.O. 1999b. The tectonic evolution of the southern part of the Aravalli Mountain belt and its environs, possible causes and time constraints. *Gondwana Res.*, v. 3, pp. 175–188.

Mandal, N., Mitra, A.K., Misra, S. and Chakrabrty, C., 2006. Is the outcrop topology of dolerite dikes of the Precambrian Singhbhum craton fractal ? *Journal Earth Syst. Sci.*, v. 115(6), pp. 643–660.

Mandal, N., Samanta, S.K. and Chakraborty, C., 2004. Problem of folding in ductile shear zones: a theoretical and experimental investigation. *Jour. Struct. Geol.*, v. 26, pp. 475–489.

Mangla Prasad, and Rao, T.C., 1999. Characteristics of a lean grade cherty–calcareous rock phosphate of Jhabua. *Indian Min. and Enginer. Journal*, v. 38(3), pp. 19–26.

Mangla Prasad, Kothyal, D.L., Misra, P., Srivastava, Rajib and Jadav P.K., 1996. Base metal mineralisation in parts of Nainital and Pithoragarh districts, Uttar Pradesh. *Geol. Surv. India Spl. Pub.*, v. 21(1), pp. 481–485.

Mangla Prasad, Dhir, N.K., Sharma, D.P., Khan, M.A, Dwivedi, G.N., Mehrotra, R.D. and Yadav, M.L., 2000. Fluid inclusion studies on Proterozoic gold prospect of Gurhar Pahar and Gulaldih, Sidhi and Sonbhadra districts, M.P and U.P. Proceedings Sem. *UNESCO–IUGS–IGCP–368* (1998), *Geol. Surv. India, Spl. Pub. 57*, pp. 250–259.

Mangla Prasad, Kothyal, D.L., Misra, S.P., Srivastava, R. and Yadav, P.K., 1996. Base metal mineralisation in parts of Nainital and Pithoragarh districts, Uttar Pradesh. *Geol. Surv. India*, Spl.Pub. 21, pp. 481–485.

Manikam, S., Victor Rajamanickam and Benjamin, R.E., 1996. Economic viability of manganese nodule mining in India: A discussion. *Jour. Geol. Soc. India*, v. 48, pp. 331–339.

Manikyamba, C., 1998. Petrology and petrochemistry of mixed oxide–silicate facies BIF from Sandur schist belt, India. *Jour. Geol. Soc. India*, v. 52, pp. 651–661.

Manikyamba, C. and Naqvi, S.M., 1995. Geochemistry of the Fe–Mn formations of the Sandur schist belt, India – mixing of clastic and chemical processes at a shallow shelf. *Precamb. Research*, v. 72, pp. 69–75.

Manikyamba, C. and Naqvi, S.M., 1997. Mineralogy and geochemistry of Archean greenstone belt–hosted Mn–formations and deposits of Dharwar craton: redox potential of proto-oceans. In K. Nicholson, J.R. Hein, B. Buhn and S. Dasgupta (Eds.), *Manganese mineralisation: Geochemistry and mineralisation of terrestrial and marine deposits. Geol. Soc.London*, Spl. Pub. 119, pp. 91–103.

Manikyamba, C., Arora, M. and Naqvi, S.M., 1991. Geochemistry of BIFs of Sandur schist belt, the effect of supergene enrichment on trace element distribution in Iron ore. *Proc. Internat. Symp. on Applied Geochemistry*, Osmania University, Hyderabad, pp. 405–416.

Manikyamba, C., Balaram, V. and Naqvi, S.M., 1993. Geochemical signatures of polygenecity of banded iron formations of the Archean Sandur greenstone belt (schist belt), Karnataka nucleus, India. *Precamb. Research*, v. 61, pp.137–164.

Manikyamba, C., Naqvi, S.M., Rammohan, M., Gnaneswar Rao, T., 2004. Gold mineralization and alteration of Penakacherla schist belt, India: constaints on Archean subduction and fluid processes. *Ore Geology Review*, v. 24, pp. 199–227.

Manikyamba, C., Naqvi, S.M., Moeen, S.I., Gnaneshwar Rao, T., Balaram, V., Ramesh, S.L. and Reddy, G.L.N., 1997. Compositional heterogeneity of graywackes from the Sandur Schist belt : implications for active plate margin processes. *Precamb. Research*, v. 84, pp. 117–138.

Mareschal, J.C. and West, G.F., 1980. A model for Archean tectonism P–T and Numerical models of vertical tectonism in greenstone belts. *Canadian Journal of Earth Sci.*, v. 17, pp. 60–71.

Marques, J.C. and Filho, C.F.F., 2003. The chromite deposit of the Ipueira–Medrado Sill, Sao Francisco craton, Bahia State, Brazil. *Econ. Geol.*, v. 98, pp. 87–108.

Martin, H and Moyen, J.F. 2002. Secular changes in tonalite–trondhjemite–granodiorite composition as markers of of progressive cooling of the Earth. *Geology*, v. 30, pp. 319–322.

Marshall, B., Vokes, F.M. and Laroque, C.L., 2000. Regional imeamorphic remobilization: upgrading and formation of ore deposits. *Rev. Econ. Geol.*, v. 11, pp. 19–38.

Marshall, B. and Gilligan, L. B., 1993. Remobilization ,syntectonic processes and massive sulfide deposits. *Ore Geol. Rev.*, v. 8, pp. 39–64.

Masterman, G.J., Cooke, D.R., Berry, R.F., Walshe, J.L., Lee, A.W. and Clark, A.H., 2005. Fluid chemistry, structural setting and emplacement history of the Rosario Cu–Mo porphyry and Cu–Ag–Au epithermal venis, Collahuasi district, Northern Chile. *Econ. Geol.*, v. 100, pp. 835–862.

Mathur, K.N., 2005. Status and development of mineral resources of Eastern and Northeastern India in the present economic scenario. In *Proc. Seminar on Mineral and Energy resources of Eastern and Northeastern India, MGMI–Calcutta Branch*, Kolkata, pp. 1–9.

Mathur, S.M. and Singh, H.N. 1971. Petrology of the Majhgawan pipe rocks. Misc. Pub., *Geol. Surv. India*, v. 19, pp. 78–85

Matveev, S. and Ballhaus, C., 2002. Origin of podiform chromitite. *Earth and Planet Sci. Letters*, v. 203, pp. 235–243.

Mazumdar, S.K., 1986 The Precambrian framework of part of Khasi Hills, Meghalaya. *Rec. Geol. Surv. India*, v. 117(2), pp 1–59.

Mazumdar, S.K., 1988. Crustal evolution of the Chhotanagpur Gneissic Complex and the Bihar Mica Belt. In D. Mukhopadhyay (Ed.), Precambrian of the Eastern Indian Shield. *Jour. Geological Society of India, Mem.* 8, pp. 49–83.

Mazumder, S.K., 1996. Precambrian geology of peninsular eastern India. *Indian Minerals*, v. 50(3), pp. 139–174.

Mc Culloch, M.T. and Bennett, V.C., 1994. Progressive growth of Earth's continental crust and depleted mantle: geochemical constraints. *Geochim. Cosmochim. Acta*, v. 58, pp. 4717–4738.

Mc Keegan, K.D., Kudryovtsev, A.B. and Schopf, J.W., 2007. Raman and ion microprobe imagery of graphitic inclusions in apatite from older than 3830 Ma Akilia supracrustal rocks, west Greenland. *Geology*, v. 35, pp. 591–594.

Mc Lellan, J.G., Oliver, N.H.S., and Schaubs, P.M., 2004. Fluid flow in extensional environments, numerical modeling with an application to Hamersley iron ores. *Journal Struct. Geol.*, v. 26, pp. 1157–1171.

Mc Lennan, S.M., 1988. Recycling of the continental crust. *Pure Appld. Geophysics*, v. 128, pp. 683–898.

McLennan, S.M. and Taylor, S.R., 1991. Sedimentary rocks and crustal evolution: Tectonic setting and secular trends. *Jour. Geol., v. 99*, pp. 1–21.

McDonald, J.A., 1965. Liquid immiscibility as one factor in chromite seam formation in the Bushveld Complex. *Econ.Geol.*, v. 60, pp. 1674–1685.

Meen, J.K., Rogers, J.J.W. and Fullagar, P.O., 1992. Lead isotope composition in Western Dharwar craton, southern India : evidence for distinct Middle Archean terrains in Late Archean cratons. *Geochim. Cosmochim. Acta*, v. 56, pp. 2455–2470.

Mehta, P.K., 1977. Rb–Sr geochronology of the Kulu–Mandi belt : Its implication for Himalayan tectogenesis. *Geol. Rundsch.*, v. 66, pp. 156–175.

Meigs, A.J., Burbank, D.W. and Beck, R.A., 1995. Middle Late Miocene (> 10 Ma) formation of the Main Boundary Thrust in the western Himalaya. *Geology*, v. 23, pp. 423–426.

Meijerink, A.M.J., Rao, D.P. and Rupke, J., 1984. Stratigraphic and structural development of the Precambrian Cuddapah basin, SE India. *Precamb. Research*, v. 26, pp. 57–104.

Meissner, B., Deters, P., Srikantappa, C., Köhler, H., 2002. Geochronological evolution of the Moyar, Bhabani and Palghat shear zones of southern India: implications for East Gondwana correlations. *Precamb. Research*, v. 114, pp. 149–175.

Metcalfe, I., 1988. Origin and assembly of Southeast Asian continental terrains. In Audley Charles, M.G. and Halm A.(Eds.), Gondwana and Tethys. *Geol.Soc.London, Spl. Pub.*, v. 37, pp. 108–118.

Mezger, K. and Cosca, M.A., 1999. The thermal history of the Eastern Ghats Belt (India) as revealed by U–Pb and ^{40}Ar–^{39}Ar dating of metamorphic and magmatic minerals: implications for the SWEAT correlation. *Precamb. Reearch*, v. 94, pp. 251–271.

Mezger, K., Cosca, M.A. and Raith, M., 1996. Thermal history of the Eastern Ghats Belt (India) deduced from U–Pb and Ar–Ar dating of metamorphic minerals. *Journal of Conference Abstracts*, v. 1, pp. 401.

Meyer, C. and Hemley, J.J., 1967. Wall rock alteration. In H.L. Barnes (Ed.), *Geochemistry of hydrothermal ore deposits*. Holt, Rinehart and Winston, New york, pp. 166–235.

Michard, A., 1989. Rare earth elements systematics in hydrothermal fluids. *Geochim. Cosmochim. Acta*, v. 53, pp. 745–750.

Middlemiss, C.S., 1915. Annual General Report of Geol. Surv. India. for the year 1914, Rec. G.S.I., 16(2), pp. 85–137.

Middlemost, E.A.K and Paul D.K., 1984. Indian kimberlites and genesis of kimberlites. *Chem. Geol.*, v.19, pp. 249–260.

Miller, C., Schuster, R., Klotzli, U., Frank, W. and Putscheller, F., 1999. Post-collisional potassic and ultrapotassic magmatism in S.W. Tibet: Geochemical and Sr–Nd–Pb–O isotopic constraints for mantle source characteristics and petrogenesis. *Journal of Petrology*, v. 40, pp. 1399–1424.

Minnitt, R.C.A., 1986. Porphyry copper–molybdenium mineralisation at Haib Rives, Namibia, South West Africa. In C.R. Anhausser and S. Maska (Eds.), Mineral deposits of Southern Africa. *Geol. Soc. South Africa*, Johannesberg, pp. 1567–1585.

Minter, W.E.L., Goedhart, M.T., Knight, J. and Frimmel, H.E., 1993. Morphology of Witwatersrand gold grains from Basal reef: Evidence for their detrital origin. *Econ. Geol.*, v. 88, pp. 237–248.

Misch, P., 1949. Metasomatic granitization of batholithic dimensions. *Amer. Journal Sci.*, v. 247, pp. 209–249.

Misch, P., 1964. Stable association of Wollastonite–anorthite and other calc–silicate assemblages in amphibolite facies crystalline schists of Nanga Parbat, Northwest Himalayas. Contrib. *Min. Pet.*, v. 10, pp. 315–356.

Mishra, B., 2008. Gold metallogeny in Dharwar craton, southern India. In *Pre-workshop(Extd. Abst.) vol., Internat. Field Workshop on Gold Metallogeny in India, Dec. 2008, NGRI – Hyderabad and Univ. of Delhi*, pp. 56–58.

Mishra, B., 2000. Evolution of the Rajpura-Dariba polymetallic sulfide deposit: constraints from sulfide– sufosalt phase equilibria and fluid inclusion studies. In M.Deb (Ed.), *Crustal evolution and metallogeny in north-western Indian shield.* Narosa, New Delhi, pp. 307–328.

Mishra, B. and Mookherjee, A., 1986. Analytical formulation of phase equilibria in two observed sulphide–sulfosalt assemblages in the Rajpura–Dariba polymetallic deposit, India. *Econ. Geol.*, v. 81, pp. 627–639.

Mishra, B. and Pal, N., 2008. Metamorphism, fluid flux and fluid evolution relative to gold mineralisation in the Hutti–Maski greenstone belt, Eastern Dharwar craton, India. *Econ. Geol.*, v. 103, pp. 801–827.

Mishra, B., Pal, D.C. and Panigrahi, M.K., 1999. Fluid evolution in quartz vein–hosted tungsten mineralisation in Chhendapathar, Bankura district, West Bengal: Evidence from fluid inclusion study. *Proc. Ind. Acad. Sci* (Earth and Planet. Scs), v. 108, pp. 23–31.

Mishra, B., Pal, N., and Ghosh, S., 2003. Fluid evolution of the Mosabani and Rakha copper deposits, Singhbhum district, Jharkhand : Evidence from fluid inclusion study of mineralized quartz venis. *Jour. Geol. Soc. India*, v. 61, pp. 51–60.

Mishra, B., Pal, N. and Basu Sarbadhikari, A., 2005. Fluid inclusion characteristics in barren and arriferous quartz veins in the Uti gold deposit, Hutti-Maski greenstone belt, southern India. *Ore Geol Rev.*, v. 25, pp. 1–16.

Mishra, M. and Rajamani, V., 2003. Geochemistry of the Archean metasedimentary rocks from the Ramagiri schist belt, Eastern Dharwar craton, India : Implications to crustal evolution. *Jour. Geol. Soc. India*, v. 62, pp. 717–738.

Mishra, R.C. and Sharma, R.P., 1975. New data on the geology of Bundelkhand Complex, Central India. *Rec. Res. Geol. 2*, Hind Publ. Corp., New Delhi, pp. 311–346.

Mishra, R.N., 1974. Sandur metallogen and its geological association with orogenesis. (Abst.) *Golden Jub. Symp., Geol. Min. Met. Soc. India*, p. 10.

Mishra, R.N., 1978. Geology and metallogenesis of Sandur basin with special reference to mining geology and exploration parameters of iron and manganese ores. *Ph.D thesis, Utkal University.*

Mishra, R.N., Padhi, R.N. and Kanungo, S.C., 1998. Graphite. In *Geology and mineral resources of Orissa.* S.G.A.T, Orissa, pp. 255–268.

Mishra, V.P., Das, N., Ramamurthy, S., Parui, P.K., Swaminarayanan, D'Souza, M. J.,Bhattacharya, D.D., Agasty, A. and Rawat. P.V.S., 1989. Stratigraphy, structures and sedimentation history of Khariar Basin, Central India. *Geol. Surv. India, Unpub. Report.*

Mishra, V.P., Datta, N. K., Kanchan, V. K., Vatsa, U.S. and Guha, K., 1984. Archaeangranulite and granite–gneiss complexes of Kondagaon area, Bastar district, M. P. *Records, Geol. Surv. India.* v. 113(6), pp. 150–158.

Mishra, V.P., Pushkar Singh and Dutta, N.K., 1988. Stratigraphy, structure andmetamorphic history of Bastar district, M.P. *Records, Geol. Surv. India*, v. 117(3 to 9), pp. 1–26.

Misra, K.S., 1998. Occurrence of polymetallic (Fe–U–Au–REE) iron oxide breccia type mineralisation around Chandil, West Singhbhum district, Bihar. *(Abst.) Sem. on Geosci. Advns. in Bihar, India in the Last Decade. Geol. Surv. India*, Patna, pp. 56.

Misra, K.S., 2002. Arterial system of lava tubes and channels within Deccan volcanics of western India. *Jour. Geol. Soc. India*, v. 59, pp. 115–124.

Misra, K.S., 2004. Deccan and associated volcanics and their relationship with the development of hydrocarbon pools around peninsular India. *Sem. Expln. Geophy. Souvenir*, pp. 13–24.

Misra, K.S., 2005. Distribution pattern, age and duration and mode of eruption of Deccan and associated volcanics. *Gondwana Geol. Mag.*, Spl. vol. 8, pp. 53–60.

Misra, K.S., Durairaju, S., Rajasekharan, P. and Das, A.K., 1997. Occurrence of U–Au–REE bearing quartz–pebble conglomerate at Syamba Taldih, Sundargarh district, Orissa. *Jour. Geol. Soc. India*, v. 50, pp. 93–94.

Misra, S. and Johnson, P.T., 2005. Geochronological constraints on evolution of Singhbhum Mobile Belt and associated basic volcanics of Eastern Indian Shield. *Gondwana Res.*, v. 8(2), pp. 129–142.

Misra, S., Deomurari, M.P., Wiedenbeck, M., Goswami, J.N., Ray, S.L., Saha, A.K., 1999. ^{207}Pb/ ^{206}Pb zircon ages and the evolution of Singhbhum craton, Eastern India: an ion microprobe study. *Precamb. Research*, v. 93, pp. 139–151.

Mitchel, C and Cox, K.G., 1988. A geological sketch map of the southern part of Deccan province. In K. V. Subbarao (Ed), Deccan flood basalts. *Jour. Geol. Soc. India*, Bangalore, pp. 27–34.

Mitchell, A.H.G., 1981. Phanerozoic plate boundarirs in mainland SE Asia, the Himalaya and Tibet. *JournalGeol. Soc. London*, v. 138, pp. 109–122.

Mitchell, A.H.G. and Garson, M.S. 1981. *Mineral deposits and global tectonic settings*. Acad. Press, London, p. 405.

Mitchell, R.H. 1986. Kimberlites: Mineralogy, geochemistry and petrology. Plenum Press, New York, p. 442.

Mitchell, R.H., 1995. Kimberlites orangeites and related rocks. Plenum Press, New York, p. 410.

Mitchell, R.H. 2004. Experimental studies at 5–12 Gpa of the Ondermatje hypabyssal kimberlite. *Lithos*, v. 76, pp. 551–564.

Mitchell, R.H., 2006. Potassic magmas derived from metasomatised lithospheric mantle: nomenclature and relevance to exploration for diamond bearing rocks. *Jour. Geol. Soc. India*, v. 67, pp. 317–327.

Mitra, G., 1997. Evolution of salients in a fold–thrust belt: the effects of sedimentary basin geometry, strain distribution and critical taper. In S.Sengupta (Ed), *Evolution of geological structures in micro to meso scales*. Chapman and Hall, London, pp. 59–90.

Mitra, S.K, 1998. Structural history of the rocks of Shillong Group around Shillong, Meghalaya. *Indian Journal Geol.*, v.70 (1and2), pp. 123–131.

Mitra, S.K., 2005. Syn–sedimentary sulphide mineralisation in Shillong Group of rocks, Meghalaya. In *Proc. Nat. Sem. on Mineral and Energy Resources of Eastern and Northeastern India, MGMI-Calcutta branch*, Kolkata, pp. 253–262.

Mitra, S.N., 1972. Metamorphic "rims" in chromite from Sukinda, Orissa, India. *Neues Jahrb. Miner. Abh.*, v. 8, pp. 360–375.

Mitra, S.K. and Mitra, S.C., 2001. Tectonic setting of the Precambrians of the North–Eastern India (Meghalaya Plateau) and age of the Shillong Group of rocks. *Geol. Surv. India Misc .Pub.*, v. 64, pp. 653–658.

Miyashiro, A., 1973. *Metamorphism and metamorphic belts*. John Wiley and Sons, Inc. New York, p. 492.

Mohan, M. and Bhoskar, K.G., 1990. Tungsten metallogeny related to acid magmatism in Sakoli Group, Central India. *Geol.Surv. India., Spl. Pub. 28*, pp. 648–657.

Mohan, M. and Bhoskar, K.G., 1990. Tungsten metallogeny related to acid magmatism in Sakoli Group, Central India. *Geol. Surv. India, Spl. Pub. 28*, pp. 648–657.

Mohan, P.M. and Victor Rajamanickam, G., 2000. Buried placer mineral deposits along east coast between Chennai and Pondichery. *Jour. Geol. Soc. India*, v. 56, pp. 1–14.

Mohanty, M. and Guha, D., 1995. Lithotectonostratigraphy of the dismembered greenstone sequence of the Mangalwar Complex around Lawa Sardargarh and Parvali areas, Raj Samand District, Rajasthan. In S. Sinha Roy and K.R. Gupta (Eds), continental crust of North Western and Central India. *Geological Society of India. Mem. 31*, pp. 141–162.

Mohanty, S., and Naha, K., 1986. Stratigraphic relations of Precambrian rocks in the Salumbar area, southeastern Rajasthan. *Jour. Geol. Soc. India*, v. 27, pp. 479–493.

Moitra, A.K., 1995. Depositional environmental history of Chhattisgarh Basin, M.P., based on stromatolites and microbiota, *Jour. Geol. Soc. India*, 46(4), pp. 359–368.

Moitra, A.K., 1999. Biostratigraphy of stromatolite and microbiota, Chhattisgarh Basin, M.P., India. *Pal. Indica*, v. 51.

Mondal, M.E. A., Sharma, K. K., Rahaman, A., and Goswami, J. G., 1998. Ion microprobe $^{207}Pb/^{206}Pb$ zircon ages for gneiss–granitoid rocks from Bundelkhand massif: Evidence for Archaean components. *Current Sci.*, v. 74(10), pp. 70–75.

Mondal, M.E.A. and Hussain, M.F., 2003. Geochemical characteristics of granitoids from Bastar craton, Central India. *Gondwana Geol. Mag.*, v. 7, pp. 193–199.

Mondal, M.E.A., Zainuddin, S.A., 1996. Evolution of the Archaean –Paleoproterozoic Bundelkhand massif, Central India – evidence from granitoid geochemistry. *Terra Nova*,v. 8, pp. 532 – 539.

Mondal, M.E.A. and Zainuddin, S.M., 1997. Geochemical characteristics of the granites of Bundelkhand massif, Central India. *Jour. Geol. Soc. India*, v. 50, pp. 69–74.

Mondal, S.K. and Baidya, T.K., 1997. Platinum-group minerals from the Nuasahi ultramafic–mafic complex, Orissa, India. *Mineral. Mag.*, v. 61, pp. 902–906.

Mondal, S.K., Frei, R. and Ripley, E.M., 2007. Os-isotope systematics of Mesoarchean chromite–PGE deposits in the Singhbhum craton (India): Implications for the evolution of lithospheric mantle. *Chem. Geol.*, v. 244, pp. 391–408.

Mondal, S.K., Ripley, E.M., Chusi Li and Frei, R., 2006. The genesis of Archean chromitites from Nuasahi and Sukinda massifs in the Singhbhum craton, India. *Precamb.Research*, v. 148, pp. 45–66.

Monrad, J.R., 1983. Evolution of sialic terranes in the vicinity of the Holenarsipur belt, Hassan District, Karnataka, India. In S.M. Naqvi and J.J.W. Rogers (Eds.), Precambrians of south India. *Geological Society of India, Mem. 4*, pp. 343–364.

Mookherjee, A., 1964. Geology of the lead–zinc mineralisation, Rajasthan, India. *Econ. Geol.*, v. 59, pp. 656–677.

Mookherjee, D., 1975. Geokhimicheskaiya cherta orudeniya v granitakh Dhabi. Abtoreferat. *Unpub. Ph.D. thesis, Moscow State Univ.* (In Russian).

Moorbath, S., Taylor, P.N., 1988. Early Precambrian crustal evolution in eastern India: the age of Singhbhum Granite and included remnants of older gneiss. *Jour. Geol. Soc. India*, v. 34(1), pp. 82–84.

Moorbath, S., Taylor, P.N., Jones, N.W., 1986. Dating the oldest terrestrial rocks – fact and fiction. *Chem. Geol.*,v. 57, pp. 63–86.

Moores, E.M., 1991. Southwest U.S. – East Antarctic (SWEAT) connection: A hypothesis. *Geology*, v. 19, pp. 425–428.

Morey, G.B.,1983. Animikie Basin, Lake Superior Region, USA, In A.F. Trendall and R.C.Morris (Eds.), *Iron formation: Facts and Problems*, Elsevier, Amsterdam, pp. 13–68.

Morey, G.B., 1999. High-grade iron ore deposits of the Mesabi Range, Minnesota – product of a continental-scale Proterozoic ground water flow system. *Econ. Geol.* v. 94, pp. 133–141.

Morgan, G.V. and London, D., 1989. Experimental reactions of amphibolite with boron-bearing aqueous fluids at 200 M Pa: implications for tourmaline stability and partial melting in mafic rocks. *Contrib. Min. Pet.*, v. 102, pp. 218–297.

Morgan, W. J., 1981. Hotspot tracks and the opening of the Atlantic and Indian Oceans. In C.Emiliani (Ed.), *The Sea*. v. 7, pp. 443–487. John Wiley, New York.

Morris, R.C., 1980. A textural and mineralogical study of the relationship of iron ore to Banded Iron Formation in the Hamersley iron province of Western Australia. *Econ. Geol.*, v. 75, p.184–209.

Morris, R.C., 1985. Genesis of iron ore in Banded Iron Formation by supergene and supergene–metamorphic processes – a conceptual model. In K. H. Wolf (Ed.), *Handbook of stratabound stratiform ore deposits*. Elsevier, Amsterdam, v. 13, pp. 73–235.

Morris, R.C., 2002. Iron ore genesis and post–ore metasomatism at Mount Tom Price. Australian Inst. of Min. Met. Pub. 8/2005, p. 3–14 (reprinted in *Transactions of Mining and Metallurgy, v. 112, sec.: Applied Earth Sci.*)

Moughal, M.Y. 1974. Uranium in Siwalik Sanstones in Sulaiman Range, Pakistan. *Formation of Uranium ore deposits, IAEA, Vienna*, pp. 383–403.

Mountain, B.W. and Wood, S.A., 1987. Solubility and transport of platinum group elements in hydrothermal solutions: thermodynamic and physical chemical constraints. In H. M. Pickard, P. J. Potts, J. F. W. Bowter and S. J. Cribbs (Eds.), *Geo–Platinum '87*, Elsevier, London, pp. 54–82.

Moyen, J.F., Jayananda, M., Nedelec, A., Martin, H., Mahabaleswar, B., and Muvray, B., 2003. From the roots to the roof of a granite: the Closepet Granite of South India. *Journal Geological Society of India*, v. 62, pp. 753–768.

Mozgova, N.N., Borodaev, yu. S. Nenasheva, S.N., Efimov, A.V., Gandhi, S.M. and Mookherjee, A., 1992. Rare minerals from Rajpura-Dariba, Rajasthan, India. *Mineral. Petrol*, v. 46, pp. 55–65.

Mukherjee, A.D., 1992. Petrotectonics of southern segment of the polymetallic sulfide-bearing Proterozoic Delhi mobile belt, Rajasthan, India, and its implications. In S.C. Sarkar (Ed.), *Metallogeny related to tectonics of the Proterozoic mobile belts*. Oxford and IBH, New Delhi, pp. 253–270.

Mukherjee, A.D., Guha–Lahiri, S. and Bhattacharya, H. N., 1991. Silver-bearing sulphosalts for Rampura Agucha massive sulphide deposit of Rajasthan. *Jour. Geol. Soc. India*, v. 37, pp. 132–135.

Mukherjee, A.D. and Bhattacharya, H.N., 1997. Stratebound pyritic copper mineralisation in the fore are setting in northwestern India– a Proterozoic Bessi type deposit. *Trans. Inst. Min. Metall., Sec B 106*, pp. B45–B51.

Mukherjee, B., 1968. Genetic significance of trace elements in certain rocks of Singhbhum, India. *Min. Mag.*, v. 36, pp. 661–670.

Mukherjee, M.M. and Natarajan, W.K., 1985. Exploration for gold in Kolar schist belt, India. In *Proc. United Nations sponsored Interregional seminar on gold exploration and development*, Bangalore. (Sec II), pp. 2.1.1–26.

Mukherjee, M.M., 1991. Deposit-scale structural control of gold mineralisation in Chigargunta area, south Kolar schist belt, India. In E.A. Ladeira (Ed.), *Brazil Gold '91*, Balkema, Rotterdam, pp. 693–698.

Mukherjee, N.K. and Dhruba Rao, B.K., 1974. Geology of the Bhotang sulfide deposit, Rangpo, Sikkim. *Jour. Geol. Soc. India,* v. 15, pp. 65–75.

Mukherjee, N.K. and Jog, R.G., 1963. Report on detailed exploration of Dikchu copper–zinc deposit, Sikkim. *Unpublished report, India Bureau of Mines*.

Mukherjee, S., 1969. Clot textures developed in the chromitites of Nausahi, Keonjhar district, Orissa. *Econ. Geol.*, v. 64, pp. 329–337.

Mukherjee, S., and Haldar, D., 1975. Sedimentary structures displayed by the ultramafic rocks of Nausahi, Keonjhar district, Orissa, India. *Mineral. Deposita*, v. 10, pp. 109–119.

Mukherji, A. and Khan, M.W.Y., 1996. Detailed facies analysis of Deodongar Member, Chhattisgarh Supergroup, Drug–Raipur districts, M.P. *Indian Journal Earth Sci.*, v. 23(3), pp. 139–146.

Mukhopadhyay, A. and Chanda, S.K., 1972. Silica diagenesis in the banded hematite–jasper and bedded chert associated with the Iron Ore Group of Jamda Koira vallcy, Orissa, India. *Jour. Geol. Soc. India,* v. 8, pp. 113–135.

Mukhopadhya, A.K., 1980. A new occurrence of scheelite, molybdenite, kowellite and gadolinite in Phalwadi-Positara area, district Sirohi, Rajasthan. *Jour. Geol. Soc. India*, v. 22, pp. 147–148.

Mukhopadhyay, A.K. and Bhattacharya, A., 1997. Tectonothermal evolution of gneiss complex at Salur in the Eastern Ghats granulite belt of India. *Journal Meta. Geol.*, v. 15, pp. 719–734.

Mukhopadhyay, D., 1984. The Singhbhum Shear Zone and its place in the evolution of the Precambrian mobile belt of North Singhbhum,. In *Proc. Sem. on Crustal Evolution of Indian Shield and its bearing on Metallogeny Indian Soc. Earth Sci*. pp. 205–212.

Mukhopadhyay, D. and Baral, M.C., 1985. Structural geology of Dharwar rocks near Chitradurga. *Jour. Geol. Soc. India,* v. 26, pp. 547–566.

Mukhopadhyay, D. and Bhattacharyya, T., 2000. Tectonic stratigraphic framework of the South Delhi fold belt in the Ajmer–Bewar region, Central Rajasthan, India: a critical review. In. M. Deb (Ed), *Crustal evolution and metallogeny in the northwestern Indian shield.* Narosa Publ. House, New Delhi, pp. 126–137.

Mukhopadhyay, D. and Dasgupta, S., 1978. Delhi– pre-Delhi relations near Badnor, Central Rajasthan. *Indian Journal Earth Sci.*, v. 5, pp. 183–190.

Mukhopadhyay, D. and Deb, G.K., 1995. Structural and textural development in Singhbhum shear zone, eastern India. *Proc. Indian Acad. Sci.* (*Earth Planet. Sci.*), v. 104(3), pp. 385–405.

Mukhopadhyay, D. and Matin, A., 1993. The structural anatomy of Sandur schist belt a greenstone belt in the Dharwar craton of south India. *Journal Structural Geol.*, v. 33, pp. 291–308.

Mukhopadhyay, D. and Sengupta, S., 1971. Structural geometry and the time–relation of metamorphic recrystallisation to deformation in the Precambrian rocks near Simulpal, Eastern India. *Geol. Soc. Amer. Bull.*, v. 82, pp. 2251–2260.

Mukhopadhyay, D. and Sengupta, S., 1979. "Eyed folds" in Precambrian marbles from southeastern Rajasthan, *Inter. Bull. Geol. Soc. Amer.*, v. 90, pp. 397–404.

Mukhopadhyay, D. and Srinivasam, R., 2003. Evolution of Dharwar Craton. In A. Roy and D.M. Mohanty (Eds), Advances in Precambrians of central India. *Gondwana Geol. Mag*, Spl. v. 7, pp. 1–7.

Mukhopadhyay, D., Bhattacharyya, T., Chattopadhyay, N., Lopez, R., and Tobisch, O.T., 2000. Anasagar Gneiss: A folded granitoid in the Proterozoic South Delhi fold belt, Central Rajasthan. *Proc. Ind. Acad. Sci. (Earth Planet Sci)*, v. 109, pp. 21–37.

Mukhopadhyay, D., Ghosh, M, Chattopadhyay, A. and Sengupta, P., 1998. Pattern of deformation and metamophism in the north Singhbhum fold belt near Dhalbhumgarh, Bihar. (Abst.) Proc. Internat. Sem. on crust in Eastern and Central India, *UNESCO–IUGS–IGCP 368, Geol. Surv. India*, Bhubaneswar, India, pp. 193–194.

Mukhopadhyay, D., Senthil Kumar, P., Srinivasam, R., Bhattacharyya, T. and Sengupta, P., 2001. Tectonics of the eastern sector of the Palghat–Cauvery lineament. *News Letters 11(1), Deep Continental Studies in India, DST*, pp. 9–13.

Mukhopadhyay, D.K. and Mookherjee, A., 1978. "Rock ball" texture from Saladipura pyrrhotite–pyrite orebody, Khetri copper belt, India. *Neues Jahrb. Miner. Abh*, pp. 106–112.

Mukhopadhyay, I., Ray, J. and Nath, S., 1995. Ductile shear zone between Shantinagar–Uppalchelka in Khammam district, Andhra Pradesh. *JournalGeol.Soc.India*, v. 46, pp. 595–601.

Mukhopadhyay, J., Beukes, N.J., Armstrong, R.A., Zimmermann, U.Ghosh, G. and Medda, R.A., 2008. Dating the oldest greenstone in India: A 3.51-Ga precise U-Pb SHRIMP zircon age for dacitic lava of the Southern Iron Ore Group Singhbhum craton. *Jour. Geol.*, v. 116, pp. 449–461.

Mukhopadhyay, J., Chaudhuri, A.K. and Chanda, S.K., 1997. Deepwater manganese deposits in the mid. to late Proterozoic Penganga Group of the Pranhita–Godavari valley. In K. Nicholson et al (Eds), Manganese mineralisation, geochemistry and mineralogy of terrestrial and marine deposits. *Geol. Soc. London, Spl. Pub. No. 119*, pp. 105–115.

Mukopadhyay, J., Ghosh, G., Beukes, N.J. and Gutzmer, J., 2007. Precambrian colluvial iron ores in the singbhum craton: Implications for origin, age of BIF-hosted high grade iron ores and stratigraphy of the Iron Ore Group. *Jour. Geol. Soc. India*, v. 70, pp. 34–42.

Mukhopadhyay, J., Ghosh, G., Nandi, A.K. and Chaudhuri, A.K., 2006. Depositional setting of the Kolhan Group: its implications for the development of a Meso to Neoproterozoic deep–water basin on the South Indian craton. *South African Journal Geol*, v. 109, pp. 183–192.

Mukhopadyay, M. and Dasgupta,S., 1988. Deep structure and tectonics of Burmese arc: construction from earthquake and gravity data. *Tectonophysics*, 149, pp. 299–322.

Muktinath, 1960. A note on the pyrite deposits of Sahabad, Bihar. *Records, Geol. Surv. India*, v. 86(4), pp. 595–612.

Muktinath, 1969. Stratigraphic sequence of the Marwar Supergroup in the Birmania basin. *Indian Minerals,* v. 23, pp. 29–42.

Muktinath, 1974. Phosphate – its present status and guides for further search in India. *Presidential address, 61st Indian Sci. Cong., Geol–Geog. Section*, pp. 91–114.

Muktinath and Wakhaloo, G.L., 1962. A note on the magnesite of Almora. *Indian Minerals,* v. 16, pp. 116–125.

Muktinath, Natarajan, W.K. and Mathur, A.L., 1968. A preliminary note on the pyrite–pyrrhotite deposit near Saladipura, Sikar district, Rajasthan. *Journal Mines Metal Fuels*, v. 6, pp. 397–399.

Mukul, M., 2000. The geometry and kinetics of the Main Boundary Thrust and related tectonics in the Darjeeling Himalayan fold–and thrust belt, West Bengal, India. *Jour. Struct. Geol*, v. 22, pp. 207–226.

Mulay, V.V., 2001. Manganese deposits of Central India – A review. *Geol. Surv. India Spl. Pub. 64*, pp. 341–352.

Muller, W.H., Schmidt, S.M. and Briegal, U., 1981. Deformation experiments on anhydrite rocks of different grain sizes: rheology and microfabric. *Tectonophysics*, v. 78, pp. 527–543.

Munshi, R.L., 1993. Petrological status of hornblende quartzite and associated rocks of Basantgarh area of Sirohi district, Rajasthan. *Jour. Geol. Soc. India,* v. 42, pp. 61–66.

Munshi, R.L. and Khan, H.H., 1981. On the geology and phosphorite deposits in parts of Jhabua district, Madhya Pradesh. *Geol. Surv. India, Spl. Pub. No. 3*, pp. 279–284.

Munshi, R.L., Khan, H.H. and Ghosh, D.B., 1974. The algal structures and phosphorite in the Aravalli rocks of Jhabua. *Current Sci.*, pp. 446–447.

Murao, S., Deb, M., Takagi, T., Seki, Y., Pringle, M. and Naito, K., 2000. Geochemical and geochronological constraints for tin–polymetallic mineralisation in Tosam area, Haryana, India. In M. Deb (Ed.), *Crustal evolution and metallogeny in northwestern Indian shield.* Narosa Publ. House, New Delhi, pp. 430–442.

Murck, B.N. and Cambpell, I.H., 1986. The effects of temperature, oxygen fugacity and composition on the behaviour of chromium in basic and ultrabasic melts. *Geochim. Cosmochim. Acta*, v. 50, pp.1871 1887.

Murphy, J.B., 2001. Flat slab subduction in the geological record: consideration of modern analogues (Abst). *Geol. Soc. Amer.*, v. 33(6), pp. A–208.

Murthy, Ch.S.N., Rao,Y. V., Aruna, M. and Khare, S., 2006. Prospects of uranium mining in India. *Proceedings 1st Asian Mining Congress (vol. I), MGMI*, Kolkata, India, pp. 175–182.

Murti, K.S., 1996. Geology, sedimentation and economic mineral potential of south-central part of Chhattisgarh Basin. *Memoir GSI, 125*, 139p.

Murti, K.S., 1987. Stratigraphy and sedimentation in Chhattisgarh Basin, In: Purana Basins of Central India. *Geological Society of India, Mem. 6.*, pp. 239–260.

Murthy, M.V.N., Vishwanathan, T.V. and Roy Chowdhury, S, 1971. The Eastern Ghats Group. *Records, Geol. Surv. India*, v. 101, pp.15–42.

Murthy, P.S.N., 1990. Origin of cyclothemic pattern in the Precambrian banded iron formation of Donimalai area in Sandur schist belt, Karnataka State, India. In *Ancient Banded Iron Formations*. Theophrastus Publications. A., Greece, pp. 327–350.

Murthy, P.S.N. and Chatterjee, A.K.,1995. The origin of the iron ore deposits of Donimalai area of Sandur schist belt, Karnataka State, India. *Jour. Geol. Soc. India,* v. 45, pp.19–31.

Murthy, P.S.N. and Reddi, K.K., 1984. 2900 My old stromatolites from Sandur greenstone belt of Karnataka craton, India. *Jour. Geol. Soc. India,* v. 25, pp. 263–266.

Murthy, P.S.N., Singh, R.Y. and Tiwari, B.S., 1983. Occurrence of Precambrian microbiota in Sandur schist belt (Abst.). *Proc. Symp. on Palaeontology and Himalayan Geology*, Chandigarh , pp. 10–11.

Nagaraja Rao, B.K., Rajurkar, S.T., Ramalingaswami, G. and Ravindra Babu, B., 1987. Stratigraphy, structure and evolution of the Cuddapah basin. *Geological Society of India Mem.* 6, pp. 33–86.

Nagaswar Rao, P., Kumar, P., Srivastava, S.K. and Sinha, R.M., 2001. Uranium mineralisation in Kurnool sub-basin, Cuddapah basin, Andhra Pradesh. *Jour. Geol. Soc. India,* v. 57., pp. 462–463.

Nagori, D.K., 1988. Palaeogeography and depositional environment of the phosphate–bearing early Proterozoic carbonate rocks around Udaipur, Rajasthan. In A.B. Roy (Ed.), Precambrians of the Aravalli Mountain, Rajasthan, India. *Geological Society of India, Mem.* 7, pp. 139–152.

Naha, K., 1965. Metamporphism in relation to stratigraphy, structure and movements in part of East Singhbhum, Eastern India. *Quart. Journal Geol. Min. Met. Soc. India*, v. 37(2), pp. 41–85.

Naha, K., 1983. Structural, stratigraphic relations of the pre–Delhi rocks of south central Rajasthan: A summary. In S. Sinha Roy (Ed), Structure and tectonics of the Precambrian rocks of India. *Recent Researches in Geology*. Hindusthan Publ., New Delhi, pp. 40–52.

Naha, K. and Ghosh, S.K., 1960. Archean palaeogeography in eastern and northern Singhbhum. *Geol. Mag.*, v. 97, pp. 463–469.

Naha, K., and Halyburton, R.V., 1974. Early Precambrian stratigraphy of central and southern Rajasthan, India. *Precamb. Research,* v. 4, pp. 55–73.

Naha, K. and Mohanty, S., 1990. Structural studies of Pre-Vindhyan rocks of Rajasthan: A summary of work on the last three decades. In K. Naha, S. Ghosh and D. Mukhopadhyay (Eds). *Proc. Ind. Acad. Sciences*, v. 99, pp. 279–290.

Naha, K. and Roy, A.B., 1983. The problem of Precambrian basement in Rajasthan, Western India. *Precamb. Res.*, v. 19, pp. 217–223.

Naha, K., Mukhopadhyay, D. Dastidar, S., and Mukhopadhyay, R.P., 1995. Basement–cover relations between a granite gneiss body and its metasedimentary envelope: A structural study from the Early Precambrian Dharwar tectonic province, Southern India. *Precamb. Research,* v. 72, pp. 283–299.

Naha, K., Mukhopadhyay, D.K. and Mohanty, R., 1988. Structural evolution of the rocks of the Delhi Group around Khetri, north eastern Rajasthan. In A.B. Roy (Ed), Precambrians of the Aravalli Mountain. *Geological Society of India., Mem.* 7, pp. 207–245.

Naik, M.S., 1989. Genesis of copper deposit at Malanjkhand, Madhya Pradesh, India. *Journal Earth Sci.*, v. 16, pp. 27–37.

Nair, K.K.K., Jain, S.C. and Yedekar, D.B., 1995: Stratigraphy, structure and geochemistry of the Mahakoshal greenstone belt, In: S. Sinha Roy and K.R., Gupta (Eds.), Continental crust of northwestern and central India. *Geological Society of India*, *Mem.* 31, pp. 403–433.

Nair, M.M. and Nair, R.V.G., 2001. Strategies for gold exploration in virgin high–grade terrains of Attapadi Valley, Kerala. *Geol. Surv. India, Spl. Pub.* 58, pp. 181–190.

Nair, M.M. and Vidyadharan, K.T., 1982. Rapakivi granite of Egimala Complex and its significance. *Jour. Geol. Soc. India,* v. 23, pp. 46–51.

Nair, N.G.K., Soman, K., Santosh, K., Santosh, M., Arkelyants, M.H. and Golubyev, V.N., 1985. K–Ar ages of three granite plutons from north Kerala. *Jour. Geol. Soc. India,* v. 26, pp. 674–676.

Nair, R.S. and Suresh Chandran, M., 1996. Gold mineralisation in Kappil-Mankada Prospect, Nilambur-Manjeri area, Malappuram district, Kerala. *Pre-workshop volume, Gold Resources of India.* Hyderabad, pp. 117–121.

Nair, R.V.G., 1993. Primary gold mineralisation at Attapadi Valley, Palakhad district, Kerala. *Jour. Geol. Soc. India,* v. 41, pp. 387.

Nair, R.V.G., Nair, M.M. and Maji, A.K.,2005. Gold mineralisation in Kottathara prospect, Attapadi valley, Kerala, India. *Gondwana Res.*, v. 8, No. 2, pp. 203–212.

Nakagawa, M., Santosh, M., Nambiar, G. and Matsubara, C., 2005. Morphology and chemistry of placer gold from Attapadi valley, South India. *Gondwana Res.*, v. 8, No. 2, pp. 213–222.

Nakata, T., 1975. On Quaternary tectonics around the Himalaya. *Tohuku University Science Reports, 7th Ser (Geography)*, v. 25, pp. 111–118.

Nakata, T., 1989. Active Faults of the Himalaya of India and Nepal. *Geol. Soc. Amer. Spec. Paper* 232, pp. 243–264.

Naldrett, A.J. and Duke, J.M., 1980. Platinum metals in magmatic sulphide ores. *Science*, v. 208, pp. 1417–1428.

Naldrett, A.J., 2004. *Magmatic sulphide deposits: Geology, geochemistry and exploration.* Springer, Berlin, p. 727.

Naldrett, A.J., Federenko, V.A., Asif, M., Lin Shushen, Kunilov, V.E., Stekhim, A.I., Lightfoot, P.C and Gorbachev, N.S., 1996. Controls on the composition on Ni–Cu sulphide deposits as illustrated by those at Norilsk, Siberia. *Econ. Geol.*, v. 91, pp. 751–773.

Nanda, J.K and Pati, U.C., 1989. Field relations and petrochemistry of the granulites and associated rocks in the Ganjam–Koraput sector of the Eastern Ghats belt. *Indian Minerals*, v. 43, pp. 247–264.

Nanda, J.K and Pati, U.C., 1998. Mafic granulites and charnockites from Berhampur – Jeypore transect in Eastern Ghats of Orissa sector: a petrochemical study. *Geol. Surv. India, Spl. Pub. No. 44*, pp. 276–285.

Nanda, J.K., Patra, R.N. and Mishra, R.N., 1996. Petrogenetic history of platinifeous magmatic breccia zone in Baula igneous complex: a conceptual model for PGM localization (Abst.). *Workshop on geology and exploration of platinum group, rare metal and rare Earth elements, GSI–JU–AMD, Calcutta, pp. 17–19.*

Naqvi, S.M., 1978. Geochemistry of Archean metasediments: evidence for prominent anorthosite-norite-troctolite (ANT) in the Archean basaftic primordial crust. It B.F. windley & S.M. Naqvi (Eds), *Archean Geochemistry*, Elsever, Amsterdan, pp. 343–360.

Naqvi, S.M., 1981. The oldest supracrustals of the Dharwar craton, India. *Jour. Geol. Soc. India,* v. 22, pp. 458–469.

Naqvi, S.M. and Hussain, S.M., 1979. Geochemistry of meta-anothorsites from a greenstone belt in karnataka, India. *Canad. Jour. Earth Sci.*, v.16. pp.1254–1264.

Naqvi, S.M., Divakar Rao, V., Hussain, S.M., Narayana, B.L., Nirmal Charan, S., Govil, P.K., Bhaskar Rao, Y.J., Jafri, S.H., Rama Rao, P., Balaram, V., Ahmed Masood, Pantulu, K.P., Gnaeswar Rao, T., and Subba Rao, D.V., 1983. Geochemistry of gneisses for Hassan district and adjoining areas, Karnataka, India. *Geological Society of India, Mem.* 4, pp. 401–413.

Naqvi, S.M., Divakara Rao, V. and Hari Narain, 1974. The protocontinental growth of Indan shield and the antiquities of its rift valleys. *Precamb. Research*, v. 1, pp. 345–398.

Naqvi, S.M. and Rogers, J.J.W., 1987. Precambrian Geology of India. *Oxford Univ. Press,* New York, p. 223.

Naqvi, S.M., Venkatachala, B.S., Manoj Sukla, Kumar, B., Natarajan, R. and Mukund Sharma, 1987. Silicified cyanobacteria from the cherts of Archean Sandur schist belt, Karnataka, India. *Jour. Geol. Soc. India,* v. 29, pp. 535–539.

Narayan Das, G.R., Sharma, D.K., Singh, G. and Singh, R., 1980. Uranium mineralisation in Sikar district, Rajasthan. *Jour. Geol. Soc. India*, v. 21, pp. 432–439.

Narayan Das, G.R., Viswanath, R.V. and Pandit, S.A., 1988. Uranium-bearing quartz–pebble conglomerates. *Geological Society of India, Mem. 9*, pp. 29–31.

Narayana K., J.R., Anantha Iyer, G.V. and Ramakrishnan, M., 1974. Co-existing aegirine and magnesio–riebeckite from the Bababudan Hills, Mysore State. *Current Sci.*, v. 43, pp. 1–3.

Narayanaswami, S., 1966 a. Tectonic problems of Precambrian rocks of peninsular India. *Symp. Tectonics. Indian Geoph. Union, Spl. Pub.*, pp. 77–94.

Narayanaswami, S., 1966 b. Tectonics of the Cuddapah basin. Jour. Geol. Soc. India, v. 7, pp. 33–50.

Narayanaswami, S., 1976. Charnockite–Khondalite and Sargur–Nellore–Khammam –Bengpal–Deogarh–Pallahara–Mahagiri rock groups, older than Dharwar type greenstone belts in the Peninsular Archaeans. *Indian Minerals*, v. 16, pp. 16–25.

Narayanaswami, S., Chakravarty, S.C., Vemban, N.A., Shukla, K.D., Subramanyam, M.R., Venkatesh, V., Rao, G.V., Anandalwar, M.A., and Nagarajaiah, R.A., 1963. The geology and manganese ore deposits of the manganese belt in Madhya Pradesh and adjoining parts of Maharashtra. *Bull. Geol. Surv. India, Ser. A., Econ. Geol.*, 22(1), p. 69.

Narayanaswami, S.,Ziauddin, M. and Ramachandra, A.V., 1960. Structural control and localisation of the gold–bearing lodes, Kolar Gold Field, India. *Econ.Geol.*, v. 55, pp. 1429–1459.

Narayanaswami and Krishnakumar, N., 1996. An overview of studies on the lateritic gold deposits (Geological, mineralogical and geochemical) of Nilambur valley, Kerala and its economic significance (Abst). *Workshop on gold resources of India, Geological Society of India, NGRI,* Hyderabad, pp. 122–127.

Naslund, H.R., Henriquez, F., Nyström, J.O., Vivallo, W. and Dobbs, F.M., 2002. Magmatic iron ores and associated mineralisation: Examples from the Chilean High Andes and coastal Cordillera. In T.M. Poter (Ed), *Hydrothermal iron oxide copper–gold and related deposits*: *A global perspective*. PGC Publ. Australia, v. 2, pp. 207–226.

Natarajan, A., Datta, D.R. and Patel, M.C., 1984. Stratigraphy and structures of Ratanpur-Pali-Chaitam area of Bilaspur district, M.P. *Unpublished report, Geol. Surv. India.*

Nayak, P.N., Choudhury, K. and Sarkar, B., 1998. A review of geophysical studies of the Eastern Ghats mobile belt. *Geol.Surv.India. Spl. Pub. No.* 44, pp. 87–94.

Neelakantam, S. 2000, Geological investigations for kimberlites / lamproites in South Indian Diamond Province: Status and scope. In *Workshop on status, complexities and challenges of diamond exploration in India,* Raipur, pp. 41–68.

Neelakantam, S. and Roy, S., 1979. Baryte deposits of Guddapah basin. *Records, Geol. Surv. India 112(5)*, pp. 51–64.

Nelson, D.R., Bhattacharya, H.N., Misra, S., Dasgupta, N. and Altermann, W., 2007. New Shrimp U–Pb dates from the Singhbhum Craton, Jharkhand–Orissa region, India.(Abst.) *Int. Conf. on Precambrian Sediments and Tectonics. 2nd GPSS meeting, IIT Bombay*, p. 47.

Neng Jiang, Xuelei Chu, Muzuta, T., Ishiyama, D. and Jiguang, Mi, 2004. A magnetite–apatite deposit in the Fanshan alkaline ultramafic complex, Northern China. *Econ. Geol.*, v. 99, pp. 397–408.

Neogi, D., 1965. Stratigraphy and structure of the area around Kishengarh, Rajasthan. *Trans. Geol. Min. Met. Inst. India (Wadia Comm. Vol.)* pp. 458–479.

Neogi, S., Dasgupta, S.N. and Fukuoka, M., 1998. High P–T polymetamorphism, dehydration melting, and generation of migmatites and granites in the Higher Himalayan Crystalline Complex, Sikkim, India. *Journal Pet.*, v. 39, pp. 61–99.

Neogi, S., Miura, H. and Hariya, Y., 1996. Geochemistry of the Dongargarh volcanic rocks, Central India: implications for the Precambrian mantle. *Precamb. Research*, v. 76, pp. 77–91.

Newton, R.C., 1990. The Late Archean high-grade terrain of South India and the deep structure of the Dharwar craton. In M.H. Salisbury and D.M. Fountain (Eds.), *Exposed cross sections of the continental crust.* Kluwer Academic Publishers, pp. 305–326.

Nicholson, D.M. and Mathez, E.A., 1991. Petrogenesis of Merensky Reef in the Rustenburg section of Bushveld Complex. *Contrib. Min. Pet.*, v. 107, pp. 293–309.

Nicholson, K., Nayak, V.K. and Nanda, J.K., 1997. Manganese ores of the Goriajhor–Monmunda area, Sundergarh district, Orissa, India: geochemical evidence for a mixed Mn-source. In K. Nicholson et al. (Eds.), Manganese mineralisation : geochemistry and mineralogy of terrestrial and marine deposits. *Geol. Soc. London, Spl. Pub. No. 119*, pp. 117–121.

Nijagunappa, R. and Naganna, C., 1983. Nuggihalli schist belt in Karnataka craton: An Archean layered complex as interpreted from chromite distribution. *Econ. Geol.*, v. 78, pp. 507–513.

Nutman, A.P., Chadwick, B., Krishna Rao, B. and Vasudev, V.N., 1996. SHRIMP U/Pb Zircon ages of acid volcanic rocks in the Chitradurga and Sandur Groups and granities adjacent to the Sandur schist belt, Karnataka. *Jour. Geol. Soc. India,* v. 47, pp. 153–164.

Nutman, A.P., Chadwick, B., Ramakrishna, K. and Viswanatha, M.N., 1992. SHRIMP U–Pb ages of detrital zircon in Sargur supracrustal rocks in Western Karnataka, South India. *Jour. Geol. Soc. India,* v. 39, pp. 367–374.

Nutman, A.P., Friend, C.R.L. and Bennett, V.C., 2001. Review of oldest (4400–3600 Ma) geological and mineralogical record: Glimpses of the beginning. *Episode*, v. 24, pp. 93–101.

Nutman, A.P., McGregor, V.R., Friend, C.R.L., Bennett, V.C. and Kinny, P.D., 1996. The Itsaq Gneiss Complex of southern West Greenland, the world's most extensive record of early crustal evolution (3900–3600 Ma). *Precamb. Research*, v. 78, pp. 1–39.

Nystrom, J.O. and Henriquez, F., 1994. Magmatic features of iron ores of Kiruna type in Chile and Sweden: Ore texture and magnetite geochemistry. *Econ. Geol.*, v. 89, pp. 820–839.

Oesterlen, M. and Vetter, U., 1986. Petrographic–geochemical characteristics and genesis of an albitized uraniferous granite in Northern Cameroon, Africa. In H. Fuchs (Ed.), *Vein type uranium deposits. Internat. Atomic Energy Agency*, Vienna, pp. 113–142.

O'Connor, J.T., 1965. A classification of quartz-rich igneous rocks based on feldspar ratios. *USGS Prof. Paper, No. 525B*, pp. 76–84.

Ohmoto, H. and Rye, R.O., 1979. Isotopes of sulfur and carbon. In H.L. Barnes (Ed.), *Geochemistry of hydrothermal ore deposits* (2nd Edn.), John Wiley and Sons, New York, p. 509–567.

Ohmoto, H., 2003. Non-redox transformation of magnetite–hematite in hydrothermal system. *Econ. Geol.*, v. 98, pp. 157–162.

Okita, P.M., 1992. Manganese carbonate mineralisation in the Molango district, Mexico. *Econ. Geol.*, v. 87, pp. 1345–1366.

Okudaira, T., Hamamoto, T., Hari Prasad, B. and Rajneesh Kumar, 2001. Sm–Nd and Rb–Sr dating of amphibolite from Nellore–Khamam schist belt, S.E. India: constraints on the collision of the Eastern Ghats terrane and Dharwar–Bastar craton. *Geol. Mag.*, v. 138, pp. 495–498.

Okuyama–Kusunose, 1994. Phase relations in andalusite–sillimanite type Fe–rich metapelites: tectonocontact metamorphic aureole, north–east Japan. *Journal Met. Geol.*, v. 12, pp. 153–168.

Oldham, R.D., 1893. Manual of Geology of India, 2nd edition, Calcutta, Supt. of Govt. Printing Press, p. 543.

Oldham, R.D., Vredenburg, E. and Datta, P.N., 1901. Geology of the Son Valley in Rewa State and of parts of the adjoining districts of Jabalpur and Mirzapur. *Geol.Surv.India, Mem.* 31(l), pp. l–178.

Oldham, T.H., 1856. Remarks on the classification of the rocks of Central India resulting from the investigation of the Geological Survey. *Journal Asiatic Soc. Bengal*, v. 25,

Oliver, N.H.S., Cleverley, J.S., Mark, G., Pollard, P.J., Bin Fu, Marshall, L.J., Rubenach, M.J., Williams, P.J. and Baker, T., 2004. Modelling the role of sodic alteration in the genesis of iron oxide–copper–gold deposits, Easter Mount Isa block, Australia. *Econ. Geol.*, v. 99, pp. 1145–1176.

Olsen, K.H. and Morgan, P., 1995. Progress in the understanding of continental rifts. In K.H. Olsen (Ed.) *Continental rifts: Evolution, structure and tectonics*. Elsevier, pp. 3–26.

Orville, P.M., 1963. Alkali ion exchange between vapour and feldspar phases. *Amer. Journal Sciences*, v. 261, pp. 201–237.

Ossandon, C.G., Freraut, R.C., Gustafson, L.B., Lindsay D.D. and Zentilli, M., 2001. Geology of Chuquikamata Mine: A progress report. *Econ. Geol.*, v. 96, pp. 249–270.

Owens, B.E. and Pasek, M.A., 2007. Kyanite quartzites in the Piedmont Province of Virginia : Evidence for a possible high–sulfidation system. *Econ.Geol.*, v. 102, pp. 495–509.

Oxburgh, E.R., 1990. Some thermal aspects of granulite history. In D. Vielzeuf and Vidal, Ph.(Eds.), *Granulites and crustal evolution. Kluwer Academic Publ.*, Netherlands, pp. 569–580. pp. 224–256.

Page, M. and Svab, M., 1985, Petrographic and geochemical variations within the carswell structure metamorphic core and their implications with respect ot uranium mineralisation. In Laine, R., Alonso, D. and Svap, M (Eds.), The carswell structure uranium deposits, Saskatchewan. *GSC Spec. Paper 29*, pp. 55–70.

Page, N.J., Banerjee, P.K. and Haffty, J., 1985. Characteristics of the Sukinda and Nuasahi ultramafic complexes, Orissa, India, by platinum group element geochemistry. *Precamb. Research*, v. 30, pp. 27–41.

Pal, D.C., Barton, M.D. and Sarangi, A.K., 2009. Deciphering a multistage history affecting U–Cu (–Fe) mineralisation in the Singhbhum shear zone, eastern India, using pyrite textures and compositions in the Turamdih U–Cu (–Fe) deposit. *Mineral. Deposita*, v. 44(1), pp. 61–80.

Pal, N. and Mishra, B., 2002. Alteration geochemistry and fluid inclusion characteristics of greenstone hosted gold deposit at Hutti, Eastern Dharwar craton, India. *Mineral. Deposita*, v. 37, pp. 722–736.

Pal, N. and Misra, B., 2003. Epigenetic nature of the banded iron formation-hosted gold mineralization at Ajjanahalli, Southern India: Evidence from ore petrography and fluid inclusion studies. *Gondwana Research*, v. 6, pp. 531–540.

Pal, T. and Bhowmik, S.K. 1998. Metamorphic history of Sausar Group of rocks. *Geol. Surv. India.,Unpub. Report*, p. 98.

Pal, T., Hi–Soo Moon and Mitra, S.N., 1994. Distribution of iron cations in natural chromites at different stages of oxidation: A 57 Fe Mossbauer investigation. *Jour. Geol. Soc. India,* v. 44, pp. 53–64.

Paliwal, H.V., Gurjar, L.K. and Craddock, P.T., 1986. Zinc and brass in ancient India. *CIM Bull.*, pp. 75–78.

Palmer, M.R. and Slack, J.F., 1989. Boron isotope composition of tourmaline from massive sulfide deposits and tourmalinites. *Contrib. Min. Pet.*, v. 103, pp. 434–451.

Panda, N.K., Rajagopalan, V. and Ravi, G.S., 2003. Rare earth element geochemistry of placer monazites from Kalingapatnam coast, Srikakulam district, Andhra Pradesh. *Jour. Geol. Soc. India,* v. 62, pp. 429–438.

Pandalai, H.S., Jadhav, G.N., Mathew, B. Panchapakesan, V., Raju, K.K., and Patil, M.L., 2003. Dissolution channels in quartz and role of pressure changes in gold and sulphide deposition in Archean greenstone hosted Hutti gold deposit, Karnataka, India. *Mineral. Deposita*, v. 38, pp. 597–624.

Pandalai, H.S., Majumdar, T. and Chanda, D., 1983. Chemistry of pyrite and black shales of Amjhore, Rhotas district, Bihar, India. *Econ. Geol.*, v. 78, pp.1505–1513.

Pande, Kanchan, 2002. Age and duration of the Deccan Traps, India: A review of radiometric and paleomagnetic constraints. *Proc. Indian Acad. Sci. (Earth Planet. Sci.)*, v. 111, No. 2, pp. 115–123.

Pande, B.K., Chabria, T. and Gupta, J.N., 1995. Geochronological characterization of the Proterozoic terrains of Peninsular India : relevance to the first order target selection for uranium exploration. *Expln. and Res. for Atomic Minerals*, AMD, India, v. 8, pp. 187–213.

Pande, B.K. Upadhyay, D.L. and Sinha, K.K., 1986. Geochronology of Jajawal-Binda-Nagnaha granitoids in relation to uranium mineralization. *Ind. Jour. Earth Sci.*, v. 13, pp. 163–168.

Pandey, B.K., Chabria, T., Veena Krishna and Krishnamurthy, P., 1994. Rb–Sr geochronology of late Proterozoic A' type granites in parts of Madurai district, Tamilnadu : Implications on uranium, rare earth and rare metal distribution. *Journal Atom. Mineral Sci.*, v. 2, pp. 79–87.

Pandey, B.K., Veena Krishna and Chabria, T., 1998. An overview of the geochronological data on the rocks of Chhotanagpur Gneiss–Granulite Complex and adjoining sedimentary sequences, eastern and central India. *Proc. Internat. Sem. on Precambrian crust in Eastern and Central India (Abst.), Geol. Surv. India, Calcutta,* pp. 131–135.

Pandey, B.K., Veena Krishna, Sastry, D.V.L.N., Chabria, T., Veerabhaskar, D., Marry, K.K. and Dhanaraju, R., 1993. Pan–African whole rock Rb–Sr isochron ages for the granites and pegmatites of Kullampathi–Suriyamalai area, Salem district, Tamilandu, India. Symp (6th vol.), *Mass spectrometry* (ISMAS), pp. 480–482.

Pandit, M.K., Khatatneh, M.K. and Saxena, R., 1996. Trondhjemite of the Alwar basin, Rajasthan : Implications of Late Proterozoic rifting in the North Delhi Fold Belh. *Current Sci.*, v. 71(8), pp. 636–641.

Panigrahi, M.K. and Mookherjee, A., 1997. The Malanjkhand copper (+ molybdenum) deposit, India: mineralisation from a low–temperature ore–fluid of granitoid affiliation. *Mineral. Deposita*, v. 32, pp. 133–148.

Panigrahi, M.K., Bream, B.R., Misra, K.C. and Naik, R.K., 2004. Age of granitic activity associated with copper–molybdenum mineralisation at Malanjkhand, Central India. *Mineral. Deposita*, v. 39, pp. 670–677.

Panigrahi, M.K., Misra, B. and Mookherjee, A., 1991. On mineralogy and fluid inclusion characteristic of different ore associations from Malanjkhand copper deposit, M.P., India. *Jour. Geol. Soc. India,* v. 37, pp. 239–256.

Panigrahi, M.K., Misra, K.C., Bream, B., Naik, R.K., 2002. Genesis of the granitoid affiliated copper–molybdenum mineralisation at Malanjkhand, central India: facts and problems. Extended abstract. in *Proceedings of the 11th Quadrennial IAGOD Symposium and Geocongress, Windhoek, Namibia* pp.563–584.

Panigrahi, M.K., Mookherjee, A., Pantulu, G.V.C. and Gopalan, K, 1993. Granitoids around the Malanjkhand deposit: types and age relationship. Proc. *Indian Acad. Sci.(Earth and Planet. Sci.)*, v. 102(2), pp. 399–413.

Pant, N.C. and Banerjee, D.M., 1990. Pattern of sedimentation in the type Bijawar basin of central India, *Geol. Surv. India Spec.* Publ., v. 28,. pp. 156–166.

Papanastassiou, D.A. and Wasserberg, G., 1971. Lunar chronology and evolution from Rb–Sr studies of Appolo 11 and 12 samples. *Earth and Planet. Sci. Letts* v.11, pp. 37–62.

Parak, T., 1975. The origin of the Kiruna iron ores, Sweden. *Geol. Undersok., Ser.* C. NR. 709, p. 209.

Pareek, H.S., 1984. Pre-Quarternary geology and mineral resources of northwestern Rajasthan. *Geol. Surv. India, Mem.*115, p. 99.

Park, A.F. and Dash, B., 1984. Charnockite and related neosome development in the Eastern Ghats, Orissa, India: petrographic evidence. *Trans. Roy. Soc. Edinburgh: Earth Sciences*, v. 75, pp. 341–352.

Park, C.F. Jr., 1972. The iron ore deposits of the Pacific Basin. *Econ. Geol.*, v. 67, pp. 339–349.

Parman, S.W.,Grove, T.L., Dann, J.C. and de Wit, M.J., 2004. A subduction origin for komatiites and cratonic lithspheric mantle. South. *African Journal Geol.*,v. 107, pp. 107–118.

Parrish, R.R. and Hodges, K.V., 1996. Isotopic constraints on the age and provenance of the Lesser and Greater Himalayan sequences, Nepal Himalaya. *Geol. Soc. Amer. Bull.*, v. 108, pp. 904–911.

Parveen, M.N. and Ghosh, B., 2007. Multiple origins of gahnite associated with hydrothermal alteration from the Bhuyari base metal prospect of Proterozoic Betul Belt, Madhya Pradesh. *Jour. Geol. Soc. India,* v. 69(2), pp. 233–241.

Parveen, M.N., Ghosh, B, Shrivastava, H.S., Dora, M.L. and Gaikwad, L.D., 2007. Sulphide mineralisation in Betul Belt: Classification and general characteristics. *Jour. Geol. Soc. India,* v. 69(1), pp. 85–91.

Pasayat, S., 1981. Precambrian ignimbrites of central India. *Geol. Surv. India, Spl. Pub.* v. 3, pp. 115–118.

Pascoe, E.H., 1928. General Report for 1927. *Records, Geol. Surv*. India, 60 p.

Pascoe, E.H., 1930. General report of 1929. *Records, Geol. Surv*. India, v. 63.

Pascoe, E.H. 1950. *A manual of the geology of India and Burma*, v. 1(*Third Edn*), reprinted in1973, *Geol. Surv.* India Pub. p. 485.

Pascoe, E.H., 1959. *A manual of the geology of India and Burma*, v. 2 (Third Edn), pp. 486–1343. Govt. of India Press, Calcutta.

Patel, S.K., Jena, B.K.,and Das, K.C., 2005. Manganiferous formations of Bonai–Kendujhar belt and a proposal for revision of its lithostratigraphy. *In Proc. Sem. on Mineral and Energy resources of Eastern and Northeastern India, MGMI – Calcutta Branch,* Kolkata, pp. 163–179.

Patil, M.L., 2008. Gold mineralisation in Hutti–Maski greenstone belt, Raichur District, Karnataka, India. In *Pre-workshop (Extd. Abst.) vol., Internat. Field Workshop on Gold Metallogeny in India, Dec. 2008,* NGRI – Hyderabad and Univ. of Delhi, pp. 73–77.

Patil, M.L. and Roy, A.B., 2000. Pattern and control of copper ore mineralsiation at Kho-Dariba, Alwar district, Rajasthan. In M. Deb (Ed.), *Crustal evolultion and metallogeny, in the northwestern Indian shield.*

Patino Douce, A.E. and Harris, N.B.W., 1998. Experimental constraints on Himalayan anatexis. *Journal of Perology*, v. 39, pp. 689–710.

Patra, R.N. and Mukherjee, M.M., 1996. Exploration for PGE in Baula–Nuasahi–Bangur ultramafic complex, Keonjhar district, Orissa. *(Abst.) Workshop on Geology and exploration of platinum group, rare metal and rare earth elements, GSI–JU–AMD,* Calcutta, pp. 7–9.

Patranabis Deb, S., 2003. Proterozoic felsic volcanism in the Pranhita–Godavari valley, India: its implication on the origin of the basin. *Journal Asian Earth Sci.*, v. 21, pp. 623–631.

Patranabis Deb, S., Bickford, M.E., Hill, B., Chaudhuri, A.K. and Basu, A., 2007. SHRIMP ages of zircon in the uppermost tuff in Chhattisgarh basin in central India require ~ 500 Ma adjustment in Indian Proterozoic stratigraphy. *Jour. Geology*, v. 115, pp. 407–415.

Pattanaik, S.C., Ghosh, S.P. and Das Gita, 1998. Base metals. In *Geology and Mineral Resources of Orissa*, S.G.A.T, Orissa, pp. 113–120.

Patwardhan, A.M. and Oka, S.S., 1988. Characteristics of the host rocks and basemetal mineralisation in the Sirohi district, southern Rajasthan. In A.B. Roy (Ed.) *Precambrians of the Aravalli Mountain, India. Mem.Geological Society of India., No. 7*, pp. 423–439.

Patwardhan, A.M. and Panchal, P.K., 1984. Some textural features of Mussoori phosphorite. In Proc. 4th Internat. Sem. on Phosphorite, Udaipur, India. *Geol. Surv. India Spl. Pub. No*. 17, pp. 147–158.

Paul, D.K., Crocket, J.H., Reddy, T.A.K. and Pant, N.C., 2007. Petrology and geochemistry including Platinum group element abundances of the Mesoproterozoic ultramafic (Lamproite) rocks of Krishna district, Southern India: Implications for source rock characteristics and petrogenesis. *Jour. Geol. Soc. India,* v. 69, pp. 577–596.

Paul, D.K., Mukhopadhyay, D., Pyne, T.K. and Bisoi, P.K.1991. Rb–Sr age of granitoid in Deo river section, Singhbhum and its relevance to the age of Iron Formation. *Indian Minerals*, v. 45, pp. 51–56.

Pearce, J.A. and Cann, J.R., 1973. Tectonic setting of basic volcanic rocks determined using trace element analysis. *Earth Planet. Sci. Lett.*, v. 19, pp. 290–300.

Pearce, J.A., Harris, N.B.W. and Tuidla, A.G. 1984. Trace element discrimination diagrams for tectonic interpretation of granitic rocks. *Journal Petrol.*, v. 25, pp. 956–983.

Peck, W.H,, Valley, J.W., Wilde, S.A. and Graham, C.M., 2001. Oxygen isotope ratios and rare earth elements in 3.3 to 4.4 Ga zircons: Ion micro-probe evidence for high $\delta^{18}O$ continental crust and oceans in the Early Archean. *Geochim. Cosmochim*. Acta, v. 65, pp. 4215–4229.

Perfenov, V.D., Chekhovskikh, M.M. and Yudin, N.I., 1979. Types and prospects of metamorphic apatite mineralisation in central Aldan. *Internat. Geol. Rev.*, v. 21, pp. 1285–1296.

Perumal, N.V.A.S., 1974. An occurrence of gadolimite near Karatupath, Madurai district, Tamilnadu. *Curr. Science*, v. 43, pp. 3–5.

Petruk, W. and Sikka, D.B., 1987. The formation of oxidized copper minerals at the Malanjkhand porphyry copper deposit in India and implications on metallurgy. In A.H. Vassilion (Ed.), *Process Mineralogy, VII, Met. Soc., AIME*, pp. 403–420.

Peucat, J.J., Bouhallier, H., Fanning, C.M. and Jayananda, M., 1995. Age of Holenarsipur greenstone belt and relationships with the surrounding gneisses (Karnataka, South India). *Journal Geol.*, v. 103, pp. 701–710.

Peucat, J.J., Mahabaleswar, M. and Jayananda, M., 1993. Age of younger tonalitic magmatism and granulite metamorphism in the amphibolite–granulite transition zone of South India (Krishnagiri area): comparison with older Peninsular Gneisses of Gorur–Hassan area. *Journal Met. Geol.*, v. 11, pp. 879–888.

Phadke, A.V. 1990: Genesis of the granitic rocks and the status of the "Tirodi Biotite Gneiss" in relation to the metamorphites of the Sausar Group and the regional tectonic setting, *Geol.Surv.India*, *Spl. Pub.*, v. 28, pp. 287–302.

Phadke, A.V. and Jhingran, A.G., 1968. On the carbonatites at Newania, Udaipur district. *Jour. Geol. Soc. India,* v. 9, pp. 165–169.

Phillips, G.N. and Law, J.D.M., 2000. Witwatersrand gold field: Geology, genesis and exploration. *Reviews in Econ. Geol.*, v. 13, pp. 439–500.

Phillips, G.N., Groves, D.I. and Martyn, J.E., 1987. Source requirements for the Golden Mile, Kalgoorlie–significance to the metamorphic replacement model for Archean gold deposit. *Cand. Journal Earth Sciences*, v. 24, pp. 1643–1651.

Philpotts, A.R. 1994. Principles of igneous and metamorphic petrology. *Prentice–Hall (India) Pvt. Ltd.*, New Delhi, p. 498.

Philpotts, A.R., 1967. Origin of certain iron–titanium oxides and apatite rocks. *Econ. Geol.*, v. 62, pp. 303–315.

Pichamuthu, C.S., 1956. The problem of the ultrabasic rocks. *Proc. Mysore Geol. Assosn*, v.15, pp.1–12.

Pillai, S.P., Peshwa, V.V., Nair, S., Sharma, M., Sukla, M. and Kale, V.S., 1999. Occurrence of a manganese-bearing horizon in the Kaladgi basin. *Jour. Geol. Soc. India*, v. 53, pp. 201–204.

Pirajno, F., 1992. Hydrothermal deposits. *Spinger,* Berlin, p. 709.

Piserevsky, S.A. , Wingate, M.T.D., Powell, C.Mc, A., Johnson, S. and Evans, D.A.D., 2003. Models of Rodinia assembly and fragmentation. In M.Yoshida, B.F. Windley and S. Dasgupta (Eds), Proterozoic East Gondwana: Supercontinent assembly and breaking up. *Geol. Soc. London, Spl. Pub.*, v. 206, pp. 35–55.

Pitchai Muthu, R., 1990. The occurrence of gabbroic anorthosites in Makrohar area, Sidhi district, Madhya Pradesh, *Geol.Surv.India. Spl. Pub.*, v. 28, pp. 320–331.

Plimer, I.R., 1987. The association of tourmalinite with stratiform scheelite deposits. *Mineral. Deposita*, v. 22, pp. 282–291.

Poddar, B.C., 1965. Lead–zinc mineralisation in the Zawar belt, *India. Econ. Geol.*, v. 60, pp. 636–638.

Poddar, B.C., 1974. Evolution of sedimentary sulphide rhythmites in the metamorphic tectonites in the base metal deposit of Rajpura, Rajasthan. *Golden Jubilee Vol., Geol. Min and Met. Soc. India*, pp. 207–222.

Pollard, P.J., 2000. Evidence of a magmatic fluid and metal source for Fe-oxide Cu–Au mineralisation. In T.M. Porter (Ed.), Hydrothermal iron oxide copper–gold and related deposits: a global perspective. *Aust. Mineral Foundation Inc.*, pp. 27–42.

Pollard, P.J., Nakapadungrat, S. and Taylor, R.G., 1995. The Phuket Supersuite, southwest Thailand: fractionated I–type granites associated with tin–tantalum mineralisation. *Econ.Geol.*, v. 90, pp. 586–602.

Poole, F.G. 1988. Stratiform barite in Palaeozoic rocks of the western United States. In Zachrisson, E (Ed)., *Proc. 7th IAGOD Quad. Symp.* E. Schweiz. Verlag., Stuttgart, pp. 309–319.

Powell, C. McA., and Conaghan, P.J., 1973. Plate tectonics and the Himalaya. *Earth Planet Sci. Letters*, v. 20, pp. 1–12.

Powell, C. McA, Li, Z. X., Pcellhinny, M.W., Meert, J.G. and Park, J.K., 1994. Paleomagnetic constraints on the timing of Neoproterzoic breakup of Rodinia and Cambrian formation of Gondwana. *Geology*, v. 22, pp. 889–892.

Powell, C. McA., Oliver, N.H.S., Li, Z.X., Martin, D. McB. And Ronaszecki, J., 1999. Synorogenic hydrothermal origin for giant Hamersley iron oxide orebodies. *Geology*, v. 27, pp. 175–178.

Powell, R. and Holland 1988. An internally consistent dataset with uncertainties and correlations.: application methods, worked examples and a computer programme. *Jour. Met. Geol.*, v. 6, pp. 173–204.

Prakash, H.S.M., Thiruvengadam, A. and Shyamala Rao, A., 1991. Investigation for gold in Penakacherla schist belt, Anantapur district, Andhra Pradesh. *Records, Geol. Surv. India*, v. 124(5), pp. 54–56.

Prakash, Narsimha, K.N., Janardan, A.S. and Mishra, V.P., 1996. A study on Kondagaon granulite belt. *Jour. South East Asian Earth Science* 14, v. 53, pp. 221–229.

Prakash, R, Swarup, P., and Srivastava, R.N. 1975. Geology and mineralisation in the southern parts of the Bundelkhand in Lalitpur district, U.P. *Jour. Geol. Soc. India*, v. 16, pp. 143–156.

Prasad, B., 1987. Geochemistry and petrogenesis of Berach Granite, Rajasthan, India. *Indian Minerals*, v. 41, pp. 1–23.

Prasad, C.V.R.K., Subba Reddy, N. and Windley, B.F., 1982. Iron formations in Archean granulite-gneiss belt with special reference to Southern India. *Jour. Geol. Soc. India*, v. 23, pp. 112–122.

Prasada Rao, G.H.S.V., Murthy, Y.G.K. and Deekshitulu, M.K., 1964. Stratigraphic relations of Precambrian Iron formations associated sedimentary sequences in parts of Keonjhar, Cuttack, Dhenkanal and Sundargarh districts, Orissa, India. *Internat. Geol. Congress, New Delhi*, Pt 10, pp. 72–87.

Proffett, J.M., 2003. Geology of the Bajo de la Alumbrera porphyry copper-gold deposit, Argentina. *Econ. Geol.*, v. 98, pp. 1535–1574.

Projnante, U.D.,Castelli, P., Benna, G., Genovese, F., Oberli, M., Meier., S. and Tonartini, S., 1990. The crystalline units of the High Himalayas in the Lahul–Zanskar region (northwest India) metamorphic–tectonic history and geochronology. *Geol. Mag.*, v. 127, pp. 101–116.

Pujari, G.N. and Shrivastava, J.P., 2003. Biogeochemical studies on some copper rich areas from Malanjkhand granitoid, Madhya Pradesh. *Jour. Geol. Soc. India*, v. 61, pp. 295–318.

Pyke, D.R., Naldrett, A.J. and Eckstrand, O.R., 1973. Archean ultramafic flows in Munro Township, Ontario. *Geol. Soc. Amer. Bull.*, v. 84, pp. 955–978.

Pyne, T.K. and Bandyopadhyay, A., 1985. Structures of Banded Gneissic Complex at and around Bandanwara, Ajmer district, Rajasthan. *Indian Journal Earth Sci.*, v. 12, pp. 9–20.

Qureshy, M.N. and Iqballuddin, 1992. A review of the geophysical constraints in modeling the Gondwana crust in India. *Tectonophysics*, v. 212, pp. 141–151.

Qureshy, M.N., Bhatia, S.C. and Subba Rao, D.V., 1972. Preliminary results of some gravity surveys in Singhbhum area and Orissa. *Jour. Geol. Soc. India*, v. 13, pp. 238–246.

Qureshy, M.N., Krishna Brahmam, M., Garde, S.C. and Mathur, B.K.,1968. Gravity anomalies and Godavari rift, India. *Geol. Soc. Amer. Bull.*, v. 79, pp. 1221–1229.

Radhakrishna, B.P. 2007. Diamond exploration in India: Retrospect and prospect. *Jour. Geol. Soc. India*, v. 69, pp. 415–418.

Radhakrishna, B.P. and Naqvi, S.M., 1986. Precambrian continental crust of India and its evolution. *Journal Geol.*, v. 94, pp. 145–166.

Raase, P., Raith, M., Ackermand, D., Viswanatha, M.N. and Lal, R.K., 1983. Mineralways of chromiferous quartizites from South India. *Jour. Geol. Soc. India*, v.24, pp.502–521.

Radhakrishna, B.P., 1983. Archean granite greenstone terrain of south India. In S.M. Naqvi and J.J.W. Rogers (Eds), Precambrian of South India. *Geological Society of India, Mem. 4*, pp. 1–46.

Radhakrishna, B.P., 1989. Suspect tectonostratigraphic terrane elements in the Indian subcontinent. *Jour. Geol. Soc. India*, v. 34, pp. 1–24.

Radhakrishna, B.P. and Sreenivasaiya, C., 1974. Bedded barytes from tghe Precambrian of Karnataka. *Jour. Geol. Soc. India*, V.15, pp. 314–317.

Radhakrishna, B.P. and Curtis, L.C., 1999. *Gold in India*. Geol. Soc. India, 307p.

Radhakrishna, B.P. and Vaidyanathan, R., 1994. Geology of Karnataka. *Geological Society of India*, Bangalore, p. 298.

Radhakrishna, K.R. Dhruba Rao, B. K. and Singhal, M. L., 1981. Explorations for copper lead and zinc ores of India. *Bull. Geol. Surv. India, Ser. A (Econ. Geol.)* v. 47, pp. 170–195.

Radhakrishna, T. and Joseph, M., 1996. Proterozoic paleomagnetism of the mafic dyke swarms in the high grade region of South India. *Precamb. Research,* v.76, pp. 31-46.

Raghunandan, K.R., 1994. Ajmer lead-sinc district, Rajasthan, India– a resume. *AMSE, Geol. Surv. India*, 10p.

Raghunandan, K.R., 1997. Goldfields of Chitradurga greenstone belt, Karnataka. In S.M. Naqvi (Ed.), *Golden Jubilee Volume, Hutti Gold Mines*, pp. 89–96.

Raghunandan, K.R., Dhruba Rao, B.K. and Singhal, M.L., 1981. Exploration for copper, lead and zinc ores in India. *Bull. Geol. Surv. India, Ser A–Econ Geol.* No., v. 47, pp. 170–195.

Raha, P.K., 1987. Stromatolites and correlation of Purana (Middle to Late Proterozoic) basins of Peninsular India. *Geological Society of India*, *Mem.6*, pp. 393–398.

Raha, P.K., Parulkar, S.N., Ghosh, S.C., Some, S., Kundu, U.S., Kumar, M., Saha, G. and Misra, I.K., 2000. Possible microsofossils from the Archean Banded Iron Formation (Bailadila Group), Madhya Pradesh, India. *Jour. Geol. Soc. India*, v. 55 pp. 663–673.

Rai Choudhury, A. and Guha, D.B., 2004. Evolution of the Great Boundary Fault : A re–evaluation. *Jour. Geol .Soc. India*, v. 64, pp. 21–31.

Rai, K.L. and Changkakoti, A. 1995. Ore genetic implication of the stable isotope geochemistry of sulfur and lead in Sargipali deposit, Orissa, India. In R. K. Srivastava and R. Chandra (Eds.), *Magmatism in relation to diverse tectonic settings. Oxford and IBH Co. Pvt. Ltd.*, New Delhi, pp. 123–134.

Rai, K.L. and Venkatesh, A.S., 1990. Malanjkhand copper deposit – a petrological and geochemical appraisal. *Geol. Surv. India , Spl. Pub. 28*, pp. 563–584.

Rai, K.L., Sarkar, S.N. and Paul, P.R., 1980. Primary depositional and diagenetic features in banded iron formation and associated iron ore deposits of Noamundi, Singhbhum district, Bihar, India. *Mineral. Deposita*, v. 15, pp. 189–200.

Rai, S.D. and Banerjee, D.C., 1995. Xenotime placers in parts of Mahan river basin, Surguja district, Madhya Pradesh. *Jour. Geol .Soc. India*, v. 45 , pp. 285–293.

Rai, S.S., Srinagesh, D. and Gour. V.K., 1993. Granulite evolution in South India—a seismic tomographic perspective. *Geological Society of India, Mem*. 25, pp. 235–263.

Raith, J.G. and Prochaska, W., 1995. Tungsten deposits in the wolfram schist, Namaqualand, South Africa: Strata–bound versus granite–related genetic concepts. *Econ.Geol.*, v. 90, pp. 1934–1954.

Raith, M.M., Srikantappa, C., Buhl, D. and Koehler, H., 1999. The Nilgiri enderbites, south India : the nature and age constraints on protolith formation, high grade metamorphism and cooling history. *Precamb. Research*, v. 98, pp. 129–150.

Rajamani, V., 1996. Gold mineralisation in the Koler schist belt. *Pre-workshop volume on Gold resources of India,* Hyderabad, pp. 255–257.

Rajamani, V., Shivakumar, K., Hanson, G.N. and Shirey, S.B., 1985. Geochemistry and petrogenesis of amphibolite, Kolar schist belt, South India: evidence from komatiitic magma derived by low percentage of melting of the mantle. *Journal Pet.*, v. 96, pp. 92–123.

Raja Rao, C.S., 1970. Sequence, structure and correlation of the metasediments and the gneissic complex of Rajasthan. *Records, Geol. Surv. India*, v. 98(2), pp. 122–131.

Raja Rao, C.S., 1976. Precambrian sequences of Rajasthan. *Geol. Surv. India*, Misc Pub. v. 23(2), pp. 497–516.

Raja Rao, C.S., Poddar, B.C. and Chatterjee, A.K. 1972. Dariba–Rajpura–Bethumni belt of zinc–lead–copper mineralisation, Udaipur district, Rajasthan. *Geol. Surv*. India Misc Pub.16(2), pp. 617–627.

Raja Rao, C.S., Poddar, B.C., Basu, K.K. and Dutta, A.K., 1971. Precambrian stratigraphy of Rajasthan : a review. *Rec. Geol. Surv. India*, v. 101(2), pp. 60–79.

Rajaiya, V., Kundu, A., Datta, D.R. and Dutta. N.K., 1990. Geology of Sambalpur–Nandghat–Baitalpur area, Durg, Bilaspur and Raipur districts, M.P., India. *Records, Geol. Surv. India*, v. 123(6), pp. 42–43.

Rajaiya. V. and Ashiya, I.D., 1993. Evolution of Sonakhan granite–greenstone bell of Raipur district, M.P. *Records, Geol. Surv. India*, v.126(6), pp. 16–17.

Rajarajan, K., 1978. Tectonics and metallogeny of the lead ore deposit near Sargipalli Orissa, India. *Geol. Surv. India, Misc. Pub.* 34, pp. 38–43.

Rajesh , H. M., Mukhopadhyay, J., Beukes, N. J., Gutzmer, J., Belyanin, G. A. and Armstrong, R. A. 2009. Evidence for an early Archaean granite from Bastar craton, India. *Journal Geol. Soc. London*, v. 166, pp. 193–196.

Rajesham, T., Bhaskara Rao, Y.J. and Murthy, K.S., 1993. The Karimnagar granulite terrain – a new sapphirine bearing granite province, south India. *Jour. Geol .Soc. India*, v. 41(1), pp. 51–59.

Raju, K.K. and Sharma, J.P., 1991. Geology and mineralisation of the Hutti gold deposit, Karnataka, India. In E.A. Ladeira (Ed.), *Brazil Gold '91*, pp. 469–476.

Rajurkar, S.T. and Ramalingaswami, G., 1975. Facies variations within the Upper Cuddapah strata in the northern parts of Cuddapah basin. *Geol. Surv. India, Misc. Pub.*, v. 23, pp. 150–157.

Rama Rao and Srirama, 1995. Geophysical studies in Central India: Barhi area, In: Geoscientific studies of the Son–Narmada–Tapti lineament zone. *Geol.Surv.India. Spl. Pub.*, v. 10, pp. 181–187.

Ramachandra, H.M., 1994. Petrological study of Precambrian granitoids in parts of Central India. *Geol.Surv. India, Unpub. Report.*

Ramachandra, H.M., 1999. Petrology of the Bhandara–Balaghat granulite belt in parts of Maharashtra and Madhya Pradesh. *Geol. Surv. India, Unpub. Report.*

Ramachandra, H.M. and Pal, R. N. 1992. Progress report on study of geochemistry and Cu–Pb–Zn mineralisation in Kherli Bazar area, Betul, M.P. *Geol. Surv. Ind., Unpub. Report.*

Ramachandra, H.M. and Roy, A, 1998. Geology and intrusive granitoids with particular reference to Dongargarh granite and their impact on tectonic evolution of the Precambrians of Central India. *Indian Minerals*, v. 52(1and2), pp. 15–33.

Ramachandra, H.M. and Roy, A., 2001. Evolution of the Bhandara–Balaghat granulite belt along the southern margin of the Sausar mobile belt of central India. *Proc. Indian Acad. Sci. (Earth Planet. Sci.)*, v. 110(4), p. 351.

Ramachandra, H.M., Mishra, V.P., Roy, A and Dutta, N.K., 1998. Evolution of the Bastar Craton – a critical review of gneiss–granitoids and supracrustal belts. (Abst.) M.S.Krishnan Comm. *Nat. Sem. on Progress in Precambrian Geology of India, Geol. Surv. India*, Calcutta, pp. 144–150.

Ramachandran, H.M., Roy, A., Mishra, V.P. amnd Dutta, N.K., 2001. A critical review of the tectonothermal evolution of the Bastar craton. *Geol. Surv. Ind., Spl. Pub., v. 55*, pp. 161–180.

Ramachandran, S. and Sinha R.P.,1992. Pegmatites of Bihar mica belt in ralation to the Chhotanagpur Granite Gneiss and their columbite–tantalite potential;. *Indian Journal Geol.*, v. 64. pp. 276–283.

Ramadurai, S., Sankaran, M., Selvan, T.A. and Windly, B.F., 1975. The stratigraphy and structure of the Sittampundi Complex, Tamilnadu, India. *Jour. Geol .Soc. India*, v. 16, pp. 409–416.

Ramakrishnan, M., 1973. Facies series in the polymetamorphic complex of Southern Bastar, Madhya Pradesh. *(Abst.)Symp. on Peninsular shield,* Hyderabad, pp. 13–14

Ramakrishnan, M., 1987. Stratigraphy, sedimentary, environment and evolution of the Late Proterozoic Indravati Basin, Cenral India. In Purana basins of Peninsular India, *Geological Society of India, Mem. 6*, pp. 139–160.

Ramakrishnan, M., 1990. Crustal development in southern Bastar, Central Indian craton. *Geol Surv. India. Spl. Pub. No. 28*. pp. 44–66.

Ramakrishnan, M. ,1994. Stratigraphic evolution of Dharwar craton. *Geo Karnataka MGD Cent.* vol., pp. 6–35.

Ramakrishnan, M., 2003. Craton–mobile belt relation in Southern Granulite terrain. In M. Ramachandran (Ed), Tectonics of Southern Granulite Terrain. *Geological Society of India, Mem. 50*, pp. 1–24.

Ramakrishnan, M. and Harinadha Babu, P., 1981. Western Ghat Belt. *Geol. Surv. India, Mem. 112*, pp. 311 – 328.

Ramakrishnan, M. and Viswanatha, M.N., 1987. Angular unconformity, structural unity argument and Sargur–Dharwar relation in Bababudan Basin. *Jour. Geol .Soc. India*, v. 29, pp. 471–482.

Ramakrishnan, M., Nanda, J.K.Augustine, P.F., 1998. Geological evolution of Proterozoic Eastern Ghats mobile belt. Proc. Workshop on Eastern Ghats Mobile Belt, *Geol. Surv. India, Spl. Pub., v. 44*, pp. 1–21.

Ramam P.K. and Murthy, V.N., 1997. Geology of Andhra Pradesh. *Geological Society of India*, p. 245.

Ramamohan Rao, T., 1969. The occurrence of ottrelite along the thrust zone in the Pakhals and Yellanpad area, Andhra Pradesh, India. *Geol. Mag.*, v. 106, pp. 452–456.

Ramamohan Rao, T., 1970. Metamorphism of the Pakhals of the Yellandlapad area, Andhra Pradesh. *W.D.West Comm. Vol.*, pp. 225–235.

Ramanathan, A., Bagchi, J., Panchapakesan, V.P. and Sahu, B.K., 1990. Sulphide mineralistion at Malanjkhand – a study. Workshop Vol, on Precambrian of Central India, *Nagpur. Geol. Surv*. India, *Spl. Pub. 28*, pp. 585–598.

Ramberg, H., 1973. Model studies of gravity–controlled tectonics by the centrifuge technique. In K.A. De Jong and R. Scholten (Eds.), *Gravity and tectonics*. Wiley, New York, pp. 49–66.

Ramesh Babu P.V., 1993. Tin and rare metal pegmatite of the Bastar–Koraput pegmatite belt, Madhya Pradesh and Orissa, India: Characterisation and classification. *Jour. Geol .Soc. India*, v. 42, pp. 180–190.

Ramesh Babu. P.V., 1999. Rare metal and rare earth pegmatites of Central India. In Special issue on rare metal and rare earth pegmatites of India. *Expln. and Res. for Atomic Minerals*, v. 12, pp. 7–52.

Ramesh Babu, P.V., Rajendran, R., Mundra, K.L., Sinha, R.P. and Banerjee, D.C., 1995. Resources of yttrium and rare earth minerals in riverine placers of parts of Madhya Pradesh and Bihar. *Proc. Sem. on Rec. Dev. Sci. Tech. Rare Earths, Cochin*, pp. 10–14.

Ranawat, P.S., Bhatnagar, S.N. and Sharma, N.K., 1988. Metamorphic character of Rampura Agucha Pb–Zn deposits, Rajasthan. In A. B. Roy (Ed.), Precambrian of the Aravalli mountains, Rajasthan, India. *Geological Society of India, Mem. 7*, pp. 397–409.

Ranawat, P.S., Rover, O., Ramboz, C. and Lakshmi, N., 2005. Native antimony float-ore from the Precambrian of Rajasthan. *Jour. Geol. Soc. India*, v. 65, pp. 353–356.

Rao, A.T., Rao, K.S.R. and Sriramadas, A., 1969. Zircon from apatite vein near Sitaramapuram, Kasipatnam, Visakhapatnam district, Andhra Pradseh. *Bull.Geol.Soc.India*, v. 6, pp. 5–8.

Rao, G.V., 1981. Sausar Group – distribution and correlation. *Geol Surv. India, Spl. Pub.* v. 3, pp. 1–7.

Rao, G.V.S. Purnachandra and Rao, J.M., 1996. Palaeomagnetic and geochemical study of Precambrian Kawar volcanic formation (Bijawar Traps), central India. *Jour. Geol. Soc.India*, v. 47, pp. 251–258.

Rao, K. Sreeramchandra, 2001. Regional surveys and exploration for gold in the greenstone granite terranes of Andhra Pradesh. *Geol.Surv. India, Spl. Pub. 58*, pp. 11–28.

Rao, K. Sreeramchandra, Roop Kumar, D., Jairam, M.S., Bhattacharya, S., Anand Murthy S. and Krishna Rao, P.V., 2001. Interpretation of geological characteristics of Dona gold deposit, Jonnagiri schist belt, Kurnool district, Andhra Pradesh. *Geol.Surv.India, Spl. Pub. 58*, pp. 217–231.

Rao, M.V., Sinha, K. K., Misra, B., Balachandran, K. Srinivasan, S. and Rajasekharan, P. 1988. Quartz–pebble conglomerate from Dhanjori – A new uranium horizon of the Singhbhum uranium province. *Geological Society of India, Mem. 9*, pp. 89–95.

Rao, N.K., 1977. Mineralogy and geochemistry of uranium prospects from parts of Singhbhum shear zone, Bihar. *Unpub. Ph.D. thesis, Banaras Hindu University*, India.

Rao, N.K. and Rao, S.V.U., 1980 a. Uraninite in the uranium deposits of Singhbhum shear zone, Bihar. *Journal Geological Society of India*. v. 24, pp. 387–397.

Rao, N.K. and Rao, S.V.U., 1980 b. Mineralogy of apatite–magnetite rocks in Singhbhum shear zone, Bihar. *Proc. 3rd Indian Geol. Cong.*, pp. 39–56.

Rao, N.K., Aggarwal, S.K. and Rao, G.V.U., 1979. Lead isotope ratios of uraninites and the age of uranium mineralisation in Singhbhum shear zone, Bihar. *Jour. Geol. Soc.India*, v. 20, pp. 124–127.

Rao, N.K., Sunil Kumar, T.S. and Narasimhan, D., 1999. Gold in copper–uranium ores of Singhbhum shear zone, Bihar. *Jour. Geol. Soc.India*, v. 54, pp. 37–42.

Rao, P. S., 1981. Inter–relationship of bauxites and laterites in Kanhangad–Kumbla area, Cannanore district, Kerala, South India. In *Lateritisation Processes*, Oxford and IBH Co. Pvt. Ltd., New Delhi, pp. 232–236.

Rao, Ramachandra, 1973. The sub–surface geology of the Indo–Gangetic plains. *Jour. Geol. Soc.India*, v. 14, pp. 217–242.

Rastogi, S.P., Prasad Mangla, Dhir, N.K., Sharma, D.P., Khan, M.A., Dwivedi, G.N., Mehrotra, R.D., Tripathi, A.K., Yadav, M.L., Afsar Hakim, and Sinha, V.P., 2001. Gold mineralisation in Son valley, Sonbhadra and Sidhi districts, U.P and M.P with special reference to its genesis. *Geol. Surv. India, Spl. Pub. 58*, pp. 83–93.

Rath, R., Rashmi and Das, S., 2005. Pebble strain analysis of Proterozoic conglomerate around Bisrampur, East Singhbhum District, Jharkhand. In *Proc. Nat. Sem. on Mineral and Energy resources of Eastern and Northeastern India. MGMI*, Kolkata, pp. 125–139.

Rath, S.C., Sahoo, K.C. and Satpathy, U.N., 1998. The Kankarakhol–Lodhajhari alkaline complex at the margin of Eastern Ghats, Deogarh district, Orissa, and India(Abst.). *Internat.Sem. on Precambrian crust of Eastern and Central India (UNESCO–IUGS–IGCP–368)*, Bhubaneswar, pp. 109–111.

Raut, P.R. and Mahakud, S.P., 2004. Geology, geochemistry and tecoinic setting of volcanosedimentary sequence of Betul belt, Madhya Pradesh and genesis of zinc and copper sulphide mineralisation. *Geol. Surv. India, Spl. Pub. 72*, pp. 133–146.

Rau, T.K., 2006. Incidence of diamonds in the beach sands of Kanyakumari coast, Tamil Nadu. *Jour. Geol. Soc. India*, v. 67, pp. 11–16.

Rau, T.K., 2007. Panna diamond belt, Madhya Pradesh – A critical review. *Jour. Geol. Soc.India*, v. 69, pp. 513–521.

Ravi Shanker, Kumar, G. and Saxena, S.P., 1989. Stratigraphy and sedimentation in Himalaya : a re–appraisal. In Ravi Sanker et al.(Eds.), *Geology and Tectonics of India, Geol. Surv. India, Spl. Pub. 26*, pp. 1–60.

Ray, A., Bhattacharyya, P.K., Mukherjee, A.D. and Roy, A.B., 1988. Sulfide mineralisation in the zone of basement-cover intersection near Anjeni, Udaipur District, Rajasthan. In A.B. Roy (Ed.), *Precamb. Arawalli Mountain, Rajasthan*, India, pp. 373–384.

Ray, K.K., 1971. Some problems on stratigraphy and tectonics of the Darjeeling and Sikkim. Himalayas. *Abst. Seminar on Recent Studies in the Himalayas, Geological Survey of India.*

Ray, K.K. 1975. Discovery of basemetal Pb–Zn mineralisation in Daling Chu. Mal Nadi area, Darjeeling district, West Bengal. *Indian Minerals*, v. 29, pp. 112–118

Ray, K.K., 1976. A review of the geology of the Darjeeling–Sikkim Himalaya. *International Himalayan Geology Seminar New Delhi*, pp. 13–17.

Ray, K.K., Ghosh Roy, A.K.,and Sengupta S., 1996. Acid volcanic rocks between the Dalma volcanic belt and the Chhotanagpur gneissic complex, East Singhbhum and Purulia districts of Bihar and West Bengal. *Indian Minerals*, v. 50, pp. 1–8.

Ray, S. and Biswas A.B., 1951. A biotite–lamprophyre from copper belt of Dhalbhum. *Proc, 39th Indian Sci. Cong.*, Pt. III, pp. 177.

Ray, S., 1947. Zonal metamorphism of Eastern Himalaya and some aspects of local geology. *Quart. Journal Geol. Min. Met. Soc. India*, v. 14, pp. 117–140.

Ray, S.K., 1974. Structural history of the Saladipura pyrite–pyrrhotite deposit and associated rocks, Khetri belt, Rajasthan. *Jour.Geol. Soc.India*, v. 15, pp. 227–238.

Ray. S.K., 1988. Structural control of copper mineralisation near Bajta, Ajmer district, Rajasthan. *Geological Society of India, Mem. 7*, pp. 383–372.

Ray, S.K., 1990. The albitite line of northern Rajasthan – a fossil intra–continental rift zone. *Jour. Geol. Soc. India*, v. 36, pp. 413–423.

Ray, S.K., 1995. Lateral variation in the geometry of thrust planes and its significance, as studied in Shumar allochthon, Lesser Himalayas, eastern Bhutan. *Tectonophysics*, v. 249, pp. 125–139.

Ray, S.K., 2004. Mineral potential of albitite line of Northern Rajasthan. *Geol. Surv. India, Spl. Pub., v. 72*, pp. 487–496.

Ray, S.K. 2006. Scaling properties of thrust fault traces in the Himalayas and inferences on the thrust fault growth. *Jour. Struct. Geol.*, v. 28, pp. 1307–1315.

Ray, S.L., Ghosh, S. and Saha, A.K., 1987. Nature of oldest known metasediments from eastern India. *Indian Minerals*, v. 41, pp. 52–60.

Raza, M., Jafri, S.H., Alvi, S.H. and Khan, M.S., 1993. Geodynamic evolution of the Indian shield during Proterozoic: Geochemical evidence from mafic rocks. *Journal Geological Society of India*, v. 41, pp. 455–469.

Reddy, P.R., 2001. Crustal seismic studies in underetsanding the collision anad subduction tectonics of Indian continent – and overview. *Geol. Surv. Ind. Spl. Pub. 64*, pp. 147–162.

Rich, R.A., Holland, H.D. and Peterson, U., 1977. *Hydrothermal uranium deposits*. Elsevier, Amsterdam, p. 265.

Richards, J.P. 2003. Tectonomagmatic precursors for porphyry Cu–(Mo–Au) deposit formation. *Econ. Geol.*, v. 98, pp. 1515–1533.

Richards, M.A., Duncan, R.A., Courtillot, V.E., 1989. Flood basalts and hotspot tracks:plume heads and tails. *Science*, 246, pp. 103–107.

Richars, J.P., Boyce, A.J. and Pringle, M.S., 2001. Geological evolution of the Escondida area, northern Chile : A model for spatial and temporal localization of Prophyry Cu–mineralisation. *Econ. Geol.* v. 96, pp. 271–305.

Rickers, K., Mezger, K and Raith, M.M., 2001. Evolution of the continental crust in the Proterozoic Eastern Ghats Belt, India and new constraints for Rodinia reconstruction: implications from Sm–Nd, Rb–Sr and Pb–Pb isotopes. *Precamb. Research*, v 112, pp. 183–20.

Rickers, K., Raith, M.M. and Dasgupta, S., 1998. Multistage reaction history of high Mg–Al granulites at Anakapalle: implications for the thermo–tectonic evolution of the Eastern Ghats belt, India (Abst.). *Int.Sem. on Precambrian crust of Eastern and Central India (UNESCO–IUGS–IGCP–368)*, Bhubaneswar, pp. 52–54.

Ridley, J.R. and Diamond, L.W., 2000. Fluid chemistry of orogenic lode gold deposits and implications for genetic models. *Rev. Econ. Geol.*, v. 13, pp. 141–162.

Ridley, J. R. Groves, D. I. And Knight, J. T. 2000. Gold deposits in amphilolite ansd granulite facies terranes of Archan Yilgarm craton, Western Australia: Evidence and implicartions of synmetamorphic mineralisation. In P. G. Spry, B. Marshall and F. M. Vokes (Eds), Metamorphosed and metamorphogenic ore deposits. *Rev. Econ. Geol.*, v. 11, pp. 265–290.

Ridley, J.R. Mikucki. E. J. and Groves, D. I. 1996. Archean lode – gold deposits: Fluid flow and chemical evolution in vertically extensive hydrothermal systems. *Ore Geol. Revs*. No. 10, pp. 279–293.

Ridley, J.R., Groves, D.I. and Knight, J.T., 2000. Gold deposits in amphilolite and granulite facies terranes of Archean Yilgarn craton, Western Australia: Evidence and implications of syn–metamorphic mineralisation. In P. G. Spry, B. Marshall and F. M. Vokes (Eds.), Metamorphosed and metamorphogenic ore deposits. *Rev. Econ. Geol.*, v. 11, pp. 265–290.

Robb. L., 2005. Introduction to ore-forming processes. *Blackwell Pub. Co.* USA and UK. p. 373.

Robert, F. and Poulsen, K.H., 2001. Vein formation and deformation in greenstone gold deposits. In J.P.Richards and R.M.Tosdal (Eds.), Structural controls on ore genesis. *Rev. Econ.Geol.*, v. 14, pp. 111–155.

Roberts, R., Palmer, M.R. and Waller. L., 2006. Sm–Nd and REE characteristics of tourmaline and scheelite from the Björkdal gold deposit, Northern Sweden : Eivdence of an intrusion–related gold deposit? *Econ. Geol.* v. 101, pp. 1415–1425.

Robinson, D., Reverdatto, V.V., Bevins, R.E., Polyansky, O.P. and Sheplev, V.S., 1999. Thermal modeling of convergent and extensional tectonic settings for the development of low grade metamorphism in Welsh Basin. *Journal Geophys. Res.* (Solid Earth), v. 104, pp. 23069–23080.

Rogers, J.J.W., 1988. The Arsikere granite of southern India : magmatism and metamorphism of previously depleted crust. *Chem. Geol.*, v. 67, pp. 155–163

Rogers, J.J.W., 1993. India and Ur. *Jour.Geol. Soc.India*, v. 42, pp. 217–222.

Rogers, J.J.W., 1996. A history of old continents in the past 3 billion years. *Journal Geology*, v. 104, pp. 91–107.

Rogers J.J.W. and Callahan, E.J., 1989. Diapiric trondhjemites of the Western Dharwar craton, southern India. *Canadian Journal Earth Sci.*, v. 26, pp. 244–256.

Rogers, J.J.W. and Santosh, M., 2002. Configuration of Columbia, a Mesoproterozoic supercontinent. *Gondwana Res.*, v. 5, pp. 5–22.

Rogers, J.J.W. and Santosh, M., 2004. *Continents and supercontinents*. Oxford University Press, New York, p. 289.

Rogers, J.J.W. and Santosh, M., 2009. Tectonics and surface effects of the supercontinent Columbia. *Gondwana Res.*, v. 15, pp. 373–380.

Rollinson, H. 1997. The Archen komatiite-related Inyala chromitite, southern Zimbabwe. *Econ. Geol.*, v. 92, pp. 98–107.

Romer, R. L., Martinson, O. and Perdahl. J. L., 1994. Geochronology of the Kiruna iron ores and hydrothermal alterations. *Econ. Geol.*, v. 89. pp. 1249–1261.

Rosiere, C.A. and Rios F.J., 2004. The origin of hematite in high grade iron ores, based on infra–red microscopy and fluid inclusion studies: The example of Conceicao mine, Quadrilatero Ferrifero, Brazil. *Econ. Geol.*, v. 99, pp. 611–624.

Rosing, M.T., Rose, N.M., Bridgewater, D. and Thompson, H.S., 1996. Earliest part of Earth's stratigraphic record: A reappraisal of the > 3.7 Ga Isua (Greenland) supracrustal sequence. *Geology*, v. 24, pp. 43–46.

Rotherham, J.F., Blake, K.L., Cartright, I. and Williams, P.J., 1998. Stable isotope evidence for the origin of the Mesoproterozoic Starra Au-Cu deposit, Choncurry district, NW Queensland. *Econ. Geol.*, v. 93, pp. 1435–1449.

Rowell, W.F. and Edger, A.D., 1986. Platinum Group element mineralisation in a hydrothermal Cu–Ni sulphide occurrence, Rathbum Lake, North Eastern Ontario. *Econ. Geol.*, v. 81, pp. 1272–1277.

Rowley, D.B. 1996. Age of initiation of collision between India and Asia: a review of the stratigraphic data. *Earth and Planetary Sci. Letter*, v. 145, pp. 1–13.

Roy, A., 1979. Polyphase folding deformation in the Hutti–Maski schist belt, Karnataka. *Jour.Geol. Soc.India*, v. 20, pp. 598–607.

Roy. A, 1983. Petrographic–mineralogic studies of Sargipalli lead–zinc sulphide deposit and their genetic implications. *Indian Minerals*, v. 37.

Roy, A., 1991. The geology of gold mineralisation at Hutti in Hutti–Maski schist belt, Karnataka, India. *Indian Minerals*, v. 45, pp. 229–250.

Roy, A. and Bandyopadhyay, B.K. 1989. Geology and geochemistry of metabasalt near Sleemanabad in the Proterozoic Mahakoshal belt of Central India. *Indian Minerals,* v. 43. pp. 303–324.

Roy, A. and Bandopadhyay, B.K., 1990. Tectonic and structural pattern of the Mahakoshal belt of Central India: A discussion. In: Precambrian of Central India, *Geol.Surv.India*, *Spl. Pub., v. 28*, pp. 226–240.

Roy, A. and Bandopadhyay, B.K., 1998. Supracrustal belts of Central India and their significance in the crustal evolution of central Indian shield. *(Abst.)National Seminar on 50 years of progress in Precambrian Geology of India*, pp. 160–164.

Roy, A. and Biswas, S.K., 1979. Metamporphic history of the Sandur schist belt, Karnataka. *Jour.Geol. Soc. India*, v. 20, pp. 179–187.

Roy, A. and Biswas, S.K., 1983. Stratigraphy and structure of Sandur schist belt, Karnataka. *Jour.Geol. Soc. India*, v. 24, pp. 19–28.

Roy, A. and Devarajan, M.K., 1999. On the occurrence of a ductile shear zone along the southern margin of the Mahakoshal supracrustal belt of central India. (Abst.). National Symposium on Developments in Geology, Mineral Deposits and Seismotectonics of Central India., *Geol. Surv. India*, Bhopal, pp. 17–18

Roy A. and Devarajan, M.K., 2000. A reappraisal of the stratigraphy and tectonics of the Palaeoproterozoic Mahakoshal Supracrustal belt, Central India. *Geol. Surv. India, Spl. Pub.*, v. 57, pp. 79–97.

Roy, A. and Hanuma Prasad, M. , 2001. Precambrian of Central India: A possible tectonic model. *Geol. Surv. India, Spl. Pub., v. 64*, pp. 177–197.

Roy, A. and Hanuma Prasad, M., 2003. Petrology and Petrogenesis of granite and charnockite–mangerite suite of rocks from northern margin of Sausar mobile belt and its implications on crustal evolution of Central Indian Tectonic Zone. *Gondwana Geol. Mag.*, Spl. Pub., v. 7, pp 241–260.

Roy A., Bandyopadhyay, B.K and Huin. A.K.1997. Geology and geochemistry of basic volcanics from the Sakoli schist belt of Central India. *Jour.Geol. Soc.India*, v. 50, pp. 209–221.

Roy,.A., Bandyopadhyay. B.K., Huin, A.K., Charles D. Mony and Saha. S.K.,1994.Geology of the Sakoli Fold Belt, Nagpur, Bhandara and Gadchiroli districts, Maharashtra, Central India. *Geol. Surv. India, Unpub. Report.*

Roy, A., Bandyopadhyay, B.K., Huin, A.K., Mony, P.C.D. and Saha, S.K., 1995. Geology of the Sakoli Fold Belt, Nagpur, Bhandara and Gadchiroli districts,Maharashtra. *Geol. Surv. India,Unpub. Report.*

Roy, A., Chore, S.A., Hanuma Prasad, M. and Sethumadhavan, M.S., 2003. Petrogenesis of mafic–ultramafic intrusives of Betul belt in central India. *Gondwana Geol. Mag., Spl. Pub., v. 7*, pp. 524–525.

Roy A., Hiroo Kagami, Masaru Yoshida, Abhinaba Roy, B.K.Bandyopadhya, A. Chattopadhyay, A.S. Khan, A.K. Huin, T. Pal, 2006. Rb–Sr and Sm–Nd dating of different metamorphic events from the Sausar mobile belt, central India: implications for Proterozoic crustal evolution. *Journal Asian Earth Sci.*, v. 26, pp. 61–76

Roy, A., Ramchandra, H.M. and Bandopadhyay, B.K., 2000. Supracrustal belts and their significance in crustal evolution of Central India. *Geol. Surv. India, Spl. Pub., v. 55*, pp. 361–380.

Roy, A., Sarkar, A., Bhattacharya, S.K., Ozaki, H and Ebihara, M., 1997. Rare earth element geochemistry of selected mafic–ultramafic units from Singhbhum craton – implications to source heterogeneity. *Jour. Geol. Soc.India*, v. 50, pp. 717–726.

Roy, A., Sarkar, A., Jeyakumar, S. Aggrawal, S.K. and Ebihara, M., 2002a. Mid–Proterozoic plume related thermal event in Eastern Indian Craton: Evidence from trace elements, REE geochemistry and Sr–Nd isotope systematics of basic–ultrabasic intrusives from Dalma volcanic belt. *Gondwana Res.*, v. 5(1), pp. 133–146.

Roy, A., Sarkar, A., Jeyakumar, S. Aggrawal, S.K. and Ebihara, M., 2002b. Sm–Nd age and mantle source characteristics of the Dhanjori volcanic rocks, Eastern India. *Geochem. Journal*, v. 36, pp. 503–518.

Roy, Abhijit, 1998. Isotopic evolution of Precambrian mantle – a case study of Singhbhum–Orissa craton. *Unpub. Ph.D. thesis, Indian School of Mines, Dhanbad, India.*

Roy Abhijit, 2002. Geochronological study in the Sausar and Sakoli belts of the Central Indian Tectonic Zone. *Records, Geol.Surv.India*, v. 136(2).

Roy, Abhijit, Sarkar, A., Jeyakumar, S., Aggarwal, S.K. and Ebihara, M., 2002a. Mid-Proterozoic plume-related thermal event in Eastern Indian craton: Evidence from trace elements, REE geochemistry and Sm-Nd isotope systematics fo basic-ultrabasic introduces from Dalma volcanic belt. *Gondwana Res.*, v. 5, pp. 133–146.

Roy, Abhijit, Sarkar, A., Jeyakumar, S., Aggarwal, S.K. and Ebihara, M., 2002b. Sm-Nd age and mantle source characteristics of the Dhanjori volcanic rocks, Eastern India. *Geochem. Jour.*, v. 36, pp. 503–518.

Roy, A.B., 1966. Interrelation of metamorphism and deformation in Central Singhbhum, Eastern India. *Geol. Mijnb.*, v. 45 c, pp. 365–374.

Roy, A.B., 1988. Stratigraphic and tectonic framework of Aravalli Mountain Range. *Geological Society of India, Mem.* 7, pp. 3–31.

Roy, A.B., 1991. Evolution of the Early Proterozoic Aravalli depositional basin. In S.K. Tandon, C. Pant and S.M. Kashyap (Eds.), *Evolution of Early Proterozoic Aravalli depositional basin. Ganodaya Prakashana*, Nainital, pp. 1–13.

Roy, A.B., 1995. Geometry and evolution of superposed folding in the Zawar lead–zinc mineralized belt, Rajasthan. *Proc. Ind. Acad. Sci.*, v. 104, pp. 349–371.

Roy, A.B., 2000. Geology of the Palaeoproterozoic Aravalli Supergroup of Rajasthan and northern Gujarat. In M. Deb (Ed), *Crustal evolution and metallogery in northwestern Indian shield.* Narosa Publ. House, New Delhi, pp. 87–114.

Roy, A.B., 2001. Tectonostratigraphy of the lead-zinc sulfide deposits in the Paleoproterozoic Arawalli supergroup. examples from Zawar and Rampura-Agueha ore deposits. In M. Deb and W.D. Goodfellow (Eds.), Pre-conf. Papers, *sediment-hosted lead-zinc sulfide deposits in the northwest Indian shield*, Delhi, Dec. 10-17, 2001.

Roy, A.B. and Bejarniya, B.R., 1990. A tectonic model for the Early Proterozoic Aravalli (Supergroup) rock from north of Udaipur, Rajasthan. In S.P.H. Sychanthavang (Ed.), *Crustal evolution and orogeny.* Oxford and IBH Pub. Co. Pvt. Ltd., New Delhi, pp. 249–273.

Roy, A.B. and Dutt, K., 1995. Tectonic evolution of the nepheline syenite and associated rocks of Kishengar district, Ajmer, Rajasthan. In S. Sinha Ray and K.R. Gupta (Eds.), *Continental crust of northwestern and central India. Geological Society of India, Mem. 31*, pp. 231–257.

Roy, A.B., and Jakhar, S.R., 2002. *Geology of Rajasthan (Northwest India) : Precambrian to recent.* Scientific Publishers (India), Jodhpur, p. 421.

Roy, A.B. and Kataria, P., 1999. Precambriam geology of Aravalli Mountains and neighbourhood: Analytical updates of recent studies. In P. Kataria (Ed.), *Proc. Seminar on geology of Rajasthan – status and perspective.* Geology Department, MLSU, Udaipur.

Roy, A.B. and Kröner, A., 1996. Single zircon evaporation ages constraning the growth of the Aravalli craton, northwestern Indian shield. *Geol. Mag.*, v. 133(3), pp. 333–342.

Roy, A.B. and Paliwal, B.S., 1981. Evolution of the Lower Proterozoic epicontinental deposits: stromatolite bearing Aravalli rocks of Udaipur, Rajasthan, India. *Precamb. Research*, v. 14, pp. 49–74.

Roy, A.B. and Patil, M.L., 1996. Re–interpretations. of structure of the Middle Proterozoic Delhi Supergroup of Kho–Dariba, Rajasthan — tectonostratigraphic implications. *Indian Minerals*, v. 50, pp. 29–40.

Roy, A.B. and Rathore, S., 1999. Tectonic setting and the model of evolution of granulites of the Sandmata Complex, the Aravalli Mountain, Rajasthan. In *Symp. Vol. on Charnockite facies rocks. Geologists Assocn.*,Tamilnadu, pp. 111–126.

Roy, A.B. and Sharma, K.K., 1999. Geology of the region around Sirohi town, Western. Rajasthan – a story of Neoproterozoic evolution of Aravalli crust. In B.S. Paliwal (Ed), *Geological evolution of Northwestern India.* Scientific Publishers (India), Jodhpur, pp. 19–33.

Roy, A.B., Kumar S., Laul, V. and Chauhan, N.K., 2004. Tectonostratigraphy of the lead–zinc-bearing metasedimentary rocks in the Rampura–Agucha mine and its neighbourhood, district Bhilwara, Rajasthan: Implications on metallogeny. In M.Deb and Goodfellow, W.D. (Eds.), *Sediment hosted lead–zinc sulphide deposits.* Narosa Publ. House, New Delhi, pp. 273–289.

Roy, A.B., Paliwal, B.S., Shekhawat, S.S., Nagori, D.K., Golani, P.R. and Bejarnia, B.R., 1988. Stratigraphy of the Arawalli Supergroup in the type area. *Geoogical Society of India, Mem. 7*, pp. 121–138.

Roy, A.B., Somain, M.K. and Sharma, N.K., 1981. Aravalli – Pre–Aravalli relationship : A study from the Bhinder region, southern Rajasthan. *Indian Journal Earth Sciences*, v. 8, pp. 119–130.

Roy, K.K., Das, L.K., Routh, P.S., Mukherjee, K.K., Das, N., Saha, D.K., Rai, M.K. and Naskar D.C., 1999. Some upper crustal information of the Singhbhum craton based on direct current geoelectrical survay. *Indian Journal Geol.* v. 71, pp. 133–142.

Roy, Minati and Dhana Raju, R., 1999. Mineragraphy and electron microscope study of the U-phases and the pyrite of the dolostone-hosted uranium deposit at Tummalapalle Cuddapah dist., Andhra Pradesh and its implications on genesis and exploitation. *Jour. Appd. Geochem.*, v. 4(2), pp. 53–75.

Roy, S., 1966. Syngenetic manganese formations of India. *Jadavpur University Publication.* Calcutta, 219 p.

Roy. S., 1981. Manganese deposits. *Academic Press,* London. p. 458.

Roy. S., 1992. Environments and processes of manganess depositon. *Econ. Geol.,* v. 87. pp. 1218–1236.

Roy, S., 2000. The manganese formation of the Neoproterozoic Penganga Group, India – Revision of an enigma – A discussion. *Econ. Geol.*, v. 95. pp. 237–238.

Roy, S., 2006. Sedimentary manganese metallogenesis in response to the evolution of the Earth System. *Earth Sci. Reviews*, v. 77, pp. 273–305.

Roy, S., Bandyopadhyay, P. C., Perseil, E.A. and Fukuoka, 1990. Late diagenetic changes in manganese ores of the Upper Proterozoic Penganga Group, India. *Ore geology Reviews*, v. 5, pp. 341–357.

Roy Chowdhury, M. K., 1958. Bauxite in Bihar, Madhya Pradesh, Vindhya Pradesh, Madhya Bharat and Bhopal, *Geol. Surv. India, Mem. 85*, p. 217.

Roy Chowdhury, M.K., 1965. Bauxite in India. *Bull Geol. Surv. India*, No. 25, p. 127.

Roy Chowdhury, M. K. and Dasgupta, S. P., 1965. Ore localization in the Khetri copper belt, Rajasthan, India. *Econ. Geol.* v. 60. pp. 69–88.

Russell, J., Chadwick, B., Krishna Rao, B. Vasudev, V.N., 1996. Whole rock Pb/Pb isotropic ages of late Archean limestones, Karnataka, India. *Precamb. Research*, v. 78, pp. 261–272.

Russel, M. J., 1988. A model for the genesis of sediments-hosted exhalative (SEDEX) deposits In E. Zachrisson (Ed.), *Proc. 7th Quad. IAGOD Symp. Schweiz. Verlag. Stuttgart*. pp. 59–66.

Russell, M.J., Hall, A.J., Boyce, A.J. and Fallick, A.E., 2005. On hydrothermal convection systems and emergence of life. *Econ. Geol.*, *100th Anniversary* vol., pp. 419–438.

Russel, M.J., Solomon, M., and Walshe, J.L., 1981. The genesis of sediment–hosted exhalative zinc + lead deposits. *Mineral Deposita*, v. 16, pp. 113–127.

Rye, R.O. and Ohmoto, H., 1974. Sulfur and carbon isotopes and ore genesis: a review. *Econ. Geol.*, v. 69. pp. 826–842.

Sachan, H.K., 1993. Early-replacement dolomitization and deep-burial modification and stabilization: a case study from the date Precambrian of the Zawar area, Rajasthan (India). *Carbonates and evaporates*, v. 8, pp. 191–198.

Safaya, H.L., 1986. Magnesite deposits of Kumaon Himalaya, Uttar Pradesh – depositonal environment, genesis and economic utility. *Geol. Surv. India, Misc. Pub., v. 41*, pp. 150–178.

Safonov, Yu. G., Genkin, A.D., Vasudev, V.N., Krishna Rao, B., Srinivasan, R., and Krishnam Raju, K. , 1988. Problems of ore gold potential of Karnataka craton. In F.V. Chukhrov, et al. (Eds.), *Zolotorudnoe pole Kolar (India)* (In Russian). Nauka Publication, Moscow, pp. 206–219.

Safonov, Yu.G., Genkin, A.D., Cheryshev, I.V., Nosik, L.P., Ananata Ayer, G.V., Krishna Rao, B. and Vasudev, V.N., 1988. Genetic model of the Kolar deposit. In F.V. Chukhrov. et al (Ed.), *Zolotorudnoe pole Kolar (India)* (In Russian), Nanka Publication Moscow, pp. 176–187.

Saha, A.K., 1948. Kolhan series – Iron Ore series boundary to the West and southwest of Chaibasa. *Sci. and Cult.*, v. 14, pp. 77–79.

Saha, A.K., 1988. Some aspects of the crustal growth of the Singhbhum–Orissa iron ore craton. Eastern India. *Indian Journal Geol.*, v. 60, pp. 270–278.

Saha, A.K., 1994. Crustal evolution of Singhbhum–North Orissa, Eastern India. *Geological Society of India.*, *Mem. 27*, p. 341.

Saha, A.K. and Ray, S.L., 1984. The structural and geochemical evolution of Singhbhum Granite batholithic complex, India. *Tectonophysics (Spl. Issue)*, v. 105, pp. 163–176.

Saha, A.K., Bose, R., Ghosh, S.N. and Roy, A.,1977. Petrology and emplacement of the Mayurbhanj Granite batholith , eastern India. Evolution of orogenic belts of India (Part–2), *Geol. Min. Met. Soc.India, Bull 49*, pp. 1–34.

Saha, A.K., Ghosal, S., and Ray, S.L., 1986. Petrochemistry and origin of the Manda–Asana–Besoi Granite in the northwestern part of Singhbhum Granite batholiths. *Quart. Journal Geol. Min. Met. Soc*. India, v. 55, pp. 181–200.

Saha, A.K., Ray, S.L. and Sarkar, S.N.,1988. Early history of the Earth: Evidence from the Eastern Indian shield. In D.Mukhopadhyay (Ed.), Precambrian of the Eastern Indian shield. *Geological Society of India*, *Mem.* 8, pp. 13–38.

Saha, Aniki, Basu, A. R., Garzione, C.N., Bandyopadhyay, P. K., Chakrabarti, A., 2004. Geochemical and petrological evidence for subduction–accretion process in Archean Eastern Indian Craton. *Earth and Planet. Sci. Letts.*, v. 220, pp. 91–106.

Saha, Ashim K., 1995. Search for gold and tungsten in the central parts of Sakoli fold belt, Bhandara district. *Records, Geol. Surv. India*, v. 128(6).

Saha, Ashim K., Bhoskar, K.G. and Chattopadhyay, S., 2001 a . Ore mineralogy, fluid inclusion and sulphur isotope studies of Parsori gold–copper deposit, Sakoli fold belt, Maharashtra and its genetic modeling. *Geol. Surv. India, Spl. Pub. 58*, pp. 163–180.

Saha, Ashim K., Chattopadhyay, S., Mahapatra, K.C. and Raut, P.K., 2001 b. Polymetallic mineralisation in Sakoli fold belt – their genetic modelling using fluid inclusion and sulphur isotope data. *Geol. Surv. India, Spl. Pub. 64*, pp. 387–398.

Saha, Ashim K. and Mohan, M., 2000. Geological and genetic aspects of tungsten mineralisation in Sakoli supracrustals. *Indian Miner.*, v. 54(1and2), pp. 35–42.

Sahai, T.N., Gupta, S.N. and Mathur, R.K., 1997. Mineral resources. In *Geol. Surv. India, Mem.* 123, pp. 171–199.

Sahoo, R.K., 1998. Chromite. In S. Acharya, N. K. Mahalik and S. Mahapatra (Eds.) *Geology and mineral resources of Orissa*. SGAT, Bhubaneswar, pp. 155–178.

Sahu, N.K. and Mukherjee, M.M., 2001. Spinifex–textured komatiite from Badampahar–Gorumahisani schist belt, Mayurbhanj district, Orissa. Jour. Geol. Soc. India, v. 57, pp. 529–534.

Sahu, N.K., Patel, K. and Ashiya, I.D., 2003. Geology and geochemistry of Precambrian rocks around Ratanpur, Bilaspur District, Chhattisgsrh. *Gondwana Geol. Mag., Spl. Pub.*, v. 7, pp. 137–151.

Samson, I.M. and Russell, M.J., 1987. Genesis of the silvermines zinc-lead-barite deposit, Ireland: fluid inclusion and stable isotoe evidence. *Econ. Geol.*, v. 82, pp. 371–394.

Sangster, D.F., 1990. Mississippi valley-type and SEDEX lead-zinc deposits: a comparative examination. *Trans. Inst. Min. Met., v. B99*, pp. B21–B42.

Sano, T., Fujii, T., Deshmukh, S.S., Fukuoka, T. and Aramaki, S., 2001. Differentiation Processes of Deccan Trap basalts: Contribution from geochemistry and experimental petrology. *Journal Pet.*, v. 42(12), pp. 2175–2195.

Santosh, M.,, 1986. Ore fluids in the auriferous Champion reef of Kolar, South India. *Econ. Geol.* v. 81, pp. 1546–1552.

Santosh, M., 1988a. The granite-molybdenite system of Ambalvayal, Kerala: Pt I. Geochemistry and Petrogenesis of granite. *Jour. Geol. Soc. India*, v. 32, pp. 83–105.

Santosh, M., 1988b. Granite-molybdenite system of Ambalvayal, Kerala: Pt II. Nature of mineralisation, sulfur isotopes, fluid characteristics and genetic model. *Jour. Geol. Soc. India*, v. 32, pp. 191–213.

Santosh, M., 1996. The Trivandrum and Nagercoil granulite blocks. In M. Santosh and M. Yoshida (Eds.), The Archean and Proterozoic terrains in Southern India within East Gondwana. *Gondwana Res. Gr.*, *Mem.* 3, pp. 243–256.

Santosh, M., Ayer, S.S., Vascoucellos, M.B.A. and Enzweiler, J., 1989. Late Precambrian alkaline plutons in Southern India: geochronologic and rare earth element constraints on Pan–African magmatism. *Lithos*, v. 24, pp. 65–79.

Santosh, M., Jackson, D.H. and Harris, N.B.W., 1993. The significance of channel and fluid inclusion of CO_2 in cordierite : Evidence from carbon isotopes. *Journal Pet.*, v. 34, pp. 233–258.

Santosh, M., Kaganir, H., Yoshida, M. and Nanda Kumar, V., 1992. Pan–African charnockite formation in East Gondwana : geochronologic (Sm–Nd and Rb–Sr) and petrogenic constraints. *Bull. Indian Geol. Assocn.*, v. 25, pp. 1–10.

Santosh, M., Nadeau, S. and Javoy, M., 1995. Stable isotopic evidence for the involvement of mantle derived fluids in Wynad gold minerlization, south India. *Journal Geol.* v. 103. pp. 718–728.

Santosh M., Yokoyama, K. and Acharyya, S.K., 2004. Geochronology and tectonic evolution of Karimnagar and Bhopalpatnam granulite belt, Central India. *Gondwana Res.*, v. 7(2), pp. 501–518.

Santosh M., Maruyama, Shingenori and Yamamoto Shinji, 2009. The making and breaking of supercontinents: some speculations based on superplumes, super downwelling and the role of tectosphere. *Gondwana Res.*, v. 15, pp. 324–341.

Sanyal, S. and Fukuoka, M., 1995. P–T–t history of granulites from Anakapalle, Eastern Ghats – evidence for polyphase (?) granulite metamorphism. *Geological Society of India, Mem. 34*, pp. 125–141.

Saraswat, A.C., Rishi, M.K., Gupta, R.K. and Bhaskar, D.V., 1977. Recognition of favourable uraniferous area in the sediments of Meghalaya, India: A case history. In *Recognition and evaluation of uraniferous areas*, IAEA, Vienna, pp. 165–182.

Sarbajna, C. and Krishnamurthy, P., 1996. The fertile granite of Allapatna, Mandya district, Karnataka : A possible parent to rare metal pegmatites of southern Karnataka. *Jour. Geol. Soc. India*, v. 46, pp. 95–98.

Sarbajna, C., Sinha, R. P. Krishnamurthy, P., Krishna, K. V. G., Viswanathan, R. and Banerjee, D. C., 1999. Mineralogy and Geochemistry of alkali beryl from the rare metal–bearing pegmatites of Marlagalla Allapatna, Mandya Distric, Karnataka. *Jour. Geol. Soc. India*, v. 54, pp. 599–608.

Sarkar, A., 1981. Structure and tectonomagmatic evolution of the Elchuru alkaline complex, Andhra Pradesh, India. *Proc.4th Regional Conf.Geol.SE* Asia, Manila, Phillipines, pp. 131–137.

Sarkar, A., 1984. Geochronology and geochemistry of the Bundelkhand granitic complex in the Jhansi – Babina – Talbehat sector, Uttar Pradesh. *Geol. Surv. India, Unpub. Report.*

Sarkar, A., 1997. Geochronology of Proterozoic mafic dykes from the Bundelkhand craton, Central India. (Abst.) *Internat. Conf. on Isotopes in Solar System*, pp. 98 – 99.

Sarkar, A. and Nanda, J.K., 1998. Tectonic segments in the Eastern Ghats Precambrian mobile belt. *(Abst.) M.S.Krishnan Comm. Nat. Sem. on Progress in Precambrian Geology of India, Calcutta*, pp. 173–74.

Sarkar, A. and Paul, D.K., 1998. Geochronology of the Eastern Ghats Precambrian mobile belt – a review. *Geol.Surv.India., Spl. Pub., v. 44*, pp. 51–86.

Sarkar, A., Bodas, M.S., Kundu, H.K., Mamgain, V.D., and Ravi Shanker, 1998. Geochronology and geochemistry of Mesoproterozoic intrusive plutonites from the eastern segment of the Mahakoshal greenstone belt, Central India. *(Abst.) Internat. Sem. on Precambrian crust in Eastern and Central India (UNESCO–IUGS–IGCP 368), Bhubaneswar. Geol. Surv. India, Spl. Pub. 57*, pp. 82–85.

Sarkar, A., Paul, D.K. and Potts, P.J., 1996: Geochronology and geochemistry of Mid –Archaean trondhjemitic gneisses from Bundelkhand craton, Central India. *Recent Researches Geol.*, v. 16, pp. 76 – 92.

Sarkar, A., Pati, U.C., Panda, P.K., Patra, P.C., Kundu, H.K. and Ghosh, S., 2000. Late Archaean charnockite rocks from the northern marginal zones of Eastern Ghat belt: A geochronological study. (Abst.) Internat. *Sem. on Precambrian crust in Eastern and Central India (UNESCO–IUGS–IGCP 368), Bhubaneswar. Geol. Surv. India, Spl. Pub. 57*, pp. 171–179.

Sarkar, A., Sarkar, G., Paul, D.K. and Mitra, N.D., 1990. Precambrian geochronology of the Central Indian shield – A Review. *Geol.Surv.India, Spl. Pub.*, v. 28, pp. 453 – 482.

Sarkar, A., Trivedi, J.R., Gopalan, K., Singh, P.N., Singh, B.K., Das, A.K. and Paul, D.K., 1984. Rb/Sr Geochronology of the Bundelkhand Granitic Complex in the Jhansi–Babina–Talbehat sector, U. P. *Ind. Jou. Earth Sci., Seminar vol. CEISM*, pp. 64 – 72.

Sarkar, A.N, 1982. Precambrian tectonic evolution of eastern India, a model of converging microplates. *Tectonophysics*, v. 86, pp. 363–397.

Sarkar, A.N., 1988. Tectonic evolution of Chhotanagpur plateau and the Gondwana basin in E. India: an interpretation based on supra–subduction geological processes. In D. Mukhopadhyay (Ed.), Precambrian of Eastern Indian Shield. *Geolological Society of India.*, *Mem.* 8, pp. 127–146.

Sarkar, A.N. and Bhattacharya, D.S., 1978. Deformation cycles and intersecting isograds : the Hesadih antiform of the Singhbhum orogenic belt. *Contrib. Min. Pet.*, v. 66, pp. 333–340.

Sarkar, G. , Corfu, F., Paul, D.K., McNaughton, N.J , Gupta, S.N. and Bishui, P.K., 1993. Early Archaean crust in Bastar craton, central India – geochemical and isotopic study. *Precamb. Research*, v. 62, pp. 127–137.

Sarkar, G., Bishui, P.K., Chattopadhyay, B., Chaudhury, S., Chaudhury, I., Saha, K.C. and Kumar, A., 1992. Geochronology of granites and felsic volcanic rocks of Delhi fold belt. *Records, Geol. Surv. India*, v. 125 (2), pp. 21–23.

Sarkar, G., Gupta, S.N. and Bishui, P.K, 1994. Rb–Sr isotopic ages and geochemistry of granite gneiss from southern Bastar: implications for crustal evolution. *Indian. Minerals*, v. 48(1and2), pp. 7–12.

Sarkar, G. , Paul, D.K., McNaughton, N.J., Delaeter, J.R. and Mishra, V.P., 1990. A geochemical and Pb, Sr isotopic study of the evolution of granite gneisses from the Bastar Craton, Central India. *Jour. Geol. Soc. India*, v. 35(5), pp. 480–496.

Sarkar, G., Roy Barman, T. and Corfu, P., 1989. Timing of continental arc type magmatism from northern India: evidence from U–Pb zircon geochronology. *Journal Geol.* v. 97, pp. 607–612.

Sarkar, N.K., Mallik, A.K., Panigrahi, D. and Ghosh, S.N., 2001. A note on the occurrence of breccia zone in Katpal chromite lode, Dhenkanal district, Orissa. *Indian Minerals*, v. 55(3–4), pp. 247–250.

Sarkar, S.C., 1995.Modelling Singhbhum uranium mineralisation in the light of Proterozoic uranium metallogeny. *Expln. Res. Atomic Minerals (AMD)*, v. 8, pp. 81–93.

Sarkar, S.C., 1966a. Structure and their control of ore mineralisation in Mainajharia–Mosabani–Surda section of the Singhbhum copper belt, Bihar. In S.Deb. (Ed.), *Contributions to the geology of Singhbhum, Jadavpur University Publication,* Calcutta, pp. 75–83.

Sarkar, S.C., 1966b. Ore deposits along the Singhbhum shear zone and their genesis. In S.Deb. (Ed.), *Contributions to the geology of Singhbhum, Jadavpur University Publication,Calcutta*, pp. 91–101.

Sarkar, S.C., 1969. Late intrusives and their relationship with sulphide ores in Mosaboni–Badia mines, Singhbhum shear zone, Bihar. *Bull.Geological Society of India*, v. 6(4), pp. 130–133

Sarkar, S.C.,1970. A study of the mineralsation of radioactive elements in the Singhbhum shear zone, Bihar. *Proc. Ind. Nat. Sci Acad.*, v. 36 A, pp. 246–261.

Sarkar, S.C., 1971. Mackinawite from the sulphide ores from the Singhbhum copper belt, India. *Amer. Mineralogist*, v. 56, pp. 1312–1318.

Sarkar, S.C.,1973. Some observations on the geology in the Khetri copper belt, Rajasthan. *Quart. Journal Geol. Min. Met. Soc. India*, v. 45, pp. 223–226.

Sarkar, S.C., 1974. Sulphide mineralisation at Sargipalli, Orissa, India. *Econ. Geol.* v. 69. pp. 206–217.

Sarkar, S.C., 1975. Some observations on the ores of a few deposits in Rajasthan. *(Abst). Sem.on recent advances in Precambrian geology and mineral deposits with special reference to Rajasthan, Udaipur*.

Sarkar, S.C., 1982. Uranium (nickel–cobalt–molybdenum) mineralisation along the Singhbhum copper belt, India and the problem of ore genesis. *Mineral. Deposita*, v. 17, pp. 257–278.

Sarkar, S.C., 1984. Geology and ore mineralisation of the Singhbhum copper–uranium belt, Eastern India. *Jadavpur University Publication*, Calcutta, p. 263.

Sarkar, S.C., 1988. Genesis and evoluation of the ore deposits in the early Precambrian greenstone belts and adjacent high–grade metamorphic terranes of peninsular India — a study in similarity and contrast. *Precamb. Research,* v. 39, pp. 107–130.

Sarkar, S.C., 1992. In search of Besshi and Kuroko type deposits in the Indian precambrians, v. 3, p. 792, *Proc. 29th Internat. Geol. Cong. Kyoto*, Japan, 1992.

Sarkar, S.C., 2000a. Crustal evolution and metallogeny in Eastern Indian Craton. *Geol. Surv. India, Spl. Pub.*, v. 55(1), pp. 169–194.

Sarkar, S.C., 2000b. Orogenic gold mineralisation: the Indian scenario. *Keynote presentation in the Symp*. pp. 11–7, *31st IGC, Rio de Janerio*, Brazil.

Sarkar, S.C., 2000c. Geological setting, characteristics, origin and evolution of the sediment-hosted sulfide ore-deposits of Rajasthan: A critique, with comments on their implications for future exploration. In M. Deb (Ed.), *Crustual evolution and metallogeny in north Indian shield*, pp. 240-292

Sarkar, S.C., 2001. Evolution of the South Indian Precambrian crust : A brief review of some existing controversies. *Geol. Surv. India, Spl. Pub.*, v. 55(2), pp. 1–14.

Sarkar, S.C., 2002. Crustal evolution and metallogeny in South Singhbhum and adjacent parts of Orissa: Results of some recent studies and their possible significance. *Indian Journal Geol.*, v. 74, pp. 27–36.

Sarkar, S.C., 2010. Gold mineralisation in India: An introduction. In and M.Deb and R.J.Goldfarb (Eds.) *Gold Metallogeny: India, and beyond.* Narosa Publ. House, New Delhi, (p. 95–122.).

Sarkar, S.C., 1964. Geological conditions of the formation of copper orebodies in Mosaboni area, Singhbhum, India. *Ph.D. thesis, Moscow State University (In Russian with a published English resume).*

Sarkar, S.C., and Banerjee, H., 1986. On mineralisation of some sulphide deposits of the Eastern Himalayas. *Geol. Surv. India, Misc. Pub.*, v. 41, pp. 127–134.

Sarkar, S.C. and Banerji, S., 2004. Carbonate-hosted lead-zinc of Zawar, Rajasthan in the context of world scenario. In M. Deb and W.D. Goodfellow (Eds.), *Sedimented-hosted lead-zinc sulfide deposits*. Narosa Pub. House, Delhi, pp. 328–349.

Sarkar, S.C. and Dasgupta, S., 1980 a. Geological setting and transformation of sulphide deposits in the northern part of the Khetri copper belt, Rajasthan, India : An outline. *Mineral. Deposita*, v. 15, pp. 117–137.

Sarkar, S.C., and Dasgupta, S. 1980 b. Ore–late intrusive contacts and the secondary hydrothermal veins at the Madan Kudan and Kolihan mines, Khetri copper belt, Rajasthan, India. *Neues Jahrb. Miner.*, v. 139, pp. 102–112.

Sarkar, S.C. and Deb, M., 1971. Dhanjori basalts and some of the related rocks. *Quart. Journal Min. Met. Soc. India*, v. 43, pp. 29–37.

Sarkar, S.C. and Deb, M., 1974. Metamorphism of the sulphides of the Singhbhum copper belt, India — the evidence from the ore fabric. *Econ. Geol.*, v. 68, pp. 1282–1193.

Sarkar, S.C. and Gupta, A. 2005. Nature and origin of the iron ores of Eastern India, a subject of scientific interest and industrial concern. *Proc. Sem. on Mineral and Energy resources of Eastern and Northeastern India. MGMI, Calcutta Branch,* Kolkata, India, pp. 79–101.

Sarkar, S.C., Bhattacharyya, P.K. and Mukherjee, A.D., 1980. Evolution of sulphide ores of Saladipura, Rajasthan, India. *Econ. Geol.*, v. 75, pp. 1152–1167.

Sarkar, S.C., Bhatacharya, S., Kabiraj, S., Das, S. and Pal, A.B., 1989. The granite–hosted early Proterozoic copper deposit of Malanjkhand, Central India – a study. In I. Haapala and Y. Kahkonen (Eds.), *(Abst.) Symp. Precamb. Granitoids: Petrogenesis, geochemistry and metallogeny.* Helsinki, p. 118.

Sarkar, S.C., Chernyshev, I.V. and Banerjee, H., 2000, Mid–Proterozoic Pb–Pb ages for some Himalayan base–metal deposits and comparison to deposits in Rajasthan, NW India. *Precamb. Research.*, v. 99, pp. 171–178.

Sarkar, S.C., Deb, M. and Roy Chowdhury, K., 1976. Sulphide ore mineralisation along Singhbhum shear zone, Bihar. *Journal Geol. Soc. Mining Japan, Spl. Issuc 3*, pp. 226–231.

Sarkar, S.C. Dwivedy, K.K. and Das, A.K., 1995. Rare earth deposits in India – an outline of their types, distribution, mineralogy, geochemistry and genesis. *Global tectonics and metallogeny*, v. 5, pp. 53–61.

Sarkar, S.C., Gupta, A. and Basu, A., 1992. North Singhbhum Proterozoic mobile belt, Eastern India: its character evaluation and metallogeny. In S.C. Sarkar (Ed.), *Metallogeny related to tectonics of Proterozoic mobile belts*, Oxford and IBH Co. Pvt. Ltd., New Delhi, India, pp. 271–305.

Sarkar, S.C. Kabiraj, S. Bhattacharyya, S. and Pal, A.B., 1996. Nature, origin and evolution of the grnitoid–hosted early Proterozoic copper–molybdenum mineralisation at Malanjkhand, Central India. *Mineral. Deposita*, v. 31, pp. 419–431.

Sarkar, S.C., Kabiraj, S., Bhattacharyya, S. and Pal, A. B., 1997 . *Reply to the comments of Changkakoti*, A. on Sarkar, *S. C. et al., 1996, in Mineral. Deposita*, 32.

Sarkar, S.C., Mukherjee, A.D. Banerjee H. And Sarkar, S. 1974. A note on the pyrite–pyrrhorite mineralisation at Saladipura, Sikar distinct, Rajasthan. *Indian. Journal Earth Sci*, v. 1, pp. 126–128.

Sarkar, S.N., 1957–1958. Stratigraphy and tectonics of the Dongargarh System, a new system in the Precambrians of Bhandara–Durg–Balaghat area, Bombay and Madhya Pradesh. *Journ. Sci. Eng. Res., I.I.T., Kharagpur, India*, v. 1(2) pp. 237–268 and v. 2(2), pp. 145–160.

Sarkar, S.N., 1968. *Pre-cambrian stratigraphy and geochronology of Peninsulas India*, Dhanbad Publications, 33p.

Sarkar, S.N., 1980. Precambrian stratigraphy and geochronology of Peninsular India: a review. *Indian Journal Earth Sci.*, v. 7(1), pp. 12–26.

Sarkar, S.N., 1994. Chronostratigraphy and tectonics of the Precambrian Dongargarh Supergroup rocks in Bhandara–Durg Region, Central India. *Indian Journ. Earth Sci.*, 21(1), pp. 19–31.

Sarkar, S.N. and Saha, A.K., 1962. A revision of Precambrian stratigraphy and tectonics of Singhbhum and adjacent region. *Quart.Journal Geol. Min. Met. Soc. India*, v. 34, pp. 97–136.

Sarkar, S.N. and Saha, A.K., 1963. On the occurrence of two intersecting Precambrian orogenic belts in Singhbhum and adjacent areas, India. *Geol. Mag.*, v. 100, pp. 69–92.

Sarkar, S.N. and Saha, A.K.,1977. The present status of Precambrian stratigraphy, tectonics and geochronology of Singhbhum–Keonjhar–Mayurbhanj region, Eastern India. *Indian Journal Earth Sci.*(S. Ray volume), pp. 37–65.

Sarkar, S.N. and Saha, A.K., 1983. Structure and tectonics of the Singhbhum–Orissa iron ore craton, Eastern India. In S. Sinha Roy (Ed.), Structure and tectonics of Precambrian rocks in India. *Recent researches in Geology*, v. 10, Hindusthan Publ. Corpn, New Delhi, pp. 1–25.

Sarkar, S.N., Basu, A. and Ghosh, J.K., 1993. Metamorphism in the Eastern part of north Singhbhum mobile Belt, India. *Journal Earth Sci.*, v. 20, pp. 173–192.

Sarkar, S.N., Gautam, K.V.V.S, and Roy, S., 1977. Structural analysis of a part of the Sausar Group rocks in Chikla, Sitekera area, Bhandara district, Maharashtra. *Jour. Geol. Soc. India*, v. 18, pp. 627–643.

Sarkar, S.N., Gerling, E.K., Polkanov, A.A. and Chukrov, F.V. 1967. Precambrian geochronology of Nagpur–Bhandara–Durg, India, *Geol. Mag.*, v. 104(6), pp. 525–549.

Sarkar, S.N., Ghosh, D., Lambert, R.J. St., 1986. Rb–Sr and lead isotope studies on the Soda granites from Mosaboni, Singhbhum copper belt, Eastern India. *Indian Journal Earth Sci.*, v. 13., pp. 101–116.

Sarkar, S.N., Gopalan, K. and Trivedi, J.R., 1981. New data on the geochronology of the Precambrians of Bhandara–Durg, Central India. *Indian Journal Earth Sci.*, v. 8(2), pp. 131–151.

Sarkar, S.N., Polkanov, A.N., Gerling, E. K., and Chukrov, R.V., 1964. Geochronologyof the Precambrian of Peninsular India, a synopsis. *Sci. Cult.*, 30, pp. 527–537.

Sarkar, S.N., Saha, A.K. and Miller, J.A.,1969. Geochronology of Precambrian rocks of Singhbhum and adjacent regions, eastern India. *Geol. Mag.*, v. 106, pp. 15–45.

Sarkar, S.N., Saha, A.K. and Sen, S. 1990. Structural pattern of Pala Lahara area, Dhenkanal district, based on ground data. *Indian Journal Earth Sci.*, v. 17, pp. 128–137.

Sarkar, S.N., Sarkar, S.S. and Ray, S.L., 1994. Geochemistry and genesis of the Dongargarh Supergroup Precambrian rocks in Bhandara–Durg region, Central India. *Indian Journal Earth Sci.*, v. 21(2), pp. 117–126.

Sarkar, S.N., Trivedi, J.R. and Gopalan K., 1986. Rb–Sr whole rock and mineral isochron age of the Tirodi gneiss, Sausar Group, Bhandara district, Maharashtra. *Jour. Geol. Soc. India*, v. 27, pp. 30–37.

Sarvothamam, H., 1995. Amphibolites of Khamam schist belt: Evidence for Precambrian Fe–tholeiite volcanism in marginal zone. *Indian Minerals*, v. 49, pp. 177–186.

Sastry, D.V.L.N., Krishna, V., Sharma, U.P., Bhatt, G., Patnaik, J.K., Kumar, M.K. and Chabra, T., 1999. Rb-Sr isochron ages on granites from basement of Bhima Basin, Gulbarga Dist., Karnataka. *Proc. 8th ISMAS Symp.*, pp. 638–641.

Sattari, P., Brenan, J.M., Horn, I. and McDonough, W.F., 2002. Experimental constraints on the sulphide–chromite–silicate melt partitioning behaviour of rhenium and platinum–group elements. *Econ. Geol.*, v. 97, pp. 385–398.

Satyanarayan, K., Sidditingam, J. and Jetty, J., 2000. Geochemistry of Archean metavolcanic rocks from Kadiri schist belt, Andhra Pradesh, India. *Gondwana Res.*, v. 3(2), pp. 235–244.

Sawarkar, A.R., 1969. Gold bearing alluvial gravels of Nilambur valley, Kozikode district, Kerala. *Indian Minerals*, v. 23, pp. 1–18.

Sawkar, R.H. and Vasudev, V.N., 2009. Gold industry in India – resources, reserves, mining, metallurgy and environment. *Jour. Geol. Soc. India,* v. 74(3), pp. 290–295.

Sawkins, F.J., 1990. Integrated tectonic-genetic model for volcanic-hosted massive sulfide deposits. *Geology*, v. 18, pp. 1061–1064.

Scaillet, B., Pichavant, M. and Roux, J., 1995. Experimental crystallisation of leucogranite magmax. *Journal of Petrology*, v. 36, pp. 663–705.

Schandl, E.S and Gorton M.P., 2004. A textural and geochemical guide to the identification of the hydrothermal monazite: Criteria for selection of samples for dating epigenetic hydrothermal ore deposits. *Econ. Geol.* v. 99, pp. 1027–1035.

Schellmann, W., 1994. Geochemical differentiation in laterite and bauxite formation. *Catena*, v. 21, pp. 131–143.

Schidlowski, M., 1998. A 3800 million year isotopic record of life from carbon in sedimentary rocks. *Nature*, v. 333, pp. 313–318.

Schidlowski, M. and Aron, P., 1992. Carbon cycle and carbon isotope record: geochemical impact of life over 3.8Ga of Earth history. In M. Schidlowski, et al (Eds.), *Early organic evolution: implications for energy resources*. Spinger-vaslay. Basin, pp. 147–175.

Schidlowski, M., Appel, P.W.U., Eichman, R. and Junge, C.E., 1979. Carbon isotope geochemistry of the 3.7 x 109 year old Isua sediments, west Greenland: Implications for the Archean carbon and oxygen cycles. *Geochim. Cosmochim. Acta*, v. 43, pp. 189–199.

Schissel, D. and Aro, P., 1992. Major Early Proterozoic sedimentary iron and manganese deposits and their tectonic settings. *Econ. Geol.* v. 87, pp. 1367–1374.

Schmidt, P. W., Prasad, V. and Raman P. K., 1983. Magnetic ages of some Indian laterites. *Palaeogeography, Palaeoecology, Palaeoclimatology*, v. 44, pp. 185–202.

Schopf, J.W. and Packer, B.M., 1987. Early Archean (3.3–3.5 billion years old) micro–fossils from Warrawoona Group, Australia. *Science*, v. 237, pp. 70–72.

Scott Smith, B. H., 2007. Lamproites and Kimberlites in India. *Journal Geological Society of India*, v. 69, pp. 443–466.

Scott Smith, B.H. and Skinner, E.M.W., 1984. A new look at Prairie Creek, Arkansas. In J. Kornprobst (Ed.), *Kimberlites and Related Rocks*, Elsevier's Development in Petrology series, No.11A, pp. 255–284.

Searle, M.P., 1983. Statigraphy structure and evolution of Tibetan Tethys zone in the Zanskar and Indus suture zone in Ladakh Himalaya. *Royal Soc. Edin.Trans. Earth Sci*, v. 73, pp. 203–217.

Searle, M.P., Waters, D.J., Dransfield, M.W. and others, 1999. Thermal and mechanical models for structurl and metamorpphic evolution of the Zhanskar High Himalaya. In McNeocaill and P.D.Ryan (Eds.),Continental Tectonics. *Geol. Soc. London Spl. Pub. 164*, pp. 139–156.

Seetharam, R., 1990. Ore mineralogy of Khobana tungsten prospect, Nagpur district, Maharashtra. *Geol. Surv. India, Spl. Pub. 28*, pp. 599–617.

Seetharam, S., 1976. Mineragraphic studies of the copper ore from Malanjkhand deposit, Balaghat district, M.P. *Geol. Surv. India, Spl. Pub. 3*, pp. 141–151.

Selvam, A. Panneer, Prasad, R.N., Dhana Raju, R. and Sinha, R.N.,1995. Rb–Sr age of the metalumious granitoids of South Khasi batholith, Meghalaya: Implications on its genesis and Pan–African activity in Northeastern India. *Geol.Soc.India, v. 46(6)*, pp 619–624.

Sen, A.K. and Guha, S., 1987. The geochemistry of the weathering sequences – present and past – in and around the Pottangi and Panchpatmali bauxite–bearing plateaus, Orissa, India. *Chem. Geol.*, v. 63, pp. 233–274.

Sen, S., 1980. Precambrian stratigraphic sequence in a part of the Aravalli Range, Rajasthan : a re–evaluation. *Quart. Journal Geol. Min. Met. Soc. India*, v. 52, pp. 67–76.

Sen, S., 1981. Proterozoic paleotectonics in the evolution of the crust and the location of metalliferous deposits, Rajasthan. *Quart. Journal Geol. Min. Met. Soc. India*, v. 53 (3–4), pp. 162–185.

Sen, S., 1983. Stratigraphy of the crystalline Precambrians of central and northern Rajasthan : A review. In S. Sinha Roy (Ed), Structure and tectonics of Precambrian rocks. *Recent Researches in Geology*, v.10, Hindusthan Publishing Corpn. Delhi, pp. 26–39.

Sen, S.K. and Bhattacharya, A,, 1990. Granulite of Satnur and Madras: A study of different behaviour of fluids. In D. Vielzeuf and Ph.Vidal(Eds.), *Granulites and crustal evolution*. Kulwer Acad. Publishers, pp. 367–384.

Sen, S.K., Bhattacharya, S. and Acharyya, A., 1995. A multi–stage pressure–temperature record in the Chilka Lake granulites: the epitome of the metamorphic evolution of the Eastern Ghats mobile belt, India? *Journal Met. Pet.*, v. 13, pp. 287–298.

Sen, S.N. and Narsinha Rao, C.H., 1967. Igneous activity in the Cuddapah basin and adjacent areas and suggestions on the palaeogeography of the basin. Proc. Symp. *Upper Mantel Project, Hyderabad, GRB and NGRI Pub.*, v. 8, pp. 261–285.

Sengor, A.M.C., 1990. A new model for the Late Mesozoic tectonic evolution of Iran and implications for Oman. In A.H.F. Robertson, M.P. Searle, and A.C. Ries (Eds.), The geology and tectonics of Oman region. *Geol. Soc. London, Spl. Pub., No. 49*, pp. 797–831.

Sengupta, N., Mukhopadhyay, D., Sengupta, P. and Hoffbauer, 2005. Tourmaline–bearing rocks in the Singhbhum shear zone, Eastern India : Evidence of boron infiltration during regional metamorphism. *Amer. Mineralogist*, v. 90, pp. 1241–1255.

Sengupta, P., Dasgupta, S., Bhattacharyya, P.K., Fukuoka, M., Chakrabarti, S. and Bhowmik, S., 1990. Petrotectonic imprints in the sapphirine granulites from Anantagiri, Eastern Ghats mobile belt, India. *Journal Pet.*, v. 31, pp. 971–996.

Sengupta, P., Karmakar, S., Dasgupta, S. And Fukuoka, M., 1991. Petrology of spinel granulites from Araku, Eastern Ghats, India, and a petrogenetic grid for sapphirine–free rocks in the system FMAS. *Journal Met.Geol.*, v. 9, pp. 451–459.

Sengupta, P., Sen, J., Dasgupta, S., Raith, M., Bhui, U.K. and Ehl, J., 1999. Ultra–high temperature metamorphism of metapelitic granulites from Kondapalle, Eastern Ghats Belt: Implications for the Indo–Antarctic correlation. *Journal Pet.*, v. 40(7), pp. 1065–1087.

Sengupta, P.R., 1972. Studies on mineralisation in the south–eastern part of the Singhbhum copper belt, Bihar. *Geol. Surv. India. Mem.101*, p. 82.

Sengupta, S. and Gangopadhyay, K.K., 1994. Preliminary investigation for wolframite around Satnala–Arhala area and extending upto Kuilapal granite body, west of Chhendapathar in parts of Bankura district, West Bengal. *Records, Geol. Surv. India*, (*Extend. Absts.of Field Season 1992–93*), pp. 114–117.

Sengupta, S. and Ghosh, S.K., 1997. The kinematic history of the Singhbhum shear zone. *Proc. Indian Acad., Sci. (Earth Planet Sci)*, v. 106, pp. 185–196.

Sengupta, S., Acharyya, S.K. and Desmeth, J.B., 1997. Geochemistry of Archean volcanic rocks from Iron Ore Supergroup, Singhbhum, Eastern India. *Proc. Ind. Acad. Sci (Earth Planet Sci.)*, v. 106, pp. 327–342.

Sengupta, S., Bandyopadhyay, P.K. and Van den Hul, H.J., 1983. Geochemistry of the Chakradharpur Granite–Gneiss Complex – A Precambrian trondhjemite body from West Singhbhum, Eastern India. *Precamb. Research*, v. 23, pp. 57–78.

Sengupta, S., Corfu, F., Mc Nutt, R.H. and Paul, D.K., 1996. Mesoarchean crustal history of the Eastern Indian Craton : Sm–Nd and U–Pb isotopic evidence. *Precamb. Research* v. 77, pp. 17–22.

Sengupta, S., Paul, D.K., Bishui, P.K.,Gupta, S.N., Chakrabarti R. and Sen, P., 1994. Geochemical and Rb–Sr isotopic study of Kuilapal and Arkasani Granophyre from the Eastern Indian craton. *Indian Minerals*, v. 48, pp. 77–88.

Sengupta, S., Paul, D.K., de Lacter, J.R., McNaughton, N.J., Bandyopadhyay, P.K. and de Smeth, J.B., 1991. Mid–Archaean evolution of Eastern Indian craton: Geochemistry and isotopic evidence from Bonai pluton. *Precamb. Research*, v. 49, pp. 23–37.

Sengupta, S., Sarkar, G., Ghosh Roy, A.K., Bhaduri, S.K., Gupta, S.N. and Mandal, A., 2000. Geochemistry and Rb–Sr geochronology of acid tuffs from the northern fringe of the Singhbhum craton and their significance in the Precambrian evolution. *Indian Minerals*, v. 54(1–2), pp. 43–56.

Sengupta, Sudipta, 1977. Deformation of pebbles in relation to associated structures in parts of Singhbhum shear zone, Eastern India. *Geol. Rundsch*., v. 66, pp. 175–192.

Sen Gupta, B., Paul, A.K. and Roy, M., 2005. A rare calcium rich uraninite from Anjangira area, Sonbhadra district. Uttar Pradesh. *Jour.Geol. Soc. India*, v. 65, pp. 296–300.

Sensarma, S. and Mukhopadhyay, D., 2003, New insight on the stratigraphy and volcanic history of Dongargarh Belt, Central India. *Gondwana Geol. Mag*., Spl. Pub., v. 7, pp. 129–136.

Seshadri, C.V., 2008. Kolar greenstone belt-geology, exploration and mining: An overview. Pre-workshop volume, *Internat Field Workshop on Gold metallogeny in India, pp. 68–72.*

Shadakshara Swamy, N., Jayananda, M. and Janardan, A.S., 1995. Geochemistry of Gundlupet Gneisses, southern Karnataka : a 2.5 Ga old reworked sialic crust. In M.Yosida, M.Santosh and A.T. Rao (Eds.), *Gondwana Res. Gr., Mem. 2*, pp. 87–97.

Shah, N., 2004. Rampura-Agucha, a remobilised SEDEX deposit, Southeastern Rajasthan, India. in M. Deb and W.D. Goodfellow (Eds.), *Sediment-hosted lead-zinc sulfide deposits*, Narosa, pp. 290–303.

Shankaran, A.V., Bhattacharyya, T.K. and Dar, K.K., 1970. Rare earth and ogher trace elements in uraninites. *Jour. Geol. Soc. India*, v. 11, pp. 205–216.

Sharma, A.K, and Kumar, B., 1999. Exploration for basemetals in Rampura-Tona-Naila-Ki Dhani area, Eastern Khetri belt, Jhunujhuna district, Rajasthan. *Records, Geol. Surv. India, v. 130(7)*, 22p.

Sharma, A.K., Shekhawat, L.S., 1999. Exploration for basemetals in Naila-Ki-Dhani, Rampura and Tonda area, Eastern Khetri belt, Jhunjhunu distric. Rajasthan. *Records, Geol. Surv. India, v. 132(7)*, pp. 19–22.

Sharma, D.S, McNaughton, N.J., Fletcher, J.R. and Groves, D.I., 2008. Timing of gold mineralisation in Dharwar craton, South India: Implications from U–Pb ages of monazite, xenotime and zircon. *Pre–workshop volume, Internat . Field Workshop on Gold Metallogeny in India*, pp. 66–67.

Sharma, K.K. and Rahman, A. 1995. Occurrence and petrogenesis of the Loda Pahar trondhjemitic gneiss from Bundelkhand craton, Central India: Remnant of an early crust. *Current Sci.*, v. 69, pp. 613 – 617.

Sharma, K.K. and Rahman, A., 2000. The early Archaean–Palaeoproterozoic crustal growth of the Bundelkhand craton, northern Indian shield. In M. Deb (Ed.): *Crustal evolution and metallogeny in the northwestern Indian Shield*, Narosa Publ. House, New Delhi, pp. 51–72.

Sharma, K.K., 1998. Geological evolution and crustal growth of the Bundelkhand craton and its relicts in the surrounding regions, N. Indian shield. In B. S. Paliwal(Ed.), *The Indian Precambrians*, pp. 33 – 43.

Sharma, K.K., 2000. Archaean–Phanerozoic crustal growth of northern Indian shield,(Abst.), *Third South Asia Geological Congress*, Lahore, Pakistan, pp. 43–44.

Sharma, M., Basu, A.R., Ray, S.L., 1994. Sm–Nd isotopic and geochemical study of Archean tonalite–amphibolite association from the eastern Indian Craton. *Contrib. Min. Pet.*, v. 117, pp. 45–55.

Sharma, M., Rai, A.K., Nagabhusana, J.C., Sinha, R.M. and Vasudeva Rao, M., 1995. Cuddapah basin and its environs as first order uranium target in the Proterozoics of India. *Explo. Res. Atomic minerals (AMD)*, v. 8, pp. 127–139.

Sharma, M., Rai, A.K., Nagabhusana, J.C., Sinha, R.M., Vasudeva Rao, M., 1995. Cuddapah Basin and its environs as first orger uranium target in the Proterozoic of India. *Expln. Res. Atomic Minerals (AMD)*, v. 8, pp. 127–139.

Sharma, R., Narayan, S. and Singh, P.N., 1996. Polymetallic mineralisation in the Tethyan sequence of Malari–Barmatiya area, Chamdu district, Uttar Pradesh. *Geol. Surv. India, Spl. Pub., v. 21*(1), pp. 515–522.

Sharma, R.K. and Kumar, H., 1969. Geology and copper occurrences of the Malanjkhand area, Balaghat district, M.P. *Indian Minerals*, v. 23, pp. 23–39.

Sharma, R.K. and Vaddadi, S., 1996. Report on lava tubes/ channels from Deccan volcanic province, Pune and Ahmadnagar district, Maharashtra. Proc. National Symp. on Deccan Flood Basalts, India. *Geol. Mag., Spl. vol. 2*, pp. 457–460

Sharma, R.S., 1977. Deformation and crystallisation history of the Precambrian rocks in north central Aravalli Mountain, Rajasthan, India. *Precamb. Research*, v. 4, pp. 133–162.

Sharma, R.S., 1988. Patterns of metamorphism in the Precambrian rocks of the Aravalli Mountain belt. In A.B. Roy (Ed), Precambrian of the Aravalli Mountain, Rajasthan, India. *Geological Society of India, Mem. 7*, pp. 33–76.

Sharma, R.S., 1995. An evolutionary model for the Precambrian crust of Rajasthan : Some petrological and geochemical considerations. *Geological Society of India, Mem. 31*, pp. 91–115.

Sharma, R.S. and Upadhyay, T.P., 1975. Multiple deformation in Precambrian rocks to the southeast of Ajmer, Rajasthan, India. *Jour.Geol. Soc. India*, v. 16, pp. 428–440.

Sharma. V.P., 1975. Note on the stratigraphic classification of the Purana rocks of the Bastar district, M.P. *Geol. Surv. India, Misc. Pub.*, v. 25(1), pp. 171–175.

Sharma, Y.C. and Nagar, R.K., 1996. Geology around Budhakedar–Ghansyalu–Chirpatyatyakhal vis–a–vis uranium resource potential, Tehri district, Garhwal Himalaya, U.P. *Geol. Surv. India, Spl. Pub.* 21(1), pp. 433–438.

Sharpe, M.R. and Irvine, T.N., 1983. Melting relations in two Bushveld chilled margin rocks and implications for the origin of chromitite. *Carneg. Inst. Washington Yearbook*, v. 82, pp. 284–288.

Shastry, C.A., 1992. Geochronology of the Precambrian rocks from Rajasthan and northern Gujarat. *Geol. Surv. India, Spl. Pub.*, v. 25, p. 96.

Shaw, R.K. and Arima, M., 1998. A corundum–quartz assemblage from the Eastern Ghats granulite belt, India: evidence for high P–T metamorphism? *Journal Met. Geol.*, v. 16., pp. 189–196.

Shaw, R.K., Arima, M., Kagami, H., Fanning, C.M., Shiraishi, K. and Motoyoshi, Y., 1997. Proterozoic events in the Eastern Ghats granulite belt, India: Evidence from Rb–Sr, Sm–Nd systematics and SHRIMP dating. *Journal Geol.*, v. 105, pp. 645 – 656.

Shaw, R.K. and Arima, M., 1996. High temperature metamorphic imprint from calcgranulites of Rayagada, Eastsern Ghats, India: implications of isobaric cooling path. *Contrib. Min. Pet.*, v. 126, pp. 169–180.

Shesadri, T.S., Chandhuri, A., Harinadha Babu and Chayapathi, N., 1981. Chitradurga Belt. *Geol .Surv. India, Mem. 112*, pp. 163–198.

Sheth, H.C.,1999a. A historical approach to continental flood basalt volcanism: insights into pre–volcanic rifting, sedimentation, and early alkaline magmatism. *Earth Planet. Sci. Lett.*, v. 168, pp. 19–26.

Sheth, H.C., 1999b. Flood basalts and large igneous provinces from deep mantle plumes: fact, fiction, and fallacy. *Tectonophysics*, v. 311, pp. 1–29.

Sheth, H.C., 2005a. From Deccan to Réunion: no trace of a mantle plume. In: Foulger, G. R., Natland, J. H., Presnall, D. C., Anderson, D. L. (Eds.), *Plates, Plumes, and Paradigms. Geol. Soc. Amer.*, Spl. Paper, v. 388, Ch. 29.

Sheth, H.C., 2005b. Were the Deccan flood basalts derived in part from ancient oceanic crust within the Indian continental lithosphere?. *Gondwana Res.*, v. 8, pp. 109–127.

Shivananda, S.R., Diwedi, K.K. and Kaul, R., 1992. Characterization of extractable uranium in Domiasiat sandstone, West Khasi Hills district, Meghalya. *Journal Geological Society of India*, v. 40(1), pp. 76–78.

Shrivastava, M. P. and Shrivastava, S.K., 1989. Geology and search for phosphorite in Gwalior basin of Madhya Pradesh. *Records, Geol. Surv. India*, v. 122(6), pp. 1–2.

Sibson, R.H., 2001. Seismogenic framework for hydrothermal transport and ore deposition. In J.P. Richards and R.M. Tosdal (Eds.), *Structural control on oregenesis. Reviews in Econ. Geol.*, v. 14, *Soc. Econ. Geol.* Inc. pp. 25–50.

Sighinolfi, G.P. and Santosh, A.M., 1976. Geochemistry of gold in Archean granulite facies terrains. *Chem. Geol.*, v. 17, pp. 113–123.

Sikka, D.B. and Nehru, C.E., 2002. Malanjkhand copper deposit, India : Is it not a porphyry type? *Jour.Geol. Soc. India*, v. 59. pp. 339–362.

Sikka, D.B., 1989. Malankkhand : Proterozoic porphyry copper deposit, M.P., India. *Jour.Geol. Soc. India*, v. 34, pp. 487–504.

Sillitoe, R.H., 1972. A plate tectonic model for the origin of porphyry copper deposits. *Econ. Geol.*, v. 67, pp. 184–197.

Sillitoe, R.H., 1991. Intrusion–related gold deposits. In R. P. Foster (Ed.). *Gold metallogeny and exploration.* Blackie and Son, Glasgow, pp. 165–209

Sillitoe, R.H., 1992. Gold and copper metallogeny of the Central Andes – past, present and future. *Econ. Geol.* v. 87, pp. 2205–2216.

Sillitoe, R.H., 1997. Characteristics and controls of the largest porphyry copper–gold and epithermal gold deposits in the circum–Pacific region. *Aust. Journal Earth Sci.*, v. 44, pp. 373–388.

Sillitoe, R.H., 2003. IOCG deposis: An Andean view. *Mineral. Deposita*, v. 38, pp. 787–812.

Silltoe, R.H., 1998. Major regional factors favouring large size, high hypogene grade– elevated gold content and supergene oxidation and enrichment of porphyry copper deposits. In T.M. Porter (Ed.), *Porphyry and hydrothermal copper and gold deposits : A global perspective. Conf. Proc. Aust. Min. Foundation*, pp. 21–34.

Sillitoe, R.H. and Burrows, D.R., 2002. New field evidence bearing on the origin of the El Laco magnetite deposit, northern Chile. *Econ. Geol.*, v. 97, pp. 1101–1109.

Sillitoe, R.H. and Thompson, J.F.H., 1998. Intrusion-related vein gold deposits: Types, tectonomagmatic setting and difficulties of distinction from orogenic gold deposits. *Resource Geology*, v. 48, pp. 237–250.

Simpson, B.M., 1987. Early silica cementation and subsequent diagenesis in arenites from four early Proterozoic iron formations in north America. *Journal Sed. Pet.*, v. 57, pp. 494–511.

Singer, D.A., 1995. World class base and precious metal deposits – a quantative analysis. *Econ. Geol.* v. 90, pp. 88–104.

Singh, A.K., Prasad,B.B. and Sinha,P.K., 1998. Integrated study for the assessment of copper mineralisation in Chhota Jamjora area, East Singhbhum district, Bihar. (Abst) *Nat. Sem. on Geosci. Advns. in Bihar, India in the Last Decade*, *Geol.Surv. India*, Patna, pp. 60–61.

Singh, B., 1995. A short note on the occurrence of gold in Sabut area, Leh district, Ladakh. *Indian Minerals*, v. 49, pp. 291–292.

Singh, G., Banerjee, D.C., Dhana Raju, R. and Saraswat, A.C., 1990. Uranium mineralisation in the Proterozoic mobile belts of India. *Exploration and Research for Atomic Minerals, AMD (India)*, v. 3, pp. 83–101.

Singh, G., Shastry, C.S., Tiwary, K.N., Sirkey, V.G. and Chatterjee, B.D., 1977. Status of exploration for atomic minerals in Purulia district, West Bengal and future possibilities. Proc. Sem. on Mineral Resources and Mineral–based Industries in Purulia district, West Bengal – Prospects and perspectives. *Mines, Metals and Fuels*, pp. 64–66.

Singh, G., Swarnakar, B.M. and Singh, R., 1995. Uncomformity proximal uranium deposits: some potential terrains in the Proterozoic of Rajasthan. *Explrn. Res. Atom. Minerals*, v. 8, pp. 73–80.

Singh, G., Singh, R., Sharma, D.K., Yadov, O.P. and Jain, R.B., 1998. Uranium and REE potential of the alkite-pyroxenite-microclinite zone of Rajasthan, India. *Expln. & Res. Atom. v. 11, pp. 1–12.*

Singh, P.P., 2005. Geology of a part of the Iron Ore Formation and associated rocks of Keonjhar. Sundergarh districts, Orissa. *Ph.D thesis, Utkal University*, 227p.

Singh, R., 2000. Uranium metallogeny in western Indian craton. In M.Deb (Ed.), *Crustal evolution and metallogeny in the north western Indian shield*, Narosa Publ. House, New Delhi, pp. 443–463.

Singh, R.N., Kumar, B. and Chaudhuri, B.K., 1998. Environment of polymetallic mineralisation in the eastern part of the Chhotanagpur Gneissic Complex with special reference to Dhawa and Biharbari deposits. (Abst.) *Proc. Nat. Sem. on Geosci. Advns. in Bihar, India in the Last Decade*, Patna, p. 68.

Singh, R.N., Saran, R.R., Kisku, S.R.,Soney Kurien, P., and Lahiri, T.C., 2005. Gold mineralisation in Singhbhum greenstone–granite terrain, Jharkhand. *Proc. Sem. Mineral and Energy Resources of Eastern and Northeastern India, MGMI–Calcutta Branch*, Kolkata, India, pp. 243–252.

Singh, S. and Singh, S.K., 2002. Petrochemical characteristics of wolframite from Balda tungsten deposit, district Sirohi, Rajasthan. *Indian Journal Geol.*, v. 74, pp. 239–248.

Singh, S.P., 1988a. Sedimentation patterns of the Proterozoic Delhi Supergroup, northeastern Rajasthan, India and their tectonic implications. *Sed.Geol.*, v. 58, pp. 79–94.

Singh, S.P., 1988b. Stratigraphy and sedimentation pattern in the Proterozoic Delhi Supergroup, northwestern India. In A.B. Ray (Ed), Precambrian Aravalli Mountain. *Geological Society of India*, *Mem. 7*, pp. 193–205.

Singh, S.P., 1998. Precambrian stratigraphy of Bihar – An overview. In B.S.Paliwal (Ed.), *The Indian Precambrian*. Scientific Publishers, Jodhpur, India, pp. 376–408.

Singh, V.K. and Verma, O.P., 1975. Controls of tungsten mineralisation, Chhendapathar, Bankura district, West Bengal. *Jour.Geol. Soc. India*, v. 16, pp. 415–427.

Singh, Y., Rai, S.D., Sinha, R.P. and Kaul, R., 1991. Lithuim pegmatites in parts of Bastar craton. *Expln. and Res. Atomic Minerals*, v. 4, pp. 93–108.

Sinha, D.K., Rai, A.K., Parihar, P.S. and Hansoti, S.K., 1998a. Khairagarh Basin, Raipur districts, Central India: its geology, geochemistry, geochronology and uranium metallogeny. *Proc. Nat. Symp.: Recent Researches on Sed. Basins*, pp. 150–171.

Sinha, D.K., Tiwari A, Verma, S.C. and Singh, R., 1998b. Geology, sedimentary environment and geochemistry of uraniferous arenite of Singhora Group, Chhattisgarh. (Abst.) *Internat. Sem. on Precambrian crust in Eastern and Central India (UNESCO–IUGS–IGCP–368), GSI,* Bhubaneswar, India., pp. 135–138.

Sinha, R.M., Shrivastave, V.K., Sarma, G.V.G. and Parthasarathi, T.N., 1995. Geologial favourability for uncomformity related uranium deposits in nothern parts of the Cuddapah Basin: evidences from Lambapur uranium occurrences, Andhra Pradesh India. *Expln. Res. Atomic Minerals (AMD)*, v. 8, pp. 111–126.

Sinha, R.P., 1999. Rare metal are rare earth pegmatites of Eastern India. In T.M. Mahadevan and R. Dhana Raju (Eds.), *Expln. and Res. Atomic Minerals*, v. 12, pp. 53–100.

Sinha, S.K., Pandalai, H.S., Panchapakesan, V. and Krishnamurthy, K.R., 1996. A note on the sulphide minerlization at Ingaldahl, Karnataka, with reference to the distribution of gold and silver in the ores. *Indian Journal Geol.*, v. 68, pp. 263–275.

Sinha Roy, S., 1984. Precambrian crustal interaction in Rajasthan, northwest India. Indian. *Journal Earth Sci., CEISM*, pp. 84–91.

Sinha Roy, S., 1985. Granite–greenstone sequence and geotectonic development of SE Rajasthan. In Proc. Symp. on Megastructures and palaeotectonics and their role as a guide to ore mineralisation. *Bull. Geol. Min. Met. Soc. India*, No. 53, pp. 115–123.

Sinha Roy, S., 1988. Proterozoic Wilson cycle in Rajasthan. In A.B. Roy (Ed), Precambrian of the Aravalli Mountain Range, Rajasthan, India. *Geological Society of India*, *Mem. 7*, pp. 95–106.

Sinha Roy, S., 1996. Stratigraphic and tectonic controls of gold mineralisation in Aravalli fold belt, Banswara district, Rajasthan. *Pre–Workshop volume on Gold Resources of India*, organized by GSI and NGRI, pp. 158–160.

Sinha Roy, S., 2004. Precambrian terranes of Rajasthan, India and their linkages with plate tectonics–controlled mineralisation types and metallogeny. In M. Deb and W.D. Goodfellow(Eds.), *Sediment-hosted lead–zinc sulphide deposits*, Narosa Publ.House, New Delhi, pp. 221–245.

Sinha Roy, S., Mohanty, M., Malhotra, G., Sharma, V.P. and Joshi, D.W., 1993. Conglomerate horizons in South-Central Rajasthan and their significance on Proterozoic stratigraphy and tectonics of the Arawalli and Delhi fold belts. *Jour. Geol. Soc. India*, v. 41, pp. 331–350.

Sinha Roy, S. and Mohanty, M., 1988. Blue schist facies metamorphism in the opliolitic mélange of the Late Proterozoic Delhi fold belt, Rajasthan, India. *Precamb. Research*, v. 42, pp. 97–105.

Sinha Roy, S., Guha, D.B. and Bhattacharya, A.K., 1992. Polymetamorphic granulite facies pelitic gneisses of the Precambrian Sandmata Complex, Rajasthan. *Indian Miner.*, v. 46, pp. 1–12.

Sinha Roy, S., Malhotra, E. and Mohanty, M., 1998. Geology of Rajasthan. *Geological Society of India*, Bangalore, p. 278.

Sinha Roy, S., Malhotra, G., and Guha, D.B., 1995. A transect across Rajasthan Precambrian terrain in relation to geology tectonic and crustal evolution of South–central Rajasthan. In S. Sinha Roy and K.R. Gupta (Eds.), Continental crust of Northwestern and Central India. *Geological Society of India*, *Mem. 31*, pp. 63–90.

Sinor, K. P.,1930. Diamond mines of Panna State in Central India. *Times India Press,* Bombay, p. 189

Siva Siddaiah, N. and Rajamani, V., 1989. The geologic setting, mineralogy, geochemistry and genesis of gold deposits of Archean Kolar schist belt, India. *Econ. Geol.*, v. 84, pp. 2155–2172.

Siva Siddhaiah, N., Hansen, G.N. and Rajamani, V., 1994. Rare earth element evidence for syngenetic origin of an Archean stratiform gold–sulphide deposit, Kolar schist belt, South India. *Econ. Geol.*, v. 89, pp. 1552–1566.

Sivaprakash, C., 1980. Mineralogy of manganese deposits of Koduru and Garbham, Andhra Pradesh, India. *Econ.Geol.*, v. 75, pp.1083–1104.

Skippen, G.B., 1974. An experimental model for low-pressure metamorphism of siliceous dolomitic marble. *Amer. Journal Sci.*, v. 274, pp. 487–509.

Skirrow, R.G. and Walshe, J.L., 2002. Reduced and oxidised Au–Cu–Bi deposits of the Tenant Creek Inlier, Australia: An integrated geologic and chemical model. *Econ. Geol.*, v. 97, pp. 1167–1202.

Slack, J.F., 1996. Tourmaline associations with hydrothermal ore deposits. In Grews, E.S. and Anovitz, L. M. (Eds), *Reviews in Mineralogy*, *Mineral Soc. Amer*, v. 33, pp. 559–643.

Sleep, N.H. and Windley,B.F.,1982. Archean plate tectonics: constraints and inferences. *Journal Geol.*, v. 90, pp. 363–379.

Smeeth, W.F., 1915. Geological map of Mysore. *Dept. Mines and Geology*, Mysore.

Smith, A.D., Lewis, C., 1999. The planet beyond the plume hypothesis. *Earth Sci. Rev.*,v. 42, pp. 135–182.

Smith, A.W., 1964. Remobilisation of sulfide orebodies. *Econ. Geol.*, v. 59, pp. 930–941.

Smith, C.B.,1983. Pb, Sr and Nd isotopic evidence for sources of Southern African Cretaceous kimberlites. *Nature*, v. 304, pp. 514–554.

Society of Economic Geologists (SEG) Newsletter, July 2009. *Exploration reviews – India* , pp. 38.

Solomon, M., Groves, D.I. and Jaques, A.J., 2000. The geology and origin of Australian mineral deposits. *Cent. Ore Depost Res.*, *University of Tasmania and Cent. Global Metallogeny, University of Western Austalia,* Perth, pp. 957

Soman, K., Santosh, M. and Golubyer, V.N., 1983. Early Palaeozoic I–type granite from central Kerala and its bearing on possible mineralisation. *Indian Journal Earth Sci.*, v. 10, op. 137–141.

Somani, O.P., Sinha, Singh, K.D.P, Babu, P.V.R. and Banerjee, D.C. ,1998. Mineralogy and geochemistry of rare metal pegmatite at Metapal, Bastar district, M.P. *Journal Atm. Min. Sci.*, v. 6, pp. 49–63.

Soni, M.K. and Jain, S.C., 2001. A review of Bundelkhand granite–greenstone belt and associated mineralisation. In *Precambrian Crustal Evolution and Mineralisation. Seminar volume of SAAEG*, Patna, pp. 25–56.

Soni, M.K., and Jha, D.K., 2001. Mahakoshal greenstone belt and associated gold mineralisation. *Geol. Surv. India, Spl. Pub. 64*, pp. 317–326.

Spear, F.S., 1993. Metamorphic phase equilibria and Pressure–Temperature–Time paths. *Monograph Min. Soc. America*, pp. 1–798.

Spencer, E., 1948. Manganese ore deposits of Jamda–Koira valley. *Trans. Mining. Geol. Met. Inst.* India (MGMI), v. 44(2).

Spencer, E. and Percival, F.G., 1952. The structure and origin of banded hematite jasper of Singhbhum, India. *Econ. Geol.*, v. 47, pp. 365–384.

Spivack, A.J., and Edmout, J.M., 1987. Boron isotopic fractionation between sea water and oceanic crust. *Geochim. Cosmochim. Acta*, v. 51, pp. 1033–1043.

Spry, P.G., 1987. The chemistry and origin of zincian spinel associated with Aggeney's Cu–Pb–Zn–Ag deposit, Namaqualand, S. Africa. *Mineral. Deposita*, v. 22, pp. 262–268.

Spry, P.G., Peter, J.M. and Slack, J.F., 2000. Meta–exhalites as exploration guides to ore. *Reviews in Econ. Geol.*, v. 11, pp. 163–201.

Sreenivasa Rao, T., 1985. A note on the stratigraphy of the upper Precambrian sediments around Ramagudam R.S., Andhra Pradesh. *Indian Minerals.*, v. 39(3), pp. 9–11.

Sreenivasa Rao, T., 2001. The Purana Formations of Godavari Valley – A conspectus. Proc. M.S.Krishnan Comm. Nat. Sem. on Progress in Precambrian Geology of India, Kolkata (1998). *Geol. Surv. India, Spl. Pub.* v. 55, pp. 67–76.

Sreenivasa Rao, T.,1987. The Pakhal basin – A perspective. *Geological Society of India, Mem.* 6, pp. 161–187.

Sreeramachandra Rao, K., Roop Kumar, D., Jairam M.S., Bhattacharjee, S., Ananda Murthy, S. and Krishna Rao, P.V., 2001. Interpretation of geological characteristics of Dona gold deposit, Jonnagiri schist belt, Kurnool district, Andhra Pradesh. *Geol. Surv. India, Spl. Pub.*, 58, pp. 217–231.

Srikantappa, C., 1993. Nature, composition and evolution of deep crustal fluids in parts of Karnataka and Tamilnadu. (Abst.) *Sem. on Fluid flow and the processes in the Earths Crust. NGRI, Hyderabad*, p. 57.

Srikantappa, C., 2001. Composition and source of deep crustal fluids and their role in the evolution of charmockitic granulites of South India. *Spec. Publn., GSI*, No 55, pp. 15–30.

Srikantia, S.V. and Bhargava, O.N., 1982. Precambrian carbonate belts of Lesser Himalaya : their geology, correlation, sedimentation and palaeogeography. *Recent researches in geology*, v. 8, Hindusthan Publ. Corp., Delhi, pp. 521–581.

Srikantia, S.V., 1995. Geology of Hutli–Maski greenstone belt. In L.C. Curtis and B.P. Radhakrishna (Eds), *Hutti Gold Mine into the 21st century. Mineral Resource Series , Geological Society of India*, pp. 8–26.

Srinivasa Rao, K., Srinivasa Rao, T. and Rajagopalan Nair, S.,1979. Stratigraphy of the Upper Precambrian Albaka belt, east of Godavari river in Andhra Pradesh and Madhya Pradesh. *Jour. Geol. Soc. India*, v. 20, pp. 205–213.

Srinivasa Sarma, D., Mc. Nanghton, N. J. Fletcher, I. R. and Groves, D. I. 2008. Timing of gold mineralisation in the Hutti gold deposit, Dharwar craton, South India. *Econ. Geol.*, v. 103, pp. 1715–1727.

Srinivasan, K.N. and Roopkumar, D., 1995. Orbicular structure from diorite body within the granitoid complex of Nellore schist belt. *Jour. Geol. Soc. India*, v. 45, pp. 277–283.

Srinivasan, K.N., 1990. Litho–stratigraphy of Veligallu schist belt, Cuddapah, Ananthapur and Chittor districts, Andhra Pradesh. *Records, Geol. Surv. India*, v. 123(5), pp. 317–318.

Srinivasan, K.N., Krishnappa, T., Katli, P.M. and Roopkumar, D., 1994. Geology of parts of Nellore schist belt, Andhra Pradesh. *Records, Geol. Surv. India*. v. 127(5), pp. 58–59.

Srinivasan, R., Naqvi, S.M., Udai Raj, B., Subba Rao, D.V., Balaram, V. and Gneneshwar Rao, T., 1989. Geochemistry of Archean greywacks from the Northwestern part of the Chitradurga schist belt, Dharwar craton, south India–evidence for granitoid upper crust in the Archean. *Jour. Geol. Soc. Ind.*, v. 34, pp. 505–516.

Srinivasan, T.V. and Odam, A.L., 1982. Zircon geochronology of Berach Granite of Chittorgarh, Rajasthan. *Journal Geological Society of India*, v. 23, pp. 575–677.

Srikantappa, C., Friend, C.R.L. and Janardhan, A.S., 1980. Petrochemical studies in chromite from Sinduvalli, Karnataka, India. *Jour. Geol. Soc. India*, v.21, pp. 473–483.

Srivastava, D.C., 1985. The deformation and strain analysis of the Gandemara conglomerate from Archaean Iron Ore Group rocks, district Singhbhum, Bihar. *Jour. Geol. Soc. India*, v. 26, pp. 233–244.

Srivastava, P., and Mitra, G., 1994. Thrust geometries and deep structure of the outer and Lesser Himalaya, Kumaon and Garhwal (India): Implications for the evolution of the Himalayan fold–and–thrust belt. *Tectonics*, v. 13, pp. 89–109.

Srivastava, P.K., Khandelwal, M.K., Deshpathi, Kirwadkar, M.P. and Dwivedy, K.K., 1990. Uranium mineralisation in the Kui-Chitrasani fault zone in Banaskantha district, Gujarat and Sirohi district, Rajasthan, India. *Expl. Res. At. Min.*, v. 3, pp. 45–56.

Srivastava, R.K. and Singh, K.K., 2003. Preccambrian mafic magmatism in Southern Bastar, Central India: Present status and future perspective. *Gondwana Geol. Mag., Spl.v. 7*, pp. 177–191.

Srivastava, R.K., 1988. Magmatism in the Aravalli Range and its environs. In: A.B. Roy (Ed), Precambrian of the Aravalli Mountain, Rajasthan, India. *Geological Society of India, Mem. 7*, pp. 78–93.

Srivastava, R.N., 1989. Bijawar phosphorites at Sonrai – geology, sedimentation, exploration strategy and origin. In D.M. Banerjee(Ed.), Phosphorites in India. *Geological Society of India, Mem. 13*, pp. 47–59.

Srivastava, S.K., Hamilton, S. and Verma, M.B., 1996. Significance of rounded uraninite and spherulites of uraniferous carbonaceous matter from umra, Udaipur district, Rajasthan, India. *Jour. Atom. Min. Sci.* v. 4, pp. 9–13.

Stacey, J.S. and Kramers, J.D., 1975. Approximation of terrestrial lead isotope evolution by a two–stage model. *Earth Planet Sci. Lett.*, v. 26, pp. 207–221.

Stachel, T.Harris., J.W. Brey, G. P., and Joswig, W., 2000. Kankan diamonds (Guinea) II: Lower mantle inclusion parageneses. *Contrib. Min. Pet.*, v. 140, pp. 16–27.

Stein, H., Hannah, J., Zimmerman, A., Markey, R., 2006. Mineralisation and deformation of the Malanjkhand terrane (2490–2440 Ma) along the southern margin of the Central Indian Tectonic Zone. *Mineral. Deposita*, v. 40, p. 755–765 (comment on the paper by Panigrahi et al., 2004).

Stein, H.J., Hanna, J.L., Zimmerman, A., Markey, R.J., Sarkar, S.C. and Pal, A.B., 2004. A 2.5 Ga porphyry Cu–Mo–Au deposit at Malanjkhand, central India : implications for late Archean continental assembly. *Precamb. Research*, v. 134, pp. 189–226.

Stein, H.J., Markey, R.J., Morgan, J.W., Hanna, J.L. and Scherstein, A., 2001. The remarkable Re–Os chronometer in molybdenite : How and why it works? *Terra Nova*, v. 13, pp. 479–486.

Stephenson, B.J., Waters, D.J. and Searle, P., 2000. Inverted metamorphism and the Main Central Thrust field relations and thermobarometric constraints from Kistwar Window, N.W. Indian Himalaya. *Jour. Met. Geol.*, v. 18, pp. 571–590.

Steven, N.M. and Moore, J.M., 1995. Tourmalinite mineralisation in the Late Proterozoic Kuiseb Formation of the Damara Orogen, Central Namibia : Evidence for a replacement origin. *Econ. Geol.*, v. 90, pp. 1098–1117.

Stoehr, E., 1870. The copper mines of Singhbhum. *Records, Geol. Surv. India*, v. 3, pp. 86–93.

Stowe, C.W., 1994. Compositions and tectonic settings of chromite deposits through time. *Econ. Geol.*, v. 89, pp. 528–546.

Straczek, J.A. and Srikantan, B., 1966. The geology of Zawar lead–zinc area, Rajasthan, India. *Geol. Surv. India, Mem. 92*, 85 p.

Straczek, J.A., Narayanswami, S., Subramanyam, M.R., Sukla, K.D., Vemban, N.A., Chakravarti, S.C. and Venkatesh, V., 1956. Manganese ore deposits of Madhya Pradesh, India. 20th Internat. *Geol., Cong. Symp. on manganese*, v. 4, pp. 63–96.

Strauss, H., 1993. The sulfur isotope record of Precambrian sulfates: new data and a critical evaluation of the existing record. *Precamb. Research*, v. 63, pp. 225–246.

Streckeisen, A.L., 1976. To each plutonic rock its proper name. *Earth Sci. Review*, v. 12, pp. 1–33.

Strens, M.R., Cann, D.L. and Cann, J.R., 1987. A thermal balance model of the formation of the sedimentary exhalative lead–zinc deposits. *Econ. Geol.*, v. 82, pp. 1192–1203.

Strong, D.F., 1988. A review and model for granite–related mineral deposits. In R.P. Taylor and D.F. Strong (Eds.), Recent advances in the geology of granite–related mineral deposits. *Canadian Inst. Min. Metal.*, Spl. vol. 39, pp. 424–445.

Subba Raju, M., Sreenivasa Rao, T., Setty, D.N. and Reddy, B.S.R., 1978. Recent advances in our knowledge of the Pakhal Supergroup with special reference to the central part of Godavari Valley. *Records, Geol. Surv. India*, v. 110(2), pp. 39–59.

Subba Rao, M.V., Rama Rao, P. and Divakara Rao, V.,1998. Advent of Proterozoic: A trigger for extensive intracrustal processes in the South Indian shield. *Gondwana Res.*,v. 1, pp. 275–283.

Subrahmaniam, V.,Viladkar, S.G. and Upendra, R .,1978. Carbonatite alkali complexof Samalpatti, Dharampuri district,Tamilnadu. *Jour. Geol.Soc. India*, v. 19, pp. 206–216.

Subrahmanian, K.S. and Mani, G., 1981. Genetic and geomorphic aspects of laterites on high and low landforms in parts of Tamilnadu, India. In *Lateritisation Processes*. Oxford and IBH Publ. Co. Pvt. Ltd., New Delhi, pp. 237–245.

Subrahmanyam, A.V., Anil Kumar, Despati, V., Deshmukh, R.D. and Viswanathan, G., 2005. Discovery of diamonds in beach placers of East Coast, Andhra Pradesh, *Current Sci.*, v. 88, pp. 1227–1228.

Subrahmanyam, A.V., Gupta, K.R., Basu, A.N. and Balakrishnan, S.P., 1997. Volcanic tuff in Lower Mahadek Formation of Meghalaya Plateau : implications on uranium source. *Atomic Mineral Sci.*, v. 5, pp. 73–79.

Subrahmanyan, C. and Verma, R.K., 1986. Gravity field, structure and tectonics of the Eastern Ghats. *Tectonophysics*, v. 126, pp. 195–212.

Sugden, T.J., Deb, M. and Windley, B.F., 1990. The tectonic setting of mineralisation in Proterozoic Aravalli–Delhi orogenic belt, N.W. India. In *Developments in Precambrian Geology, Elsevier, Amsterdam*, pp. 367–390.

Sundaram, S.M., Sinha, P.A., Ravindra Babu, B. and Muthi, V.T., 1989. Uranium mineralisation in Vempalle dolomite and Pulivendla conglomerate/quartzite of Cuddapah basin, Andhra Pradesh. *Indian Minerals*, v. 43 (2), pp. 98–103.

Sunil Kumar, T.S., Krishna Rao, N., Palrecha, M.M. Parthasarathy, R., Shah, V.L. and Sinha, K.K., 1998. Mineralogical and geochemical characteristics of the basal quartz pebble conglomerate of Dhanjori Group, Singhbhum craton, India, and their significance. *Jour. Geol. Soc. India* , v. 51, pp. 761–776.

Swaminath, J., Ramakrishnan, M. and Viswanathan, M.N., 1976. Dharwar stratigraphic model and Karnataka craton evolution. *Records, Geol. Surv. India*, v. 107(2), pp. 149–175.

Sychanthavong, S.P.H. and Desai, S.D., 1977. Proto–plate tectonics controlling the Precambrian deformations and metallogenic epochs in Northwestern India. *Mineral Sci. Eng.*, v. 9, pp. 218–236.

Sychanthavong, S.P.H. and Mehr, S.S., 1984. Proto–platetectonic – the energctic model for structural metamorphic and igneous evolution of the Precambrian rocks, N.W. Peninsular India. *Geol. Surv. India, Spl. Pub.*, v. 12, pp. 419–458.

Tahirkhali, R.A.K., Mathewner, M., Proust, F. and Tapponnier, P., 1979. The India–Eurasia suture zone in northern Pakistan: Synthesis and interpretation of recent data in plate scale. In A. Farah and K. A. DeJong (Eds.), Geodynamics of Pakistan. *Geol. Surv. Pakistan*, pp. 125–130.

Talapatra, A., 1968. Sulfide mineralisation associated with migmatisation in the southeastern part of the Singhbhum shear zone, Bihar, India. *Econ. Geol.*, v. 63, pp. 156–165.

Talapatra, A., Sarker, P. and Bandyopadhyay, K.C., 1995. Rare earth and rare metal–bearing pegmatitics along some lineaments, north of Purulia, West Bengal. *Indian Journal Earth Sci.*, v. 22, pp. 13–20.

Talluri, J.K., Pandalai, H.S. and Jadhav, G.N., 2000. Fluid chemistry and depositional mechanism of the epigenetic discordant ores of the Proterozoic, carbonate-hosted Zawar mala Pb-Zn deposit, Udaipur district, India, *Econ. Geol.*, v. 95, pp. 1505–1525.

Talusam, R.V., 2001. Possible Carlin–type disseminated gold mineralisation in the Mahakoshal fold felt, Central India. *Ore Geol. Rev.*, v. 17, pp. 241–247.

Tardy, Y.,1997. Petrology of the laterites and tropical soils. *Oxford and IBH Publ. Co. Pvt. Ltd.*, New, Delhi, 408p.

Tardy, Y. and Roquin, C., 1998. Derive des continents laterites et paleoclimates tropicaux, France. *BRGM*, 472p.

Taylor, D., Dalstra, H.J., Harding, A.E., Broadbent, G.C. and Barley, M.E., 2001. Genesis of high grade hematite bodies of the Hamersley Province, Western Australia. *Econ. Geol.*, v. 96, pp. 837–873.

Taylor, P.N., Chadwick, B., Moorbath, S., Ramakrishnan, M., and Viswanatha, M.N., 1984. Petrography, chemistry and isotopic ages of Peninsular Gneisses, Dharwar acid volcanic rocks and Chitradurga Granite with respect to the later Archean evolution of Karnataka craton, southern India. *Precamb. Research,* v. 23, pp.349–375.

Taylor, R.P. and Fryer, B.J., 1982. Rare earth element geochemistry as an aid to interpreting hydrothermal ore deposits. In: A.M. Evans (Ed.), *Mineralisation associated with acid magmatism*. John Wiley, New York, pp. 357–365.

Terziev, G.I., 1971. Geological exploration at the Mamandur copper-lead-zinc deposit. *UNDP Project Report*, 20p.

Tewari, D.N., 1973. Nesquehonite – a possible precursor in the origin of Himalayan magnesite deposits. *Him. Geol.*, v. 3, pp. 94–102.

Thayer, T.P., 1976. Metallogenic contrast in the plutonic and volcanic rocks of the ophiolite assemblage. *Min. Assocn. Canada, Spl. Paper 14*, pp. 211–219.

Thayer, T.P., 1964. Principal features and origin of podiform chromite deposits and some observations on the Guleman–Soridag district, Turkey. *Econ. Geol.*, v. 59, pp. 1497–1524.

Theodore, T.G. and Menzie, W.D., 1984. Fluorine deficient porphyry molybdenum deposits in the western North American Cordillera. In T.V. Janelidze and T.V. Tvalchrelidze (Eds.), *Proc. 6th Quad. IAGOD Symp.*, Tbilisi. Schweiz. Verlag., Stuttgart, pp. 463–470.

Thompson, J.B. and Norton, S.A., 1968. Paleozoic regional metamorphism in New England and adjacent areas. In Zen E–Au et al (Eds.), *Studies of Appalacian Geology*. Interscience Publishers, John Wiley, New York.

Thompson, J.F.H., Sillitoe, R.H., Bakar, T.L. and Mortensen, J. R., 1999. Intrusion-related gold deposit associated with tungsten–tin provinces. *Mineral. Deposita*, v. 354, pp. 323–334.

Thorat, P.K., 1996. Occurrence of lava channels and tubes in the western part of Deccan volcanic province, Proc. National Symp. on Deccan Flood Basalts, India. *Gondwana Geol. Mag., Spl. Pub.*, v. 2, pp. 449–456.

Thorat, P.K., Natarajan, A., Guha, K. and Chandra, S., 1990. Stratigraphy and sedimentation of the Precambrians in parts of Bilaspur and Rajnandgaon districts, Madhya Pradesh, Precambian of Central India. *Geol. Surv. India, Spl. Pub.*, 28, pp. 167–180.

Tiwari, A., Deb, M. and Cook, N.J., 1998. Use of pyrite microfabric as a key to tectonothermal evolution of massive sulfide deposits – and example from Deri, southern Rajasthan, India. *Mineral. Mag.*, v. 62, pp. 197–212.

Tiwari, R.C. and Casshyap, S.M.,1994. Mesozoic tectonic events including rifting in Peninsular India and their bearing on Gondwana sedimentation. Proc. 9th Internat.Gondwana Symp., Hyderabad, pp. 865–877.

Tiwary, A., 2000. Quantitative estimation of hydrothermal alteration in VMS-type deposits: examples from the western part of South Delhi fold belt, Rajasthan. In M. Deb (Ed.), *Crustal evolution and metallogeny in Northwestern Indian Shield*. Narosa, New Delhi, pp. 364–394.

Tiwary, A. and Deb, M., 1997. Geochemistry of hydrothermal alteration at the Deri massive sulphide deposit, Sirohi district, Rajasthan, NW India. *Journal Geochem. Expln.*, v. 59, pp. 99–121.

Tiwary, D.N., 1973. Nesquehonite – a possible precursor in the origin of Himalayan magnesite deposits. *Himalayan Geology*, v. 3, pp. 94–103.

Tobisch, O.T. and Kollerson, K.D., Bhattacharyya, T., and Mukhopadhyay, D., 1994. Structural relationship and Sm–Nd isotope systematics of polymetamorphic granitic gneisses and granitic rocks from central Rajasthan, India: Implications for the evolution of the Aravalli craton. *Precamb. Research*, v. 65, pp. 319–339.

Tosdal, R.M. and Richards, J.P., 2001. Magmatic and structural controls on the development of porphyry Cu ± Mo ± Au deposits. *Soc. Econ. Geol., Rev*. v. 14, pp. 157–181.

Tourn, S.M., Hermann, C.J., Ametrano, S. and de Brodtkorb, M.K., 2004. Tourmalines from the Eastern Sierra Pampeanas, Argentina. *Ore Geology Rev.*, v. 24, pp. 229–240.

Trendall, A.F., 2002. The significance of iron–formation in the Precambrian stratigraphic record. *Spl. Pub. Int. Assn. Sediment.*, v. 33, pp. 33–66.

Trendall, A.F., de Lacter, J.R., Nelson, D.R. and Bhaskar Rao, Y.J., 1997 b. Further zircon U–Pb age data for the Deginkatte Formation, Dharwar Supergroup, Karnataka craton. *Jour. Geol. Soc. India*, v. 50, pp. 25–30.

Trendall, A.F., de Lacter, J.R., Nelson, D.R. and Mukhopadhyay, D., 1997a. A precise zircon U–Pb age for the base of BIF of the Mulangiri Formation (Bababudan Group, Dharwar Supergroup) of Karnataka craton. *Jour. Geol. Soc. India*, v. 50, pp. 161–170.

Tripathi, C., 1979. The Malanjkhand-Taregaon copper deposit. Geol. Surv. India Misc. Publn. No 34(II), pp. 161–168.

Tripathi, C., 1983. Malanjkhand copper deposit. *Mining Magazine*, pp. 334–341.

Tripathi, C. and Murti, K.S., 1981, Search for source rock of alluvial diamonds in the Mahanadi Valley, Symp. on Vindhyan of Central India. *Geol. Surv. India, Misc. Pub.*, v. 50, pp. 205 –212.

Tripathi, C., Ghosh, P.K., Thambi, P.I., Rao. T.V. and Chandra, S., 1981. Elucidation of stratigraphy and structure of Chilpi Group. *Geol. Surv. India, Spl. Pub.*, v. 3, pp. 17–30.

Tripathi, Shubham, 2008. Geological characterization of the carbonate hosted polymetallic prospect at Imalia, Mahakoshal belt, Central India. *Unpub. Ph.D. Thesis, University of Delhi.*

Tripathy, C. and Kaura, S.C., 1982. Polymetallic sulphide mineralisation in Ranga Valley, Subansiri district, Arunachal Pradesh. *Indian Minerals*, v. 35, pp. 1–13.

Tsikos, H., Beukes, N.J., Moore, J.M. and Harris, C., 2003. Deposition, diagenesis and secondary enrichment of metals in the Palaeoproterozoic Hotazel iron formation, Kalahari manganese field, South Africa. *Econ. Geol.*, v. 98. pp. 1449–1462.

Tugarinov, A.I. and Voitkevitch, G.V., 1970. *Dokembrskaiya geokhronologia* materikov. Nedra, Moscow, 430p.

Turekian, K.K., and Wedepohl, K.H., 1961. Distribution of elements in some major units of the Earth. *Geol. Soc. Amer. Bull.*, v. 72, pp. 175–192.

Udai Raj, B. and Naqvi, S.M. 1995. Relics of sedimentary precursors in Archean gneisses – Melukote Paragneiss: An example from Dharwar craton, India. *Jour. Geol. Soc. India*, v. 46, pp. 497–520.

Udas, G.R. and Mahadevan, T.M., 1974. Controls and genesis of uranium mineralisation in some geological environments in India. *Formation of uranium ore deposits*, IAEA, Vienna, pp. 425–436.

Udas, G.R., 1986. Current trends in exploration for uranium in the Himalaya and future possibilities. *Geol. Surv. India, Misc. Pub. No.* 41(6), pp. 429–457.

Ulrich, T., Giinther, D. and Heinrich, C.A., 2001. The evolution of a Porphyry Cu-Au deposit, based on LA-ICP_ MS analysis of fluid inclusions: Bajo de la Alumbrera, Argentina. *Econ. Geol.*, v. 96, pp. 1743–1774.

Umamaheswar, K., Basu, H., Patnaik, J.K., Azam Ali, M., Banerjee, D.C., 2001. Uranium mineralisation in the Mesoproterozoic quartzites of Cuddapah basin in Gandi area, Cuddapah district, Andhra Pradesh: A new exploration target for uranium. *Jour. Geol. Soc. India*, v. 57, pp. 405–410.

United Nations, 1991, *Atlas of mineral resources of ESCAP region*, v.8, pp. 1–56.

Unnikrishnan–Warrier, C, Santosh, M. and Yoshida, M., 1995. First report of the Pan–African Sm–Nd and Rb–Sr mineral isochron age from regional charnockite of southern India. *Geol. Mag.*, v. 132, pp. 253–260.

Unrug, R., 1996. The assembly of Gondwanaland. *Episodes*, v. 19, pp. 11–20

Upadhyay, R., Sharma, B.L. Jr., Sharma, B.L.S. and Roy, A.B., 1992. Remnants of greenstone sequence from the Archean rocks of Rajasthan. *Current Sci.*, v. 63, pp. 87–92.

Upreti, B.N. and Le Fort, P., 1999. Lesser Himalayan crystalline nappes of Nepal: Problems of their origin. In MacFarlane, A., Sorkhabi, R.B. and Quade, J. (Eds), Himalayas and Tibet: Mountain roots to mountain tops. *Geol. Soc. Amer. Spl Paper 328*, pp. 225–238.

Valdiya, K.S., 1968. Origin of the magnesite deposits of southern Pithoragarh, Kumaon Himalaya, India. *Econ. Geol.* v. 63, pp. 924–934.

Valdiya, K.S., 1970. Simla slates, the Precambrian flysch of the lesser Himalaya: Its turbidites, sedimentary structures and paleocurrents. *Geol. Soc. Amer. Bull.*, v. 81, pp. 451–468.

Valdiya, K.S., 1976. Himalayan transverse faults and folds and their parallelism with subsurface structures of North Indian plains. *Tectonophysics*, v. 32, pp. 353–386.

Valdiya, K.S., 1987. Trans–Himadri thrust and domal upwarps immediately south of collison zone and tectonic implication, *Current Sci.*, v. 56, pp. 200–209.

Valdiya, K.S., 1995. Proterozoic sedimentation and Pan–African geodynamic development in the Himalaya. *Precamb. Research*, v. 74, pp. 35–55.

Valdiya, K.S., Joshi, D.D., Sharma, P.K. Dey, P. 1992. Active Himalayan frontal fault, Main Boundary Thrust and Ramgarh Thrust in southern Kumaon. *Jour. Geol. Soc. India*, v. 40, pp. 509–528.

Valeton, I., 1972. Bauxites. *Elsevier*, Amsterdam, pp. 106–109.

Valeton, I., 1981. Bauxites on peneplaned metamorphic and magmatic rocks on detrital sediments and on karst topography, their similarities and contrasts of genesis. In *Lateritisation Processes. Oxford and IBH Publishing Co. Pvt. Ltd.,* New Delhi, pp. 15–23.

Van Breemen, O., Bowes, D. R., Bhattacharjee, C.C. and Choudhury, P.K.,1989. Late Proterozoic–Early Palaeozoic Rb–Sr WR and mineral ages for granite and pegmatite, Goalparas, Assam, India. *Jour. Geol. Soc. India*, v. 33, pp 89–92.

Varadan, V.K.S., Narasimhan, M. and Madhukara, N., 1971. Results of proving operations in Dariba copper deposit, Alwar district, Rajasthan. *Geol. Surv. India, Misc. Pub. No. 16*(I), pp. 497–505.

Varadarajan, S., 1970. Emplacement of chromite-bearing ultramafic rocks, Mysore State India. In: *Proc-Second Symp. on Upp. Mantle,* Hyderabad, pp. 441–454.

Vasudev, V.N. and Srinivasan, R., 1979. Vanadium-bearing titaniferous magnetite deposits of Karnataka, India. *Jour. Geol. Soc. India*, v. 20, pp. 170–178.

Vasudevan, D. and Rao, T.M., 1975. The high grade schistose rocks of Nellore schist belt, Andhra Pradesh and their geologic evolution. *Indian Minerals*, v. 16, pp. 43–47.

Vasudevan, D., 1989. An Early Precambrian volcanogenic banded barite deposit near Vinjamar, Nellore district, Andhra Pradesh, India. *Jour. Geol. Soc. India*, v.33, pp. 444–446.

Vaughan, D.J. and Craig, J.R., 1978. Mineral chemistry of metal sulphides. *Camb. Univ. Press,* Cambridge. p. 493.

Veevers, J.J. ,1993. Gondwana facies of the Pangean supersequence: A review. In R.H.Findlay, R. Unrug, B. R. Banks and J.J.Veevers (Eds), *Gondwana Eight*, Balkama, Rotterdam., pp. 513–520.

Veevers, J.J., 2004. Gondwnaland from 650–500 assembly through 320 Ma merger in Pangea to 185–100 Ma break up: supercontinent tectonics via stratigraphy and radiometric dating. *Earth Science Rev.*, v. 68 pp. 1– 132.

Veizer, J.,1976. 87Sr/86 Sr evolution of sea-water during geologic history and its significance as an index of crustal evolution. In Windley,B.F.(Ed), *Early history of the Earth*. John Wiley, New York, pp. 569–578.

Venkat Rao, K., Srirama, B.V. and Ramasastry, P., 1990. A geophysical appraisal of Mahakoshal Group of upper Narmada Valley. In Precambrian of Central India, *Geol. Surv. India, Spl. Pub., v. 28*, pp. 99–117.

Venkata Dasu, S.P., Ramakrishnan, M. and Mahabales B., 1991. Sargur–Dharwar relationship around komalüte–rich Jayachamarajpura greenstone belt in Karnataka. *Jour. Geol. Soc. India* , v. 38, pp. 579–592.

Venkatachala, B.S., Shukla, M., Sharma, M., Naqvi, S.M., Srinivasan, R. and Udairaj, B., 1990. Archean microbiota from Donamalai Formation, Dharwar Supergroup, India. *Precamb. Research*, v. 47, pp. 27–34.

Venkatesan, T.R., Pande, K., Gopalan, K., 1993. Did Deccan volcanism pre–date theCretaceous / Tertiary transition?. *Earth Planet. Sci. Lett*., v. 119, pp. 181–189.

Verma, P.K., 1996. Evolution and age of the Great Boundary Fault of Rajasthan. *Geological Society of India, Mem.* 36, pp. 197–212.

Verma, R.G. and Golani, P.R., 1999. Investigation for gold in Bhukia–Jagpura area, Banswara district, Rajasthan (SE Block: Verma; E.Block: Golani). *Records, Geol. Surv. India, v. 130(7)*, pp. 23–24.

Verma, R.K. and Prasad,S.N., 1974. Palaeomagnetic study and chemistry of Newer Dolerites from Singhbhum, Bihar, India. *Candian Journal Earth Sci.*, v. 11, pp. 1043–1054.

Verma, R.K., Sharma, A.U.S. and Mukhopadhyay, M., 1984. Gravity field over Singhbhum : its relationship to geology and tectonic history. *Tectonophysics*, v. 106, pp. 86–107.

Vernon, 1975. Deformation and recrystallisation of a plagioclase grain. *Amer. Mineralogist*, v. 60, pp. 884–888.

Vidal, P. Cocherie, A., LeFort, P., 1982. Geochemical investigation of the origin of the Manaslu leucogranite (Himalaya, Nepal), *Geochim Cosmochim Acta*, v. 46, pp. 2279–2292.

Viljoen, R.P., 2006. Geological comparison between India and southern Africa: Implications for diamond exploration. *Jour. Geol. Soc. India* , v. 67, pp. 432–441.

Viljoen, R.P., Minnitt, R.C.A. and Viljoen, M.J., 1986. Porphyry copper–molybdenim mineralisation at the Lorelci, south west Africa/Nambibia. In: C.R. Anhaeusser and S. Mask (Eds.), *Mineral deposits of Southern Africa. Geol. Soc. South Africa*, Johannesberg, pp. 1559–1565.

Vinogradov, A.P., Tugarinov, A.I., Zhikov, C., Stapnikova, N., Bibikova, E. and Khores, K., 1964. *Geochronology of Indian Precambrian. 22nd Int. Geol. Congress*, New Delhi, v. 10, pp. 553–567.

Vinogradov, V. I., Reimer, T.O., Leites, A.M. and Smelov, S.D., 1976. The oldest sulfates in Archean formations of the South Africa and Aldan shield. *Lithol, Min. Res.*, v.11, pp. 407–420.

Vishwakarma, R.K., 1996. 1.66 Ga–old metamorphosed Pb–Cu deposit in Sargipali (Eastern India): Manifesations of tidal flat environment and sedex–type genesis. *Precamb. Research*, v. 77, pp. 117–130.

Vishwakarma, R.K., 2001. Isolopic characters of the Proterozoic Malanjkhand copper deposit, Central India: Implications for exhalation activity. *Journal Appld. Geochemistry*, v. 3(2), pp. 91–103.

Vishwakarma, R.K. and Ulabhaje, A.V., 1991. Sargipali galena – unusual lead isotope date from Eastern zone. *Miner. Deposita*, v. 26, pp. 26–29.

Vishwanathan, S., 1978. Early Archean volcanism and its economic significance (Abst.). *Third Indo–Soviet Symposium on Earth Sciences: Archean Geochemistry and comparative study on Deccan traps and Siberian traps.*

Viswanath, R.V., Roy, M.K., Pandit, S.A. and Narayan Das, G.R., 1988. Uranium mineralisation in quartz pebble conglomerates of Dharwar Supergroup, Karnataka. *Geological Society of India, Mem.* 9, pp. 33–41.

Viswanatha, M.N. and Ramakrishnan, M., 1981. Bababudan belt. *Geological Society of India, Mem. 112*, pp. 91–114.

Viswantha, M.N., Ramakrishnan, M. and Swaminath J., 1982. Angular unconformity between Sargur and Dharwar supracrustals in Sigegudda, Karnataka craton, South India. *Jour. Geol. Soc. India*, v. 23, pp. 85–89.

Viswanathan, S., 1974. Contemporary trends in geochemical studies of early Precambrian granite-greenstone complexes. *Jour. Geol. Soc. India*, v.15, pp. 347–379.

Viswanathiah, M.N., 1977. Lithostratigraphy of Kaladgi and Badami groups, Karnataka. In A.S. Janardhan (Ed), Proc. Sem. on Kaladgi, Badami, Bhima and Cuddapah sediments. *Indian Minerals*, v. 18, pp. 122–132.

Vohra, C.P., Dasgupta, S, Paul, D.K., Bishoi, P.K., Gupta, S.N. and Guha, S., 1991. Rb–Sr chronology and petrochemistry of granitoids from south eastern part of Singhbhum craton, Orissa. *Jour. Geol. Soc. India*, v. 38, pp. 5–22.

Vokes, F.M., 1973. "Ball texture" in sulphide ores. *Geologiska Fören. Stockholm Förhand*, v. 95, pp. 403–406.

Volpe, A.M. and Macdougall, J.D., 1990. Geochemistry and isotopic characteristics of mafic (Phulad Ophiolite) and related rocks in the Delhi Supergroup, Rajasthan, India: implications for rifting in Proterozoic. *Precamb. Research*, v. 48, pp. 167–191.

Vredenburg, E. W.,1906. Geology of the state of Panna, principally with reference to the diamond bearing deposits. *Records, Geol. Surv. India*, v. 38(4), pp. 261–314.

Wada, H. and Santosh, M., 1995. Stable isotope characterization of metamorphic fluid processes in the Kerala Khondalite belt. *Geological Society of India, Mem.* 34, pp. 161–172.

Wadia, D.N., 1932. Notes on the geology of Nanga Parvat (Mt Diamir) and adjacent portions of Chilas, Gilgit district, Kashmir. *Records, Geol. Surv. India*, v. 66, pp. 212–234.

Walker, R.J., Shiraj, S.B., Hanson, G.N. Rajamani, V., and Horan, M.F., 1989. Re–Os, Rb–Sr, and O isotope systematics of Archean Kolar schist belt, Karnataka, India. *Geochim. Cosmochim Acta*, v. 53, pp. 3005–3013.

Walker, T.L., 1900. Geological sketch of the central portion of Jeypore Zamindari, Vizagpatnam district. *Gen. Rep., Geol. Surv. India*, pp.1899–1900.

Walker, T.L., 1902. Geology of the Kalahandi state, Central Provinces. *Geol.Surv.India*, Mem. 33, pp. 1–22.

Wangdus, C., Som Nath, Tangri, S.K. Tikku, V.K. and Tiwari, G.S. 1998. Detailed geological appraisal for platinoid group of elements and gold in Kyun Tso and Niornis areas in Ladakh region, Jammu and Kashmir;. *Records, Geol. Surv. India*, v. 130(8), pp. 11–16.

Watson, J.V., 1978. Ore deposition through geologic time. *Royal Soc. Lond. Proc.*, v. A362, pp. 305–328.

Watson, J.V., 1980. Metallogenesis in relation to mantle hetrogeneity. *Phil. Trans. Royal Soc. London*, v. A97, pp. 347–352.

Weaver, B.L., 1980. Rare earth element geochemisty of Madras granulites. *Contrib. Min. Pet.*, v. 71, pp. 161–172.

Wells, P.R.A., 1980. Thermal models for the magmatic accretion and subsequent metamorphism of continental crust. *Earth and Planet Sci. Lett.*, v. 46, pp. 253–265.

Werding, G. and Schreyer, W., 1996. Experimental studies on borosilicates and selected borates. In E.S.Grew and L.M. Anovitz (Eds.), Boron: Mineralogy, petrology and geochemistry. *Mineral. Soc. Amer. Rev, Mineral.*, v. 33, pp. 117–163.

West, W.D., 1933. The origin of streaky gneisses of the Nagpur district. *Records, Geol. Surv. India*,v. 67, pp. 344–356.

West, W. D. 1936. Nappe structure in the Archaean rocks of the Nagpur District. Trans. *Nat. Inst. Sci. India*, v. 1, pp. 93–102.

White A.J.R. and Chappel, B.W., 1977. Ultrametamorphism and granitoid gneisses. *Tectonophysics*, v. 43, pp. 7–22.

White, A.J.R., Clemens, J.D., Holloway, J.R., Silver, I.T. and Chappel, B.W., 1986. S-type granites and their probable absence in the south western North America. *Geology*, v. 14, pp. 115–118.

Wickham, S.M., Oxburg, E.R., 1987. Low-pressure regional metamorphism in the Pyrenees and its implications for the thermal evolution of rifted continental crust. *Phil. Trans., Royal Soc. London* v. A321, pp. 219–242.

Widel, F., Schnell, S., Heising, S., Ehrenreich, A., Assumus, B. and Schink, B., 1993. Ferrous iron oxidation by anoxygenic phototropic bacteria. *Nature*, v. 362, pp. 834–836.

Wiedenbeck, M. and Goswami, J.N., 1994. An ion–probe single zircon 207Pb/206 Pb age from the Mewar Gneisses at Jhamarkotra, Rajasthan. *Geochim. Cosmochim Acta*, v. 58, pp. 2135–2141.

Wiedenbeck, M., Goswami, J.N. and Roy, A.B., 1996. Stabilisation of the Aravalli craton of north–western India at 2.5 Ga : An ion–microprobe zircon study. *Chem. Geol.*, v. 129, pp. 325–340.

Wilde, S.A., Valley, J.W., Peck, W.H. and Graham, C.M., 2001. Evidence from detrital zircon for the existence of continental crust and oceans of the earth 4.4 Gyr ago. *Nature*, v. 409, pp. 175–178.

Willett, S.D., 1992. Dynamic and kinematic growth and change of a Coulomb wedge. In McClay, K.R. (Ed), *Thrust Tectonics*. Chapman and Hall, London, pp. 19–31.

Williams, P.J., 1994. Iron mobility during synmetamorphic alteration in the Selwin Range area, NW Queensland: Implications for the origin of ironstone-hosted Au–Cu deposits. *Mineral. Deposita*, v. 29, pp. 250–260.

Williams, P.J.,Barton, M.D., Johnson, D.A., Fontbote, L., Haller, A.D., Mark, G., Oliver, N.H.S. and Merschik, R., 2005. Iron–oxide copper–gold deposits: geology, space–time distribution and possible modes of origin. *Econ Geol., 100th Anniversary vol.*, pp. 371–405.

Williams, W.C., Messl, E., Madrid, J. and de Machuca, B.C., 1999. The San Jorge porphyry copper deposit, Mendoza, Argentina : a combination of orthomagmatic and hydrothermal mineralisation. *Ore Geol. Rev.*, v. 14, pp. 185–201.

Wilson, A.J., Cooke, D.R. and Harper, B.L., 2003. The Ridgeway gold–copper deposit : A high grade alkalic porphyry deposit in the Lachlan fold belt, New South Wales. Australia. *Econ. Geol.*, v. 98. pp. 1637–1666.

Windley, B.F., 1995. The evolving continents (Third Edn.), *Wiley*, p. 526.

Wyllie, P.J., Cox, K.G. and Biggar, G.M., 1962. The habit of apatite in synthetic systems and igneous rocks. *Journal Petrology*, v. 3, pp. 238–243.

Wyllie. P.J. and Lee, W.J., 1999. Kimberlites, carbonatites, peridotites, and silicate–carbonate liquid immiscibility explanined parts of the system CaO (Na_2O+K_2O) – (MgO + FeO) – (SiO_2 + Al_2O_3) – CO_2. In Gurney. J. J. et. al. (Eds.) *Proc. 7th Internat. Kimberlite Conf., Cape Town*, pp. 923–932.

Xie,Q., Kerrich, R. and Fan, J., 1993. HFSE / REE fractionation recorded in the three komatiite–basalt sequences, Archean Abitibi belt: implications for plume sources and depths. *Geochim. Cosmochim. Acta*, v. 57, pp. 411–418.

Yadav, O.P., Hamilton,S., Babu,T.B. and Saxena,V.P., 2002. Signature of multimetal iron–oxide–breccia complex at Rohil, Sikar district, Rajasthan. *Indian Minerals*, v. 56 (3&4), pp. 289–294.

Yadav, R.S., Nanda, L.K. and Sen, D.B., 1996. Structure and its role in uranium mineralisation in the Lesser Himalaya of Khasa area, Simla district, Himachal Pradesh. *Geol. Surv. India, Spl. Pub.* 21(1), pp. 413–421.

Yedekar, D.B. and Jain, S.C., 1995. Geological studies along Seoni–Rajnandgaon transect, Madhya Pradesh and Maharashtra. *Records, Geol.Surv.India*, v. 128(6), pp. 205–208.

Yedekar, D.B., Jain, S.C., Nair, K.K.K., and Dutta, K.K., 1990. The Central Indian collision suture, In Precambrian of Central India. *Geol. Surv. India, Spl. Pub.,v. 28*, pp. 1–37.

Yedekar, D.B., Karmalkar, N., Pawar, N.J. and Jain, S.C., 2003. Tectonomagmatic evolution of Central Indian Terrain. *Gondwana Geol. Mag.* Spl., v. 7, pp. 67–88

Yedekar, D.B., Reddy, P.R. and Divakara Rao, V., 2000. Tectonic evolution of CIS zone Central Indian Shield. In O. P. Verma and T. M. Mahadevan (Eds.), *Research highlights in ESS, DST, Spl. v. 1*, pp. 55–75

Yoshida, M., 1995. Assembly of East Gondwanaland during the Mesoproterozoic and its rejuvination during the Pan–African period. *Geological Society of India, Mem.34*, pp. 25–46.

Yoshida, M., Funaki, M. Vannage, P.W., 1992. Proterozoic to Mesozoic East Gondwana : the juxtraposition of India, Sri Lanka and Antarctica. *Tectonics*, v. 11, pp. 381–391.

Yoshida, M., Santosh, M. and Arima, M. 2000. Pre–Pan African events in South India and their implications for Gondwana tectonics. *Proc.Internat. Sem. on Precambrian crust of Eastern and Central India (UNESCO–IUGS–IGCP–368), Geol. Surv. India, Spl. Pub. 57, pp. 9–25.*

Zachariah, J.K., Mohanta, M.K.and Rajamani, V., 1996. Accretionary evolution of Ramagiri schist belt, Eastern Dharwar craton. *Jour. Geol. Soc. India*, v. 47, pp. 279–291

Zachariah, J.K., Rajamani, V. and Hanson, G.N., 1997. Geochemistry of metabasalts from Ramagiri schist belt, south India : petrogenesis, source characteristics and implications to the origin of the eastern Dharwar craton. *Contrib. Miner. Pet.*, v. 129, pp. 87–104.

Zachariah, J.K., Rajmani, V and Hanson G.N., 1990. Geochemistry of the metavolcanics of the Ramagiri schist belt, South India, Implications to Archaean crustal genesis in Dharwar craton. (Abst.) Proc. *Sem. on Evolution of Precambrian crust in India, Geol. Surv. India*, Calcutta, pp. 62–63.

Zartman, R.E. and Doe, B.R., 1981. Plumbotectonics – the model. *Tectonophysics*, v. 75, pp. 135–162.

Zegers, T.E., de Wit, M.J., Dann,J. and White, S.H., 1998. Vaalbara, Earth's oldestassembled continent? A combined structural, geochronological , and Paleo– magnetic test. *Terra Nova*, v.10, pp. 250–259.

Zhao, G., Cawood, P.A., Wilde, S.A. and Sun, M., 2002. Review of global 2.1–1.8 Ga orogens: implications for pre–Rodinia supercontinents. *Earth Science Rev.*, v. 59, pp. 125– 162.

Zhou, H. and Murphy, M.A., 2005. Tomographic evidence for whole scale underthrusting of India beneath the entire Tibetan Plateau. *Journal Asian Earth Sci.*, v. 25, pp. 445–457.

Zhou, M.F. and Robinson, P.T., 1997. Origin and tectonic environment of podiform chromite deposits. *Econ. Geol.*, v. 92, pp. 259–262.

Ziauddin, M. and Narayanaswami, S.,1974. Gold resources of India. Bull. *Geol. Surv. India. Econ.Geol.Series*, No. 38, p. 186.

Ziauddin, M. et al., 1979. Nickel mineralisation in the Sukinda ultramafic field, Cuttack district, Orissa. *Bull. Geol. Surv. India*, 43.

Index

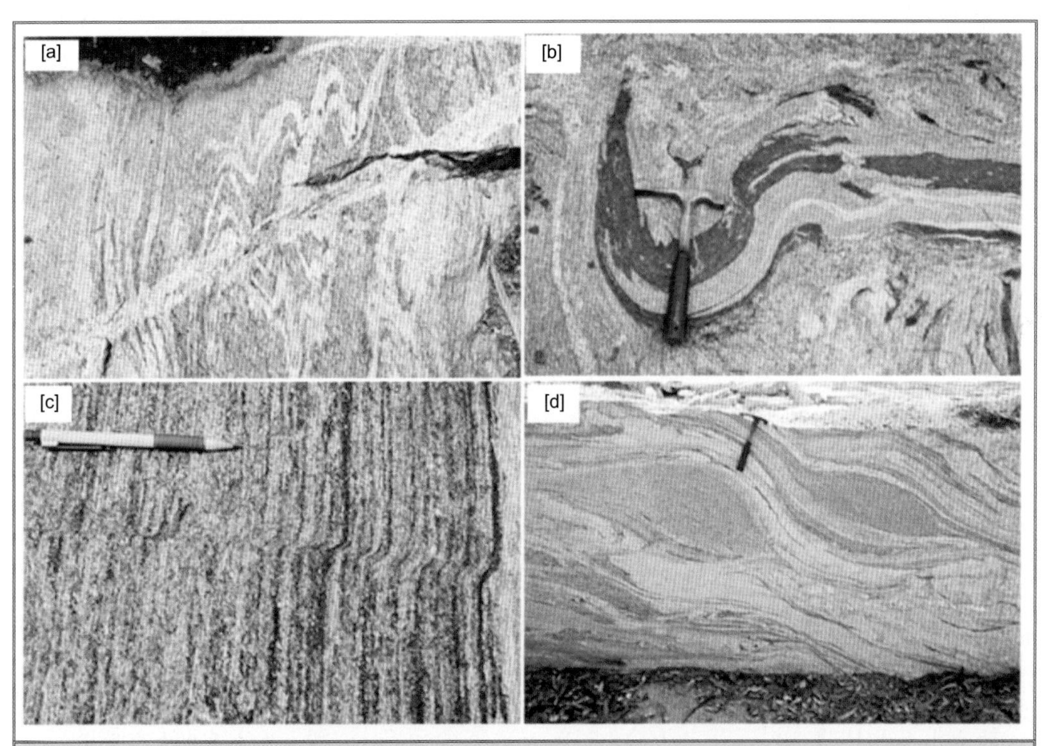

Fig. 1.1.3 (a) Intricate folding in leucosome bands in Peninsular Gneiss, west of Kolar. A later shear zone, filled up by quartzo-feldspathic material, cut across the folded bands; (b) Multiply deformed interbanded gneiss-amphibolite (dark) complex in Peninsular Gneiss, south of Bangalore. A geological hammer added to the natural sickle makes it interesting! (c) Mylonitic (proto-) banding in Peninsular Gneiss, north of Kunigal; (d) Swerving gneissosity around augen like mafic enclaves, NE of Bangalore (Photographs: courtesy V. V. Rao and A. K. Nath, GSI)

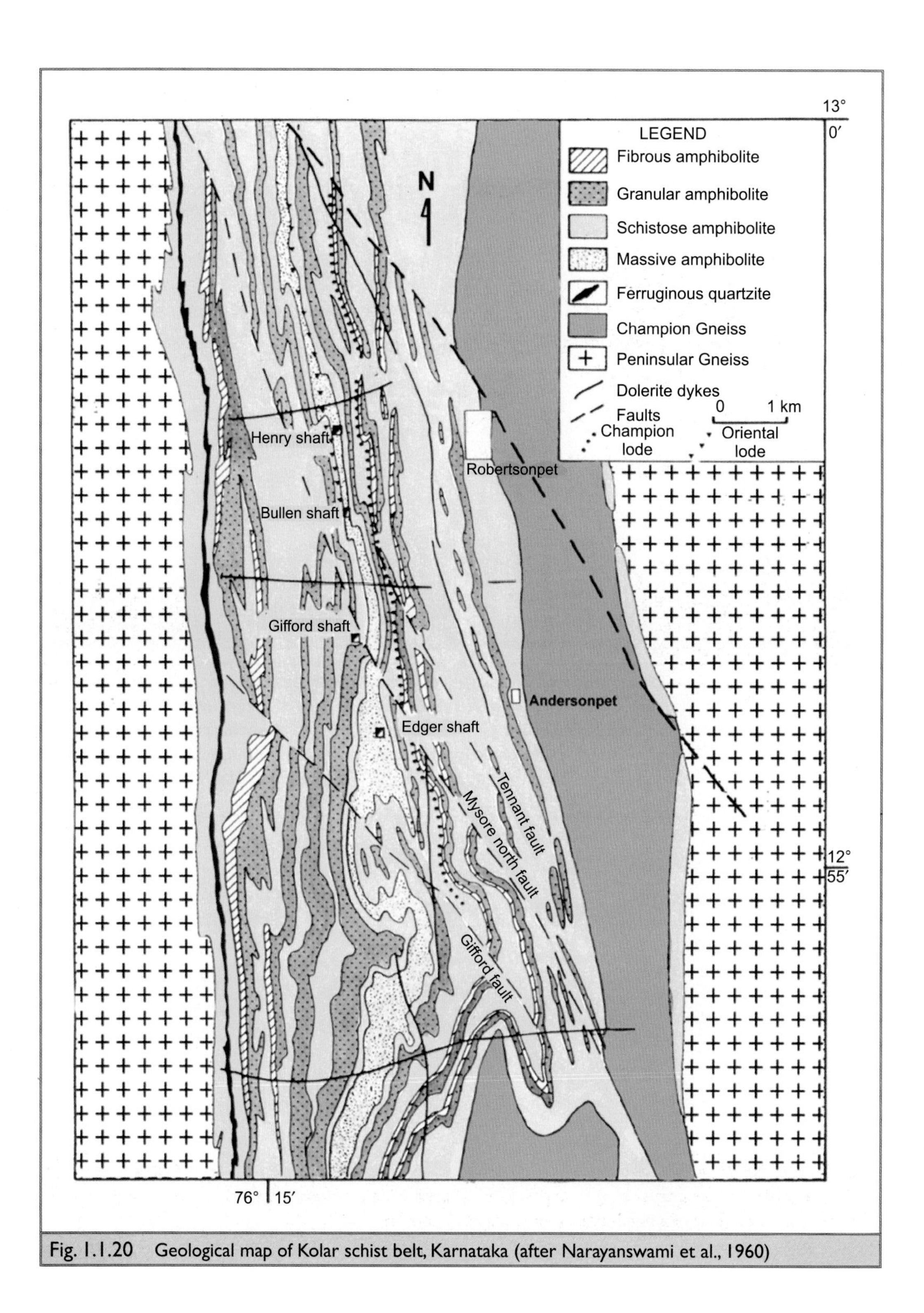

Fig. 1.1.20 Geological map of Kolar schist belt, Karnataka (after Narayanswami et al., 1960)

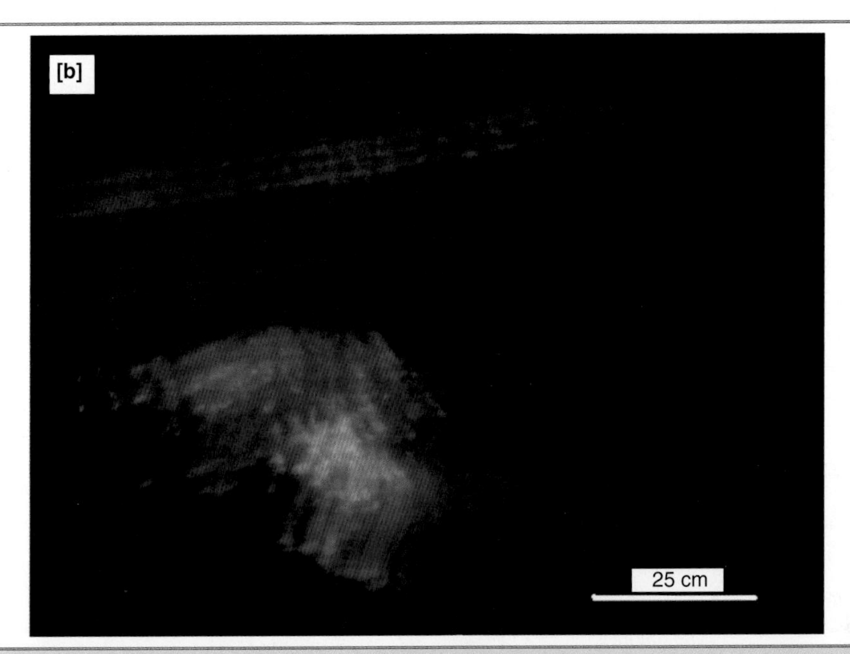

Fig. 1.2.12 Roof-views of Zone-I Reef in the drive at 24th Level (depth–2400 ft), Mallapa Shaft, Hutti mines (b) Scheelite under UV light, in a biotite-rich band and fold hinge in the wall rock of gold-bearing reef (photograph by authors)

Fig. 2.1.2 (a) Bengpal (Sukma) Gneiss exposed in Khardi river section, 3 km south of Bhanupratappur. TTG complex displays intrusion by younger coarsely crystalline granitoid followed by folding and syn–late tectonic emplacement of amphibolites (photograph by authors)

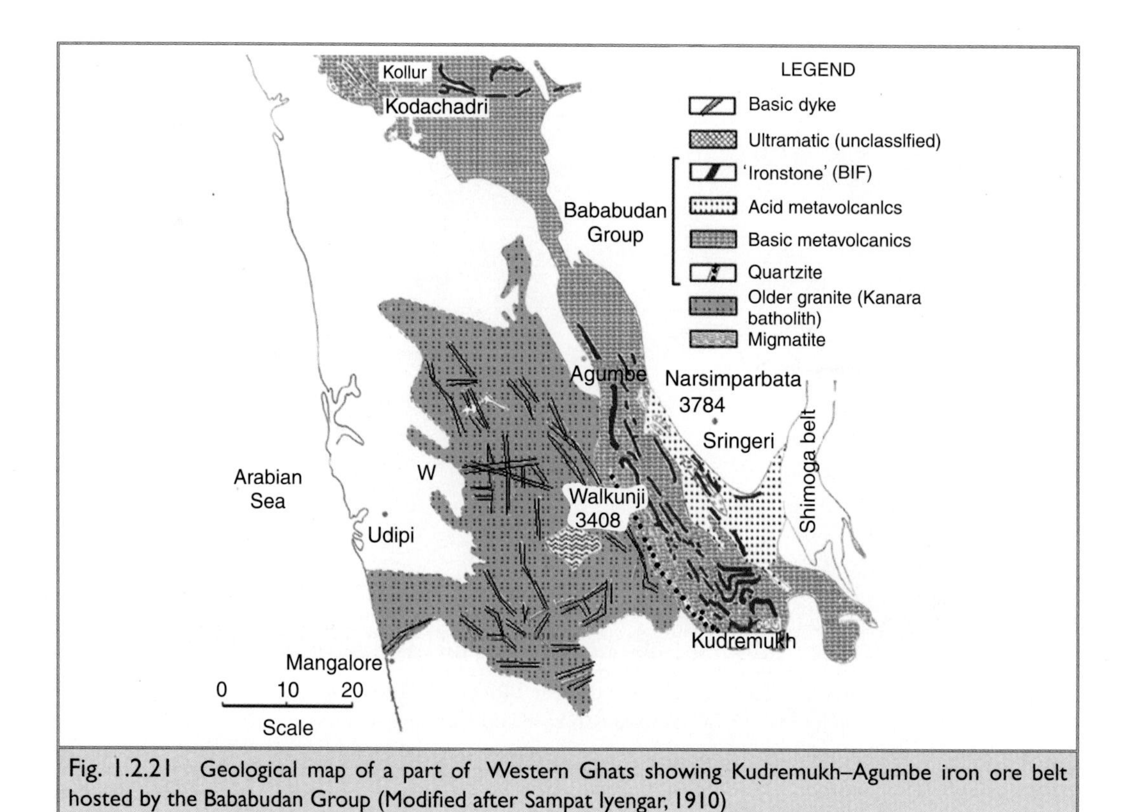

Fig. 1.2.21 Geological map of a part of Western Ghats showing Kudremukh–Agumbe iron ore belt hosted by the Bababudan Group (Modified after Sampat Iyengar, 1910)

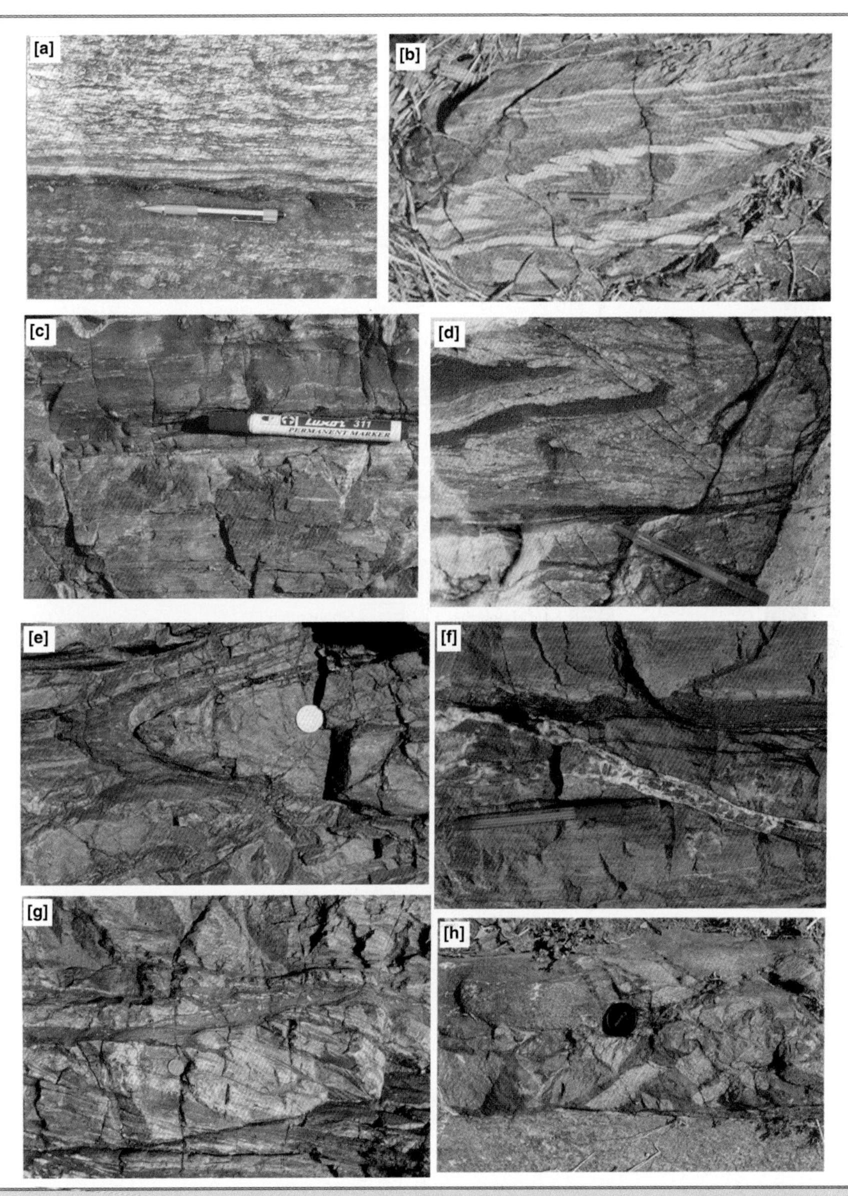

Fig. 2.1.18 (a–h) Field photographs showing (a) strongly mylonitic rock in the upper part and protomylonitic character still retained in the lower part with feldspar porphyroblasts with fish-tails, (b) axial planes of mesoscopic folds getting nearly aligned parallel to the mylonitic foliation, (c) ultramylonite (brick red) in contact with pseudotachylite (black), (d) folded cataclstite (mottled) and incipient pseudotachylite bands, (e) folded ultramylonite (brick red) and pseudotachylite (black); note the thickening of pseudotachylite band at the hinge and its flowage along axial planar fracture, (f) ultramylonite (brick red) with a foliation parallel pseudotachylite (black) band and a late discordant quartz vein (white) having included fragments of ultramylonite, (g) the planes depicting a late S–C fabric in mylonitised granite (flesh colour) invaded by pseudotachylite (dark), (h) breccia zone along a fault affecting Deccan Trap at the edge of Tan shear showing fragments of mylonite and basalt (represents last phase of movement along the zone, called Gavilgarh fault, west of Morshi). Location: Tan shear zone, Kanhan river section (Photographs: courtesy B.K. Bandyopadhyay and P.R. Golani, GSI)

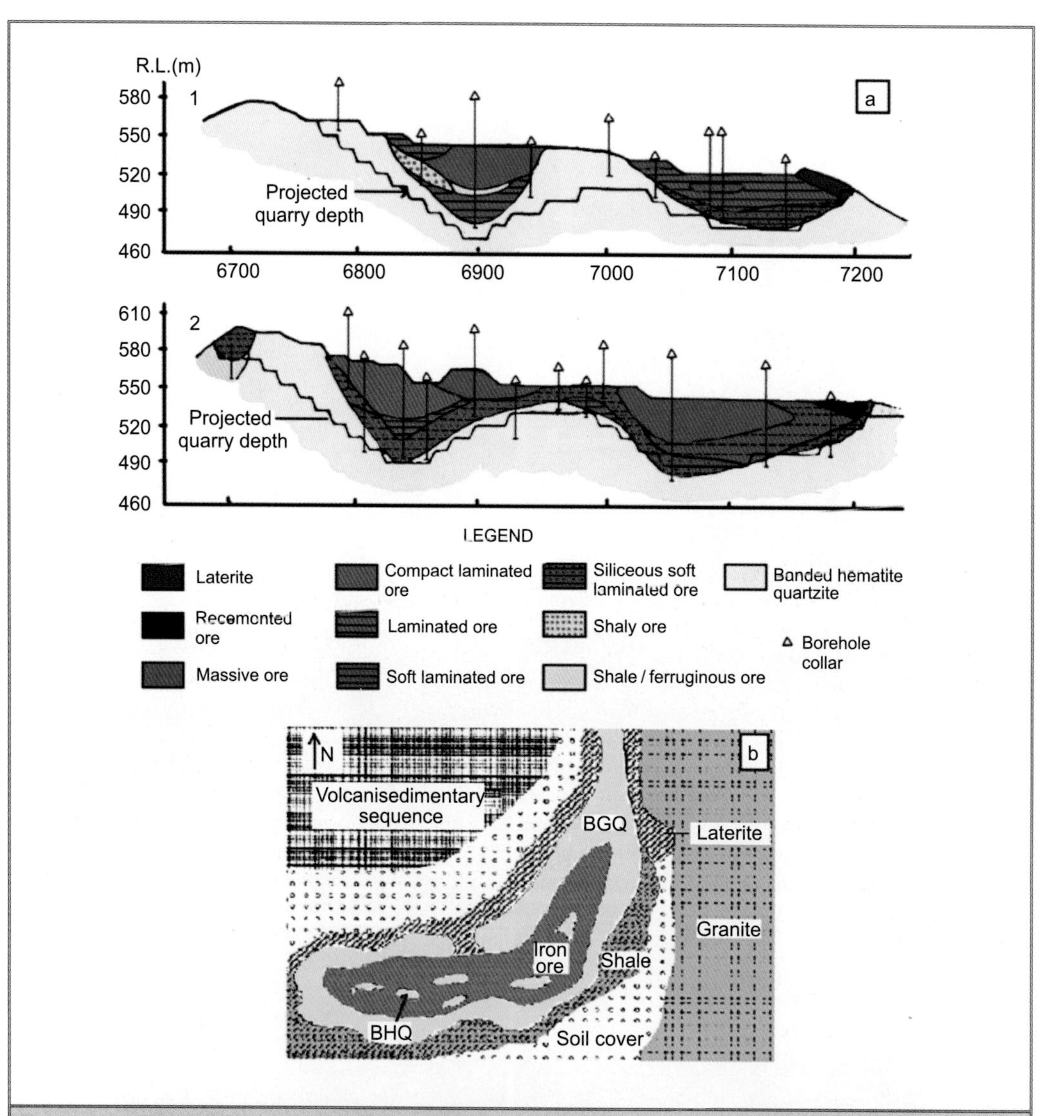

Fig. 2.2.6 (a) Cross-sections showing the disposition of iron ores with respect to BHQ (BIF) and shale in Jharandalli block, Dalli deposit (after Mines office); (b) Sketch map of Jharandalli block at 423 m RL bench level (March, 2004) (Authors)

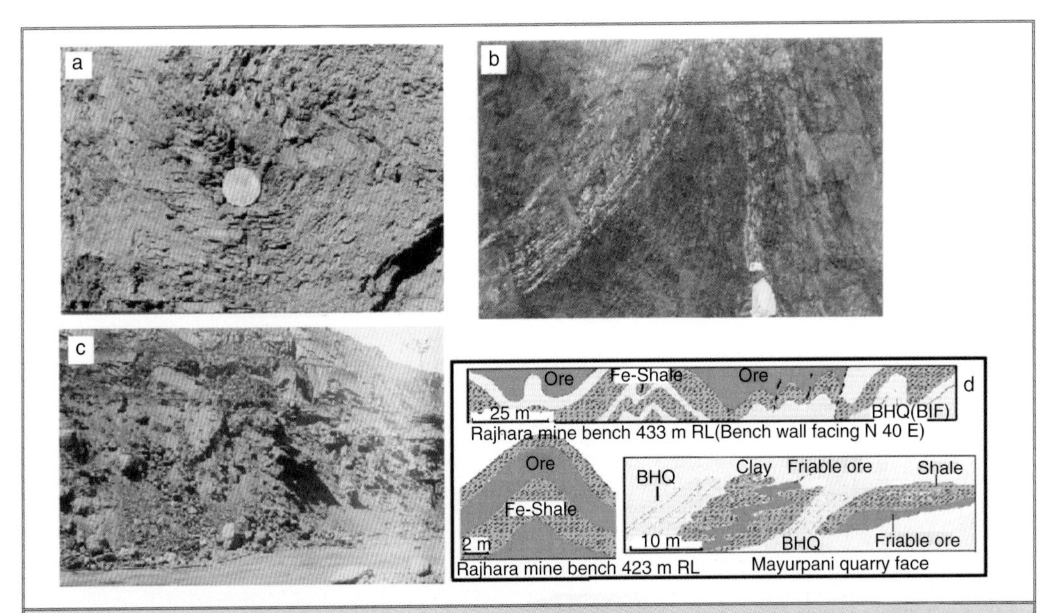

Fig. 2.2.7 (a) Folded massive iron ore in Rajhara Main block, Bench – 423 m RL, (b) Iron ore in the core of a chevron fold defined by BIF in Kandekasa (Dalli) block, Bench – 550 m RL, (c) Hard ore intimately admixed with soft ore as seen in Rajhara Mine face at 423 m RL (photographs by authors); (d) Sketch of bench faces at Rajhara and Mayurpani Mines

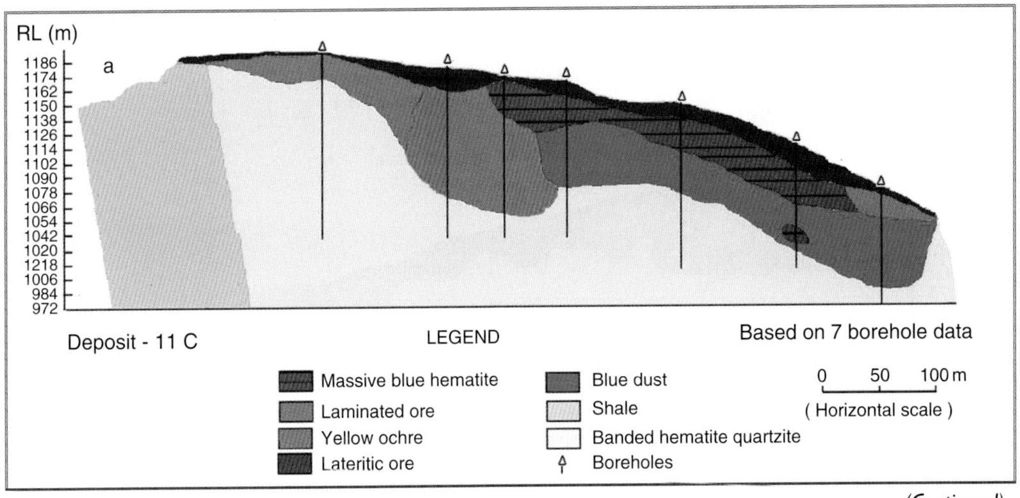

(Continued)

(Continued)

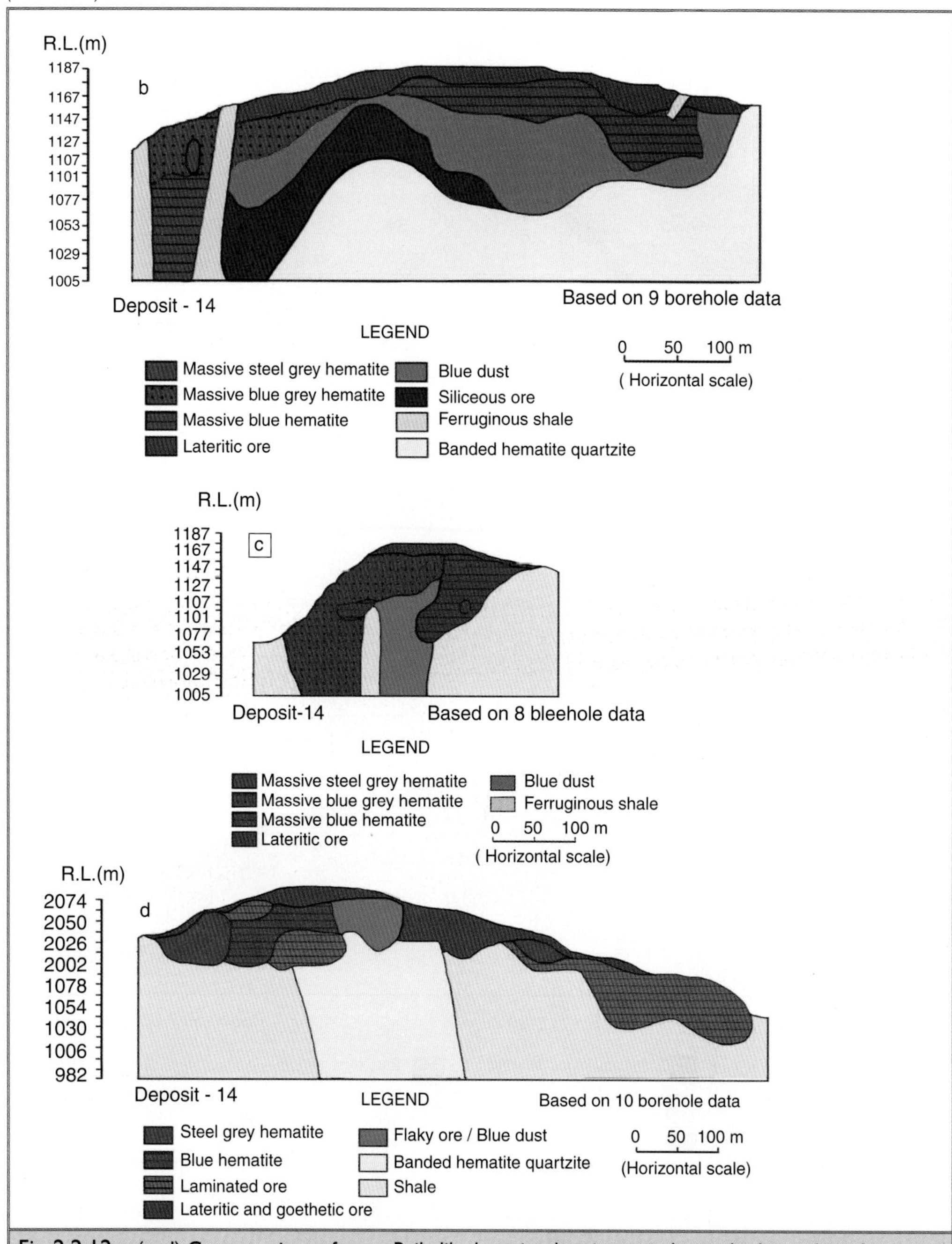

Fig. 2.2.12 (a–d) Cross sections of some Bailadila deposits showing varied spatial relationship of rich iron ore with BIF (BHQ), shale and lower grade ores (after Mines office)

Fig. 2.2.26 (a) Curvilinear bench developments in Malanjkhand open cast mine; (b) bench face in Malanjkhand mine showing fractured granite in the footwall of ore zone, overlain by subhorizontal formations of Chilpi Group. Field Photographs showing sheeted ore zones; (c) without silicification; (d) with silicification, Malanjkhand copper deposit (photographs by authors)

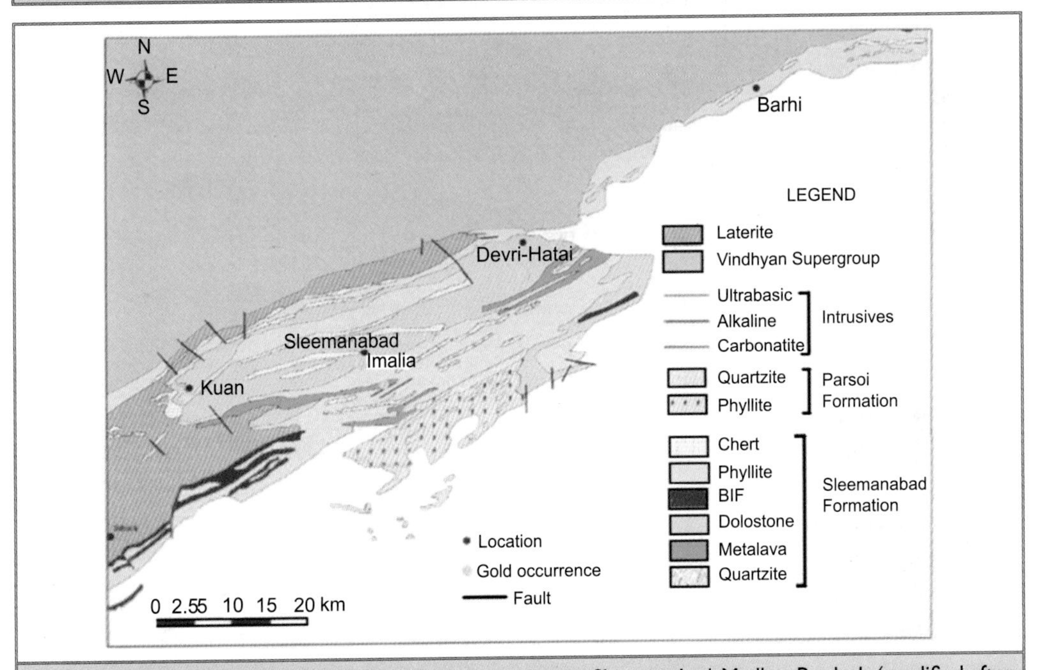

Fig. 2.2.37 Geological map of Imalia gold prospect, district Sleemanabad, Madhya Pradesh (modified after Tripathi, 2008)

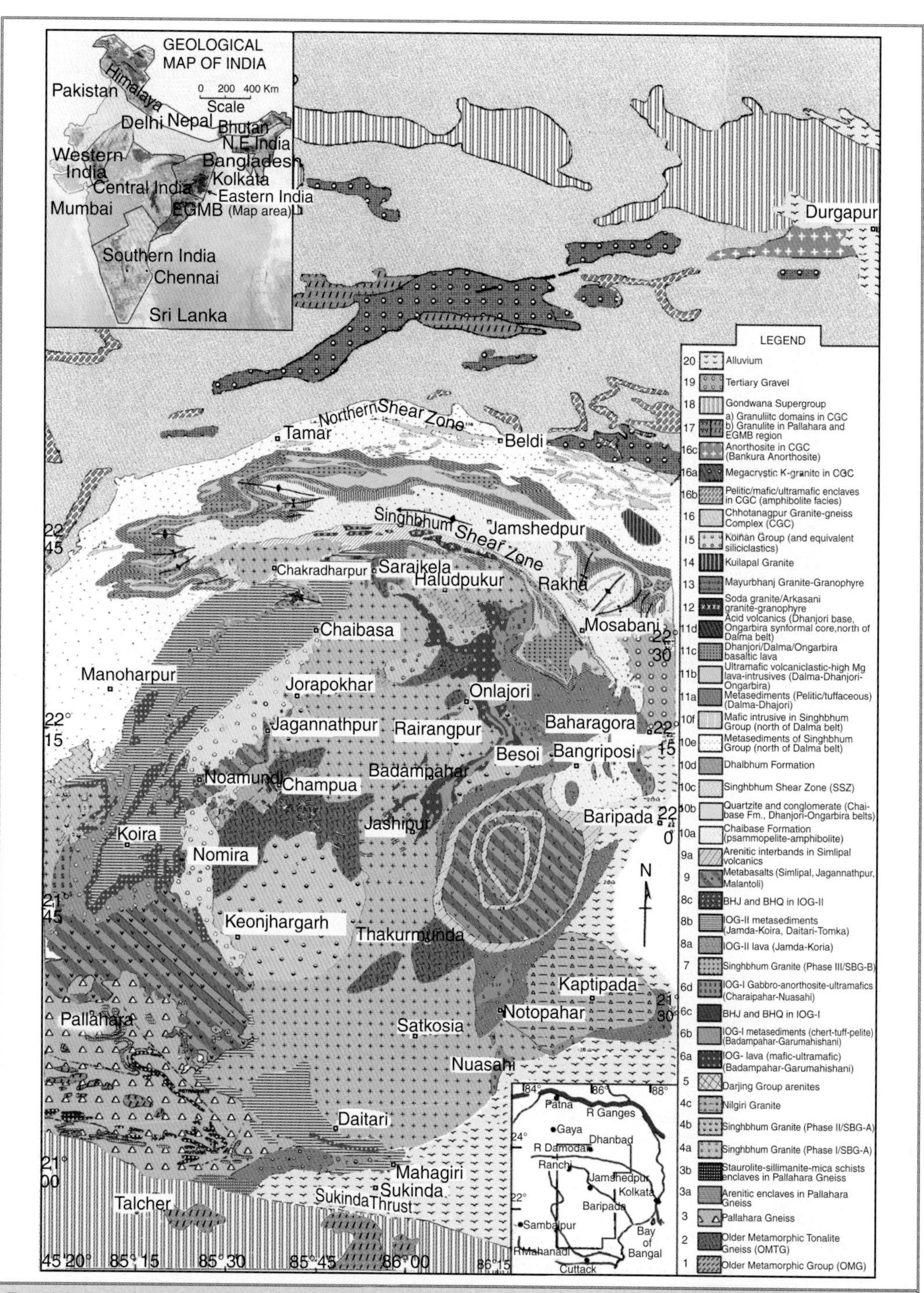

Fig. 4.1.1 Geological map of a part of the Precambrian terrain of Eastern India (Largely drawn from Dunn, 1929, 1942, Jones, 1934 Saha 1994; Sarkar and Saha, 1962; Gupta et al., 1980,1981,1985; Basu, 1985; Basu, 1993; Chattopadhyay and Ray, 1997; Blackburn and Srivastava, 1994; Ghosh and Bhattacharya (1998); Mazumder, 1996, 1988; Jena and Behera, 2000)

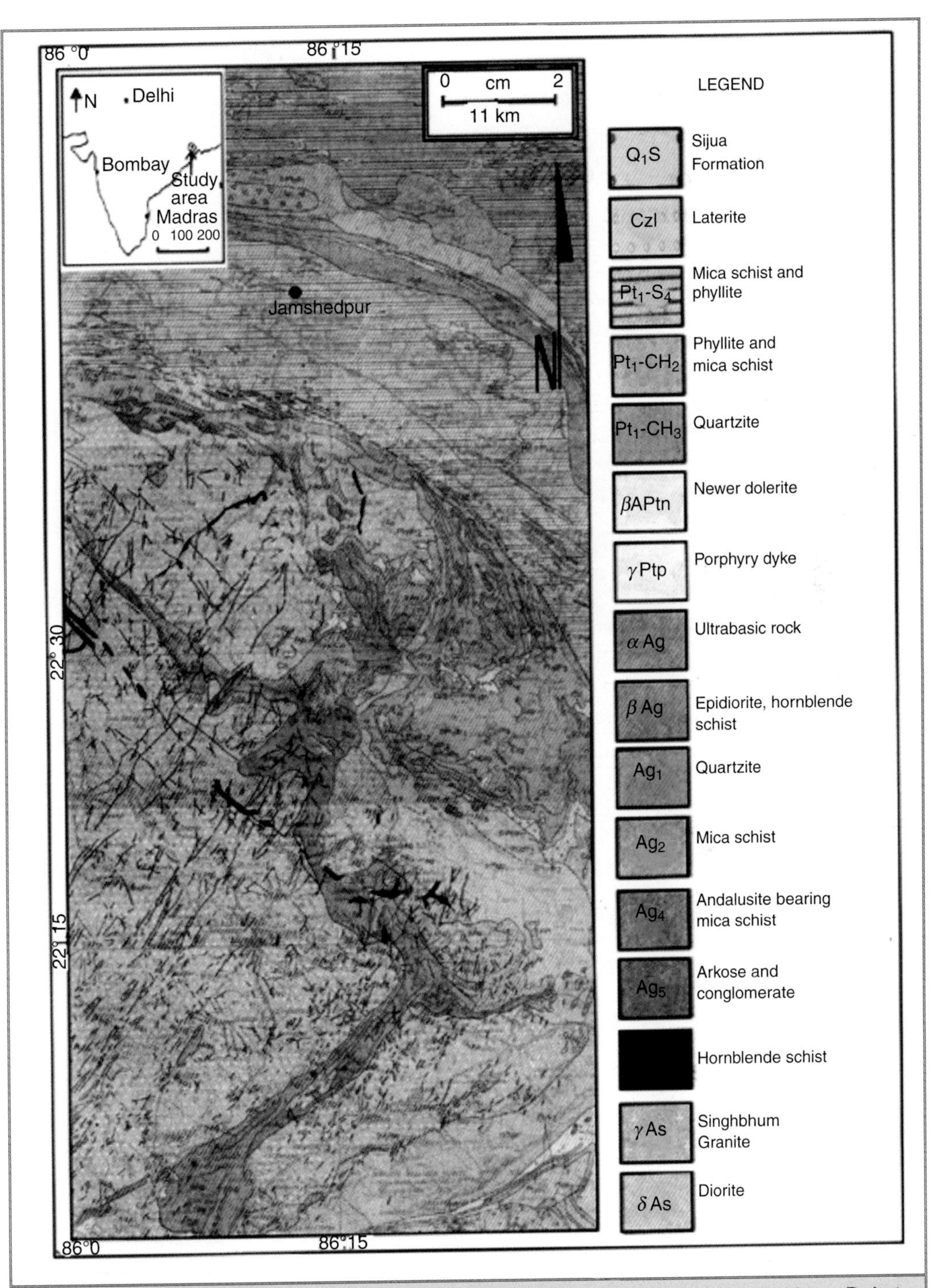

Fig. 4.1.5 Geological map of a part of Singhbhum-North Orissa cratonic nucleus showing Newer Dolerite dyke swarms (dark green) in the Singhbhum Granite country (pink) (from GSI map of Quadrangle sheet 73 J, reproduced by Mandal et al., 2006)

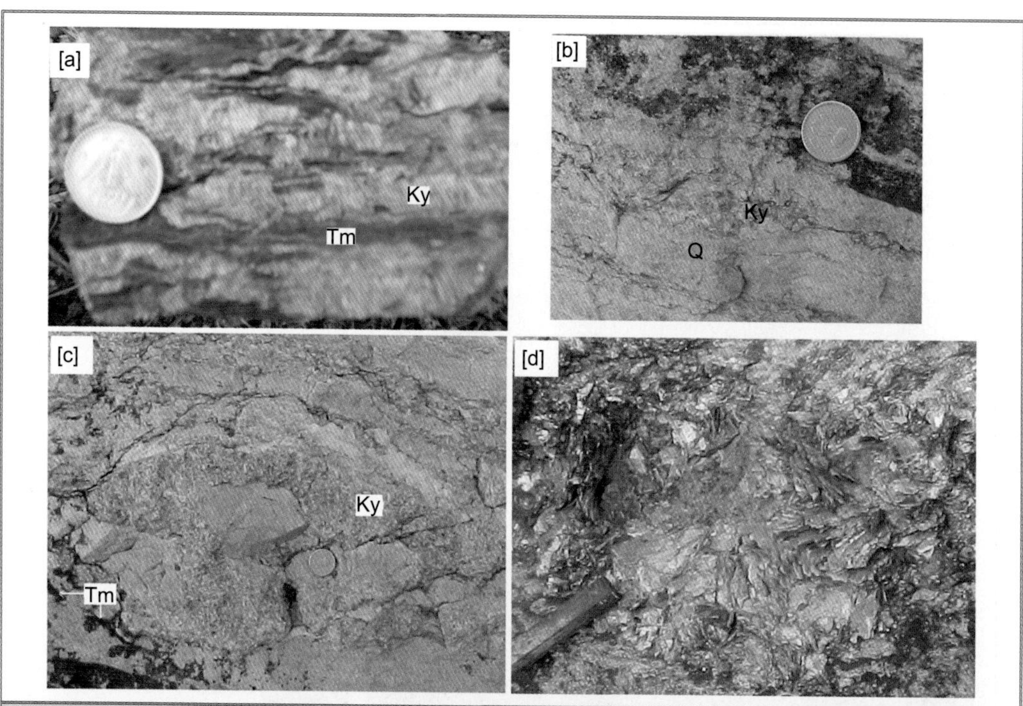

Fig. 4.1.31 (a) Kyanite (Ky)–Tourmaline (Tm) rhythmite, (b) Kyanite (Ky)-bearing bands in quartzite (Q), (c) Kyanite (Ky) in lensoid pocket within quartz-kyanite rock and tourmaline (Tm) as clots, (d) Post-tectonic kyanite blades and books of muscovite replacing kyanite (photographs by authors and P. Sengupta)

(Continued)

(Continued)

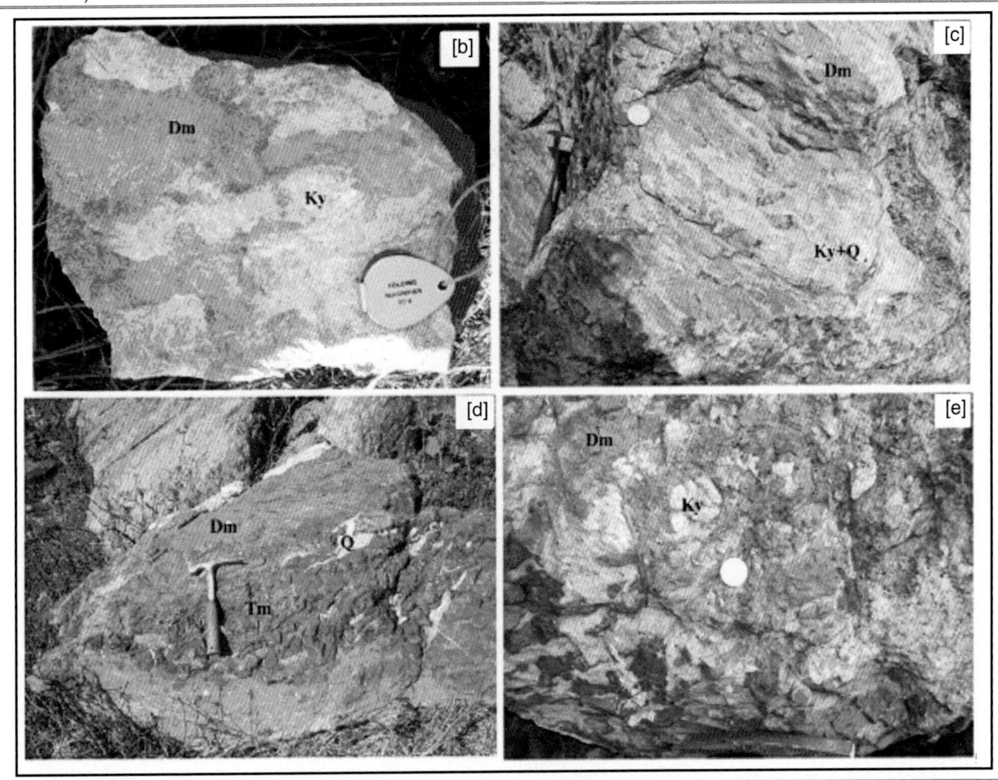

Fig. 4.1.32 (a) Dumortierite hillock at Ujanpur, west of Kharkai river; (b) Hand specimen showing dumortierite (Dm) and kyanite; (c) Schistosity parallel laths and boudins of dumortierite (Dm) in kyanite–quartz schist (Ky + Q); (d) Tourmaline pocket (Tm) surrounded by dumortierite (Dm), vein quartz (Q) in patches; (e) Kyanite (Ky) rimmed by dumortierite (Dm) (photographs by the authors)

Fig. 4.1.33 (a–d) Tourmaline mineralisation in SSZ; (a) Thinly banded tourmalinite, Surda; (b) 'Granular rock' with stringers and pockets of tourmaline; (c) Pinch-and-swell tourmalinite bands showing a cross cutting lense of apatite contained in tourmalinite; (d) Fragments of tourmalinite in lensoid body of deformed Soda granite within granular rock (photographs by S.C. Sarkar and P. Sengupta)

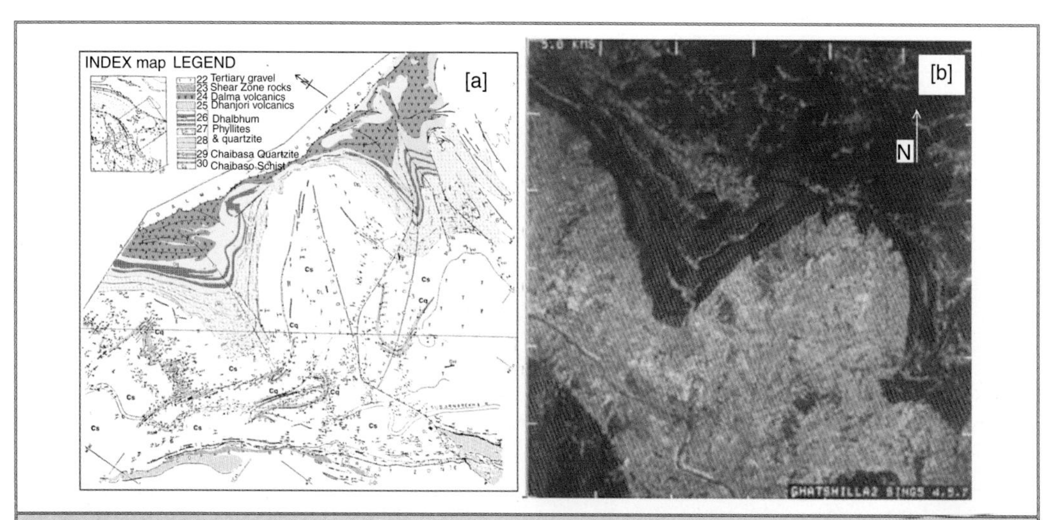

Fig. 4.1.37 (a) Geological map of Ghatsila–Galudih sector (modified after Sarkar and Saha, 1962), (b) FCC prepared from imagery for a part of the area shown in (a) (courtesy–P. Purkaet)

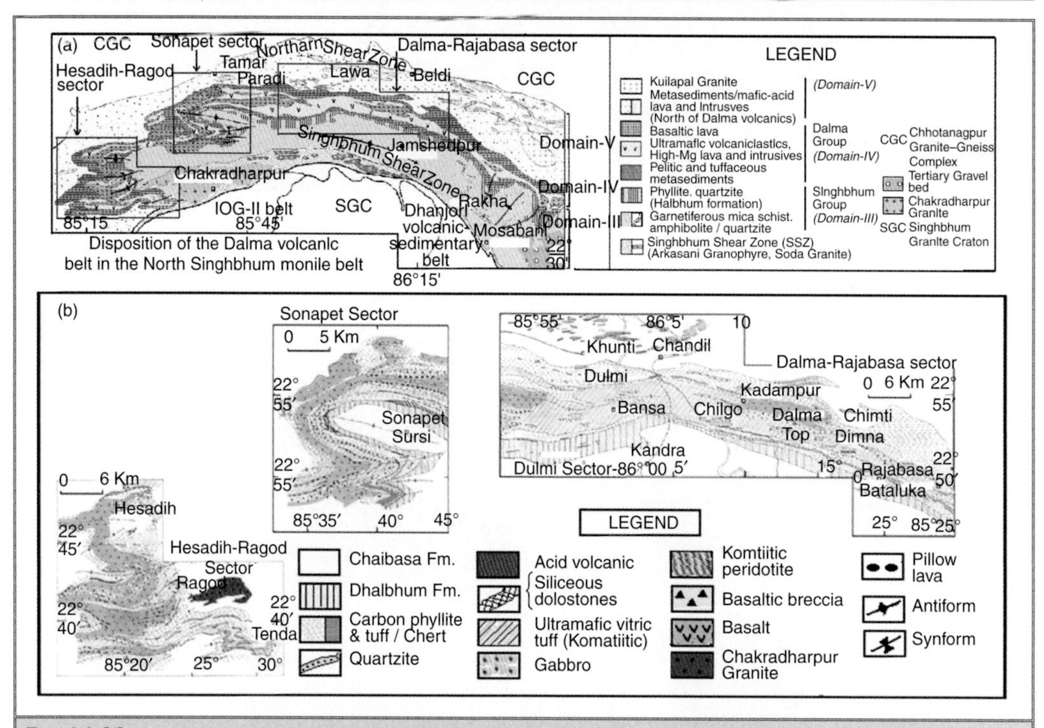

Fig. 4.1.39 (a) Map showing disposition of Dalma volcanic belt in the NSMB (excerpt from Fig.4.1.1); (b) detailed geological map of some sectors of the belt (modified after Gupta et al., 1980)

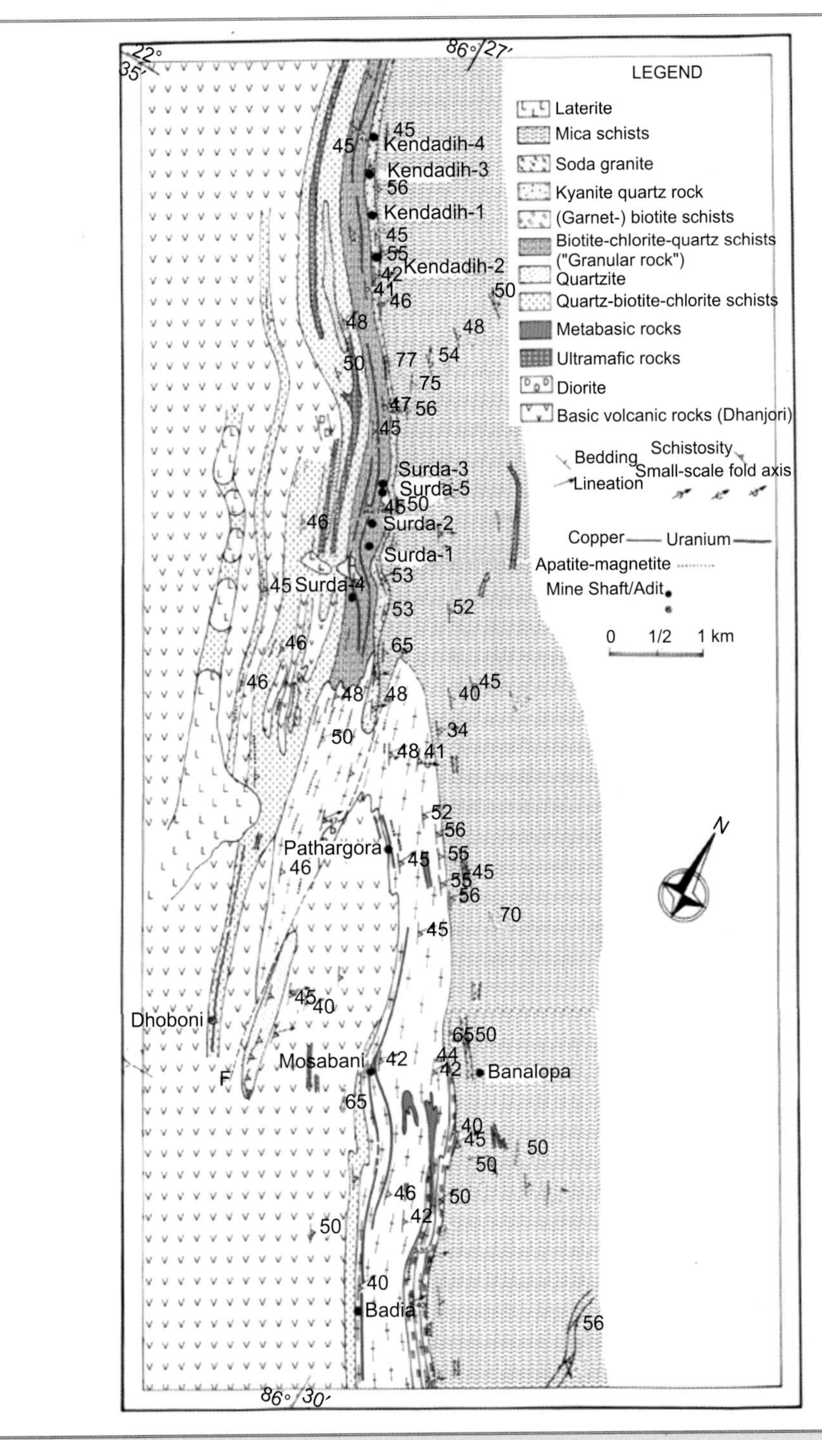

Fig. 4.2.2 Geological map of the Badia–Mosabani–Surda–Kendadih section, Singhbhum Cu-U belt, East Singhbhum district, Jharkhand (after Sarkar, 1984)

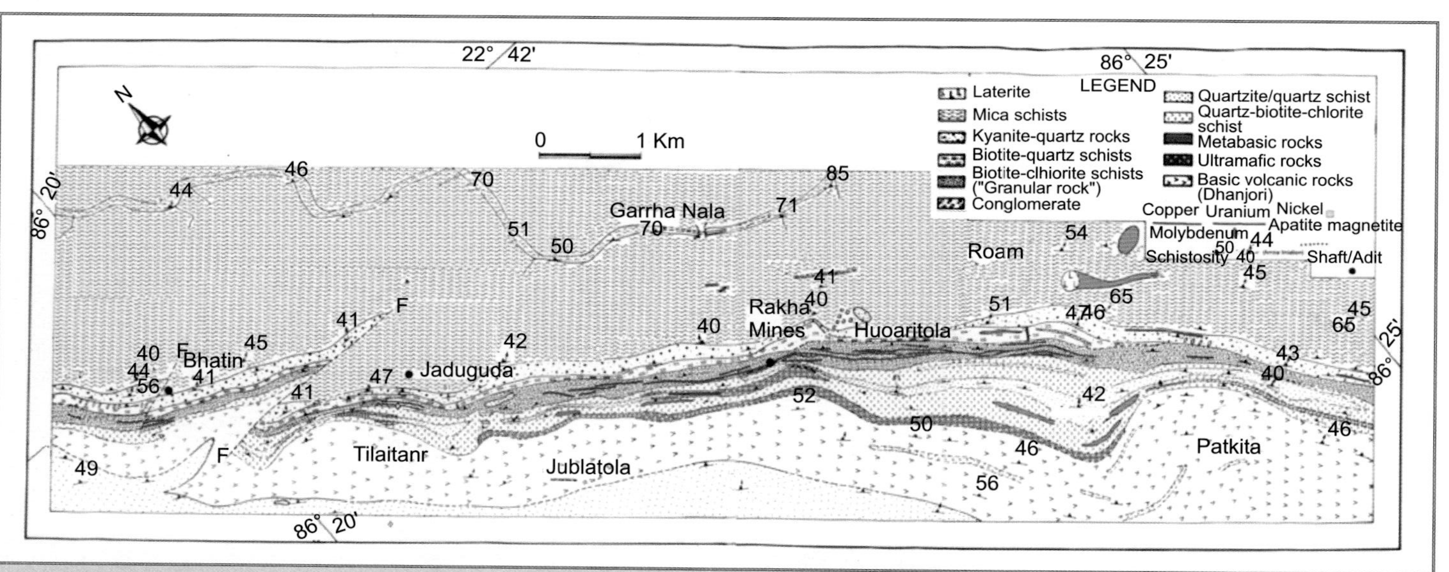

Fig. 4.2.3 Geological map of the Roam–Rakha Mines–Tamapahar–Jaduguda–Bhatin sector, Singhbhum Cu-U belt, East Singhbhum district, Jharkhand (after Deb and Sarkar, 1975)

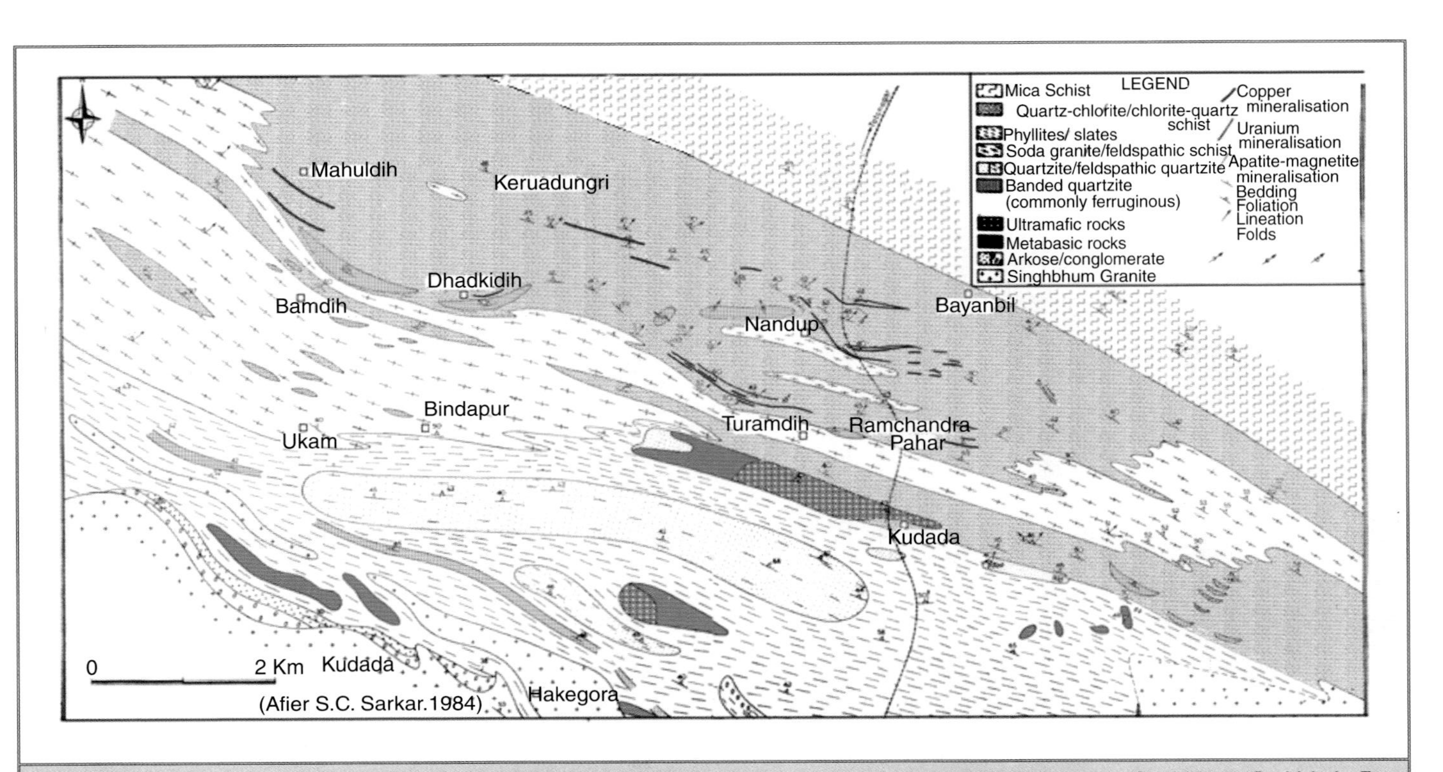

Fig. 4.2.4 Geological map of the Ramchandra Pahar–Nandup–Bayanbil–Turamdih Dhadkidih–Mahuldih sector, Singhbhum Cu-U belt, East Singhbhum district, Jharkhand (after Sarkar, 1984)

Fig. 4.2.7 (c) Chalcopyrite dominant sulphide mineralisation in chlorite-quartz schist, displaying S-C structure control (photograph by the authors)

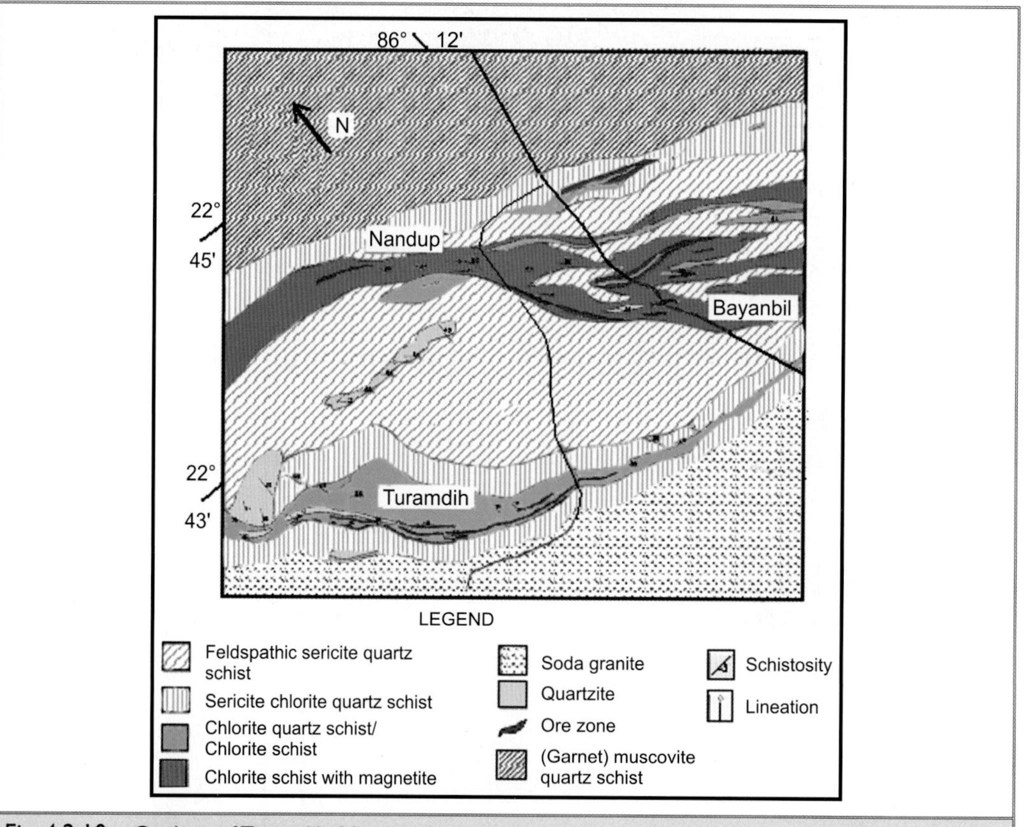

Fig. 4.2.10 Geology of Turamdih–Nandup–Bayanbil area showing Cu-sulphide ore zones (modified after Basu et al., 1979)

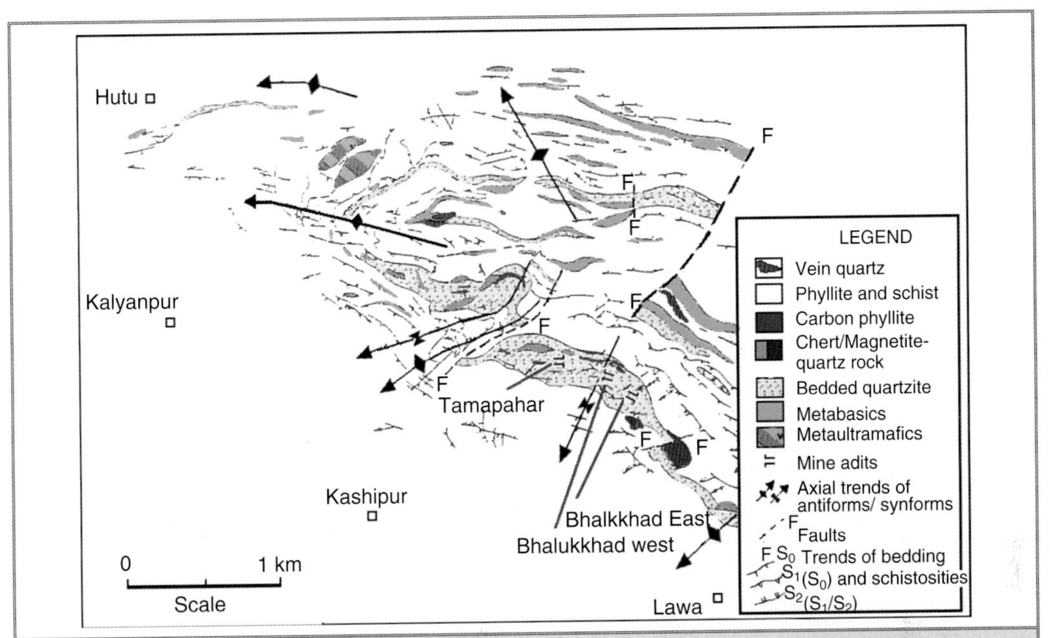

Fig. 4.2.29 Geological and structural map of Lawa area showing locations of the old gold mines at Bhalukkhad East and West and Tamapahar (Gupta, 1975, 1986)

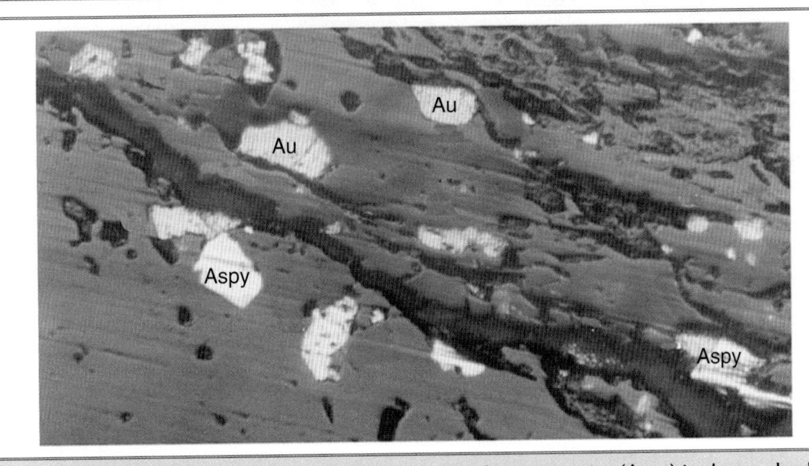

Fig. 4.2.33 Photomicrograph of disseminated gold (Au) and arsenopyrite (Aspy) in the ore body (reflected light), Parasi prospect, Jharkhand (courtesy: Dr. R.N. Singh, GSI)

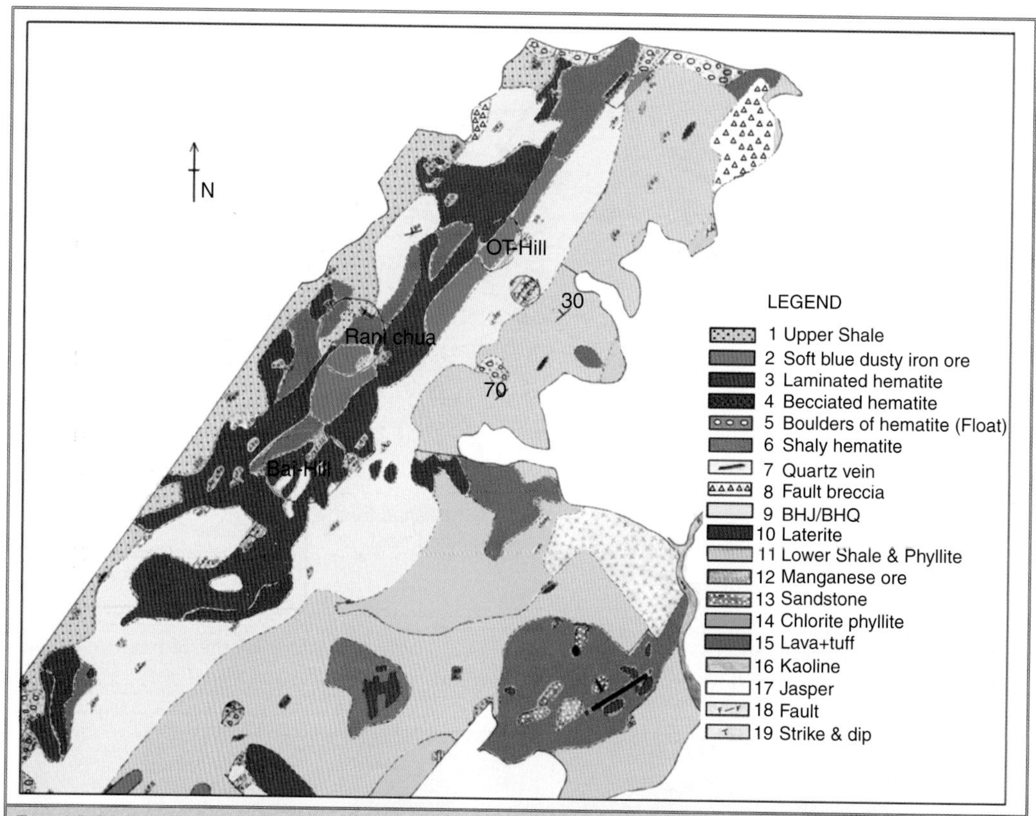

Fig. 4.2.38 Geological map of Gua iron ore deposit, West Singhbhum district, Jharkhand, showing ore body overlying BIF (BHJ/BHQ), which together are sandwiched between two shale horizons. The Lower Shale is associated with basic volcanics, represented by chloritic schists. Interfingering of ore types is notable (courtesy: mines authorities)

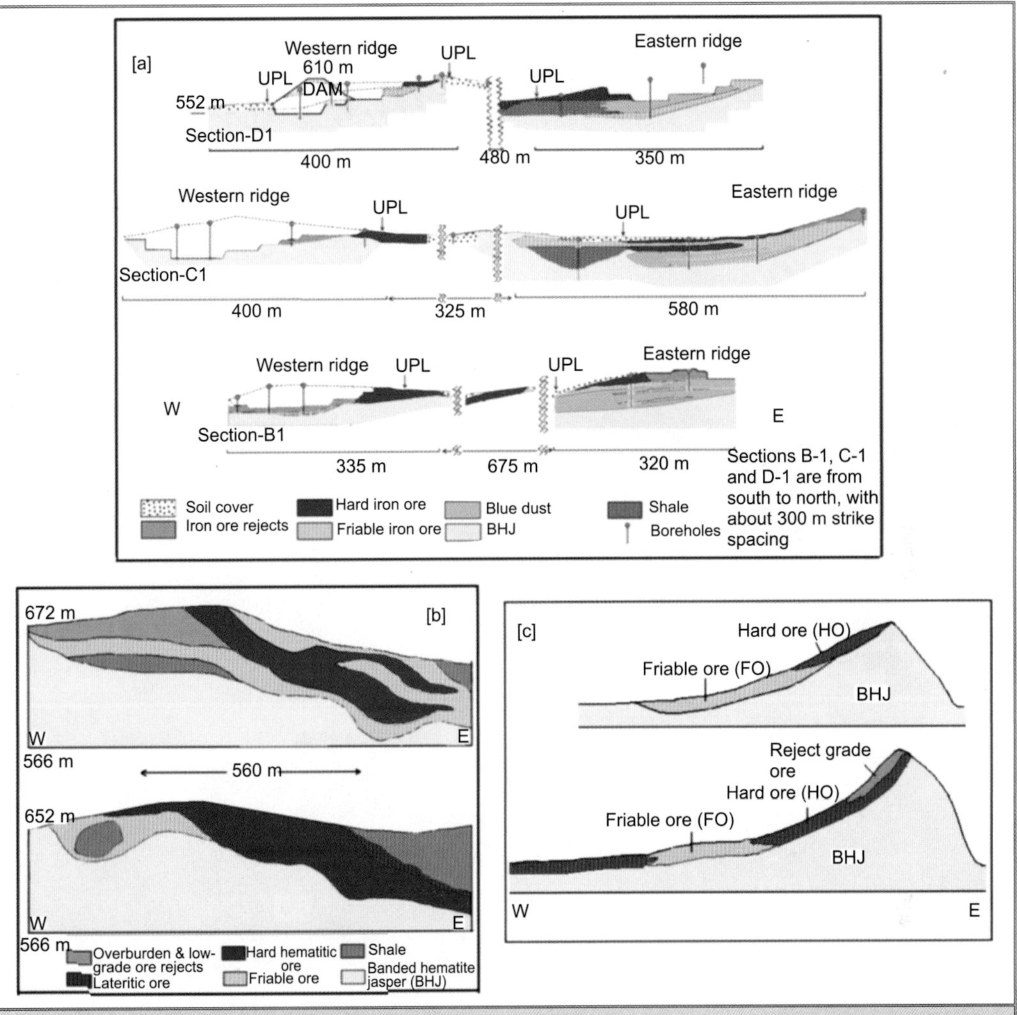

Fig. 4.2.40 (a) Three profile sections across Noamundi iron ore mines, showing low westerly dip of the ore zones and interfingering of different ore types. Borehole cross-sections of (b) Two profile sections across Khondbond deposit, Orissa, showing low easterly dip of the ore zone and (c) Two profile sections across Banspani deposit, Orissa, showing low westerly dip of the ore zone. (Data source: Tata Steel authorities)

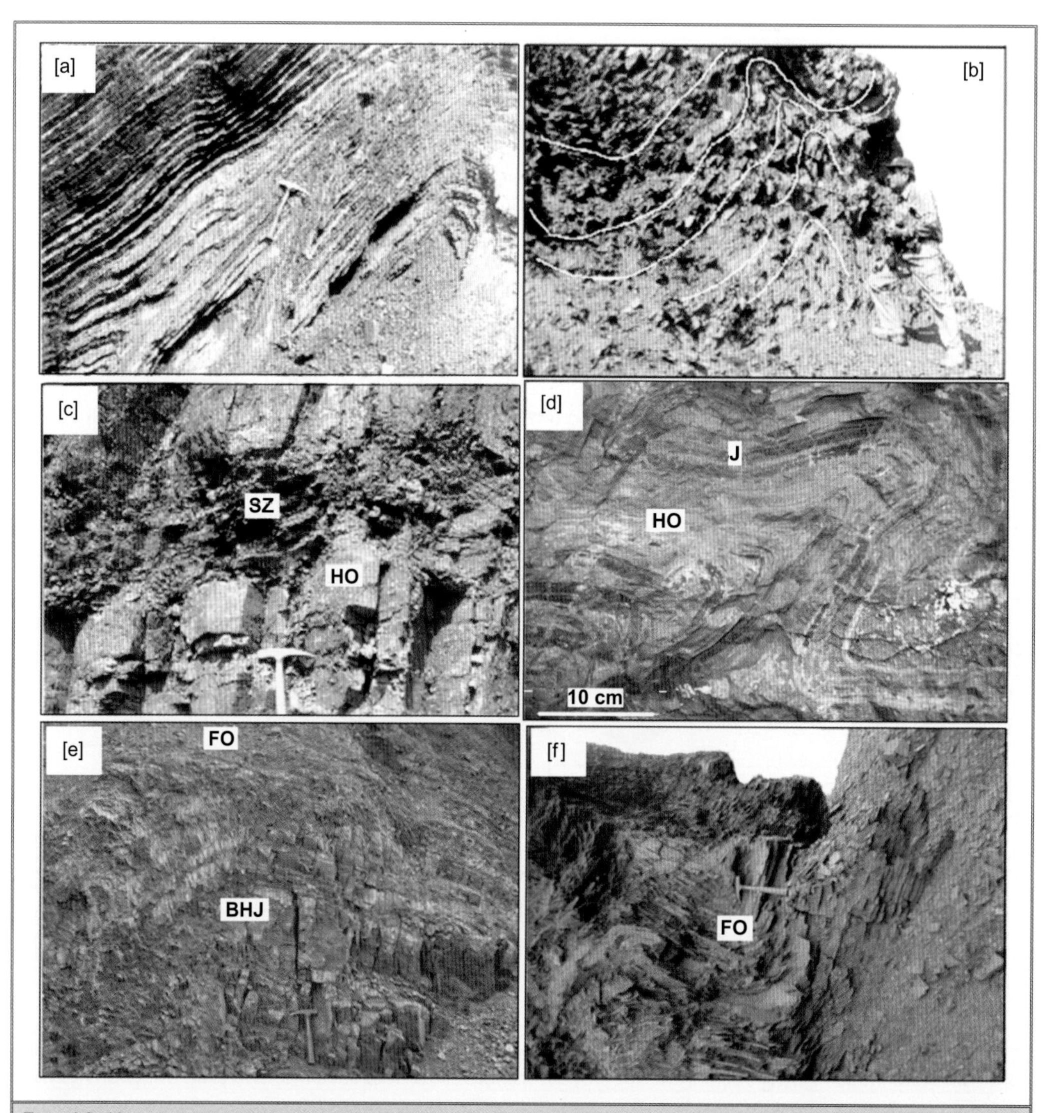

Fig. 4.2.41 Field photographs showing (a) Chevron folded BHQ (top left) changing over to Banded hematite shale, Gua mines, Singhbhum; (b) Flexure in high-grade iron ore bands with well-developed axial plane schistosity visible on close inspection, Gua mines, Singhbhum; (c) Hard and blocky high-grade iron ore (HO) with well-developed shears planes (SZ) and fractures of tectonic origin, Thakurani Pahar, Orissa; (d) Folded BHJ with alternate hard ore (dark grey-HO) and jasper (brown-J) bands in Joda mines; (e) Axial planar cleavages and joints developed in a folded BHJ band occurring within friable iron ore (FO), Hill 6, Noamundi East mine; (f) Highly folded friable iron ore (FO), Hill 4, Noamundi East mine. (Photographs by authors)

Fig. 4.2.45 Photograph and sketch showing hard hematitic iron ore (> 66% Fe) with fine undisturbed laminations alternating with non-laminated scoraceous bands. Sporadic limonite streaks/specs and minute cavities are noted in the laminated bands, Katamati mine, southern extension of Noamundi deposit

Fig. 4.2.46 (a) Near-surface limonitic ore overlying friable ore (FO) and blue dust (BD), (b) Boulders of 'Canga' derived from the surface, (c) Friable and hard ore with solution cavities parallel to the bedding surfaces, location of (a–c): Noamundi west mine, (d) Hard laminated ore alternating with contorted ore bands with cavities and solution breccia, (e) Highly scoraceous hard massive ore with goethitic patches, (f) Quartz vein in Low-grade friable ore with limonitic layers, location (d–f): Joda mines (photographs by authors)

Fig. 4.2.47 (a–b) Field photograph and representative sketch of intertwined friable ore (FO) and blue dust (BD) showing remnants of primary laminates in form of hematite plates; (c) Close-up view of a hematite plate from location of (a), showing proto-botryoid texture on the plate surface, Katamati mines, Orissa; (d) sketch showing transition of BHJ into blue dust through friable platy ore within less than a meter; note layer parallel compositional gradation along band 1 and 3, and no layer-across change to the unaltered band 2, Hill 4, Noamundi East mine (Photographs and sketches by the authors)

Fig. 4.2.48 Intricately folded BIF with quartz laminates (light coloured, high relief) interlayered with blue dust (dark grey coloured, low relief), gua mines (from Sarkar and Gupta, 2005)

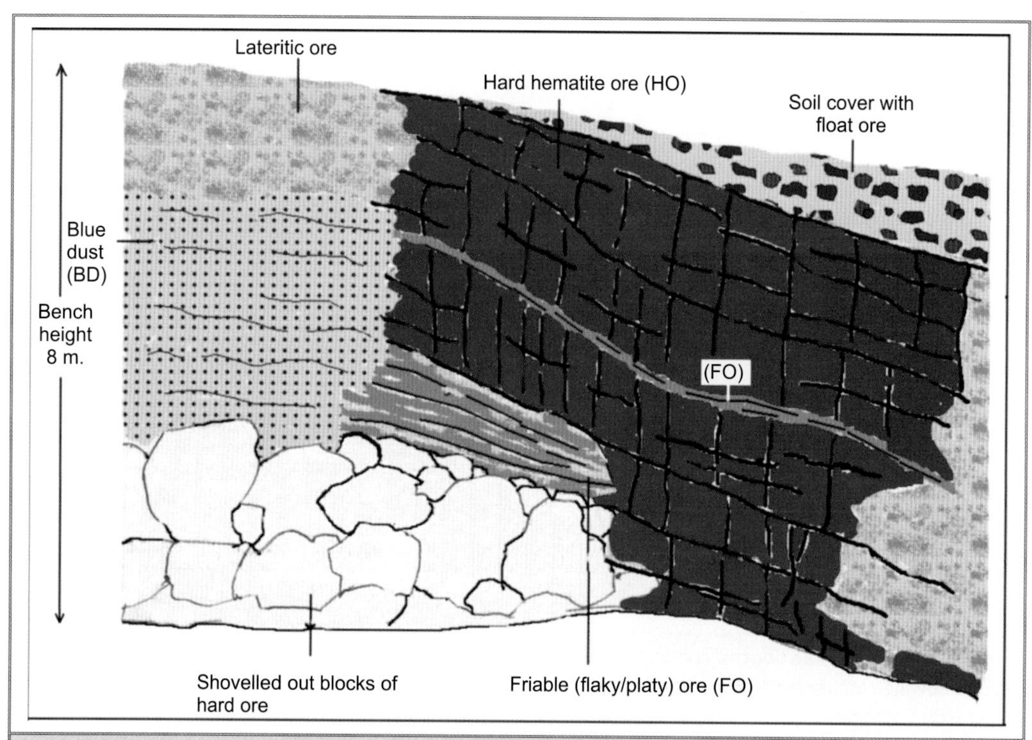

Fig. 4.2.49 Sketch of a bench face in Katamati mine showing the disposition and interrelationship between hard ore (HO) and other ore types (FO, BD, lateritic and float ore)

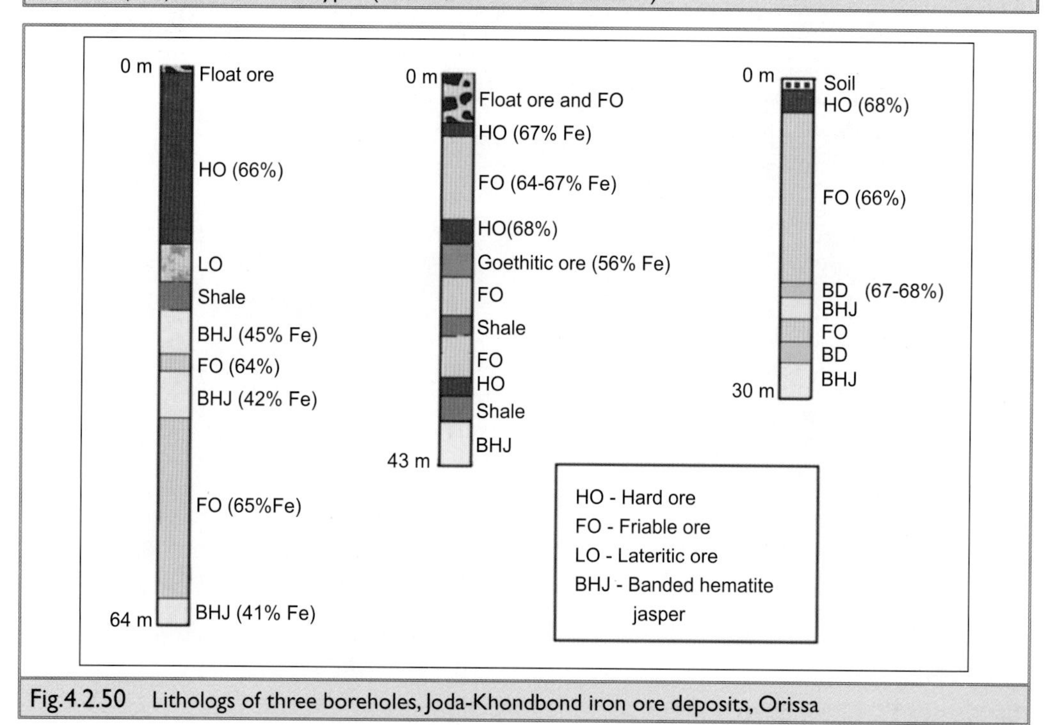

Fig.4.2.50 Lithologs of three boreholes, Joda-Khondbond iron ore deposits, Orissa

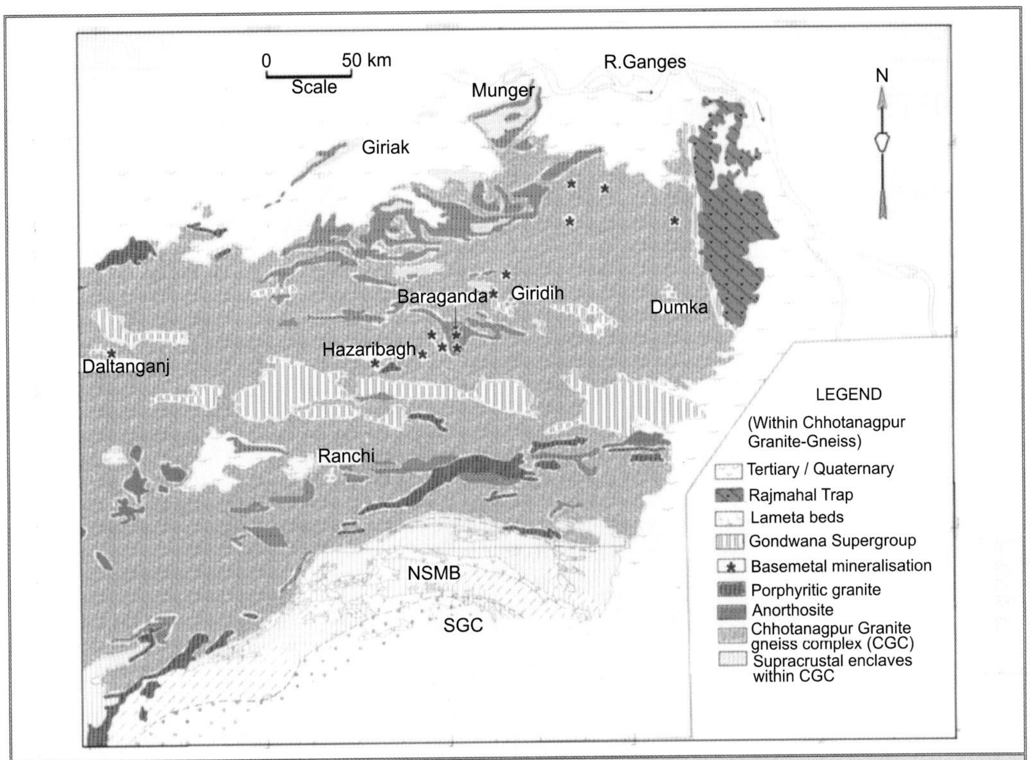

Fig. 4.2.63 Northeastern parts of Chhotanagpur Granite–Gneiss Complex (CGC) showing the locations of basemetal sulphide prospects and deposits (compiled from Fig. 4.1.1 and Mazumdar, 1988)

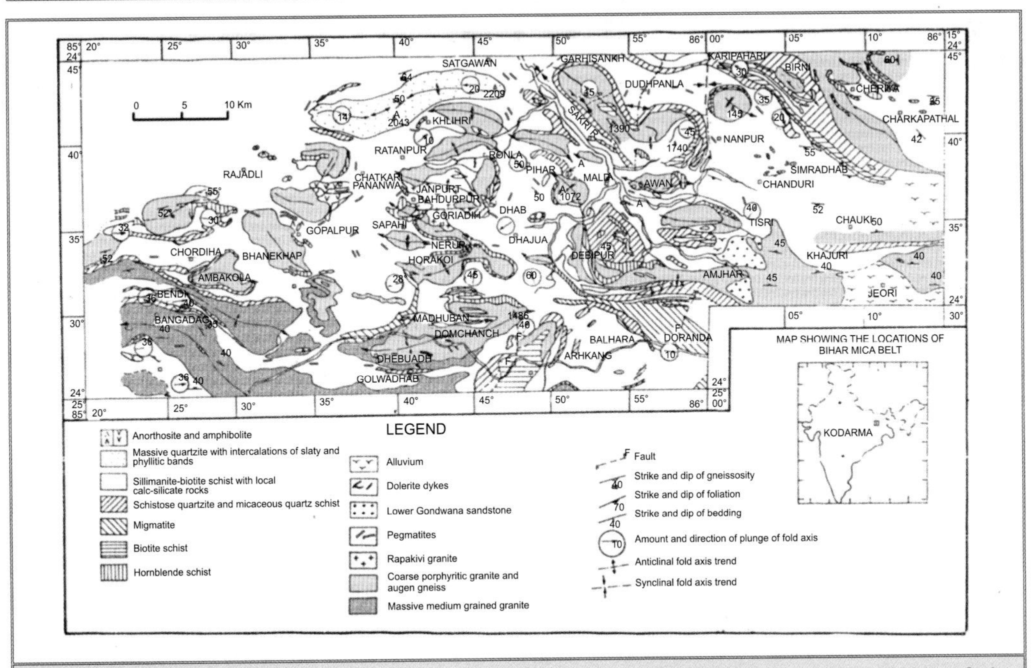

Fig. 4.2.64 Regional geological map of Bihar (Jharkhand) mica belt (after Nateswara et al., 1981, in Sinha, 1999)

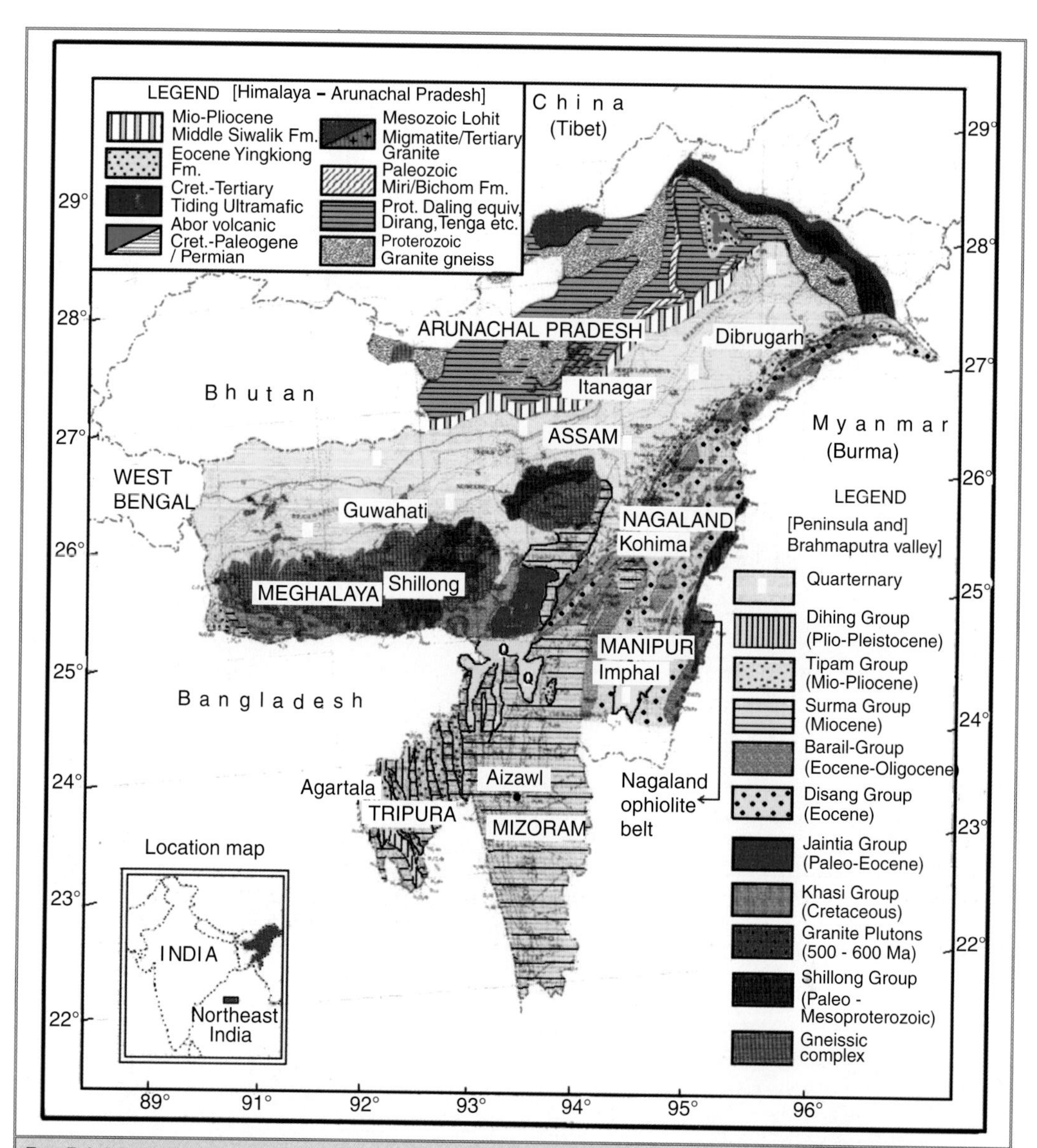

Fig. 5.1.1 Generalised geological map of the Northeastern India (partially modified from GSI's map on 1:5 million scale).

Fig. 6.1.2 Field photographs showing (a) Remnants of dark, foliated biotite gneiss within pink Untala Granite; (b) Transposition of deformed early foliation in basement gneiss; (c) Flattened sheath folds developed in basement gneiss; (d) Coaxial refolding of isoclinal fold in Banded Gneiss from Kankroli. (a–c after Roy and Jakhar, 2002)

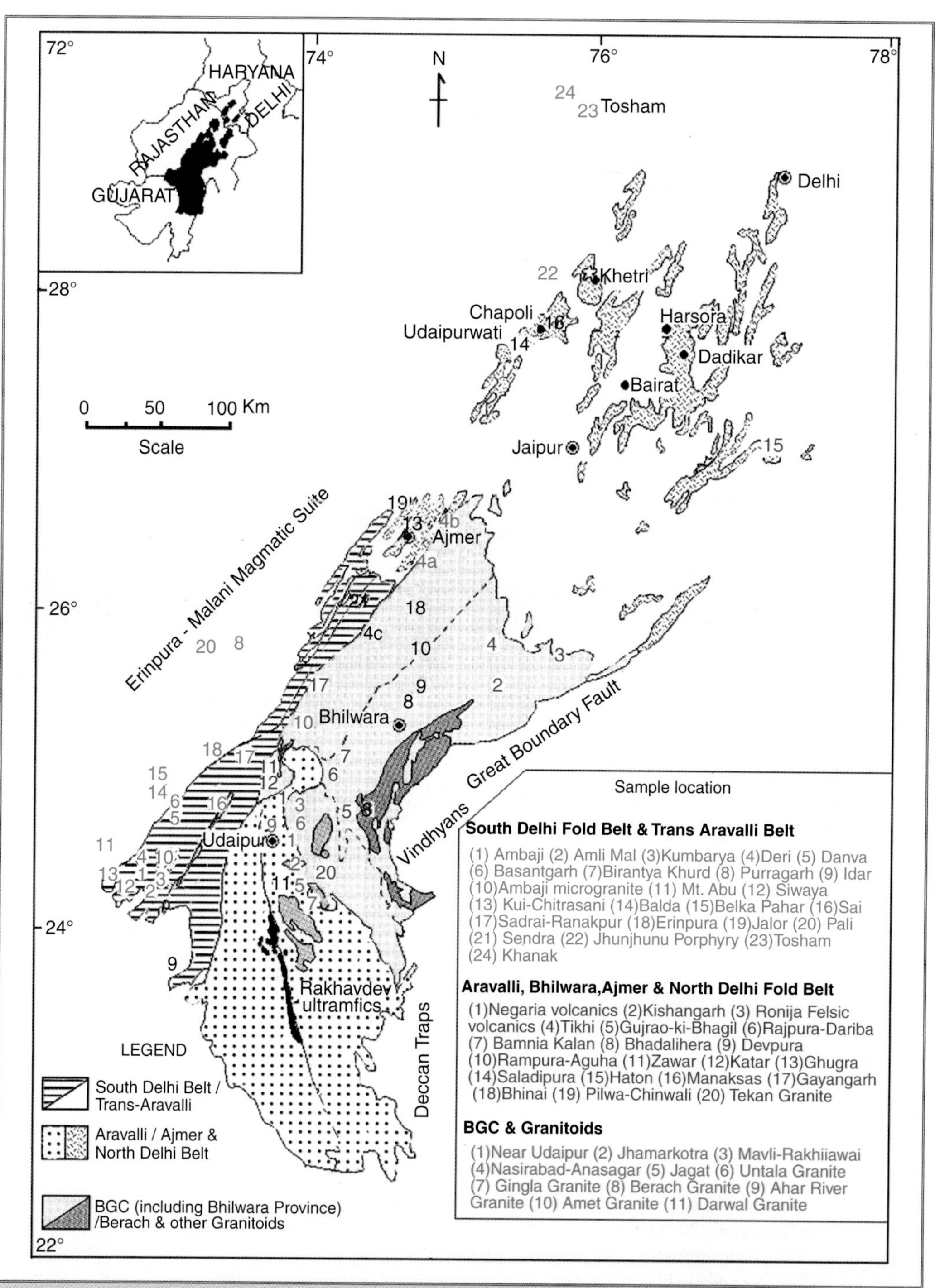

Fig. 6.1.9 Map showing locations of geochronological data available so far from different belts of Western India (Synthesised and Modified after Deb et al., 2001)

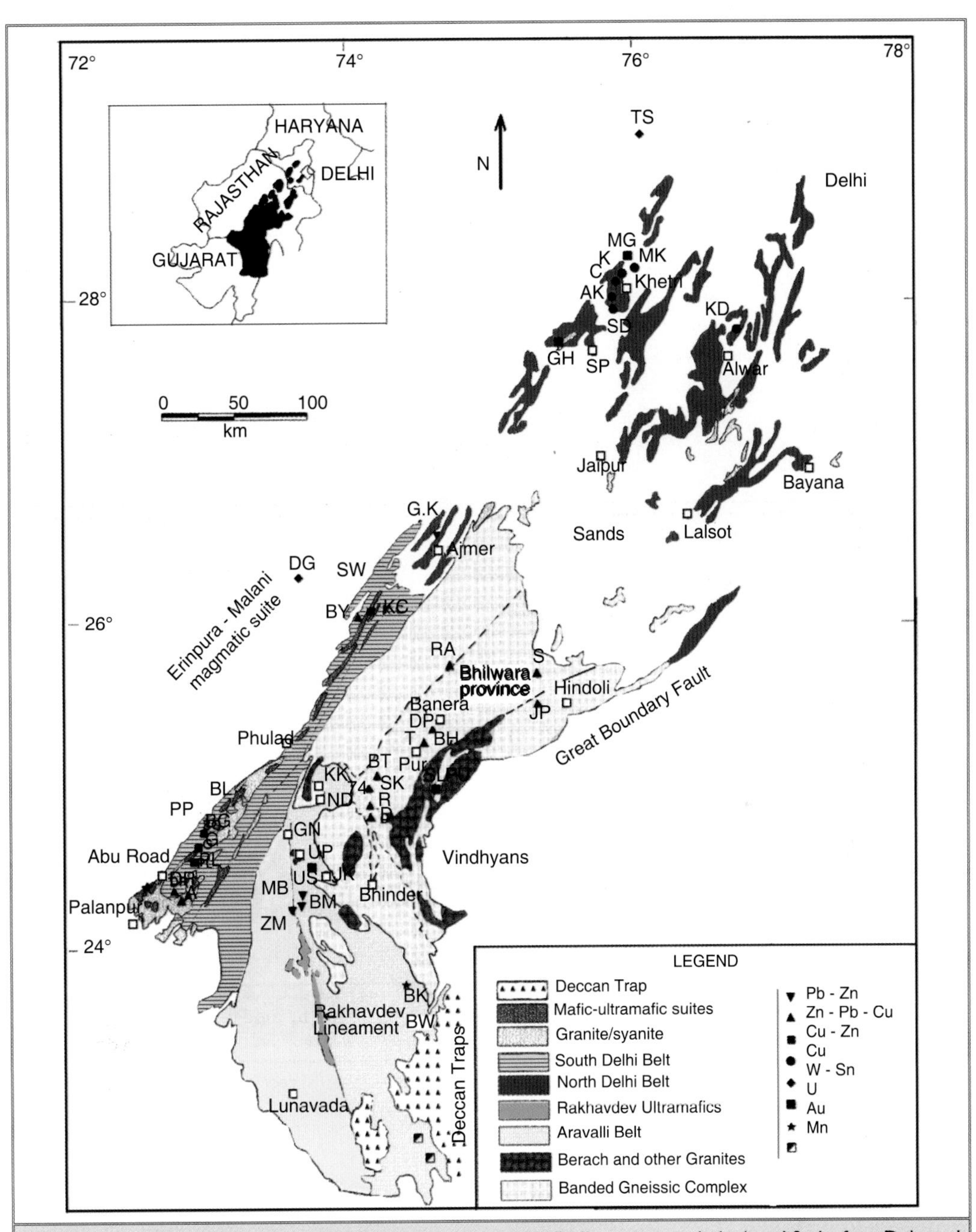

Fig. 6.2.1 Simplified regional geological map of Aravalli–Delhi orogenic belt (modified after Deb and Sarkar, 1990; Deb et al., 2001) showing locations of mineral occurrences and deposits: AK – Akwali; BG – Basantgarh; BH – Bhilwara; BK – Bhukia; BL – Balda; BM – Baroimagra; BT – Bethumni; BW – Banswara; BY – Birantiya; C – Chandmari; D – Dariba; DL – Diri; DG – Degana; DR – Deri; G – Golia; GH – Ghateswar; GK – Ghugra–Kayar; GN – Gogunda; GS – Gura Pratap Singh; JK – Jhamarkotra; JP – Jahazpur; K – Kolihan; K–C – Kalabar Chitar; KD – Kho Dariba; KK – Kankroli; MB – Mochiamagra Balaria; MG – Mewar Gujarwas; MK – Madan Kudan; ND –Nathdwara; PL – Pipela; PP – Phalwadi – Positara; R – Rajpura; RA – Rampura–Agucha; S – Sawar–Bajta; SD –Satkui–Dhanota; SL–PU – Salera–Putholi; SK – Sindesar–Kalan; SP – Saladipura; SW – Sewaria; T – Tiranga; TS – Tosham; UP – Udaipur; US – Umra–Udaisagar; ZM – Zawarmala

Fig. 6.2.8 Photograph of the gossan exposure at Dariba (courtesy: M. Deb)

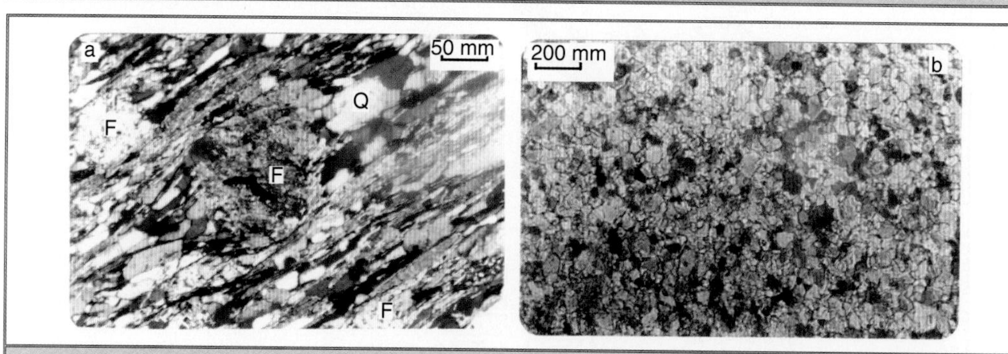

Fig. 6.2.16 Host rocks of Zawar Zn–Pb deposit (a) schistose greywake, (b) recrystallised impure dolostone (Banerjee and Sarkar, 1998). F-feldspar, Q-quartz

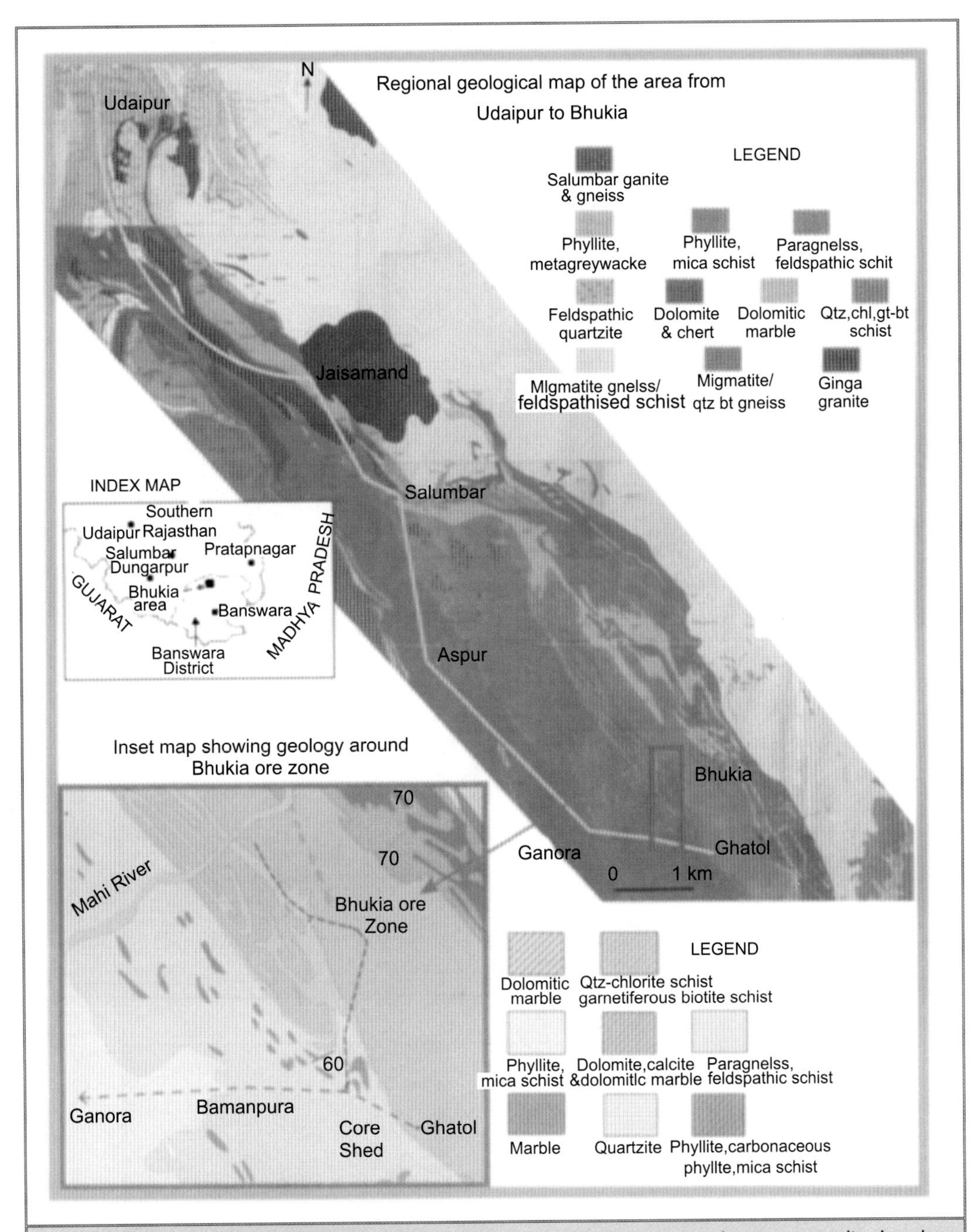

Fig. 6.2.26 Regional geological map of Udaipur–Bhukia area with inset map showing generalised geology around Bhukia Gold Prospect (modified after the GSI map)

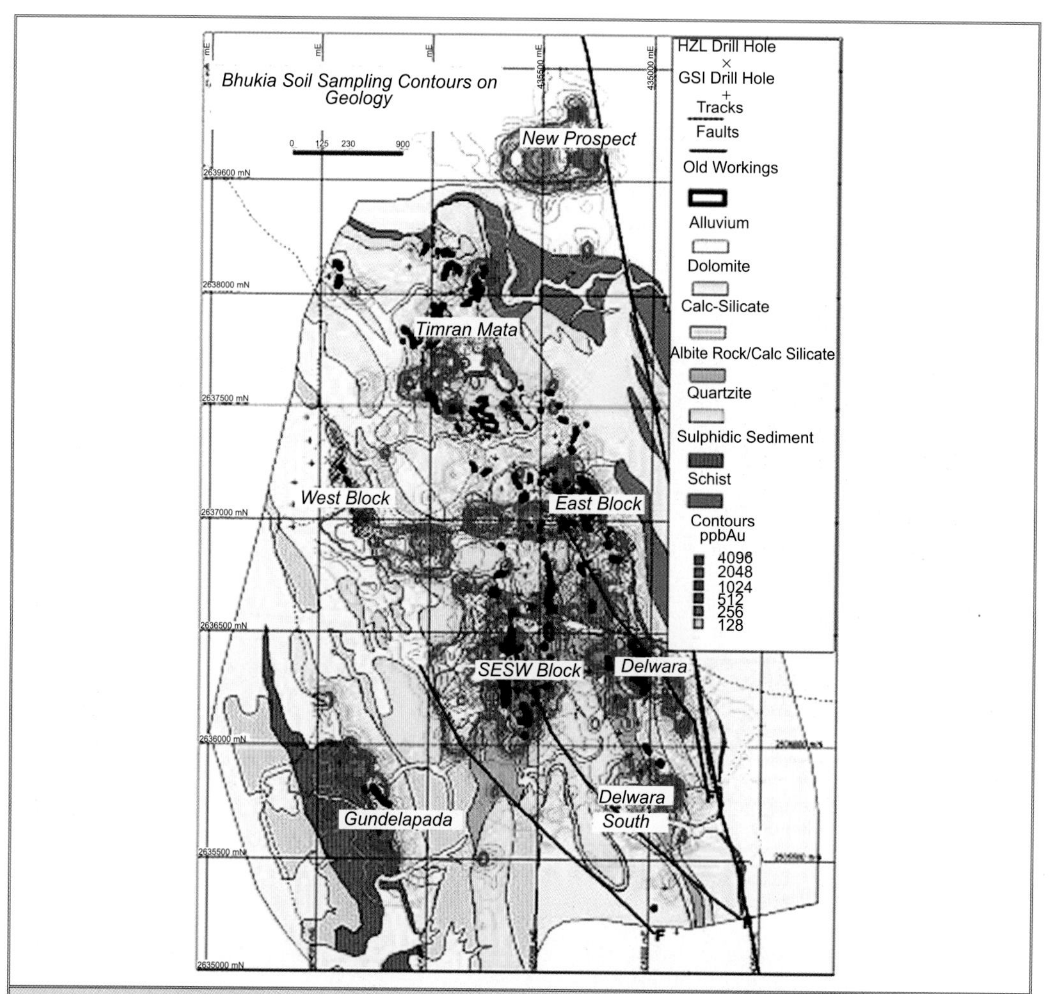

Fig. 6.2.29 Soil sample anomaly contours superposed on the geological map Bhukia–Jagpura prospect (prepared by M/s Indo Gold; www.indogold.com.au)

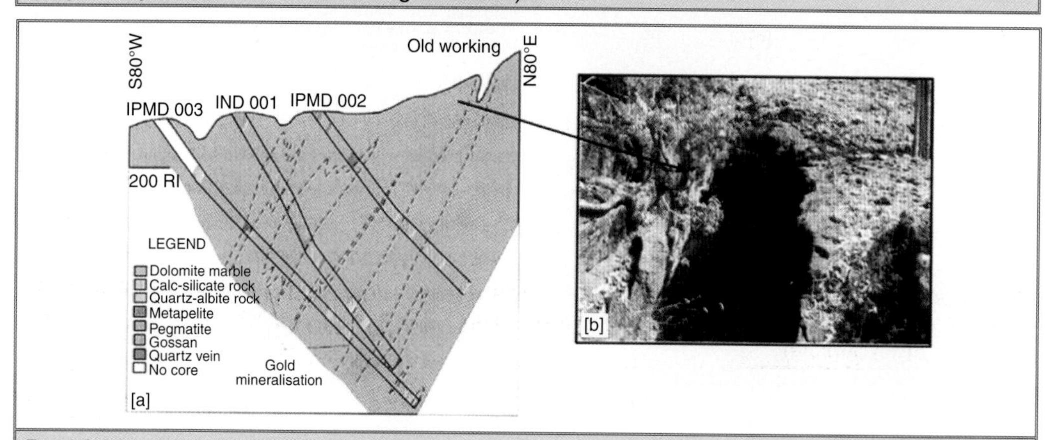

Fig. 6.2.31 (a)Transverse section showing the disposition of ore bodies at depth, (b) a major old working corresponding to the section Bhukia deposit (after Deol et al., 2008)

Fig. 6.2.32 Gold-sulphide-bearing albite-rich and calc–silicate-rich bands in carbonate units of Bhukia Formation, exposed on a hillock with several old workings. Inset-1: closer view of albititic rock exposure, Inset-2: hand specimen of sheared albite–quartz rock (photographs by authors and M. Deb)

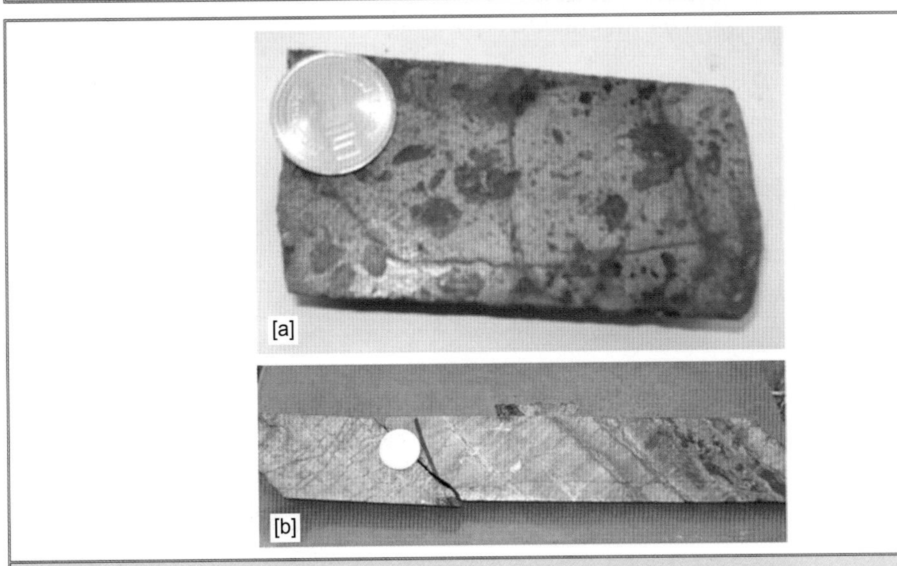

Fig. 6.2.33 Split core samples in Indo Gold core shed showing (a) 'Durchbewegung' structure displaying rounded gangue mineral aggregates in massive pyrrhotite, (b) Sulphide-bearing albite-rich veins in marble (photographs by authors)

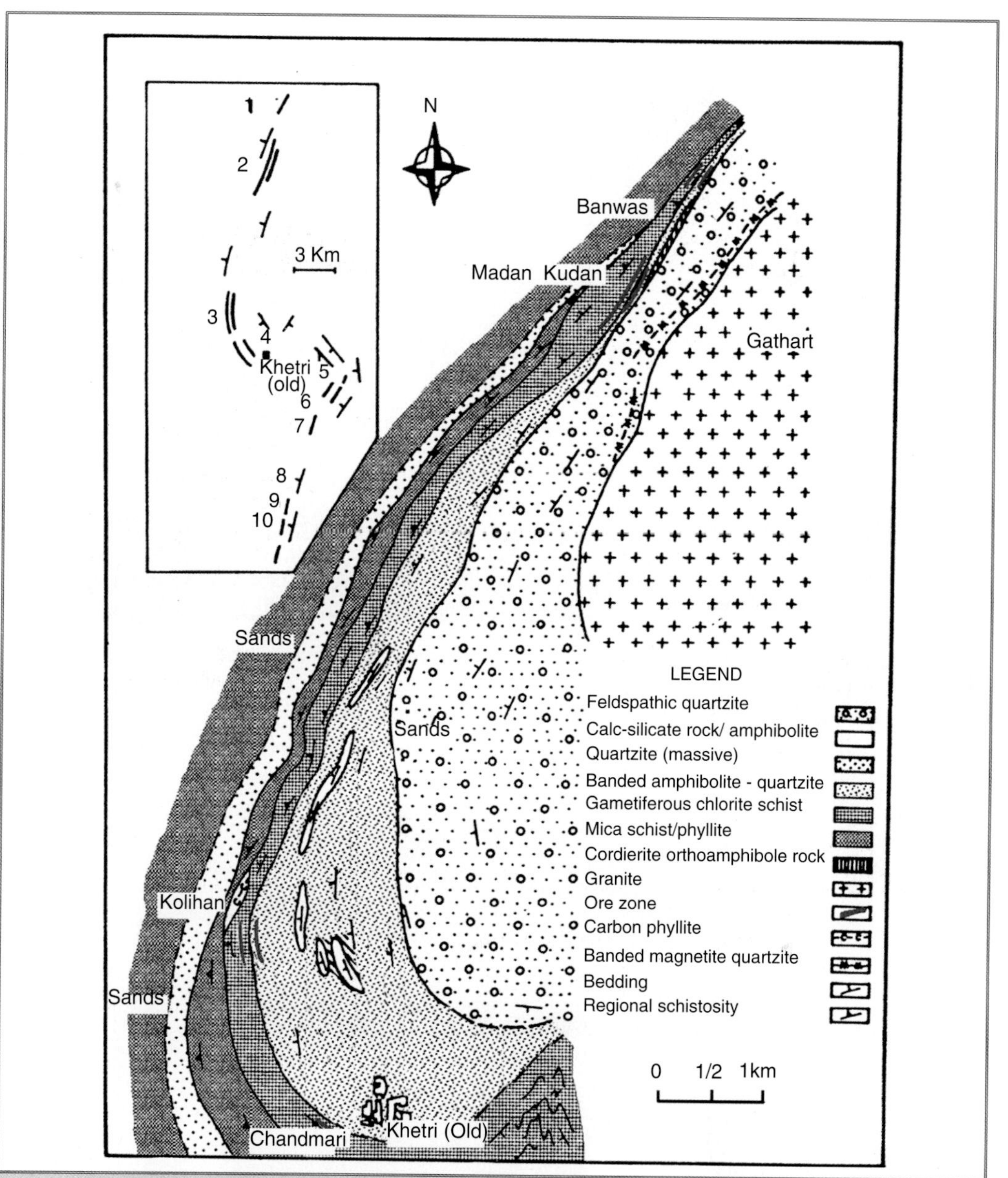

Fig. 6.2.35 Geological map of the Khetri copper belt showing the major deposits. The inset shows the folded nature of the ore belt and enclosing host rocks (after Sarkar and Dasgupta, 1980a)

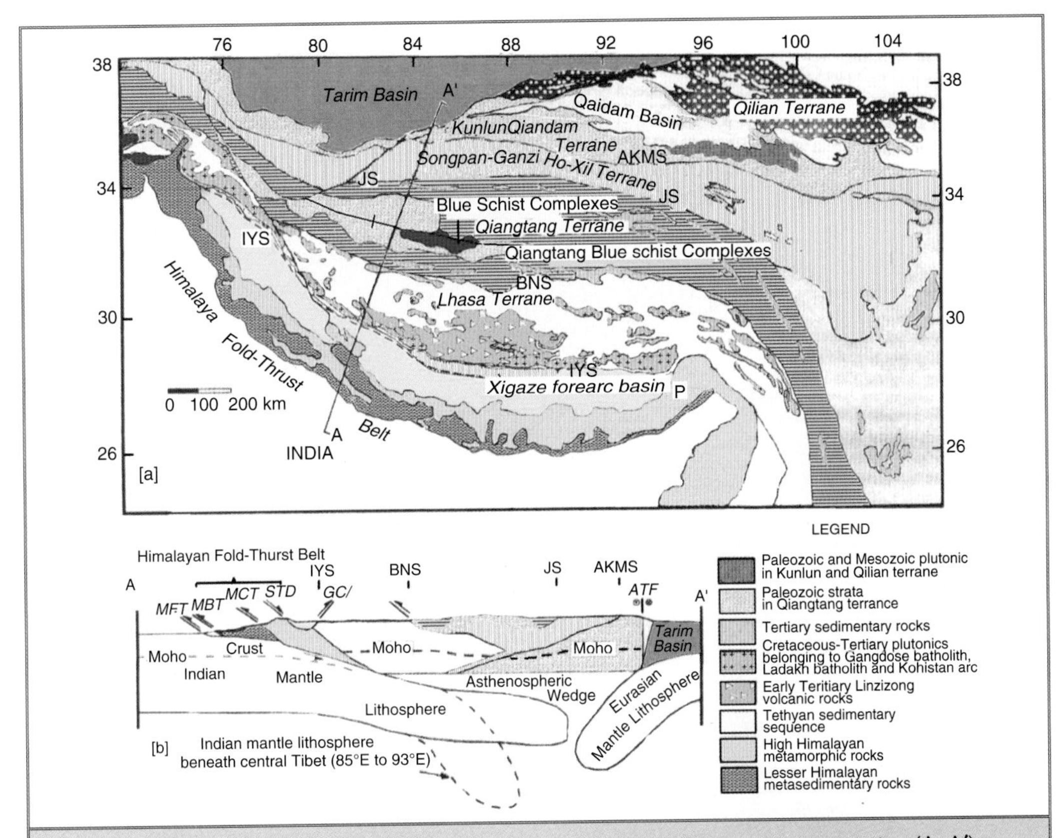

Fig.7.1.3 (a) Tectonic map of the Tibet–Himalaya collision zone (b) Schematic cross section (A–A′) across Tibet–Himalaya orogen based on the interpretation of geological and tomographic data (after Zhou and Murphy, 2005)

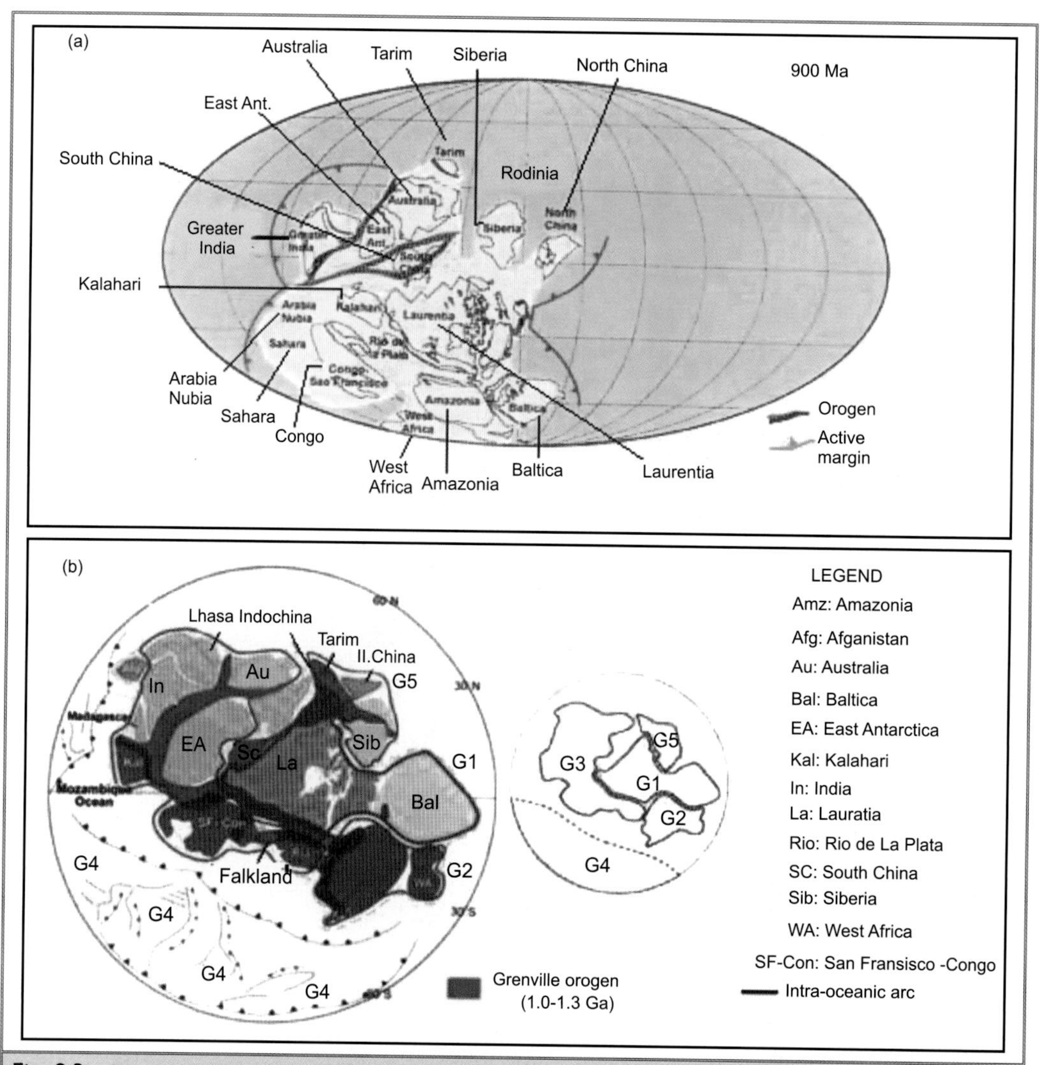

Fig. 8.3 Reconstruction of Rodinia: (a) modified after Li et al., 2008, (b) modified after Santosh et al., 2009

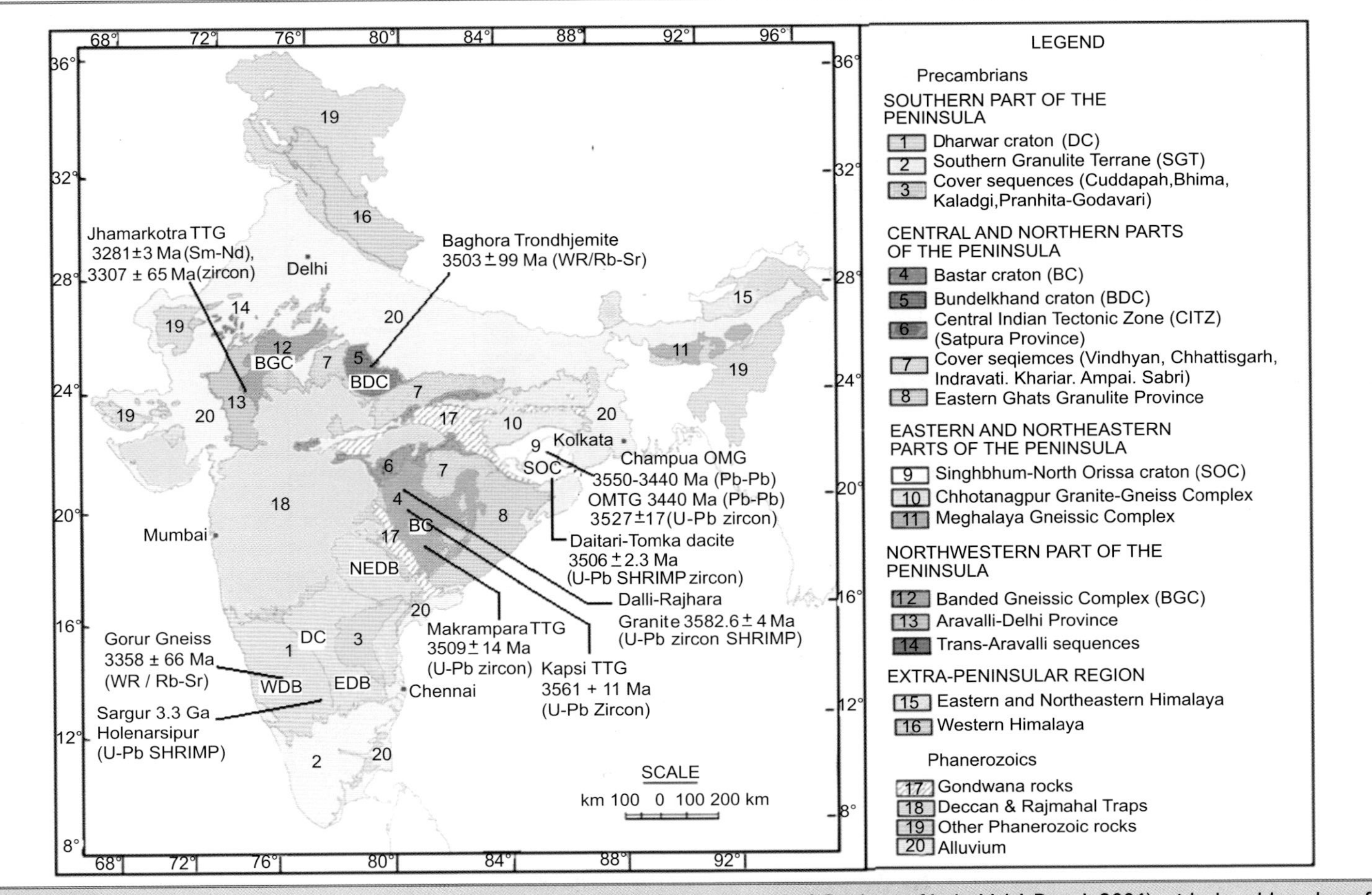

Fig. 8.7 Map showing geological provinces in India (Modified after GSI's map in Manual of Geology of India, Vol. I, Part 1, 2006) with the oldest dates from each craton: DC – Dharwar craton (WDB – Western Dharwar block, EDB – Eastern Dharwar block, NEDB – North Eastern Dharwar block), BC – Bastar craton, SOC – Singhbhum–Orissa craton, BDC – Bundelkhand craton, BGC – Banded Gneissic Complex

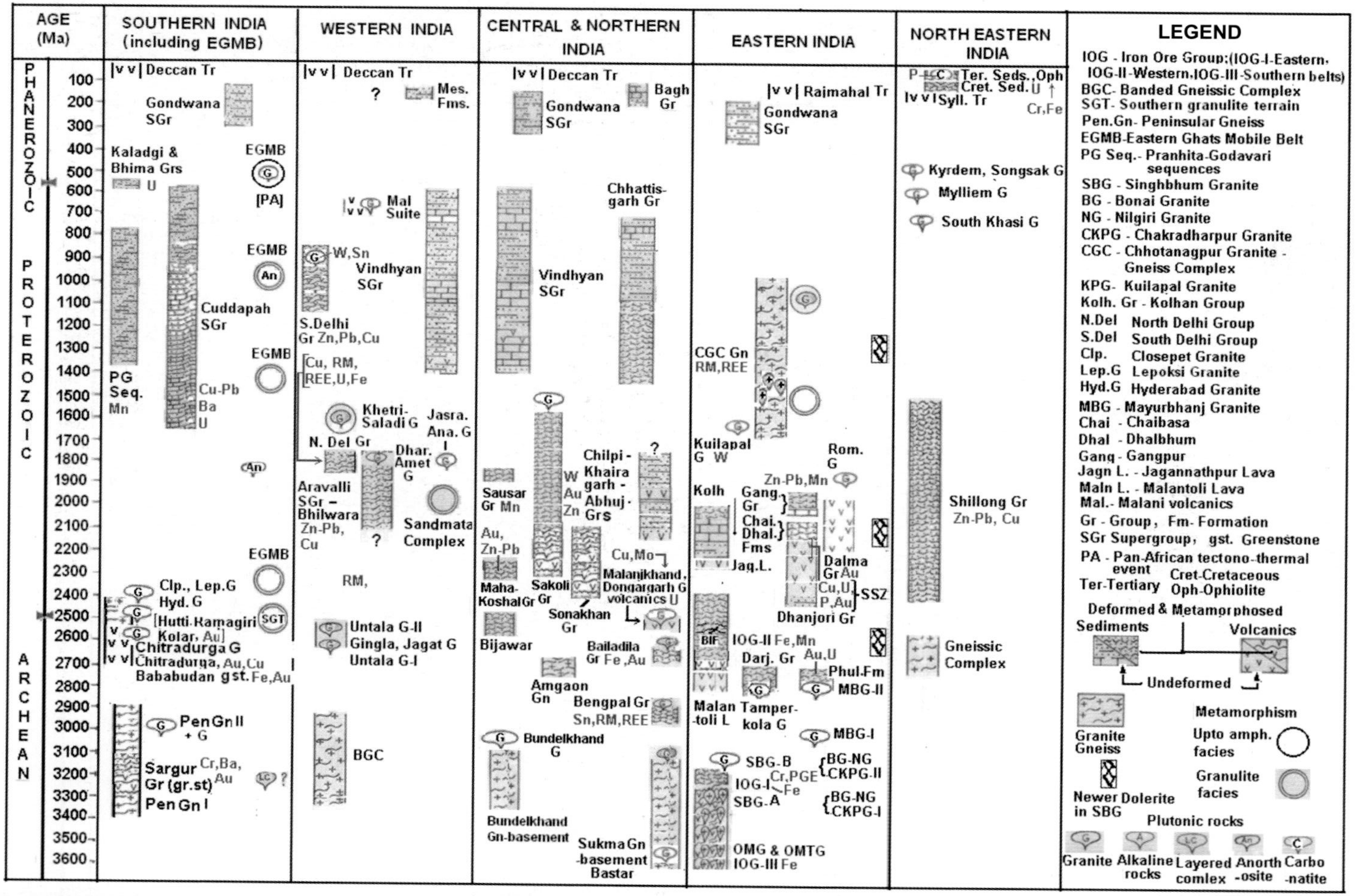

AGE (Ma)	SOUTHERN INDIA (including EGMB)	WESTERN INDIA	CENTRAL & NORTHERN INDIA	EASTERN INDIA	NORTH EASTERN INDIA
PHANEROZOIC: 100, 200, 300, 400, 500, 600; PROTEROZOIC: 700, 800, 900, 1000, 1100, 1200, 1300, 1400, 1500, 1600, 1700, 1800, 1900, 2000, 2100, 2200, 2300, 2400, 2500; ARCHEAN: 2600, 2700, 2800, 2900, 3000, 3100, 3200, 3300, 3400, 3500, 3600	v v Deccan Tr; Gondwana SGr; Kaladgi & Bhima Grs U; EGMB (G) [PA]; PG Seq. Mn; Cuddapah SGr Cu-Pb Ba U; EGMB (An); EGMB; (An); EGMB; Clp., Lep.G; Hyd. G; [Hutti-Ramagiri, Kolar, Au] (SGT); v v Chitradurga G; v v Chitradurga, Au,Cu; Bababudan gst. Fe,Au; PenGn II + G; Sargur Gr (gr.st) Cr,Ba, Au; (LC) ?; PenGn I	v v Deccan Tr; ?; Mes. Fms.; v v Mal Suite; W,Sn; Vindhyan SGr; S.Delhi Gr Zn,Pb,Cu; [Cu, RM, REE,U,Fe; Khetri-Saladi G; Jasra. Ana. G; N. Del Gr; Dhar. Amet G; Aravalli SGr – Bhilwara Zn-Pb, Cu; ?; Sandmata Complex; RM,; Untala G-II; Gingla, Jagat G; Untala G-I; BGC	v v Deccan Tr; Gondwana SGr; Bagh Gr; Chhattis-garh Gr; Vindhyan SGr; (G); ?; Chilpi - Khaira garh - Abhuj - Grs; W Au Zn; Sausar Gr Mn; Au, Zn-Pb; Cu,Mo; Malanjkhand, Dongargarh G volcanics U; Maha-Koshal Gr; Sakoli Gr; Sonakhan Gr; Bijawar; Bailadila Gr Fe ,Au; Amgaon Gn; Bengpal Gr Sn,RM,REE; Bundelkhand G; Bundelkhand Gn-basement; Sukma Gn -basement Bastar	v v Rajmahal Tr; Gondwana SGr; CGC Gn RM,REE; Kuilapal G W; Rom. G; Zn-Pb,Mn; Kolh; Gang. Gr; Chai. Dhal. Fms; Jag.L.; Dalma Gr Au; Cu,U, P,Au; SSZ; Dhanjori Gr; BIF; IOG-II Fe,Mn; Darj. Gr; Au,U; Phul-Fm; MBG-II; Malan-toli L; Tamper-kola G; MBG-I; SBG-B; BG-NG CKPG-II; IOG-I Cr,PGE Fe; SBG-A; BG-NG CKPG-I; OMG & OMTG; IOG-III Fe	P C Ter. Seds.,Oph; Cret. Sed. U; v v Syll. Tr; Cr,Fe; Kyrdem, Songsak G; Mylliem G; South Khasi G; Shillong Gr Zn-Pb, Cu; Gneissic Complex

LEGEND

IOG - Iron Ore Group:(IOG-I-Eastern. IOG-II-Western.IOG-III-Southern belts)
BGC- Banded Gneissic Complex
SGT- Southern granulite terrain
Pen.Gn- Peninsular Gneiss
EGMB-Eastern Ghats Mobile Belt
PG Seq.- Pranhita-Godavari sequences
SBG - Singhbhum Granite
BG - Bonai Granite
NG - Nilgiri Granite
CKPG - Chakradharpur Granite
CGC - Chhotanagpur Granite - Gneiss Complex
KPG- Kuilapal Granite
Kolh. Gr - Kolhan Group
N.Del North Delhi Group
S.Del South Delhi Group
Clp. Closepet Granite
Lep.G Lepoksi Granite
Hyd.G Hyderabad Granite
MBG - Mayurbhanj Granite
Chai - Chaibasa
Dhal - Dhalbhum
Gang - Gangpur
Jagn L. - Jagannathpur Lava
Maln L. - Malantoli Lava
Mal.- Malani volcanics
Gr - Group, Fm- Formation
SGr Supergroup, gst. Greenstone
PA - Pan-African tectono-thermal event
Ter-Tertiary
Cret-Cretaceous
Oph-Ophiolite

Deformed & Metamorphosed: Sediments, Volcanics
Undeformed
Granite Gneiss
Newer Dolerite in SBG
Metamorphism: Upto amph. facies; Granulite facies
Plutonic rocks: Granite, Alkaline rocks, Layered comlex, Anorth -osite, Carbo -natite

Table 8.1 Geological developments and corresponding metallogeny through time in the various crustal segments of the Indian peninsula and their age-wise correlation